electronic communication

third edition

ROBERT L. SHRADER
Former Chairman of Electronics
Laney College

McGraw-Hill Book Company
New York St. Louis Dallas San Francisco Düsseldorf Johannesburg Kuala Lumpur London
Mexico Montreal New Delhi Panama Paris São Paulo Singapore Sydney Tokyo Toronto

Library of Congress Cataloging in Publication Data

Shrader, Robert L
  Electronic communication.

  1. Electronics.   2. Radio—Examinations, questions,
etc.   3. Telecommunication—Examinations, questions,
etc.   I. Title.
TK7815.S5 1975   621.38   74-18382
ISBN 0-07-057138-4

ELECTRONIC COMMUNICATION

567890DODO783210987

The editors for this book were Gordon Rockmaker
and Alice V. Manning,
the designer was Charles A. Carson,
and the production supervisor was Phyllis D. Lemkowitz.
It was set in Helvetica Light by York Graphic Services, Inc.

contents

○ PREFACE   xv

○ CHAPTER 1

Current, Voltage, and Resistance   1

1-1 Electricity
1-2 Electrons and protons
1-3 The atom and its free electrons
1-4 The electroscope
1-5 The big three in electricity
1-6 A simple electric circuit
1-7 Current
1-8 Electromotive force
1-9 The battery in a circuit
1-10 Ionization
1-11 Types of current and voltage
1-12 Resistance
1-13 Color-coded resistors
1-14 The metric system
1-15 Wire sizes
1-16 Making low-resistance connections
1-17 Printed circuits

○ CHAPTER 2

Direct-current Circuits   21

2-1 Ohm's law
2-2 Using Ohm's law
2-3 Mathematics for Ohm's-law problems
2-4 Finding the square root
2-5 Power and energy
2-6 Using the power formulas
2-7 Power dissipation in resistors
2-8 Fuses
2-9 Meters
2-10 Types of circuits
2-11 Series circuits
2-12 Ohm's law in complex circuits
2-13 Suggestions for working complex problems
2-14 Conductance
2-15 Parallel resistances
2-16 Series and parallel batteries
2-17 Complex dc circuits
2-18 Matching load to source

○ CHAPTER 3

Magnetism   47

3-1 Magnetism and electricity
3-2 The magnetic field
3-3 Flux density, B
3-4 Mmf, $\mathscr{F}$, and field intensity, H
3-5 Permeability, μ
3-6 The atomic theory of magnetism
3-7 Ferromagnetism
3-8 The hysteresis loop
3-9 Permanent and temporary magnets
3-10 Magnetizing and demagnetizing
3-11 The magnetic circuit
3-12 English, cgs, mksa, and SI units
3-13 Permanent-magnet fields
3-14 Magnetic-field distortion
3-15 The magnetism of the earth
3-16 Using a magnetic compass in electricity
3-17 Electrons moving in a magnetic field
3-18 Generating an emf by magnetic means
3-19 Magnetostriction
3-20 Relays

○ CHAPTER 4

Alternating Current   62

4-1 Methods of producing emf
4-2 A basic concept of alternating current
4-3 The ac cycle
4-4 Peak and effective ac values
4-5 The average value of ac
4-6 Frequency
4-7 Phase
4-8 Other ac terms

○ CHAPTER 5

Inductance and Transformers   69

5-1 Inductance
5-2 Self-induction
5-3 Coiling an inductor
5-4 The time constant of an inductance
5-5 The energy in a magnetic field
5-6 Choke coils

5-7 Mutual inductance
5-8 Coefficient of coupling
5-9 Inductances in series
5-10 Inductances in parallel
5-11 Shorting a turn in a coil
5-12 Inductive reactance
5-13 Phase relations with inductance
5-14 Phase with both L and R
5-15 Transformers
5-16 Construction of transformers
5-17 Eddy currents
5-18 Hysteresis
5-19 Copper loss
5-20 External-induction loss
5-21 The voltage ratio of transformers
5-22 The power ratio of transformers
5-23 The current ratio of transformers
5-24 Transformer efficiency
5-25 Autotransformers
5-26 Practical transformer considerations

○ CHAPTER 6

Capacitance

89

6-1 The capacitor
6-2 Factors determining capacitance
6-3 Dielectric losses
6-4 Working voltage and dielectric strength
6-5 Energy stored in a capacitor
6-6 Quantity of charge in a capacitor
6-7 Electrolytic capacitors
6-8 Variable capacitors
6-9 Modern capacitors
6-10 Power factor in capacitors
6-11 Inductance in capacitors
6-12 Changing varying dc to ac
6-13 Capacitive reactance
6-14 Capacitors in parallel
6-15 Capacitors in series
6-16 Phase relations with capacitance
6-17 Considerations when selecting capacitors
6-18 Capacitance color code

○ CHAPTER 7

Alternating-current Circuits

105

7-1 Effects of inductance, capacitance, and resistance
7-2 Inductance and resistance in series
7-3 The phase angle
7-4 Voltage vector addition
7-5 Apparent and true power
7-6 Power factor
7-7 Capacitance and resistance in series
7-8 Inductance, capacitance, and resistance in series
7-9 Accuracy of computations
7-10 Proving problems
7-11 The j operator
7-12 Parallel ac circuits
7-13 Inductance and resistance in parallel
7-14 Capacitance and resistance in parallel
7-15 Capacitance, inductance, and resistance in parallel
7-16 Complex parallel-series circuit
7-17 Rectangular and polar notations

○ CHAPTER 8

Resonance and Filters

126

8-1 Resonance
8-2 Series resonance
8-3 Parallel resonance
8-4 The Q of a circuit
8-5 Logarithms and decibels
8-6 Bandwidth (BW)
8-7 Bandwidth of transformers
8-8 Filters

○ CHAPTER 9

Vacuum Tubes

146

9-1 Vacuum tubes
9-2 Cathodes
9-3 Filaments
9-4 Correct filament voltage
9-5 The anode, or plate
9-6 The plate load
9-7 Plate voltage
9-8 Reversing filament polarity
9-9 Diode tubes and rectification
9-10 The vacuum in a vacuum tube
9-11 Gaseous diodes
9-12 Cold-cathode tubes
9-13 Gaseous triodes, or thyratrons
9-14 Triode tubes
9-15 Amplification factor
9-16 How the triode amplifies
9-17 Bias voltage
9-18 Characteristic curves
9-19 Factors determining μ
9-20 Plate impedance
9-21 Mutual conductance
9-22 Power output
9-23 Plate dissipation and saturation
9-24 Classes of amplifiers
9-25 Secondary emission
9-26 Tetrode tubes

9-27 Beam-power tetrodes
9-28 Pentode tubes
9-29 Shielding tubes
9-30 Tubes with more than three grids
9-31 Multiunit tubes
9-32 Symbols used
9-33 Receiving and transmitting tubes
9-34 Causes of improper plate current
9-35 Causes of tube failure
9-36 Battery operation
9-37 High-frequency tubes
9-38 Visual-indicating tubes

○ CHAPTER 10

Basic Solid-state Devices   173

10-1 Solid-state devices
10-2 Doped semiconductors
10-3 Solid-state diodes
10-4 Light-frequency diodes
10-5 Junction transistors
10-6 Common-emitter circuits
10-7 BJT characteristic curves
10-8 Power transistors
10-9 Common-base circuits
10-10 Common-collector circuits
10-11 Junction FETs
10-12 MOSFETs
10-13 Unijunction transistors (UJTs)
10-14 Silicon-controlled rectifiers (SCRs)
10-15 Triacs and diacs
10-16 Light-activated SCRs
10-17 Integrated circuits

○ CHAPTER 11

Power Supplies   198

11-1 Power supplies
11-2 Rectifiers
11-3 Half-wave rectification
11-4 Full-wave rectification
11-5 Two voltages from one transformer
11-6 Capacitive filtering
11-7 Inductive filtering
11-8 Capacitive-input filtering
11-9 Inductive-input filtering
11-10 Some possible filter circuits
11-11 RC filters
11-12 Filter chokes
11-13 Swinging chokes
11-14 High-vacuum rectifiers
11-15 Mercury-vapor rectifiers
11-16 Inverse peak voltage
11-17 Vacuum versus mercury-vapor diodes

11-18 Ripple frequency
11-19 Rectifier filament circuits
11-20 Practical power supplies
11-21 Power-factor compensation
11-22 Voltage-multiplier circuits
11-23 The bleeder resistor
11-24 Voltage regulation
11-25 Voltage-regulator devices
11-26 Shunt-regulated power supplies
11-27 Series-regulated power supplies
11-28 Voltage dividers
11-29 Copper-oxide and selenium diodes
11-30 Dc-to-dc converters
11-31 Filter capacitors in series
11-32 Three-phase power
11-33 Three-phase power supplies
11-34 Indications of power-supply failure

○ CHAPTER 12

Measuring Devices   233

12-1 Meters
12-2 Dc meters
12-3 Linear and nonlinear scales
12-4 Dc ammeters
12-5 Computing shunt resistances
12-6 Sensitivity
12-7 Damping
12-8 The electrostatic voltmeter
12-9 Dc voltmeters
12-10 Voltmeters in high-resistance circuits
12-11 Ohmmeters
12-12 Volt-ohm-milliammeters
12-13 Dc vacuum-tube voltmeters
12-14 Ac vacuum-tube voltmeters
12-15 Ac meters
12-16 The rectifier meter
12-17 Peak-reading meters
12-18 Decibel and VU meters
12-19 Thermocouple ammeters
12-20 Hot-wire ammeters
12-21 The electrodynamometer
12-22 Repulsion-type meters
12-23 The wattmeter
12-24 Watthour meters
12-25 Frequency meters
12-26 Ampere-hour meters
12-27 Current-squared-meter scales
12-28 Bridges
12-29 The oscilloscope
12-30 Free-running oscilloscopes
12-31 Triggered oscilloscopes
12-32 Digital panel meters (DPMs)
12-33 Test equipment probes

∘ CHAPTER 13

Oscillators   264

13-1 Types of oscillators
13-2 Shock excitation
13-3 Electronic LC oscillators
13-4 Armstrong oscillator
13-5 Oscillator bias
13-6 The tuned-plate tuned-grid oscillator
13-7 Series and shunt feed
13-8 The Hartley oscillator
13-9 The Colpitts oscillator
13-10 The ultraudion oscillator
13-11 Electron-coupled oscillator
13-12 Crystal oscillators
13-13 Crystals and temperature coefficients
13-14 Crystal heater chambers
13-15 Crystal holders
13-16 Other crystal circuits
13-17 Some high-frequency oscillators
13-18 Audio oscillators
13-19 Dynatron oscillator
13-20 RC oscillators
13-21 Multivibrator oscillators
13-22 Parasitic oscillations
13-23 Indications of oscillation
13-24 Oscillator stability

∘ CHAPTER 14

Audio-frequency Amplifiers   290

14-1 Audio frequencies
14-2 Voltage versus power amplifiers
14-3 Peak-clipping distortion
14-4 μ versus stage gain
14-5 Types of coupling
14-6 Transformer coupling
14-7 Resistance coupling
14-8 Impedance coupling
14-9 Direct coupling
14-10 Types of bias used in audio amplifiers
14-11 Battery bias
14-12 Power-supply bias
14-13 Voltage-divider bias
14-14 Cathode-resistor bias
14-15 Contact-potential bias
14-16 Biasing filament tubes
14-17 Controlling volume
14-18 Tone controls
14-19 Miller effect
14-20 Decoupling, or filtering, AF stages
14-21 Classes of amplifiers
14-22 Class A audio amplifiers
14-23 Push-pull class A amplifiers
14-24 Push-pull output-transformer operation
14-25 Push-pull class A bias
14-26 If one class A push-pull tube burns out
14-27 The class AB_1 audio amplifier
14-28 The class AB_2 audio amplifier
14-29 The class B audio amplifier
14-30 Earphones
14-31 Loudspeakers
14-32 Impedance-coupled output
14-33 Matching impedances with an output transformer
14-34 Parallel-connected amplifier tubes
14-35 Types of distortion
14-36 Inverse feedback in AF amplifiers
14-37 Phase inverters
14-38 Pentode AF amplifiers
14-39 Hum in AF amplifiers
14-40 Hum reduction in push-pull stages
14-41 Miscellaneous AF amplifier items
14-42 A general-purpose vacuum-tube amplifier
14-43 A general-purpose transistor amplifier
14-44 Special transistor amplifiers
14-45 Operational amplifiers

∘ CHAPTER 15

Radio-frequency Amplifiers   329

15-1 RF amplifiers
15-2 Low-level RF amplifiers
15-3 RF power amplifiers
15-4 Battery-biased class C RF power amplifier
15-5 Tuning the plate circuit
15-6 Coupling the RF amplifier to a load
15-7 Grid-leak bias for RF amplifiers
15-8 Parallel operation
15-9 Push-pull operation
15-10 Types of feed
15-11 Methods of coupling RF amplifiers
15-12 The triode RF amplifier
15-13 Plate neutralization
15-14 Grid or Rice neutralization
15-15 Neutralizing with an RF indicator
15-16 Neutralizing by grid-current meter
15-17 Neutralizing push-pull stages
15-18 Direct neutralization
15-19 Inductive neutralization
15-20 Neutralizing tetrode and pentode tubes
15-21 Frequency multipliers
15-22 Determining C and L for resonant circuits
15-23 Center-tapping the filament
15-24 Grounded-grid RF amplifiers
15-25 Amplifiers for VHF and UHF
15-26 Transistor RF power amplifiers
15-27 Troubles in RF power amplifiers

∘ CHAPTER 16

Basic Transmitters 356

16-1 Radio transmitters
16-2 Single-stage transmitters
16-3 Keying relays
16-4 The MOPA
16-5 Tuning an MOPA transmitter
16-6 Frequency stability
16-7 Dummy antennas
16-8 Plate and cathode circuit keying
16-9 Shaping or filtering the CW signal
16-10 Primary keying
16-11 Vacuum-tube keying
16-12 Grid-block keying
16-13 Frequency-shift keying (F1)
16-14 Modulated code signals (A2)
16-15 The buffer amplifier
16-16 Grid-leak bias in CW transmitters
16-17 Frequency doublers or multipliers
16-18 Decreasing harmonic radiation
16-19 Shielding and protective devices
16-20 Transmitters today
16-21 A transistor CW transmitter
16-22 A VFO transmitter
16-23 Heterodyne transmitters
16-24 Emergency repairs
16-25 Indications of trouble

∘ CHAPTER 17

Amplitude Modulation 382

17-1 Modulation
17-2 Why the carrier is modulated
17-3 Sound
17-4 The single-button microphone
17-5 Absorption modulation
17-6 A simple series modulation
17-7 Series modulation
17-8 The modulated envelope
17-9 Basic plate modulation
17-10 Percent of modulation
17-11 Plate modulation
17-12 Plate-modulating tetrodes and pentodes
17-13 Operating power
17-14 Sidebands
17-15 Bandwidth
17-16 Heising modulation
17-17 Grid modulation
17-18 Suppressor-grid modulation
17-19 Screen-grid modulation
17-20 High-level and low-level modulation
17-21 Checking modulation with an oscilloscope
17-22 Linear RF amplifiers
17-23 Adjusting a linear RF amplifier

17-24 Doherty linear amplifiers
17-25 Carrier shift
17-26 Causes of carrier shift
17-27 A3 and A1 with the same tube
17-28 What the antenna ammeter tells
17-29 Magnetic-induction microphones
17-30 The crystal microphone
17-31 The condenser microphone
17-32 Tuning an A3 transmitter
17-33 Modulating transistor amplifiers
17-34 Single-sideband radiotelephone (A3J)
17-35 Filter-type SSB transmitters
17-36 Balanced modulators
17-37 Sideband filters
17-38 A phase-type SSB transmitter
17-39 Linear amplifiers for A3J
17-40 Special uses of SSB
17-41 Transmitter intermodulation

∘ CHAPTER 18

Amplitude-modulation Receivers 429

18-1 Receivers
18-2 Demodulating a modulated wave
18-3 Crystal detectors
18-4 Power detectors
18-5 Linear and square-law detectors
18-6 Grid-leak detectors
18-7 Regenerative and autodyne detectors
18-8 The superregenerative detector
18-9 Tuned-radio-frequency receivers
18-10 The superheterodyne
18-11 The RF amplifier
18-12 The mixer stage
18-13 The IF amplifiers
18-14 The second detector and AVC
18-15 The beat-frequency oscillator
18-16 A squelch or Q circuit
18-17 The tuning eye
18-18 S meters
18-19 Noise limiters
18-20 The audio amplifiers
18-21 The crystal filter
18-22 Wave traps
18-23 Image frequencies
18-24 A double-conversion superheterodyne
18-25 Transistor receivers
18-26 Operating a superheterodyne
18-27 Diversity reception
18-28 Transceivers
18-29 Aligning a superheterodyne
18-30 Troubleshooting in receivers
18-31 Emergency repairs
18-32 Servicing transistor equipment

∘ CHAPTER 19

Frequency Modulation 463

19-1 Purpose of frequency modulation
19-2 The four fields of FM
19-3 Basic concepts of FM
19-4 Slope detection
19-5 A stagger-tuned discriminator
19-6 The Foster-Seeley discriminator
19-7 The ratio detector
19-8 The gated-beam detector
19-9 Limiters
19-10 Pre-emphasis and de-emphasis
19-11 FM receivers
19-12 Alignment of FM receivers
19-13 Direct FM
19-14 A solid-state FM modulator
19-15 A phase-modulated transmitter
19-16 PM from the phasitron
19-17 Voice-modulated PM transmitters
19-18 FM broadcast stations
19-19 Public Safety Radio Service
19-20 A mobile FM transmitter
19-21 Noise in motor vehicles
19-22 FM stereo multiplex
19-23 A stereo multiplex transmitter system
19-24 An FM stereo multiplex receiver system
19-25 SCA and FAX

∘ CHAPTER 20

Antennas 498

20-1 Radio waves
20-2 The ionosphere
20-3 Fading
20-4 Night and day transmissions
20-5 Effect of lightning on radio reception
20-6 Polarization
20-7 The half-wave antenna
20-8 Hertz and Marconi antennas
20-9 Current and voltage in a half-wave antenna
20-10 The radiation resistance
20-11 Loading or tuning an antenna
20-12 Transmission lines
20-13 Directivity of antennas
20-14 The quarter-wave antenna
20-15 The full-wave antenna
20-16 Feeding the antenna
20-17 Collinear beam antennas
20-18 Driven arrays
20-19 Phase monitors
20-20 Parasitic arrays
20-21 Long-wire beams
20-22 Loop antennas
20-23 Top loading

20-24 Feeding antennas
20-25 Determining impedance of an antenna
20-26 Field intensity
20-27 Field intensity of harmonics
20-28 Field gain
20-29 The ground
20-30 Computing the power in an antenna
20-31 Omnidirectional antennas

∘ CHAPTER 21

Measuring Frequency 531

21-1 Means of measuring frequencies
21-2 Frequency tolerance
21-3 Absorption wavemeters
21-4 Lecher wires
21-5 Grid-dip meters
21-6 Primary standards of frequency
21-7 Secondary frequency standards
21-8 Frequency measurements with a secondary standard
21-9 Heterodyne frequency meters
21-10 A constant frequency indicator
21-11 Measuring frequency with counters
21-12 Frequency considerations for amateurs

∘ CHAPTER 22

Batteries 548

22-1 A, B, and C batteries
22-2 Primary cells
22-3 The lead-acid battery
22-4 Specific gravity of a lead-acid cell
22-5 The hydrometer
22-6 Water for batteries
22-7 Capacity of a battery
22-8 Charging batteries
22-9 Maintaining lead-acid cells
22-10 Edison batteries
22-11 Other chemical cells
22-12 Nonchemical cells

∘ CHAPTER 23

Motors and Generators 562

23-1 Electric machines
23-2 Alternators
23-3 Voltage output of an alternator
23-4 Field excitation of an alternator
23-5 Paralleling alternators
23-6 Dc generators
23-7 The externally excited dc generator
23-8 The series dc generator
23-9 The shunt dc generator
23-10 The compound dc generator

23-11 The third-brush generator
23-12 Commutating poles, or interpoles
23-13 Brush sparking
23-14 Dc motors
23-15 Series dc motors
23-16 Shunt dc motors
23-17 Compound dc motors
23-18 Ac motors
23-19 Universal motors
23-20 Synchronous motors
23-21 Squirrel-cage motors
23-22 Polyphase motors
23-23 Motor-generators
23-24 Dynamotors
23-25 Rating generators and motors
23-26 Maintenance of motors and generators

○ CHAPTER 24

Broadcast Stations   578

24-1   Standard broadcast stations
24-2   Components of a broadcast system
24-3   The broadcast console
24-4   Audio levels
24-5   Attenuator pads
24-6   Line equalizers
24-7   Peak-limiter amplifiers
24-8   Classifications of powers
24-9   Broadcast station tests
24-10 Broadcast station meters
24-11 Frequency monitoring
24-12 Modulation monitors
24-13 International broadcast stations
24-14 Logs
24-15 Broadcast operator license requirements
24-16 Remote control
24-17 Some other FCC requirements
24-18 Emergency Broadcast System (EBS)
24-19 Disk recording
24-20 Playback equipment
24-21 Tape recorders
24-22 Broadcast microphones
24-23 Radio station antenna towers

○ CHAPTER 25

Television   601

25-1   A TV broadcast system
25-2   The TV transmitter antenna system
25-3   Magnetic deflection and focusing
25-4   Interlaced scanning
25-5   The iconoscope
25-6   The image orthicon (IO)
25-7   The vidicon
25-8   The synchronizing-pulse generator

25-9   Modulation percentages of the visual carrier
25-10 Camera chains
25-11 Logs and personnel required
25-12 Motion-picture projection for TV
25-13 Television transmission requirements
25-14 The visual transmitter
25-15 The TV receiver front end
25-16 The IF strip
25-17 Detector, aural, and video circuits
25-18 Sync-pulse separation
25-19 Vertical-deflection circuits
25-20 Horizontal-deflection circuits
25-21 Horizontal output circuits
25-22 Automatic gain control
25-23 Color
25-24 Transmitting color signals
25-25 Color-TV receivers
25-26 Color-picture tubes
25-27 One-gun picture tubes

○ CHAPTER 26

Microwaves   638

26-1   Microwave range
26-2   Microwave transmission lines
26-3   Some waveguide devices
26-4   Coupling to waveguides
26-5   Detecting devices
26-6   Resonant cavities
26-7   Klystrons
26-8   Magnetrons
26-9   TWTs and BWOs
26-10 Microwave transistors and diodes
26-11 Gunn and LSA diodes
26-12 Isolators and circulators
26-13 Tunnel-diode amplifiers
26-14 YIG resonators
26-15 Microwave measurements
26-16 Installing waveguides

○ CHAPTER 27

Radar   656

27-1   Principles of radar
27-2   A basic radar system
27-3   A marine radar system
27-4   The transmitter section
27-5   Average power and duty cycle
27-6   The antenna system
27-7   TR boxes
27-8   The radar receiver
27-9   Circuits of the indicator
27-10 Antenna synchronization
27-11 Heading flash
27-12 Echo box

27-13 Operating the radar set
27-14 Radar interference
27-15 Basic radar maintenance
27-16 Nonmarine radar

∘ CHAPTER 28

Shipboard Radio Equipment   674

28-1   Radio aboard ship
28-2   Compulsory radiotelegraph installations
28-3   Compulsory radiotelephone installations
28-4   A main transmitter
28-5   A reserve transmitter
28-6   The main antenna
28-7   Medium- versus high-frequency
        communications
28-8   High-frequency transmitters
28-9   Radiotelegraph frequencies
28-10 Radiotelephone frequencies
28-11 Marine main receivers
28-12 Marine superheterodynes
28-13 Bridge-to-bridge radiotelephone
28-14 Survival radio equipment
28-15 Auto-alarm keyers
28-16 Auto-alarms
28-17 The main-antenna switch
28-18 Teleprinters
28-19 Shipboard radio operators
28-20 Standing watch
28-21 Positions and time

∘ CHAPTER 29

Radio Direction Finders   699

29-1   Basis of radio direction finding (RDF)
29-2   The DF loop
29-3   An unbalanced loop
29-4   Methods of balancing a loop
29-5   Unidirectional bearings
29-6   DF errors
29-7   Calibrating a DF
29-8   Frequencies used for marine DF
29-9   Goniometer-type direction finders
29-10 RDF receivers
29-11 Maintenance of direction finders
29-12 An automatic direction finder
29-13 Adcock antennas
29-14 Determining a position by RDF

∘ CHAPTER 30

Loran, Omega, and Facsimile   711

30-1   The loran A system
30-2   Loran A station designations
30-3   Determining time difference

30-4   Using sky-wave signals
30-5   Loran transmitters
30-6   Loran receivers
30-7   Loran C
30-8   Omega radio navigation
30-9   Facsimile transmissions
30-10 Facsimile reception

∘ CHAPTER 31

Radio Licenses and Laws: FCC

Element 1   723

31-1   Radio laws and regulations
31-2   License elements
31-3   Licenses and permits
31-4   Suspended licenses
31-5   Notices of violations
31-6   Who may operate transmitters
31-7   Who may service or adjust transmitters
31-8   Classification of communications
31-9   Logs

∘ CHAPTER 32

Voice Communications: FCC

Element 2   729

32-1   Radiotelephone operation (G)
32-2   Message priorities
32-3   Distress
32-4   Urgency signals
32-5   Safety messages
32-6   Intelligibility (G)
32-7   Phonetic alphabet (G)
32-8   Operational words (G)
32-9   Calling and working
32-10 Radiotelephone station identification
32-11 Good operating practices (G)
32-12 Calling and working frequencies
32-13 Antenna-tower lights (G)
32-14 Coast stations

∘ CHAPTER 33

Communicating by Radiotelegraph   737

33-1   Fundamentals of operating
33-2   Radiotelegraph licenses
33-3   The Morse code
33-4   Frequencies used
33-5   Calling by radiotelegraph
33-6   Answering by radiotelegraph
33-7   Tuning and testing
33-8   Station identification
33-9   Auto-alarm signal

33-10 Silence periods
33-11 Distress
33-12 The urgency signal
33-13 The safety signal
33-14 Radiotelegraph logs
33-15 Commercial radiotelegraph messages
33-16 Counting words in messages
33-17 Message charges
33-18 Different types of messages
33-19 Transmitting speed
33-20 Transmitting radiotelegraph
33-21 Call letters
33-22 Operating signals

○ CHAPTER 34

Amateur Radio   750

34-1   The amateur radio service
34-2   Amateur communications
34-3   Amateur stations
34-4   Amateur call signs
34-5   Amateur examinations

34-6   Amateur bands
34-7   The Novice class license
34-8   The Technician class license
34-9   The General class license
34-10 The Conditional class license
34-11 The Advanced class license
34-12 The Extra class license

○ APPENDIXES   758

A Greek alphabet
B Standard component values and military
  precision standard values
C Table of natural trigonometric functions
D Table of logarithms (four-place mantissas)
E Emissions
F FCC rules and regulations and field offices
G Q signals
H Radiotelegraph operating signals

○ ADDENDUM   773

○ INDEX   784

preface

The third edition of *Electronic Communication* presents in one volume a complete electronics and radio communications course, including practical information and theory required for Federal Communications Commission radio licenses of both commercial and amateur grades.

Although written primarily for the community college and technical school, the discussions in this text are well within the grasp of most senior high school radio-electronics students. The book can also be used effectively for pre-engineering courses providing a general coverage of communication electronics, and for industrial electronics courses that emphasize communications, such as those given by telephone companies. With careful independent home study this book should enable students to pass any FCC license examination satisfactorily, including the radar endorsement, radiotelephone or radiotelegraph permit, and broadcast endorsement examinations.

Proper operation, maintenance, and installation of commercial and amateur radio transmitting and receiving equipment require far more than a minimum of technical background and ability. The goal of this volume is to present, as simply as possible, the practical basic theory of electricity, electronics, and radio, as well as the many operating fundamentals and laws for personnel working in a variety of radio fields. In some cases workers in electronics areas not normally considered related to radio may find that a commercial FCC license is a requirement in some aspects of their job. In addition, a federal commercial or amateur license may form a valuable employment recommendation for many electronics jobs.

As a guide to the information necessary for commercial and amateur licenses, the FCC publishes lists of sample questions. From these, or other questions similar in content, license examinations are compiled. This book is based on these sample questions rearranged into a more useful teaching and learning order. The FCC questions for commercial licenses are included at the back of each chapter, each with a reference indicating in which section that particular question is answered. Commercial licenses are discussed in Chap. 31. Amateur license information will be found in Chap. 34. All known amateur sample questions, with either an immediate answer or a reference to the chapter and section where such material is discussed, are located just before the index.

Answering the checkup quizzes that occur every few pages throughout the book will greatly reinforce students' learning. Answers are given on the page following. The answers are somewhat abbreviated and in many cases are not the only ones possible, but they should act as a guide to thinking and as discussion points for groups. Suggestions regarding corrections and additions would be appreciated by the author.

Metric measurements have been used throughout this book except where prevailing usage makes the U.S. customary units of measurement (inch, foot, mile, etc.) still preferable.

Among those who provided considerable aid in preparing the third edition in the modern marine radio field were George F. Ong and A. J. (Bill) Tassin of the Marine Communications Department, International Telephone and Telegraph Company, Berkeley, California, and, for materials on marine facsimile systems, George F. Stafford, Vice President, Director of Communications, and Robert Dean, Engineering Technician, of Alden Electronic and Impulse Recording Equipment Company, Inc., Westborough, Maine.

The Hallicrafters Company of Rolling Meadows, Illinois, was kind enough to supply data from which the transceiver illustration was made. Donald S. Middleton, Professor of Electronics, Southern Colorado State College, Pueblo, Colorado, helped greatly with suggestions and test questions.

Materials for previous editions were made available by ITT World Communications, Inc. (formerly Mackay Radio and Telegraph Company), particularly through the courtesty of W. R. Taggart and G. G. Thommen. The RCA Corporation, particularly C. C. Pitts, G. P. Aldridge, and Ralph Scott, aided with marine radio material. The consultation activities of and the pertinent suggestions and corrections made by the late Emery L. Simpson, educator and commercial and amateur operator, are gratefully acknowledged as being of utmost aid in the preparation of the first two editions. Much credit should go to the technical reviewers of the third edition. They include the late Joseph O'Shea, for many years with the Allegheny Technical Institute, Edwin E. Pollock W6KHS, of Cabrillo College, and Oliver J. Ruel, formerly of Ranken Technical Institute.

As usual, the help and understanding of an appreciated wife Dorothy has made this and all other editions possible.

Robert L. Shrader, W6BNB

CURRENT, VOLTAGE, AND RESISTANCE

The objective of this first chapter is to familiarize the reader with some of the basic physical theories that underlie any study of electricity and electronics, specifically voltage, current, and resistance. Resistors and color codes are discussed briefly, as are some of the methods of making satisfactory electric-circuit connections. The checkup quizzes should be mastered before moving to the next section. The questions at the end of the chapter cover areas tested in Federal Communication Commission (FCC) license examinations.

1-1 ELECTRICITY

The complicated electronic systems involved in modern-day missiles, communication satellites, nuclear power plants, radio, and television, and even up-to-date automobiles, require service technicians who understand the functioning of electric and electronic circuits. It is the principal goal of this book, however, to outline the basic theory related to electronic communication, particularly such systems as require licensed personnel to operate or service them.

The term "electronic" infers circuits including either the first electronic devices—vacuum tubes—or the newer solid-state devices such as diodes and transistors, as well as integrated circuits (ICs). The term "electric" or "electrical" is usually applied to systems or circuits in which electrons flow through wires but which involve no vacuum tubes or solid-state devices. Actually, many modern electrical systems are now using electronic devices to control the electric current

that flows in them. But what is electric current? What makes such a simple thing as an electric lamp glow? It is easy to pass the problem off with the statement, "The switch connects the light to the power lines." But what does connecting the light to the power lines do? How does energy travel through solid copper wires or through space? What makes a motor turn, a radio play? What is behind the dial that allows you to pick out one radio station from thousands of others operating at the same time?

There are no simple single answers to any of these questions. Each question requires the understanding of many basic principles. By adding one basic idea to another, it is possible to answer, eventually, most of the questions that may be asked about the intriguing subjects of electricity, electronics, and radio.

When the light switch is turned on at one point in a room and the light suddenly glows, energy has found a path through the switch to the light. The paths used are usually copper wires, and the tiny particles that do the moving and carry the energy are called *electrons*. These little electrons are important to anyone studying electronics and radio, since they are usually the only particles that are considered to move in electric circuits.

To explain what is meant by an electron, it will be necessary to investigate more closely the makeup of all matter. The word "matter" means, in a general sense, anything that can be touched. It includes substances such as rubber, salt, wood, water, glass, copper, and air. The whole world is made of different kinds of matter.

Water is one of the most common of the many

items in the category of matter. If a drop of water is divided in two and then divided again and again until it can be divided no longer and still be water, this smallest particle is known as a *molecule*. The water molecule can be broken down into still smaller particles, but these new particles will not be water.

Physicists have found that there are three particles making up a molecule of water: two *atoms* of hydrogen (H) and one atom of oxygen (O) as shown in Fig. 1-1. Oxygen, at normal tempera-

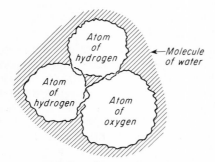

Fig. 1-1 Two atoms of hydrogen and one atom of oxygen, when chemically interlocked, form one molecule of water.

tures, is one of several gases that constitute the air we breathe. Hydrogen is also a gas in its natural state; it is found in everyday use as part of the gas used for heating or cooking. If a gaseous mixture containing 2 parts of hydrogen and 1 part of oxygen is ignited, a chemical reaction in the form of an explosion takes place. The residue of the explosion will be water (H_2O) droplets.

It has been determined that an atom is also divisible, being made up of at least two types of particles: *protons* and electrons. Both are electrical particles, and neither one is divisible. All the molecules that make up all matter of the universe are composed of these electrical proton-electron pairs.

1-2 ELECTRONS AND PROTONS

Electrons are the smallest and lightest of particles. They are said to have a *negative* charge, meaning that they are surrounded by some kind of an invisible *field of force* (Fig. 1-2) that will react in an electrically negative manner on anything

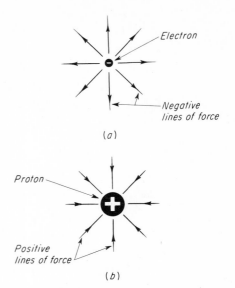

Fig. 1-2 Electrostatic lines of force shown (a) outward from a negative charge and (b) inward toward a positive charge.

that is electrically charged and brought within the limits of the field. The electric field is represented pictorially as being composed of outward-pointing *lines of force*. Whether lines of force actually exist is not known, but they are used for explanatory purposes.

Protons are about eighteen hundred times as massive as electrons and have a *positive* electric field surrounding them. The positive field is represented by inward-pointing lines. Theoretically, an electron has exactly as many outward-pointing lines as a proton has inward-pointing lines. The proton is just exactly as positive as the electron is negative; each has a unit electric charge.

In theory, negative lines of force will not join other negative lines of force. In fact, they repel each other, tending to push each electron away from every other electron (Fig. 1-3a). Positive lines of force do the same thing, as shown in Fig. 1-3b.

When an electron and a proton are far apart, as illustrated in Fig. 1-4, only a few of their lines of force join and pull together. The contracting pull between the two charges is therefore small. When brought closer together, the electron and proton are able to link more of their lines of force and will pull together with greater force. If close

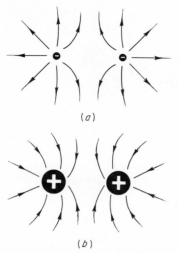

(a)

(b)

Fig. 1-3 Lines of force of like or similar direction repel one another.

enough, all the lines of force from the electron are joined to all the lines of force of the proton and there is no external field. They form a neutral, or uncharged, group. The neutral atomic particle, known as a *neutron,* exists in the nucleus of all atoms heavier than hydrogen.

The fact that electrons repel other electrons, protons repel other protons, but electrons and

Fig. 1-4 Lines of force from unlike charges attract. (*a*) If remote, there is little field interlinkage. (*b*) When close together, all lines interlink, leaving none external, resulting in a neutral-charged group.

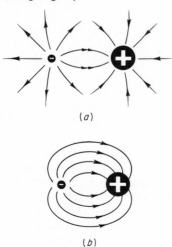

(a)

(b)

protons attract each other follows the basic physical law: *Like charges repel; unlike charges attract.*

Because the proton is about 1,800 times heavier than the electron, it seems reasonable to assume that when an electron and a proton attract each other, it will be the tiny electron that will do most of the actual moving. Such is the case. It is the electron that moves in electricity.

Regardless of the difference in apparent size and weight, the negative field of an electron is just as strong negatively as the positive field of a proton is positive. Small though it is physically, the field near the electron is quite strong.

If the field strength around an electron at a distance of one-millionth of an inch is a certain amount, at two-millionths of an inch it will be one-quarter as much; at four-millionths of an inch it will be one-sixteenth as much; and so on. If the field decreases as distance increases, the field is said to vary *inversely* with distance. Actually, it varies inversely with the distance squared.

When an increase in something produces an increase in something else, the two things are said to vary *directly* rather than inversely. Two million electrons on an object produce twice as much negative charge as one million electrons would. The charge is directly proportional to the number of electrons.

Since the electric-field strength of an electron varies inversely with the distance squared, the field strength an inch or so away might be very weak.

The fields surrounding electrons and protons are known as *electrostatic fields*. The word "static" means, in this case, "stationary," or "not caused by movement."

When electrons are made to move, the result is *dynamic electricity*. The word "dynamic" indicates that motion is involved.

To produce a movement of an electron, it will be necessary to have either a negatively charged field to push it, a positively charged field to pull it, or, as normally occurs in an electric circuit, both a negative and a positive charge (a pushing and pulling pair of forces).

1-3 THE ATOM AND ITS FREE ELECTRONS

There are more than 100 different kinds of atoms, or *elements,* from which the millions of

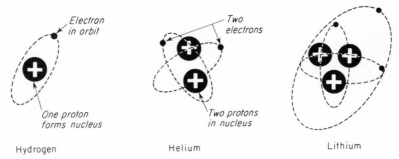

Fig. 1-5 Simplified atoms. Hydrogen has 1 proton as a nucleus and 1 orbital electron. Helium has a 2-proton nucleus with 2 orbital electrons. Lithium has a 3-proton nucleus with 2 electrons in the first layer and 1 in the second layer. (Neutrons not shown in the helium and lithium nuclei.)

different forms of matter found in the universe are composed. The heaviest elements are always radioactive and unstable, decomposing into lower-atomic-weight atoms spontaneously.

The simplest and lightest atom is hydrogen. An atom of hydrogen consists of one electron and one proton, as represented in Fig. 1-5. In one respect this atom is similar to all others: the electron whirls around the proton, or *nucleus,* of the atom, much as planets rotate around the sun. Electrons whirling around the nucleus are termed *planetary*, or *orbital*, electrons.

The next atom in terms of weight is helium, having two protons and two electrons. The third atom is lithium, with three electrons and three protons, and so on (Fig. 1-5).

Some of the common elements, in order of their atomic weights, are:

1. Hydrogen (H)	28. Nickel (Ni)
2. Helium (He)	29. Copper (Cu)
3. Lithium (Li)	30. Zinc (Zn)
6. Carbon (C)	32. Germanium (Ge)
8. Oxygen (O)	79. Gold (Au)
13. Aluminum (Al)	82. Lead (Pb)
14. Silicon (Si)	88. Radium (Ra)
26. Iron (Fe)	92. Uranium (U)

Most atoms have a nucleus consisting of all the protons of the atom and also one or more neutrons. The remainder of the electrons (always equal in number to the number of nuclear protons) are whirling around the nucleus in various layers. The first layer of electrons outside the nucleus can accommodate only two electrons. If the atom has three electrons (Fig. 1-5), two will be in the first layer and the third will be in the next layer. The second layer is completely filled when eight electrons are whirling around in it. The third is filled when it has eighteen electrons.

Some of the electrons in the outer orbit, or shell, of the atoms of many materials such as copper or silver exist in a higher *conduction level* and can be dislodged easily. These electrons travel out into the wide-open spaces between the atoms and molecules and may be termed *free* electrons. Other electrons in the outer orbit will resist dislodgment and may be called *bound* or *valence* electrons. Materials consisting of atoms (or molecules) having many free electrons will allow an easy interchange of their outer-shell electrons, while atoms with only bound electrons will hinder any electron exchange.

When a substance is heated, greater energy is developed in the free-moving electrons. The more energy they have, the more the electrons resist orderly movement through the material. The material is said to have an increased resistance to the movement of electrons through it.

1-4 THE ELECTROSCOPE

An example of electrons and electric charges acting on one another is demonstrated by the action of an *electroscope,* shown in Fig. 1-6. An electroscope consists of two very thin gold or aluminum leaves attached to the bottom of a metal rod. To prevent air currents from damaging the delicate metal-foil leaves, the rod and leaves are encased in a glass flask, the rod projecting from the top through a rubber cork.

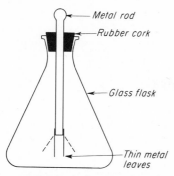

Fig. 1-6 Electroscope can indicate presence of either a negative or a positive charge.

To understand the operation of the electroscope, it is necessary to recall these facts:

1. Normally an object has a neutral charge.
2. Like charges repel; unlike charges attract.
3. Electrons are negative.
4. Metals have free electrons.

Normally the electroscope rod has a neutral charge, and the leaves hang downward parallel to each other, as shown in Fig. 1-6. Rubbing a piece of hard rubber with wool causes the wool to lose electrons to the rubber, charging the rubber negatively. When such a negatively charged object is brought near the top of the rod, some of the free electrons at the top are repelled and travel down the rod, away from the negatively charged object. Some of these electrons force themselves onto one of the leaves, and some onto the other. Now the two leaves are no longer neutral but are slightly negative and repel each other, moving outward to the position shown by the dotted lines. When the charged object is removed, electrons return up the rod to their original areas. The leaves again have a neutral charge and hang down parallel to each other.

Since the charged object did not touch the electroscope, it neither placed electrons on the rod nor took electrons from it. When electrons were driven to the bottom, making the leaves negative, these same electrons leaving the top of the rod left the top positive. The overall charge of the rod remained neutral. When the charged object was withdrawn, the positive charge at the top of the rod pulled the displaced electrons up to it. All parts of the rod were neutral again.

If a positively charged object, such as a glass rod vigorously rubbed with a piece of silk, is brought near the top of the electroscope rod, some of the free electrons in the leaves and rod will be attracted upward toward the positive object. This charges the top of the rod negatively because of the excess of free electrons there. Both leaves are left with a deficiency of free electrons and are positively charged. Since both leaves are similarly charged again, they repel each other and move outward.

If a negatively charged object is *touched* to the metal rod, a number of excess electrons will be deposited on the rod and will be immediately distributed throughout the electroscope. The leaves spread apart. When the object is taken away, an excess of electrons remains on the rod and leaves. The leaves stay spread apart. If the negatively charged electroscope is touched to any large body that can accept the excess free electrons, such as a person, a large metal object, or the earth, the excess electrons will have a path by which they can leave the electroscope and the leaves will collapse as the charge returns to neutral. The electroscope has been discharged.

If a positively charged object is touched to the metal rod, the rod will lose electrons to it and the leaves will separate. When the object is taken away, the rod and leaves still lack free electrons, are positively charged, and the leaves remain apart. A large neutral body touched to the rod will lose some of its free electrons to the electroscope, discharging it, and the leaves will hang down once more.

The electroscope demonstrates the more or less free movement of electrons that can take place through metallic objects or conductors when electric pressures, or charges, are exerted on the free electrons.

1-5 THE BIG THREE IN ELECTRICITY

Without calling them by name, the discussion thus far has touched on the three elements always present in all operating electric circuits:

◦ *Current*. A progressive movement of free electrons along a wire or other conductor produced by electrostatic lines of force.
◦ *Electromotive force*. The electron-moving force

in a circuit that pushes and pulls electrons (current) through the circuit.

  ○ *Resistance*. Any opposing effect that hinders free-electron progress through wires when an electromotive force is attempting to produce a current in the circuit.

Changes in the values of any one of these "big three" will produce a change in the value of at least one of the others. The interrelationship of these in a simple electric circuit will be discussed briefly.

1-6 A SIMPLE ELECTRIC CIRCUIT

The simplest of electric circuits consists of (1) some sort of an electron-moving force, or *source*, such as is provided by a dry cell, or battery, (2) a *load*, such as an electric light, (3) connecting wires, and (4) a control device. A method of both picturing and diagraming the connections of such a circuit is shown in Fig. 1-7.

It can be seen that the schematic diagram is simpler to draw and actually easier to read than the picture diagram. For this reason, schematic diagrams are used in electronics as much as possible. It will pay you to observe closely the diagrams that are given and practice drawing them until you can reproduce them rapidly and correctly.

Although the wires connecting the source of electromotive force to the load may have some resistance, it is usually very small in comparison with the resistance of the load and is ignored in most cases. A straight line, then, in a schematic diagram is considered to connect parts electrically

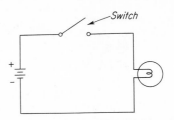

Fig. 1-8 A simple circuit consisting of a battery as the source, a lamp as the load, and a switch as the control device.

but does not represent any resistance in the circuit.

In the simple circuit shown, the cell produces the electromotive force that continually pulls electrons to its positive terminal from the lamp's filament and pushes them out the negative terminal to replace the electrons that were lost to the load by the pull of the positive terminal. The result is a continual flow of electrons through the lamp filament, connecting wires, and source. The special resistance wire of the lamp filament heats when a current of electrons flows through it. If enough current flows, the wire becomes white-hot and the lamp glows.

The addition of a switch in series with one of the connecting wires of the simple circuit affords a means of controlling the current in the circuit (Fig. 1-8). When the switch is closed, electrons find an uninterrupted path in the circuit and flow through the lamp. When the switch is opened, the electromotive force developed by the battery is normally insufficient to cause the electrons to jump the switch gap in the form of a spark and the electron flow in the circuit is interrupted. The lamp cools and no longer glows.

Since the only duty of the switch is to interrupt or close the circuit, it may be inserted anywhere in the circuit. It is shown in the upper connecting wire but will give the same results if placed in the lower. In either case it controls the flow of the electrons.

1-7 CURRENT

A stream of electrons forced into motion by an electromotive force is known as a current. The atoms in a good conducting material such as copper are more or less stationary, but one or more free electrons of the outer ring are con-

Fig. 1-7 (*a*) Lamp connected to a dry cell. (*b*) A schematic diagram of the circuit of (*a*).

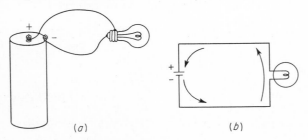

(*a*)   (*b*)

stantly flying off at a high rate of speed. Electrons from other nearby atoms fill in the gaps. Apparently there is a constant aimless movement of billions of electrons in all directions at all times in every part of any conductor.

When an electric force is impressed across the conductor (from a battery), it drives some of these aimlessly moving free electrons away from the negative force toward the positive. It is unlikely that any one electron will move more than a fraction of an inch in a second, but an energy flow takes place along the conductor at approximately 186,000 miles per second, or 300,000,000 meters per second (actually 299,792,462 m/s). A simple analogy of energy flow can be illustrated with automobiles parked in a circle (Fig. 1-9). All the automobiles are parked bumper to bumper except cars 1 and 2, which are separated by a few feet. Nothing is happening in the circuit. The driver in car 1 steps on the gas for an instant, producing mechanical energy, and his car is propelled forward, striking car 2. The force of the impact transfers energy to car 2, which in turn transfers energy to car 3, and so on. An instant later the energy is transferred to the back bumper of car 17, and it is propelled forward, striking the back of car 1. Energy has traveled completely around the circuit in a very short space of time, and yet none of the cars except cars 1 and 17 may have moved more than a fraction of an inch. In an electric circuit the electrons are somewhat similar to the cars. By moving suddenly in one direction, electrons can repel other electrons. These repel others farther along, and so on. The energy transfer (current) of a single impulse is very rapid, but the *drift* of the electrons themselves is relatively slow.

A source of electric energy does not increase the number of free electrons in a circuit; it merely produces a concerted pressure on loose, aimlessly moving electrons. If the material of the circuit is made of atoms or molecules that have no freely interchanging electrons, the source cannot produce any current in the material. Such a material is known as an *insulator*, or a *nonconductor*.

The amount of current in a circuit is basically measured in *amperes,* abbreviated A or amp. An ampere is a certain number of electrons passing or drifting past a single point in an electric circuit in one second. Therefore, an ampere is a *rate of flow,* similar to gallons per minute in a pipe.

The quantity of electrons used in determining an ampere (and other electrical units) is the *coulomb,* abbreviated C or coul. An ampere is one coulomb per second. A single coulomb is 6,250,000,000,000,000,000 electrons. This large number is more easily expressed as 6.25×10^{18}, which is read verbally as "6 point 25 times 10 to the eighteenth power." "Ten to the eighteenth power" means the decimal place in the 6.25 is moved 18 places to the *right*. This method of expressing numbers is known as the *powers of 10* and is handy to use when very large or very small numbers are involved. An example of a small number is 42.5×10^{-7}. In this case the 10^{-7} indicates the decimal point is to be moved 7 places to the *left,* making the number 0.00000425. When very large and very small numbers are multiplied, the powers of 10 are added algebraically; that is, if both are negative numbers, they are added and the sum is given a negative sign. If both are positive numbers, they are added and the sum is given a positive sign. If one is negative and the other positive, the smaller is subtracted from the larger and given the sign of the larger.

EXAMPLE: Multiply 3.2×10^{14} by 4.5×10^{-12}. When multiplied, $3.2 \times 4.5 = 14.4$. The powers of 10, that is, 10^{14} and 10^{-12}, together equal 10^2. The answer is then 14.4×10^2, or 1,440.

The unit of measurement of current is the ampere, but the unit of measurement of electrical quantity, or charge, is the coulomb.

Fig. 1-9 Explanation of how energy can travel faster than the carriers of the energy.

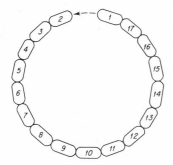

Test your understanding; answer these checkup questions.

1. What is the name of the moving particle of electricity? _____

2. What is the unit formed when atoms combine chemically? _____

3. What electric charge does a proton have? _____ An electron? _____ An atom? _____ A molecule? _____ A neutron? _____

4. What type of field forms between positive and negative bodies? _____

5. How do like charges interact? _____ Unlike charges? _____

6. What particles are found in an atomic nucleus? _____

7. Where in conductors are valence electrons? _____

8. What is the name of the device which indicates a positive or negative charge is being held near it? _____

9. What are the "big three" in electricity? _____ _____ _____

10. What are the four minimum parts of any electric circuit? _____ _____ _____ _____

11. What is the rate of electric-energy flow in meters per second? _____

12. What are materials called that have no free electrons at room temperature? _____ That have many free electrons? _____

13. Is an ampere a quantity or a rate of flow? _____

14. How many electrons are there in a coulomb? _____

15. What does 1.86×10^5 miles per second represent? _____

1-8 ELECTROMOTIVE FORCE

The electron-moving force in electricity, variously termed electromotive force (emf), electric potential, potential difference (PD), difference of potential, electric pressure, and voltage (V), is responsible for the pulling and pushing of the electric current through a circuit. The force is the result of an expenditure of some form of energy to produce an electrostatic field.

An emf exists between two objects whenever one of them has an excess of free electrons and the other has a deficiency of free electrons. This is illustrated in Fig. 1-10a. Should the two objects be connected by a conductor, a discharge current will flow from the negative body to the positive one.

An emf also exists between two objects whenever there is a difference in the number of free electrons per unit volume of the objects. This is illustrated in Fig. 1-10b. If both objects are negative, current will flow from the more negatively charged object to the less negatively charged object when they are connected together. There will also be an electron flow from a less positively charged object to a more positively charged object.

The electrostatic field, the strain of the electrons trying to reach a positive charge, or to move from a more highly negative to a less negative charge, or to move from a less positive to a more positive charge, *is* the emf in electricity. When a conducting material is placed between two points under electric strain, current flows.

The unit of measurement of electric pressure, or emf, is the *volt* (V). A single flashlight dry cell produces about 1.5 V. A wet cell of a storage battery produces about 2.1 V.

A volt can also be defined as the pressure required to force a current of one ampere through

Fig. 1-10 Direction in which energy or current will flow (a) between negative and positive bodies and (b) between highly negatively and less negatively charged bodies.

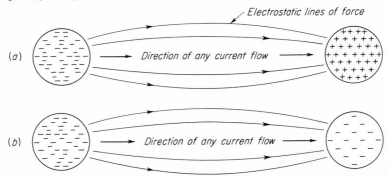

a resistance of one ohm. (The *ohm* is the unit of measurement of resistance, to be discussed in Sec. 1-12.)

An emf can be produced in many ways. Some of the more common methods, with examples of each, are:

1. Chemical (batteries)
2. Electromagnetic (generators)
3. Thermal (heating the junction of dissimilar metals)
4. Piezoelectric (mechanical vibration of certain crystals)
5. Magnetostriction (filters and special energy changers called transducers)
6. Static (laboratory static-electricity generators)
7. Photoelectric (light-sensitive cells)
8. Magnetohydrodynamic (MHD, a process that converts hot gas directly to electric power)

1-9 THE BATTERY IN A CIRCUIT

In the explanations thus far, "objects," either positively or negatively charged, have been used. A common method of producing an emf is by the chemical action in a battery. Without going into the chemical reactions that take place inside a cell, a brief outline of the operation of a battery is given here. (For a more detailed explanation, see Chap. 22.)

Consider a flashlight *cell*. Such a cell (two or more cells form a battery) is composed of a zinc can, a carbon rod down the middle of the cell, and a black, damp, pastelike *electrolyte* between them (Fig. 1-11). The zinc can is the negative terminal. The carbon rod is the positive terminal. The active chemicals in such a cell are the zinc and the electrolyte.

The materials in the cell are selected of such substances that electrons are pulled from the

Fig. 1-11 A common dry cell.

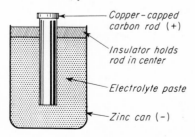

Copper-capped carbon rod (+)

Insulator holds rod in center

Electrolyte paste

Zinc can (-)

outer orbits of the molecules or atoms of the carbon terminal chemically by the electrolyte and are deposited on the zinc can. This leaves the carbon positively charged and the zinc negatively charged. The number of electrons that move is dependent upon the types of chemicals used and the relative areas of the zinc and carbon electrodes. If the cell is not connected to an electric circuit, the chemicals can pull a certain number of electrons from the rod over to the zinc. The massing of these electrons on the zinc produces a *backward* pressure of electrons, or an electric strain, equal to the chemical energy of the cell, and no more electrons can move across the electrolyte. The cell remains in this static, or stationary, 1.5-V charged condition until it is connected to some electric load.

If a wire is connected between the positive and negative terminals of the cell, the 1.5 V of emf starts a current of electrons flowing through the wire. The electrons flowing through the wire start to fill up the deficient outer orbits of the molecules of the positive rod. The electron movement away from the zinc into the wire begins to neutralize the charge of the cell. The electron pressure built up on the zinc, which held the chemical action in check, is decreased. The chemicals of the electrolyte can now force an electron stream from the positive rod through the cell to the zinc, maintaining a current of electrons through the wire and battery as long as the chemicals hold out.

Note that as soon as the wire begins to carry electrons, the electrolyte also has electrons moving through it. This motion produces an equal amount of current through the whole circuit at the same time. This point is a very important one to understand. There are no bunches of electrons moving around an electric circuit like a group of racehorses running around a track. A closed circuit is more like the racetrack with a single lane of automobiles, bumper to bumper. Either all must move at the same time, or none can move.

In an electric circuit, when electrons start flowing in one part, all parts of the circuit can be considered to have the same value of current flowing in them instantly. Most circuits are so short that the energy flow velocity, 300,000,000 meters per second, may be disregarded for the present.

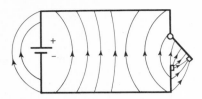

Fig. 1-12 Distribution of electrostatic lines of force across an open circuit. The greatest concentration occurs at the sharp point of the switch, but the field is present everywhere across the circuit.

When a circuit is broken by the opening of a switch, electron progress comes to a sliding halt at all points of the circuit at the same time. In this case the chemicals keep pumping electrons into the wire at the negative terminal until it attains a 1.5-V charge of electrons when compared with the positive terminal. When this occurs, the electron charge, or strain, across the open switch equals the chemical strain produced in the cell and all electron progress in the circuit ceases. The circuit is charged, and electrostatic lines of force are developed as illustrated in Fig. 1-12.

It will be noticed that more lines of force per square inch appear at the sharpest point of the switch. The sharper this point, the more concentrated the lines of force and the more likely it is that a strong field will pull free electrons from the point. Electrons pulled out at a sharp point form a *corona,* or *brush discharge.* In high-voltage circuits, care must be taken to make sure that no sharp points occur to produce such discharges. Several thousand volts is usually required to produce a corona discharge, however.

If electrons leave a sharp point, as described, in such quantities that the air is heated and becomes *ionized,* a spark of electronically heated air will be visible. A spark that holds between two points is called an *arc.*

1-10 IONIZATION

When an atom loses an electron, it lacks a negative charge and is therefore positive. The electronless atom in this condition is a *positive ion.*

In most metals the atoms are constantly losing and regaining free electrons. They may be thought of as constantly undergoing ionization. Because of this, metals are usually good electrical conductors.

Atoms in a gas are not normally ionized to any great extent, and therefore a gas is not a good conductor under low electric pressures. However, if the emf is increased across an area in which gas atoms are present, some of the outer orbiting electrons of the gas atoms will be attracted to the positive terminal of the source of emf and the remainder of the atom will be attracted toward the negative. When pressure increases enough, one or more free electrons may be torn from the atoms. The atoms are ionized. If ionization happens to enough of the atoms in the gas, a current flows through the gas. For any particular gas at any particular pressure, there is a certain voltage value that will produce ionization. Below this value, the number of ionized atoms is small. Above the critical value more atoms are ionized, producing greater current flow, which tends to hold the voltage across the gas at a constant value. In an ionized condition the gas acts as an electric conductor.

Examples of ionization of gases are lightning, neon lights, and fluorescent lights. Ionization plays an important part in electronics and radio.

1-11 TYPES OF CURRENT AND VOLTAGE

Different types of currents and voltages are dealt with in electricity. In this book the following nomenclature will be used:

○ *Direct current (dc).* There is no variation of the amplitude (strength) of the current or voltage. Obtained from batteries, dc generators, and power supplies. (See Fig. 1-13*a.*)
○ *Varying direct current (vdc).* The amplitude of the current or voltage varies but never falls to zero. Found in many transistor and vacuum tube circuits. (See Fig. 1-13*b.*)
○ *Pulsating direct current (pdc).* The amplitude

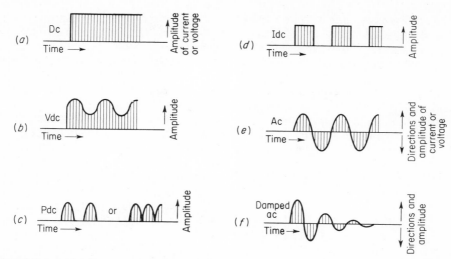

Fig. 1-13 Different forms of voltage or current. (a) Direct current. (b) Varying dc. (c) Pulsating dc. (d) Interrupted dc. (e) Alternating current. (f) Damped ac.

drops to zero periodically. Produced in rectifier circuits. (See Fig. 1-13c.)

- *Interrupted direct current.* Current or voltage starts and stops abruptly (''square wave''). Produced by vibrators, choppers, and special circuits. (See Fig. 1-13d.)
- *Alternating current* (*ac*). Electron flow reverses (alternates) periodically and usually changes amplitude in a more or less regular manner. Produced in ac generators, oscillators, some microphones, and radio in general. This is the usual house current. (See Fig. 1-13e.)
- *Damped ac.* This is alternating current which dies out in amplitude. Produced by spark-type oscillators and inadvertently in many circuits as they make and break. (See Fig. 1-13f.)

1-12 RESISTANCE

It was previously pointed out that certain metals, such as silver and copper, have many free electrons flying aimlessly, at all times, through the spaces between the atoms of the material. Other metals, such as nickel and iron, have fewer free electrons in motion. Still other materials, such as glass, rubber, porcelain, mica, and quartz, have practically no interatom free-electron movement. When an emf is applied across opposite ends of a copper or silver wire, many free electrons progress along the wire and a relatively high current

results. Copper and silver are very good conductors of electric current. When the same emf is applied across an iron wire of equivalent size, only about one-sixth as much current flows. Iron may be considered a fair conductor. When the same emf is applied across a length of rubber or glass, no electron drift results. These materials are insulators. Insulators are used between conductors when it is desired to prevent electric current from flowing between them. (*Semiconductors,* which conduct under certain circumstances, are discussed in Chap. 10.)

Silver is the best conductor, and glass is one of the best insulators. Between these two extremes are found many materials of intermediate conducting ability. While such materials can be catalogued as to their conducting ability, it is more usual to think of them by their resisting ability. Glass (when cold) completely resists the flow of current. Iron resists much less. Silver has the least resistance to current flow.

The resistance a wire or other conducting material will offer to a current depends on four physical factors:

1. The type of material from which it is made (silver, iron, etc.)
2. The length (the longer, the more the resistance)
3. Cross-sectional area of the conductor (the

more area, the more molecules with free electrons, and the less resistance)
4. Temperature (the warmer, the more resistance, except for carbon and other semiconductor materials)

A piece of silver wire of given dimensions will have less resistance than an iron wire of the same dimensions. It is reasonable to assume that if a 1-ft piece of wire has 1 ohm (the unit of measurement of resistance), 2 ft of the same wire will have 2 Ω. (The Greek letter omega, Ω, is the symbol for ohms. The Greek alphabet is shown in Appendix A.)

On the other hand, if a 1-ft piece of wire has 1 Ω, two pieces of this wire placed side by side will offer twice the cross-sectional area, will conduct current twice as well, and therefore will have half as much resistance, as shown in Fig. 1-14b.

Fig. 1-14 (a) Single wire with 1 Ω resistance. (b) Two similar wires together have twice the cross-sectional area and half the resistance. (c) Doubling the diameter quadruples the cross-sectional area and thereby reduces the resistance to one-fourth.

(The *cross-sectional area* is the area seen when a wire is cross-sectioned, or cut in two.) A wire having twice the diameter of another wire will have four times the cross-sectional area (area = $\pi \times$ radius *squared*) and therefore one-fourth the resistance. These relations are illustrated in Fig. 1-14.

A round wire of 0.001-in. diameter is said to have 1 circular mil (abbreviated cmil or cir mil) of cross-sectional area. The word "mil" means one one-thousandth of an inch. A round wire of 2-mil diameter has twice the radius and, by the formula for the area of a circle given above, has four times the cross-sectional area, or 4 cmil. A 3-mil-diameter wire has 9 cmil, and so on. The number of circular mils in any round wire is equal to the number of thousandths of an inch of diameter *squared*.

The number of circular mils is considered when determining how much current a wire may pass safely. When current flows through any wire, heat is produced in the wire. If too much heat is produced, the insulation on the wire may be set on fire or the wire may even melt.

It has been found that a copper wire having a diameter of 64 mil, or $64^2 = 4,096$ cmil, will allow 4.1 A to flow through it in a confined area without overheating. This represents about 1,000 cmil/A. Therefore it may be assumed that any copper wire may carry 1 A for every 1,000 cmil of cross-sectional area. In some applications, when a highly heat-resistant insulation is used, for example, it may be possible to use wire with 500 cmil/A or less. The wire will heat considerably more than it would with 1,000 cmil/A, but it cannot destroy the ruggedized insulation at the temperature that will be developed. A 64-mil wire that carries 4 A safely in confined spaces may carry 15 A or more in free air where it can rapidly dissipate the heat developed in it.

The unit of measurement of resistance is the ohm. For practical purposes 1 Ω may be considered to be the resistance of a round copper wire, 0.001 in. in diameter, 0.88 in. long, at 32° Fahrenheit (32°F). (Metrically, diameter = 2.54 mm, length = 2.235 cm, at 0°C. See Sec. 1-14.)

The *specific resistance* of a conductor is the number of ohms in a 1-ft-long 0.001-in.-diameter round wire of that material. The specific resistance of several common materials at room temperature is listed in Table 1-1.

An aid in recalling the order of resistance of five of the more common materials used as conductors is to remember how they go down the "scail" (misspelling of the word *scale*), where the

Table 1-1

Conductor	Specific resistance, Ω/mil ft
Silver	9.75
Copper	10.55
Aluminum	17.30
Nickel	53.00
Iron	61.00
Lead	115.00
German silver	190.00
Nichrome	660.00

letters of "scail" indicate silver, copper, aluminum, iron, and lead.

Materials such as german silver and nichrome are alloys of two or more metals and are used in the construction of resistors. When wire made of these substances is wound on a tubular ceramic form, the result is a *wire-wound resistor*, as shown in Fig. 1-15. These resistors are usually covered with a hard, vitreous protective coating.

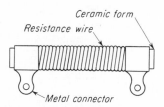

Fig. 1-15   Wire-wound fixed resistor.

A carbon resistor may be made from powdered carbon mixed with a binding material and baked into small, hard tubes with a wire attached to each end. Other carbon resistors consist of a glass or ceramic rod coated with a carbonized layer, which in turn is covered with a ceramic nonconductive coating. A connector wire projects from each end, as shown in Fig. 1-16. The value of the resistance

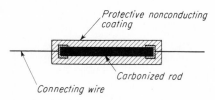

Fig. 1-16   Construction of a fixed-type carbon resistor.

depends on the percent of carbon in the mixtures used. Carbon resistors range in value from a fraction of an ohm to several million ohms. Size has no bearing on resistance value; it affects only the power (heat) or "wattage" rating of the resistor. Common ratings are 0.1, 0.25, 0.5, 1, and 2 W (watts). A 0.5-W resistor is 0.125 in. in diameter and 0.5 in. long. A 1-W resistor is 0.2 in. in diameter and 0.6 in. long. A 2-W resistor is 0.3 in. in diameter and 0.7 in. long. Older carbon resistors were slightly larger. Some newer types are considerably smaller.

The symbol used in diagrams for a *fixed*, or nonvariable, resistor, either wire-wound or carbon, is shown connected across a battery in Fig. 1-17.

Fig. 1-17   Symbol of a resistor across a battery.

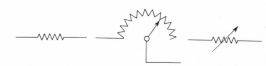

Fig. 1-18   Various symbols used to indicate a rheostat. Note that there are only two terminals on a rheostat.

The symbols used for variable types of resistors, called *rheostats,* are shown in Fig. 1-18. A rheostat has two connections, one to an end of the resistor and the other to a sliding arm that moves along the length of the resistor. Rheostats may be either wire-wound or carbon.

The symbol for a *potentiometer,* which might be considered a rheostat with connections at both ends of the resistance plus a connection to a sliding contact, is shown in Fig. 1-19. Potentiometers

Fig. 1-19   Symbol of a potentiometer.

are used in most cases to select a desired proportion of the total voltage across the potentiometer. They are used as voltage dividers.

An *adjustable* resistor is a wire-wound resistor with a sliding contact on it that can be locked into position when the desired value of resistance is determined experimentally. It can be made in the form of a rheostat but may also take the form of a cylindrical potentiometer, except that its moving contact can be tightened by a machine screw to make it immobile, as shown in Fig. 1-20. Care must be exercised when adjusting these resistors

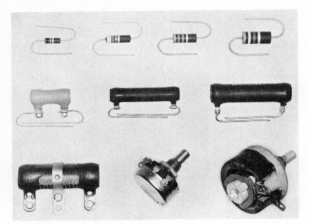

Fig. 1-20 Resistors. Top row, ½-W carbon, 1-W wire-wound, 1-W carbon, and 2-W carbon. Middle row, wire-wound 5-W, 10-W, and 20-W. Bottom row, wire-wound adjustable, carbon-type potentiometer, and wire-wound rheostat in the OFF position.

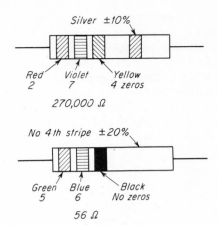

Fig. 1-21 Color-coded resistors.

to make sure that the machine screw is loose enough to allow the movable contact to be moved without damaging the fine resistance wires with which they are often wound. A resistor of this type may be partly covered with a vitreous coating, but part of the wires will be bare to allow the slider to make contact.

1-13 COLOR-CODED RESISTORS

Resistors and other components are often color-coded rather than having their resistance values printed on them. It is necessary that technicians know this code. Two methods have been used to color-code resistors. One is the three-stripe method; the other, the body-end-dot method. The body-end-dot resistors are no longer manufactured, but older equipment may contain some of them. In both methods the same color code is used. The colors and their meanings are:

Brown	1	Blue	6
Red	2	Violet	7
Orange	3	Gray	8
Yellow	4	White	9
Green	5	Black	0

To read the resistance of a color-coded resistor, as shown in Fig. 1-21, start with the stripe nearest

the end of the resistor. The first stripe is the first number. The second stripe is the second number. The third stripe, the 10 multiplier, is the number of zeros following the second number.

Resistors with values of less than 10 Ω have a multiplier, or third stripe, of gold or silver. Gold indicates that the first two numbers are to be multiplied by 0.1. (Yellow, violet, gold = 4.7 Ω.) A silver third stripe indicates multiplication by 0.01. (Green, blue, silver = 0.56 Ω.)

Three-stripe resistors have a resistance tolerance of 20%. If a silver fourth stripe is added, it indicates the resistor has a tolerance of 10%. A gold fourth stripe represents a tolerance of 5%. The ohmic values are often printed on 1% and 2% resistors.

When resistors have a double-width *first* stripe, the resistor is wire-wound rather than carbon. One is shown in Fig. 1-20.

The body-end-dot coding is read in that order. If no dot is visible, the dot and body number are assumed to be the same. Figure 1-22 illustrates a body-end-dot resistor.

Fig. 1-22 Old-fashioned body-end-dot (BED) resistor.

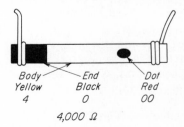

Although resistors can be specially ordered in any values, the values considered as off-the-shelf standards are shown in Appendix B.

Test your understanding; answer these checkup questions.

1. What must exist when an emf is produced between two points? _____
2. Electric current flows in what direction between a positive and a negative body? _____ Greatly positive and lesser positive bodies? _____ Greatly negative and lesser negative? _____
3. In what unit is emf measured? _____ Current? _____ Resistance? _____
4. What may develop at a sharp point in high-voltage circuits? _____
5. If an atom loses an electron, what is it then called? _____
6. Under what condition is a gas a good conductor? _____
7. List the six types of current discussed. _____ _____ _____ _____ _____ _____
8. List five common materials in order of conductivity. _____ _____ _____ _____ _____
9. Of what materials are resistors commonly made? _____
10. Does the physical size of a resistor represent power-dissipating capabilities or resistance value? _____
11. How many contacts on a carbon resistor? _____ A rheostat? _____ A potentiometer? _____ An adjustable resistor? _____
12. What resistor color coding is no longer used? _____
13. On a color-coded resistor, what does the first stripe indicate? _____ The second? _____ The third? _____ The fourth? _____
14. If the first stripe on a color-coded resistor is double width, what does this indicate? _____
15. A 40-Ω resistance is desired. What is the closest off-the-shelf value with a tolerance of 1%? _____ 5%? _____ 10%? _____
16. What is the value in ohms of a yellow-violet-brown-silver striped resistor? _____ A green-gray-gold striped resistor? _____

1-14 THE METRIC SYSTEM

Scientific measurements are more and more being given in the metric system. It is a multiple-of-10 system (similar to the United States dollar-dime-cent-mil monetary system). The basic metric units of measurement are:

○ Length: *Meter*. Approximately 39.37 in. All other units of length are multiples or submultiples of 10 of the meter.
○ Volume: *Liter*. A centimeter (cm) is $\frac{1}{100}$ meter (0.3937 in.). A cubic centimeter (cm³) is a cube with all sides 1 cm long. 1,000 cm³ is a liter (1.06 qt).
○ Weight: *Gram*. The weight of 1 cm³ of water at 4° Celsius or centigrade (4°C), or 0.353 oz, avoirdupois weight.

Volts, amperes, ohms, etc., may also use metric-based prefixes. The prefixes in general use in electronic and radio work are listed in Table 1-2. Some examples of the use of these prefixes are:

○ pF (picofarad): A trillionth of a farad, or 1 pF or 1 $\mu\mu$F
○ ns (nanosecond): A billionth of a second, or 1 nsec or 1 mμsec
○ μV (microvolt): A millionth of a volt, or 1 μV
○ mA (milliampere): $\frac{1}{1,000}$ A (.001 A)
○ mm (millimeter): $\frac{1}{1,000}$ m (.001 m)
○ cm (centimeter): $\frac{1}{100}$ m (.01 m)
○ dB (decibel): $\frac{1}{10}$ B (bel) (0.1 B)
○ kW (kilowatt): 1,000 W
○ km (kilometer): 1,000 m, 0.64 mile
○ MHz (megahertz): 1,000,000 Hz, also known as 1,000,000 cps (cycles per second)
○ GHz (Gigahertz): 1,000,000,000 Hz

It is often necessary to convert one metric-prefixed value to another. A method of doing this

Table 1-2 *METRIC-BASED PREFIXES AND POWERS OF 10*

Atto (a)	=	quintillionth of	=	10^{-18} times
Femto (f)	=	quadrillionth of	=	10^{-15} times
Pico (p), or $\mu\mu$	=	trillionth of	=	10^{-12} times
Nano (n), or mμ	=	billionth of	=	10^{-9} times
Micro (μ)	=	millionth of	=	10^{-6} times
Milli (m)	=	thousandth of	=	10^{-3} times
Centi (c)	=	hundredth of	=	10^{-2} times
Deci (d)	=	tenth of	=	10^{-1} times
		unity	=	$10^{0} = 1$
Deka (da)	=	ten times	=	10 times
Hecto (h)	=	hundred times	=	10^{2} times
Kilo (k)	=	thousand times	=	10^{3} times
Mega (M)	=	million times	=	10^{6} times
Giga (G), or kM	=	billion times	=	10^{9} times
Tera (T)	=	trillion times	=	10^{12} times

is first to arrange the two values in ratio form, larger to smaller. Then multiply this ratio by the powers-of-10 difference between the two values.

EXAMPLE: How many hertz are there in 5.23 kHz? If kilo = 10^3, the ratio of the two values would be 5.23 × 10^3 to 1, which is 5,230 Hz.

EXAMPLE: How many microvolts are there in 2.4 kV? If kilo = 10^3 and micro = 10^{-6}, in ratio form the problem would read 2.4 × 10^3 to 1 × 10^{-6}. The difference between 10^3 and 10^{-6} is 10^9. Therefore, there are 2.4 × 10^9, or 2,400,000,000 μV in 2.4 kV.

ANSWERS TO CHECKUP QUIZ ON PAGE 15

1. (*Electrostatic field*) 2. (– to +)(*Lesser + to greater +*)(*Greater – to lesser –*) 3. (*Volt*)(*Ampere*)(*Ohm*) 4. (*Corona or brush discharge*) 5. (*Ion*) 6. (*When ionized*) 7. (*Dc, vdc, pdc, idc, ac, damped ac*) 8. (*Silver, copper, aluminum, iron, lead*) 9. (*Carbon, wire*) 10. (*Power-dissipating capabilities*) 11. (*2*)(*2*)(*3*)(*3 usually*) 12. (*Body-end-dot*) 13. (*First number*)(*Second number*)(*Multiplier or number of zeros*)(*% tolerance*) 14. (*Wire-wound resistor*) 15. (*40.2*)(*39*)(*39*) 16. (*470 10%*)(*5.8*)

1-15 WIRE SIZES

Most wire used in electronics and radio is made of copper. It may be either *hard-drawn* (stiff) or *soft-drawn* (pliable). It is manufactured in various sizes, with or without an insulation coating on the wire. Some of the insulating materials used are enamel, silk, glass fibers, cotton, fiber, rubber, varnish, and various plastics.

Table 1-3 lists some of the more commonly used copper wire sizes and information regarding them. (AWG stands for American Wire Gauge.)

1-16 MAKING LOW-RESISTANCE CONNECTIONS

The addition of resistance to an electric circuit reduces the amount of current that can flow. Loose or oxidized (rusted) electrical connections may act as resistances and often result in improper operation of a circuit. When equipment is constructed or repaired, the technician must make sure that all connections are tight and have low resistance.

When a wire is scraped clean and looped around a machine screw and a nut is tightened

Table 1-3 *COPPER WIRE TABLE*

AWG* No.	Diameter, mils	Ω/1,000 ft, room temperature	Current-carrying capacity at 1,000 cmil/A, as in transformers*	Approximate current for open wiring, rubber-insulated
0	325	0.1	90 A	125 A
8	128	0.641	16.5 A	35 A
10	102	1.02	10.4 A	25 A
12	81	1.62	6.5 A	20 A
14	64	2.58	4.1 A	15 A
16	51	4.09	2.6 A	
18	40	6.51	1.6 A	
20	32	10.4	1.0 A	
22	25	16.5	640 mA*	
24	20.1	26.2	400 mA	
26	15.9	41.6	250 mA	
28	12.6	66.2	160 mA	
30	10.0	105	100 mA	
32	7.95	167	63 mA	
34	6.3	265	40 mA	

*Cmils means circular mils; mA means milliamperes; AWG means American Wire Gauge

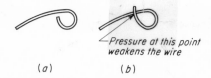

Fig. 1-23 (a) Properly looped wire to fit around a machine screw and under a nut. (b) Improperly looped wire.

down on it, a good, low-resistance electrical connection is made between the wire and the machine screw. Greater nut pressure produces more contacting surface between wire and nut and less resistance. The looped end should be fitted snugly around the screw (Fig. 1-23a) in such a direction that the tightening of the nut tends to close the loop rather than open it. The wire should not be overlapped on itself (Fig. 1-23b), as pressure exerted by the nut may squeeze the wire at the point of overlap to half thickness and weaken the joint.

Two wires may be scraped clean and then twisted tightly together, as in a Western Union splice (Fig. 1-24a). However, over a period of time

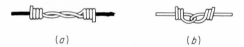

Fig. 1-24 (a) Properly made splice between two ends of wire. (b) Very poor method of connecting two wires.

such wires may corrode and a resistance joint can result. To prevent this and to give added strength, the joint may be covered with a protective layer of solder.

Solder is an alloy of tin and lead having the capability of melting at a relatively low temperature. All solders, when going from solid to liquid, change into an intermediate *plastic* state at a temperature of about 360°F. The type known as 60/40 solder, which is 60% tin and 40% lead,

becomes liquid at about 370°F. A less desirable 50/50 solder requires a temperature of about 420° before it changes from plastic to liquid and flows freely.

For electronics work solder is made up into thin, hollow wires. The hollow space is filled with rosin or some other nonacidic *flux*. The job of the flux is to melt before the solder and flow over the connection to be soldered. The hot flux reacts with the metal, cleaning it somewhat, which allows the hot solder to adhere more firmly to it. When cold, the rosin loses its ability to react with the metal and thereafter remains inert. Although acid-core solder and flux are available, they should never be used in electrical work. Acid solder joints may produce an undesirable voltage-generating cell between solder, acid, and metal that can cause noise in electronic equipment, higher resistance, and a weakening of the soldered joint. Acid flux is useful when soldering sheet metals together.

When two wires are to be soldered together, (1) the surfaces of both wires must be clean, (2) the two wires should make good mechanical contact, and (3) both wires must be heated before solder is applied. The hot soldering iron should be applied so that both wires receive heat simultaneously. Solder and flux are then held against the point where the two wires touch. As soon as hot solder flows evenly over the hot metal surfaces, the iron and solder are removed. The joint must be held motionless until the molten solder passes from liquid through plastic to solid form; otherwise, the solder may crystallize and lose its holding ability.

There are many types of soldering irons. A common type is shown in Fig. 1-25. For small, transistor-type work a 35-W iron may be satisfactory. For larger work a 50- to 100-W soldering iron or soldering gun may be more desirable. For heavy-duty work on large wires or in the open air,

Fig. 1-25 A soldering iron.

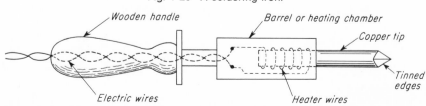

a 250-W iron may be required. It is imperative that the tip of a soldering iron be tinned. A tinned surface has a layer of fresh solder applied to a cleaned portion of the iron. The tinned area can be recognized by its shiny surface in comparison with the brown, oxidized copper of an untinned area. A tinned surface will transfer heat many times faster than an oxidized surface of a hot iron.

Solder that has been heated and cooled several times at a connection oxidizes and crystallizes. It no longer holds properly. If the solder is in this condition, it should be melted from the joint and new solder substituted.

There are other modern means of joining wires to special connector lugs in electric circuits. Crimped connections are produced with a pressure tool that physically crimps the wire and lug together. There is also a special wire-connecting tool that twists the wire around a square terminal lug so tightly that the connection is considered equal to a soldered joint and may be superior if the joint is subject to vibration.

1-17 PRINTED CIRCUITS

All methods of producing a printed circuit (PC) board result in narrow copper-strip conductors laid down on thin insulating sheets of a phenolic or Fiberglas board. The components, or parts, to be connected are then soldered to the copper strips.

One method of producing a PC board is to start with a small oblong of $\frac{1}{16}$-in.-thick phenolic insulator sheet with a thin copper plating on one side. The desired circuit wires are drawn on the copper with a special *resist* ink. The board is then immersed in a ferric chloride solution until all the copper except that under the resist ink has been etched (eaten away) chemically. After this the board is washed and the resist ink removed, leaving a series of copper conductor lines on the phenolic sheet. These may then be silver-plated or plated with solder. Holes are drilled through the copper strips and base. Connector wires from the required electronic components are then slipped into the holes and soldered to the copper lines.

Each connection can be soldered separately, which is rather slow. If all the components are fed through the holes from the insulated side of the board, all the connections can be *wave-soldered* at once by moving the copper-connection surface of the PC board over a wave of molten solder. The wave is developed over a cylinder rotating just under the surface of a small tub full of molten solder.

One of the difficulties with printed circuits stems from the difference of expansion of the copper and the phenolic boards. Temperature changes can loosen the copper strips from the phenolic backing, or hairline cracks can open up across the strips, breaking the electrical connections. This may occur during the original soldering, if heated during operation, or during unsoldering to remove and replace faulty components. Bending the boards may develop similar troubles. While resistance meters (ohmmeters) can be used to determine if a circuit has broken open, visual examination with a powerful magnifying glass will often be simpler.

When components are removed from a printed circuit, a very small soldering iron must be used to prevent overheating of the PC board. A pulling tension is applied to one end of the lead of the component being removed while a 25- to 35-W soldering iron tip is applied to the soldered contact. As the solder melts, the component wire pulls free of the PC board hole. If the hole remains plugged with solder, it may be heated and a stainless-steel wire, sharp lead pencil, or other thin device to which solder will not adhere can be run into the hole to clear it. A far better method is to use a special small soldering iron called a *desoldering tool* with a hollow tip to which is coupled a rubber bulb. The bulb is squeezed, forcing air out the hollow tip. The hot tip is then placed around the wire to be desoldered until the solder melts. When the bulb is released, solder is sucked up into the hollow iron, freeing the component lead of solder and allowing the lead to be extracted. Squeezing the bulb again forces the molten solder out of the hollow tip, readying the iron for use again.

One commercial process of producing printed circuit boards is *photoetching*. The original circuit is drawn on paper and photographed. The negative is then used to transfer, by photographic methods, an etch-resistant image of the wiring onto a photographically surfaced copper-clad phenolic board. The rest of the method is as described above. Other methods used are stenciling, chemical deposition, and vacuum distillation.

Test your understanding; answer these checkup questions.

1. How many hertz in a kilohertz? _____ In a megahertz? _____
2. How many milliamperes in 2.45 A? _____ In 0.0358 A? _____
3. How many microamperes in 29 mA? _____ In 7.2 A? _____
4. How many amperes in 450 mA? _____ In 75 μA? _____
5. How many milliwatts in 850 μW? _____ In 5.6 W? _____
6. From Table 1-3, how many ohms does 100 ft of No. 10 copper wire have? _____ No. 20? _____

7. Using 1,000 cmil/A, what size wire is required for a 4-A flow? _____
8. Using 400 cmil/A, what size wire is required for a 1-A flow? _____ A 4-A flow? _____
9. At what temperature does a good solder melt? _____
10. What is the reason for soldering a wire splice? _____
11. What is the duty of solder in PC board connections? _____
12. What is the duty of solder fluxes? _____
13. With what may circuit connections be drawn on PC boards before they are chemically treated? _____
14. What wattage iron would be used on an antenna wire outdoors? _____ On a PC board? _____ On indoor electrical wiring? _____
15. What is the coating of solder on the tip of a hot soldering iron called? _____
16. What method of soldering can make a hundred soldered connections at once on PC boards? _____
17. What is the best device to use to remove solder from PC board connections? _____

COMMERCIAL LICENSE QUESTIONS

Amateur license questions will be found in the Addendum.

An ability to answer questions similar to the following ones is required for FCC Elements 3, 4, and 6. A question followed by a bracketed number is required for that element alone. Sections in which questions are answered are shown in parentheses.

1. What is an *electron?* (1-2)
2. Explain the relationship between the physical structure of the atom and electric-current flow. (1-3, 1-7)
3. In what manner does the resistance of a copper conductor vary with variations in temperature? (1-3, 1-12)
4. Define the term *coulomb.* (1-7)
5. By what other expression may *electric-current flow* be described? (1-7)
6. By what other expression may a *difference of potential,* or *electromotive force,* be described? (1-8)
7. What is an *ion?* (1-10)
8. What is the unit of resistance? (1-12)
9. With respect to electrons, what is the difference between conductors and nonconductors? (1-12)
10. If the diameter of a conductor of given length is doubled, how will the resistance be affected? (1-12)
11. Explain the factors which affect the resistance of a conductor. (1-12)

12. What is the relationship between wire size and resistance of the wire? (1-12)
13. Name four conducting materials in their order of conductivity. (1-12)
14. What would be the value and tolerance of a resistor if color-coded, from left to right, green, brown, yellow, silver? (1-13)
15. What would be the value and tolerance of a resistor if color-coded, from left to right, red, black, orange, gold? (1-13)
16. Explain the meaning of *kilo, micro, mega, pico, micromicro, nano.* (1-14)
17. Make the following transformations: (*a*) kilohertz to hertz, (*b*) kilovolts to volts, (*c*) milliamperes to amperes. (1-14)
18. List at least two essentials for a good soldered connection. (1-16)
19. List three precautions which should be taken in soldering electrical connections to assure a permanent junction. (1-16)
20. Why is rosin used as soldering flux in radio construction work? (1-16)
21. Discuss etched-wiring printed circuits with respect to determination of wiring breaks, excessive heating, and removal or installation of components. (1-17)

NOTE: FCC license test questions are of the multiple-choice type, taking a form somewhat as follows:

22. In the metric system milli means (*a*) millionth; (*b*) million times; (*c*) thousandth; (*d*) hundredth. (1-14)
23. A component normally having three leads is a (*a*) rheostat; (*b*) carbon resistor; (*c*) dry cell; (*d*) potentiometer. (1-12)

24. The material with the highest specific resistivity is (*a*) nichrome; (*b*) copper; (*c*) iron; (*d*) silver. (1-12)

ANSWERS TO CHECKUP QUIZ ON PAGE 19

1. (*1,000*)(*1,000,000*)   **2.** (*2,450*)(*35.8*)   **3.** (*29,000*)(*7,200,000*)   **4.** (*0.45*)(*0.000075*)   **5.** (*0.85*)(*5,600*) **6.** (*0.102*)(*1.04*) **7.** (*Number 14*) **8.** (*Number 24*)(*Number 18*) **9.** (*360°F*) **10.** (*Hold wires and protect from corrosion*) **11.** (*Hold components and carry current*) **12.** (*Clean surfaces to be soldered*) **13.** (*Resist inks, adhesive tapes*) **14.** (*250 W*)(*25–35 W*)(*50–100 W*) **15.** (*Tinned surface*) **16.** (*Wave*) **17.** (*Desoldering tool*)

2
DIRECT-CURRENT CIRCUITS

The objective of this chapter is to work with Ohm's law as it applies to series, parallel, and series-parallel resistor circuits, as well as to use power formulas in such circuits.

2-1 OHM'S LAW

Wherever electric circuits are in use, whether in a simple flashlight, in motors or generators, or in radio and television circuits, voltage, current, and resistance are present. It is interesting to see how the theory of more complex circuit operation unfolds by starting with a simple circuit and slowly adding one step to another. To the beginner, each step may appear understandable enough, but remembering it and, more important, learning when to apply it are the secret to success in the study of electric circuitry. Once the reader comprehends something of the physical nature of current, voltage, and resistance, he is ready to use this knowledge to learn when, where, how, and most of all why it may be applied to electric circuits.

It was previously explained that a change in current can be produced by changing either the voltage or the resistance in the circuit. An increase in voltage will produce an increase in current. Therefore, voltage and current are *directly* proportional to each other.

An increase in resistance in a circuit produces a *decrease* in current. Therefore, resistance and current are *inversely* proportional to each other.

These two facts can be condensed into one statement, known as Ohm's law: *Current varies directly as the voltage and inversely as the resistance.*

Ohm's law is a simple statement of the functioning of an electric circuit. It can be expressed mathematically as

$$I = \frac{E}{R}$$

where I = intensity of current (in amperes, A)
E = emf (in volts, V)
R = resistance (in ohms, Ω)

Ohm's law might also be expressed as

$$\text{Amperes} = \frac{\text{volts}}{\text{ohms}}$$

By multiplying both sides of the equation $I = E/R$ by R, the Ohm's-law formula becomes

$$E = IR$$

By dividing both sides of this last equation by I, the Ohm's-law formula becomes

$$R = \frac{E}{I}$$

These three variations of the Ohm's-law formula make it possible to determine (1) the current value if the voltage and resistance are known, (2) the voltage in the circuit if the current and resistance are known, (3) how much resistance is in the circuit if the voltage and current are known.

An understanding of this law and an ability to use it are quite important. Electronic job test questions for civil service work, FCC licenses, etc., are certain to involve the application of Ohm's law in several different ways.

The ability to rearrange formulas as shown above is important. If not familiar with the indicated divisions and multiplications, pay particular attention to the section dealing with the fundamentals of this type of mathematics (Sec. 2-3). If you are unable to comprehend the briefly outlined steps, you should study a basic algebra book.

The so-called magic triangle or magic circle, shown in Fig. 2-1, may help in learning Ohm's

Fig. 2-1 Aids for remembering the three Ohm's-law formulas.

law. If the symbol for the unknown is covered, the mathematical method of solving for this letter is shown by the position of the other two symbols. For example, cover the I; it is necessary to divide E by R. Cover the E; it is necessary to multiply I by R. Cover the R; it is necessary to divide E by I.

2-2 USING OHM'S LAW

The following examples illustrate the use of Ohm's-law formulas in determining the functioning of simple electric circuits:

In the circuit of Fig. 2-2, if the voltage E of the

Fig. 2-2 Current in a circuit is directly proportional to emf and inversely proportional to resistance.

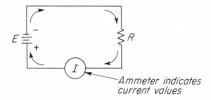

E

R

I

Ammeter indicates current values

battery is 10 V and the load resistance R is 20 Ω, what is the value of the current I in the circuit?
Solution: Using Ohm's law,

$$I = \frac{E}{R} = \frac{10}{20} = 0.5 \text{ A}$$

In the same type of circuit, if the ammeter reads 4 A and the resistance is known to be 30 Ω, what value of emf must the source have?
Solution:

$$E = IR = 4(30) = 120 \text{ V}$$

In the same type of circuit, the ammeter reads 3 A and the source voltage is known to be 150 V. What is the load-resistance value?
Solution:

$$R = \frac{E}{I} = \frac{150}{3} = 50 \text{ Ω}$$

An electric shock of more than 15 mA (15 milliamperes or 0.015 A) flowing through the body is considered dangerous to human life. What current will flow through a person having body contact resistance of 2,200 Ω across 110 V?

$$I = \frac{E}{R} = \frac{110}{2,200} = 0.05 \text{ A or 50 mA}$$

Test your understanding; answer these checkup questions.

1. What is the basic formula for Ohm's law? _____ What are the two derivations of this formula? _____
2. What current flows in a circuit with 100 V and 1,000 Ω resistance? _____
3. What voltage is required to produce 2 A of current through 60 Ω? _____
4. What resistance will limit current flow to 4 A in a circuit having a 200-V supply? _____
5. A lamp connected across 120 V is found to have 3 A flowing through it. What is its resistance? _____
6. A relay coil having 315 Ω is across 6.3 V. What current value will it draw? _____
7. A 5,000-Ω resistor in a receiver has 5 mA flowing through it. How much voltage drop is developed across it? _____
8. The resistance of a circuit remains the same, but the current through the resistor suddenly triples. What has happened to the circuit voltage? _____

9. If the voltage applied to a circuit is doubled but the resistance remains unchanged, what will the current value do? _____

10. If R triples and E doubles, what will be the new current value? _____

2-3 MATHEMATICS FOR OHM'S-LAW PROBLEMS

This is not intended to be a textbook on mathematics. It may be well, however, to point out some basic mathematical operations that can be used in working electrical problems involving formulas similar to Ohm's law. These operations are a form of simple algebra. Only a few will be given. You should undertake an outside study of algebra if you are weak in this subject.

When you work with algebraic formulas, such as the Ohm's-law equation $I = E/R$, there are certain operations to remember:

1. The sign for addition is $+$; the sign for subtraction is $-$; the sign for multiplication is $\times$ or parentheses $(\ )$.

$$+2 + 2 = +4$$
$$-3 - 4 = -7$$
$$+5 - 2 = +3$$
$$+3 - 5 = -2$$

$$3 \text{ times } 4 = 3 \times 4 = (3)(4) = 3(4)$$
$$(A)(B) = A(B) = AB$$

2. Any number (or letter) multiplied by 1 is unchanged, and therefore the 1 may be dropped.

$$1(3) = 3 \qquad 1(A) = 1A = A$$

3. Any number (or letter) divided by 1 is equal to the number (or letter), and therefore the 1 may be dropped.

$$\frac{4}{1} = 4 \qquad \frac{A}{1} = A$$

4. Any number (or letter) when multiplied by itself is equal to the number (or letter) squared.

$$2 \times 2 = 2^2 \qquad E(E) = E^2 \qquad EE = E^2$$

5. Any number (or letter) divided by itself is equal to 1.

$$\frac{4}{4} = 1 \qquad \frac{F}{F} = 1 \qquad \frac{XQZ}{ZXQ} = 1$$

6. If both sides of an equation are multiplied by the same number (or letter), equality still holds. For example,

$$2 = 2$$

If multiplied by 4:

$$4(2) = 4(2) \qquad \text{or} \qquad 8 = 8$$

Another example:

$$6 = 2(3)$$

If multiplied by 4:

$$4(6) = 4[2(3)] \qquad \text{or} \qquad 24 = 24$$

With letters:

$$X = AC$$

If multiplied by B:

$$BX = BAC$$

7. If both sides of an equation are divided by the same number (or letter), the equation will still be correct. For example,

$$6 = 2(3)$$

When divided by 2:

$$\frac{6}{2} = \frac{2(3)}{2} \qquad \text{or} \qquad 3 = 3$$

With letters:

$$E = IR$$

When divided by R:

$$\frac{E}{R} = \frac{IR}{R} \qquad \text{or} \qquad \frac{E}{R} = I$$

8. If both sides of an equation are squared, the equation will still be correct.

$$6 = 2(3) \qquad 6^2 = [2(3)]^2$$
$$6^2 = 2^2 3^2 \qquad 36 = 36$$

With letters:
$$A = BC \qquad A^2 = (BC)^2$$
$$A^2 = B^2 C^2$$

9. If the square root is taken of both sides of an equation, the equation will still be correct.

$$16 = 2(8) \quad \sqrt{16} = \sqrt{2(8)} \quad \text{or} \quad 4 = 4$$
$$A = BC \qquad \sqrt{A} = \sqrt{BC}$$
$$4^2 = 2(8) \quad \sqrt{4^2} = \sqrt{2(8)} \quad \text{or} \quad 4 = 4$$
$$E^2 = PR \quad \sqrt{E^2} = \sqrt{PR} \quad \text{or} \quad E = \sqrt{PR}$$

10. A negative number on one side of the equation becomes a positive number when moved to the other side. (No sign in front of a number or letter indicates it is positive.)

$$7 + 2 = 9$$
$$7 = 9 - 2$$

$$B + C = A$$
$$B = A - C$$

$$B - C = A$$
$$B = A + C$$

When rearranging formulas involving simple fractions, a first step may be to *cross-multiply*. This means to multiply the top of the fraction on one side of the equation by the lower number of the fraction on the opposite side of the equals sign. The same is done to the other halves of the fractions. These two answers are set as equal to each other.

$$4 = \frac{8}{2} \qquad \text{may be written} \qquad \frac{4}{1} = \frac{8}{2}$$

When cross-multiplied,

$$\frac{4}{1} \diagdown \diagup \frac{8}{2}$$

becomes $\quad 2(4) = 1(8) \quad$ or $\quad 8 = 8$

With a letter formula,

$$I = \frac{E}{R} \qquad \text{or} \qquad \frac{I}{1} = \frac{E}{R}$$

When cross-multiplied,

$$\frac{I}{1} \diagdown \diagup \frac{E}{R}$$

becomes $\quad 1(E) = I(R) \quad$ or $\quad E = IR$

By cross-multiplying it has been determined what E equals. A further possible step is to determine what R equals. To do this, divide out the unwanted letters on one side of the equation, leaving only the desired letter. For example,

$$E = IR$$

To find what R equals, divide out the I from both sides:

$$\frac{E}{I} = \frac{IR}{I}$$

The I's on the right-hand side cancel each other, leaving

$$\frac{E}{I} = R \qquad \text{or} \qquad R = \frac{E}{I}$$

These two operations, cross-multiplying and dividing out the unwanted from one side, can be used in a surprising number of electrical problems. (If trained in mathematics, you may know other methods.)

2-4 FINDING THE SQUARE ROOT

To determine the answer to many electrical problems it is necessary to find the square root of some number. Often this can be done by referring to a mathematical table of square roots,

---

ANSWERS TO CHECKUP QUIZ ON PAGE 22

1. ($I = E/R$)($R = E/I$, $\quad E = IR$)   2. (0.1 A $\quad$ or $\quad$ 100 mA)
3. (120 V)   4. (50 Ω)   5. (40 Ω)   6. (0.02 A or 20 mA)
7. (25 V)   8. (It has tripled)   9. (Double)   10. (Two-thirds the original)

by using a slide rule or a pocket calculator, or by using logarithms. If none of these are handy, it will be necessary to use the long-division-like method of finding square roots.

The square root of a given number is the number which when multiplied by itself equals the given number. The square root of 9 is 3; the square root of 16 is 4; the square root of 100 is 10; and so on. With many simple numbers such as these, the square root may be easily determined. But what is the square root of a number like 2,168? To find the square root of such a number, it should first be written out in groups of *two numbers,* starting at the decimal point.

2,168 is written as 21 68 . 00 00

Next, a line is drawn above the number.

$$\overline{21\ 68\,.\,00\ 00}$$

Above the first group of two numbers (21) a number is placed which, when multiplied by itself, will come close to equaling, but will not exceed, the number 21. This number is 4.

$$\frac{4}{21\ 68\,.\,00\ 00}$$

The number 4 multiplied by itself is 16. The 16 is placed below the 21, and the difference is indicated below the 16, as in long division.

$$\begin{array}{r} 4 \\ \hline 21\ 68\,.\,00\ 00 \\ 16 \\ \hline 5 \end{array}$$

The next two numbers, 68, are brought down next to the 5.

$$\begin{array}{r} 4 \\ \hline 21\ 68\,.\,00\ 00 \\ 16 \\ \hline 5\ 68 \end{array}$$

As the next step, the number above the main line, 4 in this problem, is doubled to 8 and placed in front of the 5 68, as shown. (A space is left for another number after the 8.)

$$\begin{array}{r} 4 \\ \hline 21\ 68\,.\,00\ 00 \\ 16 \\ 8\ {\Large /}\ 5\ 68 \end{array}$$

Above the main line and over the second group of two numbers, the 68 in this case, another number is now required. This number, when multiplied by itself with the 8 in front of it, must be equal to or slightly less than the figure 568 in the problem. By trial, the largest possible number is found to be 6. The 6 is placed above the line and also after the 8. The 86 thus produced is then multiplied by the 6, and the problem now reads

$$\begin{array}{r} 4\quad 6\,. \\ \hline 21\ 68\,.\,00\ 00 \\ 16 \\ 86\ {\Large /}\ 5\ 68 \\ 5\ 16 \\ \hline 52 \end{array}$$

The difference between 568 and 516 is determined, and the next group of two numbers, 00, is brought down. As before, the number above the line, 46, is doubled to 92 and placed in front of the 5,200.

$$\begin{array}{r} 4\quad 6\,. \\ \hline 21\ 68\,.\,00\ 00 \\ 16 \\ 86\ {\Large /}\ 5\ 68\,. \\ 5\ 16\,. \\ 92\ {\Large /}\ 52\ 00 \end{array}$$

Above the main line and above the next group of two numbers, 00, another number is required. The 92 in front of this number forms a group which when multiplied by this number must be equal to or less than 5,200. By trial the number is found to be 5. The 5 is placed above the line and also after the 92. The 925 is now multiplied by 5.

```
        4   6 . 5
   21 68 . 00  00
   16
86/  5 68
     5 16
925/    52 00
        46 25
         5 75 00
```

```
        3   1   2.        Proof: 312² = 97,344
   09 73 44.
    9
61/ 0 73
    61
622/ 12 44
     12 44
```

The number above the line, 46.5, is doubled and brought down as before. The next required number, 6, is placed above the next group of double numbers, and also placed after the 930, and multiplied.

```
        4   6 . 5   6
   21 68 . 00  00
   16
86/  5 68
     5 16
925/    52 00
        46 25
9306/    5 75 00
         5 58 36
           16 64
```

The square root of 2,168 is 46.56. For proof, the answer is multiplied by itself. It should equal the original number if the answer is carried out far enough; that is, 46.56² should equal 2,168. However, since there was a remainder in the example problem, the 46.56 when squared is only 2,167.8336, which is close enough to 2,168 for most purposes. (In most training work, figures need be correct only to the third significant figure. Thus, 2,168 might well be rounded off to 2,170.)

The important points when working square roots are: First, start marking off in double numbers from the decimal point. To find the square root of 325, it is marked off as 03 25, and not 32 50. Second, when the numbers to be multiplied are brought down from one step to the next, the number above the line is always doubled. Third, the multiplying number must be added after the last-mentioned doubled number.

Take another example. What is the square root of 97,344?

Test your understanding; answer these checkup questions.

1. If $P = EI$, then $E = ?$ _____ $I = ?$ _____
2. If $Q = X/R$, then $X = ?$ _____ $R = ?$ _____
3. If $Z = X^2/R$, then $R = ?$ _____ $X = ?$ _____
4. If $FL = 1/FC$, then $1 = ?$ _____ $C = ?$ _____ $L = ?$ _____ $F = ?$ _____
5. If $3 - A = L/X$, then $X = ?$ _____ $L = ?$ _____ $A = ?$ _____ HINT: $3 - A$ either is $+3 - A$ or can be considered as the unit $(3 - A)$.
6. If $2(B - C) = Q/Z$, then $Q = ?$ _____ $Z = ?$ _____ $B = ?$ _____ $C = ?$ _____ HINT: $2(B - C)$ is $+2B - 2C$.
7. What is the square root of 525? _____ Of 10,000? _____
8. What is the square root of 1,000? _____ Of 10? _____
9. What is the square root of 0.05? _____ Of 0.5? _____

2-5 POWER AND ENERGY

Electric pressure, or emf, by itself can do no work. A battery develops an emf, but if no load is connected across it, no current flows and no electrical work is accomplished.

When a conductor is connected across a source of emf, a current of electrons is developed. The current represents movement. The product of the pressure and the movement (volts and amperes) does accomplish work. The unit of measurement of the rate of doing work, or the unit of measurement of power, is the *watt* (W): 1 V causing 1 A to flow in a 1-Ω resistor produces 1 W of power. In formula form,

$$P = EI$$

where P = power (in watts, W)
  E = emf (in volts, V)
  I = current (in amperes, A)

EXAMPLE: What is the power input to a transmitter having a plate voltage of 2,000 V and a plate current of 0.5 A?

$$P = EI = 2,000(0.5) = 1,000 \text{ W}$$

The Ohm's-law formula states: $E = IR$. If this is true, then (IR) can be substituted for E in the power formula:

$$P = EI = (IR)I \qquad \text{or} \qquad P = I^2 R$$

where R = resistance (in ohms, Ω)

EXAMPLE: What is the heat dissipation, in watts, of a resistor of 20 Ω having a current of $\frac{1}{4}$ A passing through it?

$$P = I^2 R = 0.25^2 \times 20 = 0.0625(20) = 1.25 \text{ W}$$

From Ohm's law again, $I = E/R$. By substituting E/R for the I in the basic power formula,

$$P = EI = E\left(\frac{E}{R}\right) \qquad \text{or} \qquad P = \frac{E^2}{R}$$

EXAMPLE: What is the minimum power-dissipation rating of a resistor of 20,000 Ω to be connected across a potential of 500 V?

$$P = \frac{E^2}{R} = \frac{500^2}{20,000} = 12.5 \text{ W}$$

These three formulas to determine power in an electric circuit are undoubtedly as important to know as the three Ohm's-law formulas.

When a current of electrons flows through a conductor, the conductor always becomes warmer. Some of the power in the circuit is converted to heat and is lost. If a perfect conductor could be found, it would carry current without such a heat loss. However, even the best of conductors has some resistance, so there will always be some heat loss in electric circuits. Note that the main factors in the conversion of electric power to heat are the current and the resistance. The power formula $P = I^2 R$ can always be depended upon to give true power (heat) values.

Not all power in electricity is converted into heat. In a radio receiver, some power is converted into sound waves, but some heat will be devel-oped in the radio in the process of this conversion. With transmitters, power is changed into radio waves in the air. When current flows through the resistance-wire filament of an electric light, the filament becomes so hot that it glows brightly. The wire is hot and is radiating heat energy, but it is also radiating energy in the form of light. The power formula will give the total amount of power being consumed. The light energy, however, is a small percent of the total. Fluorescent lights, in many modern installations, utilize a method of producing light energy other than a hot filament. Less heat is required to produce the same amount of light. Such lights are more efficient because of their lower percent of heat loss.

The basic unit of measurement of power is the watt. For smaller quantities the *milliwatt*, or 0.001 W, may be used. For larger quantities, the *kilowatt*, or 1,000 W, may be used. Another unit of power is represented by 746 W, called a *horsepower*. If an electric motor were 100% efficient, 746 W of power fed to it would produce the equivalent of one mechanical horsepower of twisting force, called *torque*.

The terms "power" and "energy" have been used somewhat synonymously. Actually, these two terms do not mean the same thing, although there are occasions when they may be used interchangeably. Power is the ability to do work and is measured in watts. Energy is usually computed by multiplying the amount of power by the length of time the power is used. One watt of power working for one second is known as a *wattsecond*, or as a *joule* of energy.

If a 100-W lamp is turned on for 1 second (1 s), it uses 100 joules (100 J) of energy. During the time it is working, it is dissipating 100 W of power. If the light is left on for 10 s, it consumes 1,000 J of energy, but while it is working it is still dissipating only 100 W of power. Electric power companies may produce power, but they sell energy. Instead of using wattseconds, they use the larger basic units, the *watthour* (number of watts times the number of hours) or the *kilowatthour* (number of watts times the number of hours divided by 1,000). Every establishment buying electric energy has a kilowatthour meter measuring how many kilowatthours, abbreviated kWh, of energy flows in the power lines. If electricity costs 5 cents per kilowatthour, a 1-kW lamp may be operated for 1 hour (1 h) for 5 cents, or a 100-W lamp may be operated for 10 h for the same amount of money.

Actually, power, by the formula $P = EI$, implies time, since the ampere (I) in the formula is a coulomb per second. An ampere can be expressed as

$$A = \frac{Q}{T}$$

where A = current (in amperes, A)
  Q = electron quantity (in coulombs, C)
  T = time (in seconds, s)

For example, if 10 C moves through a circuit in 2 s, the average current is 5 A.

If power equals volts times amperes and amperes equals Q/T, then power must equal volts times Q/T. In formula form,

$$P = EI = E\left(\frac{Q}{T}\right) = \frac{EQ}{T} = EQ/T$$

This formula tells us power is equal to volts times coulombs per second.

The unit of energy was given as the watt-second, or joule, and it is usually computed as power times time. This brings up an interesting fact. If energy equals power times time and the power formula $P = EQ/T$ is multiplied by time to give energy, the equation then reads $E_n = EQT/T$ and time cancels out. Therefore, energy is actually something that may do work if given a chance and has no reference to time itself. Although normally measured for convenience as power times time, energy is timeless. The energy formula with the time canceled in $E_n = EQ$, or joules equals volt-coulombs. A power company actually sells volts of pressure times the number of coulombs, it delivers, irrespective of how long it takes the consumer to accept the energy.

Pressure times quantity, or *volt-coulombs,* equals the energy that is available. Pressure times movement, or *volt-amperes* (EI), equals work done, or power.

In practical problems involving energy, the wattsecond or watthour is usually used. For example, to determine the number of watthours of energy consumed by a radio receiver drawing 60 W of power for 20 h, the power-times-time formula is employed, using hours as the time unit. In this case, 60 W times 20 h equals 1,200 Wh of energy. This is also equal to 1.2-kWh, or 4,320,000 Ws, or joules.

EXAMPLE: It is desired to know how many kilowatthours are consumed by a receiver drawing 75 W operating for a period of 24 h. The number of watthours will be 75 × 24, or 1,800 Wh, or 1.8 kWh.

2-6 USING THE POWER FORMULAS

The power formula $P = EI$ states in mathematical form: *The power is directly proportional to both voltage and current.* Since the current increases if the voltage increases, doubling the voltage will also double the current, and the power will increase fourfold.

The power formula $P = I^2R$ states in mathematical form: *The power is directly proportional to the resistance and also to the current squared.* If the current is doubled, the power dissipation is equal to 2 squared, or 4 times as much.

The power formula $P = E^2/R$ states in mathematical form: *The power is directly proportional to the voltage squared and inversely proportional to the resistance.* If the resistance is kept constant, doubling the voltage will produce four times the power.

From the three basic power formulas, other useful formulas involving power can be derived. From the formula $P = EI$, it is possible to solve for E by dividing both sides of the equation by I:

$$\frac{P}{(I)} = \frac{EI}{(I)} \quad \text{or} \quad E = \frac{P}{I}$$

where E = emf, in V
  P = power, in W
  I = current, in A

EXAMPLE: A resistor rated at 50 W and a maximum current of 100 mA (0.1 A) will stand how much voltage without becoming excessively hot? From the formula $E = P/I$, the maximum voltage is equal to $50/0.1$, or 500 V.

From the same power formula $P = EI$, when both sides are divided by E, the result is the formula

$$\frac{P}{(E)} = \frac{EI}{(E)} \quad \text{or} \quad I = \frac{P}{E}$$

EXAMPLE: How much current will flow in a television set that is rated at 240 W when connected across mains carrying 120 V? From the formula $I = P/E$, the current is equal to $240/120$, or 2 A.

From the formula $P = I^2R$, dividing both sides by R and then taking the square root of both sides results in the formula

$$\frac{P}{(R)} = \frac{I^2R}{(R)} \quad \text{or} \quad \sqrt{\frac{P}{R}} = \sqrt{\frac{I^2R}{R}} \quad \text{or} \quad I = \sqrt{\frac{P}{R}}$$

EXAMPLE: What is the maximum rated current-carrying capacity of a resistor marked "5,000 Ω, 200 W"? From the formula $I = \sqrt{P/R}$, the current is equal to $\sqrt{200/5,000}$, or $\sqrt{0.04}$, or 0.2 A.

From the formula $P = I^2R$, dividing both sides by I^2 results in the formula

$$\frac{P}{(I^2)} = \frac{I^2R}{(I^2)} \quad \text{or} \quad R = \frac{P}{I^2}$$

EXAMPLE: A radio receiver rated at 55 W draws 2 A from the line. The effective resistance is $R = P/I^2$, or $55/2^2$, or 13.75 Ω.

From the formula $P = E^2/R$, multiplying both sides by R and then taking the square root of both sides results in the formula

$$P(R) = \frac{E^2(R)}{R} \quad \text{or} \quad E = \sqrt{PR}$$

EXAMPLE: What is the maximum voltage that may be connected across a 10-W 1,000-Ω resistor? From the formula $E = \sqrt{PR}$, the voltage is equal to $\sqrt{10(1,000)}$, or $\sqrt{10,000}$, or 100 V.

From the formula $P = E^2/R$, cross-multiplying (or multiplying both sides by R) and then dividing both sides by P results in the formula

$$R = \frac{E^2}{P}$$

EXAMPLE: What is the resistance of a 3-W 6-V lamp? The resistance is equal to 6^2 divided by 3, or $36/3$, or 12 Ω.

2-7 POWER DISSIPATION IN RESISTORS

Resistors, whether they are the carbon type or wire-wound, have a resistance and a power rating. The power rating indicates how much heat the resistor is capable of dissipating under normal circumstances. If cooled by passing air across it, the resistor may be capable of considerably greater power dissipation. If enclosed in an unventilated area, it may become excessively hot and burn out when dissipating its rated power or less. Usually, the power rating required is computed by one of the power formulas and a resistor of twice the computed power rating is employed. Thus, if it is computed that a resistor in an operating circuit must be capable of dissipating at least 5 W, a 10-W resistor would be used. If the resistor is tightly enclosed, a rating three or four times the computed value might be required.

Instead of rating resistors by power dissipation, they might be rated in current-carrying ability. In fact, some wire-wound resistors carry a resistance, a power dissipation, and a current rating. When the current rating is not given, the rearranged power formula $I = \sqrt{P/R}$ can be used. For example, a 100-Ω 1-W resistor will carry safely

$$I = \sqrt{\frac{P}{R}} = \sqrt{\frac{1}{100}} = 0.1 \text{ A or } 100 \text{ mA}$$

Test your understanding; answer these checkup questions.

1. A receiver across a 120-V line has 0.75 A flowing through it. How much power is being used? _____
2. A 420-Ω resistor has 30 mA flowing in it. How much power is producing heat in the resistor? _____
3. A 120-V electric iron has 36 Ω resistance. How much heating power does it produce? _____

4. How many milliwatts are there in a kilowatt? _____
5. How much energy is used in 30 days by a 120-V clock having an internal resistance of 5,000 Ω? _____
6. If electricity costs 4 cents a kilowatthour, how much does it cost to operate a TV set 24 h if it draws 2 A when across 120 V? _____
7. If 0.5 C passes one point in a circuit in 0.01 s, what is the average current value? _____
8. How much power is developed when 100 V forces 80 C through a point in a circuit in 0.5 s? _____ How much energy is this? _____
9. How much energy is being used by a transmitting station drawing 40 A from a 440-V power line in 2 h of operation? _____
10. Across how many volts must a 600-W heater be connected if it is drawing 5 A? _____
11. A 25-W emergency light draws how many amperes from a 6.3-V storage battery? _____
12. How much current is a 1-W 2,500-Ω resistor capable of carrying without overheating? _____
13. A 100-W lamp draws 0.9 A. What is its resistance value? _____
14. A 25-W 500-Ω resistor can be safely connected across how much voltage? _____
15. How many ohms of resistance does the ordinary 75-W house lamp have when operating at its rated 120 V? _____
16. A 1,700-Ω cathode resistor must carry 8 mA. What should the power rating of the resistor be? _____

2-8 FUSES

To protect circuits from damage caused by accidental overloads or *short circuits,* fuses are installed in series with the lines carrying current from the source to the load. The first duty of a fuse is to carry the circuit current with little or no voltage loss to the circuit. This requires a fuse with very low resistance in circuits carrying high current. Lower-current circuits may have fuses with several ohms of resistance.

A fuse is placed in such a position that all the current flowing through the circuit to be protected must flow through the fuse, as shown in Fig. 2-3. If a short circuit develops across the load, the current from the source flows through the fuse and through the low-resistance short circuit. This produces a high current. The heavy current will produce enough heat to melt the special low-melting-point fuse wire, interrupting the current flow and protecting the source from damage due to overload and excessive current flow. Without a fuse, a "short" may cause the connecting wires of a circuit to become hot enough to ignite the insulation on the wires and start a fire.

Fuses are rated for current-carrying ability and

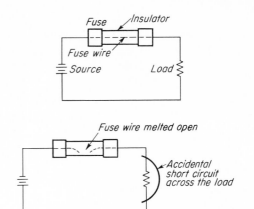

Fig. 2-3 The fuse wire melts open if excessive current is made to flow in the circuit.

also for maximum voltage of the circuit in which they are used. High-current fuses use relatively heavy fuse wire and are recognizable by their relatively large diameter. Low-current fuses may be made quite small with delicate fuse wire. Low-voltage-circuit fuses may be physically short, but fuses for high-voltage circuits are long. This prevents any high voltage that appears across the burned-out section of the fuse from jumping the gap and striking an arc of current, which would prevent the fuse from open-circuiting. The greater length results in better insulating properties of the fuse *after* it burns out.

Fuses are available in such ratings as 100 A, 30 A, 15 A, 1 A, ½ A, ¼ A, and 1/32 A. "Slow-blow fuses" are made to withstand short-duration overloads due to current surges, such as occur when a motor is started, but they will burn out after a short interval of time if the current does not decrease. They are not suitable when fast protection is important.

2-9 METERS

Five types of meters in general use are the voltmeter, ammeter, wattmeter, watthour meter, and ohmmeter. These are explained in greater detail in Chap. 12. At this point they will only be shown in their usual positions in simple electric circuits. The symbol of a meter is a circle with a letter in it to indicate the type of meter.

The *voltmeter* measures the difference of potential, or the emf, *across* a circuit. It is always

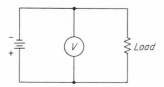

Fig. 2-4 A voltmeter is always connected across the circuit to be measured.

connected across the difference of potential to be measured, as in Fig. 2-4.

The *ammeter* is always connected *in series* with the line carrying the current to be measured. It indicates the number of electrons, in coulombs per second, flowing through it. The meter in Fig. 2-5 is measuring the total current of the circuit.

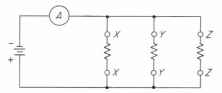

Fig. 2-5 This one ammeter reads the sum of all three branch currents.

To measure the current in any one of the three branches, the meter must be moved to points *X, Y,* or *Z* shown. Since an ampere is a relatively large current value in electronic circuits, milli-ammeters or microammeters are frequently used. (A milliampere is 0.001 A, and a microampere is 0.000001 A.)

The *wattmeter* measures electric power. It can be considered a voltmeter and ammeter combined in such a way that it gives the product of the voltage and current on its scale. It is therefore connected across the difference of potential and also in series with the line carrying the current. The wattmeter may have three or four terminals, whereas most other meters have only two. Figure 2-6 shows a three-terminal wattmeter with the

Fig. 2-6 A wattmeter is connected in series with and also across the load in the circuit.

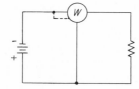

necessary connection, if a fourth terminal is used, shown in dotted lines.

In many practical applications a separate volt-meter and ammeter are used instead of a watt-meter, as in Fig. 2-7. The voltmeter value multi-

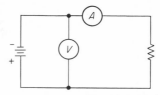

Fig. 2-7 Voltmeter times ammeter readings give power in watts demanded by a load.

plied by the ammeter value gives the power value in watts.

The *watthour meter* measures electric energy. It is actually an electric motor geared to an indicator needle similar to the hand of a clock. How far the indicator rotates depends on the current flowing through the meter and load, the voltage across the circuit, and the length of time the current flows. Figure 2-8 shows how a three-terminal watthour meter is connected in a simple circuit.

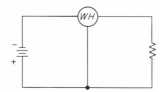

Fig. 2-8 A watthour meter is connected in series with and across a load.

The *ohmmeter* is a sensitive ammeter plus an internal battery, both contained in a small Bakelite case. It can be used only when the resistance being measured is in a dead circuit, that is, when the resistance being measured has no current flowing through it. The ohmmeter is usually a portable meter with flexible leads, as indicated in Fig. 2-9.

2-10 TYPES OF CIRCUITS

The remainder of this chapter will deal with solving for the current, voltage, resistance, or power in various types of circuits.

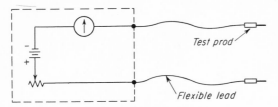

Fig. 2-9 An ohmmeter consists of a milliammeter calibrated in ohms, a battery, a calibrating resistor, and test prods.

A source of emf with a single load, as shown in Fig. 2-10, is a *simple circuit*.

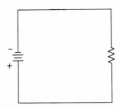

Fig. 2-10 A simple circuit.

A source of emf with two or more loads connected across it in such a way that there is only one current path through the whole circuit, as in Fig. 2-11, is called a *series circuit*.

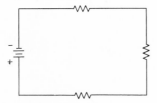

Fig. 2-11 A series circuit has loads in series.

A source of emf with two or more loads connected across it in such a way that each load has only its own current flowing through it independent of the other load or loads, as in Fig. 2-12, is termed a *parallel circuit*.

ANSWERS TO CHECKUP QUIZ ON PAGE 29

1. *(90 W)*   2. *(0.378 W)*   3. *(400 W)*   4. *(1,000,000 mW)*
5. *(2,074 Wh)*   6. *(23 cents)*   7. *(50 A)*   8. *(16,000 W)*
(8,000 J,   or   Ws)   9. *(35.2 kWh)*   10. *(120 V)*
11. *(3.97 A)*   12. *(0.02 A)*   13. *(123.4 Ω)*   14. *(112 V)*
15. *(192 Ω at 120 V)*   16. *(2 × 0.109, or in practice ¼ W)*

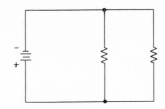

Fig. 2-12 A parallel circuit has loads in parallel.

When speaking of paralleling resistors or loads, it is sometimes said that they are connected in *shunt*. When something is connected across the terminals of something else, it may be said that the first is shunted across the second.

When a group of loads is connected in a more or less mixed and complex group of series and parallel circuits, the whole group may be said to be connected in *series-parallel*.

2-11 SERIES CIRCUITS

The diagrams in Fig. 2-13 show two series circuits. The same amount of current flows through all parts of the circuit. (More electrons can never flow into a resistance than flow out the other end. The electrons forming the current may lose energy in the form of heat while moving through a resistor, but the electrons themselves are not lost.)

In working with series circuits, the sums of the unit values are simply added. For example, Fig. 2-13 shows a series-resistance diagram with a total resistance of 150 Ω. It also shows a series of batteries with a total voltage of 600 V.

Resistors can be connected in series when it

Fig. 2-13   *(a)* These resistors in series total 150 Ω. *(b)* These batteries in series total 600 V.

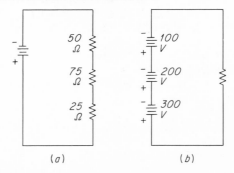

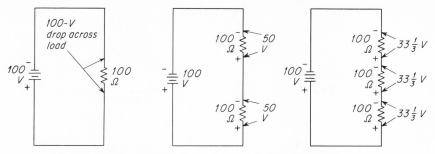

Fig. 2-14 Distribution of voltage-drops in series circuits.

is desired to have a greater resistance and thereby a smaller current.

Batteries are connected in series when it is desired to produce the highest possible voltage. However, a battery can develop only a certain value of current. By connecting batteries in series, the sum of all the voltages of all the batteries can be obtained, but the maximum current possible through the circuit is no greater than the greatest current that the weakest battery can deliver. If one of the batteries in Fig. 2-13 is capable of passing 1 A through it, another 2 A, and the third 3 A, the maximum current the three batteries in series can pass without damage to any is only 1 A. This series combination then will result in 600 V with a maximum current capability of 1 A. According to Ohm's law ($I = E/R$), if the resistor across the three batteries has 600 Ω resistance, the current in the circuit will be 1 A. If less than 600 Ω is used, the weakest battery will be overworked and may overheat and the voltage across its terminals will drop.

In Fig. 2-14, the voltage of the batteries is 100 V and each of the resistors has a value of 100 Ω. Note how the voltage is distributed in the three circuits and how the sum of the voltage-drops of the resistors always equals the battery voltage of 100 V.

If the emf across each resistor is considered to be a *voltage-drop,* or loss of voltage, then the sum of all the voltage-drops (considered as minus values) around the circuit, when added to the source voltage (a plus value), gives an algebraic sum (a plus and minus sum) of zero volts in the circuit. This is stated in Kirchhoff's voltage law: *The algebraic sum of all the voltages in a series circuit is always zero.*

The Kirchhoff current law states: *The total cur-rent flowing into a junction in a circuit will always equal the total current value flowing out of the junction.*

To find the total resistance of a group of series resistors the formula is

$$R_{\text{total}} = R_1 + R_2 + R_3 + \cdots$$

If all the resistors are equal in value, the value of any one can be multiplied by the number of resistors to give the total. Five 25-Ω resistors in series present 5(25), or 125 Ω.

2-12 OHM'S LAW IN COMPLEX SERIES CIRCUITS

The use of Ohm's law was previously discussed as it applied to a circuit made up of a source of emf and a load. If two of the three elements making up the Ohm's-law formula (*E, R,* and *I*) are known, the third can be computed.

When more complex circuits are used, Ohm's law may still be employed, but additional factors must be considered. Complex circuits, in this instance, mean parallel, series, or series-parallel types.

There are three important rules regarding the use of Ohm's law in working problems:

VOLTAGE RULE. It is possible to determine the voltage across any particular known *R* in a group of resistances, if the *current through that particular R* is known, by $E = IR$.

CURRENT RULE. It is possible to determine the current through any particular known *R* in a group of resistances, if the *voltage across that particular R* is known, by $I = E/R$.

RESISTANCE RULE. It is possible to determine the resistance of any one part of a circuit, if the

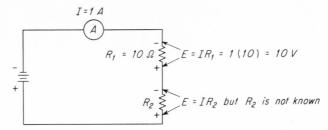

$I = 1\ A$

$R_1 = 10\ \Omega$ $E = IR_1 = 1\,(10) = 10\ V$

R_2 $E = IR_2$ but R_2 is not known

Fig. 2-15   Insufficient data given to compute unknown circuit values.

voltage across *that part* and the current flowing through *that part* are known, by $R = E/I$.

These seem simple rules, but one of the most difficult things when first starting electronics is to determine where Ohm's law can and where it cannot be used. Consider two impossible problems:

In Fig. 2-15 it is impossible to find the voltage across the second resistor because none of the three rules can be applied to any one part of the circuit.

In Fig. 2-16 it is impossible to determine the

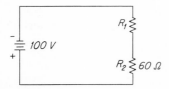

Fig. 2-16   Insufficient data given to compute unknown circuit values.

resistance of R_1 because none of the rules can be applied to any one part of the circuit. The source voltage is not across R_2 but across R_1 and R_2 in series.

A problem that can be computed is shown in Fig. 2-17. Since this is a series circuit, the current value in all parts must be equal. Since the total

Fig. 2-17   Sufficient data given to compute all circuit values.

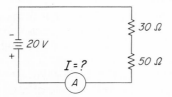

resistance is equal to the sum of all the resistors in the circuit, the total resistance across the battery is $30 + 50$, or $80\ \Omega$. (The source "sees an 80-Ω load.") Once the total resistance is known, the current value in the circuit can be found by Ohm's law:

$$I = \frac{E}{R} = \frac{20}{80} = 0.25\ A$$

The voltage drop developed across either of the resistances by the current flowing through it can be found by applying Ohm's law to that part. For the 30-Ω resistor,

$$E = IR = 0.25(30) = 7.5\ V$$

For the 50-Ω resistor,

$$E = IR = 0.25(50) = 12.5\ V$$

The voltage-drops across the resistances are considered voltage losses. The voltage of the source is considered a gain voltage. The sum of -7.5, -12.5, and $+20$ is zero volts (Kirchhoff's voltage law). This is a method of checking whether the current value has been correctly computed. The sum of the voltage-drops should equal the source emf.

In cases in which the source has internal resistance, the voltage loss across the internal resistance must be considered. Suppose a 10-V source has $1\ \Omega$ internal resistance and is connected across a 9-Ω load, as indicated in Fig. 2-18. With $10\ \Omega$ and 10 V the current is 1 A, which through the 1-Ω resistance produces a 1-V loss. Therefore the 10-V source actually produces only 9 V across its terminals and across the 9-Ω resistance load. The other volt of pressure is lost inside the source. If no current is flowing through the

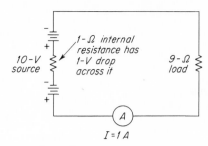

Fig. 2-18 In some cases the internal resistance in the source may have to be considered.

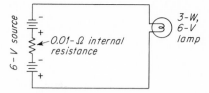

Fig. 2-20 Internal resistance in this circuit is negligible in comparison with load resistance value.

source, however, no *voltage-drop* is developed across the internal resistance and the terminal voltage is 10 V. This is important to understand!

If a circuit carrying a current of 3 A has an internal resistance of 2 Ω in the source and a 50-Ω load, what is the terminal voltage of the source? This problem can be analyzed and worked in two ways. The diagram of this circuit is shown in Fig. 2-19. The simplest solution is to consider that the

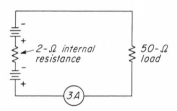

Fig. 2-19 This source has an output of 150 V with the 50-Ω load, 156 V with no load.

terminal voltage of the source is the voltage-drop across the 50-Ω resistor, since the two are directly connected. The voltage across the 50-Ω resistor when 3 A flows through it is equal to $E = IR$, or 3(50), or 150 V.

The other analysis of this problem is to consider the total resistance in the circuit as equal to 50 plus 2 Ω, or 52 Ω. The current is 3 A. From Ohm's law, the total voltage in the circuit is $E = IR$, or 3(52), or 156 V. The voltage-drop developed across the 2-Ω internal resistance is $E = IR$, or 3(2), or 6 V. This 6 V does not appear outside the source and must be subtracted from the total voltage present in the circuit. This gives 156 V less 6 V, or 150 V across the terminals of the source. If the 50-Ω resistor is disconnected, the source will have a terminal voltage of 156 V.

In some cases internal resistance may be disregarded: If a 6-V storage battery has an internal resistance of 0.01 Ω, what current will flow when a 3-W 6-V lamp is connected across it? (See Fig. 2-20.) It is assumed that the storage battery has 6 V with no load connected across it. The resistance of the lamp is determined by using the rearranged power formula $R = E^2/P$, or $6^2/3$, or 12 Ω. The total resistance of the circuit is the load plus the internal resistance of the source, or 12.01 Ω. The current, according to Ohm's law, is $I = E/R$, or 6/12.01, or 0.4996 A. By meter this would read 0.5 A. (The loss of voltage across the internal resistance, called *internal drop,* is only about 0.004996 V.)

It is possible to solve for missing values in some series problems. For example, two resistors are connected in series, as shown in Fig. 2-21. The

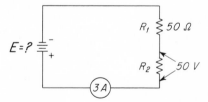

Fig. 2-21 The sum of the voltage-drops across R_1 and R_2 equals the source voltage.

current through them is 3 A. R_1 has a value of 50 Ω. R_2 is unknown but has a voltage-drop of 50 V across it. What is the total impressed emf across the whole circuit? Two factors regarding R_1 are known: $R_1 = 50$ Ω, and $I = 3$ A. The voltage across it can therefore be determined as $E = IR$, or 3(50), or 150 V. This voltage is in series with the 50 V across R_2, resulting in a total of 200 V across the two resistors, which must therefore be the source voltage.

As an example of a somewhat similar series-circuit problem, a vacuum tube has a filament

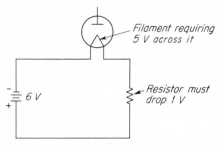

Fig. 2-22 Circuit required to drop the voltage of the source to a usable value across the filament.

rated at 0.25 A and 5 V and is to be operated from a 6-V battery. What is the value of the necessary series resistor? The circuit is illustrated in Fig. 2-22.

The filament can be considered a resistor requiring 0.25 A flowing through it to develop a 5-V drop across it. A 6-V battery directly across the filament will cause too much current to flow. To prevent this, a resistance with a value sufficient to drop 1 V across it when 0.25 A flows is required. According to Ohm's law, its resistance should be $R = E/I$, or $1/0.25$, or 4 Ω. The minimum power dissipation for this resistor is equal to $P = EI$, or $1(0.25)$, or $\frac{1}{4}$ W. In practice, a 4-Ω $\frac{1}{2}$-W resistor should be used. Actually, a 1- or 2-W resistor would be just as satisfactory but would be physically larger and more expensive.

As another example, a keying relay coil has a resistance of 500 Ω and is designed to operate on 125 mA. If the relay is to operate from a 110-V dc source, what value of resistance should be connected in series with the relay coil? The circuit is shown in Fig. 2-23.

The relay coil can be considered a resistor in this problem. A current of 125 mA equals $125/1,000$ A, or 0.125 A. The voltage required to

Fig. 2-23 Voltage-dropping resistor required to prevent excessive voltage across a relay coil.

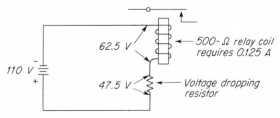

produce this current value through the coil is found by Ohm's law $E = IR$, or $0.125(500)$, or 62.5 V. The emf available is 110 V, or 47.5 V too much. To drop 47.5 V, a resistance of $R = E/I$, or $47.5/0.125$, or 380 Ω, is required. The minimum power dissipation is $P = EI$, or $47.5(0.125)$, or 5.94 W. Any 380-Ω resistor with a power rating of 10 W or more would probably be satisfactory.

Two 10-W 500-Ω resistors are connected in series. What are the power-dissipation capabilities of the combination? Since these are similar resistors, they will stand the same amount of current. With the maximum current for their rating, $I = \sqrt{P/R}$, or 0.141 A, each will produce 10 W of heat, resulting in 20-W maximum safe dissipation from the two.

On the other hand, if a 20-W 500-Ω resistor and a 10-W 500-Ω resistor are connected in series, the 10-W resistor will have a maximum safe current of 0.141 A, but the 20-W resistor will have a maximum safe current of $I = \sqrt{P/R}$, or $\sqrt{20/500}$, or 0.2 A. The limiting factor in the circuit is the 0.141 A. The 10-W resistor will dissipate 10 W, but the 20-W resistor will be held to 10 W of dissipation because only 0.141 A should be flowing through it. The total safe power dissipation for the two resistors in series is only 20 W.

When resistors are connected in series, the maximum safe current for each must be determined before the limiting current can be judged. The highest wattage rating does not always indicate the greatest safe current rating. For example, a 20-W 2,000-Ω resistor has a maximum current value of only 0.1 A, which is lower than the 0.141-A capability of a 10-W 500-Ω resistor.

2-13 SUGGESTIONS FOR WORKING COMPLEX PROBLEMS

Be neat and orderly in problem working. The tried and proved method explained here will help in thinking out the problems in an orderly, complete way.

1. Read through the whole problem carefully twice. Determine what is wanted.
2. At the top-left-hand part of the work paper, sketch a diagram of the circuit, if any is in-

volved, and label the parts with the values given in the problem. (Become proficient at freehand diagraming.)

3. Directly below the diagram, jot down any other information that may be necessary to use (such as formulas) which does not have a place on the diagram. Draw a line under this information.

4. State what is to be found at the top-right-hand part of the page.

5. Beneath the drawn line, write out the first formula used in the first operation in the problem. Do miscellaneous computations at the right-hand-side of the page rather than on another piece of paper. Keep formulas and computations close together to simplify later checking.

6. Continue solving the formulas down the left side of the page, using the right side for incidental mathematics. The last line on the left side should contain the final answer. Encircle subanswers, if any, and answers.

Test your understanding; answer these checkup questions.

1. Four 37.5-Ω resistors are in series across 120 V. What is the voltage-drop across one of them? _____

2. A 30-Ω, a 60-Ω, and a 150-Ω resistor are in series across a 24-V battery. What voltage-drop appears across the 60-Ω resistor? _____ How much current flows in the 150-Ω resistor? _____

3. A 50-Ω, a 90-Ω, and an unknown-value resistor are in series across a 60-V generator with $\frac{1}{3}$ A flowing through it. What is the voltage-drop across the unknown resistor? _____ What is its resistance? _____

4. A 12.6-V automobile battery is across a 1.5-Ω headlight lamp. If the battery has 0.14 Ω internal resistance, what is the current value? _____ What is the loaded terminal voltage of the battery? _____

5. A 500-Ω relay coil operates with 0.2 A. What resistance must be connected in series with it if operation from a 110-V line is required? _____

6. A 6-V mobile receiver draws 36 W. If it is to operate from a 12-V battery, what value resistor must be used in series with it to maintain the 36-W receiver dissipation? _____

7. A 5,000-Ω 20-W resistor and a 1,000-Ω 5-W resistor are in series. What is the maximum voltage that can be applied across them without exceeding the wattage rating of either resistor? _____

8. A 6.3-V tube filament (consider it a resistor) requires 300 mA. What resistance is required in series with it if operation is across a 110-V line? _____ What power would be dissipated in heating the resistor? _____

2-14 CONDUCTANCE

An interesting point regarding the study of electricity is the necessity, at times, of observing the same thing from two viewpoints in order to obtain a better understanding of the whole. One example of this is the resistance versus the conductance of a circuit.

In an operating electric circuit there must always be a source of emf and a load. In the simplest circuit the load may be a single resistor. As explained before, the greater the resistance in the load, the less the current in the circuit. However, the very fact that some current is flowing indicates that the resistance is not infinite (immeasurable, or endless) but is actually a conductor to some degree. It may be a very poor conductor, or it may be a fairly good conductor. In any event, the greater its conducting ability, or *conductance,* the less its resistance value. Conversely, the less its conductance, the greater its resistance. Conductance and resistance refer to the same thing but from opposite viewpoints. They are said to be reciprocals of each other.

The meaning of "reciprocal" is, roughly, "mathematically opposite." In stating that one thing is the reciprocal of another, or that it varies inversely as the other, the two are placed on opposite sides of an equals sign and one of them is expressed as a fraction by putting a 1 over it.

R stands for resistance in ohms. G stands for conductance in mhos. *Mho,* the unit of conductance, is the word *ohm* spelled backward. (Another term, *siemens, S,* may be used for *mho* and G.) Since R and G have opposite meanings, they may be expressed:

$$R = \frac{1}{G} \quad \text{or} \quad G = \frac{1}{R}$$

If a resistance has a value of 2 Ω, then its conductance value is 1 over 2, or $\frac{1}{2}$ mho. If the R value is raised to 3 Ω, the conductance value becomes $\frac{1}{3}$ mho. As the resistance is increased from 2 to 3 Ω, the conductance value decreases from $\frac{1}{2}$ to $\frac{1}{3}$ mho.

Since $R = 1/G$, Ohm's law can be expressed in terms of conductance by using $1/G$ in place of R in the three formulas:

$$E = IR = I\left(\frac{1}{G}\right) = \frac{I}{G}$$

$$I = \frac{E}{R} = \frac{E}{1/G} = E\left(\frac{G}{1}\right) = EG$$

$$R = \frac{E}{I} \qquad \frac{1}{G} = \frac{E}{I} \qquad GE = I \qquad G = \frac{I}{E}$$

EXAMPLE: What is the conductance of a circuit if 6 A flows when 12-V dc is applied to the circuit? Using the last formula above, $G = I/E$, the problem is solved:

$$G = \frac{I}{E} = \frac{6}{12} = \frac{1}{2} \text{ mho}$$

NOTE: If $R = E/I$, then G, the reciprocal, or mathematical opposite, is $G = I/E$.

2-15 PARALLEL RESISTANCES

The subject of conductance is a fitting preliminary to the subject of parallel resistors. Understanding conductance makes it possible to see that when two 10-Ω resistors are connected in parallel across a source of emf, as in Fig. 2-24,

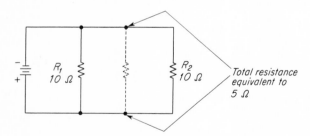

Fig. 2-24 Two 10-Ω resistors in parallel present an equivalent resistance of 5 Ω to the source.

the conductance of the circuit is greater and therefore the total resistance must be less. Two parallel 10-Ω resistors provide a conductance

---

value twice that of one resistor and therefore a resistance value of one-half of 10, or 5 Ω, and not 20 Ω as in series circuits. This apparent adding of resistances to a circuit and obtaining a resultant resistance less than any of the resistances may be confusing unless it is seen from the conductance viewpoint.

The total conductance of a circuit is equal to the sum of all the conductances connected in parallel across the circuit. This may be expressed in formula form as

$$G_t = G_1 + G_2 + G_3 + \cdots$$

Since any single conductance value is equal to the reciprocal of its resistance value, the formula for the total conductance G_t of a parallel circuit may also be expressed as

$$G_t = \frac{1}{R_1} + \frac{1}{R_2} + \frac{1}{R_3} + \cdots$$

Substituting the reciprocal of the total resistance R_t for the total conductance gives the formula

$$\frac{1}{R_t} = \frac{1}{R_1} + \frac{1}{R_2} + \frac{1}{R_3} + \cdots$$

If this equation is made into a pair of fractions by placing a 1 over both sides, it becomes

$$\frac{1}{1/R_t} = \frac{1}{1/R_1 + 1/R_2 + 1/R_3}$$

The rule for simplifying compound fractions is: Invert the lower fraction and multiply it by the upper. In this case the left-hand part of the equation,

$$\frac{1}{1/R_t} = 1 \times \frac{R_t}{1} \text{ or merely } R_t$$

The complete formula to solve for the total resistance of any number of parallel resistances is

$$R_t = \frac{1}{1/R_1 + 1/R_2 + 1/R_3}$$

This formula can be solved in one of two ways: by using fractions or by using decimals.

EXAMPLE: If resistors of 5, 3, and 15 Ω are in parallel, what is the total resistance? First, substitute known values into the formula:

$$R_t = \frac{1}{\frac{1}{5} + \frac{1}{3} + \frac{1}{15}}$$

To add the fractions $\frac{1}{5}$, $\frac{1}{3}$, and $\frac{1}{15}$, they should be expressed in their lowest common denominator: $\frac{1}{5}$ equals $\frac{3}{15}$, and $\frac{1}{3}$ equals $\frac{5}{15}$. By substituting the fractions expressed in their lowest common denominator and then adding the fractions, the problem becomes

$$R_t = \frac{1}{\frac{3}{15} + \frac{5}{15} + \frac{1}{15}} = \frac{1}{\frac{9}{15}}$$

$$= 1(\tfrac{15}{9}) = \tfrac{15}{9} = 1\tfrac{2}{3} \ \Omega$$

The same problem can be solved by expressing the original fractions $\frac{1}{5}$, $\frac{1}{3}$, and $\frac{1}{15}$ as decimal equivalents. This is accomplished by dividing 5 into 1 (0.2), then 3 into 1 (0.3333), and then 15 into 1 (0.0667). When these are substituted in the formula, it becomes

$$R_t = \frac{1}{\frac{1}{5} + \frac{1}{3} + \frac{1}{15}} = \frac{1}{0.2 + 0.3333 + 0.0667}$$

$$= \frac{1}{0.6} = 1.67 = 1\tfrac{2}{3} \ \Omega$$

When there are only two parallel resistances, the formula used above can be algebraically rearranged to read

$$R_t = \frac{R_1 R_2}{R_1 + R_2}$$

EXAMPLE: What is the total resistance of a parallel circuit consisting of one branch of 10 Ω resistance and one branch of 25 Ω resistance? By using 10 Ω as R_1 and 25 Ω as R_2, the problem is solved:

$$R_t = \frac{R_1 R_2}{R_1 + R_2} = \frac{10(25)}{10 + 25} = \frac{250}{35} = 7.14 \ \Omega$$

If there are three resistances in parallel, the total of two of them can be computed by this formula and the answer considered as R_1. By using this resistance value and the third resistance as R_2, the formula can be used to solve for the total of the three resistances. The same procedure can be employed to determine the value of any number of parallel resistances.

Another formula to compute three resistors in parallel is

$$R_t = \frac{R_1 R_2 R_3}{R_1 R_2 + R_2 R_3 + R_1 R_3}$$

When all the resistances in parallel are equal in value, such as five 100-Ω resistances, the total resistance is most easily determined by dividing the resistance value of one resistor by the number of resistors. Five 100-Ω resistors in parallel present $\frac{100}{5}$, or 20 Ω to the source.

A quick check on answers to problems involving parallel resistances is possible by noting that the answer must always be a lower value than the lowest of the parallel resistances. Also, when one resistance is about 10 times another and the two are in parallel, the total resistance will be about 10% less than the lower. If one resistance is 100 times a second resistance, it may often be possible to disregard the first entirely, since it will affect the total resistance by less than 1%.

When two 10-W 500-Ω resistors are in parallel, what are the total power-dissipation capabilities of the combination? Since they have similar power and resistance ratings, they will stand the same maximum value of voltage across them, $E = \sqrt{PR}$, or $\sqrt{10(500)}$, or 70.7 V. With the two resistors in parallel across 70.7 V, each will dissipate 10 W, giving a total 20 W of heat.

In parallel circuits the controlling and limiting factor is the voltage, whereas in series circuits it is the current. With parallel resistors it is necessary to determine what maximum voltage each will stand to produce enough current through it to make it dissipate its maximum rated power. The lowest maximum voltage of a group of resistors will determine the highest voltage that may be applied across the parallel group.

EXAMPLE: A 10,000-Ω 100-W resistor, a 40,000-Ω 50-W resistor, and a 5,000-Ω 10-W re-

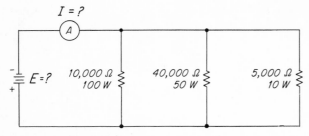

Fig. 2-25 To determine the maximum allowable current, the maximum safe source voltage must first be found.

sistor are connected in parallel. What is the maximum voltage that may be applied across the circuit? With this voltage what is the maximum total value of current through the parallel combination which will not exceed the wattage rating of any of the resistors? Figure 2-25 shows the circuit.

For the 10,000-Ω 100-W resistor:

$$E = \sqrt{PR} = \sqrt{100(10,000)}$$
$$= \sqrt{1\ 00\ 00\ 00} = 1,000\ V$$

For the 40,000-Ω 50-W resistor:

$$E = \sqrt{PR} = \sqrt{50(40,000)}$$
$$= \sqrt{2\ 00\ 00\ 00} = 1,414\ V$$

For the 5,000-Ω 10-W resistor:

$$E = \sqrt{PR} = \sqrt{10(5,000)}$$
$$= \sqrt{5\ 00\ 00} = 224\ V$$

The maximum allowable voltage across the circuit must not exceed 224 V, or the 10-W resistor will be drawing more than its rated current.
With 224 V the 10,000-Ω resistor draws

$$I = \frac{E}{R} = \frac{224}{10,000} = 0.0224\ A$$

With 224 V the 40,000-Ω resistor draws

$$I = \frac{E}{R} = \frac{224}{40,000} = 0.0056\ A$$

With 224 V the 5,000-Ω resistor draws

$$I = \frac{E}{R} = \frac{224}{5,000} = 0.0448\ A$$

The total current value in the circuit is the sum of the three separate branches, or 0.0728 A.

2-16 SERIES AND PARALLEL BATTERIES

As indicated previously, if two 45-V batteries, each capable of 0.1 A maximum safe current through it, are connected in series, as shown in Fig. 2-26, the circuit will be capable of producing 90 V at 0.1 A and therefore 9 W of power.

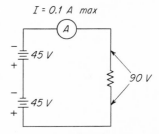

Fig. 2-26 Two 45-V 100-mA batteries in series are capable of 90 V at 100 mA.

The same two 45-V batteries connected in parallel as in Fig. 2-27, are capable of only 45 V output, but each will allow 0.1 A to flow through it, permitting the load to draw a total of 0.2 A of current from the two. The maximum safe power output is still 9 W.
To produce maximum voltage output, batteries must be connected in series, negative to positive, as shown.
To produce maximum current output, batteries must be connected in parallel, but all such batteries must have the same voltage; that is, a 40-V battery cannot be paralleled with a 45-V battery, or the higher-voltage battery will soon run down, discharging through the lower-voltage battery. Care must be taken to connect the negative ter-

Fig. 2-27 Two 45-V 100-mA batteries in parallel produce 45 V at 200 mA.

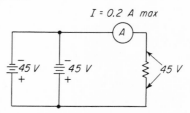

minal to negative and the positive terminal to positive when connecting in parallel.

Test your understanding; answer these checkup questions.

1. What is the conductance of a circuit having two 300-Ω resistors and one 500-Ω resistor in parallel? _____
2. What is the conductance of a circuit having 250 V and a current of 50 mA? _____
3. What is the lowest common denominator of the fractions $\frac{2}{30}$, $\frac{7}{60}$, and $\frac{11}{90}$? _____ Of $\frac{8}{25}$, $\frac{9}{15}$, and $\frac{3}{10}$? _____
4. What is the effective resistance of three parallel resistors of 1,000 Ω, 2,000 Ω, and 3,000 Ω? _____
5. A 240-Ω and a 180-Ω resistor are in parallel across a 100-V source. What is the total circuit current? _____
6. A 55-Ω resistor and a 23-Ω resistor are in parallel. The current through the source is 2.5 A. What is the source voltage? _____
7. A 4,000-Ω resistor and a 3,672-Ω resistor are in parallel across a 500-V supply. How much current flows in the 4,000-Ω resistor? _____ How much flows in the 4,000-Ω branch if the 3,672-Ω branch is disconnected? _____
8. A 75-Ω resistor and a 100-Ω resistor are in parallel. The total current through them is 3 A. How much current will flow in the 100-Ω resistor if the 75-Ω resistor is disconnected? _____
9. A 400-Ω 10-W resistor and a 1,500-Ω 50-W resistor are in parallel. What is the maximum voltage that can be applied across this circuit without exceeding the wattage rating of either resistor? _____ What is the maximum total current that can flow in the combination and not exceed the wattage rating of either resistor? _____
10. A 12.6-V tube filament and a 6.3-V 0.3-A tube filament with a series resistor are in parallel across 12.6 V. What is the value of the required resistor? _____
11. Draw a diagram by which three resistors of equal value produce a total resistance of one-third of one unit. _____
12. A power company charges 7 cents per kWh. What is the cost of operating three 120-V lamp bulbs in parallel for 24 h if each has an internal resistance of 100 Ω? _____

2-17 COMPLEX DC CIRCUITS

There are countless circuit configurations that involve resistors in series and in parallel. To attempt to give examples covering all is obviously impossible, but by applying the basic principles involved in series circuits and in parallel circuits it should be possible to solve most circuit problems. Two important points to remember are: (1) All parts connected in series have the same current flowing through them. (2) All parts in parallel have the same voltage across them.

A relatively complex circuit is shown in Fig. 2-28. By examination it can be rationalized into

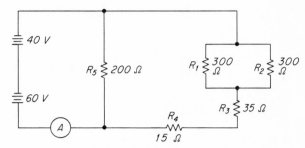

Fig. 2-28 An easily solved, relatively complex circuit.

a single resistance value. Resistors R_1 and R_2 are connected in parallel. Two 300-Ω resistors in parallel present 150 Ω of resistance. In series with this 150-Ω resistance are two other resistances, R_3 and R_4, totaling 50 Ω. To the source, this one branch presents a total of 150 plus 50, or 200 Ω.

There are two branches across the source, one made up of R_5 alone and the other of R_1, R_2, R_3, and R_4. Both branches present 200 Ω of resistance each and are in parallel with each other. Two 200-Ω parallel resistances present 100 Ω Therefore, the source of 40 plus 60 V, or 100 V, sees a total load resistance of 100 Ω. With a total emf of 100 V and a total resistance of 100 Ω, the ammeter will read $I = E/R$, or 100/100, or 1 A.

The current through the series-parallel branch is $I = E/R$, or 100/200, or 0.5 A.

With 0.5 A flowing through it, R_4 will have a voltage-drop across it of $E = IR$, or 0.5(15), or 7.5 V. Similarly, the drop across R_3 will be 17.5 V. The voltage across R_1 will be the voltage-drop across R_1 and R_2 in parallel, or that produced by $\frac{1}{2}$ A flowing through 150 Ω, or 75 V. Note that the voltage across both R_1 and R_2 is the same voltage and not merely equal voltages. The current in R_1 (or R_2) is $I = E/R$, or 75/300, or 0.25 A.

Consider the circuit shown in Fig. 2-29. Two resistors of 18 and 15 Ω are connected in parallel. In series with this combination is a 36-Ω resistor, and in parallel with the total combination is connected a 22-Ω resistor. The total current through the combination is 5 A. What is the current value in the 15-Ω resistor? The current through neither

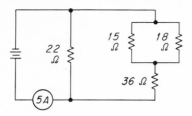

Fig. 2-29 To determine the current in the 15-Ω resistor, the source voltage must first be found.

branch is known, nor is the source voltage known. By determining the resistance of the series-parallel branch, however, both unknowns can be determined.

The 15-Ω and 18-Ω resistances are in parallel. By using a parallel-resistor formula, they are found to equal 8.18 Ω, or 8.2 Ω for simplicity. The total resistance of the series-parallel branch is 8.2 plus 36, or 44.2 Ω.

The resistances of the two branches are now known to be 22 and 44.2 Ω. When computed in parallel, they are found to equal 14.7 Ω. If 5 A flows through 14.7 Ω, the voltage across the whole combination must be $E = IR$, or 5(14.7), or 73.5 V.

With 73.5 V across the 44.2-Ω series-parallel branch, the current through it is $I = E/R$, or 73.5/44.2, or 1.66 A. If 1.66 A flows through the 8.2-Ω parallel group, the voltage across it must

be $E = IR$, or 1.66(8.2), or 13.6 V. This must be the voltage across the 15-Ω resistance also. With voltage and resistance known, the current through the 15-Ω resistor must be $I = E/R$, or 13.6/15, or 0.907 A.

In this problem it is necessary to solve almost all circuit values in order to apply Ohm's-law formula $I = E/R$ to the single resistance in question. Had the source voltage been given, the 22-Ω branch could have been disregarded, as it has no effect on the current flowing through the series-parallel branch.

Figure 2-30 is an example of an apparently highly complex circuit. Observation shows, however, that relatively simple steps can solve the total resistance. The current through the ammeter should be determinable without pencil and paper. Check these steps:

1. R_3 and R_4 can be computed as a parallel group.
2. The answer in step 1 can be computed with R_5 as a series group.
3. R_7 and R_8 can be computed as a parallel group.
4. The answers in steps 2 and 3 can be used to compute a parallel group.
5. The answer in step 4 can be added to R_6 as a series circuit.
6. The answer in step 5 can be computed with R_2 in parallel.
7. The answer in step 6 and R_1 can be computed in series, which is the effective resistance seen by the source.
8. The current is 2 A.

The diagrams in Fig. 2-31 illustrate four methods of connecting three similar resistors to

---

ANSWERS TO CHECKUP QUIZ ON PAGE 41

1. (0.00867 mho)   2. (0.0002 mho)   3. (180)(150)
4. (545 Ω) 5. (0.973 A) 6. (40.5 V) 7. (125 mA)(0.125 A)
8. (1.29 A) 9. (63.2 V)(0.201 A) 10. (21 Ω) 11. (Three in parallel) 12. (72.6 cents)

Fig. 2-30 A complex-looking circuit that may be computed mentally.

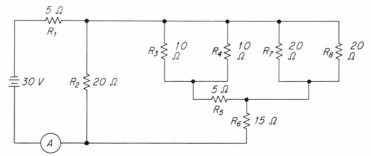

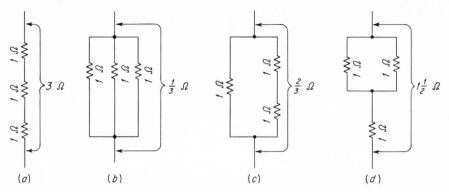

Fig. 2-31 Four possible circuit configurations for four similar resistances.

produce various total-resistance combinations. Three 1-Ω resistors in series, as in diagram *a,* result in a total resistance of 3 Ω. Three 1-Ω resistors in parallel, as shown in diagram *b,* result in a total resistance of ⅓ Ω. Two in series shunted across the third, as in diagram *c,* result in two-thirds of the resistance of one alone. Two parallel resistors in series with the third (diagram *d*) result in a total resistance of 1½ times the resistance of one resistor alone.

The way a circuit is diagramed often names the circuit. For example, the series-parallel circuit of Fig. 2-32 might be termed a *pi-type circuit* after

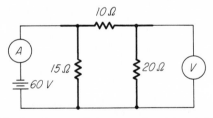

Fig. 3-32 A series-parallel group of resistors laid out as a π network.

the Greek letter π. It would be computed as a series-parallel circuit. The current in the 15-Ω leg would be $I = E/R$, or 60/15, or 4 A. The second leg consists of 10 plus 20 Ω, or 30 Ω, of resistance, and would draw $I = E/R$, or 60/30, or 2 A. The total source current is 4 A plus 2 A, or 6 A.

With 2 A flowing through the 20-Ω resistor the voltage-drop across it would be $E = IR$, or 2(20), or 40 V. Similarly, the voltage-drop would be 20 V across the 10-Ω resistor. (Does this agree with Kirchhoff's voltage law?)

The power dissipation in the 15-Ω leg would be $P = EI$, or 60(4), or 240 W. The power dissipated by the 20-Ω resistor would be 2(40), or 80 W. The power converted into heat in the 10-Ω resistor would be 40 W. The total power dissipation of this pi-type circuit would be 240 + 80 + 40 = 360 W.

2-18 MATCHING LOAD TO SOURCE

Ranking high among the important concepts for a technician to understand is matching the load to the source. To produce maximum power in any load, it is necessary that the load resistance equal the internal resistance of the source.

The diagram in Fig. 2-33 shows a 100-V source

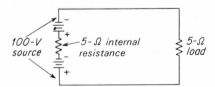

Fig. 2-33 A load resistance that matches the source resistance.

with 5 Ω internal resistance and a load with a resistance of 5 Ω also. With the load and internal resistance equal, the voltage in the circuit is 100 V, the total resistance is 10 Ω, and the current is

$$I = \frac{E}{R} = \frac{100}{10} = 10 \text{ A}$$

The power delivered to the load in this case is

$$P = I^2R = 10^2(5) = 500 \text{ W}$$

If the load resistance mismatches the source and is 15 Ω, the current in the circuit is

$$I = \frac{E}{R} = \frac{100}{20} = 5 \text{ A}$$

The power delivered to the load with this mismatch is only

$$P = I^2R = 5^2(15) = 375 \text{ W}$$

If the load resistance mismatches the source and is 3 Ω, the current in the circuit is

$$I = \frac{E}{R} = \frac{100}{8} = 12.5 \text{ A}$$

The power delivered to the load with this mismatch is only

$$P = I^2R = 12.5^2(3) = 468.75 \text{ W}$$

These figures can also be used to demonstrate an important fact. When the load matches the source, half the power is dissipated in the source and half in the load. The total in the example above is 1,000 W. The *efficiency* of the circuit, the ratio of output to input powers, is equal to 500/1,000, or 0.5, usually expressed as 50%.

With the 15-Ω load, the power dissipated in the source is 125 W. The total power is 125 W in the source and 375 W in the load, or 500 W. This results in an output efficiency of 375/500, or 75%.

With the 3-Ω load, the power dissipated in the source is 781.25 W. The total power is 781.25 W in the source and 468.75 W in the load, or 1,250 W. This results in an output efficiency of 468.75/1,250, or only 37.5%.

Matching the source to the load may produce the maximum power output in the load, but mismatching with the higher resistance in the load gives better output efficiency. Theoretically, 100% efficiency can occur only with an infinite-resistance load, in which case there is no power being fed to the load and the source is producing no load current.

When speaking of matching or mismatching a source to a load, the term impedance is usually employed rather than resistance. The word impedance is a general term indicating the opposition to current flow particularly in circuits in which ac is flowing. The term may be used properly in place of resistance in dc circuits also. Impedance is discussed in Chap. 7 where ac circuits having inductance or capacitance in series or in parallel are explained.

Test your understanding; answer these checkup questions.

1. A 200–Ω resistor and a 300-Ω resistor are in parallel. In series with them is a 180-Ω resistor. The whole combination is across a 50-V generator. What current flows through the generator? _____ Through the 180-Ω resistor? _____ Through the 200-Ω resistor? _____ What voltage appears across the 300-Ω resistor? _____
2. A 40-Ω resistor and a 60-Ω resistor are in parallel. In series with them are two other resistors of 20 Ω and 30 Ω The whole combination is connected across a 1.5-V dry cell. What is the current in the dry cell? _____ What is the current in the 60-Ω resistor? _____ What is the voltage-drop across the 20-Ω resistor? _____
3. A vacuum-tube filament requires 6.3 V across it with a current of 0.3 A. Another tube has a filament that requires 12.6 V at 0.15 A. The two tubes are connected in series. What is the value of the resistor that must be connected across the 12.6-V filament to allow it to operate properly when in series with the 6.3-V tube? _____ What is the value of the resistance that must be connected in series with the combination to allow operation from a 110-V source? _____
4. A 6BQ6 tube requires 6.3 V at 1.2 A for its filament. A 6C4 requires 6.3 V at 0.15 A. A 6SN7 requires 6.3 V at 0.6 A. A 12BE6 requires 12.6 V at 0.15 A. Draw a diagram of the most economical method of connecting these four tubes across a 12.6 V battery. _____ What are the resistance value and the wattage rating of the required resistor? _____ _____
5. A 10-Ω and a 25-Ω resistor are in parallel, and are in series with a 20-Ω and a 30-Ω parallel pair. If there is 5 V across the 10-Ω resistor, what is the current through and the voltage-drop across each component in the circuit? _____ _____ _____ _____ _____ _____ _____ _____

COMMERCIAL LICENSE QUESTIONS

Amateur license questions will be found in the Addendum.

FCC Elements 3, 4, and 6 require an ability to answer questions similar to those below. Sections in which questions are answered are shown in parentheses.

1. State the three ordinary mathematical forms of Ohm's law. (2-1)
2. If the voltage applied to a circuit is doubled and the circuit resistance increased to four times its former value, what will be the final current value? (2-2)
3. What is the unit of electric power? (2-5)
4. What is the difference between a microwatt and a kilowatt? (2-5)
5. What is the formula for determining the power in a dc circuit when the current and voltage are known? (2-5)
6. What is the formula for determining the power in a dc circuit when the current and resistance are known? (2-5)
7. What will be the heat dissipation, in watts, of a resistor of 40 Ω having a current of $\frac{1}{4}$ A passing through it? (2-5)
8. What is the formula for determining the power in a dc circuit when the voltage and resistance are known? (2-5)
9. What should be the minimum power-dissipation rating of a resistor of 20,000 Ω to be connected across a potential of 1,000 V? (2-5)
10. What is the difference between electric power and energy? (2-5)
11. What is the unit of electric energy? (2-5)
12. How much energy is consumed in 20 h by a radio receiver rated at 100 W? (2-5)
13. If the power input to a radio receiver is 75 W, how many kilowatthours does the receiver consume in 30 h of continuous operation? (2-5)
14. If the value of resistance, across which a constant emf is applied, is tripled, what will be the resultant proportional power dissipation? (2-6)
15. If the value of a resistance to which a constant emf is applied is quartered, what will be the resultant proportional power dissipation? (2-6)
16. What is the maximum rated current-carrying capacity of a resistor marked ''5,000 Ω, 100 W''? (2-6)
17. A power company charges 4 cents per kWh for electricity. How much would it cost to operate two 120-V lamp bulbs, connected in parallel, each having an internal resistance of 100 Ω, for 24 h? (2-7, 2-16)
18. What instrument measures electric power? (2-9)
19. Show by a diagram how a voltmeter and an ammeter should be connected to measure power in a dc circuit. (2-9)
20. What instrument measures electric energy? (2-9)
21. What is the sum of all the voltage-drops around a simple dc series circuit, including the source? (2-11)
22. Two resistors are connected in series. The current through these resistors is 100 mA. R_1 has a value of 50 Ω; R_2 has a voltage-drop of 50 V across it. What is the total impressed emf? (2-12)
23. If a vacuum tube having a filament rated at $\frac{1}{2}$ A and 5 V is to be operated from a 6-V battery, what is the value of the necessary series resistor? (2-12)
24. A circuit is passing a current of 5 A. The internal resistance of the source is 2 Ω. The total external resistance is 50 Ω. What is the terminal voltage of the source? (2-12)
25. A 6-V battery has an internal resistance of 1 Ω. What current will flow when a 3-W 6-V lamp is connected across it? (2-12)
26. A relay with a coil resistance of 500 Ω is designed to operate on 0.1 A. If the relay is to operate from a 100-V dc source, what value of resistance should be connected in series with the relay coil? (2-12)
27. If two 10-W 100-Ω resistors are connected in series, what is the total power-dissipation capability? (2-12)
28. What is the conductance of a circuit if 1 A flows when 12-V dc is applied to the circuit? (2-14)
29. What is the total resistance of a parallel circuit consisting of one branch of 10 Ω and one branch of 15 Ω resistance? (2-15)
30. If resistors of 8, 5, and 15 Ω are connected in parallel, what is the total resistance? (2-15)
31. If two 10-W 100-Ω resistors are connected in parallel, what are the power-dissipation capabilities of the combination? (2-15)
32. Indicate by a diagram how the total current in three branches of a parallel circuit can be measured by one ammeter. (2-15)
33. A 10-kΩ (10-kilohm) 100-W resistor, a 20-kΩ 50-W resistor, and a 2-kΩ 10-W resistor are connected in parallel. What is the maximum value of total

current through this parallel combination which will not exceed the wattage rating of any of the resistors? (2-15)

34. What method of connection should be used to obtain the maximum no-load output voltage from a group of similar cells in a storage battery? (2-16)

35. Show by a diagram how to connect battery cells in series. (2-16)

36. What method of connection should be used to obtain the maximum short-circuit current from a group of similar cells in a storage battery? (2-16)

37. Show by a diagram how to connect battery cells in parallel. (2-16)

38. Draw a circuit composed of a 12-V battery with three resistors (10, 120, and 300 Ω) arranged in a pi network. (2-17)

39. In question 38, what is the current value through each resistor, and what is the total current? (2-17)

40. In question 38, what is the voltage-drop across each resistor? (2-17)

41. In question 38, what is the power dissipated in each resistor and the total power dissipation? (2-17)

42. Draw a simple schematic diagram showing the method of connecting three resistors of equal value so that the total resistance will be three times the resistance of one unit. (2-17)

43. Draw a simple schematic diagram showing how to connect three resistors of equal value so that the total resistance will be one-third of one unit. (2-17)

44. Draw a simple schematic diagram showing the method of connecting three resistors of equal value so that the total resistance will be two-thirds of the resistance of one unit. (2-17)

45. Two resistors of 20 and 25 Ω are connected in parallel. In series with this combination is a 11-Ω resistor. In parallel with the total combination is a 20-Ω resistor. The total current through the combination is 5 A. What is the current value in the 11-Ω resistor? In the 25-Ω resistor? (2-17)

NOTE: FCC license test questions are multiple-choice types, taking a form somewhat as follows:

46. Which meter (s) have three or more terminals? (a) Ammeter; (b) watthour meter; (c) voltmeter; (d) wattmeter. (2-9)

47. The basic Ohm's-law formula states I is equal to (a) P/E; (b) Q/T; (c) E/R; (d) P/R. (2-1)

48. A 500-Ω impedance source will develop maximum power into a load with (a) 0 Ω impedance; (b) 500 Ω impedance; (c) 1,000 Ω impedance; (d) infinite Ω impedance. (2-18)

ANSWERS TO CHECKUP QUIZ ON PAGE 44

1. (0.167 A)(0.167 A)(0.1 A)(20 V)
2. (0.0203 A)(8.12 mA)(0.406 V)   3. (84 Ω)(303 Ω)
4. (12BE6 across line; 6BQ6 in series with 6SN7; 6C4 and resistor all paralleled)(14 Ω; 2 × 2.84, or 5 W, is OK)
5. (5 V, 0.5 A; 5 V, 0.2 A; 8.4 V, 0.42 A; 8.4 V, 0.28 A)

---

3 MAGNETISM

The objective of this chapter is to present the basic theories and terms regarding magnetism that will be used later in explaining inductors, transformers, ac machinery, and various other devices.

3-1 MAGNETISM AND ELECTRICITY

Any wire carrying a current of electrons is surrounded by an unseen area of force called a magnetic field. For this reason, any study of electricity or electronics must consider magnetism.

Almost everyone has had experiences with magnets or with pocket compasses at one time or another. A magnet attracts pieces of iron but has little effect on practically everything else. Why does it single out the iron? A compass, when laid on a table, swings back and forth, finally coming to rest pointing toward the North Pole of the world. Why does it always point in the same direction?

These and other questions about magnetism have puzzled scientists for hundreds of years. It is only comparatively recently that theories that seem to answer many of the perplexing questions that arise when magnetism is investigated have been developed.

Radio and electronic apparatus such as relays, circuit breakers, earphones, loudspeakers, transformers, chokes, magnetron tubes, television tubes, phonograph pickups, tape and disk recorders, microphones, meters, vibrators, motors, and generators depend on magnetic effects to make them function. Every coil in a radio receiver or transmitter is utilizing the magnetic field that surrounds it when current is flowing through it. But what is meant by the term *magnetic field?*

3-2 THE MAGNETIC FIELD

An electron at rest has a negative *electrostatic* field of force surrounding it, as explained in Chap. 1. When energy is imparted to an electron to make it move, a new type of field develops around it, at right angles to its electrostatic field. Whereas negative electrostatic lines of force are considered as radiating outward from an electron, the *electromagnetic field* of force develops as a ring around a moving electron, at right angles to the path taken, around the wire in Fig. 3-1. (A proton,

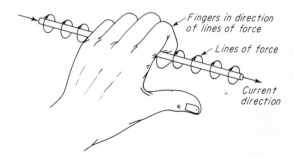

Fig. 3-1 With the thumb pointing in the direction of current, the fingers of the LEFT hand indicate the magnetic-field direction.

besides having an electrostatic positive field, also develops an electromagnetic field around it, but only when moving.)

Electrons orbiting around the nucleus of an atom or a molecule produce electromagnetic fields around their paths of motion. In most cases, these magnetic fields are either balanced or neutralized by the magnetic effect of any proton movement in the nucleus, or the movement of one orbital electron is counteracted by another orbital electron whirling in an opposite direction. In almost all substances the net result is little or no external magnetic field.

In the case of an electric conductor carrying current, the concerted movement of electrons along the wire produces a magnetic field around the conductor. The greater the current, the more intense the magnetic field.

Under normal circumstances, the field strength around a current-carrying conductor varies inversely as the distance from the conductor. At twice the distance from the conductor the magnetic-field strength is one-half as much, at five times the distance the field strength is one-fifth, and so on. At a relatively short distance from the conductor the field strength may be quite weak.

To indicate the presence of a magnetic field around a wire, circular lines may be drawn, as in Fig. 3-1. Note that the lines of force are given direction by arrowheads drawn on them. The arrowheads do not mean that the lines of force are moving in this direction but only that a relative polarity is present in them. The direction of these lines is determined in relation to the direction of the electron movement, or current flow. In the illustration the electron flow is indicated to be from left to right. If the left hand grasps a current-carrying conductor with the thumb extended in the direction of the current, the fingers will indicate the accepted field direction of the magnetic lines of force. This is known as the *left-hand magnetic-field rule*. If the current flows in the opposite direction, the left hand must be turned over so that the thumb points in the direction of the current, and the direction of the lines of force will be opposite to that shown.

If the direction of the lines of magnetic force is known, the same rule can indicate the current direction in a conductor. It is then known as the *left-hand current rule*.

When the current in a conductor is increased, more electrons flow, the magnetic-field strength increases, and the whole field extends farther outward.

In many electric circuits the current flows

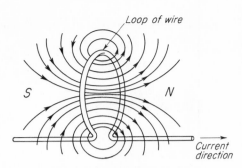

Fig. 3-2 By looping a conductor, magnetic lines of force are concentrated in the central core area of the loop.

through a coil of several turns of wire. Figure 3-2 illustrates the magnetic field as it might be set up around a single turn. Note that the indicated field direction conforms to the left-hand rule and that the greatest concentration of lines of force appears at the center, or the *core*, of the turn.

When several turns of wire are formed into a coil, the lines of force from each turn add to the fields of the other turns and a more concentrated magnetic field is produced in the core of the coil (Fig. 3-3). The more current and the more turns,

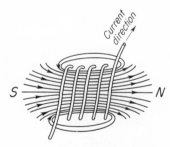

Fig. 3-3 By forming a coil with several loops, the field is concentrated and an electromagnet is produced.

the stronger the magnetic field of such an *electromagnet*. The direction of the field of force can be reversed by reversing the current direction or by reversing the winding direction.

At one end of the coil the field lines are leaving, and at the other end they are entering. When a coil or piece of metal has lines of force leaving one end of it, that end is said to have a *north* pole. The end with the lines entering is the *south* pole. Note that each line of force is actually a complete loop and has no north or south points on it.

The terms "north" and "south" indicate *magnetic* polarity, just as "negative" and "positive" indicate *electrostatic* polarity. They should not be used interchangeably. The negative end of a coil is the end connected to the negative terminal of the source and does not refer to the north or south magnetic polarity of the coil.

A different *left-hand coil rule* can be explained by Fig. 3-3. If a coil is grasped by the left hand with the fingers pointing in the direction of current flow, the thumb indicates the *north* end of the coil and also the direction of lines of force in the core. Conversely, if the direction of the lines of force through the core is known, the same left-hand coil rule can be applied to determine the current direction in the coil.

All magnetic lines of force are complete loops and may be considered somewhat similar in their action to stretched rubber bands. They will contract back into the circuit from which they came as soon as the force that produced them ceases to exist.

Magnetic lines of force never cross each other. When two lines have the same direction, they will oppose mechanically if brought near each other. This is illustrated in Fig. 3-4, which shows a

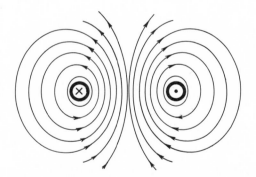

Fig. 3-4 Similar-direction magnetic lines of force always repel.

cross-sectional view of two wires with current flowing in opposite directions in them. Where they are adjacent, the lines from both wires have a similar and upward direction and repel each other. The two wires try to push apart. The little cross in the wire represents the rear view of a current-direction arrow, indicating the current to be flowing away from the viewer. A dot represents the

point of a current-direction arrow approaching the viewer.

Magnetic lines of opposite direction or polarity are attracted to each other. The loops surrounding wires carrying current in the same direction are opposite in polarity where they are adjacent, as shown in Fig. 3-5. Such loops can

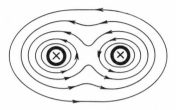

Fig. 3-5 Opposite-direction magnetic lines attract and may join.

attract each other and join into single large loops encircling both the wires carrying the current. The two wires will now be subjected to a common contracting force that tends to pull them together physically.

3-3 FLUX DENSITY, *B*

The complete magnetic field of a coil is known as the flux and is usually denoted by the Greek letter ϕ (phi). If a current flowing through a certain coil produces 100 lines of force in the core of the coil, it may be said that the core has a flux of 100 lines ($\phi = 100$ lines).

Flux density of a magnetic field is the number of lines per square inch (in English units) and is indicated by the letter *B*. In the example above, the core has a total of 100 lines. If the cross-sectional area of the core is 2 in.^{2}, the flux density *B* is 100/2, or 50 lines/in.^{2}.

3-4 Mmf, $\mathcal{F}$, and FIELD INTENSITY, *H*

The energy that produces magnetic lines of force is contained in the movement of electrons (current). The *magnetomotive force,* or mmf, is the force that produces the flux in a coil. Magnetomotive force may be computed by multiplying current flow in amperes by the number of turns (amp-turns) in a coil, or

$$\mathfrak{F} = NI$$

where $\mathfrak{F}$ = mmf, in amp-turns
N = number of turns in coil
I = current, in A

Past the ends of the coil, the magnetic field starts spreading and flux density begins decreasing. Within the core itself, however, lines of force are relatively straight and parallel and present a nearly constant flux density. A practical unit of measurement of the magnetizing force working on the column of material forming the core of the coil is determined by dividing the total mmf by the core length in inches. This reduces the total mmf to a unit for comparison and computation. Thus, a 3-in. coil with 60 amp-turns has a standardized magnetizing force, or field intensity, of 60/3, or 20 amp-turns/in. This magnetizing force is represented by H. In this case, $H = 20$ amp-turns/in.

In summary, flux, ϕ, indicates the total number of lines produced. Flux density, B, is the number of lines per square inch at some point. The total magnetizing force, $\mathfrak{F}$, is in ampere-turns, or NI. The magnetizing force, or field intensity, H, for a standard unit length of the core is in ampere-turns per inch.

3-5 PERMEABILITY, μ

When a coil of wire is wound with air as the core, a certain flux density will be developed in the core for a given value of current. If an iron core is slipped into the coil, a very much greater flux and flux density will exist in the iron core than was present when the core was air, although the current value and the number of turns have not changed.

With an air-core coil the air surrounding the turns of the coil may be thought of as pushing against the lines of force and tending to hold them close to the turns. With an iron core, however, the lines of force find a medium in which they can exist much more easily than in air. As a result, lines that were held close to the turns in the air-core coil are free to expand into the highly receptive area afforded by the iron. This allows lines of force that would have been close to the surface of the wire to expand into the iron core. Thus, the iron core produces a greater flux density, B, although no more magnetizing force (NI)

may have been developed than in the air-core coil. The iron core merely brings the lines of force out where they can be more readily used and concentrates them.

When comparing the ability of different materials to accept or allow lines of force to exist in them, air may be considered as the standard of comparison. A magnetizing force H of 20 amp-turns/in. will always produce a flux density B of 20 lines/in.^{2} in an *air-core* coil. By substituting a core of fairly pure iron, the same 20 amp-turns/in. may produce a flux density of perhaps 200,000 lines/in.^{2} in the core.

The ratio of B to H, or B/H, expresses the ability of a core material to accept, or be *permeated* by, lines of force. The permeability of most substances is very close to that of air, which may be considered as having a value of 1. A few materials, such as iron, nickel, and cobalt, are highly permeable, with permeabilities of several hundred to several thousand times that of air. (Note that the word "permeability" is a derivation of the word "permeate," meaning "to pervade or saturate," and is not related to the word "permanent.")

Permeability is represented by the Greek letter μ (mu, pronounced mew). In formula form

$$\mu = \frac{B}{H}$$

where μ = permeability (no unit)
B = flux density, in lines/in.^{2}
H = magnetizing force, in amp-turns/in.

In the example above, the iron has a permeability equal to B/H, or 200,000/20, or 10,000. In practice this is not particularly high. Pure iron approaches a permeability of 200,000. Special alloys of nickel and iron may have a permeability of 100,000. Alloying iron makes it possible to produce a wide range of permeabilities. A cast iron can be made that has almost unity (1) permeability. Most stainless steels exhibit practically no magnetic effect, although some may be magnetic.

Any substance that is not affected by magnetic lines of force and is reluctant to support a magnetic field is said to have the property of *reluctance,* the symbol of which is $\mathfrak{R}$. Air, vacuum, and most substances have unity reluctance, while iron has a very low reluctance. In formula form,

$$\mathcal{R} = \frac{\mathcal{F}}{\phi}$$

In electric circuits the reciprocal of resistance is called *conductance*. In magnetic circuits the reciprocal of reluctance is called *permeance*, symbol $\mathcal{P}$. Thus $\mathcal{P} = 1/\mathcal{R}$, and $\mathcal{R} = 1/\mathcal{P}$. Air has unity permeance, whereas iron has a high permeance.

Permeability may also be considered as permeance per unit volume. Thus permeability is lines of force (ϕ) per ampere-turn (NI) per cubic inch. When speaking of a magnetic circuit, "permeance" is used. "Permeability" is used when discussing how magnetic materials behave.

Test your understanding; answer these checkup questions.

1. Under what condition does an electron have a magnetic field? _____
2. When a preponderance of electrons are rotating around a nucleus in the same direction, what does the atom then form? _____
3. Do arrowheads on magnetic lines of force indicate movement in that direction? _____
4. Which magnetic pole has lines pointing into it? _____ Which has lines pointing out? _____
5. When the fingers of the left hand are around a coil in the direction of current flow, to what does the thumb point? _____
6. When current ceases in a coil, what happens to the magnetic-field lines? _____
7. Where in a current-carrying coil is the magnetic field most intense? _____
8. What does a little cross marked on the end of a current-carrying wire indicate? _____
9. If two parallel wires have similar-direction current flowing in them, what physical effect will be produced? _____
10. What is the symbol for magnetic flux? _____ In what English unit is it measured? _____
11. What is the symbol for flux density? _____ In what English unit is it measured? _____
12. What is the symbol for magnetomotive force? _____ In what English unit is it measured? _____
13. What is the symbol for field intensity? _____ In what English unit is it measured? _____
14. What is the permeability symbol? _____ The formula? _____
15. What is the μ of air? _____ Of pure iron? _____
16. What is reluctance? _____ What is its symbol? _____ Its formula? _____
17. What is the reciprocal of reluctance? _____ Its symbol? _____
18. Where do magnetic lines go when they collapse back into a wire? _____

3-6 THE ATOMIC THEORY OF MAGNETISM

The discussion here will be a considerably condensed version of the atomic theory of magnetism.

From atomic theory it is known that an atom is made up of a nucleus of protons surrounded by one or more electrons encircling it. The rotation of electrons and protons in most atoms is such that the magnetic forces cancel each other. Atoms or molecules of the elements iron, nickel, and cobalt arrange themselves into magnetic entities called *domains*. Each domain is completely magnetized.

Groups of domains form crystals of the magnetic material. The crystals may or may not be magnetic, depending on the arrangement of the domains in them. Investigation shows that while any single domain is fully magnetized, the external resultant of all the domains in a crystal may be a neutral field.

Each domain has three directions of magnetization: *easy* magnetization, *semihard* magnetization, and *hard* magnetization. If an iron crystal is placed in a weak field of force, the domains begin to line up in the easy direction. As the magnetizing force H is increased, the domains begin to roll over and start to align themselves in the semihard direction. Finally, as the H is increased still more, the domains are lined up in the hard direction. When all the domains have been lined up in the hard direction, the iron is said to be *saturated*. An increase in magnetizing force will then produce no more magnetic change in the material.

The result of this action can be seen in the BH curve for a piece of iron shown in Fig. 3-6. The graph shows that an increase of H produces an increase of B. At low values of magnetizing force the iron produces relatively little flux density. The permeability, or B/H, is relatively low under this condition. As the magnetization is increased, the flux density increases rapidly in respect to the increase of magnetizing force, resulting in a large B/H ratio, or a high value of permeability. As saturation is approached, further increase in H

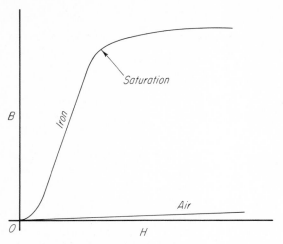

Fig. 3-6 Magnetization or *BH* curves for iron and air.

produces little increase in *B*. THis general curve form is to be expected of all magnetic materials, although the steepness of the curve changes with different permeabilities. The *BH* curve of air does not bend.

3-7 FERROMAGNETISM

Substances that can be made to form domains are said to be *ferromagnetic*, which means "iron-magnetic." The ferromagnetic elements are iron, nickel, and cobalt, but it is possible to combine some nonmagnetic elements and form a ferromagnetic substance. For example, in the proper proportions, copper, manganese, and aluminum, each by itself being nonmagnetic, produce an alloy which is similar to iron magnetically.

Materials made up of nonferromagnetic atoms, when placed in a magnetic field, may weakly attempt either to line up in the field or to turn at right angles to it. If they line themselves in the

direction of the magnetic field, they are said to be *paramagnetic*. If they try to turn from the direction of the field, they are called *diamagnetic*. There are only a few diamagnetic materials. Some of the more common are gold, silver, copper, zinc, and mercury. All materials which do not fall in the ferromagnetic or diamagnetic categories are paramagnetic. The greatest percentage of substances are paramagnetic.

Ferromagnetic substances will resist being magnetized by an external magnetic field to a certain extent. It takes some energy to rearrange even the easy-to-move domains. Once magnetized, however, ferromagnetic substances may also tend to oppose being demagnetized. They are said to have *retentivity,* or *remanence,* the ability to retain magnetism when an external field is removed.

As soon as the magnetizing force is released from a magnetized ferromagnetic substance, it tends to return at least part way back to its original unmagnetized state, but it will always retain some magnetism. This remaining magnetism is *residual magnetism*. Paramagnetic and diamagnetic materials always become completely nonmagnetic when an external magnetizing force is removed from them.

3-8 THE HYSTERESIS LOOP

When a magnetizing force *H* is applied to a piece of completely demagnetized ferromagnetic material, the flux density (*B*) rises as shown by the dotted curve *X* in Fig. 3-7. If the magnetizing force (*H*) is removed from the ferromagnetic ma-

Fig. 3-7 Magnetization of iron when subjected to a magnetizing force, dotted line *X*. Partial loss of magnetism when the magnetizing force is removed, line *Y*.

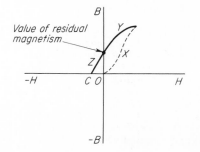

terial, the flux density decreases but drops back only part way to zero, as shown by the curve Y. The opposition of the domains to roll back to the unmagnetized state is known as *hysteresis*. The magnetism left in the metal is residual magnetism. For soft iron the residual magnetism will be small, while for steels and iron alloys it will be considerably greater. With the latter materials the B value will not drop so far toward zero.

If a magnetizing force in the opposite direction is now applied, a certain value of this −H will be necessary to counteract the residual magnetism and bring the Z portion of the curve down to point C. This point represents zero flux density, or complete demagnetization.

The opposite-direction magnetizing force, or −H, required to demagnetize the material completely is known as the *coercive force* and is indicated by the length of the line O to C.

When an iron core has a coil of wire wrapped around it and an alternating current is made to flow through the coil, a curve called a *hysteresis loop,* as shown in Fig. 3-8, may be drawn to

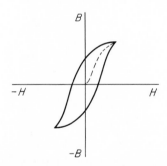

Fig. 3-8 Complete hysteresis loop in an iron core produced by one cycle of ac in the coil surrounding the core.

indicate the flux-density variations under changing magnetizing forces. The dotted line shows the magnetization when the unmagnetized iron was first subjected to the force. The remainder of the loop indicates the flux density (B) that results when the magnetizing force (H) is alternately reversed.

As current through the coil increases in one direction, the H increases, producing an increase in B. As the current decreases, the H decreases, resulting in a lessening of B. As the current reverses, the H increases in the opposite direction,

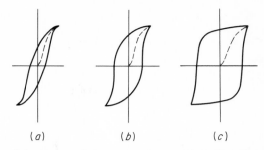

Fig. 3-9 Hysteresis loops for (a) soft iron, (b) steel, (c) Alnico-type materials.

the domains are rolled over, and the B follows in the opposite direction.

With a full cycle of ac a hysteresis loop enclosing a given area is developed. A soft iron will be represented by a slim loop (Fig. 3-9a). Steel results in a broader loop (Fig. 3-9b). Special iron alloys, such as Alnico, have a loop approaching a rectangle in shape (Fig. 3-9c).

An expenditure of a certain amount of energy is required to reverse the magnetism, or realign the domains, in a piece of ferromagnetic material. The energy lost due to hysteresis appears as heat in the magnetic material. For this reason, when transformers using ac are made with iron cores, it is necessary to select metals with narrow hysteresis loops and with little *hysteretic loss.* Even so, the core of a transformer always heats a little because of loss of energy due to hysteresis.

3-9 PERMANENT AND TEMPORARY MAGNETS

A ferromagnetic substance that holds magnetic-domain alignment well (has a high value of retentivity) and has a broad hysteresis loop is used to make permanent magnets. One of the strongest permanent magnets is made of a combination of iron, aluminum, nickel, and cobalt called *Alnico.* It is used in horseshoe magnets, electric meters, headphones, loudspeakers, radar transmitting tubes, and many other applications. Some magnetically hard, or permanent-magnet, materials are cobalt steel, nickel-aluminum steels, and special steels.

Ferromagnetic metals that lose magnetism easily (have a low value of retentivity) make temporary magnets. They find use in transformers, chokes, relays, and circuit breakers. Pure iron and Permalloys (*perm* derived from "permeable," not from "permanent") are examples of magnetically soft, temporary-magnet materials. Finely powdered iron, held together with a nonconductive binder, is used for cores in many applications. These are called *ferrite* cores.

3-10 MAGNETIZING AND DEMAGNETIZING

There are two simple methods of magnetizing a ferromagnetic material. One is to wrap a coil of wire around the material and force a direct current through the coil. If the ferromagnetic material has a high value of retentivity, it will become a permanent magnet. If the material being magnetized is heated and allowed to cool while subjected to the magnetizing force, a greater number of domains will be swung into alignment and a greater permanent flux density may result. Hammering or jarring the material while under the magnetizing force also tends to increase the number of domains that will be affected.

A less effective method of magnetizing is to stroke a high-retentivity material with a permanent magnet. This will align some of the domains of the material and induce a relatively weak permanent magnetism.

If a permanent magnet is hammered, many of its domains will be jarred out of alignment and the flux density will be lessened. If heated, it will lose its magnetism because of an increase in molecular movement that upsets the domain structure. Strong opposing magnetic fields brought near a permanent magnet may also decrease its magnetism. It is important that equipment containing permanent magnets be treated with care. The magnets must be protected from physical shocks, excessive temperatures, and strong alternating or other magnetic fields.

When tools or objects such as screwdrivers or watches become permanently magnetized, it is possible to demagnetize them by slowly moving them into and out of the core area of a many-turn coil in which a relatively strong ac is flowing. The ac produces a continually alternating magnetizing force. As the object is placed into the core area, it is alternately magnetized in one direction and then the other. As it is pulled farther away, the alternating magnetizing forces become weaker. When it is finally out of the field completely, the residual magnetism will usually be so low as to be of no consequence.

3-11 THE MAGNETIC CIRCUIT

The magnetomotive force, the reluctance of the magnetic material used, and the flux developed are often likened to the electric circuit, in which electromotive force applied across resistance produces current. The similarities are:

$E =$ electromotive force (emf) $\mathcal{F} =$ magnetomotive force (mmf)
$R =$ resistance $\mathcal{R} =$ reluctance
$I =$ current $\phi =$ flux

In the magnetic circuit the idea that the total flux is directly proportional to the mmf and inversely proportional to the reluctance of the material used can be expressed by the formula

$$\phi = \frac{\mathcal{F}}{\mathcal{R}}$$

where $\phi =$ flux, in lines
$\mathcal{F} =$ mmf, in amp-turns
$\mathcal{R} =$ reluctance (no unit)

This is sometimes referred to as *Ohm's law for magnetic circuits*.

Since the permeability graph of all magnetic materials is curved, or *nonlinear* (meaning "not a straight line"), the reluctance value of magnetic materials will also be nonlinear under varying values of H. Therefore, unless permeability curves for the metal being investigated are available, accurate magnetic-circuit computations are not possible.

3-12 ENGLISH, CGS, MKSA, AND SI UNITS

Magnetism has been explained in English terms, such as inches, square inches, and ampere-turns. However, there are two other systems that are also used when working with magnetism. The cgs, or centimeter-gram-second, was used for many years before the rationalized mksa, or meter-kilogram-second-ampere, system with its SI

Table 3-1 **MAGNETIC UNITS**

Term and symbol	English units		Cgs units		Mksa (SI) units, rationalized
Magnetic flux (ϕ)	1 LINE OF FORCE	=	1 MAXWELL (Mx)	=	10^{-8} weber (Wb)
	10^8 lines of force	=	10^8 Mx	=	1 WEBER
Magnetomotive force ($\mathcal{F}$)	1 AMPERE-TURN	=	1.26 gilbert (Gb)	=	1 AMPERE-TURN
	0.796 amp-turn	=	1 GILBERT	=	0.796 amp-turn
Magnetizing force (H)	1 AMPERE-TURN/INCH	=	0.495 oersted (Oe)	=	39.4 amp-turns/m (NI/m)
	2.02 amp-turns/in.	=	1 OERSTED, or 1 Gb/cm	=	79.6 amp-turns/m
	0.0254 amp-turn/in.	=	0.0126 Oe	=	1 AMPERE-TURN/METER
Magnetic flux density (B)	1 LINE/INCH²	=	0.155 gauss (G)	=	0.155×10^{-4} Wb/m²
	6.45 lines/in.²	=	1 GAUSS, or 1 Mx/cm²	=	10^{-4} Wb/m²
	6.45×10^4 lines/in.²	=	10^4 G	=	1 TESLA (T), or 1 Wb/m²

units (Système International d'Unités) was developed. Table 3-1 illustrates the relationship between the terms used in these three magnetic systems.

Test your understanding; answer these checkup questions.

1. What does "ferromagnetic" mean? _____ Into what basic entities do molecules of ferromagnetic substances form? _____
2. In what category is a substance that tries to turn at right angles to a magnetic field? _____ That tries to line up in the direction of a magnetic field? _____
3. When an increase of H can produce no more B, what has been reached? _____
4. Which would have no retentivity, paramagnetic, diamagnetic, or ferromagnetic materials? _____ Which would have no hysteresis loop? _____
5. Is $+H$, $-H$, $+B$, or $-B$ required to produce coercive force in a material? _____
6. What shape hysteresis loop must a material have to be a good relay core? _____ A permanent magnet? _____
7. What is finely powdered iron in a nonconductive binder called? _____ Would it have a high or low resistance value? _____
8. What happens to a permanent magnet in the core of a coil carrying a heavy alternating current? _____ Why might it heat? _____
9. What is the Ohm's-law formula for magnetic circuits? _____
10. In the English magnetic system what is the unit of measurement for ϕ? _____ For $\mathcal{F}$? _____ For H? _____ For B? _____
11. In the cgs system what is the unit of measurement for ϕ? _____ For $\mathcal{F}$? _____ For H? _____ For B? _____
12. In the mksa system what is the unit of measurement for ϕ? _____ For $\mathcal{F}$? _____ For H? _____ For B? _____

3-13 PERMANENT-MAGNET FIELDS

When a piece of magnetically hard material is subjected to a strong magnetizing force, its domains are aligned in the same direction. When the magnetizing force is removed, many of the domains remain in the aligned position and a permanent magnet results. A north pole is any place where the direction of the magnetic lines of force is outward from the magnet, as in Fig. 3-10. A

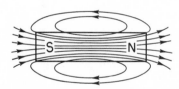

Fig. 3-10 Magnetic poles are determined by direction of lines of force: out of north, into south.

south pole is any place where the direction of the lines is inward.

It has been explained that the lines of force surrounding current-carrying wires oppose each other if the lines have the same direction and attract each other if they are of opposite direction. This is true of the fields of permanent magnets. When similar poles are held near each other, lines of similar direction are opposing, tending to push the magnets apart physically (Fig. 3-11a). Unlike, or dissimilar, poles held near each other produce a physically attracting effect because the lines of force from both magnets join as long, contracting loops (Fig. 3-11b).

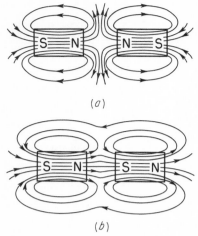

(a)

(b)

Fig. 3-11 Magnetic fields with (a) like lines repelling and pushing magnets apart, and (b) unlike lines attracting, joining, and pulling the two magnets together.

3-14 MAGNETIC-FIELD DISTORTION

Under normal circumstances a permanent magnet made into bar shape and suspended in the air will have a field surrounding it somewhat similar to that shown in Fig. 3-10.

When a piece of highly permeable material such as pure iron and a piece of low-permeability material such as copper are placed in the field, as in Fig. 3-12, the field pattern becomes distorted. The copper, having practically the same permeability as air, has almost no effect on the magnetic lines of force. The iron, being thousands of times more permeable than air, is a highly acceptable medium in which the lines of force may exist, and most of the lines on that side and some of the lines from the other side of the magnet detour into the iron. This produces a distorted field pattern and results in a highly concentrated field between the magnet and the piece of iron. This ability to

---

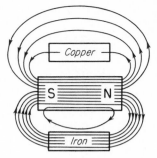

Fig. 3-12 Distortion of a bar-magnet field by iron and lack of distortion by low-permeability material.

produce a concentrated field by using iron to close the magnetic gap between north and south poles of magnets is often employed in electronics and radio.

If a magnet is completely encased in a magnetically soft iron box, all its lines of force remain in the walls of the box and there is no external field. This is known as magnetic *shielding*. Shielding may be used in the opposite manner. An object completely surrounded by an iron shield will have no external magnetic fields affecting it, as all such lines of force will remain in the permeable shield.

3-15 THE MAGNETISM OF THE EARTH

Enough of the ferromagnetic materials making up the earth have domains aligned in such a way that the earth appears to be a huge permanent magnet. The direction taken by the lines of force surrounding the surface of the earth is inward at a point near what is commonly known to be the North Pole of the world and outward near the earth's South Pole. This results in a rather confusing set of facts. The geographical North Pole of the earth is actually near its South Magnetic Pole, and the geographical South Pole is near the North Magnetic Pole, as shown in Fig. 3-13.

The familiar magnetic navigational compass consists of a small permanent magnet balanced on a pivot point. The magnetic field of the compass needle lines itself up in the earth's lines of force. As a result, the magnetic north end of the compass needle is pulled toward the earth's South Magnetic Pole, since unlike poles attract each other. This means that when the "north-

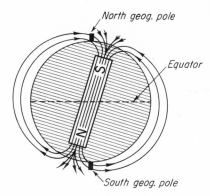

Fig. 3-13 Representation of the magnetic field of the earth.

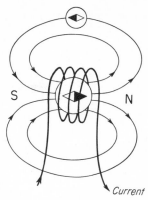

Fig. 3-15 Compass indications inside and outside a current-carrying coil.

pointing end'' of the compass is pointing toward the geographical north, this end (a magnetic north pole) is actually pointing toward a magnetic south pole.

3-16 USING A MAGNETIC COMPASS IN ELECTRICITY

A small magnetic pocket compass moved to different points in a field produced by a permanent magnet will indicate the magnet's field (Fig. 3-14). In the same manner, the direction of the

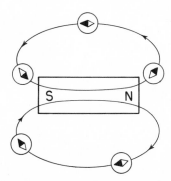

Fig. 3-14 Compass can indicate the contour of a magnetic field. (Dark ends are north.)

lines of force surrounding a coil carrying a direct current can be checked. Note the indication given by the compass needle when inside the coil (Fig. 3-15): the north end of the compass will point to the north end of the coil. Outside, the compass

lines up in the lines of force pointing toward the south end of the coil.

If the current direction in the coil is reversed, the compass will reverse. When an ac is flowing in the coil, the compass may not be able to follow rapid field reversals and will give no indication. If the ac is strong enough, the compass may be demagnetized and become useless.

The direction of current in a wire can be determined with a compass. Current flowing along a wire sets up lines of force around it. If a compass is brought near the wire, the needle will swing at right angles to the current and in the direction of the lines of force surrounding the wire. The north pole of the compass indicates the direction of the lines of force around the wire (Fig. 3-16). By

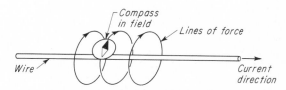

Fig. 3-16 A compass indicates the direction of the magnetic field around a current-carrying wire.

applying the left-hand current rule, the direction of current can be determined.

If the current direction is reversed, the direction of the lines of force will reverse and the compass needle will swing around 180°. As in a coil, the compass will not indicate if the current is alternating.

3-17 ELECTRONS MOVING IN A MAGNETIC FIELD

Electrons will travel in a straight line in a magnetic field provided they are moving in a direction parallel to the lines of force.

When electrons are propelled across lines of force, they will move in a curved path and at right angles to the lines of force. Figure 3-17 illustrates

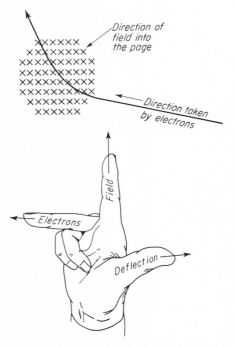

Fig. 3-17 Direction taken by electrons moving through a magnetic field and illustration of the right-hand motor rule.

the path that will be taken by electrons moving into the magnetic field shown. The crosses represent the rear view, or feathers, of arrows that indicate the direction of lines of force pointing into the page.

The direction taken by the electrons as they curve through a magnetic field can be determined by using the *right-hand motor rule,* as illustrated. This states: With the thumb, first, and second fingers of the *right* hand held at right angles to each other, the first finger pointing in the direction of the magnetic field, the second finger in the direction of the electron movement, then the

thumb will indicate the deflection of the electron. Check the electron path taken in the illustration by this rule.

If a wire is carrying the electrons through the magnetic field, an upward pressure on the electrons tends to move the wire in an upward direction. This is the principle by which electric motors operate and is discussed in Chap. 23.

Free electrons moving through magnetic deflecting fields are found in television picture tubes.

3-18 GENERATING AN EMF BY MAGNETIC MEANS

When a wire is moved parallel to lines of force, no effect is produced on free electrons in the wire. However, when a conductor is moved upward through the magnetic field as shown in Fig. 3-18, an electron flow is induced toward the right in the conductor. This current direction may be determined by using the *left-hand generator rule.* This three-finger rule is somewhat similar to the right-hand motor rule. It may be stated: With the thumb, first, and second fingers of the *left* hand held at right angles to each other, if the first finger is pointed in the direction of the magnetic field and

Fig. 3-18 Direction of induced emf when a conductor moves through a magnetic field and illustration of the left-hand generator rule.

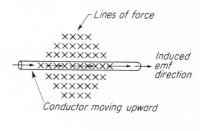

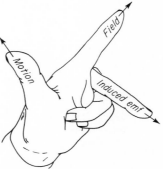

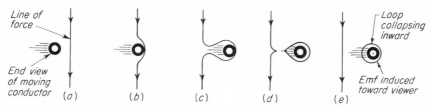

Fig. 3-19 How an emf is induced into a wire cutting through a line of force.

the thumb is pointed in the direction of the conductor's motion, then the second finger indicates the direction of induced emf in the wire and any resultant current flow.

An explanation of the induction of a current in a conductor (Fig. 3-19) may be given as follows: When a wire crosses a magnetic line of force (a), the line is stretched around the wire (b), rejoins itself behind the wire (c), and the little loop developed around the wire collapses inward (d). The free electrons in the wire move at right angles to the inward-collapsing magnetic loop (e), and are forced along the wire. The induced current in the wire due to the collapsing loop is outward from the page in the illustration. Check this with the left-hand generator rule. (HINT: The collapsing-field motion is inward, which is the same as an outward motion of the conductor!)

While the explanation is given in terms of an induced current, it is usually considered to be an emf that is induced in a conductor moving across a magnetic field. Both statements are correct, since producing a difference of potential between the two ends of the conductor requires some of the free electrons in the wire to be moved to one end from the other end. This is considered a charging current. If a load is connected across the ends of the conductor, many more electrons will be forced into movement. Induced currents always produce expanding magnetic fields around their conductors in a direction that opposes the original magnetic field. (This is called Lenz's law.) Physical or mechanical energy will now be required to move the conductor through the field. With no electrical load across the conductor, almost no energy is required to move it. This is the basic theory of the electric generator.

The factors that determine the amplitude, or strength, of the induced emf are:

1. Strength of the magnetic field (number of lines per square inch)

2. Speed of the conductor motion across the lines of force
3. Number of conductors connected in series (as in a generator)

These factors can be simply expressed by: "The greater the number of lines cut per second, the higher the voltage induced." When magnetic lines are cut at a rate of 10^8 (100,000,000) per second, an average emf of 1 V is produced.

3-19 MAGNETOSTRICTION

In some electronics and communications equipment the property of ferromagnetic materials to expand or contract when they are subjected to a magnetizing force is used. This effect is called *magnetostriction*. Nickel constricts when under a magnetizing force; iron expands. A part of the magnetic energy stored in such metals is convertible to mechanical energy in this way.

The maximum contraction or expansion occurs at magnetic saturation. The direction of the magnetizing force has no effect on the direction of strain. For example, when a piece of nickel is wrapped with a coil of wire, it will contract regardless of the direction of any current fed into the coil.

When the magnetizing force is removed, the magnetostrictive material springs back to its normal shape or size. The mechanical energy is converted back into magnetic energy and can be converted further to electric energy. If the current used as the magnetizing force is continuously pulsating or alternating, the material will continuously vibrate mechanically.

3-20 RELAYS

A relay is a relatively simple magnetic device that normally consists of a coil, a ferromagnetic

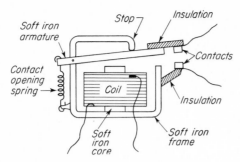

Fig. 3-20 Functional parts of a SPST NO dc relay.

core, and a movable armature on which make and break contacts are fastened. Figure 3-20 illustrates one of the simpler relays used to close a circuit when the coil is energized. It is known as a single-pole single-throw (SPST), normally open (NO) or "make-contact" relay.

The core, the U-shaped body of the relay, and the straight armature bar are all made of magnetically soft ferromagnetic materials having high permeability and little retentivity. One of the relay contacts is attached by an insulating strip to the armature, and the other to the relay body with an insulating material. The contacts are electrically separated from the operational parts of this particular relay. A spring holds the armature up and the contacts open.

When current flows in the coil, the core is magnetized and lines of force develop in the core and through the armature and the body of the relay. The gap between the core and the armature is filled with magnetic loops trying to contract. These contracting lines of force overcome the tension of the spring and pull the armature toward the core, closing the relay contacts. When the current in the coil is stopped, the magnetic circuit loses its magnetism and the spring pulls the armature up, opening the contacts.

Relays are useful in remote closing and opening of high-voltage or high-current circuits with relatively little voltage or current flow in the coil. Basic relays are indicated in Fig. 3-21.

In some cases it may be required to change the operating voltage of a relay coil.

EXAMPLE: A 6-V relay is to be used on 12 V. The magnetizing force to attract the armature is equal to the ampere-turns of the coil. If the coil has 6 Ω resistance and 300 turns, the current will be equal to $I = E/R$, or $6/6$, or 1 A. The ampere-turns required is 1 A times 300 turns, or 300 amp-turns. A resistance may be added in series with the coil to limit the current to 1 A when across 12 V. This resistance must produce a 6-V drop across it when 1 A is flowing through it. The resistance must be $R = E/I$, or $6/1$, or 6 Ω. However, the heat developed in the resistance is a complete waste. It may be more desirable to purchase a 12-V coil for the relay or to wind one. To wind one, the size of the wire in the 6-V coil must be determined. A wire having twice the resistance per unit length or 70% of the diameter must be used. This is obtained by using a wire having a gauge three units higher. For example, No. 27 wire has twice the cross-sectional area of No. 30 and therefore half the resistance. No. 30 has a diameter of 10 mils, which is 70% of the 14.2-mil diameter of No. 27 wire. When a coil is wound with the thinner wire to about the same size as the original coil, it will have four times the resistance and twice as many turns. With 12 V, $\frac{1}{2}$ A will flow through the 600 turns. This represents 300 amp-turns, the required magnetizing force.

If a 12-V relay coil is to be changed to operate on 6 V, a wire 1.4 times the diameter of the original should be used.

Relay contacts are usually made of silver or tungsten. Silver oxidizes but can be cleaned by using a very fine abrasive paper or a piece of ordinary letterhead paper rubbed between the contacts. If the contacts are pitted by heavy currents, they may be smoothed with a fine file, but the original shape of the contacts should be retained to allow a wiping action during closing to keep them clean.

Fig. 3-21 Schematic symbols of simple relays. (a) Normally open or make-contact relay. (b) Normally closed or break-contact relay. (c) Single-pole double-throw relay.

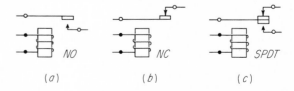

Test your understanding; answer these checkup questions.

1. Does a magnetic shield capture lines of force or repel them? _____
2. Which would shield a permanent-magnet field best, copper, lead, glass, iron, or aluminum? _____
3. The end of a compass needle that points to the geographic North Pole of the earth has what magnetic polarity? _____
4. What do the thumb and first two fingers indicate in the left-hand generator rule? _____ _____ _____ In the right-hand motor rule? _____ _____
5. "Induced currents always produce expanding magnetic fields around their conductors in a direction that opposes the original magnetic field" is an expression of what law? _____
6. How many lines of force must be cut per second to produce an average of 1 V? _____
7. What is the name of the effect whereby a ferromagnetic material changes size or shape when magnetized or demagnetized? _____
8. What would DPDT NC stand for when speaking of relays? _____
9. Why is silver a good material for relay contacts? _____ Why is tungsten? _____
10. What magnetic unit is important when determining how much pulling force a coil will have on a relay arm? _____

COMMERCIAL LICENSE QUESTIONS

Amateur license questions will be found in the Addendum.

FCC Elements 3, 4, and 6 require an ability to answer questions similar to those below. Sections in which questions are answered are shown in parentheses. A question followed by a bracketed number is required for that element alone.

1. Name at least five pieces of equipment which make use of electromagnets. (3-1)
2. What factors influence the direction of magnetic lines of force produced by an electromagnet? (3-2)
3. Compare some properties of electrostatic and electromagnetic fields. (1-8, 3-2)
4. What is meant by *ampere-turns?* (3-4, 3-12)
5. Define the term *permeability*. (3-5)
6. Define the term *reluctance*. (3-5)
7. Explain the theory of molecular alignment as it affects magnetic properties of materials. (3-6)
8. Define the term *residual magnetism*. (3-7, 3-8)
9. Define the term *hysteresis*. (3-8)
10. How can the direction of flow of dc electricity in a conductor be determined? (3-16)
11. How may a magnetic compass be affected when placed within a coil carrying an electric current? (3-16)
12. Which factors determine the amplitude of the emf induced in a conductor which is cutting magnetic lines of force? (3-18)
13. What material is frequently used for relay contacts? (3-20)
14. Explain the operation of a break-contact and a make-contact relay. (3-20)
15. Explain the method of cleaning relay contacts and why it is necessary that the original contact shape be maintained. (3-20) [4]

NOTE: FCC license test questions are multiple-choice types, taking a form somewhat as follows:

16. A magnetic domain is (*a*) without polarity; (*b*) a heterogeneously magnetized area; (*c*) an area surrounded by magnets; (*d*) an area of aligned molecules. (3-6)
17. What causes the core of a relay or transformer to heat when wire wrapped around it is fed ac? (*a*) Permeability; (*b*) reluctance; (*c*) hysteresis; (*d*) ferromagnetism. (3-8)
18. A break-contact relay (*a*) has two poles; (*b*) opens when its coil is magnetized; (*c*) is a nonmagnetic type; (*d*) is used to turn headlights on. (3-20)

ALTERNATING CURRENT

The objective of this chapter is to present some of the basic concepts and terminology regarding alternating current used in one way or another in all subsequent chapters. This includes the topics of frequency; cycles; alternations; peak, effective, and average values; and the frequency spectrum.

4-1 METHODS OF PRODUCING EMF

Some of the various methods by which an electromotive force can be produced were listed in Sec. 1-8. These can be further separated into methods of producing unidirectional pressure, called direct current (dc), and the generation of alternating-direction current (ac).

Methods of producing dc are *chemical* (in batteries, Sec. 22-2), *magnetic* (dc generators, Sec. 23-7), *thermal* (heated thermocouple junction, Sec. 12-19), *photoelectric* (conversion of light energy into movement of electrons, Sec. 25-3), *friction* (produced by rubbing two substances together, as an automobile in motion, walking across wool rugs, or stroking a cat); and *magnetohydrodynamic* (MHD, converting heat energy in hot gases directly to electric energy). Note that all these are merely methods of converting energy of one form to energy in electrical form.

Although alternating current has been men-

tioned only briefly so far, from this point on it becomes more important. Some methods by which it can be produced are *magnetic* (mechanical motion of a wire across alternate-direction magnetic lines induces alternating emf in the wire, Sec. 23-2), *magnetostrictive* (mechanical vibration of ferromagnetic materials induces an alternating emf in a wire coiled around the material, Sec. 3-19), and *piezoelectric* (mechanical vibration of quartz or rochelle salt crystals produces an alternating emf between two metal plates on opposite sides of the crystal, Secs. 13-12, 17-30).

4-2 A BASIC CONCEPT OF ALTERNATING CURRENT

A generator developing an emf that alternately forces electrons through a circuit in one direction, stops them, and then moves them in the opposite direction is called an *alternator*. It produces alternating voltages, which in turn can produce alternating currents in a circuit.

When it comes to doing electrical work, such as rotating motors, lighting lamps, and so on, either dc or ac can be used. In some cases it may be easier to do the job with one; in other cases, with the other.

Perhaps the biggest advantage of ac is the ease with which ac voltage can be stepped up or down by a transformer, in itself a relatively simple piece of equipment. When electric power is transported over long distances, much less power is lost in heating the power lines themselves if the voltage is high, because with higher voltages less current is required to produce the same amount of power

---

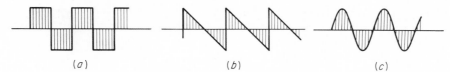

Fig. 4-1 Three possible ac waveshapes: (*a*) square wave, (*b*) sawtooth, (*c*) sine wave.

at the far end of the line. Power loss in a line is computed by the power formula $P = I^2R$. Since power loss is proportional to the current squared, anything that will lessen the required current will greatly lessen the power lost in heat. Doubling the voltage in a system reduces the required current to one-half for the same power. If power loss is proportional to the current squared, one-half squared equals one-quarter of the power loss in the resistance of the wires. This is why power companies transport electric energy over long distances at potentials of 120,000 to more than 500,000 V.

Alternating current or voltage can be produced in many forms. Figure 4-1*a* represents current flowing in one direction at a constant amplitude for a period of time, immediately reversing to the opposite direction of flow in the circuit for a period of time, again reversing to the first direction, and so on. This is a *square-wave* form of ac.

Figure 4-1*b* shows the current rising to a peak value instantaneously, slowly decreasing to zero, then slowly increasing in the opposite direction. This is a *sawtooth* form of ac.

Figure 4-1*c* shows current approaching its peak slowly. If the current increases as the sine of the angle (explained in Sec. 4-3), it is known as *sine-wave,* or *sinusoidal,* ac. Sine-wave ac is consid-

ered to be the perfect waveform. It is the type normally used by power companies and generated by radio stations. When ac is mentioned, it is considered to be sinusoidal unless specified to be otherwise.

It is also possible to have dc varying in sinusoidal, square, or sawtooth *waveform*.

4-3 THE AC CYCLE

The electron flow in an ac circuit is continually reversing. Each time it reverses, it is said to be *alternating*. Two of these alternations result in a *cycle*. Figure 4-2 shows two cycles or four alternations of ac.

The curved line represents the amount of current flow (or voltage) in a circuit. It indicates that current begins to flow at the point of time marked 0°, increases until the 90° point is reached, and then decreases in strength to the 180° point. This completes one-half of a cycle, or one *alternation*. The current now reverses direction and increases in strength or magnitude to the 270° point and then decreases to zero current at the 360° point. It has completed one cycle, 360°. The start of a new cycle has been reached. (A cycle may also be considered as starting at a 90° point and moving through to the next similar 90° point. Nor-

Fig. 4-2 Two cycles of sine-wave ac with magnitudes indicated every 30 electrical degrees.

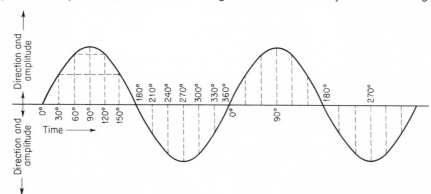

mally, however, a cycle is considered to start at 0°.)

The horizontal line in Fig. 4-2 on which the degrees are marked is known as the *time line*. It represents time progressing to the right.

A complete cycle is considered to be 360°, a half cycle to be 180°, and a quarter cycle to be 90°. A point on the time line one-third of the way from 0° to 90° is represented as 30°. A point two-thirds of the time from the zero point to the maximum is said to be 60°. The maximum points are normally at 90° and 270°.

A point 30° past 90° is 120°. The magnitude of a sinusoidal current, or voltage, at 120° will equal that of the current, or voltage, at the 60° point. This can be seen in the illustration.

If the cycle is a true sine wave, the magnitude of the current at the 30° point will be exactly one-half of the maximum value. It will have the same value at the 150°, 210°, and 330° points. Mathematically, the sine of 30° is 0.5, or one-half of the maximum value, whatever the maximum happens to be. The sine values of angles in degrees or to tenths of a degree can be found in Appendix C. At this time the sines of only four angles, 0°, 30°, 60°, and 90°, will be used.

The sine of 60° is 0.86. Therefore the amplitude of the cycle at 60° is 0.86 of the maximum value. The magnitude of the current at the 120°, 240°, and 300° points will be the same as at 60°.

The sine of 0°, 180°, and 360° is zero. The sine of 90° and 270° is 1.0, called either the maximum or the peak value.

A circle may also be used to represent a complete cycle. In electrical work, the circle is considered as starting at the right-hand side and rotating counterclockwise, as in Fig. 4-3.

Fig. 4-3 Vector method of indicating an ac cycle of 360° is a circle.

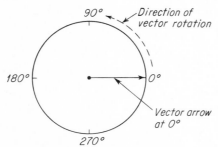

On the circle, 90° is actually one-quarter of a cycle from the start, 180° is one-half of a cycle, and so on. This method of indicating the number of electrical degrees, or the angle from 0°, is quite graphic.

The arrow from the center of the circle to the zero point in Fig. 4-3 may be termed a *vector* arrow. A quarter cycle represents 90° rotation of the vector arrow. A half cycle represents 180° rotation of the vector. The position of the vector in relation to the zero direction indicates an angle from zero. This can be called an *angular vector,* since it is indicating an angle.

An example of a vector arrow used as a *magnitude vector* is shown in the cycle of Fig. 4-4. In

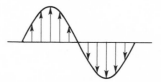

Fig. 4-4 Amplitude vectors indicating instantaneous magnitudes every 30° of one cycle.

this case the length of the vector arrows varies as the strength of the current varies. These vectors indicate not angles but the magnitude of the current, or voltage, at the angle in degrees along the time line at which they are drawn. Their direction indicates the relative direction of the current in the circuit. They show that the current is flowing in one direction in the circuit during the first half cycle and in the opposite direction during the second half cycle.

Vector arrows are handy devices to represent quantities, angles, or directions. The magnitude vectors are visually indicating the *instantaneous* values. If a vector arrow in a sine wave is drawn at the peak, the instantaneous value is the maximum value. If the vector is drawn at the 30° point on the time line, the instantaneous value is 0.5 of the maximum voltage ($0.5\,E_{max}$). The vector arrow at 60° will show an instantaneous voltage value of $0.86\,E_{max}$. An instantaneous value indicates the value of a voltage (or current) at a particular time interval from 0°.

Rotating vectors can indicate the angle of an instantaneous value of current, or voltage, in a cycle. The distance between the point of the rotat-

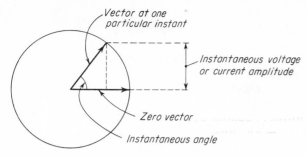

Fig. 4-5 Angle between starting point and vector determines instantaneous voltage or current value.

ing vector arrow and the zero vector line indicates the instantaneous strength, as illustrated in Fig. 4-5.

While this book uses 360° as one cycle, some engineering texts may use *radian* measure. One radian is the angle produced by a vector arrow tip rotating a distance equal to its own length. This angle is equivalent to 57.3°. From circular measure, 2π radius = circumference (π = 3.14). From this, 2π radians = 2(3.14) 57.3°, or 360°. Thus π rad = 180°, and 1 rad = 57.3°. Therefore, 3 cycles = $3(2\pi$ rad) = 6π rad, or 6(3.14) rad = 18.84 rad. Converting this from radian to degree measure, 6(3.14) 57.3 = 1080°.

For more exact pocket calculator work, π = 3.141593, and 1 radian = 57.29578.

4-4 PEAK AND EFFECTIVE AC VALUES

A 10-A dc and a 10-A *peak* ac are graphed to the same scale in Fig. 4-6. The dc represents a constant 10-A value. The ac rises to a peak of 10 A but is at this value for only an instant. Then it drops to zero, reverses, increases to 10 A in the opposite direction, and then drops to zero again. Although it is equally effective in doing work on both half cycles, during most of the cycle the ac

Fig. 4-6 Comparison of a 10-A dc with a 10-A peak ac.

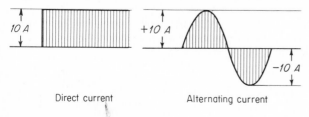

Direct current   Alternating current

has a value less than the constant dc and will be unable to produce as much heat or accomplish as much work.

Power being proportional to either E^2 or I^2 ($P = E^2/R = I^2R$), if all the instantaneous values of a half cycle of sine-wave current (or voltage) are squared and then the average, or mean, of all the squared values is found, the square root of this mean value will be 0.707 of the peak value. This *root-mean-square,* or rms, value represents how effective an ac will be in comparison with its peak value.

The *effective* value of a sine-wave ac cycle is equal to 0.707 of the peak value. Comparing dc and ac, the 10-A peak ac will be only as effective at producing heat and work as 7.07 A of dc. It may be said that the effective value of an ac is its *heating* value. Peak and effective values with dc are the same.

To determine a peak value of ac that will be as effective as a given dc, it is necessary to multiply the effective value given by the reciprocal of 0.707 (1/0.707), which is 1.414. For an ac to be as effective as 10 A of dc, it must be 10 × 1.414, or 14.14 A, at the peak.

The peak-to-peak ac value is twice the peak value. It is the value between the maximum negative and maximum positive half cycles of any ac wave.

Peak and effective factors of 1.414 and 0.707 are applicable only when the current, or voltage, is a sine wave. With square-wave ac the effective factor will be essentially 1.0. With most sawtooth waveshapes the factor is less and might approximate 0.5. The ac produced by microphones responding to certain types of noises may have an effective value of only 0.3 or 0.2 of the peak value.

When problems are given, it is assumed that the effective value is to be employed unless stated otherwise. When a power company states it is furnishing 120 V ac, it means an effective 120 V, which is 120 × 1.414, or nearly 170 V peak.

Test your understanding; answer these checkup questions.

1. What kind of voltage (ac or dc) is produced by friction? ____DC____ A crystal? ___AC___ Chemical action? ____DC____ Heat? ____DC.____
2. If all that is specified is that ac is used, what waveshape would the ac be assumed to have? ___Sine___

3. How much are power line losses decreased if power is carried at 10 times the normal voltage? _____
4. What is the main advantage of ac? _____
5. At what voltages do you think power might be carried between the mountains and cities? _____ Along a city block to homes? _____
6. How many degrees in an alternation? _____ In a cycle? _____
7. At how many points in a normal cycle will the instantaneous voltage equal the voltage at 15°? _____ At 90°? _____ At 300°? _____
8. According to Appendix C, what is the sine of 15°? _____ 88°? _____ 160°? _____ 220.5°? _____ 321.8°? _____
9. Does a vector arrow normally rotate counterclockwise or clockwise? _____
10. If sine-wave ac at maximum is 20 V, what is its amplitude at 40°? _____ At 135°? _____ At 292.3°? _____
11. What can vector arrows represent? _____ _____
12. How many degrees in one radian? _____ In 2π rad? _____
13. A current at 120° is 3 A. What is it at 265°? _____
14. A 350-V peak ac is as effective as what value of dc? _____
15. A filament of a vacuum tube requires a 0.4-A dc current to heat it. What peak ac is required? _____ What rms value? _____

4-5 THE AVERAGE VALUE OF AC

An ac cycle has a peak, an effective, and also a less frequently used value called the *average*. This is the value obtained when 180 instantaneous values, each separated by 1°, are added and then the average is found by dividing the sum by 180. The average value for a sine wave is 0.636 of the peak value.

The average value tends to be confusing to the beginner. He confuses "average" with "effective," probably because it seems that the average value should represent how much work the ac cycle should do. As pointed out previously, however, it is the effective, or rms, value that does this. To summarize, for sine-wave ac:

$$Ac\ peak\ value = 1.414 \times effective\ value$$
$$Ac\ effective\ value = 0.707 \times peak\ value$$
$$Ac\ average\ value = 0.636 \times peak\ value$$

From these values it is possible to compute the multiplying factor to change from average to peak,

1.57; the factor to change from effective to average, 0.9; and the factor to change from average to effective, 1.11.

At this point it might be well to advise the beginner to disregard the average value temporarily. Later, in Chaps. 11 and 12, in the study of meters and power supplies, when ac is changed to pulsating dc, the average value is used.

4-6 FREQUENCY

The number of times an alternating current goes through its complete cycle per second is known as its *frequency*. The International unit of measurement of frequency is the *hertz*, abbreviated Hz. The English unit is *cycles per second*, abbreviated cps (1 Hz = 1 cps). The text material will use hertz. Some problems and diagrams may use cps.

To simplify terminology, 1,000 Hz is called a *kilohertz*, abbreviated kHz, and 1,000,000 Hz is called a *megahertz*, MHz. The previous corresponding units were kc and Mc. A frequency of 3,500,000 Hz can be expressed as 3,500 kHz (previously 3,500 kc) or as 3.5 MHz (previously 3.5 Mc).

The vibration rate of sound waves in air may also use the term *frequency*. For example, when middle C is played on a musical instrument, an air disturbance with a frequency of 256 Hz is set up. The lowest tone that can be heard by human beings is about 15 Hz. The highest audible, audio, or sonic tones are usually between 15 and 20 kHz. A microphone is a device or *transducer* that can change sound waves to equivalent-frequency ac.

The list of frequencies in Table 4-1 indicates the terminology that may be used for different ac frequencies. Note the overlapping of the frequencies from 10,000 to 1,000,000 Hz. The letters in parentheses are abbreviations commonly used.

At power frequencies, materials such as fiber, cambric, cotton, some types of glass, black rubber, and impure Bakelite are satisfactory insulators. At higher radio frequencies, mica, low-loss hard rubber, special porcelains, Isolantite, Mycalex, polystyrene, steatite, plastics, and special glasses have lower losses and are to be preferred.

Table 4-1

Terms used	Frequency limits
Subaudible frequencies	0.001 to 15 Hz (cps)
Audio frequencies (AF)	15 Hz to 20 kHz (kc)
Power frequencies	10 Hz to 1 kHz
Video frequencies	0 to over 4.5 MHz (Mc)
Supersonic or ultrasonic frequencies	20 kHz to over 2 MHz
Very low radio frequencies (VLF)	10 to 30 kHz
Low radio frequencies (LF)	30 to 300 kHz
Medium radio frequencies (MF)	300 kHz to 3 MHz
High radio frequencies (HF)	3 to 30 MHz
Very high radio frequencies (VHF)	30 to 300 MHz
Ultra high radio frequencies (UHF)	300 MHz to 3 GHz (Gc)
Superhigh radio frequencies (SHF)	3 to 30 GHz
Extremely high radio frequencies (EHF)	30 to 300 GHz (3×10^{11} Hz)
Heat or infrared	3×10^{11} to 4.3×10^{14} Hz
Visible light, red to violet	4.3×10^{14} to 1×10^{15} Hz
Ultraviolet	1×10^{15} to 6×10^{16} Hz
X-rays	6×10^{16} to 3×10^{19} Hz
Gamma rays	3×10^{19} to 5×10^{20} Hz
Cosmic rays	5×10^{20} to 8×10^{21} Hz

4-7 PHASE

There are three ways in which sine-wave ac currents, emfs, or waves can differ. These are (1) amplitude, (2) frequency, and (3) phase. The *amplitude* usually means the peak height of the ac wave. The *frequency* is the number of cycles per second. The *phase* is the number of electrical degrees one wave leads or lags another. Figure 4-7a illustrates two ac waves graphed on the same time line. The solid-line wave starts upward before the dotted-line wave and is therefore leading the dotted line. The waves are out of phase, or out of step, by approximately 45°. If they were in phase, they would go to maximums at the same instant and to zero at the same instant.

The two waves shown in Fig. 4-7b start in phase, drop out of phase, and then return in phase again. To change phase this way, one or both of the waves must change frequency slightly. These waveforms may represent two voltages, two currents, or a voltage and a current. The two waves shown in Fig. 4-7c are 180° out of phase.

4-8 OTHER AC TERMS

In some electronic circuits a sinusoidal ac input signal is distorted and the output contains *harmonics* as well as the fundamental input frequency. A harmonic is a whole-number multiple of a frequency. That is, the first three harmonics of 1,000 Hz are 2 kHz, 3 kHz, and 4 kHz.

The *wavelength* of a cycle is the distance between successive positive peaks (or negative peaks). See Sec. 20-7 for further discussions of wavelength.

Fig. 4-7 Three examples of phase displacements between two waves. (a) Solid line leads dotted line by 45°. (b) Dotted line falls out of and back into phase. (c) Two waves 180° out of phase.

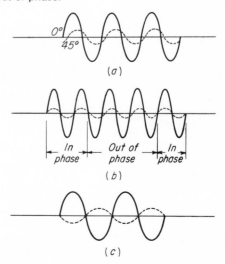

Test your understanding; answer these checkup questions.

1. What is another term for rms, or root-mean-square? _____ For hertz? _____
2. Most ac voltmeters read in rms values. What would a 50-V ac voltmeter indicate when across 28-V sinusoidal rms? _____ 50-V sinusoidal peak? _____ 100-V sinusoidal peak-to-peak? _____ 15-V square wave? _____
3. What is the average value of 50-V peak sinusoidal ac? _____ 2.5-A rms ac? _____ 8-V p-p ac? _____
4. If voltages above the time line are positive, which of the following would be negative? Voltages at 210°, 87°, 160°, 359°, 30°, 175°? _____
5. How much more than the average value is the effective value? _____

6. Which of the following materials make good insulators for power frequencies but not for high radio frequencies: mica, lead, Mycalex, polystyrene, aluminum, steatite, cotton, Isolantite, rubber? _____
7. A 10-V peak and a 5-V peak 60-Hz ac generator are in series. What will be the peak voltage if these two sources are in phase? _____ 180° out of phase? _____
8. What type of transducer changes sound waves to equivalent-frequency ac? _____ Changes ac to equivalent-frequency sound waves? _____
9. If two ac voltages do not reach their peaks at the same time, what are they said to be? _____
10. What harmonic of 1.5 MHz is 3 MHz? _____ 4.5 MHz? _____ 0.75 GHz? _____ 1,500 kHz? _____
11. What is the distance between successive negative peaks called? _____ How would this be related to frequency? _____

COMMERCIAL LICENSE QUESTIONS

Amateur license questions will be found in the Addendum.

FCC Elements 3, 4, and 6 require an ability to answer questions similar to those below. Sections in which questions are answered are shown in parentheses. A question followed by a bracketed number is required for that element alone.

1. How many degrees does a cycle represent? (4-3)
2. In ac circuits what is the relation between rms values, effective values, peak values, and peak-to-peak values? (4-4)
3. What is the relation between the effective value of an ac current and the heating value of the current? (4-4)
4. Define an average current value in an ac circuit. In what practical situation might it be used? (4-5)
5. What is the ratio of peak to average values of a sine wave? (4-5)
6. What is the frequency range associated with each of the following general frequency subdivisions: VLF, LF, MF, HF, VHF, UHF, SHF, EHF? (4-6)

7. What is an *audio frequency,* and what approximate band of frequencies is normally referred to as the audio-frequency range? (4-6) [4]
8. Name (*a*) four materials which are good insulators at radio frequencies, (*b*) four materials which are not good insulators at radio frequencies but are satisfactory for use at commercial power frequencies. (4-6)
9. What is the meaning of *phase difference?* (4-7)
10. Draw (*a*) a sine wave of voltage displaced 45° from a sine wave of current, (*b*) a wave displaced 180° from another. (4-7)

NOTE: FCC license test questions are multiple-choice types, taking a form somewhat as follows:

11. Which of the following generating methods produces ac? (*a*) Chemical; (*b*) magnetostrictive; (*c*) photoelectric; (*d*) magnetohydrodynamic. (4-1)
12. Which of the following might be associated with dc? (*a*) Square wave; (*b*) sinusoidal; (*c*) triangular; (*d*) all of these. (4-2)
13. How many degrees in an alternation? (*a*) 90°; (*b*) 360°; (*c*) 180°; (*d*) none of these. (4-3)

ANSWERS TO CHECKUP QUIZ ON PAGE 65

1. (*Dc*)(*Ac*)(*Dc*)(*Dc*) 2. (*Sinusoidal*) 3. (*100 times*)
4. (*Can be stepped up or down*) 5. (*120 kV to over 500 kV*)(*120-240 V*) 6. (*180°*)(*360°*) 7. (*4*)(*2*)(*4*)
8. (*0.25882*)(*0.99939*)(*0.34202*)(*0.64945*)(*0.61841*)
9. (*Counterclockwise*) 10. (*12.86 V*)(*14.14 V*)(*18.5 V*)
11. (*Direction, amplitude, phase*) 12. (*57.3°*)(*360°*)
13. (*3.45 A*) 14. (*247 V*) 15. (*0.567 A*)(*0.4 A*)

INDUCTANCE AND TRANSFORMERS

The objective of this chapter is to discuss self-inductance of coils and chokes and basic transformer theory in terms of mutual inductance. Power transformers and their various losses are outlined. The idea of inductive reactance and its computations are developed, along with the phase relations in inductive circuits.

5-1 INDUCTANCE

Coils of wire were mentioned in Chap. 3 when the electromagnetic effect produced by current flowing through them was considered. An equally important aspect of the operation of a coil is its property to oppose any *change* in current through it. This property is called *inductance*.

When a current of electrons starts to flow along any conductor, a magnetic field starts to expand from the center of the wire. These lines of force move outward, through the conducting material itself, and then continue into the air. As the lines of force sweep outward through the conductor, they induce an emf in the conductor itself. This induced voltage is always in a direction opposite to the direction of current flow. Because of its opposing direction it is called a *counter emf*, or a *back emf*. The direction of this self-induced counter emf can be verified by using the left-hand generator rule (Sec. 3-18).

The effect of this backward pressure built up in the conductor is to oppose the immediate establishment of maximum current. It must be understood that this is a temporary condition. When the current eventually reaches a steady value in the conductor, the lines of force will no longer be in

the process of expanding or moving and a counter emf will no longer be produced. At the instant when current begins to flow, the lines of force are expanding at the greatest rate and the greatest value of counter emf will be developed. At the starting instant the counter emf value almost

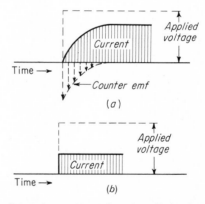

Fig. 5-1 Current, counter emf, and applied voltage plotted against a time base for (a) an inductive circuit and (b) a resistive circuit.

equals the applied, or source, voltage as in Fig. 5-1a.

The current value is small at the start of current flow. As the lines of force move outward, however, the number of lines of force cutting the conductor per second becomes progressively smaller and the counter emf becomes progressively less. After a period of time the lines of force expand to their greatest extent, the counter emf ceases to be generated, and the only emf in the circuit is that of the source. Maximum current can now flow in the wire or circuit, since the inductance is no longer reacting against the source voltage.

If it were possible to produce current by applying a voltage across a wire and not produce a counter emf, then Fig. 5-1b would represent the action of the current. The figure shows a current reaching the maximum value instantly—just as soon as the voltage is applied. This is essentially true for a purely resistive circuit.

5-2 SELF-INDUCTION

When the switch in a current-carrying circuit is suddenly opened, an action of considerable importance takes place. At the instant the switch breaks the circuit, the current due to the applied voltage would be expected to cease abruptly. With no current to support it, the magnetic field surrounding the wire should collapse back into the conductor at a tremendously high rate, inducing a high-amplitude emf in the conductor. Originally, when the field built outward, a counter emf was generated. Now, with the field collapsing inward, a voltage in the opposite direction is produced. This might be termed a *counter counter emf*, but is usually known as a *self-induced emf*. This self-induced emf is in the direction of the applied source voltage. Therefore, as the applied voltage is disconnected, the voltage due to self-induction tries to establish current flow through the circuit in the same direction and aiding the source voltage. With the switch open it would be assumed that there is no path for the current, but the induced emf immediately becomes great enough to ionize the air at the opened switch contacts and a spark of current appears between them. Arcing lasts as long as energy stored in the magnetic field exists. This energy is dissipated as heat in the arc and in the circuit itself.

ANSWERS TO CHECKUP QUIZ ON PAGE 68

1. (*Effective*)(*Cycles per second*) 2. (*28 V*)(*35.35 V*)(*35.35 V*)(*23.6 V*) 3. (*31.8 V*)(*2.25 A*)(*2.54 V*) 4. (*210°, 359°*) 5. (*1.11*) 6. (*cotton and some rubbers*) 7. (*15 V*)(*5 V*) 8. (*Microphone*)(*Loudspeaker or earphone*) 9. (*Out of phase*) 10. (*2d*)(*3d*)(*500th*)(*The first or the fundamental*) 11. (*Wavelength*)(*The longer the wavelength the fewer cycles per second*)

With circuits involving low current and short wires, the energy stored in the magnetic field will not be great and the switching spark may be insignificant. With long lines (large inductance valves) and heavy currents, inductive arcs several inches long may form between opened switch contacts on some power lines. The heat developed by arcs tends to melt the switch contacts and is a source of difficulty in high-voltage high-current switching circuits.

Note that, regardless of any change of current amplitude or direction in a conductor, the induced emfs oppose the current *change*. With a steady, unvarying direct current, there is no change of current and no opposition develops. When a varying dc is flowing, the counter emf opposes any increase of source voltage. As the source voltage decreases, the self-induced emf opposes the decrease. With alternating current the constant state of change results in a continual opposing or reacting action. From this comes the definition: *Inductance is the property of any circuit which opposes any change in current*. Another definition: *Inductance is the property of a circuit by which energy is stored in the form of an electromagnetic field*.

The unit of measurement of inductance is the *henry,* defined as the amount of inductance required to produce an average counter emf of one volt when an average current change of one ampere per second is under way in the circuit. Inductance is represented by the symbol L in electrical problems, and henrys is indicated by H.

5-3 COILING AN INDUCTOR

It has been indicated that a piece of wire has the ability of producing a counter emf and therefore has a value of inductance. Actually, a small length of wire will have an insignificant value of inductance by general electrical standards. One henry represents a relatively large inductance in many circuit applications, where millihenrys and microhenrys are more likely to be used. A straight piece of No. 22 wire one meter long has about 1.66 μH. The same wire wound onto an iron nail or other high-permeability core may produce 50 or more times that inductance.

Even without the iron core, a given length of wire will have much greater inductance if wound

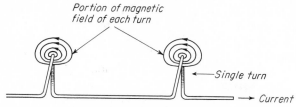

Fig. 5-2 Fields from two widely separated turns do not intercouple.

$$L = \frac{r^2 N^2}{9r + 10l}$$

where L = inductance, in microhenrys (μH)
r = radius of coil, in inches
N = number of turns
l = length of coil, in inches

Using metric measurements, essentially the same formula is

$$L = \frac{r^2 N^2}{24r + 25l}$$

where r = radius, in centimeters (cm)
l = length, in cm

into coil form. Consider Fig. 5-2, showing two loops of wire separated by enough distance so that there is essentially no interaction between their magnetic fields. If the inductance of the connecting wires is neglected, these two loops, or *turns,* have twice the inductance of a single turn.

When the two loops are wound next to each other, as in Fig. 5-3, with the same current flowing

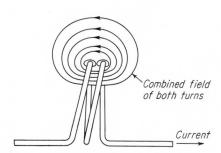

Fig. 5-3 Fields of close-wound turns intercouple.

there are twice the number of magnetic lines of force cutting each turn. With 2 turns, 4 times the counter emf is developed. With 3 turns, 3 times the number of lines of force cut 3 turns, so 9 times the counter emf is developed. The inductance of a coil varies as the number of turns squared, or as N^2. However, it can be seen that length of the coil is also going to enter into the exact computation of the inductance of a coil. If the turns are stretched out, the field intensity will be less and the inductance will be less. The larger the radius or diameter of the coil, the longer the wire used and the greater the inductance. In single-layer air-core coils with a length approximately equal to the diameter, a formula that will give the approximate inductance in microhenrys is

For more accurate and detailed formulas to compute inductance of various coils having differing permeability cores, refer to radio or electrical engineering handbooks.

The inductance of straight wires alone is encountered in antennas, in power lines, and in ultrahigh-frequency equipment. In most electronic and radio applications where inductance is required, space is limited and wire is wound into either single-layer or multilayer coils with air, powdered-iron-compound (ferrite), or laminated (many thin sheets) cores. The advantage of multilayer coil construction for high values of inductance becomes obvious when it is considered that, while 2 closely wound turns produce 4 times the inductance of 1 turn, the addition of 2 more turns closely wound on top of the first 2 will provide almost 16 times the inductance. Direction of winding has no effect on the inductance value of a given coil.

In many applications coils are constructed with ferrite cylinders that can be screwed into or out of the core space of the coil. This results in a controlled variation of inductance, maximum when the iron-core "slug" is in the coil and minimum with it out.

A special type of coil is the *toroid.* It consists of a doughnut-shaped ferrite core, either single-layer-wound as shown in Fig. 5.4, or multilayer-wound. Its advantages are high values of inductance with little wire, and therefore little resistance in the coil, and the fact that all the lines of force are in the core and none outside (provided there

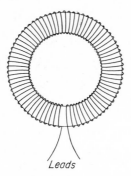

Fig. 5-4 Coil wound in toroidal form.

is no break in the core). As a result it requires no shielding to prevent its field from interfering with outside circuits and to protect it from effects of fields from outside sources. Two toroids can be mounted so close that they nearly touch and there will be almost no interaction between them.

5-4 THE TIME CONSTANT OF AN INDUCTANCE

The time required for the current to rise to its maximum value in an inductive circuit after the voltage has been applied will depend on both the inductance and the resistance in the circuit. With a constant value of resistance in a circuit, the greater the inductance, the greater the counter emf produced and the longer the time required for the current to rise to maximum.

With a constant value of inductance in a circuit, the more resistance, the less current that can flow. The less current, the less possible counter emf to oppose the source emf and the less time required to reach a maximum current value.

The time required for the current to rise to 63% of the maximum value (called the *time constant*) can be determined by

$$T_c = \frac{L}{R}$$

where T_c = time, in seconds (s)
L = inductance, in henrys (H)
R = resistance, in ohms (Ω)

According to this formula, a 10-H coil with 10-Ω resistance will allow current to rise to 63% of

maximum in one second. In the next second, the current will rise 63% of the remaining amount toward maximum, and so on. (If a coil could be produced with zero ohms of resistance, theoretically the current would never reach a maximum value.) A time equivalent to five times the time-constant formula value results in a current within 1% of maximum. This is usually considered as the maximum value. A circuit with zero inductance and only resistance will reach the maximum current value instantly (Fig. 5-1b).

The time-constant formula also indicates the time required for current to decrease by 63% from maximum in an inductive circuit.

5-5 THE ENERGY IN A MAGNETIC FIELD

Current flowing in a wire or coil produces a magnetic field around itself. If the current suddenly stops, the magnetic field held out in space by the current will collapse back into the wire or coil. Unless the moving field has induced a voltage and current into some external load circuit, all the energy taken to build up the magnetic field will be returned to the circuit in the form of electric energy as the field collapses.

The amount of energy in joules that is being stored in a magnetic field at any instant can be determined by the formula

$$E_n = \frac{LI^2}{2}$$

where E_n = energy, in joules (wattseconds)
L = inductance, in H
I = current, in A

5-6 CHOKE COILS

The ability of a coil to oppose any change of current can be used to smooth out varying or pulsating types of current. In this application an inductor is known as a *choke coil,* since it chokes out variations of amplitude. For radio-frequency (RF) ac or varying dc an air-core coil may be used. For lower-frequency circuits greater inductance is required. For this reason iron-core choke coils are found in audio-frequency and power-frequency applications. Several types of inductors are shown in Fig. 5.5.

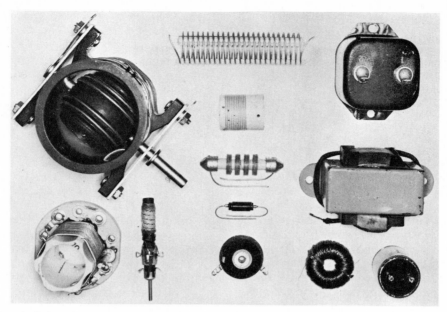

Fig. 5-5 Inductors. Left, from top, variable air-core, fixed air-core, adjustable iron-slug coil. Middle group, various air-core coils. Right group, iron-core inductors; from top, shielded audio choke, unshielded power choke, toroid, miniature high-inductance choke.

A choke coil will hold a nearly constant inductance value until the core material becomes saturated. When enough current is flowing through the coil to saturate the core magnetically, variations of current above this value can produce no appreciable counter emfs and the coil no longer acts as a high value of inductance to these variations. To prevent the core from becoming magnetically saturated, a small air gap may be left in the iron core. The air gap introduces so much reluctance in the magnetic circuit that it becomes difficult to make the core carry the number of lines of force necessary to produce saturation. The gap also decreases the inductance of the coil. An air-core coil cannot be saturated.

Fig. 5-6 (a) Air-core coil symbol. (b) Iron-core coil symbol. (c) Simplified construction of an iron-core choke coil.

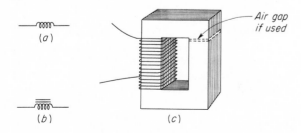

Figure 5-6 shows the symbol for an air-core coil or choke, an iron-core coil or choke, and a simplified picturization of a laminated iron-core choke.

Test your understanding; answer these checkup questions.

1. Does a 1-m wire have any inductance if ac is flowing in it? _____ If dc is flowing in it? _____
2. In a switched dc circuit, which is greater, the counter emf or the self-induced emf? _____
3. What effect causes switch or relay contacts to develop pits and mounds on them? _____
4. How much inductance is in a circuit if 2 V of counter emf is developed in 0.5 s? _____
5. What is the property of a circuit that opposes any change in current called? _____ That opposes current? _____
6. What is the advantage of using toroid coils? _____
7. What is the L of a coil having a diameter of 1.5 in., a length of 2 in., and 50 turns in μH? _____ In mH? _____
8. A 5-H coil is in series with a 100-Ω resistance, a switch, and a battery. How long will it take for the current to reach 63% of I_{max}? _____ Approximately I_{max}? _____
9. How much energy is contained in the field of a 7-H choke coil when 500 mA is flowing through it? _____

10. What relative-frequency circuit would require choke coils with air cores? _____ Iron cores? _____
11. To prevent saturation of the core of a choke coil, should the core be continuous or have an air gap? _____
12. If a 10-turn coil has a second layer of 10 turns wound over the first, what should be the total inductance? _____

5-7 MUTUAL INDUCTANCE

A single coil has a value of inductance, or self-inductance. As explained previously, a coil has 1 H of inductance if an average current change of 1 A/s produces an average counter emf of 1 V in it.

If one coil is placed near a second, it will be found that alternating or varying currents in the first produce moving magnetic fields that will induce voltages in the second coil. The farther apart the two coils are, the fewer the number of lines of force that interlink the two coils and the lower the voltage induced in the second coil (100 million lines per second cutting one turn induce 1 V).

When an average current change of 1 A/s in the first coil can produce moving fields that will induce an average of 1 V in the second, the two coils are said to have a *mutual inductance* of 1 H, regardless of the inductance values of the two coils themselves.

The mutual inductance can be increased by moving the two coils closer together or by increasing the number of turns of either coil.

In power *transformers* two coils are so arranged that almost all the lines of force of the first coil cross the turns of the second coil. A large mutual inductance value results.

When *all* the lines of force from one coil cut *all* the turns of a second coil, *unity coupling* exists, and the mutual inductance may be found by the formula

$$M = \sqrt{L_1 L_2}$$

where M = mutual inductance, in H
L_1 = inductance of coil 1, in H
L_2 = inductance of coil 2, in H

This formula assumes 100% coupling between the two coils. If all the lines from the first coil do not cut all the turns of the second, M is determined by

$$M = k\sqrt{L_1 L_2}$$

where k = percent of coupling

EXAMPLE: If a 2-H coil has 84% of its lines of force cutting a 4.5-H coil, what mutual inductance exists? By substituting in the formula given above,

$$M = k\sqrt{L_1 L_2} = 84\% \sqrt{2(4.5)}$$
$$= 0.84\sqrt{9} = 0.84(3) = 2.52 \text{ H}$$

5-8 COEFFICIENT OF COUPLING

The degree, or closeness, of coupling can be expressed as a percent, as above. Although the term percent of coupling might be used, the term *coefficient of coupling* is to be preferred. A percent of 100 is equivalent to a coefficient of coupling of 1.0, or unity; 95% is equivalent to a coefficient of coupling of 0.95; and so on.

The coefficient of coupling between two coils can be computed from the rearrangement of the mutual-inductance formula

$$k = \frac{M}{\sqrt{L_1 L_2}}$$

The answer obtained in this formula will always be a decimal, unless the coefficient is unity.

EXAMPLE: The mutual inductance between two coils is 0.1 H, and the coils have inductances of 0.2 and 0.8 H. What is the coefficient of coupling? By substituting in the formula,

$$k = \frac{M}{\sqrt{L_1 L_2}} = \frac{0.1}{\sqrt{0.2(0.8)}}$$
$$= \frac{0.1}{\sqrt{0.16}} = \frac{0.1}{0.4} = 0.25 \text{ or } 25\%$$

Coils with relatively high coefficients of coupling may be said to be *tightly* coupled. With low values of coupling they are said to be *loosely* coupled. What is tight and what is loose will vary in different applications. In power transformers the coefficient may exceed 0.98, while in some radio circuits a coefficient as low as 0.01 is all that may be required.

5-9 INANCES IN SERIES

Electric circuits often have two or more inductances in them. Whether the magnetic fields of the two coils interlink or not determines the effective amount of inductance presented to the circuit by the coils.

Figure 5-7 shows two 1-H coils and a resistor

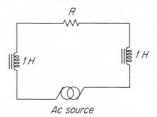

Fig. 5-7 Two uncoupled inductances in a series circuit.

in series across an ac generator, called an alternator. Since the two coils are widely separated, no interlinkage of fields occurs and the total inductance in the circuit is simply 2 H (any slight inductance in the connecting wires can be neglected). The formula for uncoupled inductances in series is

$$L_t = L_1 + L_2 + L_3 + \cdots$$

where L_t = total inductance, in H
$L_1, L_2, \ldots$ = other inductances, in H

Note the similarity of this series-inductance formula to that of series resistors.

Figure 5-8 shows two 1-H coils and a resistor in series across an alternator. The two coils are close enough that the lines of force from one coil interlink the other. Now, the mutual inductance will affect the total-inductance value. If the coils are wound in the same direction, the emf induced

Fig. 5-8 Two series inductances placed to intercouple their fields.

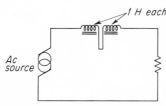

from one to the other will be *in phase,* or additive, and the total-inductance value will be more than the simple addition of the two inductance values alone. The effective inductance of two in-phase series-connected inductances can be determined by using the formula

$$L_t = L_1 + L_2 + 2M$$

where M = mutual inductance, in H

If the two coils were unity-coupled and each had 1 H of self-inductance, the total-inductance value would be 4 H.

If two coils are wound in opposite directions and coupled, the induced emf in the coils will be in opposition, or *out of phase,* and will tend to cancel each other, resulting in less effective inductance. The formula for total inductance in this case is

$$L_t = L_1 + L_2 - 2M$$

If the two coils were unity-coupled and each had 1 H of self-inductance, the total-inductance value would be zero henrys. The two coils would have completely canceled each other's inductance.

5-10 INDUCTANCES IN PARALLEL

In some circuits two or more inductors are connected in parallel as in Fig. 5-9.

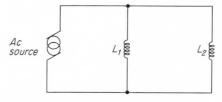

Fig. 5-9 Two parallel inductances.

If the two inductances have 1 H each, the resultant inductance (no interaction of their fields is assumed) will be ½ H. This inductance is computed by using a formula similar to the parallel-resistance formula:

$$L_t = \frac{L_1 L_2}{L_1 + L_2}$$

If three or more inductances are in parallel, a formula similar to the parallel-resistance formula may be used:

$$L_t = \frac{1}{1/L_1 + 1/L_2 + 1/L_3}$$

(Since coils are rarely connected in parallel and intercoupled, the more complicated formulas for such circuits will not be presented.)

5-11 SHORTING A TURN IN A COIL

There are several methods of reducing the inductance of a coil. Examples are taking turns off the coil, stretching the coil out until it has a greater length, using a less permeable core, or shorting one or more turns of the coil as in Fig. 5-10.

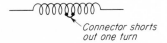

Connector shorts out one turn

Fig. 5-10 Coil with a shorted turn.

If one turn of a coil is shorted, there is one less turn in the coil and the coil has a little less inductance. If dc is flowing through the coil, there will be relatively little change in the magnetic field around the coil. However, when ac flows through the coil, the expanding and contracting fields from adjacent turns cut the shorted turn and induce an emf in it. In a shorted turn even a small emf induces a relatively high-amplitude current in the turn. Since the emf induced in the shorted turn is a counter emf, it produces a current in the shorted turn in an opposite direction to that flowing in the remainder of the coil. This results in

a counteracting field, partially canceling the field of the coil. The inductance of the coil is reduced by much more than would result from cutting off one turn. A shorted turn may become noticeably warm or hot, depending on how much current is induced in it.

In some radio applications in which a variation of inductance is desired, a shorted turn in the form of a loop of wire or a brass disk may be brought near a coil, effectively reducing the inductance of the coil. The closer the loop is to the center of the coil, the more it counteracts the inductance. A brass screw turned into the core of a coil has the same effect.

5-12 INDUCTIVE REACTANCE

It has been explained that dc flowing through an inductance produces no counter emf to oppose the current. With varying dc, as the current increases, the counter emf opposes the increase. As the current decreases, the counter counter emf opposes the decrease. Alternating current is in a constant state of change, and the effect of the magnetic fields is a continual induced voltage opposition to the current. This reacting ability of the inductance against the changing current is called *inductive reactance*.

Inductance is the property of a circuit to oppose any change in current and is measured in henrys. Inductive reactance is a measure of how much the counter emf in the circuit will oppose current variations. The opposing effect of reactance is measured in *ohms*.

It may be considered that the inductance of a coil does not change. Whether it is used in a dc circuit, a 60-Hz circuit, or a 10,000-Hz circuit or whether it lies unused on a shelf, a 1-H inductance has a value of 1 H. Its property has not changed. When dc is flowing through a 1-H coil, there will be no opposition to current flow except for any ohmic resistance in the coil. In a low-frequency 60-Hz circuit, the magnetic field builds up and collapses relatively slowly and relatively little counter emf may be developed to oppose circuit current. In a higher-frequency circuit, the magnetic field moves more rapidly and produces more counter emf, which opposes the current more. The amount of opposition, or reactance, is

---

ANSWERS TO CHECKUP QUIZ ON PAGE 73

1. (*Yes*)(*Yes, inductance is a property of the wire*) 2. (*Self-induced emf*) 3. (*Arcing*) 4. (*4 H*) 5. (*Inductance*)(*Resistance*) 6. (*No external magnetic field is developed*) 7. (*52.6 μH*)(*0.0526 mH*) 8. (*0.05 s*)(*0.25 s*) 9. (*0.875 J*) 10. (*High*)(*Low*) 11. (*Have an air gap*) 12. (*Almost 4 times original*)

directly proportional to the frequency of the current variation.

A 1-H coil will produce a certain value of opposition to a 60-Hz ac. A 2-H coil will have twice the counter-emf-producing capability and oppose the current change twice as much. Therefore, reactance is also directly proportional to the inductance value.

When the inductance in henrys is multiplied by the frequency in cycles per second and this is multiplied by $2 \times \pi$ (Greek letter pi), an inductive-reactance value results that is similar to a resistance value in ohms. The formula is

$$X_L = 2\pi f L$$

where X_L = inductive reactance, in Ω
π = 3.14
f = frequency, in Hz
L = inductance, in H

The Greek letter ω (omega) is often used to indicate $2\pi f$. (This is known as the angular velocity of the cycle and indicates how fast the current is changing.) Thus, $X_L = \omega L$.

As an example of how much inductive reactance is presented by a circuit composed of a source of ac and an inductance, consider the circuit in Fig. 5-11, in which the source has a frequency of 3,000 Hz and the inductance is 2 H. By substituting in the inductive-reactance formula,

$$X_L = 2\pi f L = 2(3.14)3,000(2)$$
$$= 6.28(6,000) = 37,680 \ \Omega$$

The 2-H coil presents 37,680 Ω of opposition to a 3,000-Hz ac and will limit the amplitude of the current in the circuit exactly as much as if a

Fig. 5-11 Current is determined by the inductive reactance, voltage, and frequency in the circuit.

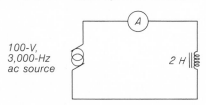

100-V,
3,000-Hz
ac source

37,680-Ω resistance were used instead. (The Greek letter ω is often used to indicate reactive ohms, with Ω used for resistive ohms.) Reactance can be substituted for resistance in Ohm's-law formulas but only if the circuits are purely reactive. Ohm's law for reactive circuits states: The current in a reactive circuit is directly proportional to the voltage and inversely proportional to the reactance. In formula form,

$$I = \frac{E}{X} \qquad E = IX \qquad X = \frac{E}{I}$$

where I = current, in A
E = emf, in V
X = reactance, in Ω

The general reactance symbol X is used in these formulas rather than X_L because there is another reactance, known as *capacitive reactance,* with the symbol X_C (explained in Chap. 6). These Ohm's-law formulas apply to either X_L or X_C. When it is desired to specify inductive reactance only, the symbol X_L should be used.

If the ac generator in Fig. 5-11 produces an effective emf of 100 V, the circuit current may be determined by Ohm's law:

$$I = \frac{E}{X} = \frac{100}{37,680} = 0.00265 \ \text{A or 2.65 mA}$$

In practice, Ohm's-law formulas for reactive circuits should be used only if the reactance value is more than 10 times the resistance. If there is an appreciable proportion of resistance in the circuit, Ohm's law for ac circuits (Sec. 7-2) must be used.

If two uncoupled inductive reactances are in series, the total reactance is

$$X_{L_T} = X_{L_1} + X_{L_2}$$

If two uncoupled inductive reactances are in parallel, the total reactance may be computed by using the formula

$$X_{L_T} = \frac{X_{L_1} X_{L_2}}{X_{L_1} + X_{L_2}}$$

*Test your understanding; answer these
checkup questions.*

1. If an average 2 A in one coil induces 0.5 V into an adjacent coil, what is the *M* value? _____
2. What is the coefficient of coupling expressed in decimals for a *k* of 3.5%? _____
3. What is the mutual inductance of two unity-coupled coils of 4 and 9 H? _____
4. What is the *M* value of a 4-H and a 5-H coil having a coefficient of coupling of 0.85? _____
5. What is the inductance of a 3-H and a 4-H coil in series, fields aiding, and 40% coupled? _____
6. What is the *L* value when a 5-H and an 8-H coil are in series, fields opposing, if the *k* value is 0.75? _____
7. What is the inductance if a 0.5-H and a 0.2-H coil are in parallel? _____
8. What happens to the *L* value if a brass slug is moved into the core area of a coil? _____ If a ferrite slug is used? _____
9. What part of a reactance formula indicates cyclic consideration? _____
10. What is the reactance of a 10-H coil to 120 Hz? _____ Of a 5-mH choke to 1,000 kHz? _____
11. What is the X_L of a 600-μH coil to 4 MHz? _____
12. What inductance is required to present 500 Ω of reactance to a 10-kHz ac? _____
13. At what frequency does a 0.04-H coil have a reactance of 3 kΩ? _____
14. How much *I* flows through a 2-H choke when connected across a 120-V 60-Hz power line? _____
15. An RF choke coil develops a 200-V drop across itself when a 5-mA 2-MHz ac flows through it. What reactance does it present to this frequency? _____
16. How much voltage-drop will occur across an audio choke coil having 5 H when 150 mA of 40-Hz ac flows through it? _____

5-13 PHASE RELATIONS WITH INDUCTANCE

When a coil is connected across an ac generator, as in Fig. 5-12, the current in the coil will not rise to a peak at the instant that the voltage attains a peak value. Theoretically, the current in the coil will lag the source voltage by 90° (provided there is no resistance in the circuit).

Fig. 5-12 A purely inductive circuit.

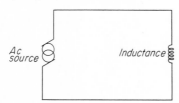

When an inductive circuit having negligible resistance is first turned on, the current and voltage start out in phase as a varying dc. (The starting operation is quite complex and will not be dealt with here.) After a few voltage cycles, however, the circuit current begins to alternate and then settles down into what is known as a *steady-state* ac and continues to operate in this manner until something in the circuit is changed. All explanations will be for steady-state conditions.

Figure 5-13 represents one cycle of steady-

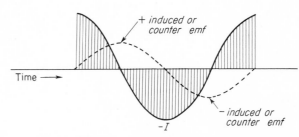

Fig. 5-13 Phase relations of the current and the induced counter emf in an inductor.

state ac flowing through a purely inductive circuit. The current cycle under consideration is from a maximum positive current value to a maximum negative value and back to a maximum positive value.

As the current increases through an inductor, the magnetic field increases, exactly in step. At the instant the current reaches its maximum positive value, its *rate of change* is zero. This means maximum field strength but zero magnetic-field *movement* and therefore zero induced or counter emf due to moving magnetic fields. The induced emf is indicated by the dotted curve. Thus, maximum current and zero induced emf occur at the same instant.

As the current diminishes toward zero, the magnetic field collapses inward toward the center of the coil. As the current nears the zero point, its rate of change is very rapid, producing maximum induced positive voltage at the zero-current instant.

The current then starts to increase in the negative direction, producing an opposite-polarity magnetic field expanding as the current increases. This reversed-polarity expanding field produces a voltage of the *same polarity* as the original contracting field produced. This is a significant point. Reversing the field polarity and at the same time reversing a contracting motion to an expanding motion constitutes a double reversal. A double reversal produces the same, or original-direction, induced emf. (If you reverse direction twice, you are still going in the original direction.) Therefore, as the current reverses, the induced emf is still in the same direction (positive) and continues to be developed as long as current is changing. As the current reaches maximum in the negative direction, its rate of change decreases to zero. At this instant there is zero induced voltage in the coil again.

As the current drops from maximum negative toward zero, its rate of change increases, developing an induced emf again, this time in the negative direction. When the current reaches zero, the induced voltage is again at a maximum but at a maximum negative value.

As the current swings up from zero to maximum in the positive direction, the induced voltage drops from maximum negative to zero once more. This completes one full cycle of current and induced voltage.

By reference to the figures it can be seen that the current and induced voltage in the coil are constantly a quarter of a cycle, or 90°, out of phase.

Since the induced voltage is a counter emf (counter to the source voltage), the source voltage must be 180° out of phase with it. The three curves in Fig. 5-14 show the phase relations of the current, the counter emf or induced voltage, and the source voltage in a purely inductive circuit. The current lags 90° behind the source voltage.

Fig. 5-14 Phase relations between current, induced or counter emf, and source emf in an inductor.

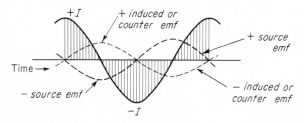

5-14 PHASE WITH BOTH *L* AND *R*

When inductance and resistance are in series in a circuit, the number of degrees the source voltage and the circuit current will be out of phase depends upon the relative resistance and reactance values. A circuit is shown in Fig. 5-15, with the resistance and the inductive reactance values equal.

Remember: The *current* in all parts of this circuit is exactly the same at any single instant, since it is a series circuit.

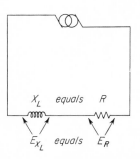

Fig. 5-15 When X_L and R are in series and equal, the voltage-drops across them are equal.

When the current through the resistance is at a maximum value, the voltage-drop across it is at a maximum, since the voltage and current across and through a pure resistance are always exactly in phase.

The current in the coil leads the induced or counter emf by 90°. The applied voltage across the coil is 180° out of phase with the induced emf and therefore leads the current by 90°.

There is only one current in the circuit, but the source sees two voltage-drops in series, one across the resistor (in phase with the current) and the other across the coil and (leading the current by 90°). Since the resistance and reactance are equal, the source sees two voltage-drops of equal magnitude but 90° out of phase. As a result, the source sees a resultant voltage-drop equal to itself but 45° out of phase and leading the circuit current. This is graphed in Fig. 5-16.

If there is more R than X_L, the phase difference will be closer to zero degrees. If there is more X_L, the phase will be closer to 90°. (This is discussed further in Sec. 7-3.)

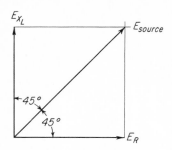

Fig. 5-16 Vector representation of the source voltage, voltage across an inductive reactance, and voltage across the resistor.

5-15 TRANSFORMERS

One of the most common components, or parts, used in electricity, electronics, and radio is the transformer. The name itself indicates that the device is used to transform, or change, something. In practice a transformer may be used to step up or step down ac voltages, to change low-voltage high-current ac to high-voltage low-current ac, or vice versa, or to change the impedance of a circuit to some other impedance in order to transfer energy better from a source to a load.

In its simplest form, a transformer consists of a *primary* wire and a *secondary* wire laid side by side as shown in Fig. 5-17. The only parts of the

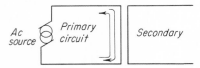

Fig. 5-17 Basic transformer with ac flowing in the primary.

primary and secondary circuits to be considered are the portions lying parallel to each other. When

the source is producing an alternating voltage, an alternating current will be developed in the primary wire, as indicated by the arrows, producing expanding and contracting alternating magnetic fields around the primary wire. These fields induce a counter emf in the primary, which attempts to counteract the source voltage and thereby limit the primary current value. In practical transformers the primary coil has a sufficient number of turns (inductance and inductive reactance) to produce a counter emf almost equal to the source voltage and therefore very little primary current.

In addition to the counter emf induced into the primary circuit by self-induction, the expanding and contracting fields from the primary cross the secondary wire and induce an ac emf in it. According to the left-hand generator rule (Sec. 3-18), if the current in the primary flows downward, the induced emf in the secondary will be upward and 180° out of step, or out of phase, with the primary emf as shown in Fig. 5-18.

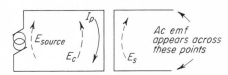

Fig. 5-18 Basic transformer voltages and currents with no load across the secondary.

There is no load shown across the secondary circuit. With no secondary load, no current flows in the circuit, although a voltage is developed across it. If the secondary has a voltage induced in it but no current, no power is developed in the secondary and the current in the primary will be the same as though there were no secondary. (Actually, some electrons flow in the turns of the secondary to produce the + and − charges that result in the secondary voltage.)

When a load resistor is connected across the secondary of a transformer (Fig. 5-19), several things occur. Step by step these are:

1. The source emf, E_{source}, produces primary current, I_p.
2. Primary current produces counter emf, E_c, in the primary.

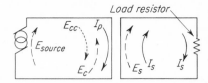

Fig. 5-19 Currents and voltages in a transformer when a load resistor is connected across the secondary.

3. Primary current also produces an induced secondary emf, E_s.
4. E_s produces current I_s through secondary and load.
5. I_s produces a magnetic field expanding outward from the secondary.
6. Expanding fields from the secondary induce a counter counter emf, E_{cc}, in the primary (opposite to the original counter emf in the primary and in the same direction as the source voltage). Mutual inductance is now present.
7. E_{cc} partially cancels the counter emf of the primary.
8. Cancellation of primary counter emf allows the source emf to send more current through the primary.

Therefore, when a load is connected across the secondary, the primary current increases. This results in feeding more electric power to the primary to be converted to magnetic energy, which is transferred in turn to the secondary and reconverted to electric energy in the secondary load.

Consider transformer loading another way. By Lenz's law, I_{sec} flows in such a direction that its magnetizing action opposes the magnetizing ac-

tion of I_{pri}. Thus, any increase in I_{sec} cancels primary flux, reduces primary cemf, and increases I_{pri}.

5-16 CONSTRUCTION OF TRANSFORMERS

Transformers are constructed with a primary coil and one or more secondary coils. A second secondary may be termed the *tertiary* (meaning "third") winding, and a third secondary may be termed the *quaternary* (meaning "fourth") winding. It is, however, more usual to designate the windings by their function. Thus, a transformer may be said to have a primary, a high-voltage secondary, a 5-V winding, etc.

Figure 5-20 illustrates two of many possible methods of constructing a power-frequency transformer and the symbol for an iron-core transformer. Winding the secondary coil over the primary coil is probably the more frequently used method.

In power transformers, there are usually several hundred turns on the primary and an equivalent number of turns on the secondary if it is desired to produce a secondary voltage equal to the voltage applied across the primary. If a greater secondary voltage is desired, more turns will be wound on the secondary than on the primary.

Transformers for higher frequencies use less iron in their cores. If the frequency is in the RF range, either air or ferrite cores are used. Figure 5-21 illustrates a possible RF transformer and its symbol used in schematic diagrams. The core is often some nonconducting material having the same permeability as air, with one or both coils "tuned" (Sec. 8-7).

Fig. 5-20 Two basic methods of constructing iron-core transformers and symbol for both.

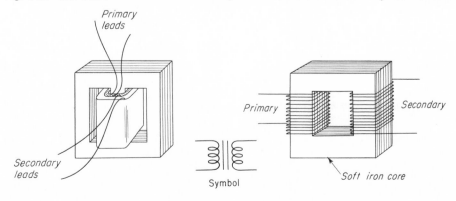

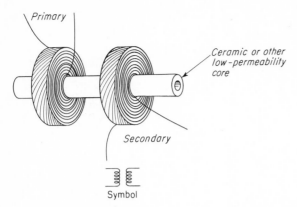

Fig. 5-21 One possible air-core transformer construction and symbol.

5-17 EDDY CURRENTS

To produce a transformer of high efficiency and with a minimum number of turns, the primary and secondary are wound on a core of iron or other high-permeability material. As a result, when a transformer is in operation, intense moving magnetic fields are produced in the core. These fields induce circulating currents in the core material because iron is a fairly good electric conductor. Such whirlpool-like currents are known as *eddy currents,* which can produce a considerable I^2R power loss in the form of heat in the core. Figure 5-22 indicates the path of eddy currents in a solid

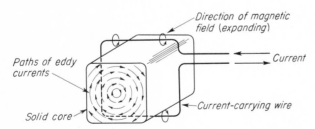

Fig. 5–22 Eddy currents induced in a solid iron core.

core wound with a single turn of wire in which the current is increasing (left-hand generator rule).

Eddy currents are decreased in strength by using many thin sheets rather than a solid block of iron. Each separate sheet must be coated with an insulating scale, varnish, or plastic coating. Any one eddy-current path is limited to the thick-

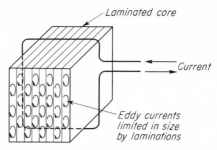

Fig. 5-23 Eddy currents are reduced in amplitude by laminating the core.

ness of the sheet (Fig. 5-23). Limiting the length of the path limits the amplitude of the eddy currents and holds the I^2R heat loss in the core to a minimum. Slicing the core into thin sheets is known as *laminating* the core. The thinner the laminations, the less the eddy-current loss. The cores in Fig. 5-20 are laminated.

As frequency is increased, magnetic flux movement and eddy-current losses both increase. It has been found advantageous to use ferrite cores in higher-frequency applications. A difficulty in constructing such iron cores is the requirement that each magnetic particle be insulated from adjoining particles to prevent eddy currents from developing should the particles touch or make electrical contact with each other.

5-18 HYSTERESIS

If iron is in an unmagnetized state, its domains are not arranged in any particular manner. When a magnetizing force is applied to them, the domains rotate into a position in line with the magnetizing force. If the magnetizing force is reversed, the domains must rotate into an opposite position. In rotating from one alignment to the opposite, the domains must overcome a frictional *hysteresis,* or resisting, effect in the substance. In some materials the resisting effect is very small; in others it is appreciable. The energy converted into heat overcoming hysteresis is known as hysteretic loss.

Hysteresis occurs in iron cores of transformers. As frequency is increased, the alternating magnetizing force will no longer be able to magnetize the core completely in either direction. Before the core becomes fully magnetized in one direction, the opposite magnetizing force will begin to be applied and start to reverse the rotation of the domains. The higher the frequency, the less fully the core magnetizes.

Transformers operated on low-frequency ac may not have much hysteresis, but the same cores used with a higher frequency have more hysteresis and may also be less efficient.

5-19 COPPER LOSS

Iron-core transformers are subject not only to eddy-current and hysteresis losses in the core but also to a *copper loss* which occurs in the copper wire of the primary and secondary. The current flowing through whatever resistance exists in these windings produces heat. The heat in either winding, in watts, can be found by the power formula $P = I^2R$. For this reason the copper loss is also known as the *I^2R loss*. The heavier the load on the transformer (the more current that is made to flow through the primary and secondary), the greater the copper loss.

With one layer of wire wound over another in a transformer, there is a greater tendency for the heat to remain in the wires than if the wires were separated and air-cooled. Increased temperature causes increased resistance of a copper wire. As a result it becomes necessary to use heavier wire to reduce resistance and heat loss in transformers than would be required for an equivalent current value if the wire were exposed to air during operation.

5-20 EXTERNAL-INDUCTION LOSS

Another loss in a transformer is due to external induction. Lines of force expanding outward from the transformer core may induce voltages and therefore currents into outside circuits. These currents flowing through any resistance in an outside circuit can produce a heating of the external resistance. The power lost in heating these outside circuits represents a power loss to the transformer, since the power is not delivered to the transformer secondary circuit. Actually, in a well-designed transformer, the amount of power lost in this fashion is usually very small. The voltages induced into nearby wires of certain types of amplifying circuits can, however, produce un-

desirable voltages in these circuits, even though the power loss to the transformer is negligible.

Any lines of force from the primary of a transformer that do not cut secondary turns and that induce no emf into the secondary are considered leakage flux, or leakage lines of force (not a power loss).

5-21 THE VOLTAGE RATIO OF TRANSFORMERS

One of the main uses of transformers is to step up a low-voltage ac to a higher voltage. This can be accomplished by having more turns on the secondary than on the primary. The transformer in Fig. 5-24 has a single wire for the primary and

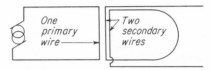

Fig. 5-24 Basic transformer with a 1:2 step-up voltage and turns ratio.

two wires in series in the secondary. Each of the secondary wires will pick up an equal voltage, since both are being cut by the same number of lines of force from the primary. If two 1-V induced emfs in the secondary are in series, this results in an output of 2 secondary volts.

As long as the coefficient of coupling is high in a transformer, the no-load voltage ratio will equal the turns ratio. If the primary is wound with 500 turns and the secondary with 1,000 turns, the secondary voltage will be twice any voltage applied across the primary. The fact that the voltage ratio is equal to the turns ratio can be expressed by

$$\frac{T_p}{T_s} = \frac{E_p}{E_s}$$

where T_p = number of primary turns
T_s = number of secondary turns
E_p = primary voltage
E_s = secondary voltage

The voltage ratio also works in reverse. If the primary has 200 turns and the secondary has 40 turns, with 100 V connected across the primary, the secondary voltage can be determined by rearranging the above formula to

$$E_s = \frac{E_p T_s}{T_p} = \frac{100(40)}{200} = \frac{4,000}{200} = 20 \text{ V}$$

5-22 THE POWER RATIO OF TRANSFORMERS

There is no step-up of power in transformers. It is possible to step voltage up or down, but the basic ratio of power into the primary to power out of the secondary is 1:1. Actually, because of the many losses in a transformer, less power will always be drawn from the secondary than goes into the primary.

Power transformers are constructed to handle a certain number of watts, or voltamperes (VA). For example, a 100-V primary, 500-W (properly 500-VA) transformer will have a primary wound with wire that will carry only enough current to produce 500 VA in the primary. By the power formula $P = EI$, or $I = P/E$, it can be computed that 5 A is all the primary wire will be required to carry, regardless of whether the transformer is going to step the secondary voltage up or step it down. The primary will be wound with the thinnest wire that will carry 5 A without excessive heating (Table 1-3). If more than 500 W is drawn by the secondary load, the primary will be called on to carry more than 5 A, will become overly hot, and may burn the insulation on the wires or melt the wire itself. The secondary also will be wound with a wire that will approach its safe heating limit when 500 W is being drawn from the secondary.

If the transformer has a step-up ratio of 1:4, the secondary voltage will be 4 times 100 V, or 400 V. Inasmuch as the limiting factor is the 500 W into the primary, the secondary can be called upon to deliver only 1.25 A at 400 V. Otherwise, the power delivered by the secondary will be more than 500 W and either the primary or the secondary winding may overheat and fail.

To protect transformers from overloads, fuses or overload relays are connected in the primary circuit. A 5-A fuse in series with the primary of the above transformer would burn out if more than the maximum safe current of 1.25 A were drawn by the transformer secondary.

Test your understanding; answer these checkup questions.

1. Where in time is the current peak in relation to the voltage peak in a purely inductive circuit? _____ In a purely resistive circuit? _____
2. At what points in a sine wave is the rate of change at maximum? _____ At zero? _____
3. A 10-Ω R is in series with a 20-Ω X_L across an ac source. Are E and I leading, lagging, or in phase in the R? _____ In the X_L? _____ In the source?
4. If a transformer secondary is unloaded, is any counter counter emf induced in the primary? _____
5. Is a ceramic-cored transformer classed as an air-core type? _____
6. What is decreased by laminating transformer cores? _____
7. What is the term that indicates the inability of a core material to reverse magnetic polarity completely? _____
8. Does an increase in load on a transformer increase the eddy-current loss? _____ The hysteretic loss? _____ The copper loss? _____ The external-induction loss? _____
9. What ratio in any transformer is always about 1:1? _____
10. What gauge wire would be used for a primary winding that must carry 5 A? _____ 10 A? _____
11. An unloaded transformer has a 4:1 step-down ratio and a k of 0.8. With 100 V across the primary, would the secondary voltage be 25 V, less than 25 V, or more than 25 V? _____
12. Short-circuiting the secondary winding of a transformer has what effect on the inductance of the primary? _____ On the current of the primary? _____
13. Would increasing the gauge size of the wire used in a transformer affect the output voltage? _____

5-23 THE CURRENT RATIO OF TRANSFORMERS

A step-up transformer may produce more voltage across the secondary than is applied across the primary, but the secondary current will have to be proportionately less than the primary current. This was indicated by the 500-W 1:4 ratio step-up transformer previously mentioned. With 100 V across the primary, the secondary produces 400 V. With a load that draws 1.25 A (a

320-Ω resistor, for example) connected across the secondary, the primary will be called upon to draw 5 A of current from the source. This represents a primary-to-secondary step-down of current from 5 to 1.25 A.

Secondary current is inversely proportional to the turns ratio. This can be expressed in formula form by

$$\frac{T_p}{T_s} = \frac{I_s}{I_p}$$

where T_p = number of primary turns
T_s = number of secondary turns
I_p = primary current
I_s = secondary current

5-24 TRANSFORMER EFFICIENCY

It will always be found that more power is fed to the primary of a transformer than is delivered by the secondary to a load. The difference between the input power and the output power is the sum of all the power *losses* in the transformer.

The ratio of the output power to the input power is the efficiency of the transformer. The factors that determine efficiency are the copper, eddy-current, hysteretic, and external-induction losses. The output/input power ratio always results in a decimal number less than 1.0. In practice, efficiency is given in percent rather than in the decimal equivalent. It is only necessary to multiply the decimal by 100 to determine the percent. The formula for percent efficiency is

$$\text{Percent efficiency} = \frac{P_o}{P_i} \times 100$$

where P_o = power output
P_i = power input

If the overall efficiency of a transformer is known, the primary power times the percent efficiency is the secondary power, or $(\%)(P_i) = P_o$

EXAMPLE: If a power transformer has a primary voltage of 4,400 V, a secondary voltage of 220 V, and an efficiency of 98% when delivering 23 A of secondary current, what is the value of primary current? In this case, 23 A at 220 V, or 5,060 W, represents 98% of the power being fed into the primary. By substituting in the formula,

$$(\%)(P_i) = P_o$$
$$P_i = \frac{P_o}{\%} = \frac{5,060}{0.98} = 5,163 \text{ W}$$

By using the power formula $P = EI$ the primary current can be found:

$$P_p = E_p I_p$$
$$I_p = \frac{P_p}{E_p} = \frac{5,163}{4,400} = 1.17 \text{ A}$$

Power transformers are always warm to the touch when operating, due to internal losses. In some cases it becomes necessary to air-cool transformers to keep them from overheating and damaging the insulation on the wires of the windings. Some transformers are built into oil-filled cases. The oil helps to insulate the internal wiring and prevent moisture from forming on the insulation, which might result in a breakdown, and also carries heat from the windings to the outer case to be dissipated into the air.

5-25 AUTOTRANSFORMERS

An *autotransformer,* or *autoformer,* consists of a single winding with one or more taps on it as shown in Fig. 5-25.

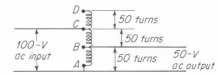

Fig. 5-25 Autotransformer.

If a 100-V source of ac is connected between points *A* and *C* and there are 100 turns between these points, an emf of 1 V will be induced in each of the turns, as well as in each of the turns from *C* to *D*. If a load is connected across points *A* and *B*, as shown, it will be across 50 V. If connected across *A* and *C*, it will be across 100 V.

If connected across *A* and *D*, it will be across 150 V. Thus an autotransformer can be used as a voltage step-down or step-up device.

If the load is connected between *A* and a tap that can be adjusted to any turn between *A* and *D*, any desired voltage up to 150 in 1-V steps can be developed across the load. Such autotransformers are made and sold under trade names such as Powerstat and Variac.

A disadvantage of autotransformers is the common connection between primary and secondary circuits. It is often desirable to have primary and secondary circuits isolated from each other electrically.

If isolation is not a factor, any common transformer can be connected as an autotransformer and more or less output voltage can be obtained than the turns ratio of the transformer would normally give.

5-26 PRACTICAL TRANSFORMER CONSIDERATIONS

In radio and electronics, there are many types of transformers in use. Three common types are power transformers, AF transformers, and RF transformers (Fig. 5-26).

A power transformer is normally made to operate across a 110- to 440-V ac line. It is a heavy iron-encased piece of equipment. The resistance in the primary winding ranges from a fraction of an ohm to possibly 5 Ω. The inductive reactance of the primary winding acts to limit the primary current to a low value when connected across an ac power line. If such a transformer is connected across a similar-voltage dc line, the low resistance will allow excessively high current to flow in the primary. The primary will overheat and burn out, or the line fuses will blow. Care must be taken

Fig. 5-26 Transformers. Left, iron-core; from top, power, variable autotransformer. Middle, unshielded, shielded, and two transistor audio transformers. Right, air-core, top two with shields removed, and transistor IF transformer.

not to connect the primary of a power transformer across a dc power line.

Power transformers are made to operate on one particular frequency, usually 60 Hz. In most cases, such transformers will operate fairly well on any frequency between 50 and 70 Hz. If the frequency is much too high, however, the inductive reactance of the primary will prevent the primary from drawing sufficient power. There will be more iron in the core than is necessary, and hysteresis and eddy-current losses will be excessive. If the frequency is too low, the primary will have insufficient reactance and too much primary current will flow, producing considerable copper loss. The transformer may start to smoke. There will not be enough iron in the core, and the transformer will not be capable of its rated power output.

If a turn of either the primary or the secondary of a power transformer shorts out for some reason, a high current will be induced in the turn, producing excessive heat in the transformer, not only because of the shorted turn's heating but also because of the cancellation of the inductance of the primary by the magnetic field set up by the shorted turn. Cancellation of the inductance materially decreases the inductive reactance of the primary, and excessive primary current flows.

Audio transformers are also iron-cored, are usually smaller than power transformers, and are connected in series with a relatively high-resistance vacuum tube or transistor across a source

ANSWERS TO CHECKUP QUIZ ON PAGE 84

1. (*Lags behind*)(*In phase*) 2. (*As time line is crossed*)(*At peaks*) 3. (*In phase*)(*I lags*)(*I lags*) 4. (*No*) 5. (*Yes*) 6. (*Eddy currents*) 7. (*Hysteresis*) 8. (*No*)(*No*)(*Yes*)(*No*) 9. (*Power*) 10. (*No. 12*)(*No. 10*) 11. (*Less than 25 V*) 12. (*Reduces L*)(*Increases I greatly*) 13. (*Theoretically not at all, but with heavy I values larger voltage-drops occur with small wires, resulting in less voltage output.*)

of dc. The resistance of the vacuum tube or transistor limits the primary current to a safe dc value and prevents the primary from burning out.

Radio-frequency transformers are normally air-core transformers and are made to operate across RF ac directly or in series with either a vacuum tube or a transistor and a source of dc.

Test your understanding; answer these checkup questions.

1. Would a voltage step-up transformer be considered a current step-up or step-down transformer? _____
2. A transformer has an E_p of 120 V, an E_s of 24 V, and an efficiency of 96% when delivering 4 A to a load. What is the I_p value? _____
3. It is desired to produce 1,000 V by using a 120-V source. How many turns should the secondary have if the primary has 240 turns? _____

4. A transformer is to supply 12.6 V from a 120-V line. How many secondary turns are required if there are 300 primary turns? _____ To supply a 4-A load at 600 cmil/A, what gauge wire should be used on the secondary? _____ On the primary? _____
5. A 120-V primary has a 240-V secondary and a 12-V tertiary. Disregarding losses, if the tertiary load draws 5 A and the secondary 0.1 A, what current flows in the primary? _____
6. If a choke coil has a tap near its center, what type of transformer might it be? _____ Would it be a step-up or a step-down type? _____
7. What might occur if 110-V dc is connected across a 120-V-primary power transformer? _____
8. What difficulties occur if the ac frequency fed to a power transformer is too low? _____ Too high? _____
9. What type of core is usually used with an RF transformer? _____ An AF transformer? _____
10. Under what condition is dc applied safely to the primary of a transformer? _____

COMMERCIAL LICENSE QUESTIONS

Amateur license questions will be found in the Addendum.

FCC Elements 3, 4, and 6 require an ability to answer questions similar to those below. Sections in which questions are answered are shown in parentheses. A question followed by a bracketed number is required for that element alone.

1. Define the term *inductance*. (5-1, 5-2)
2. What is the unit of inductance? (5-2)
3. What is the relation between the number of turns and the inductance of a coil? (5-3)
4. What is the relation between inductance of a coil and permeability of its core? (5-3)
5. Explain how values of resistance and inductance in an RL network affect its time constant. (5-4)
6. State the formula for the energy stored in the magnetic field surrounding an inductance. (5-5)
7. What does *coefficient of coupling* mean? (5-8)
8. What is meant by *unity coupling*? (5-8)
9. If the mutual inductance between two coils is 0.03 H and the coils have inductances of 0.04 and 0.09 H, what is the coefficient of coupling? (5-8) [4]
10. What is the total inductance of two coils in series but without any mutual coupling? (5-9)
11. What effect does mutual inductance have on the total inductance of two coils if they are in series? (5-9) [4]

12. When two coils of equal inductance are connected in series with unity coefficient of coupling and their fields are in phase, what is the total inductance of the two coils? (5-9) [4]
13. What is the total inductance of two coils in parallel without any mutual coupling? (5-10)
14. What will be the effect of a shorted turn in an inductance? (5-11)
15. A series inductance acting alone in an ac circuit has what properties? (5-12)
16. What is the total reactance of two inductances in series with zero mutual inductance? (5-12) [4]
17. What is the reactance of a 4-H choke coil at a frequency of 3,000 cps? (5-12)
18. What is the reactance of a 10-mH coil at a frequency of 2,000 kc? (5-12)
19. What prevents high current flow in the primary of an unloaded power transformer? (5-15)
20. Explain how self-inductance and mutual inductance produce transformer action. (5-15, 5-26)
21. What is an *eddy current*? (5-17)
22. Why are laminated-iron cores used in audio and power transformers? (5-17)
23. What factors determine the core losses in a transformer? (5-17, 5-18) [4]
24. How is power lost in an iron-core transformer? (5-17–5-20)

25. What is meant by *hysteresis* in a transformer? (5-18)
26. How is power lost in an air-core transformer? (5-18–5-20)
27. What circuit constants determine the copper losses of a transformer? (5-19)   [4]
28. In an iron-core transformer, what is the relation between the turns ratio and the primary-secondary voltage ratio? (Assume no losses.) (5-21)
29. What is the secondary voltage of a transformer which has a primary voltage of 100, primary turns of 400, and secondary turns of 150? (5-21)
30. In an iron-core transformer, what is the relation between the turns ratio and the primary-secondary current ratio? (Assume no losses.) (5-23)
31. If a power transformer having a voltage step-up ratio of 1:4 is placed under load, what will be the approximate ratio of primary to secondary current? (5-23)
32. What factors determine the efficiency of a power transformer? (5-24)
33. If a power transformer has a primary voltage of 4,800 V, a secondary voltage of 240 V, and an efficiency of 96% when delivering 20 A of secondary current, what is the value of primary current? (5-24)   [4]

34. What would be the effect if dc were applied to the primary of an ac transformer? (5-26)
35. If part of the secondary winding of a power transformer were accidentally shorted, what would be the immediate effect? (5-26)
36. What would happen if a transformer designed for operation on a 60-cps voltage were connected to a 120-cps source of the same voltage? (5-26)
37. What would happen if a transformer designed for operation on 500 cps were connected to a 60-cps source of the same voltage? (5-26)

NOTE: FCC license test questions are multiple-choice types, taking a form somewhat as follows:

38. If a circuit opposes any change in current, it is said to have (a) reactance; (b) capacitance; (c) inductance; (d) hysteresis. (5-2)
39. What are two coils said to have when there is no leakage flux between them? (a) High coefficient of coupling; (b) unity coupling; (c) mutual inductance; (d) reactance. (5-8)
40. Which factor might be the cause of a power loss in an air-core transformer? (a) Eddy-current loss; (b) leakage loss; (c) hysteretic loss; (d) copper loss. (5-19)

ANSWERS TO CHECKUP QUIZ ON PAGE 87

1. (*Step-down*) 2. (*0.833 A*) 3. (*2,000 turns*) 4. (*31.5 turns*)(*NO. 16*)(*No. 26*) 5. (*0.7 A*) 6. (*Autoformer*)(*Either*) 7. (*Burn out primary or blow fuse*) 8. (*Copper loss and low efficiency*)(*Hysteretic and eddy-current losses, low efficiency*) 9. (*Air or ferrite*)(*Iron*) 10. (*When limiting resistance is in series with primary*)

---

6 CAPACITANCE

The objective of this chapter is to explain the various types of capacitors and the factors determining capacitance and losses in capacitors. Capacitive reactance is explained, as are capacitors in series and in parallel, phase relationships, and color codes.

6-1 THE CAPACITOR

One of the most used parts in radio and electronics is the *capacitor,* previously called a *condenser.* A capacitor has the ability to hold a charge of electrons. The number of electrons it can store under a given electric pressure is a measure of its *capacitance* (sometimes called *capacity*). Two separate metallic plates with a nonconducting substance sandwiched between them, as in Fig. 6-1, form a simple capacitor. Symbols for capacitors are shown in Fig. 6-2.

It is important to understand the current and voltage changes that take place in a circuit in which a capacitor is connected. Figure 6-3 shows a simple circuit with a battery, switch, resistor, and capacitor in series and a graph of the current and

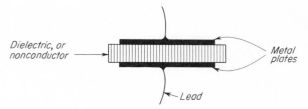

Fig. 6-1 A basic capacitor consists of two conductive plates separated by a nonconducting dielectric.

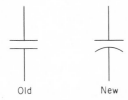

Fig. 6-2 Symbols used to denote fixed capacitors.

voltage changes that will take place when the switch is closed. With the switch open, the capacitor will be assumed to have zero charge; that is, both plates have the normal number of electrons and protons in the molecules of the metal.

Fig. 6-3 RC circuit with graph of current flowing into the capacitor and voltage developing across it after the switch is closed.

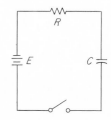
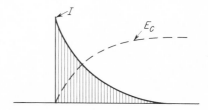

At the instant the switch is closed, the electric pressure of the battery begins to force electrons into the top plate from the negative terminal of the battery and pull others out of the bottom plate toward the positive end of the battery. As the electron difference is developed between the two plates, electrostatic lines of force appear in the region between the plates.

At the instant the switch is closed, there is no opposing emf in the capacitor and the amplitude of the current is determined only by the resistance in the circuit. As time progresses, more electrons flow into the capacitor and a greater opposing emf is developed in it. The difference between the source and the opposing emf becomes less. The opposing emf across the capacitor continually increases, and the charging current continually decreases. When the opposing emf equals the source emf, there will no longer be any charging current flowing into the capacitor. The voltage across the capacitor will be at a maximum and equal to the source voltage.

A capacitor that will store a difference of one coulomb (1 C or 6.25×10^{18} electrons) when an emf of one volt is applied across it has the capacitance value of one *farad* (1 F). A total of 2 V across this same capacitor would store 2 C. In radio and electronics 1 F is more capacitance than is normally used. Practical values are measured in microfarads (millionths of a farad) or in micromicrofarads. Microfarads are abbreviated μF, although uf, mf, μf, μfd, ufd, and mfd may also be seen at times. A micromicrofarad ($\mu\mu$F) is the same as a picofarad, abbreviated pF (incorrectly pf). It is important to be able to convert from microfarads to picofarads. For instance,

$$1 \ \mu F = 1{,}000{,}000 \ pF \ \text{or} \ \mu\mu F$$
$$0.005 \ \mu F = 5{,}000 \ pF \ \text{or} \ \mu\mu F$$
$$0.00004 \ \mu F = 40 \ pF \ \text{or} \ \mu\mu F$$
$$250 \ pF = 0.00025 \ \mu F$$
$$1 \ \mu F \times 10^{-6} = 1 \ pF \ \text{or} \ \mu\mu F$$
$$1 \ pF \times 10^{6} = 1 \ \mu F$$

The time required for a capacitor to attain a charge is proportional to the capacitance and resistance in the circuit. The *time constant* of a resistance-capacitance circuit is

$$T_c = RC$$

where T_c = time, in s
R = resistance, in Ω
C = capacitance, in farads (F)

The time in the formula is that required to attain 63% of the voltage value of the source. It is also the time required for a charged capacitor to discharge 63% when connected across the value of resistance used in the formula. (Note the difference in the time constant compared with a resistance-inductance circuit, Sec. 5-4.) The time required to bring the charge to about 99% of the source voltage is approximately five times the time computed by using the time-constant formula.

If there is no resistance in the circuit, the time required is zero and a capacitor will charge or discharge almost instantaneously.

A 4-μF capacitor and a 1-MΩ (1-megohm) resistor have a time constant of 0.000004 $\times$ 1,000,000, or 4 s.

In Chap. 5, inductance was defined as *the property of a circuit to oppose any change in current*. Capacitance may be defined as *the property of a circuit to oppose any change in voltage*. In the dc circuit discussed, it has been explained that capacitance develops an opposition emf to the change of voltage from a zero value up to the source value. When a resistor is connected across a charged capacitor that has been disconnected from the charging source, the opposing emf that had been developed in the capacitor during the charging period discharges, driving current through the resistor. In a varying dc or ac circuit, in which the emf is continually varying in amplitude, a capacitor charging and discharging will continually be opposing the source emf variations.

When a resistor and a capacitor are in series across an ac source, the voltage across the capacitor will alternate at the source frequency. However, with resistance in series, the voltage across the capacitor will always be less than the source emf. The amount less will depend on the time constant of the R and C being used. With a very long time constant the voltage across the capacitor will be very small. Except for a lower amplitude, the waveform of an ac passing through an RC circuit will remain relatively the same, although with multifrequency ac the higher frequencies will be attenuated more than the low.

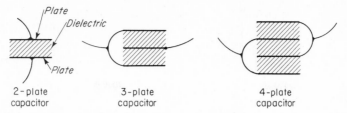

Fig. 6-4 A 3-plate capacitor has twice the capacitance of a 2-plate capacitor; a 4-plate capacitor has 3 times the capacitance of a 2-plate capacitor.

6-2 FACTORS DETERMINING CAPACITANCE

The factors that determine capacitance are the area of the plates exposed to each other, the spacing between the plates, and the composition of the nonconducting material between the plates.

Two plates, each with an area of 1 in.^{2}, when separated by 0.001 in. of air, produce a capacitance of 225 pF. If each plate area is increased to 2 in.^{2} and the spacing remains 0.001 in., the capacitance becomes twice as much, or 450 pF. Capacitance is directly proportional to the plate areas.

If the spacing of the two 1-in.^{2} plates is increased to 0.002 in., the path of the electrostatic lines of force between the negative plate and the positive plate is twice as great, resulting in only half as intense an electrostatic field and only half as much capacitance. Capacitance is inversely proportional to the spacing between plates.

The nonconducting material between the plates, called the *dielectric* material, determines the concentration of electrostatic lines of force. If the dielectric is air, a certain number of lines of force will be set up. Other materials offer less opposition to the formation of electrostatic lines of force in them. For example, with one type of paper instead of air, the number of electrostatic lines of force between the plates may be twice as great. Such a capacitor will have twice as much capacitance and will have two times as many electrons flowing into and out of it with the same applied source emf. The paper is said to have a *dielectric constant,* or *specific inductive capacity,* twice that of air. Capacitance of a capacitor is directly proportional to the dielectric constant.

A formula to determine the capacitance of a two-plate capacitor including these three factors is

$$C_{pF} = \frac{0.225\,KA}{S}$$

where C = capacitance, in pF
K = dielectric constant
A = area of one of the plates, in in.^{2}
S = spacing between plates, in inches

This formula is for a two-plate capacitor. For greater capacitance plates are stacked on top of one another and separated with strips of dielectric material. A three-plate capacitor has twice the plate area exposed, as shown in Fig. 6-4, and twice the capacitance. A four-plate capacitor has three times the plate area and capacitance. The formula for multiplate capacitors is

$$C_{pF} = \frac{0.225KA(N-1)}{S}$$

where N = number of plates in the capacitor

The approximate dielectric constant, or specific inductive capacity, of some common dielectric materials is given in Table 6-1.

Table 6-1 *DIELECTRIC CONSTANTS*

Material	Dielectric constant
Vacuum	1
Air	1.0006
Rubber	2–3
Paper	2–3
Ceramics	3–7
Glass	4–7
Quartz	4
Mica	5–7
Porcelain	6–7
Water	80
Barium titanate	7,500

The dielectric constant of solid dielectric materials may decrease with an increase in frequency. The molecules of the dielectric do not have sufficient time to conform to the rapidly changing electrostatic lines of force that they must support. If the lines of force cannot be fully developed in the molecules of the dielectric, the dielectric constant is less and the capacitance will be less. Thus, a 0.1-μF paper capacitor may have this value of capacitance at 1 MHz but will have considerably less at 100 MHz. Mica is less affected by frequency. Air and vacuum are not noticeably affected.

6-3 DIELECTRIC LOSSES

Almost all the energy stored in the electrostatic field of a capacitor is converted into some other form of energy when the capacitor is discharged. There are, however, two losses that occur in the dielectric itself.

Electrons on the negative plate of a charged capacitor may find a high-resistance path to the positive plate either through the dielectric or over the surface of the capacitor, leaking to the other plate. Therefore leakage current is a loss to a capacitor. Leakage current flowing through any high-resistance path produces heat ($P = I^2R$).

Another dielectric loss is due to hysteresis and also is indicated by heat in the capacitor. It can be considered to be caused by the friction of the molecules of the dielectric material as they are changed from one strained position to the opposite by reversing electrostatic lines of force between the plates. Hysteresis is normally significant only when an alternating current produces rapid charging and discharging of the capacitor. It increases as the frequency of the ac increases. For this reason, many capacitors operate satisfactorily at lower, but not at very high, frequencies. Vacuum, air, mica, and ceramic dielectric capacitors have little leakage or hysteresis. Paper may have considerable leakage and hysteresis, particularly if the dielectric has moisture in it.

An interesting and significant phenomenon is produced in some solid-material dielectric capacitors. If charged by a dc voltage they can be disconnected from the charging source and will remain charged. A wire touched across the capacitor terminals will discharge the capacitor, usually with an audible and visible electric spark. If the wire is disconnected for a short time and then touched across the discharged capacitor again, another spark is produced, indicating that the dielectric had not released all the stored energy on the first discharge. During the charged period dielectric molecules some distance from the capacitor plates capture some electrons. These take time to leak back to the plates through the high resistance of the dielectric after the capacitor has been discharged. This *dielectric absorption* may not be very significant when the capacitor operates in a dc circuit, but in high-frequency ac or varying dc circuits it decreases the effectiveness of the capacitor.

6-4 WORKING VOLTAGE AND DIELECTRIC STRENGTH

One rating of capacitors is the *working voltage*. This is the maximum voltage at which the capacitor will operate without leaking excessively or arcing through. Sufficient leakage through the dielectric over a period of time can produce a carbonized path across the dielectric, and the capacitor will act as a conductor. In such a case it is said to be *burned out,* or *shorted*.

A burned-out capacitor should not be confused with an *open* one. An open capacitor has lost its storage ability by the breaking off of a wire lead internally or, in the case of electrolytic capacitors, because the electrolyte has dried out.

The working voltage is usually rated as a dc value. A capacitor may be rated to work at 600 V on dc circuits; but when used in power-frequency ac circuits, its effective ac working voltage will be about one-half the dc rating. As the frequency of the ac is increased, the working voltage of the capacitor decreases, particularly when the frequency rises above a few megahertz. Any dielectric heating will decrease the breakdown voltage.

The *dielectric strength,* or number of volts that a dielectric material will stand per 0.001 in. of dielectric thickness, varies considerably with different materials. Air has a dielectric strength of about 80 V; Bakelite, about 500 V; glass 200 to 300 V; mica, about 2,000 V; untreated paper, a few hundred volts; waxed or oiled paper, 1,000 to 2,000 V; rubber, 400 V; ceramics, 80 to 200 V; Teflon, 1,500 V; practical vacuums may exceed 10,000 V.

6-5 ENERGY STORED IN A CAPACITOR

When a capacitor is charged and then disconnected from the charging source, it has a difference of electrons between the two plates and the dielectric molecules are under the stress of electrostatic lines of force. If the charged capacitor is connected across a light bulb, for example, the excess electrons on the negative plate will flow through the bulb to the positive plate until the electron inequality between the plates is balanced. When the two plates have an equal number of electrons, the capacitor will no longer have any charge and no current will flow.

While moving through the light bulb, the electrons liberate the energy of their motion in the form of heat. In this case the light bulb may flash for an instant and then go out. The amount of energy stored by the electrostatic field in the dielectric of a capacitor can be computed, using wattseconds (joules) as the unit of measurement, by the formula

$$E_n = \frac{CE^2}{2}$$

where E_n = energy, in wattseconds
C = capacitance, in F
E = voltage, in V

6-6 QUANTITY OF CHARGE IN A CAPACITOR

The *charge* of a capacitor is the difference in number of electrons on the two plates. Since this difference involves a quantity of electrons, the unit of quantity of charge is the coulomb. In formula form, the quantity of charge in a capacitor is

$$Q = CE$$

where Q = charge, in coulombs (C)
C = capacitance, in F
E = voltage, in V

EXAMPLE: If a 1-μF capacitor is across 10-V dc, the electron difference between the positive and the negative plates will be

$$Q = CE = 0.000001(10)$$
$$= 0.00001 \text{ C (or } 10^{-5} \text{ C)}$$
$$10^{-5} \text{ C} = 6.25 \times 10^{18} \times 10^{-5}$$
$$Q = 6.25 \times 10^{13} \text{ electrons}$$

It can be reasoned that if a 0.1-μF capacitor is charged by a 125-V source, an electron difference will be developed of $Q = CE$, or 0.0000001(125), or 0.0000125 C between the plates. If the charged capacitor is disconnected from the source, it still retains the electron difference on its plates (no leakage is assumed). If a similar uncharged capacitor is connected across the charged capacitor, electrons flow from the charged to the uncharged capacitor. Since both are of equal capacitance, each will have half the electron difference, or 0.00000625 C. A capacitor losing half its electron charge will have only half the voltage across it. Each of the two parallel capacitors now has 62.5 V across it. Nothing has actually been lost, however, for if the two capacitors are disconnected and reconnected in series, the total voltage-drop across them is 125 V and the same number of electrons is still in storage in the capacitors.

Test your understanding; answer these checkup questions.

1. What is the old term meaning capacitor? _____
2. If a 0.1-μF capacitor and an 8-MΩ resistor are connected across 550 V dc, how long does it take for the capacitor to charge to 63% of the source value? _____ To 99%? _____
3. What is the time constant of a 75-kΩ resistor and a 0.2-μF capacitor? _____
4. What device has the property of opposing any change in current? _____ Any change in voltage? _____
5. Is capacitance directly or inversely proportional to plate area? _____ Plate spacing? _____ Dielectric constant? _____
6. A seven-plate air-dielectric capacitor with 3.5-in.^{2} plates separated by 0.002 in. would have what capacitance in μF? _____ In pF? _____
7. What is another name for specific inductive capacity? _____
8. What dielectric material mentioned would make the high-

est-capacitance capacitor? _____ The lowest-capacitance capacitor? _____

9. What are the names of three losses present in a solid-dielectric-type capacitor? _____ _____ _____

10. What is the approximate ac working voltage of a capacitor in comparison with its dc rating? _____

11. What dielectric mentioned has the greatest dielectric strength? _____

12. A 4-μF capacitor is disconnected from a 600-V dc supply. How much electric energy will a person receive if he grabs hold of the two terminals? _____ How many electrons will pass through his flesh? _____ Might this kill him? _____

13. An uncharged 4-μF capacitor is connected across the charged capacitor in question 12. What voltage will appear across the two capacitors in parallel? _____

6-7 ELECTROLYTIC CAPACITORS

A capacitor was represented as a flat metal plate, dielectric, metal plate, dielectric, metal plate, etc. Some capacitors, mica, air, and paper particularly, are made in this manner. If the dielectric is flexible, however, capacitors are usually made of two long sheets of aluminum foil separated by a long strip of the dielectric material. The whole capacitor can then be rolled up into relatively small tubular shape with a wire lead from one end connected to one of the foil plates and a wire lead from the other end connected to the other foil plate.

An *electrolytic* capacitor consists of an aluminum-foil positive plate immersed in a solution called an *electrolyte* (ionizable solution capable of carrying current). The aluminum foil is the positive plate, and the electrolyte is the negative plate, if a liquid can be called a plate. To make an electrical connection to the liquid, another aluminum foil is placed in the solution. To prevent the two foils from touching each other, a piece of gauze is placed between them. The aluminum positive plate and negative foil in contact with the electrolyte are then subjected to an electric potential to *form* electrochemically an oxidized film on the positive plate. It is this thin oxide film between positive plate and electrolyte that is the dielectric of this capacitor. The thickness of the dielectric film depends upon the forming potential. If formed with a 450-V potential, the capacitor should be used at or slightly under this voltage. Some electrolytic capacitors formed at 450 V but used on

300-V circuits may re-form to the lower voltage, which results in a thinner film and therefore a higher capacitance value. Electrolytic capacitors are rolled into tubular shape and sealed to make them airtight.

If electrolytic capacitors are connected across a circuit with reversed polarity (positive plate of capacitor to negative terminal of circuit), the film deforms, the capacitor becomes a good conductor, high current flows, heat is developed, the electrolyte boils, and the capacitor may explode. Care must be taken when connecting *polarized* electrolytics into circuits that proper polarities are observed. This means that the average electrolytic capacitor cannot be used in ac circuits. (A special type of "nonpolarized" electrolytic capacitor can be used during starting periods for ac electric motors.)

Although physically small and relatively inexpensive, electrolytic capacitors have some disadvantages. They dry out over a period of time and lose capacitance. They have a small leakage current when in operation, which tends to raise the power factor (Sec. 6-10) of the capacitors. When used on voltages above the rated values, the leakage current increases, tending to dry out the capacitors.

Improved *tantalum* electrolytic capacitors operate at higher temperatures than are possible with the older types.

6-8 VARIABLE CAPACITORS

Two types of capacitors can be adjusted or varied. Adjustable capacitors are usually constructed of two or more flat metal plates separated by sheets of mica installed on oblong ceramic holders, as at lower left in Fig. 6-5. The plates are so bowed that they normally hold themselves apart somewhat. Turning a machine screw presses the plates together, thereby increasing the capacitance. Adjustable capacitors may be called padders or trimmers. They are commonly available in capacitances from a few to 1,000 or more pF, with a working voltage of 100 to 600 V.

Variable capacitors have *stator,* or stationary, plates and *rotor,* or rotatable, plates. When the shafts shown in Fig. 6-5 are rotated, the rotor plates mesh into the spaces between stator plates

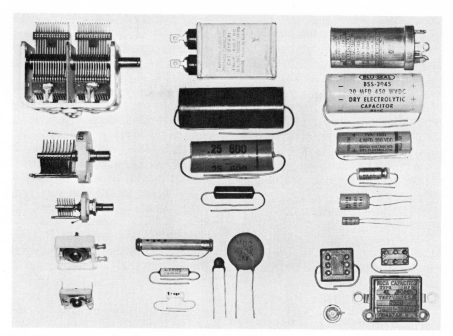

Fig. 6-5 Capacitors. Left column, variable and adjustable types. Top center, paper capacitors. Bottom center, ceramic types. Top right, electrolytics. Bottom right, flat mica capacitors.

without touching. This varies the exposed plate areas and thereby the capacitance of the capacitor. The dielectric of these capacitors is usually air, although some of them employ mica or a ceramic material. There are also special vacuum-dielectric variable capacitors. Variable capacitors in radio receivers have very little spacing between plates. Those in radio transmitters or high-power high-frequency generators may have up to 1 in. or more of spacing between plates, depending on the voltages encountered in the circuit in which they are used. Symbols for variable capacitors are shown in Fig. 6-6.

Fig. 6-6 Variable or adjustable capacitors. Circle shown dashed is used to denote a screwdriver-adjustable capacitor.

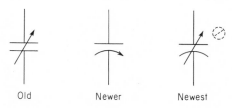

Old   Newer   Newest

6-9 MODERN CAPACITORS

There are many types of capacitors in general use in electronics and radio today. Some are listed below with a brief statement as to their relative dielectric leakage, availability in a fixed or variable type, approximate working-voltage ranges, approximate capacitance values, and frequencies with which they may be used.

1. *Vacuum-dielectric.* Practically no leakage. Made in both fixed and variable types. Used in 5,000- to 50,000-V service. Capacitances of 5 to 250 pF. Efficient to well over 1,000 MHz. Used mostly in transmitters.
2. *Air-dielectric.* Very little leakage except through insulators. Made fixed, adjustable, and variable. Used in low- and high-voltage applications, receivers and transmitters. Wide variety and capacitance range for both fixed and variable. Rarely much more than 400-pF capacitance. Used up to more than 1,000 MHz. Variable air capacitors are the usual tuning elements in receivers, transmitters, and other high-frequency equipment.

3. *Mica-dielectric*. Very little leakage except through material which encases the plates and dielectric. Both fixed and adjustable types. Working voltages range from 350 to several thousand volts. Capacitances from 1.5 pF to 0.1 μF. Used in RF circuits up to more than 300 MHz, although efficiency drops off over 10 MHz. Fixed types are used as RF bypass capacitors, etc. Adjustable types used as padders or trimmers. Silver-mica capacitors are within 5% of their rated capacitance values and hold constant capacitance under adverse conditions.

4. *Ceramic-dielectric*. Low leakage. Fixed and adjustable types. Capacitances from 1.5 pF to 0.01 μF for fixed types and up to 100 pF for adjustable types. Approximately 500 V working voltage. Useful up to more than 300 MHz. The general-purpose fixed capacitor.

5. *Paper-dielectric*. Paper impregnated with oil, wax, polychlorinated biphenyls (PCB), or ester is used. Relatively low voltage when new. When moisture seeps in, leakage becomes very high and the dielectric carbonizes at low voltages. Fixed types range from 10 pF to 10 μF. Working voltages are from 150 to several thousand volts. Efficient up to 1 or 2 MHz. Above this they rapidly become less effective because of dielectric fatigue and hysteresis.

6. *Plastic-dielectric*. Various sheet plastics used in place of paper dielectrics. Tubular forms. Better for higher frequencies than paper. Range from 0.01 μF to 2 μF and from 200 to 600 V.

7. *Electrolytic*. Considerable leakage, particularly if used on voltages over the rated value. Fixed types only. Range in capacitance from a few μF to 50,000 μF or more. Working voltages from 6 to about 700 V. Normally polarized and require that positive terminal be connected to positive terminal of the circuit. Dry out and lose

capacitance. Limited life expectancy. Useful only in dc circuits or in circuits where the dc component exceeds the ac component.

6-10 POWER FACTOR IN CAPACITORS

Capacitors may be tested to find their *power factor* (a percentage rating of the losses). With little or no leakage or losses, a capacitor has a low power factor. With no losses, the power factor is zero. As leakage develops through the capacitor, the power factor increases. A power factor of 1 denotes all leakage and no capacitive effect at all. Most capacitors when tested for power factor will indicate very close to zero. Electrolytic capacitors may read small power-factor values and still be usable. If their power factor increases above 0.1 or 0.2, it is assumed that they should be replaced.

6-11 INDUCTANCE IN CAPACITORS

In all capacitors there is a small value of inductance from the counter emf developed in the leads of the capacitor and the current traveling along the plates. In most low-frequency applications this inductance can be ignored. At frequencies over about 10 MHz the inductance of long leads may become detrimental to circuit operation.

Just as there is inductance in capacitor leads, so there is distributed capacitance in coils. There is a small value of capacitance between adjacent turns of a coil and from one end of the coil to the other. For this reason it is impossible to produce pure inductance or pure capacitance.

6-12 CHANGING VARYING DC TO AC

When a capacitor is connected in a series circuit consisting of a source of voltage, a switch, an ammeter, and a capacitor, as shown in Fig. 6-7, current will flow in the circuit under certain conditions.

If the source of emf is a battery or other dc source, when the switch is closed, electrons will flow in the circuit until the capacitor is charged to the source voltage. The ammeter will respond with a momentary indication and then fall back to zero and remain there. From then on, no cur-

---

ANSWERS TO CHECKUP QUIZ ON PAGE 93

1. (*Condenser*) 2. (*0.8 s*)(*4 s*) 3. (*0.015 s*) 4. (*Inductor*) (*Capacitor*) 5. (*Directly*)(*Inversely*)(*Directly*) 6. (*0.00236 μF*) (*2,360 pF*) 7. (*Dielectric constant*) 8. (*Barium titanate*) (*Vacuum*) 9. (*Leakage, hysteresis, absorption*) 10. (*Half*) 11. (*Vacuum*) 12. (*0.72 J*)(*1.5 $\times$ 10^{16}*)(*It might*) 13. (*300 V*)

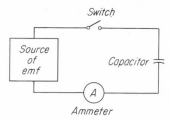

Fig. 6-7 A capacitor, ammeter, and switch in series across a source of emf (ac, dc, or varying dc).

rent will flow in the circuit. It may be said that a series capacitor blocks the flow of unvarying dc in a circuit.

If the source of emf produces ac, the emf from the source will constantly be changing, the capacitor will constantly be charging and discharging, and an alternating current will flow through the meter. A capacitor may be used to complete a circuit if the source is producing an alternating emf. It may be considered a conductor for ac, but it must be understood that electrons do not pass across the dielectric.

If the source produces a varying dc emf, the periodic increase and decrease of emf results in a continual charging and partial discharging of the capacitor. Since the charging and discharging current flows back and forth through the meter, the ammeter will have an alternating current flowing through it. A capacitor in series with a circuit having a varying dc source results in an alternating circuit current flow. The capacitor is said to block the dc but pass the ac component. Figure 6-8 shows a varying dc and the ac component of the variation. This is an important point. Most vacuum-tube and transistor circuits have a varying dc flowing in them. When a load is coupled to a varying dc circuit by a capacitor, only the variation, the ac component, is transferred to the load. Alternating current, not varying dc, flows in the load.

6-13 CAPACITIVE REACTANCE

How effective a capacitor may be in allowing ac to flow depends upon its capacitance and the frequency used. The greater the capacitance, the more electrons required to charge it to the source-voltage value. The smaller the capacitor, the fewer electrons required to bring it to full charge. It is possible to control the current flow in an ac circuit by changing the capacitance, somewhat as current can be controlled by varying the resistance in a circuit. The actual ac resistance effect of a capacitor is known as its capacitive reactance, which is measured in ohms, and can be determined by combining the value of capacitance and the frequency in the formula

$$X_c = \frac{1}{2\pi fC}$$

where X_c = reactance, in Ω
$\quad\;\; f$ = frequency, in Hz
$\quad C$ = capacitance, in F

EXAMPLE: The reactance of a 0.005-μF capacitor to a frequency of 1,000 kHz may be found by substituting in the formula:

$$X_c = \frac{1}{2\pi fC} = \frac{1}{6.28(1,000,000)0.000\ 000.005}$$

$$= \frac{1}{6.28(0.005)} = \frac{1}{0.0314} = 31.8\ \Omega$$

When C is in microfarads, the formula simplifies to

$$X_c = \frac{10^6}{2\pi fC}$$

The reactance of a capacitor is inversely proportional to the frequency. If the reactance is

Fig. 6-8 Varying dc source emf and the resultant ac component if a capacitor is in series with the circuit.

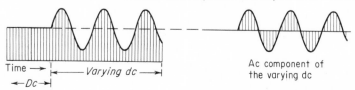

Time → |←——— Varying dc ———→|
←Dc→

Ac component of the varying dc

31.8 Ω at 1,000 kHz, it will be one-half as much, or 15.9 Ω, at 2,000 kHz and one-tenth as much at 10,000 kHz. This inverse proportion can be expressed by

$$\frac{f_1}{f_2} = \frac{X_{C_2}}{X_{C_1}}$$

where f_1 = one frequency
f_2 = another frequency
X_{C_1} = capacitive reactance at frequency f_1
X_{C_2} = capacitive reactance at frequency f_2

This formula can be used to solve problems such as the following:

What is the reactance of a capacitor at the frequency of 1,200 kHz if its reactance is 300 Ω at 680 kHz? By substituting the given values in the formula, the answer is found by

$$\frac{f_1}{f_2} = \frac{X_{C_2}}{X_{C_1}}$$

$$\frac{1,200 \text{ kHz}}{680 \text{ kHz}} = \frac{300}{X_{C_1}}$$

$$1,200 X_{C_1} = 300(680)$$

$$X_{C_1} = \frac{204,000}{1,200} = 170 \text{ Ω}$$

The same answer will be obtained if the 300-Ω reactance and 680 kHz are used in the capacitive-reactance formula rearranged to solve for the capacitance. The unknown reactance is then determined by using this capacitance at 1,200-kHz.

With the capacitive-reactance value known, it is possible to use Ohm's law for reactive circuits, as explained in Sec. 5-12. Reactance is substituted for resistance in the dc Ohm's law. The three derivations of the formula are

$$I = \frac{E}{X_C} \qquad E = IX_C \qquad X_C = \frac{E}{I}$$

6-14 CAPACITORS IN PARALLEL

It is often necessary to connect two or more capacitors in parallel to obtain greater capaci-

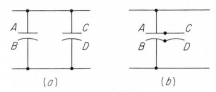

Fig. 6-9 Two similar parallel capacitors have twice the capacitance of one alone.

tance. Figure 6-9a shows two capacitors connected in parallel. Plates A and B in the first capacitor have a certain area and are separated by a dielectric. The other capacitor has similar plate areas, similar dielectric, and therefore the same capacitance. If plates A and C and plates B and D are connected, as in Fig. 6-9b, the result will be equivalent to a capacitor with twice the plate area and the same dielectric and, as a result, twice the capacitance. Whether the capacitors are connected as in diagram a or as in diagram b, the total capacitance is equal to the sum of the two capacitance values.

The formula for computing the capacitance of two or more capacitors in parallel is simply

$$C_t = C_1 + C_2 + C_3 + \cdots$$

where C_t = total capacitance
C_1 = one capacitor
C_2 = second capacitor
C_3 = third capacitor

EXAMPLE: Capacitors of 1, 3, and 5 μF are connected in parallel:

$$C_t = C_1 + C_2 + C_3 = 1 + 3 + 5 = 9 \text{ } \mu\text{F}$$

Care must be exercised when connecting two or more capacitors in parallel. The highest voltage a group of parallel capacitors will stand will be determined by the capacitor with the lowest working-voltage rating. If one capacitor has a working-voltage rating of 500 V and a second has a rating of 400 V, no more than 400 V should be used across the circuit if the two are in parallel.

Test your understanding; answer these checkup questions.

1. What is the main advantage of electrolytic capacitors? _____ What are the main disadvantages? _____

2. What would be the material of the dielectric in an electrolytic capacitor? _____ Of the nonmetallic plate? _____
3. What is the dielectric material of an improved type of electrolytic capacitor? _____
4. What are the three most likely dielectric materials of adjustable and variable capacitors? _____ _____ _____
5. Does a capacitor have any inductance? _____ An inductor any capacitance? _____ A resistor any capacitance or inductance? _____
6. Will a capacitor pass dc? _____ Ac? _____ Varying dc? _____
7. What is the capacitive reactance of a capacitor having 0.2 μF in a 15.9-kHz circuit? _____
8. What is the reactance of a 0.01-μF capacitor to a 3-kHz ac? _____
9. What is the capacitive reactance of a capacitor having 400 pF in a 3.8-MHz circuit? _____
10. If a capacitor has 1,500 Ω at a frequency of 8 MHz, what reactance will it have to 400 kHz? _____
11. If a 3-μF capacitor is across an 800-Hz signal generator producing 60 V rms, what value current flows? _____
12. If a C and an R are in series across a 100-V ac line, how much voltage will be developed across the capacitor if the X_C is 300 Ω and the current is 200 mA? _____
13. A 400-V power supply has a 450-V 40-μF capacitor and a 150-V 80-μF capacitor in parallel across its output. What is the total capacitance? _____ What will happen when the power supply is turned on? _____
14. Required, a capacitor that has 1-Ω X_C at 1 MHz. What C value is needed? _____

6-15 CAPACITORS IN SERIES

There are many cases when two or more capacitors are connected in series, as in Fig. 6-10.

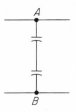

Fig. 6-10 Series capacitors increase dielectric thickness across a line and result in less total capacitance.

When two 4-μF capacitors are connected in series, the total capacitance is 2 μF. The circuit is acting as one capacitor with the top plate connected to point A and the bottom plate connected to point B. Between these two points there is twice the dielectric spacing of one capacitor. Whenever the spacing between plates is increased, the capacitance decreases, according to the formula for computing capacitance (Sec. 6-2).

The capacitance of two or more *similar* capacitors in series is equal to the capacitance of one of them divided by the total number of capacitors. Thus, four 10-μF capacitors in series present a total capacitance of 10/4, or 2.5 μF. Two equal capacitances in series total half the capacitance of one. Since capacitive reactance is inversely proportional to capacitance (from the formula $X_C = 1/2\pi fC$), the reactance of two equal series capacitors will be twice the reactance of one.

Two useful formulas for computing unequal-capacitance capacitors in series are given below. The first is for two series capacitors, and the second can be used for any number of capacitors in series. Note the similarity of these formulas to the parallel-resistance formulas (Sec. 2-15).

$$C_t = \frac{C_1 C_2}{C_1 + C_2}$$

$$C_t = \frac{1}{\dfrac{1}{C_1} + \dfrac{1}{C_2} + \dfrac{1}{C_3}}$$

EXAMPLE: The total capacitance of three capacitors in series, having 5, 3, and 7 μF, respectively, is

$$C_t = \frac{1}{\frac{1}{5} + \frac{1}{3} + \frac{1}{7}} = \frac{1}{\frac{21}{105} + \frac{35}{105} + \frac{15}{105}}$$

$$= \frac{1}{\frac{71}{105}} = \frac{105}{71} = 1.48 \ \mu F$$

The C_t of three capacitors in series can also be determined by the formula

$$C_t = \frac{C_1 C_2 C_3}{C_1 C_2 + C_2 C_3 + C_1 C_3}$$

The voltage-drop across any of the series capacitors (C_1 for example) can be found by using the formula

$$E_{C_1} = \frac{E_t C_t}{C_1}$$

Since two similar capacitors in series present twice the dielectric spacing across the circuit, the working voltage of two such capacitors in series will be doubled. If each has a rating of 500 V, the two in series will stand 1,000 V.

When capacitors in series are connected across a difference of potential, the sum of the voltage-drops across each of them will always equal the source voltage. Furthermore, the value of the voltage-drop across a particular capacitor in a series group will be inversely proportional to the ratio of its capacitance to the total, or directly proportional to its reactance. For example, a 1-μF and a 2-μF capacitor are in series across a 300-V source. There will be 200 V across the 1-μF capacitor and 100 V across the 2-μF. This will be true whether the capacitors are across ac or dc (if leakage is not present in the dc case).

It is sometimes desirable to equalize the dc voltage across two capacitors of somewhat unequal value when connected in series. This can be done by connecting resistors of equal value across them as in Fig. 6-11.

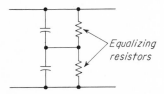

Fig. 6-11 Equalizing resistors across capacitors in series.

When an emergency arises or when a surplus of low-voltage capacitors is on hand, it is possible to connect several in series and use them across a high-voltage circuit. For example, if it is desired to have 1.5 μF of capacitance across a 1,600-V circuit and a number of capacitors rated at 400 V and 2 μF each are available, four 2-μF 400-V capacitors in series produce a capacitance of 0.5 μF capable of standing 1,600 V. Three parallel groups of four capacitors in series (twelve capacitors) will give the desired 1.5-μF capacitance suitable for the 1,600-V circuit. Equalizing resistors across the capacitors would be desirable.

As with inductive reactances, two capacitive reactances in series have a total value of

$$X_{C_t} = X_{C_1} + X_{C_2}$$

Two capacitive reactances in parallel have a total value of

$$X_{C_t} = \frac{X_{C_1} X_{C_2}}{X_{C_1} + X_{C_2}}$$

6-16 PHASE RELATIONS WITH CAPACITANCE

In Chap. 5 it was explained that alternating current and voltage will be out of phase in a circuit in which inductance is present. In an inductive circuit the current may lag the voltage by as much as 90°. In an ac circuit containing series capacitors, the current and voltage may also be out of phase by as much as 90° but the *current leads the voltage.*

The steady-state voltage and current relations of a capacitive circuit can be explained by referring to Fig. 6-12, which shows a capacitor across a source of ac emf, a graph of the emf indicated by the dashed line, and the current represented by the solid line. The voltage varies and alternates from maximum in one direction, point 1 on the

Fig. 6-12 Voltage and current phase relationships in a purely capacitive steady-state ac circuit.

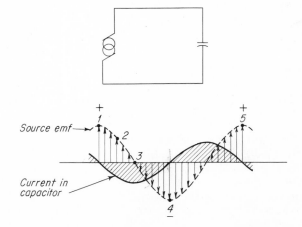

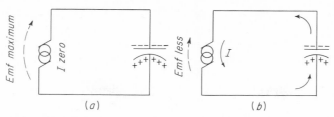

Fig. 6-13 (a) As the source emf reaches maximum, the circuit I stops. (b) As the emf decreases, I flows out of the capacitor, backward against the source E.

graph, to maximum in the opposite direction, point 4, and back to the original value, point 5.

As the source emf increases toward a maximum, more and more electrons are forced into the capacitor. At the instant of maximum pressure, point 1, the capacitor is charged; there will be no electrons moving into the capacitor or in motion at any place in the circuit. The condition that exists is maximum pressure but no current (Figs. 6-12, 6-13a).

The emf is at a maximum value for only an instant and then begins to fall off. As the electric pressure becomes less, all the electrons that had been forced into the capacitor as it was charged comprise an opposing emf greater than the decreasing source emf and current begins to move out of the negative plate of the capacitor into the source in a direction opposite to the source pressure (Fig. 6-13b and point 2 of Fig. 6-12).

The source emf continues to decrease, and the current flows out of the capacitor in a direction opposite to the emf, until the emf reaches zero and reverses. At the instant the voltage reaches zero, the number of electrons moving from one plate to the other attains a maximum value, point 3 on the graph in Fig. 6-12. From now on, as the emf increases toward maximum (in the negative direction on the graph), the capacitor is charging. When the source emf reaches maximum, the capacitor reaches maximum charge and the number of electrons moving into it will be zero. The condition that exists is again maximum source voltage and no current flow in the circuit (Fig. 6-12, point 4).

It should be noted that at point 3 on the graph the current is at a maximum in the negative direction. The voltage does not reach maximum in the negative direction until 90° later. Therefore, the current leads the voltage in a purely capacitive circuit by 90°.

The applied voltage and the current will be 90° out of phase only if there is no resistance in the circuit. With any series resistance in the circuit the difference in phase will be less than 90°. The phase difference decreases as the proportion of resistance to capacitive reactance increases. When the circuit has a small value of capacitive reactance and a proportionately large resistance value the circuit is predominantly resistive and the phase approaches 0°.

6-17 CONSIDERATIONS WHEN SELECTING CAPACITORS

When replacing or purchasing capacitors, several factors should be considered:

1. *Working voltage.* If the circuit in which the capacitor is to be used is a 350-V circuit, purchase a capacitor with a working-voltage rating at least 10 to 20% higher than 350 V.
2. *Capacitance.* Replace with a capacitor having as nearly the same capacitance as possible.
3. *Type of dielectric.* For RF, mica-, air-, vacuum-, or ceramic-dielectric capacitors are suitable. For AF ac, mica-, ceramic-, paper-, or plastic-dielectric capacitors are suitable. For dc filter circuits, electrolytic or paper capacitors are suitable.
4. *Physical size.* Ceramics are usually smaller than mica and paper equivalents. Electrolytics are much smaller than paper.
5. *Cost.* Probable cost, per microfarad, in ascending order: electrolytic, ceramic, paper, plastic, mica, air, vacuum.
6. *Variable, adjustable, or fixed.* As similar to the original as possible.
7. *Temperature.* Capacitors in confined areas may become overheated and burn out. In particular, it is not advisable to overheat paper or electrolytic capacitors. Electrolytics dry out

near hot tubes, rectifiers, transistors, or resistors.

8. *Temperature coefficient.* Some capacitors have a positive temperature coefficient (increase capacitance with an increase of heat); others have a negative temperature coefficient (decrease capacitance with an increase of heat); and still others have a zero temperature coefficient (do not change capacitance with an increase of heat). This fact is important when exact capacitance values are required, as in oscillator circuits.

6-18 CAPACITANCE COLOR CODE

Fixed capacitors may be marked with their capacitance and working voltage. The markings will be either in printed numbers or in colors, using the same number-color code employed with resistors (Sec. 1-13).

The simplest code is the three-dot code, used with 500-V mica capacitors, with a tolerance of 20% (Fig. 6-14).

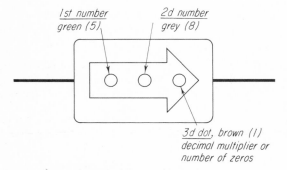

Fig. 6-14 A 3-dot EIA 580-pF mica capacitor.

In the six-dot Electronic Industries Association (EIA) code (Fig. 6-15), reading clockwise from the top left, white indicates mica; the next two dots indicate capacitance numbers; and the fourth dot is the capacitance multiplier, or number of zeros to follow the numbers. Capacitance is in picofarads. The fifth dot is the tolerance (Table 1-2, Sec. 1-13). The sixth dot is the temperature coefficient, increasing as the numbers diminish from 4. Voltage ratings are 500 V up to about 500 pF and 300 V for greater capacitances.

The military standard is similar to the six-dot

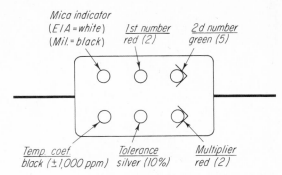

Fig. 6-15 A 6-dot EIA mica capacitor having 2,500-pF capacitance, 10% tolerance, and a temperature coefficient of ±1,000 parts per million.

EIA, except that a black first dot indicates a mica dielectric and silver a paper dielectric.

Some ceramic capacitors are cylindrical and are marked with a series of colored dots or bands. An example of a 3,800-pF 10% tolerance capacitor is shown in Fig. 6-16.

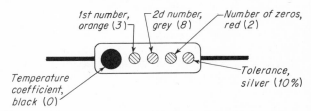

Fig. 6-16 A 3,800-pF 10% tolerance zero-temperature-coefficient ceramic capacitor.

The *temperature coefficient* is the degree by which the capacitor will change its capacitance with a change in temperature. If the capacitor does not change its capacitance at all, it has a zero coefficient. If it increases capacitance with increased temperature, it has a positive coefficient. If the coefficient is −150, the capacitance will decrease by 150 ppm per degree Celsius increase in temperature. A list of color-coded temperature coefficients for ceramic capacitors is given in Table 6-2.

Test your understanding; answer these checkup questions.

1. What is the total capacitance when six 4-μF capacitors are connected in series? _____
2. What is the total capacitance when a 2-, a 3-, and a 4-μF capacitor are in series? _____

Table 6-2 *TEMPERATURE COEFFICIENTS*

Color	TC/°C
Black	0
Brown	−33
Red	−75
Orange	−150
Yellow	−220
Green	−330
Blue	−470
Violet	−750
Gray	+30
White	+500

3. What is the voltage-drop across a 2-μF capacitor in series with a 5-μF capacitor across 100 V? _____
4. Why are resistors usually connected across capacitors in series? _____
5. What is the phase relation between E and I in a purely capacitive ac circuit? _____ Purely resistive? _____ Equally resistive and capacitively reactive circuit? _____
6. If a capacitor increases capacitance when warmed, what temperature coefficient does it have? _____
7. Would a mica-dielectric capacitor operating at 1 MHz have more, the same, or less capacitance when operating in a 100-MHz circuit? _____
8. On which plate of an electrolytic capacitor does the dielectric form? _____
9. What might be two causes of electrolytic capacitors exploding? _____
10. In a purely capacitive circuit, what is the current amplitude when the voltage amplitude is at 270°? _____ At 180°? _____
11. Why should capacitors be kept away from hot resistors, tubes, etc.? _____
12. What is the C value of a capacitor marked with dots of: Red, green, yellow? _____ Blue, gray, brown? _____ Black, green, blue, black? _____
13. A color-coded tubular capacitor has the following dots: violet, yellow, white, brown, and silver. What do you know about it? _____

COMMERCIAL LICENSE QUESTIONS

Amateur license questions will be found in the Addendum.

FCC Elements 3, 4, and 6 require an ability to answer questions similar to those below. Sections in which questions are answered are shown in parentheses. A question followed by a bracketed number is required for that element alone.

1. What is the unit of capacitance? (6-1)
2. How many picofarads (micromicrofarads) are there in 1 μF? (6-1)
3. Assuming the voltage on a capacitor is at or below the maximum allowable value, in what way does the value of the capacitor have a relation to the amount of charge it can store? (6-1)
4. What is meant by the *time constant* of an RC circuit? (6-1)   [4]
5. Explain how the values of resistance and capacitance in an RC network affect its time constant? (6-1)
6. How would the output waveform be affected by the frequency of the input in an RC network? (6-1)
7. Explain the effect of adding plates on the capacitance of a capacitor. (6-2)
8. How does the capacitance of a capacitor vary with area of plates, spacing between plates, and dielectric material between plates? (6-2)
9. What effect does a change in the dielectric constant of a capacitor's dielectric material have upon the capacitance? (6-2)
10. If the specific inductive capacity of a capacitor's dielectric material between the plates were changed from 1 to 4, what would be the resultant change in capacitance? (6-2)
11. What factors determine the breakdown-voltage rating of a capacitor? (6-4)
12. What factors determine the charge stored in a capacitor? (6-6)
13. What relation does the ability to store charges have to the total capacitance of two or more capacitors (a) in series, (b) in parallel? (6-14, 6-15)
14. Given two identical mica capacitors of 0.5 μF each. One is charged to 600 V and disconnected from the charging source. The charged capacitor is then connected in parallel with the uncharged capacitor. What voltage will now appear across the two capacitors? (6-6)
15. What is the meaning of *electrolyte?* (6-7)
16. Explain the principle of electrolytic capacitors. (6-7)
17. What precaution should be observed when connecting electrolytic capacitors in a circuit? (6-7)
18. What is a desirable feature of an electrolytic capacitor as compared with other types? (6-7)
19. What type of low-leakage capacitors are used most often in high-frequency transmitters or circuits? (6-9)

20. What is the reactance value of a capacitor of 0.005 μF at a frequency of 2,000 kHz? (6-13)
21. What is the reactance of a capacitor at the frequency of 1,200 kHz if its reactance is 100 Ω at 300 kHz? (6-13)
22. If capacitors of 4, 3, and 6 μF are connected in parallel, what is the total capacitance? (6-14)
23. What is the total reactance when two capacitors of equal value are connected in series? (6-15)
24. What formula is used to determine the total capacitance of three or more capacitors connected in series? (6-15)
25. If capacitors of 5, 8, and 10 μF are connected in series, what is the total capacitance? (6-15)
26. The voltage-drop across an individual capacitor of a group of capacitors connected in series across a potential is proportional to what factors? (6-15)
27. Having available a number of capacitors rated at 300 V and 4 μF each, how many would be necessary to obtain a combination rated at 1,200 V and 2 μF? (6-15)

28. Show by a simple graph what is meant when it is said that the current in a circuit leads the voltage. What would cause this? (6-16)　　　[4]
29. What is the value, tolerance, and voltage rating of an EIA mica capacitor whose first-row colors are (from left to right) white, brown, green and whose second-row colors are black, gold, orange? (6-18)

NOTE: FCC license test questions are multiple-choice types, taking a form somewhat as follows:

30. 500 $\mu\mu$F is the equivalent of (a) 0.0005 μF; (b) 5 $\times$ 10^{-6} F; (c) 5 $\times$ 10^{-12} F; (d) 5,000 pF. (6-1)
31. What is the capacitance of a three-plate capacitor with 2-in.^{2} plates separated by 0.003 in. of insulation having a dielectric constant of 7? (a) 315 $\mu\mu$F; (b) 2,100 pF; (c) 237 μF; (d) 105 $\mu\mu$F.
32. Which types of capacitors can be constructed with rotor and stator plates? (a) Ceramic-dielectric; (b) air-dielectric; (c) mica-dielectric; (d) all of the above. (6-8)

ANSWERS TO CHECKUP QUIZ ON PAGE 102

1. (0.667 μF)　2. (0.923 μF)　3. (71.4 V)　4. (To equalize voltages across capacitors)　5. (90°)(0°)(45°)　6. (+)　7. (Somewhat less)　8. (+ plate)　9. (Connected with incorrect polarity, used on ac)　10. (Zero)(Maximum)　11. (Change capacitance, electrolytics dry out)　12. (250,000 pF)(680 pF)(56 pF)　13. (TC = −750 ppm, 490 pF, 10% tolerance)

ALTERNATING-CURRENT CIRCUITS

The objective of this chapter is to develop methods of computing currents, voltages, impedances, phase angles, power values, and power factors for inductors, capacitors, and resistors in various series, parallel, and series-parallel circuits.

7-1 EFFECTS OF INDUCTANCE, CAPACITANCE, AND RESISTANCE

In preceding chapters it was pointed out that *inductance alone* in a circuit has the property by which it (1) opposes any change in current, (2) produces an electromagnetic field around itself, (3) tends to limit ac flow in the circuit, (4) passes dc without attenuation, and (5) produces a phase difference of 90°, with the current lagging the voltage of the circuit.

Capacitance alone in a circuit has the property by which it (1) opposes any change in voltage, (2) produces an electrostatic field between its plates, (3) tends to limit ac flow in the circuit, (4) blocks dc flow, and (5) produces a phase difference of 90°, with the current leading the voltage.

Resistance alone in a circuit limits the current that can flow at a given voltage but produces no phase difference between the current and voltage across the resistance.

7-2 INDUCTANCE AND RESISTANCE IN SERIES

When an inductance and a resistance are in series across a source of alternating emf, as in Fig. 7-1, an alternating current flows in the circuit. If resistance were alone in the circuit, the current

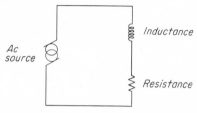

Fig. 7-1 Inductance and resistance in series across an ac source.

and voltage would be in phase, all the power in the circuit would be converted to heat, and the formula used to compute E, I, and R would be the Ohm's-law formula $I = E/R$.

If inductance were alone in the circuit, the current and voltage would be 90° out of phase, all the power in the circuit would be expended in producing a magnetic field around the coil during one half of the ac cycle, and then all the power would be returned to the circuit again during the other half of the cycle. No power would actually be lost in heat. The Ohm's-law formula used to compute E, I, and X in reactive circuits would be $I = E/X_L$.

With both resistance and inductance in a series ac circuit, it is often necessary to compute current, voltages, impedance, phase angle, volt-amperes (apparent power), true power, and power factor.

If the inductance value and the ac frequency are known, the inductive reactance of the coil in ohms can be determined by

$$X_L = 2\pi fL$$

where f = frequency, in Hz
L = inductance, in henrys (H)

Since the load in Fig. 7-1 is not purely resistive, the total opposition will not be the resistance value alone. The opposition of the load is not purely reactive either and will not be the reactance value alone. Instead, a value that is a resultant of the resistance and reactance will have to be determined. This value is the *impedance* of the circuit. The symbol assigned to it is Z, and it is measured in ohms.

While reactance acts as an opposition and in some ways may be likened to resistance, it must be considered to be setting up its opposition at right angles to the opposition of resistance rather than opposing in the same direction. This can be shown by a vector diagram as in Fig. 7-2a. The

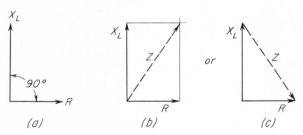

(a) *(b)* *(c)*

Fig. 7-2 *(a)* Reactance and resistance vector arrows plotted at right angles. *(b)* and *(c)* The third side of the right triangle represents the impedance value, Z.

vector diagram indicates, by the relative lengths of the vector arrows, a circuit in which the resistance value in ohms is less than the reactance value in ohms. The actual ac opposition will be something more than either X_L or R values alone, but how much?

To determine the actual value of opposition, or impedance, lines may be drawn parallel to the resistance vector and to the reactance vector, forming a parallelogram (Fig. 7-2b). The length of the dashed Z vector from the point of origin diagonally across the parallelogram indicates the impedance of the circuit.

The triangle in Fig. 7-2c, consisting of the R, the Z, and the X_L vectors, is a right triangle, since the R and X_L vectors are at $90°$. It is possible to determine the numerical value of the Z side of the right triangle, if the R and X_L lengths are known, by using the Pythagorean theorem, which states:

The square of the hypotenuse value of a right triangle is equal to the sum of the squares of the values of the other two sides. (The hypotenuse is the side opposite the right angle, the Z side in this case.) The Pythagorean theorem can be stated in formula form:

$$Z^2 = R^2 + X_L{}^2$$

By taking the square root of both sides of the equation,

$$\sqrt{Z^2} = \sqrt{R^2 + X_L{}^2}$$
$$Z = \sqrt{R^2 + X_L{}^2}$$

This formula can be used in a problem such as the following:

The resistance in a circuit is 3 Ω, and the inductive reactance is 7 Ω. What is the impedance? Figure 7-3 illustrates the circuit.

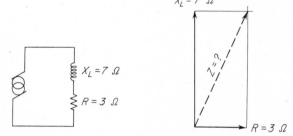

Fig. 7-3 Schematic and vector diagram of the problem: What is the impedance if R is 3 Ω and X_L is 7 Ω?

$$Z = \sqrt{R^2 + X_L{}^2}$$
$$= \sqrt{3^2 + 7^2} = \sqrt{9 + 49} = \sqrt{58} = 7.62 \ \Omega$$

With the impedance value it is possible to solve problems involving Ohm's law in ac circuits by substituting the Z value for the R in the dc Ohm's-law formulas. The formulas in Table 7-1

Table 7-1 *OHM'S-LAW FORMULAS*

For dc circuits	For purely reactive ac circuits	For ac circuits
$I = E/R$	$I = E/X$	$I = E/Z$
$E = IR$	$E = IX$	$E = IZ$
$R = E/I$	$X = E/I$	$Z = E/I$

are for use in dc circuits (or in ac circuits having negligible reactance), in reactive circuits with negligible resistance, and in ac circuits in which R and X are involved.

In the example problem used above, the impedance was computed to be 7.62 Ω. If a current of 10 A is flowing through the circuit, the voltage of the source must be

$$E = IZ = 10(7.62) = 76.2 \text{ V}$$

In problems such as "What is the impedance of a solenoid-type coil if its resistance is 5 Ω and 0.3 A flows through the winding when 110 V at 60 Hz is applied to the solenoid?" both current and voltage are given. By using Ohm's law for ac circuits, $Z = E/I$, this is solved: $Z = 110/0.3$, or 367 Ω. Neither the 5-Ω resistance nor the frequency of 60 Hz is needed to solve for the impedance.

7-3 THE PHASE ANGLE

The current in an inductance lags the voltage across it by 90°. As long as the inductance itself has zero or negligible resistance, the phase angle of the voltage and current of the coil will be 90°.

The current through a resistance will always be in phase with the voltage across it. The ac E and I phase angle is 0°.

When both resistance and reactance are present in the circuit, the phase angle of the circuit as a whole will be neither 90° nor 0° but some intermediate value. This will be the phase angle that is seen by the source as it looks into the whole circuit connected across it. The number of degrees will be equal to the angle formed between the Z and the R sides of the Z, R, X triangle of the circuit. This phase angle is usually indicated by the Greek letter θ (theta). Figure 7-4 illustrates a circuit having 10-Ω inductive reactance and

20-Ω resistance. The impedance value is found by

$$Z = \sqrt{R^2 + X^2} = \sqrt{20^2 + 10^2} = 22.4 \text{ Ω}$$

One method of determining the number of degrees of the phase angle is to draw to scale the R, X_L, Z vector diagram of the circuit and with a protractor measure the angle between the R and Z sides.

Another method of determining the phase angle is to compute the numerical value of the ratio of the R and Z sides. This ratio of R to Z, or R/Z, has been given the name *cosine*. In the problem above, the cosine of R/Z is equal to 20/22.4, or 0.8928.

Trigonometric tables (Appendix C), indicate the angle associated with the cosine value of 0.8928. By searching through the cosine (or cos) values in the tables, a number approximately equal to 0.8928 will be found. In this particular problem, a cosine value of 0.8926 is equal to 26.8°. This is close enough for general use. The angle by which the current lags the voltage in this inductive circuit is 26.8°. The circuit has a phase angle of 26.8°. (This information can be obtained from many slide rules and pocket calculators also.)

Although the phase angle can be found by using the cosine ratio R/Z, it can also be found by using the ratio of sides X/Z, which is known as the *sine* (or sin) value. The angle associated with the sine value will be the same phase angle. The ratio of sides X/R is known as the *tangent*, abbreviated tan, and may also be used to determine the θ.

In a circuit having equal values of R and X the vector arrows are equal in length, the impedance will be 1.414 times the resistance (or reactance) value, and the phase angle will be one-half of 90°, or 45°, as shown in Fig. 7-5. (Check by using cos, sin, and tan.)

Fig. 7-4 The phase angle is always formed by the resistance and impedance sides of the vector triangle.

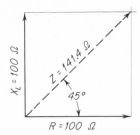

Fig. 7-5 Vector diagram of a circuit having equal resistance and inductive reactance.

7-4 VOLTAGE VECTOR ADDITION

A circuit with 10-Ω inductive reactance, 20-Ω resistance, 22.4-Ω impedance, and an ammeter indicating a 10-A current is shown in Fig. 7-6.

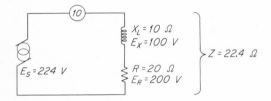

Fig. 7-6 A 224-V ac source will develop 100 V and 200 V across the components of this series circuit.

If 10 A flows through a reactance of 10 Ω, the voltage-drop across the reactance will be

$$E_{XL} = IX_L = 10(10) = 100 \text{ V}$$

If 10 A flows through a resistance of 20 Ω, the voltage-drop across the resistance will be

$$E_R = IR = 10(20) = 200 \text{ V}$$

The sum of the two voltage-drops, the reactive 100 V and the resistive 200 V, would appear to equal 300 V, but the voltage across the reactance will lead the current through it by 90°. The voltage across the resistance is in phase with the current through it. These two *voltages,* being 90° out of phase, must be added vectorially as in Fig. 7-7. This will be recognized as similar to the determination of the impedance of the circuit when reactance and resistance values in ohms were used. In this case, reactive and resistive voltages are

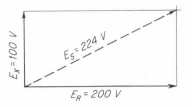

Fig. 7-7 Representation of the source voltage and the vectors of the reactive and resistive voltages.

plotted 90° out of phase, and the resultant is the source-voltage value E_s.

By the Pythagorean theorem, the formula for determining the resultant voltage is

$$E_s^2 = E_R^2 + E_X^2$$
$$E_s = \sqrt{E_R^2 + E_X^2}$$

where E_s = source voltage
E_R = voltage across R
E_X = voltage across X

By substituting the values in the problem above,

$$E_s = \sqrt{200^2 + 100^2} = \sqrt{40,000 + 10,000}$$
$$= 224 \text{ V}$$

Here is a case in which 200 and 100 V equals 224 V, but only if they are added *vectorially,* at right angles.

The total voltage may also be determined by using the Ohm's-law formula

$$E = IZ = 10 \times 22.4 = 224 \text{ V}$$

An important point here is the understanding that the source voltage represents a *resultant* of the reactive and resistive voltage-drops across the circuit. Also, in an ac circuit involving *L, C,* and *R* in series, the sum of all the voltage-drops in the circuit will add up to more than the source voltage unless they are added vectorially.

When two voltages (or currents) are in phase, they can be added vectorially, head to tail, in the same direction, as in Fig. 7-8a. This would apply when the voltage-drops across two coils in series are added. The resultant voltage is the simple sum of the voltages. If two voltages (or currents) are out of phase by 180°, they must be added vec-

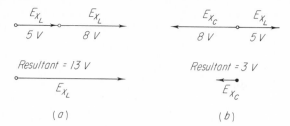

Fig. 7-8 Vector addition of (a) two voltages in phase and (b) two voltages 180° out of phase.

torially in opposite directions, starting at the same reference point, as in Fig. 7-8b. This would apply when the voltage-drop across a coil is added to the voltage-drop across a capacitor when they are in series. The resultant voltage will be the difference of the two voltages and will carry the sign of the larger.

Test your understanding; answer these checkup questions.

1. In what unit is impedance measured? _____
2. State the Pythagorean theorem in the impedance formula form. _____
3. What is probably being solved when the partial equation $\sqrt{a^2 + b^2}$ is seen? _____
4. A 100-V ac circuit has a 10-Ω R and a 12-Ω X_L in series across it. What is the Z value? _____ The I_R value? _____ The E_R value? _____
5. In question 4, what is the cosine ratio in symbol letters? _____ The cos value? _____ The phase angle of the circuit? _____
6. If a series ac circuit has an impedance of 25 Ω and 10 Ω of X_L, what is the sine ratio in symbols? _____ The sin value? _____ The θ value? _____
7. If the reactance in question 6 were capacitive, do you think the sin value would be the same? _____ The θ value? _____
8. A series ac circuit has an X_L of 50 Ω and a R of 20 Ω. What is the tangent ratio in symbols? _____ The tan value? _____ The θ? _____
9. Could the formula $Z = \sqrt{R^2 + X^2}$ be used to solve R and X in parallel? _____
10. A series ac circuit has 20 V across the resistor and 30 V across the reactor. What is the source-voltage value? _____ The tan value? _____ the θ? _____
11. Why is the Pythagorean theorem not used with I_R and I_X in series circuits to determine the source current? _____
12. What is the resultant voltage if $-E_R = 75$ V and $+E_R = 55$ V? _____ If $E_X = 75$ V and $E_R = 55$ V? _____ If $-E_{X_L} = 75$ V and $+E_{X_L} = 55$ V? _____

7-5 APPARENT AND TRUE POWER

In the simpler series-type circuits described thus far, the only part of the circuit actually using power is the resistor. The resistor loses power in the form of heat whenever current flows through it. The current flowing through the coil produces energy that is stored in the magnetic field around the coil. When the current alternates, the magnetic field collapses and returns all its energy to the circuit. Therefore, there is no loss of power in the inductance itself. This is considering the inductance to have negligible resistance in its turns and no coupling to any external circuit.

The series circuit used before but with a voltmeter included is shown in Fig. 7-9. All known

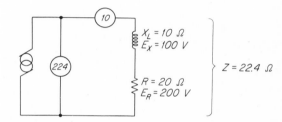

Fig. 7-9 The apparent power is the value computed when meter voltage is multiplied by the meter current.

values are indicated. The *apparent power* can be determined by using the visually apparent values shown by the voltmeter (224 V) and by the ammeter (10 A). Since power can be determined by the formula $P = EI$, the apparent power of the circuit is 224(10), or 2,240 VA (not watts).

The power lost in heat is usually computed by the formula $P = I^2R$, in this case $10^2(20)$, or 100(20), or 2,000 W. This is called the *true power* of the circuit.

The true power of this simple circuit can also be determined by using either of the formulas $P = E^2/R$ and $P = EI$ if the voltage across the resistor is used in the equation. Experience has shown, however, that the use of the formula $P = I^2R$ to solve for true power results in fewer difficulties for a beginner, and it should be employed if possible.

The use of the term *voltamperes* is preferred by many to the term "apparent power." In a purely reactive circuit, there may be volts forcing amperes to flow, but if all the energy stored in the field of the reactance is returned to the circuit, there is only an apparent power (VA) but no actual loss of power.

7-6 POWER FACTOR

When the true and apparent powers are expressed as a ratio, with the true power above and the apparent power below, a decimal number between zero and 1.0 results. The ratio true/apparent power in a circuit is known as the *power factor* of the circuit.

$$\text{pf} = \frac{\text{true power}}{\text{apparent power}} = \frac{I^2 R}{\text{VA}}$$

In the circuit used as the example thus far, this ratio is

$$\text{pf} = \frac{2,000 \text{ W}}{2,240 \text{ W}} = 0.8928$$

Note the reappearance of the number 0.8928. This will be recognized as the cosine value determined by dividing the resistance by the impedance. Besides representing the cosine of 26.8°, this decimal figure 0.8928 is the power factor of the circuit. Since either the ratio of true to apparent power or the ratio of resistance to impedance produces the same decimal figure, either ratio can be used to determine the power factor of a series circuit. Thus, three methods of determining power factor are

$$\text{pf} = \frac{I^2 R}{\text{VA}} = \frac{R}{Z} = \cos \theta$$

where θ is the EI phase angle

By algebraic rearrangement, the true power can be computed by multiplying the apparent power by the power factor:

$$P = \text{VA(pf)} \quad \text{or} \quad P = \text{VA}(\cos \theta)$$

The power factor is in reality a comparison of the amount of power a circuit is apparently using and what it is actually using. It is often expressed as a percent. Thus, a power factor of 0.8928 can be said to be a power factor of 89.28%.

Power factor is important when wiring a circuit. Consider an electric circuit having a power factor of 0.5, a source voltage of 100 V, and an impedance of 20 Ω. According to Ohm's law the current in the circuit is $I = E/Z$, or 100/20, or 5 A. An ammeter in the circuit will read 5 A. Apparent power is the product of source voltage and current through the source, in this case 100(5), or 500 VA. Since true power is apparent power times power factor, or VA($\cos \theta$), the load in the circuit must be receiving 100(5)(0.5), or 250 W. The wires of the circuit must carry the current of 5 A that is shown by the ammeter instead of only the 2.5 A normally required when 100 V produces 250 W of power in a resistive load.

Why is 5 A flowing when only 2.5 A should produce the power? The answer lies in the fields developed around the reactance of the circuit. Energy is required to build up the field of a reactance, but this energy is returned to the circuit when the field collapses. The energy is carried in the form of current from the source and is returned as current to the source. At the same time the load is constantly demanding energy and therefore current. The wires must carry the reactive field-building current too.

It must be understood that apparent power does not mean fictitious power. Apparent power has its effect in electric circuits. Engineers must so construct circuits that true power approaches apparent power (high pf), thereby enabling the use of smaller wire and lessening the cost of installation.

If the reactance causing a low power factor in a circuit is inductive reactance, it is possible to raise the power factor by adding some capacitive reactance to the circuit to balance the inductive reactance, leaving the circuit more nearly resistive (higher pf).

Power factor may be referred to as being either

leading or *lagging*. A lagging power factor indicates the *current* is lagging the voltage in the circuit (inductive circuit). A leading power factor means the *current* is leading the voltage (capacitive circuit). An example of a problem involving a power factor is the following:

If a 220-V line delivers 100 W at 80% power factor to a load, what is the phase angle between the line current and line voltage and how much current flows in the line?

First, the cosine of the phase angle is the power factor.

$$\cos \theta = pf = 0.80 = 80\%$$

From a table of trigonometric functions, 0.80 is found to be the cosine of 36.9°, which is the phase angle between the current and voltage. (It is impossible to tell whether the current is leading or lagging the voltage in this circuit from the information given.)

Second, in practice, when power is mentioned, true power is understood. Therefore the 100 W in the problem must be true power. Since true power equals voltamperes times power factor

$$P = VA(pf) = VA(\cos \theta)$$

then
$$VA = \frac{P}{\cos \theta} = \frac{100}{0.80} = 125$$

The voltampere value represents the product of what would be read by a voltmeter across the load and an ammeter in series with the load. When the apparent-power formula $P = VA$ is rearranged to read $A = P/V$, the current is

$$A = \frac{P}{V} = \frac{125}{220} = 0.568 \text{ A}$$

7-7 CAPACITANCE AND RESISTANCE IN SERIES

A circuit composed of an ac source, a capacitor, and a resistance in series is shown in Fig. 7-10. If the capacitance value and the ac frequency are known, the capacitive reactance of the capacitor in ohms can be computed by using the formula

Fig. 7-10 Series RC circuit across an ac source.

$$X_C = \frac{1}{2\pi fC}$$

where f = frequency, in Hz
C = capacitance, in F

The capacitive-reactance value in ohms can be used in conjunction with the resistance for ac computations in very much the same way as inductive reactance and resistance are used.

As with an X_L and R in series, the capacitive-reactance (X_C) vector value is also plotted at right angles to the resistance vector as shown in Fig. 7-11. Note, however, that whereas the X_L vector

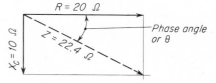

Fig. 7-11 Vector representation of resistance, capacitive reactance, and the resultant impedance.

was always shown pointing upward, the X_C vector points downward, a 180° difference in direction. The difference indicates that X_L and X_C vectors will tend to cancel each other.

The vector sum of the capacitive reactance and resistance is the impedance value and is determined by using the Pythagorean formula:

$$Z = \sqrt{R^2 + X_C^2}$$

Knowing the impedance value in ohms, voltage or current values can be computed by using Ohm's-law formulas for ac circuits:

$$I = \frac{E}{Z} \qquad E = IZ$$

In a series circuit the current is the same in all parts of the circuit at any instant. In a purely capacitive circuit the voltage lags the current by 90°. In a resistive circuit the voltage and current are in phase. In a circuit made up of both X_C and R, the phase angle between E and I will be something between 0° and 90°. The phase angle, as in inductive circuits, may be solved for graphically or by determining the angle represented by the ratio of R/Z. The ratio of the resistive voltage-drop to the source voltage (E_R/E_S) will be proportional to the ratio of resistance to impedance and may also be used to determine phase angle.

A capacitive circuit has a *leading power factor* (current leads), found by:

1. Dividing R by Z (pf $= R/Z$)
2. Dividing true power by the apparent power (pf $= I^2R/VA$)
3. Finding the cosine value of the angle between the current and voltage of the circuit (pf $= \cos\theta$)

Since the ratio of the voltage-drop across the resistance to the voltage-drop across the source is proportional to the ratio of resistance to impedance, power factor in a series circuit can be found by

$$\text{pf} = \frac{E_R}{E_S}$$

The computations of impedance, phase angle, and power factor are performed the same way, regardless of whether the circuit has a coil and resistor in series or a capacitor and resistor in series.

A leading power factor in a capacitively reactive circuit can be corrected (raised) by adding inductance to the circuit.

Test your understanding; answer these checkup questions.

1. Under what two conditions might a coil lose power? _____
2. What power is determined by multiplying ac ammeter times ac voltmeter values? _____
3. What power is registered by a wattmeter? _____
4. In what unit is apparent power measured? _____ True power? _____
5. What is the ratio of true to apparent power called? _____

6. Does the cosine value of an ac circuit indicate true power, VA, or pf? _____
7. What is the phase relation of the voltage in a circuit having a lagging pf? _____
8. A 120-V line delivers 500 W at 90% pf. What is the phase angle? _____ The line current? _____
9. What is the effective reactance in a series circuit having 12-Ω X_C and 20-Ω X_L? _____ 40-Ω X_C and 40-Ω X_L? _____ 60-Ω X_C and 35-Ω X_L? _____
10. A 300-Ω R and 100-Ω X_L are in series across a 50-V ac source. What is the Z value? _____ VA value? _____ P value? _____ Pf? _____ θ? _____ Power lost in heat? _____ I value? _____
11. A series RL circuit has a Z of 141 Ω to 500-Hz ac. An ohmmeter shows the total resistance as 100 Ω. The source voltage, E_s, is 120 V. What is the reactance value? _____ VA? _____ P? _____ θ? _____ E_L? _____ E_R? _____

7-8 INDUCTANCE, CAPACITANCE, AND RESISTANCE IN SERIES

A circuit containing a coil, capacitor, and resistor in series across a source of ac is shown in Fig. 7-12. If the inductance, capacitance, and

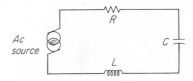

Fig. 7-12 Series RCL circuit across an ac source.

frequency are known, the reactances can be found by applying the usual reactance formulas:

$$X_L = 2\pi fL \quad \text{and} \quad X_C = \frac{1}{2\pi fC}$$

When drawing a vector diagram of resistance, inductive reactance, and capacitive reactance in series, the dissimilar reactances must be plotted in opposite directions, each 90° from resistance. Figure 7-13 illustrates a large value of inductive reactance and a smaller value of capacitive reactance. The vector diagram shows that X_C and X_L are each 90° from R but in such directions that they tend to cancel each other. If there is more X_L than X_C, as shown, the resultant, or net reactance, will be X_L. The net reactance is obtained by subtracting the smaller value from the larger. The complete impedance formula for a series ac circuit is therefore

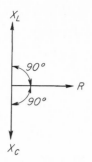

Fig. 7-13 Inductive and capacitive reactances plot 180° out of phase but 90° from resistance.

$$Z = \sqrt{R^2 + (X_L - X_C)^2}$$

In cases in which X_C is greater than X_L, the formula is

$$Z = \sqrt{R^2 + (X_C - X_L)^2}$$

EXAMPLE: What is the impedance of a series circuit consisting of a resistance of 4 Ω, an inductive reactance of 4 Ω, and a capacitive reactance of 1 Ω? The circuit and vector diagrams are shown in Fig. 7-14.

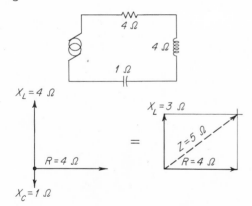

Fig. 7-14 Schematic and vector diagrams of a 4-Ω R, a 4-Ω X_L, and a 1-Ω X_C in series.

$$Z = \sqrt{R^2 + (X_L - X_C)^2} = \sqrt{4^2 + (4 - 1)^2}$$
$$= \sqrt{16 + 9} = \sqrt{25} = 5\ \Omega$$

If the generator in the preceding problem produced 50 V, the current, according to Ohm's law would be

$$I = \frac{E}{Z} = \frac{50}{5} = 10\ \text{A}$$

Conversely, if the circuit current is known to be 10 A, the voltage of the source, according to Ohm's law, is

$$E = IZ = 10(5) = 50\ \text{V}$$

In a series circuit the current is the same in all parts of the circuit. At any instant when the current reaches its peak value in the resistance, the current is also at a peak value in the inductance and in the capacitance. The voltages across the components may be out of phase with the current, but the current in all parts of the circuit, including the source, has the same value at any instant.

The voltage-drop across each part of the circuit above, with a current of 10 A, will be

$$E_R = IR = 10(4) = 40\ \text{V}$$
$$E_{X_C} = IX_C = 10(1) = 10\ \text{V}$$
$$E_{X_L} = IX_L = 10(4) = 40\ \text{V}$$

The power factor of the problem above is probably most simply determined by the ratio R/Z, or 4/5, which equals 0.80. The phase angle is determined by $\cos \theta = 0.80 = 36.9°$.

If a series circuit has capacitance and inductance but zero resistance, the impedance formula eliminates the R value and becomes

$$Z = \sqrt{(X_L - X_C)^2} = X_L - X_C$$

With no resistance, the phase angle will be 90°. The power factor will be R/Z, or O/Z, which is equal to zero. With a power factor of zero there must be no loss of power in the circuit. The circuit is completely reactive and appears as either a purely capacitive or a purely inductive circuit to the source, depending on which reactance is greater.

In the special case in which X_L equals X_C, a condition known as *resonance* occurs. When the impedance formula for a series circuit is applied to a series-resonant circuit, the impedance is

$$Z = \sqrt{R^2 + (X_L - X_C)^2} = \sqrt{R^2 + 0^2} = R$$

7-9 ACCURACY OF COMPUTATIONS

When the resistance in a series circuit is more than 10 times the net reactance, for most practical purposes the reactance could be disregarded. Consider the following:

A series circuit has a resistance of 100 Ω and a reactance of 10 Ω. The impedance is

$$Z = \sqrt{R^2 + X^2}$$
$$= \sqrt{100^2 + 10^2} = \sqrt{10{,}100} = 100.5 \ \Omega$$

The 100.5-Ω impedance is within 0.5% of 100 Ω. Most meters used for voltage, current, and resistance measurements are guaranteed to be accurate only within 1 or 2% of full scale. This is one reason why mathematical accuracy to the *third significant figure* is all that is required in general work. (15,273 = 15,300, correct to the third significant figure.)

Since the above circuit acts almost as a pure resistance, the source current and voltage will be nearly in phase and the power factor will approach unity (1).

If the values were reversed, that is, the reactance were more than 10 times the resistance, for most purposes the circuit might be considered to be completely reactive. Simplifications or generalities of this type should, however, be applied with caution.

The knowledge that any series circuit has a resistance value much larger than its total reactance value should indicate, even before a problem is worked, what the approximate impedance, phase angle, and power factor should be. If computed answers do not correspond to generalized theory, there is a good possibility that errors were made in the mathematical solving of the problem.

7-10 PROVING PROBLEMS

Occasionally it is possible to work a problem in which members of the equation are squared and an obviously incorrect answer results. This can be demonstrated by the following problem:

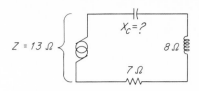

Fig. 7-15 Diagram of circuit in problem.

A series circuit contains a 7-Ω R, an 8-Ω X_L, and a 13-Ω Z. What is the X_C value? Figure 7-15 shows the diagram of the circuit. Using the series-circuit formula:

$$Z = \sqrt{R^2 + (X_L - X_C)^2}$$
$$Z^2 = R^2 + (X_L - X_C)^2$$
$$Z^2 - R^2 = (X_L - X_C)^2$$
$$\sqrt{Z^2 - R^2} = X_L - X_C$$
$$X_C + \sqrt{Z^2 - R^2} = X_L$$
$$X_C = X_L - \sqrt{Z^2 - R^2} = 8 - \sqrt{13^2 - 7^2}$$
$$X_C = 8 - \sqrt{120} = 8 - 10.95$$
$$X_C = -2.95 \ \Omega$$

When the answer, −2.95 Ω, is substituted in the original formula to prove the answer, the impedance works out to be 8.7 Ω instead of the specified 13 Ω. *Something is wrong.*

Checking back over the original formula, it is found that there are two ways of writing it:

$$Z = \sqrt{R^2 + (X_L - X_C)^2}$$

or

$$Z = \sqrt{R^2 + (X_C - X_L)^2}$$

Apparently the wrong formula was used. By the second formula the problem is solved as

$$Z = \sqrt{R^2 + (X_C - X_L)^2}$$
$$X_C = X_L + \sqrt{Z^2 - R^2} = 8 + \sqrt{13^2 - 7^2}$$
$$= 8 + 10.95 = 18.95 \ \Omega$$

When the value 18.95 is substituted in the original formula in place of X_C, the impedance value proves as 13 Ω.

A vector diagram of the problem would have shown that an X_C greater than the inductive value would be required to produce the 13-Ω impedance and that the formula with a greater X_C would be indicated as the one to use. Always draw a vector representation of an ac circuit before working it.

7-11 THE j OPERATOR

The letter j is used extensively in electrical engineering as a means of notation, or as a means of labeling the quantity in front of which it is placed. It is known as the j operator.

The j operator in front of a number indicates that the quantity is 90° (no more and no less) out of phase with something else. It can be considered a 90° vector rotation.

A positive j indicates a normal, *counterclockwise* rotation of the vector by 90°. It is possible to express "A 3.7-A current in some capacitive circuit is leading the voltage 90°" by merely writing $j3.7$ A. The notation $j2$ V indicates that some 2-V vector is leading its current vector by 90°, which would only be true in an inductive circuit.

A negative j ($-j$) indicates a *clockwise* rotation of the vector by 90°. "A 4.2-A current in some inductive circuit is lagging the voltage by 90°" may be expressed by $-j4.2$ A. The notation $-j83$ V indicates a capacitively reactive voltage of 83 V, with the voltage lagging the current by 90°.

The j indicates 90°, and the negative or positive sign prefixing it indicates the direction of vector rotation.

When drawing vectors to represent reactance values, it is standard practice to show the X_L values 90° ahead of the zero value, or upward. The X_C value is shown 90° behind the zero value, or downward. Therefore a j operator before a reactance value such as $j276\ \Omega$ indicates 276-Ω X_L, while $-j75\ \Omega$ indicates 75-Ω X_C.

A j operator is not used before a Z value because impedance assumes some resistance and therefore less than a 90° phase angle.

Test your understanding; answer these checkup questions.

1. When X_L, X_C, and R are in series, what formula is used to determine Z? _____
2. When X_L and X_C are known but R is zero, what is the formula to determine Z? _____ What is the value of θ? _____
3. What is the special term used when X_L equals X_C in a series circuit? _____ What is the formula for Z in this case? _____
4. In most practical applications, how much greater should the X value be than the R to allow R to be disregarded? _____ What percent Z error does this produce? _____ What θ is this? _____
5. A series ac circuit has 14 Ω of X_L, 6 Ω of R, and 6 Ω of X_C. What is the Z value? _____ Pf? _____ θ? _____
6. A series ac circuit of 12-Ω R, 15-Ω X_L, and 40-Ω X_C has 5 A flowing in it. What is the Z? _____ E_s? _____ Pf? _____ θ? _____ E_L? _____ E_C? _____
7. A 100-W 115-V lamp is in series with a 355-Ω X_L and a 130-Ω X_C across 220 V. What is the lamp R? _____ Z of this circuit? _____ I? _____ Pf? _____ θ? _____
8. A potential of 110 V is applied to a series circuit containing 25-ΩX_L, 10-ΩX_C, and 15-Ω R. What is the Z? _____ Pf? _____ θ? _____
9. A series circuit having 5-Ω R, 25-Ω X_C, and 12-Ω X_L has 10 A flowing. What is the voltage across the L? _____ C? _____ R? _____ What is the source voltage? _____
10. What is the Z value if a series circuit contains $j40\ \Omega$, $-j50\ \Omega$, and 15 Ω? _____ What is θ? _____

7-12 PARALLEL AC CIRCUITS

Methods for solving series ac and parallel ac circuits are quite different. Considerable confusion may result from a carry-over of procedure between the two types. To try to prevent this and at the same time to introduce two new electrical terms, an entirely different method will be employed to compute the impedance of parallel circuits. Basically, this method is similar to the method of computing resistors in parallel by converting the resistance values into their equivalent reciprocal values, their conductances, and then adding the conductances in mhos. The symbol used to denote conductance is G. The reciprocal of the total conductance is the total resistance.

$$G_t = \frac{1}{R_1} + \frac{1}{R_2} + \frac{1}{R_3} = \frac{1}{R_T}$$

and

$$R_T = \frac{1}{1/R_1 + 1/R_2 + 1/R_3}$$

This last formula is used to solve for the total resistance of any number of parallel resistors. It is valid for an ac circuit as well as a dc circuit as long as there is only resistance in the circuit. If any reactance is present in any branch of the circuit, the formula will not hold true. However, a variation of the formula can be used.

To solve for the impedance, power factor, phase angle, and so on in circuits having resistance and reactance in parallel, two new terms often employed in electric circuitry will be used. The first is *susceptance, B,* the reciprocal of reactance, or $1/X$, which is measured in mhos. The second new term is *admittance, Y,* the reciprocal of impedance, or $1/Z$, which is measured in mhos. Therefore:

$$G = \frac{1}{R} \qquad B = \frac{1}{X} \qquad Y = \frac{1}{Z}$$

The susceptance formula can be used to determine the total value of parallel reactances. What is the reactance of a 20-Ω X_L and an 8-Ω X_L in parallel? This can be solved by adding the susceptances and converting back to reactance. Thus,

$$B_t = \frac{1}{X_L} + \frac{1}{X_L} = \frac{1}{20} + \frac{1}{8}$$
$$= \frac{2}{40} + \frac{5}{40} = \frac{7}{40} \text{ mho}$$

If the total B is equal to $\frac{7}{40}$ mho, then the total X_L will be the reciprocal of B_t or $\frac{40}{7}$, or 5.71 Ω of X_L. This could also be solved by using a parallel-resistor type of formula $X_1 X_2/(X_1 + X_2)$, or $8(20)/(8 + 20) = 5.71$ Ω

If the two reactances are of different types, as in Fig. 7-16, the total B will be the *difference* between the two susceptances. If the 20 Ω is X_C and the 8 Ω is X_L, then

$$B_t = \frac{1}{8} - \frac{1}{20} = \frac{5}{40} - \frac{2}{40} = \frac{3}{40} \text{ mho}$$

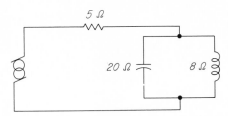

Fig. 7-16 Two pure reactances in parallel with a resistor in series.

The reciprocal of B_t is $\frac{40}{3}$, or 13.3 Ω. Since the current through the inductor would be the greater, the parallel group acts as a 13.3-Ω inductive reactance.

With the 5-Ω resistor connected in series with the parallel group, the circuit impedance can be solved as if there were only the 5-Ω resistor and the 13.3-Ω X_L, or

$$Z = \sqrt{5^2 + 13.3^2} = 14.2 \ \Omega$$

with an inductive (lagging) power factor.

7-13 INDUCTANCE AND RESISTANCE IN PARALLEL

To solve for the total impedance of a resistance and reactance in parallel, it is important to understand that the series impedance formula $Z = \sqrt{R^2 + X^2}$ cannot be applied, although the same type of formula can be used provided reciprocal values are substituted. The reciprocal of Z is Y; the reciprocal of R is G; and the reciprocal of X is B. Therefore, the formula to solve for the admittance of a parallel resistance and reactance is

$$Y = \sqrt{G^2 + B^2} \qquad \text{or} \qquad Y = G + jB$$

Note that conductance, G, plus a j operator marked susceptance, jB, is another way of expressing $\sqrt{G^2 + B^2}$. Both formulas infer that the length of the hypotenuse of a right triangle is to be solved.

Impedance, being the reciprocal of the admittance Y, can be found by dividing Y into 1. In Fig. 7-17,

$$Y = \sqrt{G^2 + B^2} = \sqrt{0.1^2 + 0.1^2}$$
$$= \sqrt{0.01 + 0.01} = \sqrt{0.02} = 0.1414 \text{ mho}$$

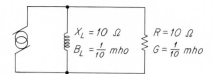

$X_L = 10 \ \Omega$ $R = 10 \ \Omega$
$B_L = \frac{1}{10} \ mho$ $G = \frac{1}{10} \ mho$

Fig. 7-17   Parallel RL circuit across an ac source.

Since $1/Y = Z$,

$$Z = \frac{1}{0.1414} = 7.07 \ \Omega$$

This formula is actually solving a right triangle having G, B, and Y sides, similar to the right triangle used in series circuits having R, X, and Z sides. A comparison of the two triangles is shown in Fig. 7-18.

Fig. 7-18   Series-circuit X, R, and Z vectors compared with parallel-circuit B, G, and Y vectors.

In series circuits the phase angle is the angle between the R and Z sides. In parallel circuits the phase angle is the angle between the $1/R$ and $1/Z$ (G and Y) sides. If G and Y are known, the phase angle of the circuit is equal to the cosine G/Y. In the circuit of Fig. 7-17, G/Y is equal to 0.100/0.1414, or 0.707. From a table of trigonometric functions, this angle, the phase angle, is found to be 45°. In this particular case, the angle should be recognized as being 45°, because both the B and the G vectors are of equal length.

With the impedance of the circuit known, Ohm's law can be applied to the circuit to solve for either current or voltage. If the source voltage in Fig. 7-17 is 20 V, the total current in the circuit is

$$I = \frac{E}{Z} = \frac{20}{7.07} = 2.83 \text{ A}$$

If each individual branch current is solved separately,

$$I_R = \frac{E}{R} = \frac{20}{10} = 2 \text{ A}$$

$$I_{X_L} = \frac{E}{X_L} = \frac{20}{10} = 2 \text{ A}$$

As with the vector diagram of B and G in a parallel circuit, it is found that a vector diagram of the branch *currents* will also plot in such a way that the phase angle and therefore the power factor can be determined from it. Figure 7-19

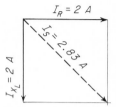

Fig. 7-19   In parallel circuits the currents of the branches can be added vectorially.

shows the current vector diagram of the circuit described above, using the resistive current and the inductive-reactance current, which together produce a resultant that represents the source current of 2.83 A.

In parallel resistance and reactance circuits, where resistance and reactance are not present in the same branch, the circuit may also be solved without the use of susceptance, conductance, and admittance by assuming a convenient source voltage and solving for the separate branch circuits. By solving the current vector triangle of I_R, I_X, and I_S, the cosine I_R/I_S will be the power factor and the angle to which it is equivalent will be the phase angle. Increasing or decreasing the assumed source voltage will have no effect on the ratio of the currents and therefore will give the same phase angle and power factor.

EXAMPLE: What is the impedance of a circuit having an inductance with a reactance of 50 Ω in parallel with a resistance of 25 Ω? If a 100-V source is assumed, Fig. 7-20 illustrates the schematic and vector diagrams.

$$I_R = \frac{E_s}{R} = \frac{100}{25} = 4 \text{ A}$$

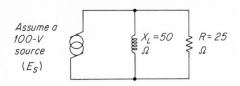

Assume a 100-V source (E_S)

$X_L = 50\ \Omega$ $R = 25\ \Omega$

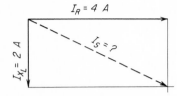

$I_R = 4\ A$

$I_{X_L} = 2\ A$

$I_S = ?$

Fig. 7-20 Any convenient source voltage can be assumed when computing impedance of a parallel circuit if branch currents are vectored.

$$I_{X_L} = \frac{E_s}{X_L} = \frac{100}{50} = 2\ A$$

$$I_S = \sqrt{I_R^2 + I_{X_L}^2} = \sqrt{4^2 + 2^2}$$
$$= \sqrt{16 + 4} = \sqrt{20} = 4.48\ A$$

Then $$Z = \frac{E_s}{I_S} = \frac{100}{4.48} = 22.2\ \Omega$$

7-14 CAPACITANCE AND RESISTANCE IN PARALLEL

Capacitance and resistance in parallel are computed in the same way as inductance and resistance, except that the susceptance vectors, or the reactive current vectors, are drawn in opposite directions.

Test your understanding; answer these checkup questions.

1. What letter symbol represents $1/R$? _____ $1/X$? _____ $1/Z$? _____
2. What term is represented by the letter symbol B? _____ Y? _____ G? _____
3. What is the X_T of a 20-Ω X and a 30-Ω X if both are similar reactances and are in series? _____ Are similar reactances and are in parallel? _____ Are opposite reactances in series? _____ Are opposite reactances in parallel? _____
4. Which two sides of an impedance triangle form the phase angle in a series circuit? _____ Which triangle sides form the θ in a parallel circuit? _____
5. A 50-V source has a parallel 40-Ω R and a 25-Ω X across it. What is the admittance of the circuit? _____ The

Z? _____ The I_L? _____ The I_R? _____ The P? _____
6. What is another way of expressing $Y = G + jB$? _____ $Y = G - jB$? _____
7. If the currents in a resistive branch and a parallel capacitive branch are known, what formula is used to solve for I_T? _____
8. If G, B, and Y are known, how is pf determined? _____
9. Voltages can be vectored in series circuits. What can be vectored in parallel circuits? _____
10. A 120-V source delivers 200 W at 90% pf to a parallel R and X_L. What is the θ? _____ I_S? _____ Z? _____ VA? _____
11. A 1-μF capacitor is across a 530-Ω electric light operating on a 110-V 60-Hz line. What is the Z_T? _____ Pf? _____ θ? _____ I_R? _____ I_S? _____

7-15 CAPACITANCE, INDUCTANCE, AND RESISTANCE IN PARALLEL

With C, L, and R in parallel, the same voltage is across all branches. Current in the inductive branch will lag the source voltage by 90°. Current in the capacitive branch will lead the source voltage by 90°. Therefore, the two reactive currents will tend to cancel each other insofar as the source is concerned. If these reactive currents happen to be equal, the only current the source must supply is to the resistive branch. This special condition, in which $X_L = X_C$, known as *parallel resonance,* is discussed in Chap. 8.

When the inductive reactance and the capacitive reactance are not equal in a parallel circuit, as in Fig. 7-21, the circuit can be solved by either

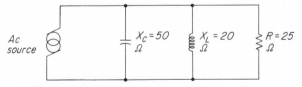

Ac source $X_C = 50\ \Omega$ $X_L = 20\ \Omega$ $R = 25\ \Omega$

Fig. 7-21 Parallel CLR circuit across an ac source.

the admittance method or the current-vector method.

By the admittance method, the reactances are converted to their susceptance values and the resistance to its conductance value. For the example circuit

$$\begin{array}{ll} X_C = 50\ \Omega & B_C = \frac{1}{50} = 0.02\ \text{mho} \\ X_L = 20\ \Omega & B_L = \frac{1}{20} = 0.05\ \text{mho} \\ R = 25\ \Omega & G = \frac{1}{25} = 0.04\ \text{mho} \end{array}$$

Since inductive and capacitive reactances are plotted 180° out of phase, the inductive and capacitive susceptances are also plotted in opposite directions. As a result, the net susceptance of the circuit will be the difference between the inductive and capacitive susceptances. In this case, the net susceptance is

$$B = 0.05 - 0.02 = 0.03 \text{ mho}$$

If the reciprocal of the difference of the *reactances* $(50 - 20 = 30)$ is used, the net value will be $\frac{1}{30}$, or 0.0333 mho, which is incorrect.

The net susceptance is 0.03 mho, while the conductance is 0.04 mho. By substituting these values in the admittance formula,

$$Y = \sqrt{G^2 + B^2} = \sqrt{0.04^2 + 0.03^2}$$
$$= \sqrt{0.0016 + 0.0009} = \sqrt{0.0025} = 0.05 \text{ mho}$$

Since $Z = 1/Y$,

$$Z = \frac{1}{0.05} = 20 \ \Omega$$

By using the impedance value and assuming a source emf of 100 V, the current of the whole circuit can be found with Ohm's law:

$$I = \frac{E}{Z} = \frac{100}{20} = 5 \text{ A}$$

Both the power and the power factor of the whole parallel circuit are determined in almost the same way as in series circuits. The true power is the power dissipated in the resistance and is usually found by the formula $P = I^2R$, where I is the current flowing through the resistance. For the circuit in Fig. 7-21,

$$I_R = \frac{E}{R} = \frac{100}{25} = 4 \text{ A}$$

$$P = I^2R = 4^2(25) = 400 \text{ W}$$

The apparent power of the whole circuit can be found by multiplying the source voltage by the source current. In this case $P_a = \text{VA} = 100(5) = 500$ W.

The power factor in series circuits is equal to R/Z. In parallel circuits it is equal to G/Y. In this circuit

$$\text{pf} = \frac{G}{Y} = \frac{0.04}{0.05} = 0.8000$$

In series circuits the power factor is the ratio of resistive to source *voltages*. In parallel circuits it is the ratio of resistive to source *currents*, or $\text{pf} = I_R/I_S$. In the example circuit, pf = 4/5, or 0.8000.

In either series or parallel circuits the power factor will be the ratio of the true to the apparent power, or $\text{pf} = P/\text{VA}$, or in this case, 400/500, or 0.8000. With parallel circuits the power factor may also be found by $\text{pf} = Z/R$.

The phase angle of the circuit is the angle represented by the cosine value of the power factor, as determined by any of the methods given above. In the example circuit it is found to be 36.9° by referring to a table of trigonometric functions.

In a parallel circuit, when the net reactance value is more than 10 times the resistance, for most practical purposes the reactance can be disregarded because it will have little effect on the value of current flowing, on the impedance of the circuit, on the phase angle, or on the power factor. For example:

A parallel circuit is made up of five branches, three of the branches being pure resistances of 7, 11, and 14 Ω. The fourth branch has an inductive-reactance value of 500 Ω. The fifth branch has a capacitive reactance of 900 Ω. What is the total impedance of this network, and which branch will dissipate the greatest amount of heat?

The net reactance of the circuit will be $1/B_T$, or $1/(B_L - B_C)$, or $1/(1/500 - 1/900)$, which is an 1124-Ω reactance, X_L. This is in parallel with a net resistance value of 3.27 Ω (the three parallel resistors). Since the net reactance value is more than 100 times the net resistance value, the impedance of the circuit can be assumed to be the resistive value of 3.27 Ω. The branch which will dissipate the greatest amount of heat is the one drawing the most current (having the lowest resistance value), which is the 7-Ω branch.

The following is an example of the use of the

current-vector method of solving a parallel ac-circuit problem:

An alternating voltage of 115 V is connected across a parallel circuit made up of a resistance of 30 Ω, an inductive reactance of 17 Ω, and a capacitive reactance of 19 Ω. What is the total circuit drain from the source? A diagram of this circuit is shown in Fig. 7-22. The current values

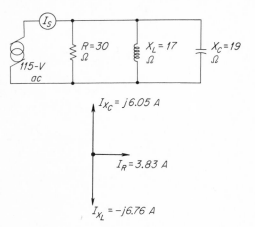

Fig. 7-22 The source current can be determined by vectoring the three branch currents.

through the three branches are

$$I_R = \frac{E}{R} = \frac{115}{30} = 3.83 \text{ A}$$

$$I_{X_L} = \frac{E}{X_L} = \frac{115}{17} = -j6.76 \text{ A}$$

$$I_{X_C} = \frac{E}{X_C} = \frac{115}{19} = j6.05 \text{ A}$$

The net current between the $-j6.76$ and the $j6.05$ A is $-j0.71$ A.

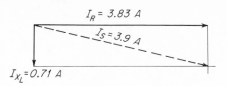

Fig. 7-23 Resultant vector diagram after the reactive-current values have been subtracted.

The resistive current is 3.83 A, and the net inductively reactive current is 0.71 A. The vector diagram of the circuit is shown in Fig. 7-23. The source current I_S is the current that flows through the parallel circuit.

By using the Pythagorean theorem to solve for the hypotenuse of the right triangle composed of sides I_R, I_{X_L}, and I_S:

$$I_S = \sqrt{I_R^2 + I_X^2} = \sqrt{3.83^2 + 0.71^2}$$
$$= \sqrt{14.7 + 0.504} = \sqrt{15.2} = 3.9 \text{ A}$$

The impedance can be found by using Ohm's law:

$$Z = \frac{E}{I} = \frac{115}{3.9} = 29.5 \text{ } \Omega$$

The power factor is the ratio of the I_R to the I_S, or pf $= I_R/I_S$, or $3.83/3.9$, or 0.9821. From a table of trigonometric functions the phase angle is found to be 10.8°.

The true power is $P = I^2R$, or $3.83^2(30)$, or 440 W.

The apparent power, or voltamperes, is 115(3.9), or 448.5 W.

7-16 COMPLEX PARALLEL-SERIES CIRCUIT

The admittance formula $Y = \sqrt{G^2 + B^2}$ can also be applied to a more complex parallel-series type of ac circuit, such as Fig. 7-24.

In branch 1, composed of pure resistance, the conductance value is simply the reciprocal of the resistance, or

$$G = \frac{1}{R}$$

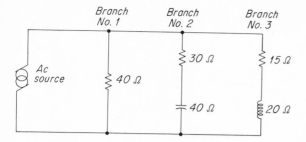

Fig. 7-24 A more complex parallel-series circuit that can be solved by the admittance method.

In the two other branches there is both resistance and reactance in series. To find the conductance component a formula that takes both resistance and reactance into consideration must be used. The special formula for this is

$$G = \frac{R}{Z^2}$$

Similarly, to compute the susceptance component of a branch that has both resistance and reactance in series the special formula is

$$B = \frac{X}{Z^2}$$

The admittance of the whole circuit can be determined by using the general formula

$$Y = \sqrt{G_T{}^2 + B_T{}^2}$$

In this formula, the total conductance (G_T) value is the sum of all the branch conductances, regardless of whether the branch is inductive or capacitive. (In a purely reactive branch the G value is zero.)

The total susceptance (B_T) is the *difference* between the inductive and capacitive susceptances of the different branches. (In a purely resistive branch the B value is zero.)

To solve for the impedance of the circuit in Fig. 7-24, the expanded admittance formula would be

$$Y = \sqrt{(G_1 + G_2 + G_3)^2 + (\pm B_1 \pm B_2 \pm B_3)^2}$$

in which inductive susceptance is assigned a plus value and capacitive susceptance a minus value.

The suggested steps are:
Branch 1:

$$G_1 = \frac{1}{R} = \frac{1}{40} = 0.025 \text{ mho}$$

$$B_1 = 0 \text{ mho}$$

Branch 2:

$$Z_2 = \sqrt{R^2 + X^2} = \sqrt{30^2 + 40^2}$$
$$= \sqrt{900 + 1{,}600} = 50 \ \Omega$$

$$G_2 = \frac{R}{R^2 + X^2} = \frac{30}{900 + 1{,}600} = \frac{30}{2{,}500}$$
$$= 0.012 \text{ mho}$$

$$B_2 = \frac{X}{R^2 + X^2} = \frac{40}{900 + 1{,}600} = \frac{40}{2{,}500}$$
$$= 0.016 \text{ mho}$$

Branch 3:

$$Z_3 = \sqrt{R^2 + X^2} = \sqrt{15^2 + 20^2}$$
$$= \sqrt{625} = 25 \ \Omega$$

$$G_3 = \frac{R}{R^2 + X^2} = \frac{15}{225 + 400}$$
$$= \frac{15}{625} = 0.024 \text{ mho}$$

$$B_3 = \frac{X}{R^2 + X^2} = \frac{20}{225 + 400}$$
$$= \frac{20}{625} = 0.032 \text{ mho}$$

The sum of all G's (G_T):

$$G_T = G_1 + G_2 + G_3$$
$$= 0.025 + 0.012 + 0.024 = 0.061 \text{ mho}$$

The sum of all B's (B_T):

$$B_T = B_1 - B_2 + B_3$$
$$= 0 - 0.016 + 0.032 = 0.016 \text{ mho}$$

Solving for impedance:

$$Y = \sqrt{G_T^2 + B_T^2} = \sqrt{0.061^2 + 0.016^2}$$

$$= \sqrt{0.003721 + 0.000256}$$

$$= \sqrt{0.003977} = 0.0631 \text{ mho}$$

$$Z = \frac{1}{Y} = \frac{1}{0.0631} = 15.8 \text{ ohms}$$

The power factor can be found by using the formula pf = P/VA and employing any convenient source voltage. The true power is then the sum of all three I^2R power values in the circuit. The apparent power is the power indicated by the product of the voltage across the circuit and the current from the source. The power factor is the cosine of the phase angle.

7-17 RECTANGULAR AND POLAR NOTATIONS

The solving of complicated ac circuit impedance, phase angle, and, from them, the currents, voltage-drops, VA, P, and pf can be accomplished by the admittance method explained above. The Pythagorean theorem formula that is used involves computations of values such as R (or G) and X (or B) at right angles, as in Fig. 7-18. This is known as *rectangular notation*. If $Z = \sqrt{R^2 + X^2}$ is a form of rectangular notation, then the *complex number* (containing a j operator) used in the expression $Z = R + jX$ has a similar meaning and is also a form of rectangular notation.

When a vector, such as a Z vector, is rotated forward, or counterclockwise (CCW), from a zero-degree direction (far right usually), the angle between its position and zero represents the phase angle of the E and I in the circuit. The length of the vector, or *phasor,* arrow indicates the value of the impedance. For example, in a series RL circuit, if $R = 3\,\Omega$ and $X_L = 4\,\Omega$, then $Z = 5\,\Omega$ (from $Z = \sqrt{R^2 + X^2}$) with a phase angle of 53.1° (by tan $\theta = X/R$). When this information is written 5/53.1° Ω, it is read as "an impedance of 5 Ω with a phase angle of 53.1°" and is said to be in *polar notation* form. For a circuit having a 10-Ω R and a 10-Ω X_C, in rectangular form $Z = \sqrt{10^2 + 10^2}$, which computes to

the polar information $Z = 14.1 \underline{/-45°}\,\Omega$. The negative sign indicates the circuit is capacitively reactive with the current ahead of, or leading the voltage, or the 0° direction. There is really nothing new in this except how the circuit Z and θ are expressed.

The trigonometric formulas, discussed briefly in Sec. 7-3, which may be utilized in converting from polar to rectangular notation, or vice versa, are

$$\cos \theta = \frac{R}{Z} \qquad \sin \theta = \frac{X}{Z} \qquad \tan \theta = \frac{X}{R}$$

For example, using the simple 3-4-5 triangle above, in which $Z = 5\underline{/53.1°}\,\Omega$, the impedance vector is 5 Ω long and is at a CCW angle of 53.1°, as shown in Fig. 7-25. To convert this to rectan-

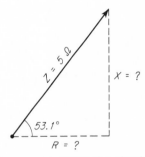

Fig. 7-25 Polar representation of a circuit having an impedance of 5 Ω and a phase angle of 53.1°.

gular notation, it is necessary to determine R and X values. To find the R value with Z known, the trigonometric formula containing the R and Z values must be used. This is cos $\theta = R/Z$. Rearranging to solve for R,

$$R = Z(\cos \theta) = 5(\cos 53.1°) = 5(0.600) = 3\,\Omega$$

To solve for the X value, the formula that contains X and Z, or sin $\theta = X/Z$, would normally be used. Rearranging to solve for X,

$$X = Z(\sin \theta) = 5(\sin 53.1°) = 5(0.800) = 4\,\Omega$$

Note below how simple it is to determine the total impedance of three series ac networks merely by adding their complex number rectangular expressions algebraically.

$$Z_1 = 10 - j30 \quad (R \text{ and } X_C)$$
$$Z_2 = 15 + j20 \quad (R \text{ and } X_L)$$
$$Z_3 = 0 + j60 \quad (\text{only } X_L)$$
$$\overline{Z_T = 25 + j50}$$

The whole series circuit has the equivalent Z of a 25-Ω R and a 50-Ω X_L in series. Try converting this to polar notation. Your answer should be $Z = 55.9 \underline{/63.4°}\ \Omega$.

It is fairly complicated to divide or multiply rectangular notation complex numbers, but polar notation value can be divided or multiplied easily. If it is desired to divide

$$\frac{15\underline{/20°}\ \Omega}{3\underline{/-25°}\ \Omega}$$

the rule is to divide the impedances but move the phase angle from below the line to above the line, change its sign, and then add the two algebraically. In this case the final polar expression is $5\underline{/20° + 25°}\ \Omega$, or $5\underline{/45°}\ \Omega$.

To multiply polar numbers, multiply the impedances and add the angles algebraically. For example, $10\underline{/40°}\ \Omega$ times $5\underline{/-10°}\ \Omega$ is simply $50\underline{/30°}\ \Omega$.

What are the parameters that can be determined from the circuit shown in Fig. 7-26a by

Fig. 7-26 (a) Schematic diagram of problem in text. (b) Block diagram of three impedances to be computed.

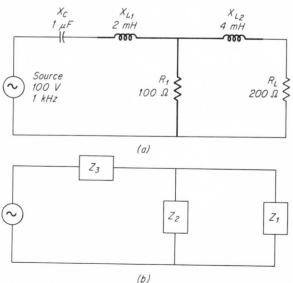

(a)

(b)

utilizing polar and rectangular notations? This circuit consists of a 100-V 1,000-Hz source coupled to a 200-Ω resistor load through a capacitor in series with an RL "T-network."

The load R_L and the 4-mH coil form a series impedance network which in turn is in parallel with the 100-Ω R_1. The capacitor and the 2-mH coil form a series impedance network. The three lumped impedances are indicated in block form in Fig. 7-26b.

Before further circuit computations can be started, it is necessary to determine the values of X_C, X_{L_1}, and X_{L_2}. By the reactance formulas (to the third significant figure only) the reactances are

$$X_C = \frac{1}{2\pi fC} = 159\ \Omega$$

$$X_{L_1} = 2\pi fL = 12.6\ \Omega$$

$$X_{L_2} = 2\pi fL = 25.1\ \Omega$$

Solving the Z_1 series network first,

$$Z_1 = \sqrt{R_L^2 + X_{L_2}^2} = \sqrt{200^2 + 25.1^2} = 202\ \Omega$$

Using the tangent formula,

$$\tan \theta = \frac{X}{R} = \frac{25.1}{200} = 0.1255$$

$$\theta = 7.2°$$

$$Z_1 = 202\underline{/7.2°}\ \Omega$$

Solving for Z_1 and Z_2 together by the general parallel impedance formula,

$$Z_T = \frac{Z_1 Z_2}{Z_1 + Z_2} = \frac{(202\underline{/7.2°})(100\underline{/0°})}{(202\underline{/7.2°}) + (100\underline{/0°})}$$

$$= \frac{(202)(100) \text{ and } 7.2° - 0°}{(200 + j25.1) + (100 + j0)}$$

$$= \frac{20{,}200\underline{/7.2°}}{300 + j25.1}$$

$$= \frac{20{,}200\underline{/7.2°}}{301\underline{/4.8°}} = 67.1\underline{/7.2° - 4.8°}$$

$$= 67.1\underline{/2.4°}\ \Omega$$

Converting to rectangular notation from $67.1\underline{/2.4°}\ \Omega$,

$X_{L_{1,2}} = Z(\sin \theta) = 67.1(\sin 2.5°) = 2.81\ \Omega$

$R_{1,2} = Z(\cos \theta) = 67.1(\cos 2.5°) = 67.0\ \Omega$

$Z_{1,2} = 67 + j2.8$

Solving for Z_3 (two pure reactances)

$$Z_3 = \begin{cases} 0 - j159 \\ \underline{0 + j12.6} \\ 0 - j146.4 \end{cases}$$

Solving for Z_T, the total of Z_3 and $Z_{1,2}$,

$$Z_T = \begin{cases} Z_3 = \ \ 0 - j146.4 \\ +Z_{1,2} = \underline{67 + j2.81} \\ \ \ \ \ \ \ \ \ \ 67 - j143.6 \end{cases}$$

$$= 157\underline{/-65°}\ \Omega$$

Solving for I_T by Ohm's law,

$$I_T = \frac{E}{Z} = \frac{100}{158} = 0.633\ \text{A}$$

Then

$E_{X_C} = I_T X_C = 0.633(158) = 100\ \text{V}$

$E_{X_1} = I_T X_1 = 0.633(12.6) = 7.98\ \text{V}$

$E_{Z_{1,2}} = I_T Z_{1,2} = 0.633(67.1) = 42.5\ \text{V}$

$I_{R_1} = \dfrac{E_{R_1}}{R_1} = \dfrac{42.5}{100} = 0.425$

$I_{R_L} = I_{Z_1} = \dfrac{E_{Z_{1,2}}}{Z_1} = \dfrac{42.5}{202} = 0.21\ \text{A}$

$E_{R_L} = I_{R_L} R_L = 0.21(200) = 42\ \text{V}$

$P_T = I_{R_1}^2 R_1 + I_{R_L}^2 R_L = 0.425^2(100) + 0.21^2(200)$

$\ \ \ \ = 18.1 + 8.82 = 26.9\ \text{W}$

$\text{VA} = 100(0.633) = 63.3\ \text{VA}$

$\text{pf} = \dfrac{P_T}{\text{VA}} \quad \text{or} \quad \cos(-65°) = 0.424 = 42.4\%$

$P_{R_1} = E_{R_1} I_{R_1} = 42.5(0.425) = 18.1\ \text{W}$

$E_{X_2} = I_{X_2} X_2 = 0.21(25.1) = 5.27\ \text{V}$

Test your understanding; answer these checkup questions.

1. When R, X_L, and X_C are in parallel, to solve for Z what is the first operation? _____ The second? _____ What formula is then used? _____
2. When R, X_L, and X_C are in parallel, to solve for source current what is the first operation? _____ The second? _____ The formula to use? _____
3. A 300-Ω X_L, a 100-Ω X_C, and a 400-Ω R are in parallel across 120 V ac. What is the Z? _____ The pf? _____ Is the pf leading or lagging? _____ What is the I_s? _____ The θ? _____
4. A 100-V ac generator has a 70-Ω X_L, a 90-Ω X_C, and a 600-Ω R and a second 400-Ω R in parallel across it. Which branch dissipates the most power? _____ What is the Z? _____ VA? _____ P? _____ Pf? _____ θ? _____
5. A 50-Ω R, a 100-Ω X_L, and a 25-Ω X_C are in parallel. Assume a 100-V source to make current computations simple. What is the I_s value? _____ By Ohm's law the Z value? _____ Pf? _____ θ? _____ If 200 V had been used, which answers would have been different? _____
6. Draw a diagram of a 50-Ω resistor across a source of 100 V with a 90-Ω X_C in series with a 40-Ω R also across the source. Using the admittance method, what is G_T? _____ B_T? _____ Y? _____ Z? _____ VA? _____ P? _____ Pf? _____ θ? _____
7. A series 20-Ω R and 30-Ω X_C are in parallel with a series 40-Ω R and 80-Ω X_L. Using $Z_T = Z_1 Z_2/(Z_1 + Z_2)$, what is the polar Z_T value? _____ The rectangular Z_T? _____ Circuit pf? _____

COMMERCIAL LICENSE QUESTIONS

Amateur license questions will be found in the Addendum.

FCC Elements 3, 4, and 6 require an ability to answer questions similar to those below. Sections in which questions are answered are shown in parentheses. A question followed by a bracketed number is required for that element alone.

1. What are the properties of (a) an inductor acting alone in an ac circuit, (b) a capacitor alone, (c) a resistor alone? (7-1)
2. What unit is used in expressing the ac impedance of a circuit? (7-2)

3. State Ohm's law for ac circuits. (7-2)
4. In what way does an inductance affect the voltage-current phase relationship of a circuit? (7-2)
5. What is the total impedance of a series ac circuit having 3-Ω R, 7-Ω X_L, and 2-Ω X_C? (7-2)
6. What is the impedance of a solenoid if its resistance is 5 Ω and 0.4 A flows through the winding when 110 V at 60 Hz is applied across it? (7-2)
7. In a circuit having 200-Ω X_L and 200-Ω R, what will be the phase angle of the current with reference to the voltage? (7-3)
8. Explain how to determine the sum of two vector quantities which have the same reference point but the directions of which are (a) out of phase by 180°, (b) in phase, (c) 90° out of phase? How does this pertain to electric currents or voltages? (7-4, 7-3)
9. What is the meaning of *power factor,* and how is it calculated? (7-6)
10. Why is the phase of a circuit important? (7-6)
11. What factors must be known in order to determine the power factor of an ac circuit? (7-6)
12. What is the meaning of the term *leading power factor?* (7-6)
13. How can low power factor in an electric power circuit be corrected? (7-6)
14. If a 220-V 60-Hz line delivers 200 W at 90% power factor to a load, what is the phase angle between line current and voltage? How much current flows in the line? (7-6)
15. A series circuit has a 4-Ω R, a 4-Ω X_L, and a 2-Ω X_C. The applied circuit ac emf is 60 V. What is the voltage-drop across (a) the resistor, (b) the inductor, (c) the capacitor? (7-8)
16. A series circuit contains a 7-Ω R and 10-Ω X_L, and the X_C is unknown. What value of reactance must the capacitor have in order that the total circuit impedance be 15 Ω? (7-10)   [4]
17. 1-A ac is flowing in a series circuit composed of a 5-Ω R, a 25-Ω X_C, and a 18-Ω X_L. What is the voltage across each component? What is the total voltage? (7-8)   [4]
18. If an ac of 2 A flows in a series circuit composed of a 12-Ω R, a 20-Ω X_L, and a 40-Ω X_C, what is the voltage across the circuit? (7-8)   [4]

19. A lamp rated at 100 W and 115 V is in series with a 300-Ω X_L and a 100-Ω X_C across a 220-V line. What is the current value through the lamp? (7-8)   [4]
20. A voltage of 110 V is applied across a series circuit of a 25-Ω X_C, a 10-Ω X_L, and a 15-Ω R. What is the phase of the E and I of this circuit? (7-8)   [4]
21. What is the relation between (a) resistance and conductance and (b) impedance and admittance? (7-12)
22. A 7-Ω R is in series with a parallel combination of a 20-Ω X_C and a 10-Ω X_L. What is the total impedance of the circuit? Is the impedance capacitive or inductive? (7-12)   [4]
23. A parallel circuit consists of a 9-, an 11-, and a 20-Ω R, a 700-Ω X_L, and a 300-Ω X_C, all in parallel. What is the impedance of the circuit? In which branch will the greatest amount of heat be dissipated? (7-15)   [4]
24. If 115-V ac is applied across a 30-Ω R, a 20-Ω X_L, and a 10-Ω X_C, all in parallel, what is the source-current value? (7-15)   [4]
25. Draw a circuit composed of a 100-V 500-Hz source and a 1-μF capacitor in series with the source followed by a T network composed of a 2-H inductor, a 100-Ω resistor, and a 4-mH inductor, with a load resistor of 200 Ω. What are the total impedance, phase angle, current, power, apparent power, and voltages and currents for the components? (7-17)

NOTE: FCC license test questions are multiple-choice types, taking a form somewhat as follows:

26. The phase angle of a parallel RL circuit can be found by (a) graphing E_X and E_R; (b) cosine θ; (c) graphing G and B; (d) all of the above. (7-4, 7-2, 7-13)
27. The E across and the I through a coil can be used to determine (a) apparent power; (b) true power; (c) resistance of the coil; (d) none of the above. (7-5)
28. Which is the most capacitively reactive series circuit? (a) $Z = 70 - j50$; (b) $Z = 50 + j50$; (c) $Z = 10 - j20$; (d) $Z = 100 + j150$. (7-17)

RESONANCE AND FILTERS

The objective of this chapter is to discuss resonant and antiresonant circuits, Q of LCR circuits, basic decibel computations, and bandwidth of LC circuits. Special filter circuits are developed by combining L, C, and R in different configurations.

8-1 RESONANCE

Resonant circuits are the basis of all transmitter, receiver, and antenna operation. Without resonant circuits there would be no radio communication.

Brief mention was made of series and parallel resonance in preceding chapters. When the inductive reactance (X_L) of a coil equals the capacitive reactance (X_C) of a capacitor in a circuit, a condition known as resonance occurs. Figure 8-1 illustrates a series-resonant and a parallel-resonant circuit.

Since resonance is the condition when X_L equals X_C, the formula for resonance is

$$X_L = X_C \quad \text{or} \quad 2\pi fL = \frac{1}{2\pi FC}$$

where X_L = inductive reactance, in Ω
X_C = capacitive reactance, in Ω

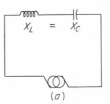 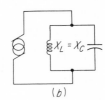

Fig. 8-1 (a) A series-resonant circuit across an ac source. (b) A parallel-resonant circuit across an ac source.

f = frequency, in Hz
L = inductance, in H
C = capacitance, in F

Examination of the formula shows that while the inductive reactance is directly proportional to frequency, the capacitive reactance is inversely proportional to frequency. With any given coil and capacitor, as the frequency increases, the reactance of the coil increases but the reactance of the capacitor decreases. At some frequency, the two reactances will be equal in value. At that one frequency the condition of resonance occurs. At all other frequencies the circuit shown in Fig. 8-1a is merely a series ac circuit, and Fig. 8-1b is merely a parallel ac circuit.

To determine the frequency at which a coil and capacitor will resonate, the resonance formula may be rearranged to solve for f:

$$2\pi fL = \frac{1}{2\pi fC}$$

$$2\pi fL(2\pi fC) = 1$$

$$4\pi^2 f^2 LC = 1$$

ANSWERS TO CHECKUP QUIZ ON PAGE 124

1. *(Convert to G and B)(Subtract B_C from B_L)($Y = G + jB$)*
2. *(Solve branch I's)(Subtract I_C from I_L)($I_T = \sqrt{I_R^2 + I_X^2}$)*
3. *(140 Ω)(0.351)(Leading)(0.857)(69.4°)* 4. *(400 Ω)(191 Ω)*
(52.4 VA)(41.7 W)(0.7955)(37.3°) 5. *(3.61 A)(27.78 Ω)*
(0.554)(56.3°)(I value only) 6. *(0.02412 mho)(0.00928 mho)*
(0.0258 mho)(38.78 Ω)(241 W)(258 VA)(0.9341)(20.9°)
7. *(41.3/−32.7° Ω)(34.75 − j223)(0.8415)*

$$f^2 = \frac{1}{4\pi^2 LC}$$

$$f = \frac{1}{2\pi \sqrt{LC}}$$

By dividing the 2π portion into the 1, the frequency formula can be simplified to

$$f = \frac{0.159}{\sqrt{LC}} \quad \text{or} \quad f = \frac{0.159}{\sqrt{L}\ \sqrt{C}}$$

These formulas can be used to determine the resonant frequency of any LC circuit.

EXAMPLE: If the inductance is 150 μH and the capacitance is 160 pF, frequency = ?

$L = 0.000150$ H, or 15×10^{-5} H

$C = 0.000000\ 000160$ F, or 16×10^{-11} F

$$F = \frac{0.159}{\sqrt{LC}} = \frac{0.159}{\sqrt{15 \times 10^{-5} \times 16 \times 10^{-11}}}$$

$$= \frac{0.159}{\sqrt{240 \times 10^{-16}}} = \frac{0.159}{15.5 \times 10^{-8}}$$

$$= \frac{0.159 \times 10^8}{15.5} = 0.01026 \times 10^8$$

$$= 1,026,000 \text{ Hz, or } 1,026 \text{ kHz,}$$

$$= 1.026 \text{ MHz}$$

According to the formula, the frequency of resonance is inversely proportional to the square root of either L or C. Thus, increasing the inductance 4 times results in a lowering of the frequency to one-half of the original. Similarly, increasing the capacitance 4 times will also result in a frequency one-half of the original.

Any time the LC product is increased 4 times, by quadrupling the inductance, by quadrupling the capacitance, by doubling both L and C, or by any other means, the frequency will be one-half of the original.

Any time the LC product is decreased to $\frac{1}{4}$ by any means, the frequency will be doubled.

As long as the LC product remains the same, the frequency will remain the same. For example, in a 1,000-kHz circuit, if the inductance is halved and the capacitance is doubled, the LC product has not changed and the frequency remains 1,000 kHz.

It is possible to rearrange the resonance formula another way to solve for the inductance needed to resonate with a given capacitance, or the capacitance needed to resonate with a given inductance, to form a resonant circuit at some desired frequency:

$$2\pi fL = \frac{1}{2\pi fC}$$

$$4\pi^2 f^2 LC = 1$$

$$C = \frac{1}{4\pi^2 f^2 L} = \frac{1}{(2\pi f)^2 L} = \frac{1}{\omega^2 L}$$

$$L = \frac{1}{4\pi^2 f^2 C} = \frac{1}{(2\pi f)^2 C} = \frac{1}{\omega^2 C}$$

For example, to determine the capacitance to shunt across a 56-μH coil to resonate at 5,000 kHz

$$C = \frac{1}{4\pi^2 f^2 L} = \frac{1}{4(9.86)(5 \times 10^6)^2(56 \times 10^{-6})}$$

$$= \frac{1}{39.4(25 \times 10^6)56} = \frac{1 \times 10^{-6}}{39.4(25)56}$$

$$= \frac{0.000001}{55,160} = 0.000\ 000\ 000\ 018 \text{ F}$$

$$= 0.000\ 018\ \mu\text{F} = 18 \text{ pF}$$

While a given coil and capacitor will resonate at essentially the same frequency whether connected in series or in parallel, a series-resonant circuit behaves in many ways opposite to a parallel-resonant circuit.

8-2 SERIES RESONANCE

The series ac circuit shown in Fig. 8-2 can be classed as series-resonant because the inductive reactance equals the capacitive reactance at the frequency of the source.

In any series circuit, whether resonant or not, the same value of current flows in all parts of the circuit at any one instant. However, the voltage across the capacitor is 90° behind the circuit

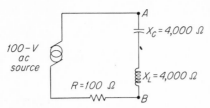

Fig. 8-2 The voltage-drop between *A* and *B* is zero. The whole source voltage appears across *R*.

current, while the voltage across the coil is 90° ahead of the current. The current through and the voltage across the resistance are in phase. Figure 8-3 is a vector diagram of the voltages and current in such a series circuit.

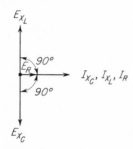

Fig. 8-3 Vector diagram of the phase and amplitude of the current and voltages in a series-resonant circuit.

The E_{X_L} and the E_{X_C} are 180° out of phase. At resonance, when the reactance values are equal, the reactive voltages, being exactly equal and opposite, cancel each other completely insofar as the source is concerned. Therefore, between points *A* and *B* in Fig. 8-2 there is zero volts, although there is a voltage-drop across the X_C equal to that across the X_L.

If the source sees the coil and capacitor together as having a zero voltage-drop across them, it sees them as a perfect conductor, or as zero impedance. If the reactances are not exactly equal (a nonresonant condition), the voltages do not exactly cancel and the source sees the two reactances as having a resultant voltage-drop across them and therefore as having some value of reactance or impedance.

Theoretically, if a series LC circuit has no resistance and is connected across a source of ac to which it is resonant, it presents zero reactance, zero resistance, and zero impedance.

The current-limiting factor in a series-resonant

circuit is the resistance. In Fig. 8-2, with a source voltage of 100 V, a resistance of 100 Ω, and 4,000-Ω reactances, the reactances cancel, leaving the source looking at the 100-Ω resistance. The current flow in the circuit is $I = E/R = 100/100 = 1$ A. Thus, a resonant circuit is considered to be purely resistive. The impedance value of a series-resonant circuit is its resistance value.

Capacitors are usually considered to have negligible series resistance in them. However, coils may have considerable resistance in the wire with which they are wound. This resistance in the coil itself is usually treated as an external resistor. In the circuit of Fig. 8-2, if the coil having 4,000 Ω of reactance also has 100 Ω of resistance in its wire, the diagram as drawn is a proper method of indicating the circuit factors.

In a common radio circuit such as Fig. 8-4

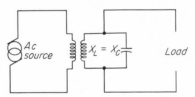

Fig. 8-4 The secondary is a series-resonant circuit.

the condition of *series* resonance is present but is not apparent. In this circuit a transformer primary is connected across a source of ac, and the secondary coil has across it a capacitor, with a reactance value equal to the reactance of the secondary coil. These reactances form a resonant circuit. At first glance it appears that the capacitor shunted across the coil forms a parallel-resonant circuit. The primary, however, is inducing an ac emf into each turn of the secondary coil. An emf is not being applied across the ends of the coil. Theoretically, the secondary may be considered to have a source of ac inserted in series with the turn, as in Fig. 8-5. Since the induced emf is apparently in series with the coil, the circuit is classed as a series-resonant circuit. Any load on the secondary, however, is in parallel with the capacitor and coil. Any emf developed across the circuit produces current in the load.

If the primary coil is brought to resonance by connecting a proper value of capacitance across it. the primary will be a parallel-resonant circuit,

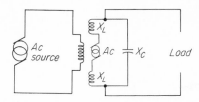

Fig. 8-5 Effectively, emf is induced in series with the secondary inductance.

since the ac voltage is being fed across it and not induced into its turns.

Tuned transformers are used in radio receivers, transmitters, and electronic circuits to select a desired frequency when many frequencies are present. Figure 8-6 shows an antenna connected

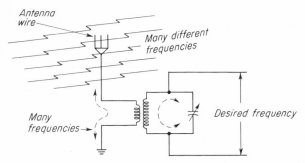

Fig. 8-6 A resonant circuit accepts signals at its resonant frequency but rejects all others.

to the primary of a transformer. The antenna circuit is completed by connecting the lower end of the primary to ground. Radio signals in the air pass across the antenna wire, inducing radio-frequency ac voltages in the antenna-to-ground circuit. Thousands of different-frequency signals are being induced into the antenna simultaneously. The problem is to pick out only the desired frequency.

A secondary coil can be loosely coupled to the primary, and a *variable* capacitor connected across its terminals. By varying the capacitance, it is possible to tune the series-resonant secondary circuit over a band of frequencies. At any frequency where the X_L of the coil equals the X_C of the capacitor, the secondary will appear as a low-impedance circuit to this frequency, and as a result this one frequency produces a significant current in the secondary. With a high-amplitude

current flowing, a relatively high-amplitude voltage at the resonant frequency will be developed across the reactances.

To any frequency other than the resonant frequency, the series-resonant circuit will have greater impedance and will oppose the flow of ac at that frequency, resulting in smaller currents and less reactive voltages for frequencies off resonance.

It has been pointed out that the impedance of a series circuit at resonance is equal to the resistance value of the circuit, but what impedance will a resonant circuit have to frequencies other than its resonant frequency?

The circuit shown in Fig. 8-7 illustrates a series-

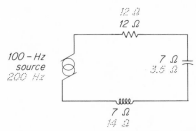

Fig. 8-7 Doubling the frequency doubles the X_L but halves the X_C.

resonant circuit with a 12-Ω resistance, a 7-Ω capacitive reactance, and a 7-Ω inductive reactance across a 100-Hz source. The impedance of the circuit is equal to the value of the resistance alone, or 12 Ω. If the frequency is increased to twice the original, or 200 Hz, what is the impedance of the circuit?

Since inductive reactance is directly proportional to frequency, when the frequency is doubled, the inductive reactance will be 2 times its original 7-Ω value, or 14 Ω. When the frequency is doubled, the capacitive reactance, being inversely proportional to frequency, becomes one-half of its original value, or 3.5 Ω. The resistance value remains the same for all frequencies. The impedance is now

$$Z = \sqrt{R^2 + (X_L - X_C)^2}$$
$$= \sqrt{12^2 + (14 - 3.5)^2} = \sqrt{144 + 110.25}$$
$$= \sqrt{254.25} = 15.9 \ \Omega \text{ inductive (lagging pf)}$$

At 50 Hz the circuit presents 15.9 Ω impedance, is capacitive, and has a leading pf.

Test your understanding; answer these checkup questions.

1. What is the basic formula for resonance? _____ To determine *f* if *L* and *C* are known? _____ To determine *L* if *f* and *C* are known? _____
2. A 2-H coil and a 0.01-μF capacitor are in series. What is the resonant frequency? _____ What is *f* if they are in parallel? _____
3. What is the resultant *f* if the *LC* product is increased 10 times? _____ Decreased to one-half? _____
4. What *C* value is needed to resonate a 2-H coil to 3 kHz? _____
5. What *L* value is needed to resonate a 70-pF *C* to 5 MHz? _____
6. Does the *Z* of a series-resonant circuit increase or decrease if the *f* is increased? _____ If the *f* is lowered? _____
7. Is a resonant circuit a resistive or a reactive load on a source? _____
8. Across what component in a resonant circuit is the source voltage-drop value developed? _____
9. What might happen if a low-resistance resonant circuit were connected across a power line? _____
10. A series circuit consisting of a 6.5-Ω *R* and equal 175-Ω inductive and capacitive reactances is across 260 V ac. What is the *Z* of the circuit? _____ *I*? _____ E_L? _____ E_C? _____ E_R? _____
11. In a series circuit having equal values of *R*, X_L, and X_C of 11 Ω, if *f* is reduced to 0.411 of its value at resonance, what is the new X_L? _____ X_C? _____ *Z*? _____

8-3 PARALLEL RESONANCE

A coil and capacitor connected as shown in Fig. 8-8 form a parallel ac circuit. If X_C and X_L have the same reactance to the frequency of the ac, the circuit is known as a *parallel-resonant* or *anti-resonant* circuit.

If the capacitor is temporarily disconnected, leaving the 100-Ω reactance coil across the 100-V source, according to Ohm's law for ac circuits the

Fig. 8-8 Parallel-resonant or antiresonant circuit.

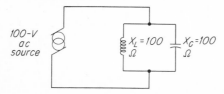

current in the coil will be

$$I = \frac{E}{X_L} = \frac{100}{100} = 1 \text{ A}$$

If the 100-Ω-reactance capacitor is reconnected across the coil, 1 A will flow in the capacitor also. Placing the capacitor across the coil in this circuit does not change the current in the coil even though a resonant circuit is formed (provided there is no series resistance in the source line).

In a parallel-resonant circuit the same voltage is across both the coil and the capacitor. In the inductive branch, however, the current lags the source voltage by 90°, and in the capacitive branch the current leads the source voltage by 90°. Since the two currents are 180° out of phase, at any instant that current is flowing down through the coil an equal current must be flowing up into the capacitor, as illustrated in Fig. 8-9.

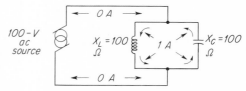

Fig. 8-9 With no resistance in a parallel-resonant circuit, current oscillates in it but the source current is zero.

As seen by the source, 1 A is flowing down through the coil and 1 A is flowing up through the capacitor at the same time. Since it is impossible for current to flow in two directions at once in either of the source lines, there must be zero current from the source but 1 A flowing from one reactance to the other. The electrons that make up the current in the reactances move from the top plate of the capacitor down through the coil and up to the lower capacitor plate. When the source voltage alternates, the electrons retrace their path back up through the coil to the top plate of the capacitor. This circulating current of 1 A flows between the reactances, but no current flows into and out of the source!

Because the source is supplying no current, it should be possible to disconnect the source and the electron current should continue to oscillate back and forth between capacitor and coil indefinitely. With no resistance or losses in the circuit, this would be true. This ability of a resonant circuit to sustain electron oscillation is known as *flywheel effect,* because of its similarity to the action of a mechanical flywheel, which, once started, tends to keep going until stopped by friction or by other losses.

Since the source voltage is across the resonant circuit and no current flows in the source, the parallel-resonant circuit impedance must be

$$Z = \frac{E_s}{I_s} = \frac{E_s}{0} = \infty \text{ (infinite) } \Omega$$

Both infinite-ohms impedance and ceaseless oscillations are impossible. The circuits shown have neglected resistance as well as inductive losses to external circuits. There is no coil or capacitor that does not have some series-resistance value. Figure 8-10 shows the circuit redrawn

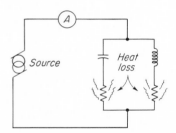

Fig. 8-10 Losses in the antiresonant circuit demand current from the source.

in a more practical form, including resistances in each branch. With the same source voltage, less current will flow in the branches because of increased impedance. The impedance of each branch can be computed as $Z = \sqrt{R^2 + X^2}$.

Current flowing through the resistances heats them. The heat represents a power loss. If the source is disconnected, the energy of the electrons will almost immediately be dissipated in heating the resistors and the electrons will cease to oscillate. With a power loss in the circuit, the source must feed power into the circuit to make up for the loss. To feed the power into the circuit, current will have to be fed to it. Therefore, enough source current must flow in all parallel-resonant circuits to make up for the losses in the circuit and maintain the voltage across the reactances.

With pure reactances and zero resistance in the circuit, the source current is zero. When resistors are added in series with either branch the source current increases and the reactive currents decrease.

As a generalization, the following applies to resonant circuits with little resistance:

SERIES RESONANCE
- The impedance across the circuit is low or zero (equals the series resistance).
- The voltage-drop across the circuit is low or zero.
- The current flow from the source is high or theoretically infinite if $R = 0$.
- The voltage-drop across either reactance is equal to $E = IX$ and may be greater than the source voltage.
- The circuit acts as a purely resistive (zero-reactance) load to the source and therefore has a power factor of 1.
- The phase of the current and voltage, as seen by the source, is 0°, or in phase.

PARALLEL RESONANCE
- The impedance across the circuit is high or infinite.
- The voltage-drop across the circuit is equal to the source voltage or less.
- The current from the source is low.
- The current through each reactance is equal to $I = E/X$ and will be greater than the source current.
- The circuit acts as a purely resistive (zero-reactance) load to the source and therefore has a power factor of 1.
- The phase of the source current and voltage is 0°, or in phase.
- The phase of the current and voltage of each reactance is 90°.
- Adding a little resistance to either branch lowers the impedance of the whole circuit.

The impedance of a parallel-resonant circuit can be determined in several ways:
1. When both the source voltage and the current are known, the impedance can be found by using Ohm's law:

$$Z_p = \frac{E_s}{I_s}$$

2. The product of the series impedance of both legs of the circuit ($Z_C = \sqrt{R^2 + X_C^2}$ multiplied by $Z_L = \sqrt{R^2 + X_L^2}$), divided by the total *series* impedance of the coil, capacitor, and any resistance in the two branches [$Z_s = \sqrt{R^2 + (X_L - X_C)^2}$], will give the parallel impedance of the circuit. The formula is

$$Z_p = \frac{Z_C Z_L}{Z_s}$$

where Z_p = impedance of parallel circuit
Z_C = impedance of capacitive leg
Z_L = impedance of inductive leg
Z_s = series impedance of the two legs

Since the impedance of a series-resonant circuit equals the resistance in the circuit, the antiresonant formula may also be given as

$$Z_p = \frac{Z_C Z_L}{R_s}$$

where R_s = total resistance of the two legs in series

3. When the reactance value is more than 10 times the resistance of the inductive branch, a simple formula that will give an approximate antiresonant impedance value is

$$Z_p = \frac{X_L^2}{R}$$

4. The circuit may be treated as a parallel ac circuit and the admittance computed by the parallel-circuit formula $Y = \sqrt{G^2 + B^2}$. The reciprocal of the admittance is the impedance.
5. The circuit may be treated as a parallel ac

circuit. Assume a convenient source voltage, compute the currents that would flow in the two legs, and then determine the circuit current by using the formula

$$I_s = \sqrt{(I_{Z_L}\cos\theta + I_{Z_C}\cos\theta)^2 \atop + (\pm I_{Z_L}\sin\theta \pm I_{Z_C}\sin\theta)^2}$$

The impedance is determined by Ohm's law, $Z_p = E_X/I_s$.

Consider the parallel-resonant circuit in Fig. 8-11, which has resistance in series with the

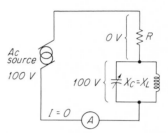

Fig. 8-11 At resonance the voltage-drop across the tuned circuit is maximum; across R, minimum.

100-V source lines. When the capacitor is varied, there should be some value of capacitance at which its reactance to the source frequency is equal to the inductive reactance offered to that frequency by the coil. At this resonant condition, the source current should drop to nearly zero (negligible resistance in the coil and capacitor is assumed). With little current through the resistor R, there will be little voltage-drop across it and almost the full 100-V source voltage will appear across the parallel circuit. The flywheel current circulating between coil and capacitor will be at a maximum.

When the capacitance is varied to any other value, the circuit will no longer be resonant. The reactive currents will no longer cancel each other. The circuit will no longer present as high an impedance to the source ac, and current will flow in the source lines. Current flowing through the resistor R will produce a voltage-drop across it. If there is a 10-V drop across the resistor, the voltage across the resonant circuit will be a little over 90 V. The further the circuit is tuned from resonance, the greater the source current and therefore the less the voltage across the parallel circuit. The less voltage appearing across the parallel circuit, the less flywheel current that will circulate in it. Maximum voltage across a parallel circuit, when resistance is in series with it, will always occur at resonance. Maximum circulating current in the LC circuit will also occur at resonance. (This explains why resonance of the plate tuning circuit of an RF amplifier stage in a transmitter is indicated by minimum plate current.)

Test your understanding; answer these checkup questions.

1. What is the basic formula for antiresonance? _____

2. If the C is removed from an antiresonant circuit, how does this change the I_{X_L} if there is zero line R? _____

3. Does maximum or minimum current flow in the line feeding a resonant circuit? _____ An antiresonant circuit? _____

4. What are electron alternations in an antiresonant circuit called? _____

5. What must be the Z of an antiresonant circuit that draws 0.2 A from a 120-V line? _____

6. If the Z of the C branch of an antiresonant circuit is 40 kΩ, the Z of the L branch is 2 kΩ, and the series Z of the two branches is 5 kΩ, what is the Z of the parallel-resonant circuit? _____

7. A parallel-resonant circuit having 380-Ω reactances and zero R is in series with a 24-Ω R across 120 V. What is the voltage-drop across R? _____ Across the LC circuit? _____ The I value in the LC circuit? _____ The X value across the C? _____

8. What is the antiresonant frequency of a 500-pF C and a 150-μH L with a 10-Ω R in it? _____

9. What C is required to tune a 40-μH coil to a frequency of 8 MHz? _____

10. What L is required to tune a circuit with 0.0005 μF to 450 kHz? _____

11. A variable capacitor has a minimum C of 20 pF and a maximum C of 300 pF. If it is connected across a 100-μH coil, what are the highest and lowest values of f to which it will resonate? _____

12. An antiresonant circuit has an 800-Ω X_L and no R. If it is across 35-V ac, what is the I value in the LC circuit? _____ In the source? _____

8-4 THE Q OF A CIRCUIT

A term often applied to ac circuits in which inductance and capacitance are involved is Q.

The symbol Q can be considered to mean "quality." That is, a coil with no resistance or other losses would be a perfect inductor and would have an infinitely high Q. Since a coil without losses is not possible, the Q of a coil will always have some finite value. The Q of a coil is merely a ratio of coil reactance to coil resistance, or losses. In formula form,

$$Q = \frac{X_L}{R} = \frac{2\pi f L}{R}$$

From the formula, an inductance should have a higher Q when used in a higher-frequency circuit (provided resistance or other losses in it do not increase proportionally).

As an example, at 6 kHz a certain coil has 1,000-Ω X_L and a 20-Ω R. Its Q is X_L/R, or 1,000/20, or 50. At 12 kHz the X_L increases to 2,000 Ω. If the resistance remains essentially the same, the Q should be 2,000/20, or 100.

At higher frequencies, however, electrons flowing in a wire or coil travel nearer the surface of the wire. Thus, only a small portion of the conductor is actually carrying current. The lessening of the usable cross-sectional area results in an effectively higher resistance of the same wire to higher-frequency ac. This increased resistance, known as *skin effect,* is one cause of lower Q in a coil. Skin effect can be decreased by (1) using larger wire, (2) silverplating the wire used, since silver is the best conducting material, (3) using fewer turns but increasing the permeability of the core by employing powdered-iron cores, and (4) employing Litzendraht ("Litz") wire, an insulated multistrand wire. Several thin strands have more surface for a given wire diameter than does a solid wire. (Use of Litz wire is effective only up to about 1 MHz.)

Capacitors also have a value of Q. As with coils, the Q is the ratio of the capacitive reactance to the effective resistance of the capacitor. This resistance may be in the leads, may be skin effect, or may be the result of internal losses of the capacitor. The formula is

$$Q = \frac{X_C}{R} = \frac{1}{2\pi f C R}$$

From the formula, the Q of a capacitor should be halved when it is used at twice as high a frequency.

When resistance is in series with any reactance, an increase in resistance produces a lower Q. When a resistor is connected across a coil or capacitor, however, the effective Q of the circuit will vary directly with the value of the resistance. A high value of shunting resistance means a high Q; a low value of shunting or parallel resistance, a low Q. Thus

$$Q_p = \frac{R}{X_L} \quad \text{or} \quad Q_p = \frac{R}{X_C}$$

The dielectric materials of capacitors normally have extremely high resistance, in the hundreds or thousands of megohms. If the dielectric leaks even a little, however, the ratio of resistance across the capacitor decreases tremendously, lowering the Q of the capacitor as well as the Q of any resonant circuit in which it might be used.

A shunt resistor is often connected across a parallel LC circuit to lower its Q. This makes the circuit less sensitive to any particular frequency. A parallel-resonant circuit with a loading resistor shunted across it to lower the Q is shown in Fig. 8-12. The lower the value of the parallel resistance, the greater the proportion of resistive current flowing in the circuit. With low resistance the

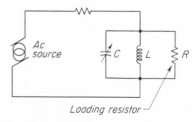

Fig. 8-12 A loading resistor across a tuned circuit lowers Q and broadens the frequency response.

---

source sees the whole LCR circuit more as a resistance than as a resonant circuit and as having a low impedance. Because of the resistance, current of all frequencies will flow from the source to the circuit. The higher the Q of a resonant circuit, the proportionally greater current that will flow in it at the resonant frequency (Sec. 8-6). Q formulas for series- and parallel-resonant circuits are

$$Q_s = \frac{\sqrt{L/C}}{R} \qquad Q_p = \frac{R}{\sqrt{L/C}}$$

Since the Q of a coil is usually much lower than the Q of a capacitor, it is the controlling factor in the Q of LC circuits. Some of the methods of attaining a higher-Q coil are (1) using larger wire, or wire of better conducting material, to reduce skin effect, (2) spacing the wires of the coil by a distance approximately equal to the diameter of the wire used, (3) using a coil with a length slightly greater than its diameter, if a single-layer coil, (4) using low-loss core materials to decrease eddy-current and hysteresis losses, and (5) covering the wire with an insulating material that has low dielectric hysteresis loss. This increases the Q of the capacitance that exists between adjacent turns in a coil, known as *distributed capacitance*.

Generally, the frequency of a tuned circuit is assumed to be determined solely by L and C values,

$$f = \frac{1}{2\pi\sqrt{LC}}$$

However, series resistance has some effect on resonant frequency, as shown by the formula

$$f = \frac{\sqrt{1 - 1/4Q^2}}{2\pi\sqrt{LC}}$$

From this formula it can be reasoned that the resonant frequency is directly proportional to circuit Q. As the Q is increased, the resonant frequency will increase somewhat.

The Q of resonant circuits used in communications may be about 5 to 15 in radio transmitters, 25 to 200 in RF tuned circuits in receivers, or several hundred in specially designed high-Q filter circuits. By using regeneration in vacuum-tube or transistor circuits, it is possible to obtain a resonant circuit with a Q well over 10,000.

8-5 LOGARITHMS AND DECIBELS

One of the important mathematical tools of communications is the *decibel* (*deci* = $\frac{1}{10}$), abbreviated dB. This is a measurement of the ratio of one power to another. An audio amplifier can increase the power of an AF ac signal. If it can increase a 5-W input to a 50-W output, it has increased the signal power 10 times, or 1 *bel,* or 10 dB. The same amplifier will also increase a 0.5-W input to a 5-W output, still a gain of 10 dB. If the volume control is turned up, the 0.5-W input signal may be amplified to a 50-W output. The hundredfold increase is a gain of two 10-times gains, or 20 dB. This will be recognized as a logarithmic increase by those familiar with logarithms. For those not familiar with them, a brief explanation is given here.

In the equation $10^2 = 100$, the 2 is the *exponent,* the 10 may be called the *base,* and 100 the *number*. The exponent 2 can also be called the *logarithm* of the number 100 to the base 10. This is written $\log_{10} 100 = 2$ and is read: "The logarithm to the base 10 of the number 100 is 2." Since the logarithms in general use always employ a base of 10, it is customary to express an equation as merely $\log 100 = 2$ and say, "The log of 100 is 2."

The equation $10^3 = 1,000$ may be expressed: $\log 1,000 = 3$. The logarithm of any number between 100 and 1,000 will have to be between 2 and 3, or 2 plus some decimal fraction. For example, $\log 500$ happens to be 2.6990. A logarithm such as 2.6990 is composed of two parts. The whole number (2.) is called the *characteristic,* and the decimal-fraction part (6990) is called the *mantissa*.

The characteristic is determined by finding between which $\log_{10}$ the number falls, as shown in Table 8-1.

Note that the characteristic is always 1 less than the number of whole digits in the number. The number 435 has a characteristic of 2; 86 has a characteristic of 1; 0.05 has a characteristic of −2; and so on.

Table 8-1

Range of numbers	Characteristic
0.001–0.009999	−3
0.01–0.09999	−2
0.1–0.9999	−1
1–9.999	0
10–99.99	1
100–999.9	2
1,000–9,999	3

The mantissa value, or decimal-fraction part of the logarithm, is found in a table of common logarithms, as in Appendix D.

EXAMPLE: $\log 8,450 = ?$ The characteristic is 1 less than the four digits in the number, or 3 in this case. Therefore, $\log 8,450 = 3. + ?$ The mantissa is found in the tables in the 84 line under the 5 column and is 9269. Thus, $\log 8,450 = 3.9269$.

Check the following logarithms for characteristic and mantissa:

$$\log 23 = 1.3617$$

$$\log 15,500 = 4.1903$$

$$\log 629 = 2.7987$$

Although $\log 0.28 = -1.4472$, the negative characteristic is not usually used. Instead, a value equal to it is employed. Thus,

$$\log 0.28 = 9.4472 - 10$$

Similarly,

$$\log 0.00862 = 7.9355 - 10$$

In this way the negative characteristic does not complicate computations.

The number of decibels change between two power values can be computed by the formula

$$dB = 10 \log_{10} \frac{P_1}{P_2}$$

$$\left(\text{from bel} = \log_{10} \frac{P_1}{P_2} \right)$$

How many decibels gain does an amplifier have if it produces 40-W output with an input of 0.016 W? By the formula,

$$dB = 10 \log \frac{40}{0.016}$$

$$= 10 \log 2{,}500 = 10(3.3979) = 33.9 \text{ dB}$$

The same formula can be used for the following problem:

How much output power will be produced by an amplifier capable of 25-dB gain if fed an input of 0.001 W? From the formula

$$dB = 10 \log \frac{P_{out}}{P_{in}} = 10 \log \frac{P_o}{0.001}$$

$$25 = 10 \log \frac{P_o}{0.001}$$

$$2.5 = \log \frac{P_o}{0.001}$$

This states that the log of the number $P_o/0.001$ is 2.5, or

$$\log \frac{P}{0.001} = 2.5000$$

The 5000 is the mantissa. By searching through the mantissas in the tables, 5000 is found in the 31 line and 6 column. The 2 is the characteristic and indicates three whole numbers in the logarithm. Therefore,

$$\frac{P}{0.001} = 316.0$$

$$P = 316(0.001) = 0.316 \text{ W}$$

For exact decibel computations the logarithmic formulas should be used, but a fairly accurate calculation is possible by applying one or more of the following ratios, particularly when logarithmic tables in books, slide rules or pocket calculators are not available:

1 dB = power gain of 1.26 (26%, approx. $\frac{1}{4}$)
3 dB = power gain of approx. 2
6 dB = power gain of approx. 4

10 dB = power gain of 10
20 dB = power gain of 100

While the power ratios given above are stated as gains, they may also be stated as losses: -3 dB indicates a loss to half power.

The 1-, 3-, and 10-dB approximations can be used in the problem involving the amplifier with 0.001-W input and 25-dB gain:

Since 10 times the power equals 10 dB, the 0.001-W input increased by 10 dB represents 0.01 W. A second 10-dB increase (total of 20 dB) represents 0.1 W. The remaining 5 dB can be computed: 3 db of this represents twice the power of 20 dB, or 0.2 W. One decibel more (total, 24 dB) represents 0.2 + $\frac{1}{4}$ of 0.2, or 0.2 + 0.05, or 0.25 W. One decibel more (total of 25 dB) represents 0.25 + $\frac{1}{4}$ of 0.25, or 0.25 + 0.063, or 0.313 W. Compare this with the 0.316 W obtained by logarithmic computation.

A simpler, "in the ball park" approximation of the last 5 dB would be to say it was slightly less than 6 dB, or a little less than 0.4 W.

Inasmuch as the decibel is used as a unit of measurement in systems in which power decreases to a threshold value before it reaches an absolute zero, it is necessary to make an arbitrary selection of some power value and assign it as "zero" dB. Power values more than the chosen zero level are considered as +dB; those less than the zero level are −dB.

In the past, several reference points have been established as zero levels. Today, 0.001 W, or 1 mW, is being accepted generally as the standard, although 0.006 W is also used. The term *dBm* signifies dB with a zero level of 1 mW. A postfix k indicates kilowatt, and μ or u indicates microvolts as the zero level, or base. The VU (volume unit) is another dB unit having 1 mW as the zero level, but in a 600-Ω impedance circuit (Secs. 12-18, 17-3).

The decibel has been explained as a ratio of two powers. Since power is proportional to both voltage (E^2/R) and current (I^2R), it is possible to compute decibel ratios between input and output voltages or currents. The basic formulas are

$$dB = 10 \log \frac{E_o^2/R_o}{E_i^2/R_i}$$

$$dB = 10 \log \frac{I_o^2 R_o}{I_i^2 R_I}$$

where E_o = output voltage
E_i = input voltage

For vacuum-tube or transistor amplifiers the power values above the fraction line would be the larger output values. For resistive pads or other power-loss networks the larger input values should be used above the fraction line.

In the special case in which the input and output resistance or impedance values are equal, these values cancel mathematically and the formulas are

$$dB = 10 \log \frac{E_o^2}{E_i^2}$$

$$dB = 10 \log \frac{I_o^2}{I_i^2}$$

Rather than work with squared values, it is possible to express the formulas

$$dB = 20 \log \frac{E_o}{E_i}$$

and $\qquad dB = 20 \log \frac{I_o}{I_i}$

As an example, the voltage across a 500-Ω input circuit of an amplifier is 6 V, and the output is 30 V across a 500-Ω load. The gain of the amplifier in decibels is

$$dB = 20 \log \frac{30}{6}$$

$$= 20 \log 5 = 20(0.6990) = 13.98 \text{ dB}$$

Test your understanding; answer these checkup questions.

1. What word does the letter Q stand for? _____ What is the formula for Q of a coil? _____ Of a capacitor? _____
2. Would a coil have a greater Q at 1 or 10 MHz? _____ If the Q is lower at 10 MHz, what is the probable cause? _____

3. Which would normally have the higher Q, a C or an L? _____ Which would have the greater effect on the Q of an LC circuit? _____
4. What is the name of the capacitance between turns of a coil? _____
5. What is the log of 420? _____ 27? _____ 42,600? _____ 0.135? _____ 0.00423? _____
6. An amplifier has a 53-dB gain and a 10-W output. What input signal power does it have? _____
7. A 2-W signal is fed to a 600-Ω resistor network which loses some of the power fed to it. How many dB must it lose to have an output of zero VU? _____
8. If a microphone is rated at −65 dBm, how many dB gain must an amplifier have to bring its output to 0.001 W? _____ 10 W? _____ 18 W? _____
9. An amplifier is fed a 2-mV signal and has a 0.025-V output. How many dB gain does it have if $Z_i = Z_o$? _____
10. An amplifier is capable of 20-dB gain. If the input signal is 0.0045 A, what is the I_o if $Z_i = Z_o$? _____
11. Is it possible to have a dB gain in a voltage step-up transformer? _____ Why? _____
12. A 5-V signal to a 200-Ω input produces what dB gain or loss if the E_o is 20 V across 10,000 Ω? _____
13. The output of an amplifier having a voltage gain of 30 dB is 25 V. What is the E_i if $Z_i = Z_o$? _____
14. What would 3 dBk indicate? _____

8-6 BANDWIDTH (BW)

In electronics there are many cases in which it is desired to limit the passage of ac to one particular frequency or, more practically, to a small band of frequencies. It has been stated (Sec. 8-2) that a series-resonant circuit in series with an ac line will pass current of the frequency of resonance very well but will attenuate (decrease or reduce) current of frequencies either higher or lower. The further from the resonant frequency, the greater the attenuation. A simple graph, or curve, of the response of such a circuit is shown in Fig. 8-13. If this curve represents the response curve of a radio receiver, when the re-

Fig. 8-13 Maximum current flows in a series circuit at resonance; less at other frequencies.

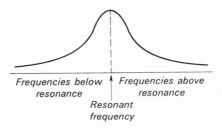

Frequencies below resonance   *Frequencies above resonance*

Resonant frequency

ceiver is tuned from station to station, the curve can be thought of as moving either up or down along the frequency spectrum.

The question arises: "How wide a band of frequencies is being passed by this series-resonant circuit?" By examination of the curve it can be seen that the band of frequencies being passed at absolute maximum may be rather small. A much wider band of frequencies is being passed between half-amplitude points. According to the curve, a very wide band of frequencies is being passed between low-amplitude points. Just where should the bandwidth be measured?

One standard method of measuring bandwidth is to measure the width of either the voltage or the current response curve between points at 0.707 of maximum. Since power is proportional to either voltage squared or current squared, dropping to a 0.707 point is also dropping to a 0.5-power point ($0.707^2 = 0.5$), equivalent to a loss of 3 dB. Thus, bandwidth is normally measured between half-power points, or -3-dB points.

The higher the Q of a resonant circuit, the narrower its bandwidth. This is expressed by the half-power bandwidth formula

$$BW = \frac{f_o}{Q}$$

where f_o = frequency of resonance

Figure 8-14 illustrates the curve of a circuit resonant at 1,000 kHz. The two points on the curve that are 0.707 of maximum are at 985 and 1,015 kHz, respectively. In this circuit the bandwidth is 30 kHz. If the same circuit were tuned to resonate at 1,020 kHz, its bandwidth would remain essentially the same, 30 kHz. To represent the circuit the curve would have to be moved to the right along the page about $\frac{3}{8}$ in. When a

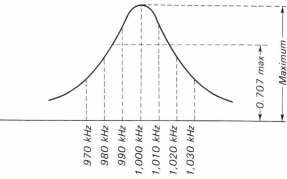
Fig. 8-14 Curve of a 30-kHz bandwidth circuit.

receiver is tuned, its response curve is being shifted up or down the frequency spectrum.

One method employed to express bandwidth, particularly when using multiple-tuned circuits, is to state the bandwidth for a certain number of decibels of loss. For example, a broad tuning receiver may be said to be 25 kHz wide 20 dB down. This means that signals $12\frac{1}{2}$ kHz higher or lower than the resonant frequency will have only one-tenth (-20 dB) as much voltage or current as signals at the resonant frequency.

8-7 BANDWIDTH OF TRANSFORMERS

An important circuit is the tuned transformer, in which both primary and secondary are tuned to the same frequency. The bandwidth of a tuned transformer will vary with degree of coupling, basic Q of both primary and secondary circuits, and load on the circuit. Determination of the bandwidth of such circuits is complex. Only a few fundamental ideas will be presented.

In Fig. 8-15 if all the lines of force from the

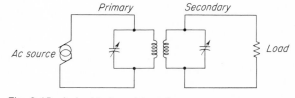
Fig. 8-15 A double-tuned transformer.

primary cut all the secondary turns, the condition of *unity coupling* exists and the coefficient of coupling k equals 1 (Sec. 5-8). In such a condition the transformer will pass practically all frequencies equally well and represents the extremity of broadness. Tuning the circuits would have very little effect on the frequency response.

In a more practical form, if the coefficient of coupling is equal to, possibly, 0.001, with each circuit tuned to the same frequency, and assuming an equal Q for both primary and secondary circuits, the frequency response of the circuit will follow the *universal resonance curve* (Fig. 8-16).

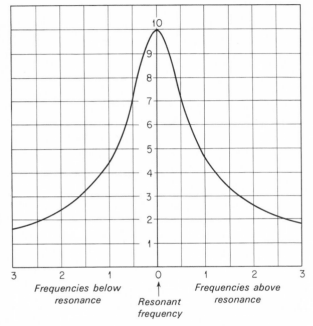

Fig. 8-16 Universal resonance curve. Current flowing in series- or parallel-tuned circuits, or impedance across a parallel circuit as resonance is approached.

This is the relative frequency response curve of series-resonant circuits (when the Q is over 5), or of parallel-resonant circuits, or of two tuned and coupled circuits.

The universal resonance curve illustrates, for example, the current amplitudes in a series-resonant circuit if a variable frequency is applied to the circuit. When the applied ac is 2 frequency units below resonance, the circuit will pass only about one-fourth the current that it will when the frequency is changed to resonance. If the frequency is raised 2 units above resonance, the current flow in the circuit will drop to about one-fourth the resonant-frequency value.

If the Q of the circuit is raised, the peak-amplitude current can be developed with less source voltage and the response at the quarter-current point will be less than 2 units from resonance. If the Q is doubled, the quarter-current frequency will be 1 unit above and below resonance for this particular circuit and the bandwidth is halved. The curve for the high-Q circuit will still have the same relative shape, however.

This same curve also indicates the relative rise in impedance for a parallel-resonant circuit when the circuit is subjected to a variable frequency. At resonance the peak impedance is reached.

A transformer with both primary and secondary tuned to the same frequency has two resonant circuits, each one attempting to select its resonant frequency and reject all other frequencies. The response curve for such a circuit, with a low coefficient of coupling, will be the product of the two curves separately. What would have been the quarter-amplitude point for one circuit alone will be $\frac{1}{4} \times \frac{1}{4}$, or $\frac{1}{16}$, the response with both circuits. The two tuned circuits (with similar Q's), when loosely coupled, produce a bandwidth approximately one-fourth that of either coil alone.

As coupling is increased, the curve of the primary current begins to dip at the resonant frequency due to a resistive effect coupled into it from the secondary (Fig. 8-17).

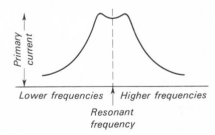

Fig. 8-17 Primary current dips at resonance if coupling is tight.

As soon as the primary-current dip occurs, the curve of the secondary current no longer continues to rise with increased coupling as sharply as when the primary had no dip and the curve of the secondary current begins to flatten at the top (Fig. 8-18).

As the coupling between primary and secondary is increased, the top of the secondary-current curve flattens more and more until an increase

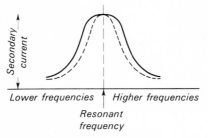

Fig. 8-18 Secondary-current curve flattens as coupling increases.

in coupling no longer raises the secondary-current peak amplitude. This value of coupling is known as *critical coupling*. The coefficient of coupling *k* may perhaps be in the region of 0.01. The bandwidth at critical coupling is wider than when the two circuits were loosely coupled, but the amplitude is maximum. If the two coils have high *Q* values, the bandwidth may still be relatively narrow. For minimum bandwidth, the coils are coupled to less than the critical degree. In many applications a flattened-peak response is desired. In these cases the two circuits will be coupled to the critical value or slightly over.

When coupling is increased past the critical point, the peak current in the secondary at resonance drops and two peaks appear, one on either side. The two circuits are now said to be over-coupled, and a secondary-current curve may result as in Fig. 8-19. The peaks of the two "shoul-

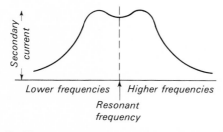

Fig. 8-19 Tight coupling develops two shoulders on the secondary-current curve, broadening the response.

ders'' remain at substantially the same amplitude as at critical coupling, but the bandwidth of the curve is considerably broader. This double-humped response, sometimes called *split tuning,* is the result of overcoupling or detuning of primary or secondary.

8-8 FILTERS

Many circuits in radio and electronics use *filters*. A filter may be considered to be a combination of capacitors, coils, and resistance that will allow certain frequencies to pass through or be impeded.

The transformer with a tuned primary and a tuned secondary forms a type of *bandpass-filter circuit*. It passes the frequency to which it is tuned and a few adjacent frequencies but attenuates those higher and lower. The *shape factor* of a bandpass filter is the ratio of its bandpass 60 dB down from the midband value to its bandpass 6 dB down. The steeper the skirts of its curve the smaller the shape factor.

A *bandstop filter* is used to attenuate one frequency or a small band of frequencies and pass all others. A *wave trap* is an elementary form of bandstop filter. It can be used in the antenna or other circuits of a receiver to prevent undesired signals from interfering with a desired signal. Figure 8-20 shows two wave traps, one a parallel-

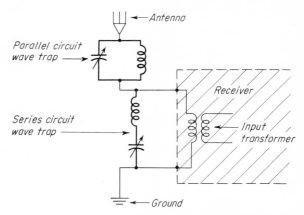

Fig. 8-20 Two possible positions for wave traps at the input (antenna-to-ground) circuit of a receiver.

resonant circuit in series with the antenna and the other a series-resonant circuit across antenna to ground of a receiver. The parallel-resonant circuit offers very high impedance to the frequency to which it is tuned and relatively low impedance to all other frequencies. This high impedance reduces current flow at the frequency of resonance into the receiver input winding. This one frequency is not received. The series-resonant circuit offers low impedance to any signals at its resonant frequency, effectively short-circuiting the antenna-to-ground circuit of the receiver at this frequency and preventing this frequency from being received. Usually only a single wave trap is used, but for greater attenuation of an interfering signal both may be used.

Two other basic types of filters are *low pass* and *high pass*.

An example of a low-pass filter using a single coil in series with a line between a source and load and a capacitor across the line is shown in Fig. 8-21. The inductive reactance of the coil will

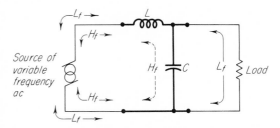

Fig. 8-21   A single-section constant-*k* low-pass filter between a source and load.

oppose higher frequencies more than it will oppose low frequencies. The capacitive reactance of the capacitor presents a better path across the line for higher frequencies than does the load. In this way, both reactances are acting to attenuate the high frequencies but allow the low frequencies to pass to the load.

This low-pass filter is one of the simplest of the group known as *constant-k filters*. They derive their name from the fact that the product of the X_L times the X_C is constant at all frequencies. For example, at a certain frequency X_L may be 400 Ω and X_C may be 100 Ω. The product of the two is 40,000. At twice the frequency X_L will be 800 Ω and X_C will be 50 Ω. The product is still 40,000. The constant k is 40,000.

A coil in series with a line will not oppose dc but will progressively decrease the ac flowing through it as the frequency is increased. However, a constant-*k* low-pass filter will pass all frequencies up to what is known as the *cutoff frequency* and then the C and L begin to attenuate all frequencies above this, as shown in Fig. 8-22.

The capacitance and inductance values for the single-section low-pass constant-*k* filter of Fig. 8-21 can be determined by the formulas

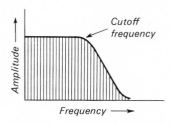

Fig. 8-22   Low-pass filter frequency response curve.

$$ L = \frac{R}{\pi f_c} \qquad C = \frac{1}{\pi f_c R} $$

where L = inductance, in H
  C = capacitance, in F
  R = impedance, in ohms, of both source and load
  f_c = cutoff frequency

Simpler filter theory assumes source and load impedances to be equal. When source and load impedance values are not equal, the filter may not have the desired frequency characteristics.

The higher the Q of the reactances used, the sharper the cutoff. When still sharper cutoff characteristics are required, two or more sections are necessary.

A simple constant-*k* high-pass filter is shown in Fig. 8-23. The capacitor in series with the line

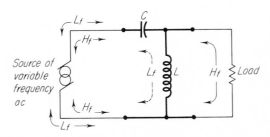

Fig. 8-23   A single-section constant-*k* high-pass filter.

passes the high frequencies to the load, while the coil across the line forms a better path than does the load for low frequencies, preventing low frequencies from appearing in the load. The result is the passage of the high frequencies only, as indicated in Fig. 8-24. The C and L values for the single-section high-pass constant-*k* filter of Fig. 8-23 can be determined by the formulas

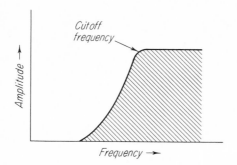

Fig. 8-24 High-pass filter frequency response curve.

$$L = \frac{R}{4\pi f_c} \qquad C = \frac{1}{4\pi f_c R}$$

where L = inductance, in H

C = capacitance, in F

R = impedance, in ohms, of both source and load when equal

f_c = cutoff frequency

The constant-k filters shown are "unbalanced." The filter in Fig. 8-25 is balanced because the

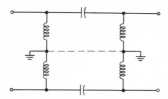

Fig. 8-25 A balanced constant-k high-pass filter circuit.

filter elements are similar on both sides of the line and because the center of the reactances across the line can be connected to ground, as indicated.

Low-pass filters are used in electronic power supplies to pass dc but not variations of current or voltage. They may be used in voice-frequency circuits where only frequencies up to perhaps 3 kHz are to be passed. They can be employed between a transmitter and an antenna to prevent frequencies higher than the desired frequencies (such as harmonics) from appearing in the antenna.

High-pass filters with the proper cutoff frequency can be used in audio-frequency circuits or between a TV receiver and its antenna to prevent nearby lower-frequency signals from interfering with TV reception.

When sharp cutoff is desired, the constant-k filter is not practical. Instead, an *m-derived* filter is used. This filter can be recognized by its parallel- or series-type resonant circuit in series with or across the line. A series-resonant wave-trap circuit across the line or a parallel-resonant wave-trap circuit in series with the line produces essentially infinite attenuation of the frequency to which it is tuned and therefore zero transmission of that frequency along the line. (Filters with series-resonant elements in *series* with a line or parallel-resonant elements *across* the line do not attenuate their frequency of resonance and would fall in the constant-k category.)

Figure 8-26 shows two simple low-pass *m*-de-

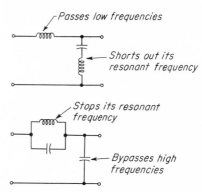

Fig. 8-26 Two possible single-section *m*-derived low-pass filter configurations.

rived filter circuits. The *m* may be considered to be a ratio of the cutoff frequency to the frequency of infinite attenuation (zero output). In a low-pass filter *m* will be something between 1 and 0. An *m*-derived low-pass filter with an *m* value of 1 has the same attenuation or transmission curve as a constant-k filter and has no point of infinite attenuation. In such a case, the tuned circuit involved has either negligible capacitance or negligible inductance. As the *m* value decreases, a point of infinite attenuation develops. As the *m* value decreases further, the point of infinite attenuation begins to approach the cutoff frequency, producing an increasingly steep attenuation (Fig. 8-27). The lower the *m* value, however, the less the attenuation of frequencies beyond the point of infinite attenuation. With an *m* value approaching zero, the untuned filter element has negligible effect and the filter operates as a wave trap, with

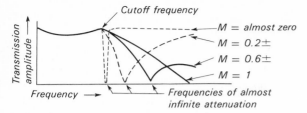

Fig. 8-27 Frequency response curves of an *m*-derived low-pass filter with different *m* values.

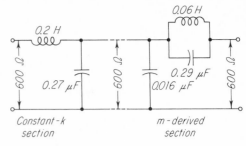

Fig. 8-30 Low-pass composite filter with one constant-*k* and one *m*-derived section.

little attenuation of frequencies except the resonant frequency. In practice an *m* value of about 0.6 is a good compromise between steep cutoff and transmission past the point of infinite attenuation.

Two single-section high-pass *m*-derived filter circuits are shown in Fig. 8-28.

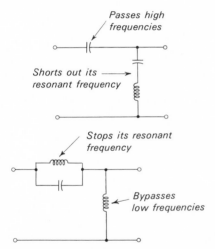

Fig. 8-28 Two single-section *m*-derived high-pass filters.

Practical filters are made up in sections. There are three basic configurations, the T, the L, and the π types, named for their appearance when diagramed (Fig. 8-29). A balanced-T type might be termed an H type.

A composite filter consists of two or more separate sections coupled together. Figure 8-30 shows a constant-*k* L section coupled to an m-derived π section to form a low-pass filter with maximum attenuation at about 1,000 Hz and operating from a 600-Ω source into a 600-Ω load. It is important that the separate sections match each other in impedance to prevent reflection of mismatches back to the opposite terminals. Figure 8-31 shows

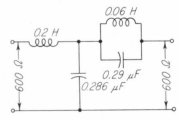

Fig. 8-31 Same low-pass filter as in Fig. 8-30 as it would actually be constructed.

the filter as it might actually be constructed. See also Sec. 17-35 for information on sideband filters.

For further information the reader is referred to the author's text *Electrical Fundamentals for Technicians* or to radio or electronic engineering handbooks.

Fig. 8-29 Basic filter-circuit configurations.

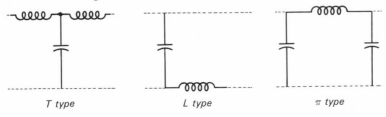

Test your understanding; answer these checkup questions.

1. At what amplitude is bandwidth (BW) of a tuned circuit usually measured? _____ What is another method of expressing BW? _____
2. Why would it be useless to try to tune a power transformer? _____
3. What would be the BW represented by Fig. 8-16 if the frequencies were in kHz? _____ In MHz? _____
4. Does the curve of Fig. 8-16 represent I versus f, E versus f, Z versus f, or all of these? _____
5. Which dips first with increased coupling, I_p or I_s? _____
6. Maximum BW without secondary dip is produced by what degree of coupling? _____
7. What are the four basic types of filters? _____ _____ _____ _____

8. Of what is a wave trap made? _____ What basic type of filter is it? _____
9. What would be the C and L values of an unbalanced single-section constant-k 500-Hz low-pass filter for use in a 1,000-Ω line? _____ _____ For a similar HP filter? _____ _____
10. What type of filter would be used between an antenna and a TV receiver to prevent interference with the picture by local lower-frequency radiations? _____ By local higher-frequency radiations? _____
11. What type of filter gives sharper cutoff response than a constant-k type? _____
12. What is the m value of a high-Q wave trap? _____ Of a constant-k LP filter? _____
13. What are the four basic configurations found in LCR filters? _____ _____ _____ _____
14. What is the shape factor of a filter with a bandpass of 2.8 kHz 6dB down if its bandpass is 6.5 kHz 60 dB down? _____

COMMERCIAL LICENSE QUESTIONS

Amateur license questions will be found in the Addendum.

FCC Elements 3, 4, and 6 require an ability to answer questions similar to those below. Sections in which questions are answered are shown in parentheses. A question followed by a bracketed number is required for that element alone.

1. State the formula for determining the resonant frequency of a circuit when the inductance and capacitance are known. (8-1)
2. What changes in circuit constants will double the resonant frequency of a resonant circuit? (8-1)
3. In a parallel circuit composed of an inductance of 200 μH and a capacitance of 200 pF, what is the resonant frequency? (8-1)   [4]
4. What value of capacitance must be shunted across a coil having an inductance of 50 μH in order that the circuit will resonate at 4,000 kHz? (8-1)   [4]
5. If a parallel circuit, resonant at 2,000 kHz, has its inductance halved and capacitance doubled, what will be the resultant resonant frequency? (8-1)
6. What is the impedance of a series circuit at resonance if it is composed of pure reactances? (8-2)
7. What is the value of total reactance in a series-resonant circuit at the resonant frequency? (8-2)
8. A series-resonant circuit has 10-Ω R and reactances of 5 Ω. What is its impedance at (a) twice the frequency and (b) half the frequency? (8-2)
9. What is the impedance of a parallel-resonant circuit if it is composed of pure reactances? (8-2)
10. What effect will a small amount of resistance in the capacitive branch of a parallel-resonant circuit have on the total impedance? (8-3)
11. Assume an inductance of 4 H in parallel with a capacitance of 2 μF. If there is no resistance in either leg of this circuit, what is the impedance of the network at resonance? (8-3)
12. What is the value of reactance across the terminals of the capacitor of a parallel-resonant circuit at the resonant frequency if zero resistance in both legs of the circuit is assumed? (8-3)
13. Under what conditions will the voltage-drop across a parallel-tuned circuit be a maximum? (8-3)
14. Given a series-resonant circuit consisting of a 10-Ω R and equal reactances of 150 Ω, what is the voltage-drop across the inductance when the applied circuit potential is 150 V? (8-3)
15. What is *skin effect?* How does it affect conductor resistance at higher radio frequencies? (8-4)
16. What is the Q of a circuit, and how is it affected by the circuit resistance? (8-4)
17. How may the Q of a parallel-resonant circuit be increased? (8-4)
18. What effect does a loading resistance have on a tuned RF circuit? (8-4, 8-7)   [4]
19. What is (a) a *bel,* (b) a *decibel?* (8-5)
20. What is the formula for determining the decibel loss or gain of power in a circuit? (8-5)
21. What is the formula for determining the decibel loss or gain of voltage in a circuit? (8-5)
22. The output of an amplifier stage having a voltage

gain of 20 dB is 15 V. What is the input-voltage level? (8-5)
23. How does the Q of a circuit affect bandwidth? (8-6)
24. What is the purpose of a wave trap in a receiver? (8-8)
25. Draw a simple schematic diagram of a wave trap in an antenna circuit for attenuating an interfering signal. (8-8)
26. In general, why are filters used? (8-8)
27. Why are bandstop, high-pass, and low-pass filters used? (8-8)
28. Draw schematic diagrams of the most commonly used filters. (8-8)
29. Explain the function of an m-derived filter. (8-8)
30. Draw a circuit diagram of a low-pass filter com-

posed of a constant-k and an m-derived section. (8-8)

NOTE: FCC license test questions are multiple-choice types, taking a form somewhat as follows:

31. Changing a 1-μF capacitor to 0.25 μF has what effect on the resonant frequency of a resonant circuit? (a) Increases it four times; (b) quarters it; (c) doubles it; (d) has no effect.
32. The Q of an antiresonant circuit is increased by (a) using a larger capacitance value; (b) decreasing resistance across it; (c) using a smaller inductance value; (d) decreasing the load on it.
33. A gain of 1,000,000 times in power is expressed by (a) 60 dB; (b) 30 dB; (c) 20 dB; (d) −30 dB.

9 VACUUM TUBES

The objective of this chapter is to outline the basic fundamentals of vacuum tubes, with special emphasis on high-power-transmitting types. Such parameters as amplification factor, bias, plate impedance, mutual conductance, power output, and plate dissipation are discussed. Triodes, tetrodes, pentodes, and special tubes, as well as some causes of tube failures, are presented.

9-1 VACUUM TUBES

Basically, there are two types of components used in electronic circuits. *Passive* devices, such as resistors, inductors (including transformers), and capacitors are considered not to alter their resistance, impedance, or reactance when ac signals are applied to them. *Active* devices, such as vacuum tubes, transistors, and diodes, do change their resistance or impedance when varying voltages are applied to them, and as a result can amplify and rectify and may modify or distort ac waveforms. Passive devices do not normally distort waveforms.

Prior to 1920, active devices in use were magnetic coherers, crystal detectors, spark oscillators, and arc transmitters, but by 1927 the vacuum tube had made them obsolete. By 1950 the

solid-state diodes and transistors began to replace vacuum tubes. In the early 1960s groups of diodes and transistors, called integrated circuits, or ICs, were constructed on a single tiny chip of silicon or germanium. This has resulted in electronic circuits continually becoming smaller and smaller. However, for amplification of high-power, high-frequency ac, vacuum tubes are still the only active devices that can be used, and no electronic training should overlook them.

In this chapter the physical construction and electrical operation of vacuum tubes (VT) will be described. Some of the many applications of vacuum tubes (and transistors) in electronic circuits will be dealt with in subsequent chapters.

9-2 CATHODES

A *cathode* is something capable of emitting electrons. For example, in a light bulb a piece of wire carrying an electric current becomes warm. If sufficient current flows, the wire becomes white-hot. Electrons whirling around the outer atoms of the hot wire move so rapidly that some of the less tightly bound ones fly outward, away from the wire into the surrounding space. This produces a cloud of free electrons around the wire as long as it is white-hot. The wire has become a cathode. The *filament* wire of a light bulb, when heated by current flowing through it, emits a radiation called light, another radiation called heat, and also the unseen cloud of electrons sprayed outward from the filament, as shown in Fig. 9-1a. If filament-wire atoms lose electrons, they are left with a positive charge. The electrons

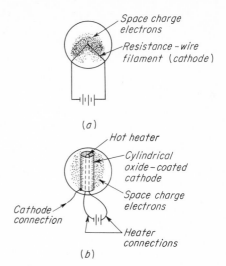

Fig. 9-1 Electron emission around (*a*) a directly heated cathode and (*b*) an indirectly heated cathode.

sprayed outward are attracted back toward these positively charged atoms. This results in a constant out-and-back movement of electrons in the area directly surrounding the wire. These temporarily loose, negatively charged electrons form a *space charge*. When the wire loses its heat, the space-charge electrons return to the wire.

The space charge has no practical use in a light bulb, but it is highly important, in vacuum tubes. The central element in vacuum tubes is a cathode. To allow space-charge electrons to travel unimpeded through a tube, all the air inside it is pumped out. In this evacuated space, with no air molecules to get in the way, the space-charge electrons can move freely to any positive element in the tube.

All early tubes had wire filaments that produced the electron emission when heated. Many high-power-transmitting tubes still use filament-type cathodes, but most tubes now employ *heater-cathodes*. A heater-cathode consists of a metal cylinder coated with special oxides that liberate great quantities of electrons when heated to a relatively low temperature. Down the center of the cylindrical, coated cathode runs an insulated heater wire, as in Fig. 9-1*b*. When current flows through the heater wire, the cylinder surrounding it heats and the external oxide coating on the outside of the cathode cylinder emits electrons.

The filament-wire type of cathode heats and

cools rather rapidly. As a result, when a 60-Hz ac is used to heat the filament, the temperature of the wire varies during each alternation, resulting in a periodic variation in the number of emitted electrons 120 times per second. Because a constant emission is usually required in vacuum tubes, this electron variation is undesirable. Since it takes 10 to 15 s for the cathode surface of a heater-cathode to heat to the point of liberating electrons, the small, rapid changes in temperature of the insulated heater wire, due to an ac flowing through it, have no effect on the surface temperature of the cathode sleeve and a constant number of electrons is emitted from the cathode surface. The symbols used to designate a filament-type ("directly heated cathode") tube and a heater-cathode type of tube are shown in Fig. 9-2. The

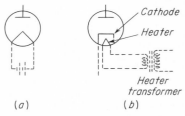

Fig. 9-2 Symbols used to indicate (*a*) directly heated and (*b*) indirectly heated cathodes.

second element in the tubes is a plate, or anode. The heater-cathode tube shows the heater circuit connected to an ac source. The cathode cylinder is indicated by the caplike symbol over the heater.

Receivers using heater-cathode tubes take 10 to 15 s to start working after being turned on because the cathode must reach an operating temperature. Receivers using filament-type tubes operate the instant the switch is turned on. Electrical insulation between the heater wires and the inner surface of the metallic cathode sleeve prevents electron emission from the heater from reaching the cathode. Such a heater-cathode current would interfere with proper operation of many circuits. If the insulation between the heater and cathode does break down, the tube is said to have a heater-cathode leak. The heater insulation will usually stand about 100 V, although some tubes are made to withstand 450 V or more without breaking down.

Alternating current is used to heat the filaments whenever possible because of its more general availability. Direct current from batteries may be quite costly.

9-3 FILAMENTS

Filaments in directly heated vacuum tubes are of three types:

1. *Nickel wires with oxide coating.* The nickel wire is heated, causing the oxide coating on the wire to emit electrons. Such filaments are heated to a red or orange-red color (880°C). When the oxide coating flakes off, the filament emission becomes very low and the tube is normally unusable.
2. *Tungsten wires.* Tungsten wire requires a relatively high, yellow or white, operating temperature (2300°C). Even at this temperature its emission is relatively small. Tungsten filament tubes require considerable power to heat them and are relatively rugged. They are used in tubes where very high plate voltages (thousands of volts) are used.
3. *Thoriated-tungsten wires.* Thorium is an excellent electron emitter, but wires made of this material are not strong enough for practical use. Mixing thorium with the more rugged tungsten produces a thoriated-tungsten wire that will emit many electrons at lower orange temperatures (1600°C). Filaments of this type are used in many transmitting tubes.

In some cases, thoriated-tungsten filaments may be rejuvenated when the outer thoriated layer loses too many thorium molecules. This is accomplished by first removing the plate voltage from the tube and then applying a filament voltage of two to three times the normal value for a few seconds. If this does not burn out the filament, the voltage is brought down to about 1½ times normal and the filament is run for 10 to 20 min. This process boils thorium atoms to the surface of the tungsten wire, and the filament may emit electrons almost as well as when new.

Generally, with plate voltages up to about 500 V, the filament or cathode will be oxide-coated. With plate voltages from about 500 to 6,000 V the filaments will usually be thoriated tungsten. Above 6,000 V pure tungsten may be most satisfactory.

When a filament wire is heated, it emits electrons; but some molecules of the wire may also boil off. These molecules fly out into the tube and deposit themselves on the inside of the glass envelope or on the other elements of the tube. Eventually, if one point in the filament wire becomes so thin that it melts apart, the tube is "burned out." Old tubes often have a dark deposit of filament molecules at the top of the glass envelope.

9-4 CORRECT FILAMENT VOLTAGE

It is important to use the correct voltage across the filament or heater of a tube to obtain optimum tube life. If a tube is rated at 6.3 V, it will have a certain life expectancy. If operated at 7 V, it will have a greater filament emission because of increased filament current, but the temperature of the filament wire will be so high that molecules of the filament wire or of the oxide coating will rapidly boil off, greatly shortening the life of the tube. On the other hand, lower filament voltage will produce less emission but may increase the life expectancy of tubes under certain circumstances.

9-5 THE ANODE, OR PLATE

The space inside the outer glass or metal shell (envelope) of a vacuum tube is evacuated until no air or gas is left inside. When a piece of metal called either a *plate* or an *anode* is placed inside the evacuated space, it will pick up electrons given off by the hot cathode. If the plate is connected to the cathode through a meter, as in Fig. 9-3, the meter will register a small *contact potential* current. When the cathode loses its heat, current in this *plate circuit* stops flowing.

If the plate is made only a few volts negative

Fig. 9-3 An uncharged plate collects some emitted electrons, and a very small plate-circuit current flows.

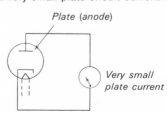

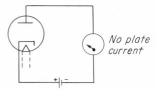

Fig. 9-4 No current flows if the plate is negatively charged in respect to the cathode.

in respect to the cathode by adding a battery in the plate circuit, as shown in Fig. 9-4, this negative charge on the plate will repel space charge electrons from the cathode and no current will flow in the circuit.

If a battery is placed in the circuit with positive polarity to the plate, as in Fig. 9-5, the plate-circuit

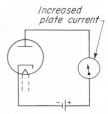

Fig. 9-5 Increasing the positive charge on the plate increases the plate current.

meter will show an increase in plate current. The higher the positive potential on the plate in respect to the cathode, the greater the plate current. The plate current in vacuum tubes is essentially directly proportional to the positive voltage applied to the plate as shown in the graph of Fig. 9-6, called an E_pI_p curve. It can be read as follows:

Fig. 9-6 Plate voltage versus plate current (E_pI_p) curve.

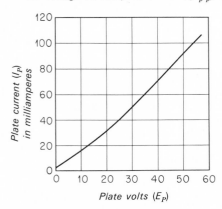

At zero volts on the plate, the current is less than 2 mA. At 10 V on the plate the current is 17 mA. At 20 V, the plate current is 31 mA. At 30 V the I_p is 50 mA. With an E_p of 40, the I_p is 71 mA, etc. The line will curve upward continually until it approaches a saturation point, when it flattens off. *Saturation* is reached when the plate is accepting all the electrons given off by the cathode.

9-6 THE PLATE LOAD

In all working circuits involving vacuum tubes a *load* will be connected somewhere in the plate circuit. The load may be a resistor, across which a varying dc voltage will be produced when the current through it varies. The load may be an inductance (a ''choke'' coil), across which an alternating voltage will be developed when the current through it is varied. The load may be the primary of a transformer, in the secondary of which an ac voltage will be induced when the primary current varies. The load may be a meter, a relay, a loudspeaker, or a pair of earphones. It is always in or across the load that the result of the operation of the tube is made apparent.

9-7 PLATE VOLTAGE

The difference in potential between the plate and the cathode of a vacuum tube is the *plate voltage*. Figure 9-7 shows a diode (two-element) tube with a plate voltage of 90 V.

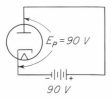

Fig. 9-7 With no resistance in the plate circuit, the plate voltage equals the supply voltage.

In Fig. 9-8, in addition to the 90-V battery, the plate circuit contains a load resistance. The current through the resistance produces a voltage-drop across it. If the voltage-drop across the load is 85 V, the plate voltage must be 5 V. In this case there is a power-supply voltage of 90 V, a load voltage of 85 V, and a plate voltage of 5 V.

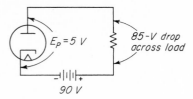

Fig. 9-8　Plate voltage is the difference between supply voltage and voltage-drop across the load.

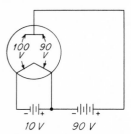

Fig. 9-9　The plate voltage is the supply voltage plus half the filament voltage.

In the filament tube of Fig. 9-9 there is a 90-V difference of potential between one side of the filament and the plate and 100 V between the other side of the filament and the plate. In this case the plate voltage will be the average of 90 and 100, or 95 V.

If the filament battery has its polarity reversed, as in Fig. 9-10, there will be 90 V between one

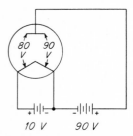

Fig. 9-10　The plate voltage is the supply voltage minus half the filament-voltage value.

end of the filament and the plate and 80 V between the other end and the plate. The average plate voltage is now 85 V.

If a 10-V peak ac is used to heat the filament as in Fig. 9-11, the average plate voltage will vary from 95 to 85 V at the frequency of the ac. This will cause any plate current to vary at this same frequency.

Usually a constant plate-circuit supply voltage

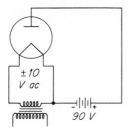

Fig. 9-11　This ac filament source would produce a varying plate voltage.

is desired in vacuum-tube circuits. With dc for both filament and plate supply there is no reason for the plate voltage to vary. To reduce plate-voltage variations and still use ac to heat the filament, the filament circuit may be center-tapped. The plate-circuit power supply is then connected to the center-tap point, as shown in Fig. 9-12.

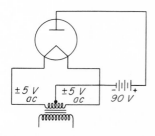

Fig. 9-12　Center-tapping the ac filament source results in a constant plate voltage and current.

If the output of the transformer is 10 V, at the instant of maximum ac voltage there will be 5 V between the center tap and either filament terminal. One terminal will be +5 V; the other will be −5 V; and the average E_p will be 90 V. The plate current will not be varied by the ac filament supply. Center-tapping the filament supply is a requirement for practically all tubes using ac on their directly heated cathodes.

An alternate method of center-tapping the filament circuit is shown in Fig. 9-13. In this case, the filament circuit has a center-tapped 10- to 50-Ω resistor connected across it.

Ac-operated filaments, however, heat and cool slightly every *alternation*. This results in a variation of electron emission as the current drops to zero and rises to maximum. With 60-Hz ac, the plate-current variation will have a "ripple" frequency of 120 Hz.

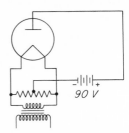

Fig. 9-13 Shunt resistor method of center-tapping an ac filament source.

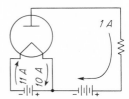

Fig. 9-14 A dc-heated filament has more emission from one end of the filament wire than from the other.

9-8 REVERSING FILAMENT POLARITY

Figure 9-14 shows a directly heated diode with dc as the heating current. For convenience, the plate current will be assigned 1 A and the filament current 10 A. The 1 A of I_p flowing through the filament battery results in 11 A at the negative end and 10 A at the positive end of the filament wire. One ampere is given off as an electron emission by the filament. Since the negative end of the filament is carrying more current, it is heated more and tends to wear out first. For this reason, high-power tube filaments heated by dc require periodic reversal of the filament-battery polarity to prolong filament life. If ac is used to heat the filament, this problem does not exist.

9-9 DIODE TUBES AND RECTIFICATION

A diode tube has a cathode that emits electrons and an anode, or plate, that collects electrons when it is positive. A diode may have a filament and plate or a heater-cathode and plate.

Diodes act as one-way conductors. If the plate is made positive with respect to the cathode, as in Fig. 9-15, it attracts space-charge electrons

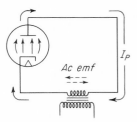

Fig. 9-15 An alternating emf in the plate circuit of a diode produces a pulsating dc plate current.

and current flows in the plate circuit. If the source of plate voltage in the plate circuit is a transformer, the plate voltage is alternating. During one half cycle the plate end of the transformer-secondary winding may be positive and the cathode end negative. On the next half cycle the plate end will be negative and the cathode end positive. While the plate is positive, electrons are attracted from cathode to plate, through the transformer secondary, and back to the cathode. Current flows throughout the plate circuit. When the plate is charged negatively, space-charge electrons are repelled and no plate current flows. (Being cold, the plate cannot emit electrons.)

A diode tube acts as a one-way gate. It is said to *rectify* ac; that is, it allows only one half cycle of the emf to produce current in the circuit. With ac plate voltage the I_p is pulsating dc, as shown in Fig. 9-16b. This one-way-gate effect is the main use of diodes. A diode is often referred to as a rectifier.

In diagrams showing some diodes, the heater circuits may not be completed. Figure 9-15 is an example of this abbreviated diagraming of a heater-cathode diode. Figure 9-17 shows an ab-

Fig. 9-16 A diode allows only one-way pulses of dc to flow if the source emf is alternating.

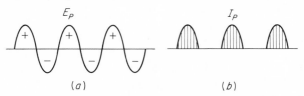

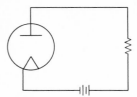

Fig. 9-17 Diagram of a filament-type diode. The filament supply is often omitted to simplify the diagram.

breviated diagram of a filament tube. It is assumed that the reader understands that the heater or filament is connected to a transformer or battery.

9-10 THE VACUUM IN A VACUUM TUBE

Any molecules of air or gas of any kind left inside a vacuum tube after it has been evacuated will interfere with the free movement of electrons from the cathode to the anode (or to any other elements in the tube). The manufacturer exhausts all the gas possible by mechanical pumping and then seals the outer glass or metal envelope of the tube. Inside are all the elements of the tube plus a small piece of chemically active metal, such as magnesium, called a *getter*. By induction, the elements and the getter are heated. The getter explodes and combines chemically with molecules of gas that are released by heating of the metal elements, or that are still in the tube, depositing part of itself in combination with the gas molecules on the inside of the envelope. From the outside this deposit appears as either a mirrorlike or a darkened area.

A high degree of vacuum is known as a *hard* vacuum. A vacuum tube which is gassy for any reason is known as a *soft* tube. Tubes become soft because of leaks through the glass or metal envelope at sealed edges or because of unintentional overheating of the elements, which may liberate some of the gas molecules that are normally present in metals.

Soft vacuum tubes usually show a faint purple or blue haze between elements. The intensity of this glow varies directly with the plate current flowing through the tube. Some tubes, when operated at high plate voltages, produce a blue glow where electrons strike the inner surface of the glass envelope. This does not indicate the pres-

ence of gas but is the conversion of electric energy into light energy as the electrons strike some of the getter or actinic glass molecules on the inner surface of the glass envelope.

Test your understanding; answer these checkup questions.

1. What are devices which change their impedance when varying voltages are applied to them called? _____
2. What are the names of two vacuum-tube elements which give off electrons? _____
3. Where in a vacuum tube does a space charge form? _____ What is its polarity? _____
4. What is the purpose of oxide coatings on filaments or cathodes in a VT? _____
5. What filament materials are used in lower-voltage tubes? _____ Medium-voltage tubes? _____ Highest-voltage tubes? _____
6. What are two names for electron-accepting elements in a diode VT? _____
7. When an element accepts all the electrons given off by a cathode, what has occurred? _____
8. What do these symbols mean: I_p? _____ I_f? _____ E_p? _____ E_f?
9. When a pulsating dc is flowing through it, what type of voltage is developed across a resistor? _____ An inductor? _____
10. In what circuit of a vacuum tube is the load always found? _____
11. What is the name of the voltage between cathode and anode? _____
12. Why are ac filaments center-tapped? _____
13. What is meant by "rectifying ac"? _____
14. Why does a plate not emit electrons? _____
15. What is a hard vacuum? _____ A soft tube? _____ A getter? _____

9-11 GASEOUS DIODES

If a diode tube is evacuated and then some liquid mercury or an inert gas is inserted, a gaseous diode is produced. Such gas-filled tubes are called soft tubes, although the word "soft" here does not indicate faulty operational characteristics as when applied to vacuum tubes.

In gaseous diodes, when the filament is heated, electrons are emitted as in vacuum tubes, but they do not go directly to the plate. They start toward the plate but soon strike a gas molecule. In the case of mercury vapor, if the difference of potential between cathode and plate exceeds about 15 V, the speed of the electrons is sufficient to dislodge one or more of them from a gas molecule. This splits the gas molecule into a positive ion (the gas molecule minus one electron) and a free electron. The relatively heavy positive ion starts to move toward the filament, and the free electron toward the positive plate. The positive ion moves only a short distance before an electron from the cathode fills in the hole produced by the freed electron, neutralizing the ion. It is no longer attracted to either a positive- or a negative-charged element.

The free electrons either travel directly to the plate or ionize another molecule as they move toward the positive element. This electron movement from ion to ion is continued, resulting in a current of electrons from filament to plate.

An advantage of gaseous over vacuum diodes is their efficiency. Since electrons are traveling only a short distance, they cannot accelerate and strike the plate with great force. Therefore, little heat dissipation is produced at the plate. Such tubes require little ventilation, carry relatively high currents, and have a relatively constant voltage-drop across them. Mercury-vapor rectifiers are discussed in Chap. 11. A gaseous diode using argon gas has a coiled filament wire and a carbon plate. It is known as a Tungar bulb and is used in battery-charging circuits. Gaseous diodes glow blue with mercury vapor, and purple, pink, or orange with other gases.

9-12 COLD-CATHODE TUBES

Tubes filled with any gas under the correct pressure will ionize and pass an electronic current if the voltage across them is greater than the ionization potential of the gas used. Provided the area of one of the electrodes in the tube is larger than the area of the other electrode, more current will flow from the large- to the small-area element than in the opposite direction. The large-area element becomes the cathode, and the small-area element becomes the anode, or plate, as indicated in Fig. 9-18. There is no heated filament or cathode. There is a small current flow in the opposite direction, but for most purposes, when an ac emf is applied across the tube, as shown, the current through the load resistor is substantially pulsating dc. The current does not start to flow until the applied voltage reaches the ionization potential and stops flowing when the applied voltage drops to the extinguishing potential of the gas used.

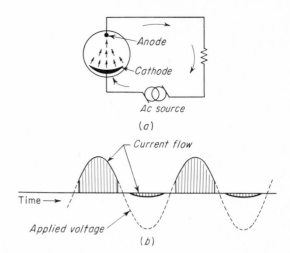

Fig. 9-18 More current flows from a large-area electrode than from a small-area one in a cold-cathode gas-filled tube.

These tubes can be used in power supplies when it is desired to change an ac to pulsating dc. In practical circuits a double-diode tube, called a *full-wave rectifier,* illustrated in Fig. 9-19, may be used. Full-wave rectification is discussed in Chap. 11.

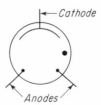

Fig. 9-19 Symbol of a two-anode gas-filled rectifier.

9-13 GASEOUS TRIODES, OR THYRATRONS

The ionizing voltage for gaseous tubes is relatively constant if the gas pressure is maintained constant. For example, if a tube ionizes at 15 V, the voltage across it can be increased from zero to 15 V. At this potential the tube ionizes and current flows. At 15 V the tube is an excellent conductor. If the voltage is now decreased by a few volts, the gas will de-ionize and no current will flow. (This feature offers electronic possibilities for maintaining a constant voltage, discussed in Chap. 11.)

If a third element is inserted between cathode

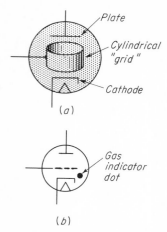

Fig. 9-20 (*a*) Cylindrical grid and (*b*) symbol of a thyratron. Dot indicates gas inside the envelope.

and plate, as in Fig. 9-20, so that an electrostatic charge on it will affect the ionization potential of the gas between cathode and anode, it is possible to control plate current in the tube. If the third element, called a *grid,* is made negative in respect to the cathode, the negative electrostatic field produced in the region between cathode and grid will counteract the positive charge of the plate, resulting in lower electron velocities and less tendency to ionize the gas. It will now be necessary to increase the positive potential on the plate to produce ionization. The value of the negative charge on the grid determines the ionization potential of the gas. Once conducting, these tubes will not de-ionize until the plate potential drops below the normal extinction voltage of the gas used.

Figure 9-20 also shows a symbol of a three-element gaseous *triode,* or *thyratron.* There are also four-element gaseous control tubes. The dot

---

---

in the symbol of a tube indicates gas. Without the dot the tube would be assumed to be a vacuum tube.

9-14 TRIODE TUBES

Vacuum-type diodes act as one-way conductors (or as one-way resistors). The resistance of a diode is very low. With only a few volts across it, enough current may flow through a diode to damage it. Practical circuits involving diodes always have some resistance as the load in the circuit, as was shown in Fig. 9-8. In this circuit an increase in plate voltage produces a proportional increase in plate current in the circuit.

When a vacuum tube with a cathode and a plate has a wire meshwork built into it between cathode and plate, the meshwork is called a grid. Such a tube can be diagramed as shown in Fig. 9-21.

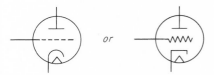

Fig. 9-21   Two symbols for triode tubes.

All electrons flowing from cathode to plate must pass through the spaces between the grid wires. If the grid has no electric charge, it will allow the electrons to pass almost as though it were not there. However, if the grid is made a little negative (in respect to the cathode), it tends to repel the negative space-charge electrons being attracted toward the plate and thereby decrease the cathode-to-plate current. A strong negative charge on the grid can stop the I_p flow entirely. A varying $-E_g$ will produce a varying I_p.

Since an increase in negative charge results in less plate current, making the grid negative increases the plate-circuit resistance. The more negative the grid, the higher the resistance between cathode and plate. The more positive the grid, the lower the resistance of the tube. The grid is not, however, made to be driven very positive. In normal operation the grid is made to vary only more or less negatively.

Because of the ability of the grid to control plate current, it is called a *control grid.* The British call tubes valves, because the grid acts like a valve controlling the flow of electrons from cathode to plate.

The dc resistance between cathode and plate can be computed by Ohm's law:

$$R_{dc} = \frac{E_p}{I_p}$$

where R_{dc} = the dc resistance of tube, in Ω
I_p = plate current, in A
E_p = plate voltage, in V

This dc resistance value is useful in some special circuits. A more important term, the *ac plate resistance,* or impedance, of the tube, will be discussed in Sec. 9-20.

9-15 AMPLIFICATION FACTOR

How effective a controlling device the grid is can be illustrated by an example. In the circuit of Fig. 9-22, the voltage in the grid circuit, $-E_g$,

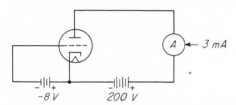

Fig. 9-22 Varying the grid voltage varies the plate current.

is -8 V. The plate-circuit voltage, E_p, is 200 V. The plate current, I_p, is 3 mA. By increasing the E_p by 40 V it is found that the I_p increases from 3 to 7 mA.

Returning to the original values, $-E_g = -8$ V, $E_p = 200$ V, and $I_p = 3$ mA, it is found that if the $-E_g$ value is reduced by 2 V, from -8 to -6 V, the I_p will again rise from 3 to 7 mA. This indicates that the same plate-current change can be produced in this tube either by changing the E_p by 40 V or by changing the $-E_g$ by 2 V. This controlling ratio of 40:2 is equal to 20. The tube is said to have a *mu* (μ) or *amplification factor* of 20. The grid is 20 times more effective in changing plate current than the plate voltage is. By formula, μ is found by

$$\mu = \frac{\Delta E_p}{\Delta E_g}$$

where Δ (Greek letter delta) indicates "a small change in."

While this tube has an amplification factor of 20, the voltage variation, or output voltage, across the plate-circuit load will not be 20 times the signal voltage applied to the grid of the tube. However, actual voltage amplifications equal to half or three-quarters of the μ value are possible. Thus, a tube with an amplification factor of 20 may produce a 10- to 15-times amplification of an ac signal fed into its grid-cathode circuit.

9-16 HOW THE TRIODE AMPLIFIES

The circuits discussed thus far have not been practical amplifiers, but the circuit in Fig. 9-23 will

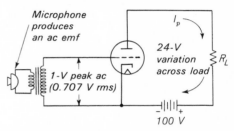

Fig. 9-23 This circuit amplifies any input-signal voltage 12 times across the plate-circuit load.

amplify an input signal. Signals that are amplified in an amplifier are either ac or varying dc voltages. In this case a microphone induces an ac voltage into the secondary of the transformer between grid and cathode.

With no signal applied to the grid and with 100 V from the plate supply, the dc voltage-drop across the load resistor R_L might be 75 V. As the input signal reaches a peak of 1 V negative, the current in the plate circuit will decrease. The voltage-drop across the plate-load resistor might decrease by 12 V, to 63 V across R_L. As the grid voltage swings to 1 V positive, the plate current will increase, until there is a voltage-drop of perhaps 87 V across the load resistor. As the grid voltage varies from -1 V to $+1$ V (a 2-V peak-to-peak variation), the voltage across the load resistance varies between 87 and 63, or 24 V. The voltage ratio of 2:24 indicates that across the plate-load resistor will appear a voltage variation that will be 12 times the voltage variation applied between grid and cathode. The true amplification,

A, or *gain* of the whole *stage,* is 12, although the tube itself may have a μ of 20.

If the resistance value of the plate load is increased, the amplification will become greater. Theoretically, if the plate-load resistance is infinite, the gain of the stage will equal the amplification factor of the tube. Resistance values much over 1,000,000 Ω, however, will decrease the plate current so much that a triode will not operate satisfactorily. In practice, plate-load-resistance values may range from 5,000 to 500,000 Ω.

In many cases, an ac signal output is desired from an amplifier stage. The signal output in Fig. 9-23 is a varying dc. If the varying plate dc is fed to a transformer primary, the signal will be converted to ac in the secondary. While this amplifier stage is capable of amplifying a signal voltage, it is still not a practical circuit. A bias voltage is needed.

9-17 BIAS VOLTAGE

In the amplifier circuit previously discussed, the grid was driven first negative and then positive. In many practical circuits, to prevent distortion of the signal, it is desirable that the grid never be allowed to become positive and thus draw *grid current* (I_g in Fig. 9-24) from the cathode. This is

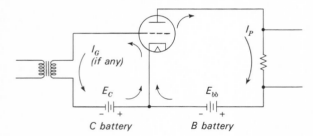

Fig. 9-24 Plate and grid (if any) current directions in a triode amplifier circuit.

accomplished by adding a C battery or other voltage supply (E_c) in series with the grid-cathode circuit. The negative terminal of the C battery is connected to the grid through the transformer, and its positive terminal is connected to the cathode. This negative dc voltage added in series with the grid circuit is known as a *bias* voltage.

If a negative 10-V bias is enough to produce plate-current cutoff with a given plate voltage,

then a possible voltage for the C battery would be half of this, or 5 V (class A operation).

With a 5-V bias applied to the grid it is possible to accommodate a peak ac emf of nearly 5 V from the secondary of the grid-circuit transformer and neither cut off the plate current at any time nor drive the grid into the positive region. NOTE: The plate supply is not applying any negative voltage to the grid circuit. It is connected between cathode and plate and is not in, or affecting, the grid circuit at all. The C battery is between cathode and grid and is not applying any voltage to the plate circuit. If the B-battery voltage E_{bb}, is 100 V and the $-E_c$ is 5 V, there may be a 105-V difference in potential between plate and grid, but this voltage is rarely considered and may be ignored.

In the basic amplifier, the grid circuit and the plate circuit are two separate circuits that happen to have a common cathode connection. The grid circuit consists of the cathode, C battery, grid-circuit signal-input device (the transformer secondary in the diagram), and the grid. The plate circuit consists of the cathode, plate, load (resistor), and B battery. The two circuits work independently of each other in this configuration.

9-18 CHARACTERISTIC CURVES

How much will a variation of the grid voltage affect the plate current in a particular triode tube? To determine this, the grid-voltage versus plate-current ($E_g I_p$) curve of Fig. 9-25 can be used. If this curve is plotted with no load resistance in the plate circuit, it is called a *static* characteristic curve. If a load is used, a *dynamic* characteristic curve results.

Fig. 9-25 Curve of plate-current versus grid-voltage values ($E_g I_p$) with a constant plate voltage.

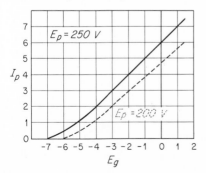

This particular graph is read as follows: The plate voltage is constant at 250 V. With zero volts on the grid (no bias), the plate current is 6 mA. With -2 V on the grid, the plate current is 4 mA. At -5 V the plate current is 1 mA, and at -6 V the plate current is 0.5 mA. This tube is completely cut off by applying a -7 V bias to its grid, provided 250 V is used on the plate. With a higher plate voltage, a higher bias voltage will be required to attain plate-current cutoff.

The negative voltage value required to produce almost complete plate-current cutoff for any triode may be found by dividing the plate voltage by the μ of the tube:

$$E_{co} = \frac{E_p}{\mu}$$

where E_{co} = cutoff bias voltage
E_p = plate voltage
μ = amplification factor of tube

The plate-current cutoff bias is the value that would be required to bring the curve to zero if it had no bend at the lower end. This would be about -6.5 V in Fig. 9-25.

By rearranging the formula above, the μ of the tube in Fig. 9-25 must be

$$\mu = \frac{E_p}{E_{co}} = \frac{250}{6.5} = 38.5$$

It is not always advisable to operate a tube in the bent portion of the curve, near cutoff. Such operation may cause distortion of the signal being amplified. In Fig. 9-26 is shown a curve with two input signals of different amplitudes. The first results in a plate-current variation having the same waveshape as the input signal because all the operation is under a straight portion of the curve. The negative half of the second, higher-amplitude input signal operates in the bent region of the curve. The input voltage operating in the bent

Fig. 9-26 Grid-voltage variation must be under straight portion of $E_g I_p$ curve to produce undistorted plate-current variations.

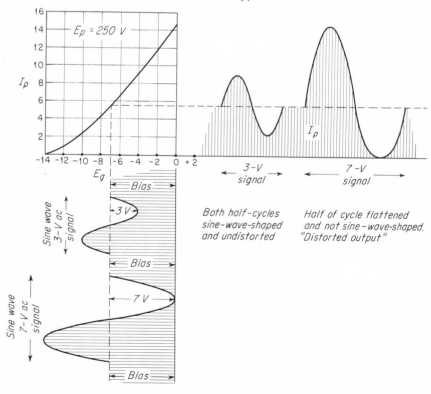

area produces a flattening of the plate-current waveshape. The voltage-drop across the plate load resistor with the higher-amplitude signal will not have the same waveshape as the voltage applied to the grid. The tube is distorting the signal.

Another type of graph is the $E_p I_p$ family of curves. This plots plate current against different plate voltages, with the grid bias held constant for each curve as in Fig. 9-27. The dashed line, called

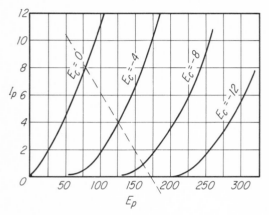

Fig. 9-27 $E_p I_p$ curves for different values of grid bias. Possible load line shown dashed.

a *load line,* is used when a transformer is the load in the plate circuit. The power-supply voltage is chosen to be 125 V. It is also decided to use -4 V as the no-signal, or quiescent bias. When a counteracting $+4$-V signal is applied to the grid circuit, the I_p should double to 8 mA. The load line is drawn from where the $E_c = 0$ curve intersects the 8-mA line, then down to $I_p = 0$ through the $E_p = 125$ V and $-E_c = 4$ mA interception. The output signal voltage peaks are indicated by the range between where $E_c = 0$ and 8 mA intercept and where the load line intercepts the $E_c = -8$ V curve. This is from $E_p = 80$ to $E_p = 160$ V, or a range of 80 V. Thus, an 8-V peak-to-peak input grid signal develops a peak-to-peak output load signal voltage of 80 V, for a stage gain, or amplification (A) or 80/8, or $A = 10$.

9-19 FACTORS DETERMINING μ

It has been indicated that different triodes have different amplification factors. The physical factors that control the μ are the relative spacing of the cathode, grid, and plate and the fineness or coarseness of the meshwork making up the grid.

A simplified cross-sectional view of a triode is shown in Fig. 9-28. When the filament is heated,

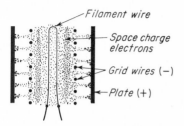

Fig. 9-28 Simplified cross section of a triode tube, showing relative element placement.

the space charge forms around it. Since the grid is biased to some negative value, it repels the electrons and holds them close to the filament. The positive charge of the plate "reaches in" through the negative field set up by the grid wires and pulls electrons out to the plate. If the grid is too highly negative, the plate potential may not be strong enough to pull electrons through the spaces between the grid wires. It may be necessary to use a higher positive potential on the plate to attract electrons through a highly negative grid area.

If the manufacturer constructs the tube with the grid wires close to the plate, the plate will be able to pull more electrons through the grid wires, since its strong positive field easily cancels the weaker repelling field of the grid. Under this condition a much higher negative grid bias is required to produce plate-current cutoff.

The greater the bias value required to produce plate-current cutoff, the greater the signal voltage required on the grid to control the plate current. Conversely, if the grid is moved closer to the filament, it takes relatively little signal voltage on the grid to vary the plate current of the tube. This produces a higher-μ tube. A cathode sleeve permits higher μ in a tube than is possible with directly heated filaments because it is possible to position the control grid closer to the cathode.

Another factor in determining the μ of the tube is how closely the grid wires are meshed. The closer the wires, the more effective they are in preventing the positive pull of the plate from reaching the space-charge electrons, the less negative the grid voltage required to produce plate-current cutoff, and the higher the μ of the tube.

9-20 PLATE IMPEDANCE

The B battery in series with the cathode-plate circuit of a tube appears as a source of energy to the plate-circuit load. When no signal is being amplified, the tube appears as a dc source to the load, but when signals are being amplified, the tube and battery appear as a varying dc source having an ac component. The value of this plate impedance can be determined by using a variation of Ohm's law. It may be found expressed in several ways:

$$Z_p = \frac{\Delta E_p}{\Delta I_p} \quad \text{or} \quad r_p = \frac{\Delta e_p}{\Delta i_p}$$

$$\text{or} \quad r_p = \frac{de_p}{di_p}$$

Note that the italic letter d has the same meaning as the Greek letter Δ, "a small change in." These formulas would be read: "The plate impedance is the ratio of a small change in plate voltage to a small change in plate current." E_p is measured in volts, I_p in amperes.

Plate impedance can be determined from an $E_p I_p$ family of curves, such as Fig. 9-27. Starting at the intersection of the load line and the $E_c = -4$ V curve, the di_p for a 25-V E_p increase is from 4 to 6.5 mA. Substituting in the formula

$$r_p = \frac{de_p}{di_p} = \frac{d25}{d0.0025} = 10,000 \ \Omega$$

Note that slightly different impedances will be obtained if other parts of the same curve or if other bias curves are used.

9-21 MUTUAL CONDUCTANCE

One of the factors by which a tube can be judged is its *mutual conductance,* also known as its *transconductance.*

In a given tube, if the grid voltage varies by some amount, the plate current will vary by a certain amount. Tubes having the ability of producing a relatively wide plate-current variation with a given grid-voltage variation have a high value of mutual conductance. Note the difference between the formulas for mutual conductance and amplification factor:

$$g_m = \frac{di_p}{de_g} \quad \mu = \frac{dE_p}{dE_g}$$

where g_m = mutual conductance, in mhos
μ = amplification factor

The transconductance formula can be remembered by its similarity to the formula for conductance $G = I/E$.

EXAMPLE: A tube changes its plate current by 5 mA with a grid-voltage variation of 1 V. What is the mutual conductance of the tube?

$$g_m = \frac{di_p}{de_g} = \frac{0.005}{1}$$

$$= 0.005 \text{ mho, or } 5,000 \text{ micromhos } (\mu\text{mhos})$$

Applying this formula to the $E_g I_p$ family of curves in Fig. 9-27, the di_p along the 100-V E_p line for a change in grid voltage from -4 V to 0 V is about 9.5 mA. Substituting in the formula

$$g_m = \frac{di_p}{de_g} = \frac{d0.0095}{d4} = 0.002375 \text{ mho}$$

$$= 2,375 \ \mu\text{mho}$$

Another formula by which mutual conductance may be determined is

$$g_m = \frac{\mu}{r_p}$$

where μ = amplification factor of tube
r_p = ac plate impedance

From this formula it is possible to derive $r_p = \mu/g_m$ and also $\mu = g_m r_p$.

If it is desired to use a tube to produce a high

value of amplification of an input voltage, a tube with a high μ should be used. If it is desired to produce a large current variation in the load, as in a transformer, a tube with a high mutual conductance should be used. Transconductance values range from about 300 to more than 12,000 μmhos.

Test your understanding; answer these checkup questions.

1. Why are gaseous diodes more efficient than vacuum diodes? _____
2. What is a gas-type triode called? _____ What does its grid control? _____ What does the grid in a vacuum triode control? _____
3. What is meant when a triode is said to have a μ of 20? _____
4. What battery is used for biasing? _____ For plate potential? _____ To heat the filament? _____
5. What occurs if signal voltages operate a tube under a bent portion of its curve? _____
6. What relative μ-value triode has a very low cutoff bias value? _____
7. What is the formula for g_m involving r_p and μ? _____
8. What is another formula for g_m? _____ For r_p? _____ For μ? _____
9. Does a high-gain tube have a high or low g_m, or is g_m unrelated to amplification? _____
10. How many volts of bias would be required to reduce the I_p of a triode to zero if E_p is 200 V and μ is 15? _____ 100? _____
11. What is the r_p of a triode if a change of 4 V E_p produces a dI_p of 0.5 mA? _____
12. Increasing the bias by 0.6 V in a triode produces a decrease of 3 mA in I_p. What is the g_m of the tube? _____
13. How is the μ affected if the grid is moved closer to the cathode? _____ If the grid wires are spread apart more? _____
14. In Fig. 9-27, using -4 and -8 V on the grid, with an E_p of 175 V, what is the g_m? _____ The μ when I_p is constant at 10 mA? _____ At 2 mA? _____

9-22 POWER OUTPUT

The consideration of amplifiers so far has centered on their ability to produce an amplified voltage variation across the plate-circuit load. Another consideration of vacuum tubes is the amount of power output obtainable from them.

A transformer-coupled amplifier circuit is shown in Fig. 9-29. With no input signal to the grid, there is a constant plate current in the primary of the plate-circuit transformer. Since there is no primary current variation, no voltage is induced into the secondary and no power is delivered to the load resistor.

When a signal voltage is fed into the grid circuit, the plate current varies and thereby induces an ac emf into the output transformer secondary and ac current in the resistance load. The amplifier is now producing power output. An increase of input signal voltage increases the power output. Since power varies as the voltage squared ($P = E^2/R$), doubling the input signal voltage quadruples the output power, provided the operation is on the straight portion of the tube's characteristic curve.

The output transformer with the resistor represents a load having a certain value of impedance. This load is connected across the source (amplifier tube and power supply). If the load impedance is equal to the ac plate impedance, the maximum output power will be produced in the load.

Computation of the power output of this circuit can be made by measuring the voltage developed across the load resistance and the current flowing through it. According to the power formula $P = EI$, the product of these two values will be the power developed in the load. Since there is always a slight power loss in a transformer, the tube will be delivering slightly more power than is computed. If the resistance of the load is known, the formulas $P = I^2R$ or $P = E^2/R$ may be used.

Fig. 9-29 Power is developed in the load resistor when an input signal is fed to the grid circuit.

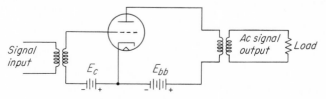

A fairly accurate formula for the output power of a triode is

$$P_o = \frac{\mu^2 E_g^2 Z_L}{(Z_L + Z_p)^2}$$

where P_o = output power, in W
μ = amplification factor
E_g = rms signal voltage on grid
Z_L = load impedance, in Ω
Z_p = plate impedance of tube, in Ω

When *maximum* power output is desired, the load impedance must match the tube impedance, and the formula is

$$P_{o,\max} = \frac{\mu^2 E_g^2}{4Z_p} \quad \text{or} \quad \frac{\mu^2 e_g^2}{8Z_p}$$

where e_g = *peak* ac signal-voltage value

Maximum *undistorted* output is obtained when Z_L equals two or three times the Z_p, depending on the requirements. A formula for undistorted output, where $Z_L = 2.5Z_p$, is

$$P_{o,\mathrm{und}} = \frac{\mu^2 E_g^2}{5Z_p}$$

9-23 PLATE DISSIPATION AND SATURATION

Electrons striking the plate surface of a tube release energy in the form of heat. The plate of a tube is made to withstand only so many watts of plate heating. If the plate current is increased beyond the value that produces the maximum rated plate dissipation, the tube may be damaged. Excessive heat may liberate gas molecules from the plate and produce a soft, or gaseous, tube or may even melt a hole in the plate.

As the plate voltage is increased in a tube, the plate current increases. Eventually a point of plate saturation is reached. Any further increase in plate voltage will not increase the plate current. If a slowly increasing plate voltage is applied to the plate of a tube, the maximum safe plate dissipation will be exceeded before plate saturation

occurs. If the plate current is made to come in pulses by applying alternate negative and positive potentials to the grid, however, it is possible to allow the plate-current pulses (if of short enough duration) to reach saturation at the pulse peaks and still have the plate operating below maximum plate dissipation, since the plate will be cooling off in between the heavy pulses of plate current. In this case the *average* plate dissipation might not exceed the maximum rated dissipation value of the tube.

9-24 CLASSES OF AMPLIFIERS

There are many ways of operating amplifier tubes. Basically, these are broken down into different classes.

Class A uses a bias value generally equal to a little more than one-half of the bias required to produce plate-current cutoff.

Class B indicates the bias is almost to the value of plate-current cutoff with no signal being applied to the grid. Between classes A and B are two intermediate classes known as AB_1 and AB_2, discussed in Chap. 14.

Class C indicates the bias is greater than the plate-current cutoff value. With no signal there is no plate current. This class is discussed in Chap. 15.

How plate dissipation affects the output from the same tube in the three basic classes can be illustrated by using the simple amplifier stage in Fig. 9-30. The tube will be assumed to have a 1-W

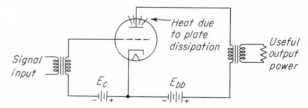

Fig. 9-30 The plate always dissipates some power in the form of heat radiation.

plate-dissipation rating. The plate, bias, and signal voltages are adjusted until the plate current produces 1 W of heat at the plate. An ac output power will be developed in the plate-circuit load because of signal being applied to the grid circuit.

CLASS A. A tube biased to a little more than half the cutoff value, operating with I_p flowing all the time and with full input signal, will have an output of about $\frac{1}{3}$ W when the plate is dissipating 1 W in heat. The total power applied to the plate circuit by the B battery is $1\frac{1}{3}$ W, of which $\frac{1}{3}$ W is usable ac power output. The remainder is only heating the plate. One-third of a watt is 25% of $1\frac{1}{3}$ W. Class A stages will be only about 25% efficient with minimal output distortion.

CLASS B. When biased almost to cutoff, the same tube will be able to stand a higher plate voltage and the grid circuit will be fed twice the signal voltage amplitude. Two tubes must be used in push-pull for undistorted class B operation. Under this condition the plate current comes in pulses and flows during only about one-half of each cycle of the signal in each tube. A total of more than 2 W can be fed into the plate circuit of a tube before the plate dissipates 1 W. As a result, the power output will be more than 1 W per tube at maximum plate dissipation. Both tubes are more than 50% efficient.

CLASS C. When biased beyond cutoff, the same tube will be able to stand even higher plate voltage. Plate current flows for only about one-third of the input-signal cycle. The grid may be fed a still greater signal voltage. A total of nearly 4 W can be developed in the plate circuit before the plate dissipates 1 W in heat. Almost 3 W is available as useful ac output. The stage is operating at about 75% efficiency.

It would seem that class C is the best way in which to operate a vacuum tube as an amplifier. In some cases it is, but such operation produces considerable distortion and cannot be used for AF amplification at all, although it can be used for certain types of RF amplifiers. Class B is also limited to certain types of AF or RF amplification. Only class A can be used in all types of amplification.

When the power delivered by the power supply ($P = E_p I_p$) is known and the useful output of the amplifier stage is also known, the plate-circuit efficiency is found by

$$\text{Percent efficiency} = \frac{P_o}{P_i} \times 100$$

where P_o = ac output power
P_i = dc plate-circuit power (as read by an ammeter and a voltmeter in the plate circuit)

EXAMPLE: What is the efficiency if the dc input power to a tube is 200 W and the output is 140 W?

$$\text{Percent efficiency} = \frac{P_o}{P_i} \times 100 = \frac{140}{200} \times 100$$
$$= 0.7 \times 100$$
$$= 70\%$$

9-25 SECONDARY EMISSION

The *primary emission* in a vacuum tube is the electron emission from the cathode.

When the plate in a vacuum tube is highly positive, it increases the velocity of electrons traveling toward it to such an extent that, when they strike, each electron may dislodge one or more electrons from the plate. These electrons moving from the plate out into the vacuum form a *secondary emission*.

The effect of secondary emission may be great enough to produce a small negative space charge (cloud of electrons) near the plate, interfering with normal electron flow to the plate.

If the control grid of a triode is driven positive, some of the secondary-emission electrons may be attracted to that element instead of to the plate. This is undesirable, since only those electrons that flow through the plate-circuit *load* can produce usable output.

The grid of a triode is normally held negative and does not attract any secondary electrons. For this reason secondary electrons in triodes are eventually returned to the plate and can flow through the load.

9-26 TETRODE TUBES

To obtain a higher amplification factor in a triode, the plate must be backed away from the cathode, or a finer grid meshwork must be used, necessitating relatively higher plate voltage to produce plate current. A disadvantage in triodes is the relatively high interelectrode capacitance that exists between the plate and the grid because of their proximity to each other. This capacitance interferes with the ability of the tube to amplify properly at higher frequencies.

To overcome some of the difficulties that arise when the μ of a triode tube is increased and, possibly most important, to reduce the grid-plate capacitance, the *tetrode,* or four-element tube, was developed. In the tetrode, the grid is kept close to the cathode and the plate is moved outward from the cathode, increasing the μ. Then, between grid and plate, a second *screen* grid is placed (Fig. 9-31).

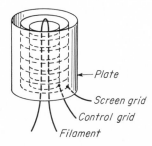

Fig. 9-31 Illustration of the relative position of the elements in a filament-type tetrode tube.

The screen grid is connected to a static, or unvarying, positive dc potential of 100 V or more. The positive field of the screen grid draws space-charge electrons from the cathode through the control grid wires. This puts the space-charge electrons in such a position that the plate potential can attract them. Thus, relatively high plate current and high amplification or g_m are possible. The screen grid is dynamically (ac) connected to the cathode with a *bypass* capacitor as in Fig. 9-32. The diagram of the tetrode circuit is the same as that of the triode except for the addition of the screen grid and the bypass capacitor.

Since the control grid and plate are much farther apart, the interelectrode capacitance between them is greatly reduced. More important,

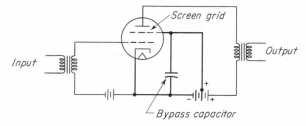

Fig. 9-32 Screen-grid circuit shown in heavy lines.

bypassing the screen grid to cathode effectively places an electrostatic shield between plate and grid. With a higher g_m and less plate-to-grid interaction due to capacitance, the tetrode tube produces more stable operation and higher gain. Amplifications up to several hundred times are possible.

Triode tubes are rated by their μ values, but since the voltage applied to the screen grid can control the amplification of the tube, tetrodes carry no μ rating, only g_m.

The positive screen grid has the effect of accelerating the electrons on their way to the plate. As a result, the secondary-emission current from plate to screen becomes rather high. In fact, if the plate voltage falls below that of the screen grid, the secondary-emission current from the plate may exceed the electron current to the plate.

Tetrodes have high plate-impedance values, ranging from 50,000 to 100,000 Ω. This presents difficulties in load-impedance matching if it is desired to use a transformer as the load.

As long as the plate voltage is greater than the screen voltage, the plate current in a tetrode is nearly independent of the plate voltage. Doubling the plate voltage will raise the plate current very little, whereas in a triode doubling the plate voltage will more than double the plate current. However, grid-voltage variations, either control grid or screen grid, will control the plate current. The control grid is about fifty times as effective as the screen in controlling I_p in tetrodes.

9-27 BEAM-POWER TETRODES

One of the later developments in tubes is the beam-power tetrode. This tube is used not as a voltage amplifier but as a power amplifier.

The disadvantage of the secondary-emission

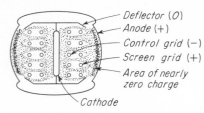

Fig. 9-33 Basic idea of placement of cathode, grids, deflector, and anode in a beam-power tetrode.

electrons returning to the screen grid in the original tetrodes is overcome in two ways. First, the screen and control grids are so placed in the tube that the screen-grid wires are in the electron shadow of the control grid, as shown in Fig. 9-33. Most of the electrons leaving the space charge are made to deviate outward from the negative control grid and do not converge again until they have passed the screen grid. This tends to form strong streams of electrons flowing to the plate. Second, the tube has two deflector plates held at cathode potential, which further restrict the travel of the electrons into a beam toward the plate.

The beaming results in many electrons hitting the plate and some secondary emission. Secondary-emission electrons, however, are swept backward toward the plate again by the advancing beam of electrons. The net result is an area next to the highly positive plate where there is virtually no charge. This area of zero charge tends to lower the speed of the electrons flowing to the plate and thereby reduce secondary emission.

The shading of the screen grid is not complete. Some I_{sg} flows, but it is usually less than one-tenth of the I_p value. When the control grid is at a small negative value, the screen grid is no longer shadowed as well as with a more highly negative control grid and screen-grid current increases. If the positive plate potential is reduced, the electrons that would have gone to the plate now go to the screen and may melt the thin screen-grid wires. Beam-power tubes should not be operated without applying plate potential unless a highly negative control-grid bias is used.

Beam-power tubes made for audio circuits may have rather poor grid-plate shielding. Other beam-power tubes, made for operation in RF circuits, are specially designed to reduce interelec-

trode capacitance and may require little or no *neutralization* (Sec. 15-13).

Connecting tetrode control and screen grids together produces a high-μ triode. Connecting screen grid and plate together forms a low-μ triode.

9-28 PENTODE TUBES

Secondary-emission electrons moving from plate to screen grid in the original tetrodes is undesirable. The addition of a zero-charged third grid, called a *suppressor grid,* between screen grid and plate decreases the velocity of the electrons approaching the plate. Any secondary emission that does occur has insufficient energy to move back across the zero-charged suppressor grid field. In this way secondary-emission current from plate to screen grid is stopped. This five-element tube is known as a *pentode*. Placement of the elements in a pentode is shown in Fig. 9-34.

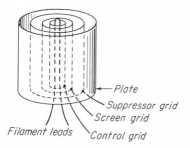

Fig. 9-34 Relative positions of pentode elements.

The only difference between diagraming pentode and tetrode amplifiers is the inclusion of the suppressor grid. The suppressor is normally connected directly to the cathode. In fact, in many pentode tubes, the suppressor grid is connected to the cathode internally by the manufacturer. A pentode tube in an amplifier circuit is shown in Fig. 9-35a.

Besides having higher gain than tetrodes, pentodes have better shielding between plate and control grid, requiring no neutralization when used in high-frequency circuits.

The original type of tetrode is no longer found, but beam-power tetrodes are employed in both AF and RF power amplifiers. Pentodes find use

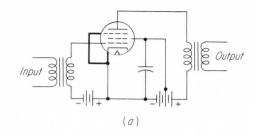

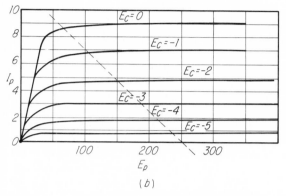

Fig. 9-35 (a) Suppressor grid circuit in heavy lines. (b) E_pI_p curves for a pentode. Possible load line shown dashed.

where voltage amplification is desired, as in receivers. Some high-power transmitting pentodes are also used.

Pentode plate impedances vary from about 100,000 to over 1,000,000 Ω, making load matching difficult except when resistors or resonant transformers form the load. Compare the pentode E_pI_p curves of Fig. 9-35b with those of a triode in Fig. 9-27. The dashed line on Fig. 9-35b is a possible load line centered on $-E_c = -2$ V and an E_p of 250 V.

For some circuits a tube is required that does not drop its plate current sharply to cutoff, as is normal for most tubes. This behavior can be produced by spacing the control-grid wires close together at the top and bottom, but far apart at the center as in Fig. 9-36. With no bias, electrons move freely to the plate between all grid wires. With a little negative bias, the closely spaced wires cut off the plate current through them, but the widely spaced wires continue to allow electrons to flow between them. A very high bias is required to cut off plate current completely in such remote-cutoff, variable-μ, or supercontrol tubes. The

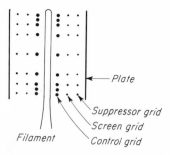

Fig. 9-36 Spacing of control and other grid wires in a remote-cutoff pentode tube.

E_gI_p curves for the usual sharp-cutoff tube and for a remote-cutoff tube are shown in Fig. 9-37. Using remote-cutoff pentodes for AVC (automatic volume control) is described in Sec. 18-14.

Test your understanding; answer these checkup questions.

1. Under what condition does a VT produce maximum power output to a load? _____
2. A 2-V rms signal fed to the grid of a 20-μ triode with a 10-kΩ r_p will produce how much power in a 10-kΩ load? _____ In a 20-kΩ load? _____ Which should have the lesser distortion? _____
3. What is the advantage of class A operation? _____ Of class C? _____
4. A triode with a 42-W dc input has a measured 12-W output. How much heat must the plate be dissipating? _____
5. If $E_p = 300$ V, $I_p = 200$ mA, and the load is dissipating 15 W, what is the efficiency of the stage? _____
6. From cathode, what is the name of the first grid? _____ Second grid? _____ Third grid? _____
7. What dc polarity does the first grid usually have? _____ Second grid? _____ Third grid? _____ Plate? _____ Cathode? _____
8. What component is always connected between screen grid and cathode in tetrodes? _____ Is this true

Fig. 9-37 Comparision of E_gI_p curves of sharp- and remote-cutoff pentodes.

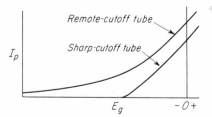

of pentodes? _____ What does this component do? _____

9. What type of tetrode is in general use today? _____
10. What is the name of the grid which reduces secondary emission? _____
11. What type of tube(s) can be used in high-frequency amplifiers without neutralization? _____
12. What kind of VT has irregularly spaced control-grid wires? _____ What are its two other names? _____

_____

9-29 SHIELDING TUBES

In the earlier days of radio the tubes were rather large, and in some circuits one tube interacted with others because of the capacitance between them. It became necessary to construct a metal shield around each tube to prevent this interaction. Later, the metal tube was developed and the metal envelope was used as the shield. More recently miniature tubes are being made so small that they may not interact in many circuits. If they do, small metal shields are placed around them and grounded as in earlier days.

9-30 TUBES WITH MORE THAN THREE GRIDS

The pentode has the greatest number of grids used in normal amplifying tubes. However, there are tubes having more than three grids. A hexode, or six-element tube, has four grids. A heptode, or seven-element tube, has five grids. An octode, or eight-element tube, has six grids. These tubes are used in special circuits having functions other than mere amplification. The pentagrid tube is shown in symbol form in Fig. 9-38. The cathode, G_1, and G_2 are used in conjunction with one signal frequency. The cathode and G_4 are used in conjunction with a second signal frequency. The current flowing from cathode to plate will have components of both frequencies in it, producing a third, a *difference*-frequency component

Fig. 9-38 Symbol of a pentagrid converter.

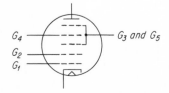

and a fourth, a *sum*-frequency component, in the plate-circuit current.

9-31 MULTIUNIT TUBES

To conserve space in equipment, it is often possible to employ multiunit tubes as in Fig. 9-39.

- *Twin, or duodiodes.* Two diodes in one envelope with common or separate cathodes, such as a seven-pin miniature type 6AL5.
- *Twin triodes.* Two triodes with either a common cathode or separate cathodes, such as a 12AT7 with a nine-pin noval base.
- *Triode-pentodes.* A triode and a pentode with a common cathode, such as a 6F7 with a grid connection at the top of the tube.
- *Duplex-diode-triodes.* Two diodes and one triode with a common cathode, such as a 6SQ7 with an octal base.
- *Twin tetrodes.* Beam-power tetrodes with common cathodes and screen grids, such as an 829 with plate leads coming out of the top.
- *Diode-triode-power pentode.* A diode, triode, and pentode using a common filament, such a 1D8, battery-portable tube. There are many other multiunit tubes, triple diodes, triple triodes, etc.

9-32 SYMBOLS USED

Symbols are used throughout electronics and radio. *E* stands for voltage, *I* for current, *R* for resistance, and so on. A capital letter usually indicates a dc, an average dc, or an effective ac value. A lowercase letter indicates either an instantaneous value of an ac or its peak value. In general, vacuum-tube circuitry symbols follow this same line of thinking:

E_f = dc or effective ac filament voltage
E_c = dc bias voltage (as from a C battery)
E_g = effective grid signal voltage
e_g = instantaneous or peak grid signal voltage
E_p = dc or average dc plate voltage, measured cathode to plate
e_p = instantaneous plate-to-cathode voltage during signal cycle
E_{bb} = plate-circuit battery or power-supply dc voltage

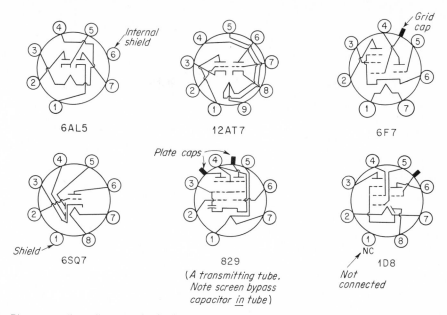

6AL5   12AT7   6F7

6SQ7   829   1D8

(A transmitting tube. Note screen bypass capacitor in tube)

Fig. 9-39 Pin connections (bottom view) of several types of directly and indirectly heated cathode tubes.

E_{sg} = dc screen-grid voltage
E_{sp} = dc suppressor-grid voltage

The same idea is used to indicate currents:

I_g = grid current as read by meter
I_p = plate current as read by meter
i_p = instantaneous plate current during signal cycle

9-33 RECEIVING AND TRANSMITTING TUBES

Diode, triode, tetrode, and pentode tubes are used in both transmitters and receivers. The difference between a transmitting and a receiving tube is generally in physical size and ruggedness of element construction. Receiving tubes are small and usually have all the element connections attached to metal pins at the base of the tube. (In the past the control-grid connection was sometimes at the top of the tube.)

Transmitting tubes are larger than receiving tubes and have heavier filaments for greater electron emission. The larger plates accommodate heavier plate current and dissipate greater power in heat. The plate connection of larger trans-

mitting tubes is usually brought out the top of the tube to provide lower grid-plate capacitance and maximum insulation between the plate and other elements. Voltages applied to the plate may be 500 to 10,000 V or more.

In some transmitting tubes the plate forms part of the outside envelope of the tube. When this is the case, fins may be attached to the outer surface of the plate, as shown in Fig. 9-40. A forced draft of air rapidly dissipates heat developed on the plate. Such tubes are said to be

Fig. 9-40 Air flowing over metal fins attached to external plate removes heat dissipated by plate.

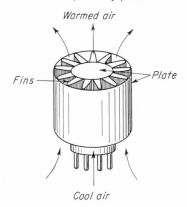

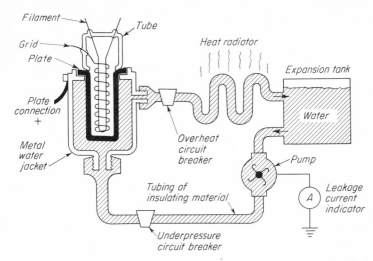

Fig. 9-41 Essential components of a water-cooling system for a water-cooled triode.

air-cooled. Heat on the plate of the usual smaller vacuum tubes must be *radiated* from the plate, resulting in much less rapid dissipation of the heat.

In some high-powered transmitting tubes the plate is made the outer shell, as shown in Fig. 9-41. The plate connection is made by contact with the metal socket or jacket. Grid and filament leads come out through the glass part of the envelope. A stream of pure water is pumped over the outer surface of the plate, keeping it cool. A possible plumbing system to accompany a water-cooled-tube installation is shown. The water passing over the plate of the tube is warmed, led off to the radiator, cooled, and then pumped back through the system again.

If the water leaving the tube is too hot because of excessive plate current, an overheat circuit breaker may be activated and the plate current is turned off automatically. If the pump is unable to supply sufficient water to the plate, the tube may become too hot, and an underpressure circuit breaker can automatically turn off the plate

current until proper water pressure has been restored.

All water lines are made of rubber, plastic, porcelain, or other insulating materials. The water must be pure in order that no current will flow through it from the ungrounded pump to the highly positive plate. Any small leakage current that does flow through the water will be indicated by the meter between pump and ground. If the water becomes impure, it will pass more current. The greater the meter reading, the more impure the water. In a few areas the water supplied by the local water company is pure enough for use directly into the system, but usually distilled water is required.

9-34 CAUSES OF IMPROPER PLATE CURRENT

When vacuum-tube circuits are operating improperly, the plate current may be found to be either excessive or insufficient. A few causes are listed below:

EXCESSIVE PLATE CURRENT. Indicated by excessive heating of resistors and parts in the plate circuit or by reddening of the plate of glass-envelope tubes. May be caused by excessive screen voltage, excessive plate voltage, insufficient negative bias, positive grid bias, a positively charged suppressor grid, or a gaseous tube.

INSUFFICIENT PLATE CURRENT. Indicated

ANSWERS TO CHECKUP QUIZ ON PAGE 165

1. $(Z_L = r_p)$ 2. (0.04 W)(0.035 W)(20-kΩ load) 3. (Low distortion)(High efficiency) 4. (30 W) 5. (25%) 6. (Control) (Screen)(Suppressor) 7. (−)(+)(0)(+)(0) 8. (Bypass C)(Yes)(Holds grid potential constant) 9. (Beam power) 10. (Suppressor) 11. (RF pentode or tetrode) 12. (Remote cutoff)(Variable-μ, supercontrol)

---

by poor or no response of the circuits. May be caused by insufficient plate voltage, insufficient screen-grid voltage, excessive negative grid bias, disconnected control or screen grids, low emission from the cathode due to aging or to low filament current, or a gaseous tube.

9-35 CAUSES OF TUBE FAILURE

There are several reasons why vacuum tubes become inoperative. Some of the more common are:

FILAMENT FAILURE. Filament wires gradually lose molecules, weaken at one point, and burn out. Excessive jarring may rupture the filament. Too much filament current may burn out the wire. Filaments lose their ability to give off sufficient electrons.

TUBE BECOMES GASSY. If the envelope leaks, air is drawn into the tube. Internal elements give off gas when overheated by excessive current.

SHORTED ELEMENTS. Heated elements or wires in the envelope may sag and touch. Heater-to-cathode insulation may break down.

LOOSE ELEMENTS. When elements are not welded properly, they vibrate, causing opens or short circuits in the tubes and noises in circuits in which the tubes are used. Jarring or heating may loosen internal-element welds.

9-36 BATTERY OPERATION

When batteries are employed to power equipment, the A battery produces the filament-heating current. It must produce relatively high current at voltages between 1.5 and 12 V.

The plate circuit, or B battery, has a relatively high voltage and low current drain. The screen grid is also connected to the B battery, usually to a lower-voltage tap. B batteries range from $22\frac{1}{2}$ to 135 V.

The bias battery is the C battery. Since there is no grid current in low-power equipment, a C battery has almost zero current drain and long life. It ranges from 1.5 to 22.5 V.

9-37 HIGH-FREQUENCY TUBES

Ordinary diode, triode, tetrode, and pentode tubes will operate with frequencies up to well over 20,000,000 Hz (20 MHz). Many of the newer miniature and subminiature tubes will operate satisfactorily up to 500 or 1,000 MHz. When higher frequencies are to be used, however, special design is required in the manufacture of tubes.

In the SHF (superhigh frequency) range, 3,000 to 30,000 MHz, magnetrons and klystrons are often used. These do not operate on the same principles as the tubes discussed in this chapter. (See Chap. 26.)

In the UHF (ultrahigh frequency) range, 300 to 3,000 MHz, magnetrons and klystrons are used, but specially constructed triodes such as the *lighthouse* tube shown in Fig. 9-42 may also be

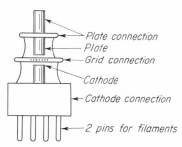

Fig. 9-42 Construction of a lighthouse high-frequency triode.

employed. It operates at ultrahigh frequencies because its leads are very short and therefore have a small value of inductance. The spacing between cathode and plate is very small, allowing *transit time,* the time taken by electrons in moving from the cathode to the plate, to be cut to a minimum.

The requirements for all tubes operating at higher frequencies are to have minimum inductance in the leads to the elements, short transit time, and as little interelectrode capacitance as possible.

One of the first successful high-frequency tubes was the *acorn* type, which looked like a glass acorn. The elements were very small. The leads were brought out directly through the sides of the glass envelope to shorten them as much as possible and to reduce interelectrode capacitance. It operated up to several hundred megacycles.

There are other special VHF and UHF tubes, such as the *nuvistors,* useful up to 1,000 MHz.

A few of the thousands of different tubes that have been manufactured are shown in Fig. 9-43.

Fig. 9-43 Electron tubes. Top row, from left, three octal (8-pin) metal tubes, a loctal glass tube, three 7- or 9-pin glass miniature tubes, two subminiature tubes, and a nuvistor high-frequency tube. Bottom row, vacuum phototube, glass octal amplifier, gaseous VR tube, two low-power lighthouse tubes, air-cooled anode lighthouse tube, and air-cooled tetrode transmitting tube.

9-38 VISUAL-INDICATING TUBES

There are some types of tubes that give visual indications. One is the *tuning-eye* tube, or electron-ray tube, used in receivers and discussed in Chap. 18. Another is the cathode-ray tube, used in oscilloscopes, television receivers, radar, loran, and measuring devices. The basic operation of the cathode-ray tube is discussed in Chap. 12, and its applications in other chapters.

Test your understanding; answer these checkup questions.

1. What are two methods used to shield tubes? _____ _____

2. How many grids does an octode tube have? _____ Hexode? _____ Heptode? _____ Pentagrid? _____

3. What is the difference between E_g and e_g? _____

4. What does E_{bb} mean? _____ E_{cc}? _____ E_{ff}? _____

5. What element connection is usually brought out at the top of transmitting tubes? _____ Why? _____

6. What are the three methods used to cool VT plates? _____ _____ _____

7. What does the meter between pump and ground indicate in a water-cooled tube system? _____

8. What are the two most likely causes of excessive I_p? _____ _____

9. What are the three most likely causes of insufficient I_p? _____ _____ _____

10. What are three possible causes of filament failure? _____ _____ _____

11. What are two reasons why tubes become gassy? _____

12. What may cause internal shorts in a VT? _____

13. Which VT battery has least current drain? _____ Most? _____

14. What are the names of three specially constructed VHF or UHF tubes? _____ _____ _____

15. What are the three effects that must be reduced in high-frequency tubes? _____ _____

16. What do you think would be the best test when checking receiving-type tubes, I_p, μ, I_f, g_m, or I_g? _____ The next best? _____ The most practical for transmitting tubes? _____ The next best? _____

COMMERCIAL LICENSE QUESTIONS

Amateur license questions will be found in the Addendum.

FCC elements 3 and 6 require an ability to answer questions similar to those below. Sections in which questions are answered are shown in parentheses.

1. What is the meaning of *electron emission?* (9-2)
2. What is a *cathode* in a vacuum tube? (9-2)
3. Why is it desirable to use an ac filament supply for vacuum tubes? (9-2, 9-8)
4. Why should the cathode of an indirectly heated type of vacuum tube be maintained at nearly the same potential as the heater circuit? (9-2)
5. Are there any advantages or disadvantages of filament-type vacuum tubes when compared with the indirectly heated types? (9-2, 9-19)
6. What is meant by *space charge?* (9-2)
7. What is the composition of filaments, heaters, and cathodes in vacuum tubes? (9-3)
8. Is a tungsten filament operated at higher or lower temperatures than a thoriated filament? (9-3)
9. What kind of vacuum tube responds to filament reactivation, and how is reactivation accomplished? (9-3)
10. Is it important to maintain transmitting-tube filaments at recommended voltages? (9-4)
11. What is the meaning of *plate saturation?* (9-5, 9-23)
12. What is meant by the *load* on a vacuum tube? (9-6)
13. When an ac filament supply is used, why is a filament center tap usually provided for the vacuum-tube plate and grid return circuits? (9-7)
14. Why is it advisable to reverse periodically the polarity of the filament potential of a high-power vacuum tube with a dc filament supply? (9-8)
15. What are the physical characteristics and common usage of diode tubes? (9-9)
16. What is the *getter* in a vacuum tube? (9-10)
17. What is meant by a *soft* vacuum tube? (9-10)
18. What does a blue haze in the space between filament and plate of a high-vacuum rectifier tube indicate? (9-10)
19. What are some of the indications of a defective vacuum tube? (9-10, 9-35)
20. What are the primary characteristics of a gas-filled rectifier tube? (9-11)
21. What are the advantages of mercury-vapor as compared with high-vacuum rectifier tubes? (9-11)
22. Describe the construction and characteristics of a battery-charger rectifier tube. (9-11)

23. What are the physical characteristics and common usage of cold-cathode tubes? (9-12)
24. Explain the principle of operation of cold-cathode gaseous rectifying diodes. (9-12)
25. What are the physical characteristics and common usage of thyratron tubes? (9-13)
26. What is the primary purpose of the control grid of a triode? (9-14)
27. Discuss the physical characteristics and common usage of triode tubes. (9-14, 9-19)
28. What is meant by the *amplification factor* or μ of a triode vacuum tube? (9-15)
29. Under what conditions would amplifier gain approach the value of the mu (μ) of the tube? (9-16)
30. Draw a diagram of a resistance load in the plate circuit of a vacuum tube. Indicate the direction of electronic flow in this load. (Fig. 9-24)
31. Explain the operation of a triode as an amplifier. (9-16, 9-17)
32. What is meant by the voltage *gain* of a vacuum-tube amplifier? (9-16)
33. What is the purpose of a bias voltage on the grid of an amplifier tube? (9-17, 9-24)
34. What is the direction of electronic flow in the grid circuit of a vacuum tube? (Fig. 9-24)
35. What occurs in the grid circuit when the grid is driven positive? (9-17)
36. What is meant by *plate-current cutoff?* (9-18)
37. Draw a graph indicating how the plate current in a vacuum tube varies with plate voltage, grid bias remaining constant. (9-18)
38. Draw a graph of plate-current versus grid-voltage ($E_g I_p$) curves for various plate voltages on a triode tube. (9-18)
39. How would I_p vary with input signal voltage in a class A amplifier? (9-18)
40. What is the relationship between distortion in the output-current waveform and (*a*) the class of operation, (*b*) the portion of the transfer characteristic curve over which the signal is operating, (*c*) amplitude of input signal? (9-18)
41. Draw a graph of plate-current versus plate-voltage curves for different bias voltages on a typical triode vacuum tube. (Fig. 9-27)
42. Operation over which portion of the $E_g I_p$ curve produces the least distortion? What effect does a resistive plate load have on the $E_g I_p$ curve? (9-18)
43. What is meant by *plate resistance* of a vacuum tube and on what does its value depend? (9-20, 9-21)

44. What is the meaning of the term *maximum plate dissipation?* (9-23)
45. Why is the efficiency of an amplifier operated in class C higher than if operation were in class A or B? (9-24)
46. What is meant by *secondary emission?* (9-25)
47. What are the physical characteristics and common usage of tetrode tubes? (9-26)
48. What is the primary purpose of a screen grid in a tube? (9-26)
49. Compare tetrode tubes with triode tubes in reference to high plate current and interelectrode capacitance. (9-26, 9-27)
50. What are the physical characteristics and common usage of beam-power tubes? (9-27)
51. What are the physical characteristics and common usage of pentode tubes? (9-28)
52. What are the physical characteristics and common usage of remote-cutoff tubes? (9-28)
53. What is the primary purpose of a suppressor grid in a multielement vacuum tube? (9-28)
54. Draw a diagram of a grounded-cathode AF pentode amplifier with battery biasing. (Fig. 9-36)
55. What are the physical characteristics and common usage of duodiode tubes? (9-31)
56. What are some possible causes of overheating vacuum-tube plates? (9-35, 9-36)
57. What factors may cause low plate current in an amplifier tube? (9-36)
58. Name at least three abnormal conditions which would tend to shorten the life of a vacuum tube and the probable causes. (9-36)
59. What is an *A battery?* A *B battery?* A *C battery?* (9-36)
60. What are *lighthouse* and *acorn* tubes and for what frequency range were they designed? (9-37)
61. Why are special tubes sometimes required at UHF and above? (9-37)

ANSWERS TO CHECKUP QUIZ ON PAGE 170

1. (*Metal shields, metal envelopes*) 2. (*6)(4)(5)(5*) 3. (E_g *is a dc or rms value; e_g is an instantaneous value*) 4. (B *battery voltage)(C battery voltage)(A battery voltage*) 5. (*Plate)(Insulation and minimum g-p capacitance*) 6. (*Radiation, air, water*) 7. (*Degree of water inpurity*) 8. (*Low or + bias on grid, gas in tube*) 9. (*Low E_p, low filament emission, high negative bias*) 10. (*Jarring, wearing thin and opening, losing emission*) 11. (*Air through envelope, gas from overheated elements*) 12. (*Loose element welds, heated elements expand and touch*) 13. (*C)(A*) 14. (*Acorn, lighthouse, nuvistor*) 15. (*Lead inductance, interelectrode C, transit time*) 16. ($G_m)(I_p)(I_p)(I_g$)

10

BASIC SOLID-STATE DEVICES

The objective of this chapter is to provide a basic coverage of the most generally employed solid-state devices, such as diodes of various types, bipolar junction transistors (BJTs), FETs, MOSFETs, SCRs, and ICs, as well as a few of the characteristic circuits in which they are used.

10-1 SOLID-STATE DEVICES

An electronic semiconductor component capable of controlling electron flow through it in some way is known as a solid-state device. Vacuum-tube devices control the flow of electrons through a vacuum. Solid-state devices control the flow of electrons along a solid piece of semiconductor material. A representative list of some vacuum devices might be capacitors, diodes, photodiodes, triodes, tetrodes, and pentodes. Gaseous control devices include diodes, neon or other glow tubes, and thyratrons. Solid-state devices include voltage-variable capacitors, rectifier diodes, zener diodes, tunnel diodes, photosensitive diodes, light-emitting diodes, a variety of junction transistors, a variety of field-effect transistors, silicon-controlled rectifiers, triacs, diacs, light-activated SCRs, and thermistors, to name a few.

Solid-state devices are much smaller physically, more rugged, and lighter than vacuum types but can not withstand heat as well, changing their characteristics as they warm. However, many newer special solid-state devices will operate at much higher frequencies than any vacuum devices can.

10-2 DOPED SEMICONDUCTORS

The outer-orbit, or valence, electrons of some atoms, such as metals, can be detached with relative ease at almost any temperature and may be called *free electrons*. These valence electrons are able to move outward from a normal outer-orbit level into a *conduction level* or *band,* from which they can be dislodged easily. Such materials make good electrical conductors. Other substances, such as glass, rubber, and plastics, have no free valence electrons in their outer conduction bands at room temperatures and are good insulators. A few materials have a limited number of electrons in the conduction level at room temperatures and are called *semiconductors*. Applying energy in the form of *photons* (small packets of light or heat energy) to the valence electrons moves some of them up into the conduction band, and the semiconductors then become better electrical conductors. Energy of some form is required to raise semiconductor electrons to a conduction level. Conversely, if an electron drops to the valence level from a higher conduction level, it will radiate energy in some high-frequency form such as heat, light, infrared, ultraviolet, and, if the fall is great enough, x-rays.

Two semiconductors, germanium and silicon, have four outer-ring electrons. Crystals of these can be laboratory-grown in pure (intrinsic, I) form. A perfectly formed intrinsic semiconductor crystal lattice, as illustrated in Fig. 10-1*a,* acts more like an insulator than a conductor at room temperature. However, if during manufacture of a germa-

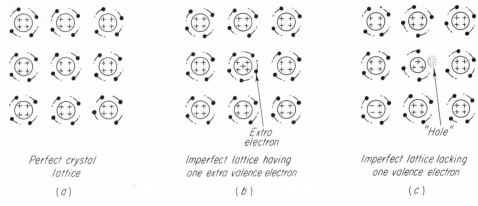

Fig. 10-1 (*a*) I germanium. (*b*) N germanium. (*c*) P germanium.

nium crystal about one in a million atoms of an impurity such as arsenic, with five outer-ring electrons, is added, the resulting crystal is imperfect, as in Fig. 10-1*b*. It has one in a million atoms in the lattice with an apparently excess outer-ring electron not being tightly held. When an electrostatic field is developed across such *arsenic-doped* germanium, the semiconductor is about 1,000 times better as a conductor. Doped germanium with such relatively free electrons is known as *N germanium,* and it is a reasonably good conductor. (To form *N silicon,* phosphorus can be used as the dopant for silicon.)

When germanium is doped with gallium, which has three valence electrons, the crystal lattice is again imperfect, as in Fig. 10-1*c*. This time there is an area, or *hole,* in the lattice that apparently lacks an electron. While the hole may not be actually positive, at least it is an area in which electrons might not be repelled by a negative charge. This positive-appearing semiconductor material is called *P germanium*. When an electrostatic field is impressed across a P-type semiconductor, the hole areas act as stepping stones for electron travel through the material. It can be said that *hole current* flows in a direction opposite to the electron flow. Note that both N germanium and P germanium have zero electric charge because both have an equal number of electrons and protons in all of their atoms. (One dopant used to produce P silicon is boron.)

Doped silicon has considerably more resistance than germanium, but it is useful in higher-voltage applications, does not change its resistance as

much when heated, and can withstand greater temperatures without its crystalline structure being destroyed.

10-3 SOLID-STATE DIODES

When N-doped and P-doped semiconductor materials are grown together to form a single long crystal, as in Fig. 10-2*a,* a solid-state diode results. The area in which the N and P substances join is called the junction. Some of the relatively free electrons in the N-type material move into the more or less positively charged holes in the adjacent P-type material, developing an area at the junction which is actually slightly negative on the P side of the junction and slightly positive on the N side. This produces a barrier to any further electron flow of about 0.2 V with germanium and 0.6 V with silicon diodes.

The bias battery in Fig. 10-2*b* has its negative terminal to the N-type semiconductor and its positive terminal to the P-type. The negative-to-positive electrostatic field developed by the battery across the junction overcomes the junction barrier voltage, and current flows through the *forward-biased* junction. It is necessary to add a current-limiting resistor to prevent excessive current flow in the semiconductor because of its low resistance in the forward-biased condition. Because of this low resistance, the voltage-drop across a conducting solid-state diode is very small in comparison with that across a conducting vacuum diode.

The circuit in Fig. 10-2*c* has the bias connected

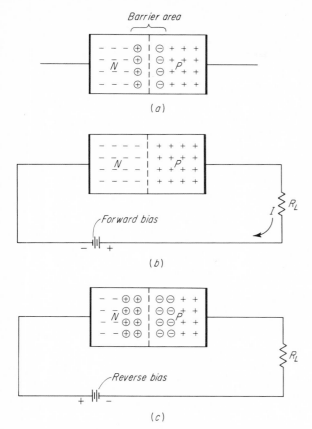

Fig. 10-2 (*a*) Solid-state diode develops barrier area with no bias. (*b*) Forward bias neutralizes barrier. (*c*) Reverse bias increases barrier area.

in reverse. The negative pole of the battery drives electrons into the holes of the P-type material, while the positive of the battery pulls electrons from the N-type material. The barrier area of the junction widens and no current flows. In this condition the diode has very high resistance. (There are always a few free electrons in P-type materials and holes in N-type materials, particularly at a junction. As a result of these *minority carriers* there is always a small backward leakage current in solid-state devices.) If the reverse voltage is excessively high, the barrier may break down and reverse current will flow. This is called *zener effect* if the emf value at breakdown is less than about 5 V and *avalanche* if it is more than about 5 V. This is not a normal operating condition for most semiconductor diodes and may cause lattice damage, ruining the diode.

The reverse-voltage breakdown effect, however, is used in special *zener diodes*. Curves for three diodes are shown in Fig. 10-3*a*. The solid line represents the actions of a silicon diode that has a low reverse-voltage breakdown [$V_{(BR)r}$]. In the forward-biased direction current starts flowing when V_f exceeds 0.6 V. In the reverse direction, V_r, the diode breaks down at about 4 V. If V_r were increased a few volts more, the diode might be destroyed unless manufactured to operate at high reverse currents or unless a resistor were in the circuit to limit current flow. The circuit across which the diode is connected will not increase in voltage over the 4 to 5 V of the zener effect. For this reason, zener diodes are used as shunt (parallel) voltage-regulating devices. Zener diodes with less than a 5-V breakdown will have a negative temperature coefficient (TC) of resistance, but with breakdown values over 5 V have a positive TC. (In the forward-biased direction all diodes have a negative TC.) The dashed lines might also represent characteristic operation of normal solid-state diodes as well as zeners. Figure 10-3*b*

Fig. 10-3 (*a*) Zener and avalanche curves for zener diodes. (*b*) Symbols used for zener diodes.

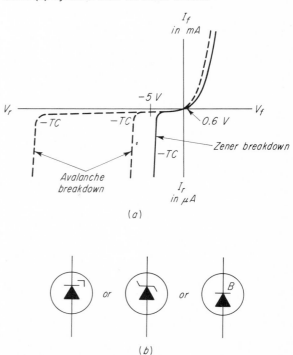

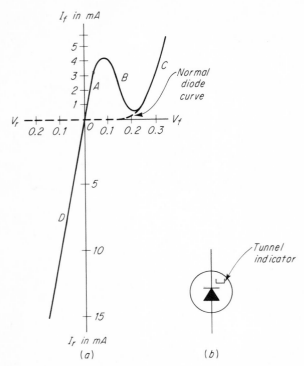

Fig. 10-4 (a) Tunnel-diode curve. (b) Symbol of tunnel diode.

shows symbols used to designate zener or break-down diodes.

A *voltage-variable capacitor,* called a *varactor,* utilizes the variation of barrier width in a reverse-biased diode. Since the barrier of a diode acts as a nonconductor, a diode forms a capacitor when reverse-biased, with the N material as one plate, the P material as the second plate, and the junction as the dielectric. If the reverse-bias voltage is increased, the barrier (dielectric) widens, effectively separating the two capacitor plates and reducing the capacitance. The frequency of resonance of an LC circuit can be varied by using a varactor across a coil, discussed in Chap. 19.

A *tunnel diode* is a heavily doped germanium or gallium arsenide diode with radically different VI characteristics. Because of the heavy doping, the diode acts as a good conductor when reverse-biased and has no zener effect, as shown by the solid line in Fig. 10-4a. The heavy doping allows the valence and conduction levels of the N- and P-area atoms to overlap under zero-bias conditions. With a small forward-bias voltage the

diode acts as a conductor and electrons "tunnel" through the barrier area with the speed of light. As V_f increases, I_f increases linearly up to about 0.05 V. At this emf value the electrostatic field across the junction begins to develop a barrier. For a further small increase in forward bias voltage the I_f decreases sharply. This part of the operation curve represents a *negative-resistance* effect (opposite to Ohm's law) in which an increase of E results in less I. At a little more than 0.2 V the junction begins to behave as a normal forward-biased junction, shown by the dashed curve. The negative-resistance effect is used in oscillators (Chap. 13) and in microwave amplifiers (Chap. 26).

When a doped semiconductor crystal is formed against a metal conductor, the heavily occupied conduction band of the metal and the lightly occupied conduction band of the semiconductor average out to be a diode junction which requires almost no voltage to produce conduction but which does not conduct under reverse-voltage conditions. Furthermore, since there are no minority carriers in the junction, current turn-on and turn-off can occur instantaneously. In any all-solid-state diodes, minority carriers must be swept out of the junction before complete turn-off occurs. These fast-acting diodes are called *Schottky* or, because they require so little energy to carry electrons across the junction, *hot-carrier diodes* (HCDs).

10-4 LIGHT-FREQUENCY DIODES

A *light-emitting diode* (LED), also known as a solid-state lamp (SSL), utilizes the fall of an electron from the conduction level to the valence level to develop an energy release in the form of heat or light. An electron moving across any PN junction moves to a hole area. This can allow a nearby conduction electron to fall to its valence level, radiating energy. In common diodes and transistors made from germanium, silicon, or gallium arsenide, this electromagnetic radiation is usually at a heat frequency, which is lower than light frequencies. With gallium arsenide phosphide the radiation occurs at red light frequencies. Gallium phosphides produce still higher frequency yellow through green radiations. Gallium nitride radiates blue light. An LED in a biased circuit is shown

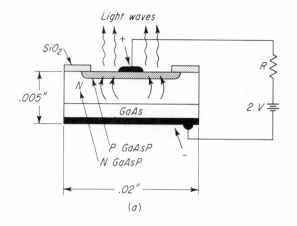

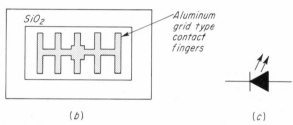

Fig. 10-5 LED construction: (a) cross section and (b) top view looking into light-emitting surface. (c) Symbol.

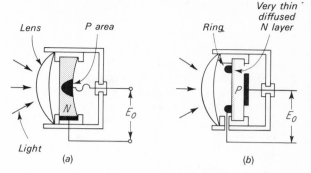

Fig. 10-6 Two forms of photovoltaic diodes.

in Fig. 10.5a. Starting with the bottom gold contact, there is a gallium arsenide layer, an N-type gallium arsenide phosphide layer, and then a very thin P-type GaAsP layer. Light passing through the thin P GaAsP layer develops red light radiations. About 5 mA of current flow in the direction shown produces a weak red glow. A current of 20 mA results in quite a bright glow. Maximum safe current may be in the 100-mA region. At about 20 mA the life expectancy is about 100,000 h. To prevent a maximum glow at only the tiny positive electrode, this contact is made in the form of fingers of aluminum spread out across the P GaAsP surface, as shown in Fig. 10-5b, resulting in a broader more visible area of red glow. Since this is a diode, reversing the bias battery will produce a barrier at the junction, no current will flow, and no glow will appear.

There are several photodiodes and photosensitive devices. The photodiodes convert photons to electric emf. Two of these are illustrated in Fig. 10-6. In (a) an N silicon chip has a P Si diffused into it until the front of the N Si is so thin that

light can penetrate to the PN junction. Photon energy lifts electrons from the valence level of the P Si and injects them above the conduction level of the N Si. As long as light strikes the junction, the diode converts photon energy to emf, up to about 0.5 V, across the diode. Figure 10-6b indicates a metal ring connector held against a thin diffused N layer on a P substrate with a contact against the back of it. Light striking the PN junction develops a movement of electrons into the P Si. These devices are called photovoltaic cells or diodes.

A *photojunction,* or *photoresistive,* device is not a diode; it is a semiconductor material that increases its conductance (reduces its resistance) when struck by photons of light or heat. Photojunction devices are usually slow-acting, whereas the photovoltaic diodes act instantaneously.

A PIN-type silicon photodiode is shown in Fig. 10-7. From the bottom, it consists of a gold electrode, a thin N Si layer, a thick I Si layer, a thin P Si layer, and finally a gold electrode making contact with the P Si. The silicon dioxide is an

Fig. 10-7 PIN photodiode.

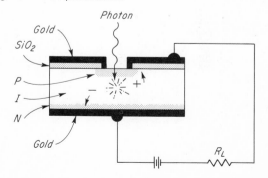

insulating or passivating (protecting) material. Photons of heat or light passing through the thin P Si layer strike atoms in the intrinsic layer and produce free electrons and holes. The electrons move to the positive bias potential, and electrons fill the holes from the negative potential. The stronger the light the more electron-hole pairs developed and the more current that can flow. These are not true diodes, but are very fast acting photoresistors.

Mounting an LED facing a photodiode cadmium sulfide cell in a tiny lighttight enclosure produces an *opto-isolator*. When current flows in the LED, it illuminates the CdS cell and produces current in it. An opto-isolator allows signal transfer without coupling wires, capacitors, or transformers. It can couple digital (on-off) or analog (variable) signals.

A variety of diodes are shown in Fig. 10-8.

Test your understanding; answer these checkup questions.

1. Where must valence electrons be in conductors? _____

2. What are small packets of light or heat energy called? _____

3. When impurities are added to a semiconductor crystal lattice, what is the process called? _____

4. Is P germanium positively, negatively, or neutrally charged? _____ N Si? _____

5. Heating a semiconductor has what effect on its conductance? _____

6. Is the P material of a diode positive, negative, or neutral? _____

7. What is the junction barrier potential difference using Ge? _____ Si? _____

8. What current do minority carriers produce? _____

9. How does forward-biasing affect junction resistance?

10. What is another name for a varactor? _____ Would it normally be forward- or reverse-biased? _____

11. What are two materials used with tunnel diodes? _____ _____

12. What is different about the characteristic curve of a tunnel-type diode? _____

13. What makes Schottky diodes high-speed types? _____

14. What is an LED? _____ Would it rectify? _____

15. Which determines color in an LED, the material used or the current amplitude? _____

16. What frequency radiation is produced by a forward-biased Ge or Si diode? _____

17. Which are fast-acting devices: photodiodes, photoresistive devices; PIN devices? _____

18. What might be considered as the "majority carrier" in N Si? _____ In P Ge? _____

Fig. 10-8 Various types of diodes. The cathode end may be marked with a wide stripe or a dot, may be pointed to by an arrowhead, or may be the threaded end.

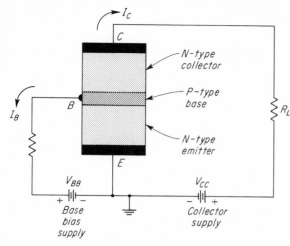

Fig. 10-9 NPN-transistor circuit showing currents flowing out of both base and collector.

10-5 JUNCTION TRANSISTORS

The basic transistor can be thought of as two diode PN junctions constructed in series, as shown in Fig. 10-9. From the bottom there is a contact plate against an N-type *emitter* element, E. Next to this is a thin P-type *base* element, B, with a metal electrode connected to it, forming the first PN junction. A second N-type *collector* element, C, is added, with a contact on it, forming the second PN junction. This produces an NPN transistor.

With the base element disconnected, regardless of the polarity of the collector supply, no current can flow in the emitter-collector (EC) circuit (disregarding a small minority current) because one of the PN junctions would always be reverse-biased. However, with the circuit connected as shown in Fig. 10-9, the emitter-base junction is forward-biased and electrons move from emitter to base. Since a semiconductor has few usable carriers, electrons must detour throughout the thin base layer to find a path to the base contact. In so doing some of them move through the base-collector junction area, aided by the higher collector voltage. With carriers in this junction it becomes a conductor, allowing emitter-collector current to flow. Increasing and decreasing EB current results in increasing and decreasing EC current. The significant feature about a transistor

is that a small variation of EB current can control 50 to 150 times as much EC current. Thus, the transistor is an ideal control and amplifying device. A junction transistor of this type can be called a *bipolar junction transistor* (BJT) to differentiate it from a field-effect transistor (FET) discussed later. Note that $I_E = I_B + I_C$ at all times.

Transistors are manufactured in a variety of ways, as indicated in Fig. 10-10. The original transistors were *point-contact* types, Fig. 10-10a.

Fig. 10-10 Methods of constructing transistors.

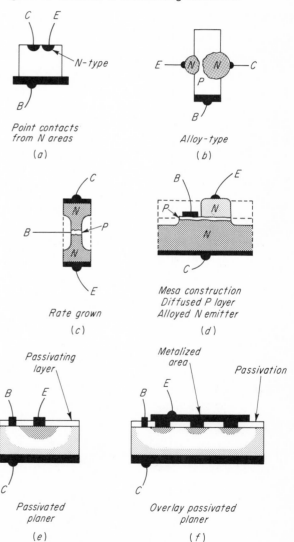

The P germanium chip had two wires, C and E, attached to its top. *Forming currents* developed N Ge pockets in the P Ge, leaving a small P Ge layer between them as the base. This type of transistor is no longer in general use.

An old method, still used in power transistors, has N-type pellets alloyed (thermally fused) into both sides of a P-type substrate, as shown in Fig. 10-10*b*, to form an NPN transistor (or P-type pellets alloyed into an N-type chip to form a PNP transistor).

During the growing of an N-type crystal it is possible to stop the process, add P-type liquid, slowly pull the crystal upward, and grow a thin P-type layer on it. Then, by changing to N-type liquid again, another N-type section can be grown on the crystal, as in Fig. 10-10*c*. The crystal is hollowed to give a small base area, and connections are made to E, B, and C to form a *grown* transistor.

An *epitaxial* layer is produced by starting with an N-type substrate, heating it, and diffusing it with hot P-type gas, resulting in a top layer of P-type material. By diffusing twice, first with P-type material to a given depth and then again to a lesser depth with N-type material, a *planar epitaxial* transistor is formed.

A *mesa* (tablelike) transistor, as shown in Fig. 10-10*d*, has an N-type substrate with a P-type diffused top layer and an alloyed N-type area over it. To reduce junction capacitances, much of the crystal is etched away (dashed lines show original size), leaving the collector as the largest volume element, since it must dissipate the greatest amount of heat. Power mesa transistors are also fabricated in a *homotaxial* manner, diffusing both sides of a chip with the same gas dopant and etching away unwanted areas.

The transistor shown in Fig. 10-10*e* is a *passivated planar* type. The passivating or protecting layer consists of an insulating material such as silicon dioxide, which protects the PN junctions and prevents unwanted surface leakages across them.

To obtain high-frequency operation, the turn-on and turn-off times must be reduced by increasing the edge-to-area ratio of the emitter. An *overlay* transistor does this. The emitter is developed into many tiny islands (or fingers) that spread out over the base region, illustrated in Fig. 10-10*f*.

A *tetrode* transistor has a second base connection at the side opposite from the first, as shown in Fig. 10-11. A dc emf, negative to B_2 and posi-

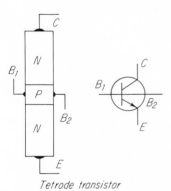

Tetrode transistor

Fig. 10-11 A tetrode or NPNP transistor.

tive to B_1, produces an electrostatic field across the base that forces emitter-collector current to flow only near B_1. Restricting active base area in this way results in a smaller active area of the base, less base current, less interelement capacitance, and thus better high-frequency operation.

A *phototransistor* is a very sensitive photoresistor. When light is allowed to fall on the EB and BC junctions, EC current flows. The brighter the light the greater the EC current.

10-6 COMMON-EMITTER CIRCUITS

The basic transistor amplifier circuit is the common-emitter, shown in Fig. 10-12*a* with an NPN transistor and in Fig. 10-12*b* with a PNP BJT. The notable differences between the two circuits are reversed arrow directions in the symbols and reversed currents due to reversed bias and collector supply polarities. (Note similarities of: cathode-emitter, grid-base, plate-collector.)

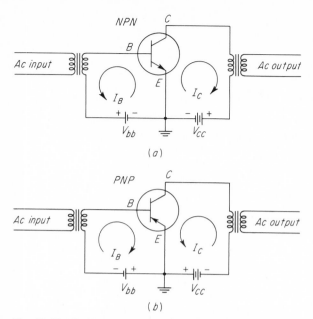

Fig. 10-12 Fundamental transistor amplifiers. Note polarities (a) with NPN and (b) with PNP transistors.

With a medium value of bias voltage, a medium value of I_B will flow in the input circuit and a medium I_C will flow in the output or collector circuit (class A operation). In a representative transistor, a current variation of 20 μA in the base circuit might produce a 2,000 μA (2 mA) change in the collector circuit, or a current gain of 100 times. A small audio ac coupled into the base circuit becomes a much larger audio current variation in the output.

Current gain is the ratio of collector current change occurring in the collector circuit load compared with the change in base current that is causing the I_C change, or

$$A_i = \frac{\Delta I_C}{\Delta I_B}$$

Beta (Greek letter β) is an expression of the change in the collector current of a transistor compared with the change in the base current when there is no load in the collector circuit and the emitter-collector voltage is held constant, or

$$\beta_{\mathrm{ac}} = h_{fe} = \frac{\Delta I_C}{\Delta I_B}\bigg|_{V_{EC}\text{ constant}}$$

The *dc beta* is the ratio of I_C compared with the I_B value that produces it, and it is roughly similar to the ac beta above. In formula form it is

$$\beta_{\mathrm{dc}} = h_{FE} = \frac{I_C}{I_B}\bigg|_{V_{EC}\text{ constant}}$$

Producible current gains in amplifiers will range around 70% of the published beta values with load resistors up to perhaps 10 kΩ.

The *alpha cutoff frequency* of a transistor is the frequency at which its current gain drops 3 dB from its gain at 1,000 Hz.

It is important to understand that a transistor must have the base-emitter circuit forward-biased to allow any collector current to flow. With no bias, or with a reverse bias, no I_C can flow. This is not true with vacuum tubes. With no bias a heavy plate current will flow. In a vacuum tube a reverse (negative) bias is always used to reduce static I_p. Note that a reverse bias is a negative voltage for an NPN transistor, similarly to the vacuum tube. Conversely, a PNP transistor is reverse-biased with a positive potential on its base. To simplify explanations, most circuits will be shown with NPN transistors, due to their similarity to vacuum-tube biasing, with the understanding that the circuits will operate with PNP transistors if the supply polarities are reversed.

The circuits shown are not practical because the battery bias voltage required for forward-biasing a germanium transistor is slightly over 0.2 V and slightly over 0.6 V for silicon. No common batteries have such emf values. The bias voltages for transistors are quite critical and must usually be held correct to small parts of one volt. With transistors it is easier to work with bias currents flowing through high-value resistors. Thus a certain low-power transistor might have a bias of 50 μA to produce a medium I_B and I_C flow. Another, higher-power transistor might require 10 mA of I_B to produce a medium value of I_C for that transistor.

One method of biasing a base circuit, called *fixed* bias, is shown in Fig. 10-13. The dashed line indicates the collector current path through transistor, load, and power supply. The dotted line represents the base-biasing current path into the emitter, out of the base, through the base-biasing

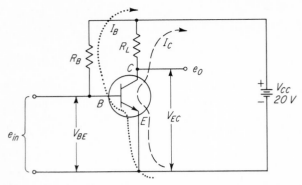

Fig. 10-13 Fixed bias in a common-emitter amplifier, showing base and collector current paths.

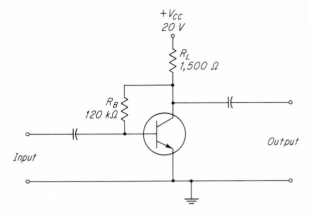

Fig. 10-14 BJT amplifier with self-bias.

resistor, and to the power supply. The value of biasing resistance is varied until the collector-emitter voltage, V_{EC}, is just half of the power-supply voltage, or 10 V in this case. An ac input signal to the base circuit can now vary the voltage-drop across the load, R_L, from a quiescent (no-signal) 10 V up to almost 20 V when the base is heavily forward-biased (collector-current-saturated) by the signal and down to 0 V when the base voltage drops below the EB junction barrier voltage of 0.2 V for germanium and 0.6 V for silicon. The R_B value is usually in the range of 200,000 Ω for low-power transistors and a few thousand ohms for power transistors.

Unfortunately, this simple biasing arrangement is quite unstable thermally. If the transistor warms for any reason, due to a rise in ambient (surrounding) temperature or due to current flow through it, the I_C value increases. The higher the current gain of the transistor the greater the instability of the circuit. Obtaining base bias current by connecting R_B to the bottom of the load resistor, as in Fig. 10-14, improves thermal stability by a factor of 2 or more. The value of R_B is then about half of the fixed-bias value. This is called *self-bias,* and it produces a collector form of *degeneration.*

When another resistor is connected emitter-to-ground, as in Fig. 10-15, a positive-going signal on the base increases not only I_B and I_C through R_L but also I_E through R_E. The increased voltage-drop across R_E reduces V_{EC}, and the gain of the stage is reduced. This is called degeneration, and it results in less output from the stage. However, the greater the degeneration the greater the ther-

mal stability, and usually the less distortion produced in the stage. To provide long-period thermal stability but to allow minimal ac signal degeneration, the bypass capacitor C_{bp} is added across R_E. If this capacitor is large enough (X_C is less than $\frac{1}{10} R_E$ at the lowest signal frequency), rapid signal variations do not change its charge materially and no degeneration of the signal from this part of the circuit is produced. Note the labeling of the various voltages in this stage. Other symbols often seen are V_{CBO}, the voltage between collector and base with the other element (emitter) open-circuited, and I_{CBO}, the minority carrier or leakage current between C and B with E open.

By far, the most used bias system with transistors is the voltage-divider type, shown with a

Fig. 10-15 Adding R_E improves dc thermal stabilization and adds degeneration.

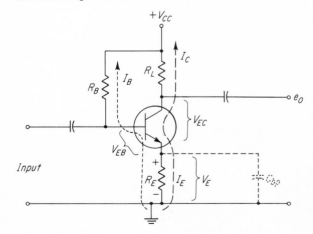

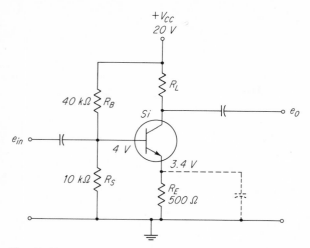

Fig. 10-16 Voltage-divider bias.

silicon transistor in Fig. 10-16. By ratio, the voltage-drop across R_S must be $^{10}/_{50}$ of 20 V, or 4 V. (Since I_B in R_B will be only $^1/_{100}$ of I_C in a 100-β transistor, it will be disregarded to simplify this explanation.) Because the V_{EB} of a silicon transistor is about 0.6 V, when the base is at $+4$ V, the emitter must be at about $+3.4$ V above ground (or $-V_{CC}$). With 3.4 V across 500 Ω, I_{R_E} must be $I = E/R,$ or 3.4/500, or 6.8 mA. Disregarding the small I_B again, I_C must be equal to I_E, or also 6.8 mA. Since R_L is 1,500 Ω and I_C is 6.8 mA, $E_{R_L} = IR,$ or 1,500(0.0068), or 10.2 V. The emitter-to-collector voltage must be 20 − (10.2 + 3.4), or 6.4 V. If equal V_{EC} and E_{R_L} values are desired, R_L must be decreased in value somewhat.

The addition of R_S increases thermal stability. The lower the resistance of the voltage-divider network the better the thermal stability but the lower the input impedance and the greater the loss of signal due to power dissipated in these resistors. The dashed bypass capacitor is used to prevent signal degeneration while maintaining long-term or dc thermal stability.

A special heat-sensitive resistor made of selenium, silicon, or metal oxides, having a negative TC and called a *thermistor,* can be physically attached to the transistor and electrically connected in series with R_S for further thermal stability. As transistor and thermistor warm, the transistor I_C tends to increase, but the thermistor

resistance decreases. This reduces the forward bias on the transistor, lowering the increase in I_C. The negative-TC characteristic of a forward-biased diode can be used in this circuit by adding the diode, connected in a forward-biased direction, in series with R_S. Some solid-state diodes can be used as thermistors.

Test your understanding; answer these checkup questions.

1. Of what type material is the base of an NPN transistor made? _____ The collector of a PNP? _____
2. How is the EC resistance affected when the EB junction is forward-biased? _____
3. What does BJT stand for? _____
4. What method of fabrication was used with the first transistors? _____ Is this a junction transistor? _____
5. What is the layer which protects junctions in transistors called? _____
6. What are the names of two types of high-frequency transistors? _____ _____
7. In a class A amplifier what value of I_C flows? _____ I_B? _____
8. What is the formula for current gain in a BJT? _____
9. What is the formula for beta? _____
10. What is the frequency at which the gain of a transistor drops 3 dB from its 1,000-Hz gain called? _____
11. Name three types of biasing circuits used in transistor amplifiers. _____ _____ _____
12. Increasing degeneration decreases what undesirable effects in a transistor amplifier circuit? _____
13. What symbol indicates minority carrier current between B and C with E open? _____
14. What type of biasing arrangement is probably the most common with transistor amplifiers? _____
15. What is a thermistor? _____ Must a diode be forward- or reverse-biased to act as a thermistor? _____
16. Would a thermistor be used in series with R_B, R_E, or R_S? _____

10-7 BJT CHARACTERISTIC CURVES

The diagram in Fig. 10-17a is a simple bipolar junction transistor resistance-load amplifier. The characteristic curves shown in (b) could be supplied by the manufacturer of the transistor or could be plotted by hand. The lowest solid line represents the collector current as the V_{EC} (or since these curves normally would be developed with no load, the V_{CC}) is varied from 0 to 25 V,

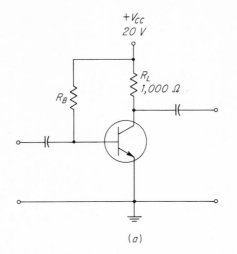

+V_{CC}
20 V

R_L
1,000 Ω

R_B

(a)

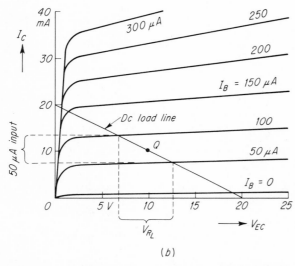

(b)

Fig. 10-17 (a) CE amplifier. (b) Load line and Q point with a 50-μA peak-to-peak input signal. V_{R_L} is output voltage.

with the I_B at the point of practically zero. The next curve is the I_C with a V_{EC} variation with I_B held at 50 μA, and so on.

What bias should be used when using an R_L of 1,000 Ω and a V_{CC} of 20 V? With these conditions known, two points can be marked on the graph. (1) 20 V for V_{EC} when the transistor is at zero bias and no I_C flows. (2) Maximum $I_C = E/R$, or 20/1,000, or 20 mA when the transistor is forward-biased to I_C saturation. A line drawn between these two points is the *load line*. A no-signal or quiescent point, Q, has been chosen close to the center of the load line, in this case at 75 μA. Resistor R_B can be found by Ohm's law, considering the voltage-drop across it to be 20 − 0.6, or 19.4 V. Thus $R_B = E/I$, or 19.4/0.000075, or 260,000 Ω.

A sinusoidal input signal of ±25 μA (Δ50 μA) peak value produces a sinusoidal I_C variation of about 5 mA peak, or a current gain of about 100. The voltage variation across the load with this input signal will be about 5.5 V. (The V_{EB} variation would be only a few hundredths of a volt.) The impedance of the transistor is found by $Z = \Delta E/\Delta I$, or 5.5/0.005, or 1,100 Ω. Since the Q point is approximately half of the V_{CC} value, it might have been expected that the transistor and the load impedances would be nearly equal.

10-8 POWER TRANSISTORS

Most transistors are tiny, low-power devices operating with fractions to a few milliamperes of I_C, as shown in Fig. 10-18. When power is required to drive loudspeakers or radiate RF energy, the transistors employed must be capable of at least hundreds of milliamperes, and usually many amperes, of I_C. Audio-frequency and radio-frequency transistors are different devices. The lower-frequency transistors are more or less standard types but made larger, resulting in more junction capacitance. Since C_j prevents high-frequency circuits from working properly, RF transistors must be specially designed overlay types or perhaps of the double-diffused or triple-diffused low-capacitance types.

A basic requirement in power transistors is to remove the heat developed in the device when I_C flows. The EB junction is forward-biased and therefore has a low resistance or impedance

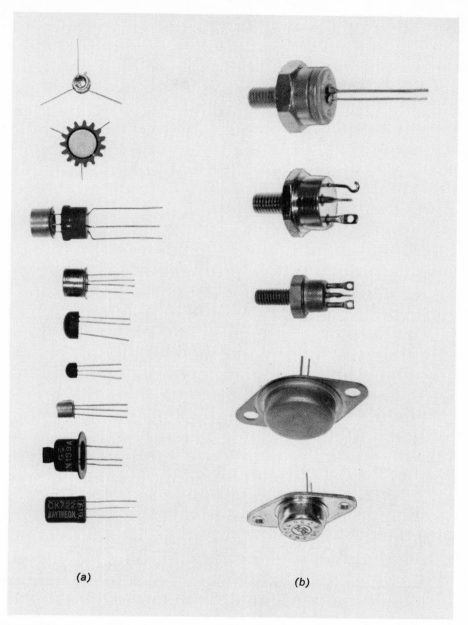

(a)   (b)

Fig. 10-18 Left, from top, phototransistor, transistor with a heat sink around it, transistor in a socket, and six varieties of transistors. Right, from top, silicon-controlled rectifiers (SCRs) and power transistors.

value. The CB junction, on the other hand, is reverse-biased and has considerably greater resistance. Thus, with I_E and I_C essentially equal (except for a small I_B in the I_E), from the power formula, $P = I^2R$, the higher-resistance collector junction will develop greater power in heat. For this reason the collector in power transistors is always constructed with a means of leading heat to the outside air to allow it to be dissipated. The collector is usually fastened to a relatively large

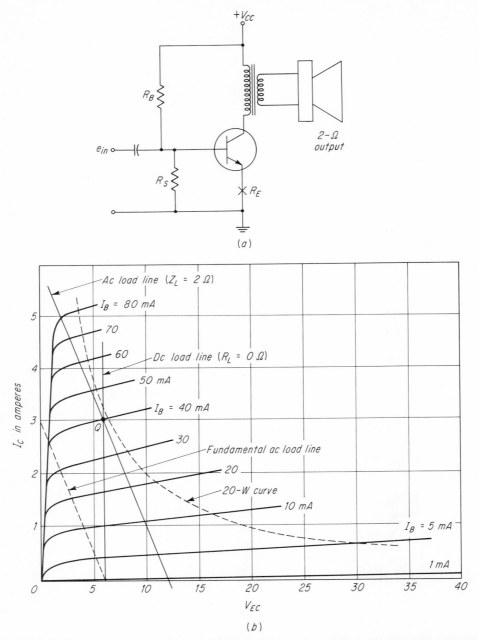

Fig. 10-19 (a) Power transistor in an AF amplifier circuit with an output transformer. (b) Dc and ac load lines for one type of power transistor.

metal body that can radiate heat developed in the collector junction. Any metal cooling device, such as fins clamped to the collector connection, is called a *heat sink*. A transistor that can safely dissipate only 1 W of power without a heat sink might safely dissipate 10 W with an efficient heat sink.

The circuit shown in Fig. 10-19a is a simple BJT

power amplifier. It is similar to previous amplifier circuits except that it has a transformer and loudspeaker as the load in the collector circuit. If additional degeneration is required, a 2- to perhaps 10-Ω emitter resistor might be added at point X in the diagram.

The characteristic curves shown in Fig. 10-19b represent a transistor with a 20-W collector dissipation rating. The dashed curve indicates that operations must be to the left side of this line, since it follows a 20-W value. (At $V_{EC} = 20$, $I_C = 1$, and $P = 20$ W. At $V_{EC} = 5$, $I_C = 4$, $P = 20$ W, and so on.)

Normally a dc load line is drawn from V_{EC} to I_C at its maximum possible value. With a low-resistance transformer primary as the load, the maximum I_C would be very high, resulting in an almost vertical load line, as indicated at 6-V V_{EC}.

When an ac signal is fed to the base-emitter circuit, the collector current will vary in the transformer primary (which has a relatively high ac impedance). If the transformer has a 1:1 turns ratio, it also has a 1:1 impedance ratio. If a 2-Ω load is connected across the secondary, the primary impedance will look into the 2-Ω load and present this impedance to the transistor. Thus, "ac" current flowing in the transformer should be $I = E/Z$, or 6/2, or 3 A. Using this value of current and 6 V as the V_{EC}, a fundamental ac load line can be drawn, as shown with long dashes.

The curves show that a bias of 40-mA I_B will produce approximately 3 A with a 6-V V_{EC}, which is safely to the left of the 20-W curve. This is a possible quiescent or Q value. An ac load line (solid line) is drawn through Q exactly parallel to the dashed fundamental load line.

A sinusoidal input signal that drives the base from its 40-mA quiescent value down to zero, up to 80, and back to 40 mA is an 80-mA peak-to-peak varying input signal. From the load line, a change of I_B from 80 to 0 mA produces a I_C change from 5 to 0 A. The voltage-drop across the transistor is from 2 to 12 (the 12 is the sum of the 6 V V_{EC} plus the inductive transformer primary voltage that is developed) or 10 V. The impedance of the transistor must be $Z = E/I$, or 10/5, or 2 Ω.

Since the load impedance matches the 2 Ω of the transistor, maximum power output will result. The actual power being produced across the 2-Ω

load can be found by $P = EI$, but effective values must be used, not these peak-to-peak values. If the p-p load voltage is 10 V, the peak value is 5 V and the effective value is 0.707(5), or 3.5 V. The effective value of the p-p 5 A is 0.707(2.5), or 1.77 A. Using these values, the output ac power is

$$P_o = EI = 3.5(1.77) = 6.2 \text{ W}$$

The power being dissipated in the collector of the transistor with no signal is the dc input power, or

$$P_{dc} = I_C V_{EC} = 3(6) = 18 \text{ W}$$

The efficiency of the stage, the ratio of the ac output to the dc input power is

$$\text{Eff} = \frac{P_o}{P_{in}} = \frac{P_{ac}}{P_{dc}} = \frac{6.2}{18} = 0.34 \text{ or } 34\%$$

Note the distortion being produced. An increase of 40 mA of I_B increases I_C by 2 A, but a decrease of 40 mA decreases I_C by 3 A. Had the curves been parallel and equally spaced, there would have been none of this distortion and the efficiency of the stage operating over the center of the load line (class A operation) would have been 50%, the theoretical class A maximum.

When the Q point is about half of maximum rated I_B, as in the transformer output circuit, without a very efficient heat sink internal transistor heating may occur. This can result in increasing I_C, an upward shift of the Q point (crossing the maximum dissipation curve), and a devastating *thermal runaway*: more heat, less semiconductor resistance, more current, more heat, and so on. When the Q point is about half of V_{CC} in resistance-load stages, self-heating is usually self-correcting if dc degeneration is used. Thermal runaway is not likely to happen at normal ambient temperatures.

10-9 COMMON-BASE CIRCUITS

The previously discussed *common-emitter*, or *grounded-emitter*, circuit might be considered the standard transistor amplifier. However, there are two other basic circuits. One is the *common-base*, or *grounded-base* amplifier. A simplified version

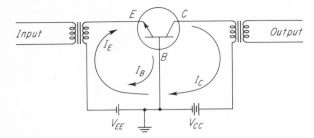

Fig. 10-20 Common-base amplifier.

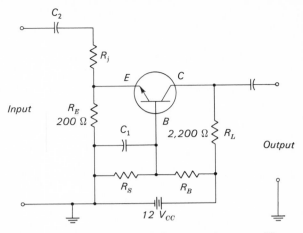

Fig. 10-21 Resistance-coupled common-base amplifier.

using batteries and transformers is shown in Fig. 10-20. The two batteries are in series and produce the emitter-collector current that flows through input and output circuits. The base, tapped up the battery, is forward-biased, allowing I_C to flow. The input circuit current, I_E, is composed of I_C plus I_B. The ratio of common-base I_C to I_E is known as the *alpha* (Greek letter α), and in formula form it is

$$\alpha = \frac{I_C}{I_E}$$

Since I_E is always greater than I_C, the alpha value is always less than unity in junction transistors. Alpha ranges from about 0.95 to 0.98. The smaller the I_B required to control the I_C the more nearly equal I_E and I_C will be and the higher the value of alpha. The alpha cutoff frequency of a transistor in a common-base circuit is the frequency at which the gain drops 3 dB from its gain at 1 kHz.

An input signal between E and B that forward-biases the emitter-base junction will increase both I_E and I_C further. The opposite-polarity input signal decreases the forward bias and I_E and I_C decrease. The phase characteristics of common-base amplifiers have input and output currents in phase, preventing self-oscillation (Chap. 14) of such amplifier stages, even at high frequencies. The input impedance is normally quite low, less than 200 Ω. The output load impedance can be anything from a few ohms to over 50 kΩ.

Although the CB amplifier has no current gain, it can have a voltage gain. Assume a 100-Ω input impedance and a 1,000-Ω output impedance. With essentially the same current flowing through both circuits the voltage-drop across the output

load will be 10 times that fed into the input. There is also considerable power gain in a CB amplifier. The I_C versus V_{CB} curves of a CB circuit are nearly parallel and equally spaced, producing a minimum of distortion.

In the resistance-coupled common-base amplifier shown in Fig. 10-21, the base is forward-biased with a voltage-divider network. The bias voltage is held constant by C_1. Since the emitter-base voltage varies very little during operation of a transistor, the voltage-drop across R_E remains essentially constant. The input signal current, via C_2 and R_i, takes a path through emitter to base and collector to ground. What small voltage difference develops across the EB junction produces the amplifying action in the transistor.

10-10 COMMON-COLLECTOR CIRCUITS

The third basic transistor amplifier circuit is the *common-collector,* or *emitter-follower,* shown in Fig. 10-22. This circuit is often used as either an impedance converter or an isolation stage. Although it has both current and power gain, an emitter follower always has less voltage output than input. The input impedance seen by the stage ahead is approximately R_L times beta, shunted by R_B, since R_B goes to ground through the power supply. (With a 100-β transistor in the circuit shown this would be 60,000 Ω shunted by 54,000 Ω, or 28,200 Ω input impedance.) The output impedance is usually slightly less than R_L,

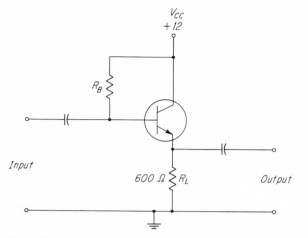

V_{CC}
+12

R_B

Input

600 Ω R_L

Output

Fig. 10-22 Common-collector or emitter-follower amplifier.

although if a voltage-divider biasing network is used with values of resistances in the few thousand ohms, the output impedance may appear in the 20- to 100-Ω range.

The emitter, or output, voltage follows the input signal almost exactly. The stage has nearly 100% degeneration, and there is little or no possibility of unwanted oscillations and distortion in the stage.

Table 10-1 shows some rough approximations

Table 10-1 PARAMETER APPROXIMATIONS

Parameter	Type of circuit		
	CE	CB	CC
Input Z	1,000 Ω	60 Ω	40,000 Ω
Output Z	40,000 Ω	200,000 Ω	1,000 Ω
Voltage gain	500	800	0.96
Current gain	20	0.95	50
Power gain	10,000	760	48
Phase in/out	180°	0°	0°

of impedances, gains, and phase relationships of the three basic amplifiers, the common-emitter, the common-base, and the common-collector.

Test your understanding; answer these checkup questions.

1. Where is the Q point on a load line for class A operation? _____

2. Why does the BC junction heat more than the EB? _____

3. What would cooling fins on a transistor be called? _____

4. What is the rms value of a 100-V p-p? _____

5. Is the power output from a transistor amplifier normally ac or dc? _____ The power input? _____

6. What is the formula for percent efficiency of an amplifier? _____

7. What is the maximum possible class A amplifier efficiency? _____

8. What will be the result of unchecked thermal runaway? _____

9. What is another term meaning the same as common emitter? _____ Common base? _____ Common collector? _____

10. Is the alpha of a transistor always more or less than unity? _____ The beta? _____

11. Does a CE amplifier have voltage gain? _____ Current gain? _____ Power gain? _____

12. Does a CB amplifier have V gain? _____ I gain? _____ P gain? _____

13. Does a CC amplifier have V gain? _____ I gain? _____ P gain? _____

14. If CB circuit I_C versus V_{CB} curves are parallel and equally spaced, what does this indicate? _____

15. What are two main uses of CC circuits? _____

10-11 JUNCTION FETs

Field-effect and bipolar junction transistors are entirely different devices. In fact, the FETs are solid-state amplifying devices that have operating characteristics somewhat similar to those of triode VTs. There are three types of FETs, the junction (JFET), and two *metal oxide semiconductor* types known as MOSFETs. One is a *depletion* MOSFET; the other is an *enhancement* MOSFET.

The essentials of a JFET are illustrated in Fig. 10-23a. A P-type substrate has an N-type area diffused into it, on top of which is diffused another P-type area. Normal barrier or depletion areas develop at the two PN junctions. The operating current of an N-channel JFET amplifier, as in Fig. 10-23b, is driven by the supply, V_{DD}, into the source end, S, of the N-type channel, out the drain end, D, and through the load. The two P-type areas are connected together and are called the *gate*, G. With no potentials connected to the gate, the source-drain current, I_D, would be relatively high.

By connecting a reverse-bias voltage, V_{GG}, between gate and source, the depletion areas are increased, reducing the volume of the N channel and tending to pinch off the drain current. If sufficient reverse gate bias is used, the I_D can be

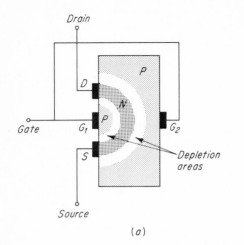

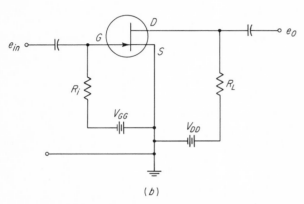

Fig. 10-23 (a) Diffused N-channel JFET. (b) Simple battery-biased JFET amplifier.

pinched off completely. Half of the pinch-off bias value results in a medium I_D and class A operation. A signal voltage (not current as in BJTs) added across gate-to-source varies the effective bias, resulting in a variation in I_D through the load and an amplified voltage-drop across the load resistor. The field-effect transistor is a voltage-operated device similar to the vacuum tube.

Because no current flows in an FET input (GS) circuit, there is no electron-current "shot effect" noise due to random electron motion, resulting in more noise-free amplification than is possible with BJTs or vacuum tubes.

One evaluation of a JFET is its *amplification factor*, or μ. This is the ratio of the change in gate-source voltage necessary to produce a constant drain current if the drain-source voltage is changed, or

$$\mu = \frac{\Delta V_{DS}}{\Delta V_{GS}}\bigg|_{I_D \text{ constant}}$$

It must be understood that the μ is not the voltage gain that can be achieved in any practical circuit. A JFET with a μ of 50 might operate with voltage amplifications (A_e) of 20 to 30. Either a higher load resistance value in the drain circuit or a greater V_{CC} increases the possible A_e.

Another measurement of a JFET's operation is its *dynamic drain resistance*, or r_d. The dc or ohmic resistance of the N channel of the JFET shown might read from a few hundred ohms with no bias to essentially infinite ohms with a pinch-off bias value. The impedance from source to drain to varying signal voltages is more significant. Dynamic drain resistance or impedance (Z_d) is the ratio of a change in drain-source voltage compared with the change in drain current that is produced by the voltage change, assuming the gate bias voltage is held constant, or

$$r_d = \frac{\Delta v_{ds}}{\Delta i_d}\bigg|_{V_{GS} \text{ constant}}$$

Values of r_d of different JFETs will range from a few thousand to several hundred thousand ohms. Matching this drain impedance to the load impedance produces maximum output power in the load.

The *transconductance*, or g_m, of a JFET is its ability to vary the output circuit current, I_D, when an input-voltage variation is applied. This is also called the forward transadmittance, y_f or g_f. In formula form

$$g_m = \frac{\Delta i_d}{\Delta v_{gs}}\bigg|_{V_{DS} \text{ constant}}$$

This can be expressed in mhos, or since it is always a small decimal number, in micromhos (μmhos).

The relation between μ, r_d, and g_m are

$$\mu = r_d g_m \qquad g_m = \frac{\mu}{r_d} \qquad r_d = \frac{\mu}{g_m}$$

Although only an N-channel JFET has been discussed, P-channel JFETs are also used. Supply polarities are reversed, as is the arrow direction of the gate in the schematic diagram symbol.

10-12 MOSFETs

When an FET is formed as indicated in Fig. 10-24a, with the gate insulated from the slim N channel by a thin layer of silicon dioxide, an insulated-gate FET, IGFET, or MOSFET (C-MOS,

Fig. 10-24 (a) N-channel depletion MOSFET amplifier. (b) Symbols of N-channel MOSFETs. (B indicates bulk or base contact, sometimes shown as U.)

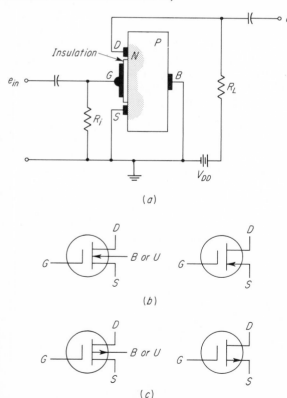

(a)

(b)

(c)

COS/MOS, V-MOS, etc., when used in complementary circuitry forms) is produced. The P-type base substrate leads may or may not be connected to source or ground.

When a voltage is applied, negative to gate and positive to source, its electrostatic field extends into the N channel and reduces its ability to carry current by repelling channel electrons, depleting the channel of its carriers. This is a *depletion* MOSFET. Sufficient reverse bias can pinch off the I_D completely. Conversely, a forward bias can increase I_D up to a certain point. As a result, this device can operate with medium I_D (class A) with no bias at all, which simplifies circuitry. Unlike vacuum tubes and JFETS, a forward bias does not produce any input circuit current (cathode to grid in VTs and source to gate in JFETS) due to the insulation between gate and channel. MOSFETs have extremely high impedance input circuits.

A set of transfer characteristic curves for a depletion MOSFET is shown in Fig. 10-25a. The same general information is contained in the V_{GS} versus I_D curve in (b). Zero bias is close to the center of both sets of curves.

When a MOSFET is formed as illustrated in Fig. 10-26a, N areas almost touching, with no bias there will be essentially no source-drain current flowing. If a forward bias (+ with this N-channel device) is used, it attracts negative minority carriers from the P area into the channel, providing carriers to support a source-to-drain current. The more forward bias the more I_D, as indicated in Fig. 10-26b. This type of MOSFET is known as an *enhancement* MOSFET because the channel requires carrier enhancement to produce current flow. Even with forward bias, no gate current can flow through the insulation. The input impedance may be in the hundreds or thousands of megohms. Note that in the symbols of MOSFETs the spacing between gate and channel represents the gate insulation.

Although only N-channel MOSFETs have been discussed, P-channel devices are also used. Their supply polarities must be reversed, and the arrowheads on the symbols are also reversed.

If the gate contact of a MOSFET is split down the middle and each half has a lead fastened to it, a *dual-gate* MOSFET results. Both gates can affect drain current amplitude.

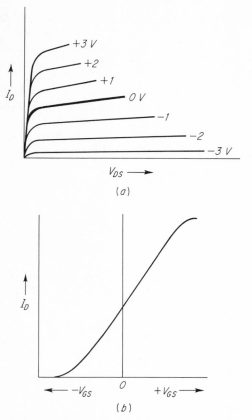

10-13 UNIJUNCTION TRANSISTORS (UJTs)

A bar of N-type semiconductor with contacts on both ends and a small P-type area alloyed into it, as shown in Fig. 10-27a, forms a double-contact diode which is called a *unijunction transistor*. The ohmic resistance between the *cathode*,

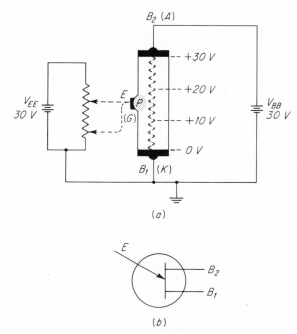

Fig. 10-25 (a) V_{DS} versus I_D transfer characteristic curves for N-channel depletion MOSFET. (b) V_{GS} versus I_D curve for the same MOSFET.

Fig. 10-27 (a) Unijunction transistor in test circuit. (b) Symbol for UJT.

Fig. 10-26 (a) N-channel enhancement MOSFET in amplifier circuit. (b) V_{GS} versus I_D curve. (c) Symbol. (d) Symbol of P-channel enhancement MOSFET.

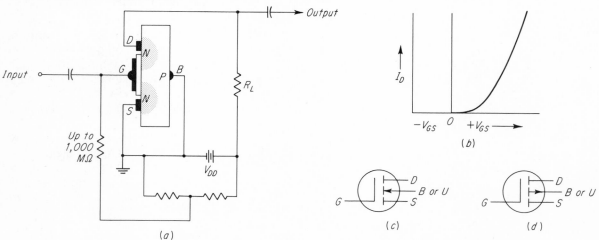

K, or base 1, and the *anode, A,* or base 2, may be about 10,000 Ω. The emitter or gate, E or G, is about 60% up the bar from base 1. The emitter acts as a voltage-divider tap on a fixed resistor, with E in the diagram at a potential of 60% of 30 V, or at 18 V. When V_{EE} is increased by moving the potentiometer arm upward, at about 18.6 V (silicon UJT) the PN diode becomes forward-biased and emitter current flows. As I_E starts, holes are injected into the N area between B_1 and E and this area becomes a good conductor. As I_E increases, the voltage-drop between B_1 and E decreases, which is opposite to Ohm's law. For this reason the UJT is said to exhibit negative resistance, an effect similar to regeneration (discussed in Chap. 13), greatly speeding the increase of I_E to a maximum and the fall of the B_1-E resistance to minimum.

When V_E is decreased to a low value, a point is reached at which the PN diode reverse-biases, and I_E decreases to zero as fast as it increased to maximum. While this transistor is not used as an amplifier, it finds use in certain types of oscillator and pulse circuits, as well as a means of firing SCRs.

10-14 SILICON-CONTROLLED RECTIFIERS (SCRs)

The device shown in Fig. 10-28 illustrates one of several NPNP, or multiple-layer semiconductor devices. This one is called a *silicon-controlled rectifier* (SCR). Since there are three junctions,

Fig. 10-28 SCR in a dc circuit.

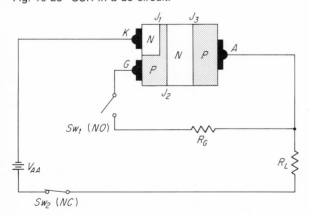

regardless of any reasonable potential connected across it, there will be at least one junction which will be reverse-biased, limiting current flow to some negligible value. The basic current path is from V_{AA} to the cathode, K, through the device to the anode, A, through the load, and back to the source.

Assuming V_{AA} is some value below the forward breakover voltage, $V_{f(BO)}$, of the device, negligible current should flow. If the normally-open (NO) gate-electrode switch, Sw_1, is closed, the first P-type segment, acting like the base of an NPN BJT, is forward-biased through R_G and R_L, allowing current to flow through J_2. Since J_3 is forward-biased already, current flows through the whole device, K to A, and through the load. The gate switch can now be opened and current will continue to flow because J_2 loses control of its current carriers. Even reverse-biasing the gate will not stop cathode-anode current. It is necessary to reduce V_{AA} to almost zero or open Sw_2 to stop current flow. No current will flow if the source potential is reversed, even with Sw_1 closed. Therefore, with an ac source the SCR acts as a rectifier (allows current to flow in one direction only) and is in effect a solid-state thyratron.

An ac lamp-dimming circuit operated by controlling the triggering voltage phase of an SCR is shown in Fig. 10-29. During the half cycle in which the SCR can fire, capacitor C charges through R_L and R_G until it builds up to enough voltage to fire the SCR. If R_G is a high resistance, C charges slowly and the SCR is fired late, resulting in only a few degrees of current (cycle1). If R_G is a medium value, C charges faster and current flows for perhaps 90° (cycle 2). With a low value of R_G, the capacitor charges almost instantaneously, the SCR fires immediately, and almost 180° of current flow is produced. The diode, D, prevents a reverse voltage from being applied to G when the ac polarity reverses.

One rating given SCRs is the maximum safe peak forward voltage (PFV). This is somewhat greater than the forward breakover voltage, and if exceeded, will usually damage the SCR. If reverse-direction voltage exceeds the *reverse breakdown voltage,* the SCR will go into thermal runaway and be destroyed.

If two similar SCRs with similar triggering circuits are connected in reverse, back to back, the

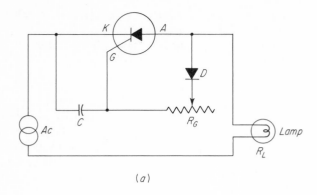

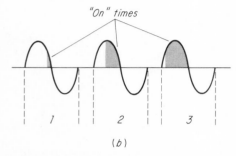

Fig. 10-29 (a) RC phase-control circuit produces (b) almost 180° of controlled ON time.

circuit will conduct in both directions and both halves of each cycle will add power to the load circuit.

10-15 TRIACS AND DIACS

Another breakdown semiconductor, or *thyristor* device, is the *triac*. It is a two-way SCR, shown in Fig. 10-30. Across the triac, the layers from terminals T_2 to T_1 are PNPN. When the source emf attempts to produce current flow downward through R_L, it is opposed by junction J_1, which is reverse-biased. When the emf reverses, J_2 prevents current flow. As T_2 starts to become positive, C begins to charge (shown dotted) through R and R_2. The voltage across C produces a forward bias for the NP junction next to T_1. The gate acts as a P-type base of an NPN transistor. A positive gate injects carriers across J_2, and current can flow across the top of the triac from T_1 to T_2. Once conducting, there is essentially no voltage-drop across the triac, and C discharges. The higher the resistance of R, or the greater the capacitance of C, the longer the delay of voltage

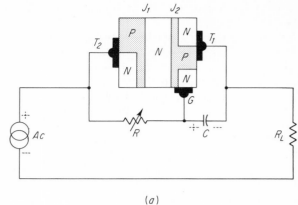

Fig. 10-30 (a) Triac and full-wave phase-control circuit. (b) Triac symbol.

buildup across C to fire the triac, controlling the average current flow.

When the source ac reverses, current through the triac ceases, and C starts charging with negative polarity toward the gate. When the gate NP junction is biased with a negative voltage, carriers are injected into the adjacent junction and current flows across the bottom of the triac. The triac acts as two SCRs back to back, in reverse direction, both directions being under the control of the one gate, in what can be termed a *full-wave* control circuit, since it operates on both half cycles of the ac.

Another thyristor device, called a *diac,* is a three-layer semiconductor, as shown in Fig. 10-31, with equal doping in both N layers (BJTs have heavy doping of the emitter). As an ac voltage develops across it, the diac reaches an ava-

Fig. 10-31 (a) Diac. (b) Symbol.

lanche point and breaks into conduction. The avalanche voltage is between 25 and 35 V in either direction. Diacs are usually used for delaying the firing of an SCR until the control voltage rises to about 30 V.

10-16 LIGHT-ACTIVATED SCRs

A light-activated SCR, also called an LASCR, is similar to a light-activated NPN transistor. It is constructed with a glass lens on one side of the metal case surrounding the semiconductor layers, as in Fig. 10-32a. When light falls on the junc-

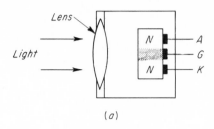

(a)

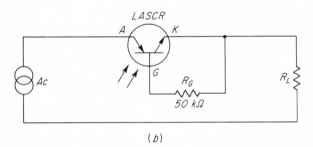

(b)

Fig. 10-32 (a) LASCR. (b) Light-operated LASCR circuit.

tions, the device latches ON and cathode-anode current can flow through the device. The diagram of Fig. 10-32b is a possible circuit that will turn on a LASCR when it is light-activated. The two arrows pointing to the symbol indicate light and are always part of LASCR or other light-activated device symbols.

10-17 INTEGRATED CIRCUITS

The advent of transistors greatly reduced electronic circuit size. When several transistors and diodes were developed on a single $\frac{1}{8}$- by $\frac{1}{8}$-in. substrate chip into what is called an *integrated circuit,* in some cases size was reduced to the almost unbelievable point.

An integrated circuit (IC) can be formed on a tiny P-type foundation chip. Into the top surface can be diffused an epitaxial N-type layer. (If the junction thus formed is reverse-biased, the N layer can be considered to be floating above ground potential.) If tiny spots of P-type material are diffused into the top of the epitaxial layer, and if leads are attached to them, diodes are formed. If still smaller N-type spots are diffused into the top of the P-spots, NPN transistors are formed. Another method of fabricating an IC might be to lay down four alternate P, N, P, and N layers, etch away undesired areas from the top layers, and add metallic connections to the required exposed areas to produce diodes and transistors.

The top of an IC may be insulated with a layer of silicon dioxide. The SiO_2 layer can be used as the dielectric for tiny capacitors by depositing a metallic cover layer over part of it to form one plate and utilizing a conductive P or N layer below it as the other plate. Channels diffused in an IC chip can also be used as resistors. Thus, one IC may contain many diodes, transistors, resistors, and small-value capacitors interconnected into various circuits. Because the separate devices on a chip are so small (they can only be seen with a microscope), one method of fabrication is to develop the desired circuitry in large form on a drawing board and then, by photographic means, to reduce the circuit and chemically reproduce the desired spots on the IC chip. By using electron beams in an evacuated chamber and programing the beam movement over a semiconductor chip by a computer, ten times as many transistors or diodes can be fabricated on the same size chip as is possible by photographic means. Electrical leads must then be attached to the required areas and be brought out to enable soldered connections to be made to the components of the IC. The number of leads from a single IC may range from three or four up to forty or more. Several complete amplifying circuits, plus power supply circuitry, plus separate transistors and diodes may be incorporated in a single IC.

One of the advantages of PNP and NPN transistors is the possibility of using them in *complementary* circuitry—where the same negative-going signal, for instance, increases the output current of the PNP but decreases the NPN output. Complementary, or *complementary symmetry,* FET circuits can be developed on a single base chip easily (difficult with BJTs). These can provide very useful COS/MOS (complementary symmetry MOSFET) integrated circuits.

There are two basic types of ICs. One is a *digital* type used in computers and other digital circuits, in which all devices must be able to change from fully on to fully off in fractions of a microsecond (actually, in a few nanoseconds). The other type is the *linear* IC, having transistors with characteristic curves that can be used in audio amplifiers. The number of circuit possibilities with ICs are endless, although IC handbooks illustrate some of the more standard combinations of circuits available off the shelf.

Test your understanding; answer these checkup questions.

1. A JFET is similar in operation to what other device? _____
2. Why is the JFET supply labeled V_{DD}? _____
3. In N-channel JFETs, at what forward emf and polarity would I_G begin to flow in silicon devices? _____ In germanium P-channel JFETs? _____
4. Under class A operation does I_G flow in a VT? _____ Does I_B flow in a BJT? _____ Does I_G flow in a JFET? _____
5. To what is the JFET pinch-off bias value analogous in VT operation? _____
6. What is the formula for μ in triodes? _____ In JFETs? _____
7. What is the formula for r_p in triodes? _____ For r_d in JFETs? _____ For r_d in MOSFETs? _____
8. What is the formula for g_m in triodes? _____ For g_f in JFETs? _____
9. In JFETs what is increased to pinch off the channel? _____
10. Why would most MOSFETs have a relatively low insulation-layer voltage tolerance? _____
11. For class A operation, which FET requires forward bias? _____ No bias? _____ Reverse bias? _____
12. Which FET would have the lowest Z_{in}? _____
13. In UJTs, what is produced once the emitter voltage is raised to the start of I_B? _____ What effect does this have? _____
14. For what are UJTs used? _____
15. How many layers are there in an SCR? _____

What must be done to start current in an SCR? _____
16. Why are SCRs usually used in ac circuits? _____
17. What is the name of a single-unit solid-state two-way SCR? _____
18. Name three thyristor devices. _____ _____
19. Which SCR is turned on by high-frequency radiations? _____
20. What are the four types of devices listed as being built into ICs? _____ _____ _____ _____
21. Could MOSFETs be built into ICs? _____ FETs?
22. What are the two types of ICs? _____
23. How many microseconds (μs) in 45 nanoseconds (ns)? _____
24. Identify the symbols shown in Fig. 10-33 on a separate piece of paper.

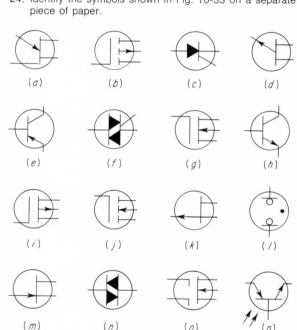

Fig. 10-33 See if you can identify these symbols.

25. Draw diagrams of the following circuits and check against those shown in the chapter. NPN CE amplifier with fixed bias. NPN CE power amplifier with voltage-divider bias. NPN CB resistance-coupled amplifier. NPN emitter-follower. N-channel JFET amplifier. N-channel depletion MOSFET amplifier using symbol. N-channel enhancement MOSFET amplifier using symbol. SCR with RC phase control in ac circuit.

COMMERCIAL LICENSE QUESTIONS

Amateur license questions will be found in the Addendum.

FCC Elements 3, 4, and 6 require an ability to answer questions similar to those below. Sections in which questions are answered are shown in parentheses. A question followed by a bracketed number is required for that element alone.

1. What are the main disadvantages of using transistors in circuits rather than vacuum tubes, assuming cost is the same for both? (10-1)
2. Describe the difference between P- and N-type semiconductors with respect to the internal resistance. (10-2)
3. What is the direction of current flow when an external emf is applied to (a) an N-type semiconductor? (b) A P-type? (10-2)
4. Discuss silicon and germanium rectifying diodes. (10-3)
5. Describe the physical structure of transistors. (10-5)
6. What is a junction-tetrode transistor? How and why does it differ from other transistors in operating frequency? (10-5)
7. What is the gain factor of a transistor? (10-6, 10-9)
8. What effect does biasing have on the performance of a PNP or NPN transistor? (10-6)
9. Explain how transistors operate as an amplifier. (10-6)
10. Show connections of external batteries, resistance load, and signal source as they would appear in a properly biased common-emitter transistor amplifier. (10-6, Fig. 10-15)
11. What is the difference between forward- and reverse-biasing of transistors? (10-6)
12. Explain the cutoff frequency of a common-emitter transistor amplifier. (10-6)
13. Why is stabilization of a transistor amplifier usually necessary? How would a thermistor be used for this? (10-6)
14. Draw a diagram of a method of obtaining self-bias, without current feedback, in a common-emitter amplifier. Explain the voltage-drops across the resistors. (10-6, Fig. 10-16)
15. Draw a circuit diagram of a common-emitter amplifier with emitter bias and explain its operation. (10-6, Fig. 10-15)
16. What is ambient temperature? (10-6)
17. What is meant by V_{CE}, V_{EB}, I_C, and V_{CBO}? (10-6)
18. What is the function of a heat sink? (10-8)
19. What is meant by transistor dissipation? (10-8)
20. What is alpha and what is alpha cutoff frequency in a transistor? (10-9)   [4]
21. Draw a transistor amplifier circuit that would be analogous to that of a vacuum-type cathode-follower amplifier. (10-10, Fig. 10-20)
22. Name some of the common types of transistors and draw their schematic symbols. (10-17, Fig. 10-31)

NOTE: FCC license test questions are multiple-choice types, taking a form somewhat as follows:

23. A solid-state device containing diodes, transistors, resistors and capacitors is called (a) an IC; (b) an SCR; (c) a LASCR; (d) A MOSFET. (10-17)
24. The most frequently used transistor amplifier circuit is the (a) grounded base? (b) common base? (c) common collector; (d) grounded emitter. (10-9)
25. The output circuit current in class A operation of an amplifier is normally a (a) zero value; (b) saturated value; (c) medium value; (d) high value. (10-6)

11 POWER SUPPLIES

The objective of this chapter is to study the basic half-wave and full-wave rectifier circuits and the use of C and L as filters. Solid-state, high-vacuum, and mercury-vapor rectifiers are considered. Voltage-multiplying circuits, as well as shunt and series regulating circuits are explained, as are dc-to-dc converters and three-phase power systems.

11-1 POWER SUPPLIES

Power to operate electronic equipment may be obtained from a variety of sources. Batteries can produce a dc emf by chemical action. Photons of heat or light from the sun can be converted to dc electric energy by photocells. Fuel cells combine hydrogen and oxygen gases in an electrolyte to produce a dc emf. A fossil-fuel motor or a fall of water can rotate dc or ac generators.

The dc sources are often able to operate electronic equipment directly, although some means of regulating or maintaining a constant emf under changing-load conditions may be necessary. The most generally available energy, alternating current, must be changed (rectified) to a pulsating dc, which in turn must be smoothed (filtered) to a nonvarying voltage. The resultant dc may also require voltage regulation to operate some electronic circuits properly.

Generally speaking, in electronics the term "power supply" is usually considered to mean a rectifier-filter system that converts ac to pure dc. There are many different power-supply circuits that may be employed to do this. The basic components used for the simpler circuits are transformers, rectifiers, resistors, capacitors, and inductors. More complex regulated supplies may add transistors or triodes as voltage-sensing and controlling devices, plus zener diodes or VR tubes to establish reference voltages.

11-2 RECTIFIERS

Various types of rectifiers are used in power supplies. High-vacuum diodes, mercury-vapor diodes, and silicon diodes have been described in preceding chapters. There are some terms regarding these devices which should be understood.

A vacuum diode may have either a filament or a heater-cathode as the electron emitter. The electron collector is called either a plate or an anode. When the plate is made positive in respect to the cathode, electrons flow to the anode. The result is a relatively low resistance to current when the plate potential has the proper polarity. When the plate potential is reversed, no electrons flow through the tube. The voltage-drop across a vacuum diode operating in the conduction direction may be only a fraction of a volt under very light loads, or it may be hundreds of volts under high-current loads. Thus, the output voltage of a vacuum-diode rectifier power supply may vary considerably when the load demand varies.

The voltage that may be applied across a vacuum diode when the plate is made negative in respect to the cathode may vary from one or two hundred volts to many thousands of volts, depending on the internal insulation capabilities and the spacing between cathode and anode elements. If this *inverse voltage* is exceeded, a spark, called a flashback, or arc-back, from plate to anode may occur. This can damage the cathode and ruin the tube. The amount of current a vacuum diode can carry depends on the cathode emission capabilities, the area of the plate, and the ability of the plate to lose heat developed in it by the plate current. This heat is usually dissipated by direct radiation through a glass envelope, often aided by radiation fins attached to the plate.

The mercury-vapor diode is the most commonly used gaseous rectifier tube. When the plate becomes about $+15$ V in respect to the hot cathode, the mercury gas ionizes suddenly and current begins to flow to the plate. When the plate potential decreases to a value below about -14 V, the plate current ceases abruptly. With an inverse voltage no ionization occurs and no current flows to or from the plate.

Silicon diodes, and other solid-state diodes such as germanium, selenium, and copper-oxide, require no cathode-heating supply and operate cooler than do the vacuum and mercury-vapor diodes when passing a given current value. However, the solid-state diodes do require heat sinks attached to their anodes when they carry heavy currents. The solid-state diodes have relatively constant-value voltage-drops across them, about 0.3 V for germanium and 0.6 V for silicon, for example. As the current through them increases,

the voltage-drop values also increase somewhat. Solid-state diodes are physically quite rugged and may be operated in any position.

Although ac power may be converted to dc by using an ac motor to turn a dc generator, this is not considered to be rectification.

11-3 HALF-WAVE RECTIFICATION

The simplest rectifier circuit is the half-wave circuit shown in Fig. 11-1. The ac input produces

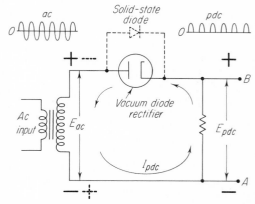

Fig. 11-1 Vacuum-tube or solid-state half-wave rectifier circuits produce pulsating dc.

an alternating emf in the secondary of the transformer, which attempts to force current through the secondary circuit, first in one direction and then in the opposite, alternately. If the rectifier were not in the circuit, an ac would flow through the load resistance, but the characteristic of a rectifier is such that it will allow current to pass through it in one direction only. Although the transformer-secondary voltage may be alternating, the current can flow in the resistor only during one-half of each cycle. This produces a pulsating dc in the circuit, as shown. The voltage-drop across points A and B is pulsating and has essentially the same voltage value as the positive peak of the ac from the transformer secondary.

The half-wave circuit may be the simplest rectifier circuit, but it has some serious disadvantages. One of them is that only half of each ac cycle is used. The other half cycle, when a negative potential is being applied to the anode of the rectifier, is powerless to cause current to flow backward through the rectifier and circuit.

The average current of the half-wave pulses is equal to only 0.318 (half of 0.636, Sec. 4-5) of the peak-current-amplitude value.

11-4 FULL-WAVE RECTIFICATION

Two circuits that can be used in rectifier power supplies to utilize both halves of the ac cycle are the *bridge* and the *full-wave center-tapped* rectifier circuits.

Figure 11-2 shows the currents in a bridge

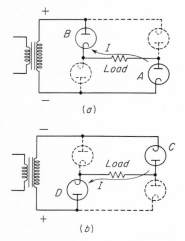

(a)

(b)

Fig. 11-2 The path of current flow through the load in a bridge-rectifier circuit during two alternate half cycles.

circuit during two alternate half cycles of ac. Note the polarity markings on the transformer secondaries. During the half cycle when the top of the transformer is positive, rectifiers A and B pass current through the load resistor from right to left. During the half cycle when the bottom of the transformer is positive, rectifiers C and D pass current through the load resistor, again from right to left. On both half cycles the electrons find a path between the negative end of the transformer and the positive end, each time producing current in the same direction through the load resistor.

Figure 11-3 shows the bridge rectifier with the load at the right, as it usually is drawn in diagrams. Note that the bridge-rectifier system needs three separate filament sources. It is possible to parallel the filaments of tubes B and D and use one source for them, but tubes A and C must have separate filament sources of their own. (Try sketching a solid-state bridge-rectifier circuit.)

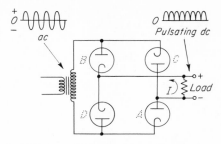

Fig. 11-3 Bridge-rectifier circuit as often drawn.

Current pulses from a half-wave rectifier circuit and from a full-wave rectifier are shown in Fig. 11-4. The average value of the pulses with full-

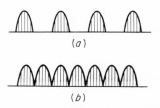

(a)

(b)

Fig. 11-4 Comparison of rectified current pulses: (a) half-wave, (b) full-wave.

wave rectification of a sine-wave cycle is 0.636 of the peak value.

The other full-wave rectifier circuit is the center-tapped transformer type (Fig. 11-5). During the half cycle when the top of the transformer is positive, current flows out of the center tap, up

Fig. 11-5 Current-flow path through the load in a center-tap full-wave circuit during two alternate half cycles.

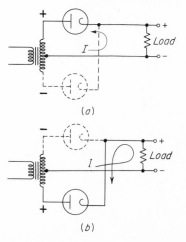

(a)

(b)

through the load resistor, and through the upper rectifier to the positive end of the transformer. During the half cycle when the bottom of the transformer is positive, current flows out of the center tap, up through the load resistor, and through the lower rectifier to the positive end of the transformer. During both halves of the cycle current flows upward through the resistor. The two filaments can be connected in parallel and operated from one filament source. (Try drawing a solid-state center-tap full-wave circuit.)

Whereas the bridge circuit uses the full output voltage of the transformer, the center-tap circuit uses only one-half of the total transformer voltage at one time. A 2,000-V center-tapped transformer produces only about 1,000-V dc output with the center-tap circuit. In some cases this may be a disadvantage. In others, the three filament sources and the requirement of four tubes in a bridge circuit may be disadvantages.

Using the same power transformer, a bridge-rectifier circuit will produce an output voltage twice that obtainable from a center-tapped full-wave circuit but at only half the current. Thus, the power capabilities ($P = EI$) of both circuits are essentially the same.

11-5 TWO VOLTAGES FROM ONE TRANSFORMER

By using a center-tapped transformer, the bridge-rectifier circuit can produce two separate full-wave-rectified voltages, one with half the voltage of the other (Fig. 11-6).

If the secondary voltage of the transformer is 2,000 V, the output voltage across the load R_{L_1}

will be essentially 2,000 V. By using the center tap on the secondary and the negative lead of the 2,000-V circuit, a full-wave-rectified 1,000 V will be obtained across the second load R_{L_2}. The 2,000-V circuit is a normal bridge rectifier. The 1,000-V circuit uses rectifiers A and B as a two-filament-source full-wave center-tapped rectification system. With this circuit configuration the two voltages have a common negative lead.

Tubes A and B are being used in both power supplies and will therefore be carrying more current than the other two rectifiers. (Try drawing this circuit using solid-state diodes.)

11-6 CAPACITIVE FILTERING

The pulsating current through the load resistor of the half-wave rectifier and therefore the voltage developed across it are not at all smooth. Since the voltage required in most applications must have an unvarying characteristic, it will be necessary to smooth the pulsations by *filtering*. One method is capacitive filtering.

Figure 11-7 shows a capacitor connected across the output of a half-wave rectifier circuit with no other load. During one-half of the ac cycle, when the rectifier refuses to pass current, nothing happens in the circuit insofar as current

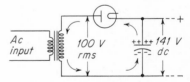

Fig. 11-7 A filter capacitor across the output charges to the peak ac voltage value.

Fig. 11-6 Two output-voltage values from a single bridge-rectifier circuit.

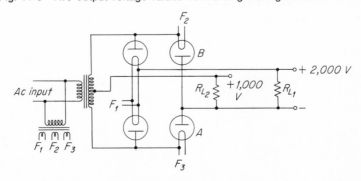

is concerned. On the other half cycle, the top of the transformer becomes positive, pulls electrons off the top plate of the capacitor through the rectifier, and drives electrons onto the lower plate of the capacitor. This charges the capacitor to the peak voltage of the ac. For example, if the secondary voltage is 100 V effective, the capacitor will be charged to a voltage equal to 1.414 × 100, or 141 V dc.

On the next half cycle, the current cannot push back through the rectifier; so the capacitor remains charged at 141 V.

The next current-moving half cycle finds the potential again in position to push or pull electrons through the rectifier, but since the capacitor is already charged to the peak value, nothing happens in the circuit. The voltage across the filter capacitor is an unvarying dc.

During the charging half cycle, the current in the transformer secondary has two components, illustrated in Fig. 11.8a. One component charges

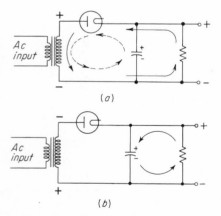

(a)

(b)

Fig. 11-8 (a) During the conduction half cycle the load's current comes from the transformer. (b) During the nonconduction half cycle the load's current comes from the charged filter capacitor.

the capacitor; the other flows through the load resistor.

During the noncharging half cycle, the transformer cannot push electrons backward through the rectifier, so any current in the circuit during this time must come from the electrons that have been stored on the plates of the capacitor. During this half cycle, the capacitor discharges through the resistor (Fig. 11-8b). The current direction

through the resistor is still upward. If the capacitor is large, it can hold sufficient electrons to keep current flowing through the resistor during all the noncharging half cycle. As it discharges, however, the potential difference across capacitor and resistor decreases and the current decreases in amplitude. The voltage across the resistor drops off slowly during the noncharging half cycle. During the next charging half cycle, the capacitor is recharged to full voltage and the transformer drives the peak value of current through the resistor again. This charge and discharge action continues as long as ac is fed to the transformer primary. The voltage across the load resistor will vary as shown in Fig. 11-9.

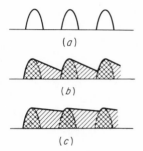

(a)

(b)

(c)

Fig. 11-9 Adding filter capacitance smooths the varying dc waveform. (a) No filter. (b) Some filter capacitance. (c) Still more filter.

A load having a very high resistance will discharge the capacitor very slowly. It is comparatively simple and inexpensive to filter such a light load. If a heavy load (low-value resistance) is connected across the output, however, the capacitor will be discharged rapidly by the load, causing considerable variation of voltage between cycles. Additional capacitance will be required to counteract the rapid drop-off. In many cases it is not practical to keep adding capacitance. Instead, another type of filtering, called *inductive filtering,* can be added.

11-7 INDUCTIVE FILTERING

When an inductance is connected in series with a rectifier circuit (Fig. 11-10), a filtering, or smoothing, action results.

By definition, inductance has the property to oppose any change in current. Pulses of current

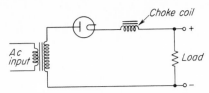

Fig. 11-10 A choke coil in series will filter pulsating dc.

through the choke coil build up a magnetic field around it, taking energy from the circuit to produce the field. As the pulse tries to decrease in amplitude, the magnetic field collapses and returns energy in the form of current to the circuit, thereby tending to hold the current constant. When pulsating dc from a half-wave rectifier passes through the circuit in which a choke coil is placed, the pulse amplitude will be lessened and the dropping off of the pulse will result in a lengthening of the pulse duration (Fig. 11-11). It

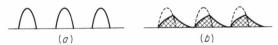

Fig. 11-11 (a) Pulses without filtering and (b) effect of inductive filtering on the pulses.

is impractical to produce steady dc by the use of inductance only, but it is possible to use inductive and capacitive filtering together and produce substantially pure dc with reasonably inexpensive parts.

Test your understanding; answer these checkup questions.

1. What polarity does a VT diode anode have if there is a 200-V inverse voltage across the tube? _____
2. Which of the three common types of rectifiers has the lowest forward resistance? _____ The highest? _____
3. How do anodes of tube-type rectifiers dissipate heat? _____ Solid-state rectifiers? _____
4. What is the average value of a 200-V rms sine-wave ac if half-wave-rectified? _____ If full-wave-rectified? _____
5. Which rectification tends to permanently magnetize the core of the supply transformer? _____ Why? _____
6. What two circuits produce full-wave rectification? _____ _____ Which may require three filament windings? _____
7. Why are most modern rectifier circuits bridge type? _____

8. What is required with a bridge rectifier circuit to obtain two voltages, one twice the other? _____
9. What components produce shunt filtering? _____ Series filtering? _____
10. What type of filtering smooths the leading edge of pulses? _____ The trailing edge? _____
11. In a half-wave filtered supply what feeds current to the load during the conduction half cycle? _____ During the nonconduction half cycle? _____
12. A 100-V rms secondary feeds a rectifier and a filter capacitor with no load. What is the voltage across C if the rectifier is a VT? _____ A silicon diode? _____ A mercury-vapor diode? _____
13. Draw diagrams and waveforms of a half-wave solid-state rectifier, a full-wave bridge VT rectifier, and a full-wave center-tap rectifier.

11-8 CAPACITIVE-INPUT FILTERING

While there are many variations of filter circuits, one used often in power supplies is the capacitive-input type in Fig. 11-12.

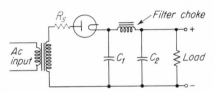

Fig. 11-12 Common π-type capacitive-input power-supply filter circuit.

This filter circuit can be recognized as a low-pass pi-type filter (Sec. 8-8). Used to filter the output of a rectifier, it discriminates against any voltage and current variations but passes steady dc (zero frequency) without attenuation.

When looking at the filter from the source of power, the transformer in this case, the first part of the filter seen is the capacitor C_1. This is known as the *input capacitor* of the filter. It charges to the peak-voltage value of the transformer ac and discharges somewhat during the noncharging half cycle. During the noncharging interval, the input-capacitor current flows through the load resistor and through the choke coil. The choke opposes any dropping off or change in this current, resulting in a fairly smooth dc flow through the load. Added to this, the output capacitor C_2 further tends to hold the voltage constant across the load. The net result is a smooth dc suitable for most vacuum-tube circuits.

Practical values for a half-wave 300-V power

supply might be $C_1 = 10$–$20 \ \mu F$, $C_2 = 20$–$40 \ \mu F$, and choke coil $= 10$–30 H. With full-wave rectification circuits, the values of the filter components might be $C_1 = 5$–$10 \ \mu F$, $C_2 = 10$–$20 \ \mu F$, and choke coil $= 5$–15 H. Full-wave rectification requires only about half as much filter as half-wave. In better communications equipment half-wave rectification is rarely used in power supplies. Note that the values of the output capacitor are double those of the input. An input capacitor tends to determine the output voltage of the supply, but an output capacitor acts more as storage tank of electrons for the load.

In practical applications, under medium load, the output of a capacitive-input filter system is roughly 85% of the ac peak-voltage value. Power supplies having outputs of more than 450 V use paper-dielectric capacitors. Below this voltage either paper or electrolytic capacitors are suitable. Below 100 V, electrolytics are used almost exclusively.

Low-voltage high-current supplies employed to power transistorized equipment may use filter capacitors of 500 to 20,000 μF. When the supply is turned on, the first surge of charging current with such high-value capacitors may damage a solid-state rectifier. To prevent this a few ohms of series resistance (R_s in Fig. 11-12) may be added to limit surge current. High-voltage transformers used with vacuum diodes generally have sufficient internal resistance in the secondary winding. Inductances used in low-voltage high-current filters are usually a fraction of a henry.

11-9 INDUCTIVE-INPUT FILTERING

The first parts seen by the source when looking into the filter of Fig. 11-13 are a coil L_1 and a capacitor C_1 in series. To differentiate between this circuit and the capacitive-input filter, this one is known as an *inductive-input* filter. If the load

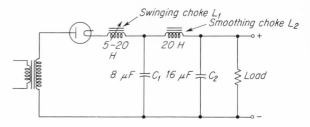

Fig. 11-13 A practical inductive-input high-voltage power-supply filter circuit for VT circuits.

is disconnected, the pulsating dc from the rectifier will charge C_1 and C_2 to the peak value of the transformer after a few pulses. When a load is connected, however, current flows through L_1 and L_2. Since the current through L_1 will be pulsating and may be said to have an ac component, a reactive voltage-drop will occur across this choke coil ($E = IX_L$). C_1 can no longer charge to the peak value but will usually fall to about 65% of the peak. As a result, the voltage across C_2 may never be greater than this value and may be somewhat less. Because the current through L_2 is essentially unvarying dc, there is very little reactive voltage-drop across it, although there will be a resistive voltage-drop across the resistance of the wire in the choke ($E = IR$). The inductive-input filter circuit will usually have better voltage regulation, but it will have less output voltage than a capacitive-input filter with the same power transformer. Technically, a choke-input filter may have a small value of input capacitance and still be considered inductive input. The input choke is usually a swinging choke, explained in Sec. 11-13. Any other chokes are smoothing chokes.

When an inductive-input filter is used, no surge resistor R_s is needed with any type of rectifier.

11-10 SOME POSSIBLE FILTER CIRCUITS

In general, the more inductance and capacitance used in a filter circuit, the smoother the resulting dc. Some of the many possible filters are shown in Fig. 11-14.

The first filter (1) is a low-pass capacitive-input π-section filter. It is a single-section filter, whereas the second (2) is a two-section capacitive-input filter. The third (3) is another capacitive-input filter, adding a choke in the lower lead.

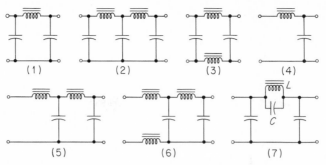

Fig. 11-14 Some possible power-supply filter configurations.

Filters 4, 5, and 6 are choke-input types. It is usually considered undesirable to use an input choke in the grounded lead of a power supply, as in filter 6, because a ripple voltage of the source frequency can be developed across it, owing to primary-secondary capacitance in the power transformer.

Filter 7 is a resonant filter. If C and L resonate at the pulsating dc frequency (120 Hz for full-wave 60-Hz ac), they form a parallel-resonant circuit and a high impedance to this frequency. This opposes variations at this frequency and results in better filtering.

Note that a power-supply filter always has a capacitor across the output.

11-11 RC FILTERS

When the load is constant (as in equipment having all class A amplifiers), adequate filtering may be produced by using a resistor in place of the choke coil of the usual π-type filter circuit (Fig. 11-15). Many solid-state low-voltage supplies use this filter when it is followed by a voltage-regulating circuit.

Two disadvantages of this type of filtering alone are the voltage-drop across the resistor used in the circuit (10 to 100,000 Ω, depending on the load), and the variation of the output voltage when the load changes.

11-12 FILTER CHOKES

Some important factors to be considered regarding filter chokes are insulation, shielding, resistance, inductance, distributed capacitance, current-carrying ability, losses, and saturation.

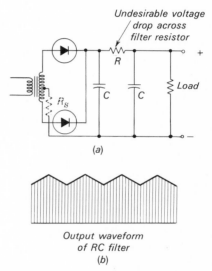

Fig. 11-15 (a) Full-wave power supply with an RC filter. (b) Output waveform under heavy load.

It is necessary that the winding of a choke coil be insulated from the core to withstand the peak voltages developed between winding and the usually grounded core.

To prevent ac from being induced in turns of a choke by external alternating magnetic fields, it may be necessary to shield the choke in a sheet-iron case. Shielding will also prevent fluctuating fields of the choke from inducing an ac emf in nearby wires in which such voltages would be detrimental (AF amplifier leads, for example).

If the windings of chokes are of small-diameter wire and many turns, the inductance may be great but the resistance of the coil may also be high. There will be a dc voltage-drop across the choke equal to $E = IR$. Increasing the current increases the voltage-drop across the choke, resulting in poor power-supply voltage regulation. The greater the inductance in a choke, the more reactance to varying currents and therefore the greater choking action it will have. It becomes a problem of how much inductance, how much resistance, and how much bulk.

Since capacitance acts to pass ac or pulsating dc, a choke with a high value of distributed capacitance across it will choke down the variations by its inductance but pass them by its distributed capacitance. One counteracts the other. Chokes must have low distributed capacitance.

The amount of current that can be carried by a choke is determined primarily by the size of the wire used. The ventilation of the choke windings determines the current-carrying ability also. With poor ventilation a relatively small current may cause enough heat accumulation to produce deterioration of the insulation on the wires. Heating the windings also increases the resistance of the wires, which increases voltage-regulation difficulties.

If the variation of current in a choke is of high amplitude, eddy currents may develop in the core, resulting in a waste of energy heating the core. The cores of chokes are laminated with low-retentivity steel to decrease hysteresis and eddy-current losses. If the laminations are not securely bolted together, they may vibrate when varying dc flows through the windings. Except for an audible buzzing sound, the operation of the choke is not impaired.

Choke coils are capable of reacting against current variations only if their cores are not magnetically saturated. To prevent magnetic saturation at a relatively low current, the core is constructed with an interruption in the continuity of the magnetic-iron circuit. To increase the reluctance of the core, an air gap will be found in all cores of *smoothing*-type filter chokes. This gap will be only a small fraction of an inch, perhaps with a piece of cardboard in it, but the gap prevents the core from becoming saturated unless an excessively high current flows through the windings.

Whenever possible, the metal cases of chokes and transformers are connected to ground potential or to the metal chassis. This prevents personnel from receiving electric shocks and the metal cases from picking up an electrostatic charge. Undesired interaction between the supply and other parts of the equipment is avoided.

Smoothing chokes in high-voltage 60-Hz vacuum-tube supplies may range from 2 to 30 H. Low-voltage high-current (or high-frequency) supplies will use chokes ranging from millihenrys to perhaps 1 H.

11-13 SWINGING CHOKES

Vacuum-tube-rectifier power supplies may use a *swinging* choke as the input inductor. A swinging choke has little or no air gap in its core. Because of this the magnetic-iron core begins to saturate at a medium-current value. When low current is flowing through it, the choke has high inductance and filters effectively. With high current it saturates and has less inductance. Thus, with a light load and little current the swinging choke has a high reactance and develops a large voltage-drop across it. When the load increases, the choke saturates and has less reactance and less voltage-drop across it. The voltage of the power supply tends to remain constant under varying loads, improving the voltage regulation of the supply.

It is usual to employ an input choke with twice the computed critical value at minimum load. By formula

$$L_c = \frac{R}{1,000}$$

where L_c = critical inductance in H
R = load resistance in Ω (includes bleeder resistor, Sec. 11-23)

A swinging choke might have inductance values that swing from 5 H with a heavy load (high current) to 20 H with a light load.

11-14 HIGH-VACUUM RECTIFIERS

High-vacuum diodes were discussed in Chap. 9. These tubes are used in vacuum-tube-receiver power supplies as well as in high-voltage supplies in high-powered transmitters.

High-voltage vacuum rectifiers are usually made in single units, with filament leads coming out at the base and the plate connection at the top of the tube.

Low-voltage vacuum rectifiers are usually constructed as duodiodes, with two plates and two filaments in one envelope. The filament and plate leads all terminate in pins at the base of the tube.

Some of the significant points regarding high-vacuum rectifiers are:

1. The voltage-drop across the tubes will vary directly with the current flowing through them.
2. Vacuum diodes dissipate considerable energy in heat, both from high filament temperatures and by radiation from the plate. Adequate ventilation must be provided, particularly in high-power applications.
3. They can be used with either inductive- or capacitive-input filter circuits.
4. Low-power filament-type diodes can be operated as soon as the filament is turned on, with no warm-up period required. When the cathode is indirectly heated, there is a cathode warm-up period of about 10 s before sufficient electrons are emitted from the cathode.
5. As with all high-power tubes having heavy, directly heated filaments, tube life is materially increased by slowly increasing the filament voltage and current when the tube is first turned on each day, to prevent rapid heating and expansion from fracturing the filament wires.
6. Any light blue or purple haze between filament and plate of a high-vacuum rectifier indicates unwanted gas. In some cases the tube may operate for a long period of time with some gas in it. In other cases, when the gas leak is more rapid, the tube may require immediate replacement. Any gassy tubes bear watching.

11-15 MERCURY-VAPOR RECTIFIERS

Mercury-vapor diodes were mentioned in Chap. 9. There are duodiode types of mercury-vapor diodes, but most are of the single-diode, high-voltage, high-current type. They are primarily used in transmitters and other high-power applications.

The ionization potential of a gaseous tube depends on several factors: type of gas used, pressure of the gas, size of electrodes, and whether a filament is one of the electrodes. As the pressure is reduced, the ionization-voltage value increases until in a vacuum tube (with practically no gas present) ionization does not occur under normal operating conditions. On the other hand, a gas with a certain medium gas pressure will ionize and pass current at relatively low voltages. Should the gas be compressed further, the voltage required to produce ionization will again increase.

Ionized mercury vapor affords a low-resistance path between cathode and plate and will support relatively high-current flow with little heating. Such a tube depends on ionization of the gas rather than on electron emission from a relatively hot filament to produce current flow. This results in greater current through the tube, less power required to heat the filament, less heating of the plate, and greater overall efficiency of operation.

Mercury-vapor rectifier tubes are usually constructed with a plate and an oxide-coated filament (Fig. 11-16). The glass envelope is evacuated

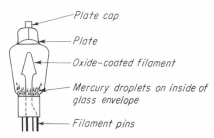

Fig. 11-16 A mercury-vapor rectifier tube (type 866A) used in many transmitters.

except for some liquid mercury that forms in droplets at the bottom of the envelope. When the filament is turned on, the liquid mercury is vaporized by heat radiated from the hot filament and the gas pressure in the tube rises. This requires a little time. Electron emission from the filament partially ionizes the mercury vapor. When the tube is at its operating temperature, the gas pressure rises until about 15 V between the hot filament and the plate will cause heavy ionization and allow a large current flow through the tube. Over a rather wide range of gas pressure this 15-V ionization potential remains approximately the same. The potential required to produce an undesired current backward through the tube (plate to heated filament) is several hundred or thousand volts. By holding the gas pressure to a certain value, it is possible to have a rectifier tube with a 15-V ionization emf one way and an *inverse* ionization voltage of several thousand volts in the other direction.

If overheated, gas pressure may increase to the point where the inverse ionization voltage will fall below the peak voltage of the transformer to which the tube is connected and a flashback, or arc-back, will occur.

If the mercury vapor is too cold, insufficient gas may be present. The forward ionization voltage may increase to more than 15 V. If it rises to about 22 V, a phenomenon known as *double ionization* will occur. Double ionization means that two electrons are torn from molecules of the mercury. This results in the positive ions having twice the attraction toward the filament. High-speed ion bombardment of the filament will disintegrate the electron-emitting surface of the filament wire and ruin the tube. Excessive current through a mercury-vapor diode due to too heavy a load may also cause double ionization and deterioration of the filament.

Mercury-vapor tubes must be warmed up for a period of 15 to 20 s and must not be allowed to become overheated. They operate satisfactorily between about 20 and 70°C.

Mercury-vapor tubes must not work into a capacitive-input filter. In such a circuit, when the alternating emf of the transformer increases from zero to +15 V, current suddenly starts flowing. This sudden, steep-sided wave of current represents the equivalent of a very-high-frequency pulse. The reactance of the capacitor is very low,

practically a short circuit, for such a high frequency. As a result, during this instant, a relatively high current flows in the transformer-tube-capacitor circuit, producing double ionization and deterioration of the filament. A mercury-vapor tube may operate only a few minutes or hours into a capacitive-input filter.

When a choke is added in series with the input-filter capacitor (inductive-input filtering), it opposes any sudden change in current and prevents any heavy instantaneous current flow, protecting the tube.

Even with an input choke, a slight current surge due to distributed capacitance across both choke coil and transformer may be developed at every instant of ionization. This surge in the mercury *plasma* (ionized gas) sets up a damped oscillatory wave, similar to a spark-transmitter emission, in the transformer-tube-filter circuit. A low-intensity wideband signal can be picked up by nearby receivers as a disturbing *hash,* or buzzing sound, over much of the usable RF spectrum. It may also be heard in high-gain AF amplifiers, tape recorders, etc. To stop it, 5- to 20-mH RF chokes may be connected in series with each plate lead of the rectifier tube and 0.001- to 0.005-μF mica RF filter capacitors may be added, as in Fig. 11-17.

Mercury-vapor tubes pass relatively high current for a given filament rating. For example, an 866A tube may pass a peak current of more than 2 A although an average of only 0.5 A. When more current must be passed, two or more tubes may be connected in parallel, as in Fig. 11-18. The voltage-drop across a mercury-vapor tube remains at approximately 15 V regardless of the

Fig. 11-17 Power supply with full-wave rectification, mercury-vapor tubes, inductive-input filter, hash filters, and one means of heating filaments before high voltage is applied.

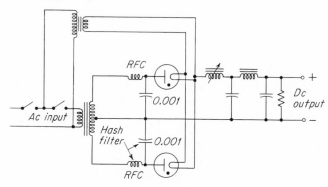

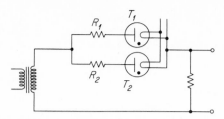

Fig. 11-18 Equalizing resistors are required when mercury-vapor tubes are paralleled.

value of current passing through it. Two parallel tubes of the same type may have slightly different ionization voltages, however. Suppose tube T_1, in Fig. 11-18, will ionize at 14.9 V and tube T_2 at 15.1 V. With these tubes in parallel the voltage builds up across both at the same time. Tube T_1 ionizes as soon as the emf across it reaches 14.9 V. With no more than 14.9 V across T_2, it will never ionize. If R_1 and R_2 are in the circuit, the voltage across T_2 will be the 14.9 V across T_1 plus any voltage-drop across R_1 when current flows through it. Under load, this will be more than the necessary 15.1 V, and T_2 ionizes, carrying half the circuit current. These *equalizing resistors* may range from 0.5 to 5 Ω.

When mercury-vapor tubes are *first* plugged into a power supply, to assure vaporization of droplets of mercury on the plate the filament of the tube must be run for about 15 min before plate voltage is applied. Thereafter, a 15- to 20-s warm-up period is all that is necessary.

A newer xenon gas rectifier tube produces no hash and requires no hash filter. It features high current, high efficiency, a constant 10-V drop, and inverse peak voltages from 4.5 to 10 kV.

Test your understanding; answer these checkup questions.

1. How long a warm-up period should mercury-vapor tubes have when first plugged into equipment? _____ When equipment is turned on? _____
2. What is the usual input-output capacitance ratio of π-type power-supply filters? _____
3. What is the approximate loaded output voltage of a C-input filter and a full-wave power supply if the peak ac is 1,000 V? _____ If the filter is L-input? _____
4. With what type(s) of rectifier may surge resistors be necessary? _____ Equalizing resistors? _____

5. What might be the advantages of using a resistor in place of a choke in a π-type filter? _____
6. What kind of filter C is used in 1,000-V supplies? _____ In 10-V supplies? _____
7. Which input filtering should result in the best voltage regulation? _____
8. Why does a core of a choke coil have a gap? _____
9. What is the relative gap size in the cores of smoothing chokes? _____ Swinging chokes? _____
10. What might be the L values with full-wave 60-Hz power-supply filter chokes? _____ With 600 Hz? _____
11. A 500- to 1,000-Ω changing load on a power supply should have an input choke with what minimum value of L? _____
12. What type cathode is usually used in mercury-vapor diodes? _____ High-voltage high-vacuum diodes? _____
13. What causes cathode bombardment in a mercury-vapor diode? _____
14. What filter configuration is never used with mercury-vapor diodes? _____
15. Draw diagrams of a half-wave VT rectifier with π-type filter and a full-wave mercury-vapor rectifier with filter.

11-16 INVERSE PEAK VOLTAGE

The inverse peak values of high-vacuum diodes are equal to the insulation properties of the internal elements of the tube, since there is little or no gas in the tubes to ionize and arc back.

The inverse peak voltage that a gas diode will stand depends upon gas pressure. The higher the pressure, the lower the reverse voltage needed to produce an arc-back in the tube. The only control on gas pressure is mercury temperature. If the mercury temperature increases, gas pressure increases and the tube is more likely to arc back.

The inverse peak voltage being applied to a rectifier tube by the power transformer in a full-wave center-tap circuit can be determined by multiplying the ac rms (voltmeter) voltage across the entire secondary of the transformer by 1.414, which gives the peak voltage. In Fig. 11-19 the ac voltage across the secondary is 500 plus 500, or 1,000 V rms. Therefore, 1,000 × 1.414 = 1,414 V peak develops across the secondary. When the top of the transformer is positive, T_1 is conducting and there is a 15-V drop across it. At that instant, T_2 has a negative potential on its plate. The potential difference between the plate and filament of tube T_2 is 1,414 V less the 15 V across T_1, or 1,399 V. If the tubes had inverse-peak-voltage ratings of 1,200 V, they would arc back. If they had inverse peak ratings of more

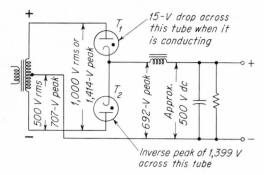

+

15-V drop across
this tube when it
is conducting

T_1

1,000 V rms or
1,414-V peak

500 V rms
707-V peak

T_2

692-V peak

Approx.
500 V dc

+

−

Inverse peak of 1,399 V
across this tube

−

Fig. 11-19 Voltages appearing across various parts of a full-wave rectifier with inductive-input filtering and mercury-vapor tubes.

than 1,400 V, they would operate without any arc-back. For safety, the transformer peak voltage should be less than 90% of the peak-inverse-voltage rating of the tube to guard against line-voltage surges.

The inverse peak voltage across any mercury-vapor tube in a bridge-rectifier circuit is also equal to the peak ac voltage minus the 15-V drop across the other tube operating in series with it at the time.

The inverse peak voltage of a half-wave rectifier circuit with capacitive-input filtering would be practically *twice* the peak ac voltage of the secondary. For example, if the plate is driven to a peak voltage of 1,400 V by the transformer, the first filter capacitor charges to nearly 1,400 V. The capacitor holds this charge for a time. Since the plate is charged 1,400 V negative when the ac cycle reverses, there is nearly 1,400 V positive on the cathode and 1,400 V negative on the plate, or a difference of about 2,800 V across the tube.

11-17 VACUUM VERSUS MERCURY-VAPOR DIODES

A summary of the advantages and disadvantages of high-vacuum and mercury-vapor diode tubes is listed in Table 11-1.

11-18 RIPPLE FREQUENCY

When filters are designed for power supplies, the pulse frequency being filtered is important. The higher the ripple frequency, the easier it is

Table 11-1

High-vacuum diodes	Mercury-vapor diodes
Advantages	Advantages
No warm-up time required for low-power tubes.	Constant voltage-drop of 15 V across the tube.
May be used with any type of input-filter circuit.	Better voltage regulation possible.
Momentary overloads do little damage to the tube.	Cooler in operation.
Can utilize capacitive-input filter; therefore, the output voltage can be greater for a given transformer.	Can use oxide-coated filaments in high-voltage applications, which are efficient electron emitters.
	High current for size and filament power.
	High efficiency.
Disadvantages	Disadvantages
May require ventilation in high-power applications.	Must be warmed 15 to 20 s.
Require tungsten or thoriated-tungsten filaments for high-voltage applications: greater filament heating power required.	Low inverse peak voltage if operating overly warm.
Less efficient.	Insufficient gas pressure or momentary overloads cause double ionization and filament damage.
Varying voltage-drop across the tube under heavy load conditions.	Usually require hash filters.
Radiate considerable heat.	Usually not desirable for receivers.
	Must use inductive-input filter.
	Inductive-filter requirement decreases voltage output.
	Require equalizing resistors with tubes in parallel.

to filter. For example, the usual power frequency of ac is 60 Hz. A half-wave 60-Hz rectifier produces 60 pulses, or ripples, per second and requires a certain amount of inductance and capacitance to filter the pulses adequately. A full-wave 60-Hz rectifier produces a ripple frequency of 120 pulses/s and requires only about half as much C and L to filter it. In aircraft, frequencies up to 800 Hz produce a full-wave ripple frequency of 1,600 Hz, requiring much lighter and smaller filter components.

A full-wave 60-Hz rectifier produces a 120-Hz ripple frequency (Fig. 11-20a), but if one of the

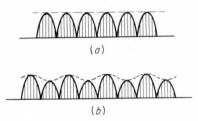

(a)

(b)

Fig. 11-20 (a) Pulses when both rectifiers are good. (b) Pulses with one weak rectifier tube showing resultant half-frequency hum component.

two rectifier tubes is weak, the alternate pulses will be of different amplitudes and a 60-Hz component (Fig. 11-20b) which is more difficult to filter will occur.

When three-phase ac is investigated (Sec. 11-31), it is found that a half-wave 60-Hz three-phase rectifier has a ripple frequency of 180 Hz. A full-wave 60-Hz three-phase rectifier has a ripple frequency of 360 Hz. It will also be found that the rectified three-phase current is in the form of varying instead of pulsating dc and is therefore

much easier to filter. For this reason three-phase ac is very often used in broadcasting stations and other high-powered applications.

11-19 RECTIFIER FILAMENT CIRCUITS

The electric power to heat the filament or cathode of rectifier diodes in power supplies is obtained in one of four ways:

1. In low-power VT equipment, receivers, small amplifiers, etc., the power transformer consists of a (a) primary, (b) high-voltage center-tapped secondary, (c) filament winding for the tubes of the receiver or amplifier, and (d) separate filament winding for the rectifier tube or tubes. An example of this type of power supply is shown in Fig. 11-21.
2. In older equipment operating from 6- or 12-V batteries, the rectifier tube usually has an indirectly heated cathode, the heater being connected to the battery. The heater-to-cathode insulation of such tubes must be capable of withstanding half the transformer-secondary peak-voltage value. Many of these rectifiers are being replaced by solid-state diodes.
3. In higher-power equipment, the filament of the rectifier (and amplifier) tubes must be heated for an appreciable time before the high voltage is applied across the tubes. As a result, a separate filament and high-voltage power transformer are used (Fig. 11-22). After the filament circuit has been energized, the high-voltage transformer can receive no voltage until the time-delay relay has closed and the primary circuit of the high-voltage transformer has been completed.

Fig. 11-21 Power supply used in many VT receivers.

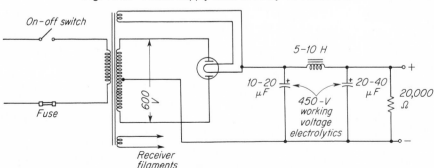

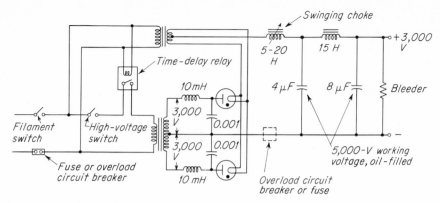

Fig. 11-22 A complete full-wave center-tap high-voltage power supply, including possible component values.

4. In ac/dc radio and TV equipment, in which there is no power transformer, the heaters of all the tubes in the equipment are connected in series across the 120-V source. All the tubes in this type of equipment must have the same heater-current ratings.

In Sec. 9-7 center-tapping ac filament leads was discussed. Failure to center-tap results in a periodic variation of the current through the tube. In power-supply rectifier tubes the same thing occurs, except that the rectifier circuit is followed by a filter the chief function of which is to filter out hum, or current variations. Therefore, in power supplies center-tapping the filament may not be so necessary as in other circuits such as amplifiers or oscillators. When an absolute minimum of hum is required, the filaments of the rectifiers should be center-tapped, as in Fig. 11-22.

11-20 PRACTICAL POWER SUPPLIES

The diagram in Fig. 11-22 is an illustration of a possible high-voltage transmitter supply utilizing full-wave center-tapped rectification, choke-input filtering, a bleeder resistor, a time-delay relay to allow the filaments to heat before applying high voltage, and two possible places for fuses or overload relays.

The diagram in Fig. 11-23 illustrates a possible high-voltage power supply employing a bridge rectifier with capacitive-input filtering. Three secondary windings are required on the filament transformer (same power transformer as in Fig. 11-22).

The diagram in Fig. 11-21 represents a power supply used in many receivers and amplifiers to provide a dc voltage of perhaps 200 to 400 V. The filament winding for the rectifier is a tertiary wind-

Fig. 11-23 A practical high-voltage power supply using the same power transformer as in Fig. 11-22. The output voltage is doubled by employing bridge rectification.

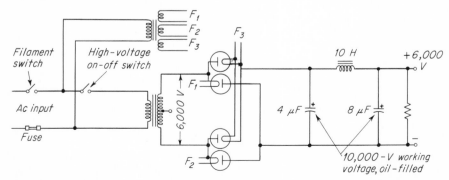

ing on the power transformer. The filament voltage for the other tubes of the receiver is taken from the fourth winding.

11-21 POWER-FACTOR COMPENSATION

All power-transformer primaries have considerable inductance. When inductance is operating in an ac circuit, current and voltage are out of phase. Whenever the current lags or leads the voltage, there is a lessening of the power factor in the circuit. A low power factor means less secondary power output per ampere of current flowing in the primary. For 60-Hz ac, power factor may be disregarded in most power supplies. When 400- to 800-Hz ac is used, as in aircraft, the power factor may decrease considerably. To overcome the inductive effect of the transformer, an 8- to 20-μF capacitor can be connected in series with the primary. The current lead of the capacitor counteracts the current lag of the inductance, raising the power factor.

11-22 VOLTAGE-MULTIPLIER CIRCUITS

It is possible to rectify an ac voltage and produce a dc voltage underload almost twice the ac value by using a *voltage-doubler* circuit.

In Fig. 11-24a, when the top of the transformer secondary is positive, current flows through the rectifier and charges capacitor C_1, as shown. When the cycle reverses and the bottom of the secondary is positive, C_2 is charged by a second rectifier, as shown in Fig. 11-24b. The whole voltage-doubling circuit is drawn out in Fig. 11-24c, with the output dc across the two charged capacitors in series. The larger the value of the two capacitors, the better the regulation and the higher the output voltage of the power supply under loaded conditions. Values in the range of 40 μF to more than 100 μF for each capacitor are used. The output voltage of the voltage-doubler circuit ranges from twice the peak value of the applied ac at no load to about the peak value under heavy-load conditions. With 120-V rms ac input, this would produce a no-load voltage of almost 340 V (from 1.414 $\times$ 240 V) but a heavy-load output of perhaps only 150 to 200 V.

The inverse peak voltage on the rectifiers is equal to approximately twice the peak ac voltage of the secondary of the transformer.

If vacuum-type diodes are used, two filament windings are required. A single filament winding may be used with duodiode heater-cathode rectifiers, such as a type 6X5. When solid-state rectifiers are used in voltage doublers, a 5- to 50-Ω surge resistor (R_s) should be connected in series with the rectifiers to limit surge currents.

Another doubler circuit is shown in Fig. 11-25.

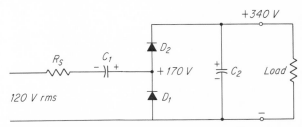

Fig. 11-25 A solid-state half-wave or cascade voltage-doubler circuit.

When the top power line is negative, the input capacitor charges through D_1 to the voltage of the line, as indicated. When the line voltage reverses, the output capacitor, through diode D_2, is connected to the positive line voltage in series with the charged input capacitor. The output capacitor charges to almost twice the line-voltage peak value. Both capacitors can be electrolytics with polarities as shown.

Fig. 11-24 Voltage doubling. (a) and (b) Capacitors C_1 and C_2 charge on alternate half cycles and discharge in series into the load. (c) Actual circuit.

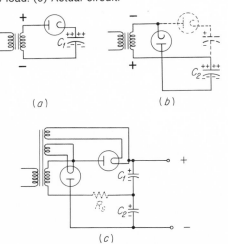

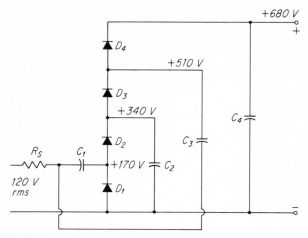

Fig. 11-26 Voltage-quadrupler circuit.

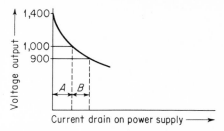

Fig. 11-27 Illustrating the advantage of using a bleeder resistor.

The same basic circuit with the addition of D_3 and C_3, as shown in Fig. 11-26, produces a voltage tripler. A fourth diode and capacitor results in a quadrupler. As in Fig. 11-25, D_1, C_1, D_2, and C_2 act to charge C_2 to nearly 340 V. When the upper input line is negative, the third capacitor is in series with D_3 and the charged C_2. When C_3 charges, it adds another 170 V, for a total of nearly 510 V. When the upper line reverses to positive, C_4 charges through D_4, adding another 170 V, for a total unloaded voltage of nearly 680 V. With any load at all the output voltage drops drastically. Voltage reduction depends on the capacitance values used and the current demand of the load that tends to discharge the capacitors.

11-23 THE BLEEDER RESISTOR

High-voltage power supplies use a *bleeder* resistor across the output of the supply. It serves two purposes. First, it bleeds off the charge of the capacitors when the power supply is turned off. (It is possible to turn off a transmitter with an open bleeder and several hours later receive a lethal shock from the charge still left in unbled filter capacitors.) Second, the bleeder resistor aids in holding the voltage output more nearly constant, as indicated in the simplified graph in Fig. 11-27.

According to the graph, with no current drain on it, the power supply has an output of 1,400 V. If a current drain of the magnitude indicated by line *A* is drawn from the supply, the output voltage is reduced to 1,000 V, a 40% voltage variation. By connecting a bleeder resistor which draws a current of the magnitude of *A* across a power supply, any load equal to *A* drops the output voltage to 900 V. This is only an 11% load voltage variation.

A power supply must have sufficient output power to accommodate not only the load demand but also the bleeder-resistor demand. The bleeder takes between 10 and 25% of the total power-supply output current. Within limits, the lower the resistance value of the bleeder, the better the regulation. If the bleeder has too low a resistance, however, it may discharge the output capacitor and increase the ripple.

A power supply to operate at 1,000 V and 200 mA might have a bleeder current of 40 mA. The resistance of the bleeder, by Ohm's law, would be $R = E/I$, or 1,000/0.04, or 25,000 Ω. The power rating of the bleeder should be at least twice the computed power dissipation by the formula

$$P = EI = 1,000(0.04) = 40 \text{ W}$$

A 100-W 25,000-Ω resistor would probably be used.

11-24 VOLTAGE REGULATION

When a power supply is operating into a load, it will have a certain output voltage. If the load is removed, the output voltage will increase. The percent of voltage increase is considered the *regulation* of the power supply. For example, a power supply delivers 1,000 V to a telegraph transmitter with the key down and 1,200 V with the key up. The regulation is the ratio of the difference of voltages to the full-load voltage. The percent of voltage regulation can be found by the formula

$$\% \text{ Reg} = \frac{(E_{nl} - E_{fl})100}{E_{fl}}$$

$$= \frac{(1{,}200 - 1{,}000)100}{1{,}000} = \frac{200(100)}{1{,}000} = 20\%$$

where E_{nl} = no-load voltage
E_{fl} = full-load voltage

The same formula can be applied to batteries, motor-generators, dynamotors, transformers, etc., to determine regulation. The lower the percent, the better the regulation.

The regulation formula can be rearranged to compute the full-load voltage, as in the following problem:

The no-load voltage is 140 V, and the regulation is 15%. What is the full-load voltage?

$$\% = \frac{(E_{nl} - E_{fl})100}{E_{fl}}$$

$$\% E_{fl} = 100E_{nl} - 100E_{fl}$$

$$\% E_{fl} + 100E_{fl} = 100E_{nl}$$

$$E_{fl}(\% + 100) = 100E_{nl}$$

$$E_{fl} = \frac{100E_{nl}}{\% + 100} = \frac{14{,}000}{115} = 121.7 \text{ V}$$

The regulation formula can be rearranged to compute the no-load voltage, as follows:

Full-load voltage is 240 V, and regulation is 11%. What is the no-load voltage?

$$\frac{(E_{nl} - E_{fl})100}{E_{fl}} = \%$$

$$(E_{nl} - E_{fl})100 = \%(E_{fl})$$

$$E_{nl} - E_{fl} = \frac{\%(E_{fl})}{100}$$

$$E_{nl} = \frac{\%(E_{fl})}{100} + E_{fl} = \frac{11(240)}{100} + 240$$

$$= 26.4 + 240 = 266.4 \text{ V}$$

If a power-supply voltage decreases when the load across it demands more current, it is acting as a high-impedance device. A varying load cur- rent results in a varying power-supply output voltage. Thus, variations of current of one circuit connected to a common supply produce variations of voltage in other stages connected to the same supply. This is one form of interstage feedback and can result in weakened output, distorted amplification, or oscillation of the system. Well-regulated power supplies are said to have low impedance, from

$$Z = \frac{dE}{dI}$$

where d = a change in

If E does not change when I changes, the Z value must be low.

The following conditions in a power supply may cause an undesirable increase in percent of voltage regulation:

1. Resistance in the wires of the choke coils and transformer. When the output current is increased, the voltage-drops across all resistances increase, decreasing the output voltage of the supply.
2. The voltage-drop across a vacuum diode increases when the current through it increases, because of the internal resistance of such tubes. This is not true of mercury-vapor diodes, which have a nearly constant 15-V drop. Semiconductor diodes also have nearly constant voltage-drops across them.
3. Use of capacitive-input filtering instead of inductive input.
4. Failure to use a swinging choke as the input-filter element.
5. Insufficient capacitance in the filter circuit. The load bleeds off the charge in the capacitors too rapidly.
6. Too high a bleeder-resistance value.
7. Use of half-wave rectification instead of full-wave.

Test your understanding; answer these checkup questions.

1. If a mercury-vapor tube heats excessively, what effect does this have on its inverse peak rating? _____
2. What is the inverse peak voltage on mercury-vapor diodes with 1,000-V peak ac using a bridge rectifier?

_____ Using center-tap full-wave? _____
Using half-wave? _____

3. What is the ripple frequency of single-phase 600-Hz ac if half-wave-rectified? _____ If bridge-rectified? _____ If bridge-rectified with three-phase ac? _____

4. What type of diode radiates the most heat energy? _____

5. What is reduced by center-tapping the filament winding in a power supply? _____

6. What are two good places to have overload circuit breakers in a center-tapped full-wave power supply? _____ _____

7. When is power-factor compensation usually required for power-supply transformers? _____

8. An unloaded voltage doubler is used with 100-V rms ac. What is the output voltage? _____ If a tripler circuit is used? _____

9. What are two reasons for using bleeder resistors on power supplies? _____ _____

10. What resistance bleeder might be used across a 5,000-V 0.5-A output power supply? _____ What should be its power rating? _____

11. A power supply has 12 V when loaded. What is its regulation if the no-load E_o is 13 V? _____ If it is 12.1 V? _____

12. A transformer has 11.1% regulation with 900 V as the loaded output. What is the unloaded output? _____

13. What is the voltage-drop value across a current-carrying mercury-vapor diode? _____ Germanium diode? _____ Silicon diode? _____

14. Draw diagrams of a 3,000-V center-tap mercury-vapor supply with a filament warm-up circuit, a 6,000-V bridge VT supply with a filament warm-up circuit, a solid-state voltage doubler, and a solid-state voltage tripler.

11-25 VOLTAGE-REGULATOR DEVICES

The constant voltage-drop across a gas-filled tube when the gas is ionized, or across a zener diode, can be used as a means of regulating the output of a power supply. If a tube is constructed with two electrodes and is filled with neon gas at the proper pressure, the voltage across it is maintained at about 75 V with any value of current flowing through it from about 5 to 40 mA. (Five mA is required to keep the gas ionized properly.) The tube can be used to regulate the voltage across a load drawing up to 35 mA of current.

Figure 11-28 shows a *voltage-regulator* (VR) *tube* connected to a 300-V power supply to provide a regulated 75 V for a load. In the diagram assume the load draws 20 mA but is expected to increase 5 mA and to decrease to zero. The maximum current through resistor R will be 25 mA for the load, plus 5 mA to keep the tube ionized

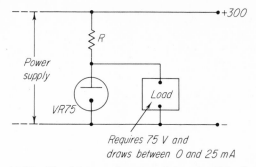

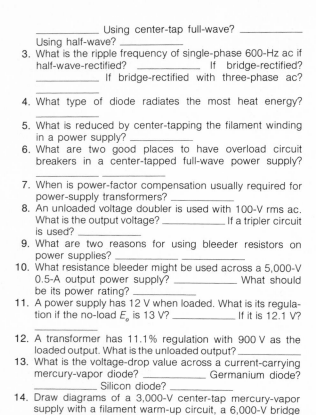

Requires 75 V and draws between 0 and 25 mA

Fig. 11-28 Shunt-type voltage-regulator circuit to feed a constant voltage to a varying load.

properly, a total of 0.030 A. The voltage-drop developed across resistor R will be 300 − 75, or 225 V. By Ohm's law, $R = E/I$, or 225/0.03, or 7,500 Ω. When the power supply is turned on, the voltage across the VR tube rises until it reaches about 90 V, or the striking potential of the gas in the tube. As soon as the gas ionizes, the voltage across the tube drops to 75 V. As long as the load draws no more than 25 mA, the voltage across the tube and load will hold very close to 75 V. From zero load to 25 mA load the voltage will only vary about 3 V. Because the regulator device is in parallel with the load, it may be called a shunt-type regulator.

If the line voltage varies, the VR tube keeps the voltage across the load constant. The VR tube acts as a filter, feeding constant-voltage dc to the load even if the power supply produces a varying dc voltage.

Some commonly used VR tubes are the VR75 (or OA3), the VR90 (OB3), the VR105 (OC3), and the VR150 (OD3). Two VR150 tubes in series may be used to regulate to 300 V, or a VR150 in series with a VR90 will produce a regulated 240 V. The unregulated supply should have an output voltage about 50% greater than the regulated-voltage value.

The anode of a VR tube is a thin wire down the center of the tube. The cathode is a large-area concentric metal cylinder around the anode. To obtain better ionization, a radioactive substance may be painted on a portion of the anode of some regulator tubes, making them potential health hazards if broken.

If the dc potential across a VR tube is reversed, less current will flow through the tube, although it will ionize and may regulate small load currents. The difference in front-to-back current is produced by the difference in area of the two electrodes. The smaller an electrode, the greater the proportion of current that will flow to it. If the anode is made small enough, more than 10 times as much current can be made to flow toward it in comparison to the current from it. This essentially unidirectional behavior was used in helium-filled type BH (also OZ4) rectifiers, which had a large-area dish-shaped cathode and two short thin-wire anode electrodes. This produced a full-wave *cold-cathode* rectifier tube.

Solid-state zener diodes are used in simple shunt regulation circuits similar to VR tubes. Zeners range in breakdown (operating) voltages from about 2.4 V to 200 V. They may be used to regulate currents from a few milliamperes to several amperes at the lower voltages.

11-26 SHUNT-REGULATED POWER SUPPLIES

Whereas a volt of power-supply variation may not seriously degrade the operation of many high-voltage VT circuits, most transistor circuits require low-voltage, high-current power supplies that must reduce output variations to less than 20 mV at worst. More often, the variations or ripple must be reduced to less than 1 mV. To do this, more complex circuits than simple VR tube or zener diode circuits may be necessary.

Figure 11-29a represents one shunt-type regulator with the regulating transistor and the load in parallel (shunt). If the load increases for any reason, the output voltage sags somewhat. This reduces the forward bias of Q_1 (provided by voltage divider $R_B R_S$), and its emitter-collector current decreases, allowing this amount of current to go to the load. This tends to return the output voltage to its original value.

If the shunt regulator can be made more sensitive to output voltage changes by amplifying the variations, the regulation will be better. The circuit in Fig. 11-29b uses a zener to hold the base-collector of Q_1 constant. If the load increases and the output voltage sags, the base voltage of Q_1 becomes less positive due to the decreased voltage-drop across R_Z. This decreases the I_C of amplifier Q_1, which decreases the I_C of amplifier

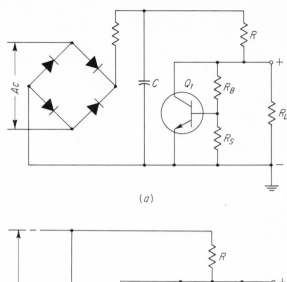

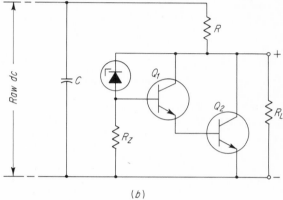

Fig. 11-29 Simple transistor shunt regulators.

Q_2. This decreases the current through R, decreasing the voltage-drop across it, increasing the voltage output. Such a circuit might have 100 times the sensitivity of the single-transistor shunt regulator.

11-27 SERIES-REGULATED POWER SUPPLIES

The current-regulating device that results in a regulated voltage output can be connected in series with one lead of a power supply, as in Fig. 11-30. The circuit to the left of the dashed line is a full-wave center-tap unregulated 400-V power supply with capacitive filter (C_1). Assume the load requires a regulated 180 V. The series tube V_1 acts as a dropping resistor to drop the voltage to 180 V for the load. The resistance of V_1 can be varied by varying the bias on its grid. The

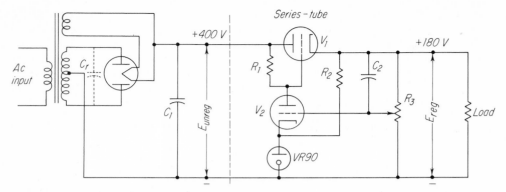

Fig. 11-30 Series-tube voltage-regulator circuit. Output voltage value is controlled by R_3.

difference between the voltage-drop across amplifier V_2 and the voltage-drop across R_2 is the cathode-grid bias voltage for V_1.

The VR90 tube produces a constant +90-V reference voltage. The bias control R_3 is adjusted until the voltage across the load is the desired 180 V. If the voltage at the arm contact of R_3 is now +85 V, the grid bias on V_2 is −5 V.

Plate current for V_2 flows up through R_1. Ionizing current for the VR tube flows up through R_2 and through V_2.

If the load suddenly demands more current, the voltage-drop across V_1 increases and the load voltage might drop to +178 V. This drops the voltage at the contact on R_3 to about +84 V, representing an increase of 1 V of bias for V_2. Higher negative bias on V_2 reduces its plate current, the voltage-drop across R_1 decreases, and the plate of V_2 becomes slightly more positive. This means the grid of V_1 becomes slightly more positive (or less negative). Less negative bias on V_1 decreases its cathode-plate resistance. The voltage-drop across V_1 decreases, increasing the voltage across the load back to almost 180 V.

Capacitor C_2 has a value of about 0.1 μF. It aids in reducing any residual ripple that might appear in the output.

If the load current decreases, the opposite voltage changes occur in the circuit to reduce the load voltage to 180 V. Should the line voltage vary, the 400-V supply will vary but the output voltage holds constant. By paralleling two or more series tubes, load-current capabilities increase. Beam-power tetrodes rather than triodes are usually employed as series tubes.

If C_r (dotted) is chosen to resonate with the transformer secondary, strong circulating resonant currents tend to saturate the core. That results in voltage regulation of the transformer output.

One of the simpler transistor series-regulator circuits is shown in Fig. 11-31a. The bridge rectifier and capacitor, C, produce a reasonably well filtered 20-V dc. With no load connected, no I_c flows through Q_1. Its base is held at +12 V above ground by the zener diode. When a load, R_L, is connected, collector current, I_c, flows through Q_1. Since Q_1 is normally a silicon device, under a light load there is a voltage-drop of 0.6 V between emitter and base, which leaves +11.4 V as the load voltage. If the load increases, I_c increases and the EB voltage increases slightly. This represents a greater forward bias for Q_1, it passes more current to the load, and the output voltage tends to rise. The change in voltage caused by a change in load current can be called an error voltage or signal. The correction of the error voltage in most circuits does not quite return the output voltage to its former value. If this error voltage can be amplified, the voltage regulation can be improved. Note that Q_1 is functioning as an *emitter-follower* amplifier, with R_L as its load, with the base bias being held constant by the zener.

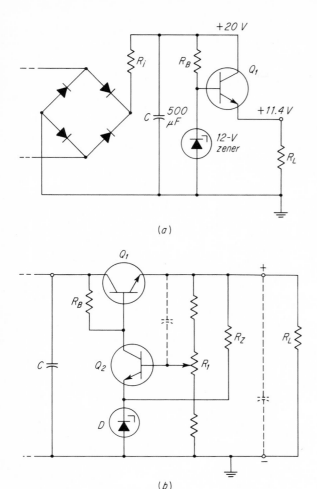

(a)

(b)

Fig. 11-31 Simple series regulators.

The circuit in Fig. 11-31b uses a second amplifier transistor, Q_2, to sense the error voltage and feed an amplified error signal to the control transistor, Q_1. Resistor R_B with Q_2 and D in series forms a voltage-divider bias circuit for Q_1. If transistor Q_1 has its bias changed, its conduction also will change. Potentiometer R_1 and the resistors in series with it form a voltage-divider biasing circuit for Q_2. Moving the arm on R_1 will vary the emitter-collector resistance of Q_2, which in turn will vary the bias on Q_1. Assume R_1 is adjusted to produce an output voltage of 15 V. The zener diode holds the Q_2 emitter at a constant voltage value. If the load is then increased, the output voltage should decrease slightly. With the error voltage becoming

less positive (less forward bias for Q_2), the EC resistance of Q_2 increases. This allows the base of Q_1 to become more positive (more forward-biased), and the series transistor conducts more, reestablishing the output voltage to nearly the original value. Since Q_2 is acting as an amplifier for the error voltage, the regulation of this circuit is quite good. (More complex regulators may include more amplification as well as circuits to counteract temperature variations in the circuit.)

Since short-circuiting the output of a power supply may damage the rectifiers and the series control device, some form of overload protection should be provided. This might take the form of a fuse or overload relay. An electronic overload circuit is shown in Fig. 11-32. Q_1 is a silicon

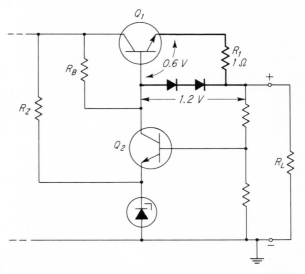

Fig. 11-32 Overload protection for series-regulated power supply.

transistor (EB = 0.6 V). Two silicon diodes in series and a 1-Ω resistor form the overload circuit. The diodes are forward-biased whenever load current flows through R_1. However, the voltage-drop across them must exceed 1.2 V before they will conduct. A load current of 0.5 A produces 0.5 V across R_1, which, in series with the 0.6-V EB, is still not enough to make them conduct. As soon as the load current reaches 0.6 A the diodes begin to conduct, holding the base of Q_1 at 1.2 V above the output voltage. The diodes in series with R_B form a voltage-divider bias circuit. As the

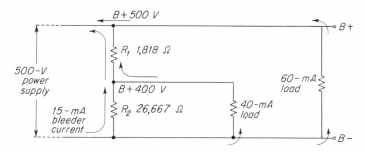

Fig. 11-33 Voltage-divider circuit to feed different voltages to two loads.

load increases, the Q_1 bias ($E_{R_1} - 1.2$ V) decreases, and not enough current can flow through Q_1 to damage it.

11-28 VOLTAGE DIVIDERS

In many applications it is necessary to have two load voltages, one higher than the other. It is possible to use a *voltage-divider* network of resistors to develop the lower voltage from the higher voltage of the power supply. The voltage divider may also serve as a bleeder resistor. Regulation is poor with voltage dividers.

EXAMPLE: A 500-V power supply is used in a two-tube transmitter. One tube requires 500 V on the plate at a current of 60 mA. The other tube requires 400 V at 40 mA. The bleeder current is to be 15 mA. What resistors will the voltage-divider circuit require? See Fig. 11-33.

R_1 and R_2 in series form both the voltage-divider and the bleeder circuit.

The voltage across R_2 is 400 V at a current of 0.015 A. According to Ohm's law, $R_2 = E/I$, or 400/0.015, or 26,667 Ω.

The current through R_1 is the sum of the 0.015-A bleeder current plus the 0.04 A of the 400-V load circuit, or 0.055 A. The voltage-drop across this resistor is 100 V. The resistance of R_1, according to Ohm's law, is $R_1 = E/I$, or 100/0.055, or 1,818 Ω. The wattage rating for the resistors is usually twice the computed I^2R, the EI, or the E^2/R values.

11-29 COPPER-OXIDE AND SELENIUM DIODES

If a sheet of lead and a sheet of copper covered with copper oxide are pressed together, current

will flow more readily from copper to oxide than in the other direction, resulting in a rectifying action. Such a unit is called a *copper-oxide rectifier*. Units of this type will stand only a limited inverse peak voltage. When used in high-voltage applications, many units must be connected in series to give the desired inverse-voltage capabilities.

When iron is coated with selenium, a selenium rectifier is produced. Both copper-oxide and selenium rectifiers were used to rectify ac in power supplies, in battery chargers, and in meters, but in most cases silicon rectifiers are supplanting copper-oxide and selenium units.

Copper-oxide and selenium rectifiers are sometimes manufactured in a full-wave bridge circuit. Four rectifiers are mounted on a single rod and interconnected in such a manner that when ac is fed to the proper pair of leads, dc is obtained from the other two leads. Figure 11-34 shows,

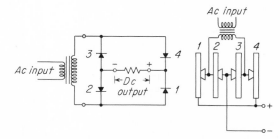

Fig. 11-34 Bridge circuit used with copper-oxide rectifiers and method of stacking the units.

first, a bridge-rectifier circuit with each rectifier numbered and, second, the physical interconnections of such a rectifier group correspondingly numbered.

11-30 DC-TO-DC CONVERTERS

Up to this point power supplies have converted ac to dc. There are many cases in which low-voltage dc, 6 or 12 V from an automobile battery, for example, must be converted to 20- to 80-V dc for power transistors or to several hundred volts for vacuum-tube circuits.

The dc-to-dc converter in Fig. 11-35 uses two

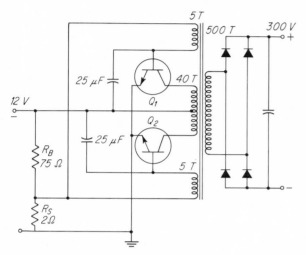

Fig. 11-35 Dc-to-dc converter.

transistors which alternately pulse current through the 40-turn primary. This is stepped up by the transformer, bridge-rectified, and filtered to provide, in this case, 300 V for a VT circuit.

When 12 V is applied to the forward-biased BJTs, one of them, perhaps Q_1, begins to draw more I_C than the other. As a result, greater current would start flowing downward through the 40-turn primary. This induces voltages into both 5-turn secondaries. The upper secondary voltage forward-biases Q_1 still more and its I_C rapidly rises to saturation. At the same time the voltage induced into the lower secondary reverse-biases Q_2 and no I_C flows in Q_2. The heavy increasing I_C of Q_1 in the primary induces a high voltage into the high-voltage secondary.

Very shortly after the I_C of Q_1 reaches saturation, the core of the transformer no longer has a magnetic flux building up in it and the induced voltages in the 5-turn secondaries drop to zero. The 25-μF capacitor of Q_1, which has been positively charged by the upper 5-turn secondary,

begins to discharge through the $R_B R_S$ bias resistors. This produces a reversed current in the upper 5-turn secondary, and by induction Q_2 becomes forward-biased. This drives an I_C pulse upward through the lower half of the 40-turn primary, reverse-biasing Q_1 and cutting off its I_C, but driving Q_2 into I_C saturation. The Q_2 upward-flowing I_C pulse induces a high voltage into the high-voltage secondary of opposite polarity from the voltage induced when Q_1 was functioning. The net result is a continuing oscillation of saturated I_C pulses up and down in the primary inducing a square-wave ac into the secondary. Since the induced ac is flat-topped, it is easier to filter when rectified than rectified sine-wave ac would be. The whole BJT circuit is a push-pull Armstrong oscillator (Sec. 13-4). The frequency of the ac developed depends on the values of the two capacitors, the biasing resistances, and to some extent the inductance of the coils involved. The higher the frequency of the ac the less filter necessary. For this reason these circuits usually are engineered to produce 500- to 1,000-Hz ac. The transformers are often iron-cored toroids.

Another method of converting low-voltage dc to high-voltage dc is a vibrator-rectifier-filter system. The circuit shown in Fig. 11-36 is a *nonsyn-*

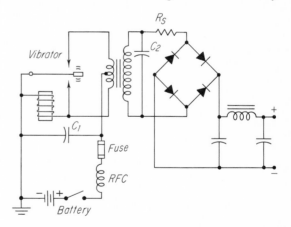

Fig. 11-36 Nonsynchronous-vibrator power supply with bridge rectification.

chronous vibrator supply. When the switch is closed, the high-resistance vibrator magnet coil is energized, pulling the spring steel vibrator arm downward. When it strikes the lower contact, the arm shorts out the magnet coil and a heavy surge of current flows downward in the lower half of the primary of the transformer. With the magnet coil shorted there is nothing to hold the arm down, so it springs back up. The iron weight on its end allows the arm to swing up far enough to strike the upper contact, producing a heavy current downward in the top part of the primary. Once the magnet coil is no longer shorted, it attracts the arm again. After the arm strikes the top contact, it returns to the bottom contact, and so on. The result is an alternating series of pulses in the primary, which induces an ac into the secondary. This is rectified and filtered to provide the desired higher-voltage dc. The ac frequency is usually several hundred hertz.

The few-turn-coil radio-frequency choke (RFC) and capacitor (C_1) reduce sparking at the vibrator contacts and prevent radiation of undesired broadband radio-frequency interference (RFI) or hash that often results from sparking circuits. The RFI can be picked up as unwanted buzzing noises in radio receivers, amplifiers, and other electronic equipment. These supplies must always be completely shielded (enclosed in a metal case). The *buffer* capacitor, C_2, smooths any sharp spikes that might develop on the ac and also reduces vibrator contact sparking. The value of this capacitance is critical. If the buffer capacitor becomes faulty, the vibrator will usually ruin its contacts and will have to be replaced. Whenever a vibrator is replaced, the buffer capacitor should also be replaced with one of similar value, as it might have been the cause of the vibrator trouble.

When vacuum-tube rectifiers are used in this type of supply they must have a high heater-to-cathode insulation rating, as in 6X4- or 6X5-type tubes.

A *synchronous vibrator* supply is shown in Fig. 11-37. It uses no diodes, being a mechanical rectifier. Its primary circuit is similar to that of a nonsynchronous vibrator. Another set of contacts is struck by the vibrating reed or arm.

The reed is pulled down by magnetic attraction, making a contact that drives a pulse of dc upward through the lower half of the primary, at the same time connecting the lower end of the secondary to ground (B−) through the second lower contact. The high-voltage emf developed in the secondary by the primary pulse forces current through the load R_L in an upward direction, as shown by the arrow.

The reed springs back, momentarily closing the two circuits through the upper contacts. This sends a dc pulse downward through the upper half of the primary, inducing a high voltage upward in the secondary. It also connects the upper end of the secondary to ground and forces current through the load R_L, again in the upward direction.

The current that flows in the load is always in the upward direction and is equivalent to a full-wave pulsating dc. The low-pass capacitive-input filter smooths the pulses to a relatively pure dc.

As an aid to waveshaping, damping resistors of 50 to 100 Ω may be connected across the primary (not shown). A resistance having a few thousand ohms is often connected in series with the buffer capacitor C_2.

Test your understanding; answer these checkup questions.

1. What is the minimum *I* value that should flow through a VR tube? _____ The maximum? _____
2. What change in *I* occurs through the dropping *R* to a VR tube when I_L increases? _____
3. What is responsible for the ability of a cold-cathode gas tube to rectify? _____
4. What maximum-value dropping *R* would be used to feed a 12-V zener in shunt with a 100-mA load from a 20-V source? _____ To allow 10 mA for the zener, what *R* value would be used? _____ Its *P* rating? _____
5. Does electron current flow in the arrowhead direction on a zener diode symbol? _____ On a normal rectifying diode? _____
6. Would a load fed from a voltage divider have a low or high E_{reg} percent? _____
7. What solid-state rectifiers were used prior to silicon and germanium types? _____ _____
8. What type of current does an oscillator produce? _____

Fig. 11-37 Synchronous-vibrator power supply.

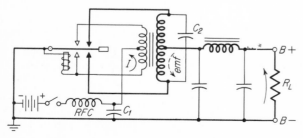

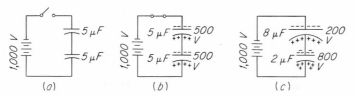

Fig. 11-38 Equal capacitors in series divide the voltage-drop equally. Unequal capacitors divide the voltage inversely as the ratio of capacitances.

9. Why would VTs not be used in 12- to 300-V dc-to-dc converter oscillators? _____
10. What is the advantage of a synchronous over a nonsynchronous vibrator supply? _____
11. What else should be replaced whenever a power-supply vibrator is replaced? _____
12. Will reversing the battery feeding a vibrator supply reverse the output polarity in a nonsynchronous vibrator circuit? _____ In a synchronous vibrator–type circuit? _____
13. Draw diagrams of a transistor shunt regulator, a series-tube voltage regulator, a transistor series regulator with overload protection, a dc-to-dc converter, a nonsynchronous vibrator supply, and a synchronous vibrator supply.

11-31 FILTER CAPACITORS IN SERIES

In an emergency it is sometimes necessary to replace a high-voltage filter capacitor. If the only available capacitors have voltage ratings less than the value required, two or more may be connected in series. Two capacitors with equal capacitance will divide the total voltage across them equally. If one has more capacitance than the other, the one with the lesser capacitance will have the greater voltage-drop developed across it (Fig. 11-38).

When the switch in Fig. 11-38b is closed, the same number of electrons flow in the whole circuit until the capacitors are charged. Since both are of 5-μF capacitance, they charge with the same number of electrons per square centimeter of plate area. The same number of electrostatic lines of force are developed across each square centimeter of dielectric, resulting in an equal voltage-drop across each capacitor, in this case 500 V.

In the third circuit, the same charging current flows into both capacitors. The smaller must now hold more electrons per square centimeter of plate area. This produces more electrostatic lines of force for each square centimeter of dielectric, resulting in a higher voltage-drop across this one.

The voltage-drop across two capacitances in series is inversely proportional to their respective capacitances, or directly proportional to their reactances.

If two 500-V 5-μF paper-dielectric or oil-filled-dielectric capacitors are connected across 1,000 V, they will operate satisfactorily. If one has less capacitance than the other, more than 500 V will be developed across it and it may short out. This will place the full 1,000 V on the other one, shorting it out also. Equalizing resistors should be connected across series capacitors, as in Fig. 11-39. Equal-value resistors tend to hold the

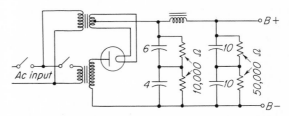

Fig. 11-39 Equalizing resistors across unequal and equal capacitors in series.

voltage-drop across each equal to half the total voltage. The more nearly equal the capacitances, the higher the resistance value the equalizers may have (possibly 100 kΩ). When the two capacitors are not of the same value, the resistors must be lower in value (10 to 50 kΩ). Generally, the higher the resistance values used the the better.

When electrolytic capacitors are connected in series, polarity must be considered, connecting positive leads to negative.

If it is necessary that a 500-V paper and a 500-V electrolytic capacitor be connected in series, equalizing resistors must be used. Such a circuit is not recommended except in emergencies.

Since the leakage currents in two equal-capacitance electrolytic capacitors will normally vary

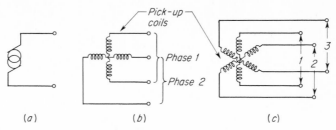

Fig. 11-40 Two terminals from a single-phase alternator, four from a two-phase alternator, and six from a three-phase alternator.

somewhat, equalizing resistors should always be used across such capacitors when in series. The polarity (positive and negative) of electrolytic capacitors must be considered when connecting them in series.

Equalizing resistances can also serve as bleeder resistors, but they must not be used as a voltage-divider circuit.

11-32 THREE-PHASE POWER

For high-power high-voltage supplies in AF or RF stages of radio transmitters, three-phase (3-φ) power has several advantages over single-phase:

1. The ripple frequency is three times that of single-phase. A 60-Hz single-phase full-wave rectifier has a ripple frequency of 120 Hz. A 60-Hz 3-φ full-wave rectifier has a ripple frequency of 360 Hz. Half-wave 3-φ with its 180-Hz ripple is easier to filter than full-wave single-phase.
2. The pulses of rectified current in either full- or half-wave 3-φ circuits overlap. The current value never drops to an instantaneous value of zero, making the 3-φ rectified current still easier to filter. Power is present in the unfiltered circuit at all times.
3. The output voltage of a 3-φ system may be 73% higher than the turns ratio of the transformers would indicate.

Three-phase ac is produced by special ac generators, properly termed *alternators*. In effect, they are three single-phase alternators in one. A single-phase alternator has two leads coming from it (Fig. 11-40). A 2-φ alternator would have four leads, and a 3-φ alternator six leads.

A simple 3-φ alternator may have three separate pickup coils that rotate between electromagnetic-field poles. In each pickup coil a single-phase ac is induced. These coils are set on the revolving part, called a *rotor*, or *armature*, in such a way that the voltages induced in the different coils are 120° apart. The 3-φ alternator shown in Fig. 11-40 would be capable of supplying three separate single-phase circuits, or the outputs can be combined in one of two methods to produce the three-wire three-phase circuit shown in Fig. 11-41. The first diagram shows the Y connection

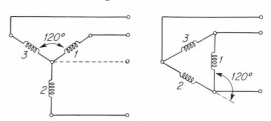

Fig. 11-41 Alternator pickup coils or transformer secondaries connected in Y and in delta (Δ).

(also known as *wye*, or *star*). All three pickup coils are connected together at one point. The other ends of the coils form the leads that are brought out of the alternator. Any two of the output leads will carry voltages from two of the coils in series. Since the voltages in the coils are 120° out of phase, there will never be a time when the two voltages will be at a maximum together. With 100 V as the peak in all coils, there will never be a time when 200 peak V will appear across any two leads. By trigonometry, the peak voltage can be computed to be 173 V between any two legs of such a Y-connected, 100-V per coil, 3-φ alternator. The voltage output across any leg of a Y-connected alternator can be found by multiplying the output voltage of one of its pickup coils by the factor 1.73.

In the Δ-connected (delta) circuits of Figs. 11-41 and 11-42, any two lines leading from the ma-

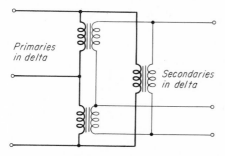

Fig. 11-42 Transformers connected Δ-Δ.

chine are directly connected across a pickup coil. If the coil has 100 V induced in it, there will be 100 V across the line. The other two coils have voltages induced in them, but these will be out of phase, with a resultant of 100 V at the instant the first coil attains its peak value. This parallels all voltages at all times and in such phases as to produce 1.73 times as much current as any one pickup coil alone could carry.

A Y-connected alternator produces higher voltage at a lower current. The same generator Δ-connected produces lower voltage output but greater current. The power output is the same either way.

Three-phase power lines usually have three wires, although the center, or *neutral,* connection on a Y circuit may be grounded as shown dashed in Fig. 11-43.

When either a single-core 3-φ transformer or three separate transformers terminate in a three-wire line, there are five methods of connecting the windings: (1) Δ primary, Δ secondary—termed Δ-Δ; (2) Y-Y; (3) Δ-Y; (4) Y-Δ; and (5) open Δ, which requires only two single-phase transformers.

Figure 11-42 illustrates three transformers with primaries connected in Δ and the secondaries also in Δ. Figure 11-43 illustrates three transform-

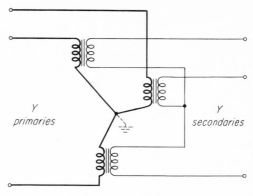

Fig. 11-43 Transformers connected Y-Y.

ers with primaries in Y and the secondaries also in Y.

If the transformers used have a 1:1 ratio, there will be no step-up or step-down in either the Δ-Δ or the Y-Y connections. With the same transformers connected with primaries in Δ and secondaries in Y, as in Fig. 11-44, each output phase

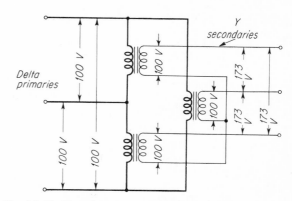

Fig. 11-44 Transformers connected Δ-Y.

will have 1.73 times the voltage of the primary phases. If the transformers have a 1:10 step-up ratio, the secondary-line voltage will be 17.3 times the primary voltage.

Using the same transformers connected with primaries in Y and secondaries in Δ, as in Fig. 11-45, each output phase will have a voltage equal to the reciprocal of 1.73, or 0.578, times the input- or primary-phase voltage. However, the current for each phase of the output can be 1.73 times the current in a primary phase.

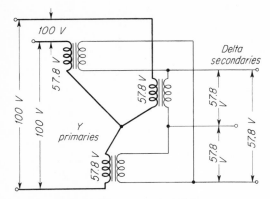

Fig. 11-45 Transformers connected Y-Δ.

With balanced loads on a 1:1 3-ϕ system the following apply:

Delta-connected:

$$E_{\text{line}} = E_{\text{phase}}$$
$$I_{\text{line}} = 1.73 I_{\text{phase}}$$
$$I_{\text{phase}} = I_{\text{line}}/1.73$$

Wye-connected:

$$E_{\text{line}} = 1.73 E_{\text{phase}}$$
$$E_{\text{phase}} = E_{\text{line}}/1.73$$
$$I_{\text{line}} = I_{\text{phase}}$$

For either delta or wye:

$$VA = 1.73 E_{\text{line}}(I_{\text{line}})$$
$$W = 1.73 E_{\text{line}}(I_{\text{line}}) \text{ pf}$$
$$\text{pf} = W/VA$$

It is possible to use only two transformers in a 3-ϕ system. This configuration is known as an open-Δ circuit. It produces 3-ϕ ac by using the three secondary wires, as indicated in Fig. 11-46. It is equivalent to a three-transformer-Δ circuit with

Fig. 11-46 Two transformers connected open-Δ.

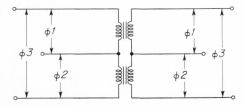

one of the transformers open-circuited in either or both primary and secondary, or with one transformer disconnected. In an emergency it is possible to change a Δ power system to an open Δ and operate at 58% of the full-load capabilities of the three-transformer system.

A Y-connected output circuit can power one, two, or three single-phase circuits and still operate 3-ϕ motors, etc., from the same transformers at the same time (Fig. 11-47). (Only the secondaries of the transformers are shown.)

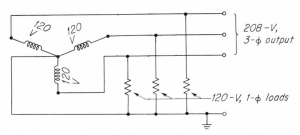

Fig. 11-47 A 208-V three-phase Y with grounded neutral can also supply three 120-V single-phase lines.

In some cases a phase winding of a 240-V Δ-connected power transformer may be center-tapped and used as the neutral for two relatively low-power 120-V single-phase circuits.

Computations of power become quite involved in 3-ϕ circuits. However, if reasonably well-balanced, with each phase taking approximately one-third of the load, the total power can be determined by measuring the power in any two phases separately. If the power factor is more than 0.5, as is usually the case, the total power of the three phases is equal to the *sum* of the two power readings taken. If the power factor is less than 0.5, the total power is equal to the *difference* between the two readings. Figure 11-48 shows how two single-phase wattmeters can be connected to read the total power in a

Fig. 11-48 Two single-phase wattmeters connected to read the power in a three-phase system.

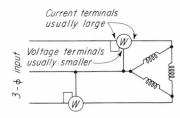

three-phase system. There are also three-phase wattmeters that consider all phases and indicate the true power of the whole circuit.

Proper phasing (connections) of transformers in either Δ or Y circuits is absolutely necessary.

11-33 THREE-PHASE POWER SUPPLIES

When the three phases of 3-φ ac are plotted against time, they form the pattern shown in Fig. 11-49. This can be resolved into three separate

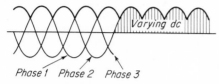

Fig. 11-49 Three-phase ac and varying dc ripple when half-wave-rectified.

single-phase sine waves separated by 120°. The shaded lines indicate the resultant voltage or current when a three-phase ac is half-wave-rectified. The pulses do not drop to zero at any time, making the waveform a varying rather than a pulsating dc.

A half-wave-rectified 3-φ power-supply circuit is shown in Fig. 11-50. The filament transformer is single-phase and may be connected to any of the primary phases. The 1:10 ratio transformers are connected Δ-Y. This circuit does not give the 1.73-voltage gain as a half-wave rectifier, because no two phases are in series across the load at any time.

A full-wave 3-φ bridge-rectifier circuit is shown in Fig. 11-51. It requires four filament windings (not shown). The power transformers are connected Δ-Y. The neutral connection cannot be grounded if B− is. The dashed lines indicate one

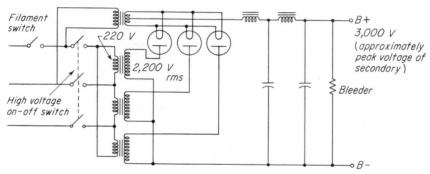

Fig. 11-50   Three-phase half-wave Y-connected power supply.

Fig. 11-51   Three-phase full-wave Y-connected power supply showing current path during the peak of one of the phases.

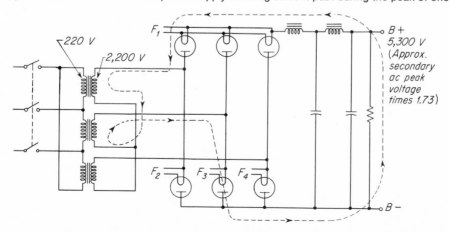

current path at one instant of maximum voltage in one phase. Two of the secondary windings are in series across the load. This results in a 1.73-voltage step-up over the turns ratio of the transformers used.

The waveform of full-wave-rectified 3-ϕ ac is shown by the shaded lines in Fig. 11-52. Because

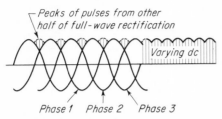

Fig. 11-52 Three-phase ac and varying dc ripple when full-wave-rectified.

of the small amplitude variation, these pulses are very easily filtered.

11-34 INDICATIONS OF POWER-SUPPLY FAILURE

One of the important phases of a technician's duties is to be able to locate and repair failures in equipment when they occur or are about to occur.

There are several indications of power-supply failures. Light gray wisps of smoke may curl up from under a chassis. Sparks may be seen in a rectifier tube. The hissing of an electric arc-over may be heard. Rectifier plates may turn red-hot. Mercury-vapor tubes may turn a brilliant blue-white. A purple glow may appear between filament and plate in a vacuum diode. Meters in associated equipment drop to zero or indicate excessive values. Transformers or chokes may hum ominously or start to smoke. Circuit breakers or overload relays may snap open; fuses may burn out; lights may appear on panels; alarm bells may ring.

Depending on the circumstances, the operator should (1) immediately shut down all the equipment or (2) note the stages that indicate normal operation and then shut down the equipment and start looking for the trouble in the first stage that indicates improper operation.

In power supplies, tubes age and lose emission rather slowly. This condition is usually indicated by a slowly decreasing output current. It is also possible for a filament of a tube to burn out without warning. Vacuum tubes may become slightly gassy and develop a purple glow between filament and plate even without excessive current flow through them. If the tube develops a rapid gas leak, it will suddenly turn milky and the filament will burn out immediately, depositing a white oxide layer on the inside of the envelope. If the trouble is due to a weakened tube, it is unlikely that a fuse will burn out. Subnormal current in the associated equipment is the usual indication. Visual examination of the power supply may tell the operator which tube is at fault.

Smaller receiving-type rectifier tubes can be tested on a tube tester. Transmitting tubes can be tested by heating the filament of a *good tube* to a normal degree and connecting a dc source, an ammeter, and a variable resistance in series with the plate circuit of the tube. The resistance is adjusted until the maximum rated operating plate current flows through the tube. When a suspected tube is tested in place of the good tube, the plate current should be at least 80% of that of the tube known to be good.

A short circuit in a power supply or in associated equipment usually produces the more dramatic indications mentioned above (red plates, internal sparking in the tube, blown fuses). A short circuit is frequently produced when a filter-capacitor dielectric arcs over and carbonizes. This results in a low resistance across the capacitor and a heavy current in the transformer-rectifier-filter circuit. Since chokes are usually installed in the positive high-voltage lead with their cores and metal cases grounded, a short circuit between winding and core will cause heavy current in the transformer-rectifier-filter circuit. A similar effect may be produced if the insulation of the filament-transformer winding breaks down, or if the bleeder resistor shorts out, or if any of the positive-terminal wiring touches the chassis.

If any of the above-mentioned parts short to ground, an ohmmeter reading between ground and the positive terminal will show a very low resistance. If there is no short circuit, the ohmmeter reading should be the value of the bleeder resistor, usually between 10,000 and 40,000 Ω.

(Be sure the equipment is off and capacitors are discharged before using an ohmmeter for any tests.)

Sometimes a filter capacitor does not completely short out but develops a high resistance across it. It is said to have developed a *leak*. A leaking capacitor can be checked by connecting an ohmmeter across it. Any good capacitor of 0.1 μF or more will produce a kick of the ohmmeter needle when first tested. This is the charging current flowing into the capacitor from the ohmmeter battery. If a paper capacitor registers less than about 10 MΩ (10 megohms) after a second or two, it may be considered as leaking. An electrolytic will read low resistance the first instant it is tested but should register more than 100,000 Ω after a few seconds. If an electrolytic capacitor is tested with the ohmmeter leads reversed, a different resistance value will usually be obtained. This is normal.

If one is available, a *capacitor analyzer* will give a reading of the capacitance, leakage, and power factor and show whether the capacitor is open or shorted. This is always better than an ohmmeter test.

To test a capacitor alone in a power supply, one of its leads must be disconnected from the circuit and a meter connected across it; otherwise, the whole filter circuit is being tested.

It is also possible that a short in the load circuit will produce the same symptoms in the power supply as a short in the power supply itself. Disconnecting the power supply from the load and checking both with an ohmmeter will make it possible to localize the trouble more closely. If a B + -to-ground reading across the load shows low resistance with the power supply disconnected, the short is in the load, not in the power supply.

Sometimes one or more turns in the power transformer short together because of insulation breakdown. Heavy primary current will flow whether the shorted turns are in the primary or in the secondary, and fuses or overload relays will go out. Often the transformer will smoke. If it continues to heat or smoke when all parts are disconnected from it, a new transformer is required.

If one-half of the secondary of a center-tapped transformer used in a full-wave rectification system burns out, the equipment may operate but at reduced power because of lower voltage output when the circuit operates as a half-wave power supply. A 60-Hz hum will usually be audible when the output of the equipment is checked.

Sometimes a short circuit is an instantaneous thing caused by arcing across a moistened surface or by an insect crawling between two points of high potential in the equipment. A sudden line surge can produce an overload that will burn out fuses or open circuit breakers. Sometimes fuses themselves oxidize and burn out for no apparent reason. As a result, when a piece of equipment suddenly ceases operating and it is desirable that it be placed in operation as soon as possible, it may be expedient to replace the fuse or close the circuit breaker and try momentarily turning on the equipment again. If there is nothing seriously wrong, the piece of equipment will operate normally. If not, the fuse or circuit breaker will go out again and it will be necessary to shut down to locate the trouble. In the first case, it would be well to examine the equipment closely for burns or signs of arcing the next time it is turned off, cleaning it thoroughly at the same time.

After the equipment has been turned off for servicing, the operator should *always* touch the positive terminals of all power supplies with a grounded flexible wire before servicing the equipment. If such a wire is not available, an insulated-handle screwdriver held across the filter-capacitor terminals will discharge such capacitors. *The operator should make sure he is not grounded* when he does this.

There have been occasions when a piece of equipment using a power transformer in the power supply has been plugged into a power line carrying 110-V dc instead of the required 110-V ac. The inductance of the primary limits the current flow with ac, but it has no opposition to dc and an excessively high current flows through the primary, burning out the primary or a fuse immediately.

The metallic shields and cores of power transformers and power-supply choke coils should be grounded to prevent them from picking up a static-electricity charge which may be dangerous to personnel. Ungrounded equipment in transmitters may allow high-frequency RF ac to find its way into it and cause damage to internal insulation.

When testing solid-state power supplies, remember that an ohmmeter will read low resistance across many good solid-state devices, but if the ohmmeter leads are reversed the resistance indication will be high or infinite.

Test your understanding; answer these checkup questions.

1. Under what conditions are equalizing resistors required with series capacitors? _____
2. Why is a high ripple frequency an advantage in power supplies? _____
3. Why is full-wave 3-ϕ ac simple to filter? _____
4. What is the proper name for any ac generator? _____
5. The primaries of three 1:10 step-up transformers across 220-V 3-ϕ ac are in Y. What is the output voltage across any output lines if the secondaries are in Y? _____ If in delta? _____
6. If the primaries in question 5 are connected in Δ, what is the output if the secondaries are in Y? _____ If in Δ? _____
7. What is another name for Y? _____
8. What is the advantage of using 208-V 3-ϕ Y with a grounded neutral? _____
9. If one secondary of a Δ-Δ transformer system open-circuited, what would be the result on the output voltage of the three legs? _____ On the output current capabilities? _____
10. How many 1-ϕ wattmeters are required to measure the power in 3-ϕ lines? _____
11. Before servicing power supplies, what two things should an operator be sure to do? _____ _____
12. Under what conditions does a smoking power transformer not indicate trouble in the transformer? _____
13. Why might 120-V ac lines sometimes cause a breakdown in 200-V circuits? _____
14. Draw diagrams of (a) transformers connected Y-Y, Δ-Y, Y-Δ, Δ-Δ, and open Δ, (b) a half-wave 3-ϕ power supply, (c) a full-wave 3-ϕ power supply.

COMMERCIAL LICENSE QUESTIONS

Amateur license questions will be found in the Addendum.

FCC Elements 3, 4, and 6 require an ability to answer questions similar to those below. Sections in which questions are answered are shown in parentheses. A question followed by a bracketed number is required for that element alone.

1. Discuss high-vacuum, mercury-vapor, selenium, and copper-oxide diodes or rectifiers. (11-2, 11-14, 11-15, 11-29)
2. Explain the operation of power-supply rectifiers, waveshapes, and filters. (11-3 through 11-13)
3. What are the advantages of full-wave versus half-wave rectification? (11-4, 11-8, 11-18, 11-24)
4. What advantage may a bridge-rectifier circuit have over a conventional full-wave rectifier? (11-4)
5. Draw a diagram of a rectifier system supplying two voltages, one twice the other, using one high-voltage transformer with a single center-tapped secondary and such filament supplies as are needed. (Fig. 11-6)
6. What is the principal function of the filter in a power supply? (11-6)
7. What is the purpose of a choke coil? (11-7)
8. What are the characteristics and relative advantages of capacitor-input and choke-input filters when used with rectifiers? (11-8, 11-9)
9. What are the approximate values of power-supply filter inductors and capacitors? (11-8, 11-9, 11-12, 11-20)
10. Draw a diagram and explain the operation of a vacuum-tube-diode half-wave rectifier with a capacitive-input pi-section filter. (11-8, Fig. 11-12)
11. What is the effect upon a filter choke if a large value of dc flows through it? (11-8, 11-9)
12. How may a capacitor be added to a choke-input filter system to increase the full-load voltage? (11-9)
13. Draw a diagram and explain the operation of a full-wave rectifier with RC filter. (11-11, Fig. 11-15)
14. If the reluctance of an iron-core choke is increased by increasing the air gap of the magnetic path, in what other way does this affect the properties of the choke? (11-12)
15. What is the effect of loose laminations in a filter choke? (11-12)
16. What effect does the resistance of filter chokes have on the regulation of a power supply in which they are used? (11-12, 11-24)
17. What are swinging chokes, and where are they normally used? (11-13)
18. Why should the operating temperature of mercury-vapor tubes be kept within specified limits? (11-15)

19. Why should the temperature of the filament or heater in a mercury-vapor rectifier tube reach normal operating temperature before the plate voltage is applied? (11-15)
20. What is the voltage-drop across a mercury-vapor rectifier under normal conditions? (11-15)
21. What is arc-back or flashback in a rectifier tube? (11-15)
22. What factors permit high-conduction currents in a hot-cathode mercury-vapor tube? (11-15)
23. Why is a time-delay relay arranged to apply the high voltage to anodes of mercury-vapor rectifier tubes some time after the application of filament voltage? (11-15, 11-19)
24. Why are resistors sometimes placed in series with each plate of mercury-vapor tubes connected in parallel? (11-15)
25. What is meant by the peak-inverse-voltage rating of a diode, and how can it be computed for a full-wave power supply? (11-16)
26. What is the maximum allowable secondary voltage of a transformer with a center-tapped full-wave circuit using rectifier tubes having a 10,000-V peak-inverse-voltage rating? (11-16)
27. Compare advantages and disadvantages of high-vacuum and hot-cathode mercury-vapor rectifier tubes. (11-17)
28. What is the ripple frequency in the output of a single-phase full-wave rectifier when the primary source is 60 Hz? (11-18)
29. What is the ratio of the frequencies of the output and input circuits of a single-phase full-wave rectifier? (11-18)
30. Draw a schematic diagram of a full-wave rectifier and filter for supplying plate voltage to a radio receiver. (Fig. 11-21)
31. Draw a diagram of a full-wave bridge rectifier and indicate polarity of output. (Figs. 11-3, 11-34, 11-36)
32. Why is a capacitor sometimes placed in series with the primary of a power transformer? (11-21)
33. Draw a diagram of a voltage-doubling supply using two half-wave rectifiers. (Figs. 11-24c, 11-25)
34. What are the purposes of bleeder resistors in power supplies? (11-23)
35. Would varying the value of the bleeder resistor have any effect on the ripple voltage? (11-23)
36. What is voltage regulation as applied to power supplies? (11-24)
37. What effect does the amount of current required by the load have upon the voltage regulation of a power supply? Why is voltage regulation an important factor? (11-24)
38. If a power supply has an output of 200 V at no load and the regulation is 15%, what is the output voltage at full load? (11-24)
39. If a power supply has a regulation of 10% when the output voltage at full load is 250 V, what is the output voltage at no load? (11-24)
40. What is the percent regulation of a power supply with a no-load voltage of 120 V and a full-load voltage of 115 V? (11-24)
41. What causes poor voltage regulation? (11-24)
42. Explain the action of a VR tube. (11-25)
43. How is the value of the resistor in series with a zener diode (or VR tube) computed to form a voltage-regulating circuit? (11-25)
44. Describe the theory of conduction and rectification of cold-cathode gas-diode tubes. (11-25)
45. Draw a simple diagram of a cold-cathode electron tube voltage regulator. (Fig. 11-28)
46. Show a method of obtaining two voltages from one power supply. (Fig. 11-33)
47. A power supply is to furnish 500 V at 50 mA to one circuit and 400 V at 20 mA to another. The bleeder current value is 10 mA. What value of resistance should be between the 500- and 400-V taps of the voltage divider? (11-28)
48. Draw a diagram and explain the operation of a nonsynchronous-vibrator power supply with silicon-diode bridge-circuit rectifiers and capacitive-input π-section filtering. (11-30, Fig. 11-36)
49. Draw a diagram and explain the operation of a synchronous-vibrator power supply with capacitive-input π-type filtering. (11-30, Fig. 11-37)
50. When filter capacitors are in series, why are high-value resistors often connected across them? (11-31)
51. May a 500-V electrolytic and a 500-V paper capacitor be used successfully in series across a potential of 1,000 V? Explain your answer. (11-31)
52. What precaution should be observed when connecting electrolytic capacitors in series? (11-31)
53. In what circuits of a radio station are three-phase circuits sometimes employed? (11-32)
54. Show by diagrams the delta method and the wye method of connecting transformers in a power system. (Figs. 11-41, 11-42, 11-43)   [4]
55. Show by diagrams how various output voltages may be obtained from wye- and delta-connected transformers. (Figs. 11-44, 11-45)   [4]
56. Draw a schematic wiring diagram of a three-phase transformer with delta-connected primary and Y-connected secondary. (Fig. 11-44)   [4]
57. What system of connections for a three-phase three-transformer bank will provide maximum secondary voltage? (11-32, 11-33)
58. Three single-phase transformers, each with a ratio of 220 to 4,400 V, are connected across a 220-V

three-phase line with primaries in delta. If the secondaries are connected in Y, what is the secondary-line voltage? (11-32)

59. By diagram, show how only two transformers can be connected for operation on a three-phase circuit. (Fig. 11-46)

60. What does a blue haze in the space between the filament and plate of a high-vacuum rectifier indicate? (11-34, 11-14)

61. If part of the secondary winding of the power-supply transformer of a transmitter were accidentally shorted, what would be the immediate effect? (11-34)

62. If a high-vacuum rectifier suddenly shows severe internal sparking and then fails to operate, what elements of the rectifier-filter system should be checked for possible failure before installing a new rectifier tube? (11-34)

63. If the plate or plates of a rectifier tube suddenly became red-hot, what might be the cause and how could remedies be effected? (11-34)

NOTE: FCC license test questions are multiple-choice types, taking a form somewhat as follows:

64. A filter capacitor may be checked for leakage by using (a) a voltmeter and battery; (b) an ohmmeter; (c) a capacitor analyzer; (d) any of these. (11-34)

65. A power-supply transformer is designed to provide 250 V when operating from a 120-V 60-Hz line. If connected across a 100-V dc line, will it (a) produce 208 V; (b) produce 4.1 V; (c) burn out; (d) hum audibly? (11-34)

66. The metallic case of a high-voltage transformer should (a) be grounded; (b) be made of nonferrous material; (c) be insulated; (d) have cooling fins. (11-34)

ANSWERS TO CHECKUP QUIZ ON PAGE 230

1. (*Unequal C values, or if paper and electrolytic types in series*) 2. (*Easy to filter*) 3. (*Unfiltered dc has little ripple amplitude*) 4. (*Alternator*) 5. (*2,200 V*)(*1,272 V*) 6. (*3,806 V*)(*2,200 V*) 7. (*Star or wye*) 8. (*Gives three 120-V lines to neutral*) 9. (*None*)(*Reduced to 57.8%*) 10. (*Two*) 11. (*Open power lines, hang shorting wire across power-supply capacitors*) 12. (*When a load is still connected to secondary*) 13. (*Line voltage surges*) 14. (*See chapter illustrations*)

12 MEASURING DEVICES

The objective of this chapter is to present the information a technician should know about dc and ac ammeters and voltmeters of various types, as well as about ohmmeters, bridges, and some digital indicating meters. The fundamentals of free-running and triggered oscilloscopes are also presented.

12-1 METERS

In radio and electronics, meters are employed to measure or indicate current, voltage, power, resistance, frequency, decibels, volume units, watthours, ampere-hours, etc.

Since alternating and direct current behave differently, it is necessary to develop special meters for dc and others for ac, although a few meters can be used on both.

Almost all meters in general use are current-type meters; that is, they depend upon a current of electrons flowing through them to move the indicator pointer, or needle, across the scale. The greater the current, the farther the pointer moves. The pointer will be pulled back to the zero reading when current ceases by either a taut band or by thin spiral springs attached to it.

Meters vary in size from the tiny 1-in. type used in portable equipment, through the more frequently found 2- to 4-in. *panel* meters, up to *switchboard* meters measuring as much as 1 ft across, made for distant viewing. Most meters are expected to be read at distances of 1 to 6 ft.

12-2 DC METERS

Practically all the dc meters used in commercial radio applications are of the same general type, known as a *moving-coil*, a *galvanometer*, or a *D'Arsonval* meter. This one type of meter can be used as a dc ammeter, milliammeter, micro-ammeter, voltmeter, or ohmmeter, and with recti-fiers it can indicate alternating currents and voltages.

The moving-coil meter is an electromagnetic device consisting of the following parts:

1. A horseshoe-shaped permanent magnet
2. A round iron core between the magnet poles
3. A rotatable mechanism, which includes
   a. A lightweight coil
   b. A pointer affixed to the coil
   c. Two delicate spiral springs to return the pointer to zero position
   d. Two precisely ground bearings
4. A calibrated paper or metal scale
5. A metal or plastic case

A simplified drawing of the working parts of a moving-coil meter is shown in Fig. 12-1. The coil and pointer are shown away from their normal position in the circular slots between the magnet poles and the iron-core piece. The iron core does not rotate, but the coil assembly rotates in the spaces between the core and the magnet.

The distance between the magnet and the soft-iron core is made quite small to reduce the reluctance of the path of the lines of force from

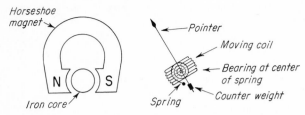

Fig. 12-1 Component parts of a D'Arsonval meter. The moving-coil assembly has been removed from its place between the magnet poles.

north to south pole. This provides a strong field in which the coil moves, reduces leakage lines of force from the magnet, and increases the sensitivity of the meter. It also decreases interaction between the lines of force and other outside magnetic fields that might change the field strength of the magnet.

Each of the two spiral bronze springs is connected electrically to an end of the rotating coil. The other ends of the springs are brought to physical and electrical terminals on the magnet and pole-piece assembly. Besides being used to zero the indicating needle, the springs provide the only path by which current is fed to and from the moving coil.

When current flows through it, the coil becomes an electromagnet with a north pole at one end and a south pole at the other. If the current is in such a direction as to develop an electromagnetic north pole at the upper end of the coil in the illustration, there will be a magnetic attraction between the north end of the coil and the south pole of the magnet. At the same time the south end of the coil will be attracted to the north pole of the magnet. This rotates the coil assembly (and the pointer) clockwise against the spring tension. If the current is small, the springs will not allow much rotation. The greater the current flow, the farther the pointer and coil assembly will rotate.

Since the coil is constructed of many turns of very fine wire, care must be taken never to feed excessive current through it. The full-scale deflection current will not injure the coil, but a 100% or greater overload may burn out the coil or take the temper out of the springs or burn them out, or the very delicate aluminum pointer may be bent against the bumper at the end of the scale.

When current through the coil ceases, the springs return the pointer to the zero setting. Reversing the current through the coil moves the pointer backward, off the scale to the left. This may not damage the meter mechanism, but it may bend the pointer.

On most meters an adjustment screw is brought out at the front. By rotating this screw, more or less torque, or twisting effort, can be placed on one of the centering springs. With this adjustment the pointer can be accurately set to the zero point on the scale when no current is flowing through the meter.

Well-balanced meters will read the same whether held horizontally or vertically. Others may require a zero readjustment if their operating position is changed. This can sometimes be corrected by adjusting the position of the counterweight (Fig. 12-1) to balance the coil and pointer assembly more accurately. Additional *quadrantal* counterweights are often used at right angles to the indicator needle.

Meters are delicate instruments and must be handled gently. They should not be subjected to strong external magnetic fields. When treated properly, meters may hold satisfactory accuracy for 60 years or more.

Meters are made to be mounted on either steel (iron) panels or nonmagnetic panels (aluminum, plastic, etc.). If mounted on an improper panel, the permanent-magnet-field strength will be affected and incorrect indications may result.

Most general-purpose meters are accurate to 2% or better of their full-scale value. They are usually considered most accurate above half-scale. Many ac meters are difficult to read below one-third scale.

There are other dc meters operating on different electromagnetic pulling principles, but they are not in wide use. Some meters use a thin steel *taut band* instead of a pivot, jewel, and springs, which results in a very fine movement.

12-3 LINEAR AND NONLINEAR SCALES

A moving-coil meter usually has good scale *linearity;* that is, if a current flow of 10 A causes the pointer tip to move through an arc of 4 cm, 5 A will move the pointer tip through an arc of 2 cm, 2½ A will result in a 1-cm deflection, and so on.

Meters other than D'Arsonval types often have a nonlinear scale. The scale divisions near the zero value (on the left side of the scale) are usually crowded together and widen toward the higher values. Alternating-current and current-squared meters (Sec. 12-27) are examples of such scales. Figure 12-2 shows a linear meter scale and a possible nonlinear meter scale.

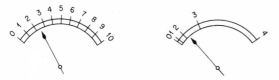

Fig. 12-2 Examples of a linear and a possible nonlinear meter scale.

12-4 DC AMMETERS

In dc meters calibrated to read higher values of current, a 0–1-mA meter might be used. By connecting a resistor across the meter, the current in the circuit divides, part going through the meter and part going through the shunt resistor, as in Fig. 12-3.

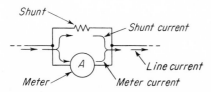

Fig. 12-3 Part of the line current flows through the shunt of an ammeter.

If the coil of a 0–1-mA meter has 25 Ω resistance and a 25-Ω resistor is connected across it, half of any current flowing through this parallel circuit will pass through the meter and the other half will flow through the resistance. In this case, if 1 mA is flowing in the circuit, ½ mA flows through the meter. Only half as much electromagnetic effect will be developed in the meter coil, and only half deflection of the meter pointer will result. The meter will now read full deflection when 2 mA flows in the line. For correct meter indications it will be necessary to replace the 0–1-mA scale with a 0–2-mA scale.

By using the correct value of shunt resistance and scale calibration, it is possible to make a 0–1-mA meter read full-scale deflection with 50 mA, 500 mA, 1 A, 10 A, or any desired value above its basic 0–1-mA reading.

While the resistor shunt can be connected to the external contacts of the meter, it is usually found inside the case. The shunt is considered an integral part of the meter and is not indicated in diagrams unless externally connected.

An ammeter (ampere, mA, or μA meter) is always connected in series with one of the current-carrying wires of a circuit, as in Fig. 12-4. It indi-

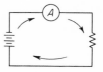

Fig. 12-4 An ammeter is connected in series with the line.

cates how much current is passing through that particular point in the circuit.

In a parallel circuit, one ammeter can be used to measure the total line current, or ammeters can be inserted in the parallel branches, as shown in Fig. 12-5. The sum of the branch currents will equal the line current.

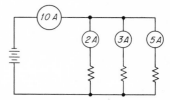

Fig. 12-5 Placement of an ammeter to read the total current of three branches.

Most metals heat when current passes through them, and their resistance values usually increase. To retain accuracy, an ammeter must use metal shunts that do not change resistance under a change of temperature. Special alloys with *zero temperature coefficient of resistance,* such as constantan or manganin, are used.

If the current flowing is more than the full-scale value of an ammeter connected in a circuit, it is possible to connect another ammeter in parallel with the first. The readings may not be equal if the sensitivities of the meters are not the same. However, the circuit current will be the sum of the two readings.

If the accuracy of the ammeters on hand is not

known, a more accurate value results from connecting two ammeters in series and using the average of the two readings. If both meters are accurate, they will both read the same current value, regardless of their sensitivities.

12-5 COMPUTING SHUNT RESISTANCES

If a milliammeter with a 0–1-mA movement is available, it is possible to convert it to a full-scale 0–10- or 0–100-mA meter by connecting different shunt resistors across it.

EXAMPLE: It is desired to make a 0–1-mA meter with 25-Ω internal resistance read 0–10 mA full-scale. The shunt will have to carry 0.009 A, and the meter 0.001 A. Being in parallel, the meter and the shunt will have the same voltage across them. The current in any leg of a parallel circuit is inversely proportional to the resistance. Therefore, $\frac{1}{9}$ of 25 Ω will be required for the shunt, or 2.78 Ω. Whatever the meter reads must now be multiplied by 10. A reading of 0.6 on the meter now indicates 6 mA.

If the meter is to be used as a 0–100-mA meter, the shunt must be $\frac{1}{99}$ of 25 Ω, or 0.253 Ω. The scale graduations must now be multiplied by 100. A reading of 0.74 indicates 74 mA.

Ammeters have low resistance and must not be connected across a source of potential, or excessive current may flow through them.

It is possible to determine the current in a circuit by connecting a meter across a resistance in series with the circuit.

EXAMPLE: A 0–0.001-A meter with 25 Ω internal resistance is connected across a 4-Ω resistor in a circuit (Fig. 12-6) and reads 0.0004 A. The total current in the line can be determined by Ohm's law. The voltage-drop across the meter (and the 4-Ω shunt) is equal to $E = IR$, or 0.0004(25), or

Fig. 12-6 Circuit of meter across resistor in problem.

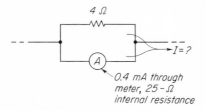

4 Ω

$I = ?$

0.4 mA through meter, 25 - Ω internal resistance

0.01 V. The current through the shunt is $I = E/R$, or 0.01/4, or 0.0025 A. The shunt current is 0.0025, and the meter current is 0.0004, a total of 2.9 mA, which is the line-current value.

12-6 SENSITIVITY

A sensitive meter is one that requires very little current to produce full-scale deflection of the pointer. A meter that will swing to full scale with 1 mA is more sensitive than a meter that requires 2 mA to produce full-scale deflection.

If anything is done to a meter that makes it necessary to use more than normal current to obtain full-scale deflection (by shunting a resistor across it, for example), the meter is said to be desensitized. Actually, shunting a resistor across a meter may make the meter operate as if it were less sensitive, but the *movement sensitivity,* the coil assembly and the magnet's ability to produce full-scale deflection with a given current, has not been changed.

The movement sensitivity of meters in electronic equipment and testing apparatus varies widely. Some of the more common dc meters have sensitivities of 0–50 μA, 0–200 μA, 0–500 μA, 0–1 mA, and 0–5 mA, to mention a few.

Another means of expressing the sensitivity of a meter is in *ohms per volt* (Sec. 12-9).

12-7 DAMPING

Meters with no shunt resistors across them sometimes have very lively moving pointers. Those with very-low-resistance shunts may have slower-moving pointers. This slowing of the pointer movement is known as *damping.* Some damping is desirable to prevent the pointer from oscillating back and forth when the current through the meter is changed a little. If a current is suddenly fed through a meter and the pointer moves up past the correct reading, the meter is underdamped. If it comes to the correct reading rapidly but does not overshoot, it is *critically damped.* If overly damped, it will rise slowly and will not indicate short pulse peaks adequately.

Damping can be produced electromagnetically. In ammeters, the coil is usually across the low-resistance shunt of the meter. When current flows, the coil moves. As it is driven across the lines of force of the magnet, a counter emf is developed in it. This induced voltage develops a counter-current through coil and shunt. The counter-current bucks the current flow through the meter coil, preventing the coil and pointer from swinging upward as rapidly as they normally might. As the coil slows, the counter emf drops off. The meter is damped in both upward and downward motions. The lower the resistance of the shunt, the greater the damping effect.

Damping is aided by using an aluminum form for the moving coil. The metal form acts as a shorted turn. When the coil moves in the magnetic field, it induces a current in the shorted turn, setting up a counter field that tends to oppose the movement of the shorted turn and coil.

A third method of producing damping utilizes small aluminum paddles attached to the coil assembly. The motion of the paddles through an enclosed air chamber prevents rapid movement of the coil and pointer.

12-8 THE ELECTROSTATIC VOLTMETER

A voltmeter is an instrument that will indicate the difference of potential across a circuit. It is always connected across the difference of potential, as in Fig. 12-7. Note that the ammeter is in

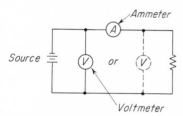

Fig. 12-7 A voltmeter is connected across the source or load.

series with the circuit; the voltmeter is across the circuit.

It might be said that the only real emf-indicating instruments are the oscilloscope (Sec. 12-29) and the electrostatic voltmeter. The electrostatic voltmeter is constructed in the form of a variable capacitor with a pair of stationary metal plates and a pair of light, balanced metal plates that rotate on the central axis (Fig. 12-8). A pointer is attached to the rotating plates. A spiral spring returns the pointer to zero.

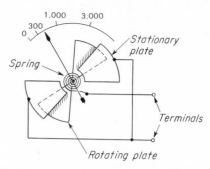

Fig. 12-8 An electrostatic voltmeter. When oppositely charged (+ and −), the rotor plates are pulled toward the stator plates and move the pointer.

When the plates of the meter are connected across a positive and a negative source of voltage, the positive- and negative-charged plates are attracted to each other. The rotating plates swing toward the stationary plates, overcoming the tension of the spring, and rotate the pointer across the scale.

Electrostatic voltmeters can be used to measure either dc or ac emf values from about 50 V to several thousand volts. The scale is not linear at the lower voltage readings but can be made reasonably linear at the higher-scale readings by shaping the plates. These meters are used to measure high-potential circuit voltages, particularly when little or no current drain on the circuit is required. The electrostatic voltmeter requires no current flow through it to produce deflection, although it does require electrons as an initial charging current. When used to measure ac, its capacitive reactance results in an apparent alternating current, although no electrons actually flow through the meter. The electrostatic voltmeter is usually confined to laboratory-type measurements.

Test your understanding; answer these checkup questions.

1. How are D'Arsonval meters adjusted to exact zero? _____
2. How is current led into and out of the moving coil of a D'Arsonval meter movement? _____
3. What must be adjusted if a meter changes its zero position when moved? _____
4. Can external magnetic fields affect the indications of moving-coil meters? _____
5. Of what are meter pointers usually made? _____

6. At what part of their scale are meters assumed to be most accurate? _____

7. To what might quadrantal counterweights be attached? _____

8. What returns a taut-band meter pointer to zero? _____

9. Which would usually have the more linear scales, dc or ac meters? _____

10. What is it that constantan and manganin have that makes them good shunt metals? _____

11. Why should ammeters never be connected across a line? _____

12. What is the shunt value needed to make a 50-Ω 0-1-mA meter into a 0–50-mA meter? _____

13. What are two methods of expressing the sensitivity of a meter? _____ _____

14. What do pointers on underdamped meters do? _____ On overdamped meters? _____

15. What are three ways of producing damping in meters? _____ _____ _____

16. What meter requires no current flow through it to produce deflection? _____ Does it read ac, dc, or both? _____

12-9 DC VOLTMETERS

The usual dc voltmeter is composed of a sensitive moving-coil milliammeter with a resistance in *series* with the meter. Notice that the dc ammeter usually has a low-value shunt resistor, while the voltmeter has a high-value series resistor, called a *multiplier,* normally installed inside the voltmeter case. In some special cases the multiplier may be externally connected.

A common meter movement used in voltmeters is a 0–1-mA meter with an internal resistance of about 25 Ω. If this meter is connected across a 100-V dc source, the current through it will be $I = E/R$, or $100/25$, or 4 A. Since the moving coil is wound for a maximum current of 0.001 A, it would promptly burn out if across 100 V. To limit the current to 0.001 A, the meter must have a resistance in series with it of $R = E/I$, or $100/0.001$, or 100,000 Ω. The meter has approximately 25 Ω resistance in the coil. To this must be added 99,975 Ω (nominally 100,000 Ω). Such a meter with this value of resistance connected in series with it will give full-scale deflection when across 100 V, half-scale deflection across 50 V, quarter-scale deflection across 25 V, and so on.

If it takes 100 V to produce a full-scale deflection with a 100,000-Ω multiplier resistor, a 200,000-Ω multiplier will require 200 V to produce full-scale deflection. For this reason, a 0–1-mA

meter is said to have a sensitivity of 1,000 Ω/V. A 0–50-μA meter (0.00005 A), being 20 times as sensitive as a 0–1-mA meter, has a sensitivity of 20,000 Ω/V. Since the resistance of the moving coil of the meter is such a small percent of the total multiplier-resistance value, it would be disregarded. To produce a 0–150-V meter, either a 0-1-mA meter with a 150,000-Ω resistance or a 0–50-μA meter with 3,000,000 Ω as the multiplier could be used.

The ohms/volt sensitivity of a meter can be determined by computing the reciprocal of its full-scale *current* reading in μA. Thus, for a 100-μA meter

$$\text{Sensitivity} = \frac{1}{100 \times 10^{-6}}$$

$$= \frac{1}{10^{-4}} = 10,000 \ \Omega/V$$

To measure the voltage across a circuit that is known to be more than the full-scale voltage of any of the meters on hand, it is possible to connect two voltmeters in series. The sum of the two voltage readings will be the voltage across the circuit. The readings may not be equal if the sensitivity of the meters is not the same. Each meter is actually indicating the voltage-drop existing across itself.

If the accuracy of the voltmeters on hand is not known, two voltmeters can be connected in parallel across the circuit. The average of the two readings should produce a fairly accurate value.

12-10 VOLTMETERS IN HIGH-RESISTANCE CIRCUITS

A low-sensitivity meter may give correct readings when measuring circuits having low-resistance values but may give very inaccurate indications when used to measure voltages in high-resistance circuits. For example, consider the circuit shown in Fig. 12-9.

When a 0–100-V 1,000-Ω/V meter is connected across the source of emf, *A* to *B,* it will read 100 V. Since no current is flowing through the 100,000-Ω resistance *R,* there is no voltage-drop across it and there must be 100 V between points *A* and

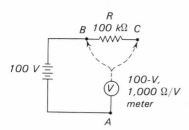

Fig. 12-9 Voltmeter reads 100 V between *A* and *B* but only 50 V between *A* and *C*.

C. If the meter is connected between points *A* and *C*, however, the meter will indicate only 50 V. In this case, current is flowing through the 100,000-Ω resistor *R* and through the meter with its multiplier resistance of 100,000 Ω. With a source of 100 V across 200,000 Ω, the current will be $I = E/R$, or 100/200,000, or 0.0005 A, and the meter reads only half-scale. The meter is actually reading the correct voltage across its terminals. There is now a 50-V drop across resistor *R*. As soon as the meter is disconnected, the voltage across points *A* and *C* rises to 100 V again.

If a 100-V 20,000-Ω/V meter is substituted for the 1,000-Ω/V meter in the same circuit, it will give a different reading. Across points *A* and *B* it will show 100 V, but when connected across points *A* and *C,* it will read more nearly the actual voltage that exists across these points when no meter is connected between them. The multiplier resistance of this 100-V meter is 2,000,000 Ω. The multiplier plus the resistance *R* has a total of 2,100,000 Ω. The current through the combination is found by $I = E/R$, or 100/2,100,000, or 0.0000476 A. This is 47.6 μA. The meter, being a 50-μA meter, will read 47.6/50 × 100 V, or 95.2 V, an error of only about 5%.

It can be seen that the more sensitive the meter used, the more accurate the voltage readings will be when high-resistance circuits such as the automatic volume control (AVC) circuit in receivers and the plate circuits of resistance-coupled amplifiers are measured. It is possible that even the 20,000-Ω/V meter might not give accurate enough readings in very high resistance circuits.

12-11 OHMMETERS

A standard piece of test equipment used in radio and electronics is an ohmmeter. With this

meter it is possible to read directly the value of a resistor, the amount of resistance in a coil, or the value of resistance in a circuit or to make continuity tests on filaments of vacuum tubes and on transistors, capacitors, transformers, or entire circuits.

An ohmmeter can be a relatively simple piece of equipment, as in Fig. 12-10, composed of a

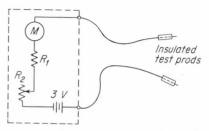

Fig. 12-10 Component parts of a simple ohmmeter.

moving-coil meter, a low-voltage battery (1½ or 3 V), a fixed resistor R_1, and a rheostat R_2.

The meter and the two resistors R_1 and R_2 form a 3-V voltmeter. When the two test probes are held together, the meter is connected across the battery and reads full-scale. The probes touching each other represent a zero-resistance connection. Therefore, this point is marked 0 Ω on the meter scale. The zero-ohms point is on the far-right end of the scale, as in Fig. 12-11.

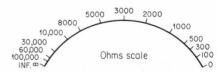

Fig. 12-11 Ohmmeter scale with zero ohms at one end and infinite ohms at the other.

If the meter is a 0–1-mA meter with negligible resistance, the total of R_1 and R_2 will equal 3,000 Ω. If the probes are touched across a 3,000-Ω resistor, there are 6,000 Ω in the circuit and the current is ½ mA. The meter deflects to a half-scale reading. If the probes are held across 1,000 Ω, the total resistance in the circuit is 4,000 Ω and the meter will deflect to three-quarters scale. Deflection can be computed by

$$D = \frac{R_m(100)}{R_m + R_x}$$

where D = percent of deflection
R_m = resistance of multiplier
R_x = resistance of unknown resistance

If a 60,000-Ω resistance is measured, the meter deflects only one-twentieth of the full scale. Thus, resistance values crowd together at the high-resistance end, with infinite resistance being equal to the zero-deflection setting in Fig. 12-11. This particular meter will not read values above about 60,000 Ω accurately, nor will it give satisfactory readings of resistances under 100 Ω.

The rheostat R_2 is made variable to compensate for battery aging. Dry cells have about 1.5 V when new but about 1.3 V when they age. As a result, it is necessary to hold the test probes together and adjust the rheostat until the meter reads exactly zero ohms before taking a resistance reading. This calibration to 0 Ω assures correct resistance readings.

A multirange ohmmeter is shown in Fig. 12-12.

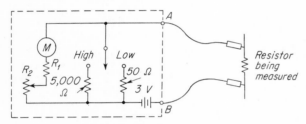

Fig. 12-12 Circuit of a multirange ohmmeter.

The meter and resistors R_1 and R_2 form a 3-V meter as before. Test probes are connected to terminals A and B. When the probes are held together, the voltmeter is across the 3-V battery. The rheostat R_2 is adjusted to an accurate zero-ohms setting. If the probes are across a 50-Ω resistor, as shown, and the high-low switch is in the low position, a voltage-divider circuit is

formed. The meter is across only 1.5 V and will read half-scale. Fairly accurate readings can be obtained from zero to about 500 Ω with this setting. If the switch is in the high position, a 5,000-Ω resistance across the probes will give approximately half-scale deflection. Reasonably accurate readings can be expected up to about 100,000 Ω.

By using a 50-μA meter instead of the 0–1-mA meter and several values of calibrating resistors instead of only two, fairly accurate readings can be obtained from a few ohms to 1,000,000 Ω or more.

CAUTION. The current through ohmmeters, particularly on the low-resistance settings, may be a hundred or more milliamperes. Milliammeters, microammeters, transistors, germanium diodes, or circuits which will not stand this much current through them must not be measured or tested with an ohmmeter on its low ranges. When using an ohmmeter, it is also essential that the circuit being tested have no current flowing in it. If the circuit is not completely dead, the sensitive meter in the ohmmeter may be burned out by inadvertently connecting the ohmmeter across a relatively high voltage.

12-12 VOLT-OHM-MILLIAMMETERS

A handy piece of equipment is a *volt-ohm-milliammeter* (VOM). This is usually a relatively sensitive dc meter in a small box with a battery, switch, and several terminals. If connection is made across the proper terminals and the switch is set to the correct position, the single meter can be used as a voltmeter, an ohmmeter, or a milliammeter. The meter face will be marked with three separate scales. Figure 12-13 shows a simple VOM circuit.

Volt-ohm-milliammeters may have bridge-type solid-state rectifiers to enable them to indicate ac voltages. A separate ac volts scale will appear on the face of the meter. Usually three or four voltmeter ranges, three or four milliammeter or ammeter ranges, and two or more ohmmeter ranges can be selected.

12-13 DC VACUUM-TUBE VOLTMETERS

A voltmeter that may load a high-resistance circuit even less than a 20,000-Ω/V meter does

---

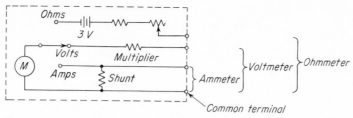

Fig. 12-13 Essentials of a simple volt-ohm-milliampere meter (VOM).

is the *vacuum-tube voltmeter* (VTVM), also called *electronic voltmeter*. It utilizes the fact that a small variation in grid voltage can produce a relatively large variation of the plate current.

The simplest type of VTVM is shown in Fig. 12-14. It consists of a high-μ triode with a 0–1-mA

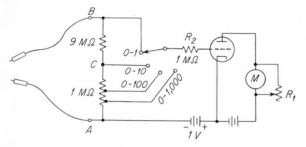

Fig. 12-14 Rudimentary vacuum-tube voltmeter.

meter in the plate circuit. If the probes are touched to a 1-V source, positive to *B* and negative to *A*, the −1-V grid bias will be neutralized and the plate current will increase to several milliamperes. If the shunt resistance across the meter is adjusted until the meter reads full-scale deflection, the meter can be calibrated to read from zero up to 1 V at full deflection. Unfortunately, because of the bend in the $E_g I_p$ curve of a tube, the lower-scale indications will be crowded together, resulting in nonlinearity at the low-voltage end.

Resistor R_2 prevents excessive current flow through the meter. If any positive voltage applied to the grid is greater than the 1-V bias, the grid will draw current, producing a voltage-drop across R_2, which maintains the grid-to-cathode voltage at essentially 1 V. If the meter is inadvertently connected across a high voltage, the meter will not be damaged.

If the resistance between the bias battery and point *B* is 10 MΩ (10 megohms), the meter sensi-

tivity is 10,000,000 Ω/V. This is 500 times better than the sensitivity of a 20,000-Ω/V meter at 1 V.

To change from a 0–1-V to a 0–10-V meter, the 10-MΩ input resistance is tapped to form 9- and 1-MΩ resistances, forming a voltage divider in the grid circuit, as shown. The grid is connected to the 1-MΩ resistor, and the meter now deflects to full scale when 10 V is applied to the terminals *A* and *B*. The meter face can carry a double calibration, one from 0 to 1 V and the other from 0 to 10 V. The operator must note the voltage range being used to know which scale to read. The sensitivity of the meter is now 1,000,000 Ω/V. If the 1-MΩ resistance is also made into a voltage divider, the meter can be made into a 0–100-V meter. The sensitivity is then 100,000 Ω/V. If the meter is made into a 0–1,000-V meter, the sensitivity will be only 10,000 Ω/V, which is less than the 20,000 Ω/V of a 50-μA meter used as a voltmeter. Above 500 V it may be better to use a standard 20,000-Ω/V meter unless the VTVM has special high-voltage circuits built into it. (Note that the VTVM input impedance is constant at 10,000,000 Ω on all ranges.)

Practical VTVMs use two tubes in a balanced circuit similar to the diagram shown in Fig. 12-15. Triodes are shown for simplicity, although pentodes may be used. Many modern meters use dc amplifiers ahead of the balanced circuit to produce greater sensitivity.

In the circuit shown, if no voltage is being applied to the input terminals, the grids of both tubes are at the B− potential. Both tubes are conducting; the currents through R_1 and R_2 should be equal; and the voltage-drop across each should be the same. The meter is therefore connected across two points having no difference in potential (both plates are the same number of volts below the B+ value), and no current flows through the meter.

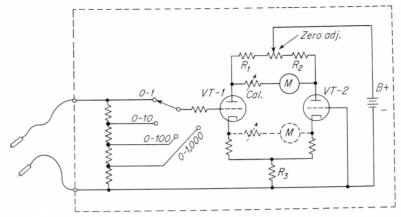

Fig. 12-15 A practical VTVM using a balanced tube circuit.

If the voltage-drop across R_1 and R_2 is not exactly the same with no voltage applied to the input terminals, the zero-adjust potentiometer, mounted on the front panel, can be adjusted to correct this error and bring the indicator to zero.

When a positive voltage is applied to the input circuit, the grid of VT-1 becomes less negative. Plate current through VT-1, R_1, and R_3 increases, increasing the voltage-drop across R_1 and R_3. The increase in voltage across R_3 increases the bias on VT-2, and the plate current for this tube decreases, decreasing the voltage across R_2. The meter is now connected across a voltage difference. If the voltage-drop across R_1 and R_2 was 10 V with no input, now there may be 11 V across R_1 and 9 V across R_2. The meter is across a 2-V differential; current flows through it, and it deflects. (The resistor in series with the meter is used to calibrate the meter when new tubes are installed. When the meter has once been calibrated correctly, this control need not be adjusted again.)

The $E_g I_p$ curves of all tubes bend, but by using two opposing tubes the effect of their bent curves is canceled and a linear scale results. Furthermore, the cathode resistors produce a degenerative effect that further flattens any tendency toward nonlinearity.

In some VTVM circuits, the meter may be connected between the two cathodes, as indicated by the dotted lines in the diagram. The theory of operation is essentially the same.

One advantage of a VTVM is its polarity-

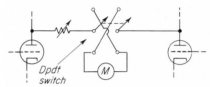

Fig. 12-16 Double-pole double-throw polarity-reversing circuit used in a VTVM.

reversing switch, shown in Fig. 12-16. It reverses the connections of the indicator meter. If the VTVM reads backward during a test, the test probes need not be reversed. It is necessary only to throw the polarity-reversing switch.

Many vacuum-tube voltmeters have a switch that allows the meter to function as an ohmmeter. The meter face carries another scale calibrated in ohms. When set to read resistance, a battery and two (or more) calibrating resistors, R_a and R_b, are connected as shown in Fig. 12-17. As soon as the switch is thrown to the ohms position, the meter swings to full deflection, being connected across the battery (R_a and R_b have no current flowing through them.) If the probes are held together, the pointer drops back to a zero reading. If the resistance-range switch connects R_a in the circuit, when a resistance equal in value to R_a is connected across the probes, as shown, the pointer indicates half-scale, as in previously explained ohmmeters.

Note that this VTVM ohmmeter scale will read 0 Ω at the left side of the scale and infinite resistance at the right side, the opposite of ohmmeters previously explained.

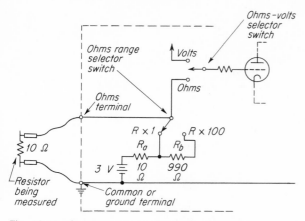

Fig. 12-17 Ohmmeter connections for a VTVM.

When transistors are used instead of vacuum tubes, a transistor voltmeter, or TVM, results. The use of BJTs alone is not particularly successful due to their relatively low input impedance. This can be overcome by the addition of a high-input-impedance amplifier ahead of the BJTs. The normally high input impedance of JFETs (or MOSFETs) makes them ideal for even battery-powered portable TVMs, as in Fig. 12-18. Note the similarity of VTVM and JFET TVM circuitry. The addition of the two diodes in parallel across the meter is a protective circuit. If the voltage-drop across the meter ever exceeds 0.6 V (silicon diodes), the diodes conduct and prevent any greater current flow through the meter. Such a double-diode protection circuit is often used across ammeters or across the movement (coil) of a voltmeter.

A difficulty develops when using the conventional ohmmeter circuit shown, with 1.5 or 3 V as the battery voltage, to measure resistances in solid-state circuits. Since the barrier voltages of solid-state junctions are either 0.2 or 0.6 V, having as little as 1.5 V in the ohmmeter will produce conduction and a low reading on the ohmmeter if any solid-state junctions are in the circuit being measured. By using a *low-power ohms* circuit, in which the battery voltage is reduced to about 0.05 V, circuit resistance measurements become possible in BJT, FET, and IC circuitry without removing the devices from the circuit. It is important that the 0.05-V power source have low impedance, such as produced by some form of

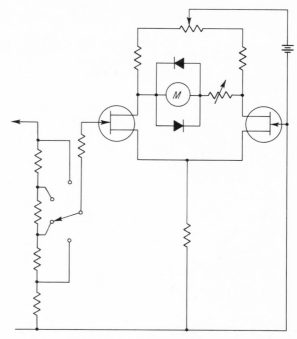

Fig. 12-18 Electronic voltmeter (EVM) using FETs in place of VTs of a VTVM.

series regulation (Sec. 11-27). Because of the low voltage involved, even such things as sensitive meter movements can be measured directly without producing enough current to damage them.

Test your understanding; answer these checkup questions.

1. What is the name of the resistor connected in series with voltmeter movements? _____
2. Which is more sensitive, a 0-50-μA or a 0-1-mA meter? _____
3. What is the Ω/V sensitivity of a 0–2-mA meter? _____ Of a 0–25-μA meter? _____
4. What would be the value of the multiplier used with a 50-μA-movement 300-V meter? _____
5. Could a 0–1-mA-movement 100-V voltmeter and a 0–50-μA-movement 100-V voltmeter be used in series across 125 V? _____ Why? _____
6. A 100-V 0–1-mA voltmeter reads 100 V when across one of two 100,000-Ω series resistors across a power supply. What is the power-supply voltage? _____
7. If a 20,000-Ω/V meter with 5 $k\Omega$ internal resistance is used in an ohmmeter with a 3-V battery, what internal resistance is required in the meter to produce proper zeroing? _____
8. On a simple ohmmeter where is the 0-Ω mark? _____ On a multirange ohmmeter? _____

9. On what range on multirange ohmmeters is the meter most likely to be dangerous to equipment being tested? _____
10. What does VOM mean? _____
11. Does switching to a higher full-scale range produce more, less, or the same sensitivity with a VTVM? _____ With a D'Arsonval voltmeter? _____
12. What component prevents meter damage in a VTVM? _____
13. What is gained by using balanced tubes in a VTVM? _____
14. Is the calibrate or the zero-adjust control on the front panel of a VTVM? _____
15. In what kind of circuits are "low-power" ohmmeters particularly useful? _____

12-14 AC VACUUM-TUBE VOLTMETERS

Most VTVMs will measure dc voltages up to 1,000 V or more, resistance from 1 Ω to possibly 1,000 MΩ, and ac voltage up to about 1,000 V. In some cases, for ac voltage measurements the sensitive meter in the VTVM, in conjunction with a solid-state bridge rectifier and multiplier resistors, is used. The vacuum tubes are not used at all. This results in a meter that may not respond accurately to ac voltages above the AF range of about 15,000 Hz. Other VTVMs may have diode rectifiers inside an ac voltage probe. These meters may satisfactorily measure RF voltages.

It would seem that a single-diode tube could be used to rectify the ac being measured. The half-wave-rectified pulses could be filtered with a resistance and capacitance filter (D_1, R_1, and C_1 in Fig. 12-19), and the resulting E_{dc} across C_1 could be measured by the dc VTVM. Essentially this is correct. However, as soon as the cathode of the probe diode D_1 is heated, electrons flow to its plate, forming a 1- to 2-V negative *contact potential* on it. This voltage indicates on the VTVM, making it impossible to zero the meter properly.

By using twin diodes, one to provide the rectified ac voltage to be read by the VTVM and the other to develop an equal value of contact potential to be applied to the other VTVM tube grid, it is possible to balance out the contact-potential indication. The contact potential of one tube equals the contact potential fed to the other tube, resulting in an equal decrease in both plate currents. This allows the meter to be brought to a zero reading with no voltage applied.

A VTVM having an ac voltage function may have a separate scale for rms and peak ac voltages. The probe-to-ground capacitance varies in different meters from perhaps 3 to 20 pF. When measuring tuned circuits, the circuits may be detuned slightly by this capacitance. The input impedance is somewhat lower than when the meters are used to measure dc voltages.

Some ac voltage probes contain a single solid-state diode (no contact potential) which can be used for practically any frequency but is limited

Fig. 12-19 Ac probe for a VTVM with a contact-potential-canceling diode.

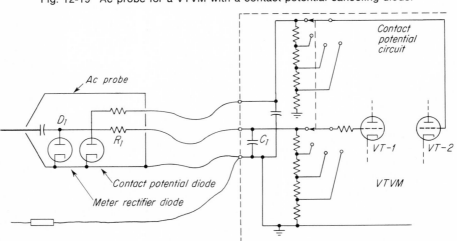

to about 100 V. They have low probe-to-ground capacitance and require no filament wires in the probe cable. If the diode is germanium, it will not measure voltages below 0.2 V.

12-15 AC METERS

The D'Arsonval, or moving-coil, meter will deflect in one direction or the other, depending on the direction of the current flowing through it. With a 1-Hz ac, it will swing back and forth at a 1-Hz rate. With 10-Hz ac, it may attempt to swing back and forth 10 times per second, but because of damping and inertia the pointer cannot move fast enough and will only vibrate a little above and below the zero reading. With any frequency higher than about 20 Hz, the needle will not move at all. It is possible to increase the ac through such a meter until the meter burns out and still have no deflection or indication at any time. Thus, the D'Arsonval meter alone is not suitable for ac-circuit measurements.

As previously mentioned, the electrostatic meter will give an indication with either dc or ac. The ac calibrations are usually in effective values (0.707 of the peak value if the waveform is sinusoidal).

12-16 THE RECTIFIER METER

In radio and electronic circuits, where the frequency involved is not much above the audible-frequency range (20,000 Hz), a moving-coil meter with a bridge rectifier is frequently used. Four rectifiers are connected in a full-wave bridge-rectifying circuit to form an ac voltmeter, as shown in Fig. 12-20. The bridge rectifier converts ac into pulsating dc and produces a deflection

Fig. 12-20 Component parts of a rectifier-type ac voltmeter.

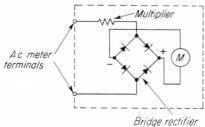

equivalent to 0.636 (the average value) of the ac peak.

It is standard practice to calibrate the face scales of ac meters in effective values. Thus, the peak-voltage value of any sine-wave ac being shown is the meter reading times 1.414.

The rotation of the moving coil in a dc meter is always proportional to the *average* value of the current or voltage being measured. When a voltmeter is connected across a 6.3-V battery, the meter reads 6.3 V. If a 2-V peak ac is added in series with the 6.3-V dc, the voltage varies from 4.3 to 8.3 V alternately but the average is still 6.3 V. This is the value the meter will indicate.

As explained in Chap. 4, the effective value of a sine-wave cycle of ac is 0.707 of the peak value. The average value is 0.636 of the peak. To convert from the effective value to the peak value, the multiplying factor 1.414 (reciprocal of 0.707) is used. To convert from average to peak, the factor 1.57 (or 1/0.636) is used.

If a 10-V *dc* meter is to be used to read *ac* voltage by employing a full-wave bridge rectifier, an indication of 10 V on the dc scale will actually be the average value of the dc pulses flowing through it, or 0.636 of some peak-voltage value. To find the peak value when the average is known, the multiplying factor 1.57 is used: 1.57 × 10 V = 15.7 V. Therefore, the 10-V dc scale reading indicates 15.7-V peak ac. The effective value of a 15.7-V peak is 0.707 of it, or 11.1 V. This is 1.11 times the average 10-V value shown on the meter. Thus, when using a dc-calibrated meter and a full-wave rectifier to measure ac, the dc scale reading must be multiplied by 1.11 if the effective ac value is desired.

If a half-wave rectifier is used, the average value is half the full-wave value of 0.636, or 0.318. Now a 10-V reading on the meter indicates twice the full-wave peak, or 31.4 V peak. Therefore, to determine the effective ac value, the conversion factor to use if a half-wave rectifier is employed with a dc meter is 2.22 times the dc scale reading. (All rectifiers are nonlinear near their zero-voltage points. As a result, neither of these theoretical factors, 1.11 for full-wave and 2.22 for half-wave rectifiers, will give exact values. Special scale calibrations are required for low-voltage ac meters.)

Meters using selenium or copper-oxide rectifiers may not be accurate at higher frequencies because of the capacitance across the rectifier units. This can be partially overcome by using low-capacitance germanium diodes, but the inductive reactance of the moving coil increases with frequency and the meter reads lower as the frequency increases.

D'Arsonval meters with self-contained rectifier units are normally calibrated in effective ac values. A 10-V reading on their scales indicates a 10-V effective value of ac. This is 0.707 of the peak only when the waveform is sinusoidal.

12-17 PEAK-READING METERS

In audio work, in broadcasting and recording, the waveform of the ac used is rarely sine-wave. As a result, the effective values shown on the meters are rarely 0.707 of the peak values. If the peak value is desired, the normal ac meter is not satisfactory. A *peak-reading* voltmeter that rectifies the ac being measured, charges a capacitor with the pulsating dc obtained, and then measures the voltage across the charged capacitor (Fig. 12-21) is required.

The meter used may be either a sensitive moving-coil meter or a vacuum-tube voltmeter. The

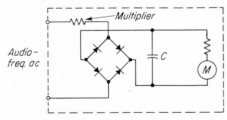

Fig. 12-21 A peak-reading ac voltmeter. Meter *M* is a sensitive dc or vacuum-tube voltmeter to indicate voltage across the capacitor.

ANSWERS TO CHECKUP QUIZ ON PAGE 243

1. (*Multiplier*)   2. (*0-50-µA*)   3. (*40 kΩ/V*)(*400 kΩ/V*)
4. (*6 MΩ*)   5. (*No*)(*60 µA flows, 50-µA meter off scale*)
6. (*300 V*)   7. (*60 kΩ*)   8. (*Far right*)(*Same*)   9. (*Low R*)
10. (*Volt-ohm-milliampere meter*)   11. (*More*)(*The same*)
12. (*R at grid*)   13. (*Scale linearity*)   14. (*Zero adjust*)
15. (*Solid-state*)

---

high resistance of the meter circuit slowly leaks off the charge of the capacitor. The indicator rises rapidly as the capacitor charges but hangs at or near the peak reading for a period of time. The less sensitive the meter, the less resistance it has and the quicker the capacitor discharges through the meter.

Most ac vacuum-tube voltmeters actually respond to the peak values of the ac they are measuring because of the capacitance across the output of the ac probe circuit. The scale on the meter is calibrated in effective values (0.707 of the peak). As a result, readings on such scales, even when measuring speech or music signals, if multiplied by 1.414, approximate the actual peak values.

12-18 DECIBEL AND VU METERS

The VU (volume-unit) meter is used in broadcast work as well as in other audio applications. It is an ac peak-reading voltmeter of the rectifier type used to monitor the amplitude of the AF program signals in different parts of the broadcast or other audio circuits. It must be connected across a 600-Ω impedance line, since it can be used to read *power* levels when across such a line.

In the past, different services used different values of power as the reference, or 0-dB value. Some used 500 Ω, and others 600 Ω as standard line impedances. Some used 0.006 W as the 0 reference point, some 0.0125 W, and others 0.001 W. All such meters are calibrated in decibel units (Sec. 8-6), with the zero-power-level indication at about center scale. If the power in the line is more than the reference value, it produces a higher voltage across the line and the meter reads to the right of the zero mark. If the power is less than the reference value, the voltage is less and the indicator needle moves to less than center scale.

If a reference power of 0.006 W is used, a +10-dB reading on the meter indicates that 10 times the reference power, or 0.06 W, is being delivered to the 600-Ω load. A 20-dB reading indicates 100 × 0.006, or 0.6 W, etc. On the other hand, a −10-dB reading on the meter indicates $\frac{1}{10}$ of 0.006, or 0.0006 W. A −20-dB reading indicates 0.00006 W, etc.

In 1940, a standard type of power-level meter was decided upon. It is a rectifier-type meter having specified needle damping ("ballistics") and scale calibrations. It is calibrated with 0.001 W as the zero reference level and must be connected across a 600-Ω impedance line. This is the standard VU meter in all broadcasting services. All VU meters are built with identical characteristics, whereas dB meters may not have the same scale markings, damping, or zero levels. The VU meter has a fairly high degree of damping, whereas the ordinary dB meter usually has less. The dB meter is used in applications where the power level being measured is not changing rapidly. In circuits carrying speech and music the power level is always changing and VU meters are used.

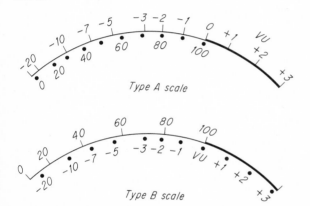

Fig. 12-22 VU meter scales graduated in plus or minus volume units and 0% to 100% modulation.

There are two scale markings used in VU meters, as shown in Fig. 12-22. The VU meter is constructed to operate with a 3,600-Ω external series multiplier resistor (Fig. 12-23). When the power in the line being measured is higher than the +3 VU at the high end of the meter scale, three properly selected resistors forming a T-type *attenuation pad* can be inserted in the meter circuit to desensitize the meter by a predetermined number of volume units (decibels) to prevent it from indicating off scale. If the attenuation pad is made to desensitize the meter by 8 VU, any reading given by the meter will be 8 VU lower than the true value. An indication of −11 VU represents an actual line power of −3 VU with an 8-VU attenuator pad in the circuit.

The 3,600-Ω external resistor is not installed inside the meter in order that the user may include an attenuation pad in the meter circuit if it is required.

12-19 THERMOCOUPLE AMMETERS

A thermocouple ammeter consists of a dc moving-coil meter connected across a thermocouple junction as in Fig. 12-24. Current flowing from A to B or from B to A produces heat at the junction. The junction is heated whether the current through it is ac or dc. The heat produced is independent of frequency.

The junction is composed of two dissimilar metals welded together. When the joint is heated, different values of electron activity are developed

Fig. 12-23 Circuits for one VU meter connected across a 600-Ω line and another VU meter with an attenuation pad to desensitize it.

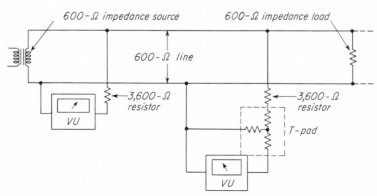

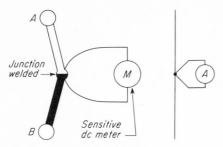

Fig. 12-24 Thermocouple ammeter consists of a sensitive dc meter and a thermocouple junction. Symbol of thermocouple ammeter at right.

in the two metals. This results in a dc emf between them and an electron movement from one to the other. The heated junction becomes a thermal dc generator. The current developed is small, but with a sensitive meter and high enough junction temperature satisfactory deflections are produced. The greater the current through the resistance of the junction, the warmer it becomes and the greater the dc developed across the meter. A thermocouple ammeter's indicator needle moves rather slowly because of the time required to heat the junction.

Thermocouple meters are normally calibrated to read effective ac values. If calibrated at 60 Hz, the calibrations will be accurate up to 20 MHz or more. If used in dc circuits, some meters may read either slightly high or slightly low, depending on the direction of the current through them, due to the small dc voltage-drop developed across the junction by the line current. This emf is in series with that produced by the heated junction and may either aid or buck it. The average of two readings, one with the current flowing in one direction and one with the current in the other, will give a correct value.

The range of a thermocouple meter can be varied in several ways. Different junction metals will have different electron-activity capabilities and will produce different dc emfs with the same heat. A resistance in series with the meter lead will reduce the current through the meter, decreasing the sensitivity of the meter as a whole. A shunt across the junction will reduce the current flowing through the junction and will also reduce heating. Deflection is essentially current squared with most heat-actuated meters.

There are other methods of producing a ther-

mocouple junction. For example, the two dissimilar metals may be welded together and to the surface of a current-carrying wire.

Since they operate on practically any frequency, thermocouple meters can be used to measure dc, AF, and RF currents. They are customarily used as antenna RF ammeters in radio transmitters.

12-20 HOT-WIRE AMMETERS

A hot-wire ammeter is a relatively simple meter and indicates equally well for dc or ac of any frequency.

Figure 12-25 illustrates the basic principle of

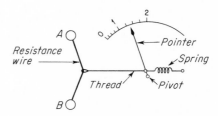

Fig. 12-25 Essentials of a hot-wire ammeter.

the hot-wire ammeter. When current flows from A to B or from B to A, the resistance wire becomes warm and expands. When the wire expands, the spring tension pulls the center of the wire toward the right, at the same time pulling the pointer over. The greater the current flow, the greater the heat developed, the more the wire expands, and the greater the movement of the pointer. Deflection is essentially current squared.

Because of high internal resistance, the effect of variations of air temperature, failure to return to zero always, and slow pointer movement, it is rarely seen today.

12-21 THE ELECTRODYNAMOMETER

The electrodynamometer, or dynamometer, is somewhat similar to the dc moving-coil meter except that it has no permanent-magnet field. Instead, it has a pair of air-core field *coils* that produce an electromagnetic field when current flows through them. The pointer is attached to a moving coil that is returned to the zero position

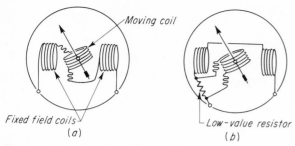

Fig. 12-26 (a) Low-current and (b) high-current dynamometer connections.

by a pair of spiral springs (represented by the wavy lines, Fig. 12-26).

Two methods of connecting the field and moving coils of an electrodynamometer are shown. The series connection is used for voltmeters and low-current ammeters. The circuit with the moving coil across a low-value series resistance is used in high-current ammeters. In either case, the currents in the moving and stationary coils are in phase. When current flows through one, it also flows through the others, producing a magnetic pulling effect that rotates the pointer against the tension of the springs. Because the polarity of both the fixed- and the rotary-coil magnetic fields reverses with a reversal of current, the rotational pull is always in the same direction. For this reason, the meter can be used to indicate on either ac or dc. Although electrodynamometers can be made with sensitivities of a few milliamperes, they are generally rather insensitive (a few ohms per volt). These meters find their greatest use in power-frequency circuits where power drain is of little importance.

Some electrodynamometers can be calibrated on dc and will hold calibration reasonably well for ac up to about 500 Hz. At higher frequencies increasing reactance of the coil decreases the current flow through the meter and introduces errors. The electrodynamometer has no iron as a core and should not be operated near iron masses.

The electrodynamometer, as well as other power-frequency ac meters, may use a step-down transformer when measuring potentials higher than 1,000 V. When measuring heavy currents, a step-down *current transformer* can be connected in series with the line. The meter is connected across the secondary.

Because of their low sensitivity, low-range dynamometer voltmeters may require as much as ½ A to produce full-scale deflection.

The deflection of the pointer against the spring tension requires current flow in both the moving coil and the field coil. As a result, the deflection of the needle is small with low currents and increases as the square of the current, producing a current-squared scale.

Test your understanding; answer these checkup questions.

1. What is the approximate value of contact potential? _____ Could this affect biasing of vacuum tubes? _____
2. What do you know about the diode in an ac probe that measures nothing below 0.6 V? _____
3. How could an ac voltage function be added to a VOM? _____
4. Under what condition would 10 V rms shown on a meter not indicate 14.14 V peak? _____
5. What is the average value if an ac voltmeter reads 10 V effective? _____
6. Why do many ac voltmeters read low at frequencies above about 20 kHz? _____
7. What might be the effect if a very sensitive meter were used as a peak-reading meter? _____
8. Across what impedance line will VU meters indicate accurately? _____ Is this true of dB meters? _____
9. What power is in the line if a VU meter indicates +30? _____ −20? _____ How is it possible for a VU meter to read +30? _____
10. Are VU meters dB meters? _____
11. What do the two scales of VU meters measure? _____
12. If the thermocouple part of a thermocouple meter were used as a thermometer, what would have to be done to the meter scale? _____
13. Would a thermocouple ammeter read accurately on 5 Hz? _____ Would the pointer vary at this rate? _____ Why? _____
14. What thermal ammeter is rarely seen anymore? _____
15. Basically, how does a dynamometer differ from a D'Arsonval movement? _____
16. What meter(s) might normally have current-squared calibrations? _____

12-22 REPULSION-TYPE METERS

The repulsion-type *moving-vane* meter consists of a coil of wire and, inside the coil, two *vanes* made of thin sheets of highly permeable, low-retentivity soft iron that can magnetize and demagnetize easily. One of the vanes is stationary and the other rotary, somewhat as indicated in Fig. 12-27. The pointer is attached to the moving

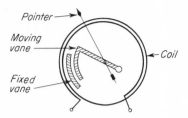

Fig. 12-27 Iron-vane type of repulsion ac meter.

vane and is returned to the zero setting by a spiral spring (not shown).

When the coil is energized, it magnetizes both vanes by induction. Since they are lying in the same plane, the two vanes are magnetized with like polarity and repel each other. The moving vane is repelled from the fixed vane and against the tension of the restoring spring. Regardless of the direction of the current, the vanes continually repel each other as long as current flows in the coil.

A moving-vane meter has low sensitivity and is normally used as a voltmeter or ammeter for power-frequency ac, although it can be used to measure dc. If the meter is used as an ammeter, the coil is made of a few turns of heavy wire. As a voltmeter, the coil has many turns of fine wire. By proper shaping of the vanes it is possible to produce fairly linear graduations over most of the scale.

An *inclined-vane* (Thompson) meter has its iron vane set on an inclined plane inside the coil and fastened to the pointer. When magnetized, the vane tries to line itself in the lines of force, rotating

the indicator needle against a spiral-spring tension. It will operate with either ac or dc.

A purely ac meter is the *inclined-coil,* or inclined-loop, meter. Moving fields due to ac flowing in the external coil induce a current into the internal shorted coil, or loop. The induced current in the loop produces a magnetic field of its own, in opposition to the external-coil field, rotating the loop and the indicator attached to it against spring tension.

12-23 THE WATTMETER

The value of power in a circuit can be determined by multiplying the voltage by the current. If an electrodynamometer type of meter is connected with its field coils in series with the line, as shown in Fig. 12-28, all the current to the load

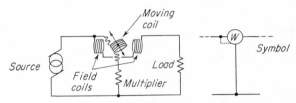

Fig. 12-28 Wattmeter and symbol. Dotted connection is used if meter has a fourth terminal on it.

passes through the field coils and produces a magnetic field proportional to the current. If the moving coil and a resistor are connected as a voltmeter across the line, the magnetic field around the moving coil is developed proportional to the voltage across the circuit. In one meter are represented both current and voltage effects. Increasing the current value increases the deflection of the pointer. Increasing the voltage across the line results in a greater current through the moving coil, a stronger magnetic field around it, and a greater scale deflection. Increases of either current or voltage, or both, increase the power in the circuit and the scale deflection of the meter. The meter may be calibrated in watts or kilowatts. It may be used with dc or low-frequency ac.

A wattmeter always indicates the *true power* in an ac circuit. If the current and voltage are out of phase, the current-carrying field coils and the voltage coil automatically allow for this and it is

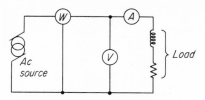

Fig. 12-29 Circuit by which the power factor of a load can be determined (pf = P/VA).

not necessary to multiply the indication by the power factor.

In Fig. 12-29 the product of the current times the voltage gives the apparent power. The wattmeter reads true power. By dividing the true-power reading by the computed apparent power, the power factor of the load can be determined. The formula is

$$pf = \frac{P}{VA}$$

where pf = power factor
 P = true power
 VA = apparent power

The power used by a load in a dc circuit can be determined by a wattmeter alone or by connecting a voltmeter and ammeter in the circuit and multiplying E by I.

12-24 WATTHOUR METERS

Electric energy is measured in watthours, watt-seconds, or joules. Energy-measuring meters are known as watthour meters or kilowatthour meters. They operate on a principle somewhat similar to that of a wattmeter. Instead of moving an indicator needle, however, the current and voltage fields rotate the armature of an electric motor. The motor rotation gear-drives indicator hands that rotate like clock hands. A thousand watts operating for a minute will rotate the motor for a minute, gear-driving the hands through a small arc. The same power operating for an hour will move the indicator hand through an arc 60 times as great. There are usually four indicator hands, each reduction-geared 10 times the preceding indicator. The first reads kilowatthours; the second, 10s of kilowatthours; the third, 100s of kilo-

Fig. 12-30 Symbol of a watthour meter.

watthours; and the fourth, 1,000s of kilowatt-hours. The symbol of a watthour meter is shown in Fig. 12-30.

12-25 FREQUENCY METERS

There are five basic *frequency meters:* the vibrating-reed type, the induction type, the electrodynamic type, the beat-frequency type (described in Chap. 21), and the digital counter with 1-s (1-second) gating.

Except for the beat-frequency and digital types, frequency meters can measure only a narrow band of frequencies. The vibrating-reed type may indicate from only 58 to 62 Hz, for example. It might be composed of nine vibrating steel reeds, having natural periods of vibration of 58 Hz, 58.5 Hz, 59 Hz, and so on. An electromagnet excited from the circuit being measured produces an alternating field at the frequency of the current in the circuit. The reeds are placed in this alternating magnetic field. If the frequency is 60 Hz, the reed tuned to 60 Hz falls into resonance with this frequency and vibrates with considerable amplitude, while adjacent reeds vibrate less. Observation of the reeds indicates the frequency of the ac. If two adjacent reeds vibrate at the same amplitude, the frequency of the ac is halfway between the frequency of vibration of the two reeds.

The induction and electrodynamic types of frequency meters utilize a principle of balancing an indicator needle at the center of the scale by using the magnetic field of a resistive circuit and the magnetic field of an inductively reactive circuit to oppose each other. If the frequency increases, the current through the reactive circuit decreases while the current through the resistive circuit remains the same. This pulls the needle toward the resistive-circuit side of the meter. A lower frequency produces greater field strength of the reactive circuit, pulling the indicator toward the reactive-circuit side of the scale. These frequency meters are connected across the line similarly to a voltmeter.

The digital-type frequency meter employs an electronic counter that can register with neon tubes or other digital readout devices the number of pulses fed into it. Between the decimal-counting register in Fig. 12-31 and the input circuit is

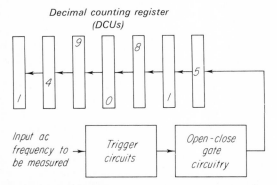

Fig. 12-31 Block diagram of a digital counter used as a frequency-measuring device. Gate opens for precisely 1 s.

a gate circuit that opens for precisely 1 s and then closes. During the open-gate time any pulses being fed into the machine are registered on the decimal-counting units (DCUs). If a 25-Hz ac is being fed into the machine, the trigger circuits rectify the ac to 25 pulses of dc and shape the pulses to a square wave for best operation of the counter circuits. During the 1 s of open-gate condition 25 pulses pass through the gating circuit. The first DCU registers the first nine pulses. The tenth pulse registers on the second DCU as a 1, and the first DCU returns to a zero indication. The next nine pulses again run up the first DCU. The twentieth pulse registers as a 2 on the second DCU. The last five pulses run the first DCU up to 5, and then the gate shuts. The first DCU is left with a 5, and the second with a 2. The DCUs are so arranged that the viewer sees the numbers as 0000025 on the register. The DCUs can be set to hold the reading or to sample the frequency again after a desired period of time. In the block diagram of the meter the register is reading 1,490,815 Hz.

If it happens that the first half cycle of the ac to be counted has the improper polarity to be half-wave-rectified, the counter will read one digit lower than the correct frequency. A one-cycle error is always a possibility with digital counters.

12-26 AMPERE-HOUR METERS

Ampere-hour meters are specialized energy-indicating instruments. In radio the dc type is limited almost exclusively to battery-charging circuits.

Briefly, when a battery is being *discharged,* current flows out of its negative terminal and into its positive. To *recharge* a battery, it is necessary to reverse the current flow through it by using an emf greater than that of the battery. This forces current into the negative and out of the positive terminals, as shown in Fig. 12-32. If an ampere-

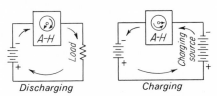

Fig. 12-32 Ampere-hour meters in a circuit in which a battery is discharging and in which the battery is being charged.

hour meter is connected in series with a discharging battery, the current flows through the meter in one direction. When the battery is charging, the current flows through the meter in the opposite direction.

The ampere-hour meter is a small mercury-pool motor. The direction of motor rotation depends on the current direction through the meter. The speed of rotation depends on current strength. The motor is geared to a rotating indicator needle. While the battery is discharging, the indicator needle rotates clockwise. While the battery is charging, the indicator needle reverses its motion, moving counterclockwise. The more current flowing, the farther the indicator needle rotates in a given time.

12-27 CURRENT-SQUARED-METER SCALES

In the electrodynamometer, the same current flows through the field and moving coils. Doubling the current in one coil results in a doubling of the current in the other and a quadrupled magnetic effect. Thus, deflection is proportional to the *current squared*. For example, if the pointer deflects 1 cm when 1 mA flows, it will deflect 4 cm with 2 mA, 9 cm with 3 mA, etc.

If the current values are squared and the squared markings are also placed on the meter scale, it will be found that the squared markings are linear, as in Fig. 12-33

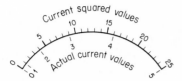

Fig. 12-33 Current-squared-meter face. I^2 and actual current values indicated.

12-28 BRIDGES

Any discussion of measuring instruments must include some mention of bridge-type devices. The Wheatstone bridge of Fig. 12-34 can be used to

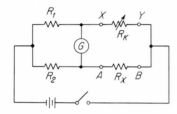

Fig. 12-34 Resistance-bridge circuit for visual null indications.

measure resistance values accurately from a fraction of an ohm to millions of ohms. The unknown resistance R_x is connected between points A and B. A known variable resistance R_k is connected between points X and Y. If the four resistances are proportional so that R_1 is to R_k as R_2 is to R_x, the voltage-drops across R_k and R_x will be equal. With no difference of potential across the meter, it will read zero. If the resistances are not proportional, the meter will indicate some value, plus or minus. The formula for the balanced circuit is

$$\frac{R_1}{R_k} = \frac{R_2}{R_x} \quad \text{or} \quad R_x = \frac{R_k R_2}{R_1}$$

In Fig. 12-34, if it is known that R_1 is 5 Ω, R_2 is 10 Ω, and R_k is 50 Ω, R_x must be

$$R_x = \frac{50(10)}{5} = 100 \ \Omega$$

The bridge circuit requires a sensitive meter with a zero setting in the *center* of the scale. Such a meter is called a *galvanometer*.

The theory of using a proportional balance to produce a null indication is also employed in other bridges. There are many different types capable of measuring impedance, reactance, frequency, capacitance, and inductance. Space limits this discussion to one other type, an inductance bridge (Fig. 12-35). This bridge uses an ac

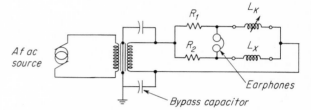

Fig. 12-35 Inductance bridge giving aural indication of the null.

source, such as an audio-frequency oscillator, transformer-coupled to the bridge. When R_1 is to L_k as R_2 is to L_x, the difference of ac potential across the earphones will be zero and no tone will be heard. If the proportions are not correct, a signal from the oscillator will be heard as a tone in the earphones.

The formula for determining the unknown inductance is the same as used for the resistance in the Wheatstone bridge. Other variations are possible. For example, it might be more practical to use L_k as a fixed value and R_1 as the variable element, since varying large values of inductance is difficult. In measuring inductors, any resistance in the unknown inductance results in some error.

The variable resistances (inductances or capacitances) used to balance bridge circuits are available in *decade boxes* ("deca" means "ten"). A decade box may have five or more rotary switches; the first switch selects, for example, one of nine resistances from 0.1 to 0.9 Ω. The second switch adds resistance in single-ohm units up to 9 Ω. The third switch adds resistance in 10-Ω units up to 90 Ω. The fourth switch adds resistance in

100-Ω units up to 900 Ω. The fifth switch adds resistance in 1,000-Ω units up to 9,000 Ω. By proper selection of the various switches, any value of resistance within 0.1 Ω from 0.1 to 9,999.9 Ω can be selected.

Resistor decades are very accurate. Capacitor decades are slightly less accurate. Inductance decades may be accurate only to within 5%. Try to diagram a resistance decade box.

Test your understanding; answer these checkup questions.

1. Name five types of ac voltmeters. _____ _____ _____ _____ _____
2. Which of the meters of question 1 will not indicate with dc applied to it? _____
3. In a wattmeter to what is moving-coil current proportional? _____ Field-coil current? _____
4. How is power factor determined if volts, amperes, and watts of a circuit are known? _____
5. What kind of device is a watthour meter? _____
6. Name five types of frequency meters. _____ _____ _____ _____ _____
7. Which frequency meter(s) can measure over a wide band of frequencies? _____
8. What does DCU mean? _____
9. Why might a 10-s gate opening result in a more accurate frequency count than a 1-s opening with digital counters measuring low frequencies? _____
10. What kind of a device is an ampere-hour meter? _____
11. For what are ampere-hour meters usually used? _____
12. Are the graduations of a current-squared meter linear? _____ Are the graduations of squared current-squared values linear? _____
13. For what is a Wheatstone bridge used? _____
14. What is a center-zero meter called? _____
15. What is used in a Wheatstone bridge to serve as the known resistance value? _____

16. Is a VTVM a bridge circuit? _____
17. Is a VOM ohmmeter a bridge circuit? _____
18. Why might AF ac and earphones make a good null indicator when used in bridges? _____

12-29 THE OSCILLOSCOPE

An extremely useful measuring device is the *oscilloscope,* or *oscillograph.* It presents an instantaneous visual indication of voltage excursions that no other measuring device can show. The indicator is known as a cathode-ray tube (CRT), which is somewhat similar to the picture tube in a television receiver, the basic difference being in the type of deflection used.

A CRT consists of an *electron gun,* four deflection plates, a flourescent screen, an Aquadag coating, and a glass envelope.

The electron gun, shown in Fig. 12-36, consists of a heater cathode to emit electrons, a small metal cylinder with a hole in one end called the grid, and two other metal cylinders, one named the focusing anode and the other the accelerating anode. The grid is negatively charged in respect to the cathode by being connected to a more negative point on a voltage-divider resistor across a power supply of 600 to 1,500 V or more.

The negative charge on the grid would normally prevent electrons from passing through the one small hole in it, except that the focusing anode on the other side is at a relatively high positive potential and attracts them. Some of these electrons strike the focusing anode and move to the power supply, but most of them pass on through the focusing anode and into the *electrostatic lens* that is formed at the open end of this anode.

Because of the difference in potential between

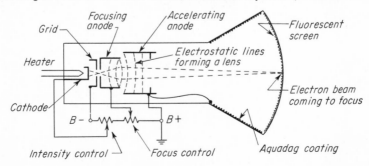

Fig. 12-36 Electron gun of an electrostatic-deflection cathode-ray tube. (Deflection plates not shown.)

the two anodes, the electrostatic lines of force in this area bend. These lines of *equipotential* across the opening between the anodes form an electrostatic lens. Electrons moving through this lens are made to converge at the screen, as light beams can be made to converge and focus when passed through a glass lens. By varying the voltage difference between the first and second anodes (focus control), the configuration of the lens can be changed, allowing the electron beam to focus as a spot smaller than the head of a pin on the fluorescent screen.

The second anode, besides aiding in focusing the beam, increases the speed of the electrons by its high potential and is known as the accelerating anode. The more rapidly the electrons travel, the brighter the spot they produce when they strike the fluorescent screen at the face of the tube.

The intensity of the spot produced on the screen is basically controlled by changing the potential on the grid. The more negative the grid, the fewer electrons that can be pulled through the hole in the grid and the less intense a spot produced on the screen. Control of the intensity can be accomplished by making the return lead from the cathode to the voltage-divider resistor a variable control. This may also have a slight effect on the focus.

The electron beam can be deflected up or down by including two *vertical deflection plates* in the neck of the tube, above and below the beam, as in Fig. 12-37.

If the lower deflection plate is connected to the

Fig. 12-37 Deflection plates of an electrostatic-deflection CRT.

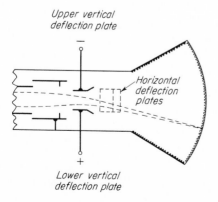

Upper vertical
deflection plate

Horizontal
deflection
plates

Lower vertical
deflection plate

second anode (ground) and a positive potential in respect to it is applied to the upper deflection plate, the electrons in the beam will be attracted toward the upper plate and the spot on the screen will move upward. How high the spot moves is determined by the magnitude of the voltage applied to the deflection plate. If the upper plate is made negative in respect to the lower plate, the electron beam will be deflected downward a distance proportional to the voltage applied. Applying a voltage to be measured to the two plates gives an indication on the screen of the peak-voltage value. A transparent graph with millimeter spacing between lines can be laid over the face of the tube and the length of the deflection measured. If the sensitivity of the deflection plate with a particular anode voltage is 0.33 mm/V, for example, and the deflection of a pulsating dc is 6 mm, the peak voltage applied to the plates must be 18.2 V. The direction of deflection from the center of the screen indicates the polarity of the voltage if dc is being measured. Since there is no horizontal deflection, the indication will be a solid vertical line. With an ac voltage applied to the deflection plates the line on the screen will extend equally above and below the center of the screen. The oscilloscope is an excellent peak-to-peak indicating device for voltages applied to its deflection plates.

The cathode-ray tube also has a pair of *horizontal deflection plates,* shown dashed, placed at right angles to the vertical deflection plates. Since these plates are closer to the screen, they will deflect the beam less and as a result have a slightly lower sensitivity. Voltages applied to the horizontal plates deflect the spot horizontally.

The *Aquadag* coating, a conductive material sprayed on the inside of the tube, is connected to the accelerating anode. Electrons striking the phosphorescent-fluorescent painted surface, or screen, on the inner face of the tube cause a bright spot wherever they hit. These electrons bounce back as a secondary emission, are attracted to the positively charged Aquadag coating, and from there move to the positive terminal of the power supply.

The connections to the elements are usually made to pins at the cathode end of the tube. These pins fit into a 7- to 14-pin socket, depending on the tube type. There are many different types of cathode-ray tubes, having different face sizes and shapes, sensitivities, persistence of illumination, colors of screens, numbers of anodes, filament voltages, etc. Oscilloscope tubes may have round or oblong screens, of 1-in. to over 12-in. diameter or width.

There are many uses for oscilloscopes. In practically any electronic or radio circuit the oscilloscope can be made to picture what is occurring to circuit voltages.

12-30 FREE-RUNNING OSCILLOSCOPES

The original and many of the simpler modern oscilloscopes use a free-running oscillator (generator of a continuous running sawtooth ac or dc) for the horizontal-deflection voltage. When the sawtooth voltage is applied to the horizontal-deflection plates of the cathode-ray tube, the dot is moved slowly (relatively) across the screen, left to right, and then is snapped back so fast that no trace is left. A 1-Hz sawtooth ac will produce a dot that moves across the screen in 1 s, disappears, and reappears immediately to move across the screen again. If this sweep ac is over 20 Hz, the dot appears as a line, due only to the persistence of vision of the human eye. The persistence of illumination of most CRT tube screens is less than 1 ms (1 millisecond) and may be in the range of 40 μs (40 microseconds).

The usual application of an oscilloscope is to view some signal voltage waveform. A simple oscilloscope system that will do this is shown in Fig. 12-38 in *block diagram* form. A horizontal triangle block represents an amplifier, the apex of the triangle indicating the direction of signal progression through the amplifier (usually shown left to right). The label on an oblong block indi-

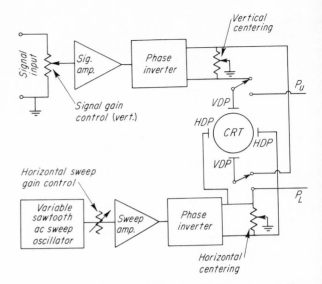

Fig. 12-38 Block diagram of a free-running sweep oscilloscope.

cates the function of that particular circuit. The cathode-ray tube is represented as a round screen with the horizontal- and vertical-deflection plates (HDPs and VDPs) shown outside the CRT for better clarity.

There are four separate sections of this type of oscilloscope. One is the CRT. The second is the sweep circuitry. The third is the signal amplification circuitry. The fourth is the power supply, not indicated for simplicity.

The phase-inverter circuits convert the normal varying dc output signal from a resistance-coupled amplifier to equal-amplitude positive-going and negative-going ac voltages. This allows an amplified ac signal voltage to be centered on the CRT, rather than being displayed above or below the center line. The centering controls allow both vertical and horizontal centering, or any other movement of the display desired. The gain controls are adjusted by the operator to the desired signal size and sweep width.

Consider what is displayed when a 100-Hz sine-wave ac is applied to the vertical-deflection plates and a 100-Hz sawtooth ac is applied to the horizontal-deflection plates. The signal voltage will be considered as starting with zero volts on the VDPs and the sawtooth sweep at the instant of maximum positive on the left-hand HDP (maxi-

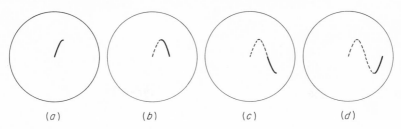

Fig. 12-39 Developing a single cycle on an oscillograph screen.

mum negative on the right-hand HDP). This places the spot at the far-left-hand edge of the screen at the start.

1. In $\frac{1}{400}$ s the horizontal sweep moves the dot one-fourth of the distance to the right. At the same time the sine wave has risen to a maximum potential (positive on top VDP) and has pulled the dot upward, perhaps an inch, depending on the magnitude of the voltage, tracing the first quarter of a sine wave on the screen (Fig. 12-39a).
2. During the next $\frac{1}{400}$ s the horizontal sweep moves the dot to a position one-half of the total distance to the right. At the same time, the sine wave has fallen to zero volts, allowing the dot to fall back to the starting level. The second quarter of a sine wave has been traced (Fig. 12-39b).
3. During the next $\frac{1}{400}$ s the horizontal sweep moves the dot to a position three-fourths of the total distance to the right. At the same time, the sine wave increases to a maximum negative and the dot moves in a negative (downward) direction, tracing the third quarter of a sine wave on the screen (Fig. 12-39c).
4. In the next $\frac{1}{400}$ s the horizontal sweep moves as far to the right as it is going and then snaps back to the starting point. During this time, the sine-wave voltage drops back to the starting level, tracing the final quarter of the sine wave on the screen (Fig. 12-39d). One cycle is completed, and the next retraces immediately.

With the cycles occurring 100 times per second, to the eye a sine wave appears to be standing still on the screen. If the ac applied to the vertical plates is nonsinusoidal, the display will be nonsinusoidal. In this way, the waveshape of any ac or varying dc applied to the vertical plates is made visible. It is only necessary to synchronize the horizontal-sweep frequency with the frequency of the wave being applied to the vertical plate to make the pattern stand still on the screen. It is also possible to use the sweep at some submultiple of the signal frequency and stop the motion of the figure displayed. For example, if the sweep is 50 Hz and the signal is 100 Hz, the sweep makes only one-half of its total excursion by the time the signal completes its first cycle. As a result, two cycles of the signal voltage will be shown across the screen. If the signal and sweep voltages are not synchronized, the display moves to the right or left across the screen.

If the sweep frequency (or the sweep rate) is known accurately, it is possible to measure the frequency of an ac by stopping the display of it on an oscilloscope, counting the number of cycles shown, and multiplying the cycles by the sweep frequency.

Some oscilloscopes have switches to permit direct connections to the VDPs (P_U and P_L on the diagram). This allows signals to be fed directly to the deflection plates without passing through the amplifiers, which may not be linear or amplify at all at very high frequencies. Signals up to several hundred megahertz can be displayed with such direct connections, provided they have a high enough voltage amplitude to give adequate deflection. Since there are so many cycles per second, the display is not of separate cycles, but an illuminated band of closely packed waveforms across the screen. However, variations of the signal strength will show as variations of the band height, which is one way of monitoring amplitude modulation in radiotelephone transmissions discussed in Chap 17.

Patterns developed by two different ac cycles fed to the VDP and HDP and shown on a CRT are known as *lissajous figures*. If two sine waves

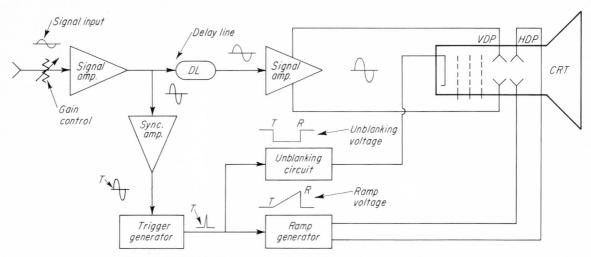

Fig. 12-40   Block diagram of a simple triggered-sweep oscilloscope.

of equal frequency are used, a stationary single circle, oval, or loop will be formed on the screen. If the ratio of frequencies is exactly 2 : 1, there will be two stationary loops formed, etc.

12-31 TRIGGERED OSCILLOSCOPES

A more sophisticated type of oscilloscope uses a triggered sweep. It is usually accurate enough to measure voltages and times. With no input signal no sweep voltage is developed. The cathode of the CRT is maintained so highly positive in respect to the grid that no electrons can form a beam to the screen. When a signal is applied to the input amplifier, it is amplified and fed in two directions: to a synchronization amplifier and to a delay line, as in Fig. 12-40.

The input signal voltage is further increased by the sync amplifier. Either the rising positive or the rising negative portion of the signal cycle is used to key a circuit to develop a sharp trigger pulse of voltage. The trigger pulse is fed to two circuits. One produces a slowly increasing dc *ramp* voltage, time *T* on the diagram, which is used to sweep the spot horizontally across the display screen; the other produces an *unblanking* bias voltage at time *T* to bring the cathode of the CRT to a potential that allows the electron beam to be formed through the electron gun. The ramp generator and the unblanking circuit are interconnected in such a way that when the ramp

returns to zero volts, time *R* on the diagram, the unblanking circuit again produces a blanking voltage on the cathode that stops the electron beam.

During the generation of the trigger, ramp, and unblanking voltages, the input signal is also being fed through a delay line that delays the signal about 150 ns (150 nanoseconds) before it is further amplified and fed to the vertical-deflection plates. This short delay is time enough to allow the developing of the trigger and the starting of the ramp voltage so that the beginning of the signal cycle that produced the trigger can be displayed on the screen of the CRT.

If the amplifiers are single-ended (one of the two output lines is at ground potential), it will be necessary to use phase inversion to feed the deflection plates. In many cases, phase inversion is produced at the beginning of the amplification system and all signals are then fed through direct-coupled double-ended (push-pull) amplifiers to prevent distortion of high-frequency and very low frequency signals. These direct-coupled amplifiers are usually low-impedance transistor types. It is not unusual for such oscilloscopes to show 200-MHz signals as well as be able to show variations that take several seconds for a single cycle. Vacuum-tube scope amplifiers rarely are used at more than 5 to perhaps 10 MHz. The duration of the ramp voltage generated may be either continuously variable or variable in switched steps, such as 1, 10, 100 μs, 1, 10, 100 ms, 1, 10, 100 s, etc.

There are a variety of other possibilities for these scopes. Dual-trace scopes show two displays on one screen by alternating a sweep line above the center and a sweep line below the center and applying two different signals to the two lines to allow comparison of two signals (such as an input and output signal of some amplifier) at the same time. Some scopes feature variable-persistence or *storage* CRTs that retain a selected display for fractions of a second, seconds, hours, or days if need be. When the frequency to be viewed is higher than the capabilities of the vertical amplifiers, use is made of a *sampling* system whereby an additional unit ahead of the scope input samples repetitive waveforms and delivers the samples at a frequency that the amplifiers can handle and display. Sampling scopes can display signals well up into the gigahertz range.

Test your understanding; answer these checkup questions.

1. List the contents of a CRT. _____
2. What is an advantage of using a high potential on the last anode of a CRT? _____
3. What is controlled by varying the potentials between anodes? _____ By varying grid bias? _____
4. Which deflection plates control the beam more? _____
5. To which plates are signal voltages usually applied? _____ What is applied to the other plates? _____
6. What is the function of the Aquadag? _____ Its potential in respect to ground? _____ In respect to the cathode? _____
7. What determines the number of cycles shown on a scope for a given frequency? _____
8. What is a block diagram symbol of an amplifier? _____ Of an oscillator? _____
9. Why is phase inversion used in scopes? _____ What type of amplifiers do they follow? _____
10. What will be seen with sawtooth 500 Hz on the HDPs and sine-wave 2 kHz on the VDPs? _____ If the sawtooth is 501 Hz? _____ If 499 Hz? _____
11. How can high-frequency ac be displayed if the amplifiers can not handle the frequency? _____
12. In a triggered scope, what starts the sweep voltage? _____ What determines sweep speed? _____ Why must the HDPs use a blanking voltage? _____ Why is a delay line used? _____
13. Why are transistor amplifiers desirable in scopes? _____
14. What is a CRT that can hold a display called? _____

12-32 DIGITAL PANEL METERS (DPMs)

A new type of meter indicates voltage, current, resistance, etc., in digital fashion on a series of decimal counting assemblies (DCAs) similar to those used in counters (Sec. 12-25). D'Arsonval meters may be superior for monitoring varying signals and may have long, reliable lives, but digital displays can give better repeatability, readability, and accuracy than a simple analog meter. The cost of digital panel meters has been steadily decreasing. Such digital displays can be produced by several methods; single-ramp, dual-ramp, and voltage-to-frequency are examples.

A single-ramp meter has a circuit that generates a dc ramp voltage, as in Fig. 12-41, a few times

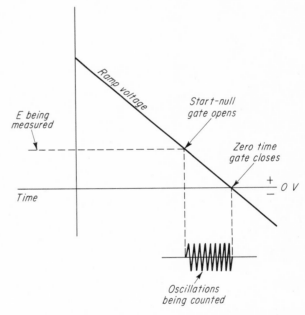

Fig. 12-41 Digital voltmeter (DVM) ramp voltage. While counting gates are open, internal oscillations are counted.

a second. Both the ramp voltage and the voltage to be measured are fed to a null detector circuit. When the ramp voltage decreases to a value equal to the voltage being measured, the null detector triggers a high-frequency oscillator into operation. Every cycle of the oscillator is counted on a series of decimal counting assemblies and displayed in digital form by neon Nixie tubes, seven-segment light-emitting diode figures, or liquid crystal figures. When the ramp voltage reaches zero, another null detector circuit stops the oscillator and the last digital number displayed represents the voltage being measured. Small voltages allow the oscillator to run for only a few cycles. Higher voltages allow more cycles to be generated and counted.

A dual-ramp DPM uses an *operational amplifier* (op amp) with an amplifying capability of several hundred thousand; but by feedback of output to input circuits, the net amplification is reduced to perhaps 10. Any distortion produced in such amplifiers is reduced by almost the feedback ratio. As a result, the op amp output current will be a faithful replica of the input voltage (to be measured). In this way the op amp acts as a voltage-to-current converter. The current, which is proportional to the input voltage, is fed to an integrator circuit which is gated on for a short period of time. The integrator circuit current builds up at a rate determined by the current fed to it by the converter, time T in Fig. 12-42. At the end

of time T, the integrator circuit cuts off and (1) the current starts to decay, I_1 and I_2, and (2) a high-frequency oscillator is fed into a decimal counting assembly. When the current decays to zero, a null detector stops the DCAs and the numbers shown represent the voltage. A low voltage produces a small-rise current, I_1. A higher voltage produces a high-rise current, I_2. Since both decay at the same rate, I_1 shunts off the DCAs before I_2 and therefore indicates less voltage.

A voltage-to-frequency DPM also employs an op amp as a voltage-to-current converter. The output current from the op amp is used to charge a small capacitor, C in Fig. 12-43. When the

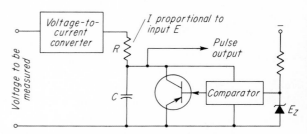

Fig. 12-43 Voltage-to-frequency conversion for a DPM.

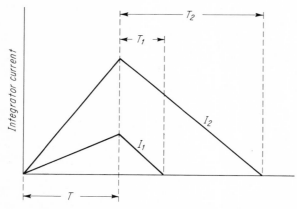

Fig. 12-42 Dual slopes of two different voltages being measured in a DPM.

capacitor voltage builds up to the zener voltage E_z, a comparator or null detector circuit forward-biases the transistor across C, discharging C immediately. The bias drops off the transistor and C starts recharging again. The greater the voltage being measured the greater the charging current and therefore the more times per second the charge-discharge pulses are generated. These pulses are fed to DCAs and counted. By gating the comparator circuit on an off over the proper interval, the digits shown on the DCAs indicate the proper voltage.

If a DPM has two DCAs, it can theoretically count up to 99. If it has three DCAs, it can count to 999. But what is a $2\frac{1}{2}$-digit DPM? If a meter can show three digits but has only the capability of counting to 99 plus perhaps 100% overranging (essentially 200 counts before it must start recounting), it is usually considered as a $2\frac{1}{2}$-digit meter. If its basic ability is 99 counts plus perhaps 300% overranging, it can count to essentially 400 and it may be called a $2\frac{3}{4}$ digit meter. (Different manufacturers may use different meanings for these terms.) The price of DPMs nearly doubles for each added full digit.

Most DPMs have a hold circuit that allows the voltage display to be held from a fraction of a second to any desired time before it recycles again. One advantage of these meters is the binary pulse output information which allows recording of readings or the feeding of any number of remote displays. For current measurements the voltage across a shunt resistor in a circuit is read. For resistance readings the DPM is used in place of the analog meter in the usual ohmmeter circuits (Sec. 12-11). For added sensitivity, dc amplifiers (op amps usually) may be used ahead of the analog section (circuits that convert voltage to pulses). Only by using ICs can such complicated systems be developed in cases smaller than about 6 cm high, 12 cm wide, and 10 cm deep. To enable the meters to operate in circuits that are above ground potential, the analog portion must be isolated by transformer coupling from the counting and power-supply circuits.

12-33 TEST EQUIPMENT PROBES

Electronic test equipment (oscilloscopes, VTVMs, digital voltmeters, etc.) usually has a ground lead that attaches to the ground circuit on the chassis being tested and a "hot lead" wire or probe that is touched to the component under observation in the chassis.

Probes may range from simple wires up to highly complex compensated, amplifying devices. One simple type consists of an insulated wire terminated with an insulating-material handle with a contact tip at the end, as shown in Fig. 12-44a. Inside the handle a 10–100-kΩ isolation resistor may be inserted. These probes are adequate for relatively low-impedance circuits involving frequencies up through the AF range (20 kHz).

For higher frequencies a simple compensated probe, such as that shown in Fig. 12-44b, may be used. The internal capacitance of the flexible coaxial cable used, represented by the dotted capacitor, has a continually lowering reactance as frequency increases. The result would be a constantly decreasing signal reaching the oscilloscope as higher frequencies were tested. To compensate for this, a 3–30-pF capacitor is connected across a series input resistor. At dc the

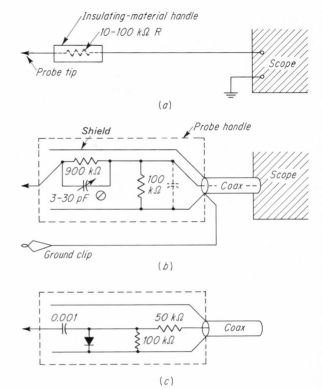

Fig. 12-44 Three simple probes: (a) low frequency; (b) medium HF; (c) HF through UHF.

two capacitances are not effective, and the probe is a 10:1[(900 + 100):100] reduction probe. The coaxial capacitance bypasses the ac signal more as frequency increases, but the reactance of the compensating capacitor is also decreased, feeding more signal to the coaxial line, maintaining the signal to the test equipment at the same amplitude. The circle with a line across it is the symbol of a screwdriver adjustment for the capacitor. The probe is calibrated by watching a square-wave ac on a scope. The capacitor is varied to show the sharpest corners possible on the square-wave signal. A ground clip may be added at the base of the handle. To prevent hand capacitance effect, the inner wall of the handle may be metallic and grounded to the outer sheath of the coaxial cable (shielded).

The probe in Fig. 12-44c is a half-wave rectifier that converts RF ac to pulsating dc, which is filtered to dc by the capacitance of the coaxial cable. This probe presents a dc voltage to the test instrument that varies as the circuit ac varies in amplitude. Either bridge rectification or a voltage-doubling diode arrangement can be used to increase the amplitude of the voltage being presented to the test equipment.

Test your understanding; answer these checkup questions.

1. Could digital meters be used as bench meters as well as panel meters? _____
2. What are the three methods used for developing digital displays? _____ _____ _____
3. What are the three types of digital display devices listed? _____ _____ _____
4. What digital meter(s) count(s) cycles? _____
5. What digital meter(s) use(s) an op amp? _____
6. What are the zero-feedback amplification possibilities of op amps? _____ With feedback? _____ What is the advantage of using feedback? _____
7. What circuit can be used as a linear voltage-to-current converter? _____
8. How many counts would a three-digit meter display? _____ A 4½ digit meter? _____ A 4¾ digit meter? _____
9. How can digital meters be used to display amperes? _____ Ohms? _____
10. If the analog section of a digital meter is isolated by a transformer from the display and power-supply section, how might the analog section obtain its operating power? _____
11. What does DVM mean? _____
12. What appears to be the main disadvantage of compensated probes? _____

COMMERCIAL LICENSE QUESTIONS

Amateur license questions will be found in the Addendum.

FCC Elements 3, 4, and 6 require an ability to answer questions similar to those below. Sections in which questions are answered are shown in parentheses. A question followed by a bracketed number is required for that element alone.

1. Make a sketch showing the construction of a D'Arsonval meter. Label parts. (12-2)
2. What is the purpose of a shunt as used with an ammeter? (12-4)
3. If two ammeters are in parallel, how may the total current in the line be determined? (12-4)
4. If two ammeters are in series, how may the total current in the line be determined? (12-4)
5. Indicate by a diagram how the total current in three branches of a parallel circuit can be measured by one ammeter. (12-4)
6. A 0–1-mA meter having a resistance of 30 Ω is used to measure an unknown current by shunting the meter with a 3-Ω resistor. It then reads 0.5 mA. What is the unknown-current value? (12-5)
7. Which voltmeter absorbs no power from the circuit under test? (12-8) [4]
8. Why is a multiplier resistance used with a voltmeter? (12-9)
9. How may a dc milliammeter, in an emergency, be used to indicate voltage? (12-9)
10. If two voltmeters are connected in series, how would you be able to determine the total drop across both instruments? (12-9)
11. If a 0-to-1 dc milliammeter is to be converted into a voltmeter with a full-scale calibration of 200 V,

what value of resistance should be connected in series with the meter? (12-9)
12. What type of meter is suitable for measuring the AVC voltage in a receiver? (12-10, 12-13) [4]
13. Draw a diagram of an ohmmeter and explain its operation. (12-11)
14. Draw a circuit diagram of a VTVM. (12-13)
15. Does an ac ammeter indicate peak, average, or effective values? Explain. (12-15)
16. By what factor must the voltage indicated on the scale of an ac voltmeter be multiplied in order to obtain the peak value? (12-16)
17. If a dc voltmeter is used to measure effective ac voltages by the use of a bridge-type full-wave rectifier, by what factor must the meter readings be multiplied to give corrected readings? (12-16) [6]
18. By what factor must the voltage indicated on the scale of an ac voltmeter be multiplied to obtain the average-voltage value? (12-16) [6]
19. Why are copper-oxide rectifiers, associated with dc voltmeters for the purpose of measuring ac, not suitable for the measurement of voltages at radio frequencies? (12-16)
20. What type of meter is suitable for measuring peak ac voltages? (12-17, 12-18) [4]
21. What unit has been adopted by program-transmission organizations as a volume unit and to what power is it equivalent? (12-18) [4]
22. Audio frequency VU meters should be used across transmission lines of what impedance value? (12-18) [4]

23. What is a use of a T-pad attenuator? (12-18) [4]
24. Show by a circuit diagram a method of desensitizing a VU meter to make it read lower than the normal value. (12-18) [4]
25. Describe the construction and characteristics of a thermocouple meter. (12-19)
26. What type of indicating instrument is best suited for use in measuring RF currents? (12-19)
27. How may the range of a thermocouple ammeter be increased? (12-19) [6]
28. Which meters may be used to measure RF currents? (12-9, 12-20)
29. Describe the construction and characteristics of a dynamometer type of meter. (12-21)
30. Is the angular-scale deflection of a repulsion iron-vane ammeter proportional to the square or the square root of the current? (12-22)
31. What instrument measures electric power? (12-23)
32. Draw a circuit diagram of a wattmeter. (12-23)
33. Show by a diagram how a voltmeter and an ammeter should be connected to measure power in a dc circuit. (12-23)
34. What instrument measures electric energy? (12-24)
35. Explain the operation of a resistance bridge. (12-28)
36. If the known resistors in a resistance bridge are 5 and 10 Ω and if adjusting the third resistance to 820 Ω produces a perfect balance, what is the unknown resistance? (12-28) [4]
37. Describe how horizontal and vertical deflection takes place in a cathode-ray oscilloscope. (12-29)
38. Discuss the waveforms involved in a cathode-ray oscilloscope. (12-29)

13
OSCILLATORS

The objective of this chapter is to explain the operation of some of the many different types of oscillator circuits used today with semiconductor devices and vacuum tubes. They include Hartley, Colpitts, TPTG, ECO, crystal, and RC oscillator circuits. (Some instructors might prefer to teach the audio-frequency amplifier chapter before this oscillator chapter.)

13-1 TYPES OF OSCILLATORS

An oscillator circuit produces alternating current from a fraction of a watt to sometimes thousands of watts. When high-power ac for power-frequency transmission (30 to 1,000 Hz) is required, various types of electromagnetic alternators are used. For higher frequencies, in the audio- and radio-frequency ranges, transistor or tube oscillator circuits are employed.

The first type of oscillator used to generate HF (high-frequency) ac was the spark circuit. It produced a damped (decaying) ac waveform up to more than 2 MHz. Another early HF oscillator was

the Poulsen arc, which produced a constant-amplitude ac up to about 500 kHz. Still other types of early RF generators were the Alexanderson and Goldsmith alternators for frequencies below 50 kHz. Then came the vacuum tube, now capable of generating ac up to more than 3 GHz.

Since the 1930s, the magnetron tube, klystron tube, traveling-wave tube (TWT), and backward-wave-oscillator (BWO) tube have extended the vacuum-device oscillating frequency up to 50 GHz and more. Recent transistors can produce ac into the 3- to 5-GHz range, while tunnel and other special diodes can produce ac at frequencies over 100 GHz.

13-2 SHOCK EXCITATION

As a means of explaining the operation of oscillator circuits involving coils and capacitors, the shock-excitation, or flywheel, theory will be used.

If the switch in Fig. 13-1 is closed for an instant and then opened, electrons from the battery (1) flow to the top plate of the capacitor and charge it negative and (2) are pulled from the bottom

---

Fig. 13-1 Basic LC oscillator. Quick closing and opening of the switch charges the capacitor.

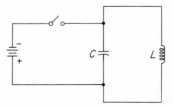

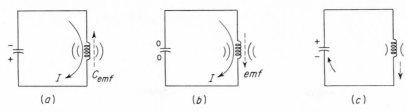

Fig. 13-2 (*a*) Capacitor discharges through the coil, producing a magnetic field. (*b, c*) Collapsing field recharges capacitor to opposite polarity.

plate, charging it positive. The inductance of the coil prevents any material current flow through it for the instant that the switch is closed. As the switch is opened, electrons deposited on the top plate of the charged capacitor start to move toward the positive plate, downward through the coil. The battery has *shock-excited* the coil-capacitor circuit, and the circuit starts into operation, using energy obtained from the battery as the motivating power.

The current of electrons through the coil causes a magnetic field to expand outward (Fig. 13-2*a*), inducing a counter emf in the coil which prevents the capacitor from discharging immediately.

As the capacitor discharges, it eventually reaches a point where there are the same number of electrons on both plates (Fig. 13-2*b*). With no emf across the coil, there should be no current in it and nothing to hold the field out around the coil. The field collapses back inward, inducing a downward-direction voltage in the coil. This forces free electrons in the wire of the coil and from the top plate of the capacitor down through the coil to the bottom plate, charging the bottom plate negative and the top plate positive (Fig. 13-2*c*).

Now the capacitor is charged again but this time with an opposite polarity. The amplitude of the charge is exactly the same as that of the original. This charged condition will cause the current to reverse itself in the circuit, swing back up through the coil, and recharge the capacitor to a polarity the same as originally developed in it by the shock excitation of the battery. One cycle of ac has been produced.

If there were no losses in the circuit, the current would *oscillate* back and forth, indefinitely, with a constant amplitude for each cycle. This would be a perpetual ac generator. Because there are always losses in circuits, each succeeding half

cycle has an amplitude less than that of the one before. It is not long before the ac is damped out entirely (Fig. 13.3). The less the resistance or

Fig. 13-3 Alternating current damps out (*a*) rapidly in low-*Q* high-resistance circuits and (*b*) slowly in low-resistance high-*Q* circuits.

losses, the longer it takes the current to damp out.

The oscillation of electrons back and forth in an LC (inductance-capacitance) circuit is known as *flywheel effect*. Most sine-wave-generating oscillators utilize this effect.

This explanation of shock excitation assumed that the capacitor was the part receiving the original shock of energy. It is also possible to shock-excite an LC circuit by inducing a pulse of current into the coil, which would also produce an oscillation in the circuit.

If analyzed, electrostatic energy in the capacitor is seen to be alternately converted into electromagnetic energy around the coil and then back to electrostatic energy in the capacitor, and so on. The energy in an LC circuit oscillates from an electrostatic form to an electromagnetic form.

The spark oscillator originally used for radio communication inserted a spark gap in series with an LC circuit and then fed a low-frequency ac voltage across the capacitor, as shown in Fig. 13-4. When the emf reached a voltage sufficient to ionize the air between the electrodes of the gap, a spark flashed across, heating the air. The ionized hot air acted as a conductor, allowing the charge that had built up in the capacitor to produce oscillations of a damped type at the natural frequency of resonance of the LC circuit.

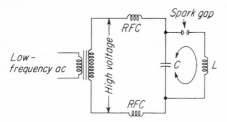

Fig. 13-4 Spark-gap oscillator. Low-frequency ac charges *C*, ionizes air of gap, and allows damped high-frequency ac to oscillate in the LC circuit.

13-3 ELECTRONIC LC OSCILLATORS

The basic theory of the generation of sine-wave ac of constant amplitude and frequency in vacuum-tube and transistor oscillator circuits might be outlined as follows:

1. A tuned LC circuit is shock-excited into oscillation by any small voltage.
2. The ac voltage from this oscillating circuit is amplified by vacuum tube or transistor.
3. Amplified ac energy is fed back to the original tuned circuit by either inductive or capacitive coupling.
4. The energy must be fed back in such a phase as to aid the oscillating energy in the tuned circuit and be strong enough to overcome any LC circuit losses.
5. The feedback keeps the shock-excited LC circuit oscillating at its resonant frequency.
6. The oscillator circuit draws all its operating energy from the dc power supply of the tube or transistor.
7. Ac power produced in the oscillating LC circuit can be taken from it by either inductive or capacitive coupling (Fig. 13-5).

13-4 ARMSTRONG OSCILLATOR

The Armstrong, or inductive-feedback, oscillator circuit is shown in Fig. 13-5a and b.

When the switch is closed in Fig. 13-5a, a surge of electrons begins to flow through the *tickler coil* producing an expanding magnetic field around it. The field cuts across the turns of the coil of the LC circuit, inducing a voltage in it. This induced voltage shock-excites the tuned circuit, which starts to oscillate at the LC frequency.

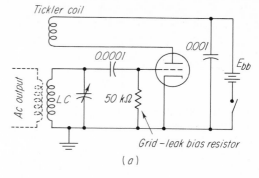

(a)

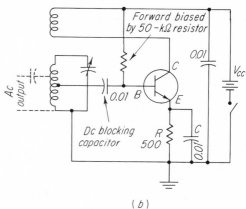

(b)

Fig. 13-5 Armstrong oscillator circuits. (a) Triode with inductive-output coupling. (b) NPN transistor with capacitive-output coupling.

As the LC circuit starts to oscillate, the ac voltages developed across it are fed to the amplifying device (tube or transistor), producing a relatively high-amplitude *variation* of tickler current. As a result, expanding and contracting magnetic fields around the tickler coil induce an ac back into the LC circuit, keeping the circuit oscillating.

Note that the tickler coil and the LC circuit form a transformer, with the tickler coil as the primary. All the ac energy in the LC circuit is the result of induction from the tickler.

As long as the two coils are oriented in such a manner as to produce an aiding effect between them, the oscillation of the LC circuit will continue to increase in strength until maximum oscillation power is reached (almost immediately). The maximum amplitude is determined by several factors—power-supply voltage, degree of coupling between the primary and secondary coils, the Q of the LC circuit, the amount of bias developed, and the value of capacitance from tickler coil to ground.

If the tickler coil were wound in the opposite direction, the voltages induced into the LC circuit by the tickler-coil current would be out of phase with oscillations in the LC circuit, the circuit would not produce sustained oscillations, and no ac would be generated.

All LC circuits have some losses because of resistance in them or induction losses into some external circuit. The energy fed back from tickler to LC circuit must be sufficient to overcome any losses to allow the LC circuit to oscillate at a constant amplitude and not damp out. Actually, considerably more feedback than this minimum value is used.

The in-phase feedback effect capable of producing oscillation is known as *regeneration*. Out-of-phase feedback will prevent oscillation and is known as *degeneration*. If it is desired to make an electronic circuit oscillate, it will be necessary to introduce regeneration in it. If it is desired to prevent it from going into oscillation, it may be necessary to introduce degeneration in it, or *neutralize* it.

In the Armstrong circuit there are several methods by which the quantity of tickler-to-LC circuit feedback can be controlled: (1) by varying the tickler-coil coupling to the LC circuit, (2) by varying the dc power-supply voltage, thereby varying the amplitude of the current flowing through the tickler coil, and (3) by varying the tickler-to-ground capacitance, which completes the ac path from tickler circuit to cathode. The greater this capacitance, the greater the ac component flowing in the tickler coil.

The operation of the NPN Armstrong circuit (Fig. 13-5b) is essentially the same as with the triode. Note, however, that the LC circuit is tapped down to a lower impedance point on the coil to match the relatively low base-emitter impedance of the transistor more closely. Note also that the lower-impedance transistor circuits use larger circuit capacitors than the higher-impedance vacuum-tube circuitry does. Changes in V_{cc} voltage will affect the frequency of the oscillator considerably.

The parallel RC network in the emitter circuit is a stabilizing circuit to prevent *thermal runaway*. When current flows through the emitter or collector junctions, it heats the junctions. Since transistors have a negative temperature coefficient, increasing their temperature decreases their resistance and more emitter-collector current flows, heating the junctions still more. The action can progress until the transistor burns out. Increasing emitter current increases the voltage-drop across the emitter resistor. This voltage is a reverse bias for the base and decreases the conductivity of the transistor, preventing thermal runaway. A stabilizing network is found in many transistor circuits.

13-5 OSCILLATOR BIAS

A tube in an oscillator circuit requires a relatively high negative dc bias voltage to allow it to operate efficiently. If a battery were used to supply the bias voltage, the circuit would oscillate but might not be self-starting. A satisfactory bias voltage can be developed with *grid-leak* bias. The oscillator will always be self-starting because no bias voltage is developed until the circuit actually begins to oscillate. Figure 13-5a is a shunt-resistor grid-leak bias circuit. Figure 13-6 shows series-resistor bias.

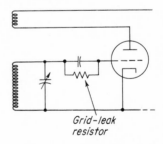

Fig. 13-6 Series-type grid-leak bias circuit.

When the power-supply switch is closed in Fig. 13-5a, plate current begins to flow through the tickler coil. A magnetic field expands, inducing a voltage into the grid-circuit coil, and electrons start oscillating in the LC circuit. On the positive half of the first induced LC cycle the grid is driven positive, further increasing the plate current, which still further increases the voltage induced into the LC circuit. This regenerative action continues until I_p reaches maximum. When I_p reaches saturation, the magnetic field of the tickler coil can no longer expand and induction into the LC circuit ceases. Electrons in the LC circuit now begin to move in the opposite direction. The positive voltage at the grid drops off and is driven negative during the negative half of the cycle.

While the grid was being driven positive, it picked up a large number of electrons from the space charge near the cathode. These electrons could not leak back to the cathode immediately because of the presence of the 0.0001-μF grid-leak blocking capacitor and the high resistance of the grid-leak resistor (25 to 100 kΩ). The electrons that are trapped on the grid form a highly negative charge that does not leak off completely even during the negative half cycle of LC oscillation because of the relatively long RC time constant of the grid circuit. The electron charge on the grid over a whole cycle of oscillation averages enough to produce the desired high (class C) bias value. During a small part of the positive cycle on the grid, I_p pulses flow and the oscillator operates quite efficiently (50 to 70%).

The higher the grid-leak resistance, the higher the average bias voltage and the greater the efficiency of circuit operation. However, if the bias is too high, the average plate-current pulses become too low in amplitude and the oscillator will not deliver much ac power output. The size of the grid-leak capacitor may vary from 0.00005 μF for high-frequency circuits to perhaps 0.001 μF for low-frequency oscillators. Too great a capacitance may cause low-frequency RC oscillations to occur at the same time the LC circuit is oscillating at a high frequency. Such a low-frequency *parasitic* oscillation is normally undesirable.

A transistor Armstrong oscillator without any forward bias can oscillate but would not be self-starting. Without forward bias, closing the switch in the collector circuit does not produce any current (I_C) flow in the tickler coil to excite the LC circuit into oscillation. Forward-biased as in Fig. 13-5b with a 50-kΩ dropping resistor, the circuit will start oscillating as soon as the switch is closed. Compare the bias resistor connections for tube and transistor circuits.

Test your understanding; answer these checkup questions.

1. List the three early types of high-power HF ac generators. _____ _____ _____
2. In what type of circuit does flywheel effect occur? _____
3. What are trains of ac that decay called? _____
4. When electrons rotate back and forth in an LC circuit, what is it termed? _____
5. In an Armstrong oscillator, what type of feedback is used? _____ What is the plate or collector coil called? _____
6. Why must a transistor Armstrong oscillator be forward-biased? _____ Why is this not true of VT Armstrongs? _____
7. In Fig. 13-5, what kind of BJT is used? _____
8. If the plate coil of an Armstrong oscillator could be rotated 180°, what would be controlled? _____
9. What term is used to indicate in-phase feedback? _____ Out-of-phase feedback? _____
10. Why are LC tank circuits often tapped down when BJTs are the active devices? _____ What other method might produce the same effect? _____
11. How may thermal runaway in transistor oscillators be prevented? _____
12. If an oscillator oscillates at two frequencies at the same time, what is the undesired oscillation called? _____
13. With grid-leak bias does grid current flow when circuit is oscillating? _____ If circuit is not oscillating because of excessive coupling to a load? _____
14. Would a JFET Armstrong circuit resemble a BJT or a VT circuit? _____

13-6 THE TUNED-PLATE TUNED-GRID OSCILLATOR

The Armstrong oscillator is an example of an inductively coupled oscillator. Energy from the plate circuit is induced into the grid circuit by transformer action of two inductively coupled coils. The tuned-plate tuned-grid (TPTG) oscillator circuit is an example of a capacitively coupled oscillator.

The reason the TPTG oscillator works is the reason why many high-frequency amplifiers may *not* operate satisfactorily. When it is understood how this circuit oscillates, it will also be evident why an amplifier using the same general circuitry will sometimes oscillate instead of amplifying as it should.

In Fig. 13-7a, a capacitor C_{gp}, shown in dotted lines, is usually not needed. (C_{cb} in Fig. 13-7b may

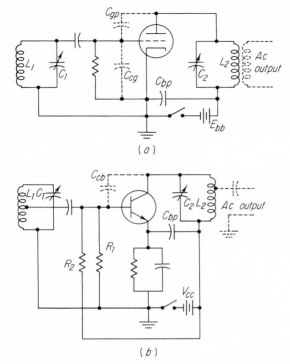

(a)

(b)

Fig. 13-7 TPTG oscillators: (a) with a triode and (b) a transistor form.

be required.) The grid-to-plate capacitance of the tube elements themselves is normally sufficient to couple back enough ac energy capacitively from the plate circuit to the grid circuit to produce sustained oscillations of electrons in the tuned L_1C_1 circuit, which in turn keeps L_2C_2 oscillating.

The ac grid circuit of Fig. 13-7a consists of the grid, the grid-leak capacitor, the LC circuit, and the cathode. The dc grid circuit consists of the grid, the grid-leak resistor, and the cathode.

The ac plate circuit consists of the plate, the LC circuit, the bypass capacitor C_{bp}, and the cathode. The dc plate circuit consists of the plate, the LC circuit, the B battery, the switch, and the cathode. When drawing diagrams, it is always wise to check for complete grid and plate (base and emitter) circuits for both ac and dc continuity.

The TPTG oscillator must have its grid and plate LC circuits tuned to approximately the same frequency. When the switch is closed (Fig. 13-7a), a sudden surge of plate current begins to flow in the plate circuit. This shock-excites the plate

LC circuit into oscillation at a frequency determined by the values of its inductance and capacitance.

When L_2C_2 starts oscillating, an ac voltage is developed across it. This ac voltage is divided across the bypass capacitor C_{bp}, the cathode-grid capacitance of the tube C_{cg}, and the grid-plate capacitance of the tube C_{gp}.

The bypass capacitor is relatively large (has low reactance) and has very little of the total ac voltage-drop across it. As a result, practically all the ac voltage across the plate LC circuit is developed between cathode and plate in the tube. If C_{cg} and C_{gp} were equal, about half the L_2C_2 ac would be developed across the grid-cathode circuit. In Fig. 13-7a, the grid LC circuit is connected between cathode and grid through the grid-leak capacitor and is therefore subjected to a considerable fraction of the ac emf developed across the plate LC circuit. This voltage, fed to the grid LC circuit, forces it into oscillation at the frequency of the plate circuit. Both LC circuits should oscillate.

If the two circuits are not adjusted to approximately the same frequency, the two ac voltages may be sufficiently out of phase to prevent oscillation.

To change frequency in a TPTG oscillator, the grid and plate circuits should be tuned together to keep them in proper phase and the circuit in optimum oscillation. Actually, the plate circuit must be tuned to a frequency slightly higher than that of the grid to prevent both circuits from being completely resistive.

Since there are two tuned circuits in a TPTG oscillator, there are two possible frequencies of oscillation. The circuit will normally operate on the frequency of the LC circuit having the higher Q.

When used to generate lower-frequency ac, the TPTG circuit may not have enough energy fed back from plate to grid circuit by the interelectrode capacitances of the tube. A capacitor (10 to 50 pF) may have to be connected between grid and plate (C_{gp} in Fig. 13-7a).

Because of the low impedance of transistor junctions, an output-to-input capacitor (C_{cb} in Fig. 13-7b) may be needed in transistor TCTB (TPTG) circuits. Note also, in the circuit shown, how a voltage-divider network (R_1 and R_2) is used to forward-bias the base instead of using a simple voltage-dropping resistor from $+V_{cc}$ to base.

13-7 SERIES AND SHUNT FEED

The route taken by the dc in either the plate or the grid circuit determines the type of *feed* used in the circuit. There are two possible means of feeding, *series* and *shunt* (parallel). (1) If dc flows through a tube or transistor, an LC circuit, and a dc source in series, the circuit is series-fed. (2) If a tube or transistor, a tank circuit, and a dc source are all in parallel, the circuit is shunt-fed.

Figure 13-8 illustrates series-fed plate and grid circuits. Figure 13-9 illustrates shunt-fed plate and

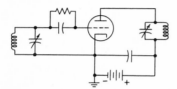

Fig. 13-8 TPTG oscillator, series-fed plate and grid circuits.

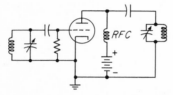

Fig. 13-9 TPTG oscillator, shunt-fed plate and grid circuits (RFC required).

grid circuits. It is not necessary to use the same type of feed in both grid and plate circuits. When a vacuum-tube circuit is said to be series-fed, it usually means that the plate circuit is series-fed.

Figures 13-10 to 13-12 illustrate shunt-fed Armstrong oscillators. The radio-frequency choke (RFC) coils shown in the shunt-fed circuits have

---

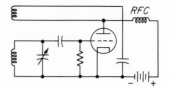

Fig. 13-10 Armstrong oscillator, shunt-fed plate and grid circuits (RFC required).

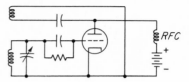

Fig. 13-11 Armstrong oscillator, shunt-fed, with both grid and plate coils at ground or B– potential.

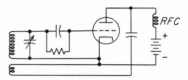

Fig. 13-12 Armstrong oscillator with tickler coil below the LC circuit. Note that the electrical connections have not changed in any way from Fig. 13-11.

sufficient inductance and little enough distributed capacitance to exhibit high impedance to RF ac or RF varying dc. The higher this impedance value, the greater the RF ac emf that can be developed across the RFC and its associated tuned LC circuit. Note that an RFC is not needed in series-fed circuits, although it is sometimes used to keep RF ac out of the power supply.

Series feed of plate-tank (LC) circuits puts them at high dc voltages with attendant insulation and safety problems.

13-8 THE HARTLEY OSCILLATOR

A widely used variable-frequency oscillator is the Hartley (Figs. 13-13 to 13-16). In its usual form it is an inductively coupled circuit; that is, plate-current variations in the plate half of the tank coil produce induced voltages in the grid half of the coil, which are in phase, or regenerative, and produce sustained oscillations of the tank circuit. However, the circuit will oscillate if the grid and plate halves of the tank circuit are isolated from each other, indicating that part of the regenerative

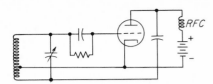

Fig. 13-13 With tickler incorporated in LC circuit the whole coil, not only the grid coil, is tuned. The Armstrong has become a Hartley oscillator.

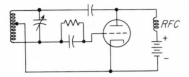

Fig. 13-14 Same circuit as Fig. 13-13 except positions of parts are rearranged, making the top of the coil the plate instead of the grid end.

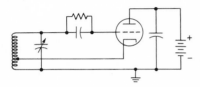

Fig. 13-15 In this series-fed Hartley circuit, one side of the tuning capacitor is grounded (B−). The cathode is above ground potential.

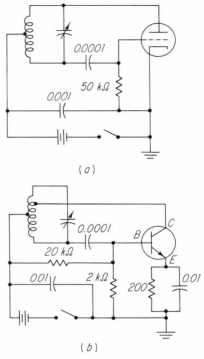

(a)

(b)

Fig. 13-16 Hartley oscillators, series-fed: (a) using triode and (b) transistor. For better frequency stability, collector is tapped down the LC circuit.

effect must also result from the capacitive coupling through the tuning capacitor in the LC circuit.

The Hartley oscillator may be series- or shunt-fed. It has only one center-tapped coil and one tuning capacitor. The tap will be closer to the grid end of the coil for maximum power output and to the plate end for best frequency stability.

Figures 13-10 to 13-16 show the evolution of several forms of the Hartley oscillator from an Armstrong oscillator. All these diagrams are practical oscillator circuits and may be found in use. The circuit to be used will depend on the requirements. If it is desired to keep the rotor of the tuning capacitor at ground potential to reduce *hand capacitance* (detuning the oscillator frequency when any object is brought near the tuning capacitor), the circuit of Fig. 13-15 can be used.

Triodes, tetrodes, pentodes, BJTs, or JFETs can be used in any of these circuits. Compare Fig. 13-17 with Fig. 13-15. The bypass capacitor C_{bp} holds the screen grid and cathode at the same

ac potential above ground. To prevent the dc supply from ac-grounding the cathode, a high-impedance RFC (or resistor) must be added between screen grid and the dc supply.

In all vacuum-tube (or transistor) Hartley circuits the grid (or base) couples to one end of the LC circuit, the plate (or collector) to the other end, and the cathode (or emitter) to the center tap on the coil. Check for this in the Colpitts oscillators described next.

Fig. 13-17 Tetrode-tube Hartley circuit (pentode if suppressor grid is used).

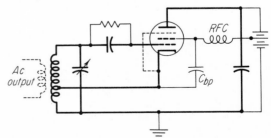

13-9 THE COLPITTS OSCILLATOR

A frequently used variable-frequency oscillator is the *Colpitts* circuit. It is similar to a shunt-fed Hartley except that, instead of the tank coil being center-tapped, two series capacitors are employed to center-tap the LC circuit as in Fig. 13-18.

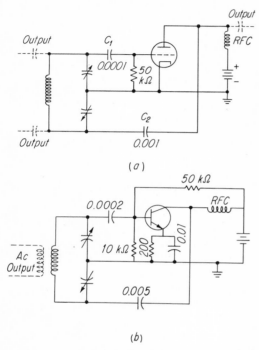

(a)

(b)

Fig. 13-18 Colpitts oscillators: (a) with triode and (b) with transistor. (Capacitors shown in dotted lines in tube circuit indicate three points from which ac can be capacitively coupled to another circuit.)

With vacuum tubes the Colpitts oscillator is normally shunt-fed in both grid and plate circuits. While both C_1 and C_2 are shown in the diagram, the circuit will oscillate with either one shorted out but operates better with both in use. C_1 is the grid-leak capacitor, and C_2 the plate-voltage blocking capacitor. Without the latter, the dc plate voltage would be across the tuning capacitors. Should any plates short while the capacitor is being varied, the plate supply would be shorted and damage would result to power supply or RFC.

Since the grid-end and the plate-end tuning capacitors in the Colpitts circuit are in series, it is necessary that each of them be twice the value of the tuning capacitor used in an equivalent LC circuit of a Hartley oscillator. These large capacitances across the cathode-to-grid and cathode-to-plate circuits of the tube mean that small changes in interelectrode capacitances that may occur in the tube as it warms up will have little effect on the frequency of oscillation of the LC circuit. Tube capacitances have only about half as much effect on the frequency of oscillation in the Colpitts as they have in the Hartley with any given tank-circuit coil.

In any of the variable-frequency oscillators, the greatest frequency stability is usually obtained at the lower-frequency end of the tuning range, because at this end the shunting capacitance of the tuning circuit is greatest and the ratio of possible interelectrode-capacitance change to tuning capacitance is at its highest.

As in the Hartley, the center tap of the tank circuit is not exactly in the center. If the grid-end tuning capacitor has the greater capacitance, the greater voltage-drop will appear across the plate-end capacitor and the greater power output will be produced. Better frequency stability may result if the plate-end capacitor has the greater capacitance.

Tetrodes or pentodes can also be used in a Colpitts circuit. Try diagraming a pentode Colpitts circuit and then a JFET or MOSFET circuit.

13-10 THE ULTRAUDION OSCILLATOR

The ultraudion oscillator circuit in its usual form is actually a series-fed Colpitts circuit. It is used in very high frequency (VHF) or ultrahigh frequency (UHF) applications. It is a simple circuit operating on the same principle as the Colpitts, where the cathode is brought to the approximate center of the LC circuit capacitance. In the case of the ultraudion, these capacitances are the interelectrode capacitances of the tube, shown in dotted lines in Fig. 13-19.

Since the circuit is used mostly at frequencies of over 100 MHz, the required coils and capacitors are quite small. As a result, the interelectrode capacitances are large enough to serve as the voltage-dividing network across the LC circuit. Although shown connected to the "cold" (center) point on the coil, the circuit can operate with the

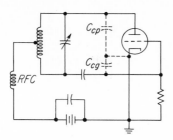

Fig. 13-19 Ultraudion oscillator circuit.

RFC connected to either the grid or plate end of the coil.

13-11 ELECTRON-COUPLED OSCILLATOR

The electron-coupled oscillator (ECO) is in reality a combination of one of the previously described variable-frequency oscillators plus the amplification possible in the plate circuit of a tetrode or pentode tube (Figs. 13-20, 13-21). It can

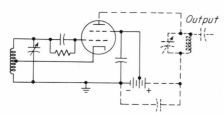

Fig. 13-20 Hartley-type ECO. Output circuit shown dashed.

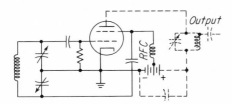

Fig. 13-21 Colpitts-type ECO. Output circuit dashed.

be considered similar to both an oscillator and an amplifier stage in one tube. The addition of the amplifier-stage effect produces greater power output and, more important, tends to isolate the oscillator from external circuits (prevents frequency changes due to variations in any circuits following the oscillator).

Although the Armstrong or TPTG oscillator could be used in an ECO circuit, usually a Hartley, Colpitts, or crystal circuit is employed. The os-

cillating section will oscillate even if the output circuit, in dotted lines, is disconnected completely.

In the ECO the oscillating circuit is made up of the cathode, grid, and screen grid of the tube. The screen grid acts as the anode of the oscillator circuit. When the oscillator is oscillating, the grid voltage varies at the oscillation frequency, varying the plate-circuit current at the same frequency. The only connection between the output coil in the plate circuit and the actual oscillating circuit is by electron pulses passing the wires of the screen grid on their way to the plate; hence the name *electron-coupled oscillator*. The oscillator itself does not rely upon electron coupling, but the amplifying section does. Changing the output LC circuit tuning or loading should produce little effect on the frequency of oscillation. This is a highly desirable characteristic.

The plate-circuit load is shown as a tuned LC tank circuit. It may be found that when output is desired on the fundamental frequency of the oscillator, this circuit will be replaced with an RFC, or possibly a 2,000- to perhaps 10,000-Ω resistor. If maximum output is desired on a *harmonic* frequency (two, three, four times, etc., the oscillator frequency), the plate tank should be tuned to the desired harmonic for maximum output.

For simplicity, tetrode tubes are shown in the diagrams, but pentodes or other multigrid tubes are usually used. The suppressor grid is connected either directly to ground for better isolation of the oscillator section or to the cathode of the tube for greater power output.

Notice that the bottom of the *output* tuned circuit is always bypassed to ground to complete the ac plate circuit.

Test your understanding; answer these checkup questions.

1. What produces the feedback in a TPTG circuit? _____
2. From which LC tank of a TPTG is ac power taken? _____ What are the two methods used? _____ _____
3. To provide proper phasing for oscillation, which tank of a TPTG should be at the higher frequency? _____
4. Why does the grid tank usually determine the oscillation frequency of a TPTG? _____
5. What type of feed has current flowing through the tank coil? _____ Through an RFC? _____

6. In what way is a Hartley oscillator similar to a TPTG? _____ To an Armstrong? _____
7. Where on a VT Hartley tank is the tap for greatest power output? _____ Why? _____
8. If you reach toward an oscillating LC circuit and it changes frequency, what causes this? _____
9. Basically, how does a Colpitts differ from a Hartley? _____
10. Why might Colpitts stability be better than Hartley? _____
11. Is it possible to have a series-fed Hartley? _____ A series-fed Colpitts? _____
12. What is the name of the VHF oscillator that uses inter-electrode capacitances as the ac voltage divider? _____
13. What does ECO mean? _____ Can an ECO be used with BJTs? _____ Triodes? _____ Tetrodes? _____
14. What are three advantages of an ECO over simple Hartley or Colpitts circuits? _____ _____ _____

13-12 CRYSTAL OSCILLATORS

Most modern commercial transmitters, either telegraph or telephone, use *crystal* oscillators because they will not drift more than a few hertz from the frequency for which they are ground. A variable frequency "self-excited" oscillator (VFO) tends to drift considerably more.

An oscillating quartz crystal usually looks like a piece of frosted window glass cut into $\frac{1}{2}$- to 1-in. squares and then ground smooth on all surfaces. Actually, glass does not have the required properties to operate as an oscillator. Instead, special crystalline quartz is cut in thin slices and ground smooth. Such quartz crystals have peculiar properties. If a crystal is held between two flat metal plates and the plates are pressed together, a small emf will be developed between the two plates, as if the crystal became a battery for an instant. When the plates are released, the crystal springs back to its original shape and an opposite-polarity emf is developed between the two plates. In this way, mechanical energy is converted to electric energy by the crystal. Furthermore, if an electric emf is applied across the two plates of a crystal, the crystal will distort its normal shape. If an opposite-polarity emf is applied, the crystal will reverse its physical distortion. In this way, electric energy is converted to mechanical energy by the crystal. These two reciprocal effects in a crystal are known as *piezoelectric effect*. A crystal oscillator circuit can be called a piezoelectric oscillator. Man-made lithium tantalate crystals may be superior to quartz crystals.

If a crystal between metal plates is shock-excited by either a physical stress or an electric charge, it will vibrate mechanically at its natural frequency for a short while and at the same time produce an ac emf between the plates. This is somewhat similar to the damped electron oscillation of a shock-excited LC circuit. Actually, a vibrating crystal will produce an alternating emf longer than an LC circuit will when shock-excited, because the crystal has a much higher Q (fewer losses) than ordinary LC circuits.

A crystal TPTG oscillator substitutes a crystal, held between two plates, for the LC tank in the grid circuit, as shown in Fig. 13-22. In this circuit

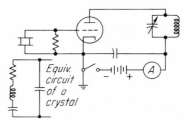

Fig. 13-22 Crystal oscillator, TPTG type. An equivalent circuit of a crystal in holder is shown.

the crystal is operating as a high-Q parallel-resonant circuit. No grid-leak blocking capacitor is needed, since the crystal blocks any electron flow from grid to cathode.

When the switch is closed, the LC tank in the plate circuit is shock-excited into oscillation by the sudden surge of plate current. The ac developed across this LC circuit is fed back to the top crystal plate through the grid-plate interelectrode capacitance and to the bottom plate of the crystal through the bypass capacitor from the LC circuit. The crystal starts vibrating and working as an ac generator on its own. The emf generated by the crystal, applied to the grid and cathode, produces plate-current variations in the plate LC circuit. With both crystal and LC circuit oscillating and feeding each other in proper phase, the whole circuit continues in sustained oscillation and acts as a very stable ac source.

The plate LC circuit must be tuned slightly higher than the frequency of the crystal to produce the required phase relationship between the two circuits to sustain oscillations.

With no oscillation of the circuit, no grid bias will be developed across the grid-leak resistor and the plate current will rise to a high value. Therefore, it is possible to tune this type of oscillator by watching the action of a milliammeter in the plate circuit. A decrease of plate current, as the plate LC circuit is tuned, is an indication that the circuit is oscillating and developing grid bias. The harder the crystal oscillates, the greater the grid-leak bias developed and the lower the plate-current indication.

When tuning a crystal stage, as the plate LC tank is *increased* in frequency, the circuit breaks into strong oscillation and the plate current suddenly drops to minimum when the resonant frequency of the crystal is reached. However, if the plate circuit is *decreased* in frequency while tuning, the plate current gradually decreases to the minimum value and then pops up to a maximum as the circuit stops oscillating. This is a tuning characteristic of all TPTG circuits.

The minimum plate-current reading may indicate strongest oscillation of a crystal-oscillator stage (also greatest crystal heating), but it does not necessarily indicate the optimum operating condition. For most satisfactory operation the plate-current value should be brought down about three-quarters of the way from maximum toward minimum. This will also allow the circuit to be immediately self-starting (very important in radio-telegraph transmitters).

The frequency of vibration, or oscillation, of a crystal is determined primarily by its thickness, physical size, angle of cut, and temperature. It is possible to vary the frequency of oscillation by a few hertz by connecting a small variable capacitor across it or a small variable inductance in series with it. Too large a capacitance or inductance may stop the crystal from oscillating.

While the plate LC circuit is being adjusted to the proper tuning point, the frequency of oscillation may vary as much as a kilohertz. Replacement of the oscillator tube with a new one often changes the oscillation frequency by a few hertz. Variation of the pressure on the crystal by the plates will shift the frequency of oscillation also. Even if the manufacturer indicates a certain frequency of oscillation for a given crystal, it is pos-

sible that it may oscillate a few hertz one way or the other depending on the oscillator circuit used.

Low-frequency crystals may require a small capacitor between plate and grid of the tube to increase feedback sufficiently to produce oscillation. Excessive feedback capacitance may result in crystal fracture.

Although *thickness* vibration of thin crystals has been used in the explanations, a crystal, if cut from quartz at the correct angle, will vibrate corner to corner (*shear*) or end to end (*longitudinally*). The last-named mode results in a much lower frequency of oscillation when using the same-sized crystal. Crystals are often silver-plated on the two flat surfaces, with a connecting wire soldered to the middle of each silvered surface. This allows shear or longitudinal oscillations but damps out thickness vibrations. Silvered crystals with connections made to the edges of the plates will produce thickness oscillations. Crystals may also vibrate *flexurally* (by bending back and forth), *torsionally* (in a twisting movement), or in two modes at the same time.

Although crystals are quite stable in frequency, changes in plate potential may shift the oscillation frequency. Because of this, some form of separate oscillator power supply is advisable in a multistage transmitter.

13-13 CRYSTALS AND TEMPERATURE COEFFICIENTS

The manner in which a crystal is sliced from the raw quartz will determine its natural oscillating frequency, its frequency stability, and its temperature coefficient and even name the type of crystal.

Oscillating crystals are cut from six-sided quartz crystals that are found in nature. An end view of two such quartz crystals is shown in Fig. 13-23.

Fig. 13-23 Angles at which Y-cut and X-cut crystals are sliced from the raw crystalline quartz.

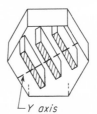
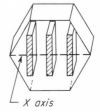

└ Y axis   └ X axis

The dotted lines indicate the X and the Y axes. The Y axis is from one flat face to the opposite flat face of the crystal. The X axis is from one corner to the opposite corner. A crystal sliced out of the quartz at right angles to the Y axis is called a Y-cut, crystal. If cut at right angles to the X axis, it is called an X-cut crystal, as shown.

When crystals are cut at angles other than X and Y cuts, they are given other names. Some other cuts are the AT, BT, CT, and Z. Some cuts operate more satisfactorily on high frequencies; others, on lower frequencies. Different cuts have different temperature coefficients and oscillating characteristics.

The following should make clear the meaning of temperature coefficient of a crystal.

1. If a change in temperature produces a relatively large variation in its oscillation frequency, a crystal has a high temperature coefficient (TC).
2. If a change in temperature produces a relatively small variation in its oscillation frequency, a crystal has a low TC.
3. If a change in temperature produces no variation in its oscillation frequency, a crystal has a zero TC. (Actually no crystal has a true zero coefficient.)
4. If an increase in temperature produces an increase in its oscillating frequency, a crystal has a positive TC.
5. If an increase in temperature produces a decrease in its oscillating frequency, a crystal has a negative TC.

Some crystals have a +TC when operated at about 70°F, nearly zero at about 110°, and a −TC at temperatures above 140°. Such crystals, if their temperature can be held at about 110°, will operate as though they had a zero TC.

Y-cut crystals have a TC range of about −25 to +100 cycles/°C/MHz. They also have the disadvantage of having a second frequency at which they may oscillate. A frequency meter (Chap. 21) must be used when tuning such a crystal stage to make sure that the frequency of oscillation is the desired one.

X-cut crystals have a TC range of about −10 to −25 cycles/°C/MHz. Compare these ranges with those of a GT-cut, which has a −1 to +1-cycle/°C/MHz range. The GT-cut is almost a zero-TC crystal from freezing to boiling, but it is useful at frequencies only up to a few hundred kilohertz.

A TC example:

A 600-kHz X-cut crystal, calibrated at 50°C and having a TC of −20 parts per million (ppm) per degree Celsius, will oscillate at what frequency when its temperature is 60°C?

First, the "−20 parts per million per degree" can be read as "a −TC of 20 cycles/°C/MHz." Since 600 kHz is 0.6 MHz, the expression above can be stated "−20 cycles/°C/MHz × 0.6." The change in temperature is 10°. The change in hertz is then

$$-20 \times 10 \times 0.6 = -120 \text{ Hz}$$

By increasing 10° in temperature, the crystal loses 120 Hz from its 600,000-Hz calibration, or oscillates at a frequency of 599,880 Hz, or 599.88 kHz. (If the crystal had a similar but +TC, the 120 Hz would have been added instead of subtracted.)

In many transmitters, the oscillator may be operating on a certain frequency and the output may be on some integral multiple, or harmonic, of the fundamental frequency.

For example: If a transmitter uses a 1,000-kHz crystal with a TC of −4 cycles/°C/MHz and the crystal temperature increases 6°, the crystal will decrease its frequency of oscillation −4 × 6, or −24 Hz. If the output of the transmitter is to be 5 times the fundamental, or 5,000 kHz, the output frequency will decrease 5 × −24, or −120 Hz. The output frequency will be 5,000,000 less 120 Hz, or 4,999.88 kHz.

13-14 CRYSTAL HEATER CHAMBERS

Crystals warm when operating and change frequency. To prevent this they may be operated in special insulated chambers held at perhaps 140°F, which is higher than ambient or room temperature. The chamber contains the crystal, a heating resistor, and some form of thermal transducer, such as a special thermometer, a bimetallic element, or a thermocouple. The chamber takes time to reach a stable operating temperature, and so it may be left running even if the equipment in which it is used operates only part of each day.

Figure 13-24 illustrates an electronic circuit

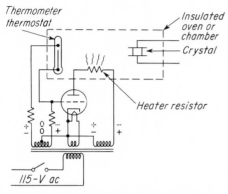

Fig. 13-24 Circuit to control chamber or oven temperature using a mercury thermostat.

used to control chamber, or "oven," temperature. The important point about this circuit is the center-tapped transformer winding in the grid circuit. The center tap, because it is connected directly to the cathode, is always considered to be at zero potential. At any instant that the right side of the grid winding is positive, the left side will be negative in respect to the cathode. During the other half cycle the polarities reverse.

The grid and plate windings of the transformer are so connected that, when the plate is positive and the thermometer contacts are open, the grid is connected through the grid resistor to the positive side of the grid winding. This places a positive voltage on the grid whenever a positive voltage is applied to the plate. If both grid and plate are positive at the same time, plate current flows and the resistor heats. The chamber starts to warm,

and the thermometer column starts to rise. When both grid and plate are negative, no plate current can flow. Therefore, the resistor is heating on only one half cycle.

When the thermometer contact is closed by the rising mercury, the grid is connected through a resistor to the other end of the grid winding. Now when the plate potential is positive, the grid is negative and no plate current flows. With the thermostat open the chamber heats, but with the thermostat circuit closed the chamber cools. The heating and cooling cycle may take 5 to 30 s. A lamp can be included in the plate circuit. It will glow during the heating period and be dark during the cooling period, giving a visual indication of chamber operation.

A bimetallic element is often used as a thermostat. It is composed of two thin strips of different metals welded together. The metals are so chosen that one expands considerably when heated and the other expands very little. When current flows through the resistance wire in Fig. 13-25, the air

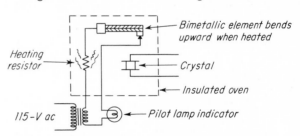

Fig. 13-25 Temperature-controlled chamber using a bimetallic element.

in the chamber, the crystal, and the bimetallic element all heat. If the bottom strip expands more than the top, the element will be forced to bend upward at the free end, opening the two contacts and breaking the heater circuit. When the chamber cools, the strip will bend down again and reestablish contact, starting the heating cycle again. An inert gas in the chamber prevents oxidation of the contacts due to sparking on the make and break of the circuit.

There are many possible types of controls to hold the temperature of a crystal chamber constant. Consider the circuit in Fig. 13-26. A thermocouple (Sec. 12-19) produces a dc voltage and current when heated. If the chamber is cold, there

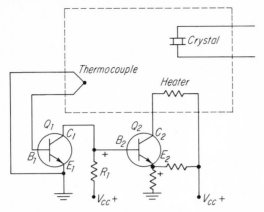

Fig. 13-26 Thermocouple-controlled crystal oven. Variation of R_1 will determine operating temperature.

is no voltage from the thermocouple, B_1 of Q_1 has no bias, and the resistance between C_1 and E_1 is very high. Thus, B_2 is forward-biased through R_1, and collector current flows through the heater resistor, warming the chamber. As it heats, the thermocouple warms, producing a voltage which forward-biases B_1. With forward bias the C_1E_1 resistance is reduced, bringing B_2 closer to ground potential. This reduces the forward bias on Q_2, and its collector current decreases, reducing the heating of the chamber. The amplifying ability of the transistors produces a highly sensitive thermostat.

13-15 CRYSTAL HOLDERS

Several types of holders are used for crystals. The oldest type consists of a hollowed-out block of Bakelite or other insulating material in which two flat metal plates and the crystal are mounted, as in Fig. 13-27. A spring against the top plate holds together the sandwich formed by the crystal between the two plates. An electrical connection is made to each plate.

Fig. 13-27 One type of crystal holder.

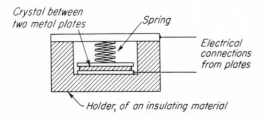

Another type of holder uses metal plates with raised corners touching the crystal. These air-gap mountings are often used with higher-frequency crystals.

Crystals will not oscillate if they become dirty. There are two methods of cleaning them: (1) wash with warm soap and water and rinse in clean water or (2) wash with a cleaning solvent. The crystals should not be touched with the bare hands but should be handled with clean, lint-free cloths and by the edges rather than by the flat surfaces. The metal plates must also be cleaned.

Crystals are very fragile. They will fracture easily if dropped. If excessive regeneration is used in an oscillator circuit, the crystal may oscillate so hard that it will fracture itself. If the edge of a crystal becomes chipped, the crystal can sometimes be ground down on that edge, but the frequency of oscillation will increase. Grinding is done on a flat surface such as a piece of plate glass, using a mixture of very fine carborundum dust and water as the grinding compound. The frequency of a crystal can be raised by grinding its surfaces. The frequency can be lowered a few hundred hertz by marking an X from corner to corner on one or both flat surfaces with solder or lead.

Various types of holders are employed with *plated* crystals. Instead of pressing plates against its flat surfaces, a crystal may be plated with a thin layer of metal on both flat surfaces. The crystal is often suspended in a thin sheet-metal holder by the two steel wires that make connections to the two plated surfaces.

13-16 OTHER CRYSTAL CIRCUITS

Besides the basic TPTG-type crystal oscillator described, crystals can be used in other circuits. The *Pierce* circuit of Fig. 13-28 uses a crystal as

Fig. 13-28 Triode Pierce crystal oscillator.

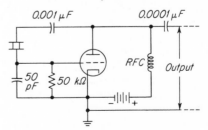

the LC circuit in a shunt-fed ultraudion oscillator. A Pierce oscillator has the advantage of not requiring any tuned circuits. Relatively low plate voltages must be used, limiting the output power from the circuit. In vacuum-tube circuits the 0.001-μF capacitor is inserted in series with the crystal to prevent dc voltage strain across the crystal, but it is not needed with BJTs or FETs.

In Fig. 13-29, the crystal is used in place of the

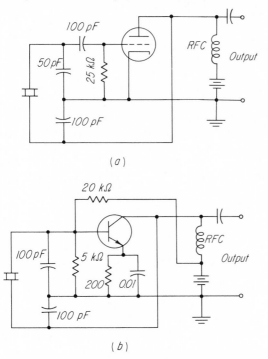

(a)

(b)

Fig. 13-29 Crystal Colpitts oscillators: (a) triode and (b) transistor.

coil in Colpitts oscillators. The two capacitors across the crystal control the degree of coupling between plate and grid (collector and base), and they should be as small as possible.

A very common transistor crystal oscillator uses either a Hartley or a Colpitts circuit tapped down to produce very little feedback and with the crystal connected between base and emitter (grid and cathode). When the LC tank is tuned close enough to the crystal frequency, feedback is sufficient to start the crystal oscillating, and it controls the oscillation frequency of the stage.

Crystals are extremely thin when ground to resonate at frequencies above 10 MHz. For higher

frequencies it may be desirable to use a lower-fundamental-frequency crystal and employ circuits such as the ECO with its output plate-tank circuit tuned to a harmonic of the crystal frequency. Figure 13-30 illustrates a crystal ECO

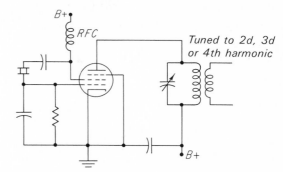

Fig. 13-30 Pierce ECO crystal oscillator capable of even- or odd-harmonic output.

employing the first three elements as a triode Pierce oscillator. The harmonic output tank can be tuned to two, three, or four times the crystal frequency. The output will be an exact multiple of the oscillation frequency of the crystal.

If a crystal is used as a coupling device from the output of a two-stage amplifier to the input of the first stage, the two stages will operate at the series-resonant frequency of the crystal, which is slightly lower than the parallel-resonant frequency produced in crystal oscillator circuits such as the TPTG.

It is possible to make almost any crystal oscillate in three layers. This results in a *third-mode* oscillation, which produces a harmonic output of the *series*-resonant frequency of the crystal. These *overtone* oscillator circuits resemble some self-excited LC oscillators with the crystal acting as the coupling device to complete the oscillator circuit. In Fig. 13-31a the crystal is completing the circuit to the feedback (grid tickler) coil in a tuned-plate-circuit Armstrong oscillator. The LC circuit must be tuned to the third, fifth, or seventh overtone frequency. No fundamental frequency will be generated at any place in the circuit. Thus, overtone and harmonic output crystal oscillators such as an ECO are not the same. Figure 13-31b shows a Colpitts-type overtone oscillator used with a transistor.

The TPTG crystal circuit depends on plate-grid

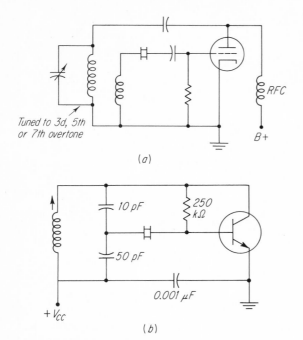

Tuned to 3d, 5th
or 7th overtone

(a)

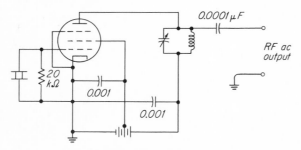

(b)

Fig. 13-31 Odd-harmonic overtone oscillators: (a) Armstrong VT type and (b) Colpitts BJT type.

capacitance to feed energy from the plate circuit back to the grid to support sustained oscillations. It would seem that the pentode crystal circuit of Fig. 13-32 would not be practical, because of low

Fig. 13-32 Pentode crystal oscillator.

C_{gp}, but the gain of a pentode tube is so great and the Q of a crystal so high (10,000 to 100,000) that enough energy will feed back to produce oscillation of the circuit. Since so little RF ac excitation is required to keep the crystal oscillating in a pentode (or tetrode) circuit, it is possible to use relatively high plate voltage and obtain higher power output than might be possible with a triode but not overheat or fracture the crystal.

Test your understanding; answer these checkup questions.

1. What kind of oscillators are not VFOs? _____
2. To what does a quartz crystal convert mechanical energy? _____ What is this effect called? _____
3. To what is an oscillating crystal equivalent electrically? _____
4. What is the relative I_p value when a crystal oscillator is oscillating? _____ The $-E_g$? _____
5. Why is it advantageous to have a separate regulated oscillator power supply? _____
6. What TC does a crystal have if an increase in temperature increases its frequency? _____ Has no effect on frequency? _____
7. What type crystals are sliced parallel to the sides of the raw crystal? _____
8. A 4-MHz crystal at 50°C with a +10 Hz/°C/MHz has what frequency at 60°C? _____
9. Is a harmonic ever anything other than a whole-number multiple of the fundamental? _____
10. Why are crystals often kept in temperature-controlled chambers? _____ Would this benefit LC tanks also? _____
11. What are three devices used to keep crystal chambers at constant temperature? _____ _____ _____
12. To what LC oscillator circuit is a Pierce similar? _____
13. What are two advantages of using a crystal ECO? _____ _____
14. Is an overtone the same as a harmonic? _____
15. Which overtones can be used in crystal oscillators? _____
16. What is the first overtone frequency of a 10-MHz crystal? _____

13-17 SOME HIGH-FREQUENCY OSCILLATORS

Generating low-frequency ac by vacuum tube or transistor presents no serious difficulties. The higher the frequency, however, the less power output and the poorer the frequency stability of oscillation. Audio frequencies, up to about 20 kHz, may use laminated-iron-core coils and capacitors for the LC tank circuit. For frequencies up to several megahertz either air-core or powdered-iron-core coils may be used. Up to several hundred megahertz air-core coils, sometimes with brass slugs in the core area, are employed. In the UHF and superhigh frequency (SHF) range resonant-line tanks, hairpin tanks, coaxial tanks, and resonant cavities are required. The higher the frequency, the less inductance and capacitance required to produce a resonant circuit. At about

300 MHz a 4-in. piece of wire bent into a hairpin and connected between grid and plate leads of a triode to form a Colpitts-type oscillator may provide all the *L* and *C* needed to act as the resonant circuit. Above this frequency the usual LC-circuit idea may not be practical.

An electric impulse travels at 186,000 miles, or 300,000,000 meters, per second. A 300-MHz ac completes its cycle in $\frac{1}{300,000,000}$ s. In this time, from the formula $\lambda = v/f$, or *wavelength = velocity/frequency*, the 300-MHz ac travels 300,000,000/300,000,000, or 1 m (39.37 in.).

If a 300-MHz generator is connected to a load by a long pair of wires as in Fig. 13-33, the instan-

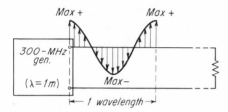

Fig. 13-33 Instantaneous voltage distribution of an ac having a frequency of 300 MHz.

taneous voltage values along the line at one instant might be as shown. Note that points one-half of a wavelength apart are 180° out of phase. This is the end-to-end phase relationship of a resonant LC circuit. Therefore, it should be expected that a half-wavelength wire connected between grid and plate of a tube should form a resonant colpitts or Hartley tank at 300 MHz. When bent into a hairpin shape, however, the end-to-end capacitance of the wire, plus the interelectrode capacitance of the tube added to the inductance of the tube leads, requires a shorter hairpin to resonate at this frequency. If a small tuning capacitor is added to tune the circuit, the hairpin length must be still shorter.

Figure 13-34 shows a TPTG-type hairpin tank oscillator. It should be understood that the quarter-wavelength indications represent the electrical, not the physical, length of the grid- and plate-circuit hairpins.

Hairpin tanks are not practical at low frequencies. One wavelength at 5 MHz is *v/f,* or 300/5, or 60 m (6,000 cm, 196.85 ft).

A variation of the hairpin form of tank circuit is the *coaxial* tank, a thin conductor running up

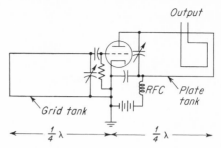

Fig. 13-34 UHF TPTG-type quarter-wavelength hairpin-tank oscillator.

the center of a piece of copper tubing. The conductor is connected to the sealed end. A TPTG circuit using coaxial tanks is shown in Fig. 13-35.

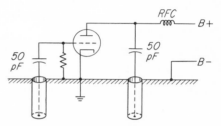

Fig. 13-35 Quarter-wavelength coaxial tanks in a TPTG oscillator.

Such tank circuits isolate the oscillating electrons from outside effects. Electrons oscillate up and down the central wire and on the *inner* surface of the copper tubing, not on the outside. RF ac output is obtained through a hole in the coaxial tank.

At still higher frequencies (UHF and SHF) the *resonant cavity* is used. As a simplified explanation, many hairpin circuits of the same dimensions laid side by side and soldered together form a single wide hairpin. If the open sides of the hairpin are closed over with a sheet of metal, the metallic cavity shown in Fig. 13-36 is produced. Electrons oscillate back and forth inside the cavity, setting up alternating magnetic fields in the cavity. A loop or hook acts as a means of extracting energy or of feeding energy to the cavity. Cavity resonators are discussed in Chap. 27.

The higher the frequency, the greater the skin effect and the higher the effective resistance of a conductor. As a result, high-frequency circuits use large wires or large-area components such as resonant cavities. Since the current travels only

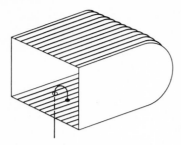

Fig. 13-36 Development of a resonant cavity from many parallel hairpin tanks.

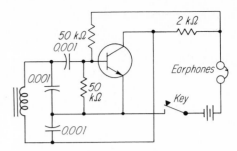

Fig. 13-38 Code-practice circuit using a BJT in a Colpitts-type oscillator circuit.

on the skin of conductors at high frequencies, it is preferable to silver-plate conductors to increase their Q.

13-18 AUDIO OSCILLATORS

The oscillators described thus far have been RF types. The ac generated is presumably well above the audible frequencies. An audible-frequency ac may be desired for a code-practice circuit or as an audible-frequency source. An Armstrong oscillator using an iron-core transformer is shown in Fig. 13-37. The grid-leak resistor and capacitor

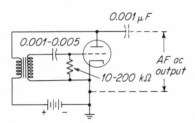

Fig. 13-37 Armstrong VT audio oscillator.

help to determine oscillation frequency by their RC time constant.

A Colpitts-circuit AF oscillator using a choke coil instead of a transformer is shown in Fig. 13-38. Note the use of a resistor instead of an

AF choke in the collector circuit. As long as only a little audio power is desired, it is much more economical to use the resistance to obtain the required impedance across which the AF ac voltage-drop is developed.

13-19 DYNATRON OSCILLATOR

In all the oscillator circuits described thus far, inductive or capacitive feedback of energy from plate to grid circuit is used to produce sustained oscillations. Dynatron oscillators operate on an entirely different *negative resistance* principle. According to Ohm's law, an increase in emf in a circuit produces an increase in current. If an increase of emf in a circuit results in less current, the circuit is said to be exhibiting a negative resistance effect.

A negative resistance effect can be developed by operating the grid of a triode at a positive potential a little greater than the plate potential. With the grid at a constant positive value, the plate current will increase as the plate voltage is brought up from zero to a certain point, X in Fig. 13-39. Then, as the plate voltage is increased further, the plate current begins to decrease because of secondary-emission electrons from the plate traveling to the grid. (If more electrons are leaving the plate as secondary emission than arrive on it, current flows from plate to grid in the tube.) If the plate voltage is increased further, the plate is able to attract the secondary-emission electrons more than the grid and the plate current will increase again. The downward slope of the $E_p I_p$ curve, X to Z, indicates the range of plate voltages over which a negative resistance effect is exhibited by the tube.

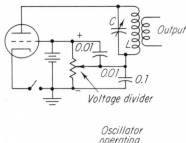

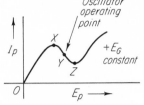

Fig. 13-39 Basic dynatron oscillator. Below, E_pI_p curve, showing portion exhibiting negative resistance, X to Z, and bias point to produce LC-circuit oscillation.

When the switch is closed, the plate LC circuit is shock-excited and the ac voltages developed across the tuned circuit are added to the plate voltage. If the plate voltage is adjusted to a value near Y on the curve, the negative resistance effect will be operating in such a phase as to oppose any resistance effect of the LC circuit and the circuit will function as a generator of sustained oscillations.

Compare the basic dynatron circuit with the basic solid-state tunnel-diode oscillator shown in Fig. 13-40, in which an LC circuit is across a

Fig. 13-40 Basic tunnel-diode oscillator. Below, V_bI_b curve, showing portion of curve exhibiting negative resistance, and bias point to produce oscillation.

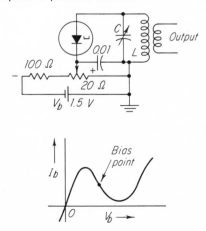

diode biased to the midpoint of the negative resistance portion of its curve. As long as an LC circuit is across a source exhibiting negative resistance (provided tank circuit losses are not excessive), LC oscillations should result. The bias-voltage point for germanium tunnel diodes is about 0.15 V. The tunnel diode may be tapped down the LC circuit instead of being across the whole coil as shown.

Practical dynatron oscillators, such as Fig. 13-41, use tetrode tubes because control grids

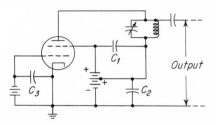

Fig. 13-41 Common dynatron uses a tetrode.

of triodes are not sufficiently rugged to carry enough current. Bypass capacitor C_1 affords a low-reactance ac path from the LC circuit to the screen grid. C_2 brings one end of the LC circuit to ac ground potential. C_3 holds the control grid to a constant value. Frequency stability of dynatron oscillators is good, but power output is low. The circuit is rarely used, but the dynatron principle occurs in many electronic devices.

13-20 RC OSCILLATORS

The oscillator circuits discussed thus far generate essentially sine-wave ac, since this is the waveshape produced by flywheel effect in an oscillating LC circuit. There are many applications for which other waveshapes, such as sawtooth or square-wave, are required. The circuits used to produce these waveshapes usually rely on charge and discharge times of capacitors in conjunction with resistors.

A simple RC oscillator, shown in Fig. 13-42, employs a resistor, capacitor, and neon bulb in a *relaxation* oscillator circuit. When the switch is closed, the time required to develop a charge in the capacitor C depends on the E_b and R values in series with C. When the voltage across C rises to about 75 V, the ionization potential of neon, the

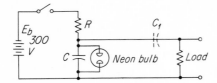

Fig. 13-42 Neon bulb relaxation oscillator.

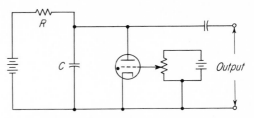

Fig. 13-44 Thyratron relaxation oscillator. Frequency can be controlled by *R*, *C*, or bias voltage.

neon acts as a good conductor and discharges *C* rapidly to about 50 V. At this voltage the neon de-ionizes, and the capacitor begins to charge slowly again. Thus the voltage across *C* repeatedly rises slowly from 50 to 75 V and falls rapidly to 50 V, producing a sawtooth varying dc waveshape across capacitor and load, as shown in Fig. 13-43*a*. If a capacitor C_1 is added in series with the load resistor, the waveshape remains the same but the charge-discharge action of the output capacitor converts the current in the load to sawtooth ac, illustrated in Fig. 13-43*b*. The power-supply voltage should be several times the ionization potential of the neon or other gas used.

Gas-filled thyratrons can be used in the relaxation oscillator circuit of Fig. 13-44. Variation of the grid bias controls the ionization potential of the tube and therefore the frequency of the output wave. Oscillators of this type are limited to a few thousand hertz. For higher-frequency sawtooth waves other types of circuits must be used. SCRs can be used in somewhat similar circuits.

True sine-wave ac has no harmonic energy output, but sawtooth ac is rich in harmonics. A 100-Hz sawtooth wave may contain weak but usable harmonic energy every 100 Hz up to several hundred thousand hertz.

13-21 MULTIVIBRATOR OSCILLATORS

One of the many possible relaxation oscillators is the *multivibrator*. It consists of two vacuum tubes or transistors connected in RC-coupled circuits such as those in Fig. 13-45.

Multistage amplifiers have 180° phase change from the grid of one stage to the plate of that stage, or from the grid of the first stage to the grid of the second. If the grid of the first tube suddenly goes more negative, the grid of the second goes more positive. As the grid of the second tube goes more positive, the plate of that tube is going more negative (less positive). If the output of the plate of the second tube is fed back to the grid of the first, the feedback voltage will be in phase (180 + 180 = 360° = 0°) and will increase the effect of the originating voltage change

Fig. 13-43 (*a*) RC charging curve and varying dc produced across bulb in oscillator. (*b*) Sawtooth ac wave if C_1 is used.

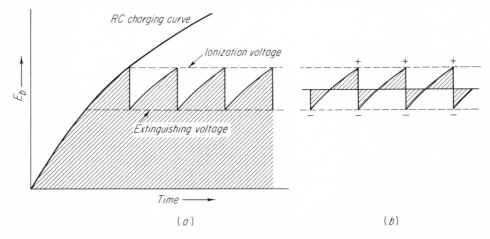

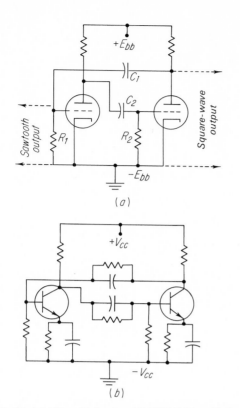

Fig. 13-45 (*a*) VT multivibrator. (*b*) NPN BJT multivibrator.

forward-bias the bases. *Switching* transistors with low storage times, and abruptly rising collector-current curves make them suitable for oscillators and other class C biased circuits.

Multivibrators are used to produce sawtooth or square-wave ac, to generate a fundamental frequency with many harmonics, and to develop voltages to gate or switch electronic circuits on and off.

13-22 PARASITIC OSCILLATIONS

Parasitic oscillations are unwanted (parasite) oscillations that may occur in almost any type of circuit, oscillator, amplifier, power supply, receiver, transmitter, etc. As an example, a transmitter is operating on 6 MHz, but nearby receivers can hear signals from it every 100 kHz from about 4 to 8 MHz. Furthermore, the transmitter tubes may overheat from overload, or tuning is found to be erratic. In all probability one or more stages are oscillating at 100 kHz in some way. This is a low-frequency parasitic oscillation.

In another case a transmitter is operating, but tuning is erratic, the plates of the tubes show excessive color, plate current is not normal, and perhaps capacitors spark over for no apparent reason. It may be found that one or more of the stages are oscillating at possibly 100 MHz or some other very high frequency. This would be a high-frequency parasitic oscillation.

Sometimes, in audio amplifiers, undesired feedback effects will produce an oscillation at an audible frequency, or at some subaudible frequency, or at a supersonic frequency. These are all forms of parasitic oscillations.

One possible method of producing a parasitic oscillation at a frequency lower than the operating frequency is illustrated in Fig. 13-46, showing a TPTG oscillator (which is similar also to a basic RF amplifier stage).

In this diagram, if L_1 and C_1 happen to resonate at the same frequency as L_2 and C_2, the circuit may oscillate at this particular frequency and at the same time work at the frequency of the two tank circuits. This would be a lower frequency parasitic. RF chokes have greater inductance than the coils in the tuned circuits have, and the bypass capacitors have far greater capacitance than the tuning capacitors of the LC circuits have.

on the grid of the first tube. This in-phase feedback produces a slow discharging of coupling capacitors C_1 and C_2 through grid resistors R_1 and R_2 and then a rapid charging through the low-resistance plate resistors when the tube ahead starts to conduct. A sustained relaxation oscillation is produced. The voltage waveform developed across the grid resistors, as they discharge the coupling capacitor, is sawtooth. The voltage waveform across the plate resistors, as their tubes are alternately held at cutoff and then full conduction, is square-wave. Frequency of oscillation depends on the RC time constant of the coupling capacitors and their associated grid resistors. Unequal RC constants produce cycles of ac in which one alternation is longer than the other. If an LC circuit is substituted for one of the plate load resistors, an essentially sine-wave voltage can be developed across the LC circuit.

Two transistors connected in a multivibrator circuit are shown in Fig. 13-45*b*. The resistors across the coupling capacitors are required to

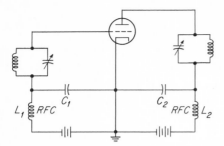

Fig. 13-46 TPTG oscillator (or amplifier) circuit in which low-frequency parasitic oscillations are possible.

The parasitic oscillation can be stopped by using a resistance instead of the grid RFC, by using a smaller-inductance RF choke in the plate circuit, or a smaller value bypass capacitor in the grid circuit.

TPTG-type parasitic oscillations are not the only ones. It is possible to have chance Hartley, RC, Colpitts, and even dynatron parasitic oscillations in electronic circuits.

High-frequency parasitic oscillations may exist in an amplifier or oscillator and not be detected. TV receivers make excellent detectors for any parasitic oscillations that fall near TV channels.

The two diagrams in Fig. 13-47 show the same

Fig. 13-47 (a) TPTG oscillating at the LC circuit frequency may form (b) a quarter-wave hairpin-tank VHF parasitic oscillator at the same time.

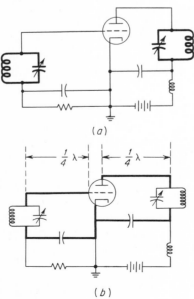

(a)

(b)

circuit, first, as normal TPTG oscillations are concerned and, second, as seen by the high-frequency parasitic oscillations. Figure 13-47a shows the elements of the stage which determine its normal frequency of operation in heavy lines. Figure 13-47b shows a parasitic oscillation circuit at a frequency determined by two quarter-wave hairpin tanks; the tuning capacitors act as bypass capacitors across the ends of the quarter-wave lines. Should both the grid and plate circuits happen to have approximately the same-length leads, the circuit may produce oscillations at a high frequency at the same time that it is operating on the desired frequency. The coils of the tuned circuits act as RF chokes for the high frequencies and are effectively out of the circuit.

High-frequency parasitics may be stopped by winding a half-dozen turns of wire around a 50-Ω 1-W carbon resistor and inserting this *parasitic choke* in either the grid (base) or the plate (collector) lead of a circuit, as in Fig. 13-48. Parasitic

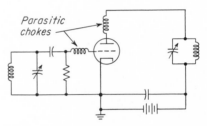

Fig. 13-48 Parasitic chokes may be placed in plate or grid (base or collector, gate or drain) leads in a circuit.

chokes should be installed as close to the terminals of the tube or transistor as possible. Sometimes a 20- to 300-Ω resistor alone will produce the same result as a parasitic choke.

13-23 INDICATIONS OF OSCILLATION

When a vacuum tube or transistor RF oscillator is turned on, several methods can be used to determine whether the circuit is oscillating.

RECEIVER. A nearby radio receiver will indicate by a change of sound when an oscillator is tuned across the frequency to which it is set. With the beat-frequency oscillator circuit of the receiver turned on, the oscillator will be heard as a whistle.

BIAS VOLTAGE. The presence of grid bias voltage in a vacuum-tube oscillator or reverse base bias in a transistor oscillator is a good indication that the circuit is oscillating. A change of bias voltage will be produced if the oscillator is stopped from oscillating by shorting the capacitor of the LC circuit in either tube or transistor circuits.

PLATE OR COLLECTOR CURRENT. When a vacuum-tube or transistor oscillator is oscillating, bias voltage is developed, reducing the plate or collector current. High current may be an indication of nonoscillation of a circuit. (Transistor circuits sometimes increase collector current when in oscillation.)

RF INDICATOR. A flashlight lamp in series with a loop of insulated wire will glow when the loop is coupled to the oscillator tank coil if the oscillator produces a watt or more of RF power. A sensitive RF thermogalvanometer can be used instead of the lamp. A 0- to 1-mA meter with a solid-state diode across it and in series with a pickup loop will act as a sensitive RF indicator suitable even for low-power oscillators.

OSCILLOSCOPE. An insulated-wire loop coupled to the oscillator tank coil and to the vertical input of the oscilloscope will show the presence of RF energy as a horizontal band on the screen. With low-frequency oscilloscopes it is necessary that HF or UHF RF be fed directly to the CRT plates, since the vertical-signal amplifiers are unable to amplify such high frequencies.

NEON LAMP. A neon lamp will glow if touched against the plate or grid lead of a low-power oscillating circuit, provided the oscillator is generating 80 V peak or more in the tank circuit.

LEAD PENCIL. A dangerous method with high-voltage vacuum-tube circuits is the use of a wooden pencil with soft lead. When touched to an oscillating tank circuit, it will produce a spark if the circuit is generating more than a few watts of RF energy. This method should not be used with circuits generating more than approximately 50 W of RF.

For audio-frequency oscillators. The presence of bias is a reliable indication. A tone in earphones in series with two 0.0005-μF capacitors connected from ground to either grid or plate (base or collector) is a good indication. The output of the oscillator can be connected to an oscilloscope for a visual indication.

13-24 OSCILLATOR STABILITY

The main requirement for an oscillator circuit whether crystal or variable-frequency (VFO) types is usually frequency stability. Factors for this are:

1. *Constant dc supply voltage*. Use a separate oscillator power supply or use regulated voltages in a multistage system.
2. *Low plate or collector current*. The less current the less heating of the devices involved and of the components in the circuits.
3. *Low power output*. Loose coupling keeps plate and collector currents low and tank Q high, tending to hold frequency constant.
4. *Rigid mechanical structure*. Vibration of almost any oscillator part produces frequency variation.
5. *Buffer stage*. Oscillators should be followed by a buffer stage to prevent output stages from affecting the oscillator.
6. *Heavy coil wire*. Heavy wire for the oscillator coil results in higher Q (particularly if silver-plated), less contraction and expansion due to heating, and less chance of vibration.
7. *Drafts*. Changing temperature of oscillator-stage parts because of air drafts produces a drift in the frequency.
8. *Temperature-control chamber for crystals*.
9. *Proper tapping of coils*. For the Hartley, move tap to point on coil where best stability results. For the Colpitts, proper tuning-capacitor ratios are required.
10. *High C/L ratio*. The more capacitance in the tuning circuit, the less effect external and internal parameters have on frequency.
11. *Tapping down coil*. Tapping down the coil, as in Fig. 13-5b, produces higher Q, resulting in circuit changes having less effect on the frequency of resonance.
12. *Shielding parts*. The shielding of parts prevents air currents, hand capacitance, and humidity from affecting tubes, transistors, coils, resistors, and capacitors.
13. *Grid leak*. Highly important factors are the correct values of grid-leak or biasing resistances and capacitances, usually found by trial and test.
14. *Active device used*. Some vacuum tubes or transistors change too much when in operation and will never make stable oscillators.

1. Name four types of tanks used in UHF and SHF. _____ _____ _____ _____
2. An LC tank represents what wavelength? _____
3. What type of VT makes the best dynatron oscillator? _____ Could BJTs or FETs be used? _____
4. What solid-state device mentioned has a negative resistance effect? _____
5. What is the bias point for all negative resistance oscillators? _____
6. What waveform(s) is produced by neon lamp RC oscillators? _____
7. From what part of a multivibrator is a square wave available? _____ A sawtooth wave? _____
8. If the grid resistance of one VT of a multivibrator is 10 times the second R_g, what is the output wave at the first VT plate? _____
9. Why is the output of a multivibrator rich in harmonics? _____
10. What are two possible methods of producing a lower-frequency parasitic oscillation? _____ _____
11. What can be done to stop higher-frequency parasitic oscillations? _____
12. What are some signs of parasitic oscillation? _____
13. List seven devices that can be used to indicate oscillation in an RF oscillator. _____ _____ _____ _____ _____ _____ _____
14. List 10 factors that aid frequency stability of an oscillator. _____ _____ _____ _____ _____ _____ _____ _____ _____ _____

COMMERCIAL LICENSE QUESTIONS

Amateur license questions will be found in the Addendum.

FCC Elements 3, 4, and 6 require an ability to answer questions similar to those below. Sections in which questions are answered are shown in parentheses. A question followed by a bracketed number is required for that element alone.

1. What is meant by *shock excitation* of a circuit? (13-2)
2. What is meant by *flywheel effect* of a tank circuit? (13-2)
3. Draw a diagram and explain the principles of an Armstrong oscillator. (13-4)
4. Explain how grid-bias voltage is developed by the grid-leak resistor in an oscillator. (13-5)
5. Draw a diagram and explain the principles of a series-fed TPTG oscillator. (13-6, 13-7)
6. By what means is feedback coupling obtained in a TPTG oscillator? (13-6)
7. Draw a diagram and explain the principles of a shunt-fed TPTG oscillator. (13-7)
8. Draw a diagram of a tuned-grid Armstrong oscillator with shunt-fed plate. (13-7)
9. What is the purpose of an RF choke? (13-7)
10. Draw a diagram and explain the principles of a vacuum-tube Hartley oscillator. (13-8)
11. Draw a diagram and explain the principles of a transistor Hartley oscillator. (13-8)
12. How do Colpitts and Hartley oscillators differ? (13-9)
13. Why is a high ratio of capacitance to inductance employed in the grid circuit of some oscillators? (13-9, 13-24)
14. Draw a diagram and explain the principles of a shunt-fed Colpitts vacuum-tube oscillator. (13-9)
15. Draw a schematic diagram and explain the principles of a transistor Colpitts oscillator. (13-9)
16. Draw a diagram and explain the principles of an ECO oscillator. (13-11)
17. What are the characteristics of an ECO? (13-11)
18. What is meant by a *harmonic*? (13-11)
19. For what is a quartz crystal used in a radio transmitter? (13-12)
20. What will result if a dc potential is applied between the two parallel surfaces of a quartz crystal? (13-12)
21. What are the principal advantages of crystal control over tuned-circuit oscillators? (13-12)
22. What factors affect the resonant frequency of a crystal? (13-12)
23. For maximum stability, should the tuned circuit of a crystal oscillator be tuned to exact crystal frequency? (13-12)
24. Draw a diagram and explain the principles of operation of a series-fed and a shunt-fed TPTG crystal oscillator. (13-12)
25. Draw the approximate equivalent circuit of a quartz crystal. (13-12)   [4]
26. What may result if a high degree of feedback coupling exists between the plate and grid of a crystal-controlled oscillator? (13-12)
27. Why is a separate source of plate power desirable for a crystal-oscillator stage in a radio transmitter? (13-12)
28. Why is an additional plate-grid feedback capacitor

sometimes necessary in a crystal oscillator? (13-12)

29. What is the maximum temperature variation at the crystal from the normal operating temperature when using (a) X- or Y-cut crystals, (b) low-temperature-coefficient crystals? (13-13)   [4]

30. What is the purpose of maintaining the temperature of a quartz crystal as nearly constant as possible? (13-13, 13-14)

31. What is meant by the *temperature coefficient* of a crystal? (13-13)

32. What is the approximate range of temperature coefficients to be encountered with X-cut crystals? (13-13)

33. A 500-kHz crystal, calibrated at 50°C and having a temperature coefficient of −25 ppm/°C, will oscillate at what frequency when its temperature is 60°C? (13-13)

34. A transmitter is operating on 4 MHz, using a 1-MHz crystal with a temperature coefficient of −5 cycles/°C/MHz. If the crystal temperature increases 7°, what is the change in the output frequency of the transmitter? (13-13)

35. Why are quartz crystals sometimes operated in temperature-controlled ovens? (13-14)

36. Why are crystal heaters left on even when a broadcast station is not on the air? (13-14)   [4]

37. Explain by the use of simple drawings the physical construction and operation of mercury-thermometer, bimetallic-element, and thermocouple types of crystal heater controls. (13-14)   [4]

38. Why must the surfaces of a quartz crystal be clean? (13-15)

39. What cleaning agents may be used to clean a quartz crystal? (13-15)

40. Draw a diagram and explain the principles of operation of a Pierce oscillator. (13-16)

41. What is a *third-mode* crystal? (13-16)

42. What are the characteristics and possible uses of an overtone crystal? (13-16)

43. Draw a schematic diagram of a pentode-tube crystal oscillator. (13-16)

44. What is the wavelength in meters and centimeters of a 6-MHz ac? (13-17)

45. Would a cavity resonator be associated with higher or lower frequencies? (13-17)

46. Explain the principles of operation of a cavity resonator. (13-17)

47. Draw a circuit diagram of a vacuum-tube audio oscillator using an iron-core choke. (13-18)

48. Draw a diagram and explain the characteristics of a dynatron oscillator. (13-19)

49. What type of oscillator depends upon secondary emission from the anode for its operation? (13-19)

50. What is a *multivibrator,* and what are its uses? (13-21)

51. Show by simple diagrams at least two ways of obtaining a sawtooth wave. Explain how the wave is formed. (13-20, 21)   [4]

52. What determines the fundamental operating frequency of a multivibrator? (13-21)

53. How do multivibrator oscillators differ from a Hartley oscillator? (13-21)

54. Draw a diagram and explain the principles of operation of a multivibrator oscillator. (13-21)

55. What is a *parasitic oscillation?* (13-22)

56. What may be the result of parasitic oscillations? (13-22)

57. How are parasitic oscillations detected and prevented? (13-22)

58. Name four devices that could be used to indicate oscillation in a crystal oscillator. (13-23)

59. Explain some methods of determining if oscillation is occurring in an oscillator circuit. (13-23)

60. What may be the effects of shielding applied to RF inductances? (13-24)

61. Explain some of the factors involved in the stability of a crystal- or LC-tank oscillator. (13-24)

62. What is the meaning of VFO? (13-24)

14
AUDIO-FREQUENCY AMPLIFIERS

The objective of this chapter is to describe the operation of amplifiers that amplify audible-frequency signals. These may be voltage or power amplifiers, solid-state or vacuum-tube, single-ended or push-pull, classes A, AB, or B. The various types of coupling methods and distortions developed in these circuits are discussed. The effects of inverse feedback are described. Special solid-state circuits, such as the Darlington, differential, cascode, complementary symmetry, and operational amplifiers, are explained.

14-1 AUDIO FREQUENCIES

The chapters discussing active devices, vacuum tubes, BJTs, FETs, etc., outlined the basic principles of amplification. Variations of voltage or current from small fractions of a cycle per second up to thousands of millions of hertz can be amplified. For variations below the physiological limit of audibility (about 16 Hz) direct-coupled stages are usually required. Direct, capacitive, or transformer coupling can be used between amplifier stages for frequencies up to 1,000 MHz or so. Above this frequency (microwaves) specialized

equipment is used, to be discussed in later chapters.

The frequencies that can be heard when applied to earphones or loudspeakers are called audible or audio frequencies (AF). These are generally considered to be from 20 to 20,000 Hz. The basic terminology for frequencies above 20 kHz is radio frequency (RF), or supersonic.

If a system can amplify only the frequencies between about 250 and 3,500 Hz, it will do a reasonably good job of transmitting voltages developed in a microphone by the human voice. This is the range of frequencies that telephone companies must amplify. Music signals transmitted through such speech-type systems leave something to be desired. If the amplifiers are engineered to pass all frequencies from 50 to 5,000 Hz equally well, music (and voice) is reproduced quite well. But for true high fidelity (low distortion), audio amplifiers must be able to amplify equally well all frequencies, 20 to 20,000 Hz. In this chapter the fundamentals of amplifiers of only audible-frequency signals will be discussed.

14-2 VOLTAGE VERSUS POWER AMPLIFIERS

The output ac signals from microphones, turntables, detector stages of receivers, etc., may be in the microvolt to millivolt range. To enable such weak signals to drive the input of a power amplifier adequately, one or more voltage, or low-level, amplifiers may be required. Generally, the output-circuit current of such low-level amplifiers may be only one or two milliamperes.

A power amplifier operates into some load that demands more than a few milliwatts of power. The load may be a loudspeaker, a tape or disk recorder, or the modulator circuit of a broadcast transmitter. Power amplifiers range in output from a fraction of a watt to many kilowatts. The higher-power amplifiers may require many watts of power just to drive their input circuits, which then require a power amplifier to drive them.

In general, voltage or low-level amplifiers will be biased to operate in the class A region (discussed later). Audio power amplifiers may also operate as class A stages, or they may be class AB or class B types. (Consider AF amplifiers as never being class C.)

Physically, voltage-amplifying devices are relatively small and operate at very low current and power levels. Power-amplifying devices are relatively larger, may carry amperes of current, and may heat enough to require special cooling systems for the plates or collectors. Many low-level stages may be operated from the same power supply. A power amplifier stage often requires a separate power supply of its own.

14-3 PEAK-CLIPPING DISTORTION

Consider the simple class A transformer-coupled input and output triode AF amplifier in Fig. 14-1. The triode tube has a μ of 30. The E_{bb}

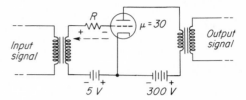

Fig. 14-1 A degenerative voltage develops across any resistance in a grid circuit when I_g flows through it.

is 300 V. Since there will be negligible voltage-drop in the primary of the output transformer, the E_p also will be 300 V. From the cutoff-bias formula (Sec. 9-18)

$$E_{co} = \frac{E_p}{\mu} = \frac{300}{30} = 10 \text{ V}$$

The $E_g I_p$ curve of this triode is shown in Fig. 14-2. With -10 V on the grid the I_p is at essen-

Fig. 14-2 Flattening of plate-current pulse peaks due to overdriving.

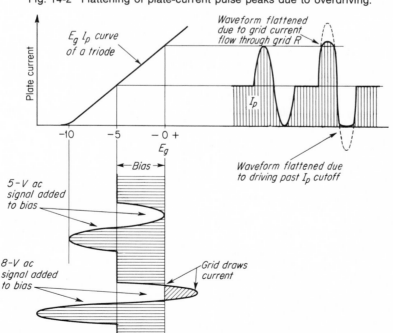

tially zero. If a bias voltage of half of the cutoff value (-5 V) is used, a medium I_p value is developed. Now, if a 5-V peak ac signal is added to the -5-V bias, the acting grid circuit voltage varies from -5 to 0 to -10 and back to -5 V. The resulting I_p varies from a medium to a high to zero and back to the medium value. Since the waveform of the input grid signal and the output plate current are the same, it would be assumed that the amplifier is not distorting the signal. Since the grid is never being driven positive, it would also be assumed that the grid will not pick up electrons from the cathode and no grid current will flow through the resistor R that has been added in this particular circuit.

If an 8-V peak ac signal is fed to the grid, during part of the negative half cycle (-5 to -8 V) the plate current is completely cut off, resulting in serious distortion to the peak of this half cycle.

During the positive half cycle, the grid is driven into the positive region and collects electrons from the space charge, and grid current flows backward, through the bias battery. When grid current flows through any resistance in the grid circuit, it develops a voltage-drop across the resistor. Because of the direction of current flow, the grid end of the resistor is negative and the transformer end is positive. Thus, as the grid is driven positive by the transformer (which would tend to increase plate current), a negative voltage-drop is added in series with the grid circuit (which would tend to decrease plate current). As a result, the net grid voltage never reaches the 8-V peak value but remains at essentially 5 V. This produces a plate-current waveshape with a flattened or clipped positive (and negative) peak, as illustrated.

This explanation has neglected two facts. (1) Varying plate current in the load impedance produces a varying voltage-drop across the plate load, lowering the voltage on the plate during the positive half cycle and raising it during the negative. Raising the plate voltage results in a greater cutoff voltage value. With an E_{co} of 10 V an amplifier would use a bias of $0.66E_{co}$ and could accommodate a greater input peak signal. (2) The contact potential on a VT grid approximates 1.5 V. Thus, when the grid is driven within 1.5 V of the 0 point, it begins to pick up grid current, resulting in positive-peak distortion if there is any resist-

ance in the grid circuit. Not only must a tube be biased correctly, but also the input ac amplitude must be held within certain limits, or serious distortion of the input waveshape will be present in the output circuit. This is true of transistors also.

In some cases it may be possible to change circuit voltages to enable a circuit to handle input signals which might otherwise produce distortion. For example, overdriving the grid produces positive and negative peak clipping. If the plate voltage is increased, the cutoff point will be increased. It will now require a greater value of negative grid voltage to reach the proper class A bias value, but the input voltage peaks may not fall into the cutoff region or into the positive grid region and the tube will operate without distorting. However, there are practical limits on plate voltages.

If a given signal voltage is overdriving a tube, it may be possible to use a lower-μ tube. Low-μ tubes require higher bias voltages to produce plate-current cutoff, thereby making it possible for them to accommodate greater grid-voltage amplitudes.

14-4 μ VERSUS STAGE GAIN

The μ, or amplification factor, of a triode tube (β of a transistor) is built into the device by the manufacturer. In general, the voltage amplification of a triode using a resistor as the plate load depends on the μ of the tube. It usually varies from about one-half to two-thirds of the μ value. With a μ of 30 a voltage amplification of gain between 15 and 20 times can be expected.

The amount an input ac voltage is amplified in passing through an amplifier stage is known as the *stage gain* (A). Low-μ tubes produce low stage gains, while high-μ tubes produce higher stage gains. The μ of a tube cannot be changed, but the stage gain can be varied within limits.

One factor in the voltage gain of a triode stage is the plate supply voltage. The higher the E_{bb} the greater the possible gain. For example, a tube with a μ of 20 may have a stage gain of 12 with 90 V E_{bb} but a gain of 14 if the voltage is raised to 300 V.

Another factor is the plate-load resistance. A tube may have a gain of 11 with a plate-load resistance of 50,000 Ω but a gain of 14 with

500,000 Ω. The higher the R_L the greater the stage gain.

In an amplifier stage the resistance of the tube and the load resistance are in series across the power supply. The ratio of the load resistance to the total resistance of the tube and load, when multiplied by the μ of the tube, gives the stage gain. In formula form,

$$A = \frac{\mu R_L}{R_L + R_p}$$

where A = stage gain
R_p = plate resistance of tube
R_L = resistance of load
μ = amplification factor of tube

EXAMPLE: An amplifier has a μ of 30, an R_p of 5,000 Ω, and an R_L of 10,000 Ω. The stage gain is

$$A = \frac{\mu R_L}{R_L + R_p}$$

$$= \frac{30(10,000)}{10,000 + 5,000} = \frac{300,000}{15,000} = \text{gain of 20}$$

If a step-up coupling transformer is used in an amplifier stage, the stage gain times the step-up ratio may exceed the μ of the tube.

Note that each time the μ of a tube was mentioned the tube was specified as a triode. Tetrode or pentode tubes are also used as audio amplifiers, but the values of screen- or suppressor-grid voltages will affect the actual amplification of the tube. For these tubes it is best to refer to tube-characteristics charts for resistances and capacitances to be used in circuits. The amplification of pentodes can be several hundred, making them highly desirable in some applications.

14-5 TYPES OF COUPLING

Four types of coupling circuits are used between AF stages:

1. Transformer, or inductive, coupling
2. Resistance, or capacitive, coupling
3. Impedance (another form of capacitive) coupling
4. Direct coupling

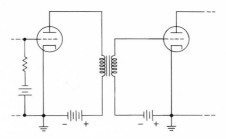

Fig. 14-3 Basic transformer, or inductive, coupling.

14-6 TRANSFORMER COUPLING

Transformer, or inductive, coupling between two tubes is diagramed in Fig. 14-3. With inductive coupling, the output voltage of the first stage can be stepped up by the use of a step-up ratio transformer, and additional gain per stage is possible. The low resistance of the primary of a transformer results in a plate voltage almost equal to the supply voltage, an advantage in power amplifiers.

Transformer coupling is more costly than other forms, will weigh more, will be more bulky, and, unless well shielded, will be prone to pick up hum, or unwanted signal voltages, from nearby wiring. Less expensive transformers often introduce distortion into the signal.

If the signal voltage in an input transformer is amplified by one or two other stages and is inadvertently coupled back into the transformer by nearby fields, it may be induced in a degenerative phase and cause a weakening of the output or be induced in a regenerative phase and cause the stages to break into self-oscillation at some audible or supersonic frequency. The stages may *motorboat* (oscillate at a few hertz rate). To prevent induction due to stray fields, AF transformers are encased in iron shields. Magnetic fields from nearby wires remain in the iron shields rather than pass through them and across the turns inside.

Some transformers may also have *electrostatic* (Faraday) shielding between primary and secondary. This may take the form of a single layer of insulated wire or a single turn of sheet brass (insulated to prevent forming a shorted turn) placed between primary and secondary windings. The shield is grounded or connected to the core. An electrostatic shield reduces capacitive coupling between primary and secondary coils. Since high frequencies are passed better by a capacitance than low frequencies are, any capacitive feed increases the transfer of higher frequencies into the secondary and also produces a difference in phase between the capacitive and inductive coupling. With this shield, any signal attempting to get from primary to secondary by capacitance effects encounters the shield and is led to ground. The only signal induced into the secondary is that produced by magnetic induction. Magnetic fields are not affected by the nonferrous electrostatic shield.

Transformer coupling can be used in either voltage or power amplifiers.

In Sec. 9-22 it was pointed out that *maximum power output* is produced when the impedance of the load matches the plate impedance of the tube. With triode tubes, slightly less output power but also less distortion is produced by increasing the load impedance until it is two or three times the plate impedance. This produces *maximum undistorted power output*.

Power output of a triode can be determined by laying out a load line on a family of E_g curves on an $E_p I_p$ graph (Sec. 9-18) and then applying the formula

$$P_o = \frac{(dI_L)(dE_L)}{8}$$

where dI_L = change in amperes in the load
dE_L = change in voltage across the load

The no-signal $E_p I_p$ product must not exceed the plate dissipation rating of the tube or tube life will be shortened.

Tetrode and pentode tubes have very high plate impedances. The primary impedance of a coupling transformer for these tubes is usually between one-fifth and one-tenth of the plate impedance. Transformers with primary impedances of much more than about 20,000 Ω are difficult to produce. Transistors, with their low impedances, make transformer coupling quite practical.

Matching impedances is important only in *power* amplifiers. Resistance-coupled amplifiers produce essentially undistorted output-signal *voltages* regardless of the plate-load-resistance value.

The general rule that dc must not be fed to the primary of a transformer does not hold true with transformers used as loads for tubes or transistors. The resistance of active devices in series with a transformer across a dc power supply limits the current to a safe value for the primary wires. Dc flowing through the primary may tend to magnetize the core, although the results of such magnetization may not always be significant. A form of inductive coupling that prevents dc from flowing through the primary of the coupling transformer and saturating the core is shown in Fig. 14-4. Isolating the transformer from the dc component of the plate current in this manner can result in less distortion of the signal.

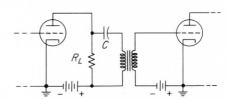

Fig. 14-4 Inductive coupling with no plate current flowing in the transformer.

14-7 RESISTANCE COUPLING

A common method of coupling signals from the output circuit of one voltage amplifier to the input of the next is by *resistance coupling* as shown in Fig. 14-5.

In Fig. 14-5a, changes of grid voltage on the first tube produce a varying plate current through the plate load resistor R_L. Varying currents through this resistor produce varying voltage-drops across it. The grid resistor R_g of the next stage is dynamically connected across R_L through two capacitors, the coupling capacitor C_c, at the top, and the bypass capacitor C_b, at the bottom. These two capacitors are large enough in value to have a low reactance to the lowest audio frequencies to be used in the circuit. The grid resistor R_g is chosen to have at least twice the resistance of the plate load R_L. Varying dc voltages produced across the plate-load resistor appear as similar-frequency ac voltages across the grid resistor. If R_g has a resistance value at least 10 times the reactance of C_c at the lowest frequency to be amplified, practically all the signal voltage will be developed across the grid resistor. (Actually, R_L

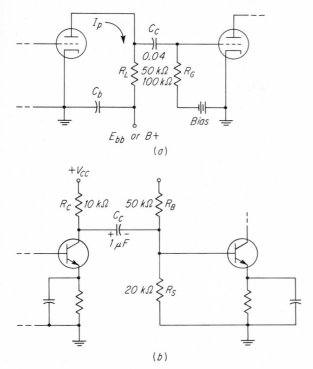

Fig. 14-5 Resistance coupling between (a) VTs and (b) BJTs.

in parallel with R_g is the ac plate-load impedance.) Any voltages across R_g are in series with the grid circuit and will cause the plate current of this tube to vary.

The bypass capacitor C_b may not be included in some circuits. Instead, the output-filter capacitor of the power supply may do the bypassing.

To keep the stage gain high, the value of the plate-load resistor is made several times the impedance of the tube. The higher the plate-load and grid resistances the smaller the value of coupling capacitor needed. Two sets of possible component values for a resistance-coupled circuit using the same plate supply voltage and a tube with a μ of 20 are:

$R_L = 50,000\ \Omega$	$R_L = 250,000\ \Omega$
$R_g = 100,000\ \Omega$	$R_g = 500,000\ \Omega$
$C_c = 0.04\ \mu F$	$C_c = 0.008\ \mu F$
Stage gain $= 11$	Stage gain $= 14$

Figure 14-5b shows resistance coupling between two transistors. Note the increased capaci-

tance used in C_c in low-impedance circuits. Whereas VT amplifiers may use paper- or plastic-dielectric capacitors, transistor circuits use electrolytics and polarity must be observed.

Some of the important points regarding resistance coupling are: The input resistance to the next stage should be at least twice the output load resistance. The larger the coupling capacitor, the better the coupling of low-frequency signals to the next grid. Too much coupling capacitance may cause cascaded stages (several stages, one after another) to break into self-oscillation. Resistance coupling in vacuum tubes is considered a method of coupling voltages, not power. The larger the value of the plate-load resistance, the greater the stage gain. The higher the supply voltage, the greater the stage gain. The higher the grid resistance, the more grid-cathode interelectrode capacitance (and Miller effect, Sec. 14-19) bypasses higher-frequency signals.

14-8 IMPEDANCE COUPLING

Impedance coupling is somewhat similar to resistance coupling, using an iron-core AF choke coil in place of the plate resistor as in Fig. 14-6.

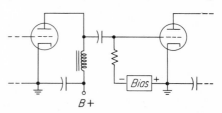

Fig. 14-6 Impedance coupling between two tubes.

With a plate-circuit choke as the impedance, the plate voltage on the tube will be nearly the full voltage of the power supply instead of less than half the power-supply voltage, as is usual with resistance coupling. This results in considerably greater output signal voltages. However, the choke may not present the same impedance to all audio frequencies, thereby introducing a form of frequency distortion not as appreciable in resistance coupling.

Choke coils may pick up stray fields and hum voltages, as do transformers. It is possible to use impedance coupling in either voltage or power amplifiers when a high-impedance output circuit is desired.

1. What are the upper and lower limits required to transmit speech? _____ High-fidelity music? _____
2. Would an amplifier be a voltage or a power amplifier if it drives a pair of earphones? _____ If it drives a class A amplifier? _____
3. A triode has $E_p = 200$ V, $\mu = 50$. What is the bias value for I_p cutoff? _____ For class A? _____
4. What produces peak clipping in a class A amplifier? _____
5. When is resistance in an AF grid circuit undesirable? _____
6. With a 100-μ triode, about what gain can be expected from a resistance-coupled amplifier? _____
7. Contact potential can produce about how much bias? _____
8. What is the gain of a triode stage if $\mu = 60$, $R_p = 50$ kΩ, and $R_L = 100$ kΩ? _____ If $R_L = 50$ kΩ? _____
9. What are two advantages of inductive coupling? _____ _____
10. Is motorboating a high- or low-frequency oscillation? _____
11. Why are AF transformers shielded? _____
12. In the formula $P_o = (dI_L)(dE_L)/8$, should the values be rms, peak-to-peak, peak, average, or any of these? _____
13. What is gained by matching load to device impedance? _____ What is the disadvantage in AF amplifiers? _____
14. In RC coupling, what should be the ratio of R_L to the resistance in the input circuit of the next device? _____ The ratio of coupling capacitance X_C to input circuit R? _____
15. What type of device might be used as the impedance in RF amplifier impedance-coupling circuits? _____
16. What are the advantages of using impedance coupling over RC coupling in AF stages? _____ _____ Disadvantages? _____ _____

14-9 DIRECT COUPLING

The capacitors used as the coupling element in resistance and impedance coupling and the transformers used in inductive coupling may not pass frequencies below 50 Hz too well. To improve low-frequency response, a direct-coupled *Loftin-White* circuit can be used. A simplified version is shown in Fig. 14-7.

Plate current flowing out of the negative terminal of the power supply divides, the major portion flowing through resistor R and a little flowing through the first tube and resistor R_L. These two

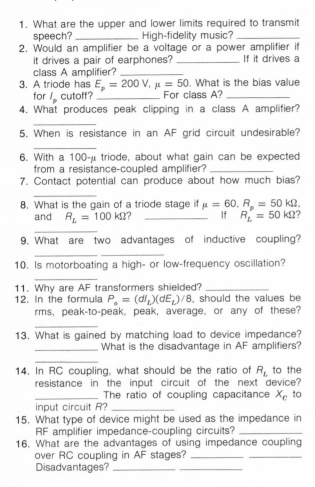

Fig. 14-7 Direct coupling between two tubes.

currents join and as one current flow through the second tube, through the primary of the output transformer, and to the power supply.

The plate current of the first tube, flowing through R_L, produces a voltage-drop across it. Since the negative end of the resistor is at the top and the positive is at the bottom, this potential acts as bias.

When a signal voltage is fed to the grid of the first tube, the plate current varies, producing a varying potential across the plate-load resistor. This varying dc potential is directly coupled to the grid-cathode circuit of the second tube. The varying potential in its grid circuit causes the plate current of the second tube to vary. Varying current in the primary of the output transformer produces ac in the secondary.

The capacitor C_1 is large enough to hold the voltage across R constant. The capacitor C_2 is large and tends to hold the voltage constant from the cathode of the second tube to B+. Effectively, there are two power supplies in series. The one actual power supply must supply enough voltage for the two stages in series. With two separate power supplies, or by using a zener diode in place of R, this circuit is capable of coupling between the two stages with little distortion over the audio spectrum and down to dc. Thus, such a circuit is also called a *dc amplifier*.

14-10 TYPES OF BIAS USED IN AUDIO AMPLIFIERS

There are several methods of obtaining bias voltage for VT AF amplifier stages. The following terminology will be used: (1) battery bias, (2) power-supply bias, (3) voltage-divider bias, (4) cathode-resistor bias, (5) contact-potential bias. (Grid-leak bias is not used in AF amplifiers.)

14-11 BATTERY BIAS

The use of dry-cell C batteries for biasing vacuum-tube AF amplifiers is usually limited to portable equipment. Wet, or storage, cells were sometimes used in the past. (See Figs. 14-4, 14-5.)

14-12 POWER-SUPPLY BIAS

In high-powered transmitting audio equipment electronic power supplies are usually used to supply the relatively high bias voltages that may be required by the high-powered tubes employed.

14-13 VOLTAGE-DIVIDER BIAS

It is possible to connect a tapped resistor across a power supply, as in Fig. 14-8, and obtain

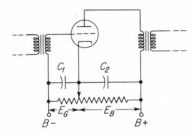

Fig. 14-8   Voltage-divider bias.

both plate-potential and grid-bias voltage from the same supply. As far as the right-hand, or B+, end of the voltage-dividing resistor is concerned, the cathode is more negative. The return of the grid circuit is to a still more negative point. The amount by which the grid is more negative than the cathode is the bias value. It can be adjusted to the desired value by moving the tap up or down the resistor. The greater the grid-bias voltage, the less the plate supply voltage. The capacitors C_1 and C_2 are filter capacitors that hold the plate supply and the bias voltages constant when signals produce a varying plate current.

14-14 CATHODE-RESISTOR BIAS

The most common type of bias in audio amplifiers is *cathode-resistor, automatic,* or *self-bias,* shown in Fig. 14-9. This is a form of voltage-divider bias. The cathode resistor, the tube, and the plate load are in series across the power sup-

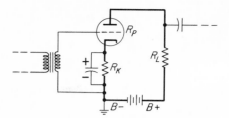

Fig. 14-9   Cathode-resistor or self-bias.

ply. Starting from the positive end of the power supply, there is a voltage-drop across the load resistance, another across the tube, and another across the cathode resistor. The sum of these voltage-drops equals the supply voltage. Since the cathode is connected part way up this circuit toward the positive, it is more positive than the grid, which is returned to the bottom of R_k. The difference of potential across R_k is the bias voltage.

When a signal ac is applied between grid and cathode, the plate (and cathode) current varies, producing a varying voltage-drop across the biasing resistor. When the grid is driven negative by a signal voltage, the plate current decreases. This decreases the voltage-drop across the cathode resistor and produces less negative bias. Therefore, as the grid signal goes more negative, the grid bias becomes less negative. Actually, the bias voltage does not drop as much as the signal increases, so that the tube does function as an amplifier but at low gain.

If a large value of capacitance is placed across the cathode resistor, the bias voltage is held constant, resulting in full output from the tube. Resistor R_k tends to discharge the capacitor, The lower the resistance value, the larger the capacitance required to maintain a nearly constant voltage. If the capacitor is not large enough, the resistance will almost completely discharge it when low-frequency signals are being amplified but may be able to maintain the voltage relatively constant for high-frequency signals. This action will result in a loss of amplification (degeneration) of the low frequencies but full amplification of the high. The biasing resistance can be computed by Ohm's law, $R = E/I$.

EXAMPLE: A vacuum-tube handbook gives this information on a triode tube: plate voltage =

250 V, plate current = 9 mA, bias voltage = −8 V, μ = 20. What is the value of cathode resistance to produce the required −8 V bias?

$$R = \frac{E}{I} = \frac{8}{0.009} = 889\ \Omega$$

where R = cathode resistance
E = bias voltage
I = cathode current in *amperes*

In practice any resistor between 820 and 910 Ω would be satisfactory. Note that this method of determining the bias resistance is used only in transformer or impedance coupling. For pentodes and tetrodes the screen current plus the plate current flows through the biasing resistor. Use the sum of these currents as the cathode current when computing the cathode-bias-resistor value.

For component values in resistance-coupled stages refer to resistance-coupled amplifier tables in vacuum-tube handbooks. If no chart is available, a usable approximation for the cathode resistor of a triode is

$$R_k = \frac{R_L}{\mu}$$

where R_L = plate-load resistor in parallel with grid resistor of next stage
μ = amplification factor of tube

Determining the value of the capacitor to connect across the cathode resistor for reasonable response down to 100 Hz is rather complex. A rough approximation for this capacitor can be obtained by using two times the RC time-constant formula

$$T = 2RC \qquad \text{or} \qquad \frac{T}{2R_k} = C_k$$

where C_k = capacitance in farads (F)
T = duration of the lowest frequency cycle, seconds, or $1/f$
R = resistance in Ω

For a 500-Ω cathode resistor the capacitor might be

$$C_k = \frac{T}{2R_k} = \frac{0.01}{2(500)} = 10\ \mu\text{F}$$

For operation down to 50 Hz, twice this capacitance would be used, or 20 μF. For operation down to 500 Hz, a 2-μF capacitor should suffice.

14-15 CONTACT-POTENTIAL BIAS

This type of bias is used only with high-μ tubes operating with very small input-signal voltages. In Fig. 14-10 there is apparently no bias on the grid.

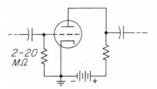

Fig. 14-10 Contact-potential bias with high-μ triode.

If the tube has a high μ (70 or more) and the resistance grid-to-cathode is several megohms, a negative voltage of 1 to 2 V will be produced by electrons from the cathode striking the closely spaced grid wires of the high-μ tube. These electrons are unable to leak back to the cathode through the high resistance fast enough to hold the grid at zero potential. This piling up of electrons on the grid forms the negative *contact-potential* bias voltage. (The diagram resembles grid-leak biasing used in oscillators and RF amplifiers, but the theory of operation is not the same.)

Contact-potential or grid-resistor bias is used only in the first stage of audio amplifiers, following microphones or detector circuits. It can decrease the hum that is sometimes developed when cathode-resistor bias is employed with indirectly heated tubes using ac as the heater current. The beginner should not use this form of biasing when drawing AF amplifier diagrams, as its sim-

---

ANSWERS TO CHECKUP QUIZ ON PAGE 296

1. (*250 and 3,500 Hz*)(*20 and 20,000 Hz*)   2. (*P*)(*E*)
3. (*4 V*)(*2.64 V*)   4. (*Overdriving input*)   5. (*When I_g flows*)
6. (*50–70*)   7. (*1.5 V*)   8. (*40*)(*30*)   9. (*Increased gain, couple power*)   10. (*Low*)   11. (*Prevent signal induction*)
12. (*Any, as long as both are the same*)   13. (*Maximum P_o*)(*Distorts at high gain*)   14. (*1:2*)(*10:1*)   15. (*RFC*)
16. (*Added gain, delivers power*)(*Costly, signal induction*)

plicity tempts him to use it where it may not be applicable.

14-16 BIASING FILAMENT TUBES

Thus far, only tubes with indirectly heated cathodes have been considered. When bias is applied to directly heated (filament-type) tubes using ac as the heater current, the center tap of the filament circuit must be used as the plate and grid return point. Figure 14-11 shows two methods of

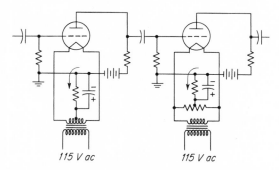

Fig. 14-11 Two methods of returning grid and plate circuits to the center tap of the filament for cathode-resistor biasing.

returning to the center of the filament circuit. Either the center tap of the filament transformer or the center tap of a resistor across the filament circuit may be used. (The center-tapped resistor adds one-quarter of its total value as additional cathode-resistance bias.)

The voltage developed across a resistor in series with the power supply can be filtered with a capacitor and be used as a bias voltage, as in Fig. 14-12.

When dc is used to heat a filament-type tube, the filament circuit is not center-tapped.

Fig. 14-12 Bias for a dc-heated filament-type tube.

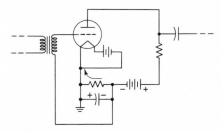

14-17 CONTROLLING VOLUME

The usual method of controlling volume in AF amplifiers is to use a voltage divider in the grid circuit. This takes the form of a potentiometer as in Fig. 14-13. In resistance-coupled circuits the

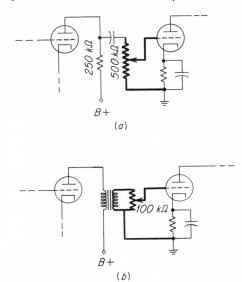

Fig. 14-13 Volume controls for (a) resistance-coupled stages and (b) transformer-coupled stages.

potentiometer in the grid circuit should have twice the plate-load resistance of the tube ahead. When the sliding contact is at the top of the potentiometer, the grid receives full output voltage from the first stage. When halfway down the potentiometer, it receives half voltage, and at the bottom, no signal.

The potentiometer across the secondary of the coupling transformer should have a resistance high enough to reflect the proper impedance back to the primary. It may range from 1,000 to 500,000 Ω.

Potentiometers have different resistance tapers. A *liner-taper* potentiometer has the same resistance change for a given rotation angle of the arm regardless of whether the arm is at the lower end, the middle, or the top end. It will be found, however, that when starting from zero signal and increasing the control slightly, the audio signal will rapidly increase at first. Above the midpoint there seems to be relatively little increase in volume. To overcome this undesirable feature, potentiometers are available with logarithmic *audio tapers*. A small angle of rotation at the low end of the potentiometer moves the arm over a relatively small resistance change. As the arm is continued in rotation, more and more resistance change occurs for the same angle of rotation. This results in a volume control that has a less critical adjustment at the low-volume end.

Varying the filament voltage, screen voltage, plate voltage, or bias voltage is not a satisfactory means of controlling volume in AF amplifiers. However, bias- and screen-voltage variations may be used to control the gain of *radio*-frequency (RF) amplifiers.

In transistor circuits the base connection to the biasing resistors must not be varied or the bias value would change. For this reason the volume-control potentiometer is placed on the other side of the coupling capacitor, as in Fig. 14-14.

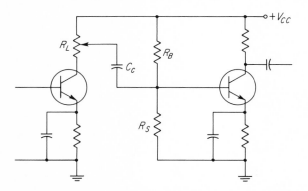

Fig. 14-14 A common volume control used in transistor AF amplifiers.

14-18 TONE CONTROLS

Theoretically, an AF amplifier should have high fidelity. It should amplify all frequencies equally well. It should amplify 50 Hz just as much as it amplifies 500 or 15,000 Hz. When correctly designed, modern amplifiers actually have such a response. They are said to have a *linear* (straight-line), or *flat,* response. If any capacitance is connected across the input or output circuits, however, the reactance of such a capacitance will form a path to ground for the signal voltage. Since the reactance of a capacitor varies inversely as the frequency, the same capacitor will bypass the high frequencies to ground more than it will the low frequencies. The larger the capacitance across a circuit, the more it will attenuate high frequencies.

The higher the impedance of the circuit across which the capacitor is placed, the more effect it will have on the loss of high frequencies. For example, a 0.0002-μF capacitor across a 100,000-Ω circuit will effectively decrease the higher audio frequencies. It may take a 0.001-μF capacitor to produce the same effect across a 20,000-Ω circuit.

An amplifier in which the high frequencies are attenuated will sound to the listener as though the low frequencies are being highly amplified. Many people prefer an increased *bass* (low-frequency) response.

Random noises in amplifiers originate from thermal agitation of electrons in the first-amplifier-stage wires, in the uneven streams of electrons from cathode to plate in tubes, or in electron-hole recombinations in transistors. These and other noises contain more energy at frequencies above about 5 kHz. A capacitor shunted across an amplifier stage can materially reduce such noises.

While a variable capacitor across the grid-cathode circuit would effectively decrease treble (high-frequency) response, it is not very practical. Possible treble-attenuating (high-frequency-reducing) tone controls are shown in Fig. 14-15.

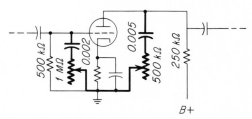

Fig. 14-15 Simple shunt-capacitance tone controls for VTs or JFETs. Multiply C and divide R values by about 100 for BJTs.

With maximum resistance in series with the capacitor, there is practically no capacitive effect. With zero resistance, the capacitor is directly across the circuit, attenuating the high frequencies. Since the plate-circuit impedances are less than grid, capacitance values in this circuit must be greater.

The dual tone controls of Fig. 14-16 are capable of boosting or attenuating either the bass or

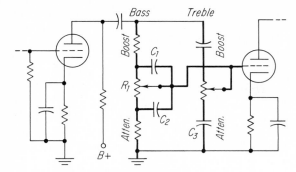

Fig. 14-16 A bass-treble tone-control circuit.

the treble to the grid of the second tube. When the arm of the bass control is at the bottom of R_1 (attenuate), high frequencies are coupled to the grid by C_1. With the arm up (bass boost), C_2 shunts the high frequencies. When the treble-control arm is down, the grid is shunted by C_3, which attenuates the high frequencies at the grid. With the arm up, C_3 is not as effective, and maximum high frequencies are coupled to the grid.

Low- or high-pass LC filters are often used in amplifiers. They are not adjustable but have sharp cutoff characteristics.

14-19 MILLER EFFECT

When a high-gain resistance-coupled triode amplifier is constructed, it is sometimes found that the high-frequency response is poor, even with no capacitors across the grid-cathode circuit. This may be due to *Miller effect,* an electronic capacitance between grid and cathode in a tube. The value of this capacitive effect is approximately

$$C_m = C_{gk} + (A)C_{gp}$$

where C_m = effective grid-to-ground capacitance in pF
C_{gk} = grid-cathode tube capacitance in pF
A = amplification of stage
C_{gp} = grid-plate tube capacitance in pF

As an example of the effective capacitance across the input circuit of a tube in which C_{gk} is 4.2 pF, A is 12, and C_{gp} is 3.8 pF.

$$C_m = C_{gk} + (A)C_{gp}$$
$$= 4.2 + (12)3.8 = 49.8 \text{ pF}$$

The tube itself may have only 4.2 pF grid-cathode interelectrode capacitance, but when it is operated in a circuit, it may operate as though more than 50 pF is across the grid circuit. If the grid circuit has a high impedance, this effective capacitance provides a low-reactance path for high-frequency signals to ground.

Miller effect varies with the stage gain. High-amplification stages produce high effective capacitive effects, while low-gain stages have proportionately less shunting capacitance. This is one reason why high-fidelity VT amplifiers may use lower-μ triodes. With their low grid-plate capacitance, pentode tubes have little Miller effect even with relatively high stage gain. Inverse feedback (Sec. 14-36) decreases Miller effect.

Miller effect is also present in common-emitter transistor circuits and is computed the same way, substituting C_{EB} for C_{gk}, C_{BC} for C_{gp}, and A_v for A. Since the gain of transistors is so very dependent on voltages, currents, etc., Miller effect can become a problem, particularly in circuits which must be held to a certain frequency.

Test your understanding; answer these checkup questions.

1. What is a Loftin-White circuit? _____
2. What is another name for cathode-resistor bias? _____
3. A power-amplifier tube requires -7-V bias, $I_p = 50$ mA, $I_{sg} = 10$ mA. What value should the cathode resistor be? _____ What power rating? _____
4. What value of R_k might be used with a triode having a μ of 100 and a 250-kΩ R_L? _____
5. For operation down to 100 Hz what value C_k might be used in question 3? _____ In question 4?
6. What is another name for contact-potential bias? _____ What value resistor is used? _____
7. What results if an ac filament circuit is not center-tapped? _____ Must VT heaters be center-tapped?
8. What taper potentiometer is recommended for AF amplifiers? _____ Would other tapers work?
9. What type of amplifiers vary bias or screen voltage to control gain? _____
10. Why are gain controls not included in the base circuit of BJT AF amplifiers? _____
11. Why might a 5-kHz low-pass filter be an advantage on AF amplifiers or receivers? _____
12. Could the bass-treble tone control shown in Fig. 14-16

be used with JFETs? _____ What might have to be done to it to have it work with BJTs? _____

13. Why might high-μ tubes have high values of Miller effect? _____

14. What effect does degeneration have on Miller effect? _____

14-20 DECOUPLING, OR FILTERING, AF STAGES

The phase of any signal passing through a normal amplifier stage is changed 180°. As the grid of one tube in an amplifier is becoming less negative, the grid of the next tube is becoming more negative. This may be seen by following the step-by-step action of the circuit in Fig. 14-17.

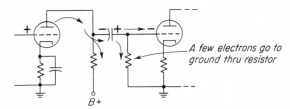

A few electrons go to ground thru resistor

Fig. 14-17 As the first grid goes positive, the second is driven negative.

1. Grid 1 goes less negative (more positive).
2. The plate current in this tube increases.
3. The voltage-drop across the plate-load resistance increases because the resistance of the tube has decreased.
4. The top of the plate-load resistor is now closer to ground potential, or more negative than it was previously.
5. The plate side of the coupling capacitor is now more negative.
6. When the left side of the capacitor becomes more negative, it drives electrons out of the right side onto the grid of tube 2.
7. Electrons forced onto the grid charge it negatively.

The plate currents of two tubes in cascade are also 180° out of phase. As the first current increases, the other decreases. Any coupling, either capacitive or inductive, between grids of two adjacent stages, or between plates of two adjacent stages, will tend to cancel each other. This decreases the overall amplification of the cascade

amplifier but also decreases any tendency for the stage to oscillate. The latter is usually desirable. On the other hand, any feedback from the third- to the first-stage plate circuits in a cascade amplifier is twice 180°, or 360°, and is in phase, or *regenerative*. The stages may form an oscillatory circuit.

The *degenerative,* or out-of-phase, type of feedback is usually not bothersome. In fact, many high-fidelity amplifiers incorporate circuits utilizing some *inverse feedback* to stabilize the circuits (prevent self-oscillations) and to decrease certain types of distortion.

The regenerative type of feedback is rarely desirable in audio amplifiers, since it increases noise in the circuit, increases distortion, and may cause erratic operation of the stages or produce oscillations. To decrease this effect, it has been found that decoupling the plate load of each stage using a common power supply with poor voltage regulation will prevent current variations in one stage from producing plate-voltage variations in other tubes.

Figure 14-18 shows an AF amplifier stage em-

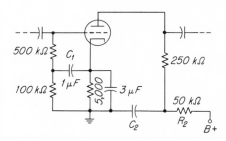

Fig. 14-18 Grid-circuit and plate-circuit decoupling.

ploying both plate and grid RC decoupling, or filtering. The plate and grid circuits are decoupled from the power supply and from other stages insofar as audio voltage variations are concerned.

The long-time-constant circuit, R_2 and C_2, holds the bottom of the 250,000-Ω plate-load resistor at a constant voltage above ground, regardless of any instantaneous-voltage variations of the power supply.

The 100-kΩ grid decoupling resistor and C_1 form an RC grid-circuit filter. The 5,000-Ω cathode resistor is partially filtered by the 3-μF capacitor across it. Any voltage variations across the cathode resistor will be smoothed out by the action of C_1 and the 100-kΩ resistor because of their long time constant. The voltage across C_1 will vary very little even with considerable variation of plate current through the cathode resistor.

Filter circuits decrease hum if it is due to an inadequately filtered power supply. Output circuit decoupling is used in at least the first stage of practically all audio amplifiers. Such decoupling is very important in many transistor circuits.

14-21 CLASSES OF AMPLIFIERS

The final amplifier in a series of audio-amplifier stages is normally a *power amplifier*. It is so called because its output is used to do some form of work, such as operating a loudspeaker or modulating a transmitter.

A convenient method of partially describing the many conditions under which vacuum tubes can be used as AF amplifiers is to classify them by their approximate bias values, as in Table 14-1.

Table 14-1

Class	Approximate bias used	Number of tubes	Circuit
A	0.5–0.66 × cutoff	1 or 2	Single-ended, parallel, or push-pull
AB_1	0.85 × cutoff	2	Push-pull
AB_2	0.90 × cutoff	2	Push-pull
B	1.00 × cutoff	2	Push-pull

The cutoff bias value is computed by the formula $E_{co} = E_p/\mu$. (Parallel and push-pull operations are discussed in Secs. 14-23 and 14-34.)

The bias values given are only approximate. There is considerable latitude in the selection of the bias value for an amplifier. There are other factors in determining the class in which the tube is operating. If no grid current is drawn during any portion of the input signal, the tube is considered as operating as class A, AB_1, or B_1. If the grid is driven positive so that grid current flows during a portion of the input cycle, the tube is operating either as a class AB_2 or as a class B (or B_2) amplifier.

Class C is used in RF power amplifiers and oscillators. Its bias value is $1\frac{1}{2}$ to 4 times cutoff.

During more than half of its input cycle no plate current flows, which would result in distortion in audio applications.

In AF amplifiers, BJTs must be forward-biased, resulting in base current in all classes (except class C). Table 14-2 lists the various

Table 14-2

Class	Biased to	Number of transistors required	Circuit
A	Medium I_C	1 or 2	Single-ended, parallel, or push-pull
AB	Medium-low I_C	2	Push-pull
B	Low I_C	2	Push-pull

classes for transistors in AF circuits by the relative collector current that flows under no-signal input conditions. This listing is also valid for vacuum tubes if I_p is substituted for I_C.

14-22 CLASS A AUDIO AMPLIFIERS

Vacuum-tube voltage amplifiers are always biased to class A. Bias can be obtained by any of the methods mentioned previously. Theoretically, in power amplifiers the plate-circuit efficiency can be 50%, but they are usually operated at about 20%. If the plate circuit draws 10 W from its power supply, only about 2 W of the undistorted audio power can be produced. If the stage is resistance-coupled, power output is extremely low and is not considered.

Figure 14-19 shows a class A resistance-coupled voltage amplifier driving a class A

Fig. 14-19 Class A triode, with plate decoupling, and a class A power amplifier with current feedback.

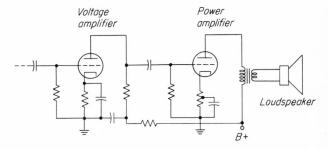

power amplifier, inductively coupled to a loud-speaker. A low-voltage AF signal fed through the coupling capacitor to the grid of the first tube is amplified and appears across the grid circuit of the second tube. The ac voltage on the grid of this power amplifier produces variation of a relatively high-amplitude plate current in the primary of the output transformer. The heavy varying current produces a strong varying magnetic field, which transfers considerable ac energy into the secondary. This causes the loudspeaker diaphragm to vibrate, producing airwaves, or sound. Note plate-current decoupling and current feedback (Sec. 14-36).

Class A power amplifiers may use a single tube or two tubes in either parallel or push-pull. All other classes of audio power amplifiers must use two tubes in push-pull.

Class A stages should have no grid-current flow. Therefore, the stages ahead of them can be voltage amplifiers. If a class A stage draws grid current, it indicates insufficient grid bias, excessive signal voltage, or A_2 operation (which requires a low-impedance input circuit). If a power-amplifier stage draws any grid current, the stage driving it must also be a power amplifier.

A class A VT stage is limited to a peak signal input approximately equal to the bias value. It must operate on the straight portion of the E_gI_p curve. This results in low plate-circuit efficiency but little distortion. Incorrect bias values result in serious output distortion.

To minimize distortion the output transformer for a class A stage should have a primary impedance two or three times the plate impedance of the output tube. It must have enough iron in the core that dc plate current flowing through the primary will not saturate the core. Magnetic saturation of the core prevents the voltage induced into the secondary from having the same waveform as the current flowing in the primary. This produces distortion in the output.

ANSWERS TO CHECKUP QUIZ ON PAGE 301

1. (*Direct-coupled*)   2. (*Self*)   3. (*120 Ω*)(*Twice 0.43 or 1 W*)   4. (*2.5 kΩ*)   5. (*40 μF*)(*2 μF*)   6. (*Grid resistor*) (*2–20 MΩ*)   7. (*Hum*)(*No*)   8. (*Logarithmic or audio*)(*Yes*) 9. (*RF*)   10. (*Change bias*)   11. (*Reduce HF noise*) 12. (*Yes*)(*Reduce R's, increase C's, C-couple to base*) 13. (*High gain or A*)   14. (*Reduces*)

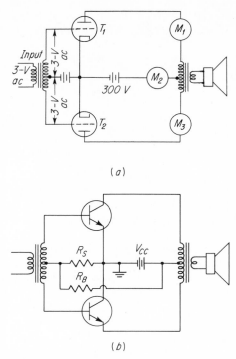

Fig. 14-20   Push-pull AF amplifiers: (*a*) VT and (*b*) BJT.

14-23 PUSH-PULL CLASS A AMPLIFIERS

Figure 14-20*a* shows a push-pull VT amplifier. If the bias battery voltage is about 0.66 of cutoff, the stage is a class A amplifier. If the bias is adjusted to about 0.95 of cutoff, the stage will be a class B amplifier. In other words, the *diagram* of class A push-pull and all other classes will be identical if battery bias is used. Furthermore, the theory of operation is similar in many respects. At this time the circuit will be assumed to be biased for class A operation. Although the explanations are in terms of vacuum tubes, circuit operation is essentially the same with transistors.

NO-SIGNAL CONDITION. With no signal to the grids, 300 V E_{bb}, and a bias voltage of 66% of cutoff, a possible plate current for each tube might be 30 mA. M_1 and M_3 should both read 30 mA. If they read different values, the tubes should be changed until a matched pair is obtained to reduce even-order harmonic generation. M_2 should read 60 mA.

POSITIVE SIGNAL CONDITION. When an input signal of 3 V is fed to the 2:1 step-up ratio input transformer, on one-half of the input cycle the grid of T_1 will become 3 V more positive (actually 3 V less negative) than it was with bias alone and no signal. This reduction in negative voltage on the grid of T_1 will increase its plate current, possibly by 6 mA. Now M_1 should read 36 mA. As the T_1 grid becomes less negative, the T_2 grid will become 3 V more negative than with no signal. The increase in negative charge on T_2 will produce a reduction in plate current through this tube of 6 mA. Now M_3 should read 24 mA. Note that as one tube draws more current, the other draws an equal amount less. The net result, insofar as M_2 is concerned, is no change in current.

NEGATIVE SIGNAL CONDITION. On the other half of the input-signal cycle, T_1 is driven more negative than with no signal, and its plate current drops. T_2 is driven less negative than with no signal, and its plate current increases. Again the total plate current in M_2 remains the same.

The signal voltages used have been considered to be ac but of such a low frequency that the meters M_1 and M_3 could follow their variations. Actually, AF signals produce very rapid variations in plate current in each tube. Any increase in current in either tube is followed so soon by the decrease due to the other half of the ac cycle that the meter can read only the average of these high and low values. For this reason, under normal operating conditions, M_1, M_2, and M_3 should hold constant when a signal is applied to the input circuit. If a meter in the position of M_2 does change when signals are suddenly applied to a class A stage, whether push-pull or single-ended, it indicates that some form of distortion is being developed in the circuit. In the push-pull circuit the trouble might be in mismatched tubes. In the single-ended circuit the trouble might be in a faulty tube, incorrect bias value, or excessive signal voltage.

A push-pull transistor stage is shown in Fig. 14-20b. Note how R_S plus R_B provides forward bias with a transformer.

An advantage of using a push-pull circuit instead of a single-ended circuit (either single or parallel tubes) is the tendency to cancel the even-order harmonics (2d, 4th, etc.) that may be developed in the stage.

The dc plate current of a single-ended amplifier tends to magnetize the core of the output trans-former, resulting in distortion. In push-pull stages the plate currents are flowing in opposite directions, thereby preventing core magnetization or saturation. Thus, push-pull produces less distortion.

Push-pull stages can be operated in the more efficient classes of AB and B. This is another reason why push-pull is preferable to single-ended operation when a large value of audio power output is required.

Still another advantage is the ability of a push-pull stage to cancel hum due to any power-supply ripple (Sec. 14-39).

14-24 PUSH-PULL OUTPUT-TRANSFORMER OPERATION

The operation of the currents and induced voltages in a push-pull output transformer should be understood. Figure 14-21 indicates the direction taken by the plate currents of the two tubes.

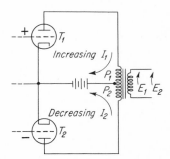

Fig. 14-21 As current increases in one half of primary, it decreases in the other, but voltages induced in secondary are additive.

As the grid of T_1 becomes more positive, the current flowing down through its primary P_1 increases, producing an expanding magnetic field of an assumed north polarity, which induces an upward voltage E_1 in the secondary. At the same time the grid of T_2 is becoming more negative, and the plate current, flowing upward in P_2, decreases, producing a contracting magnetic field of south polarity. This also induces a voltage in the secondary in an upward direction. The expanding north field produces the same voltage effect as the contracting south field.

When the T_1 grid is going more positive, the T_1 plate current is increasing, inducing a counter emf in the upper half of the primary. Since this emf is in a direction opposite to the power-supply voltage, while it exists, the plate voltage of T_1 is less than the power-supply voltage. At the same time, the increase in T_1 plate current also induces an emf in the other half of the primary that is in series with the power-supply voltage for T_2. Therefore, the T_2 plate voltage is considerably higher than the power-supply voltage as its grid is going negative.

14-25 PUSH-PULL CLASS A BIAS

One of the advantages of the class A amplifier is the ease of biasing. The single-ended circuit was described in Sec. 14-14. How cathode-resistor bias is produced in a push-pull class A stage is shown in Fig. 14-22. The plate current

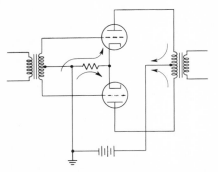

Fig. 14-22 In class A push-pull, cathode resistor requires no filter capacitor, as current is constant.

for both tubes flows through the cathode resistor. When a signal is impressed on the grid circuits, the plate current of one tube increases and the plate current of the other decreases by a like amount at the same time. This results in a constant value of current through the cathode-biasing resistor at all times. A constant current produces a constant voltage-drop. Therefore, no filter capacitor is needed across the cathode resistor in a push-pull class A stage, although one is often used in case the two tubes are unbalanced. Note which end of the bias resistor is grounded.

14-26 IF ONE CLASS A PUSH-PULL TUBE BURNS OUT

A push-pull class A amplifier stage is actually two class A stages *back to back;* that is, the upper

tube alone is a complete class A stage, as is the lower-tube circuit. If one of the two tubes suffers a filament burnout and ceases to function or is pulled out of its socket, the other tube continues to function as a single-ended class A amplifier. It will produce only about half the normal power output of the two tubes (-3 dB), but the output will remain relatively undistorted. When only a small percentage of the available audible output power of an amplifier is actually used, the amplifier may operate for long periods of time with one tube inoperative and the difference may not be noted. Other classes of amplifiers will produce considerable distortion if one tube ceases to function, since only about half the input cycle will be properly amplified.

14-27 THE CLASS AB$_1$ AUDIO AMPLIFIER

A push-pull audio amplifier biased a little higher than class A, but not driven to the point where grid current will flow, is considered to be a class AB$_1$ stage. On the negative half of the input cycle the peak signal usually adds enough negative voltage to the bias voltage to cut off the plate current for perhaps one-quarter of the input cycle (Fig. 14-23).

The positive peak of the signal voltage does not

Fig. 14-23 Operating class AB$_1$ and relative bias voltages for classes A, AB$_2$, B, and C.

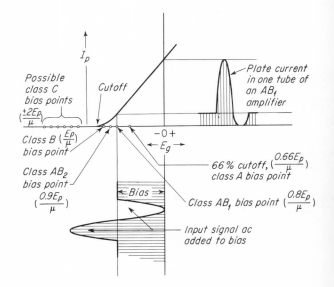

quite reach the positive grid region during any portion of the cycle fed to the grid.

Because a signal strong enough to reach zero grid potential with the amplifier biased to class AB_1 will result in cutoff of plate current on the negative half cycle, class AB_1 amplifiers must use two tubes in push-pull. When one tube is cutting off for a short time, the other is reaching its high plate-current value. The output transformer has plate current flowing in it at all times to reproduce the signal voltage in the secondary. The output signal remains relatively undistorted.

A pair of triode tubes capable of producing about 8 W of audio in class A may produce about 16 W of audio operating as class AB_1. With the higher bias it is possible to use higher plate voltage without causing too much average plate current to flow. Increasing the plate voltage increases the cutoff voltage point. This allows a greater input-signal voltage without driving into the positive grid region. With a greater grid-voltage swing, greater plate-current variation takes place, increasing the power output. During high negative signal peaks, the plate current ceases to flow for a time and the plate has a chance to cool. Because the plate cools or rests during part of the input cycle, the tube can be worked much harder without increasing the average plate heat. As long as the tube is so operated that its plate does not exceed its rated plate (heat) dissipation, it should give satisfactory operation for long periods. When its plate dissipation is exceeded, its life expectancy is shortened.

The plate-circuit efficiency (dc power input to ac power output) for a class AB_1 amplifier is in the 25 to 50% region.

Battery, power-supply, and voltage-divider biasing methods are satisfactory for class AB_1 stages. If the bias voltage is only a little more than class A values, cathode-resistor bias with a large filter capacitor across the resistor is satisfactory.

14-28 THE CLASS AB_2 AUDIO AMPLIFIER

The class AB_2 amplifier is always a push-pull stage and is usually biased nearer the plate-current cutoff point than is a class AB_1 amplifier. The increased bias allows still higher plate voltage to be used on the tubes and therefore greater grid-voltage swing than in class AB_1. The grid-circuit transformer must be designed to withstand grid-current flow as the high-amplitude input signals drive into the positive grid region on most peak signal voltages and also cut off the plate current for a considerable portion of the negative half cycle. The cutting off of the plate current allows the plate to rest and cool. These factors produce greater efficiency and power output.

A pair of tubes capable of about 8-W output in class A may be capable of about 16-W output in class AB_1 and possibly 24-W output in class AB_2.

In the class AB_2 amplifier the *no-signal plate current* is relatively low. If a milliammeter is connected in series with the plate-circuit power supply, it will read a low value with no signals but will increase whenever medium or strong signals are fed into the grid circuit.

Battery, power-supply, and voltage-divider bias can be used with class AB_2 stages, but not cathode-resistor bias. Good power-supply regulation is required.

14-29 THE CLASS B AUDIO AMPLIFIER

When two tubes connected in push-pull are biased almost to cutoff, they are considered to be operating in class B. The $E_g I_p$ curve, input-signal voltage, and resulting plate-current pulse for one of the two tubes of a class B stage are indicated in Fig. 14-24.

Normal class B operation assumes signal driv-

Fig. 14-24 Plate-current versus input-signal voltage for one tube of a class B stage. With no resistance in grid circuit the plate pulse is not flattened at peak.

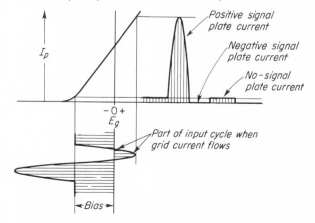

ing, or excitation, voltages applied to the grids of the tubes that are high enough to drive well up into the grid-current region. Considered alone, one of the tubes will operate with plate current for about 200° of the input-signal voltage cycle. No plate current flows during about 160° of the negative half cycle. The tube rests for almost the full half cycle. When it does work, a class B tube may be worked quite hard before its average plate-dissipation rating is exceeded. Since each tube in a class B AF amplifier handles only one half cycle, considerable distortion is produced if only one tube is operating. Grid current may flow for as much as 30° of the positive input cycle.

The bias for class B stages requires separate bias supplies of low impedance (good voltage regulation) to maintain the same bias voltage with and without grid current.

Both primary and secondary of the driver transformer of a class B stage should have low resistance to prevent formation of degenerative voltages in them during the portion of the cycle in which the grids are driven positive and draw current. The stage ahead of the class B amplifier should be a power stage to develop the necessary power during grid-current times.

A milliammeter in the plate circuit of a class B stage shows very little current with no signal, but its reading increases as soon as a signal is applied to the grids. A rough estimate of how much power the stage is producing is indicated by the meter pointer.

It is possible to use special class B high-μ triodes having μ values of 100 to 150 and use no bias voltage at all, because high-μ tubes with no bias limit the plate current to very low values. The whole positive half cycle of the input signal will cause grid current to flow in each tube. Most of the negative half cycle drives the grid beyond cutoff. With two such tubes in push-pull, grid current flows for 360° of the input ac cycle and considerable power is required from the driving stage.

Beam-power tetrodes and pentodes are not operated in class B. They are usually limited to class A, AB_1, or AB_2.

The plate-circuit efficiency of class B audio amplifiers is usually between 50 and 60%, depending upon the distortion that can be tolerated.

Since a class B amplifier is biased almost to cutoff, a vacuum tube having a plate voltage of 1,000 V and a μ of 25 will require approximately 40 V of negative bias (Sec. 14-21).

Test your understanding; answer these checkup questions.

1. What phase angle feedback is produced by a capacitor from one grid to the grid of the preceding stage? _____

2. What classes are expected to operate with I_g flowing? _____ Must I_g flow to operate without distortion? _____

3. When is plate (collector) circuit decoupling required? _____

4. Besides bias value, what else helps determine the class in which a VT will be operating? _____

5. In reference to E_{co}, what bias does a class A stage have? _____ Class AB_1? _____ AB_2? _____ B? _____ C? _____

6. What is the approximate efficiency of a class A stage with no signal input? _____ With full signal input? _____

7. In what class do stages operate if they have only R and C components? _____

8. The I_p meter in a push-pull class A amplifier rises with a medium input signal. Is the bias too high or too low?

9. What polarity applied to the grid forces the E_p of a VT with an R_L to drop nearer to cathode potential? _____ Is this true with PNP BJTs? _____

10. What is the only class of push-pull operation in which I_p (I_c) flows during the whole input cycle? _____

11. A single VT with $E_{co} = 10$ V is biased to -8 V but is never fed more than 1 V of signal. What class is it? _____ Will it distort? _____ Would it operate similarly if biased to -3 V? _____

12. What classes of operation must use push-pull? _____

13. What operating class has almost no I_p with no signal? _____

14. What operating class has highest efficiency? _____

14-30 EARPHONES

Converting varying-amplitude dc or AF ac into sound waves is accomplished by earphones or loudspeakers. Such devices may be given the general name of *transducers*. They usually operate on an electromagnetic principle whereby a varying current produces a varying-strength magnetic field. The varying magnetic field attracts and releases a thin iron diaphragm. Vibration of the diaphragm sets up airwaves which are recognized as sound by the human ear.

Figure 14-25 illustrates the basic principle of

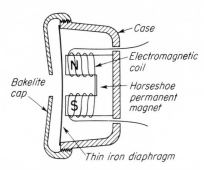

Fig. 14-25 Construction details of an earphone.

the earphone. When no current is flowing through the coils, the permanent magnet attracts the diaphragm and holds it in a strained position, bent slightly inward. If current flows through the coils, it will either add to the magnetism of the magnet and pull the diaphragm farther inward or counteract the magnetic field and allow the diaphragm to swing outward a little, depending on the current direction. If the current is varying or alternating, the diaphragm will swing back and forth, or vibrate, at the frequency of the current variations or alternations. The diaphragm vibrations cause airwaves or sound to come out through the hole in the plastic cap.

High-impedance (2,000- to 10,000-Ω) earphones have many turns of fine wire around the core. When connected directly in the output circuit of an amplifier (Fig. 14-26a), care must be

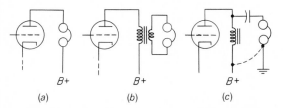

Fig. 14-26 (a) Direct, (b) inductive, and (c) impedance coupling earphones to the plate circuit of a triode (JFET or BJT).

taken to connect them in such a way that current will flow in a direction that will increase the magnetic pull of the core and keep the permanent magnets from becoming demagnetized. Earphone cords are usually marked with a red tracer to indicate which wire should be connected to the positive end of the power supply. When earphones are connected across the secondary of

an output transformer, or impedance-coupled, they are fed ac and the polarity of the leads does not matter. If earphones lose their permanent magnetism, each half cycle of AF ac applied to them will produce one vibration, an extreme form of distortion.

Low-impedance (5- to 600-Ω) earphones have fewer turns of heavier wire and operate with higher currents. This is readily obtainable by the use of a step-down output transformer (Fig. 14-26b). Low-impedance phones are not made to operate directly in the plate circuit of an amplifier tube, although they will produce weak signals if connected in such a manner. Figure 14-26c shows earphones impedance-coupled to a triode.

A cathode-follower (or emitter-follower) circuit may be used to convert high impedance to low impedance (Fig. 14-27). With no input signal the I_p is at a low value. Signals vary the cathode-plate

Fig. 14-27 Coupling low-impedance earphones to (a) a VT cathode-follower and (b) an NPN emitter-follower circuit.

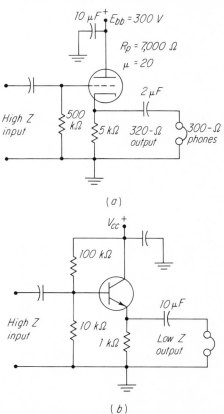

current, but the voltage-drop across the load is always less than the signal voltage. The gain can be found from

$$A = \frac{\mu R_k}{R_p + R_k(1 + \mu)}$$

The output impedance Z_0 is approximately $R_p R_k / (R_p + \mu R_k)$, or about 320 Ω in Fig. 14-27a. Sometimes the earphones can be connected in place of the cathode resistor. Unlike an amplifier, which changes the signal phase 180°, the cathode-follower produces no phase change. This type of circuit is useful as an isolating circuit because any changes that may occur in the load do not affect the input circuit.

Earphones should be handled carefully. A dented diaphragm will cause weak signal output or distortion if the diaphragm happens to touch the core during its vibrations. A mechanical shock, such as dropping on the floor, demagnetizes the magnet core, causing weakened signal output and distortion.

Besides electromagnetic earphones, there are piezoelectric, or crystal, earphones. These are very high impedance types, ranging in the region of 20,000 Ω. They operate from a high-impedance output transformer or with impedance coupling. Inside the earphone is a piezoelectric crystal attached firmly to a thin diaphragm. The two sides of the flat crystal are metal-coated. When an audio voltage is applied to the two opposite metalized surfaces of the crystal, the crystal bends, pushing the diaphragm inward or outward, depending on the polarity of the audio voltages being applied. Piezoelectric earphones are quite sensitive and have excellent fidelity.

14-31 LOUDSPEAKERS

In early-day radio, loudspeakers were similar to high-impedance electromagnetic earphones but were larger and had heavier wire in them to handle greater plate currents. A horn, or large diaphragm area, was used to allow a larger mass of air to vibrate, producing louder sounds.

Today, most loudspeakers operate by either the permanent-magnetic or the electromagnetic field principle. The former is known as a p-m, or *dynamic,* speaker; the latter, as an *electrodynamic* speaker. Figure 14-28 illustrates the basic principle of the p-m type.

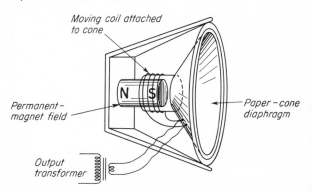

Fig. 14-28 Components of a p-m loudspeaker.

When current flows through the coil attached to the diaphragm assembly, the coil becomes an electromagnet. The coil will now be either attracted inward toward the magnet or repelled outward away from it, depending upon the direction of the current in the coil and its magnetic polarity. Since the coil is attached to the diaphragm, any movement of the coil carries the diaphragm back and forth with it, producing the air vibrations necessary to make sound. The ends of the coil are attached to two points on the paper diaphragm. Flexible leads from these points are connected across the output-transformer secondary. In this way ac signal currents are carried to the coil from the transformer.

The impedance of loudspeakers of this type varies from 2 or 3 to 100 or more Ω. They are made to operate from high-ratio step-down output transformers.

Electromagnetic-field loudspeakers are similar in construction to p-m speakers except that the central core is an iron temporary magnet. On the rear of this core is a fixed field coil, which, when energized, produces the strong field that was produced by the permanent magnet in the p-m speaker.

Electrostatic loudspeakers are very high impedance devices and might be likened to an air-dielectric capacitor with one plate free to move. Varying voltages attract and release the movable plate-diaphragm.

14-32 IMPEDANCE-COUPLED OUTPUT

To prevent dc from flowing through high-impedance types of loudspeakers, an impedance-coupling circuit can be used, as in Fig. 14-29.

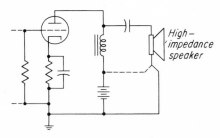

Fig. 14-29 Impedance-coupling a high-impedance type of speaker to a VT plate circuit.

The varying plate current flowing through the choke coil develops an ac voltage across the choke, which is fed to the loudspeaker through the capacitor C.

The lower connection on the speaker in the diagram could be connected to the lower end of the choke coil, shown in dotted lines, instead of to ground

14-33 MATCHING IMPEDANCES WITH AN OUTPUT TRANSFORMER

When the maximum power output is desired from an ac source, the load impedance must equal the source impedance. While this goal cannot always be attained in practice, it can be approached. If a 4-Ω loudspeaker is connected directly into a 4,000-Ω plate circuit, quite a mismatch will be produced. In such a case the speaker will produce almost no signal output.

Transformers can step up or step down voltages, depending on the turns ratio of the primary and the secondary. A 3:1 ratio transformer is assumed either to step up or to step down the voltage three times.

Transformers can also be used to convert from one impedance value to another. As might be expected, a 1:1 ratio transformer will have the same impedance in the secondary as in the primary. When the turns ratio is anything other than unity, however, the impedance ratio of primary and secondary will be equal to the turns ratio squared:

$$\left(\frac{T_p}{T_s}\right)^2 = \frac{Z_p}{Z_s} \quad \text{or} \quad \frac{T_p}{T_s} = \sqrt{\frac{Z_p}{Z_s}}$$

where T_p = number of primary turns
T_s = number of secondary turns
Z_p = primary impedance
Z_s = secondary impedance

The turns ratio (T_p/T_s) of a transformer to match a source impedance of 500 Ω to a load of 10 Ω is

$$\frac{T_p}{T_s} = \sqrt{\frac{Z_p}{Z_s}} = \sqrt{\frac{500}{10}} = \sqrt{50} = 7.07:1$$

If a transformer has a turns ratio of 4:1, the impedance ratio can be found applying the formula.

$$\frac{Z_p}{Z_s} = \left(\frac{T_p}{T_s}\right)^2 = \left(\frac{4}{1}\right)^2 = \frac{16}{1}$$

The answer obtained is the impedance ratio between primary and secondary. If the primary is looking into an impedance of 16,000 Ω, the secondary will appear as a 1,000-Ω source for a load.

To match a 4-Ω speaker to a 4,000-Ω power tube, what would be the transformer turns ratio for maximum power output?

$$\frac{T_p}{T_s} = \sqrt{\frac{4,000}{4}} = \sqrt{1,000} = 31.6:1$$

Unfortunately, in AF amplifiers maximum power output produces some distortion. It is usually better for the load not to match but to have an impedance at least twice the plate impedance of the tube. With an 8,000-Ω primary and 4-Ω secondary the turns ratio of the output transformer would be

$$\frac{T_s}{T_s} = \sqrt{\frac{Z_p}{Z_s}} = \sqrt{\frac{8,000}{4}} = \sqrt{2,000} = 44.7:1$$

Proper matching of primary, secondary, and load impedances sometimes requires additional loading of the secondary to utilize a transformer that does not have the desired turns ratio. This may be accomplished by connecting a second load resistor, R, across the secondary, as shown in Fig. 14-30.

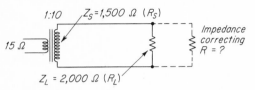

Fig. 14-30. Matching a 2,000-Ω load to a 1,500-Ω transformer secondary.

A transformer has a step-up ratio of 10:1, with a 15-Ω primary impedance. The secondary impedance is equal to the turns ratio squared, or 100 times the primary. The secondary impedance is therefore 1,500 Ω. If the load across the secondary is 2,000 Ω, there is a mismatch between secondary and load. If a second resistor is paralleled across the 2,000-Ω load however, it is possible to match impedances. It is possible to compute the required value of the loading resistance, R_x, by rearranging the parallel-resistance formula:

$$R_s = \frac{1}{1/R_L + 1/R_x}$$

$$\frac{1}{R_L} + \frac{1}{R_x} = \frac{1}{R_s}$$

$$\frac{1}{R_x} = \frac{1}{R_s} - \frac{1}{R_L}$$

$$R_x = \frac{1}{1/R_s - 1/R_L} = \frac{1}{1/1,500 - 1/2,000}$$

$$= \frac{1}{4/6,000 - 3/6,000} = \frac{1}{1/6,000}$$

$$= 6,000 \ \Omega$$

Also,
$$R_s = \frac{R_1 R_2}{R_1 + R_2} = \frac{R_L R_x}{R_L + R_x}$$

from which
$$R_x = \frac{R_L R_s}{R_L - R_s} = 6,000 \ \Omega$$

The addition of the 6,000-Ω loading resistor

across the 2,000-Ω load results in a total load of 1,500 Ω, which matches the secondary impedance of the transformer.

14-34 PARALLEL-CONNECTED AMPLIFIER TUBES

To double the output power of a class A amplifier, two similar-type tubes can be used. They can be connected in either a push-pull or a parallel circuit. The diagram for two triode tubes in a class A parallel circuit is shown in Fig. 14-31.

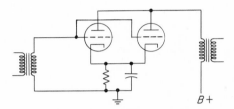

Fig. 14-31 A parallel-tube class A triode amplifier.

Two tubes in parallel act in the same way as one larger tube of twice the power rating. With battery bias, if one tube burns out, the other tube continues to operate, delivering about half the power output and with little change in percent of distortion. With cathode-resistor bias, when one tube burns out, the current through the cathode resistor decreases, decreasing the bias on the remaining tube. This causes the single tube to operate at a higher-than-normal plate current, at half the power output of the two tubes.

A disadvantage of parallel-tube operation is the doubling of the total plate current. This requires a transformer wound with heavier wire in the primary. Parallel-connected tubes do not cancel even-order harmonics, or power-supply hum, as push-pull operation does.

While transformer coupling is shown in the grid circuit, it is also possible to use resistance coupling to any class A stage. It is possible to use four tubes in a push-pull-parallel circuit biased to class A, AB_1, AB_2, or B.

Transistors may also be connected in parallel, push-pull, or push-pull-parallel.

14-35 TYPES OF DISTORTION

When the waveshape of the output signal from an amplifier varies in any respect other than in amplitude from the waveshape of the signal fed into the amplifier, the amplifier is distorting the signal. There are four basic forms of distortion:

1. *Amplitude.* If a 3-V sine-wave signal is fed into an amplifier and produces a 30-V output signal but a 6-V signal produces only a 40-V output signal, the amplifier is not amplifying all signal amplitudes in the correct proportion. This is amplitude distortion. The output waveshape of the 6-V input signal must be flattened on the peaks. This same flattening of the peaks can be produced by adding to the fundamental signal just the correct number and amplitudes of harmonics of the signal frequency. Since the amplifier is emitting the fundamental plus all the harmonics, amplitude distortion is also known as *harmonic distortion.* It is usually caused by overdriving the amplifier tube or working on a nonlinear part of the curve of the tube.
2. *Frequency.* If a 3-V 100-Hz signal is fed to an amplifier and produces a 30-V output voltage but a 3-V 5,000-Hz signal produces only a 20-V output voltage, the amplifier is not amplifying all frequencies equally well. This type of distortion is known as *frequency distortion.* In this case, if a curve of frequency-versus-output amplitude is drawn, it will vary up and down. It is said to be nonlinear (not a straight line), or not flat. A perfectly flat amplifier will amplify all frequencies equally well.
3. *Phase.* If a 500-Hz signal shifts phase 180° in passing through an amplifier but a 5,000-Hz signal shifts 175°, the amplifier is said to be producing *phase-shift distortion.* This differing amount of phase changing for different frequencies is caused by inductances and capacitances in the amplifier circuit working in conjunction with resistances. Phase distortion is not usually considered particularly bothersome in audio amplifiers since the ear does not distinguish this distortion easily. In circuits amplifying TV, video, square, or complex waves, phase distortion produces highly undesirable effects.
4. *Intermodulation.* When two different-frequency signals are added in a nonlinear circuit, they intermodulate, producing sum and difference frequencies of the original two frequencies.

These products appear as dissonant high- or low-frequency sounds or noises in audio amplifiers. Nonlinearity may be produced by a signal driving into a nonlinear portion of an $E_g I_p$ or $V_B I_C$ curve.

Some possible causes of distortion in AF amplifiers are:

1. Loss of cathode emission due to aging or to low filament current
2. Overdriving an input circuit with too high a signal amplitude
3. Tube becoming gaseous
4. Leaking coupling or bypass capacitors
5. Open or shorted resistors or capacitors
6. Improper bias values
7. Improper matching of load impedances
8. Filament circuit not grounded at some point
9. Poor-quality transformer or shorted turns in a transformer
10. Hum introduced into amplifier circuits
11. Undesired interstage coupling
12. Poorly regulated power supply
13. Core saturation of the output transformer
14. Malfunctioning vacuum tubes or transistors

14-36 INVERSE FEEDBACK IN AF AMPLIFIERS

Signals fed into audio amplifiers are usually distorted somewhat as they pass through each stage. This is particularly true when audio transformers are used. It is possible to decrease the distortion produced in a stage by taking a small part of the output signal and feeding it back into the input circuit ±180° out of phase from the signal being applied to the input. This may be called out-of-phase, inverse, degenerative, or negative feedback. Any variations of the signal waveform produced in the stage and fed back out of phase will be materially reduced when reamplified out of phase. This results in a less distorted but weaker output signal. However, the input signal can now be increased, allowing the output of the stage to increase.

A VT circuit with *inverse-voltage feedback* is shown in Fig. 14-32a. The inverse circuit itself is shown in darker lines.

A signal applied to the grid circuit is amplified and appears across the primary of the output

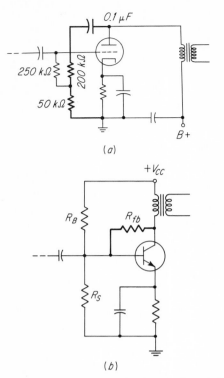

Fig. 14-32 Inverse-voltage feedback: (a) VT and (b) BJT.

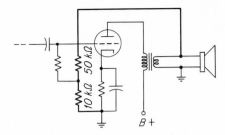

Fig. 14-33 Another inverse-voltage feedback system.

transformer. Since the bottom of the transformer primary is bypassed to ground, the output signal also appears between plate and ground. Connected across this output signal is a dc blocking capacitor and the voltage-divider network made up of the 200-kΩ resistor and the 50-kΩ resistor. Approximately one-fifth of the output-signal ac will appear across the 50-kΩ resistor. This small portion of the output is being developed in series with the grid circuit. Since the output-voltage signal of an amplifier stage is always 180° out of phase with the signal on the grid, that part of the output signal across the 50-kΩ resistor represents an out-of-phase voltage being fed back from plate to grid circuit.

The transistor circuit uses a high resistance, R_{fb}, as the feedback element. R_B is usually not included when R_{fb} is used.

Another inverse-voltage feedback circuit takes the feedback voltage from the secondary of the output transformer (Fig. 14-33). If the voltage fed back to the grid is found to be regenerative and raises the gain, or causes the stage to oscillate, it can be made degenerative by reversing either the primary or the secondary connections of the output transformer.

It is also possible to use a feedback loop from the output back to the input of the stage ahead. This decreases distortion developed in both stages. Such a circuit is particularly adaptable when resistance coupling is employed between a driver and final amplifier. With transformer *interstage* coupling, a phase reversal may occur at some frequency in the audio range, and increased distortion due to regeneration at the critical frequency may result.

Current feedback is shown in Fig. 14-34a. It

Fig. 14-34 Inverse-current feedback: (a) VT and (b) BJT.

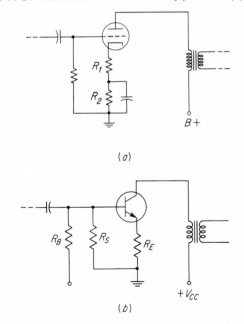

utilizes the inverse voltage developed across the cathode-biasing resistor to produce the degeneration. In this circuit, the sum of R_1 and R_2 equals the necessary resistance to bias the stage to class A. Any signal emf developed across the unfiltered resistor will be 180° out of phase with the signal applied to the grid and in series with the grid-to-cathode circuit. If all the cathode resistor is unfiltered, the degeneration developed may be more than desired. For this reason, only about half the resistance is usually filtered or bypassed. With transistors the whole emitter resistance may be unbypassed, as in Fig. 14-34b.

Mathematically, the voltage gain of an amplifier at the midfrequency of its response curve with inverse feedback can be found by

$$A_f = \frac{A}{1 + (-\beta)A}$$

where A_f = amplification with feedback
A = amplification without feedback
$-\beta$ = decimal fraction of output voltage fed back to input

As an example, an amplifier has a gain of 20. If 1 V is fed to its grid, 20 V should appear across the plate load. If a $-\beta$ of 0.05 of the output is added back into the grid circuit, the total gain of the amplifier is

$$A_f = \frac{20}{1 + 0.05(20)} = \frac{20}{2} = \text{gain of 10}$$

Without feedback this amplifier would have dropped to half-voltage output at some lower frequency. With this value of feedback, at the same frequency the output would have dropped by only 10%. Thus, inverse feedback widens the bandwidth of an amplifier. Also, instead of developing a 45° phase shift at the original half-voltage frequency, the resultant phase shift would have been only 27°. Therefore, inverse feedback reduces phase shift in an amplifier. Voltage feedback increases the input impedance of the stage and reduces the output impedance. Current feedback reduces input and increases output impedances.

If the gain of an amplifier without feedback is 10 and with feedback is 8, the feedback factor

β can be determined by rearranging the formula given above to

$$\beta = \frac{1}{A_f} - \frac{1}{A} = \frac{1}{8} - \frac{1}{10} = \frac{1}{40} = 0.025$$

Test your understanding; answer these checkup questions.

1. What are the two different types of earphones used? _____ _____ Which always has high impedance? _____
2. The magnet in an earphone is completely demagnetized. 500-Hz ac is fed to it. What tone is heard? _____ Why? _____
3. If $Z_o = R_p R_k/(R_p + R_k)$, what is the formula to determine R_k? _____
4. Name two basic types of loudspeakers? _____ _____ What relative impedances do they have? _____
5. What ratio transformer is required to convert a 2-kΩ primary to a 16-Ω secondary? _____
6. By paralleling an additional active device, what gain in dB is produced? _____ Does this cancel harmonic distortion? _____
7. Why is inductive input coupling not required with class A stages? _____
8. What is another name for amplitude distortion? _____
9. If an amplifier gain-versus-frequency graph is not flat, what distortion is present? _____
10. The ear is not sensitive to which distortion? _____
11. Two frequencies mix in a nonlinear circuit. What kind of distortion is produced? _____
12. What are three possible advantages that result from inverse feedback? _____ _____ _____
13. What negative feedback results if a cathode (collector) resistor is not bypassed? _____
14. What negative feedback increases Z_{in} and reduces Z_o? _____
15. Stage gain is 20 with no feedback and 12 with feedback. What is the β value? _____

14-37 PHASE INVERTERS

A push-pull stage must be fed two signals of equal amplitude but opposite phase. As one grid is driven a few volts positive by the signal, the other must be driven an equal number of volts negative. The simplest form of a *phase inverter* is a center-tapped transformer connected as shown in Fig. 14-35. If the center tap is in the electrical center of the secondary and is at ground potential, when one end of the secondary is more positive, the other is an equal value more nega-

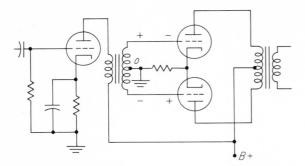

Fig. 14-35 Phase inversion using center-tapped transformer from single-ended stage to push-pull stage.

tive. Transformers are particularly useful for class AB_2 and B_1 or for push-pull transistors, when signal *currents* flow in the input circuit.

Audio transformers are relatively expensive and may pass certain frequencies better than others. To produce phase inversion at less cost and with a flat frequency response, vacuum-tube phase inverters were developed. These vacuum-tube phase inverters can be used only with circuits that draw no grid current, which limits them to classes A or AB_1.

Figure 14-36a is a single-tube *paraphase*

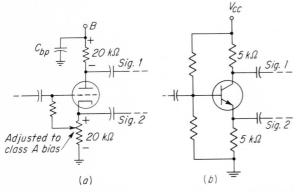

Fig. 14-36 Split-load, or paraphase, phase inverters: (a) VT and (b) BJT.

ANSWERS TO CHECKUP QUIZ ON PAGE 315

1. (*Electromagnetic, crystal*)(*Crystal*) 2. (*1 kHz*)(*Diaphragm pulled each half cycle*) 3. [$R_k = R_p Z_o/(R_p - Z_o)$] 4. (*Dynamic, electrodynamic*)(*Usually 3 to 16+ Ω*) 5. (*11.2:1*) 6. (*+3 dB*)(*No*) 7. (*No I$_g$*) 8. (*Harmonic*) 9. (*Frequency*) 10. (*Phase*) 11. (*Intermodulation*) 12. (*Less distortion, less Miller effect, change of input or output impedances*) 13. (*Current*) 14. (*Voltage*) 15. (*0.0333*)

phase inverter. The cathode resistor, tube, and plate resistor are all in series. The same current flows through them all. If both resistances are equal, the voltage-drop across them will always have to be equal. B+ can be considered to be at ground potential insofar as ac is concerned because of the bypass capacitor C_{bp}. Note the relative potentials across the resistors. The cathode is positive in respect to ground. The plate is negative in respect to B+. If the grid becomes a little more negative, (1) the current through both resistors decreases, (2) the cathode becomes less positive in respect to ground, and (3) the plate becomes less negative in respect to ground. Therefore, the signals taken from the cathode and from the plate will be equal in amplitude but opposite in phase. The lower resistor is actually a cathode-follower, which means the circuit will have less than unity gain. (A 1-V input produces about 0.8 V output.) The transistor version is shown in Fig. 14-36b.

In Fig. 14-37, T_1 is a normal resistance-coupled

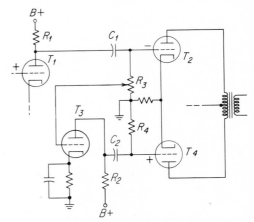

Fig. 14-37 Using an additional amplifier stage for phase inversion.

amplifier feeding the grid of T_2. If T_3 is capable of amplifying 10 times, its grid will be fed from a point one-tenth of the way up the voltage-divider grid resistor R_3. The 10-times-amplified voltage across the plate-load resistor of T_3 will be equal in amplitude to the voltage applied to the grid of T_2, but because it passes through the T_3 amplifier stage, it is 180° out of phase. Therefore, T_2 and T_4 are being fed equal-amplitude but 180° out-of-phase signals. R_1 and R_2 must be equal in value; C_1 and C_2 must be similar; and R_3 and R_4 must be similar for equal signal voltages at the push-pull grids.

14-38 PENTODE AF AMPLIFIERS

The AF-circuit explanations have been in terms of triodes or transistors. In vacuum-tube amplifying systems, however, voltage amplifiers are often pentode resistance-coupled stages and power-output tubes are beam-power tetrodes, single-ended or in push-pull. Figure 14-38 illustrates

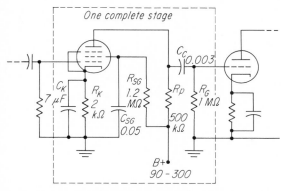

Fig. 14-38 Pentode resistance-coupled amplifier with representative component values.

a class A pentode resistance-coupled voltage amplifier. Values of plate-load resistance R_p may range from 25 kΩ to as much as 1 MΩ. The higher the plate-load resistance, the higher the gain of the stage. With a 6AU6 tube, a 250-kΩ load produces a gain of 280 times, a 500-kΩ load gives a gain of 360 times, and a 1-MΩ load gives a gain of about 450 times.

The grid resistor R_g is usually twice the value of R_L. The screen-grid dropping resistor R_{sg} is as much as twice the plate-load-resistance value. The cathode resistor R_k ranges from 1 to 10 kΩ, depending on plate voltages and plate-load-resistance values.

The value of the coupling capacitor C_c depends on the plate-load and grid resistances. With a 250-kΩ load a capacitor of 0.01 μF may be required. With a 1-MΩ load a 0.002-μF capacitor may suffice. The cathode capacitor ranges from about 10 μF across a 1-kΩ resistor to about 1.5 μF across a 10-kΩ resistor. Screen-grid bypass or

filter capacitors are usually between 0.03 and 0.5 μF.

The capacitance values quoted are for reasonably linear amplification down to 100 Hz. If linear results are required to 50 Hz, double all the capacitance values suggested above.

14-39 HUM IN AF AMPLIFIERS

Hum in AF amplifiers may fall into one of five general categories: (1) improperly filtered power supply, (2) improper functioning of some part in the amplifier, (3) capacitive coupling of hum voltages, (4) inductive coupling of hum voltages, or (5) low-frequency oscillation of the amplifier.

If an amplifier has been operating properly but slowly develops a constant low-frequency tone, or *hum*, the probability is that the power-supply capacitors are losing their effectiveness by drying out or a tube is developing a filament-cathode leak. If the signal strength drops at the same time, a filter capacitor may be leaking or shorted or a coupling capacitor may be leaking. In the latter case, considerable distortion will be present.

If a VT amplifier is newly constructed or has been revamped in some way, it may hum. The frequency of the hum should give a clue to the trouble. With 60-Hz hum, the trouble may be due to capacitive or inductive coupling from the filament leads to the grid leads of the first amplifier stage. Any single wire carrying heavy ac will induce that frequency in any other wire near it. If two filament wires are twisted together, however, an equal value of current is flowing in each wire but in opposite directions. The fields from the two wires induce equal and opposite voltages in any nearby wires, which cancel. Relocation of grid or the filament wiring may help.

If a power-supply transformer is too close to the input-amplifier stage, a 60-Hz hum may be induced into the input leads of this stage. Power-supply transformers may induce a 60-Hz hum voltage into coupling transformers of low-level stages that are less than 2 or 3 ft away. Special double iron shields completely surrounding the transformers will allow closer operation. Often it is possible to rotate the coupling transformer to find a position of minimum hum induction.

If the cathode-biasing resistor in the first stages of an AF amplifier does not have a sufficiently large capacitor across it, hum may be produced because of the difference in potentials or leakage between cathode and filament. Some tubes are better than others of the same type and make. Selection of a quiet tube is necessary in the first stage of high-gain amplifiers.

Grounding the center tap of the filament transformer is usually effective in decreasing hum. If a filament transformer has no center tap, ground the center tap of a 20- to 50-Ω resistor connected across the filament circuit.

If the frequency of the hum is 120 Hz, insufficient power-supply filtering or excessive load is indicated. If an amplifier is drawing too much current, the reason must be determined by test. To increase the filtering, more capacitance or inductance can be added to the power supply. It may be that only the first amplifier stages need more filtering. Plate-circuit decoupling networks help to filter such stages. The degree of filtering required of the power-supply voltage decreases as the level of the amplified signal increases; that is, the first amplifier requires the most filtering and the last stage the least.

To prevent hum voltages induced in metal chassis from being included in amplifier circuits, all ground connections of each stage should be brought to a common ground point.

When a long shielded input lead is used, as from a microphone or turntable pickup, the shield should be grounded at the output end, not at the microphone or pickup end, and never at both ends.

Transistor amplifiers do not have the filament difficulties of VTs, and with their regulated power supplies have far fewer hum troubles.

14-40 HUM REDUCTION IN PUSH-PULL STAGES

One of the advantages of using push-pull, particularly in class A and AB$_1$ stages, is the ability of the circuit to cancel hum produced by power-supply ripple. In the push-pull circuit of Fig. 14-39, the plate currents from both tubes are flowing in opposite directions in the primary of the output transformer. If the power-supply voltage varies at 120 Hz (hum), both primary currents decrease at the same time, producing two equal but opposite voltages in the secondary that cancel

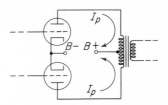

Fig. 14-39 Power-supply variations in primary induce voltages in secondary that cancel each other.

each other. Signal voltages are amplified by the stage, but the hum voltages are canceled in the output.

14-41 MISCELLANEOUS AF AMPLIFIER ITEMS

The interstage coupling capacitor of resistance-coupled stages is often troublesome. The voltage difference across this capacitor is relatively high. As the capacitor ages, it may become *intermittent,* operating correctly part of the time and either shorting or opening during the remainder of the time. This produces an amplifier which works normally part of the time and then either distorts or decreases output. If it begins to leak (dielectric breaks down), an improper bias is applied to the stage, which produces distortion. If this capacitor opens, the capacitance decreases drastically and the signal transferred across it drops to almost zero. In transistor amplifiers the coupling capacitor is an electrolytic type and may dry out.

Another capacitor that often shorts is the screen-grid bypass. If it shorts, the screen voltage drops to zero and plate current ceases. A resistor in the screen circuit will heat and possibly burn out in this case. If a resistance in an amplifier begins to heat, it will rarely be the fault of the resistance. Look for a shorted capacitor, a shorted tube, or two wires touching and shorting. Charred resistors should be replaced, as heat may change their resistance or cause intermittent or noisy operation.

Electrolytic capacitors dry and lose their capacitance. If power-supply capacitors lose their capacitance, an increase in hum will occur and an audible oscillation may result because of poor power-supply regulation. When a power-supply filter capacitor shorts, the plate voltage drops to zero on all stages. The power-supply transformer, rectifiers, and chokes may overheat. The tar in the transformer and choke may boil, smoke, and smell. Rectifier tube plates may become red; mercury-vapor tubes may glow intensely blue; but solid-state diodes give no visual indications.

If an electrolytic cathode bypass capacitor loses capacitance by drying out, the low-frequency response of the amplifier will decrease. Signals sound thin, or tinny. If this capacitor shorts, the tube loses its bias, high plate current flows, and distortion is excessive.

It is important to understand the difference between *plate supply voltage* and *plate voltage*. The plate supply voltage is the voltage reading when a voltmeter is placed across the power supply. The plate voltage is the voltage reading when the voltmeter is connected from cathode to plate at these terminals of the tube. These two voltages will differ by the amount equal to the voltage-drop across any cathode-resistance bias circuit plus the voltage-drop across the plate-circuit load resistance. It is possible, if battery bias is used and the plate-load transformer primary has almost no resistance, that the plate voltage and the plate supply voltage will be essentially the same. There is always a great difference between E_p and E_{bb} in resistance-coupled stages (V_{cc} and V_C or V_{EC} in transistor circuits).

Test your understanding; answer these checkup questions.

1. What is another name for a paraphase phase inverter? _____ How many loads does it have? _____
2. Why might pentode RC amplifiers be preferred over triodes? _____

3. List the general categories that might produce hum in transistor amplifiers. _____ _____
4. Why are input stages more susceptible to hum pickup than higher-level stages? _____
5. What type of bias may prevent heater-cathode leakage from producing hum in an input stage? _____
6. How can electromagnetic hum induction be decreased? _____ Electrostatic induction of hum? _____
7. What might cause 60-Hz hum? _____ What usually causes 120-Hz hum? _____
8. Which end(s) of a long shielded microphone cable should be grounded? _____
9. Is hum due to insufficient power-supply filter decreased by single-ended operation? _____ Parallel? _____ Push-pull? _____
10. What should be done if an amplifier resistor suddenly chars and burns out? _____
11. Why might a dried-out power-supply filter cause amplifier oscillation? _____
12. If a transformer begins to smoke, is this always an indication of a faulty transformer? _____
13. In Fig. 14-40a and b find the grid-bias voltage _____ _____ Power-supply voltage _____ _____ Bleeder-resistor voltage _____ _____ Plate voltage _____ _____ Plate load voltage _____ _____

14-42 A GENERAL-PURPOSE VACUUM-TUBE AMPLIFIER

Figure 14-41 represents a possible four-stage VT amplifier capable of amplifying weak signals from a source such as a microphone up to 10 W of undistorted power, or adequate audio power for a 40- by 80-ft room.

The first stage is a contact-potential-biased 6AU6 pentode *preamplifier* (stage before the volume control) with a tone control across the output circuit and a 10-μF 20-kΩ plate-circuit decoupling network.

The second stage is one-half of a 12AU7 twin-

Fig. 14-40 Practice-problem circuits.

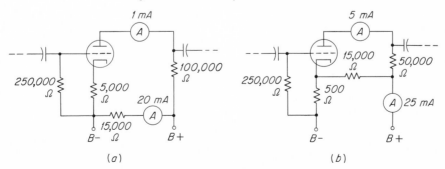

(a)

(b)

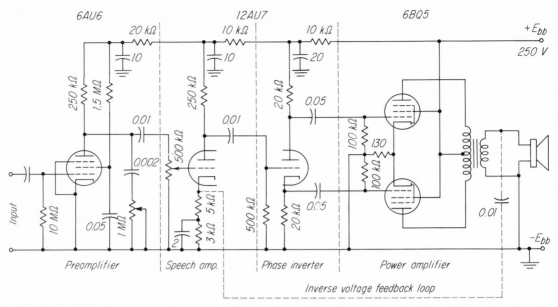

Fig. 14-41 A four-stage 10-W AF amplifier including a pentode preamplifier, triode speech amplifier, phase inverter, and push-pull class A power pentode.

triode amplifier with a volume or gain control in the input circuit, with inverse-current feedback in the cathode circuit, and with plate-circuit decoupling. If the dotted circuit is included, the output stage, the phase inverter, and the speech amplifier are all included in an inverse-voltage feedback loop.

The third stage is a split-load phase inverter with plate-circuit decoupling, using the other half of the dual-triode 12AU7.

The fourth stage is a push-pull 6BQ5 class A power amplifier capable of 10 W of undistorted output. If the E_{bb} is raised to 300 V, the output should exceed 15 W (a gain of only about 2 dB).

The resistance-coupled stages should draw

about 1 mA of plate current each, the phase inverter about 2 mA, and the power amplifiers about 62 mA of plate current and 7 mA of screen current with no signal, rising to a total of about 75 plus 15, or 90 mA under full output. At 6.3 filament V, the 6AU6 and the 12AU7 draw 0.3 A each and the 6BQ5s draw 0.75 A each. Thus the power supply must furnish 250-V dc at 94 mA as well as 6.3-V ac at 2.1 A.

14-43 A GENERAL-PURPOSE TRANSISTOR AMPLIFIER

Figure 14-42 illustrates a possible transistor amplifier capable of perhaps 10 W if the power transistors have adequate heat sinks but only about 1 W if they do not.

The first two stages are directly coupled and use an NPN and a PNP transistor. The resistors in these stages must be chosen carefully so that both transistors are forward-biased to their optimum class A points. A positive-going incoming signal to the NPN base increases the stages' collector current. This produces an amplified negative-going (180° out-of-phase) potential at the NPN collector. Since the base of the PNP is directly coupled to the NPN collector, the negative-

---

ANSWERS TO CHECKUP QUIZ ON PAGE 319

1. (Split-phase)(Two)   2. (Higher gain)   3. (Improper power-supply filter, low-frequency oscillation)   4. (All following stages amplify any hum picked up)   5. (Contact potential)   6. (Iron shield, separate physically)(Metal shield, separate)   7. (Filament circuits, half-wave rectification) (Insufficient filter in power supply)   8. (Amplifier end)   9. (No)(No)(Yes)   10. (Determine component at fault and replace it, replace R)   11. (Poor power-supply regulation, interstage feedback)   12. (Not necessarily, shorted component may cause high currents)   13. (5 V, 12.5 V)(300V, 312.5 V)(300 V, 300 V)(100 V, 250 V)(195 V, 50 V)

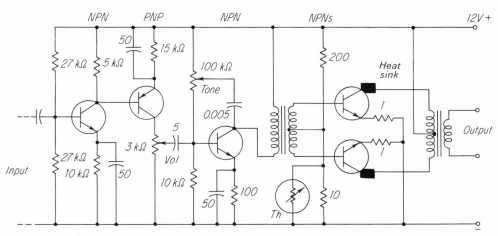

Fig. 14-42 A four-stage transistor AF amplifier including two direct-coupled stages, a volume and tone control in driver stage, and class B output stage.

going potential forward-biases the PNP base, and it also increases its collector current. The increase of current through the 3-kΩ resistor develops an amplified positive-going signal at the top of this resistor. A negative-going input signal reduces the currents in both stages.

The third stage is an NPN common-emitter power amplifier that should be capable of at least 0.5 W output to drive the following push-pull stage adequately. The volume control is at the input of this driver stage, which also has a simple tone control incorporated in its voltage-divider bias network. The stage has an emitter-stabilization RC circuit to protect it against thermal runaway if I_C flow heats the transistor. The current in this class A stage should be approximately 100 mA.

The fourth stage uses two push-pull NPNs biased to class AB or B for high power output. Note the low value of emitter-stabilizing resistors to allow maximum emitter-collector current to flow in the transistors. These resistors are so low in value that it is not feasible to try to bypass them. A dc milliammeter in the center-tap lead of the output transformer should read only a few milliamperes with no signal and perhaps 1,000 mA or more with full-power output.

With zero bias on the base of a transistor, collector current should be negligible, which might be considered as being biased to cutoff or class B. In the output stage, however, some forward bias must be provided to prevent *crossover distortion* when the signal ac switches conduction

from one transistor to the other. The 10- and 200-Ω voltage-divider resistors provide this very small forward bias.

When transistors become warm, they become better conductors and their collector current increases. The *thermistor* (Th) is a negative-temperature-coefficient resistor. As the temperature of the circuitry increases, the thermistor resistance decreases, thereby decreasing the forward bias, reducing collector current, and preventing thermal runaway. The thermistor must be physically attached to the transistors.

14-44 SPECIAL TRANSISTOR AMPLIFIERS

There are many electronic circuits that could use vacuum tubes but do not because of the difficulty of heating the filaments. In such circuits transistors find considerable application. The first two stages in Fig. 14-42 are an example; in them an NPN and a PNP amplifier are direct-coupled. Using an above-ground-positive power supply, the PNP has to be connected "upside down" to allow I_C to flow through it.

Another direct-coupled or dc amplifier is shown in Fig. 14-43. Q_2 receives forward bias through Q_1. A positive-going input signal on the base of Q_1 decreases its EC resistance in proportion to the beta of Q_1. This increases the forward bias on Q_2, and it amplifies the input change still further. The gain is approximately $\beta\beta R_L$.

An extensively used amplifier is the *Darlington*

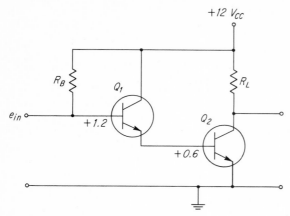

Fig. 14-43   Two-transistor dc amplifier.

pair, shown in Fig. 14-44. It is the same circuit as Fig. 14-43, except that the collector lead of Q_1 is connected to the opposite side of R_L. This introduces degeneration, and the circuit is thermally more stable as the devices warm during operation.

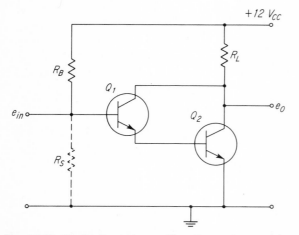

Fig. 14-44   Darlington pair amplifier. R_s can be used for greater stability.

The *differential amplifier* of Fig. 14-45 responds to the difference between two signals. Assume a gain of 10 for each JFET. If input 1 goes $+1$ V and input 2 goes $+1$ V also, the currents in both R_{L_1} and R_{L_2} increase like amounts and there is no voltage difference across the output terminals. If input 1 goes $+1$ V, but input 2 goes to only $+0.5$ V, the current in R_{L_2} increases only half as

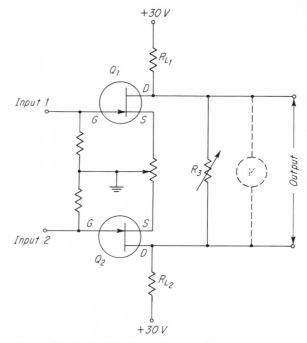

Fig. 14-45   Basic FET differential amplifier.

much as that in R_{L_1}, and the difference of the two voltage-drops is $10 - 5$, or 5 V. The differential amplifier amplified the 0.5-V difference to 5 V. The potentiometer is used to balance the two circuits to zero dc output voltage with no input voltage. The gain of the amplifier is controlled by R_3. With zero resistance the output must be zero; with high resistance the gain is high. For better thermal stability differential amplifiers use some form of inverse feedback.

A series-connected low-noise amplifier pair that has the first device sensitive to input voltages and the second sensitive to current variations in the first is called a *cascode* amplifier. A JFET cascode circuit is shown in Fig. 14-46a. The Q_2 gate is biased to a fixed positive value by the power supplies, with its source at a higher positive potential (the gate is negatively biased). If a positive-going signal is fed to the Q_1 gate, current increases through Q_1, R_{L_1}, Q_2, R_{L_2} and the power supplies. The positive-going Q_1 gate pulls its drain voltage down to a less positive value. If the source of Q_2 is more negative, its fixed potential gate must now be relatively more positive, further in-

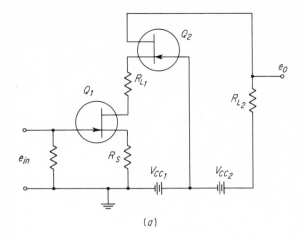

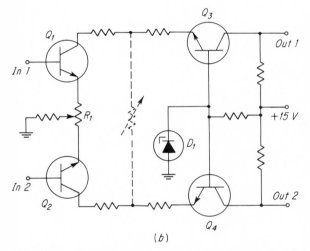

Fig. 14-46 Cascode amplifiers. (a) Single-ended with JFETs. (b) BJT form, thermally balanced.

creasing the current flow through R_{L_2}. Being unbypassed, R_{L_1} introduces degeneration into the circuit, thermally stabilizing the whole stage. The amplitude stability and bandwidth are good up to about 150 MHz with JFET and BJTs and to about 400 MHz with ICs.

The balanced cascode circuit of Fig. 14-46b has better thermal stability, assuming similar devices on both sides of the circuit. If the base of Q_2 is grounded, the amplifier can be used as either a balanced single-ended cascode amplifier or a cascode phase inverter if both outputs are used. When voltages are applied to the bases of both Q_1 and Q_2, the circuit operates as either a differential or a push-pull cascode amplifier. The zener diode (D_1) holds the bases of Q_3 and Q_4 at a constant potential. The dashed rheostat would act as a gain control. Potentiometer R_1 is used to balance the circuit.

The three diagrams in Fig. 14-47 illustrate three solid-state cascaded amplifiers. The first is RC-coupled; the others are direct-coupled.

A complementary-symmetry power-output amplifier is shown in Fig. 14-48. The NPN and PNP power transistors are picked with identical characteristics. R_{B_1} and R_{S_1} bias Q_1. R_{B_2} and R_{S_2} bias Q_2. If both biases are equal, similar I_C values try to flow through R_L, resulting in zero current in the load. When a positive half cycle of ac is applied to the input, I_C increases in the NPN but decreases in the PNP and current flows to the left in the load. On the negative half cycle, the PNP is forward-biased and current flows to the right through R_L. An ac input produces an ac in the load. The difficulties with this circuit are the two power supplies and deciding on what point to ground.

Another complementary-symmetry amplifier using only one power supply is shown in Fig. 14-49. Again matched NPN and PNP power transistors are used. If properly biased, the midpoint in the circuit, M, will be at $+20$ V. Capacitor C charges to this potential through R_L. A positive-going signal to Q_1, as indicated, decreases the EC resistance of Q_1, increasing the forward bias on the PNP. This decreases the PNP EC resistance also, discharging C downward through R_L. As the collector of Q_1 becomes less positive, the base of Q_2 is made more negative, tending to cut off Q_2. Point M can decrease almost to zero if the positive input signal is strong enough. When the input signal reverses polarity, Q_3 is driven toward cutoff and Q_2 is forward-biased. Point M increases in potential and capacitor C charges toward $+40$ V, pulling current up through R_L.

14-45 OPERATIONAL AMPLIFIERS

One of the most versatile of all electronic circuits is the operational amplifier, or op amp. It consists of (1) a very high gain ($A_e = 10^6$) dc amplifier that reverses signal phase $180°$, (2) a feedback impedance, and (3) a series input im-

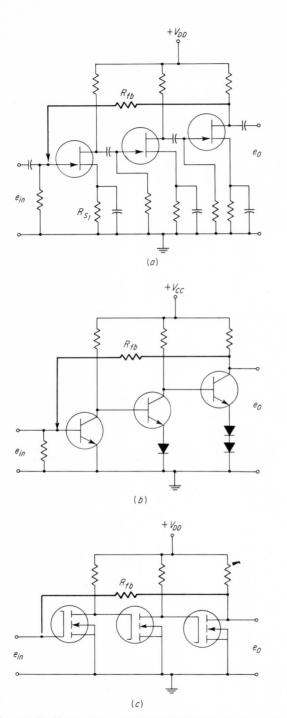

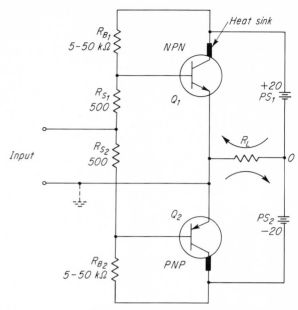

Fig. 14-48 Complementary-symmetry amplifier requiring two power supplies.

Fig. 14-49 Complementary-symmetry amplifier with one power supply.

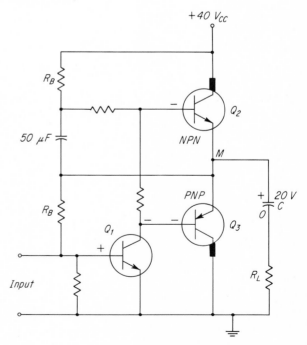

Fig. 14-47 Three-stage amplifier with feedback. (a) JFET with RC coupling. (b) Direct-coupled BJTs. (c) Direct-coupled enhancement MOSFETs.

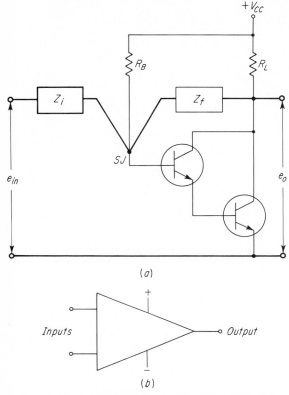

$+V_{CC}$

R_B

R_L

Z_i

Z_f

SJ

e_{in}

e_o

(a)

Inputs

Output

+

−

(b)

Fig. 14-50 (a) Basic operational amplifier using a single Darlington pair. (b) Symbol of a common op amp.

pedance, as shown in Fig. 14-50. For simplicity, only a single Darlington pair is shown, which alone would not produce sufficient gain. The point where Z_i, Z_f, and the base of Q_1 meet is called the *summing junction* (SJ).

With an A_e of 10^6, a 1-μV ac signal should be able to produce a 1-V output with perhaps 3% distortion. The feedback impedance, Z_f, usually holds the gain to less than 1,000. This would reduce distortion to less than 0.003%, which is negligible. Because of the large feedback factor, the ac voltage-drop from SJ to ground is so small that it may be neglected. Therefore, the input is the voltage-drop developed across Z_i. Also, the gain of the whole circuit will be equal to e_o/e_{in}, which is the ratio of Z_f/Z_i. Thus, if both Z_i and Z_f are 500 Ω, the gain of the system is 1, or unity. if Z_i is 1 kΩ and Z_f is 10 kΩ, the gain is 10. If Z_i is 1 kΩ and Z_f is a 100-Ω resistor plus a 100-kΩ rheostat in series, an amplifier with a variable gain

from 0.1 to 100 results. The peak output voltage is limited to about one-quarter of the V_{CC} value.

To mention a few of the many uses of op amps: If two separate input impedances are connected to SJ, the output voltage will be the algebraic sum of the two inputs. Thus, the circuit can be used as an analog (nondigital) computer.

If Z_f is a capacitor and Z_i is a resistor, the circuit becomes a precision integrator and develops a linearly rising voltage. If Z_f is a resistor and Z_i is a capacitor, the circuit can be used as a precision mathematical differentiator or to develop sharp peaks from square-wave input signals. If Z_i is a resistor and Z_f is a diode, the circuit becomes a logarithmic amplifier (one that amplifies according to a logarithmic rather than a linear curve).

One of the common types of op amps consists of two cascaded differential amplifiers followed by a single-ended output amplifier. Either or both of the differential amplifier inputs (inverting and non-inverting) can be used. Op amps are manufactured in IC form and are made up in a variety of packaging types. The symbol of a simple op amp is shown in Fig. 14-50b.

Test your understanding; answer these checkup questions.

1. In Fig. 14-41, which stage has greatest voltage gain? _____ Least? _____ What type bias is used in the input stage? _____ What class is used in the output stage? _____ Which stage(s) has (have) current feedback? _____ In which stage is the gain control? _____
2. In Fig. 14-42, why the high-value emitter R_s in the first two stages? _____ Why no bypasses across emitter R_s in the output stage? _____ What is the purpose of the thermistor? _____ Why must the power supply have good regulation? _____ What three methods of coupling are used? _____
3. In the Darlington circuit with germanium BJTs, what would be the voltage value at the bottom of R_B? _____ At the bottom of R_L? _____
4. Why are JFETs not used in a Darlington circuit? _____
5. In Fig. 14-45, what is producing the bias? _____ If R_3 were 0 Ω, could the circuit amplify as a single-ended amplifier? _____
6. Could VTs be used in place of the JFETs in the single-ended cascode amplifier? _____ What produces bias for Q_1? _____ For Q_2? _____
7. To simplify the circuit in the thermally balanced cascode amplifier, what has been left out? _____

8. How does a cascode amplifier differ from a cascade? _____

9. Is the load current ac or varying dc in Fig. 14-48? _____ In Fig. 14-49? _____

10. If the op-amp BJTs were silicon, what would be the dc SJ voltage value? _____ Ac? _____

11. How many stages would be required in an op amp if each had a gain of 100? _____ What would be the maximum p-p output voltage if its power supply were 10 V? _____

COMMERCIAL LICENSE QUESTIONS

Amateur license questions will be found in the Addendum.

FCC Elements 3, 4, and 6 require an ability to answer questions similar to those below. Sections in which questions are answered are shown in parenthesis. A question followed by a bracketed number is required for that element alone.

1. Is the dc bias voltage normally positive or negative in a vacuum-tube amplifier? (14-3)
2. Why is bias voltage used on the grid of an AF amplifier tube? (14-3).
3. Does dc grid current normally flow in a class A amplifier? (14-3)
4. What is the cutoff bias voltage E_{co} of a triode with a μ of 20 and a plate voltage of 250 V? (14-3)
5. What is the maximum rms value of AF voltage which can be applied to the grid of a class A AF amplifier with a grid-bias value of -8 V? (14-3)
6. What is the stage gain of a grounded-cathode triode with a μ of 20, plate impedance of 8,000 Ω, and load impedance of 50,000 Ω? (14-4)
7. What is the most desirable factor in the choice of a tube to be used as a voltage amplifier? (14-4)
8. What would be the effect if dc were applied to the primary of an AF transformer? (14-6)
9. What may cause self-oscillation in an AF amplifier? (14-6, 14-7, 14-20)
10. Why is it preferable to isolate the dc from the primary winding of an AF transformer? (14-6)
11. What is the advantage of transformer over resistance coupling in AF amplifiers? (14-6)
12. If the capacitance of a coupling capacitor in a resistance-coupled AF amplifier is increased, what effect may be noted? (14-7).
13. Draw a diagram of a resistance load connected in the plate circuit of a vacuum tube and indicate direction of electron flow. (Fig. 14-5)
14. Draw a diagram of impedance coupling between two tubes. (Fig. 14-6)
15. Draw a diagram of direct coupling between two triodes. (Fig. 14-7)
16. Draw schematic diagrams illustrating battery, voltage-divider, and cathode-resistor biasing. (Figs. 14-6, 14-8, 14-9)

17. How is the approximate value of cathode-bias resistance necessary to provide correct grid bias for any particular tube determined? (14-14)
18. If the desired bias voltage of a triode is -7 V and the cathode current is 40 mA, what cathode-resistance value is required? (14-14)
19. What size capacitor should be used to bypass sufficiently a 500-Ω cathode resistor if the lowest approximate frequency desired is 100 Hz? (14-14)
20. A 5-kΩ cathode resistor is bypassed to ground with a capacitor. If it is desired to operate this amplifier at a minimum frequency of 5 kHz, what size capacitor should be used? (14-14) [4]
21. What would probably be the effect on the output amplitude and waveform if the cathode-resistor bypass capacitor in an AF stage were removed? (14-14)
22. What is the purpose of bypass capacitors connected across a cathode-bias resistor? (14-14)
23. Is grid-leak biasing practical in AF amplifiers? (14-15)
24. Draw a diagram showing a method of obtaining grid bias for a filament-type vacuum tube by use of resistance in the plate circuit. (Figs. 14-11, 14-12)
25. What is the purpose of a center-tap connection on a filament transformer? (14-16)
26. What is meant by the *fidelity* of an AF amplifier? (14-18) [4]
27. What causes resistance noise in electrical conductors and shot-effect noise in diodes? (14-18) [4]
28. Why do vacuum tubes produce random noise? (14-18)
29. Draw circuit diagrams and explain the operation of two common tone-control circuits. (Figs. 14-15, 14-16)
30. Why are decoupling resistors and capacitors used in stages having a common power supply? (14-20)
31. What means, besides shielding and parts placement, is used to prevent interaction between the stages of a multistage AF amplifier (14-20)
32. What is a class A amplifier? (14-21, 14-22)

33. Compare the operating characteristics of class A, B, and C amplifiers. (14-21, 14-27–14-29)
34. What factors should be taken into consideration when ordering class A or class B AF output transformers to feed a speaker of known ohmic value? (14-22)
35. Would saturation of an output transformer create distortion? (14-22)
36. What is the effect of incorrect bias in a class A amplifier? (14-22)
37. Draw a diagram of a triode inductively coupled to a loudspeaker. (Fig. 14-19)
38. Draw a diagram and explain the operation of a class A push-pull amplifier. (14-23)
39. What are the advantages of push-pull as compared with single-ended operation? (14-23)
40. What is the advantage of using two tubes in push-pull over the two tubes in parallel? (14-23, 14-34)
41. When a signal is impressed on the grid of a properly adjusted and operated class A AF amplifier, what change in average value of plate current will take place? (14-23)
42. How may even-order harmonic energy be reduced in the output of an AF amplifier? (14-23)
43. Draw the $E_g I_p$ characteristic curve of a vacuum tube and indicate points for class A, B, and C operation. (Fig. 14-23)
44. Why does a class B AF amplifier stage require considerably greater driving power than a class A? (14-29)
45. In a class B AF amplifier should the plate current fluctuate or remain steady? (14-29)
46. During what portion of the excitation-voltage cycle does plate current flow when a tube is used as a class B amplifier? (14-29).
47. Why is it necessary to use two tubes in a class B AF amplifier? (14-29).
48. How is class B bias-voltage value determined? (14-29)
49. Why are permanent magnets used in earphones? (14-30)
50. Why should polarity be observed in connecting earphones directly in a circuit carrying varying dc? (14-30)
51. If low-impedance earphones are to be connected to the output of a vacuum-tube amplifier, how may this be done? (14-30)
52. What is the voltage gain of a cathode-follower amplifier? (14-30)
53. Draw a diagram and explain the principles of operation of a cathode-follower amplifier. (14-30)
54. Draw a transistor amplifier circuit that would be analogous to that of a vacuum-tube cathode-follower amplifier. (Fig. 14-27)

55. Draw a diagram illustrating a method of coupling a high-impedance loudspeaker to an AF tube without flow of plate current through the speaker windings and without the use of a transformer. (Fig. 14-29)
56. Why is impedance matching between electrical devices an important factor? Is it always to be desired? Can it always be attained in practice? (14-33)
57. An AF transformer has a resistive load connected across its secondary. What is the relation between this resistance, the turns ratio, and the input impedance at the primary terminals? (14-33)
58. A loudspeaker with an impedance of 4 Ω is working in a plate circuit which has an impedance of 4,000 Ω. What is the impedance ratio of an output transformer used to match the plate circuit to the speaker? What is the turns ratio? (14-33)
59. In a transformer having a turns ratio of 4:1, working into a load impedance of 320 Ω and out of a circuit having an impedance of 10 Ω, what value of resistance may be connected across the load for an impedance match? (14-33)   [4]
60. Draw a diagram and explain the operation of two tubes in parallel operated as a class A amplifier. (14-34)
61. List four causes of distortion in a class A AF amplifier. (14-35)
62. Why is noise often produced when an audio signal is distorted? (14-35)
63. In a low-level amplifier using degenerative feedback, at a nominal midfrequency, what is the phase relation between the feedback voltage and the input voltage? (14-36)   [4]
64. What is the purpose of deliberately introduced degenerative feedback in AF amplifiers? (14-36)   [4]
65. What is the technical requirement for negative feedback? (14-36)   [4]
66. Draw diagrams of a circuit having negative voltage feedback and of one having negative current feedback. (Figs. 14-32, 14-34)   [4]
67. What is the formula to determine amount of feedback if gain with and gain without feedback are known? Solve a sample problem illustrating this. (14-36)   [4]
68. Draw a diagram and explain the principles of operation of two types of phase inverters for feeding push-pull amplifiers from a single-ended stage. (14-37)
69. Draw a diagram illustrating resistance coupling between a pentode and a triode. (Fig. 14-38)
70. Why are pairs of wires carrying ac heater currents in AF amplifiers preferably twisted together? (14-39)

71. Name some causes of hum in audio amplifiers and the methods of reducing it. (14-39)
72. What would be the effect of a leaking or short-circuited coupling capacitor in a resistance-coupled AF amplifier? (14-41)
73. What is the dc plate voltage of a resistance coupled amplifier with a plate supply voltage of 300 V, a plate current of 2 mA, a plate-load resistance of 50,000 Ω, and a cathode resistor of 2,000 Ω? (14-41)

74. Draw a diagram and explain the principle of operation of a PNP transistor directly coupled to an NPN type. (14-43)
75. Draw a diagram and explain the principles of a class B push-pull AF amplifier using transistors. (14-43)
76. How is a thermistor used in stabilization of a transistor amplifier? (14-43)

ANSWERS TO CHECKUP QUIZ ON PAGE 325

1. (*The first*)(*Fourth*)(*Contact potential*)(*A*)(*Second*)(*Second*)
2. (*Produce proper PNP bias*)(*Too low R values*)(*Heat compensation*)(*Class B output*)(*Direct, RC, inductive*) 3. (*About +0.5 V*)(*About 6 V*) 4. (*No S to G conduction*) 5. (*Potentiometer*)(*Yes*) 6. (*Yes*)(*R from S to ground*)(*Difference in voltage-drop between R_1-Q_1-R_S and V_{CC_1}*) 7. (*Input circuit biasing*) 8. (*Cascode has devices in series*) 9. (*Ac*)(*Ac*) 10. (*About 1.2 V*)(*0 V*) 11. (*Three*)(*About 5 V*)

15

RADIO-FREQUENCY AMPLIFIERS

The objective of this chapter is to discuss briefly low-level radio-receiver, type RF amplifiers and then, at greater length, transmitter RF power amplifiers, both vacuum-tube and transistor. Unneutralized tetrode and pentode stages are described first, followed by neutralization methods with vacuum and semiconductor triode-type devices. Parallel, push-pull, and push-push amplifiers with various forms of coupling are considered. Special requirements for high-power amplifier stages are discussed, as are amplifiers for VHF and UHF. A section on amplifier troubleshooting practice is included.

15-1 RF AMPLIFIERS

In this chapter two basic types of high-frequency radio-frequency (RF) amplifiers will be discussed: (1) RF voltage, low-level, or small-signal amplifiers and (2) RF power amplifiers. The low-level amplifiers are found in radio receivers or in small-signal applications in transmitters. RF power amplifiers are used in transmitters or special applications where large values of RF ac power are required.

The RF spectrum is considered to be from 10 kHz to 300 GHz, although communications are usually limited to frequencies below 60 GHz.

The fundamental difference between AF and RF amplifiers is the bandwidth they are expected to amplify. AF amplifiers amplify a major portion of the AF spectrum (20 to 20,000 Hz) equally well throughout. RF amplifiers amplify only a relatively narrow portion of the RF spectrum, attenuating all other frequencies. Radiotelegraphic code

transmissions (called CW or A1) require receiver amplifiers of 200- to perhaps 1,000-Hz bandwidth. Single-sideband (SSB or A3J) voice transmissions require about a 3-kHz bandwidth. RF carrier waves amplitude-modulated (AM or A3) by voice frequencies require 3 kHz each side of the carrier, or a 6-kHz bandwidth. Standard broadcast band AM for voice and up to 5,000 Hz for music requires at least 5 kHz on each side of the carrier, or 10 kHz. Frequency-modulated (FM or F3) broadcast carriers are operated to utilize a bandwidth of 200 kHz. Television transmissions (TV or A5C) use a 6-MHz bandwidth. Radar may use a bandwidth up to 10 MHz. Even the bandwidth of radar emissions represents only a small portion of the whole RF spectrum.

RF amplifiers are tuned to a desired frequency, amplifying only that frequency and a relatively few frequencies on both sides. AF amplifiers may amplify frequencies from perhaps 20 to 100 Hz up to perhaps 800 Hz for CW, 3 kHz for voice, and up to 5 to 20 kHz for music. (Special *wideband,* or *video,* amplifiers are usually directly coupled types, not unlike basic AF amplifiers, but may be flat from dc to 5 MHz or possibly 100 MHz in some applications. These are used in oscilloscopes, TV, and radar; they are discussed in later chapters.)

In general, amplifiers with tuned LC circuits accept the resonant frequency and a few frequencies above and below. At 50 kHz LC circuits tend to have high Q and narrow bandwidth, perhaps as low as 3 kHz. At 500 kHz the bandwidth may be 5 to 30 kHz, again depending on the Q of the LC circuits. At 5 MHz the bandwidth may be in the hundreds of kilohertz. Thus, for applications requiring amplification of a wide band of frequencies, signals can be converted to high-frequency RF and then amplified. If only the frequencies in a narrow band are to be amplified, signals may be converted to lower RF frequencies and be amplified at those frequencies. However, by utilizing the high-Q capabilities of piezoelectric crystals, quite narrow bandwidths can be obtained for any radio frequencies up to about 15 MHz with crystal filters. Radio receivers and low-level stages of transmitters take advantage of the passband of LC and other filter circuits. This is discussed in Chaps. 16 through 19.

At high frequencies the Q of coils decreases because of skin effect and increased core losses. The dielectric losses in capacitors increase. Interaction between stages in receivers or transmitters increases. Shielding becomes more necessary and more difficult to achieve. Wiring between parts of a circuit must be made shorter. Equipment must be made more compact. Insulating materials, such as Bakelite, paper, cambric, black rubber, and cotton, usually satisfactory at lower frequencies are found to have too much loss at higher RF frequencies. Special insulating materials, such as mica, steatite, Isolantite, ceramics, and special plastics, must be used. In general, at higher frequencies circuits become more cranky, more difficult to manage. New, smaller parts and more advanced circuitry, however, are gradually widening the usable spectrum space. Before 1940, little commercial use was being made of frequencies above 100 MHz. At present, frequencies up to at least 60 GHz are in daily commercial use.

15-2 LOW-LEVEL RF AMPLIFIERS

A low-level, or small-input-signal RF amplifier can use either vacuum tubes or transistors, is biased to class A, and may be involved with ac signals in the microvolt or microampere range. Both input and output loads may be resonant circuits in small-input-signal RF amplifiers, but in RF power amplifiers the output circuit is always tuned.

Pentodes are the vacuum tubes most commonly used in low-level RF amplifiers. Both BJTs and FETs are commonly used solid-state devices

in small-signal circuits. While the basic explanations will be given in vacuum-tube terms, there should be little trouble in converting to transistor circuitry as long as it is remembered that vacuum tubes and FETs are high-impedance input devices, but BJTs have low input impedance. This is illustrated in Fig. 15-1, in which the grid-

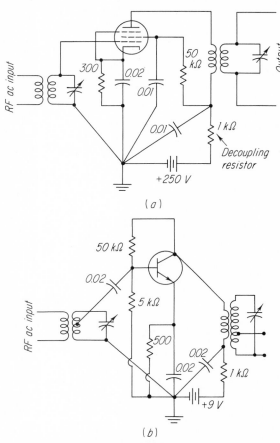

Fig. 15-1 Representative small-signal-input RF amplifiers. (a) Pentode VT. (b) NPN BJT with degenerative voltage-divider bias and emitter stabilization.

cathode connections are made across the whole tuned circuit, while the base-emitter circuit of the transistor is connected across only a small fraction (low-impedance section) of the tuned circuit.

In the vacuum-tube diagram, screen-grid current flowing through the 50-kΩ screen-grid dropping resistor produces a voltage-drop across it, presenting a lower voltage to the screen than that applied to the plate. If the screen requires 3 mA at 100 V, for example, but the power supply is 250 V, the screen dropping resistor must lose 150 V across it. The resistance required to drop the voltage 150 V when 0.003 A flows through it is found by $R = E/I$, or $150/0.003$, or $50,000\ \Omega$. Different pentodes or tetrodes use other values of resistance, of course. When the power supply produces 100 V or less, the screen-grid dropping resistor may be eliminated, allowing the screen to operate at the same voltage as the plate.

The 0.01-μF screen bypass capacitor and 1-kΩ resistor shown in Fig. 15-1a form an RC filter. They hold the screen-grid voltage essentially constant even when the current flowing through the tube varies. Without the filter capacitor, a positive signal voltage on the control grid produces an increased plate and screen current. The increase in screen current results in a voltage-drop of more than 150 V across the dropping resistor and therefore less than 100 V between screen grid and cathode. When the screen-grid voltage decreases, plate current also decreases. Therefore, the screen grid will tend to decrease I_p when the signal on the control grid should be increasing I_p. This is a degenerative effect and results in a loss of amplification. For the short duration of one RF cycle, the charged 0.01-μF capacitor can maintain the screen voltage at approximately the 100 V required. For an AF cycle, which lasts a much longer time, a 0.1-μF capacitor might be required to hold the voltage steady enough.

The 300-Ω cathode resistor biases the tube to class A. Since the function of the receiver RF amplifier is to produce voltage amplification, a class A amplifier using a pentode tube with high transconductance fulfills the requirements. The 0.02-μF capacitor across the cathode-bias resistor holds the bias voltage essentially constant even when signal voltages on the control grid produce a variation in plate current through the resistor. It may also be considered as forming a low-reactance or low-impedance path from cathode to ground, resulting in a decrease in degenerative effect that would otherwise decrease the amplification of the stage.

The 0.01-μF bypass capacitor connected from the bottom of the plate-circuit load to ground tends to hold the voltage across it constant. It may also be considered an ac signal completion circuit from plate load to cathode.

Notice that all leads of the RF amplifier are brought to one point. This is important in high-frequency construction. Using a single ground point for audio amplifiers is desirable in most cases, but in RF amplifiers it becomes a necessity if minimum interaction between stages is required. Each stage should have a separate ground point. The higher the frequency being amplified, the greater the necessity of using single ground points. All ac grid, plate, and ground return leads must be as physically short as possible. Instead of amplifying, a stage with long leads may produce self-sustained oscillation, or it may interact with one or more other stages. Note the RF amplifier coils and transformers have air cores. AF amplifiers have iron cores.

In Fig. 15-1 the input circuit is tuned and the output is untuned. Although the output is not tuned, it is coupled close enough to the next stage's tuned circuit so that a tuned-plate tuned-grid (TPTG) oscillator may be produced if any capacitive feedback between output and input circuits exists. These leads must be kept short and well separated to minimize stray capacitive coupling across a tube or transistor.

To prevent inductive coupling due to the fields of one tuned circuit inducing voltages into the other, both tuned circuits may be encased in metal shield cans. The best shielding material is the best conductor, silver, but aluminum or copper is generally used. Since RF currents induced into the inner walls of the shield tend to travel on the inner surface only (because of skin effect), an excellent shield is produced by silver-plating the inside of a thin copper can. Practically all communications receivers shield all RF and IF (intermediate-radio-frequency) coils to prevent erratic tuning, unwanted oscillations, or reception of unwanted signals (Chap. 18).

Figure 15-1b illustrates low-signal RF amplification using a BJT.

15-3 RF POWER AMPLIFIERS

The remainder of this chapter pertains to RF amplifiers as used in high-power high-frequency applications such as radio transmitters. These amplifiers are designed to take a relatively high-amplitude RF ac input signal, amplify it, and produce enough power output to drive another, higher-power RF amplifier, excite a transmitting antenna, or operate other loads (nuclear research, etc.). Since present transistors are not capable of high power output (over 500 W) at high frequencies, the discussion will be mostly in vacuum tube terms.

While the small-signal RF amplifier is basically class A in operation, distorting the signal very little at any time, the RF power amplifier is usually biased from $1\frac{1}{2}$ to 4 times the plate-current cutoff, allowing it to operate as a high-efficiency class C amplifier. Ten watts applied by the power supply to a class A stage may produce only about 2.5 W of ac output (25% efficiency). The same tube, biased to class C, may be able to produce 6 W output (60% efficiency). To do this, however, it is necessary to change the sine-wave input signal to pulses of plate current in the output circuit and then rely on the flywheel effect of the plate-circuit tuned coil and capacitor to restore the sine-wave shape to the output signal ac.

In many cases in which class C amplifiers are used, either class A or class B could be used but with lower efficiency. When it is desired to amplify RF ac signals without distortion, class A, AB, or B is used.

Bias for the different classes can be determined by the formulas

$$\text{Class A (0.66 cutoff)} = \frac{0.66E_p}{\mu}$$

$$\text{Class B (cutoff)} = \frac{E_p}{\mu}$$

$$\text{Class C (1}\tfrac{1}{2}\text{–4} \times \text{cutoff)} = \frac{1.5E_p}{\mu} \text{ to } \frac{4E_p}{\mu}$$

EXAMPLE: For a tube with a plate voltage of 1,250 and a μ of 25,

$$\text{Class A} = \frac{0.66(1,250)}{25} = 33 \text{ V}$$

$$\text{Class B} = \frac{1,250}{25} = 50 \text{ V}$$

$$\text{Class C} = \frac{1.5(1,250)}{25} = 75 \text{ V}$$

$$= \frac{4(1,250)}{25} = 200 \text{ V}$$

For transmitting tetrodes, the cutoff bias voltage is approximately one-fifth of the screen-grid voltage, or $E_{co} = 0.2E_{sg}$. Class C bias ranges from $0.3E_{sg}$ to $0.8E_{sg}$.

A common-emitter transistor circuit with no forward bias has no emitter-collector current and is therefore in class C. Any RC emitter-stabilization circuit produces reverse bias for the base when a signal is applied and emitter current begins to flow through it. This places the stage in class C operation when driven. Care must, however, be taken not to overdrive the base-emitter circuit.

15-4 BATTERY-BIASED CLASS C RF POWER AMPLIFIER

Shown in Fig. 15-2a is a possible beam-power tetrode RF power-amplifier circuit with power-supply bias. The plate-current cutoff bias is

Fig. 15-2 RF power amplifiers: (a) tetrode VT and (b) NPN BJT. Direct lines indicate wiring that must be as short as possible.

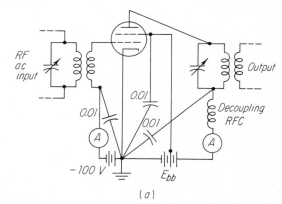

(a)

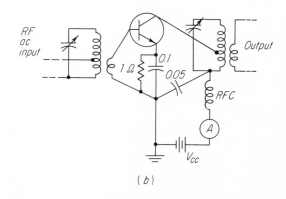

(b)

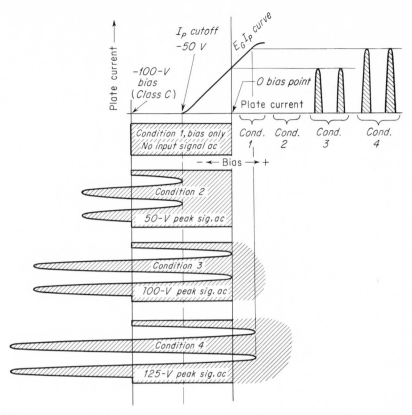

Fig. 15-3 E_gI_p curve and four possible plate-current conditions with class C amplifier operation.

−50 V. Since the grid is biased to −100 V, the tube is biased to class C. With no signal voltage being applied from the RF ac source, the bias holds the plate and screen-grid currents to zero. This condition is indicated on the E_gI_p curve of Fig. 15-3 as *condition* 1. The curve shows the cutoff bias point as −50 V, the working bias value of −100 V, and the zero bias point. The resultant plate current during condition 1 is zero.

In *condition 2,* the RF ac source induces a 50-V peak in the grid-circuit coil. The ac is added to the −100-V bias, producing, in effect, a varying dc bias. This varying bias ranges alternately from −50 to −150 V, but still no plate current flows in the tube because the grid voltage is still not driven to a lower value than cutoff.

In *condition 3,* the RF ac source induces a 100-V peak into the grid coil. This, added to the −100-V bias, produces a dc that varies from zero to −200 V. During the portion of the input-signal

cycle when the bias is actually effective under the E_gI_p curve, plate current flows as a steep, narrow pulse. Note that the width of the pulse is less than a half cycle. The plate current flows for less than 180° of the input cycle, for 120° in this case. Thus far, the grid has not been driven into the positive region, and therefore grid current does not flow in the grid circuit. This is a possible condition for a class C stage but is not the optimum or normal condition.

Condition 4 is the usually accepted condition for class C RF amplifiers. The RF source induces a voltage with a peak value considerably greater than the bias value. This drives the grid into the positive region for a portion of each cycle, causing a pulsating dc grid current to flow from the cathode through the tube to the grid, down through the grid coil and the ammeter, and through the bias battery in a direction opposite to normal current flow for a battery. The plate-current peaks are considerably higher in amplitude and exist for more than 120°.

Plate current heats the plate of a tube. The plate-dissipation rating of a tube expresses how much average heat the plate is capable of dissipating without overheating dangerously. The use of high bias voltages results in narrow plate-current pulses. Thus, class C amplifiers have a relatively long duration between pulses, allowing the tube to rest for a major portion of each cycle. During the time the tube is working it can be driven very hard, and still its average plate dissipation may not exceed the plate-dissipation rating. For this reason class C amplifiers are high in efficiency. Class B amplifiers, biased to cutoff, have plate current flowing for a full half of the input cycle, rest proportionately less, and are therefore less efficient. Class A amplifiers have plate current flowing all the time and tend to overheat their plates unless the plate voltage is held to a lower value than can be used with class B and C amplifiers.

It is notable that an RF power amplifier may be biased to the class A or AB_1 value but be driven into the positive grid region and will draw grid current. While not the accepted class A or AB_1 operation, it can be used in RF amplifiers and results in a considerable increase in stage efficiency.

As in low-level RF stages, screen-grid, grid, plate, and cathode circuit bypass capacitors should return to a single ground point. Some special high-frequency vacuum tubes may have two (or more) leads to the cathode. By grounding both of them, the cathode-to-ground inductance is halved, which helps to stabilize the circuit.

Current varies in the screen-grid circuit at the same time as does the plate current, although at a considerably lower amplitude. Two important points to remember about the screen grid in an RF amplifier are: (1) Use short leads on the bypass capacitor. (2) Do not apply higher screen-grid voltage than recommended in tube manuals or excessive plate and screen currents may decrease tube life.

The radio-frequency choke (RFC) coil in the plate circuit may not be needed. If the bypass capacitor can maintain the bottom of the plate-circuit coil at essentially RF ac ground potential,

the additional filtering or decoupling of the RFC may be unnecessary.

Radio-frequency chokes may be constructed as long single-layer coils, but may exhibit parallel-resonant (high-impedance) or series-resonant (low-impedance) effects to certain frequencies. The resonant peaks are harmonically related. To overcome such undesirable effects, RF chokes may be made up of three, four, or five separate, often *universal-wound* (layers wound on top of other layers) *pies,* with all pies connected in series as shown in Fig. 15-4. This produces a high

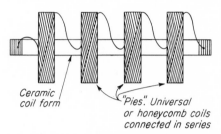

Ceramic coil form

"Pies". Universal or honeycomb coils connected in series

Fig. 15-4 Radio-frequency chokes are usually constructed in pies.

impedance to practically all frequencies over the relatively wide band for which the choke is manufactured and greatly reduces end-to-end distributed capacitance. The lead nearest the core should be connected to any high-impedance RF ac point. Circuits sometimes operate better if the RFC is reversed.

A BJT counterpart of the VT RF power amplifier is shown in Fig. 15-2b.

Test your understanding; answer these checkup questions.

1. Three RF amplifier systems are available, a 50-kHz, a 500-kHz, and a 40-MHz. To which of these would radar signals be converted? _____ AM broadcasts? _____ CW? _____
2. In what way do input circuits of BJTs differ from those of VTs or FETs? _____
3. In Fig. 15-1a, what might be removed to produce degeneration? _____
4. In Fig. 15-1b, what produces degeneration? _____
5. How can electromagnetic feedback from output to input of an amplifier be decreased? _____ Electrostatic? _____
6. What would be the output waveform if pulses of dc were fed to a class C AF amplifier? _____ To a normal class C RF amplifier? _____

7. How is bias value for a class C RF triode amplifier computed? _____ For a tetrode? _____
8. With zero forward bias, to what class is a BJT biased? _____ A depletion MOSFET? _____ An enhancement MOSFET? _____
9. Does control-grid bias affect the E_{sg} in the same way it does I_p? _____
10. In a power-supply-biased class C stage is it possible to have an input signal and no output signal? _____
11. What is the approximate angle of I_p flow in a properly operating class C stage? _____
12. Will I_g flowing in a bias battery charge or discharge the battery? _____
13. Can active devices operate with higher power-supply voltage if in class C? _____
14. What is sometimes added to RF amplifier tubes to improve HF operation? _____

15-5 TUNING THE PLATE CIRCUIT

The plate current flowing through a class C tube occurs as short-duration pulses of dc which appear to the plate LC circuit more as a distorted form of ac than as a dc. Consequently, it is reasonable to consider that the tuned circuit sees the tube and power supply as a source of quasi ac energy. The LC circuit is a parallel-resonant circuit connected across an ac source, presenting a high impedance to the pulses. Little plate current flows. The plate-circuit meter will read a minimum value.

If the plate LC circuit is detuned from the frequency of the plate-current pulses, it presents a lower impedance to the pulses and allows a greater pulsating current to flow through the source and the LC circuit. The ammeter in the plate circuit will read a higher value. As a result, the plate-current ammeter makes a good indicator to determine when the plate LC circuit is tuned to resonance. At resonance the plate-current meter reads minimum. The farther from resonance the greater the plate-current value. The same is true in transistor collector circuits.

In a parallel circuit at resonance, a maximum number of electrons oscillate back and forth between the coil and capacitor of the LC circuit. Because they are oscillating at the same frequency as the incoming pulses from the source, little energy is required from the source pulses to maintain oscillations and the ac voltage across the LC circuit is nearly equal to the voltage of the source. (Since the LC circuit is in parallel with the

source, the peak RF ac should approximate the source-voltage value.) At resonance it takes little energy and therefore little plate current to produce this ac voltage value in the LC circuit. Thus maximum RF current is circulating in the LC circuit when dc plate current is at a minimum.

Because the plate is taking a minimum number of electrons from the space charge of the tube when the plate load is tuned to the resonant frequency, the grid is free to attract a maximum number during its positive swing. As a result, the grid current, as read on a grid-circuit milliammeter, rises somewhat when the plate circuit is tuned to resonance. The grid current falls off as the plate circuit is detuned or if the plate current increases for any reason. As a result, a grid-current meter may indicate plate-circuit resonance. I_{sg} also increases when I_p decreases, but meters are rarely included in the screen-grid circuit.

15-6 COUPLING THE RF AMPLIFIER TO A LOAD

The RF ac energy oscillating in the plate LC circuit is coupled into an antenna, into the grid circuit of another RF amplifier, or into some other kind of load circuit. If the coupling between the plate LC circuit and the circuit that follows is very loose, little energy will be drawn from the plate-tank circuit. In this case, the plate-current dip, when the LC circuit is tuned to resonance, will be very deep. For example, if the off-resonance plate-current maximum is 200 mA, the resonant value may dip to perhaps 20 or 30 mA (and the screen current may rise dangerously).

If the coupling between the plate tank and the following circuit is increased, the tank circuit will have more energy drawn from it. This lowers the parallel-circuit impedance value of the tank circuit, increasing the pulsating current flowing through the LC circuit and the source. The plate-current dip will no longer be as great. The off-resonance value may still be 200 mA, but the value at resonance may now be 100 mA. If the load is too tightly coupled to the tank circuit, it may be found that no plate-current dip will be noticeable.

How much coupling should be used? The best answer to this question is to consult a manual giving the operating characteristics of the tube being used. Adjust the plate and screen voltages to the recommended values. Adjust the grid bias and grid drive to produce the recommended value of grid current. Then, starting with loose coupling, keep increasing the coupling and dipping the plate current by tuning the plate circuit until the plate-current minimum at the dip is the value recommended in the manual. Recheck the E_p, E_{sg}, and $-E_{cc}$ voltages and the I_g and readjust to the recommended values if necessary.

In an emergency adjusting the coupling until the plate current at the dip is about 75% of the off-resonance maximum should result in fairly efficient operation. If the off-resonance value is 200 mA, the coupling should be increased until the maximum dip reads 150 mA. Proper operation of an RF amplifier rarely results with less than a 20% dip in plate current from the off-resonance value.

A more accurate method of determining the proper coupling to the output circuit is to use some means of measuring the output power or of obtaining a relative indication of it. An RF ammeter in series with the load circuit, as in Fig. 15-5, will serve as an indicator of the relative power output of an amplifier circuit. An RF wattmeter may also be used.

The plate-power *input* to the amplifier is equal

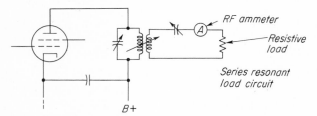

Fig. 15-5 RF power output can be determined by multiplying the load resistance in ohms by the RF current squared ($P = I^2R$).

to the product of the power-supply voltage times the plate-current meter indication, or $P_i = E_{bb}I_p$ (or E_pI_p, since E_{bb} and E_p are the same). With loose coupling to the load circuit, both the dc plate input power at resonance and the RF ac output-power value will be low. As the coupling is increased, both dc input and RF output values will increase. A degree of coupling at which the RF output power will show no further increase will be reached, but the dc input power will continue to increase as coupling is increased. Further increase of coupling will usually result in a decrease in RF output and a still further increased input power. This indicates that optimum coupling has been passed. The coupling should be reduced to the lowest value of dc power that will give the greatest RF output. At this degree of coupling a condition of approximate impedance match between the plate impedance of the tube and the tank-circuit impedance occurs, resulting in a maximum power being developed in the load circuit.

It is usually possible to obtain a still greater power output by increasing the capacitance and decreasing the inductance of the tank circuit, or vice versa, and running through the tests again. When maximum RF output results with as little dc power as possible being used to produce it, optimum coupling has been attained. Care must be taken not to operate a tube (or transistor) at plate (collector) current values higher than those recommended by the manufacturer of the device.

Exact values for capacitance and inductance of the plate tank, degree of coupling, and load impedance are not easily determined. Fortunately, a rather wide variation of L and C values from the optimum will still produce satisfactory output.

The plate-circuit efficiency is the ratio of RF power output ($P = I^2R$) to the dc plate-power input ($P = E_pI_p$), or

$$\text{Percent efficiency} = \frac{P_{\text{rf}}}{P_{\text{dc}}} \times 100$$

Of the total dc plate-power input, some is dissipated as heat by the plate of the tube, a little is dissipated as heat in the LC tank and other wiring of the circuit, and the remainder is the RF power output. For example, if the plate-power input to a tube is 1,000 W and the RF power output is 700 W, plate dissipation is the remainder, or approximately 300 W. The efficiency of the stage, from the formula given above, must be (700/1,000) × 100, or 70%.

Maximum output power is obtained when the load impedance matches the source impedance, but the higher the load impedance in comparison with the plate impedance of the tube the greater the plate-circuit efficiency. However, above the 1:1 matching ratio the higher the ratio of load impedance to plate impedance the higher the efficiency but the lower the power output. A transmitter should not be adjusted with highest plate-circuit efficiency as the only consideration.

15-7 GRID-LEAK BIAS FOR RF AMPLIFIERS

Grid-leak bias (Sec. 13-5) can be used only if grid current flows. It cannot be used in receiver or other small-signal RF amplifiers because no grid current flows in these circuits.

Grid-leak bias depends upon the flow of grid current through a grid-leak resistor when the grid is driven into the positive region by the signal from the preceding stage. This requires a power amplifier as the driver stage to produce the required grid current.

Figure 15-6 shows a grid-leak-biased class C

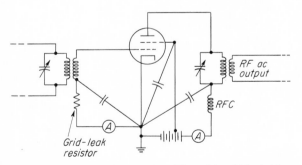

Fig. 15-6 Tetrode RF power amplifier with grid-leak bias.

tetrode RF amplifier stage. The grid-leak resistor is selected to produce the proper average class C bias voltage. The voltage developed by grid-leak bias varies in amplitude due to the positive and negative peaks of the input RF ac signal voltage, but the average value (held by the grid-circuit capacitor) is more than that required to

produce plate-current cutoff. Therefore, the stage is biased to class C.

The value of the bias voltage can be determined by Ohm's law

$$E = IR$$

where I = grid current in A
R = grid-leak resistance in Ω

If the average grid current, as read on the meter, is 12 mA and the resistance is 10,000 Ω, the average grid bias is 0.012(10,000), or 120 V. If this bias voltage is less than is desired, the coupling from the RF ac source can be tightened or the value of the grid-leak resistance may have to be increased.

There is one danger with grid-leak bias. It is normally used in class C amplifiers with relatively high plate voltage. If the driver stage suddenly ceases operating for any reason, the RF amplifier is left with no input signal, zero bias, and high plate voltage. This will produce dangerously high plate and screen currents that may damage the amplifier tube unless some precaution is taken. Because of this danger, many class C amplifiers use either an external bias supply plus grid-leak bias or some cathode-resistance bias (Sec. 14-14) in addition to the grid-leak bias. In this way, if the drive is suddenly interrupted, the fixed bias will be sufficient to limit the plate and screen current of the amplifier to some safe value.

Another safety device is a plate-circuit *overload relay*. If excessive plate current flows through the coil of an electromagnetic relay in series with the plate circuit, the relay arm is pulled in, trips a latch, and shuts off the plate and screen-grid power supplies.

15-8 PARALLEL OPERATION

To produce twice the RF power output possible from one tube or transistor, two or more tubes or transistors may be connected in parallel or in push-pull.

A parallel RF amplifier circuit is shown in Fig. 15-7. Parallel-connected tubes have their similar elements connected together, cathode to cathode, grid to grid, etc. Parasitic chokes are connected as close as possible to the plate terminals

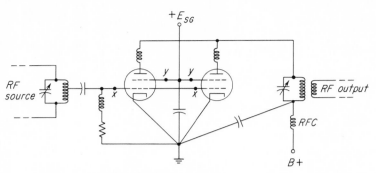

Fig. 15-7 Parallel tetrode tubes in an RF power amplifier.

to discourage high-frequency parasitic oscillations (Sec. 13-22).

To test an amplifier stage for parasitic oscillations, the RF drive is first removed, leaving the stage with grid-leak bias only. The plate and screen voltages are adjusted low enough to limit the plate and screen currents to values that will not exceed the plate- and screen-dissipation ratings of the tubes. Any material value of grid current indicates parasitic oscillations. To keep parasitic oscillations from occurring at higher frequencies than the operating frequency, parasitic chokes or 10- to 100-Ω noninductive resistors may also have to be added to the control grids (X) or the screen grids (Y) or both. If the oscillation is occurring near the operating frequency, neutralization (Sec. 15-13) is required. If lower than the operating frequency, RF chokes in the grid and plate circuits may be causing the parasitic oscillation.

Transistors made for VHF or UHF operation may give parasitic trouble when used in HF systems.

Two tubes in parallel lower the tube impedance as seen by the plate-tank circuit. This may require lowering the value of inductance and raising the capacitance of the tuned circuit in order to match tube and load impedances. More oscillatory current will be flowing through the coil, which may require a coil made with larger-diameter wire.

Two tubes in parallel will have twice the plate-to-cathode, plate-to-grid, and grid-to-cathode interelectrode capacitances, which may be a disadvantage for very high frequency (VHF) operation.

15-9 PUSH-PULL OPERATION

RF amplifier tubes or transistors can be connected in push-pull. The output power from two such devices will be twice that obtainable from one alone in a single-ended circuit.

Figure 15-8 shows a pair of tetrode tubes connected in push-pull. An advantage of push-pull stages is their tendency to cancel even-order (2d, 4th, etc.) harmonic energy in their output. To do this effectively, each tube must be fed the same-amplitude signal voltage, the plate-tank cir-

Fig. 15-8 Push-pull tetrode RF power amplifier.

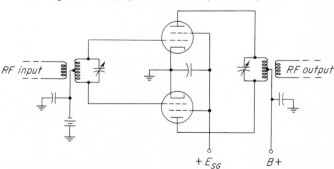

cuit must be accurately center-tapped, the coupling to the output should be taken equally from both halves of the plate-tank circuit, and the two tubes must be evenly matched.

The plate-to-cathode and grid-to-cathode interelectrode capacitances across the tuned circuits are in series. This results in only half the capacitance across the tuned circuits that would be present with one tube alone and one-quarter the capacitance of parallel operation, a decided advantage at higher frequencies.

Note that multiple ground indications simplify diagraming. This does not indicate that a single ground point is not necessary in the stage. It is assumed that the reader now understands that each stage has a single grounding point.

15-10 TYPES OF FEED

As mentioned in Sec. 13-7, it is possible to feed plate current from the tube through the coil of the tuned plate tank circuit (series feed, Fig. 15-9a),

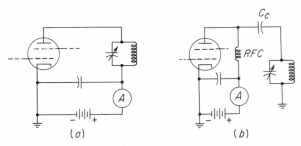

Fig. 15-9 (a) Series-fed and (b) shunt-fed RF amplifier plate circuits.

or else the plate current can be led off through an RF choke. In the latter case, the reactive ac voltage-drop across the RFC developed by pulsating dc flowing through it is coupled to a tuned circuit by a coupling capacitor C_c (shunt feed, Fig. 15-9b).

In either series- or shunt-fed circuits, as the plate LC circuit is tuned to resonate at the frequency applied to the grid, the plate-current value will dip. With shunt feed, one side of the tuning capacitor is grounded. This is a mechanical advantage to the manufacturer of equipment in some cases. Also, there are no high dc potentials on the coil or capacitor. Either series or shunt

feed can be used in single-ended, parallel, or push-pull stages.

Test your understanding; answer these checkup questions.

1. What is a simple way to tell when an RF power amplifier output LC tank is at resonance? _____
2. With a tuned and unloaded LC circuit in a class C amplifier with E_{bb} of 1 kV, what voltage values might be developed between plate and cathode? _____ If heavily loaded? _____
3. What, if anything, happens to I_g when I_p increases? _____ Why? _____
4. Why might it be dangerous to operate power tetrode stages with no load? _____ Power BJT stages? _____
5. Should class C BJT power amplifiers show an I_c dip if the LC tank is resonated? _____
6. Does "efficiency = P_{rf}/P_{dc}" apply equally to VT and BJT circuits? _____
7. Is a class C power amplifier as efficient with grid-leak bias as with power-supply bias? _____ What is the main danger of using grid-leak bias? _____
8. What might be the advantage of using a 50-kΩ grid-leak resistor over a 10-kΩ? _____
9. What is probably the main disadvantage of using parallel tubes over push-pull? _____ The main advantage? _____
10. An undriven, unloaded grid-leak-biased RF amplifier has its E_p reduced to one-tenth of normal. Why would some I_g be indicated? _____ If I_g reduces as a load is coupled, what is indicated? _____

15-11 METHODS OF COUPLING RF AMPLIFIERS

The two stages in Fig. 15-10 are inductively or transformer coupled. With tuned inductive coupling between stages, as shown, the circuits will require very loose coupling. They may have to be placed so far apart physically that more space may be required to construct them than is desired. It may be preferable to close-couple the two coils and tune only one of them.

Fig. 15-10 Inductive or transformer coupling between two RF amplifiers.

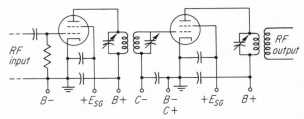

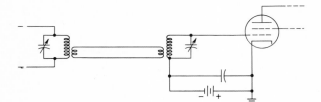

Fig. 15-11 Link coupling between two tuned circuits.

The two-transformer method of coupling shown in Fig. 15-11 is known as *link* coupling. The plate coil is the primary of a step-down transformer, and the grid coil is the secondary winding of a step-up transformer. The first transformer ratio reduces the impedance of the link line to a low value. The second steps it up again. Energy can be carried long distances with little loss, particularly if coaxial cable is used between the terminating loops. As a result, the intercoupled plate and grid circuits need not be physically close together, as is necessary with other forms of coupling. Link coupling reduces stray capacitive coupling, thereby reducing transmission of unwanted harmonic frequencies that might be generated in the plate circuit. This is aided by grounding one side of the link line.

A form of capacitive or impedance coupling is shown in Fig. 15-12. The degree of coupling can

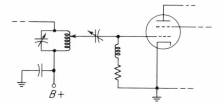

Fig. 15-12 Capacitive or impedance coupling in an RF amplifier.

be determined either by adjustment of the coupling capacitor or by the tap on the coil. The closer the tap to the ground end of the coil, the

---

---

less the coupling. The RF choke may not be necessary in the grid circuit. This form of coupling finds wide use, but it tends to transfer harmonics generated in the plate of the driving stage to the next grid circuit.

Two forms of direct coupling, inductive and capacitive, are shown in Fig. 15-13. The coupling

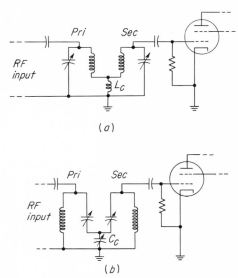

Fig. 15-13 (*a*) Direct-inductive and (*b*) direct-capacitive coupling circuits.

coil L_c and the coupling capacitor C_c are common elements in these coupling systems. The reactive voltage developed across these coupling reactors by current in the primary circuits produces current in the secondary circuit. The smaller the inductance L_c and the larger the coupling capacitance C_c, the less the coupling between circuits. There is no mutual inductance between coils.

15-12 THE TRIODE RF AMPLIFIER

Triode tubes are also used in RF power amplifiers, although tetrodes and pentodes have the advantage of greater sensitivity, require less RF drive to their grids, and do not require neutralization. Thus far, tetrode or pentode tubes have been used in the explanations of RF amplifiers because such tubes usually require no out-of-phase feedback, or neutralization. A triode-type RF amplifier (including BJTs and FETs), with plate and grid circuit tuned to the same frequency, becomes a TPTG oscillator because of feedback from plate to grid circuits via the grid-plate interelectrode capacitance. If an oscillator is desired, such a circuit might be used, but when the stage is supposed to be an RF amplifier, it must not break into self-oscillation. Introduction of the proper amount of neutralizing voltage into the circuit will prevent self-oscillation. If an RF amplifier in a transmitter is not properly neutralized, it may generate many spurious signals, interfering with other radio services. The plate current may not dip smoothly as the plate circuit is tuned through resonance (a common indication of an improperly neutralized stage), or there may be RF output when the driving ac ceases.

There are several ways by which neutralization may be introduced into a stage to prevent it from oscillating. In vacuum-tube terms, some of these are (1) plate neutralization, (2) grid neutralization, (3) direct neutralization, (4) inductive neutralization, and (5) use of a losser resistor.

The general theory of the first four methods is somewhat the same. Any ac emf fed back to the grid from the plate circuit by interelectrode capacitance is counteracted by feeding back another equal but opposite phase voltage, resulting in a neutralizing of the feedback effect. Although the collector-base capacitance of transistors is relatively low, as is the base-emitter impedance, transistor RF amplifiers must also be neutralized.

The losser resistor consists of a resistor in series with the grid lead as in Fig. 15-14. The loss

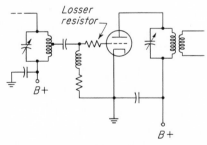

Fig. 15-14 A losser resistor in the grid circuit of an RF amplifier.

of energy in the grid resistor plus the out-of-phase, or degenerative, voltage developed across the resistor whenever current flows through it adds enough loss to the input circuit to counteract small values of regenerated energy. A losser re-

sistor prevents HF parasitic oscillations and also produces partial neutralization at the frequency to which the amplifier is tuned. The losser-resistor value should be as low as possible to produce the desired suppression of oscillation without too much loss of amplification. Values used may range from 50 to more than 1,000 Ω. Losser resistors may also be included in screen-grid and plate leads in place of parasitic chokes to stabilize the operation of the stage. An unbypassed cathode-biasing resistor also discourages self-oscillation of an amplifier and acts as a losser resistor.

15-13 PLATE NEUTRALIZATION

Plate neutralization, also known as a *Hazeltine balance* circuit, is shown in Fig. 15-15. Interelec-

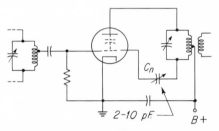

Fig. 15-15 A plate-neutralized triode RF amplifier.

trode capacitance between grid and plate (dotted capacitor inside tube) can act as a conductor of RF energy from the plate-tank circuit back to the grid circuit. As explained in Chap. 13, this will produce a sustained oscillation in the amplifier stage if the plate and grid circuits are tuned to approximately the same frequency. If they are not tuned to the same frequency, the stage may not oscillate, but neither will it effectively amplify the frequency fed to the grid by the source of RF ac.

Notice that the plate-tank-circuit coil is center-tapped with the center tap bypassed to the cathode (ground). This develops a zero RF-potential, or ground-potential, point at the center of the coil. With the center of the coil at zero RF potential at all times and with the tank circuit oscillating, one end of the LC circuit will be RF-positive at the same instant the other end is RF-negative, and vice versa. Insofar as the coil is concerned, the two ends always have equal-amplitude voltages but opposite RF polarities.

The top end of the tank circuit is feeding RF energy back to the grid via the grid-plate capacitance of the tube and is attempting to make the stage oscillate. When a *neutralizing capacitor* C_n, having the same capacitance as the interelectrode capacitance of the tube, is connected between the bottom of the tank circuit and the grid of the tube, the grid is fed two equal and opposite voltages at the same time, one through the tube and one through the neutralizing capacitor. The net result is zero effective feedback to the grid from the plate circuit, and the stage will not oscillate. It is neutralized.

In practice, the neutralizing capacitance is a variable capacitor in order that the proper capacitance can be obtained by trial.

Actually, it is not necessary to center-tap the plate coil exactly. If the tap is closer to the bottom of the coil, a smaller neutralizing voltage will be developed across the lower part of the coil. This can be overcome by adjusting the neutralizing capacitor to a higher value of capacitance.

Instead of center-tapping the plate coil. a *split-stator* capacitor may be connected across the tuning coil as in Fig. 15-16. This is a dual variable

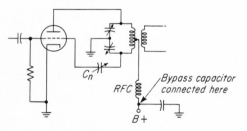

Fig. 15-16 Plate-neutralized triode RF amplifier with a split-stator capacitor.

capacitor with a common rotor connection for both stator sections. It is used to center-tap the capacitive half of the LC circuit. Center-tapping the tuning circuit in this manner again establishes a central ground-potential point, allowing the opposite ends of the coil to have equal potentials but opposite polarities. Both methods of center-tapping the tuned circuit may work equally well, but center-tap only the coil or the capacitor, not both. If split-stator capacitors are used, the coil should not be bypassed to ground at the center, or the circuit forms two tuned circuits, one made up of the top capacitor and top half of the coil

and the other of the bottom capacitor and bottom half of the coil. The center of the coil should be connected to the power supply through an RF choke, as shown.

15-14 GRID OR RICE NEUTRALIZATION

Another way of neutralizing a triode RF amplifier is to center-tap the grid circuit as in Fig. 15-17. Amplified energy in the plate circuit will

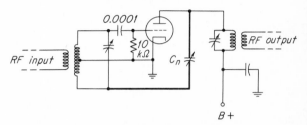

Fig. 15-17 A grid-neutralized triode RF amplifier.

feed back to the top of the grid-circuit coil via the interelectrode capacitance C_{gp}. The neutralizing capacitor C_n, if properly adjusted, will feed the same value of energy back to the bottom of the grid-circuit coil. Now the same voltage is being fed into both ends of the grid coil at the same time, canceling any effective feedback of energy. The stage is neutralized. As with plate neutralization, the tuned grid circuit may use a split-stator capacitor instead of a center-tapped coil.

15-15 NEUTRALIZING WITH AN RF INDICATOR

A method of neutralizing a low-powered triode RF amplifier stage is:

1. Turn off the plate-power-supply voltage. (Neutralization cannot be accomplished with an RF indicator if plate power is on.)
2. Couple to the plate-tank circuit some form of RF device that will indicate the presence of RF ac in the tank circuit (a loop of insulated wire in series with a sensitive RF thermogalvanometer, a loop and flashlight lamp, a loop and an oscilloscope, a loop and a milliammeter with a germanium diode across it, any form of wavemeter having an RF-indicating device in it, or a neon lamp held to the plate end of the tank coil).

3. Make sure the grid circuit is being excited by RF ac. (The presence of grid current is an indication.)
4. Rotate the neutralizing capacitor to either minimum or maximum capacitance to completely deneutralize the circuit.
5. Tune the plate circuit for maximum RF ac indication. (The plate circuit is now being fed RF via the neutralizing capacitor or interelectrode capacitance of the tube and should give a strong RF indication when the plate LC circuit is tuned to resonance.)
6. Rotate the neutralizing capacitance to zero RF ac indication.

The stage should now be neutralized and ready to operate as soon as plate voltage is applied. The RF indicator must be removed before plate voltage is applied; otherwise, the indicator may burn out.

This method of neutralizing can be applied to either plate- or grid-neutralized vacuum-tube or equivalent circuit transistor RF power amplifiers.

15-16 NEUTRALIZING BY GRID-CURRENT METER

Another method of neutralizing a triode RF amplifier uses the grid-current meter only.

1. Turn off the plate supply voltage of the stage to be neutralized.
2. Tune the plate-circuit tank capacitor through the resonance point and watch the grid-current meter. If the grid current fluctuates at all during the tuning, the stage is not neutralized.
3. Keep tuning the plate circuit back and forth through resonance and adjusting the neutral-izing capacitor until a position is found for the neutralizing capacitor where the *grid current* does not change as the plate circuit is tuned past the resonant frequency. The stage is now neutralized. This method of neutralizing can be applied to either plate-neutralized or grid-neutralized RF amplifiers.

A sensitive RF voltmeter can be connected across the input circuit of either vacuum tubes or transistors if there is no milliammeter in the input circuit.

15-17 NEUTRALIZING PUSH-PULL STAGES

The neutralizing circuit used with push-pull RF amplifiers, shown in Fig. 15-18, is a combination of both grid and plate neutralization. Note that C_{n1} is acting as a plate-neutralizing capacitor for the top tube and a grid-neutralizing capacitor for the bottom tube. C_{n2} acts as grid neutralizer for the top tube and plate neutralizer for the bottom. Both neutralizing capacitors are adjusted at the same time to maintain equal-capacitance values for each.

Because both grid and plate circuits are center-tapped and are assumed to be evenly balanced electrically, neutralization of push-pull stages tends to produce a neutralized condition over a relatively wide band of frequencies. Single-ended stages may require reneutralization if the frequency of operation is changed materially.

15-18 DIRECT NEUTRALIZATION

A completely different method of producing neutralization is the *direct* method. It operates on the theory that a parallel-tuned circuit has a very

Fig. 15-18 Neutralization circuit when triodes are in push-pull, using a Marconi-type antenna as the load.

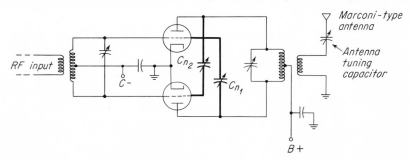

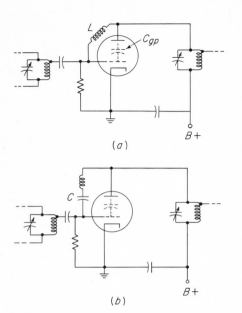

Fig. 15-19 Direct neutralization. (*a*) Basic circuit. (*b*) Actual circuit.

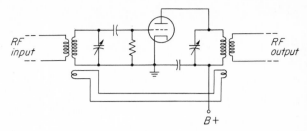

Fig. 15-20 Inductive neutralization of an RF amplifier stage.

high impedance to the frequency to which it is tuned. If the correct value of inductance is selected and connected as L in Fig. 15-19*a*, a parallel-resonant circuit is formed by L and C_{gp}. If this circuit resonates at the desired frequency of operation, it forms a very high impedance to this frequency and no ac energy is fed back from plate to grid circuit.

If the coil L were connected as shown in Fig. 15-19*a*, it would place B+ on the the grid and the stage would not operate. To prevent this, a blocking capacitor C is connected in series with the coil as in Fig. 15-19*b*. This capacitor is selected to have a very low reactance to the operating frequency, so that the L and C_{gp} circuit resonates as if it were not there. The dc of the power supply will be blocked and will find no path to the grid. Direct neutralization is effective only if the L and C_{gp} tune naturally to the frequency of operation. A new value of inductance is required if another frequency of operation is desired.

15-19 INDUCTIVE NEUTRALIZATION

Another method of neutralizing an RF amplifier is *inductive neutralization,* shown in Fig. 15-20.

To neutralize plate-to-grid energy transfer, it is possible to couple a pickup loop loosely to the plate circuit. The energy picked up by this loop can be inductively coupled into the grid-tank circuit by a second loop. If either loop is turned over, the phase of the inductive feedback will be reversed. With the loops coupled one way, the whole circuit may be regenerative and will oscillate. If one loop is then turned over, the feedback is degenerative. Varying the coupling of one or both loops can produce neutralization.

The loops should be coupled to the *cold,* or ground-potential, ends of the tuned circuits, as shown. This is generally true; a load should always be coupled to the cold end rather than to the *hot,* or plate (collector), end of any LC circuit to prevent detuning it.

15-20 NEUTRALIZING TETRODE AND PENTODE TUBES

In general, tetrode and pentode tubes made for use in RF circuits will need no neutralization at frequencies below 20 to 50 MHz. Above this, particularly if the input and output circuits are not adequately isolated and shielded, neutralization may be necessary. The very small grid-plate capacitance in such tubes makes necessary very little neutralizing capacitance. For example, with the grid circuit below the chassis level and the plate circuit above, it may only be necessary to run a stiff wire from the bottom of a center-tapped grid coil up through a hole in the chassis to a position near the plate of the tube to produce enough capacitance to neutralize the tube.

Some high-frequency tetrodes are constructed in such a way that they may operate up to 500 MHz without neutralization.

15-21 FREQUENCY MULTIPLIERS

Oscillator circuits may have good frequency stability when operated on lower frequencies but at higher frequencies the stability may not be good.

It is possible to operate an oscillator at a relatively low frequency and feed its output to frequency-doubler or -multiplier stages. Such stages will produce a frequency in their output circuit two, three, four, or five times the frequency fed to their grid circuits. The stability, as well as the frequency of the output, is an exact multiple of that of the oscillator. This results in an output stability considerably better than if the oscillator itself were used to generate the higher frequency directly. As a result, it is common to find multiplier stages in HF applications.

Basically, a multiplier stage is a normal class C RF power amplifier, but with its plate circuit tuned to resonate at a frequency some whole-number multiple of that fed to its grid circuit. Tetrode or pentode multipliers closely resemble normal RF amplifiers, insofar as diagraming is concerned. It is necessary to label the plate circuit as being tuned to a multiple of the grid-circuit frequency to indicate that the stage is acting as a multiplier.

If a triode tube is used in a multiplier circuit, as in Fig. 15-21, it is recognized as a frequency-multiplier stage because it lacks any neutralizing circuit. Since the plate is not tuned to the same frequency as the grid circuit, the stage cannot break into self-oscillation and therefore needs no neutralizing.

Fig. 15-21 A doubler or frequency-multiplier stage indicating grid and plate signal voltages and currents.

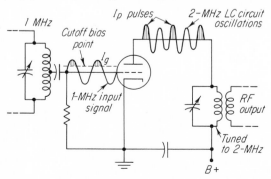

A tube operating as a frequency doubler usually has an RF output power about half of what might be expected from it if it were used as a fundamental-frequency or straight-through amplifier. Its plate efficiency may range between 30 and 40%.

When the plate circuit is tuned to resonate at a frequency three times that applied to the grid, the stage operates as a *tripler,* emitting energy at a frequency exactly three times that of the exciting frequency. The power output from a tripler is considerably less than that from a doubler. Still higher multiplications produce progressively less power output. For this reason it is rather unusual to have a frequency multiplier operating at more than the third, or at the most the fifth, harmonic.

Maximum output is produced in a doubler by using a high value of class C bias on the stage, usually two to four times the cutoff value, and driving well up into the positive grid region. This produces high-amplitude, narrow-width plate-current pulses. Each pulse excites the plate LC circuit. The electrons are driven down through the LC circuit but oscillate back up, then down, and then up again by flywheel effect before the next dc plate-current pulse arrives. In this way each pulse produces two cycles of nearly sine-wave ac in the plate-tank circuit. In the case of a tripler, the LC circuit must use the flywheel effect of a high-Q circuit to allow three complete cycles to be formed for each plate-current pulse. Each succeeding unexcited oscillation, or ac cycle, decreases in amplitude, resulting in a lowered average output power from the doubler and still less from triplers.

The conditions for high power output on a harmonic frequency in a multiplier stage can be summarized as: High bias voltage, high driving voltage, high plate voltage, high-Q tank circuit, and loose coupling to the load.

The opposite conditions are generally required to reduce harmonic output from an RF amplifier: Use as little bias as possible and still operate efficiently. Use as little drive as possible and still obtain good efficiency. Use loose coupling between the output tank circuit and the load. Use a low L/C ratio (low L and high C) in the output tank circuit.

A frequency doubler with high efficiency is the *push-push* circuit, shown in Fig. 15-22. The grid

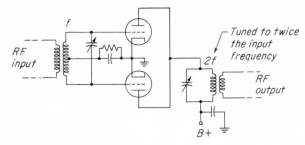

Fig. 15-22 A push-push frequency-doubler.

circuits are connected in push-pull, and the plates are in parallel. The stage is biased to class C. The plate circuit is tuned to twice the frequency of the grid circuit. The two grids are being excited 180° out of phase.

When the grid of the top tube is positive, a plate-current pulse flows downward in the plate LC circuit, starting an oscillation at the LC circuit frequency. During this time the grid of the lower tube is highly negative, and this tube is inoperative. Between half cycles of the excitation voltage there is a period of time when neither tube is passing current and the flywheel effect of the tank circuit allows electrons to oscillate back up through the tank coil. When the cycle in the grid circuit reverses, the grid of the bottom tube becomes positive and a plate-current pulse again flows downward in the plate LC circuit, exciting the LC circuit again. In the push-push circuit only one half cycle of each plate-circuit oscillation depends on flywheel effect. In single-tube doublers, the flywheel effect must operate by itself for three half cycles before another excitation pulse occurs. This is the reason the push-push circuit can produce RF output at twice the grid frequency with almost the same efficiency as a single-ended RF amplifier.

The push-push amplifier needs no neutralization. It will not operate as a tripler but will produce power output on even-order (2d, 4th, etc.) harmonics. The push-push stage has a relatively high plate-to-cathode capacitance because the two plates are connected in parallel, making VHF and UHF operation difficult. Transistors in push-push operate up into the microwave region.

Unlike the push-push amplifier, a *push-pull* class C stage (Fig. 15-18, but without the neutralizing capacitors) will not operate as a doubler but makes an excellent tripler, producing power output on odd harmonics, such as the 3d, 5th, etc.

It should be pointed out that the first harmonic is the fundamental frequency. A harmonic is a whole-number multiple of some frequency. When a fundamental is multiplied by 2, the answer is the second harmonic. Thus, the second harmonic of 3 MHz is 6 MHz. The second harmonic of 790 kHz is 1,580 kHz, the third harmonic is 2,370 kHz, etc.

To develop output power on high frequencies, several frequency multipliers can be used in cascade (succession). The output frequency when three doublers are connected in cascade is $2 \times 2 \times 2$, or 8 times the fundamental. If the output frequency of three doublers in cascade is 16,840 kHz, the input frequency to the first doubler is 16,840/8, or 2,105 kHz.

If a tripler and two doublers are used with an input of 1 MHz, the output will be $3 \times 2 \times 2$, or 12 MHz. The same output will result if the arrangement is doubler, doubler, tripler, or doubler, tripler, doubler.

The term *octave* is often used in electronics. Octave represents twice (or half) the frequency. Up one octave from 1 MHz is to 2 MHz. Two octaves above 1 MHz is up to twice 2, or to 4 MHz. Three octaves above 1 MHz is 8 MHz. This is not the same as the third harmonic of 1 MHz, which is 3 MHz.

Test your understanding; answer these checkup questions.

1. What type of coupling would be used to transfer RF energy from one building to another? _____
2. What type of coupling transfers harmonic energy better than fundamental frequency energy? _____
3. What are two disadvantages if triode RF amplifiers are used? _____ _____ What is an advantage? _____
4. Plate neutralization in VTs is equivalent to what neutralization in BJTs? _____ FETs? _____
5. Grid neutralization is equivalent to what in BJTs? _____ In FETs? _____
6. What phase displacement always exists between grid and plate of an amplifier? _____ What phase difference must a neutralizing circuit feed back to the input? _____ What is the total? _____
7. What are five devices to use as an RF indicator for neutralizing? _____ _____ _____ _____ _____

8. Does a maximum RF indication indicate best plate neutralization? _____ Best grid neutralization? _____

9. Why is it undesirable to couple to the hot end of an RF tank circuit? _____

10. What type of neutralization is produced by running a stiff wire from a grid coil up next to the plate of a tube? _____

11. Would it be possible to multiply a 2-MHz signal to 8.5 MHz? _____

12. From the diagram how can you tell if a triode stage is a multiplier or an amplifier? _____ A tetrode stage? _____

13. What is the ratio of I_p pulses to input frequency in a frequency tripler? _____

14. What circuit has a push-pull input and parallel output circuit? _____ Will it triple its input f? _____

15-22 DETERMINING C AND L FOR RESONANT CIRCUITS

When tuned circuits are described, the reader may wonder what size of coil and what value of capacitance should be used. Actually, there is only one value of inductance and capacitance that will give maximum output for a given set of circuit conditions. So many of the factors in a circuit are variable, however, that an accurate determination is rather involved. An old rule of thumb for HF circuits can be used to determine a workable capacitance. With the resonance formula an inductance value can be computed. In the earlier days of radio, frequency of RF ac was not employed as much as it is today. *Wavelength* was used instead. A wavelength is the distance that an electric impulse, or wave, will travel in the time it takes to complete one cycle of the ac being considered (Fig. 15-23). For example, consider a frequency of 1,000,000 Hz. The period $(1/f)$ of one such cycle is 0.000001 s. It takes $\frac{1}{1,000,000}$ s to complete one cycle of this alternating current.

Fig. 15-23 One wavelength on a two-cycle ac wave can be from 0° to 360° or from peak-to-peak. The time to complete a cycle is the period.

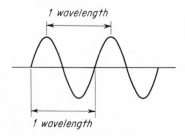

1 wavelength

1 wavelength

In that time, an electric impulse, or wave, will travel 300 m (300 meters) through space. The velocity of electric impulses, or radio waves, is 300,000,000 m/s. A wave having a frequency of 1 MHz has a wavelength of 300 m.

The formulas to convert wavelength to frequency or frequency to wavelength are

$$\lambda = \frac{V}{f} = \frac{300,000,000}{f} = \frac{300,000}{f_{kHz}} = \frac{300}{f_{MHz}}$$

$$f_{Hz} = \frac{V}{\lambda} = \frac{300,000,000}{\lambda}$$

where V = velocity of propagation in m/s
f = frequency in Hz
λ = wavelength in m

The rule of thumb mentioned above is: For a resonant circuit, use 1 pF of capacitance for each meter of wavelength. For a 1-MHz resonant circuit, 300 pF is a possible capacitance to be used in it.

What capacitance could be used in a relatively high impedance tuned circuit operating on a frequency of 6,000 kHz? The wavelength is

$$\lambda = \frac{300,000,000}{6,000,000} = 50 \text{ m}$$

A possible value of capacitance to use is 50 pF. It must be pointed out that this is unlikely to be the value of capacitance for maximum efficiency of the tuned circuit. It is merely a simple method of arriving at a starting approximation. In some circuits twice the value may be better; in others half may be better. For lower-impedance transistor RF circuits about four times the rule of thumb might be best.

To determine the inductance required to resonate the 50-pF capacitance to 6,000 kHz, the resonance formula (Sec. 8-1) can be used:

$$L = \frac{1}{4\pi^2 F^2 C}$$

$$= \frac{1}{4(9.86)(36 \times 10^{12})(50 \times 10^{-12})}$$

$$= \frac{1}{70,900} = 14 \text{ } \mu\text{H}$$

If the coil must have 14 μH, what size and how many turns should it have? Assume an RF coil with diameter and length approximately equal. In Sec. 5-3, a formula is given for such a coil:

$$L = \frac{r^2 n^2}{9r + 10l}$$

The required coil might be made in many dimensions. For this problem, a 1-in. diameter and a 1-in. length will be tried. How many turns will be needed? Rearranging to solve for turns, n,

$$n = \sqrt{\frac{L(9r + 10l)}{r^2}} = \sqrt{\frac{14[9(0.5) + 10(1)]}{0.5^2}}$$

$$= \sqrt{\frac{203}{0.25}} = \sqrt{812} = 28.5 \text{ turns}$$

On examination of a copper-wire table in a handbook, it is found that a No. 19 enamel-insulated wire will wind approximately 27 turns to the inch. Therefore, a 1-in.-long coil wound with No. 19 enameled copper wire on a 1-in. diameter will produce an inductance of approximately 14 μH. This coil, when shunted with a 50-pF capacitor, should form a resonant circuit at about 6,000 kHz. (Distributed capacitance and other circuit capacitances may affect the frequency of resonance greatly.)

If the diameter of No. 19 wire is too small for the current that will be flowing, it will be necessary to increase the dimensions of the coil until the desired wire size can be accommodated.

The capacitance to resonate a coil of known inductance to a given frequency can be determined as follows:

If a frequency-doubler stage has an input frequency of 1,000 kHz and the plate-circuit induct-

ance is 60 μH, what value of plate capacitance is necessary for resonance? Since the stage is a doubler, the plate-circuit frequency is 2,000 kHz. By using 60 μH and solving for the capacitance with the resonance formula rearranged,

$$C = \frac{1}{4\pi^2 F^2 L} = \frac{1}{4(9.86)(4 \times 10^{12})(60 \times 10^{-6})}$$

$$= \frac{10^{-6}}{39.4(240)} = 0.000106 \ \mu\text{F or } 106 \text{ pF}$$

This is the total capacitance, including tube or transistor and incidental circuit capacitances needed to resonate the 60-μH coil to 2,000 kHz.

15-23 CENTER-TAPPING THE FILAMENT

Many high-power transmitting tubes use directly heated cathodes. The heating of these filaments is usually by 60-Hz ac. As explained in Sec. 9-7, to produce as little variation of plate current, or hum modulation, as possible, the filament circuit is center-tapped and both grid- and plate-circuit returns are made to the center-tap point, which is grounded as in Fig. 15-24. It is important

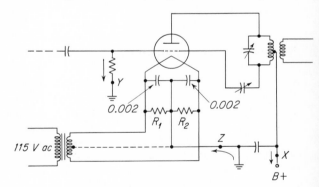

Fig. 15-24 Center-tapping the filament circuit of an RF amplifier triode.

that the two bypass capacitors shown be connected from the filament terminals to ground with leads as short as possible. The center tap of the filament can be developed by using two equal-value resistors, R_1 and R_2 (10 to 50 Ω), or the center of the filament winding can be grounded as shown by the dotted line. In either case the capacitors are required. Since the capacitors complete the RF ac circuit from filament to ground, the length of the filament wires to the transformer is not critical.

A milliammeter inserted at point X will read the plate current. If it is inserted at point Y, it will read the grid current. If it is inserted at point Z, it will read the sum of both plate and grid currents, called the *cathode current,* since both currents flow to the filament center tap. In a tetrode it would read the sum of the plate, grid, and screen currents. In a pentode it would also read any suppressor current.

15-24 GROUNDED-GRID RF AMPLIFIERS

When the input signal is fed to the grid-cathode circuit with the grid at ground potential, as in Fig. 15-25, a grounded-grid circuit is produced. As the

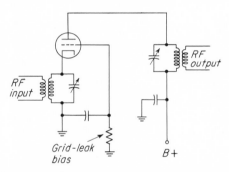

Fig. 15-25 A grounded-grid class C RF power amplifier.

cathode is driven more positive by the input signal, the grid at the other end of the input tank becomes more negative than the cathode. As the grid becomes negative in respect to the cathode, the plate current decreases and the plate itself becomes more positive. Thus, the top of the input and the output tanks are positive at the same time and are in phase, not 180° out of phase, as with grounded-cathode amplifiers. With this phase relation no regenerative feedback occurs and no neutralization is required, an advantage in HF, VHF, and UHF circuits.

The input impedance of a grounded-grid circuit is relatively low, requiring a high C/L ratio in the input tank. Such a stage requires several times as much power to drive it, but most of the driving power is available as output because input and output circuits are in series. Since the input and output are in series, the output signal will be pro-

portional to the input voltage plus the μ of the tube $(\mu + 1)$. The gain of a grounded-grid stage is:

$$A = \frac{(\mu + 1)R_L}{R_L + R_p}$$

With the grid grounded by the grid-leak resistor and capacitor, during one-half of the input cycle the grid is driven more positive than the cathode, and grid current flows. Current flowing down through the grid-leak resistor results in enough bias to produce class C operation.

In some RF power amplifiers it is necessary that the output circuit reproduce any variations in RF amplitude that are fed to the input circuit without changing, or distorting, the variations. Under this condition the amplifier must be a linear device and must be biased to class A, class AB, or, more likely, class B with its higher efficiency. To make the grounded-grid class C amplifier of Fig. 15-25 into a class B stage, substitute a bias supply for the grid-leak resistor. The bias supply must have low impedance (good regulation).

Although an indirectly heated cathode tube is shown for simplicity in the basic diagram, high-power tubes use directly heated cathodes. One method of feeding filament current to the filament without allowing RF ac to appear in the filament transformer is to make the cathode coil of hollow copper tubing and feed both filament leads through the hollow center, bypassing the filament terminals to the top of the cathode coil, as in Fig. 15-26. Another possibility is to use an RF choke

Fig. 15-26 A method of feeding the filament in a grounded-grid power amplifier.

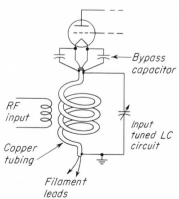

in each filament lead and bypass both filament terminals to the top of the cathode coil.

A class A grounded-grid amplifier is sometimes used in low-signal-input VHF circuits, TV receivers, etc. A circuit similar to that shown in Fig. 15-27 may be used.

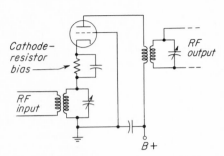

Fig. 15-27 Class A grounded-grid voltage amplifier used in VHF circuits.

15-25 AMPLIFIERS FOR VHF AND UHF

For audio-frequency work the length of component leads, placement of parts, shielding, and physical size are not overly important. In the HF range, from 3 to 30 MHz, these factors begin to affect operation of the stages. Capacitor leads must be short; coils and parts must be placed in positions that will not induce unwanted voltages from one to the other; interstage shielding becomes important; grid and plate leads must be kept apart; and neutralization of pentodes may be required.

In the VHF range, from 30 to 300 MHz the factors mentioned above become increasingly important. Interelectrode capacitances of tubes shunting the input or output circuits may lower the frequency of tuned circuits so that they will not resonate at the desired frequency. *Transit time,* the time required for electrons to move from the grid area to the plate, may be longer than one-half of a cycle of the signal frequency, preventing operation of circuits. Special, close-element tubes must be used. LC circuits are supplanted by hairpin, coaxial, or cavity tanks. Grounded-grid circuits are favored over grounded-cathode. Connecting wires are larger to obtain higher Q and less loss in the circuitry, and they may be silver-plated to reduce skin effect. Parts are placed as close together as possible to cut down lead length. Wiring goes as directly as possible, component to component, with no length loss in right-angled lead dress. Capacitors must be mica or other low-loss dielectric types. Because of reduced effective capacitance, push-pull circuits are better than single-ended ones.

15-26 TRANSISTOR RF POWER AMPLIFIERS

Whereas vacuum-tube RF power amplifiers may produce hundreds of kilowatts of RF power, HF transistor RF amplifiers rarely exceed a few hundred watts. VHF and UHF amplifiers are usually of less than 100 W. However, there are far more transistorized transmitters and receivers being built than vacuum-tube types.

Generally speaking, transistor RF amplifier circuitry resembles vacuum-tube RF circuitry (single-ended, push-pull, etc.). As with tubes, there is an unlimited number of possible circuits. Figure 15-28 shows four circuits cascaded to illustrate some of the variations possible in high-frequency transistor circuits.

The first stage is a common-base circuit (similar to a grounded-grid circuit in vacuum tubes) with

Fig. 15-28 Various possible forms of NPN BJT RF power amplifier and multiplier circuits.

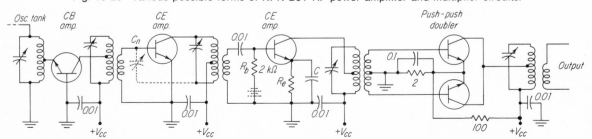

no base-emitter bias at all; it therefore operates as class C. It should not need neutralization.

The second stage is a common-emitter circuit, again with no bias. The neutralizing circuit in dotted lines should be recognized as being similar to triode plate neutralization.

The third stage is also a common-emitter circuit, again with no forward bias in the base circuit. When the base is driven with a signal, emitter current flows through R_e and a voltage-drop develops across it and capacitor C. If C is large enough to store it, this voltage represents a reverse bias for the base through R_b, biasing the stage well into class C. A bias supply in series with R_b (dotted) would also bias the stage to class C. Positive half cycles of the input signal will draw electrons from the emitter. As these electrons leak down through R_b, they develop a reverse bias voltage that biases the stage to class C.

The fourth stage is a push-push frequency doubler, also biased to class C by the 2- and 100-Ω resistors. Push-push and push-pull stages require transistors with very short storage times, or they will not function at high frequencies.

Variations in the output-circuit tuning or load impedance of transistor amplifiers reflect considerable impedance variation back into the input circuit. This in turn tends to shift the tuning of the driving stage. Tuning capacitors with large values of variation may be required.

With no input signal, collector current is essentially zero in these circuits. With signal drive, I_c increases. When the collector loads are tuned to resonance, I_c values should dip to a minimum.

The C/L ratio is always high in transistor tuned circuits, being three to five times that of tube circuits unless the coil is tapped. When using a high-impedance LC circuit, the collector tap point may be from one-tenth to one-half of the way up the coil.

The π-network output circuit shown in Fig. 15-29 is often used with transistors. C_1 and C_2 are in series across L and form the resonant circuit. The ratio of C_1 to C_2 determines what part of the total impedance across L is coupled to the 50-Ω load. C_2 is the output-impedance-matching device and C_1 is used to resonate the LC circuit. The greater the capacitance of C_2 the lower the output impedance. Many circuits that are not

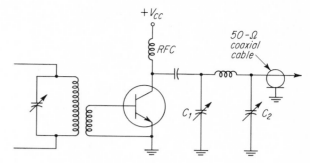

Fig. 15-29 A π-network output circuit.

diagramed as shown can be analyzed to be derived from this basic circuit. The π-network can be used for interstage coupling and for impedance transformation either up or down, and it can be used with common-base or emitter-follower circuits, as well as with VT circuits.

Test your understanding; answer these checkup questions.

1. How much greater is a fourth octave than a fourth harmonic? _____
2. What is the wavelength of a 5-MHz wave or cycle? _____ What C value might be used for such a frequency LC circuit? _____ What L value? _____
3. What is reduced by center-tapping the filament circuit of directly heated RF amplifier tubes? _____
4. What currents make up the cathode current in an RF beam-power tetrode? _____
5. Why do grounded-grid input circuits have large values of tuning capacitance? _____
6. How does the RF power driving a grounded-grid stage differ from that of a grounded-cathode type? _____
7. What is the main electrical difficulty when constructing grounded-grid RF power amplifiers? _____
8. Would transit time affect operation more in 1,000-kHz or in 500-MHz circuits? _____
9. Why are UHF and microwave inductors and tanks often silver-plated? _____
10. Why do two tubes in push-pull require less tuning C than a single-ended stage? _____
11. In Fig. 15-28, what kind of RF circuit must the third stage be? _____ Is the fourth stage forward-biased? _____
12. What advantages might π-network coupling have over transformer coupling in RF power amplifiers? _____
13. Would π-network coupling be practical for interstage coupling between BJTs? _____ VTs? _____

15-27 TROUBLES IN RF POWER AMPLIFIERS

When something goes wrong with an RF amplifier, servicing personnel must consider all the evidence given by meters, smoke, heat, and whatever else can be seen or heard. The meters alone can help to localize the difficulty.

The output-circuit milliammeter indicates normal or improper operation by its reading. If it reads *low,* there are several possibilities (check the reasoning behind these): (1) The power-supply voltage has dropped off. (2) The *load* circuit is not tuned or is not functioning properly. (3) The bias voltage may be too high. (4) There may be a loss of RF drive if the circuit is power-supply-biased. (5) With vacuum tubes, the filament voltage may have decreased or the tube may have lost its emission. (6) The screen potential may be low.

If the output-circuit milliammeter indicates *higher* current than normal, some possibilities are: (1) The output circuit is detuned from the minimum current value. (2) The load is too tightly coupled. (3) There is loss of bias or insufficient bias. (4) With vacuum tubes, screen or plate volt-

age is too high. (5) There is lack of RF drive if the stage has grid-leak bias only.

The grid-circuit milliammeter is also a good indicator of the operation of VT stages. If it reads abnormally *low,* it may indicate: (1) Too much bias. (2) Low filament voltage, or the tube is losing its emission. (3) Lack of RF drive to the grid. (4) An excessive plate current.

If the grid-circuit milliammeter indicates *higher* than normal, some possibilities are: (1) Loss of, or decrease in, bias voltage. (2) Too much driving voltage. (3) Plate and/or screen circuits open or not drawing current or shorted to ground. (4) The load is decoupled from the plate circuit.

Troubleshooting practice

The TPTG oscillator, link-coupled to a neutralized triode amplifier (Fig. 15-30), represents a possible transmitter. What would the different meters read (VH for very high, H for higher than normal, N for normal, L for lower than normal, VL for very low, R for reversed reading, and 0 for zero) if the faults listed below occurred? (Answers on page 354.)

	M_1	M_2	M_3	M_4	M_5
1. If C_1 shorted out (example):[1]	VL	H	VL	VH	L
2. If C_2 shorted out:	___	___	___	___	___

[1] M_1, no oscillation, contact-potential current only; M_2, no bias, high I_p; M_3, no drive, contact current only; M_4, no bias, high I_p; M_5, heavily loaded power supply, voltage drops.

ANSWERS TO CHECKUP QUIZ ON PAGE 351

1. *(Four times)* 2. *(60 m)(60 pF)(17 μH)* 3. *(Hum modulation)* 4. *(I_p, I_g, I_{sg})* 5. *(Low-Z circuits)* 6. *(Appears in output)* 7. *(Feeding filaments)* 8. *(500-MHz)* 9. *(Reduce skin-effect losses)* 10. *(C_{gp} of two tubes in series across L)* 11. *(Multiplier)(No)* 12. *(Simple to match Z's, output C tends to bypass harmonics)* 13. *(Yes)(Yes)*

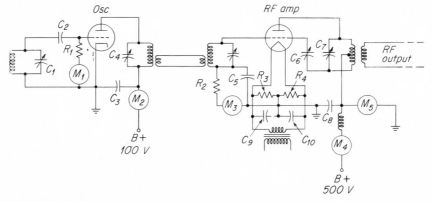

Fig. 15-30 Practice quiz TPTG oscillator link-coupled to a triode RF amplifier.

	M_1	M_2	M_3	M_4	M_5			M_1	M_2	M_3	M_4	M_5
3. If C_3 shorted out:	___	___	___	___	___		13. If M_4 burned out:	___	___	___	___	___
4. If C_4 shorted out:	___	___	___	___	___		14. If M_5 burned out:	___	___	___	___	___
5. If C_5 shorted out:	___	___	___	___	___		15. If RFC shorted out:	___	___	___	___	___
6. If C_6 shorted out:	___	___	___	___	___		16. If R_1 burned out:	___	___	___	___	___
7. If C_7 shorted out:	___	___	___	___	___		17. If R_2 burned out:	___	___	___	___	___
8. If C_8 shorted out:	___	___	___	___	___		18. If R_3 burned out:	___	___	___	___	___
9. If C_9 shorted out:	___	___	___	___	___		19. If oscillator filament burned out:					
10. If M_1 burned out:	___	___	___	___	___		20. If amplifier filament burned out:	___	___	___	___	___
11. If M_2 burned out:	___	___	___	___	___							
12. If M_3 burned out:	___	___	___	___	___							

COMMERCIAL LICENSE QUESTIONS

Amateur license questions will be found in the Addendum.

FCC Elements 3, 4, and 6 require an ability to answer questions similar to those below. Sections in which questions are answered are shown in parentheses. A question followed by a bracketed number is required for that element alone.

1. What type of tube is generally employed in RF voltage amplifiers? (15-2)
2. Why are bypass capacitors used across the cathode-bias resistors of an RF amplifier? (15-2, 15-7)
3. What is the purpose of shielding between RF amplifier stages? (15-2)
4. What materials may be used in shields to prevent stray magnetic fields in the vicinity of RF circuits? (15-2)
5. What is the difference between RF voltage amplifiers and RF power amplifiers in regard to applied bias? (15-3)
6. What is the principal advantage of a class C amplifier? (15-3)
7. What are the approximate efficiencies of class A, B, and C amplifiers? (15-3)
8. What value of control-grid bias must be used for class C operation of a triode with 1,000 V on the plate, 200-mA plate current, and a μ of 15? (15-3)
9. Draw a diagram and explain operation of a class C RF power amplifier with battery bias. (Fig. 15-2)
10. Draw a $E_g I_p$ curve of a triode and indicate the operating point for class C operation. (15-4)
11. Does grid current flow in a vacuum-tube amplifier when properly operated as class C? (15-4)
12. During what approximate portion of the excitation-voltage cycle does plate current flow in a class C amplifier? (15-4)
13. Why is the efficiency of an amplifier operated as class C higher than one operated as class A or B? (15-4)
14. What is an *RFC*, and why is it used? (15-4)
15. Why do some tubes have three prongs connected to the cathode? (15-4)
16. If upon tuning the plate circuit of a triode RF amplifier the grid current undergoes variations, what defect, if any, is indicated? (15-5, 15-16)
17. When adjusting the plate-tank circuit of an RF amplifier, does minimum or maximum plate current indicate resonance? (15-5)
18. In a class C RF amplifier, what ratio of load impedance to dynamic plate impedance will give the greatest plate efficiency? (15-6)
19. What load conditions must be satisfied in order to obtain maximum power in the load? (15-6)
20. What is the primary purpose of a grid leak in a class C amplifier? (15-7)
21. Given $E_p = 1,000$ V, $I_p = 150$ mA, $I_g = 10$ mA, and grid leak $= 5,000$ Ω, what would be the value of dc grid bias? (15-7)
22. Draw a diagram and explain the operation of a tetrode RF power amplifier with grid-leak bias. (Fig. 15-6)

23. What are the advantages of using a resistor in series with the cathode of a class C RF amplifier tube to provide bias? (15-7)
24. Draw a diagram and explain the operation of an RF power amplifier with two tetrode tubes in parallel. (Fig. 15-7)   [4]
25. Draw a diagram and explain the operation of an RF power amplifier with two tetrode tubes in push-pull. (Fig. 15-8)   [4]
26. How may the generation of even-harmonic energy in an RF amplifier stage be minimized? (15-9, 15-21)
27. Draw a diagram of a series-fed plate circuit of an RF amplifier. (Fig. 15-9)
28. Draw a diagram of a shunt-fed plate circuit of an RF amplifier. (Fig. 15-9)
29. Draw a diagram of a method of coupling between two tetrode RF amplifier stages. (Fig. 15-10)
30. Draw a diagram of inductive coupling between two tuned RF circuits. (15-11)
31. Draw a diagram of impedance (capacitive) coupling between two RF amplifier stages. (Fig. 15-12)
32. What is *link coupling,* and for what purposes is it used? (15-11)
33. Under what circumstances is neutralization of a triode RF amplifier not required? (15-12, 15-21, 15-24)
34. Does a pentode usually require neutralization when used as an RF amplifier? (15-12)
35. What is the principal advantage of a tetrode tube over a triode as an RF amplifier? (15-12)

ANSWERS TO PRACTICE QUIZ ON PAGE 352

	M_1	M_2	M_3	M_4	M_5
2.	O	H	VL	VH	L
3.	L	VH	VL	VH	L
4.	VL	H	VL	VH	L
5.	L	H	O	H	L
6.	H	L	R	VH	L
7.	H	L	L	VH	L
8.	L	H	H	VH	O
9.	N	N	N	N	N
10.	O	O	VL	VH	L
11.	L	O	VL	VH	L
12.	H	L	O	O	H
13.	L	H	H	O	O
14.	N	N	N	N	O
15.	N	N	N	N	N
16.	O	O	VL	VH	L
17.	H	L	O	O	H
18.	N	N	L	L	N
19.	O	O	VL	VH	L
20.	H	L	O	O	H

36. Why must some RF amplifiers be neutralized? (15-12)
37. What will determine if an RF power-amplifier stage is properly neutralized? (15-12, 15-16)
38. Draw a diagram and explain the operation of a plate-neutralized triode RF amplifier. (Fig. 15-15)
39. Explain the methods of neutralization in RF amplifiers. (15-13, 15-14, 15-18)
40. Draw a diagram and explain the operation of a grid-neutralized triode RF amplifier. (Fig. 15-17)   [4]
41. Is it necessary to remove the plate voltage from the tube being neutralized? (15-15)
42. What instruments or devices may be used to adjust an amplifier stage and determine that it is properly neutralized? (15-15)
43. Explain, step by step, a procedure for neutralizing an RF amplifier stage. (15-15, 15-16)
44. Draw a diagram of a push-pull neutralized RF amplifier coupled to an antenna. (Fig. 15-18)
45. Why should grid wiring be separated as far as possible from plate wiring? (15-20)
46. What class of amplifier is used to generate harmonic frequencies? (15-21)
47. Draw a diagram and explain the operation of a triode frequency-doubler stage, indicating points which distinguish it as a doubler. (Fig. 15-21)
48. How may harmonic-energy generation in an RF amplifier be minimized? (15-21)
49. What factors are important when operating a vacuum tube as a frequency multiplier? (15-21)
50. If a circuit has two identical tubes, the grids in push-pull and the plates in parallel, what relation will hold between input and output frequencies? (15-21)
51. Draw a diagram and explain the operation of a push-push frequency-doubler stage. (Fig. 15-22)   [4]
52. Explain the principle of a push-pull frequency multiplier. (15-21)
53. Push-pull frequency multipliers normally produce what order of harmonics, even or odd? (15-21)
54. What is a *harmonic?* (15-21)
55. List the fundamental frequency and the first 10 harmonic frequencies of a broadcast station licensed to operate at 800 kHz. (15-21)
56. What is the crystal frequency of a transmitter having three doubler stages and an output of 16,000 kHz? (15-21)
57. Draw two cycles of an RF wave and indicate one wavelength on it. (Fig. 15-23)
58. What is the formula to determine wavelength when frequency in kilohertz is known? (15-22)
59. If the period of one complete cycle of a radio wave is 0.000002 s, what is the wavelength? (15-22)

60. What is the velocity of propagation of RF waves in space? (15-22)

61. If a frequency-doubler stage has an input of 2 MHz and the plate inductance is 40 μH, what value of plate capacitance is necessary for resonance? (15-22) [4]

62. When an ac filament supply is used, why is a filament center tap usually provided for the vacuum-tube plate and grid return circuits? (15-23)

63. What currents will be indicated by a milliammeter between the center tap of the filament transformer of a tetrode and negative high voltage (ground)? (15-23)

64. Draw a diagram of a grounded-grid RF amplifier and explain its operation. (Figs. 15-25, 15-27)

65. Why are grounded-grid amplifiers sometimes used at VHF? (15-24)

66. What are the advantages and disadvantages of grounded-grid amplifiers? (15-24)

67. What factors are significant in computing the gain of a grounded-grid amplifier? (15-24)

68. Draw a diagram of a grounded-grid triode RF amplifier for class B operation. (15-24)

69. Name at least three circuit factors (not including tube types and component values) in a one-stage amplifier circuit that should be considered at VHF but would not be of concern at VLF (very low frequency). (15-25)

70. What are some factors regarding wire diameter, lead length, wiring configuration (placement and bending), coaxial cables and transmissions lines, and capacitor types that should be considered at VHF but would not be of concern at MF (medium frequency) or below? (15-25)

71. In a shunt-fed plate circuit of an amplifier, what would be the result of a short circuit of the plate RFC? (15-27)

72. In a shunt-fed plate circuit of an amplifier, what would result if an open circuit occurred in the plate RF choke? (15-27)

73. In a series-fed plate circuit of an amplifier, what would be the result of a short circuit of the plate bypass capacitor? (15-27)

74. Why might an RF amplifier tube have excessive plate current? (15-27)

75. What might be the result of a short circuit of the plate RFC in an RF amplifier? (15-27)

76. Draw a diagram of link coupling between a TPTG oscillator and a neutralized radio-frequency amplifier. (15-27)

16

BASIC TRANSMITTERS

The objective of this chapter is to outline the basic systems used to provide the carrier frequency and power for all types of transmitters—keyed radiotelegraph, amplitude-modulated, and frequency-modulated. It deals with some of the many on-off and frequency-shift keying methods by which a carrier can be controlled.

16-1 RADIO TRANSMITTERS

Thus far, study has been confined to basic devices and circuits, such as vacuum tubes, transistors, oscillators, AF amplifiers, RF amplifiers, power supplies, and meters. This and subsequent chapters will be concerned with complete electronic systems. For example, a radiotelegraph transmitter may consist of an oscillator, one or more low-level RF amplifiers or multiplier stages, an RF power amplifier, a power supply, a means of keying the system, and an antenna. Such transmitters may operate in the frequency range of 10 kHz to well over 200 MHz.

The transmission of intelligence in a radiotelegraph system is accomplished by *keying* the system. For example, if a telegraph key is inserted in the plate supply circuit of an oscillator, as in Fig. 16-1, the oscillator generates an ac when the key is held down and none when the key is up. By operating the key according to some dot-dash or other code system, it is possible to transmit words and information from one point to another. The most familiar code is the Morse code, shown in Chap. 33. Another often-used code is the teleprinter, or teletype, code, consisting of accurately timed intervals of on-off groups controlled by a

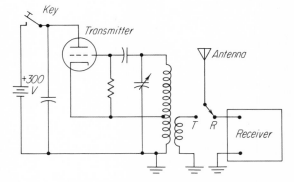

Fig. 16-1 Single-stage transmitter with changeover (transmit-receive, or TR) antenna switch.

sending keyboard. A remote receiver-keyboard system detects the code and prints out the data being transmitted.

Historically, the first radio transmitters were *spark* oscillators (Sec. 13-1) coupled to long-wire antennas. They produced trains of damped high-frequency ac (type B) that could be radiated into space by the antenna. Manual operation of a telegraph key was the only method of sending used with these transmitters. Their damped emission had a basic carrier frequency but contained many energy components called *sidebands* on both sides of the carrier, resulting in an extremely wide transmission bandwidth.

Another method of transmitting radio frequencies was by the use of high-frequency generators, called Alexanderson or Goldsmith alternators. These machines could be made to generate sinusoidal ac (type A1) up to about 50 kHz but were extremely bulky and could not be keyed rapidly.

For many years *arc* transmitters were used on land and sea. The electric arc has a negative-resistance effect and, if in series with an LC circuit, can produce sinusoidal RF ac energy. They generated a much purer A1 waveform than the spark, and were used for high-power RF transmissions for many years. However, they were limited to an upper frequency of about 500 kHz. Few if any of these transmitters are now in use.

As the ability to use higher and higher frequencies increased, the more adaptable vacuum tube outmoded all other types of transmitters. Today, all high-power transmitters use vacuum tubes. However, in low-power applications, transistor transmitters are in general use. In many cases, transistors are used for low-power circuitry of a transmitter and only the high-power stages will use vacuum tubes.

16-2 SINGLE-STAGE TRANSMITTERS

The simple Hartley oscillator coupled to an antenna wire does produce a workable transmitter. By using high-powered tubes, such a single-stage transmitter could produce as much as 100,000 W of output power and at high efficiency. However, its frequency stability would not be good. That is, when the key was pressed, the plate (or collector) potential would rise from zero volts to the voltage of the power supply. As the plate voltage on an oscillator changes, the frequency of the oscillator will always change somewhat. Therefore, during the make and break of keying, the oscillator would not be frequency-stable, producing *chirping* sounds in a receiver and sidebands on both sides of the oscillator's operating carrier frequency. The result is a wide-frequency-spectrum signal, which is undesirable.

The instability of an oscillator increases as the power and frequency of oscillation increase, so that even if a transmitter might produce a reasonably stable emission at a frequency of 500 kHz, at 5 MHz the chirp produced might be intolerable. From previous discussions on oscillators, a crystal oscillator would be an improvement over self-excited oscillator circuits.

The closer, or tighter, the coupling between any oscillator and the antenna the less stability the oscillator will have. Variations of antenna characteristics, such as tuning the antenna, varying the effective height, swinging of the antenna wire, etc., will also affect the frequency of oscillation.

16-3 KEYING RELAYS

The single-stage transmitter of Fig. 16-1 represents a *simplex* method of transmission and reception. The same antenna is used for both transmitting and receiving. The operator sends his message, then throws the antenna changeover switch manually from the transmit position to the receive position and listens for the answering information from the distant station.

In this single-stage transmitter circuit, if the operator allows his fingers to fall across the key connections, he will receive essentially the full voltage of the power supply across his fingers. To prevent this, a *keying relay,* as in Fig. 16-2, may be used.

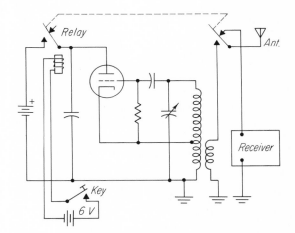

Fig. 16-2 Single-stage transmitter with break-in relay.

A keying relay is a fast-acting double-pole double-throw type of switch. With the key up, the relay coil is not energized and both relay arms are in the up position, as shown. This connects the antenna to the receiver, and incoming signals can be heard. When the operator presses the key to the down position, the relay coil becomes a magnet and attracts the relay arms downward. This disconnects the antenna from the receiver and connects it to the transmitter. At the same time the power supply is connected to the oscillator, producing a carrier-frequency signal that energizes the antenna and transmits energy.

This type of simplex operation is known as *break-in* because the transmitting operator can hear the other station whenever his key is in the up position. If the distant operator wishes to break in on the sending, he merely holds his own key down for a second and the local operator hears the solid tone between his dots and dashes and stops sending. The distant operator then explains why he has broken in. Most manual commercial radiotelegraphic communication systems use break-in, although *duplex,* or simultaneous transmission in two directions, is made possible by using two different frequencies.

While the key is pressed, the amplitude of the emission, or wave, is continuously the same. Hence the common term for such code transmission is CW (continuous wave). This is also indicated by the FCC letter-number designation A1.

Relay contacts may be silver, nickel, or tungsten. The relay arms are usually so made that the contacts slide or wipe across each other slightly, by having a resiliency in the relay arm or by spring-loading the contacts. The self-wiping motion assures better contact by continually polishing the contacting areas.

Silver makes an excellent contact material but tends to pit if the contacts are used in circuits where an arc is produced when the contacts open. Such arcing leaves a small depression in one contact and a small mound on the other. The mound may form to such an extent that eventually the contacts may not open at all. Relay contacts should be burnished, or filed with a fine, flat file

every so often. Over a long period of time they may have to be replaced with new contacts.

It is advantageous to use vacuum relays which have all their working parts in an evacuated bulb. Their tungsten contacts will stand extremely high voltages and currents without sparking or deteriorating.

16-4 THE MOPA

To develop a high-power ac at high frequency and with good frequency stability it is necessary to go to a system of two or more stages. The oscillator stage determines the frequency of operation, and the RF amplifier stage or stages that follow produce the high-power output. For maximum power output, high-efficiency class C operation of the amplifier is employed. Such a two-stage transmitter is known as a master oscillator power amplifier (MOPA). A possible MOPA is shown in Fig. 16-3. It consists of a Hartley oscillator capacitively coupled to a neutralized triode power amplifier which is inductively coupled to a tunable Marconi antenna.

Since the B− is being keyed, both oscillator and amplifier are keyed on and off at the same time. No provisions are made for break-in.

If the oscillator is left running at all times and only the B+ to the amplifier is keyed, a greatly improved frequency stability will result. However, the oscillator signal may block out a local receiver tuned to its frequency. If the amplifier is properly neutralized, the oscillator-signal leak-through (called a *backwave*) will be minimized.

Fig. 16-3 A two-stage MOPA transmitter.

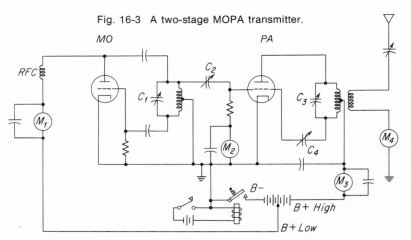

Dc meters in a transmitter should be bypassed with 0.001- to 0.01-μF mica- or ceramic-dielectric capacitors. This prevents RF ac from flowing through the thin wires and hairsprings of the meter and burning them out. Although dc meters are never in hot parts of RF circuits (between the plate and the RF choke or between the grid and the grid-leak resistor), there is still the possibility that some RF ac may flow through them.

The RF ammeter in the antenna circuit must not be bypassed, of course, or it will not read at all.

Neutralization of the amplifier is important. If it is improperly neutralized, not only will there be a strong local signal when the key is open, but the oscillator frequency will pull (shift) as the amplifier or antenna is tuned and the stage may break into strong self-oscillation and emit signals on frequencies off the operating frequency. Often, if the grid drive from the MO is sufficiently high, the PA may stay in synchronism and operate satisfactorily if only slightly out of neutralization. If the drive is low or drops to zero, the PA may then break into self-oscillation. With no grid drive to an amplifier, self-oscillation is indicated if grid current flows at some setting of the plate-tuning capacitor. The frequency of the ac generated can be determined with a receiver or a wavemeter.

16-5 TUNING AN MOPA TRANSMITTER

The tuning of an MOPA transmitter, such as Fig. 16-3, to operate on a desired frequency involves several steps. The procedure outlined below is a possible one.

1. Disconnect the B+ lead of the amplifier. Decouple the antenna coil. Adjust the interstage coupling capacitor to minimum capacitance.
2. Turn on the filaments and, after a few minutes warm-up time, close the key, energizing the oscillator. The oscillator-plate-current meter M_1 will usually read between 5 and 20 mA. The amplifier's grid-current meter M_2 should read very little. The amplifier-plate-current meter M_3 and the antenna meter M_4 will both read zero.
3. Select the desired frequency on a receiver or frequency meter and rotate the oscillator tuning capacitor C_1 until the signal is heard.

4. Couple energy to the grid of the amplifier by increasing C_2 until the grid current on meter M_2 is about 10% more than required for normal operation. The oscillator-plate current should increase by 10 to 50%. Retune the oscillator if necessary.
5. Neutralize the amplifier stage as explained in Sec. 15-15.
6. Connect the amplifier B+ lead to the power supply and immediately tune the plate-tank-circuit capacitor C_3 for minimum amplifier-plate current.
7. With *very loose* coupling between antenna and amplifier tank, tune the antenna by adjusting the variable capacitor in the antenna for a maximum peak indication of the amplifier-plate current, or for maximum indication of RF current in the antenna meter M_4. (The antenna is now tuned to resonance and usually needs no further adjustment.) Note that it is necessary to tune the final amplifier before the antenna circuit can be tuned.
8. Increase antenna coupling. This will increase antenna current and amplifier-plate current at the same time. Both are indications that the antenna is taking RF power from the amplifier-plate circuit.
9. Redip the amplifier-plate current by adjusting the plate-tank capacitor C_3 and readjust the antenna coupling until the desired plate current occurs with the tank circuit at resonance (plate-current dip).
10. Recheck the grid current, adjusting capacitor C_2 if necessary to bring the grid current to the desired value.
11. Recheck the amplifier-plate tuning for desired plate current.
12. Recheck the frequency once more, and the transmitter is ready to operate.

Note that it is necessary to watch the oscillator frequency closely to prevent off-frequency operation of the transmitter. If the oscillator stage uses a crystal, no frequency adjustments may be necessary.

16-6 FREQUENCY STABILITY

Self-excited oscillator circuits, such as Hartley and Colpitts, may have poor frequency stability and may not hold a constant frequency of oscillation. Some possible causes of an *abrupt* frequency variation in an oscillator are (1) power-supply voltage changes due to power-line voltage variations, (2) loose connections in the oscillator, amplifier, or antenna circuits, (3) poor soldered connections in the oscillator or in the following stage, (4) poor connections between tube or transistor base pins and sockets, (5) faulty fixed capacitors, (6) faulty resistors, (7) faulty tubes or transistors, (8) poor electric contact between the bearing surfaces of variable capacitors, (9) loose shield cans, and (10) parts which, when heated, expand and either make or break a contact.

Some causes of a *slow drift* of frequency in a self-excited oscillator are: (1) Heating and expansion of the oscillator coil due to RF ac flowing in it. (2) Heating of any capacitors in the oscillator or the stage following it. (3) Heating of the oscillator tube or transistor causing its elements to expand and change their electrical characteristics. (4) Dc and RF ac heating of resistors, causing them to change values slightly (grid-leak resistor, for example). (5) As a tube or transistor ages, its characteristics may change, resulting in a slight frequency change.

Practically all these slow frequency changes can be effectively decreased or eliminated by using a quartz crystal oscillator instead of a self-excited oscillator. Where the *frequency tolerance* (number of cycles per second that a transmitter may operate off the assigned frequency) is small, it may be necessary not only to use crystal oscillators but to keep the crystals in temperature-controlled chambers (Sec. 13-14).

A source of frequency variation, present even in crystal oscillators to some extent, is variation of the plate voltage. This difficulty can be minimized by using a separate power supply for the oscillator stage. In this way, the voltage of the oscillator will be independent of all the other stages of the transmitter, regardless of how they may be tuned or operated. It is common to use voltage regulation on oscillator stage power supplies for this reason.

Transistor oscillators are very sensitive to frequency change if the collector voltage changes. They always require voltage regulation.

All oscillators will have a warm-up period, during which some frequency variation will be present. With temperature-controlled crystals, the period is the time required for the oven and crystal to attain their operating temperature. With self-excited oscillators, half an hour to several hours may be required before no frequency drift is noticeable. These oscillators can be more quickly stabilized if they utilize correct-value temperature-coefficient capacitors to counteract the effect of heating in the circuit.

A sudden cold draft striking a self-excited-oscillator stage may cause a change of several hundred hertz in frequency, but only a few hertz with a crystal.

16-7 DUMMY ANTENNAS

To prevent a transmitter from emitting a signal and interfering with someone else on the frequency during tests, it should be disconnected from the antenna and coupled to a *dummy* (*artificial, phantom*) antenna, as in Fig. 16-4.

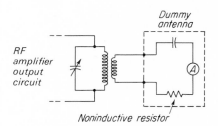

Fig. 16-4 A dummy-antenna circuit.

The capacitor of the dummy antenna should have a reactance value similar to the reactance of the coupling coil being used. The noninductive resistor (not made in coil form) should have a resistance similar to the impedance expected to be exhibited by the antenna when it is connected to the transmitter. With the dummy antenna coupled to the amplifier-tank circuit, the resistor should dissipate, in heat, the amount of energy the antenna will dissipate in radiated radio waves.

The power output of the transmitter can be approximated by using the power formula $P = I^2R$, employing the current value shown by the ammeter in the dummy antenna and the resistance value of the noninductive resistor. When the antenna is connected in place of the dummy (and with a similar amplifier-plate-current value), the antenna should be emitting a power approximately that computed for the dummy circuit.

For low-powered transmitters an electric light bulb and a capacitor can be used as a dummy antenna. The RF power heats the filament of the bulb. By comparing the brilliance of the bulb with its normal brilliance when across a 120-V circuit, a rough approximation of the power output can be obtained.

Preliminary tuning of any transmitter, either with or without a dummy antenna, should be done at reduced power, usually obtained by using one-half to one-quarter of the operating plate (and screen-grid) voltage for the stage. This not only helps to protect the equipment from inadvertent overloads during tuning operations but decreases possible interference to other stations operating on the same or adjacent frequencies.

16-8 PLATE AND CATHODE CIRCUIT KEYING

One method of sending dots and dashes with a radio transmitter is to open and close the plate circuit with a telegraph key or keying relay. The key can be inserted in series with the negative end of the plate power supply, at point X in Fig. 16-5, or in series with the positive terminal, at point Y.

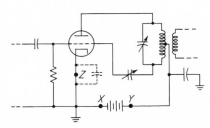

Fig. 16-5 Keying at X or Y is plate-circuit keying; at Z, cathode keying.

If the key is connected in series with the plate and grid circuits, between cathode and ground, point Z, the stage is *cathode-keyed*. The cathode must be bypassed to ground.

Simple keying methods such as these may produce key clicks in receivers, at frequencies near and even far removed from the carrier frequency. This is a serious fault in a CW transmitter and must be corrected.

Test your understanding; answer these checkup questions.

1. In what frequency range might radiotelegraph transmitters operate? _____
2. What were the three methods of generating carriers before vacuum tubes were used? _____ _____ _____
3. What is the main disadvantage of a single-tube transmitter? _____
4. Why would a swinging antenna change the frequency of a simple transmitter?
5. What is the name of the operation that uses an antenna changeover switch? _____ What is required to make this break-in? _____
6. If A1 is CW, what would AØ (A-zero) be? _____ When might AØ be used? _____
7. What is the main advantage of an MOPA over a single-tube transmitter?
8. What are two methods of neutralization that might be used with the MOPA? _____ _____
9. Plate tank resonance of a PA is shown by what I_p indication? _____ What I_g indication? _____
10. Why might heating a fixed capacitor change its capacitance? _____ A variable capacitor? _____
11. What are two other names for dummy antennas? _____ _____
12. Why are dummy antennas used? _____
13. What three formulas might be used with dummy antennas to determine power output of a transmitter? _____ _____ _____
14. If both grid and plate circuits are keyed simultaneously, what is this keying called? _____ What would a comparative transistor stage keying be called? _____

16-9 SHAPING OR FILTERING THE CW SIGNAL

When the key of a code transmitter is closed, RF ac is radiated. The RF ac of the letter S (three dots) can be represented as in Fig. 16-6. The

Fig. 16-6 Three keyed bursts of RF ac forming the telegraphic letter S.

waveshape of the letter S, as shown, is square. Figure 16-7 shows the sharp corners of the square wave, as it looks at the receiver.

Transmission of this type of square-wave signal will produce sidebands, or frequencies on both sides of the carrier frequency. The reason for this

Fig. 16-7 How the letter *S* appears in the detector of a receiver.

can be understood by considering what constitutes a square wave.

Figure 16-8 shows, first, a sine wave, second, the result of adding the third harmonic, and finally, the result of adding up to the ninth harmonic.

By starting with a fundamental frequency, 10 Hz, for example, and adding the first 9 harmonics in proper phase and amplitude, it is possible to produce a fairly square 10-Hz waveform. If the first 50 harmonics are added in the proper proportions, an excellent square wave can be produced. The more harmonics involved, the squarer the corners of the square-wave signal. By working backward with this theory, it can be seen that if a fundamental plus harmonics can produce a square wave, then a square wave must be made up of a fundamental plus many harmonics of that fundamental frequency.

If a series of square-wave dots are transmitted, the signal must be made up of the dots plus many harmonics of the *dot frequency* (not to be confused with possible harmonics of the RF carrier frequency). If the dots are made at a speed of 10 per second, they form a 10-Hz fundamental

keying frequency. If the square waveshape of the dots is good, there may be up to 50 or more strong harmonics present. This means that harmonics will be present for 500 Hz on both sides of the carrier (10 Hz × 50 = 500 Hz). If there are 20 dots per second, harmonics will be generated out to 1,000 Hz or more on each side of the carrier. The amount of RF spectrum space required is directly proportional to the speed of the sending. High-speed sending takes up more spectrum space than does low-speed. Shaping the transmitted square-wave signals by rounding them off represents the presence of fewer harmonics and produces good readable signals up to 25 or 35 words per minute without involving a spectrum band of more than 100 to 200 Hz in width.

The generation of sidebands by keying may be considerably exaggerated by variations of the power-supply voltage of the transmitter. When the key is pressed, a heavy load is suddenly applied to the power supply and its output voltage drops. Actually, the voltage sags and rises several times in a fraction of a second. This *transient* variation of the power-supply voltage imposes an added waveshape on the transmitted signal. The mixture of the transient waveshape with the normal waveshape can produce high-order harmonics and objectionable sideband energy hundreds of kilohertz on both sides of the carrier. These key clicks produce spurious clicking sounds in receivers tuned far from the fundamental frequency of the transmitter.

The rounding of the corners of a CW signal is said to produce *soft* keying. *Hard* keying has a sharp-cornered square waveform. Hard keying can be softened by the use of a *key-click filter,* shown in heavy lines in Fig. 16-9. When the key is closed, the inductance *L* opposes the rise of

---

ANSWERS TO CHECKUP QUIZ ON PAGE 361

1. (*10 kHz to 200 MHz+*)   2. (*Spark, arc, alternator*)
3. (*Frequency instability*)   4. (*Reflects changing C*)
5. (*Simplex*) (*Relay*)   6. (*Carrier only*) (*When tuning*)
7. (*Frequency stability*)   8. (*RF indicator, I_g meter*)   9. (*Dip*)
(*Slight peaking*)   10. (*Dielectric material expands*) (*Plate supports expand*)   11. (*Artificial, phantom*)   12. (*Tuning without transmitting a signal, determine P_o*)   13. (*$P = EI$, $P = I^2R$, $P = E^2/R$*)   14. (*Cathode*) (*Emitter*)

Fig. 16-8 Formation of a square-wave cycle. (*a*) Fundamental. (*b*) Addition of third harmonic. (*c*) Addition of many harmonics.

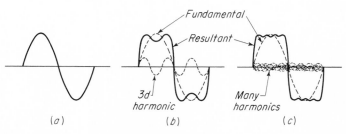

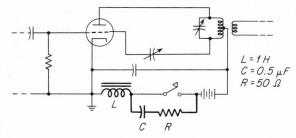

Fig. 16-9 LCR key-click filter in plate circuit of an amplifier.

plate current, rounding off the sharp *make* corner of the signal. When the key is opened, the capacitor allows plate current to flow until the capacitor becomes charged to the power-supply potential. The resistance *R* decreases sparking at the key contacts when the key is closed again. The values of *L, C,* and *R* vary with plate-current values. A characteristic waveshape with such a filter is shown in Fig. 16-10.

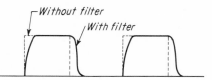

Fig. 16-10 Shape of square-wave plate-current pulses when a key-click filter is in the circuit, showing the rounding off of the make and break parts of pulses.

If sparking is excessive at any switch contacts, a capacitor and resistor may be connected in series across the switch to decrease it.

16-10 PRIMARY KEYING

An old method of producing radiotelegraph transmissions with little or no key click was to use *primary keying* (Fig. 16-11). This circuit has a full-wave center-tapped rectifier with inductive-input filtering. The key makes and breaks the power to the primary of the power transformer. The choke prevents plate voltage from building up rapidly, rounding off the make corners. The capacitance of the capacitor determines how long the transmitter will continue to operate after the key is opened, preventing a rapid decrease in plate current, resulting in a rounded characteristic of the break corner of the transmitted dot or dash. The values of *L* and *C* vary with plate-current requirements as well as the frequency of the power ac. For 60-Hz power, the inductance may be between 1 and 3 H and the capacitance between 1 and 3 μF. Higher-frequency power ac requires smaller component values.

16-11 VACUUM-TUBE KEYING

One method of controlling the emission of a telegraph transmitter is by using *vacuum-tube keying* as in Fig. 16-12. The circuit to be keyed is opened, and a low-μ triode tube is connected in series with the cathode circuit. In the circuit shown, the keying tube is in series with the plate supply voltage of the RF amplifier. If the keyer tube is biased to stop all current flow through it, the RF amplifier stage must also have no current flow and no RF output.

Fig. 16-11 A possible primary-keying circuit.

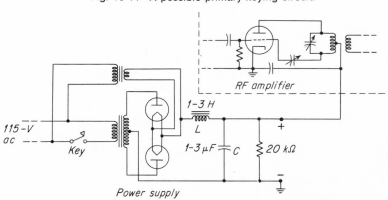

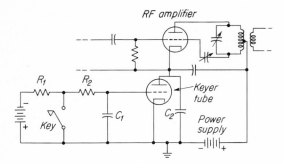

Fig. 16-12 Vacuum-tube-keying circuit.

With the key open, full bias is applied to the keyer grid-cathode circuit, biasing the tube past cutoff, preventing the RF amplifier stage from operating. C_1 charges to the full bias voltage.

When the key is closed, three things happen: (1) All the bias voltage is developed across R_1, preventing a short circuit of the battery. (2) Capacitor C_1 discharges through R_2 at a rate dependent upon the R_2 and C_1 values, lowering the negative bias value on the keyer grid to zero. (3) With no bias, current can flow through the keyer tube and the RF amplifier, allowing the amplifier to operate.

When the key is opened, C_1 charges at a rate dependent upon the values of C_1 and of R_1 plus R_2. Since C_1 is connected between grid and cathode, when the voltage across C_1 builds up to the cutoff value of the tube, the keyer tube stops passing current and the RF amplifier stops operating.

C_2 is a bypass capacitor, keeping the cathode of the RF amplifier at ground potential insofar as RF ac is concerned. It has little effect on shaping the transmitted wave.

Two or more keyer tubes may be paralleled when keying a high-power RF amplifier. This keying is capable of very fast operation and high-speed transmissions.

The transistor counterpart of this circuit is simple. Since a BJT with zero bias has no conductance, keying is produced merely by applying a forward bias to the keying BJT.

16-12 GRID-BLOCK KEYING

If the bias of an RF amplifier is made sufficiently negative to prevent plate-current flow in the tube,

regardless of how high the driving RF ac voltage may be, the amplifier stage cannot operate. The blocking of the plate current by applying a high negative charge to the grid is known as *grid-block, blocked-grid,* or *grid-bias* keying. One such keying circuit requires a separate bias supply, as shown in Fig. 16-13.

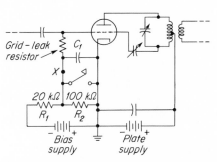

Fig. 16-13 Blocked-grid keying with bias supply.

When the key is open, R_1 and R_2 form a voltage divider across the bias supply. The voltage-drop across R_2 must be high enough to cut off the plate current of the stage, preventing any RF output.

When the key is closed, all the bias supply voltage is across R_1 with none across the key or in series with the grid circuit. The only bias in the grid circuit is that developed by the grid-leak resistor. The stage will now operate normally, emitting an RF signal.

Capacitor C_1 is an RF bypass, holding the bottom of the grid-leak resistor at cathode potential insofar as RF ac is concerned. If key clicks are produced, a resistance inserted in the circuit at point X may reduce them.

Should the bias supply voltage fail for any reason, the stage will continue to emit a signal even with the key open. The emission may be only slightly less when the key is up. It may appear to the operator that the key is shorted, since the local signal will be extremely strong under both conditions. Pitted contacts that refuse to open on a keying relay would also produce an emission with the key contacts apparently open. A faulty relay will give the same indication.

Another circuit in which the grid is blocked with the key open but which requires no separate bias supply is shown in Fig. 16-14. With the key open, the 20- and 40-kΩ resistors form a voltage divider

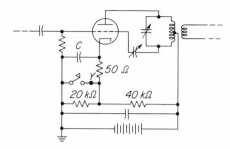

Fig. 16-14 Blocked-grid keying requiring no bias supply.

across the power supply. One-third of the power-supply voltage is placed on the grid to block it. When the key is closed, the 20-kΩ resistor is shorted out and the only bias in the grid circuit is that developed across the grid-leak resistor. The full voltage of the power supply is now connected between cathode and plate.

In some installations it might be more advantageous to ground the cathode instead of the negative terminal of the plate supply.

Screen-grid and suppressor-grid keying are possible, but are rarely used.

Another type of emission can be produced by connecting a chopper wheel (Fig. 16-15) into Fig.

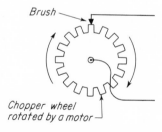

Fig. 16-15 A chopper wheel electrically makes and breaks at the brush.

16-14 at point Y. If the metal chopper makes and breaks contact 500 times per second and the key is held down for $1/10$ s to transmit a dot, the carrier will be turned on and off 50 times. This results in an interrupted CW (A2) signal, composed of pulses of square-wave RF ac, only slightly rounded off by the filtering effect of capacitor C and the 50-Ω resistor across key and chopper. This produces a rather broad emission that can be detected by any type of receiver tuned to the frequency. For this reason it would be desirable for distress transmissions.

16-13 FREQUENCY-SHIFT KEYING (F1)

The usual radio receiver is sensitive to changes in signal amplitude. When a carrier of a transmitter is turned on and off, as when sending dots and dashes, the receiver should produce a sound for each dot or dash and be quiet between times. Unfortunately, both man-made and natural static occur as changes in amplitude, producing noises in the receiver when the transmitter is off. If the received signal is strong enough, it will be well above this noise level and the dots and dashes can be easily distinguished. When signals are only slightly stronger than the noise level, however, instantaneous crashes of noise between dots and dashes confuse reception of the code.

As long as there is a carrier, the receiver is quieted somewhat. Therefore, if the carrier can be left on at all times, some of the effect of background noise can be overcome. If the carrier is to be left on all the time, how can dots and dashes be sent? One method is to shift the carrier back and forth from one frequency to another nearby. When no code is being sent, the carrier is left on one frequency, called the *space* frequency. When a dot or dash is transmitted, the carrier is shifted to the other *mark* frequency. This is *frequency-shift keying* (FSK or F1). If the mark and space frequencies are relatively close (60 to 800 Hz), the receiver remains in a quieted condition, since it can receive both signals on one setting of the dial. By making the receiver sensitive to frequency changes only, rather than to amplitude changes, an improvement of reception results.

Figure 16-16 shows a basic keying circuit by which mark and space signals can be produced. The closer the coupling of the loop, the greater the frequency variation when the relay shorts the

Fig. 16-16 FSK using a shorted loop to change LC circuit inductance.

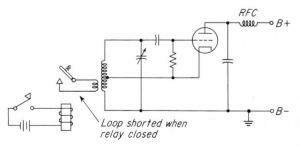

loop. A shorted loop coupled to an inductance lowers the effective inductance of the associated *oscillator* LC circuit.

Another method of shifting the frequency is to connect and disconnect a small capacitor across part of the tuned circuit of an oscillator. The capacitance lowers the frequency. A possible circuit for this is shown in Fig. 16-17. Adjusting the

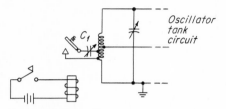

Fig. 16-17 FSK by keying a capacitor across the LC tank circuit of an oscillator.

capacitance C_1 controls the amount of frequency shift.

An FSK circuit using two diode tubes biased *off* by having their plates more negative than their cathodes is shown in Fig. 16-18. With the two

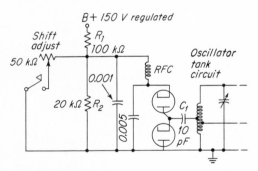

Fig. 16-18 FSK circuit using biased diodes.

tubes biased to nonconductance (determined by the voltage divider composed of R_1 and R_2) capacitor C_1 is not effective. When the key is closed, the bias voltage is decreased enough that some of the peaks of both halves of the RF ac cycles produced in the oscillator are greater than the bias. During these portions of the RF cycle the capacitor is effectively coupled across part of the tuned circuit, lowering the frequency of oscillation. The lower the bias value on the diodes, the

greater the proportion of the cycle during which the capacitor is effective and the greater the frequency variation. With this circuit remote frequency control is possible.

It is possible to frequency-shift a crystal oscillator a little. This can be done by keying a small inductance in and out of the crystal circuit itself or by keying a small capacitor across the crystal with a double-diode circuit. Crystals at 500 kHz may shift only a few hertz, but in the 4-MHz region they may shift 400 Hz or more and still give good mark-space output.

A reactance-tube modulator circuit or a voltage-variable diode (Sec. 19-13) across the tank circuit of the master oscillator are two other methods of producing FSK.

16-14 MODULATED CODE SIGNALS (A2)

Short-wave signals, either radiotelegraph or radiotelephone, are often subject to very rapid and deep fading. Two frequencies separated by only a few hundred hertz may fade at different times. To overcome this, code signals are sometimes *tone-modulated*. This may be accomplished in several ways. One method consists of adding an AF ac voltage in series with the dc plate voltage of the final RF amplifier stage, resulting in a varying dc plate voltage for this tube. The strength of the RF ac output will now vary up and down in amplitude at the ac rate. The RF carrier is said to be *amplitude-modulated* at the AF rate. The chopper wheel, mentioned in Sec. 16-12, produces a similar effect.

If an 800-Hz ac is added in series with the dc plate voltage of an RF amplifier, the output RF power will increase and decrease in amplitude 800 times per second. The addition of this modulation to the carrier wave produces two other RF signals, one having a frequency 800 Hz higher and the other a frequency 800 Hz lower than the carrier frequency. These new RF signals are *sidebands*.

Under conditions of severe fading the carrier frequency may fade out completely but one or the other sideband may remain strong. As a result, with the usual CW receiver, a continually readable signal may be received, another reason why all distress transmissions should use A2 emissions.

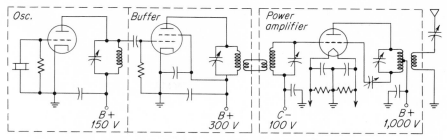

Fig. 16-19 Transmitter with crystal oscillator, tetrode buffer, and triode power amplifier.

16-15 THE BUFFER AMPLIFIER

The MOPA transmitter represents a decided improvement over a simple oscillator in both power output and frequency stability. Even so, varying the amplifier tuning or antenna coupling causes some frequency pulling. More nearly perfect operation is possible if a transmitter consists of an oscillator, an intermediate-powered amplifier, and a high-powered final amplifier, as in Fig. 16-19. The intermediate stage is called a *buffer* amplifier. It has two duties: (1) to amplify the weak output of the oscillator enough to excite the grid of the final amplifier adequately and (2) to prevent interaction between the oscillator and any power amplifiers which might result in frequency instability.

A buffer amplifier can be a standard triode, tetrode, pentode, or transistor RF amplifier. It is relatively low powered. The power-amplifier stage of commercial-type CW transmitters, may have an output ranging from 200 to more than 10,000 W. Amateur transmitters are limited to 1 kW input to the final amplifier.

When both the oscillator and the buffer stages are properly shielded and the buffer has been neutralized (if necessary), tuning power amplifiers or antenna circuits should have little or no effect on the oscillator frequency. If the oscillator and buffer operate from a power supply separate from the final-amplifier supply, keying the final amplifier should produce excellent keying characteristics. Fairly good keying can be produced by keying the buffer, particularly if it has a power supply separate from that of the oscillator. However, keying an intermediate driver amplifier between the buffer and the final amplifier is to be preferred. In general, the oscillator and buffer stages should

be considered as a frequency-determining pair and should not be keyed in radiotelegraph or be amplitude-modulated in radiotelephone transmitters.

The three-stage transmitter shown consists of a triode crystal oscillator, capacitively coupled, for simplicity, to a tetrode buffer stage, link-coupled to a plate-neutralized triode final amplifier, inductively coupled to an antenna. A possible power input to the final-amplifier plate circuit ($E_p I_p$) is about 200 W, with about 140 W of RF power output.

Buffer amplifiers may be biased to class C, although class A or B is often used. Transistor stages are usually unbiased (class B or C). A procedure for tuning multistage transmitters might be:

1. Start with the plate voltage off on all stages.
2. Connect the plate voltage to the oscillator and tune this stage. (Neutralize next stage, if required.)
3. Connect the plate voltage to the next stage and tune it. Recheck the previously tuned stage.
4. Progress stage by stage as in steps 2 and 3 to the final amplifier.
5. Tune the final amplifier to a plate-current dip with very loose antenna coupling and low plate voltage.
6. Tune the antenna to resonate with the final amplifier (if a tunable antenna is used), indicated by a peaking of the final plate current.
7. Increase the antenna coupling to the desired final-amplifier plate-current value or required antenna-current value.

It is good procedure to recheck each stage again, starting at the oscillator and working up to the final amplifier and antenna.

All triode-type stages (except multiplier or grounded-grid amplifiers) must be neutralized, before they are tuned, to prevent self-oscillation, generation of spurious oscillation frequencies, erratic tuning, excessive plate current, or poor isolation of the stages before or after them. Once neutralized, a stage may require no reneutralization for a long period of time.

16-16 GRID-LEAK BIAS IN CW TRANSMITTERS

A grid-leak-biased amplifier following a keyed stage in a CW transmitter has no bias when there is no RF excitation to the grid. Without bias, the stage may draw too much plate current and damage the tube.

There are three methods of protecting such a grid-leak-biased stage.

1. The amplifier may be biased with a bias supply that limits the plate current to some safe value when the key is up.
2. The stage may employ cathode-resistor bias to produce a safety bias when no RF excitation is applied.
3. A clamp tube may be used in screen-grid amplifier tubes, as shown in Fig. 16-20.

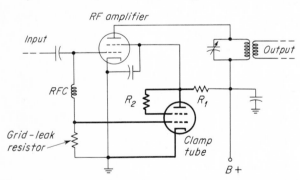

Fig. 16-20 Clamp tube used when a preceding stage is keyed.

The clamp tube should have a high transconductance, a low cutoff-bias value, and a large plate, such as a type 6Y6 beam-power tetrode. With no excitation to the RF amplifier, there is no grid-leak bias and the clamp tube, which also derives its bias from the grid leak, has a low plate resistance, passing a high value of current through itself and the screen dropping resistor R_1.

This produces a large voltage-drop across R_1, a low voltage across the clamp tube, a low screen voltage for the RF amplifier, and therefore a low plate current.

When RF excitation is applied to the grid of the amplifier, grid-leak bias is produced, biasing the clamp tube to cutoff and stopping its plate current. This reestablishes the normal screen-grid voltage, and the amplifier stage operates as though the clamp tube were not in the circuit. Resistor R_2 prevents excessive clamp tube screen-grid current.

Clamping circuits are forms of limiter or clipper circuits and may be found in many applications in radio and electronics.

16-17 FREQUENCY DOUBLERS OR MULTIPLIERS

The lower the frequency of oscillation, the more stable the oscillator and the less it is likely to drift (Secs. 13-9, 13-24). To take advantage of this better frequency stability, the oscillator is often operated at a lower frequency than the frequency of operation of the final amplifier. For example, it is desired to transmit on a frequency of 4,000 kHz. If the oscillator operates on 1,000 kHz stably, the 1,000-kHz signal can be fed into a pair of frequency-doubler stages. The output of these stages will drive the final amplifier at the desired 4,000-kHz frequency with greater stability than if the oscillator had been on 4,000 kHz. A block diagram of such a transmitter is shown in Fig. 16-21.

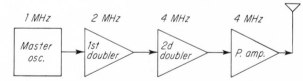

Fig. 16-21 Block diagram of a transmitter with an output on 4,000 kHz and an oscillator on 1,000 kHz.

Since the final-amplifier frequency is not the same as the frequency of the oscillator, there is much less chance of interaction between the two stages. Shielding between multiplier stages may not be required.

Transmitters operating in the VHF band, 30 to 300 MHz, usually use three or more multiplier

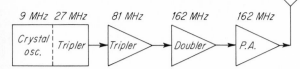

Fig. 16-22 Block diagram of a system to produce 162-MHz output.

stages. For example, to transmit a signal on 162 MHz, a possible arrangement is shown in the block diagram in Fig. 16-22. A crystal is the only commercially acceptable oscillator to maintain frequency stability. Frequency synthesizers (Sec. 28-12) may be used in more sophisticated equipment.

If the frequency of the oscillator stage in a multistage transmitter changes for any reason, the multiplier stages multiply the change.

EXAMPLE: A 1,000-kHz crystal is used in a transmitter operating on an assigned frequency of 8,000 kHz. The transmitter might have three doubler stages. If the crystal changes 5 Hz in frequency, the first doubler changes 10 Hz, the second doubler changes 20 Hz, and the third doubler changes 40 Hz. The 5-Hz oscillator change produces a 40-Hz change in the output signal.

If the crystal being used has a temperature coefficient of -4 Hz/MHz/°C and the temperature of the crystal increases 6°C, the output frequency will decrease by 4 (-4 Hz indicates a negative temperature coefficient) times 1 (1,000 kHz equals 1 MHz) times 2 times 2 times 2 (three doubler stages) times 6 (temperature increase in degrees), or 192 Hz. If the output frequency had been 8,000,000 Hz, increasing the temperature of the crystal 6° would have decreased the frequency of the output by 192 Hz, to 7,999,808 Hz, or 7,999.808 kHz. If the crystal had a positive temperature coefficient of $+4$ Hz/MHz/°C, the rise in temperature would have increased the frequency to 8,000,192 Hz.

Test your understanding; answer these check-up questions.

1. Is emitted bandwidth directly or inversely proportional to keying speed? _____
2. What is reduced by rounding off square-wave emissions? _____ _____

3. Does a key-click filter result in soft or hard keying? _____
4. What components make up a key-click filter? _____
5. Why does vacuum-tube keying require no inductor for soft keying? _____
6. Why might the counterpart of blocked-grid keying not be used in a transistor CW stage? _____
7. Why should A2 be used for distress transmissions? _____ _____
8. Is static an amplitude- or frequency-shifting impulse? _____
9. What are the two frequencies called with FSK? _____ _____
10. Why would the diode FSK system be better than a relay type? _____ _____
11. What are the two duties of a buffer amplifier? _____
12. Why might pentodes be preferable to triodes as buffer amplifiers? _____ _____
13. Would the clamp tube circuit shown operate in a triode RF amplifier? _____
14. What is gained by operating an oscillator on some subharmonic of the transmission frequency? _____
15. A 1-MHz oscillator has 100-Hz FSK. What is the output shift if this signal is fed through three doublers? _____

16-18 DECREASING HARMONIC RADIATION

No RF amplifier will emit a pure sine-wave RF ac. Inasmuch as any deviation from the pure sine wave represents harmonic content, RF amplifiers can be expected to transmit some harmonic energy as well as the fundamental frequency. In many cases, harmonic energy from a transmitter will fall on frequencies used by other radio services, causing them interference. When a harmonic or spurious signal falls on TV frequencies, it is known as television interference (TVI). This must be eliminated.

There are many methods of decreasing harmonic generation and radiation. Figure 16-23 illustrates some of them. Each numbered point will be briefly described.

1. The use of link or any other form of inductive coupling discriminates against the passing of higher frequencies, whereas most forms of capacitive coupling will couple higher (harmonic) frequencies better than they will the lower fundamental frequency.
2. A high-Q circuit tuned to the fundamental in the grid circuit decreases the possibility of

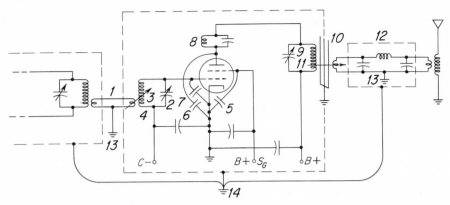

Fig. 16-23 Methods of decreasing harmonic radiation from a power amplifier.

transferring harmonics generated in the driving stage to the final-amplifier grid.

3. The coupling to the grid circuit should be held to a low enough value to prevent driving the grid to the point of plate-current saturation at the peak of the RF voltage, which always results in a high percentage of harmonic output.

4. The bias voltage should be as low as possible, no more than $1\frac{1}{2}$ times the cutoff value for class C amplifiers, to produce plate-current flow as near 180° of the RF cycle as possible without materially decreasing the efficiency of the stage.

5. Also 6 and 7. Small mica- or air-dielectric capacitors (10–50 pF) may be connected to ground from grid, screen, and plate terminals of the tube to bypass harmonic frequencies that fall in the VHF range.

8. A small coil and capacitor tuned to any VHF being interfered with tends to decrease harmonic transfer through the tube to the plate-tank circuit and also functions as a parasitic-oscillation-stopping device.

ANSWERS TO CHECKUP QUIZ ON PAGE 369

ANSWERS TO CHECKUP QUIZ ON PAGE 369

1. (*Directly*) 2. (*Harmonics, bandwidth*) 3. (*Soft*) 4. (*L, C, R*) 5. (*Slow charge-discharge of C_1 through R*) 6. (*Too much voltage between E and B*) 7. (*Any receiver can receive; signals with selective fading*) 8. (*Amplitude*) 9. (*Mark, space*) 10. (*No moving parts to jar oscillator; remote frequency-shift adjustment possible*) 11. (*Amplify oscillator; prevent oscillator frequency pulling*) 12. (*Require less power from oscillator; no neutralization required*) 13. (*No*) 14. (*Frequency stability*) 15. (*800 Hz*)

9. The tank circuit should have a high Q to produce good flywheel effect.

10. A *Faraday shield* may be installed between the final tank and the antenna. It is a gridwork of thin parallel wires, not connected together at the top but soldered together at the bottom and grounded. This *electrostatic shield* materially decreases capacitive coupling, thereby decreasing the possibility of harmonic transfer to the antenna circuit.

11. As loose a coupling as possible should be used between final-tank circuit and the antenna and still maintain output power. Over-coupling lowers tank-circuit Q.

12. A low-pass filter with a cutoff frequency slightly below the second-harmonic frequency of the transmitter may be added to the antenna feed line. All frequencies above its cutoff frequency will be attenuated, will not be delivered to the antenna, and therefore will not be transmitted. (A more complex filter is preferable to the simple constant-*k* type shown.)

13. One side of a link-coupling circuit should always be grounded.

14. All stages, as well as the low-pass filter, should be separately shielded and then grounded to a common ground point on the chassis. A water pipe or a copper rod in damp earth makes a good ground.

Not shown in this illustration, but particularly effective in reducing even-order harmonics, is a well-balanced push-pull stage.

The presence and strength of harmonic radiation from a transmitter can be determined by use of an accurately calibrated field-strength meter, by a receiver with an accurately calibrated strength-of-signal meter, or by some types of calibrated wavemeters. Any indication of harmonic signal detected must be below the limit of tolerance of harmonic radiation for that transmitter.

16-19 SHIELDING AND PROTECTIVE DEVICES

The advisability of shielding stages, circuits, or parts has been mentioned. Shields must do two things: (1) shield one circuit electromagnetically from other circuits and (2) shield one circuit electrostatically from other circuits.

As a simple case of shielding, consider the two coils in Fig. 16-24. The dotted lines surrounding

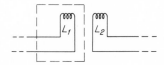

Fig. 16-24 Coil L_1 is shielded from L_2.

the first coil indicate that this coil is encased in a metal container or shield.

Assume, first, that there is no shield present. Coils 1 and 2 are mutually inductive. Ac in the first will induce ac into the second because of expanding and contracting magnetic fields. If a metallic box is placed around the first coil, almost no current is induced into the second coil. The expanding and contracting magnetic fields induce an alternating voltage and current in the metal shield. This induced current produces a field, according to Lenz's law, opposite in phase to that of the field that produced it. If the shield were a perfect conductor, the strength of the opposing field would equal the inducing field and there would be zero resulting field to move outward toward the second coil. The shielding would be perfect.

Unfortunately, there is no perfect conductor. Therefore shielding of magnetic fields can never be perfect. However, silver is near enough to a perfect conductor to produce very good shielding. Copper is only a little less effective. Then, in order, come gold, aluminum, zinc, iron, and lead,

to mention some common metals. Because of high cost most shielding is aluminum, except in circuits operating at frequencies over about 100 MHz, when silver-plated copper or brass is used.

Once again, consider the two coils without the shield. Each coil is a conductor, and the air between them is a dielectric. They form a capacitor. Ac energy can travel from the first coil to the second via the electrostatic lines of force of capacitance. Actually, energy is transferred from one coil to the other by both electromagnetic and electrostatic lines of force.

When the metallic shield is placed around the first coil, there is capacitance between the first coil and the shield and also between the shield and the second coil. If the shield is connected to ground, the electrostatic energy flows back and forth between the first coil, the shield, and ground and none reaches the second coil. Electrostatic shielding can be considered complete with most metals.

What is true of shielding between the two coils is also true between two or more RF stages, between an RF amplifier and an antenna, and between an RF and an AF stage.

Proper shielding can decrease or prevent degeneration, regeneration, oscillation, distortion, parasitics, instability, hand capacitance (detuning), and other undesired effects. Communications equipment is encased in metal to provide solid physical construction as well as good shielding.

Properly constructed transmitters and receivers have no high-voltage-carrying leads external to the equipment. All exposed metal parts are engineered in such a way that they can be connected to the chassis and grounded, thereby preventing operating personnel from receiving shocks. Grounding the metal parts of a transmitter also prevents the accumulation of a static charge on them. As a further protection to personnel, *interlock switches* may be incorporated in doors to interrupt all high-voltage circuits automatically if the doors of the transmitter are opened. All possible controls to adjustable parts are brought outside, making it unnecessary for the operator to reach inside the cabinet, or housing, to tune or adjust the circuits. When fuses are used, they are usually available from outside the equipment.

When electromagnetic overload relays or circuit breakers are used instead of fuses, the restore mechanisms are usually available from outside. The operator determines the operation of the equipment by meter readings alone. The meters to be found on a VT transmitter are power-line voltmeter, filament voltmeter, plate-current meter or meters, plate-voltage meter or meters, grid-current meter or meters, and either an antenna ammeter or an antenna voltmeter.

16-20 TRANSMITTERS TODAY

Modern transmitters might be catalogued as to whether they operate on one assigned frequency, on several assigned frequencies, or in bands of frequencies.

AM broadcast, TV, FM broadcast and some fixed point-to-point stations are examples of transmitters assigned to operate on one frequency only. Mobile police, fire, etc., may operate on one, two, or three relatively nearby frequencies. They all use crystal-controlled oscillators. The RF sections of such transmitters are tunable but, once tuned, are not expected to be adjusted for long periods of time. Mobile equipment is constructed to operate from 6- or 12-V dc sources, is as light and rugged as possible, and has a minimum power drain per watt of output. Mobiles usually operate on assigned frequencies from 30 to 500 MHz.

Shipboard radiotelegraph or radiotelephone, aircraft, and many point-to-point transmitters are examples of transmitters that operate on certain assigned frequencies during one time of the day and on other frequencies at other times of the day to communicate over the same or different distances. Crystal-controlled oscillators are used. There may be a separate crystal for each assigned frequency, or the frequencies may be harmonically related, and frequency-multiplier stages can be utilized. The use of crystals in these transmitters simplifies selection of the correct frequency, as well as assuring good frequency stability. The desired crystal is chosen by rotating a selector switch. The same switch may shift the RF stages to pretuned settings, requiring only antenna and final-amplifier trimming to assure proper operation of the transmitter. In some equipment the tuning is done automatically and electronically. In others, each buffer, multiplier, and amplifier requires individual retuning by an operator.

In the early days of radio, all transmitters used variable-frequency oscillators. Today, amateur and some marine equipment may be the only transmitters with variable-frequency oscillators. Commercial equipment uses crystals or electronic frequency synthesizers to assure operation within the strict frequency tolerances set by national and international communications agreements.

Since the amateur has been assigned bands of frequencies, such as 3.5 to 4 MHz, 7 to 7.3 MHz, and 14 to 14.35 MHz, his transmitter may have a variable-frequency oscillator to allow him to operate on any frequency within the band limits. Crystal-controlled oscillators may be used, but they restrict his ability to move away from interfering signals or to shift to any particularly desirable frequency. The amateur transmitter should be capable of operating on several bands at the choice of the operator and on any desired frequency in these bands.

The single-frequency type of transmitter is discussed in Chap. 24. The multifrequency type of transmitter is discussed in Chap. 28. A variable-frequency-oscillator radiotelegraph transmitter circuit suitable for amateur-band CW operation is discussed in Sec. 16-22.

16-21 A TRANSISTOR CW TRANSMITTER

Transistorized CW transmitters are usually capable of 1 to 100 W, depending on the transistors used. Circuitry is simple. Figure 16-25 illustrates a possible crystal oscillator transmitter capable of 10 W by using heat sinks and a 12-V battery or of more than 50 W output with a V_{cc} of 30 V or more.

Capacitive coupling is shown in low-level stages and inductive coupling to the push-pull power stage. Many dual-transistor stages are parallel, although push-pull is shown. Keying can be accomplished by inserting a key into the closed-circuit jack in the oscillator emitter circuit. Common-base circuits are used for stability. If one of the stages is to operate as a frequency multiplier, it should be changed to a common-emitter configuration.

Capacitors C_2 and C_4 are variable because

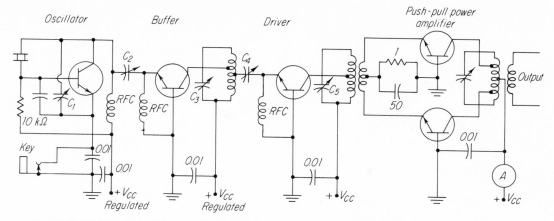

Fig. 16-25 A transistor CW transmitter capable of 1 to 50 W, depending on transistors and V_{cc} used.

coupling between stages may be critical. C_1 should be adjusted for best keying characteristics.

While the collector ammeter should dip at resonance, some form of RF indicator such as an oscilloscope may be preferable to tune the stages.

If the push-pull stage requires neutralization, the two-capacitor circuit shown in Fig. 15-18 could be used.

16-22 A VFO TRANSMITTER

The *variable-frequency oscillator* (VFO) radiotelegraph transmitter diagram of Fig. 16-26 represents a possible amateur-band CW transmitter capable of 200- to 400-W input. By adjusting the ganged three-pole four-position band-selector switch (Sw), operation on any one of four amateur

Fig. 16-26 Schematic diagram of a possible four-band VFO amateur band transmitter.

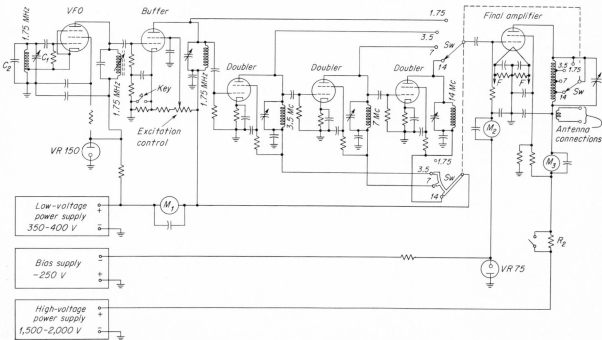

bands is possible. The switch is shown in the 14.0- to 14.35-MHz band position.

The electron-coupled oscillator operates on the "160-meter," (1.8- to 2.0-MHz) amateur band. C_1 is the tuning capacitor, and C_2 is a padding capacitor to add capacitance to the tuned circuit. The output plate circuit of the oscillator is broadly tuned by using a tightly coupled, low-Q resonant circuit which is tuned by adjusting the powdered-iron core, or slug, in the inductance. The oscillator tube is a receiving-type pentode.

The second stage is a buffer amplifier operating on the same frequency as the oscillator. The screen-grid voltage is variable to control excitation to the stages that follow. With the switch in the position shown, the output of the buffer feeds the grid circuit of a beam-power tetrode doubler.

The third stage is a doubler, with the plate circuit tuned to the "80-meter," or 3.5- to 4.0-MHz band. The cathode-resistor bias holds the plate current to some safe value while no signal is applied to the grid of the tube. When the key is closed, the grid is excited and bias is developed across the grid-leak resistor. The tube now acts as a doubler, feeding a 3.5- to 4-MHz signal to the next stage.

The fourth stage is similar to the third except that the plate circuit is tuned to cover the "40-meter," or 7.0- to 7.3-MHz amateur band. Note that 7.3 to 8.0 MHz is not included in the amateur band. The transmitter must not be operated outside assigned band limits.

The fifth stage is similar to the fourth except that the plate circuit is tuned to cover the "20-meter," or 14.0- to 14.35-MHz amateur band.

The final amplifier feeds RF power to the antenna. It is a beam-tetrode power amplifier biased by a separate power supply to more than plate-current cutoff.

TUNING. First, the high-voltage power supply is turned off and the antenna coupling reduced to minimum. The remainder of the transmitter is turned on. After a warm-up period of a few minutes, the desired frequency of operation is selected on the oscillator tuning dial.

When the key is pressed, the buffer begins amplifying and feeding RF ac to the first doubler, which excites the second doubler, and so on. The final-amplifier grid circuit draws current when excited by the last doubler. The buffer and dou-

bler stages can be tuned by watching either M_1 or M_2. Meter M_1 will show lower current values as the buffer or multipliers are tuned to resonance, while meter M_2 will show an increase as the stages are brought to resonance. When the buffer and doublers have been tuned properly, the *excitation control* in the screen-grid circuit of the buffer can be adjusted to the required grid current for the final amplifier.

Now the high-voltage power supply can be turned on at a low-power setting by reducing the plate voltage, either with a variable autotransformer (Fig. 16-27) or by inserting a resistor in

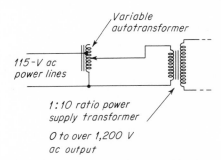

Fig. 16-27 Autotransformer connected between a power-supply transformer of a power amplifier and the power line.

series with the high-voltage lead, R_2 in Fig. 16-26. The plate-tank circuit is tuned to a plate-current minimum. The antenna circuit is tuned to resonance, indicated by a slight rise in plate current. The antenna is coupled a little tighter, and the plate tank redipped, as coupling the antenna usually detunes the final plate tank a little. The full plate voltage is applied to the final amplifier, and the antenna coupling is adjusted until the desired plate current, as specified in a tube manual, is attained. The grid current must be rechecked to produce the specified value. (Power output from a CW transmitter can be controlled by varying E_p or E_{sg}, by grid excitation, or by antenna coupling.)

It is possible to have the plate LC circuit of the last multiplier resonate from 22 to 14 MHz and operate the last multiplier as a tripler. If the final amplifier has a plate LC circuit capable of tuning over this same range, the transmitter can operate on the "15-meter," or 21- to 21.45-MHz band. Care must be exercised not to tune, inadvertently, to the wrong harmonic frequency.

To operate on the 7.0-MHz band, the band-selector switch is set to the 7-MHz position. This disconnects the plate supply from the last multiplier, connects the final-amplifier grid to the 7-MHz stage, and shorts out less of the final-amplifier plate coil, allowing it to resonate at a lower frequency. Operation on the 3.5- and 1.8-MHz bands is accomplished in a similar manner.

Shifting the fundamental frequency of the transmitter a few kilohertz may not require retuning of all the stages. If the frequency is changed by 50 or more kilohertz, however, the tuned circuits will have to be retuned. It is possible to gang all tuned circuits so that satisfactory operation can be obtained on four bands with a single tuning control for oscillator buffer and multipliers. Separate controls are always used with the final plate and antenna circuits.

16-23 HETERODYNE TRANSMITTERS

Many modern transmitters utilize the heterodyne principle. If two frequencies are fed into a nonlinear circuit, both a sum and a difference frequency will be developed. A block diagram of a possible heterodyne CW system is shown in Fig. 16-28. Two oscillators on separate frequencies

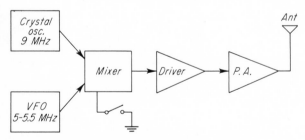

Fig. 16-28 Heterodyne CW transmitter system.

feed into a mixer or heterodyne circuit. In this circuit the mixing of 9- and 5-MHz ac produces either $5 + 9 = 14$-MHz, or $9 - 5 = 4$-MHz ac. The one frequency to which the output LC circuit of the mixer is tuned determines which frequency will be fed to the driver and then the power amplifier. These last two stages will also have to be tuned to the desired frequency.

The advantages of this system are: (1) Frequency stability is excellent, since both oscillators run continuously. (2) Keying characteristics are excellent, since only the mixer stage is keyed, not an oscillator. (3) There is no oscillator running on the transmitting frequency, allowing break-in keying. (4) There are no frequency multiplying stages, thereby reducing harmonic production.

The output frequency can be shifted over a band of 500 kHz in this system by adjusting the self-excited variable-frequency oscillator. This may also require retuning or trimming of all following stages. For amateur transmitters it is possible to change from one band to another by changing crystals in the crystal oscillator stage, rather than by using frequency multipliers. For single-frequency commercial, marine, or citizen band operation both oscillators would be crystal types.

The mixer circuit can be either special mixer tubes or transistor circuits, discussed in later chapters. Mixer stage current flows only when the key is closed. However, even with the key open there may be a production of noise due to random electron motion in the mixer device. Such "white" (all-frequency) noise can be amplified and heard by local receivers as a hissing sound on all frequencies near which the transmitter LC circuits are tuned unless the driver and PA are biased beyond cutoff.

Applications of transmitters are discussed further in Chaps. 17, 19, 24, 25, and particularly 28. The subject of *transceivers* is discussed in Sec. 18-28.

Test your understanding; answer these checkup questions.

1. What type of coupling discriminates against harmonic transmission? _____
2. Why does a high-Q circuit discriminate against harmonics? _____
3. What class of bias should produce least harmonics? _____
4. What type of field does a Faraday shield stop? _____
5. What type of circuit reduces even-order harmonics? _____
6. Which is more difficult to shield, electromagnetic or electrostatic fields? _____
7. What is the purpose of an interlock switch? _____
8. What are the only types of transmitters that now use VFOs? _____
9. In Fig. 16-25, what type of crystal oscillator is used? _____ In what class is the buffer and driver operating? _____ What is the purpose of the 1-Ω resistor in the PA? _____

10. In Fig. 16-26, what type of ECO is shown? _____ What kind of keying is being used? _____ In what class would the doublers operate with adequate drive? _____ With inadequate drive? _____
11. In Fig. 16-26, what does M_1 read? _____ M_2? _____ M_3? _____ What bias value is applied to the PA? _____ What is the oscillator E_p value? _____ To change from 14,100- to 14,140-MHz output, how many hertz must the VFO be changed? _____
12. All-frequency noise is called what? _____
13. What are four possible output frequencies if f_1 and f_2 are fed to a mixer? _____ _____ _____ _____
14. What are the advantages of heterodyne CW transmitters? _____ _____ _____ _____
15. Why should the two frequencies fed to a mixer not overdrive the mixer circuit? _____

16-24 EMERGENCY REPAIRS

There are many possible sources of trouble in a transmitter. Any part can become faulty and prevent proper operation. When something does go wrong, it is up to a licensed operator to determine the fault and make the necessary repairs. When a faulty part is found, it is best to use an identical replacement. Unfortunately, identical replacements are not always immediately available, and emergency repairs must often be made. This sometimes tests the ingenuity of the operator.

Some of the many possible troubles encountered in transmitter operation will be discussed briefly. In most radio stations there will be at least a volt-ohm-milliammeter with which equipment can be tested. Between voltmeter readings and continuity tests with an ohmmeter, most troubles can be located.

TUBES. Vacuum tubes wear out. Excessive filament voltage and current shorten their lives. An indication of loss of filament emission is a permanent decrease of grid and plate current. Excitation to the next stage decreases. If the tube has a glass envelope, a blue or purple haze may be seen between filament and plate, indicating gas. The grid, screen grid, or plate may become red-hot. If the envelope develops a crack in it, air will rush in and a white coating forms on the inside of the glass as the filament wire burns out. Loose elements may sag and short to adjacent elements, causing sparks and high-current indications on meters. Often the terminals of tubes become oxidized and develop a poor contact. If the tube is pulled out and then put back, the equipment may work normally again. The terminals of the tube and the socket contacts should be lightly sanded with fine sandpaper. A continuity check with an ohmmeter will tell if a filament is burned out. An ohmmeter may also be used to test a tube for element-to-element shorts. Receiving-type tubes can be checked in a tube tester, if one is available, but transmitting tubes must have identical replacements to determine if they are faulty. In some cases, tubes of somewhat similar construction and similar socket-pin connections can be substituted temporarily. (A 6L6 might be used in place of a 6V6, or vice versa.)

TUBE SOCKETS. These may short out internally or open-circuit. Because they so seldom become faulty, they may be overlooked when they do. Ohmmeter tests will usually indicate the difficulty.

TRANSFORMERS. Power transformers often give trouble. If they become damp, the insulation around the internal wires may ionize and spark across, sometimes starting a fire. Radio-frequency ac sometimes finds its way into a transformer and burns the insulation or short-circuits the transformer. If charred insulation is found, it is sometimes possible to scrape all the blackened parts away and coat the wires with an insulating material, such as clear lacquer, insulating paint, or household cement. The transmitter may operate under lowered power for long periods of time with such a temporary repair. If RF is suspected in a transformer, bypass capacitors should be connected across both primary and secondary windings. It is not unusual to find transformer wires corroded to the extent that they finally part, open-circuiting the transformer. If the wires are within reach, it is sometimes possible to scrape them clean and solder them together again. Sometimes a transformer with several taps for different voltages will open-circuit. It is often possible to jumper the open part and use the remainder of the winding, operating at reduced power. An ohmmeter can be used to check continuity of windings in suspected transformers. A transformer can be checked by putting a low-voltage ac across the primary and measuring the secondary voltage. If there is no voltage output, the transformer is probably burned out or shorted. If burned out, it will remain cool; if shorted internally, it will soon become hot.

RESISTORS. Resistors often burn out. In some cases they may become excessively heated over a long period of time and increase resistance to a high value, or open. In many cases, they overheat because some other part is faulty. For example, a screen-grid dropping resistor suddenly becomes hot and burns out. It will usually be found that the screen bypass capacitor is shorted, connecting the resistor from ground to B+, increasing the current through it, and burning it out. In many cases, as in grid or screen-grid circuits, replacement of resistors with others of half or twice the resistance of the burned-out part will produce a temporary repair that will be satisfactory. Sometimes it will not. In some cases, wire-wound resistors can be temporarily repaired by holding aluminum foil against the burned-out section with a wrapping of wire.

CAPACITORS. When capacitors are at work, the dielectric is under constant strain. If there is a weak spot in the dielectric, eventually it will break down and short the capacitor. Sometimes the short circuit is intermittent, making it very hard to find. Electrolytic capacitors can dry out and lose their capacitance but still not short out. When tested with an ohmmeter, these capacitors will not show the charging current that a good capacitor will. They will not show low resistance as a shorted capacitor would. If a mica capacitor in an RF circuit shorts out, it may be possible to use paper capacitors, although paper capacitors are not usually considered satisfactory in transmitters, particularly at frequencies of over 2 or 3 MHz. The bearings, or sliding contacts, on variable capacitors often become clogged with dirt, producing a poor connection. They can be cleaned with alcohol, lacquer thinner, or other solvents. In many cases, 100% or more variation in the capacitance value of bypass capacitors will not have any noticeable effect on the operation of a transmitter. In the case of filter capacitors in power supplies, more than the original capacitance is almost always satisfactory. In an emergency, particularly in CW transmitters, a power supply can be operated with little or no filter.

CHOKE COILS. Radio-frequency choke coils sometimes oxidize and burn out. If the faulty layer, or pie, can be found with an ohmmeter, it can be jumpered and the choke will usually work satisfactorily. In many cases, a handmade RFC can be fabricated by winding several hundred turns of wire onto a pencil-sized piece of insulating material or even dry wood. A choke coil in a power supply may burn out or short to the core. If burned out, it may be jumpered, or the primary or secondary of a spare power transformer may be used in its place. If shorted to the core, the core can be loosened from the metal chassis of the power supply, insulated from the chassis by some form of sheet insulation, and operated temporarily. The core will be "hot" electrically when the power supply is on and must not be touched.

METERS. Direct-current ammeters or milliammeters usually have shunt resistors in them. The moving coil can burn out or oxidize apart, and the meter will still carry current through the shunt. When this happens, the transmitter continues to operate satisfactorily but with no current indication in the one meter. If a shunt burns out, the moving-coil element always burns out too. In a voltmeter, the series-multiplier resistor can oxidize and open, but the moving part may not be damaged. Installation of a new series resistor of approximately the same resistance in the meter will result in fairly accurate readings by the voltmeter. Since the plate- and grid-current values are functions of the filament temperature, if the filament circuit has a voltmeter across it and the meter burns out, it is possible to arrive at a good approximation of the correct filament voltage by adjusting the filament rheostat until normal grid and plate current is indicated on the grid and plate meters or until the color of the hot filament seems normal. In an RF thermocouple meter, if the thermocouple circuit becomes faulty, RF ac may still flow through the thermocouple, but no dc will be fed to the meter and no indication of RF will be given. If the RF ammeter is used as the only means of indicating when the antenna and final amplifier are correctly tuned, it is possible to substitute a 150- to 300-W electric light for the meter if it burns out. Care must be taken not to operate the equipment at a power level that will burn out the light. After the transmitter is tuned at low power, the light can be shorted out and the transmitter can be run at full-power output. A 6-V pilot light may be shunted across 6 to 10 in. of the antenna wire and act as a tuning indicator.

ANTENNAS. Any kind of wire will work as an antenna in an emergency. (Even wet rope soaked in salt water has been used for transmission and reception for short periods.) While copper wire is better, iron wire will radiate almost as well, and the difference in reception is not noticeable. No wire is too thick, and any wire that does not burn out is satisfactory for emergency communications. Dry string, rope, most plastic materials, and dry, oiled, or waxed wood make satisfactory temporary antenna insulators.

RELAYS. If a relay coil burns out, it can often be taken apart and rewound (Sec. 3-20). If the relay contacts become so badly worn that they are inoperative, it is sometimes possible to substitute pieces of silver coins or sheet silver for the worn contacts.

16-25 INDICATIONS OF TROUBLE

Obvious indications of trouble in transmitters are the hiss or crackle of an electric spark jumping, the blue curl of smoke, the flicker of a flame, the smell of burning insulation, a loud humming sound from a transformer, lack of filament glow in tubes, failure of the keying relay to close when the key is pressed, and the clatter of an overload circuit breaker opening. Less obvious but just as significant are the meters on the panel of a VT transmitter. (Many items also apply to transistor transmitters.) Some possible meter indications of circuit malfunctions are outlined below.

OSCILLATOR. *Plate Current Meter Reading Increases*. Stage probably not oscillating because of inoperative crystal, detuned plate circuit, or shorted grid circuit or tube. Unlikely that the power-supply voltage will increase materially.

Plate Current Meter Reading Decreases. The next stage may not be accepting RF power, the oscillator tube may be faulty, or the filament or plate voltage may be low.

Plate Current Meter Reading Zero. The power supply may not be functioning, the meter may be burned out, the oscillator tube filament may be burned out, there may be a poor connection in the plate circuit, a fuse may be burned out, or the screen or control grid may be open.

RF AMPLIFIERS. *Grid Current Meter Reading Increases*. Loss of bias supply voltage, excessive RF drive from preceding stage, shorted grid leak, or low plate current. When the plate is taking no electrons, the grid is free to collect more of them than normal. Thus, low plate or screen voltages, decreased coupling to next stage, or open plate or screen circuits may all result in increased grid current.

Grid Current Meter Reading Decreases. Decreased excitation to grid, driving circuit detuned, low filament voltage, low filament emission from tube, or increased plate current. (As plate current increases, grid current decreases. Therefore, anything that produces high plate current, detuned plate circuit, high screen voltage, or tight coupling to next stage produces a lowered grid current.)

Grid Current Meter Reading Zero. A burned-out filament, no filament voltage, resistor or RF choke in grid circuit burned out, insufficient excitation if power-supply bias used, no excitation if grid-leak bias used, or burned-out meter.

Plate Current Meter Reading Increases. Loss of power-supply bias voltage, loss of RF excitation from previous stage if grid-leak bias is used, a detuned plate circuit, an increased coupling to the next stage, a soft, or gassy, tube, or a high screen-grid voltage.

Plate Current Meter Reading Decreases. May be caused by low plate supply voltage, decreased coupling to plate-circuit load, low filament emission, low filament voltage, low screen-grid voltage, excessive bias on the grid, or an open in the antenna or other load circuit.

Plate Current Meter Reading Zero. No plate supply voltage because of burned-out fuse in power supply, shorted filter capacitor, open RFC or power-supply choke, open overload relay in plate circuit, no filament voltage, burned-out filament in tube, zero screen-grid voltage, or an open in the plate circuit.

ANTENNA. *Ammeter Reading Increases*. Plate voltage inadvertently increased, or faulty ammeter. If an antenna wire falls to the ground, the antenna meter usually reads zero. In some cases if the antenna can be retuned, the meter may register more than normal current.

Antenna Ammeter Reading Decreases. This is the usual indication when something is wrong in the transmitter. It may be caused by a low or excessive degree of coupling to the antenna, a detuned plate circuit, low plate voltage, low grid excitation, an excessive grid bias on the final amplifier, low filament emission from the tube, low filament voltage, insufficient or excessive screen-grid voltage, or loss of battery bias on the amplifier.

Antenna Ammeter Reading Zero. Either there is no RF output from final amplifier, the antenna system or coupling circuit is faulty, or the meter is burned out. It is possible to have the antenna-meter moving coil burn out and still have the transmitter emitting normally. The meters in the final amplifier will read properly in this case.

A PINNED METER. When a meter is suddenly *pinned* (the pointer driven to the stop pin past the maximum scale indication), there is usually a shorted part in the stage involved. A bypass capacitor may be shorted, an RFC may short to ground, a variable capacitor may short to ground, or the tube may develop a short circuit by the loosening and touching of internal elements. If there is no fuse, circuit breaker, or overload relay in the circuit, the meter may be burned out before the transmitter and power supply can be turned off. With the transmitter completely off, an ohm-meter can be used to determine what is shorted.

Some transmitters incorporate overload *recycling* relays that restore a certain number of times automatically before staying open. These relays are designed for transmitters in which sometimes, for no accountable reason, a spark will jump a gap and momentarily cause a surge of high current in the circuit. The overload relay opens, the spark extinguishes, and the overload closes again. If the spark occurs again, or if the fault does not clear itself, the overload falls out again. The relay may be constructed to hold open after the second or third restoration until manually restored when the fault has been found. Lightning striking on or near a transmitting antenna will often cause an instantaneous interruption of this type.

If the RF excitation to an amplifier is removed but the antenna meter continues to indicate the presence of RF, either the amplifier is improperly neutralized and has broken into self-oscillation on some frequency near the assigned frequency or the meter needle may be stuck.

An overload relay carrying the plate current of an amplifier through its coil can be made to latch open if the current becomes excessive, thereby protecting the stage.

Some relays are *time delay* types, opening only after a greater-than-normal current has been flowing for a period of time. The time delay can be by some mechanical means, such as an oil or air dashpot, or a thermal device which snaps open when heated by an excessive current through it.

Test your understanding; answer these checkup questions.

1. What might a decrease in PA I_g indicate? _____
2. What does a blue haze between filament and plate indicate? _____
3. Does a cherry red plate indicate excessive I_p? _____
4. How is it determined if a high-power-transmitting tube is faulty? _____ A receiving-type tube? _____
5. What might go wrong with a tube socket? _____
6. What may cause overheating of a power transformer? _____
7. Why are bypass capacitors often connected across power or audio transformers in a transmitter? _____
8. What about a transformer can be tested with an ohm-meter? _____ _____ _____
9. What should be looked for if a resistor burns out? _____
10. If an ammeter burns out, what can be done in an emergency to get the transmitter on the air? _____
11. If a voltmeter burns out, will the ability of the transmitter to function be affected? _____
12. What is the usual indication if a VT crystal circuit stops oscillating? _____
13. If the PA I_p meter suddenly changes but other meters in the transmitter remain normal, where might the trouble be? _____
14. The PA I_p meter reads normally but the antenna ammeter drops to zero. What is probably wrong? _____
15. Buffer and amplifier stage meters in a transistorized transmitter drop to zero. What are two possibilities? _____ _____
16. If the drive to the PA ceases but the antenna ammeter continues to read, what is the probable trouble? _____

COMMERCIAL LICENSE QUESTIONS

Amateur license questions will be found in the Addendum.

FCC Elements 3, 4, and 6 require an ability to answer questions similar to those below. Sections in which questions are answered are shown in parentheses. A question followed by a bracketed number is required for that element alone.

1. What are the lowest radio frequencies useful in radio communication? (16-1)
2. What is meant by *carrier frequency?* (16-1)
3. Define *type B, A1, A2,* and *F1 emissions*. (16-1, 16-3, 16-13, 16-14)
4. What are the disadvantages of using a self-excited-oscillator type of transmitter? (16-2)
5. What is the effect of a swinging antenna on the output of a simple oscillator? (16-2)
6. What is the effect of excessive coupling between the output circuit of a simple oscillator and an antenna? (16-2)
7. Draw a diagram showing a method of coupling the RF output of a final power amplifier to an antenna. (16-2, 16-3, 16-15, 16-18)
8. Draw a circuit diagram showing the principle of operation of a telegraph keying relay. (16-3) [6]
9. What is meant by *break-in operation* at a radiotelegraph station, and how is it accomplished? (16-3) [5, 6]
10. What materials are used for relay contacts and why? (16-3)
11. What is meant by *self-wiping contacts* as used in connection with relays? (16-3)
12. What is the primary function of the power-amplifier stage of a radiotelegraph transmitter? (16-4)
13. What are possible causes of and methods of reducing spurious emissions other than harmonics from transmitters? (16-4, 16-9)

ANSWERS TO CHECKUP QUIZ ON PAGE 379

1. (*Low drive, weakening tube, high I_p, high bias*) 2. (*Gas in tube*) 3. (*Not necessarily; some tubes normally operate with plates nearly white hot*) 4. (*Meter readings; substitute a new one and compare*) (*Tester*) 5. (*Shorts or leakage between pins or open circuits*) 6. (*Heavy load, high primary voltage, shorted turns*) 7. (*Keep RF ac out*) 8. (*Winding continuity, shorts to case, interwinding shorts*) 9. (*Shorted part or tube causing overload*) 10. (*Jumper between terminals*) 11. (*No*) 12. (*High I_p in oscillator*) 13. (*Antenna circuit*) 14. (*Coil of meter is open*) 15. (*Oscillator stopped, power-supply output zero*) 16. (*PA oscillating*)

14. How are meters in transmitters protected against damage due to stray RF? (16-4)
15. What are the various points in a radiotelegraph transmitter where keying can be accomplished? (16-4, 16-10–16-13, 16-22) [6]
16. What are the advantages of an MOPA as compared with a simple oscillator transmitter? (16-4)
17. What class of amplifier should be employed in the final stage of a radiotelegraph transmitter for maximum plate efficiency? (16-4)
18. Why is a separate source of plate power desirable for a crystal oscillator in a transmitter? (16-6)
19. What are the principal advantages of crystal control over tuned-circuit oscillators in transmitters? (16-6)
20. Why is a *dummy antenna* sometimes used in testing a transmitter? By what other names is this device known? (16-7)
21. Why should a transmitter be tuned initially at reduced power? (16-7, 16-22)
22. Does code speed (number of words per minute) have any effect on the bandwidth of emission from a radiotelegraph transmitter? (16-9) [6]
23. Diagram a key-click filter. (16-9) [6]
24. Diagram primary keying of a radiotelegraph transmitter. (16-10) [6]
25. What are the general characteristics of the emission of a transmitter which uses a chopper to obtain A2 emission? (16-12) [6]
26. What is meant by a *blocked grid?* (16-12)
27. Draw a diagram of grid-block keying in an RF amplifier. (16-12) [6]
28. In a radiotelegraph transmitter with grid-bias keying, when the key contacts are open, the emission continues. What could be the trouble? (16-12) [6]
29. Explain briefly the principles involved in FSK transmissions. (16-13)
30. Diagram capacitive coupling between an oscillator and a buffer amplifier. (16-15)
31. What is the purpose of a buffer-amplifier stage in a transmitter? (16-15)
32. In what order should circuits be adjusted in placing a transmitter in operation. (16-15)
33. What effect on the plate current of the final amplifier will be observed as the antenna circuit is brought into resonance? (16-15)
34. Should the antenna circuit of an MOPA-type transmitter be adjusted to resonance before the

plate-tank circuit of the final stage? Why? (16-15)

35. Draw a block diagram of a transmitter with master oscillator operating on 2,017.5 kHz and output on 8,070 kHz. (16-17) [6]
36. Diagram link coupling of a final RF stage to an antenna. Include a low-pass filter. (16-18)
37. Describe the usual method and equipment employed for measuring the harmonic attenuation of a transmitter. (16-18)
38. Discuss the following with respect to harmonic-attenuation properties in a transmitter: link coupling, tuned circuits, degree of coupling, bias voltage, decoupling circuits, and shielding. (16-18)
39. Explain several frequently used methods of attenuating harmonics in transmitters. (16-18)
40. What is a Faraday screen? (16-18)
41. What is an electrostatic shield? (16-18)
42. Why must transmitters remain on frequency and their harmonics be attenuated? (16-18)
43. In relation to safety, explain the function of interlocks and circuit breakers. (16-19)
44. Why should all exposed metal transmitter parts be grounded? (16-19)

45. Explain overload relays. (16-19, 16-25)
46. How is the power output of a VT telegraph transmitter ordinarily adjusted? (16-22) [6]
47. What emergency repairs may be made to an inductance coil having burned or charred insulation? (16-24) [6]
48. What are some possible indications of a defective transmitting tube? (16-24)
49. What would indicate subnormal filament emission of a tube in a transmitter? (16-24)
50. If the plate current of the final RF amplifier in a transmitter suddenly increased and radiation decreased although the antenna circuit was in good order, what are possible causes? (16-25) [6]
51. An MOPA-type transmitter has been operating normally. Suddenly the antenna ammeter reads zero, although all filaments are burning and plate and grid meters are indicating normal voltages and currents. What are possible causes? (16-25) [6]
52. Explain overload, time delay, and recycling relays. (16-24)

17 AMPLITUDE MODULATION

The objective of this chapter is to discuss the various important facets of amplitude modulation, starting with the historic and simplest forms, progressing through the types of circuits used in standard broadcast transmitters and in the picture information of television, up to the use of single-sideband suppressed-carrier (SSSC or SSB) generators used in many modern two-way voice communication systems.

17-1 MODULATION

The term modulation may be considered to imply or mean variation or shaping. If a direct current in a circuit is being made to vary in amplitude (strength) 500 times per second, it is being *amplitude-modulated* at a 500-Hz rate. In this case, the dc is the carrier current, and the variations impressed on it represent modulation. This is the basic principle of the telephone.

If a transmitter emits an RF ac carrier wave at some frequency and this carrier is made to vary in amplitude at a 500-Hz rate, amplitude modulation (AM or A3 emission) is being produced. Here, ac is the carrier current, and the variations impressed on it represent the modulation. This is the basic principle of AM radiotelephone, or AM broadcast transmissions.

If the RF carrier is made to vary in frequency (not amplitude) 500 times per second, it is being *frequency-modulated* (FM or F3, Chap. 19). A complete listing of emissions is included in Appendix E.

17-2 WHY THE CARRIER IS MODULATED

A normal radio receiver can be tuned to the carrier frequency of a radio transmitter but will reproduce no sound from this carrier. However, if such an inaudible carrier wave is made to vary in amplitude at speech or musical tone rates, it is said to be modulated by the speech or music. The detector stage in the receiver can demodulate the modulation on the carrier and reproduce the speech or music variations, producing intelligible speech or music sounds in its loudspeaker.

There are many methods by which speech or music frequency ac can be made to amplitude-modulate a carrier, but first there are some basic concepts regarding sound and microphones that should be understood.

17-3 SOUND

When a firecracker is lighted, thrown up into the air, and explodes, a sound is heard. If the firecracker is near, it sounds loud. If it explodes far away, it sounds weak. When the firecracker explodes, it suddenly and violently pushes air outward, creating an expanding ball of compressed air all around itself. Compressing molecules of air as it goes, the ball of compressed air moves outward in all directions, at a speed of approximately 1,100 ft/s. If the ball were examined closely, it would be found that just inside, and next to, the compressed-air molecules is a rarefied area of air molecules. This compression with its attendant rarefaction forms a *sound wave*.

As the sound wave travels outward, it is constantly expanding. The mechanical energy it contained when it was confined in a small area near the firecracker rapidly becomes less for any given square inch of the wave front. If the wave travels only a short distance, it strikes the eardrum with considerable energy. If it travels a long distance, it strikes the eardrum with very little energy.

A sound wave of mechanical energy striking the eardrum causes the outer-ear diaphragm to vibrate. This transmits a mechanical vibration into the inner ear, in turn actuating delicate nerve endings that relay an electrochemical impulse to the brain. If the person is "conscious," he is made aware of the "sound." By past experience, this particular nerve-to-brain impulse from an explosion is recognized as a "bang."

Any musical instrument producing a constant, single tone sets up a continuous series of waves at a definite frequency. These waves, striking the diaphragm of the ear, are transferred to the inner ear and energize the nerve endings that are resonant to this particular frequency, making the listener aware of this tone only. Higher-frequency sound waves energize high-frequency nerve endings, and the *pitch,* or tone, the listener hears, is correspondingly higher.

The greater the energy content of a wave striking the eardrum, the stronger the nerve impulse generated and the louder the sound to the listener.

A sound wave, then, has both amplitude (loudness) and frequency (pitch). It also has what may be termed *quality,* or purity (great or little harmonic content). A sine-wave ac signal generator adjusted to 256 Hz (middle C) coupled to a loudspeaker produces a pure 256-Hz tone. A musical instrument playing a fundamental 256-Hz tone will sound different because it cannot produce a pure vibration. It may produce a fundamental 256-Hz vibration, but many harmonics, or *overtones,* are generated at the same time. The number and amplitudes of the different harmonics produced identify the type of instrument being played. Thus, a piano, an oboe, and a saxophone all sound different even though they may all be playing the same fundamental tone. Two singers singing the same tone will also sound different because their vocal cords have different dimensions and produce different harmonics.

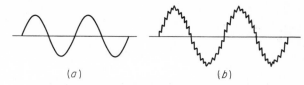

Fig. 17-1 Representation of two signals of the same fundamental frequency. (*a*) A pure tone. (*b*) A fundamental with harmonics, or overtones.

Most sounds are highly complex, being made up of fundamentals and many harmonics. Figure 17-1 shows, first, a representation of a pure tone and, second, a possible musical or speech tone having the same fundamental frequency plus harmonics.

Sound waves are often represented by sine waves for pure tones and jagged waveforms for speech or musical tones. The half of the drawing above the zero, or time line, may indicate the compression of air molecules, in which case that shown below the line will represent the rarefaction part of the sound wave. The jagged form indicates high-frequency harmonics (or other tones) added to a fundamental frequency.

More sine-wave ac power is required to modulate a carrier to a maximum extent than is required by other ac waveshapes. All natural-sound waveshapes produce a given peak value with a lesser average power. Therefore, a transmitter engineered to handle sine-wave audio-power values will be capable of handling any waveforms of normal sounds. For this reason, explanations are made in terms of sine-wave-modulating ac. It takes about half as much average audio power to modulate a transmitter with speech sounds as it does to modulate one with sine-wave tones.

The amount of sound power transmitted to a diaphragm can be measured in microwatts per square centimeter. For example, the lowest audible intensity, or threshold value, of sound for the human ear is approximately $1 \times 10^{-10} \ \mu W/cm^2$. On the loudness scale, this power level is given a *zero* value. Ten times this sound power is said to be one *bel* louder. Ten times more sound (100 times the threshold value) is again 1 bel louder, or has an absolute value of 2 bels. This is a logarithmic increase. The logarithmic ratio is used because the human ear responds closely to this ratio; that is, an increase in a weak sound of one decibel (one-tenth of a bel, abbreviated dB) is barely noticeable to a listener. Actually, this is a power increase of 26% (Sec. 8-5). A very loud sound (containing considerably more power), when increased by 26%, will also result in a just barely discernible change in sound intensity to a listener.

The ear responds best to sounds between 1,000 and 4,000 Hz. The response is down about 20 dB at a frequency of 200 Hz and down about 40 dB at a frequency of 100 Hz. The older a person is the less response his ear produces at the higher frequencies. For teen-agers, sounds of 15,000 Hz may be down 20 dB, and 18,000 Hz may be down 40 dB. For elderly persons, there may be little or no response above 12,000 Hz.

17-4 THE SINGLE-BUTTON MICROPHONE

A microphone is a device used to convert mechanical sound energy into electric energy of equivalent frequency and relative amplitude.

The original *single-button,* or *carbon-button,* microphone is still used in telephones and in applications where fine quality of reproduction is not essential. It will convert voice sounds into varying electric currents fairly well but will not faithfully reproduce over a wide enough frequency range to be used for high-fidelity music. It is in reality a sound-variable resistor. When its resistance is changed, the current flowing through it varies accordingly.

The construction of a single-button microphone is shown in Fig. 17-2. The carbon or metal button

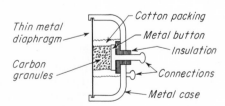

Fig. 17-2 Cross-sectional picturization of a single-button carbon microphone.

is electrically insulated from both the diaphragm and the frame of the microphone. One electrical connection is made to the frame and diaphragm, and another to the metal button. Sand-sized particles of carbon are held between diaphragm and

button by fluffy cotton washers. This light cotton packing allows the diaphragm to vibrate to and from the stationary button without allowing the carbon particles to drop away from between them. If the diaphragm is moved toward the button, the granules are compressed and the diaphragm-to-button resistance is decreased.

When sound waves beat against the diaphragm, compressions push it inward; rarefactions pull it outward. Thus sound waves cause the resistance of the microphone to vary at the rate of the compressions and rarefactions. High-frequency tones vibrate the diaphragm rapidly; low frequencies vibrate it slowly. Weak sounds cause little vibration. Loud sounds produce wide movements of the diaphragm and large resistance changes. Pure tones produce an even, smooth inward and outward swing. Harmonic-containing sounds produce a jerky inward and outward swing of the diaphragm.

The single-button-microphone circuit in Fig. 17-3 shows a microphone, battery, switch, ca-

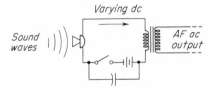

Fig. 17-3 A single-button-microphone circuit.

pacitor, and microphone transformer. The battery keeps current flowing through the microphone and transformer primary at all times. Resistance variations, due to sound waves, produce corresponding current variations (varying dc) in the microphone circuit. The varying dc in the primary produces an ac in the secondary with a frequency and amplitude proportional to the variations of the current in the primary. Therefore, the sound waves are represented in the secondary of the transformer by an ac whose frequency and amplitude vary in proportion to the frequency and amplitude of the sound itself.

If earphones are connected either in place of the primary of the transformer or across the secondary, the electrical variations will reproduce the original sound with recognizable clarity, although the sound will not be very loud. This represents a simple one-way telephone circuit.

The single-button microphone has several advantages. It is relatively inexpensive, withstands rough handling, and is not subject to deterioration due to heat or cold. It has a relatively high output power in comparison with many other types. It can be spoken into at close range, allowing the voice to override background noises. It is usually made most sensitive to voice frequencies, between 200 and 3,000 Hz.

Once in a while the carbon granules will either pick up moisture or weld together and *pack*. The microphone may have to be shaken or jarred lightly to free the granules. A disadvantage is the requirement of a battery in the microphone circuit. This microphone distorts sounds more than most other types of microphones. When direct current flows through the granules, they move very slightly, producing a random variation of current resulting in a constant, weak hissing sound in the output. A current of 5 to 20 mA in the microphone circuit produces satisfactory results. The microphone has an impedance of about 200 Ω.

17-5 ABSORPTION MODULATION

The oldest and simplest method of modulating a radio carrier wave with voice frequencies is called *absorption,* or *loop,* modulation. An RF oscillator is necessary to provide a constant RF ac carrier to the antenna. In Fig. 17-4, a Hartley oscillator is used.

A single-button microphone is inductively coupled to the antenna circuit. Most of the RF carrier energy being fed into the antenna circuit by the oscillator is radiated, but some of the energy of the oscillator is fed into the loop and is dissipated by the resistance of the microphone as heat.

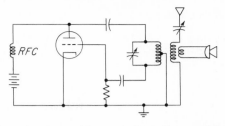

Fig. 17-4 One form of loop, or absorption, modulation.

The resistance of the microphone varies at the audio rate of any sound causing its diaphragm to vibrate. As a result, the amount of RF energy being absorbed and dissipated into heat by the microphone is varying.

If the power output of the oscillator is essentially constant but the microphone is absorbing alternately more and less power from the oscillator, the antenna finds itself with alternately less and more power to be radiated. This variation of RF power output at an audio rate from the antenna produces the modulated carrier of the system. Since the power output is being varied in accordance with the sound waves, the modulation is known as *amplitude modulation* or AM.

The effect of the microphone on the output RF carrier power is shown in Fig. 17-5. This form of modulation is impractical by present standards and is rarely used. (However, variations of the absorption principle using biased PIN diodes across microwave waveguides are in use.)

The absorption loop acts as a partly shorted turn coupled to the oscillator tank. As the microphone resistance changes, the effective inductance of the shorted turn changes, which also modulates the *frequency* of the oscillator. Simultaneous FM and AM produces serious unwanted distortion products.

Fig. 17-5 Antenna current (*a*) with no modulation and (*b*) when the microphone is alternately absorbing less and then more power from the circuit.

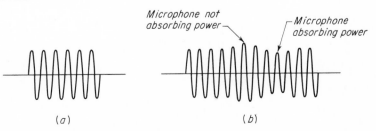

17-6 A SIMPLE SERIES MODULATION

A system of modulation which will be called *simple series modulation* utilizes the single-button microphone in another manner. In Fig. 17-6 the plate current flows through the microphone.

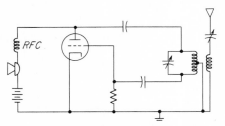

Fig. 17-6 Carbon microphone in series with the plate circuit produces a form of plate modulation.

When spoken into, the microphone varies its resistance, in turn varying the plate current of the tube and thereby the output power of the oscillator. If a 500-Hz sound wave strikes the diaphragm of the microphone, the output RF carrier will be amplitude-modulated at a 500-Hz rate because of the variation of the resistance of the

microphone at this frequency. Only a small amount of modulation can be produced in this circuit.

Simultaneous FM and AM is always produced by varying the plate voltage or current of an oscillator. For this reason oscillators are rarely amplitude-modulated. It is necessary to modulate an RF amplifier stage, as in Fig. 17-7.

If the plate circuit of the RF power amplifier is modulated, there will be much less frequency shift, or FM. Any such frequency shift is called *dynamic instability*. The term refers to the carrier and indicates that the carrier frequency, for some reason, is not stable under operating conditions. Some of the possible causes of dynamic instability, or FM, in an MOPA transmitter are Miller effect, imperfect power-supply regulation, imperfect neutralization of the amplifier, unwanted feedback effects, vibration of parts of the oscillator if the transmitter is jarred, faulty parts causing sudden frequency changes, and loose connections.

To prevent FM entirely, a separate power supply is used for the oscillator, and one or more buffer

Fig. 17-7 A transmitter circuit using series modulation.

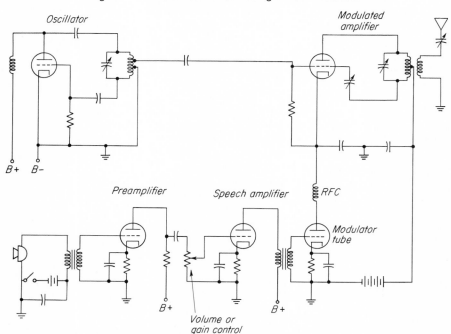

stages are inserted between the oscillator and the modulated stage, with all stages adequately shielded and neutralized.

17-7 SERIES MODULATION

One satisfactory method of producing A3 is the *series modulation* circuit shown in Fig. 17-7. A continuous RF carrier is provided by an oscillator and RF amplifier. The output of the class C amplifier is fed to an antenna and is radiated.

The RF stage directly involved in the modulation process is the *modulated amplifier*.

The final AF stage coupled to the modulated amplifier is the *modulator*.

The stage or stages preceding the modulator are known as the *speech amplifiers*. They amplify the weak AF emf from the microphone to a value that will drive the grid of the modulator tube.

In series modulation, the plate circuit of the modulator tube, the plate circuit of the modulated stage, and the power supply are all in series. A variation of the resistance in any part of this series circuit will affect the dc flowing in the RF amplifier and therefore the power output of that stage.

The modulator tube in series modulation might be termed an electronic variable resistor, since an AF voltage variation applied to its grid varies the dc plate resistance of the tube. (Compare with the vacuum-tube-keying circuit in Fig. 16-12.)

When a sound strikes the microphone, a weak AF ac will be developed across the microphone-transformer secondary. This is amplified by the speech amplifiers and applied to the grid of the modulator tube. When the modulator grid is driven *less negative* than normal, its plate resistance decreases, the voltage-drop across the modulator decreases, and the RF amplifier is across a greater proportion of the plate supply voltage. More current flows through the RF amplifier, and its RF output increases.

When the modulator grid is driven *more negative,* its plate resistance increases and a greater voltage-drop is developed across the modulator. Less of the total power-supply voltage is across the RF amplifier, and its output decreases. In this way, AF variations of the grid voltage of the modulator vary the power output of the RF amplifier.

Series modulation has the advantage of not requiring any modulation transformer to limit frequency response. It has the disadvantage of requiring approximately twice the plate voltage from the power supply, since both tubes, modulator and modulated amplifier, are in series across the supply. If the cathode of the modulator is at ground potential, the cathode of the RF amplifier will be at a high dc potential above ground, producing an insulation problem in high-powered equipment. Series modulation is useful in modulating transistors.

The audio voltage applied to the grid of the modulator must never be high enough in amplitude to produce plate-current cutoff or grid-current flow in the stage. This places the operation of the modulator in the class A category. It can use cathode-resistor bias, as shown, or a bias supply.

The RF amplifier is a class C stage. The grid-leak bias value remains essentially constant with or without modulation of the plate circuit.

17-8 THE MODULATED ENVELOPE

All modulated RF stages are biased to class C. As a result, the plate current will always be narrow pulses of dc that occur during the positive half cycle of the RF ac grid excitation. These pulses of plate current produce a flywheel effect in the plate-tank circuit, resulting in a very nearly sinusoidal RF ac output waveform as illustrated in Fig. 17-8.

I_p pulses ⟶ By flywheel effect produce ⟶ Ac in tank circuit

Fig. 17-8 Constant-amplitude pulses of plate current by flywheel effect produce constant-amplitude sinusoidal RF ac in the LC tank circuit.

If the modulator suddenly allows more plate current to flow through the RF amplifier, the plate pulses increase in amplitude and the RF flywheel amplitude also increases. Note that both the positive and the negative halves of each RF ac cycle are increased. Figure 17-9 illustrates the result of

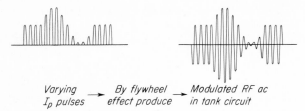

Varying → By flywheel → Modulated RF ac
I_p pulses   effect produce   in tank circuit

Fig. 17-9 Varying-amplitude plate-current pulses produce a modulated RF ac in the tank circuit.

flywheel effect on the output ac when varying-amplitude plate-current pulses are flowing in the plate circuit. The diagram of the RF varying in amplitude during modulation illustrates what is known as the *modulation envelope.*

It is important to understand that in single-ended circuits, flywheel action alone is responsible for the reproduction of the second half of the RF ac cycle. Each cyle of RF of the modulated envelope produced by flywheel action has an almost perfect sine waveshape if a tank-circuit *Q* of 10 or more is used. If the RF amplifier is push-pull, plate-current pulses flow in opposite directions alternately in the tank circuit and aid in producing the second half of the RF cycle.

Test your understanding; answer these checkup questions.

1. What is the letter-number designation of AM? _____ FM? _____
2. If a broadcast station is transmitting but not modulating, what is heard in a receiver tuned to its frequency? _____
3. What is a radio wave? _____ What is a sound wave? _____
4. Does audible "pitch" refer to frequency, amplitude, purity, or harmonics? _____
5. What is the frequency of middle C? _____ Of one octave above middle C? _____
6. What is the threshold power value for the human ear? _____ What power is 1 bel louder? _____
7. What frequencies are heard best by the human ear? _____
8. What is another name for a single-button microphone? _____ Where are these used most? _____
9. How is a packed single-button microphone cleared? _____
10. What was the first type of modulation, using a SB microphone, called? _____
11. What is dynamic instability? _____
12. In series modulation what name is given to the last AF stage? _____ To the RF stage to which it is coupled? _____

13. What are three advantages of series modulation? _____ _____ _____
14. What is another name for an AF volume control? _____
15. What device can be used to make a modulated envelope visible? _____

17-9 BASIC PLATE MODULATION

It is possible to produce modulation of a carrier by adding an AF ac in series with the plate circuit, as in Fig. 17-10. This is basic plate modulation.

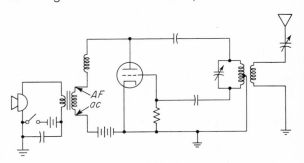

Fig. 17-10 Basic plate modulation of an oscillator.

Sound waves striking the diaphragm of the microphone produce an AF ac voltage in the secondary of the microphone transformer.

On the positive half cycle of the AF ac, the audio emf will be in the same direction as the plate-supply emf, and the two will be additive. As the plate voltage increases, the RF output increases.

On the negative half cycle the audio emf will be opposing the plate-supply emf, and the plate-circuit voltage will be the difference between the two. The resultant plate voltage is decreased, the plate current decreases, and the RF output decreases.

During the positive half cycle of the modulating AF ac, the *positive peak of modulation* of the carrier is produced, as shown in Fig. 17-11. During this half cycle, both positive and negative half cycles of the RF ac output are greater than the unmodulated, or carrier, values.

During the negative half cycle of the modulating ac, the *negative peak of modulation* is produced, also shown in Fig. 17-11. In this half cycle both positive and negative half cycles of RF ac output are less than the carrier value. (It may seem peculiar that the negative *peak* should actually be the

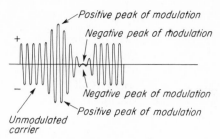

Fig. 17-11 Illustrating positive and negative peaks of modulation.

lowest point as far as the RF ac voltage in the emitted wave is concerned.)

The explanations have been given in terms of voltage. The illustrations representing carrier voltages are valid illustrations of the carrier currents also, since current is directly proportional to voltage. They are not, however, correct representations of power. If the carrier voltage is doubled, the carrier current is also doubled. At the instant the voltage and current are doubled, the output power is four times the carrier value. The illustration of power at the positive peak of modulation would have to be drawn four times as high as the carrier value.

If sufficient ac voltage is used to modulate a self-excited oscillator to a high degree, the variation of plate voltage will produce considerable FM. Furthermore, during high-amplitude negative peaks of the modulating ac, the plate voltage may drop so low that the oscillator will stop oscillating completely for a short period of time, producing extreme distortion. For these reasons, full modulation of an oscillator is always undesirable.

The voltage output of a microphone alone will not be sufficient to produce much modulation. However, by adding audio amplifiers, it is possible to increase the audio power and voltage to produce adequate modulation.

17-10 PERCENT OF MODULATION

The percent of modulation of a carrier wave is determined by how much the carrier voltage or current varies in amplitude. A strong carrier with a low percent of variation may give a weaker response in a receiver than a weaker carrier with a greater percent of variation.

The variation value is expressed as a *percent of modulation,* with 100% being the highest possible undistorted modulation.

Figure 17-12 illustrates several percents of sine-wave modulation of an RF carrier. The unmodulated carrier (Fig. 17-12a) represents 0% modulation.

A carrier is modulated 50% if its positive peak voltage rises to a value 50% greater than the unmodulated-carrier-voltage maximum and drops 50% at the negative peak.

A carrier is modulated 100% if its positive peak rises to a value twice the unmodulated-carrier maximum and also drops 100% (to zero) at the negative peak.

If too much modulating voltage is applied to the modulated stage, overmodulation occurs. The positive peak rises to more than twice the carrier level, and the negative peak drops to zero and remains at zero for a time. The 50 and 100% illustrations show sine-wave modulation. The overmodulated signal is not a sine wave at the negative peak and therefore will produce many undesirable harmonics of the modulating frequency, known as *splatter, buckshot,* or *spurious emissions.*

The results of overmodulation are interference to other radio services on frequencies above and below the carrier frequency and a distorted, harsh-sounding transmitted signal. As a result, overmodulation is illegal. Care must be exercised that no more than 100% modulation of the nega-

Fig. 17-12 (*a*) An unmodulated carrier. (*b*) A 50% modulated carrier. (*c*) A 100% modulated carrier. (*d*) An overmodulated carrier.

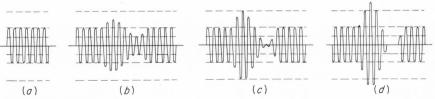

tive peaks occurs in radiotelephone transmitters. As long as positive peaks of overmodulation retain an undistorted waveform, they are not responsible for distortion. It is the abrupt arrival of the negative peak at zero output and its abrupt rise from zero that produce the undesirable byproducts of overmodulation.

Because the percent of modulation is directly related to the loudness and intelligibility of the signal produced in a receiver, the higher the average percent of modulation that can be maintained without producing distortion, the more effective the communication can be. To maintain as high a percent as possible, an *audio peak limiter* is usually incorporated in one of the speech amplifiers of commercial radiotelephone equipment. It clips off signal-voltage peaks passing through it which rise above a predetermined value. While this clipping distorts the peaks and generates AF harmonics, the latter can be filtered with 3-kHz low-pass filters. The distortion becomes hardly noticeable, and the emission is not broadened. Excessive limiting, or clipping, of the waveform does produce noticeable distortion.

ANSWERS TO CHECKUP QUIZ ON PAGE 388

1. (*A3*) (*F3*) 2. (*Nothing*) 3. (*Alternate NS electromagnetic and + − electrostatic waves*) (*Alternate air molecule compressions and rarefactions*) 4. (*Frequency*) 5. (*256 Hz*) (*512 Hz*) 6. ($10^{-10}\ \mu W/cm^2$) ($10^{-9}\ \mu W/cm^2$) 7. (*1–4 kHz*) 8. (*Carbon-button*) (*Telephones*) 9. (*Jarred*) 10. (*Absorption or loop*) 11. (*Carrier frequency variation*) 12. (*Modulator*) (*Modulated amplifier*) 13. (*Simple, no transformer, modulates any frequency*) 14. (*Gain*) 15. (*Oscilloscope*)

17-11 PLATE MODULATION

The standard method of amplitude modulating is by transformer-type plate modulation. A diagram of a simple, workable A3 circuit is shown in Fig. 17-13. With this circuit most of the basic theory of a plate-modulation system can be explained.

In the diagram, a triode RF amplifier stage is being modulated by adding an AF ac voltage in series with the plate power-supply voltage of the RF amplifier. The power-supply voltage of the RF amplifier is 2,000 V. The value of the AF voltage output of the modulation transformer must be 2,000 peak V, positive and negative, to produce 100% modulation. Under this condition, the plate voltage will be forced to vary alternately from 4,000 to 0 V. This produces positive voltage peaks of modulation twice the value of the carrier voltage and negative peaks of zero output.

The RF amplifier stage is grid-leak-biased to class C by a constant-amplitude RF ac applied to the grid by an oscillator, buffer, doubler, or driver stage. The output of the RF amplifier is coupled to a Marconi-type antenna (grounded at one end).

The modulator shown is a class B pentode stage which develops the relatively high audio power required to produce the desired modulation and at the same time reduce even-order AF harmonics.

An important factor in producing plate modulation is determining how much audio power is required to produce a desired percent of modulation (usually 100%). Assume that a modulated RF

Fig. 17-13 A practical plate-modulated RF power amplifier with a class B push-pull modulator.

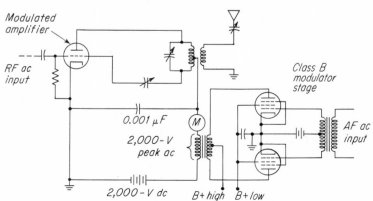

amplifier is operating under the following conditions:

$$E_p = 2,000 \text{ V}$$

$$I_p = 0.5 \text{ A}$$

where E_p = dc plate supply voltage
I_p = plate current in A, as read on plate-circuit ammeter M

To determine the power required to modulate the RF amplifier to 100%:

1. According to the power formula $P = EI$, the dc power input to the RF stage is $2,000 \times 0.5$, or 1,000 W.
2. According to Ohm's law $R = E/I$, the resistance of the plate circuit of the RF stage is $2,000/0.5$, or 4,000 Ω.
3. To produce 100% plate modulation, the sum of the instantaneous AF ac peak voltage, plus the dc plate voltage, must equal twice the unmodulated RF amplifier-plate voltage and zero volts, alternately.
4. Twice the voltage on the plate of the RF tube will produce twice the plate current, or 1 A.
5. Twice the voltage and twice the current produce a *peak* power of 4 times the dc power input, or $4 \times 1,000$, or 4,000 W.
6. Even at 100% modulation, the plate-current meter M will not visibly vary. Therefore the *average* power being drawn from the RF amplifier power supply must still be 1,000 W. It would seem that the modulator must furnish 3,000 W to produce the 4,000-W peak power. However, consider what the modulator is actually doing.
7. The modulator tube is feeding its power output, through the modulation transformer, into the 4,000-Ω plate circuit of the RF amplifier.
8. The modulator must develop a 2,000-V *peak* ac voltage into the 4,000-Ω resistive RF amplifier-plate circuit to produce 100% modulation. How much power will it take to produce 2,000-V peak ac across 4,000 Ω?
9. The power formula $P = E^2/R$ assumes effective-voltage values. To use this formula the peak-voltage value must be converted to the effective value by multiplying the peak by

0.707. In this case, $2,000 \times 0.707 = 1,414$-V effective AF ac.
10. By substituting these figures in the formula for power,

$$P = \frac{E^2}{R} = \frac{1,414^2}{4,000} = \frac{2,000,000}{4,000} = 500 \text{ W}$$

11. It requires 500 W of sinusoidal AF power from the modulator to produce a 2,000-V peak and 100% plate-modulate an RF amplifier with a plate-power input of 1,000 W. This 2:1 ratio holds for sinusoidal modulation only.

Note that the AF power must be delivered into the RF amplifier-plate circuit. Any AF power lost, because of inefficiency of the modulation transformer, will not reach the RF amplifier. As a result, the modulator must feed slightly more than one-half of the dc plate power into its output-transformer primary to produce the required secondary power.

It is possible to apply the same reasoning and formulas to determine the required AF power to produce any percent of modulation. For example:

A peak AF modulating ac equal to 100% of the dc E_p produces 100% modulation.

A peak AF modulating ac equal to 75% of the dc E_p produces 75% modulation.

A peak AF modulating ac equal to 50% of the dc E_p produces 50% modulation.

To find the required audio power to plate-modulate an RF amplifier with 2,000-V plate voltage and 0.5-A plate current ($R_p = 4,000$ Ω) to 50%:

$$P = \frac{E^2}{R} = \frac{(1,000 \times 0.707)^2}{4,000}$$

$$= \frac{707^2}{4,000} = \frac{500,000}{4,000} = 125 \text{ W}$$

Therefore 50% modulation requires only one-quarter of the power that is required for 100% modulation. This represents a 75% decrease in required audio power between 100 and 50% modulation.

A simpler method of determining the required power to produce a given percent of modulation is to use the formula

$$P_{af} = \frac{m^2 P_{dc}}{2}$$

where P_{af} = audio power needed
P_{dc} = dc plate-power input
m = modulation percent as a decimal (0.5 = 50%)

EXAMPLE: To modulate an amplifier 100% when the dc power input is 1,000 W, the AF power is

$$P_{af} = \frac{m^2 P_{dc}}{2} = \frac{1^2(1,000)}{2} = 500 \text{ W}$$

For 50% modulation of the same amplifier,

$$P_{af} = \frac{m^2 P_{dc}}{2}$$
$$= \frac{0.5^2(1,000)}{2} = \frac{0.25(1,000)}{2} = 125 \text{ W}$$

If a given amount of AF power is available, it is possible to determine the dc power which can be modulated to a desired percent by rearranging the formula above:

$$P_{af} = \frac{m^2 P_{dc}}{2}$$
$$2P_{af} = m^2 P_{dc}$$
$$\frac{2P_{af}}{m^2} = P_{dc}$$

EXAMPLE: An AF power of 500 W will modulate what RF amplifier dc power input to 50%?

$$P_{dc} = \frac{2P_{af}}{m^2} = \frac{2(500)}{0.5^2} = \frac{1,000}{0.25} = 4,000 \text{ W}$$

How much modulator dc power input is required to produce a given AF ac power? If the modulator stage is class A, it will be only about 25% efficient. If class B, it may be more than 60% efficient. If the transmitter has an output power of 100 W and the final RF amplifier is only 50% efficient, the plate input power must be 200 W.

It will require 100 W of sine-wave AF ac to modulate the 200 W of dc plate input. The AF power output from the modulator stage will be equal to its plate-power input times the percent of efficiency of the stage, or

$$P_o = P_{in}(\%)$$

where P_o = ac power output
P_{in} = dc power input
% = efficiency of stage

This formula can be rearranged to

$$P_{in} = \frac{P_o}{\%}$$

If the modulator is a class B audio amplifier with an efficiency of 66%, the dc power input to the modulator required to produce the needed 100 W of audio is

$$P_{in} = \frac{P_o}{\%} = \frac{100}{0.66} = 151 \text{ W}$$

In these figures the modulation is considered as being sinusoidal. Because of the jagged, high-peaked characteristics of speech sounds, their average power is far below that required to produce a given peak value of sine-wave voltage. It is generally considered that if 500 W of audio will produce 100% sine-wave modulation, half of 500 W, or 250 W of voice-type audio power will produce 100% modulation on peaks. However, with some voices 250 average watts of audio may not have AF voltage peaks high enough to produce 100% modulation at any instant, while other voices may have many peaks overmodulating the transmitter with an average of less than 250 W.

In the circuit of Fig. 17-13, at the positive peak of modulation the plate potential rises from the 2,000-V carrier value to 4,000 V. This should produce exactly twice the plate current and therefore four times the RF power output. However, as the plate voltage is doubled, the plate current may not quite double. As a result, the output power may not be quite four times, and the positive peak will not rise as much as it should. If power-supply bias alone is used in the RF amplifier, this distortion of the positive peak will always occur. With grid-leak bias, when the plate current increases, fewer electrons will be available to flow to the grid, reducing the grid current and therefore the bias value. Reducing the bias increases the plate current, tending to increase the positive peak of modulation. For this reason, plate-modulated stages should either use power-supply *and* some grid-leak bias or use grid-leak bias alone.

The value of the RF grid excitation is quite important. A grid-leak-biased amplifier with no RF grid excitation has no bias, and excessively high plate current will flow. With little RF grid excitation there will be only a little bias and the signal voltage will operate over only a small portion of the $E_g I_p$ curve of the tube. This will result in low efficiency and low RF ac power output. If the plate circuit is modulated, the grid is not being driven positive enough to lower the impedance of the tube and the positive peaks of modulation will not be developed, although the negative peaks may be. As excitation is increased (without modulation), the RF power output will increase up to a point. The tube is approaching the point of saturation. A further increase of excitation will not materially increase the RF output.

If modulation is applied with low drive, low percent of modulation is produced but high positive peaks of modulation will not be. The excitation must be increased to allow high-percent positive peaks to be produced linearly. The correct RF excitation will be the minimum value required to produce a positive voltage peak of modulation twice the carrier value and a negative peak just to zero, as indicated on an oscilloscope (Sec. 17-21). If an oscilloscope is not available, the approximate values can be approached by adjusting the stage according to the operating data provided by the manufacturer of the tube being used.

As explained in Chap. 14, the primary of a modulation (audio) transformer should have an impedance value between two and three times the impedance of a triode modulator tube (or tubes) for undistorted output. The secondary impedance of the transformer must match the plate resistance of the class C modulated stage. As an example:

A certain class C amplifier has an E_p of 1,000 V and an I_p of 0.15 A. The modulator tube has a plate impedance of 15,000 Ω. What turns-ratio transformer will match the modulator to the modulated tube?

If the modulator Z_p is 15,000 Ω, about 2½ times this value, or a 37,500-Ω primary, will be suitable. The secondary Z must match the Z of the class C plate circuit and is computed

$$Z_p = \frac{E_p}{I_p} = \frac{1,000}{0.15} = 6,670 \ \Omega$$

From the turns-ratio formula in Sec. 14-33,

$$\text{Turns ratio} = \sqrt{\frac{Z_1}{Z_2}}$$
$$= \sqrt{\frac{37,500}{6,670}} = \sqrt{5.62} = 2.37:1$$

The primary should have approximately 2.37 times as many turns as the secondary. In practice, any ratio between 2:1 and 2.5:1 would operate satisfactorily. Excessive mismatch results in distortion of the modulation or inability to produce high positive peaks.

17-12 PLATE-MODULATING TETRODES AND PENTODES

Tetrode or pentode tubes are often used as the modulated RF stage. Triodes must be neutralized, whereas tetrodes and pentodes may not require neutralization below 30 MHz and usually have higher-power sensitivities, requiring less RF driving power and less bias voltage.

If a modulating voltage is applied to the plate circuit of a tetrode or pentode but a constant voltage is applied to the screen grid, 100% modulation is not possible because plate current is fairly independent of the plate voltage. It is necessary to modulate both plate and screen-grid circuits simultaneously to produce high percent modulation.

The modulation transformer may have a secondary and a tertiary winding, producing a high AF ac voltage for the plate circuit and a lower modulating voltage for the screen-grid circuit as in Fig. 17-14. (The plate- and screen-modulating voltages must be in phase.)

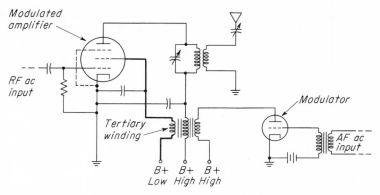

Fig. 17-14 Plate modulation of a tetrode (or pentode) using a modulation transformer with a tertiary winding.

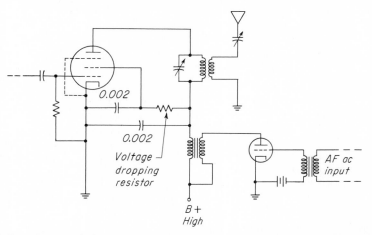

Fig. 17-15 Plate modulation of a tetrode (or pentode) using a voltage-dropping resistor and one power supply.

The screen grid may be fed a modulating voltage through a voltage-dropping resistor connected to the plate end of the modulation-transformer secondary, as in Fig. 17-15. If the resistor is connected to the power-supply end of the transformer, no modulating AF ac will be fed to the screen grid.

A high-inductance, low-resistance choke coil in series with the screen-grid supply lead, as in Fig. 17-16, will produce self-modulation of the screen-grid voltage. Increasing positive plate potential on the plate causes increasing plate current and decreasing screen-grid current. Decreasing screen current allows the magnetic field of the choke to collapse, inducing a more positive voltage on the screen. Thus the screen and plate become more positive at the same time.

Fig. 17-16 Self-modulating the screen grid of a tetrode while plate-modulating.

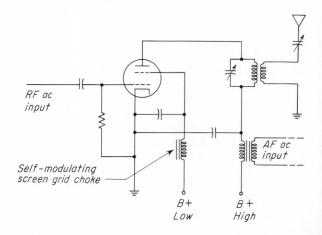

Test your understanding; answer these checkup questions.

1. In a single-tube class C amplifier, why is the radiated modulated envelope not made up of varying amplitude dc pulses? _____
2. In a modulated wave is the power in the positive peak equal to the power in the negative peak? _____
3. Why is simple plate modulation not practical? _____
4. What is the greatest possible modulation value without distortion? _____
5. A plate-modulated RF stage has $E_p = 1$ kV. What modulation is produced if an AF of 2,000 V p-p is added in the plate circuit? _____ If 500 V peak is added? _____ 100 V peak? _____ 700 V rms? _____ 1,000 V rms? _____
6. What does overmodulation produce? _____
7. Which produces the worst interference, overmodulated positive or negative peaks? _____
8. Does using a transmitter audio peak limiter result in louder or weaker received signals? _____ Why? _____
9. How much AF power is required to sine-wave plate-modulate a 50,000-W transmitter to 100%? _____ To 50%? _____ To 10%? _____
10. In question 9, if the plate modulator were 50% efficient, what would be the dc input to this stage when producing 100% sinusoidal modulation? _____ What class would it probably be? _____
11. What kind of bias should be used on a plate-modulated stage? _____ Why? _____
12. What is the result of low RF excitation to a plate-modulated RF stage? _____
13. What are the three ways of simultaneously modulating the screen and plate circuits of pentode or tetrode tubes? _____ _____ _____

17-13 OPERATING POWER

A plate-modulated RF amplifier with 2,000-V plate potential and 0.5-A plate current has a 1,000-W input to the plate circuit. How much of this 1,000-W input is actually radiated? It is usually assumed that the maximum RF output from a class C RF amplifier will be about 70% of the input. For 1,000-W input, about 700 W of output, or *operating,* power will be produced.

For A3 broadcast transmitters licensed at more than 5,000 W, the 70% factor may be raised to 85%. A transmitter licensed for 8,500-W operating power will require an input power of 10,000 W.

The efficiency of an RF amplifier should remain constant regardless of the modulation percent,

provided the tubes are operated within their rated values. A 1,000-W input transmitter has a 700-W carrier-power rating whether modulated or not. When it is 100% sine-wave-modulated, however, the dc plus the audio ac power input is 1,000 plus 500 W, or 1,500 W. The total RF power output with 100% modulation is 70% of 1,500 W, or 1,050 W. The carrier is still only 700 W. The other 350 W of RF power is in *sidebands,* two other RF signals being generated by the modulation process and emitted at the same time the carrier is transmitted.

17-14 SIDEBANDS

The sine wave is the perfect ac waveform. An RF carrier consisting of sine-wave ac will have no harmonics or any other frequency components in it. If such a carrier has a frequency of 1,500 kHz, the only possible emission from it would be a 1,500-kHz signal.

When the illustration of the modulated envelope in Fig. 17-11 is examined, during the time the carrier is constant in amplitude and not modulated, the RF ac might have a substantially sine waveshape. When modulation is applied, however, each succeeding cycle is higher or lower in amplitude than the one before it. Any progressive increase or decrease in amplitude such as this changes the sine waveshape slightly. The change that occurs when a 2,000-Hz sine-wave modulating signal is applied is such that two other RF frequencies, called sideband signals, are developed along with the carrier. One of the sideband signals occurs 2 kHz above the carrier frequency and the other 2 kHz below the carrier, as in Fig. 17-17.

The explanation of sidebands deals with the

Fig. 17-17 (a) Representation of a 1,500-kHz carrier signal and (b) the sideband signals formed when the carrier is modulated by a 2-kHz audio signal.

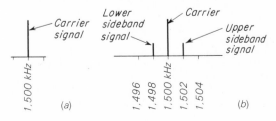

theory of *mixing, beating,* or *heterodyning* one frequency with another in a nonlinear circuit, such as in the plate circuit of an RF amplifier. The result is always at least four output frequencies: (1) one of the original frequencies, (2) the other original frequency, (3) the *sum* of the two frequencies, and (4) the *difference* between the two frequencies.

In a modulated RF amplifier-plate circuit, the four frequencies might be a 1,500-kHz carrier and a 2-kHz AF from the modulator. When mixed, they result in the following four frequencies: (1) 1,500,000 Hz, (2) 2,000 Hz, (3) 1,502,000 Hz, and (4) 1,498,000 Hz.

Since the RF amplifier-tank circuit and the antenna circuit have practically zero impedance to the 2,000-Hz frequency, no voltage of this frequency can develop in them and the 2,000-Hz frequency is lost.

The other frequencies, the carrier and the two sidebands, are close enough in frequency that the amplifier-tank circuit and the antenna will accept and radiate them.

Any transmitted intelligence is in the sidebands produced during modulation. If 350 W of RF power is radiated in the sidebands, each sideband will have half this total, or 175 W. An extremely sharp, or narrow-bandwidth, receiver can tune to the carrier or to either sideband signal alone. With a 2-kHz tone modulation, the receiver will produce no tone from any of the three signals *alone*. A broader receiver will accept all three signals at once and produce a 2-kHz tone when the carrier and sideband frequencies recombine in the non-linear receiver circuits, mixing or demodulating, to reproduce the missing 2-kHz resultant. It is also possible to have a sharp receiver pick up the carrier and only one sideband. These two frequencies can combine in the receiver to produce the 2-kHz, but at only half the power (-3 dB).

If the 2-kHz tone modulation is distorted for any reason, harmonics of 2 kHz will be present in the modulated envelope. Instead of confining all the radiated energy within 2 kHz on either side of the carrier, spurious harmonics of 2 kHz will be produced as additional sidebands far out, above and below the carrier frequency. This results in a broad emission which produces interference to other radio services as well as distorted-sounding signals.

In plate modulation, the power output of the modulator produces the sidebands. Modulator output power is converted to RF power by the RF amplifier,

$$P_{sb} = P_{af} \times \text{RF amplifier efficiency}$$

The sideband power present in a modulated signal can be determined by using the same formula as is used to determine the amount of audio power required to produce a given percent of modulation:

$$P_{af} = \frac{m^2 P_{dc}}{2}$$

where P_{af} = sideband power produced by AF modulator
P_{dc} = carrier power produced by dc power supply of RF amplifier
m = percent of modulation divided by 100

EXAMPLE: If the carrier output of a transmitter is 1,000 W, how much power is in the sidebands when the carrier is modulated 80% by sine-wave AF?

$$P_{af} = \frac{m^2 P_{dc}}{2} = \frac{0.8^2(1,000)}{2}$$

$$= \frac{0.64(1,000)}{2} = \frac{640}{2} = 320 \text{ W in sidebands}$$

17-15 BANDWIDTH

The *bandwidth* of an AM transmitter is determined by the highest-frequency audio ac being transmitted. The bandwidth is the difference in frequency between the furthest removed upper and lower sideband signals produced. A carrier modulated with an 800-Hz audio tone has a bandwidth of 1,600 Hz. If modulated with a 3-kHz tone, the bandwidth is 6 kHz.

If the modulating frequency is 3 kHz but the audio is distorted, as from overmodulation, the bandwidth will be determined by the number of harmonics of 3 kHz that are significantly strong. If the fifth harmonic is still relatively strong, the bandwidth is at least $2 \times 15 = 30$ kHz. Near the transmitter, where even weak harmonics are receivable, the signal will appear to be even wider.

For radiotelegraph transmitters keying at 30 to 40 words per minute, the bandwidth should not be more than about 250 Hz. For speech transmission, in which 3,000 Hz is the highest frequency to be transmitted, the bandwidth should not be much more than 6 kHz. For music transmission, in which the highest frequency to be transmitted is 5,000 Hz, the bandwidth should not be much more than 10 kHz. For high-fidelity transmissions, in which the highest frequency is 15,000 Hz, the bandwidth should be 30 kHz.

17-16 HEISING MODULATION

A simple plate modulation is the Heising system shown in Fig. 17-18. The plate currents of both modulator and RF amplifier flow down through the modulation choke. The same plate voltage is being applied to modulator and modulated ampli-

fier. The modulator is a class A stage and the RF amplifier is class C.

The modulation choke has enough inductance (30 to 100 H) to present a very high impedance to any AF current variations that attempt to flow through it. With no modulation, there is a steady average dc flowing through the choke, part being the modulator plate current and part being the RF amplifier plate current.

When the grid of the modulator is driven less negative, the plate current of the modulator tube increases. The current increase (downward) through the choke develops a counter emf (upward voltage) across the choke. This counter emf is in series with the power-supply voltage and the plate circuit of the RF amplifier. However, the direction of the counter emf is opposite to the power-supply voltage and partially cancels the plate voltage on the RF amplifier. This decreases the RF amplifier plate current and the RF ac output. The RF amplifier plate current decreases just as much as the modulator plate current increases.

When the modulator grid is driven more negative, the modulator plate current decreases, producing a collapsing magnetic field and an induced emf (downward) in the choke which adds to the power-supply voltage for the RF amplifier. This increases the amplifier plate current and the RF ac output. The RF amplifier plate current increases as much as modulator plate current decreases. This results in a constant current value

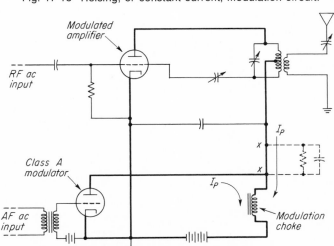

Fig. 17-18 Heising, or constant-current, modulation circuit.

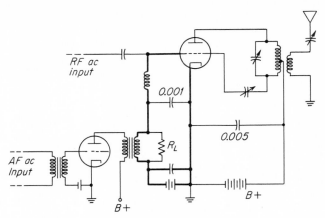

Fig. 17-19 A control-grid, or grid-bias, modulation circuit.

in the choke and power supply and is why Heising modulation is also known as *constant-current modulation*.

The peak AF ac voltages developed across the choke coil in a Heising class A modulation system will never equal much more than 80% of the plate supply voltage. If the RF amplifier plate voltage cannot be doubled at the peaks of modulation, 100% modulation cannot be obtained in the Heising circuit.

If a resistor, with an AF bypass capacitor across it, is connected in series with the plate circuit at the points marked X in the diagram, the voltage-drop across the resistor will lower the dc plate voltage on the RF amplifier plate but not on the modulator plate. If the power supply is 1,000 V and the modulator can develop 800-V peak AF ac across the choke, the series resistor should have a value large enough to provide a voltage-drop of at least 200 V to produce 100% modulation.

The capacitor across the resistor should have a low reactance to all audio frequencies in order not to attenuate them. The resistor alone will produce a dc voltage-drop across itself, but the modulating AF will also produce a voltage-drop across it. The capacitor passes the AF voltages to the RF amplifier plate.

17-17 GRID MODULATION

In plate modulation, an AF ac is added to the dc plate supply voltage, forming a varying dc plate voltage that modulates the output wave.

In grid (grid-bias) modulation an AF voltage is added in series with the bias supply of the modulated amplifier as in Fig. 17-19. Since the bias voltage can control the output power of the amplifier, variations of the bias voltage can produce A3 in the output. In plate modulation, the modulating ac works into a plate circuit having a definite resistance value ($R = E_p/I_p$). In grid modulation if the bias and modulating-signal values are so adjusted that the grid draws no current at any time during modulation, the modulator tube is then working into an infinite-ohm load resistance, and almost no power is required to produce modulation. This low audio power requirement is one of the advantages of grid modulation.

A grid-modulated stage is biased to class C and always uses some type of bias supply, never grid-leak bias. Grid-leak bias requires grid current flowing through the grid-leak resistance to produce the bias voltage. With no grid current in grid modulation there would be no grid-leak bias voltage.

Figure 17-20 illustrates the $E_g I_p$ curve of a grid-modulated amplifier in which no grid current flows. In condition *A*, the bias is adjusted to about $1\frac{1}{2}$ times the plate-current cutoff value. There are no RF excitation or modulating voltages and no output.

In condition *B*, the RF excitation voltage is increased to the point where the peaks are at the midpoint of the $E_g I_p$ curve (between zero grid volts and the cutoff value).

In condition *C* the bias voltage is modulated. The RF excitation peaks are made to move toward

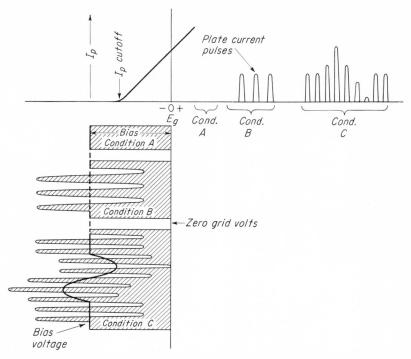

Fig. 17-20 How modulation of the grid-bias voltage produces modulated-amplitude plate-current pulses.

and away from the zero grid-voltage line. The plate-current pulses vary in amplitude as the modulating voltage changes. Note that the RF excitation voltage does not vary in amplitude, but the whole excitation signal is swung back and forth under the curve by varying the bias voltage, thereby producing the varying-amplitude plate-current pulses.

There seems to be an inconsistency in grid modulation. The plate power-supply voltage is constant. At the point of 100% modulation, the amplitude of the plate-current pulses is twice what it was with the carrier alone. With the same plate voltage and twice the current there should be only twice the power output from the tube at 100% modulation peaks. But the positive peak of modulation requires four times the power output (Sec. 17-9). During modulation the excitation voltages are operated over a greater portion of the curve, and the efficiency of operation of the tube increases. When the bias is swung sufficiently to produce plate-current pulses of double amplitude, twice the length of $E_g I_p$ curve is being used and

the efficiency of stage operation doubles. As a result, at 100% modulation the tube is twice as efficient, and the peak current is twice as much. This causes a fourfold increase of RF power output at the 100% positive peak of modulation. Because of the variation of efficiency, grid modulation is also known as *efficiency modulation*.

During grid (or plate) modulation, the plate-current pulses increase as much as they decrease, resulting in a constant average plate current. A milliammeter in the plate circuit shows no variation during modulation, unless distortion is present.

Operating a grid-modulated amplifier as explained will produce reasonably undistorted modulation up to about 95%. The stage will operate at only about 20% efficiency when producing the carrier alone, rising to a peak of about 40% under 100% modulation conditions. To increase the efficiency of the amplifier stage, the RF excitation may be raised to the point where the peaks of the RF grid voltage approach the zero grid-voltage point. This operates the tube over a greater proportion of its curve, at about 35% efficiency. When modulated to 100%, the positive peak of modulation will be produced while the tube is operating at about 70% efficiency. Under these conditions, grid current flows during most of the positive peaks of modulation, tending to flatten the peaks. Partially to overcome this type of distortion, the resistor R_L, in Fig. 17-19, maintains a fixed load on the modulator stage at all times. A similar loading resistor may be connected across any grid LC circuit to load both halves of an RF excitation cycle. The amount of AF power required for grid modulation is quite small, rarely being more than 1 or 2 W.

To produce the highest possible power output and the best linearity of modulation, the plate voltage on the modulated stage should be as high as the tube will stand safely. When resistance-coupled, grid modulation can be linear from a few hertz to over 4 MHz.

Among the disadvantages of grid modulation are the rather critical adjustments for proper operation. The degree of coupling to the antenna, the correct L/C ratio in the plate-tank circuit, the bias voltage, the excitation voltage, and the modulating voltage are all more critical than in plate modulation.

Grid modulation requires less audio power than plate modulation, has simpler circuits, is more critical to tune and keep in adjustment, requires a higher plate voltage, and is easily overmodulated. Actually, considering efficiences, cost, size, and weight, one type of modulation may be very little better than the other.

17-18 SUPPRESSOR-GRID MODULATION

The suppressor grid of a pentode may be modulated. RF excitation voltage is applied to the control grid, which may be either power-supply or grid-leak-biased. The AF modulating voltage is added in series with the suppressor-grid circuit, which must be class A-biased. A suppressor-grid-modulated pentode RF amplifier is shown in Fig. 17-21. The modulating voltage value for suppressor-grid modulation is considerably greater than for control-grid modulation because of the lower transconductance of the suppressor grid. Very little modulating power is required.

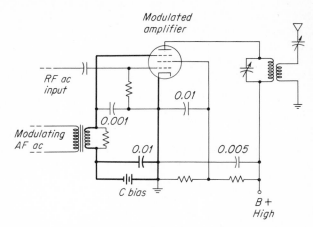

Fig. 17-21 A suppressor-grid modulation circuit.

17-19 SCREEN-GRID MODULATION

Modulating the screen grid of a beam-power tetrode or a pentode is similar in some respects to both grid modulation and plate modulation. The audio power requirement is considerably lower than for plate modulation and greater than for grid. The AF ac is added to a positive potential from a dc power supply, as shown in Fig. 17-22.

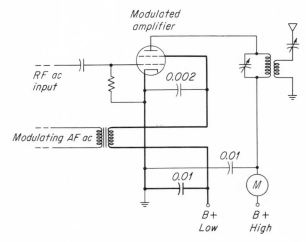

Fig. 17-22 A screen-grid modulation circuit.

The plate voltage should be as high as possible for maximum output. The screen voltage must be lowered to about two-thirds of its normal operating value and varied from that point by the modulating voltage. To increase linearity at higher percent of modulation, the modulating voltage can be applied to the screen and control grids simultaneously. The plate-current meter usually rises slightly during modulation.

17-20 HIGH-LEVEL AND LOW-LEVEL MODULATION

Modulated stages can be divided into three separate categories. These are:

1. High-level modulated stages, in which the modulated RF stage is plate-modulated and its output feeds the antenna
2. Low-level modulated stages, in which the modulated RF stage is either plate- or grid-modulated followed by a linear amplifier which feeds the amplified modulated signal to the antenna
3. Grid modulation, in which the final RF amplifier is grid-modulated

Test your understanding; answer these checkup questions.

1. What is the approximate dc input power of a broadcast station licensed for 2-kW operating power? _____ If licensed for 7.5 kW? _____
2. What percent of the total radiated power is in the sidebands with 100% sinusoidal modulation? _____ 50%? _____
3. What are the four output frequencies when 4 MHz and 5 MHz are mixed? _____ When 500 Hz and 8 MHz are mixed? _____
4. If one of two signals being heterodyned is distorted, what is the result? _____
5. What is the approximate bandwidth of a voice-modulated A3 transmitter? _____ Hi-fi A3? _____
6. What are two names of the A3 modulation which uses an AF choke? _____ _____ Why is an RC network used in series with the choke? _____
7. What do you think is the main advantage of grid over plate modulation? _____ Disadvantage? _____
8. Under what conditions can grid current be allowed in grid modulation? _____ _____
9. Why is grid modulation called efficiency modulation? _____
10. Is any substantial AF power required for grid modulation? _____ Suppressor-grid modulation? _____ Screen-grid modulation? _____
11. What is meant by high-level modulation? _____ Low-level? _____ What other type is there? _____

17-21 CHECKING MODULATION WITH AN OSCILLOSCOPE

The oscilloscope was described in Chap. 12. This piece of equipment provides one of the most satisfactory methods of determining the percent of modulation on a carrier, as well as indicating the presence of certain types of distortion. There are many methods of using an oscilloscope. One shows the modulation envelope. Another produces a trapezoidal modulation figure on the screen.

The modulation envelope is displayed by using a sawtooth ac of 20 to 200 Hz on the horizontal-deflection plates. This is supplied by the sawtooth ac-generating circuit incorporated in the oscilloscope. Radio-frequency ac is picked up inductively by loose-coupling a two- or three-turn loop to the antenna or final-amplifier coil as shown in Fig. 17-23. (A short antenna with an accompanying tuned LC circuit can also be used as the RF pickup device.)

With no modulation, the RF ac drives the electron beam up and down as the sawtooth ac drives it left and right. This results in a wide band across the face of the cathode-ray tube. When modulation is applied, the band is modulated, the positive peaks increasing the height of the band and the negative peaks decreasing the band to a narrow line. If the sawtooth sweep voltage has a frequency of 200 Hz and the modulation is a sinusoidal 400-Hz signal, the modulated carrier will appear as sine-wave variations on the top and also on the bottom of the carrier band. If the carrier is modulated 100%, the negative peaks will show as a spot on the screen and the positive peaks will have twice the vertical amplitude of the carrier alone, as shown. The oscilloscope shows the modulation envelope as it was explained at the beginning of this chapter. If the transmitter is being modulated by a sine-wave signal, any deviation from the sine waveshape on the modulated envelope indicates that distortion is being produced in the modulation system.

It is possible to check stage by stage with an oscilloscope to determine where distortion first appears. To do this, a 0.01- to 0.1-μF capacitor can be connected to the ungrounded vertical plate, where the RF pickup loop was previously

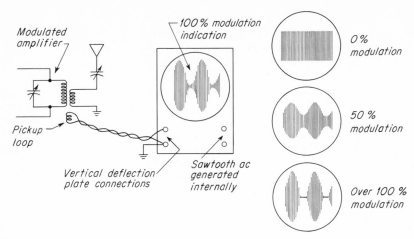

Fig. 17-23 The circuit used to produce a modulated-envelope display on an oscilloscope and four possible displays.

connected. By touching the capacitor to the plate or grid terminal of each tube in the audio section of the transmitter, an indication will be given of the waveshape at that point. Care must be taken when touching plate-circuit terminals because dangerously high voltages may exist at those points. Modulator AF ac may be too high in amplitude for the plates of the cathode-ray tube. A voltage-divider circuit (Fig. 17-24) can be used to decrease the voltage applied to the oscillo-

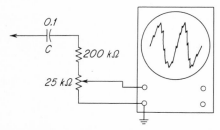

Fig. 17-24 Method of displaying an AF voltage on an oscilloscope.

ANSWERS TO CHECKUP QUIZ ON PAGE 401

1. (2,860 W) (8.8 kW) 2. (33%) (20%) 3. (4, 5, 9, and 1 MHz) (500 Hz, 8, 8.0005, and 7,9995 MHz) 4. (Harmonics of one signal also beat against other signal) 5. (6 kHz) (30 kHz) 6. (Heising, constant I) (Attain 100%) 7. (Low AF power) (Low power efficiency, difficult to adjust) 8. (High drive, low-R grid circuit) 9. (0 efficiency at 0%, ±60% at + peaks) 10. (No) (No) (Yes) 11. (Plate-modulated PA) (Modulation other than PA) (Grid-modulated PA)

scope vertical plates. WARNING: Turn off the transmitter, fasten the capacitor to the desired point, and then turn on the transmitter to make this test.

The pattern on the oscilloscope will stand still only if the modulating frequency is some exact multiple of the sawtooth sweep frequency. Voice and music produce a jagged, jumping series of waveforms on the face of the tube. If the positive or negative peaks are badly flattened (distorted), this will be apparent to the trained observer.

If bright spots are developed on the modulated envelope at the negative peaks of modulation, this is an indication of overmodulation and consequently of distortion and broadening of the bandwidth of the emission.

Precise neutralization of the modulated stage can be accomplished by overmodulating the negative peaks and then adjusting the neutralizing capacitor to minimum line width.

The trapezoidal display is produced by feeding modulated RF directly to the vertical plates of the oscilloscope, as before, and by feeding a small fraction of the modulating ac from the modulation-transformer secondary to the horizontal plates as in Fig. 17-25.

With no modulation, the carrier produces a thin vertical line on the screen of the oscilloscope. When modulation is applied, the AF drives the electron beam back and forth and the line thickens to the right and left sides. The negative peaks of modulation are indicated by the side that

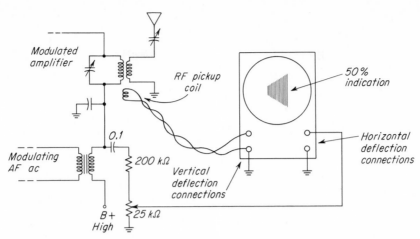

Fig. 17-25 Circuit to produce a trapezoidal modulation display on an oscilloscope.

drops off in amplitude, and the positive peaks by the side that increases in amplitude. At 100% modulation, the trapezoid forms a point at one side and is twice the carrier amplitude on the other side (a triangle). Figure 17-26 shows (*a*) an unmodulated carrier, (*b*) a 50% modulated carrier, (*c*), a 100% modulated carrier, and (*d*) an overmodulated carrier. If the slanting sides of the trapezoid are not perfectly straight, distortion is indicated. If the positive peaks show a flattening (Fig. 17-26*d*) instead of rising to a sharp point, negative carrier shift (Sec. 17-24) is indicated.

Envelope displays on an oscilloscope show the results of adding sidebands to the carrier but do not show the sidebands themselves. The beginner often believes that the part of the display above the carrier level is a sideband, which is not true. The oscilloscope represents only a carrier and how the resultant voltage of the modulated carrier and sidebands varies from instant to instant.

17-22 LINEAR RF AMPLIFIERS

A linear amplifier is one which will amplify without distortion.

Linearity can be obtained only by operating an amplifier tube over the straight portion of its $E_g I_p$ curve. If operated over a nonlinear section, distorted amplification will result. The only amplifiers that produce undistorted signals are the class A, AB, and B. Class C amplifiers are considered nonlinear because they are not operated on the straight portion of the $E_g I_p$ curve for the whole of the input cycle.

In a high-level A3 transmitter, where the output of the modulated stage feeds into an antenna, no linear amplifiers are needed. If the modulated RF stage is not the final RF amplifier, *all stages* after the modulated stage must be linear amplifiers to enable them to amplify the modulated RF carrier without distorting it. A simple block diagram of a

Fig. 17-26 Four possible trapezoidal displays of modulation on an oscilloscope: (*a*) 0%, (*b*) 50%, (*c*) 100%, and (*d*) overmodulation.

   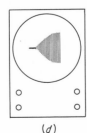

(*a*)   (*b*)   (*c*)   (*d*)

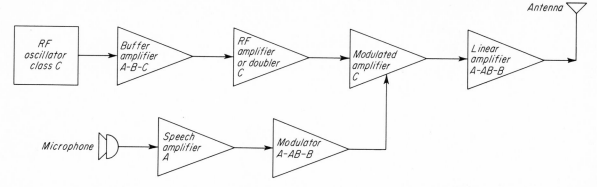

Fig. 17-27 Block diagram of the stages in a radiotelephone transmitter using low-level modulation and a linear amplifier.

low-level modulated transmitter is shown in Fig. 17-27. The letters indicate possible classes of operation of the stages.

Class A amplifiers are quite low in efficiency. Therefore, classes AB and B are more likely to be used as linear RF power amplifiers.

In audio, a class B amplifier must be a push-pull type to reproduce both halves of the input cycle. In RF amplifiers, however, a class B amplifier can operate with only one tube. The missing half of each RF cycle is reproduced by the flywheel action of the tuned plate-tank circuit of the stage. However, class B radio-frequency amplifiers may use two tubes in a push-pull circuit.

Figure 17-28 shows a 100% modulated enve-

Fig. 17-28 When the grid excitation to a linear amplifier is modulated, the plate-current pulses are varied in amplitude.

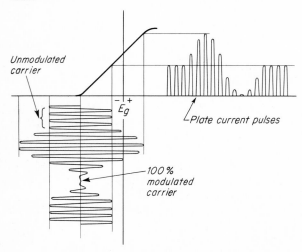

lope impressed on the $E_g I_p$ curve of a class B linear amplifier and the resulting plate-current pulses. The amplifier is shown excited almost to zero grid volts by the carrier signal. High positive peaks of modulation fall into the positive grid region and produce grid-current flow. The amplifier requires a bias supply with good regulation as well as low dc resistance between grid and cathode.

To produce undistorted modulation from any modulated amplifier it is necessary for it to work into a constant load, such as is presented by an antenna in high-level modulation systems. To produce an unvarying impedance load on the modulated stage, the linear-amplifier grid circuit may be loaded with a noninductive resistance of such a value as to reflect the desired impedance on the plate circuit of the modulated stage, as in Fig. 17-29.

To take advantage of the characteristic higher efficiency of a class C stage, a linear amplifier is sometimes biased to class C, but the modulated input signal is made to vary only over the linear portion of the $E_g I_p$ curve. The signal fed to the grid must not be modulated to a high percent. This is an advantage, because it is not difficult to produce low percent of essentially undistorted modulation with most systems. In Fig. 17-30 the linear amplifier is biased to about $1\frac{1}{2}$ times cutoff and requires an excitation carrier modulated to only about 70% to produce 100% modulated output RF ac. Since the stage is biased to class C but modulates over the same portion of the curve as would be used if it were biased to class B, this is known as *class BC*.

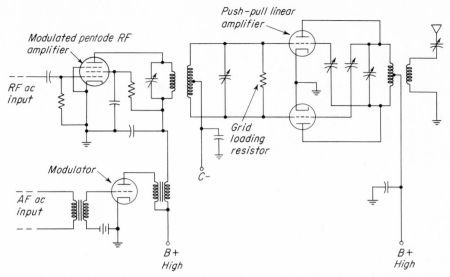

Fig. 17-29   A low-level plate-modulated pentode stage and a push-pull linear RF amplifier.

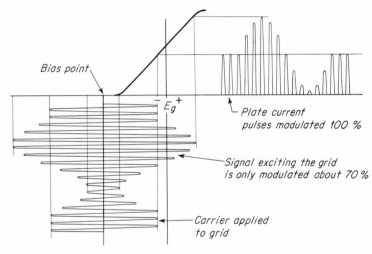

Fig. 17-30   Biasing the linear amplifier to class BC produces 100% plate-current-pulse modulation with only a 70% modulated input.

Notice that in Fig. 17-30 the positive peaks of modulation do not operate into the saturation, or bent, portion of the $E_g I_p$ curve. If excitation due to the carrier alone is increased to the point of saturation, positive peaks of modulation can not produce higher plate-current pulses than the carrier does. Negative peak variations can be produced in the output, but not the positive peaks. This is a serious form of distortion. In such a case, the milliammeter in the linear-amplifier plate circuit

will decrease sharply when the stage is amplifying a modulated input signal, whereas it should remain steady or vary only slightly as the percent of modulation changes.

To hold distortion to a minimum in a linear amplifier, the plate, screen-grid, and grid bias supplies must be well regulated.

A tube used in a linear-amplifier circuit can operate with higher plate voltage than if it were plate-modulated. A tube limited to 2,000 V on the plate as a plate-modulated amplifier may use 2,500 V as a linear amplifier.

According to the FCC, the dc plate power input times 0.35 equals the operating power output of a class B linear amplifier, an efficiency of approximately 35%. Compare this with the factor of 0.7 to 0.85 for plate-modulated stages. (However, linear amplifiers used with single-sideband suppressed-carrier emissions may be more than 60% efficient.)

A linear amplifier in an A3 transmitter is similar to a grid-modulated stage in many respects. The dc plate-current meter remains constant. When the excitation voltage is doubled, the plate-current pulses double in amplitude, efficiency doubles, and the output increases four times. The output power is therefore proportional to the excitation voltage squared. With an increased percent of modulation, the plate current varies over a wider portion of the $E_g I_p$ curve and the efficiency of the amplifier increases. Whereas the plate of a plate-modulated tube may turn red-hot at a high percent of modulation, a grid-modulated or linear-amplifier plate will cool under modulated conditions because of lessened plate dissipation. (If the dc input is constant, the plate dissipation must decrease as the RF output increases.)

17-23 ADJUSTING A LINEAR RF AMPLIFIER

It is possible to tune a class B linear RF amplifier of an A3 transmitter by these steps:

1. Bias the stage to cutoff by applying full operating plate voltage and observing the plate current as the bias voltage is decreased from a value known to be more than is necessary for plate-current cutoff. (Use no RF excitation while determining the cutoff-bias value.) When the plate current begins to read a few milliamperes, the bias value is correct.
2. Apply a weak unmodulated RF excitation to the grid, and tune the plate circuit to minimum plate current. (Neutralize if necessary.)
3. Increase the RF excitation until some grid current begins to flow. At this value of RF excitation note the plate current. (For example, it might be 300 mA.)
4. Decrease the RF excitation until the plate-current value is one-half of the above-noted value (150 mA). This should be approximately the correct unmodulated carrier-excitation value.
5. Couple the antenna to the amplifier. Check the modulation percent and linearity as shown on an oscilloscope, with a sine-wave signal generator feeding into the speech amplifier, making adjustments on bias, RF excitation, and antenna coupling until essentially undistorted 100% modulation is produced at the desired power output.

17-24 DOHERTY LINEAR AMPLIFIERS

The usual linear amplifier of an A3 signal operates at only about 30 to 35% efficiency. The two-tube Doherty linear, shown in Fig. 17-31, may be 60% efficient. The class B stage operates at all times, but the class C stage operates only during the positive half cycles of modulation. With the carrier alone, and during negative halves of the modulated cycle, the class C tube does nothing.

The class B tube is excited just to saturation by the carrier, producing a wide swing of plate current, operating at about 60% efficiency, but into a plate tank load of twice the tube impedance. This mismatch decreases the power output somewhat, but gives good efficiency. The class B stage produces all the output power of the carrier alone plus the negative half cycle of modulation.

During the positive half cycle of modulation the class B stage is in a saturated condition and assumably does not increase its output. The class C stage can now go into operation. It feeds RF power into the output circuit. This reflects a lowered impedance back on the tank circuit of the class B stage, forcing this tube and load impedance to match better, and additional power comes from the class B stage. As a result, at the positive peak of modulation the necessary RF peak power of four times the carrier value is produced by the action of the two tubes.

To produce the necessary impedance matching in the output circuit, the equivalent of a quarter-wave line must be inserted in the output of the class B tube. This shifts the phase of the output RF by 90°, necessitating an opposite change of phase, either in the grid of the class B stage, as shown, or in the grid circuit of the class C stage,

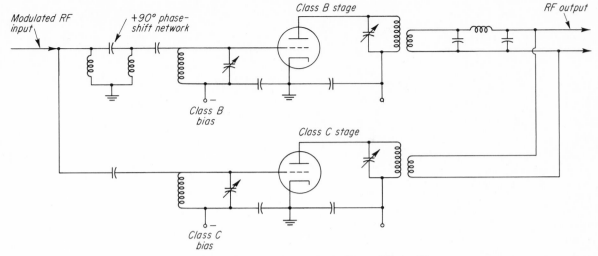

Fig. 17-31 Doherty high-efficiency linear RF amplifier.

in order that the RF output of both stages be in phase.

These amplifiers are used in broadcast stations where operation is on only a single frequency. Tuning from one frequency to another becomes complicated because the phase-shift networks must also be tuned.

17-25 CARRIER SHIFT

Distortion of modulation in an A3 transmitter may result in *carrier shift,* either *positive* or *negative.*

A meter that will indicate carrier shift is shown in Fig. 17-32. A dc voltmeter (50-μA meter and

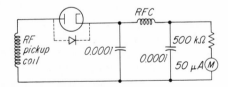

Fig. 17-32 A carrier-shift meter.

resistor) is connected across the output of an RF pickup coil and a half-wave rectifier. The output of the rectifier is dc, pulsating at the RF rate. The RFC and the two capacitors form a low-pass filter for RF but have no smoothing effect on audio-frequency variations. The pickup coil is coupled closely enough to the modulated RF output of a

transmitter to give approximately half-scale reading on the meter.

When undistorted modulation is produced by the transmitter, the increased positive peaks and decreased negative peaks of modulation should be equal. Since the meter cannot swing fast enough to follow the audio variations of the modulated envelope and since the average remains constant, the meter needle does not change with or without modulation.

If during the positive peak of modulation the RF voltage increases more than it decreases during the negative peak, the result is a shifting of the average carrier voltage upward, in the positive peak direction, as in Fig. 17-33a. Positive carrier shift is present. As modulation is applied, the meter swings upward.

If negative peaks of modulation decrease more than positive peaks increase, the result is a shifting of the average carrier voltage downward, in the negative peak direction, as in Fig. 17-33b. Negative carrier shift is present. The average carrier voltage being lower during modulation, the carrier-shift meter swings lower as modulation is applied. Negative carrier shift is known as *downward modulation.*

With carrier shift, even if the original AF voltage were sinusoidal, the modulated envelope voltage is no longer sine-wave-shaped. This indicates distorted or *asymmetrical* (nonsymmetrical) modulation.

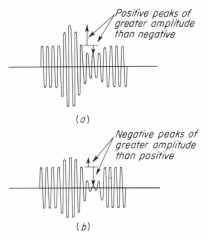

Positive peaks of greater amplitude than negative

(a)

Negative peaks of greater amplitude than positive

(b)

Fig. 17-33 Illustration of a carrier with (a) positive carrier shift and (b) negative carrier shift.

When carrier shift is present, there is distortion of the signal. Up to about 5% of carrier shift is probably not noticeable to the average listener. More than this may be. If considerable negative carrier shift is present, the signals have a compressed, choked, and weak sound, as if the percent of modulation were low, which it is. With negative carrier shift the antenna ammeter will not increase normally or may even decrease. With positive carrier shift the antenna ammeter will increase more than normally.

There is a tendency, when seeing the words "carrier shift," to infer a shifting of the frequency of the carrier. This is not correct. Any frequency variation may be termed frequency shift, dynamic instability, or frequency modulation. Carrier shift refers to amplitude variations of the carrier, not frequency.

For voice communication, some positive carrier shift in the transmission is an advantage, provided the negative peaks are not materially distorted in shape. A greater proportion of sideband power can be radiated for a given carrier value, and the receiver produces louder signals.

17-26 CAUSES OF CARRIER SHIFT

There are many types of distortion that will indicate on a carrier-shift meter:

PLATE-MODULATED STAGES. Insufficient RF excitation to the grid is one of the most frequent causes of negative carrier shift. Battery bias alone on the modulated stage, distorted audio from the modulator, improper impedance match between the modulator and the modulated stage, improper antenna coupling, low filament voltage or an old tube in the RF amplifier, excessive screen voltage, or failure to modulate the screen of a tetrode or pentode tube will result in negative carrier shift. If one of the two tubes of a class AB or B push-pull modulator burns out, severe distortion and carrier shift will occur.

Insufficient or no plate voltage on the modulated RF amplifier will result in an overmodulated condition and considerable positive carrier shift, as will too high a setting of the gain control in the speech amplifier.

GRID-MODULATED STAGES. Either form of carrier shift is easily produced if the grid-modulated stage and the modulator are not carefully adjusted. Too high a bias can produce positive carrier shift, as can too little RF grid excitation. Too low a bias can produce negative carrier shift, as can too much RF grid excitation. Excessive audio modulating voltage will usually result in positive carrier shift. A distorted audio waveform may produce either positive or negative carrier shift. A weak tube or low filament voltage in the modulated stage can produce negative carrier shift. Improper antenna coupling can produce either form of carrier shift, depending upon other circuit factors.

LINEAR AMPLIFIERS OF A3 SIGNALS. These stages are similar to grid-modulated stages. Proper bias and RF excitation values are very important. An overexcited grid produces severe negative carrier shift. If the modulated stage has a certain type of carrier shift in its output, the output of the linear amplifier normally has the same. It is possible to balance out small values of carrier shift as the signal passes through a linear amplifier. If the original modulated signal has a slight negative carrier shift, the linear amplifier can be adjusted to produce a slight positive carrier shift.

If an RF amplifier tuning circuit or the antenna circuit insulation breaks down and sparks across on positive peaks of modulation, the indication given by the carrier-shift meter will be downward.

17-27 A3 AND A1 WITH THE SAME TUBE

If a tube in a transmitter is to be used for both plate-modulated A3 and A1 emissions, the carrier output for radiotelephone will usually have to be held between 65 and 75% of that used for A1. The differential in the output is necessary to allow for the additional power being handled by the tube under modulated conditions. If a tube is capable of 1,000 W in radiotelegraph service, it should be operated at approximately 700 W if modulated, to allow for the 350 W of sideband power that will be added to the carrier under 100% sine-wave-modulation conditions. Running at 700-W carrier, the tube is actually being overdriven if continually operated at full modulation. However, it is unusual to have a transmitter operating under this particular condition for any period of time. The average modulation for speech and music will produce far less than 350 W of average sideband power for a 700-W carrier.

If the same tube is grid-modulated, instead of a 1,000-W carrier output, only about 200 to 250 W can be expected as the carrier power.

Test your understanding; answer these checkup questions.

1. What sweep frequencies would be used on a scope to stop an envelope pattern of 600-Hz modulation? _____
2. Can voice patterns be stopped on a scope? _____
3. Why would trapezoidal patterns be better than envelope for checking speech modulation? _____
4. What percent of modulation must be used when neutralizing an operating transmitter? _____
5. What is the shape of a trapezoidal pattern at 100% modulation? _____
6. What class amplifier is usually used in linear RF amplifiers with A3? _____
7. What is the advantage of biasing an RF linear amplifier to class BC? _____
8. Why are linear amplifier grid circuits often resistor-loaded? _____
9. Does the efficiency of a linear amplifier increase with percent of modulation? _____ Why? _____
10. What is the advantage of a Doherty linear? _____ To what class is the carrier amplifier biased? _____ The + peak amplifier? _____
11. Does carrier shift refer to amplitude, phase, or frequency? _____ What is another term meaning the same thing? _____
12. What carrier shift results with insufficient grid excitation

to a high-level modulated stage? _____ Grid-modulated stage? _____
13. Would AF peak-limiting produce negative carrier shift? _____
14. Why must a plate-modulated A3 RF amplifier use less E_p than with A1? _____

17-28 WHAT THE ANTENNA AMMETER TELLS

If a transmitter is turned on, the antenna ammeter will rise to some value and remain there until the transmitter is turned off again.

If the transmitter is turned on and then modulated, the antenna meter will rise to a higher value than with the carrier alone. The modulation being applied to the RF amplifier appears in the radiated wave as sideband power. The added power is responsible for the increase in the antenna-current reading.

How much will the antenna current rise for 100% sine-wave modulation? Assume a carrier power of 100 W in the antenna and an antenna resistance of 1 Ω. (The antenna resistance will not vary with changes of power.)

1. The current in the antenna, from the power formula $P = I^2R$, is

$$I = \sqrt{\frac{P}{R}} = \sqrt{\frac{100}{1}} = 10 \text{ A}$$

2. With 100% modulation, the sideband power will be equal to one-half of the carrier power (Sec. 17-11). The total power in the antenna at this time is 100-W carrier plus 50-W sideband, or 150 W.
3. The antenna current with 150 W is:

$$I = \sqrt{\frac{P}{R}} = \sqrt{\frac{150}{1}} = 12.25 \text{ A}$$

4. The antenna current increases from 10 A unmodulated to 12.25 A when 100% modulated. This is a 22.5% increase.

Whenever an antenna ammeter indicates an increase of 22.5% over the unmodulated value, provided a sine-wave-modulation signal is used and provided no distortion is present, the transmitter can be assumed to be modulated 100%. An 8-A unmodulated carrier should read 8 + (8 × 0.225), or 9.8 A, when modulated 100%.

For 50% sine-wave modulation, assuming the same 100-W carrier and antenna as above:

1. A 50% modulation produces only one-fourth as much sideband power as is developed at 100% (Sec. 17-11), or 12.5 W.
2. At 50% modulation the total power in the antenna is 112.5 W.
3. The antenna current will be

$$I = \sqrt{\frac{P}{R}} = \sqrt{\frac{112.5}{1}} = 10.6 \text{ A}$$

4. The increase from 10 to 10.6 is 0.06 greater than the original carrier value. Therefore, the antenna ammeter will rise only 6% with 50% modulation.

Note that sine-wave-modulating voltages are specified. Pure tones of music are somewhat sinusoidal, but normal speech and music are far from sine-wave-shaped. When a transmitter is voice-modulated, the antenna ammeter may hardly move at all, although some of the peaks of modulation may be 100% or more.

If the antenna meter decreases when modulation is applied, downward modulation, or severe negative carrier shift, is present. If the meter does not move when modulation is applied, this may indicate low percent of modulation, some negative carrier shift, or both. Sometimes the ammeter will move upward with low percentages of modulation but decrease with an increase of modulating voltage. This may indicate nearly normal modulation to a certain percent. Above this value negative carrier shift sets in.

ANSWERS TO CHECKUP QUIZ ON PAGE 409

1. (*Any submultiple of 600; 300, 150, etc.*) 2. (*Not on the usual scope*) 3. (*Slope of trapezoid indicates distortion*) 4. (*100% or over*) 5. (*Triangular*) 6. (*B or AB₂*) 7. (*100% modulation output from lower percent modulation input*) 8. (*Constant load on modulator stage, help stabilize amplifier*) 9. (*Yes*) (*Signal operating over more of E_g/I_p curve*) 10. (*High efficiency*) (*B*) (*C*) 11. (*Amplitude*) (*Asymmetrical*) 12. (*Negative*) (*Positive*) 13. (*No*) 14. (*Allow for additional E from modulator*)

17-29 MAGNETIC-INDUCTION MICROPHONES

There are many types of microphones that fall into the general category of magnetic-induction microphones. They all operate on the principle that sound waves striking a diaphragm produce a relative movement between a magnetic field and a conductor, thereby inducing a voltage in the conductor. A *dynamic microphone* is similar in general construction to a permanent-magnet (p-m) dynamic loudspeaker, described in Sec. 14-31. It has an essentially flat frequency range from perhaps 60 to well over 10,000 Hz in the better models. It requires no battery, being a form of ac generator in itself. It can be built in a light, small, and rugged form at a comparatively low cost. It has an output power of perhaps −30 to −80 dBm (decibels using 1 milliwatt as a zero reference), depending on sound amplitude.

Figure 17-34 shows a simplified construction of

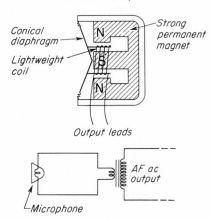

Fig. 17-34 Cross section of a dynamic microphone and the circuit in which it is used.

a dynamic microphone and its circuit. The conical diaphragm has a small coil attached to it but is free to move in and out of the space between north and south poles of the magnet. Vibration of the diaphragm by sound waves moves the coil back and forth across the lines of force, inducing an ac in the turns of the coil. The ac has a frequency equal to the diaphragm vibration frequency and an amplitude proportional to the extent of the diaphragm vibration.

The moving-coil impedance is only a few ohms. A step-up transformer, usually built into the microphone case, is required to enable the output of the microphone to match the 50-, 250-, or 600-Ω or high-impedance input circuits of amplifiers with which it is to be used.

Sound-powered microphones, used for voice communications in special telephone lines, are dynamic-microphone types. Since high fidelity is not an important factor for voice transmission, it is possible to obtain relatively high-power output (0 dBm) by close talking into the microphone. When earphones are connected across them, the resulting signal is quite loud in the earphones. Such a microphone can also be used as an earphone, operating as a p-m loudspeaker.

Dynamic microphones must be kept away from ac fields. If the coil is held near a power transformer, for example, a hum voltage will be induced in the microphone coil. It is possible to minimize such hum by turning the microphone in a different direction.

Other microphones operating on the magnetic-induction principle have various names: magnetic, variable-reluctance, ribbon, velocity (Fig. 17-35), etc. Most of these can be used in broadcast stations.

When it is necessary to use long microphone leads, the lower the impedance of the lines, the less extraneous noise picked up and the less high-frequency attenuation due to capacitance

between the conductors making up the line will be present. This rule of low impedance for long lines is true for all types of microphones and audio lines in general.

17-30 THE CRYSTAL MICROPHONE

The crystal microphone is found mostly in public address, amateur communication, and home recording systems, but rarely in broadcasting. It has good frequency response, from about 50 to 10,000 Hz, and relatively good sensitivity, -40 to -60 dBm. It is fairly rugged mechanically, is satisfactory when used as a hand microphone, and may be spoken into at close range.

The crystals used in older microphones are affected by moisture or a temperature of more than 120°F (easily produced by summer sun striking the microphone). Newer developments with ceramics have overcome these difficulties.

The rochelle salt crystals of these microphones have piezoelectric properties similar to those of quartz crystals used in oscillators. When the crystal is vibrated mechanically, an ac emf is developed between any two metal plates held against opposite surfaces of the crystal. This ac will be proportional to both the vibration frequency and amplitude.

Sound waves striking the conical aluminum diaphragm of the crystal microphone in Fig. 17-36 will cause it to vibrate. These vibrations are mechanically transmitted by a small driving rod to two of the opposite corners of a thin, square crystal. The mechanical vibrations bend the crystal alternately in and out, producing an AF ac between the two foil plates cemented to the two flat sides of the crystal. The output leads from the microphone are connected to the two plates.

The output impedance of the crystal microphone is high, in the range of several million ohms, necessitating relatively short leads to the input circuit of the microphone amplifier to prevent loss of the higher audio frequencies, or hum pickup. The leads must be well shielded and not more than about 20 ft long.

17-31 THE CONDENSER MICROPHONE

Since the late 1920s, one of the best broadcasting microphones has been the electrostatic,

Fig. 17-35 Basic construction of a ribbon-type microphone. Sound waves vibrate the ribbon across the magnetic field, inducing ac into the ribbon.

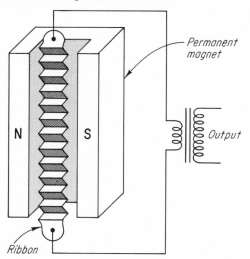

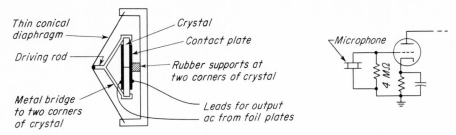

Fig. 17-36   Cross section of a crystal microphone and the circuit in which it is used.

or *condenser* (capacitor), microphone. Its frequency response is good, from about 50 to 15,000 Hz. It has an exceptionally high impedance, however, and its leads must be very short. The first-amplifier stage is constructed in the microphone case for this reason. The cable to the microphone contains not only the output lines from the microphone unit but also the required power-supply lines for the tubes or FETs, and condenser head. The diaphragm is tightly stretched to prevent it from resonating at any audible frequency.

The cross-sectional illustration and diagram in Fig. 17-37 show the elements of a condenser-microphone head, accompanied by a possible amplifier using a JFET. When sound waves strike the tightly stretched diaphragm, it vibrates, changing the capacitance between front and back plates. This capacitance changes in accordance with the frequency and amplitude of the sound waves. In this way, sound waves produce a variation of capacitance.

When the circuit is operating, electrons move down through the resistor R until there is a 200-V charge across the microphone. Then the current ceases to move, but an excess of electrons remains on the front plate and a deficiency on the back. If the charged capacitor suddenly changes to a greater capacitance by the plates being forced closer together, a current of electrons will flow through the resistor until the charge on the capacitor becomes 200 V again. This capacitor-charging current flowing through R produces a voltage-drop across it, the value being equal to $E = IR$. As the capacitance varies, because of diaphragm vibration, electrons are forced to flow back and forth through R and an ac emf is developed across the resistor. The emf will vary in frequency and amplitude in accordance with the sound waves striking the diaphragm. The condenser microphone has a sensitivity of about −90 dBm.

The case in which a microphone is placed plays an important part in its directional characteristics. If the diaphragm is pointed upward, it can receive signals equally well from all directions (omnidirectional). By mounting the diaphragm in a vertical direction and closing the back of the case, the microphone will pick up signals best from the front (unidirectional). By leaving both front and back of the diaphragm open (as in a ribbon microphone), the microphone receives signals approaching from both front and back (bidirectional).

Fig. 17-37   Cross section of a condenser microphone and the circuit in which it is used.

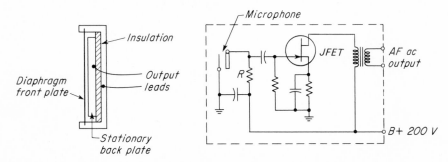

When two or more microphones are fed into the same amplifier, it is possible that signals from a sound source unequal in distance from the microphones may arrive at the microphones 180° out of phase, canceling each other. Therefore, it is important that microphones be phased properly (with polarity of microphone leads correct and placement of microphones proper) when they are used to pick up large orchestras, etc.

17-32 TUNING AN A3 TRANSMITTER

A simple but workable radiotelephone transmitter is shown in Fig. 17-38. A short description of the operation of each stage is given. At the start, the filaments of all tubes are on, but all plate voltage supplies are off.

The crystal-oscillator power supply is turned on. The plate-tank circuit of this stage is tuned to a plate-current dip, as indicated by meter M_1. The grid-current meter M_2 will usually peak when the M_1 reading is minimum. If the M_2 reading is too low, the capacitance of coupling capacitor C_1 is increased.

The buffer is neutralized, and then its power supply is turned on. The buffer-plate tank is tuned to a dip on meter M_3. The grid tank LC_3 is tuned to maximum grid current on meter M_4. If the grid current is not the desired value, the link coupling to LC_3 can be tightened or loosened.

The modulated-amplifier power supply is set for low-voltage output and turned on. The final plate tank is tuned to a dip on M_5.

The antenna is loosely coupled to the final-tank

Fig. 17-38 Schematic diagram of a radiotelephone transmitter and meters that might be used to indicate its operation.

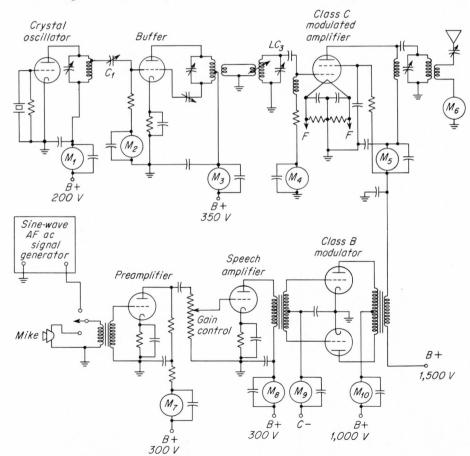

circuit and tuned to resonance, indicated by a peak reading of the RF current meter M_6, or a peaking of the plate current on meter M_5. The modulated amplifier-plate voltage is raised to the operating value, and the coupling is increased until the desired antenna current is produced or until the desired plate-current value is shown by M_5.

The *preamplifier stage* is turned on and a sine-wave signal is fed to the input transformer of the preamplifier. The signal-generator output is increased until a change of plate current is noted in meter M_7 and is then decreased slightly. The class A preamplifier is probably not distorting at this signal-input level.

The *speech-amplifier* gain control is turned down, and the stage is turned on. The gain control can be increased up to the point where plate current begins to change, as shown by meter M_8. This is as high as the gain control may be turned with the signal generator set at its present level.

The bias-voltage value of the modulator stage should be checked. If the modulator is to be operated as push-pull class A, there should be no indication of grid current in meter M_9. It may be necessary to reduce the gain control to a point where no grid current flows. If the modulator is biased for operation as class AB_2 or class B, grid current should flow with the gain control set to its maximum allowable point determined previously.

With the gain control turned down, the modulator is turned on. If the modulators are biased to class A, the plate current in meter M_{10} should read the value indicated in the tube manufacturer's data sheet for the type of tubes used in the modulator stage. If the modulators are biased to class AB_2, there will be considerably less *static* (no input signal) plate current. If they are biased to class B, there will be very little static plate current.

When the gain control is turned up, the RF ammeter should rise and all other meters, except possibly the modulator-stage grid and plate meters, should remain steady. If the modulator is biased to class A, neither of its meters should move. However, if it is biased to class AB_2 or B, at a low gain control setting, no grid current will flow but plate current will increase with modula-

tion. At a high gain-control setting, both the plate current and the grid current will rise.

The average plate current of the modulated amplifier should remain constant, with or without modulation. Any material change of meter M_5 indicates the presence of carrier shift and distortion. Actually, at high percentages of modulation this current value will usually drop a little, but not more than 1 or 2%.

If an oscilloscope is available, a check on percent of modulation and linearity of the modulated waveform should now be made.

The microphone can be switched on and spoken into. The antenna current should rise slightly. It will probably be necessary to raise the gain-control setting to produce a high percent of modulation because of the low-amplitude output of microphones. In fact, a second preamplifier or speech-amplifier stage would undoubtedly be needed to bring the weak output of the microphone up to the required level to modulate the transmitter completely. The percent of modulation of speech is best checked with an oscilloscope.

If the modulation transformer suddenly develops one or more shorted turns in either the primary or the secondary, all meter readings will be normal with no modulation. When modulation is applied, a heavy current will be induced in the shorted part, lowering the impedance of the primary, increasing the modulator plate current, and distorting the audio signal. The indication in meter M_{10} will increase excessively, the percent of modulation will drop off, negative carrier shift will be produced, and the antenna current will not increase as much as normally. The shorted transformer may buzz audibly, may become excessively warm, may smoke after a short time, and may burn up. The plates of the modulator tubes may become hotter than normal.

With the modulator stage biased to class B, the plate-current indication varies directly with the signal output of the stage. Meter M_{10} can be used as a relative indication of the percent of modulation. A small plate-current increase indicates a low percent of modulation; a large plate-current increase indicates a high percent.

When tuning a transmitter, keep in mind that the equipment contains lethal voltages. As a general rule, while tuning, keep one hand in a pocket. Do not touch the microphone. Touching the antenna leads on high-powered transmitters can result in severe burns or worse. Most commercial transmitters have built-in safety devices to protect personnel and equipment, such as door interlocks that disconnect high-voltage supplies if the doors of the transmitter are opened, and overload relays that automatically shut down the equipment if excessive current flows. In some cases lights flash or bells ring if trouble occurs in any circuit. Relays can be set to sound an alarm if either more or less than normal current flows.

If the operator knows the circuits in his equipment, he can tell by the various meter readings if and where any trouble occurs. A transmitter should require only periodic retuning unless it is repaired or retubed.

17-33 MODULATING TRANSISTOR AMPLIFIERS

A BJT transistor stage can be modulated in very much the same way as a VT triode. Figure 17-39

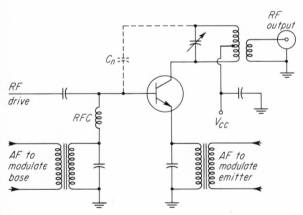

Fig. 17-39 Where AF might be connected to base and emitter to modulate a BJT stage.

illustrates how the modulating AF might be coupled to the base circuit to produce base modulation, as well as how AF ac might be fed into the emitter circuit to develop emitter modulation.

To produce 100% collector modulation of a transistor, it is usually necessary to modulate the driver and modulated amplifier collectors simultaneously, as in Fig. 17-40.

A common form of output coupling used with

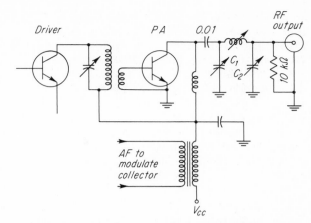

Fig. 17-40 Collector modulation of a BJT for high percent modulation.

transistors (as well as VT) stages is the pi-network circuit. The 10-kΩ resistor is intended only to drain off any dc static charge that might build up on the antenna; an RFC might be used instead. The first capacitor, C_1, is basically to tune the LC circuit, while C_2 is adjusted to match the impedance of the LC circuit in parallel with the BJT to the antenna impedance. Maximum RF output occurs when these two parameters are matched.

17-34 SINGLE-SIDEBAND RADIOTELEPHONE (A3J)

The information to this point has dealt with amplitude modulation of a carrier which results in a carrier and two sidebands (A3). This emission is used in standard broadcast and shortwave international broadcast stations, aircraft, some industrial communications, some citizens band, and some amateur equipment. The picture information of television transmissions is also amplitude-modulated, but it is designated A5.

Originally, all radiotelephonic communications used A3. Many short-distance communication systems, such as police and fire, have converted to frequency modulation (FM or F3), particularly with frequencies above 27 MHz. Below this, most radiotelephonic transmissions are now being made with single-sideband suppressed-carrier (SSSC, SSB, or A3J) emissions. Two immediately obvious advantages of SSSC over A3 are (1) power is saved by not transmitting a carrier and (2) with only one sideband the required bandwidth is halved, allowing twice as many stations to use the available spectrum space. As an example, a 1-kW A3 carrier when 100% modulated emits 1.5 kW of RF. If the carrier is deleted, the same information can be transmitted with 500 W of sidebands, a saving of 67% in power. Since each SB carries the same information, deleting one of them is a further saving (84%), as well as requiring half the spectrum space. A 110-W SSB transmitter will do essentially as well for voice communication as a 1,000-W A3 transmitter will. Why are all A3 transmissions not converted to A3J? The difficulty lies mostly in the complexity of the required receivers (Chap. 18).

Fading signals have always been a problem in long-distance radio communications. Not only do A3 signals fade up and down in strength, but the carrier may fade at different rates and times from either sideband. This results in varying signal strengths and a characteristic rolling distortion effect. The result of this may be voice transmissions that are completely unintelligible at times, regardless of how loud they are. The difficulty is due to carrier and sidebands fading at different rates. This can be eliminated by balancing out the carrier at the transmitter, filtering out one sideband, and transmitting only the remaining sideband. Figure 17-41 shows (a) a normal double-sideband A3 signal. (b) the same emission with one sideband filtered out (A3H), and (c) the same signal with one sideband and the carrier removed (A3J). If it is desired to transmit a pilot carrier for receivers to lock onto, instead of canceling the carrier it may be reduced to about 10% of its normal value. This is called an A3A emission.

A3J signals produce a muffled noise in the usual A3 receiver and cannot be understood because there is no carrier for the sideband signals to beat against to produce audible heterodyne, or beat, frequencies in the receiver. At the receiver it is possible to use an oscillator adjusted to the frequency that the carrier would have had if it had been transmitted. By feeding this frequency into the receiver at the same time as the sidebands are received, the sideband signals can beat against a correct-frequency (local) carrier to produce the audio tones of the original intelligence or sounds. Substituting the nonfading carrier signal at the receiver represents quite a saving for the transmitter.

Voice transmissions on A3J may fade up and down in strength, and at different instants high or low audio frequencies may predominate, but the rolling distortion fade of distant signals is eliminated.

Test your understanding; answer these checkup questions.

1. An antenna ammeter reads 6 A with carrier alone. What should it read with 100% sine A3 modulation? _____ 50%? _____
2. If an antenna ammeter decreases with A3 modulation, what is indicated? _____
3. What types of microphones have permanent magnets in them? _____ _____ _____ What types do not? _____ _____ _____
4. What types of microphones have high Z output? _____ _____ _____ Low Z output? _____ _____ _____
5. What microphone has lowest power output? _____
6. In what order would a three-stage transmitter be tuned? _____
7. What is a preamplifier? _____
8. What two meters in Fig. 17-38 might give indications that modulation was in progress? _____ _____
9. How does basic plate modulation differ from basic collector modulation? _____

Fig. 17-41 (a) Carrier and sidebands produced by multitone A3 modulation. (b) Single-sideband A3H with full carrier. (c) Single-sideband A3J with suppressed carrier.

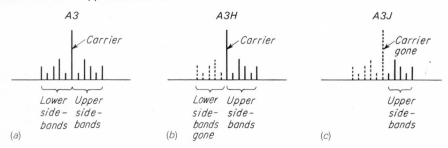

10. Would series modulation work with transistors? _____

11. What are two obvious advantages of A3J over A3? _____ _____

12. Why do fading A3J signals distort less than fading A3 signals? _____

13. What are the letter-number designations of carrier and 1 SB? _____ 1 SB and no carrier? _____

14. What are five uses of A3-type transmissions now? _____ _____ _____ _____ _____

15. About how many decibels of transmitting power are gained by using A3J over A3? _____

17-35 FILTER-TYPE SSB TRANSMITTERS

Figure 17-42 is a block diagram of a possible single-sideband transmitter to operate on a (suppressed) carrier frequency of 8,000 kHz. A 3,000-kHz crystal oscillator provides the RF carrier to be modulated. The microphone and speech amplifier provide the AF modulating signal (200 to 3,000 Hz). The RF and AF are mixed in a special circuit, called a *balanced modulator* (Sec. 17-36), which amplitude-modulates the 3-MHz RF, producing two sets of normal AM sidebands, but the balanced modulator circuit cancels or balances out the carrier frequency signal completely. The output of the balanced modulator in this case is a set of upper sidebands consisting of signals from 3,000,200 to 3,003,000 Hz, and a set of lower sidebands from 2,999,800 to 2,997,000 Hz.

In the diagram the switches are in the USB position. Only upper-sideband frequencies can pass through this filter. (LSB signals should be attenuated 50 dB or more.) Thus, no carrier and only upper-sideband signals are fed to the mixer. At the same time, another crystal oscillator operating at 5 MHz is fed to the mixer stage. The sum frequency of 3 + 5 MHz is the 8-MHz carrier frequency to be transmitted. Actually, the 5-MHz oscillator is providing a carrier and the 3-MHz sidebands modulate (mix with) this carrier frequency, producing an output that includes the 5-MHz carrier, USBs 3 MHz above 5 MHz (8 MHz), and LSBs 3 MHz below 5 MHz (2 MHz). The tuned output circuits of the mixer are selective enough to act as a bandpass filter to prevent the undesired products of mixing (the 5-MHz carrier and the 2-MHz signal) from being fed to the linear amplifier. Since the 3-MHz carrier was canceled by the balanced modulator, there can be no 8-MHz output from the transmitter without modulation, but when modulation is being produced, upper-sideband information mixes with the 5-MHz oscillator output to produce A3J USB signals having an 8-MHz carrier as their base.

If the filter switch is moved to the LSB position, only the lower-sideband information will pass to the mixer and only a LSB A3J emission will result, still based on an 8-MHz carrier frequency.

By changing the 5-MHz crystal to 4 MHz, the output frequency of the transmitter would be 3 + 4 = 7 MHz (or 4 − 3 = 1 MHz). Only the output circuits of the mixer and linear amplifier would have to be tuned to the new frequency.

Fig. 17-42 Block diagram of an SSB transmitter using 3-MHz filters.

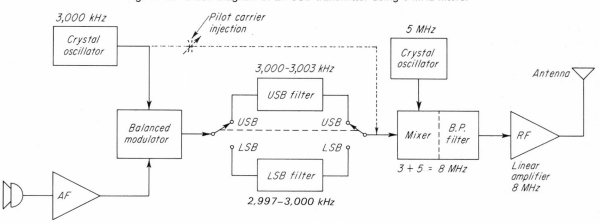

Either USB or LSB signals at any desired carrier frequency could be produced by this system with the proper mixer crystal. By using a VFO in place of the mixer-crystal oscillator, transmission can be made on a band of frequencies.

In commercial equipment, a *vestigial,* or *pilot,* carrier may be transmitted (A3A). Although the carrier is canceled at the balanced modulator, a small amount of RF ac from the carrier oscillator can be fed into the mixer input (dashed lines), equivalent to about 10% modulation. At the receiver the pilot carrier is used to lock in the receiver's carrier oscillator on the exact frequency required to produce correct-frequency sideband mixing.

Before high-frequency crystal filters were generally available, SSB transmitters used LC filters. To obtain the desired steep-skirt falloff and a narrow bandpass of 3 kHz, it was necessary to do the filtering at low radio frequencies, in the range of 20 to 100 kHz, in order to obtain coils with high enough *Q* values. Figure 17-43 illustrates in block diagram form an LC-filter system that might be used to develop an 8-MHz SSB emission. The 50-kHz crystal oscillator and the AF mix in the balanced modulator, developing no 50-kHz carrier, but both USB and LSB information. By switching to the desired filter either USBs or LSBs are fed to the first mixer. Assuming the use of the USB filter, USB signals mixing with the 500-kHz crystal oscillator results in an output of a 500-kHz carrier and two sidebands 550 to 553 kHz from it in each direction. The tuned circuits in the output of the first mixer are made selective enough that, when tuned to the upper 550–553-kHz sidebands, the carrier and the lower set of sidebands are not fed to the second mixer. The 3-MHz crystal oscillator is modulated by, or mixes with, the 550–553-kHz sidebands, producing 3,550–3,553-kHz sidebands. Again, the tuned output circuits of the mixer stage are selective enough to prevent the other products of this mixing, the carrier and a 2,550–2,547-kHz set of sidebands, from being fed to the third mixer. The output of the second mixer is an SSSC emission capable of either USB or LSB transmission based on a carrier of 3.55 MHz. From this 3.55-MHz SSB emission the third mixer oscillator is chosen to translate to any desired frequency. To transmit on 8 MHz a 4.45-MHz (or 11.55-MHz) crystal would be employed. Use of a VFO would allow trans-

ANSWERS TO CHECKUP QUIZ ON PAGE 416

1. (*7.35 A*)(*6.35 A*) 2. (*Negative carrier shift*) 3. (*Dynamic, ribbon*) (*Carbon, crystal, condenser*) 4. (*Crystal, condenser*)(*Dynamic, ribbon, carbon*) 5. (*Condenser*) 6. (*Oscillator, driver, PA*) 7. (*AF amplifier ahead of gain control*) 8. (M_6, M_{10}) 9. (*Driver must be modulated too*) 10. (*Yes*) 11. (*Power saving, less spectrum*) 12. (*SB not beating against fading carrier*) 13. (*A3H*)(*A3J*) 14. (*BC, CB, amateur, aircraft, industrial, TV picture*) 15. (*About 9 dB, from 84% saving*)

Fig. 17-43 Filter-type SSB system using LC filters.

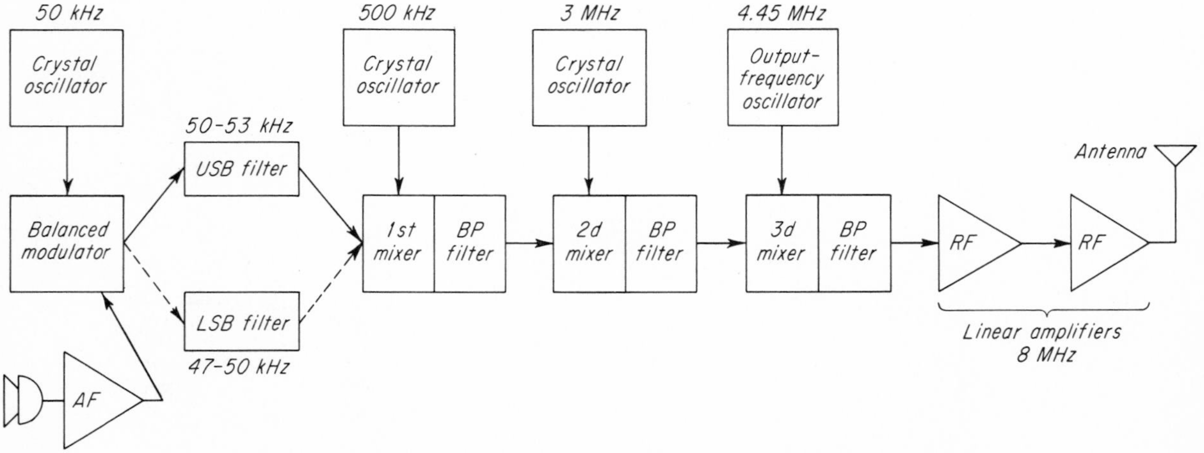

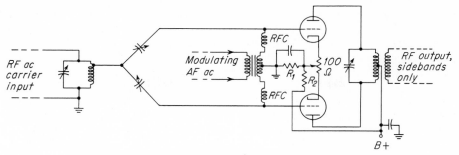

Fig. 17-44 A balanced-modulator circuit using triodes.

mission on a band of frequencies. These systems only translate the same 3-kHz sidebands from a low frequency up to a high frequency where they are transmitted.

17-36 BALANCED MODULATORS

Figure 17-44 illustrates a possible two-triode balanced modulator. In this circuit the RF is fed to the two grids in parallel and the AF modulating voltage (control-grid modulation in this case) is fed to the two grids in push-pull. The RF carrier frequency balances out in the plate circuit because both grids are driven RF-positive at the same time, producing equal and opposite current flows in the plate coil. Since the tuned circuit has zero impedance to AF, there can be no audio output in the plate circuit either. The results of modulation, namely, the upper and lower sidebands, are not canceled, however. The output from the tuned circuit consists of two sets of

sidebands only, provided the circuit is properly balanced. Balancing may be accomplished by adjustment of the two variable capacitors feeding the grids until the carrier output (as seen on an oscilloscope) becomes zero. Balance may also be accomplished by adjusting the 100-Ω potentiometer in the cathode circuit. Adjusting the center tap on the plate tank circuit will also balance out the carrier. Selection of R_1 and R_2 values determines the class of the stage, class A, AB, B, or C.

Balanced-modulator circuits can be developed with transistors and solid-state diodes. Transistor circuits would be similar to triode circuits. Figure 17-45 shows three diode-type balanced modulators. The first, a shunt type, requires closely balanced diodes to cancel the carrier adequately. The others have potentiometers to balance out the carrier. The second is called a ring-type balanced modulator, and the third is a modified ring. In each case, AF is fed across the mixer or modulator circuit and the RF carrier is connected, not 180°

Fig. 17-45 Diode-type balanced modulators: (*a*) shunt type, (*b*) ring type, and (*c*) modified ring.

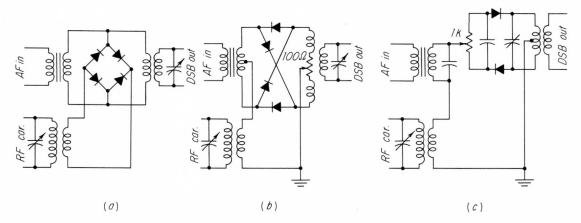

(*a*) (*b*) (*c*)

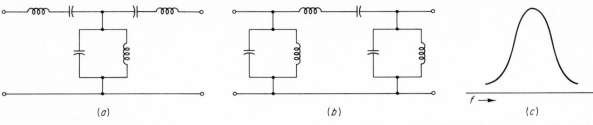

Fig. 17-46   Constant-*k* bandpass filters: (*a*) T type and (*b*) π type. (*c*) Possible passband curve (less steep on the high-frequency side).

out of phase, but in what appears to be *quadrature* (90°). This results in cancellation of the carrier and a double-sideband (DSB) output.

17-37 SIDEBAND FILTERS

Bandpass filters were described briefly in Sec. 8-8. The circuit shown in Fig. 17-46*a* is called a T-type, constant-*k* bandpass filter. It consists of two series-resonant arms of the T with a shunt-resonant circuit across the line. All three circuits attempt to pass along the frequency to which they are resonant. Figure 17-46*b* is a π type, also made up of resonant circuits. These filters are rather simple, and the skirts drop off rather slowly. They are said to have a large shape factor (Sec. 8-8). A 100-*Q* constant-*k* filter may have a shape factor of 15:1. By using cascaded (a series of) sections, the shape factor becomes less (steeper skirts) but the insertion loss (losses from input to output in dB) increases.

Filters with better shape factors can be produced by using a series-resonant circuit across the line or a parallel-resonant circuit in series with the line to provide frequencies of infinite attenuation. These are *m*-derived filters. They may have steeper skirts, but the response pops up again past the point of infinite attenuation. Figure 17-47 is a series-type, *m*-derived bandpass filter with the series arms passing the desired frequencies and the two midsection circuits resonant to frequencies just past the desired passband frequencies. Note the pop-up past the points of infinite attenuation.

LC filters become too broad to be useful as SSB filters at higher frequencies because of the relatively low *Q* of the coils. Piezoelectric crystals have *Q* values in the thousands, even up in the 10-MHz range. When used as elements in a

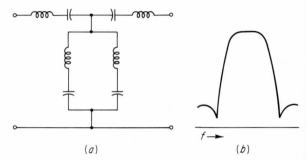

Fig. 17-47   (*a*) *M*-derived T-type bandpass filter. (*b*) Possible bandpass curve showing pop-up past points of coattenuation.

filter, they can produce narrow passbands and steep skirts with good shape factors, but they will have some pop-up also. Figure 17-48*a* is a half-lattice crystal filter which uses two crystals of only 2.8-kHz frequency difference for a 3-kHz passband. It may have a shape factor of 4:1. By

Fig. 17-48   (*a*) Half-lattice crystal filter. (*b*) Full-lattice circuit.

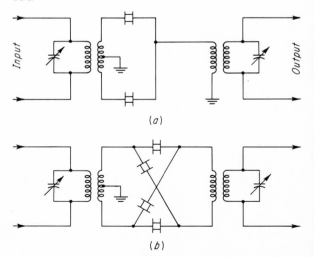

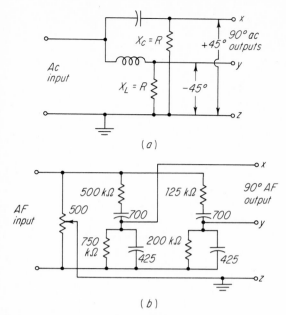

(a)

(b)

Fig. 17-49 Two 90° phase-shift networks. (a) Simple network used for a single-frequency RF. (b) AF network for a frequency band from 300 to 3,000 Hz (approximate values).

using a full-lattice, four-crystal circuit, shown in Fig. 17-48b, a shape factor of 3:1 may be attained.

Mechanical filters, consisting of tiny thin, machined nickel disks mechanically coupled with wires welded to their edges and with electromechanical coupling devices at both ends, can produce 3-kHz filters in the 100- to 500-kHz range with shape factors of 2:1.

17-38 A PHASE-TYPE SSB TRANSMITTER

An entirely different method of producing an A3J emission employs two 90° phasing networks. A simple 90° phase-splitting network is shown in Fig. 17-49a. With the reactances equal to the resistances, the current will lag 45° in the inductive circuit and will lead 45° in the capacitive circuit, totaling 90°. The same ac is available as two separate signals (X to Z and Y to Z) out of time by 90°. However, this phase relationship holds true for only a single frequency. It can be used to shift the frequency of an RF carrier 90° by selecting and tuning the components.

A practical AF phase network, using only resistors and capacitors but capable of holding a constant 90° phase difference from 300 to 3,000 Hz, is shown in Fig. 17-49b.

In the block diagram of Fig. 17-50, AF from a microphone is amplified and fed into the 90° AF phase-shifting network and then to two balanced modulators. An RF carrier is shifted in phase 90° and is also fed into the same two balanced modulators. The carrier frequency is completely canceled, but each modulator has upper- and lower-sideband signals developed. These are in such phase that, when added in a common circuit, the upper-sideband signals, for example, are in phase but the other sideband signals are now 180° out of phase (+90° and −90°) and cancel. This leaves only a single-sideband signal to be amplified by the linear amplifier.

By reversing either the AF or the RF leads to the balanced modulators, the opposite sidebands

Fig. 17-50 Block diagram of a phasing system that produces a single-sideband suppressed-carrier A3J emission.

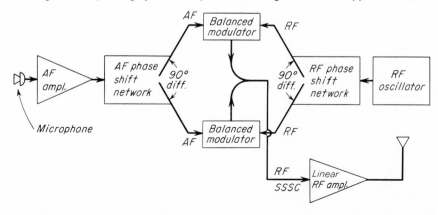

will add and cancel, providing a simple method of switching sidebands.

The phasing type of system has the advantage of not requiring any heterodyning to operate on any one given high frequency. Its carrier frequency may be the frequency generated by the RF oscillator, or frequency multipliers may be used between the oscillator and the RF phase-shifting circuits. The RF phase-shifting network will, however, have to be changed if the frequency of operation is changed.

It is not very difficult to produce audio phase-shifting networks that will hold close enough to 90° phase difference over a range of 300 to 3,000 Hz, which is the voice-frequency range. It becomes difficult to extend further the AF range and maintain a precise 90° phase difference to produce proper balancing and cancellation of one sideband.

17-39 LINEAR AMPLIFIERS FOR A3J

The SSB signal may be generated and converted to the desired transmission frequency at a level well under 1 W. At the transmission frequency it must then be amplified without distortion to the desired power level. Power pentodes, tetrodes, and triodes are used as high-power linear amplifiers in either grounded-cathode or grounded-grid circuits.

A3 transmissions are rated at either RF carrier-power output (operating power) or dc plate-power input to the final amplifier. A 1-kW dc input to an A3 modulated amplifier can produce about 700 W of carrier output with plate modulation, about 350 W if being used as a linear amplifier with an A3 signal input, and about 250 W if grid-modulated.

Linear amplifiers for A3J emissions are essentially the same as those used to amplify A3 emissions, except that with SSB there is zero watts output when there is no modulation. The output power will be proportional to the sideband signals applied to the grid of the amplifier tube. If biased to class B, there will be almost no plate current with no SSB drive, and plate current will increase according to the amplitude of the SSB excitation. With class AB_2, there will be more resting current and somewhat less average plate current variation

when SSB excitation is applied. In class AB_1 there will be still more resting current and little average plate current increase when excitation is applied. With class A operation the resting current will be high and the average I_p will not vary with SB excitation.

If the $E_p I_p$ meter readings of the usual class AB_2 type linear RF amplifier indicate 1,000-W dc input with modulation produced by a single sinusoidal AF tone (assume 2 kHz), the amplifier should produce a single, constant-amplitude RF sideband 2 kHz from the carrier frequency. The sideband should have a constant-amplitude peak envelope power (PEP) of RF output equal to 700 W (assuming 70% amplifier efficiency). With no modulation the $E_p I_p$ values may indicate 100 W or so of dc input, but the RF output will be zero.

When the same amplifier is voice-modulated to an apparent 1-kW dc input (difficult to determine because of bouncing meter pointers), the actual peaks of the input PEP may be two or more times the meter-indicated value and the output PEP may be greater than 2×700 W, although this depends on the characteristics of the voice doing the modulating.

To determine the voice PEP output values, it is possible to single-tone-modulate to a given visual value on an oscilloscope. When voice peaks are produced, they can be compared with the single-tone peak value. If they are twice the single-tone value, they represent twice the peak voltage, or a 4-times peak power. An increase of 1.4 times indicates a 2-times peak power.

The presentation of single-tone A3J modulation on an oscilloscope using a sampling of antenna RF to the vertical plates and a low-frequency horizontal sawtooth sweep (about 30 Hz) is a horizontal band, as shown in Fig. 17-51a. Proper

Fig. 17-51 (a) Single-tone SSB emission as presented on a scope. (b) Voice modulation. (c) Overdriven voice modulation.

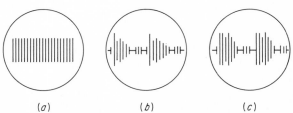

(a)   (b)   (c)

voice modulation shows clear, sharp peaks, as in Fig. 17-51*b*. Overdrive flattens the peaks and indicates severe distortion, as in Fig. 17-51*c*. Underdrive produces a voice pattern that looks normal but is reduced in amplitude.

A preferred presentation is a two-tone test. This is produced by feeding two constant-frequency sinusoidal AF tones, perhaps 1,000 Hz apart, into the microphone input. If there is no distortion, the presentation should appear as in Fig. 17-52*a*. The

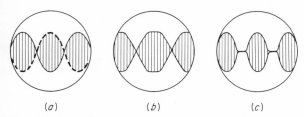

| (a) | (b) | (c) |

Fig. 17-52 (*a*) Two-tone modulation test signal. (*b*) Overdriven two-tone modulation. (*c*) Result of excessive bias on linear amplifier.

dashed outline should be sinusoidal. Assuming the SSB generator is operating properly, overdrive of the linear amplifier flattens the peaks, as in Fig. 17-52*b*. If the pattern separates at the middle as in Fig. 17-52*c*, it indicates the linear amplifier is operating in the class C region. Parasitics in the linear amplifier produce a fuzzy or ghostlike pattern. The degree of loading on the linear amplifier has a decided effect on the linearity of the output. Insufficient loading flattens the peaks, whereas excessive loading narrows the peaks too much and lowers the output peak values. With two-tone input the dc PEP is twice the $E_p I_p$ meter indication. With single tone the ratio is 1:1. With voice it is usually more than 2:1.

With commercial SSB equipment it is usually required that the carrier be reduced 40 dB below the maximum PEP value. Thus, a 2-kW PEP transmitter must reduce its carrier to less than 0.2 W. The highest audio frequency to be passed by the internal AF low-pass filter is 2.8 kHz. All limiting must occur before the filter to decrease radiation of distortion products.

It is interesting to note that the clamp-tube circuit, back in Fig. 16-20, can also be used as a linear amplifier for SSB. With no signal there is no drive to the tetrode or pentode amplifier and no grid-leak bias. The clamp tube clamps the screen voltage to nearly zero. With weak SSB signals applied to the amplifier grid, a little grid-leak bias is produced. This unclamps the clamp tube a little, and the amplifier begins to amplify the input. With strong input signals, the clamp tube is completely unclamped and the amplifier is free to amplify normally, using grid-leak bias. The clamp tube follows the SSB signal, acting as a *gating* circuit for the amplifier. This linear amplifier does not require well-regulated screen and bias supplies, unlike all other types of RF linear stages.

17-40 SPECIAL USES OF SSB

In some commercial sideband communications the voice is inverted or scrambled by heterodyning against some frequency such as 5 kHz. The result will be a 5-kHz carrier with USB (5 to 8 kHz) and LSB (2 to 5 kHz). Since LF AF becomes HF AF, the 2- to 5-kHz signals would be unintelligible to a listener. Such an inverted signal can be used as a scrambled modulating signal and be fed into either an A3 or an A3J transmitter.

One voice signal can be transmitted using the USB and a second voice signal can be transmitted simultaneously as the LSB of a single suppressed-carrier emission. This is called A3B, and it is a form of *multiplexing*. Either two separate receivers (one tuned for USB, the other for LSB) or a single receiver having two sideband filters, each feeding a separate detector and AF amplifier, can be used.

Multiplexing is the transmission of two or more voice channels on a single carrier ("mux"), or with telephone systems, into a single pair of wires or into a single coaxial transmission line. Multiplexing may be accomplished by *frequency division* (FDM) or by *time division* (TDM).

Telephone companies usually use FDM. One system results in a basic 12-channel group, each channel 4 kHz wide, starting at 60 kHz and ending at 108 kHz (a band of 48 kHz). Harmonics of a 4-kHz oscillator feed each channel as a carrier. Subscriber 1 is made to produce a filter-type LSB emission based on 64 kHz (16th harmonic of 4 kHz). Subscriber 2 produces LSB signals based on 68 kHz, and so on. All 12 LSB signals can be fed in parallel into a coaxial cable which connects to a distant point. At the receiving end 12 separate sideband filters feed separate detectors and AF amplifiers which produce the signals heard by receiving subscribers. A similar system is needed in the reverse direction for answering signals. To provide more channels on a single coaxial line, one whole 12-channel group can be translated up to 302–350 kHz (48 kHz), another to 350–398 kHz, and so on, until five 12-channel groups are stacked to form a 60-channel supergroup using the band of 302–552 kHz. Supergroups can be stacked into 600-channel mastergroups, and six stacked mastergroups provide 3,600 channels (jumbo group) into a single coaxial line. Instead of being fed into coaxial cables, mastergroups are often used to modulate microwave transmitters.

· A simple frequency-division radio-multiplexing system is illustrated in block form in Fig. 17-53. Voice 1 signals are fed through a 0- to 3-kHz LP filter. Voice 2 signals are first translated up 5 kHz by a balanced modulator and a 5- to 8-kHz USB filter. Both voice signals are then translated up 50 kHz by a balanced modulator and passed through a 50- to 58-kHz USB filter. Finally, this signal is translated up to the desired transmitting frequency. At the receiving end, voice 1 would be audible with any USB receiver (using a 0- to 3-kHz LP filter after the detector to remove the 5- to 8-kHz voice 2 signals). A 5- to 8-kHz filter passes voice 2 voltages to a mixer where they beat with a 5-kHz oscillator to reproduce the voice 2 information. Each channel requires its own AF amplifying system at the receiver. Instead of using voice signals, either channel might have been used for FSK teleprinter signals, for facsimile (Sec. 30-9), for binary computer data, and so on. This FDM type of emission is known as A9B.

Time-division multiplex is another means of transmitting two or more voice signals without requiring filters. TDM is a synchronized system, usually involving pulse-code modulation (PCM). If the electrical voice waveform is chopped at a rate equal to twice the highest expected voice frequency ($2 \times 4,000$ Hz = 8,000 pps) and these varying amplitude pulses are transmitted along a line, at the far end a listener will hear the 8,000/s varying amplitude pulses almost as the original voice sounded. If the chopping were done at twice this frequency, the received signals would only be slightly better. Furthermore, if chopping were at the 16,000-pps rate and only every other pulse were allowed to be transmitted, the reception would still be good. By chopping another voice channel at the same rate, but accurately sandwiching it in between the first voice pulses, at the receiving end either voice could be heard alone if the receiver circuit were opened and closed in synchronism with the desired channel. A chopping frequency of 32,000 pps would allow four voices to be transmitted with only slight degradation. One of the difficulties with this system is maintaining a sufficiently accurate clock pulse at transmitting and receiving ends to hold the channels in synchronism. Since the pulses are square wave, the bandwidth to transmit TDM is much

Fig. 17-53 Simple FDM system transmits two voice signals simultaneously.

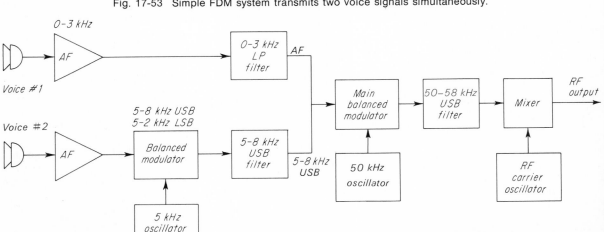

greater than with FDM. Rather than using true AF (analog) variations of the various pulses, the variations may be *quantisized* or converted into one of 128 steps or amplitudes. Each step has a binary (on-off, or 0-1) coded value which is the actual information transmitted. At the receiving end the binary code is decoded and the quantisized value is fed to the receiver, reproducing essentially the original voice sounds or tones.

A circuit used in many SSB systems is voice-operated transmission (VOX). A simplified VOX diagram is shown in Fig. 17-54. When the micro-

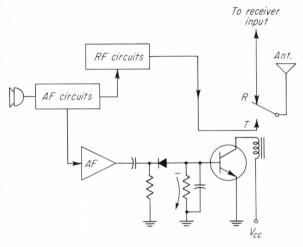

Fig. 17-54 Essentials of a VOX circuit.

phone is being spoken into, some of the AF ac is led off, amplified, and rectified and the resulting dc is used to bias ON a BJT (or VT). When collector current flows in the BJT circuit, its relay is energized and the antenna contacts are made to shift from receiver to transmitter. A second set of relay contacts (not shown) would be used to mute the receiver input or AF circuits while the transmitter is on. As soon as the speaking stops, the BJT becomes unbiased and the relay transfers the antenna back to the receiver. (See also Sec. 18-28.)

17-41 TRANSMITTER INTERMODULATION

A difficulty that often arises is the effect one transmitter has on another when the two are in close proximity. For example, transmitter A radiates a modulated signal from antenna A. A few

yards away transmitter B radiates a modulated signal from antenna B on some other frequency. If these two transmitters are on the air at the same time, the B signal may induce some of the B energy into the A antenna. The final-amplifier circuit of A now has both A and B energy mixing together, or *intermodulating*. The result is sum and difference frequencies of the B and A carriers, plus sum and difference frequencies of B and all harmonics of A. Many of these frequencies may be radiated by antenna A and cause interference to other services operating on these frequencies. The same type of intermodulation may be going on simultaneously in transmitter B.

This type of interference can be stopped if a wave trap (Sec. 8-8) tuned to the B frequency is installed in the connecting line between the A transmitter and its antenna. This prevents B energy from appearing and mixing in the nonlinear RF amplifier-plate circuit of A, and no intermodulation can occur. The antennas themselves may have both A and B energy in them, but antennas are linear *R, L,* and *C* circuits and cannot mix or beat two frequencies to produce sum and difference output. If there is a poor, or oxidized, connection in one of the antennas, however, this nonlinear joint can mix and produce beat output.

Test your understanding; answer these checkup questions.

1. What is the designation of SSB with no carrier? _____ With reduced carrier? _____ With full carrier? _____
2. What is the advantage of using crystal filters over LC and mechanical filters? _____
3. Why would it not be possible to use a doubler stage to change an SSB signal from a carrier frequency of 3 to 6 MHz? _____
4. What is the result if a balanced modulator is not perfectly balanced? _____
5. If two signals are in quadrature, what is their phase relation? _____
6. With what types of filters does a pop-up occur past the first skirt minimum? _____
7. Which is more desirable, a large or small shape factor? _____
8. In Fig. 17-42 how would it be possible to obtain both USB and LSB emissions on the same carrier frequency with only the USB filter being used? _____
9. What is an advantage in using a phase system for A3J? _____ A disadvantage? _____
10. In a filter system of A3J, how many circuits must

be balanced? _____ In a phase system?
_____

11. Why might it be difficult to produce music transmissions with phase-type A3J? _____

12. Why might linear amplifiers with A3 excitation be more stable than with A3J operation? _____

13. Why might two-tone tests be better than single-tone or voice if checking with a scope? _____

14. What is A3B emission? _____ A9B? _____ VOX? _____

15. How is transmitter intermodulation stopped or decreased? _____

COMMERCIAL LICENSE QUESTIONS

Amateur license questions will be found in the Addendum.

FCC Elements 3, 4, and 6 require an ability to answer questions similar to those below. Sections in which questions are answered are shown in parentheses. A question followed by a bracketed number is required for that element alone.

1. What is meant by the emission designations *A3, A3J, A3A, F3, F5,* and *P3D?* (17-1, 17-34, 17-35, Appendix E)

2. What is a *carrier frequency?* (17-1)

3. Explain how radio signals are transmitted by using amplitude modulation. (17-1)

4. What causes sound, and how is sound transmitted in air? (17-3)

5. What form of energy is contained in a sound wave? (17-3)

6. What characteristic determines the pitch of a sound? (17-3)

7. Define the term *decibel.* (17-3)

8. What was the first microphone that is still in use? (17-4)

9. Sketch the physical construction of a single-button microphone. List advantages and disadvantages. (17-4) [4]

10. Draw a diagram of a single-button-microphone circuit. (17-4)

11. What may cause packing of granules in a single-button microphone? (17-4)

12. What is meant by *dynamic instability* with reference to a modulated RF emission? (17-6)

13. What might cause FM in an AM radiotelephone transmitter? (17-6)

14. Draw a diagram of a microphone with two stages of amplification. (17-7, 17-31)

15. What is the last AF amplifier stage which modulates the RF stage termed? (17-7)

16. What is a speech amplifier? (17-7)

17. What is the ratio of unmodulated carrier power to instantaneous peak power at 100% modulation? (17-9, 17-11)

18. What is the relationship between percentage of modulation and the shape of the waveform envelope relative to carrier amplitude? (17-10)

19. Why is a high percent of modulation desirable in AM transmitters? (17-10)

20. What undesirable effects result from overmodulation of an AM carrier? (17-10)

21. What are possible causes of and methods of reducing transmitter spurious emissions other than harmonics? (17-10, 17-40)

22. What is the relation between the amount of power in the sidebands and the intelligibility of the signal at the receiver? (17-10, 17-14)

23. What are the results of using an audio peak limiter on the output signal of an A3 transmitter? (17-10) [4]

24. What is meant by *plate modulation?* (17-11)

25. Draw a diagram showing plate modulation of an RF amplifier. (17-11)

26. Draw a diagram of a class B push-pull pentode AF amplifier. (17-11)

27. Draw a diagram and explain operation of a triode class C RF amplifier modulated by a push-pull modulator. (17-11, 17-31) [4]

28. What is the relation between the average ac power output of the modulator and the dc plate-circuit input of the modulated amplifier with 100% sinusoidal plate modulation? How does this differ when voice modulation is employed? (17-11)

29. Discuss the characteristics of a modulated class C amplifier. (17-11, 17-17) [4]

30. How is the load impedance of a modulator determined if it is modulating the plate circuit of a class C RF stage? (17-11) [4]

31. What percent increase in average RF output power is obtained under 100% sinusoidal modulation? (17-11)

32. If the percent of modulation is decreased from 100 to 50, by what percent is the power in the sidebands decreased? (17-11, 17-14)

33. The dc plate input to a modulated class C amplifier with an efficiency of 60% is 200 W. What value of sinusoidal AF power is required to produce (a) 100% modulation and (b) 50%? (17-11)

34. A class C amplifier with E_p of 1,000 V and I_p of 150 mA is modulated by a class A amplifier with E_p of 2,000 V, I_p of 200 mA, and plate impedance of 15,000 Ω. What is the proper turns ratio for the coupling transformer? (17-11)

35. A transmitter has an output of 100 W and an RF amplifier efficiency of 50%. If the modulator is 66% efficient, what dc plate input to the modulator is necessary for 100% sinusoidal modulation? (17-11)

36. Draw a diagram showing how to plate-modulate a tetrode or pentode RF amplifier. (17-12)

37. When a carrier is amplitude-modulated, what produces the sideband frequencies? (17-14)

38. If a 1,500-kHz radio wave is modulated by a 2,000-Hz sine-wave tone, what frequencies are contained in the modulated wave? (17-14)

39. In AM, what is the relation of sideband power, output carrier power, and percent of modulation? Give an example. (17-14)

40. What is meant by and what determines the *band-width* of an emission? (17-15)

41. What is the bandwidth of a transmitter using A2 emission, a modulating frequency of 800 Hz, and a carrier frequency of 500 kHz? Using A1? (17-15)

42. Why does exceeding 100% modulation in an AM transmission cause excessive bandwidth? (17-15)

43. Draw a diagram of a Heising modulation system capable of 100% modulation. (17-16)

44. What is the purpose of the plate choke in Heising modulation? (17-16)

45. Why is a series resistor used in the dc plate supply of a modulated RF amplifier, between amplifier and modulator, in Heising modulation? (17-16)

46. What is meant by *grid modulation?* (17-17)

47. How should the bias of a grid-modulated RF stage be adjusted? (17-17)

48. Does grid current flow in a grid-bias-modulated stage under modulated conditions? (17-17)

49. Draw a diagram of grid modulation of an RF amplifier. (17-17)

50. Compare advantages and disadvantages of grid modulation and plate modulation. (17-17)

51. In a properly adjusted grid-bias-modulated RF amplifier, under what circumstances will the plate dc meter current vary? (17-17)

52. Is the efficiency of a grid-bias-modulated stage maximum at 100% or zero modulation? (17-17)

53. Draw a diagram showing suppressor-grid modulation of an RF amplifier. (17-18)

54. What is meant by *low-level* and *high-level* modulation? (17-20)

55. Sketch oscilloscope displays for 0, 50, 100, and 120% amplitude modulation. (17-21)

56. How does a linear RF power amplifier differ from other types? (17-22)

57. Under what class of amplification are the vacuum tubes in a linear-amplifier stage, following a modulated stage, operated? (17-22)

58. Why are tubes used in linear RF amplifiers not normally biased to class A? (17-22)   [4]

59. Draw a diagram of a low-level plate-modulated pentode RF amplifier coupled to a push-pull linear RF amplifier and explain its operation. (17-22)

60. Doubling the excitation voltage of a class B linear RF amplifier gives what increase in RF power output? (17-22)

61. What do variations in the final-amplifier plate current of a transmitter employing low-level modulation indicate? (17-22)

62. How may distortion effects of class B operation of an RF amplifier be minimized? (17-22)

63. If a final RF amplifier operated as class B linear were excited to saturation with no modulation, what would be the effects when it was undergoing modulation? (17-22)

64. What may be the cause of a decrease in plate and antenna current during modulation if a class B linear RF amplifier is used? (17-22)

65. What is a Doherty amplifier and for what is it used? (17-24)   [4]

66. What is *carrier shift* or *asymmetrical modulation* and how is it measured? (17-25, 17-26)

67. What type of carrier shift might cause a dip in antenna current during modulation? (17-25)

68. What may cause positive carrier shift in a linear RF amplifier output? (17-26)

69. What is the effect of insufficient excitation to a modulated class C RF amplifier? (17-26)

70. What are causes of downward fluctuation of the antenna current when modulating? (17-26)

71. What are the effects of overexcitation of a class B RF amplifier-grid circuit? (17-26)

72. What are possible causes of negative carrier shift in a linear RF amplifier? (17-26)

73. If an amplifier is adjusted for maximum power output for telegraph operation, why must the E_p be reduced if amplitude-modulated? (17-27)

74. What percent of antenna-current increase would be expected between unmodulated conditions and 100% sinusoidal modulation? (17-28)

75. Sketch the physical construction of dynamic-type microphones. List some advantages, disadvantages, and uses. (17-29)   [4]

76. Sketch the physical construction of a crystal- or ceramic-type microphone. What are some advantages and disadvantages? (17-30)

77. What precaution should be observed when using and storing crystal microphones? (17-30)

78. What types of microphones have a high-impedance output? (17-30, 17-31)

79. Why are the diaphragms of some microphones stretched? (17-31)

80. What is meant by *unidirectional, bidirectional,* and *omnidirectional* microphones? (17-31)   [4]

81. What is meant by the *phasing* of microphones, and when is this necessary? (17-31)   [4]

82. Draw a diagram and explain operation of a beam-power tetrode class C amplifier coupled to a push-pull class B power amplifier, showing modulating-signal input, RF exciting-voltage input, and modulated output. (17-32)   [4]

83. Draw a circuit diagram of a complete radiotelephone transmitter showing microphone input, preamplifier, speech amplifier, class B modulator, crystal oscillator, buffer amplifier, class C modulated amplifier, and antenna output connection. Insert meters where necessary and explain how the transmitter is tuned. (17-32)   [4]

84. What might be the cause of variation in plate current of a class B type of modulator? (17-32)

85. Should the plate current of a modulated class C amplifier stage vary or remain constant under modulation conditions? (17-32)

86. What is a *preamplifier,* and where is it normally used in a broadcast station? (17-32)   [4]

87. What would be the effect of a shorted turn in a modulation transformer? (17-32)

88. What precautions should be observed in the adjustment of a radiotelephone transmitter? (17-32)

89. When should transmitters be retuned? (17-32)

90. During 100% modulation, what percent of the output power is in the sidebands in (*a*) an A3 emission and (*b*) an A3J emission? (17-34)

91. Explain the principles involved in a single-sideband suppressed-carrier (SSSC) emission. (17-34)

92. How does bandwidth of an SSSC (SSB) emission and required power compare with that of full carrier and sidebands? (17-34)

93. Draw a block diagram of an SSB transmitter (filter type) with a 50-kHz oscillator and emission frequencies in the range of 8 MHz. Explain the function of each stage. (17-35)

94. Explain briefly the principles involved in multiplexing. (17-40)

95. Define *transmitter intermodulation* and list the causes, effects, and steps that can reduce it. (17-41)

ANSWERS TO CHECKUP QUIZ ON PAGE 425

1. (*A3J*) (*A3A*) (*A3H*) 2. (*Sharp at HF*) 3. (*Doubles SSBs*) 4. (*Carrier transmitted*) 5. (*90°*) 6. (*m-derived and crystal*) 7. (*Small*) 8. (*Use carrier crystal at 5,003 and mixer crystal at 2,997 kHz*) 9. (*No beating required*) (*RF phase network must be tuned if frequency changed*) 10. (*1*) (*2*) 11. (*Difficult to produce 90° AF phase network to 5 or 10 kHz*) 12. (*Continual carrier holds them from oscillating on some other frequency*) 13. (*Steady modulation pattern can be evaluated*) 14. (*Two separate SSB emissions, one carrier*) (*Multiplexed signals*) (*Voice-operated transmission*) 15. (*Wave traps*)

18

AMPLITUDE-MODULATION RECEIVERS

The objective of this chapter is to discuss the various types of detectors used to demodulate amplitude-modulated signals and then to explain the TRF and superheterodyne systems used in AM receivers. The variations of system requirements to receive A1, A2, A3, and A3J emissions are developed in preparation for the FM and TV superheterodynes outlined in Chaps. 19 and 25. Some alignment and troubleshooting techniques are included. More sophisticated receivers are discussed in the chapters on radar, television, shipboard radio equipment, and frequency modulation.

18-1 RECEIVERS

A receiving system consists of an antenna, in which all passing radio waves induce an emf, a means of selecting the desired signal, a means of detecting, or *demodulating,* the modulation in the signal, and a means of making the detected electrical signal audible.

A wire may serve as the antenna. A particular signal can be selected by using a resonant circuit. A detector can change the modulated RF ac to an AF ac or varying dc. Earphones or a loudspeaker can produce an audible response when fed the AF.

It might be thought that earphones across a tuned circuit, as in Fig. 18-1, would operate as a receiver. If the resonant circuit is tuned to perhaps 1,000 kHz, a voltage of that frequency is being developed across the tuned circuit and a current of that frequency flows through the earphones. However, even if the earphone dia-

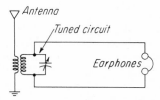

Fig. 18-1 Earphones alone across a tuned circuit cannot detect radio signals even though signals may be present in the LC circuit.

phragms could vibrate at that rate and produce airwaves at a frequency of 1,000,000 hertz, nothing would be heard. The ear is not sensitive to waves in air higher than about 20,000 Hz. If the carrier is modulated, the carrier frequency is still 1,000 kHz, the sidebands are a little higher and lower, and all are inaudible. To change the inaudible RF ac to an audible-frequency ac, a *detector* is needed.

18-2 DEMODULATING A MODULATED WAVE

If the received RF carrier is amplitude-modulated by voice or music, it is possible to detect the variation of the carrier amplitude with a simple half-wave rectifier and filter circuit, as shown in Fig. 18-2.

When the transformer secondary is tuned to the frequency of a local modulated transmitter, RF ac at this frequency is developed in it. The modulated RF is illustrated. The high-impedance earphones and the rectifier form a load on the tuned circuit. Because of the rectifier, only unidirectional pulses

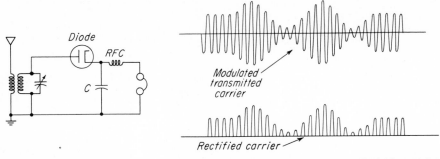

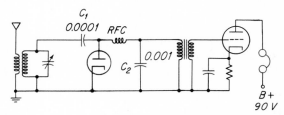

Fig. 18-2 A series-diode detector circuit, a modulated carrier wave, and the rectified (demodulated) carrier.

can flow through the earphones. The resultant pulsating dc is also shown.

With an unmodulated carrier, constant-amplitude dc pulses flow through the earphones. This pulls the diaphragm inward a little and holds it there. When the carrier is modulated, the average current of the pulses varies with the modulation and the earphone diaphragm is pulled farther inward or is released to move outward. This controlled vibration of the earphone diaphragm generates airwaves (sound).

The average amplitude of the sinusoidal half-wave pulses is equal to one-half of the average value (one-half of 0.636 maximum), or roughly one-third of the peak value. The average value can be increased by using a capacitive-input type LC filter circuit as shown in Fig. 18-2. A 0.0005- to 0.002-μF capacitor charges to the peak value and tends to hold this value until the next RF pulse arrives. As a result, the average current through the earphones is increased and a louder signal is heard.

If capacitor C is too large, it will not discharge rapidly enough and the waveform of a cycle of high-frequency modulation may drop and rise again before the capacitor has had time to discharge. As a result, current fed to the load will not vary at high-frequency audio rates and those frequencies will be lost. With correct capacitance, this detector can reproduce AM with reasonable fidelity. It is being used in almost all AM broadcast receivers.

In this detector there is no power supply. The incoming signal supplies all the power that appears in the earphones. Only relatively strong or local signals will be audible.

Fig. 18-3 A shunt-type diode detector with a stage of audio amplification.

The diode detector described is a series-type half-wave rectifier. A shunt-rectifier circuit as shown in Fig. 18-3 will demodulate equally well. In this circuit an RF emf from the tuned circuit is applied across the diode. When the plate is made positive, electrons flow and are stored on the right-hand plate of C_1. On the next half cycle the plate is driven negative and does not draw electrons. Those that had been drawn to it may now discharge through the RFC, charging capacitor C_2. This capacitor discharges through the primary of the audio transformer. If the carrier is modulated, the charge across capacitor C_2 varies with the amplitude of the modulation, producing a varying dc through the transformer primary and an ac in the secondary. The ac is amplified by the triode, and relatively loud signals are now audible in the earphones. The RFC is required in the shunt circuit to prevent the coupling capacitor C_1 and the filter capacitor C_2 from being in series across the tuned circuit and detuning it.

18-3 CRYSTAL DETECTORS

One of the first rectifying demodulators was the *crystal detector*. Rectifier crystals were metallic types, such as galena, iron pyrites, silicon, and carborundum (not to be confused with quartz and rochelle-salt crystals used in oscillator circuits and microphones).

The crystal was embedded in a lead mounting. A thin, pointed wire, called a *cat whisker,* was pressed against the surface of the crystal. When a sensitive spot was found, considerably more current flowed in one direction than in the opposite and rectification resulted. Crystal detectors were the first junction rectifiers. They were used in a manner similar to a diode tube, as shown in Fig. 18-4. If the diode and earphones are tapped

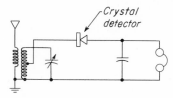

Fig. 18-4 A crystal detector.

part of the way down the tuned circuit, less loading of the tuned circuit results. The circuit has a higher Q, is more selective (has narrower bandwidth), and is more sensitive (produces more signal voltage from weak signals).

18-4 POWER DETECTORS

With rectifying detectors the power of the signal in the earphones is always somewhat less than the RF power picked up by the antenna wire. With a triode tube or transistor it is possible to obtain detection as well as amplification of the input signal.

One such circuit is the *plate,* or *power, detector* shown in Fig. 18-5. The rectification effect responsible for the demodulation of the signal takes place in the plate (or collector) circuit. It is called a power detector because with a high power-supply voltage and a relatively high-amplitude signal applied to the input circuit a relatively high output audio power can be produced.

The power detector can be considered a class B-biased RF input circuit and an AF amplifier output circuit. The vacuum-tube circuit grid is biased almost to cutoff, as illustrated in the $E_g I_p$ graph. With no signal input there is almost no plate current. When a carrier is received, the positive half cycles of the signal produce plate-current pulses and the negative half cycles produce no plate current. This is rectification in the plate circuit. When the carrier is amplitude-modulated, the pulses in the plate circuit vary in amplitude. The voltage across C varies with the modulation, and

Fig. 18-5 A battery-biased power-detector circuit and the rectification effect on the plate current. An NPN transistor circuit which would operate in a similar manner.

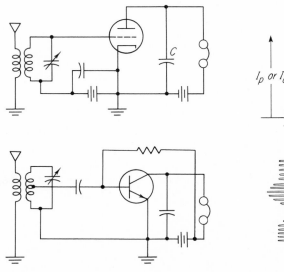
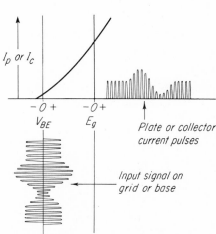

the current flowing through the earphones varies similarly.

The grid circuit of the power detector has relatively high bias, and grid current does not flow under normal signal conditions. As a result, the tuned circuit operates as though it had no load on it. With no load the Q can be quite high, allowing a relatively high signal voltage to be developed with a given input signal.

Instead of using a bias battery, a cathode resistor of about 20,000 to 50,000 Ω with a 1- or 2-μF capacitor across it, as in Fig. 18-6, will pro-

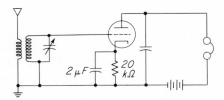

Fig. 18-6   A cathode-resistor-biased power detector.

duce the necessary bias for the stage. With a transistor a high resistance to V_{CC} produces the necessary forward bias to bring the stage to class B. Each positive half cycle of the signal forward-biases the base more and greater collector current flows.

18-5 LINEAR AND SQUARE-LAW DETECTORS

The graph in Fig. 18-7 illustrates (a) a plate current that is relatively linear and (b) a plate current that graphs according to the *square law*. In the linear case, each time the voltage is doubled, the plate current is doubled. In the square-law case, each time the voltage is doubled, the plate current increases four times, or as the square of the voltage increase.

A linear-detector circuit will produce a current waveshape essentially the same as the voltage waveshape. With a square-law rectifier the waveshape of the output current will differ from the input-voltage waveshape, and distortion is produced. The same signal input will produce considerably more output from a square-law detector.

A diode is considered a fairly linear device unless operated close to the zero current point by weak signals, where it rounds off in a more or less square-law manner.

18-6 GRID-LEAK DETECTORS

The grid-leak detector, shown in Fig. 18-8, was popular in the past because it had better sensitivity than a diode. It has a diode rectifier grid

Fig. 18-8   A grid-leak detector circuit.

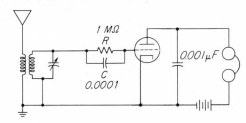

Fig. 18-7   Voltage versus current in (a) a linear rectifying circuit and (b) a square-law rectifying circuit.

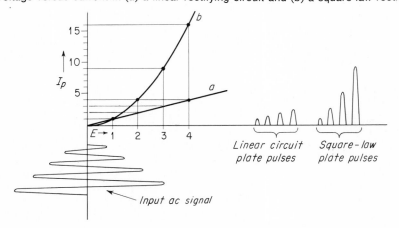

circuit and an AF amplifier plate circuit. The grid circuit consists of a tuned circuit (the RF ac source), a rectifier (the cathode and grid), and a resistance as a load with a filter capacitor C across it. Received signals produce a varying dc voltage across the grid-leak resistor R and the grid-leak capacitor C. Rectified current flows from grid, through the resistor, to cathode, making the grid end of R negative, biasing the triode. With no signal input, there is almost no grid current and almost no bias, plate current is relatively high, and therefore a low plate voltage (15 to 30 V) is used. With an unmodulated carrier, a bias voltage is developed and the plate current decreases. When the carrier is modulated, the bias varies with the modulation waveform, producing a relatively large plate-current variation in the earphones. As a result, the listener hears loud sounds.

18-7 REGENERATIVE AND AUTODYNE DETECTORS

To increase the sensitivity and selectivity of the grid-leak detector, a plate-circuit *tickler coil* can be used to regenerate, or couple, energy from the plate circuit into the grid circuit, as in Fig. 18-9.

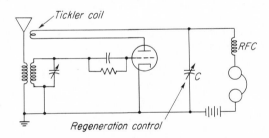

Fig. 18-9 An Armstrong-type regenerative detector.

The tuned circuit of a grid-leak detector should have a fairly high Q except that there are always losses in the LC circuit plus grid-leak resistor loading. When an RF signal voltage appears in the grid circuit, it is amplified and fed back *in phase* to the LC circuit by the tickler coil. Some of the signal energy lost due to grid-circuit loading is partially made up. The result is an apparent lessening of the losses and a higher Q. With no regeneration (in-phase feedback) the stage operates as a straight grid-leak detector. As regeneration is increased, the detection efficiency and the

signal output increase. If enough regeneration is present, all the losses in the grid circuit will be made up and the LC circuit will begin to oscillate at its natural frequency of resonance. The regenerative detector is now an Armstrong oscillator. The variable bypass capacitor C completes the ac plate circuit and controls regeneration.

The higher the resistance of the grid leak, the more sensitive and selective the detector. Maximum sensitivity and selectivity, for detection of A2- or A3-modulated signals, occur when the regeneration is just under the value required to produce oscillation. If the stage oscillates, it is no longer satisfactory as a detector for A3-modulated signals.

As the regeneration and the Q of the LC circuit increase, the coupling requirement from antenna to tuned circuit decreases markedly.

As soon as the regenerative detector breaks into oscillation, it becomes an *autodyne* (self-heterodyning) detector. In this condition it can be used to detect A1, A2, A3J, or F1 emissions. The basis of detection changes from rectification to heterodyne.

As discussed previously, when two different-frequency ac signals are mixed, or heterodyned, in a nonlinear circuit, four frequencies will appear: (1) the first frequency, (2) the second frequency, (3) the sum of the first and second frequencies, and (4) the difference between the first and second frequencies.

If an autodyne detector such as that of Fig. 18-9 is oscillating at 1,002,000 Hz and a signal is fed into it at 1,000,000 Hz, the difference between these two is 2,000 Hz. This 2,000-Hz frequency and the three other heterodyne signals appear in the plate circuit. Since the plate circuit has an RF filter composed of capacitor C and the RFC, the three higher-frequency signals do not pass through the plate-circuit load. Only the 2,000-Hz beat flows through the earphones. The beat tone can be changed by tuning the autodyne LC circuit. If the 1,000,000-Hz incoming signals are in dots and dashes, the audible beat will reproduce the dots and dashes as 2,000-Hz tone signals in the plate load. (Compare the theories of heterodyne detection and sideband generation in Sec. 17-14.)

The autodyne detector is most sensitive to weak A1 or A2 signals when oscillating weakly. Maximum beat response is produced when the two signals being heterodyned have the same amplitude. Since two separate frequencies operating in the same circuit tend to synchronize into one frequency of oscillation, a weakly oscillating LC circuit may be forced to synchronize with a strong incoming signal and no beat note will be produced. To prevent such *frequency pulling* it may be necessary to loosen the coupling to the LC circuit, or to increase regeneration until the autodyne circuit is in strong oscillation. To copy weak signals through strong signal interference, antenna coupling should be loosened, increasing the Q of the LC circuit, and the detector should be set into fairly strong oscillation to prevent frequency pulling. The degree of regeneration and oscillation can be controlled by varying the value of capacitance C, the degree of coupling between tickler and LC circuit, or the plate-supply-voltage value.

As regeneration is increased, a faint swishing sound marks the point of oscillation. If a regenerative detector is oscillating and it is tuned across the frequency of a station to which another nearby receiver is set, the detector will produce a whistle on the receiver. The ability of an autodyne to radiate a signal is one of its disadvantages. If coupled directly to an antenna, such a detector may radiate a receivable signal for miles.

Test your understanding; answer these checkup questions.

1. What is the circuit that changes modulated RF to AF called? _____
2. Name two transducers that change AF ac to sound waves. _____ _____
3. Why is a diode called an envelope-type detector? _____ Why might it also be considered a heterodyne type? _____
4. If the bypass capacitor across the earphones in a diode detector is too large, what is the result? _____
5. Where do crystal receivers obtain the power that vibrates the earphone diaphragms? _____ Where do power detectors? _____
6. Would shunt-type grid-leak bias be possible in a regenerative detector? _____
7. What kind of crystals were used in crystal detectors? _____ In oscillators? _____
8. Why is a VT power detector more selective than a BJT power detector or a diode detector? _____
9. What is a plate detector? _____

10. Which produces the louder signal, a linear or a square-law detector? _____ The more distorted signal?
11. If earphones were substituted for the grid-leak resistor in a grid-leak detector, why would modulation be audible? _____
12. Why does some regeneration in a grid-leak detector result in louder A3 signals? _____ Narrower bandwidth?
13. What happens if a regenerative detector starts oscillating while tuned to an A3 signal? _____ What detector is it now? _____
14. How can the beat tone in an oscillating regenerative detector be changed? _____ Why does it change if the signal fades? _____
15. How is a regenerative detector adjusted to receive weak A1 or A2 signals? _____ A3? _____ Why is it not satisfactory in detecting A3J signals?

18-8 THE SUPERREGENERATIVE DETECTOR

A demodulator used in the past in the VHF range is the superregenerative detector shown in Fig. 18-10. It is the most sensitive detector ever

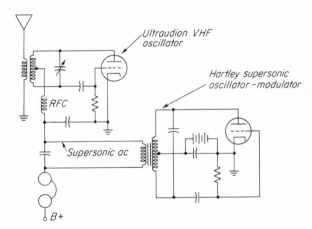

Fig. 18-10 An externally quenched superregenerative detector.

developed, detecting signals in the low-microvolt region. Basically, it is an autodyne detector with either its plate or its grid circuit overmodulated by a supersonic ac emf. As a result, the autodyne circuit is effectively turned on and quenched, possibly 20,000 or more times per second.

Each time the supersonic quenching voltage allows the ultraudion circuit to start oscillating, it produces a burst or pulse of RF ac at the very high frequency to which the LC circuit is tuned. With no signal being received, noise in the grid circuit of the VHF oscillator circuit allows random starting times for the VHF ac bursts and therefore random-length bursts. Random burst lengths result in a loud hissing noise in the output of the detector.

When a carrier signal is received, RF voltages from the carrier force all the bursts to start at a constant time interval, quieting the noise output. When modulation appears on the carrier, the variation of the *amplitude* of the signal voltage changes the *time* that it takes to start the VHF pulse, resulting in modulated burst widths and therefore a modulated average energy output. This is an application of *pulse-width demodulation*.

A relatively weak carrier signal will suppress the background noise, and its modulation will produce almost as much output-signal strength as the modulation of a strong carrier. High-amplitude impulse noises, usually very strong in the VHF range, are effectively limited in amplitude. They will not appear at all if they occur during an RF burst.

The superregenerative detector is extremely broad-tuning even with a relatively high-Q LC circuit. When coupled to an antenna, the super-regenerator radiates an equally broad signal. It is rarely used.

A simpler superregenerative detector is the self-quenching circuit shown in Fig. 18-11. The

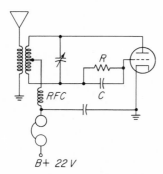

Fig. 18-11 A self-quenching superregenerative detector.

one tube oscillates at the very high RF signal frequency, while the grid-leak resistor and capacitor sets up a supersonic RC relaxation oscillation at the same time, which produces the quenching

effect. This type of superregeneration is often produced in RF amplifiers and is characterized by a wide band of spurious signals that it generates. Changing RC circuit constants usually cures the difficulty.

18-9 TUNED-RADIO-FREQUENCY RECEIVERS

The disadvantages of detectors when used alone as receivers are lack of sensitivity, selectivity, and output audio power.

Selectivity, the ability to tune out all frequencies except the one desired (measured in kilohertz of bandwidth), is readily obtainable by using a series of loosely coupled tuned circuits as in Fig. 18-12.

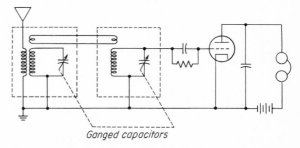

Fig. 18-12 One method of increasing selectivity is by cascading tuned circuits.

These circuits can be gang-tuned so they will tune to identical frequencies, or *track,* at all times.

In place of link coupling, as shown, a vacuum tube (or transistor) can be used to couple between two tuned circuits. These devices also amplify, so that any loss in the signal as it passes through the tuned circuits is more than made up by the amplification.

Figure 18-13 illustrates a block diagram of a three-tube *tuned-radio-frequency* amplifier (TRF) receiver. TRF receivers may use two or three RF amplifiers ahead of the detector. However, the

Fig. 18-13 A block diagram of a three-stage TRF receiver.

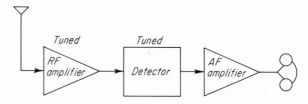

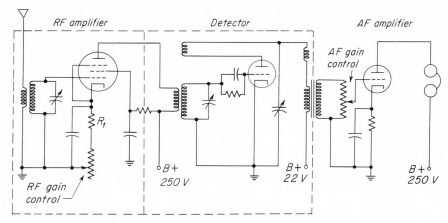

Fig. 18-14 Schematic diagram of a three-tube TRF receiver.

more stages, the greater difficulty experienced in making them track properly and in preventing them from breaking into oscillation. Only two or three audio stages are needed to bring the detector output up to loudspeaker volume.

The three-tube TRF receiver in Fig. 18-14 represents the basic requirements for a broadcast or shortwave receiver. It consists of a pentode RF amplifier stage with a variable bias resistor to control the gain or sensitivity. R_1 is the minimum value of resistance required to bias the tube to class A. Any additional resistance added in series with R_1 increases the bias and decreases the sensitivity of the stage. The detector stage is a regenerative type, permitting rectification detection for A2- and A3-modulated signals when not oscillating and autodyne detection for A1 or A2 radiotelegraphic signals or A3J signals when in oscillation. The audio stage is a standard type of volume-controlled amplifier.

ANSWERS TO CHECKUP QUIZ ON PAGE 434

1. (*Detector, demodulator*)　2. (*Earphones, loudspeaker*)　3. (*Envelope pattern produced on scope*) (*Carrier and sidebands mix in the nonlinear diode*)　4. (*Loss of high AF*)　5. (*From transmitter*) (*From its power supply*)　6. (*Yes, same output*)　7. (*Galena, iron pyrites, silicon, carborundum*) (*Quartz*)　8. (*LC circuit works into ∞-Ω load*)　9. (*Envelope rectification occurs in plate circuit*)　10. (*Square-law*) (*Square-law*)　11. (*Grid circuit would now be a diode detector*)　12. (*Raises Q of LC circuit*) (*Higher Q narrower BW*)　13. (*Beat tone usually heard, radiates a signal*) (*Autodyne*)　14. (*Tune LC circuit*) (*Oscillator frequency pulled by incoming carrier ac*)　15. (*Just in oscillator*) (*Just under oscillator*) (*SBs pull oscillator frequency*)

Note that A2 (tone-modulated radiotelegraphic code) can be received either as a modulated signal or by beating it with the detector in oscillation. Regardless of how any detector is adjusted, A2 signals can be detected and copied. For this reason SOS, or telegraphic distress, signals are transmitted with an A2 emission if possible.

The detector stage should be completely shielded in a grounded metal box to prevent signals from reaching it directly. It is desirable to shield RF amplifiers also. The dashed lines indicate shielding.

The use of a well-shielded and neutralized RF amplifier ahead of the autodyne or superregenerative detectors also prevents oscillations in the detector stage from reaching the antenna and being radiated. Transistorized TRF receivers use JFET or MOSFET RF amplifiers which may require neutralization.

TRF receivers are still used in simple low-frequency (±500 kHz) receivers. Above this both the sensitivity and selectivity fall off because of lowered tuned-circuit Q. Shipboard emergency receivers and auto-alarm receivers are examples of modern use of the TRF. There are some fixed-frequency microwave TRF receivers also.

18-10 THE SUPERHETERODYNE

Practically all radio receivers today are superheterodynes. While this receiver system is more complicated than the TRF, the ability to operate satisfactorily on any frequency with a constant value of selectivity and good sensitivity makes it highly desirable. A block diagram of a simple communications superheterodyne receiver capable of detecting A1, A2, A3, or A3J emission signals is shown in Fig. 18-15.

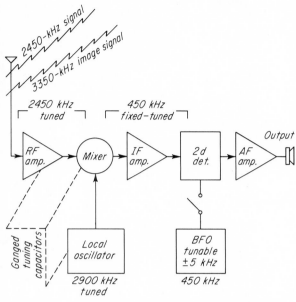

Fig. 18-15 Block diagram of a relatively simple short-wave-communications superheterodyne receiver.

The basic operation of the stages of the superheterodyne is as follows:

The receiver is tuned to 2,450 kHz and to any station on that frequency.

The RF amplifier is tuned to 2,450 kHz and amplifies the received signal. The output from this stage is fed to a heterodyne detector, known as the *first-detector, mixer,* or *converter* stage.

The mixer stage is also tuned to 2,450 kHz but is being fed two signals. One is the received signal from the RF amplifier, and the other is a constant-amplitude oscillation from the *local-oscillator* (LO) stage. These two frequencies are heterodyned in the mixer stage, where the sum and the difference frequencies are developed in the mixer-output circuit. If the local oscillator is tuned to 2,900 kHz and the input-signal frequency is 2,450 kHz, the difference frequency is 450 kHz. In the superheterodyne the difference frequency is known as the *intermediate frequency* (IF). The IF signal is fed into an IF amplifier stage. (If the sum fre-

quency is used, the IF is said to be *up-converted*.)

The IF stage has both input and output circuits tuned to 450 kHz to amplify only the difference frequency. It rejects the RF signal, the local-oscillator signal, and the sum frequency, all of which are present in the output circuit of the mixer stage. The selectivity of a simple superheterodyne is determined primarily by the number and Q of the tuned circuits in the IF amplifier stage or stages. The IF amplifier amplifies the difference frequency of 450 kHz and feeds it to the second detector.

The second detector may consist of only a simple diode detector to rectify the 450-kHz IF signal. If the signal contains modulation, the output of the diode rectifier is an AF varying dc that is fed to the AF amplifier and then to earphones or loudspeaker.

If the signal is an A1 emission, it is necessary to change the second detector from a rectifying detector to a heterodyne detector by switching on the *beat-frequency oscillator,* or BFO. The BFO is often tunable a few kilohertz above and below the 450-kHz IF. By tuning either the LO or the BFO, it is possible to produce any beat tone desired.

To prevent interaction between all the stages and to prevent interference from undesired signals, all RF, IF, and oscillator stages should be completely shielded.

18-11 THE RF AMPLIFIER

The first stage in a communications superheterodyne usually consists of a JFET or MOSFET or of a remote-cutoff, or variable-μ, RF pentode tube such as a 6BJ6 or a 6BA6 in a circuit similar to that of Fig. 18-16. The grid circuit is tuned to the signal frequency. The plate circuit is untuned. The screen-grid voltage is obtained from a voltage-divider network to give good voltage regulation. The stage obtains its class A bias from the cathode resistor and additional bias from the automatic-volume-control, or AVC, voltage (explained in Sec. 18-14). A sensitivity or RF gain control may also be incorporated in this stage, as in Fig. 18-14, or gain can be controlled by varying the screen voltage.

Capacitor C prevents the dc AVC bias voltage from grounding but allows completion of the coil-

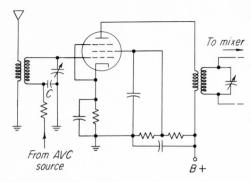

Fig. 18-16 One possible RF amplifier circuit.

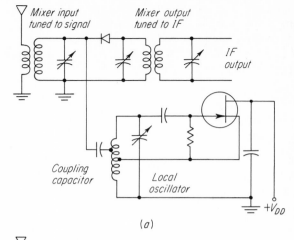

(a)

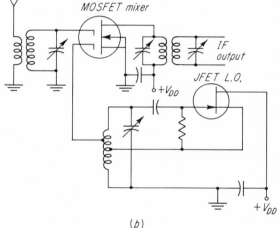

(b)

Fig. 18-17 (a) Diode or crystal mixer circuit with capacitive coupling from a JFET local oscillator. (b) Dual-gate MOSFET mixer.

to-ground circuit for RF ac. Bias voltage developed across the cathode resistor reaches the grid through the AVC circuit (not shown), one end of which is grounded.

Since all the stages following the RF amplifier amplify any noise generated in it, this tube and its circuit must be selected for minimum noise output, or highest signal-to-noise ratio. Signal-to-noise ratio (S/N) is a relation of signal power to noise power developed at the input of an amplifier. If the signal power is 10 times the power generated by thermal action in the input circuit resistance, the S/N is 10.

The noise figure (NF) of an amplifier is the ratio of the input S/N to the output S/N, or NF $= (S_i/N_i)/(S_o/N_o)$. This can be expressed in decibels (dB). An NF of 6 dB indicates the S/N power ratio is 4 times worse at the output than at the input. This is quite good; the amplifier has only two times the noise voltage ratio over that present in the input circuit. The receiver has added very little noise to the signal.

The output of the RF amplifier is fed to the mixer stage.

18-12 THE MIXER STAGE

The mixer, converter, or first-detector stage is a heterodyne detector in which the difference frequency is a relatively low RF rather than an AF. A simple mixer is shown in Fig. 18-17a. In this circuit either a VT or a solid-state diode can be used. Both the signal and the local oscillator are coupled to the signal LC circuit. They mix or beat in the nonlinear rectifier, producing the intermediate frequency. The disadvantage of a diode

mixing circuit is the lack of amplification of the IF signal produced. Mixing can be accomplished in the input circuit of a VT, BJT, or FET, and amplification of the difference frequency can be produced in the output.

Since the drain current is affected by both gates, a dual-gate MOSFET is a natural mixer device, as shown in Fig. 18-17b.

Special tubes have been developed to mix two frequencies efficiently in the electron stream between cathode and plate. One such tube is shown in Fig. 18-18, and another is shown in Fig. 18-33. The *pentagrid converter* in Fig. 18-18 (using a 6SA7, 6BE6, etc.) employs the cathode, the first

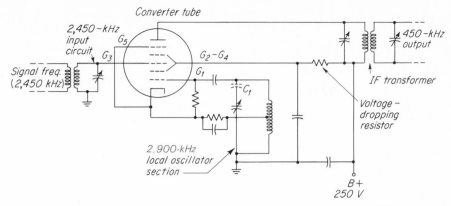

Fig. 18-18   A pentagrid converter circuit used to produce the IF.

grid, and the second grid as a cathode, control grid, and plate in a Hartley-type triode oscillator circuit.

If the frequency of the IF amplifier is to be 450 kHz and the signal frequency is 2,450 kHz, the local oscillator will have to be on either 2,000 or 2,900 kHz. In receivers operating up to about 20 MHz, it is standard practice to use an oscillator operating on a frequency *higher* than the signal frequency to simplify tracking. With a 450-kHz IF, suppose the local oscillator is operated at a frequency *lower* than the signal frequency. The oscillator would have to be on 450 kHz to receive a station on 900 kHz. The strength of the LO signal would completely block the 450-kHz IF amplifier, and nothing would be heard when the receiver was tuned to 900 kHz.

The oscillator circuit is usually a Hartley or an Armstrong in tunable receivers. For fixed-frequency reception, the local oscillator may be a crystal oscillator, making off-frequency operation unlikely.

It is necessary to track the tuning of the oscillator, the mixer, and the RF amplifier(s) so that, when the mixer and RF amplifier are tuned to the desired frequency, the oscillator will always be exactly 450 kHz higher. To do this the capacitor plates of the oscillator stage can be manufactured specially to provide proper tracking. For all-frequency operation, however, the tracking may be accomplished by inserting a tracking capacitor C_1. This allows three similar, ganged variable capacitors to be used for RF, mixer, and LO circuits. The higher the frequency band being tuned,

the greater the capacitance required of the tracking capacitors. Above 20 MHz no capacitor may be necessary. Small trimmer capacitors are connected across all tuned coils, and some method of varying the inductance of the coils, such as adjustable brass or powdered-iron cores, is also used. In more elaborate receivers having two RF amplifiers, four ganged circuits must all track properly.

The stability of the oscillator is important in communications superheterodynes. All insulating materials used in the oscillator should be ceramic or other low-loss types. Mica or ceramic rather than paper capacitors should be used for bypassing. All oscillator components should be rigidly mounted to prevent frequency wobble when the receiver is physically jarred.

Although signal gain is produced in the mixer stage, it is not as great as in an RF or IF amplifier. RF pentodes have transconductance values from 2,000 to 12,000 μmhos, whereas the conversion transconductance of mixer circuits ranges from about 400 to 900.

The conversion efficiency (gain) of a mixer is the ratio of its IF output voltage to its signal input voltage.

Since mixers always have a high noise figure, it is an advantage to have an RF amplifier ahead of the mixer.

18-13 THE IF AMPLIFIERS

The superheterodyne is the most practical tunable receiver chiefly because the IF amplifiers do

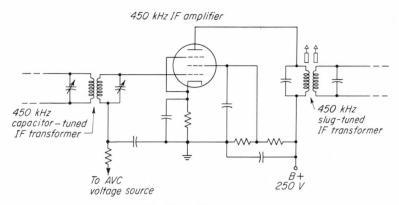

Fig. 18-19 A possible IF amplifier circuit using series-fed AVC.

not require tracking. They can be adjusted to the IF and remain that way, regardless of the frequency to which the *front end* (RF amplifier and converter) is tuned. Since the IF is usually a relatively low RF, pentode and many transistor amplifier circuits are stable and require no neutralization. Both input and output circuits are tuned, resulting in high gain for the amplifiers and an effective rejection of all frequencies except those on the IF and a few kilohertz on each side. Figure 18-19 is a diagram of an IF amplifier. Except for the tuned plate circuit, it is essentially the same circuit as an RF amplifier.

IF transformers usually consist of two multilayer coils wound on thin cylinders of insulating material. The coils may be tuned with two adjustable mica or air capacitors, all enclosed in an aluminum shield can. Most IF transformer coils have fixed mica capacitors across them. Powdered-iron *slugs* are screwed into or out of the cores of the coils to tune them by variation of their inductance. By using low-loss, high-permeability core materials, the coils require only a fraction of the number of turns needed for air-core coils. This can result in higher-Q circuits with narrower bandwidth (Secs. 8-6, 8-7) and higher output voltages.

Several methods are used to produce either a narrower or a variable IF bandwidth. The coupling between primary and secondary of the transformers can be varied from less than critical coupling for a narrow bandwidth to an overcoupled condition for a broad bandwidth. Resistances can be switched in series with some of the tuned circuits of the IF amplifiers to lower the Q of the circuits and broaden the bandwidth. In general, the

greater the number of tuned circuits in the IF strip, the narrower the passband. Two top-coupled transformers may be used between stages as shown in Fig. 18-20. Tapping down the secondary

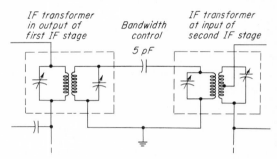

Fig. 18-20 IF selectivity can be increased by cascading loosely top-coupled and tapped-down tuned circuits.

coil raises the Q of the output circuit, further narrowing the bandwidth. Bandpass filters of various bandwidths can be switched into the IF strip.

If a crystal lattice or a high-Q *m*-derived bandpass LC filter can develop a narrow bandpass response by itself, the amplifiers that follow it need not be narrow-bandwidth circuits. Some receivers use mechanical filters which have IF response curves with extremely steep-sided skirts and narrow, flat tops. Such IF filters are available with passbands from 400 Hz for A1 reception to more than 30 kHz for FM communications.

Shielding is required between IF amplifier stages to reduce any interaction of electromagnetic fields from one IF transformer to another. Input and output connecting leads must be kept apart physically and be as short as possible.

Many smaller receivers use only one IF amplifier. Most communication receivers have two IF stages, and some three. It is often found that the addition of a third high-gain stage may produce self-oscillation or bring up the background noise level without improving the signal-to-noise ratio.

A rheostat may be connected in series with the cathode-bias resistor, or the screen-grid voltage may be made variable to act as a manual gain control for either or both RF and IF amplifier stages, particularly when receiving A1 or single-sideband signals.

18-14 THE SECOND DETECTOR AND AVC

The signal amplified by the last IF stage is fed to the second detector, often a diode *envelope* detector as shown in Fig. 18-21.

The usual tube for the second-detector first-audio-amplifier AVC circuit is a twin-diode high-μ triode, such as a 6AT6. In the diagram shown, the two diode plates are connected together to form a single diode. The 250-kΩ potentiometer is the load on the diode rectifier circuit. Two 0.0001-μF capacitors and a 50,000-Ω resistor form an RC-type RF filter, resulting in only an audio varying dc across the potentiometer. The setting of the potentiometer (volume control) determines how much AF signal is coupled to the grid of the triode amplifier.

Since the rectified signal current flows from left to right through the potentiometer, the left end is negatively charged. The stronger the signal received, the more negative the left end becomes. This negative voltage, which varies with the strength and modulation of the incoming signal, is taken off through a long-time-constant RC filter (1-MΩ R and 0.05-μF C) that filters out any AF variation of the dc. The voltage across this capacitor will vary only with average carrier variations; it is the AVC voltage for the RF and IF amplifier-grid circuits. A strong signal produces a high AVC voltage and biases the remote-cutoff amplifier tubes to a point on their $E_g I_p$ curves where the slope is slight and they amplify very little. With a weak signal there is very little AVC bias, and the tubes operate on the steep part of their curves at full gain. This tends to make all signals nearly equal in amplitude at the second detector. Thus it is not necessary to adjust the audio volume control when tuning from a stronger A3 station to a weaker one.

When receiving A1, A2, and A3J signals, the AVC voltage may be grounded. A PHONE-CW switch shorts out the AVC voltage, leaving the RF and IF amplifiers with only the class A bias developed by their respective cathode-resistor bias circuits. Receiver volume is then controlled manually by the RF gain control.

For A1, A2, and A3J signals an automatic-gain-control (AGC) voltage may be used. AGC differs from AVC in that it has a fast attack time and a slow decay time. For example, if the diode

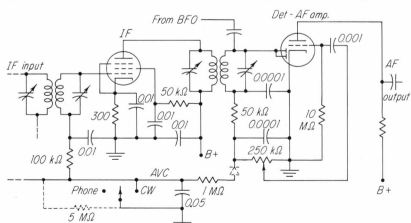

Fig. 18-21 Last IF, second detector, first AF amplifier, and AVC (or AGC) circuits.

and resistor shown dashed on the AVC line are used, the 0.05-μF capacitor could charge rapidly through the 1-MΩ (1-megohm) resistor, but it could not discharge backward through the diode. It would have to discharge slowly through the 5-MΩ resistor. With A1 and A2 signals this prevents a bias pumping effect between dots and dashes, allowing the gain to ride along with the signal if it fades slowly. The same is true with A3J. The AGC voltage rises to the highest sideband peaks and tends to hang there for a time. If no more equally high peaks appear, the bias slowly decreases, raising the sensitivity of the receiver. The gain variation is so slow that no pumping action should be noticed. When AF ac is rectified and filtered in a long-time-constant circuit, this can be used as an AGC voltage for RF, IF, or AF amplifiers when receiving A1, A2, A3J, and F1 signals.

Even on weak signals, the AVC circuit in Fig. 18-21 produces bias, thereby weakening such signals. To prevent this, an AVC circuit that does not go into operation until the signal reaches a predetermined amplitude may be used. This is known as *delayed AVC,* or DAVC. Using the AGC diode and resistor in Fig. 18-21, AVC voltage can not begin to be produced until the carrier voltage rises to about 0.3 V with a germanium diode and 0.6 V with a silicon diode, which is a form of DAVC.

In Fig. 18-22 the delayed voltage level can be adjusted from nearly zero to a volt or two by adjusting the 1-MΩ potentiometer. In this circuit the top diode plate, the IF transformer secondary, and the 250-kΩ "pot" form a normal diode detector. Triode plate current flowing through the cathode resistor develops a bias of about 1 V across it. This bias voltage is applied to both the triode grid and the lower diode plate. Until the lower plate receives a signal greater than 1 V positive, current cannot flow to it from the cathode. If the signal strength is over 1 V, current flows to this diode plate and down through the 1-MΩ resistor, developing a negative biasing voltage across it. The DAVC voltage is filtered and fed to the grid circuits of the RF and IF amplifiers. Weak signals receive full amplification, and only stronger signals developing more than a 1-V peak at the second detector have AVC applied to them.

In general, AVC is not fed to the mixer stage of a communication receiver, since a changing bias on the grid of such tubes tends to pull or vary the oscillator frequency as the signal fades up and down. Small broadcast receivers having no RF amplifiers usually do have the AVC fed to the mixer. Their IF passband is so broad that a little detuning is not noticeable.

Test your understanding; answer these checkup questions.

1. Regarding superregenerative detectors, what are two advantages? _____ _____ Two disadvantages? _____ _____ Used in what part of the spectrum? _____
2. Where is superregeneration found most today? _____
3. What is meant when it is said that two circuits track?
4. What is the term that expresses frequency versus attenuation of a circuit? _____
5. Of what does a TRF receiver consist? _____ Could AVC be used in one? _____
6. List the seven stages in a CW superheterodyne from antenna to loudspeaker. _____ _____ _____ _____ _____ _____
7. Why might a triode be better than a pentode as an input RF amplifier? _____ What is the formula for NF? _____
8. Which is better, an NF of 20 dB or one of 10 dB? _____
9. What are two other names for the first detector? _____
10. What separate circuit usually accompanies a first detector? _____
11. What circuits are usually used in the LO? _____
12. What part of a superheterodyne is responsible for the bandpass of the receiver? _____
13. What are three types of filters used in IF strips? _____ _____ _____

Fig. 18-22 A delayed AVC (DAVC) circuit.

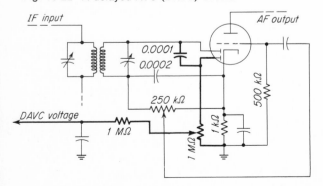

14. What type of second detector is used for A3? _____ For A3J? _____ For A1? _____

15. Is AVC voltage directly or inversely proportional to carrier amplitude? _____ Is it − or + with VTs? _____ With NPN BJTs? _____ With PNP BJTs? _____

16. What is the basic difference between AVC and AGC? _____ Which would be used with A3 signals? _____ A1? _____ A3J? _____

17. In DAVC, what is delayed? _____

18-15 THE BEAT-FREQUENCY OSCILLATOR

To change the second detector from a rectifying or envelope detector to a heterodyne detector to receive A1, A2, A3J, or F1 (FSK), the beat-frequency oscillator (BFO) is turned on. It is a variable-frequency oscillator using a Hartley, Colpitts, or Armstrong circuit. It is often tunable to the IF and a few kilohertz higher and lower. The BFO heterodynes with any signal coming through the IF strip, producing an audible beat frequency in the detector. Both BFO and LO must have good frequency stability to prevent drifting of the beat tone.

The BFO may be capacitively coupled to the diode plate of the second detector, as in Fig. 18-21, or to the plate of the last IF amplifier, as in Fig. 18-32.

Since many BFO circuits are simple oscillators, strong signals delivered to the second detector will tend to pull their frequency, resulting in a changing beat tone with fading signals. Reception of single-sideband signals will be very poor under such conditions. It is preferable to use an ECO as in Fig. 18-23 to prevent such pulling. To isolate the BFO and the second detector, a buffer amplifier may be connected between the two stages.

Fig. 18-23 An ECO-type BFO circuit.

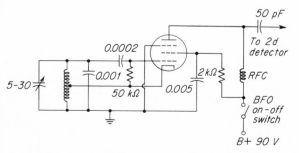

More effective detectors for single sideband are pentagrid converters or other special mixing circuits termed *product detectors* in this application.

In receivers made for A3J only, the BFO may be called the *carrier oscillator.*

18-16 A SQUELCH OR Q CIRCUIT

One circuit that can operate from the AVC voltage is the *squelch,* or Q (quieting), circuit. Modern communication receivers have high gain. When no signal is tuned in, a loud buzzing and crackling background noise is developed. To quiet the receiver until a signal appears on the frequency to which the set is tuned, a squelch circuit as in Fig. 18-24 can be used. With no signal there is no

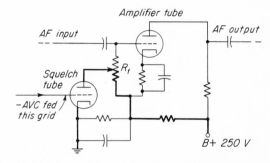

Fig. 18-24 A squelch or interstation quieting (Q) circuit.

negative AVC voltage, and the squelch tube allows current to flow through it, through part of R_1, to the +250-V point. Current flow through R_1 produces a dc voltage-drop across it, more negative at the midpoint than at the bottom. Being in series with the amplifier-grid circuit, this voltage-drop biases the high-μ amplifier tube past cutoff, preventing it from functioning.

When a signal is received, AVC voltage biases the high-μ *squelch tube* to cutoff, stopping its plate-current flow. The voltage-drop across R_1 ceases, allowing the amplifier to operate as a normal resistance-coupled stage. The position of the tap on R_1 controls the quieting point.

18-17 THE TUNING EYE

A device that operates from AVC voltage is the *electron-ray tube,* or the *tuning eye tube.* In one of its forms it is a special double-triode tube, as

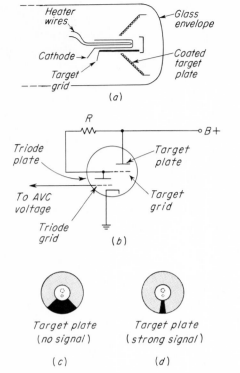

Fig. 18-25 6E5-type tuning indicator tube. (a) Side view of target elements. (b) Double-triode circuit. (c) Target with no received signal. (d) Target with signal.

shown in Fig. 18-25a. One triode is a normal amplifier, but the other has a specially formed, circular, shallow, cone-shaped target plate, with a hole in the center. The common cathode and the target grid wire extend through the hole. The target plate is coated with a substance that glows bright green when electrons strike it.

When the cathode is hot, it emits electrons that

ANSWERS TO CHECKUP QUIZ ON PAGE 442

1. (*Very sensitive, limit noise pulses*) (*Broad tuning, radiate signals*) (*VHF or UHF*) 2. (*Circuit malfunction*) 3. (*Tune together over same portion of spectrum*) 4. (*Selectivity*) 5. (*RF amplifier detector, AF amplifier*) (*Yes, if diode detector is used*) 6. (*RF amplifier, mixer, LO, IF, second detector, BFO, AF amplifier*) 7. (*Lower NF*) $[(S_i/N_i)/(S_o/N_o)]$ 8. (*10 dB*) 9. (*Mixer, converter*) 10. (*LO*) 11. (*Hartley, Armstrong*) 12. (*IF or filter in IF*) 13. (*Crystal, mechanical, LC*) 14. (*Diode*) (*Heterodyne*) (*Heterodyne*) 15. (*Directly*) $(-)(-)(+)$ 16. (*Fast attack time with AGC*) (*Either*) (*AGC*) (*AGC*) 17. (*Development of AVC voltage until carrier develops more than delay voltage value*)

---

are attracted to the positively charged target plate. If the target grid has no charge, the whole target-plate area glows green. If the target-grid wire is charged negatively, it repels the electrons from the cathode, effectively shielding a portion of the plate area from electron bombardment. This segment of the plate area appears dark.

When there is no AVC voltage, the amplifier triode grid has no bias, and plate current flows through resistor R in (b). Because of the current direction through R, the triode plate and therefore the target grid are more negative than the target plate, shielding a segment of the target plate, shown dark in (c).

When a signal is received, AVC voltage applies a negative charge to the triode grid and decreases the voltage-drop across R, resulting in a smaller shielded or dark target area, (d). If the signal is strong enough, all the target area may glow. The width of the shaded area is a visual indication of signal intensity. When the receiver is tuned exactly on the carrier frequency of a station, maximum AVC voltage is developed and the eye closes to the greatest degree.

If the dc voltage fed to the triode grid is taken directly from the detector load instead of after the AVC RC filter, the closing and opening of the shaded area can be made to indicate the relative percent of modulation on the received carrier.

18-18 S METERS

A visual indicator of signal strength is the S meter. The simplest S meter consists of only a 10-mA meter in series with an RF or IF amplifier plate (collector) circuit. With no signal, there is no AVC bias voltage and maximum plate current flows. With a signal, the AVC biases the tube, reducing the plate current and the indication on the meter. The stronger the signal, the less current the meter indicates. The meter face can be calibrated in relative signal strengths, usually 6 dB per calibration unit up to S9, which is equivalent to a received signal of about 250 μV at the antenna terminals of the receiver (assuming 1 μV as S1 and being just above the receiver noise level).

A more representative S meter is shown in Fig. 18-26. If, with no signal being received, the resistance of the tube (BJT, FET) equals the resistance of R_1, and R_2 equals R_3, the four-resistance

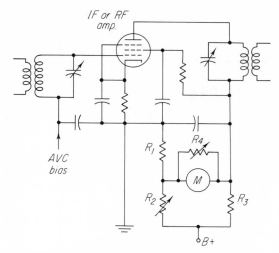

Fig. 18-26 A bridge-type S meter in a VT stage.

bridge-type circuit is balanced and no voltage is developed across the meter. It reads zero. When a signal is being received, AVC is developed, the bias on the tube increases, its plate resistance increases, and the bridge is unbalanced. The meter is now across a difference of potential, and current flows through it. If R_2 is variable, it can be used to balance the bridge circuit to a zero reading when no signal is being received. R_4 can be varied to control the sensitivity of the meter movement.

A separate S meter amplifier may be used to indicate AVC voltage values similar to the rudimentary VTVM of Fig. 12-14.

For A1 and A3J signals rectified AF from a point ahead of the volume control and applied to a dc voltmeter will produce S-meter indications.

18-19 NOISE LIMITERS

Impulse noises, such as automobile ignition, lightning, static, sparking of motors, switching on and off of heavy-current circuits, and power leaks, can completely cover signals that might otherwise be perfectly readable. All these represent an amplitude-modulated form of signal and are detected by the usual AM communications receiver. (Since an FM receiver does not detect amplitude variations, such noises may cause little or no difficulty with FM reception.) While noise impulses may be of extremely short duration, they may

have amplitudes 10 to 1,000 times the strength of the signal it is desired to hear.

A circuit used to clip off, or limit, such pulses is the *series noise limiter*. It uses a diode that conducts signals to the audio amplifier as long as they represent less than about 85% modulation. When impulses of greater amplitude than this come through the detector, the diode stops conducting and no signal is passed to the audio amplifier until the amplitude has decreased to the 85% value again. Figure 18-27 shows a series

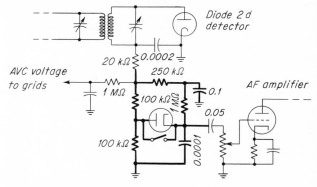

Fig. 18-27 A series-type audio noise-limiter circuit.

noise limiter between the second detector and the first audio amplifier.

The varying dc audio signal of the second detector is developed across the two 100-kΩ resistors in series that form the load for the detector. If the carrier signal produces 10 V negative at the AVC takeoff point, the plate of the limiter diode is then 5 V positive (5 V less negative) in respect to this point. The cathode of the diode is connected to the −10-V point through the 250-kΩ and 1-MΩ resistors. The 0.1-μF capacitor between these two resistors holds an average −10-V charge whether the carrier is modulated or not. (Actually it is not quite −10 V because of the current flowing through the resistors.) Since the plate is more positive than its cathode, the diode conducts. The AF signals developed across the lower 100-kΩ resistor are passed through the conducting diode to the 0.05-μF coupling capacitor to the volume control in the grid circuit of the AF amplifier.

A peak of 100% modulation will produce a −20-V peak of dc at the top of the two 100-kΩ resistors. At this instant the plate of the limiter diode will be −10 V, the same voltage as its cathode, since the cathode is obtaining its charge from the 0.1-μF capacitor. The diode ceases to conduct, and no impulse signal of this amplitude or higher can be transferred to the audio stage. A very small signal can be passed across the diode by the cathode-plate interelectrode capacitance. The 0.0001-μF capacitor bypasses any such leakage signal.

A switch across the limiter diode stops the limiting action. This limiter decreases the possible signal-voltage amplitude from the second detector to one-half of what it would be without the limiter. Distortion is produced on all signals exceeding about 85% modulation with the limiter in the circuit. This limiter will not operate with the BFO on, since the BFO produces so much rectified voltage that the diode conducts all the time. As a result, the limiter will not function for code or SSB signals.

A shunt noise limiter that will limit impulses over a set voltage value can be useful for A1 reception. Two zener diodes can be connected in series, in opposite polarity, as in Fig. 18-28a, across an

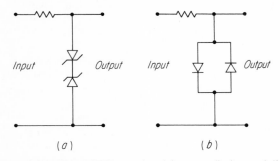

(a)   (b)

Fig. 18-28   Shunt limiters using (a) zener diodes and (b) silicon diodes.

audio-amplifier grid circuit. Any noise impulse above the zener voltage causes breakdown of one of the diodes, limiting the grid voltage for the duration of the impulse. An opposite-polarity impulse breaks down the other zener. (Zener diodes conduct in the forward direction.) Zeners range in voltage ratings from 2.5 V to about 200 V.

Two silicon diodes in parallel and with polarities reversed, as in Fig. 18-28b, are useful as limiters across low-voltage circuits, such as the primary or secondary of the IF transformer, or across earphones. Silicon diodes require about 0.6 V

before they conduct. When they begin to pass current, they effectively short-circuit the line across which they are connected for any voltage peaks above 0.6 V.

To limit an amplifier when copying code signals, the power-supply voltage of the first audio amplifier can be reduced, which limits the possible output voltage of the stage.

Noise impulse limiters in the IF strip are called noise blankers. A block diagram of a noise blanker circuit is shown in Fig. 18-29. The IF signal from

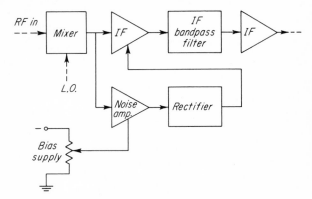

Fig. 18-29   A noise blanker system.

the mixer is fed to the IF amplifier and also to another, somewhat similar, stage, which is called a noise amplifier. If it is desired to clip all incoming pulses over some particular voltage value, the noise-amplifier threshold control is set to bias the amplifier so that no signal up to that value will produce any noise-amplifier output current. Pulses over that value will be amplified and rectified, and these pulses will be fed back into the IF amplifier 180° out of phase to cancel the pulse as it is being developed in this stage. It is important that the blanking be accomplished ahead of the IF bandpass filter, since blanking produces spurious signals off the IF frequency. These will then be attenuated by the filter and will not reach the second detector.

18-20 THE AUDIO AMPLIFIERS

The audio amplifiers in a receiver should have a relatively flat response if high-fidelity music is to be received. Communication receivers may limit the audio response of all frequencies above approximately 3,000 Hz to decrease noise response. Where code or other single-tone reception is required, the audio stages may incorporate a tuned filter to accentuate a narrow band of frequencies. A tone control is often helpful.

When multiplex transmissions are being received, special bandpass filters may be required in the audio circuits.

Most receivers are constructed to operate into loudspeakers having 3- to 16-Ω impedances. When it is desired to use earphones, they may be connected across the loudspeaker, or provision may be made to plug them into the output of the stage before the final amplifier. When the earphones are plugged in, the loudspeaker may be automatically disconnected.

18-21 THE CRYSTAL FILTER

Included in some communication receivers is a crystal filter, the basic element being a quartz crystal ground to series-resonate at the IF. Characteristics of quartz crystals are high Q and narrow bandwidth. The crystal circuit is inserted between the mixer and the first IF amplifier, as in Fig. 18-30.

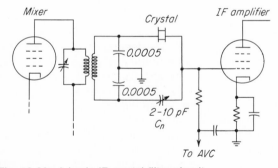

Fig. 18-30 A basic IF crystal-filter circuit.

A signal at the resonant frequency of the crystal passes through it with practically no attenuation. Signals on frequencies only a few hundred hertz removed from the crystal frequency may be greatly attenuated. Unfortunately, the capacitance between the metal plates that hold the crystal will pass all frequencies near the IF equally well. As a result, not only is the frequency of resonance passed by the crystal but other adjacent frequencies are passed by the capacitance of the crystal

holder. To neutralize this capacitive signal transfer, a similar capacitive signal from the opposite end of the input transformer, 180° out of phase, is also fed to the IF amplifier. If the two signals are of equal amplitude (crystal holder capacitance equals C_n), the only signal transfer will be by the crystal. A narrow, sharply peaked passband is developed. About 500 Hz above the resonant peak is the antiresonant notch that greatly attenuates signals on its frequency. Simple crystal circuits of this type are not useful for A3, A3J, or F3 reception. A switch across the crystal removes it from the circuit, broadening the IF bandpass.

A narrow, flat-topped, crystal-lattice bandpass filter is developed by substituting a second crystal for the neutralizing capacitor C_n. Bandwidth is about 3 kHz if the crystals are resonant to frequencies about 2.5 kHz apart. Neutralization should not be required with similar crystal holders. These and full-lattice crystal filters (Sec. 17-37) are effective for A3, A3J, and F3 emissions.

18-22 WAVE TRAPS

When undesired local signals produce a response in a receiver, either series- or parallel-resonant-circuit wave traps may be used to advantage (Fig. 18-31).

A *series-resonant* wave trap utilizes the *low* impedance of a series LC circuit to its resonant frequency. A series trap connected across a circuit will effectively reduce the formation of any resonant-frequency voltage across it. Such a wave trap may be connected across the input (antenna and ground) circuit of a receiver, from grid to ground or from plate to ground, and sometimes across the power lines where they enter the receiver. As long as the desired signal is not too close to the wave-trap frequency, the wave trap will not reduce the desired signal materially.

A *parallel-resonant* wave trap utilizes the very *high* impedance developed across a parallel LC circuit at its resonant frequency. A parallel wave trap may be connected in series with the antenna input terminal, in series with a grid or plate circuit lead in an amplifier, or in series with one or both power lines as they enter the receiver. This introduces so much impedance to the frequency to which the trap is tuned that little energy at this frequency will flow in the circuit containing the

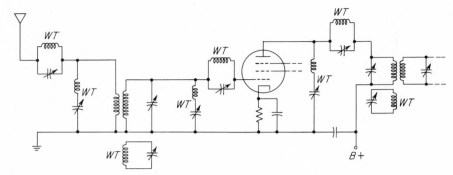

Fig. 18-31 Possible placement of series- and parallel-tuned wave traps (WT) in an RF amplifier and antenna circuit.

trap. A trap in a circuit may detune it somewhat.

A parallel-resonant wave trap may be inductively coupled to any tuned circuit. Signals of the wave-trap frequency are induced into the trap, produce an oscillation in it, and reinduce into the tuned circuit of the receiver a voltage of the same frequency but 180° out of phase (Lenz's law). This effectively cancels or attenuates the signal to which the wave trap is tuned.

18-23 IMAGE FREQUENCIES

Any frequency besides the desired signal that beats with the local oscillator of a superheterodyne and produces a difference frequency equal to the IF is an *image*. In most cases only one image is considered. As an example, if the IF is 450 kHz and the local oscillator is oscillating on 8,000 kHz, a signal on either 8,450 or 7,550 kHz will beat against the oscillator and produce the IF.

The relations between the signal, the oscillator, the IF, and the image frequencies can be expressed in formula form as:

$$f_i = f_o + f_{if} \qquad f_i = f_s + 2f_{if}$$
$$f_i = f_o - f_{if} \qquad f_i = f_s - 2f_{if}$$

where f_i = image frequency
f_o = oscillator frequency
f_{if} = intermediate frequency
f_s = signal frequency

If the receiver above is tuned to a frequency of 7,550 kHz, the mixer tuned circuit usually has a low Q and is quite broad. Alone, it may not

reject a strong signal on the image frequency of 8,450 kHz to any great extent. The addition of a tuned RF amplifier before the mixer produces considerable image rejection. A second RF amplifier will reject the image even more. However, at frequencies in the 30-MHz range even two RF amplifiers may not reject images satisfactorily.

The IF of 450 kHz is too low for image-free high-frequency reception. The image signal is only 2 times 450, or 900 kHz, from the signal frequency. By using an IF of 2,000 kHz, the image will be twice 2,000 kHz, or 4 MHz from the signal frequency. Even a relatively low-Q circuit will reject a signal 4 MHz removed.

The problem of image rejection can be solved by using *double-conversion* superheterodyne circuits. The first IF may be in the 2.5-MHz region. After an IF stage at this frequency, a second mixer stage is used, with a crystal-controlled oscillator to convert to a second, much lower IF, usually in the 50- to 100-kHz region. It is relatively simple to produce a high-Q tuned circuit at these low frequencies. As a result two low-frequency IF stages may produce a very narrow passband. (Narrow bandwidth and image rejection may also be obtained by using a crystal lattice filter at the 2.5-Mhz frequency.)

If the local crystal oscillator used in conjunction with the second conversion stage is not adequately shielded and isolated, harmonics of this oscillation may produce images or signals on higher-frequency bands. This is somewhat similar to the way in which many broadcast receivers with unshielded oscillators hear higher-frequency amateur stations apparently in the broadcast band. Actually, the stations are beating against

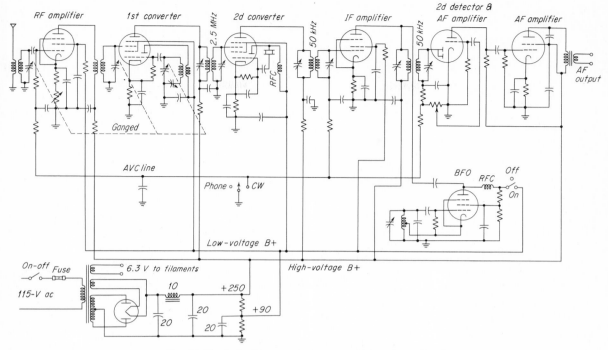

Fig. 18-32 A double-conversion superheterodyne circuit.

18-24 A DOUBLE-CONVERSION SUPERHETERODYNE

The diagram of the VT receiver in Fig. 18-32 represents a double-conversion (two mixer, or converter, stages) or double-superheterodyne communication receiver. Only single RF and IF stages are used. No limiters, crystal filters, audio filters, band switching (switching in different coils to receive different parts of the RF spectrum), squelch, or S-meter circuits have been included.

The receiver consists of a remote-cutoff RF amplifier with an RF gain control in the cathode. This is followed by the first triode-hexode converter stage, using an Armstrong local oscillator to produce a first IF or 2.5 MHz. This in turn is followed by the second converter stage, using a Pierce-type local crystal oscillator circuit. (To isolate the first and second conversion stages better, a stage of 2.5-MHz IF amplification might be

added between them.) The second converter is followed by an IF stage of 50 kHz. This feeds into a second-detector stage in which is developed the AVC voltage. The triode part of this duplex diode-triode tube is also the first AF amplifier, which is resistance-coupled to a beam-power tetrode AF power amplifier. An ECO-type BFO is coupled into the plate circuit of the last IF amplifier. A full-wave rectifier having a single-section pi filter with a voltage divider makes up the power supply.

If the second converter stage is deleted, the diagram is of the simpler and more common single-conversion superheterodyne.

The first IF must be a frequency that is not included in the tuning range of the receiver, since signals near the IF will tune broadly. Furthermore, the IF section may break into oscillation as the front end is tuned close to the IF frequency.

18-25 TRANSISTOR RECEIVERS

For practically every vacuum-tube circuit there is a transistorized counterpart. There are transistor TRF receivers, but most transistor receivers

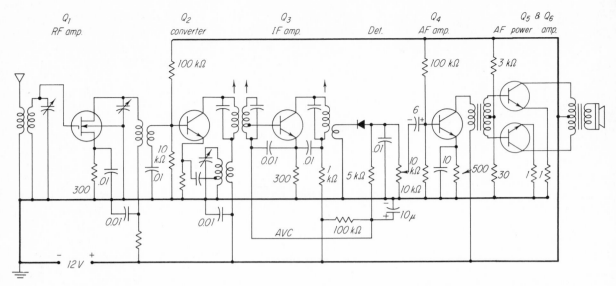

Fig. 18-33 A simple transistor superheterodyne receiver circuit using a MOSFET RF amplifier and BJTs.

are superheterodynes. Figure 18-33 represents a basic superheterodyne without BFO, limiter, S meter, or squelch, etc., all of which could be added. The circuit shows only one IF stage, although two would be preferable in a communication receiver. Small broadcast receivers would not use the RF amplifier, employing instead a high-Q loop stick as the antenna and tuned circuit of the converter. Automobile broadcast receivers would use the RF amplifier stage.

The input tuned circuit feeds the RF signal to the gate-source circuit of the amplifier Q_1. Source-resistor biasing is used. The drain circuit is tuned and acts as the tuned input for the base-emitter circuit of the converter stage Q_2. The converter acts as both a mixer and an Armstrong oscillator through the use of the tuned circuit in the emitter and the tickler coil in series with the IF transformer primary coupled to it. The secondary of the first IF transformer is tapped to match the low impedance of the base of Q_3. The second IF transformer may be tapped to match impedances better and to allow a higher Q of the tuned circuits, and also if neutralization is required. The diode detector has a 10-kΩ potentiometer for a load.

The IF amplifier is forward-biased by the voltage-divider network made up of the 100-kΩ,

the 5-kΩ, and the 10-kΩ resistances. Current flows upward in the potentiometer and charges the 10-μF capacitor. When a signal is present, the diode rectifies it and a downward emf is developed across the potentiometer. This lowers the positive charge on the 10-μF capacitor and therefore the forward bias of the IF amplifier. The gain of the amplifier is reduced by the decrease of this forward-biasing AVC current. To apply AVC to the RF amplifier the standard *VT-type* circuit would be used.

The first AF amplifier Q_4 is a class A stage driving the class B push-pull power-amplifier output stage Q_5 and Q_6. If the 30-Ω resistor were increased in value, the push-pull stage would be biased to class AB or A.

Test your understanding; answer these checkup questions.

1. How could a TRF receiver be designed to receive A3J signals adequately? _____
2. If the BFO is tunable, over what range of frequencies do you think it should operate? _____
3. What is the BFO in A3J receivers called? _____
4. What does a squelch circuit eliminate? _____ In what part of a receiver is it normally found? _____
5. What are two possible uses of an electron-ray tube? _____ _____

6. For what is an S meter used? _____ Will it vary if the receiver is tuned to a local station transmitting A3? _____ A3J? _____ A1? _____
7. How many dB gain represents one S unit? _____
8. What are circuits that clip off high AF noise peaks called? _____ That clip pulses in IF or RF stages? _____
9. What are the two desirable frequencies present when a single-crystal IF filter is used? _____ _____ Are these present in half-lattice or full-lattice filters? _____
10. Will signals be nulled or peaked if a series wave trap is across a circuit? _____ Parallel WT shunts a circuit? _____ Series WT is in series with a circuit? _____ Parallel WT is inductively coupled to a circuit? _____
11. A receiver with an IF of 450 kHz is tuned to 1.5 MHz. What is the probable image frequency? _____ Why might it receive a local signal that is on 3.450 or 4.350? _____
12. What are two ways in which image response is reduced? _____ _____
13. Does narrowing the IF bandpass reduce images? _____
14. In Fig. 18-33, what type of FET is used? _____ Why use an FET rather than a BJT? _____ What would have to be changed if P-channel and PNP transistors were used? _____

18-26 OPERATING A SUPERHETERODYNE

The small broadcast-band superheterodyne is quite simple to operate. It has an on-off switch, usually on the same shaft as the volume control, and a tuning knob. Some superheterodynes have tone controls. There is usually one band of operation, the *standard broadcast band,* from 540 to 1,640 kHz, and one type of emission, A3. The tuning knob is rotated to the desired station, the volume is adjusted to the desired level, and the tone control may be varied.

Receivers manufactured to monitor a single frequency, such as police, fire, taxicab, and other communication services, are crystal-controlled, having an on-off switch, volume control, and usually a squelch-circuit on-off switch with an accompanying noise-threshold adjustment. Operation is as simple as possible.

Communication receivers are another matter. They may have an on-off switch, a band-selector switch, a wideband tuning control, a vernier control to tune over a narrow band of frequencies, a crystal or other filter on-off switch, a variable IF bandwidth control, an AVC on-off switch, an RF gain control, an audio volume control, a tone control or switchable bandpass filters, a BFO on-off switch, a BFO tuning control, a limiter for use with A3 signals, a limiter for use with A1 signals, an audio filter switch, an earphone or loudspeaker switch or jacks, and a standby on-off switch which is used when a local transmitter is in operation but which leaves the oscillators (and filaments) of the receiver on continuously.

As the band-selector switch is turned, different-sized coils are simultaneously connected across the RF amplifier, mixer, and local-oscillator tuning capacitors, allowing the receiver to tune over different portions of the radio spectrum. *Low-frequency* receivers will tune from about 15 to 550 kHz in possibly four bands. *All-wave* receivers usually tune from 560 kHz to 30 or 40 MHz in four to seven bands, depending on how wide a frequency coverage is desired on each band. In some receivers, band changing is accomplished by shorting out part of each tuning coil simultaneously. This is not considered particularly good engineering, but it is better than merely tapping down the coils to change frequency, because the unused upper portions may fall into parallel resonance at some frequency and act as a wave trap at that frequency. *Shortwave* receivers tune from about 1.5 to 30 or 55 MHz in four to eight bands.

To tune in a broadcast signal when relatively high fidelity is desired, the crystal filter is turned off, the bandwidth is set to 10 to 15 kHz, the AVC is turned on, the RF gain control is turned up full, the audio volume control is advanced to the desired sound level, the tone control is set to maximum high-frequency response, the noise limiters are normally left off, and the BFO is off.

To tune in an A3 voice signal on the higher-frequency bands where sidebands up to only 3,000 Hz are desired, the crystal filter is off, the bandwidth should be set to about 6 kHz, AVC is on, RF gain is on full, tone control is set to reduce the high-frequency audio response, the noise limiter is on or off as required, CW audio filter is off, and the BFO is off. Loudness is controlled with the AF volume control. If interference appears on one sideband of the desired signal, it is possible to reduce the bandwidth or set the crystal filter to a bandpass of 3 kHz. The receiver is tuned to pick up the carrier and the set of sidebands that are not being interfered with. This is single-sideband-with-carrier reception and is possible on an A3 transmission with some loss of signal strength.

To tune in an A3J voice signal (SSSC), the bandwidth of the receiver is adjusted to 3 kHz. The AVC is switched off (although long-time constant AGC could be left on, and volume would be controlled with the AF volume control). The audio volume control is set near maximum, the tone control is set to decrease high-frequency response, and the noise limiter and CW filter are turned off. The RF gain control, *not* the audio volume control, is used to control the volume of the received signal. The BFO is turned on and set to a frequency 1.5 kHz higher or lower than the IF center frequency. The BFO is now inserting a carrier in place of the missing carrier of the transmitted signal, and the 3-kHz sideband signals are passing through the IF amplifiers. The receiver is tuned very slowly and carefully until the voice is readable. If the voice cannot be detected, the BFO is tuned 1.5 kHz on the other side of the IF center frequency and the receiver is again tuned. It will be noted that as much as 10-Hz variation in the setting of the vernier tuning control or of the BFO can make the received signal sound unnatural. Two hundred hertz off, and the signal is difficult to understand. For this reason receivers used for SSSC reception must have exceptional BFO and LO frequency stability.

To tune in A1 or A2 radiotelegraph signals the crystal filter may or may not be used, depending on the interference from adjacent frequency signals. The bandwidth control may be adjusted to as narrow a setting as possible. The AVC is turned

ANSWERS TO CHECKUP QUIZ ON PAGE 450

1. (*Use detector with BFO*) 2. (*IF center* ± 3 *kHz*)
3. (*Carrier oscillator*) 4. (*Background noise when carrier not present*) (*AF*) 5. (*Tuning indicator, modulation percent indicator*) 6. (*Tuning indicator*) (*No*) (*Yes*) (*Yes, unless long-delay AGC is used*) 7. (*6 dB*) 8. (*Limiters, also clippers*) (*Blankers*) 9. (*Series resonant to peak, parallel resonant to null*) (*No*) 10. (*Nulled*) (*Peaked*) (*Peaked*) (*Nulled*) 11. (*1.95 MHz*) (*Images of second harmonic of LO*) 12. (*Add RF stages, use high-frequency IF*) 13. (*No*) 14. (*Depletion MOSFET*) (*Less load on LC, higher Q, less noisy*) (*Power supply polarity, diode polarity, polarity of three electrolytic capacitors*)

off; the audio volume is set near maximum; the BFO is turned on and adjusted to a frequency about 800 Hz from the center of the IF channel; the tone control is adjusted to remove as many high audio frequencies as possible; the A3 noise limiter is turned off, and the A1 noise limiter may be turned on if necessary. The audio filter may be used if interference is bothersome or if the signal is extremely weak. The RF gain control is used to control the volume of the received signal. If adjacent frequency signals interfere, the IF filter can be placed in a sharper condition.

Frequency-shift keying (FSK) is a form of code transmission in which a dot or dash is transmitted on one frequency and the carrier is then shifted 60, 170, or 850 Hz up or down for the intervening spaces (Sec. 16-13). With the BFO on in an AM receiver, a shifting carrier produces two different beat tones. By using two sharp audio filters, it is possible to separate the dot and dash tones from the space tones and reproduce the transmitted code at the receiver. Since the carrier is on all the time, the receiver hears none of the background noises that are present when MAB (make and break) keying is used. FSK is used with RTTY (radio teleprinters) as well as with Morse code.

When trying to copy through interference with a receiver, as low a degree of coupling as possible should be used between antenna and receiver to increase the Q of the input circuit. The RF gain control should be operated as low as possible, and any interstage coupling that can be varied should be at a minimum to prevent overloading of the tuned circuits. The narrower the bandwidth of the IF strip, the less noise present in the output.

To relieve listening fatigue due to continued listening to a single tone while copying code signals, the operator can tune either the BFO or the main tuning dial very slightly to change the beat tone of the receiver signal, perferably about half an octave.

Strong CW signals are prevented from blocking the receiver by reducing the RF gain control.

See also Secs. 28-11 and 28-12 for additional information on modern communication receivers.

18-27 DIVERSITY RECEPTION

To overcome the effects of fading, the AVC circuit was developed. While this is partially effective, signals can still fade in and then out completely. It has been found that two antennas a few hundred feet apart may have two entirely different amplitude signals in them from the same distant station at any particular instant. It is possible to connect two widely spaced antennas to separate front ends and IF strips, as in Fig. 18-34, and feed

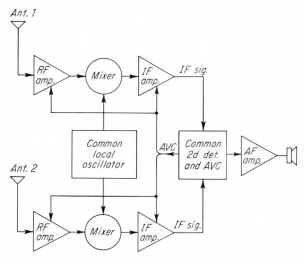

Fig. 18-34 One type of diversity-reception receiver system.

them to a common second detector, AVC, and AF system. The front end with the stronger signal develops the most AVC, which biases the other one off. Signals that might have been fading severely on a single antenna and receiver may have almost no signal-strength variation using space diversity reception. Antennas directly adjacent will give a certain amount of diversity reception, but the farther they are apart the better the diversity action. To assure that both front ends are tuned to the same station, a common local oscillator is employed, as shown.

18-28 TRANSCEIVERS

A transceiver is a combination transmitter-receiver in one enclosure. Although both units might be tuned separately and be operated on different frequencies, the two sections are usually made to operate on the same frequency at the same time. Citizen-band and commercial transceivers will be either single- or multiple-channel systems with frequencies controlled by

crystal oscillators. Amateur-band transceivers use a VFO to allow any frequency in a given band to be used. If the receiver is tuned to 4,000 kHz, the transmitter will transmit on that frequency also. The circuitry to produce a system that does this is not simple. There are a variety of methods by which this may be accomplished. One such SSB system is shown in block form in Fig. 18-35.

The system shown in heavy lines is a double-conversion superheterodyne receiver. For operation in the 3.5- to 4-MHz amateur band the incoming signal is amplified by an RF amplifier and mixed with a 10-MHz crystal oscillator which converts it to some frequency between 6 and 6.5 MHz. This signal is amplified by the 6- to 6.5-MHz broadband IF amplifier and is then fed to the second receiver mixer. Tuning is accomplished by varying the MASTER VFO feeding the second mixer. The output of this mixer is 1.65 MHz, which is fed to an IF amplifier tuned to this frequency. A single 3-kHz crystal filter (1.650 to 1.653 MHz) can be used for both upper- and lower-sideband reception by selecting one of two crystals in the carrier oscillator.

When transmitting, some of the receiver circuits are used. Starting at the microphone, the voice signals are amplified and then mixed with the desired carrier crystal frequency in a balanced modulator. The output of this stage is two sidebands but no carrier. Both sets of sidebands are amplified by the 1.65-MHz IF amplifier, but only one will pass through the crystal filter. The sideband that is passed by the filter is fed to the first transmitting mixer, where it is then heterodyned against the 4.35–4.85-MHz MASTER VFO. When the resulting 6–6.5- and the 2.7–3.2-MHz signals are fed to the 6–6.5-MHz IF amplifier, only the 6–6.5-MHz SSB signals are amplified and fed to the second transmitting mixer. This signal is mixed with the 10-MHz heterodyne crystal oscillator and is translated to 3.5–4 MHz. The output circuit of the second transmitting mixer is tunable from 3.5–4 MHz, as are the driver amplifier and the power amplifier. The transmitter output is coupled to the antenna by the relay that transfers the antenna from the receive to the transmit position when the equipment is switched to transmit.

The change from receive to transmit may be accomplished by a manual switch that shifts the antenna from receive to transmit and also con-

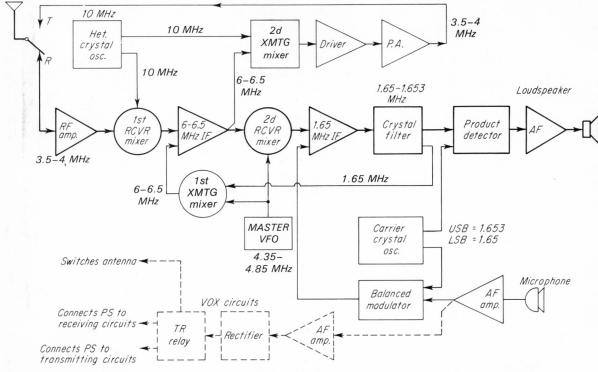

Fig. 18-35 Block diagram of amateur A3J transceiver. (Simplified from Hallicrafters model SR-400A)

nects the power supply from receiver circuitry to transmitter circuitry. This may also be accomplished by a voice-operated transmitting (VOX) system. AF picked up by the microphone is amplified, rectified, and made to throw a TR relay that switches the antenna and power-supply circuits. When voice signals cease, the relay falls back to the receive position and the operator can hear incoming signals. To prevent received sounds from tripping the VOX circuit, an antitrip circuit (not shown) rectifies some of the received AF and uses this dc to buck the VOX dc, thus preventing the VOX TR relay from changing over. Since local voice sounds do not emanate from the receiver loudspeaker, they are not bucked out and the VOX circuits operate.

The frequency selection is accomplished by varying the MASTER VFO, but the receiver RF amplifier, the second transmitter mixer, and the driver are gang-tuned to allow trimming them to

frequency for optimum operation. The PA has its own output tuning and coupling controls.

Test your understanding; answer these checkup questions.

1. What services would use crystal LOs? _____
2. What services would use variable LOs? _____
3. What services might use narrowband AF filters? _____
4. What does a band switch do? _____
5. With what type of emission does mistuning by 200 Hz materially degrade the received signal? _____
6. Why does FSK not require an FM receiver? _____
7. What system is used to give even better protection against fading than is provided by AVC? _____
8. Would a separate transmitter and receiver in one enclosure be considered a transceiver? _____
9. In the transceiver system described, how many crystal oscillators are required to tune a 500-kHz band? _____ Which stages would have to be trimmed to frequency? _____
10. In the transceiver, how is upper- or lower-sideband operation selected? _____ What stages would be fed

AGC? _____ Would an antitrip circuit be required if earphones were used? _____ What must be the MASTER VFO frequency range to produce 3.5 to 4 MHz? _____

11. What does VOX mean? _____ What might MOX mean? _____ What is a TR switch or relay? _____

12. In the transceiver, to change to another band, what six things would have to be changed? _____ _____ _____ _____ _____

13. What does "translated" mean when referring to SBs? _____

14. What are the common circuits in the space diversity receiver shown? _____ What would frequency diversity be? _____ Could a single antenna be used for frequency diversity? _____

18-29 ALIGNING A SUPERHETERODYNE

Tuning the RF, mixer, oscillator, IF, and detector stages of a superheterodyne to their correct frequencies is known as *aligning* the receiver. Fortunately, superheterodynes rarely need realignment. When tubes are changed, only on the higher-frequency bands may the local oscillator be detuned enough to require realignment. In many cases, inexpert attempts to realign will result in greater misalignment. However, once in a while a do-it-yourself mechanic will tighten all the screws in the receiver, including the RF and IF trimmer capacitors. In this case no signals will be audible, even if the receiver circuits are electrically perfect. Sometimes one or more stages in a receiver are changed to include new parts, tubes, transistors, and so on. In these cases realignment will be necessary.

To align a superheterodyne, the technician should have at least a tone-modulated RF signal generator capable of being tuned across the IF frequency and across that part of the radio spectrum the receiver is to tune, a small screwdriver made of insulating material, and two 0.0001- to 0.0005-μF fixed capacitors. A VTVM is the best instrument, although a 20,000-Ω/V meter on the 10-V range can be used to measure the AVC voltage during alignment. An oscilloscope can be connected across the loudspeaker terminals to give a visual indication of strength of signals.

The basic procedure is to work from the second-detector input circuit back to the RF input circuits, checking each stage in turn.

To align a small broadcast receiver, one procedure might be as follows:

The tone-modulated RF signal generator is adjusted to the desired IF, usually 455 to 465 kHz, and is connected to the grid of the last IF stage and to ground (or negative terminal of the power supply), using the 0.0001-μF capacitors in series with the signal-generator leads. The detector stage should demodulate this signal, a tone should be heard in the loudspeaker, and a pattern will be seen on the oscilloscope. A VTVM connected between the AVC line and ground will read the AVC voltage developed across the diode detector load. The RF signal generator should always be adjusted to an output as weak as possible and still obtain an audible or visual tuning indication.

The primary and the secondary of the IF transformer between the last IF amplifier and the detector are tuned for maximum audible or visual indication, using the insulating-material screwdriver to turn the adjusting screws of the IF transformer.

After this transformer has been tuned, the signal generator is moved to the grid of the mixer stage (or to the grid of the IF stage ahead, if there is one). A wire is connected across the local-oscillator tuning capacitor to prevent the LO from oscillating. The IF transformer primary and secondary between the mixer and IF stages (or between any two IF stages) are tuned for maximum audible or visual indication. This completes the alignment of the IF section.

The alignment of the mixer and oscillator stages requires tracking them. The various capacitors involved in such circuits are shown and labeled in Fig. 18-36. After the shorting wire has been removed from the oscillator capacitor, the receiver dial is set to some high-frequency point, such as 1,400 kHz. The signal generator is set to the same frequency and coupled to the external antenna terminal. If the signal generator is not heard on the receiver, the oscillator *trimmer* capacitor can be varied until the signal is heard. Then the input trimmer capacitor of the mixer is tuned until a peak is indicated audibly or visually. The receiver is next tuned to a calibrated point at the low-frequency end of the dial, such as 600 kHz. The signal generator is then set to 600 kHz. If the

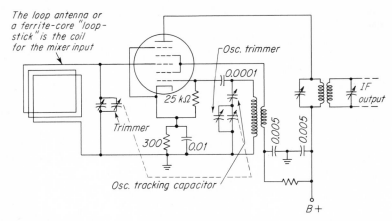

The loop antenna or a ferrite-core "loop-stick" is the coil for the mixer input

Osc. trimmer

0.0001

25 kΩ

IF output

Trimmer

300

0.01

0.005

0.005

Osc. tracking capacitor

B+

Fig. 18-36 A VT converter stage of a loop-antenna broadcast-band receiver.

signal generator cannot be heard in the receiver, the oscillator is not tracking with the dial markings. At the low-frequency end of the dial the oscillator *tracking* capacitor is adjusted until the signal is heard. Then the receiver and signal generator are returned to the 1,400-kHz settings, and the oscillator *trimmer* capacitor is readjusted for maximum indication. The high- and low-frequency points are rechecked until the signals track with the dial readings, adjusting the oscillator trimmer capacitor on the high-frequency point and the oscillator tracking capacitor on the low-frequency point. The receiver is now aligned to the dial markings.

Note that it may not be necessary to attempt to track the mixer input circuit of the smaller receivers. In many receivers there is no way to

track this circuit, and it is assumed that the dial has been calibrated to follow the tuning characteristics of this input circuit. If the input trimmer is adjusted at the high-frequency end and found to be off tune at the low-frequency end, the receiver may be set to a frequency of about 1,200 kHz and peaked. The input circuit will probably be tuned close enough on all other frequencies to produce satisfactory operation. If the input-circuit tuning capacitor has slotted outer plates, it is possible to bend out one or more slotted sections at one end or the other, reducing the capacitance at either the high- or the low-frequency end as desired, to produce better tracking. This will require more trimmer adjustments.

In the more expensive receivers, there are trimmer capacitors across each coil and a means of changing the inductance of each coil, usually by running either brass or powdered-iron-compound slugs into or out of the core area of the coils. (Brass slugs lower inductance, acting as shorted turns, whereas iron increases inductance when used as a core.) After the oscillator stage has been tuned to conform to the dial markings, the mixer input stage is made to tune over the same band of frequencies, adjusting the *inductance* at the *low-frequency end* of the band and the trimmer *capacitance* at the *high-frequency end*. With receivers having RF amplifiers, the signal is finally fed into the antenna and the RF amplifier is tuned to cover the same band of frequencies, using the same method as was used with the mixer input

---

ANSWERS TO CHECKUP QUIZ ON PAGE 454

1. (*Police, fire, aircraft, ship*) 2. (*Amateur, ship, AM, FM, TV, shortwave listeners*) 3. (*CW, RTTY*) 4. (*Switches different coils across the tuning capacitors*) 5. (*A3J*) 6. (*Frequency shift causes two tones in receiver*) 7. (*Diversity reception*) 8. (*Yes*) 9. (*One*) (*RF, first receiver mixer input, second transmitting mixer output, driver, PA*) 10. (*Changing carrier crystals*) (*RF, first receiver mixer*) (*No*) (*4.35 to 4.85 MHz*) 11. (*Voice-operated transmission*) (*Manually operated transmission*) (*Transmit-receive*) 12. (*RF coils, first receiver mixer coils, heterodyne crystal, second transmitting mixer output, driver output, PA output*) 13. (*SBs transferred to another part of spectrum without changing relation of low- and high-frequency SBs*) 14. (*LO, AVC, second detector*) (*Front ends tuned to two stations transmitting the same information but on different frequencies*) (*Yes*)

circuit. The basic steps are tune oscillator, tune mixer input, and then tune RF amplifiers.

While the audible-output method is satisfactory for small broadcast receivers, an unmodulated-signal-generator output is preferable for narrow-bandwidth communication receivers, which are always aligned in their minimum-bandwidth position. With such receivers, the sidebands of a modulated signal can give confusing indications both audibly and visually. A visual peaking of AVC voltage on a VTVM and a peaking of the S meter are both good indications when aligning.

When a crystal filter is the input circuit of the IF section, it should be adjusted for minimum bandwidth and the signal generator fed into the mixer stage and varied until a maximum visual indication is obtained. The signal generator is now on the crystal-filter frequency. All IF amplifiers are then trimmed to this frequency.

The aligning of transistor superheterodynes is accomplished in basically the same way as that of vacuum-tube circuits.

18-30 TROUBLESHOOTING IN RECEIVERS

One of the requirements of most technical jobs is servicing equipment when it ceases to function properly. Many excellent books have been written on the subject. Only a few basic ideas can be included here. Experience with previous breakdowns is probably as important as anything else, but if the operator has not had experience, what is he to do? First, what indicates that the equipment is not operating properly? If it is smoke, the equipment must be turned off immediately. If the signals are weak, distorted, or nonexistent, the required steps may vary considerably. In any case, first try to see what is not normal.

Regardless of the number of jokes made on the subject, in an emergency one of the first steps taken when a receiver suddenly stops working is to deliver a sharp blow with the heel of the clenched fist to each side of the receiver, to the top, and to the front panel. In a remarkable number of cases the receiver will start operating again, but it is not fixed. As soon as possible the equipment should be checked thoroughly. Something may be loose; there may be an oxidized pin connection in a tube or other part; or some screw or solder connection may have loosened. Each tube should be worked back and forth gently in

its socket with the receiver on, to see if the cause of the intermittent operation can be localized. Tapping the parts will often help to localize the trouble. Transistor receivers usually have parts mounted on printed-circuit (PC) boards, sometimes with transistors in sockets. The usual difficulty with PC boards is a hairline crack in one of the interpart leads. A magnifying glass is used to locate the trouble.

Fuses should be checked, as should line-cord connections if there is no heating of filaments or dial lights. It is discouraging to find, after fruitless testing of components and circuits, that the line cord has a broken connection or that the fuse blew out when the receiver was turned on, as it often will. An old fuse may burn out for no apparent reason. In the case of a blown fuse, it is usually a good practice to renew the fuse and turn on the equipment. If there is something wrong with the equipment, the fuse will burn out again. Always check the line cord and plug.

Failure of a tube is more often the reason for improper operation of a VT receiver than any other cause. See if all filaments are lit. A dark filament usually means a burned-out tube. A tube with a burned-out filament will feel cold. Power transistors also should feel warm. If possible, test each tube, noting the physical condition of the tube pins and the socket and surrounding chassis. Clean out any dust. If a tube or transistor tester is not available, substitute each tube or transistor singly with another, similar device, taking care to replace all devices to their original sockets if not proved faulty. Mixing up the arrangement of even the same types of devices in a communication or TV receiver may cause improper operation because of slight variations between supposedly similar devices.

If all the steps outlined above have been taken and the equipment still does not operate, something may have burned out, opened, or shorted. The receiver should be removed from its cabinet and given the eyeball test. Careful scrutiny of the parts may show a burned resistor, signs of excessive heat, bulging fixed capacitors, or loose wires.

In a surprising number of cases, the trouble in a receiver can be localized by using nothing more complicated than a fairly sensitive VOM. The voltmeter is used if the receiver is turned on and the power supply is found to be operating. The ohmmeter is used only when the receiver is turned off and the ac line cord is pulled out of the service outlet. An ohmmeter touched across loudspeaker terminals should produce a loud click. Shorted transistors can often be located with an ohmmeter. They read the same low resistance when the test probes are reversed across them.

In a VT receiver, if the normal 100 to 300 V is being produced by the power supply, the voltage between the plate pin of each tube and B− can be tested. The plate of every triode, tetrode, and pentode should read a positive voltage. If not, there may be an open in that circuit or the plate circuit may be shorted to ground. The plate voltage of resistance-coupled amplifiers will read much lower than other stages. Screen-grid terminals should also show a voltage reading. Touching the voltmeter to the plates and grids of audio-amplifier tubes may result in an audible click in the loudspeaker, telling that the audio stages are operating. Each cathode of an amplifier stage can be checked for voltage. If the tube is not passing current for some reason, no bias voltage will be developed, although there may be voltage at the plate terminal. An open cathode resistor may produce a high positive voltage reading at the cathode. Each control grid can be tested and should read zero volts to ground. A positive voltage on a control grid may indicate a leaking coupling capacitor. A small negative voltage may mean an open grid circuit if the tubes are lit. In general, similar tests can be made in transistor circuits, but the voltages are very low. The bias-voltage variation between proper and improper operation may be only a fraction of a volt.

To check whether an oscillator stage is oscillating, an RF choke or a 100,000-Ω carbon resistor can be connected to the end of the ungrounded voltmeter probe and held against the control grid of the oscillator tube. If there is plate voltage on the tube but no negative dc grid voltage, the oscillator must not be oscillating. The bias for tube and transistor oscillators should indicate class C operation (negative for VTs and normally reverse bias for transistors).

If all measured voltages seem within reason and still no signals can be heard, an audio signal generator can be connected across each audio amplifier in turn, starting at the final amplifier and working back to the second detector. If this does not localize trouble to be in an AF stage, a tone-modulated RF signal generator can be connected to the IF amplifiers and RF circuits, following a procedure of circuit checking similar to alignment, each stage being checked until one is found producing little or no output when fed an input signal.

Distortion in the audio amplifiers can be checked with an oscilloscope or even with earphones in series with a 0.005-μF capacitor. When the stage in which the distortion is produced has been reached, it will be visually or audibly evident. It is then necessary to determine which part is faulty. Open grid resistors, leaking capacitors, and shorted transformer turns are all possibilities.

Sometimes the AVC line becomes grounded, and overdriving of the IF stages occurs, distorting all but the weakest signals. Sometimes the diode load resistor opens, and all signals may be weak and distorted. Volume-control potentiometers often become faulty (sound scratchy when turned) and may open (very distorted AF).

If the power supply does not have any output voltage, the power-supply switch should be left on and the power-line plug pulled out. An ohmmeter test across the plug and line cord should give a reading of 3 to 10 Ω. If a very high resistance reading is indicated, something in the primary circuit (fuse, line cord, or the primary winding) is open.

The secondary of VT power transformers should show continuity with resistance values of 25 to 350 Ω (often less than an ohm in transistor supplies). The B+-to-ground circuit should read several thousand ohms, or whatever the bleeder resistance is. If it reads nearly zero, a filter capacitor in the power supply, or a bypass capacitor in the receiver, may be the trouble. The output lead of the power supply can be unsoldered from the receiver circuitry, and the B+ to ground of the power supply checked. If it shows a relatively high resistance, the receiver B+ lead must show the low-resistance reading. The section that reads low resistance has the shorted capacitor or part in it. It will be necessary to continue disconnecting circuits from the B+ line until the low-resistance circuit is found. Always reverse ohmmeter probes when testing for possible shorts in transistor circuits. (Semiconductor junctions are diodes.)

Noisy operation of a receiver can be caused by many things. Sometimes it arises from poor connections and vibration of the equipment. It can be due to intermittent breaking down of the dielectric of coupling or bypass capacitors. It can be due to faulty resistors, transistors, or tubes. If the equipment is battery-operated, noise can be produced by polarization of old, worn-out batteries.

18-31 EMERGENCY REPAIRS

Sometimes there are no spare parts for a receiver that burns out. The operator's problem is to put the receiver into working condition, possibly with a minimum of test equipment.

If the first RF amplifier stage or tube becomes faulty, it is possible to capacitively couple the antenna directly to the next RF amplifier grid, or to the mixer input grid, by wrapping an insulated wire around any exposed wiring in the grid circuit. This forms a *gimmick,* which has a few picofarads of capacitance to the wire around which it is connected. It is sometimes possible to remove the faulty tube or transistor and couple the antenna through a 50- to 100-pF capacitor into the plate or collector hole in the socket. The primary of the output RF transformer then acts as an antenna coil. It is still connected to a dc potential, however, and must be treated with caution.

If an RF or IF amplifier tube or transistor becomes inoperative and there are no replacements, the plate and grid (base-collector, gate-drain) connections of this stage can be connected together with a gimmick or small capacitor and signals may be heard.

If the second-detector tube fails, audible signals may be heard by connecting earphones between the cathode and the AVC end of the second, or even the first, IF amplifier transformer.

If the local oscillator ceases to oscillate, earphones can be connected in series with the mixer output circuit or in series with the last RF stage. These circuits will act as detectors, and strong modulated signals can be demodulated.

If one of the audio amplifiers becomes faulty, it may be possible to jump the signal over this stage with a capacitor of any value between 0.001 and 0.1 μF, or earphones can be connected ahead of the faulty stage.

If test equipment is not available, a continuity checker can be rigged by connecting a $1\frac{1}{2}$-V battery in series with a pair of earphones. When leads from these are connected and disconnected across a coil or the primary or secondary of a transformer, for example, clicks will be heard. If the winding is open, little or no click will be audible. Capacitors of more than 0.002-μF capacitance can also be tested. A capacitor such as a 0.1-μF or larger will click loudly on the first contact, charge to the battery voltage, and produce almost no click if the connection is made to it again immediately. If it continues to click, it is possibly leaking or shorted. The higher the resistance tested with this continuity checker, the lower the amplitude of the click. Tubes can be tested for filament continuity or for shorts between elements.

If the earphones have 1,000 Ω resistance or more, emitter-base and base-collector sections of BJTs can be checked. A click will be heard one way, but not if the test leads are reversed.

A pair of earphones in series with a 0.0001- to 0.01-μF capacitor can be used to test whether AF signals are present in different stages. When connected from cathode to the AVC connection of an RF or IF transformer, modulated signals should be audible if the receiver is operating up to that point. From the second detector on, the audio signal can be traced to each grid and plate circuit of the different audio stages, indicating where a loss of signal occurs.

18-32 SERVICING TRANSISTOR EQUIPMENT

In most transistorized equipment the transistors are soldered into printed-circuit boards (Sec. 1-17), which complicates servicing. First, it is difficult to loosen the transistors. Second, if the soldering iron is left on the wire lead of a transistor, heat travels up the wire and deforms the junction to which it is connected and the transistor is destroyed. To help prevent this, the transistor wire lead should be gripped, with a copper heat-sink clip or clamp, between the soldered joint and the transistor so that heat traveling up the wire will be transferred to the clip rather than to the transistor. Sockets are used in some equipment into which the transistors are plugged. When servicing is being done, the transistors can be removed easily and tested. Always pull transistors from their sockets when soldering to the socket terminals.

Care must be taken that an electrostatic difference of potential does not build up between transistors in a circuit and the soldering iron. The difference of potential can be reduced by touching the iron tip to a grounded point on the equipment just before soldering to another point in the circuit. Never solder transistorized equipment while it is operating.

Some soldering guns have ac flowing in a hairpin-shaped tip. The magnetic field from this hairpin can induce voltages into low-impedance transistor circuits that can burn out the transistors.

Printed-circuit boards are fairly delicate. The printed patterns will pull loose from the backing if they are overheated. Faulty parts can be carefully unsoldered and new units placed in the holes in the board. Often it is quicker and better to cut off a faulty part, leaving $1/4$ in. of its two leads sticking out from the board. Then solder the replacement part to these two leads.

When printed patterns crack apart, they can be repaired by coating them with solder, although it may be advisable to lay a fine copper wire across the joint and solder it into place.

Transistors soldered into PC boards must be unsoldered $1/16$ in. of lead at a time, moving from one lead to the next, using heat-sink clamps and preferably a flow of air across the transistor. Special desoldering irons remove solder from a connection by suction and should be used on PC boards. When the new transistor is replaced, leave a lead length of about $1/4$ in. above the board connection. Hold leads with a heat-sink clip while soldering them in place.

Power transistors may have their cases bolted to a heat-sink metal, but they often require insulation between case and heat sink. A thin piece of mica coated with special silicon grease to improve heat conduction through it may be used. An ohmmeter check should always be made to make sure the case and heat sink are insulated from each other.

Test your understanding; answer these checkup questions.

1. What sections of a superheterodyne are involved in a complete alignment? _____
2. What tools and equipment should be used when aligning? _____
3. What stage is aligned first? _____ Last? _____
4. Why is a tone generator best for broadcast receivers when aligning? _____ Why may an A0 (unmodulated RF) signal generator be better when aligning a communication receiver? _____
5. When aligning the first IF transformer, what circuit should be stopped from operating? _____
6. When aligning the front end, is L or C varied at the high-frequency end of the dial? _____ At the low-frequency end? _____
7. When aligning using AVC, do you tune for a dip or a peak of the voltage? _____ With an S meter? _____
8. What is the first technique to use when servicing an inoperative receiver at the bench? _____ The second? _____ The third? _____ The fourth? _____
9. What polarity and approximate voltage should be read on a VTVM between chassis (ground or B−) and a VT plate? _____ Grid? _____ Cathode? _____ Screen grid? _____ AVC bus? _____
10. What might be approximate VTVM readings between a PNP BJT emitter and its base? _____ Ground? _____ Collector? _____
11. What is the best indication of oscillation of an oscillator using a VTVM? _____ How else can it be determined? _____
12. Why are 100-W soldering irons never used on transistorized circuits? _____
13. Why is silicon grease used on power BJT insulators? _____

COMMERCIAL LICENSE QUESTIONS

Amateur license questions will be found in the Addendum.

FCC Elements 3, 4, and 6 require an ability to answer questions similar to those below. Sections in which questions are answered are shown in parentheses. A question followed by a bracketed number is required for that element alone.

1. Explain operation of a diode detector. (18-2)
2. Draw a diagram of a diode detector coupled to an AF amplifier. (18-2)
3. Draw a diagram of a crystal detector and explain its operation. (18-3)

4. Name four materials which can be used as crystal detectors. (18-3) **[6]**
5. What is meant by the *sensitivity* of a receiver? In what unit is it measured? (18-3, 18-8)
6. What is meant by the *selectivity* of a receiver? In what unit is it measured? (18-3, 18-9)
7. Explain operation of a *power detector*. (18-4)
8. Draw a diagram of a grid-leak detector and explain its operation. (18-6) **[6]**
9. What effect does the reception of a signal have on (a) the plate current of a grid-leak detector, (b) a power detector? (18-6, 18-4)
10. Draw a diagram of a regenerative detector and explain its operation. (18-7) **[6]**
11. In a regenerative-type receiver, how is oscillation of the detector indicated? (18-7) **[6]**
12. What are the objections to the operation of regenerative detectors when directly coupled to an antenna? (18-7) **[6]**
13. How may a regenerative receiver be adjusted for maximum sensitivity? (18-7) **[6]**
14. What adjustment should be made to a radiotelegraph receiver if the receiver blocks on the reception of strong signals? (18-7, 18-26) **[5]**
15. Do oscillators operating on adjacent frequencies have a tendency to synchronize or drift apart in frequency? (18-7, 18-13) **[6]**
16. Why is an oscillating detector used for reception of an unmodulated carrier? (18-7)
17. What feedback conditions must be satisfied in a regenerative detector to obtain oscillations? (18-7)
18. Describe the principle of operation of a super-regenerative receiver. (18-8) **[6]**
19. Draw a block diagram and a schematic diagram of a TRF receiver. (18-9)
20. What type of radiotelephone receiver using vacuum tubes does not require an oscillator? (18-9)
21. What is the purpose of shielding in a multistage radio receiver? (18-9, 18-23)
22. Draw a block diagram of a single-conversion superheterodyne receiver capable of receiving both AM and CW. Indicate the frequencies present in the various stages. Explain what occurs in each stage. (18-10)
23. What is the purpose of the first detector in a superheterodyne receiver? (18-10, 18-12)
24. What is a mixer in a superheterodyne? (18-10, 18-14)
25. What type of radio receiver has an IF transformer? (18-10)
26. What is meant by *double detection* in a receiver? (18-10)
27. What is the purpose of an oscillator in a receiver operating near the IF of the receiver? (18-10, 18-15)

28. If a superheterodyne receiver is receiving a signal on 1,000 kHz and the mixer oscillator is tuned to 1,500 kHz, what is the IF? (18-10)
29. Why do some superheterodyne receivers employ a crystal oscillator in the first detector? (18-12)
30. Why are iron cores of special construction used in RF or IF transformers? (18-13)
31. What is meant by the *bandwidth* of an amplifier? (18-13)
32. In what way are electrical properties of circuit elements affected by electromagnetic fields? Are interstage connecting leads susceptible to these fields? (18-13)
33. Draw a diagram of an AM second detector and an AF amplifier (in one envelope) showing AVC circuitry. Explain principles of operation. Show coupling to adjacent stages. (18-14)
34. Draw a diagram showing a second detector with a DAVC circuit. (18-14)
35. How should the AVC switch be set for reception of (a) CW signals (b) SSSC signals? (18-14, 18-26) **[5]**
36. Draw a BFO circuit diagram and explain its use in detection. (18-15)
37. What type of radio receiver is little affected by static interference? (18-19, 18-8)
38. What type of modulation is largely contained in static and lightning radio waves? (18-19)
39. What are advantages and disadvantages of using a bandpass switch on a receiver? (18-20)
40. What is the main advantage of a tuned AF amplifier in a CW receiver? (18-20) **[6]**
41. What is the purpose of an IF crystal filter in a superheterodyne communications receiver? When is it used? (18-21) **[6]**
42. Explain the use of the crystal-filter switch on a communications receiver. (18-21, 18-26) **[6]**
43. Show by a diagram how to connect a wave trap in the antenna circuit of a radio receiver to attenuate an interfering signal. (18-22)
44. If broadcast signals interfere with reception of signals on 500 kHz aboard ship, how can this interference be reduced or eliminated? (18-22) **[6]**
45. What type of radio receiver is subject to image interference? (18-23)
46. Explain the relation between signal, oscillator, and image frequencies in a superheterodyne. (18-23)
47. What are three ways of reducing image response in a superheterodyne receiver? (18-23) **[6]**
48. If IF and signal frequency in a superheterodyne are known, how can the probable image frequency be determined? (18-23) **[6]**
49. If a superheterodyne is tuned to 1,000 kHz and its conversion oscillator is operating at 1,300 kHz,

what frequency would cause an image? (18-23)

50. What is the chief advantage in using high IFs in a superheterodyne? (18-23) [6]

51. Why should a superheterodyne used for A1 signals have at least one stage of RF amplification ahead of the first detector? (18-23) [6]

52. Sometimes a station can be heard at more than one place on the tuning dial of a receiver. Is this always an indication that the station is transmitting on more than one frequency? (18-23) [5]

53. Draw a diagram of a superheterodyne receiver with AVC and explain the principle of operation. (18-24) [6]

54. Why is a superheterodyne receiver not successfully used for reception of frequencies near its IF? (18-24) [6]

55. Explain briefly how an SSSC emission is detected. (18-26)

56. Describe how to adjust a communications receiver for reception of (a) CW (A1) signals and (b) A2 signals. (18-26). [5]

57. Discuss methods whereby interference in radio reception can be reduced. (18-26)

58. Why are the unused portions of inductances in receivers sometimes shorted? (18-26) [6]

59. After long periods of listening to CW signals of constant tone, what adjustment can the operator make to relieve hearing fatigue? (18-26) [5]

60. Explain briefly the principles involved in detecting an FSK signal. (18-26)

61. What is the purpose of a diversity-antenna receiving system? (18-27)

62. Explain, step by step, how to align an AM receiver by using a signal generator with a speaker, oscilloscope, and VTVM. What is occurring during each step? (18-30)

63. Explain how spurious signals can be received or created in a receiver. How could this be reduced in sets having sealed untunable filters? (18-30, 18-23)

64. State some conditions under which readings of AVC voltage would be helpful in troubleshooting a receiver. (18-30)

65. Explain how to test components in a receiver. (18-30) [6]

66. What may be the cause of noisy operation of a receiver? (18-30)

67. If a tube in the RF stage of a receiver burned out, how could temporary repairs or modifications be made to permit operation of the receiver if no spare tubes were available? (18-31) [6]

68. List some precautions to be observed when soldering transistors and repairing printed circuits. (18-32)

ANSWERS TO CHECKUP QUIZ ON PAGE 460

1. (*IF strip and front end*) 2. (*Signal generator, insulated screwdriver, VTVM*) 3. (*Second detector*) (*First* RF) 4. (*Simple, audible, fast*) (*No SBs to confuse indications*) 5. (*LO*) 6. (*C*) (*L*) 7. (*Peak*) (*Peak*) 8. (*Observe carefully*) (*Check cord and input to power supply*) (*Check tubes or transistors*) (*Measure voltages or resistances, starting at last AF*) 9. (+ *high*) (*0*) (+ *few volts*) (*60–100 V +*) (− *few volts with signal input*) 10. (*0.3–0.6 V*) (*Perhaps* $\frac{1}{10}$ *PS voltage if resistor in circuit*) (*Half to full PS voltage*) 11. (− *grid bias for VTs, reverse bias for transistors*) (*With another receiver, with a scope*) 12. (*Overheating junctions damages them*) 13. (*Improve thermal conductivity*)

19 FREQUENCY MODULATION

The objective of this chapter is to present simple theories of frequency and phase modulation, to discuss circuits used to generate and to detect FM signals, and to apply these circuits to previously discussed basic systems to produce wideband broadcast FM and narrowband communication FM transmitters and receivers. Stereophonic, SCA, and FAX multiplex systems that may be applied to an FM broadcast carrier are briefly outlined.

19-1 PURPOSE OF FREQUENCY MODULATION

Since its inception there has been a constant search for better methods of utilizing radio as a means of communication. By 1930 the superheterodyne had been developed, as had high-powered CW and AM transmitters. Since that time communication equipment has become less bulky and better insofar as sensitivity and selectivity are concerned, but difficulty with man-made and natural static still exists. Radio broadcasters are usually interested in a primary coverage of only a few miles, so the standard broadcast band of 535 to 1,605 kHz with its 1- to 50-kW stations, is satisfactory. However, even in the primary coverage area a lightning storm can make reception of a 50-kW station unpleasant. In order to prevent noises that vary in amplitude from interfering with reception, it is necessary to make the receiver unresponsive to amplitude variations.

As previously stated, a sinusoidal ac can be changed in only three ways: (1) amplitude, (2) frequency, and (3) phase.

The study thus far has been mostly about transmitters and receivers that produce and demodulate amplitude changes. The short discussions on frequency-shift keying (FSK) indicated that it is possible to transmit intelligence by changing the carrier frequency without changing the carrier amplitude. Frequency modulation (FM) is similar to FSK in that the carrier is made to swing back and forth in frequency, although at an audio rate rather than at a code rate. If the carrier sweeps back and forth 1,000 times per second, the carrier is being frequency-modulated at a 1,000-Hz rate. When the audio voltage that produces the FM of the carrier is no longer present, the carrier comes to rest at a *center frequency* and remains there until another modulating voltage is applied.

An FM receiver is similar in many respects to an AM receiver. It is a superheterodyne but has a special second detector that demodulates frequency changes instead of amplitude variations. The IF stages are amplifiers, but the last two may also act as amplitude limiters. If the FM receiver is not sensitive to amplitude variations, static crashes and other impulse signals are not demodulated and are therefore not audible.

19-2 THE FOUR FIELDS OF FM

There are four major fields in which FM is in use. One is in the FM broadcast band, from 88 to 108 MHz, in which FM stations broadcast programs to the everyday radio listener. While FM transmissions may have greater fidelity of transmission than those on the standard AM broadcast band, it is only because the AM audio is fed through low-pass filters that usually prevent frequencies higher than 5,000, or perhaps 7,500, Hz from modulating the transmitter to prevent interfering with adjacent-channel transmissions on the AM band. FM transmitters attenuate no frequencies under 15,000 Hz.

A second use of FM is in television. The video, or visible, signals are amplitude-modulated, but the sound is transmitted on a separate transmitter and is frequency-modulated. Thus, a TV receiver must be an AM and an FM receiver at the same time.

A third use of FM is in the mobile or emergency services, such as taxicabs, police, and fire communications, which are interested in transmitting voice frequencies only, up to about 3,000 Hz.

A fourth use of FM is in the amateur bands. Here again, only voice frequencies are transmitted.

19-3 BASIC CONCEPTS OF FM

In AM, the louder the sound striking the microphone, the greater the variation of the strength of the carrier and the stronger the signal developed in the detector stage of the receiver. In FM, the louder the sound, the greater the variation of the carrier frequency from the assigned or center frequency and the stronger the signal developed in the FM detector stage.

If a carrier is made to deviate 75 kHz on each side of a center, or resting, frequency (a total swing of 150 kHz), a signal will be developed at the detector of a receiver that is well above other normal noises that might also be received. This is the maximum carrier excursion set by the FCC for FM broadcast stations, and it is considered *100% modulation*. (Actually, any excursion in frequency could have been selected as 100%.) If the carrier is 50% modulated, it swings 37.5 kHz on each side of the center frequency; 60% modulation produces a *deviation* of 45 kHz; 80% modulation gives a 60-kHz deviation. Doubling the modulating AF voltage doubles the frequency swing. As long as the percent of modulation remains the same, the frequency swing remains the same, regardless of the frequency of the modulating voltage (disregarding pre-emphasis).

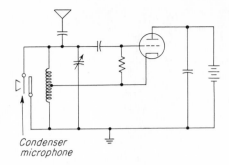

Fig. 19-1 A possible method of producing FM.

A simple system by which FM could be developed is shown in Fig. 19-1. The circuit shows a series-fed Hartley oscillator with a condenser microphone connected across the tuned circuit. When sounds strike the diaphragm, it swings in and out, changing the capacitance across the microphone, which changes the frequency of the oscillator. The *amplitude* of the RF ac generated by the oscillator will not change, however.

The louder the sounds striking the microphone, the greater its capacitance change and the farther the carrier swings from its center frequency. If the tone of the sound is 500 Hz, the microphone diaphragm vibrates 500 times per second and the frequency of the oscillator swings higher and lower 500 times per second. If the same-strength sound, but one with a frequency of 1,000 Hz, strikes the microphone, the diaphragm will vibrate twice as rapidly, producing a carrier that increases and decreases frequency 1,000 times per second. Since the strength of the tone is the same, the deviation of the carrier from the center frequency is the same. If the sound striking the microphone is weaker, the deviation (and percent of modulation) will be less.

In AM, modulation of the carrier produces sidebands. Any single tone used to modulate the carrier produces a sideband on each side of the carrier frequency. During modulation each succeeding RF ac cycle is slightly greater or less in *amplitude* than the one before. This distortion of the RF wave can be considered as producing the sidebands.

In FM, swinging a carrier from one frequency to another means that no RF ac cycle can be a pure sine wave, since each succeeding cycle is at a slightly higher or slightly lower *frequency* than the one preceding it. This results in sidebands, but instead of only one pair of sidebands for any one tone (as in AM), the number of significantly strong sidebands produced by FM will depend on how far the carrier is swung. The greater the swing, the greater the number of sidebands. Theoretically, an infinite number of sidebands are produced by FM, but only a few may be strong enough to be significant.

In broadcast FM, 15 kHz is the highest required AF to be transmitted, and 75 kHz is the widest frequency deviation allowed. The ratio of the greatest allowable deviation to the highest modulating frequency is therefore 75:15. This is known as the *deviation ratio* of an FM broadcast transmitter.

The ratio of the maximum deviation allowed (75 kHz) to a specific modulating frequency being used (1 kHz, 5 kHz, etc.) is known as the *modulation index*. (Deviation ratio and modulation index are the same only with 75-kHz deviation and a modulating frequency of 15 kHz.)

The deviation ratio of 5 that is used in FM broadcasting will produce eight significant sidebands above and eight below the center frequency with a 15-kHz modulating tone. (The number of significant sidebands is mathematically determined by Bessel functions, which are beyond the scope of this book.) A 15-kHz modulating tone, forcing the carrier to deviate 75 kHz, produces eight significant sidebands, each one 15 kHz from the adjacent sidebands (Fig. 19-2). Such a modulating signal will produce significant sidebands 120 kHz each side of the center fre-

Fig. 19-2 Significant sidebands developed by a 15-kHz tone swinging the carrier 75 kHz above and below the center frequency.

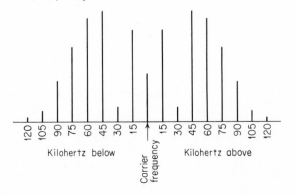

quency, (a bandwidth of 240 kHz). Since the fundamental tones of all musical instruments are below 5 kHz, however, only the relatively weaker overtones, or harmonics, are in the range of 5 to 15 kHz. It is unlikely a strong signal will ever be transmitted at as high an audio frequency as 15 kHz. With weaker signals, only a few sidebands may be developed, requiring perhaps 120 to 150 kHz of bandwidth.

With a 10-kHz tone the modulation index will be 75/10, or 7.5. This will produce about eleven significant sidebands, or a total bandwidth of $2 \times 10 \times 11$, or 220 kHz. So high a frequency is probably greater than will be transmitted at full volume, and therefore a 200-kHz bandwidth will still suffice.

When a 5-kHz tone is transmitted, its modulation index is 75/5, or 15. A modulation index of 15 has 20 significant sidebands and will require a bandwidth of $2 \times 5 \times 20$, or 200 kHz, provided the carrier is deviated a full 75 kHz by the 5,000-Hz tone. While this is a high frequency, it is possible that it *may* be applied to the transmitter with enough amplitude to produce a full 75-kHz deviation (if pre-emphasis is being used).

An FM broadcast channel is 200 kHz wide, whereas a standard AM broadcast channel is 10 kHz wide. By using a part of the spectrum (88 to 108 MHz) where radio waves travel only a little more than line of sight (usually less than 100 miles) daytime or nighttime, FM stations can use wide channels economically.

The FM band is 20 MHz wide, with each channel being 200 kHz wide. (This includes the 75 kHz each side of center frequency for 100% modulation, plus 25 kHz on each side as "guard bands.") The first channel has a center frequency of 88.1 MHz; the second, 88.3 MHz, etc., providing 100 FM channels. In one locality, however, stations are never assigned adjacent channels, to prevent interference due to overlapping of outer sidebands. This reduces the possible number of stations in one local area to 50.

Since only the frequency of the oscillator is modulated in FM, only a fraction of a watt of audio power is required to produce 100% modulation. Compare this with the 500 W of audio power required to amplitude-modulate 1,000 W of dc plate-power input to an RF amplifier. In AM, the RF sidebands are added to the carrier power. In

FM, the carrier power is used to form the sidebands. As a result, the antenna current of an FM transmitter should not change when the carrier is modulated. However, if the tuned circuits of the RF amplifiers of the transmitter or of the antenna have too high a Q, the bandpass of these circuits may be too narrow and sidebands widely displaced from the center frequency may be attenuated. In this case, the antenna current may drop off slightly at high percentages of modulation. This may also occur if the stages are not properly tuned.

The various frequency-shift or frequency-modulated types of emissions are as follows:

○ F1. A carrier that is shifted or keyed in frequency according to some code, such as International Morse or radioteleprinter.
○ F2. An on-off keyed carrier that is shifting in frequency at some audible rate, or a carrier that has AF code signals frequency modulating it.
○ F3. Frequency-modulated voice or music transmissions.
○ F4. A carrier shifted in frequency in accordance with still-picture elements. Facsimile transmissions.
○ F5. A carrier shifted in frequency in accordance with moving-picture elements. Television transmissions. (Used only experimentally, not on commercial TV channels.)

19-4 SLOPE DETECTION

FM transmissions can be detected with AM receivers by tuning the receiver a few kilohertz off the transmitter carrier frequency. Figure 19-3 illustrates a possible response curve of an AM receiver with a bandwidth of 6 kHz and centered on 460 kHz. By tuning the receiver so that the FM carrier is at 464 kHz in the IF strip, variations of the carrier up and down in frequency (jagged waveform) will produce similarly shaped amplitude variations in the detector output. (The FM carrier could be tuned to the other slope, 456 kHz, and be detected as well.) If the FM deviation is greater than the linear slope of the response curve, distortion is produced. For this reason, only narrowband FM can be detected with this receiver. Furthermore, the main advan-

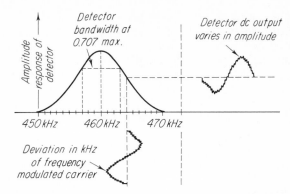

Fig. 19-3 How frequency variations along the slope of the response curve of an AM receiver can produce amplitude variations of the detector output.

tage of FM, limiting of impulse noise, does not occur with slope detection. Tuning the FM carrier to the center of the IF strip produces almost no audible output, as FM sidebands cancel each other in an AM detector.

19-5 A STAGGER-TUNED DISCRIMINATOR

An FM detector that operates on a theory of bucking voltages is shown in Fig. 19-4a. The IF signal, usually centered on 10.7 MHz, is fed equally to two LC circuits, one tuned 200 kHz above the carrier frequency (10.9 MHz) and the other 200 kHz below (10.5 MHz). When a signal comes through the IF strip on 10.7 MHz, the signals induced in both LC circuits and rectified by D_1 and D_2 are equal. The currents flow as shown, developing equal bucking voltages E_1 and E_2 across the equal resistors R_1 and R_2. Therefore, the output voltage E_0 is zero.

If the carrier deviates 50 kHz higher in frequency, the 10.9-MHz LC circuit has greater current induced into it and E_1 becomes greater than E_2. From the current directions shown, this results in a negative E_0. If the carrier deviates 50 kHz low, E_2 exceeds E_1 and E_0 becomes positive. In this way frequency variations are converted to negative-positive voltage alternations. FM is converted to AF ac in this manner.

The heavy-line curve of Fig. 19-4b illustrates a normal discriminator curve. As long as the FM of the carrier, f_0, is held beneath the straight portion of the curve, the output AF ac will be an exact

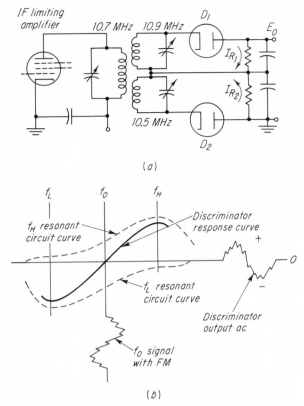

Fig. 19-4 (*a*) A stagger-tuned or balanced discriminator capable of converting frequency variations to amplitude alternations. (*b*) Resultant characteristic discriminator S curve.

replica of the FM. The dashed lines in this case are the 10.9-MHz (f_H) and the 10.5-MHz (f_L) LC circuit resonant curves, dropping off at the extremities due to the resonance effect of the primary. The primary Q should be half of that of the secondaries.

Since variations in *strength* of the FM carrier will also produce amplitude variations in E_o, all signals fed to any discriminator should first be limited to a constant amplitude.

The negative- or positive-going output from a discriminator can be used as a detector for FSK signals. A negative-going output signal can forward-bias a PNP transistor into conduction, and the positive-going output signal can forward-bias an NPN transistor to actuate relays or circuits to key a teleprinter, for example.

1. Why was FM first developed? _____
2. To what feature of the modulating tone is FM deviation proportional? _____
3. What are the four basic fields of FM? _____ _____
4. Under what condition would a 1-kHz AF signal produce a single pair of FM sidebands? _____
5. What deviation is considered 100% for FM BC stations? _____ What receiver bandwidth is required to receive this? _____
6. In FM broadcasting, what is the highest required modulating frequency? _____ Lowest? _____
7. What is the ratio of the greatest allowable deviation to the highest modulating frequency called? _____
8. What is the ratio of maximum deviation allowed to the modulating frequency being used called? _____
9. How much AF power is required to plate-modulate a 10-kW PA of an AM transmitter? _____ To modulate a 10-kW FM station? _____
10. What is the letter-number designation for voice or music FM? _____ FM facsimile? _____ FSK? _____ A keyed FM-tone carrier? _____
11. What is the disadvantage of slope detection of FM? _____
12. Why should discriminators be tuned for a straight characteristic S curve? _____ What would be the advantage of a steep curve? _____
13. Would high-Q coils be more desirable in wide- or narrow-band stagger-tuned-type discriminator transformers? _____ Why? _____

19-6 THE FOSTER-SEELEY DISCRIMINATOR

Besides its use as an FM detector, this well-known circuit has many applications in electronics where a change in frequency must produce a negative or positive control voltage.

To understand how the Foster-Seeley circuit operates it is necessary to examine basic transformer operation. In an unloaded transformer, as in Fig. 19-5a, when primary current I_p increases, a counter-direction voltage E_i is induced in the secondary. This produces a difference of potential E_s across the secondary. The phase relations are shown to the right, E_i and E_s in phase and I_p 180° out of phase. No secondary current flows.

In Fig. 19-5b, the load has a low resistance value, and a high current I_s flows in the secondary. Since the inductance of the secondary and the resistor are in series, I_s lags the induced voltage E_i and by a relatively large angle. The reactive

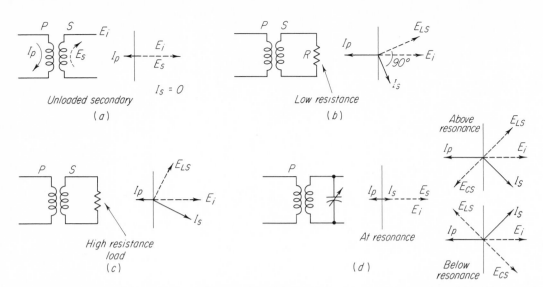

Fig. 19-5 Voltage-current relations in a transformer when (a) unloaded, (b) heavily loaded, and (c) lightly loaded and (d) when the secondary is a series-resonant circuit.

voltage developed in the coil E_{Ls} is always 90° ahead of I_s, as indicated. If R were zero ohms, I_s would be 90° out of phase with the induced voltage E_i.

When the load resistance is high as in Fig. 19-5c, little current flows in the secondary and I_s is more nearly in phase with the induced voltage E_i. With infinite resistance, E_i and I_s would be in phase, theoretically, as in (a).

When the load is a resonant circuit as in Fig. 19-5d, the secondary-circuit reactances cancel. I_s and E_i are in phase. (A resonant circuit is always a resistive load.)

When the secondary LC circuit is tuned above the resonant frequency, the X_L of the coil increases, X_C decreases, and the LC circuit current is determined more by inductance. As a result, I_s lags E_i.

When the LC circuit is tuned lower than the resonant frequency, X_C predominates and I_s leads

E_i. The farther from resonance, the greater the lag or lead of I_s.

If a resonant secondary is center-tapped and connected in a circuit such as Fig. 19-6a, the rectified dc will be in the direction shown by the arrows. From the center tap CT, the two voltages E_{R_1} and E_{R_2} are 180° out of phase, as indicated by the vector arrows. At the resonant or any frequency the circuit is balanced, and E_0 is zero.

The circuit in Fig. 19-6b is a Foster-Seeley discriminator. When the secondary LC circuit is tuned to the signal in the IF amplifier, E_{R_1} and E_{R_2} are equal and out of phase and the output voltage E_0 is zero.

The heavy lines show that L_4 is essentially in parallel with L_1 insofar as ac is concerned. Therefore, any voltage across L_1 must be in phase with the voltage across L_4. Voltage E_{L_4} can be used as a reference voltage.

The phase relationships that are present in the circuit when the signal is above the resonant frequency are shown in Fig. 19-6c. The LC circuit appears inductive, and its current I_s lags. The reactive voltages E_{L_2} and E_{L_3} must each be 90° from this current, as shown. Diode D_1 and load resistor R_1 are across E_{L_2} and E_{L_4}. Therefore, the current that flows in this circuit will be proportional to the *resultant* of these two voltages and is labeled E_{D_1}. Similarly, the current in the second

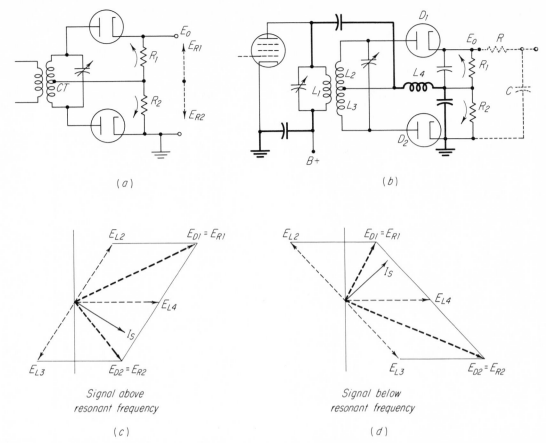

Fig. 19-6 Foster-Seeley discriminator. (*a*) Preliminary circuit. (*b*) Foster-Seeley circuit. (*c*) Phase relations with signal above the resonant frequency. (*d*) Phase relations with signal below the resonant frequency.

diode is proportional to the voltages across L_3 and L_4 and is labeled E_{D_2}. As a result, when the received signal deviates upward in frequency, E_{R_1} (or E_{D_1}) is greater than E_{R_2} (or E_{D_2}) and the output E_o is positive.

When the received signal deviates below the resonant frequency of the secondary, the phase relationships are as illustrated in Fig. 19-6*d*. E_{R_2} exceeds E_{R_1}, and the output voltage E_o is negative. In this way, a received carrier deviating up and down in frequency produces an ac output voltage E_o.

There are many variations of this basic circuit which can produce similar results. They may be used with solid-state diodes in transistor receivers.

All discriminators are sensitive to amplitude variations and must be fed signals that have been limited in amplitude.

The R and C, shown in dotted lines, form a de-emphasis circuit, used in FM broadcast receivers (Sec. 19-10).

19-7 THE RATIO DETECTOR

The FM detector shown in Fig. 19-7 demodulates FM signals and suppresses amplitude noise impulses without limiter stages ahead of it.

In the ratio detector, the tuned circuit, the two diodes, and the resistor R_1 are all in series. (Note that the top diode is reversed from the Foster-Seeley circuit.) This allows the voltages that are built up across C_1 and C_2 to be in series, instead of opposing as in the discriminators discussed

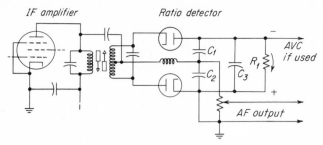

Fig. 19-7 A basic ratio-detector circuit with slug-tuned primary and secondary.

above. The current through R_1 produces a voltage-drop across it proportional to the average carrier strength being received. The capacitor C_3 has sufficient capacitance (10 μF) to enable it to hold the voltage across R_1 constant even if there are instantaneous amplitude variations. The voltage across C_1 plus the voltage across C_2 must always equal the voltage across C_3, from which the circuit derives its name. (Without C_3 the circuit is an AM detector.)

When the received signal is on the center frequency of the tuned secondary, the dc voltages across C_1 and C_2 are equal, as in the Foster-Seeley circuit. If the carrier is deviated lower in frequency, the vector-sum voltage across C_1 will increase and that across C_2 will decrease. When the carrier deviates higher in frequency, C_2 volt-

age will increase. Thus, frequency modulation at an audio rate produces a varying dc audio signal across C_2 (or C_1).

Note that the top of C_3 is negative in respect to ground. The stronger the input signal, the more negative it becomes. The voltage is used as AVC and is fed to the grids of RF and IF stages. Ratio detectors are often used in transistor receivers.

19-8 THE GATED-BEAM DETECTOR

The gated-beam FM detector is used in many TV receivers, which always use a 4.5-MHz IF to demodulate the sound portion of the program. Its operation depends upon a special three-grid tube, the 6BN6 (6DT6, 6HZ6, etc.), and circuit (Fig. 19-8a). The limited grid produces plate-current

Fig. 19-8 A gated-beam FM detector. (a) Circuit. (b) Limiting or clipping effect. (c) Result of two out-of-phase gating voltages on plate-current pulses.

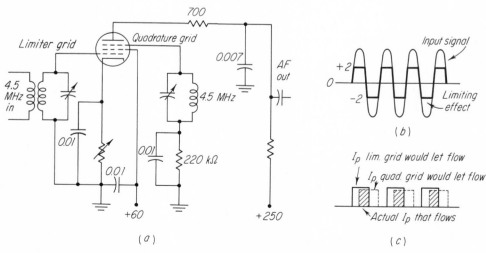

cutoff at approximately -2 V and plate-current saturation at about $+2$ V. As a result, any signal with an amplitude over 2 V will be limited in the output (Fig. 19-8b). The quadrature grid has a similar characteristic, producing I_p cutoff at -3 V and I_p saturation at $+2$ V.

Signal voltages at the limiter grid produce pulses of plate current. As these bunches of electrons flow past the quadrature grid, some strike it, setting its 4.5-MHz LC circuit into oscillation. It is characteristic of oscillating circuits at resonance that the current flowing through them and the voltage across them is $90°$ out of phase. Thus, the quadrature grid has a voltage similar to the limiter-grid signal voltage, except that it is $90°$ out of phase. Both grids are effective in stopping or starting (gating) plate current, as indicated in Fig. 19-8c. With both operating, I_p flows for approximately one-quarter of the whole cycle.

If the input signal deviates up in frequency, the phase relationship of the 4.5-MHz resonant circuit changes and the gated plate-current pulses become narrower, representing less power in the output. As the carrier deviates lower in frequency, the phase shift reverses and the plate-current pulses lengthen, representing more power in the output. A deviating carrier produces a variation of power output. This produces a pulse-width demodulation somewhat similar to superregenerative reception (Sec. 18-8).

19-9 LIMITERS

Except for the gated-beam and the ratio detectors, FM detectors are inherently susceptible to amplitude variations and noise. It is necessary to use one or, preferably, two *limiter stages* ahead of the detector. These must limit the amplitude of the carrier being received to a constant value, regardless of the strength of the signal. They are actually low-gain IF amplifiers using either grid- or plate-circuit limiting or both, as in Fig. 19-9. The limiter shown uses impedance coupling. Transformer and resistance coupling may be used. The grid leak R_1 produces increased bias if the signal increases, decreasing the plate current and holding the output signal to a constant value. The voltage-divider network of R_2, R_3, and R_4 lowers the plate and screen voltages to such a value that the tube cannot have very much signal-voltage output. In most cases it is preferable to have two limiter stages, since a single stage of this type cannot limit a wide variation of signal-amplitude change. When signals are very weak, the limiters act as low-gain amplifiers without limiting and the discriminator will have noise in its output.

Solid-state diodes or zener diodes acting as shunt-type limiters can be connected across the grid circuit of an IF stage.

While FM receivers using limiter stages may not

Fig. 19-9 IF limiter amplifier with both grid- and plate-circuit limiting.

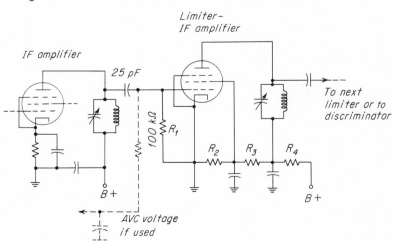

require it, AVC may be used to prevent front-end overloading by strong local signals. An AVC voltage can be taken from the grid of the *first* limiter stage through a long-time-constant RC network, shown in dotted lines in Fig. 19-9.

19-10 PRE-EMPHASIS AND DE-EMPHASIS

Unwanted amplitude variations in electronic circuits are classified as noise. Random electron motion, whether in tubes, transistors, diodes, wires, resistors, or capacitors, produces some noise. The warmer the circuits the greater the noise produced. In the low AF range thermal-noise content is low, but at higher frequencies noise increases. Such noise degrades signals coming through any amplifier. To overcome this partially, pre-emphasis and de-emphasis are used. The FM broadcast transmitter amplifies or emphasizes the higher audio-frequency program signals more than it does the lower. A *pre-emphasis* circuit consists of a series capacitor and a shunt resistor with a time constant of 75 μs ($T = RC$, or $RC = 75 \times 10^{-6}$) in some audio circuit. At 1,000 Hz the pre-emphasis is about 1 dB, at 5 kHz about 8 dB, and at 15 kHz about 17 dB. Therefore, all FM broadcast receivers must have a similar 75-μs time-constant *de-emphasis* circuit (a series R and shunt C) as was shown

in Fig. 19-6, following any discriminator or other detector.

19-11 FM RECEIVERS

FM broadcast-band receivers may be constructed as shown in block form in Fig. 19-10. These receivers tune from 88 to 108 MHz and have an IF bandwidth of 200 kHz or more. Less-expensive receivers may leave out the RF amplifier, an IF amplifier, or a limiter stage.

Although single-tube pentagrid converters or triode-hexode converters may be used as the mixer and local oscillator, in VHF receivers it may be preferable to use a separate oscillator tube to reduce frequency pulling and detuning of the oscillator when the mixer is tuned during alignment. The local oscillator is operated on a lower frequency than the mixer to take advantage of better oscillator frequency stability at lower frequencies.

An FM communication receiver suitable for VHF or UHF mobile or fixed-station operation is shown in block form in Fig. 19-11. It is a double-conversion superheterodyne using crystal oscillators in the mixer stages.

FM receivers produce a loud noise when no carrier is being received to quiet them. To prevent this, a negative AVC-type voltage from the first

Fig. 19-10 Block diagram of FM broadcast receiver. Automatic-frequency-control (AFC) circuit shown in dashed lines.

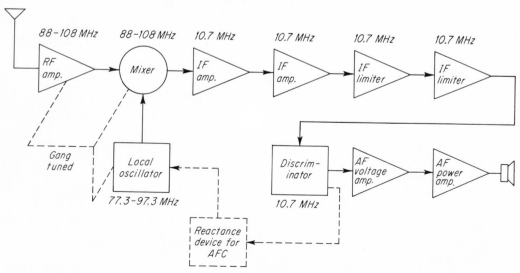

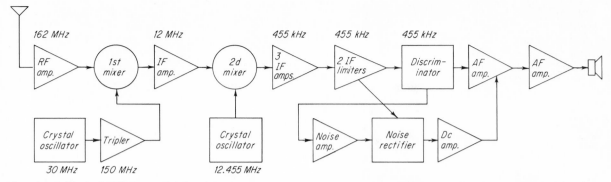

Fig. 19-11 Block diagram of a VHF FM communication receiver with squelch voltage that rides just above the noise level.

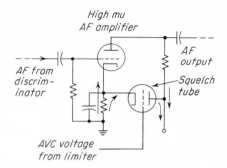

Fig. 19-12 A possible squelch circuit.

limiter grid can be used to actuate the squelch circuit that was discussed in Sec. 18-16. Another possible squelch circuit is shown in Fig. 19-12. With no signal there is no AVC voltage and the squelch tube conducts heavily, developing a high current through the cathode resistor and a high bias voltage for the high-μ AF amplifier tube, cutting it off. When a signal is received, AVC voltage is developed, cutting off the squelch tube plate current, and the AF stage operates normally. In mobile equipment the noise level is constantly changing, which would require constant adjustment of the squelch-circuit level. For mobile receivers a McMurphy differential squelch (Fig. 19-13) can be used. The output of the discriminator, after passing through a de-emphasis circuit, is fed in two directions: (1) to the AF amplifier V_1, (2) to a noise amplifier through a high-pass filter C_1 and R_1.

With no carrier being received, noise appearing at the discriminator is amplified by the noise amplifier and is fed to two diode rectifiers. One charges C_2 positively, and the other charges C_3

negatively. The negative charge of C_3 is fed to the limiter grid, which reduces its gain and the noise in the system. The positive charge on C_2 is fed to the dc amplifier or squelch tube Q, making plate current flow in this tube, down through the 100-kΩ resistor in the AF amplifier grid, biasing this tube to cutoff, and quieting the AF amplifier.

With a carrier, there is no noise, only audio frequencies of 3 kHz and lower due to modulation will be heard. This energy has little charging effect on C_2 or C_3, and the positive charge on C_2 drops off. The carrier produces a negative bias across the limiter-grid resistor which is fed to the grid of the squelch tube, cutting its plate current off and removing the cutoff bias on the AF amplifier V_1.

With no carrier being received, whenever the noise level rises, the bias on the limiter grid is increased and the squelch tube cuts off more. This results in a squelch action that rides just above the noise level at all times.

An example of a solid-state squelch system is shown in Fig. 19-14. A limiter is feeding a discriminator, which in turn feeds its output to an AF amplifier. First, consider the circuit when no carrier is being received. High-amplitude receiver-developed noise is present at the output of the limiter and also from the discriminator. D_1C_1 together act as an AM detector, producing a short-time-constant positive voltage where indicated. (AF to 3 kHz represents low-time-constant signals, whereas noise consists mostly of frequency components above 5 kHz.) If the squelch control is turned down to ground, the positive signals feed through the 0.03 capacitor to another detector, D_2C_2, this time producing a negative potential that

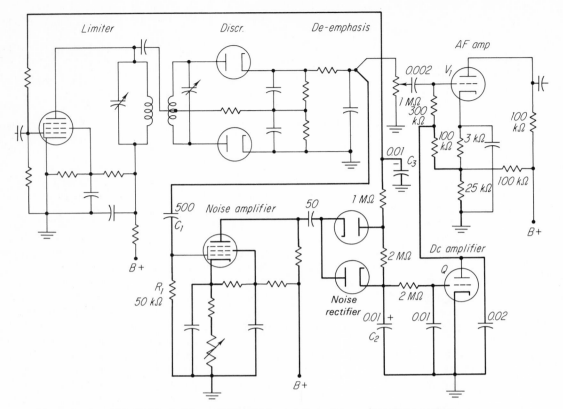

Fig. 19-13 A differential-squelch circuit used in mobile FM receivers.

Fig. 19-14 A solid-state squelch system.

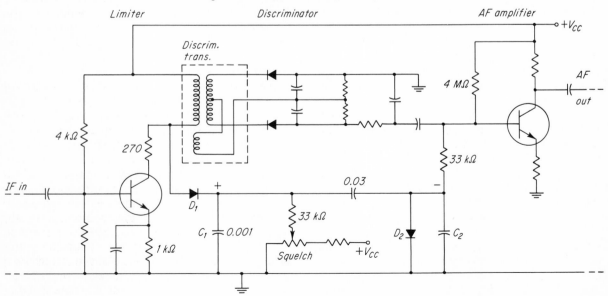

can reverse-bias the AF amplifier, completely silencing it. If the squelch control is moved up, it produces a bucking positive voltage that prevents D_1C_1 from rectifying noise, resulting in no noise signal fed to D_2C_2. With no reverse bias the AF BJT is free to amplify the high noise level that comes from the discriminator. There will be a threshold squelch setting that will just stop all noise amplification.

When a carrier is received, it quiets the discriminator and the limiter. With no signal for D_1C_1 to rectify, no reverse bias is developed for the AF BJT. Any modulation detected by the discriminator is amplified by the now normally biased AF BJT. Due to the short time constant of D_1C_1, AF modulation on the carrier will not be detected with sufficient signal strength to develop reverse bias in D_2C_2.

VHF reception is subject to much man-made impulse-noise interference from automobile ignition systems, motors, and electric circuits being made and broken. For a given signal strength, the limiter stages in an FM receiver are more effective in eliminating such interference than audio peak limiters in AM receivers. Probably of more importance, in the VHF region signals are subject to multipath transmission. Signals transmitted from a car may reach the fixed-station receiver from two or more different directions because of the susceptibility of such frequency signals to reflection by wires and large metal objects. If the multipath signals arrive in phase, the signal is strong. If they are out of phase, the signal is weakened. If the car is moving, the signals may fade up and down rapidly, producing a fluttering effect in the receiver. The AVC circuits in AM receivers can not follow such a rapid fade, and reception suffers. The FM receiver depends on its limiter stages to keep the signal to the discriminator at a constant level. FM reception of fading signals is not subject to flutter until the signal becomes so weak as not to operate the limiters. Many voice communications at frequencies over 100 MHz are FM; between 30 and 100 MHz they may be F3, A3J, or A3; and below 30 MHz most are A3J or A3.

Although FM is authorized in the amateur bands, it is not popular except on the VHF bands. It is relatively simple to modulate an amateur CW transmitter by direct FM or by PM (phase modulation), but few amateurs have a satisfactory narrowband FM receiver to receive such signals properly.

Each of two amplitude-modulated stations transmitting on the same frequency will produce sideband signals in an AM detector. Both can be heard, even if one signal is 10 times the strength of the other. In FM, if one signal is twice as strong as another on the same frequency, the stronger will capture the oscillation frequency of the discriminator circuit and the weaker signal will not be able to produce any output in the receiver. This capture effect in amateur bands is a disadvantage and the reason why aircraft usually use A3 or A3J.

19-12 ALIGNMENT OF FM RECEIVERS

FM broadcast receivers use a high IF, 10.7 MHz, to produce a relatively broad, low-Q, 200-kHz bandwidth. On the other hand, FM communication receivers, such as Fig. 19-11, operating in the 25- to 1,000-MHz range use double conversion, with the first IF at high frequency, 2 to 30 MHz, to reject image signals. The second IF may be near 450 kHz to produce the 20-kHz IF bandwidth required for ±5-kHz deviation voice transmissions.

The alignment of the RF amplifier, mixer, and IF amplifiers in FM receivers is similar to the procedure used in AM receivers. In Foster-Seeley receivers the discriminator and limiters are aligned first. An unmodulated signal generator is adjusted to the IF and fed to the input of the last limiter. This feeds a signal to the discriminator transformer. A sensitive dc voltmeter with an RF choke, a 1-MΩ resistor in the ungrounded probe, or a vacuum-tube voltmeter is connected between ground and the center tap of the two series resistors in the discriminator circuit. The secondary of the discriminator is detuned until an indication of voltage is obtained on the meter. The primary of the discriminator transformer is then tuned for a maximum indication on the meter.

The meter is next connected to read the voltage across both series resistors. The secondary of the discriminator is then tuned to zero volts across the load resistors. The signal generator is shifted 100 kHz above and then 100 kHz below the IF to check the linearity of the discriminator. The same-amplitude voltage, but of opposite polarity, should appear 100 kHz from the carrier in both directions.

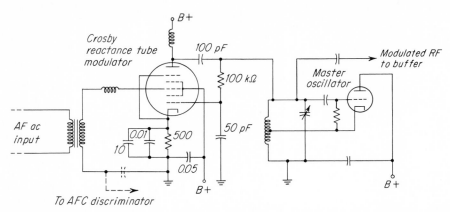

Fig. 19-15 Crosby reactance-tube modulator across the tuned circuit of a self-excited oscillator.

The signal generator is next moved to the grid of the first limiter. The voltmeter is connected across the grid leak of the second limiter, and the plate circuit of the first limiter is tuned for maximum voltage indication. If transformer coupling is used between limiters, the grid circuit of the second limiter is next tuned, again for maximum voltage indication. The weakest possible signal from the signal generator must be used.

The signal generator is then moved to the grid of the last IF amplifier; the voltmeter is moved to the grid of the first limiter; and the plate circuit of the last IF amplifier is tuned for maximum voltage. If transformer coupling is used between stages, the grid circuit of the first limiter is tuned next.

The voltmeter can be left in its last position, and the remainder of the alignment of the IF, mixer, oscillator, and RF stages will follow the pattern explained for AM receivers (Sec. 18-28).

To align a ratio detector the signal generator is fed to the last IF amplifier input. Both the primary and the secondary of the ratio-detector transformer are tuned for maximum voltage across the large capacitor in the detector circuit. The voltmeter may be left across the capacitor to tune the IF transformers also.

Test your understanding; answer these checkup questions.

1. In the Foster-Seeley circuit shown, for what is L_4 used? _____ Could a resistor be used instead? _____

2. Is the AF output voltage varying dc or ac in a Foster-Seeley circuit? _____ Stagger-tuned discriminator? _____ Ratio detector? _____ Gated-beam detector? _____

3. In question 2, which circuits require limiters ahead of them? _____ Which provide an AVC voltage? _____ Which has its diodes in series? _____

4. To what frequency must the gated-beam quadrature circuit be tuned in a TV receiver? _____ In an FM BC (broadcast) receiver? _____

5. In what way is the gated-beam detector similar to a superregenerative detector? _____

6. What are the three methods of limiting that were mentioned? _____ _____ Are they all applicable to transistorized circuits? _____

7. With what FM detector(s) would AVC be an advantage? _____

8. What is the reason for using pre-emphasis? _____ How much is produced at 1, 5, and 15 kHz? _____

9. What are the two types of stages in an FM receiver that differ from those in an AM receiver? _____

10. What special circuits are used in a squelch system that can follow changing noise levels? _____

11. Why might FM be better than AM for mobiles? _____ _____

12. What is the order of circuit alignment in an FM receiver? _____

19-13 DIRECT FM

There are many methods of producing FM. One direct FM method, the Crosby, utilizes a vacuum tube that is made to appear as an inductive or a capacitive reactance across a self-excited oscillator LC circuit, as shown in Fig. 19-15.

The 100-kΩ resistor and the 50-pF capacitor across the oscillator LC circuit feed oscillator fre-

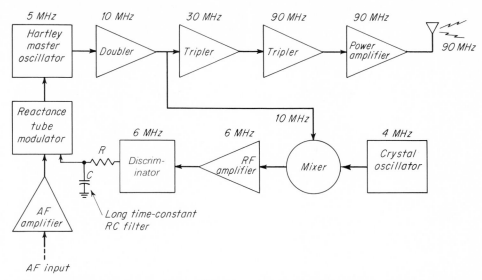

Fig. 19-16 Block diagram of a reactance-tube-modulated FM transmitter with an AFC system.

quency voltage to the modulator grid, but at nearly 90° phase *lead* due to the low reactance of the capacitor and the high resistance of the resistor. This grid voltage produces a modulator plate-current variation through the RFC nearly 90° out of phase with the oscillator tank voltage. The oscillator tank circuit is coupled through a 100-pF capacitor to a circuit that appears to have the same frequency ac but *lagging* nearly 90° (remember the 180° phase shift between grid and plate of an amplifier tube). The 90° current lag makes the modulator look like an inductive reactance to the oscillator tank. Paralleling an inductance across an LC circuit raises the resonant frequency, so the oscillator operates on some higher frequency. When audio voltages are applied to the third grid of the modulator, they control the gain of this tube, the vector sum of oscillator and modulator RF voltages, and therefore the apparent reactance of the modulator. Varying the reactance varies the frequency of oscillation of the LC circuit. Thus, AF voltages on the third modulator grid produce FM of the oscillator.

Since any voltage change in any part of the oscillator or modulator will shift the center frequency, it is necessary to use some method of assuring that the resting-carrier frequency remain constant and on the assigned frequency. A discriminator circuit can be used to return the resting

carrier to the center frequency. A block diagram of an FM transmitter using automatic frequency control (AFC) is shown in Fig. 19-16. The Hartley oscillator has a center frequency of 5 MHz. The doubler output is 10 MHz. The first tripler has a 30-MHz output, and the second tripler has a 90-MHz output which drives the power amplifier. Some of the 10-MHz output is fed to a mixer stage. A 4-MHz crystal oscillator is also fed into the mixer. The difference frequency between 10 and 4 MHz is fed to a 6-MHz amplifier and to a discriminator tuned to exactly 6 MHz.

As long as the oscillator remains on 5 MHz, its multiplied output, 10 MHz, will beat against the 4-MHz crystal to produce 6-MHz ac and zero voltage at the discriminator output. If the center frequency of the master oscillator drifts upward, the mixer will be fed a frequency higher than 10 MHz. This beats against the 4-MHz crystal and produces a difference frequency higher than 6 MHz and develops a voltage at the discriminator output. This voltage is fed to the grid of the reactance modulator, changing its reactance and shifting the oscillator frequency down again until no voltage is present at the discriminator. If the oscillator drifts lower in frequency, an opposite-polarity dc voltage is developed at the discriminator, shifting the carrier back to the center frequency again. A reactance tube or transistor AFC system, as shown in Fig. 19-10, is used in FM, AM, and TV receivers to keep them tuned to the desired station.

A discriminator alternately develops positive and negative voltages at the modulation frequency. If fed back to a reactance tube in the AFC circuit of a transmitter, they would cancel the modulation. To prevent this, a long-time-constant RC circuit is connected between the discriminator output voltage and the modulator grid. The average voltage no longer follows the deviation of the modulation, although it will follow any slow drift of the carrier.

A multiplication of 12 to 24 times the oscillator frequency is usually adequate to produce an undistorted 75-kHz swing with reactance-tube modulators. A multiplication of 18 can be attained with a doubler and two triplers. Thus, a transmitter with a center frequency of 88.1 MHz requires an oscillator with a center frequency of 4,894.3 kHz. A deviation of 2 kHz at the oscillator produces a 36-kHz deviation of the final amplifier.

Actually, a single-ended Crosby circuit produces a somewhat distorted FM, which is usually overcome by using a push-pull-type reactance modulator.

19-14 A SOLID-STATE FM MODULATOR

A *voltage-variable capacitor,* a special solid-state diode (Varactor, Varicap, etc.), when reverse-biased, will vary its junction capacitance with a variation of bias (Sec. 10-3). A circuit employing these to produce frequency modulation (or FSK) of an oscillator is shown in Fig. 19-17. The two diodes are reverse-biased by the +20 V through the RFC. Any AF ac added in series with the bias will modulate the bias, change the capacitance of the diodes, and shift the frequency

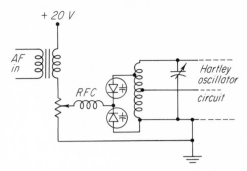

Fig. 19-17 Reverse-biased voltage-variable capacitors (Varactors) used to frequency-modulate an oscillator.

of the oscillator. The diodes are tapped down the LC circuit so that the RF ac peaks will not be greater than the bias on them. Most general-purpose solid-state diodes will work in this circuit but have so little capacitance that the deviation produced is not sufficient.

19-15 A PHASE-MODULATED TRANSMITTER

Holding a self-excited oscillator on an assigned center frequency is difficult. A temperature-controlled crystal oscillator would be a help, but the frequency deviation produced by reactance-tube-modulating a crystal is quite small, not linear, and requires many multiplications of the oscillator frequency.

If the crystal-oscillator stage itself is not modulated but modulation is applied to following stages, the frequency of the emission cannot be directly varied. However, the modulating voltage can shift the phase of the carrier voltage or current. This results in indirect FM called *phase modulation,* or PM.

A modulating voltage applied to the grid of a reactance tube shifts the frequency of the oscillator and holds it at the new frequency until the modulating voltage has been changed. In phase modulation the result of shifting the phase will be an instantaneous frequency change, but the carrier cannot be held at any frequency except the center frequency.

An early system of PM, called the Armstrong, used a crystal oscillator for frequency stability and phase-modulated a buffer stage. It employed a balanced modulator to cancel the carrier and produce amplitude-modulated sidebands only.

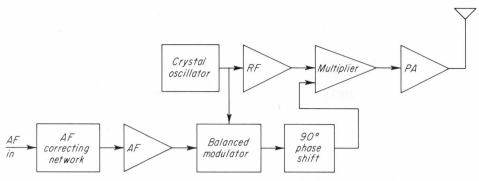

Fig. 19-18 Block diagram of an Armstrong PM system of producing FM.

These were shifted in phase 90° and then fed back to the output of the stage from which the carrier was taken (Fig. 19-18). With no modulation there are no sidebands, and the buffer output is that of the crystal. When audio is applied, its amplitude determines the amount of 90° phase-shift energy fed to the output of the buffer. The grid of the next stage is fed (1) a carrier and (2) a 90° carrier frequency of varying amplitude. This stage responds to the vector sum of the two phased voltages. With a weak audio signal, little sideband energy is fed to the third stage, there is little phase shift, and the PM is small. With more audio there is greater sideband energy, greater phase shift, and greater PM. Any AM produced in this or other systems is flattened out by the limiting action of the class C multiplier stages operating into the plate-current saturation part of their curves.

With any method of PM a constant-amplitude, variable-frequency tone capable of producing a 25-Hz deviation of the carrier with an AF of 50 Hz will produce a 7,500-Hz deviation with an AF of 15,000 Hz. To produce equal deviation for all modulating frequencies, it is necessary to use a low-pass RC network in one of the AF amplifier stages. This makes the audio output amplitude inversely proportional to output frequency. Such a network may consist, for example, of a 500,000-Ω resistor in series with the grid circuit in a resistance-coupled audio amplifier with a 0.1-μF capacitor from grid to cathode. Then, if pre-emphasis is desired, another network is required to raise the higher-frequency response according to the pre-emphasis curve.

If the maximum low-frequency deviation that can be developed at the modulated stage is only about 25 Hz, it is necessary to multiply this 3,000 times to produce 75,000-Hz deviation. For a carrier frequency of 90 MHz, this would require an oscillator frequency of 30 kHz. In actual practice an oscillator frequency of about 200 kHz can be used. It is fed through a series of multipliers until the frequency is in the 30-MHz region. This is heterodyned against a second crystal oscillator and is translated down to about 5 MHz (without losing any frequency deviation) and then multiplied up to the 90-MHz region. This requires many multiplier circuits.

19-16 PM FROM THE PHASITRON

Another method of producing PM is by the use of a specially designed *phasitron* tube. Essentially, this tube develops a thin wheel, or disk, of electrons that radiate outward from the cathode toward the eventual plate. Before striking this plate, the electrons must pass through holes punched in an inner plate. The flat wheel of electrons is ruffled by three-phase RF ac voltages applied to a series of grids placed between the cathode and the first plate. The ruffled edges of the electron wheel are made to sweep around, sometimes striking the holes in the first plate and allowing current to flow to the second plate and sometimes striking the first plate and stopping all current to the second plate. The on-off current flow to the second plate is the current that excites the output RF LC circuit into oscillation. A crystal oscillator is used to develop the three-phase ac that rotates the ruffled wheel of electrons. As a result, the output plate-current pulses are directly

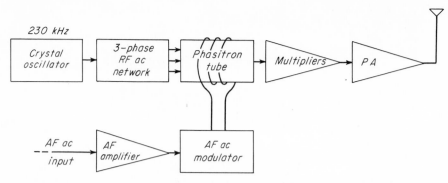

Fig. 19-19 Block diagram of the essentials of a phasitron tube FM transmitter.

proportional to the crystal frequency, assuring constant center frequency of the transmitter. A block diagram of a phasitron system is shown in Fig. 19-19.

A coil of wire, wrapped around the tube and excited by an AF ac, produces a magnetic field in the tube that tends to deflect and thereby either increase the speed of the ruffles in their sweep around the tube or slow them slightly, depending on the magnetic polarity developed in the coil by the AF ac modulating signal. This produces the PM of the plate-current pulses that flow to the second plate. A deviation of about 175 Hz is possible with this system, requiring a multiplication of 424 times to produce a 75-kHz deviation.

19-17 VOICE-MODULATED PM TRANSMITTERS

While the Crosby reactance-tube system, the Armstrong phase system, and the phasitron system represent some of the FM methods used in broadcast stations, by far the greatest number of frequency-modulated transmitters in use today are found in the thousands of taxicabs, police and fire cars, and other communication services. The designers of these systems are interested only in voice transmissions. The requirement is to transmit only audio frequencies in the range of 200 to about 3,000 Hz. By reducing the deviation ratio and modulating 200 to 3,000 Hz only, 5-kHz deviation can be produced using phase modulators with multiplications of only about twelve times. The reduction of the deviation ratio results in a lessened audio output from the receivers, but this is partially made up by the lessening of noise with

weak signals when the receiver bandwidth is narrowed. Narrowband FM in amateur work is limited to 3-kHz deviation, although above 26 MHz, wideband FM (75-kHz deviation) is permitted in certain portions of the bands.

A PM circuit used in some mobile transmitters is shown in Fig. 19-20a. The oscillator RF ac signal is fed across the $C_1R_1C_2$ network and, at the same time, across the C_3L_2 network. The frequency of the ac across R_1 and across L_2 will be the same, but because one circuit is an RC circuit and the other an LC circuit, the voltage between the top of L_2 and ground will be leading the voltage across R_1. If the phase-modulator tube is pulled out, the phase of the voltage developed across L_2 is solely dependent upon C_3L_2. When the tube is in the circuit, the out-of-phase voltage on the grid of the modulator produces an amplified out-of-phase current in the plate circuit through L_2. The current that flows through L_2 has two out-of-phase components, one due to the voltage fed to the grid and one due to the voltage from the oscillator. The resultant voltage-drop across L_2 is therefore not in phase with either voltage but is at some resultant phase. When an AF ac is added to the grid circuit of the phase modulator, the plate current still retains the steady oscillator component but the grid-circuit component varies in amplitude, producing a resultant phase shift that varies in accordance with the AF ac. In this way, PM is developed across L_2 and is fed to R_2 and the multiplier stages. Figure 19.20b is a solid-state counterpart PM system.

Communications transmitters are required to have a means of limiting the percent of modulation with AM or a means of limiting the frequency

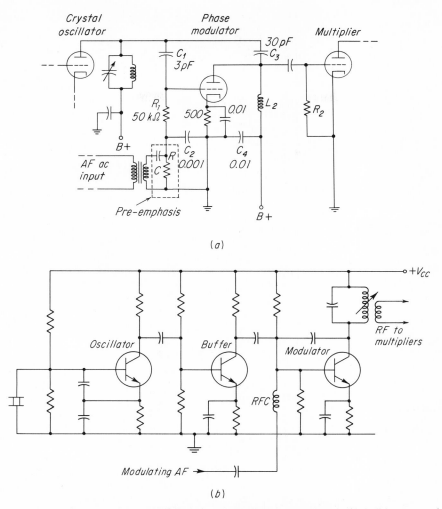

Fig. 19-20 (a) A phase-modulation system used in some mobile VT FM transmitters. (b) Solid-state equivalent simplified.

deviation with FM. In the limiter used in some mobile FM transmitters, shown in Fig. 19-21a, current flows through R_2 and R all the time. Current also flows through R_1 and R. When a positive audio signal is applied across R_1, the cathode of this half of the tube approaches the plate potential and current flow decreases. This decreases the current through R, and less voltage-drop is developed across it, while more is developed across R_2. Thus, AF ac from V_1 is passed to the filter network. When the cathode of the first half of the diode tube reaches or exceeds its plate potential (because of high-amplitude audio peak voltages), the diode ceases to conduct and any audio sig-

nals above this value do not affect the current through R and therefore through R_2. The signal is limited, or the peak is clipped off.

On the negative half cycle of audio, the cathode of the first diode is made more negative (plate relatively more positive), increasing the current through R, developing a greater voltage-drop across it, and decreasing the voltage-drop across R_2. When the negative signal reaches a certain value, the current through the diode approaches the maximum. Any further increase in the negative signal produces almost no increase in current in the first diode, nearly full voltage-drop across R, and therefore almost no signal variation across R_2.

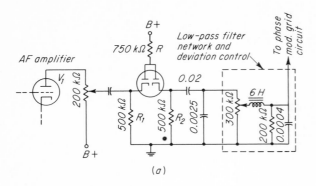

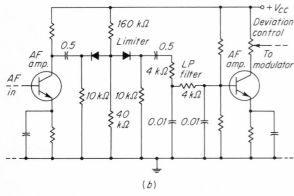

Fig. 19-21 (*a*) A VT audio-limiter, or clipper, circuit with deviation control. (*b*) Solid-state counterpart.

In this way, both the positive and the negative halves of high-amplitude audio signals are limited.

The audio developed across R_2 is fed through a low-pass filter network that does two things. It tends to round off or filter any square-wave-shaped signals developed in the clipping process, thereby reducing higher-frequency audio harmonics that might otherwise be transmitted. It also makes the audio output inversely proportional to the frequency required in PM systems to make them act as direct FM. The setting of the potentiometer arm determines the deviation or percent of modulation of the transmitter.

A semiconductor version of the same type of clipper-filter is shown in Fig. 19-21*b*.

Crystal-controlled transmitters of this type may gradually drift lower in frequency with age. In a few months they may need to be readjusted to the proper center frequency because the silver plating on the crystal increases in weight when oxidized. To allow the crystal to be *warped* to the

desired frequency, variable capacitors and inductors may be included in the crystal circuit, by which the crystal frequency can be warped several hundred hertz on its fundamental. This produces several kilohertz of control at the assigned carrier frequency.

Test your understanding; answer these checkup questions.

1. What is another name for the reactance-tube modulator? _____ Would it operate with a triode? _____ A transistor? _____
2. What effect would be produced if a small L were used in place of the 50-pF C in the reactance-tube modulator? _____ If the 50-pF C and 100-kΩ R were interchanged? _____
3. What are the five significant circuits in the AFC system shown? _____ _____ _____
4. Besides the reactance-tube modulator, what is another method of producing direct FM? _____ Could AFC be used with it? _____ Would it produce linear FM? _____
5. Back in Fig. 16-18, where might AF ac be applied to produce direct FM? _____
6. In a PM system, where is the PM changed to FM? _____
7. How does the output of a PM stage differ from that of an FM stage? _____
8. What is the advantage of PM? _____ The disadvantage? _____
9. Is AM produced in a reactance modulator? _____ A PM modulator? _____ How is AM eliminated in FM transmitters? _____
10. What was the first PM system called? _____
11. Why are limiters used in FM transmitters? _____ In FM receivers? _____
12. In Fig. 19-20, which control would reduce excessive limiting? _____ Which would set maximum deviation? _____
13. Why might it be difficult to produce FM at frequencies much above 15,000 kHz with a phasitron? _____

19-18 FM BROADCAST STATIONS

The block diagram of an FM transmitter and station shown in Fig. 19-22 represents a station operating with a minimum of facilities. Most stations have more studios, turntables, and control positions. This control room houses the FM transmitter, a frequency and modulation monitor, and the *master* audio-control console, with which the operator can connect the audio output of the studios, turntables, tape recorders, or his own local microphone to the transmitter.

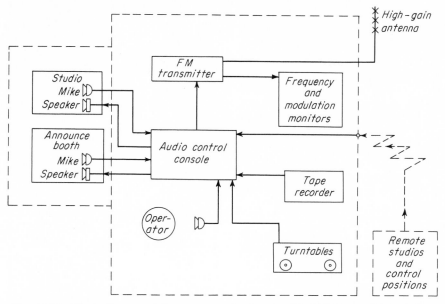

Fig. 19-22 Block diagram of the components of a simplified FM transmitting station.

Since most FM transmitters are located on top of hills for best radiation of signals, programs may originate in remote downtown studios and be controlled at that point. The program is then fed to the transmitter station by telephone lines or by a microwave radio link.

Studios or announce booths contain one or more microphones and a monitoring loudspeaker. When the studio is activated, the output of the microphone is fed to the console and the monitor speaker in the studio is cut off to prevent an audio feedback, or howl. As soon as the studio microphone is switched off by the control operator, the monitor speaker in the studio goes on, and the program now on the air is fed into the studio, notifying the studio personnel that their microphone is no longer alive. The speaker may also be used as part of an intercom system to the control position.

The transmitting antenna is usually a high-gain, multielement, horizontally polarized, omnidirectional, on a tower, 100 to 500 ft high, installed on top of a tall building or hill near the populated area it is desired to service. The more elements in the antenna, the higher the gain, the more the transmitted signal is radiated downward toward the horizon, and the stronger the field strength of the signals received on the ground.

Except for transmitter and antenna, an FM broadcast system may be identical with a standard AM broadcast station. In fact, some of the same programs may be fed to an AM and an FM broadcast station simultaneously, the only difference being AF response. The FM station uses an AF of 50 to 15,000 Hz, whereas most AM stations attenuate frequencies above 5 or 7.5 kHz.

The operating power of FM transmitters can be determined by either the direct or the indirect method. The *direct method* substitutes a dummy antenna (Sec. 16-7) for the actual antenna. When the desired power is developed in the dummy, the plate current is noted. The transmitter is then connected to the antenna and tuned for the same plate-current value. The *indirect method* computes operating power by the product of plate voltage (E_p) and plate current (I_p) of the final RF amplifier stage, times an efficiency factor (F):

$$\text{Operating power} = E_p \times I_p \times F$$

The factor F is determined by the transmitter manufacturer or by the FCC (usually 0.65 to 0.8 for class C amplifiers). The operating power must not exceed 105% or fall below 90% of the authorized operating power.

Accurate meters must be used to measure

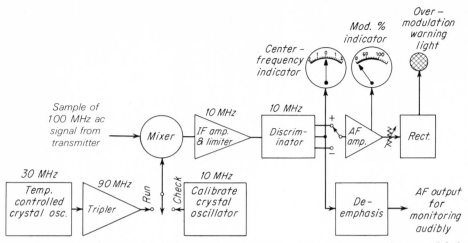

Fig. 19-23 Block diagram of an FM station frequency monitor, modulation indicator, and overmodulation warning.

final-amplifier plate voltage and current. A meter must also be provided to measure the RF current or voltage of the main transmission line to the antenna. Accuracy of the meters must be at least 2% of the full-scale reading.

Commercial and educational FM broadcast stations are required to keep their center frequency within a *frequency tolerance* of $\pm2,000$ Hz of the assigned frequency. This requires a continuously operating *frequency monitor,* such as shown in Fig. 19-23. Its meter face must have a zero reading at center scale and frequency calibrations up to at least 2,000 Hz on each side of zero in graduations of 100 Hz. The frequency monitor is first calibrated by switching in the 10-MHz-calibration crystal oscillator. This signal is fed through the mixer stage, the 10-MHz IF amplifier, and the limiter to the 10-MHz discriminator. The discriminator is trimmed to read zero on the meter with the calibrating signal. The 30-MHz running crystal

is then switched in and fed through a tripler to the mixer. When the transmitter is turned on (assume a 100-MHz transmitter), the 90- and 100-MHz signals mix to produce a 10-MHz resultant. If the transmitter is exactly on frequency, the 10-MHz resultant will read zero on the discriminator meter. If the carrier is off frequency, the meter will read to the left or right of zero, indicating in hertz the carrier-frequency error. The frequency of the calibration crystal can be checked against station WWV or some other primary frequency standard (Sec. 21-6). A different running-crystal frequency is required if the transmitter operates on any frequency other than 100 MHz.

Since the discriminator is an FM detector, its output can (1) be used as an audible monitor signal after passing through a de-emphasis circuit or (2) operate a percent-of-modulation or deviation meter. The percent-of-modulation meter shown is a VU (volume-unit) type meter (Sec. 12-18) so calibrated that 75-kHz deviation (whether produced by FM or PM) reads 100% modulation.

If the frequency meter shows the carrier is within tolerance but modulation causes the meter to vary, it indicates that the carrier is deviating in one direction more than in the other. Some possible causes of such nonlinearity are detuned multiplier or amplifier LC tank circuits, detuned antenna, nonlinear modulation of the oscillator, improper neutralization, an improperly operating AFC system, overmodulation, or audio distortion.

When the frequency monitor indicates improper operation, the oscillator and each stage after it should be retuned, one by one, to an accurate peaking of the grid-current meter in the final amplifier. Then the final-amplifier plate circuit should be tuned to a plate-current dip. With a direct FM transmitter the AFC crystal oscillator, IF, and discriminator stages should be tuned for proper indications when modulation is applied.

LICENSE REQUIREMENTS AT FM BROADCAST STATIONS. *If over 25 kW.* Chief or Acting Chief Operator, First Class radiotelephone license. For routine operation, First or Second Class radiotelephone or Third Class permit with BC (Element 9) endorsement is required when on duty at transmitter or control point.

If under 25 kW. A First Class radiotelephone license holder must be available at all times. First or Second Class radiotelephone license holder or BC-endorsed permittee required to be on duty at transmitter or control point.

First Class license holders are the only ones who may make technical adjustment to transmitters and monitoring equipment except to:

1. Turn transmitter on or off.
2. Compensate for line voltage fluctuations.
3. Maintain proper modulation levels.
4. Make routine authorized power changes.

Daily (5 days a week) inspections of the transmitting system and monitoring equipment must be made by a First Class license holder.

LICENSE REQUIREMENTS FOR NONCOMMERCIAL EDUCATIONAL FM STATIONS. For routine operation, operator in charge may hold First or Second Class radiotelephone license or permit.

For adjustments, inspections and maintenance of transmitting system:

If over 25 kW. Same as for FM BC stations.

If more than 1 to 25 kW: First Class radiotelephone license holder.

If More Than 10 W to 1 kW. First or Second Class radiotelephone license holder.

Not More Than 10 W. First or Second Class radiotelephone or radiotelegraph license holders.

If an unqualified operator notes improper operation, he should shut down the transmitter immediately, calling a First Class license holder to correct the difficulty.

If it is noted that the frequency or modulation monitors become defective, the fact should be logged in the maintenance log (Sec. 24-14) and the FCC Engineer-in-Charge of the radio district advised, but the station may continue to operate for a period of 60 days. Log entries are to be made when the meter is placed back in service.

All FM broadcast stations must make equipment-performance measurements annually. These are AF response measurements of frequencies between 50 and 15,000 Hz; AF harmonic-distortion measurements at 25, 50, and 100% modulation with de-emphasis; output noise of frequency-modulated type between 50 and 15,000 Hz; and output noise of amplitude type between 50 and 15,000 Hz.

All tests of FM (only) broadcast stations for testing or maintenance purposes must be made within the *experimental period,* midnight to 6 A.M. local time.

19-19 PUBLIC SAFETY RADIO SERVICE

This is one example of a service using narrowband FM to enable more stations to operate in a given band of frequencies. Whereas FM broadcast channels are 200 kHz wide, Public Safety communication FM channels are only 20 kHz wide.

The Public Safety Radio Service includes developmental-operation, local government, police, fire, highway-maintenance, forestry-conservation, special-emergency, and state guard radio services. As examples, some of these services may operate between 1.6 and 2.5 MHz with 8-kHz-bandwidth A3 emissions; from 2.8 to 8 MHz with 0.25-kHz-bandwidth A1 emissions; and from 25 to 500 MHz with 5-kHz-deviation, 20-kHz-bandwidth F3. Transmitters are type-accepted (manufactured equipment acceptable to the FCC for the service intended), often using from 10 to 60 W for mobiles. Base-station powers range from 60 to 2,000 W. All transmitting and receiving antennas are vertically polarized for mobile work (horizontal for FM broadcasting).

So many systems are licensed that adjacent channel interference becomes a problem and maintaining required frequency tolerances (Table 19-1) becomes important.

Before establishing an FM communication sys-

Table 19-1 FREQUENCY TOLERANCES

Frequency range, MHz	All fixed and base stations, %	All mobile stations	
		Over 3 W, %	3 W or less, %
Below 25	0.01	0.01	0.02
25–50	0.002	0.002	0.005
50–450	0.0005	0.0005	0.005
450–512	0.00025	0.0005	0.0005
Above 950	*	*	*

*To be specified in the station authorization.

tem, it is necessary to submit an application for a *radio station authorization* from the FCC. The standard forms for this are obtained from either the Washington, D.C., office or any field office of the FCC. After the station has been authorized by a *construction permit* and before the installation has been completed, a *station license* must be applied for. The Engineer-in-Charge of the local field office must be notified before tests are made.

While normally an authorization must be obtained before operating a new station, in the case of a mobile unit, in emergencies telephonic authorization may be given, or a mobile may be used as a base station for a period of less than 10 days.

Authorized bandwidth means the maximum bandwidth occupied by the emission. This bandwidth of frequencies must contain at least 99% of the emitted energy of the transmitter.

Each communication system must have some *control point* under the supervision of the licensee, where the monitoring facilities for the system are located and where persons responsible for operation are stationed. In the Public Safety Service the base station and the mobiles may either transmit and receive on the same frequency or have separate frequencies. Identification must be made at the end of each transmission or once every 30 min, whichever is selected. Identification can be by the assigned call signs or by special identifiers, provided the FCC is notified. Transmitter measurements required are (1) measurement of power input to the final stage, (2) modulation percent, and (3) frequency of the carrier. Measurements must be made when the station is first licensed, whenever a change is made in the transmitter, and at least once a year. The results of these measurements must be entered in the log or records of the station by a responsible operator who signs the records for any maintenance. Operators must sign on and off duty, indicate the calls made and the times antenna lights were checked, and note any failures of equipment.

When Public Safety radio transmitters use amplitude modulation, the peaks of modulation should be maintained above 70%, but the negative peaks of modulation shall not exceed 100%.

FM sideband energy past the authorized 20-kHz bandwidth must be attenuated by 25 dB at 100% past the bandwidth, by 35 dB up to 250%, and by $43 + 10 \log_{10}$ (mean output watts) dB past this point. This can be measured with special spectrum-analyzer equipment, usually with the transmitter operating into a dummy antenna.

A complete FM installation consists of a power supply (battery in a mobile), possibly an auxiliary supply, the transmitter, the antenna transmission line, and the antenna. To measure frequency, besides the frequency monitor, a counter, a secondary frequency standard, or a heterodyne-frequency meter are required. To measure power, a dummy antenna, an RF wattmeter, or a standing-wave bridge is needed. Besides hand tools, maintenance equipment might include a VOM, a VTVM, an oscilloscope, and an AF signal generator.

19-20 A MOBILE FM TRANSMITTER

The block diagram of Fig. 19-24 represents a possible mobile transmitter for operation in the 150-MHz region. The crystal oscillator is fed to a buffer. The output of this stage is fed to a phase modulator and to the first doubler. The signal is doubled in frequency again and then tripled before driving the power amplifier on 155 MHz. The microphone has a switch mounted on it. When the operator picks up the microphone and pushes the button, the receiver is muted, the antenna is shifted from receiver to transmitter, the transmitter power supply is turned on, and the transmitter feeds energy to the antenna. RF circuitry might be similar to the basic transmitter in Sec. 16-10.

A possible tuning procedure for a transmitter of this type is outlined below. It will be assumed that all stages are detuned and that each stage

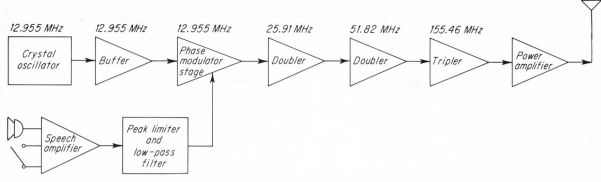

12.955 MHz	12.955 MHz	12.955 MHz	25.91 MHz	51.82 MHz	155.46 MHz

Crystal oscillator → Buffer → Phase modulator stage → Doubler → Doubler → Tripler → Power amplifier

Speech amplifier → Peak limiter and low-pass filter →

Fig. 19-24 Block diagram of a mobile PM-type FM transmitter.

has a meter jack in the grid and cathode circuit to measure grid and cathode (plate) current.

1. An open plug should be inserted in the cathode jack of the final amplifier to prevent high plate current during tune-up of the low-level stages.
2. A milliammeter is plugged into the grid stage of the buffer, and the press-to-talk switch is closed. If the oscillator is a Pierce circuit, grid current should flow. If a tunable crystal oscillator is used, its tank circuit should be tuned to maximum buffer grid current.
3. Move the meter to the buffer cathode, press the button, and tune the buffer plate circuit for minimum cathode current. To assure operation on the proper frequency, an absorption wavemeter (Sec. 21-3) could be coupled to the buffer tank circuit to check the frequency of the RF ac in the tank circuit.
4. Maximum grid and minimum cathode currents of each of the doubler and tripler stages should be checked, following the method used with the buffer stage.
5. To tune the final amplifier, remove the open plug from the cathode jack. This will allow grid and cathode current to flow.
6. Before tuning the final amplifier, to prevent interference to communications, make sure by listening on a receiver that the operating frequency is not being used, or better, use a dummy antenna.
7. Decouple the antenna from the final-tank circuit a little. Peak the final-amplifier grid current by retuning the tripler plate-tank circuit. Insufficient grid drive will decrease output power,

and detuned LC circuits cause distorted sidebands. Dip the cathode current by tuning the final-amplifier tank circuit. This should produce a plate current lower than the recommended current for the tube (determined by reference to a tube manual).
8. If the antenna is tunable, now tune the antenna for a rise in final-amplifier cathode current, which should still be less than the desired operating value. To bring the cathode current to this value, slowly increase the antenna-to-amplifier coupling. Retune the final tank for accurate cathode-current minimum. (A nearby field-strength meter should register maximum as the final tank is dipped.) Identify the transmission as a test.

The power input to the final stage is determined by the power formula $P_{in} = E_p I_p$. The plate voltage is determined by a voltmeter. The plate current for a tetrode or pentode is the cathode current minus the screen-grid and the control-grid currents. The control-grid current is read with the milliammeter, but the screen current must be computed. Determine the resistor value between B+ and the screen grid. With the stage operating, determine the voltage-drop across this resistor (the difference between B+ and the voltage at the screen grid). The screen current, by Ohm's law, is then

$$I = \frac{E}{R}$$

where E = voltage across resistor
R = resistor value in Ω

A simple check to determine if the final amplifier is self-oscillating is to remove the drive to the final stage by inserting an open plug into the tripler cathode jack and pressing the press-to-talk switch. If grid current flows in the final amplifier, the stage must be oscillating. Adjust the neutralizing control, until oscillation ceases.

The deviation, or modulation, can be checked by using a calibrated deviation meter. When the microphone is spoken into, the deviation should not exceed the licensed limit of ±5 kHz. Soft speaking should produce about 3-kHz deviation and loud speaking or whistling only 5-kHz deviation, indicating that the audio limiter is operating. If the deviation is greater than 5 kHz with a whistle, the AF level should be reduced (move arm of potentiometer in Fig. 19-20 toward ground).

After the transmitter has been tuned, its operating frequency should be checked with a frequency meter. In the frequency region of this transmitter the frequency must be within 0.0005% of 155.46 MHz (which is 777 Hz), or between 155,460,777 and 155,459,223 Hz. However, if the frequency meter is guaranteed accurate only to 0.0002% (which is 311 Hz) to assure transmission within licensed tolerance, the meter should indicate the frequency to be within $777 - 311$, or ±466 Hz. The meter should indicate the frequency to be ±0.0003%, or between 155,460,466 and 155,459,534 Hz.

The frequency can also be measured with a heterodyne frequency meter (Sec. 21-9). If such a meter has crystal checkpoints at every megahertz and is continuously tunable and calibrated to any frequency over a range of 20 to 40 MHz, its harmonics can be used to determine the frequency of the transmitter. For example, the fourth harmonic of 38.865 MHz is 155.46 MHz. If a zero beat is heard in the frequency meter when the dial reads 38.865 MHz, the transmitter should be on frequency. To assure operation on the correct harmonic, an absorption wavemeter may be coupled to the final-tank circuit.

If the frequency meter has no audio output, it is also possible to measure the frequency of the transmitter by tuning it in on a receiver and feeding the output of the heterodyne frequency meter to the receiver simultaneously. The receiver should respond with a zero beat when the hetero-dyne oscillator is set to 38.865 MHz. If the zero beat is heard at 38,865,100 Hz on the dial of the frequency meter, it means that the transmitter is 4×100, or 400 Hz high, which is within either the 777- or the 466-Hz tolerance above.

Licensed transmitting equipment is said to have *type approval* if the FCC tests its power and frequency characteristics and *type acceptance* if manufacturers' test data show the equipment to be acceptable. Generally, changes that might affect the power output, frequency stability, or percent of modulation of a transmitter must be approved by the FCC. Other types of changes may be made without Commission approval.

Transmitting equipment should be tested when first installed, when technical changes have been made, and at least once a year thereafter.

Test your understanding; answer these checkup questions.

1. What signals might feed into an FM broadcast station audio control console? _____ Feed out of it? _____
2. What method of measuring FM transmitter power uses a dummy antenna? _____ The formula $E_p I_p F$? _____ Who sets F? _____
3. What accuracy must meters in the output stage of an FM transmitter have? _____
4. What is the frequency tolerance of FM BC stations? _____ Educational FM stations? _____
5. What information can be obtained from FM BC station frequency monitors? _____ _____
6. What is indicated if the frequency meter deviates when FM is applied? _____ How might it be corrected? _____
7. What adjustments may holders of licenses or permits lower than first class make on FM BC stations? _____ _____ _____
8. For how long after a modulation monitor becomes defective may an FM station be operated? _____
9. How often must equipment performance tests be made on an FM BC transmitter system? _____
10. What is the experimental period? _____
11. What antenna polarization is used with Public Safety Service equipment? _____ FM BC stations? _____
12. For Public Safety Service transmitters, what is the FM deviation? _____ The three required measurements? _____ _____ _____ Maximum and minimum modulation percent using AM? _____
13. Should a mobile transmitter be tuned on the air or on

a dummy antenna first? _____ How is its power determined? _____

14. What are the upper and lower frequency limits if a 160-MHz transmitter has a tolerance of 0.0005? _____ Of 0.05%? _____

19-21 NOISE IN MOTOR VEHICLES

Most communication receivers will satisfactorily amplify and detect a received signal as low as 1 μV, whether AM or FM. However, if the *noise* being received has an average value of more than the signal, neither AM nor FM will reproduce the signal satisfactorily. It is important to reduce local noise as much as possible.

There are some general methods of reducing noise generated in land-based electrical equipment. One is to ground the equipment with as short a ground wire as possible. Another is to bypass the input and output electrical lines to ground with 0.001- to 0.01-μF capacitors, using series RF chokes in the lines if necessary. Another aid is to shield the equipment and lines with metallic housings or braid and ground the shield. Still another aid is to connect a bypass capacitor across any making and breaking electric contact or switch.

In motor vehicles the major source of noise is the ignition system. The popping noise developed by the spark plugs increases in frequency with increased engine speed. It can usually be decreased by using resistor spark plugs and resistor ignition cables, making sure all ignition leads are tight, connecting suppressor resistors in the lead from the distributor to the ignition coil, and bypassing the primary of the ignition coil to ground. Transistor ignitions are sometimes noisy.

Another source of noise in mobile receivers is the battery-charging generator. When the engine changes speed, the whining noise changes. This noise can be reduced by bypassing the armature terminal of the generator (not the field terminal) with a 0.1- to 1-μF capacitor. It may help to bypass the battery terminal and the armature terminal of the voltage regulator to ground. Alternators should cause no noise.

Gasoline, temperature, and oil gauges can produce clicking noises. The leads to these gauges can be bypassed with a 0.25-μF capacitor at the source and at the dashboard.

An irregular clicking noise that disappears when the brakes are applied is known as *wheel static*. The front wheels riding on an insulating layer of grease may build up a static charge which sparks across when the voltage increases sufficiently. This can be stopped by using wiping-type front-wheel-static eliminators. The use of antistatic powder in inner tubes will decrease tire static.

If the metal parts of the chassis of a vehicle become loose, a voltage difference may be developed between two parts and sparking may occur when the parts work together. Adequate bonding of the major parts of the car, hood, body, chassis, motor, cables, rear axles, brake and speedometer cables, rear-axle assembly, doors, and fenders with flexible-braid conductors is often necessary.

All electrical leads coming through the fire wall may have to be bypassed at the point where they leave the engine compartment.

19-22 FM STEREO MULTIPLEX

In 1961 the FCC authorized a compatible system of stereophonic FM broadcasting called *stereo multiplex*. It is compatible because it can be received on stereo-type receivers or on standard monaural FM receivers. Many FM broadcast stations now use stereo multiplex.

Stereo is a means of making music or sounds two-dimensional to a listener. It makes sounds that would have approached a listener from the left in a concert hall approach from his left in his home and sounds that would have approached from the right in a concert hall approach from the right in the home. This effect can be accomplished by having one microphone on the left side of the concert-hall stage and a second microphone on the right side of the stage. Then, by transmitting both left and right signals separately, by detecting them separately, and by feeding them to two separate loudspeakers placed at the left and right, the listener is enabled to hear sounds in very much the same way as though he were at the concert. In the past such a system actually employed two transmitters and two receivers. In FM stereo multiplex this is accomplished by using two microphones, one transmitter, one receiver having a stereo matrix, two AF amplifiers, and two loudspeakers.

FM multiplex transmission is the means by

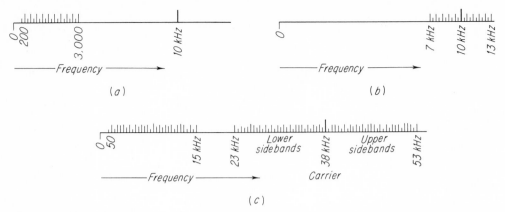

Fig. 19-25 (a) Voice frequencies and a 10-kHz tone. (b) 10-kHz tone carrier modulated by voice frequencies. (c) High-fidelity sounds and a 38-kHz carrier modulated by hi-fi sounds.

which separate left-side (L) and right-side (R) signals can be made to frequency-modulate a single radio carrier. FM multiplex reception is the means by which the L and R signals can be separated in the receiving system to feed separate loudspeakers.

If voice-type audio ac signals having frequencies from 200 to 3,000 Hz are fed to a loudspeaker, they produce intelligible sound to a listener. These voice frequencies can be represented as in Fig. 19-25a. If a 10-kHz steady tone is also fed to the loudspeaker, the listener will hear a high-pitched whistle. This is also represented.

If the voice frequencies could be made to amplitude-modulate the 10-kHz carrier frequency, the resultant signals would consist of the 10-kHz carrier plus upper and lower sidebands extending from 7 to 13 kHz (Fig. 19-25b). Although a loudspeaker could reproduce all these frequencies, the listener would hear them not as under-

standable sounds but as a conglomerate group of high-pitched squeaks—sounds in the range of 7 to 13 kHz. This is an example of a carrier and sidebands, both in the AF spectrum, but unintelligible because they have not been detected. If this carrier and its sidebands were fed to a 10-kHz receiver and detected, the result should be the 200- to 3,000-Hz signals that were used to modulate the 10-kHz carrier.

By raising the modulated carrier to a supersonic 38 kHz, a listener could hear nothing because this is higher than humans can hear. If this 38-kHz carrier frequency is amplitude-modulated by high-fidelity AF ac (50 to 15,000 Hz), the sidebands produced would extend 15 kHz above and below 38 kHz, or from 23 to 53 kHz (Fig. 19-25c). None of these supersonic frequencies are audible either, but they could be detected with a 38-kHz low-frequency receiver. This is essentially one of the signals that are used to modulate the FM stereo multiplex transmitter.

19-23 A STEREO MULTIPLEX TRANSMITTER SYSTEM

A possible system to produce FM stereo multiplex transmissions is shown in Fig. 19-26 in block form. Most circuits have been discussed previously in other applications.

The transmitter is reactance-tube-modulated, but could be phase-modulated. Both L and R microphone signals are fed into a linear AF *adder* circuit and are used to produce a composite

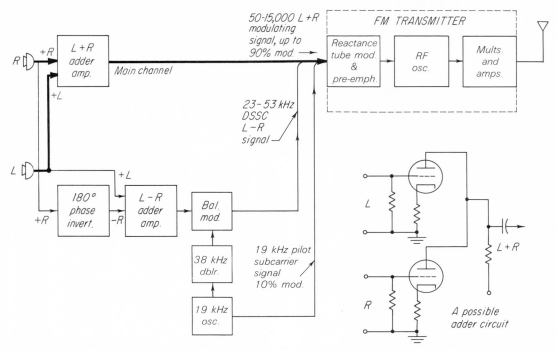

Fig. 19-26 Block diagram of a possible FM stereo multiplex transmitting system. Right, a simple adder to combine L and R signals to an L + R signal.

L + R main-channel signal to frequency-modulate the oscillator of the transmitter. This 50- to 15,000-Hz signal produces modulation that can be detected by standard monaural FM receivers.

Note the polarity indications in the illustration. The L + R amplifier is being fed the L and R signals in one phase. The L − R amplifier is being fed both L and R signals, but by passing the R signal through a phase inverter it reverses phase and is designated −R. Thus this adder is fed a +L and a −R, producing L − R output. To a listener the result of L + R sounds exactly the same as L − R. The resultant in both cases contains the same frequency components and in the same amplitudes and would therefore sound the same to the ear.

The 50- to 15,000-Hz L − R signal is used to amplitude-modulate a 38-kHz carrier signal (the second harmonic of the 19-kHz pilot subcarrier shown). Since this modulation occurs in a balanced modulator, the output from this stage is a double-sideband suppressed 38-kHz carrier set of supersonic ac signals having frequency components ranging from 23 to 53 kHz. If the 38-kHz

carrier were transmitted, it would produce an undesirable beat interference in receivers. For this reason, the L − R signal is converted to a double-sideband, without-carrier (DSSC) signal, which is used to frequency-modulate the transmitter.

The whole multiplex modulating signal for an FM stereo broadcast consists of the monaural L + R audio ac signals, a supersonic 19-kHz (±2-Hz) ac pilot subcarrier, and the L − R stereophonic ac signal consisting of sidebands from 23 to 53 kHz, but without their 38-kHz carrier, as shown in Fig. 19-27.

Theoretically there should be no *cross talk* between the monaural and the stereophonic channels, but there always is some. By FCC regulations the cross talk must be held at a level of more than 40 dB between channels.

Some other stereophonic transmission standards are: The stereophonic subcarrier must be suppressed to a level of less than 1% modulation of the main carrier. Pre-emphasis for the stereophonic subchannel must be identical with that of the main channel. Peak modulation of either

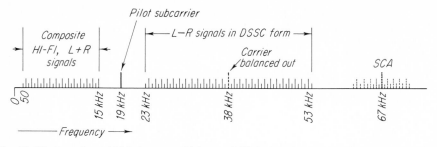

Fig. 19-27   Location of FM stereo multiplex and SCA signals in the frequency spectrum.

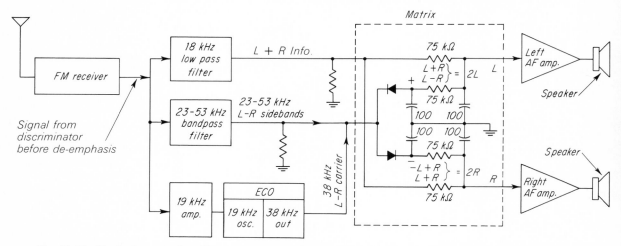

Fig. 19-28   Block diagram of an FM stereo multiplex receiving system, showing the matrix in schematic form.

channel alone is 45% of the total (pilot subcarrier is 10%). A positive left signal deviates the carrier to a higher frequency.

19-24 AN FM STEREO MULTIPLEX RECEIVER SYSTEM

A possible stereo multiplex receiving system is shown in block form in Fig. 19-28. A normal FM receiver system is employed up to, but not including, the de-emphasis circuit (which is added later in the matrix). With the exception of the matrix, all circuitry shown has been explained elsewhere.

The FM receiver discriminator detects from the received FM stereo signal (1) the 50- to 15,000-Hz composite L + R AF signals, (2) the 19-kHz pilot-subcarrier frequency, and (3) the 23- to 53-kHz L − R sideband signals.

This complex group of audio and supersonic ac signals is fed to three circuits. One is a 0- to 18-kHz low-pass filter. The output of this filter contains only the composite 50- to 15,000-Hz L + R AF signal. (With monaural transmissions, this signal is de-emphasized in the matrix, amplified by both amplifiers, and fed to both loudspeakers as monaural sound.)

The output of the discriminator is also fed to a 23- to 53-kHz bandpass filter (or 53-kHz low-pass filter). The output signal of this bandpass filter is the L − R sidebands (to which the 38-kHz carrier signal will be added in the matrix).

The third circuit amplifies the 19-kHz pilot sub-carrier. This signal is fed to a self-excited 19-kHz ECO (electron-coupled oscillator) with its output circuit tuned to the second harmonic of 19 kHz, or 38 kHz. The received 19-kHz signal synchronizes the ECO circuit, forcing it to oscillate at exactly 19 kHz. This produces the correct 38-kHz carrier, which when added to the L − R sidebands forms a 38-kHz A3 signal.

The matrix has two diode detectors that detect the amplitude-modulated 38-kHz L − R signals simultaneously but in opposite polarities. That is, L − R signal information is detected to a varying dc with positive polarity and also to a varying dc with a negative polarity.

The positive-polarity L − R signal (now detected into its original 50- to 15,000-kHz AF form) is added to the composite L + R AF signal in the matrix. Combining the positive L − R and the L + R results in (L + R) + (L − R), or 2L, a strong left output signal from the top terminal of the matrix. Combining the negative L − R and the L + R results in (L + R) + (−L + R), or 2R, a strong right output signal from the bottom terminal of the matrix. Thus, the output of the matrix is the two audio signals that originated in the L and R microphones in the broadcast studio. Each signal is fed to its own amplifier and loudspeaker, forming the stereo reproduction. Theoretically there is no cross talk.

Note the four de-emphasis circuits in the matrix, four 75-kΩ resistors and four 100-pF capacitors to give a 75-μs de-emphasis—and how both positive and negative L − R detectors mix with the same L + R signal in the de-emphasis circuits.

19-25 SCA AND FAX

Besides the multiplexing of stereo information into an FM broadcast transmission, there are two other signals that may be multiplexed. One is called *subsidiary communication authorization* (SCA), used to transmit background music, weather, time signals, educational information, etc. The other is *facsimile* (FAX), a method of transmitting photographs, maps, etc., which print out on special paper at the receiving terminal.

Whereas stereophonic multiplex signals consist of AF ac translated up to a subcarrier at 38 kHz with AM sidebands 15 kHz on both sides, SCA information is AF ac translated up somewhere between 20 and 75 kHz. SCA modulation is always by 7.5-kHz deviation FM of the subcarrier frequency used. Since the maximum SCA sideband power must never exceed 30% of the total modulation of the main FM carrier, the SCA will be a narrowband-type FM. If the main carrier is also being modulated by stereo multiplex, the SCA multiplex must not exceed 10% of the total

main-carrier modulation. When main channel, stereo multiplex, and SCA multiplex are all in operation simultaneously, the SCA information must be in the range of 53 to 75 kHz, usually with a carrier frequency of 67 kHz, as indicated in Fig. 19-26. To produce the SCA signal at the transmitter, an SCA carrier (67 kHz, for example) is frequency-modulated, and the resulting carrier and FM sidebands are fed to the reactance or other modulator without pre-emphasis at the same time that the pre-emphasized 50- to 15,000-Hz FM program material is fed to the modulator.

An SCA receiver is simpler than a stereo multiplex. The output of an FM broadcast receiver discriminator is fed to two circuits. One is the normal de-emphasis circuit and AF amplifiers for the main channel program. The other is to a 67-kHz (or whatever SCA carrier frequency is being used) bandpass filter and discriminator, which extracts the SCA information. This is amplified and fed to a loudspeaker. The listener has the choice of the main-channel program, the SCA program, or both simultaneously.

The third possible multiplex signal on FM broadcast emissions is FAX, or photograph transmission. A picture is photoelectrically scanned from left to right in thin horizontal lines. Where the picture is black, a maximum signal is developed. Where the picture is white, a minimum signal is developed. The variations of scanned black, grays, and white produce the modulating signal that is transmitted. The following are the standards for rectilinear multiplex FAX picture information. The scanning spot moves from left to right and from top to bottom, using 105 lines per vertical inch of picture, 360 lines per minute. Line use is seven-eighths, or 315°. The one-eighth not included in picture transmission is transmitted 15° at white level, another 15° at black level, and the third 15° at white again. In multiplex operation the modulation of the FM carrier by the modulated subcarrier must not exceed 5%.

The band used for FAX FM multiplex is from 22 to 28 kHz, with the subcarrier at 25 kHz. The FAX signal may be either amplitude-modulated or frequency-shifted. If amplitude-modulated, maximum subcarrier strength produces the blackest area on the paper at the receiver. If frequency-modulated, the highest-frequency excursion of the subcarrier produces the blackest spot on the paper at the receiver.

A FAX receiver consists of a normal FM broadcast receiver with the discriminator output, before de-emphasis, being fed to either a 25-kHz diode envelope detector, if the FAX is in AM form, or to a 25-kHz discriminator if the FAX is in FM form. The demodulated FAX signals are fed to a printer that has a left-to-right moving stylus that passes over a special paper. When the FAX signal is at a maximum, the stylus electrochemically turns the paper dark. When the signal is at a minimum, the stylus voltage does not affect the paper and a white portion remains at that point. Medium-level FAX signals darken the paper slightly, producing gray tones. The 45° of white-and-black transmission can be used to synchronize the receiver stylus with the photoelectric spot scanner at the transmitter. FAX is discussed further in Sec. 30-9.

Test your understanding; answer these checkup questions.

1. What signal sensitivity do most good receivers have? _____

2. Name five sources of radio noise in automobiles. _____ _____ _____ _____ _____
3. How does an adder differ from a mixer? _____
4. With stereo FM transmissions, does a monaural receiver produce all the sounds that a stereo does? _____
5. Why do you think 19 kHz was picked for the pilot subcarrier in stereo FM? _____
6. When fed to the stereo FM reactance modulator, in what form are the L + R signals? _____ The L − R signals? _____
7. In stereo FM, of the total modulation, what percent is main channel? _____ Stereo subchannel? _____ Pilot subcarrier? _____
8. Into what three signals is the discriminator output separated in an FM stereo matrix? _____ _____ _____ What are the two matrix outputs? _____ _____
9. Where is de-emphasis added in a stereo FM receiver? _____
10. What types of modulating signals are fed to the transmitter modulator with: Stereo multiplex? _____ SCA? _____ FAX? _____
11. Can a stereo multiplex transmitter also transmit SCA? _____ FAX? _____
12. Why must the SCA and FAX discriminator output be taken before de-emphasis? _____
13. With FAX, how many lines per vertical inch of picture? _____ How many lines per minute? _____ How much of a line is picture usage? _____ Would a black signal be a high or low amplitude if using AM? _____

COMMERCIAL LICENSE QUESTIONS

Amateur license questions will be found in the Addendum.

FCC Elements 3, 4, and 6 require an ability to answer questions similar to those below. Sections in which questions are answered are shown in parentheses. A question followed by a bracketed number is required for that element alone.

1. What type of modulation is contained in static? (19-1)
2. What types of radio receivers do not respond to static interference? (19-1)
3. To what is deviation proportional in a PM or FM transmitter? (19-3)
4. Discuss FM sidebands. (19-3)
5. What are the criteria for determining bandwidth of emissions in FM? (19-3)
6. What is the relation between number of sidebands and bandwidth of FM emissions. (19-3)
7. What is the relation between the spacing of the sidebands and the modulating frequency in FM? (19-3)
8. What is the relation between modulation index or deviation ratio and the number of sidebands in FM? (19-3)
9. What is the relation between percent of modulation and the number of sidebands in FM? (19-3)
10. What is the relation between the number of sidebands and amplitude of the modulating voltage in FM? (19-3)
11. What is the relation between the number of sidebands and the modulating frequency in FM? (19-3)
12. Explain what is meant by the emissions *F1, F2, F3, F4,* and *F5.* (19-3)
13. What effect does modulation have on the antenna current of an FM transmitter? (19-3)
14. In FM, what is the meaning of (a) *modulation*

index and (*b*) *deviation ratio?* What values of deviation ratio are used in FM? (19-3)

15. What are the merits of an FM system compared with an AM system? (19-3, 19-19)

16. How wide is an FM broadcast channel? (19-3) [4]

17. What characteristic of an audio tone determines the percent of modulation of an FM broadcast transmitter? (19-3)

18. Explain in a general way how radio signals are transmitted through the use of frequency modulation. (19-3, 19-18)

19. What AF range must an FM broadcast station be capable of transmitting? (19-3, 19-18)

20. What is *center frequency* in FM? (19-3)

21. What is *frequency swing* in FM? (19-3)

22. What frequency swing represents 100% modulation in broadcast FM? (19-3)

23. An FM broadcast transmitter is modulated 40% by a 5-kHz tone. When the percent of modulation is doubled, what is the frequency swing of the transmitter? (19-3) [4]

24. If the transmission-line current of an FM broadcast transmitter is 8.5 A without modulation, what is the current at 90% modulation? (19-3) [4]

25. Explain in a general way how radio signals are received when using FM. (19-4, 19-11) [4]

26. Draw a diagram and explain how a discriminator of an FM receiver is sensitive to frequency changes. (19-5, 19-6)

27. Explain briefly the principles involved in detecting FSK. (19-5)

28. Draw a diagram of a radio detector and explain its operation. (19-7)

29. Draw a diagram of a limiter circuit in an FM receiver and explain its operation. (19-9)

30. What are the reasons for pre-emphasis and de-emphasis in FM? (19-10)

31. Draw a diagram of a method of obtaining (*a*) de-emphasis, (*b*) pre-emphasis. (19-6, 19-19) [4]

32. Draw a block diagram of an FM superheterodyne receiver and explain its operation. (19-11)

33. Draw a diagram of a mixer for an FM receiver and explain its operation. (19-11)

34. How wide a frequency band must the IF amplifier of an FM broadcast receiver pass? (19-11)

35. Draw a diagram of a differential-squelch circuit in an FM receiver and explain it. (19-11)

36. Describe, step by step, a proper procedure for aligning an FM double-conversion superheterodyne receiver. (19-12)

37. Draw a block diagram of an FM transmitter, indi-

cating center frequency of radiated signal and frequency of the oscillator. (19-13)

38. Draw a diagram of a frequency-modulated oscillator using a reactance-tube modulator and explain the principle of operation. (19-13)

39. How is good stability of a reactance-tube modulator achieved? (19-13)

40. If an FM transmitter has a doubler and two triplers, what is the carrier swing when the oscillator frequency swing is 2 kHz? (19-13)

41. Could the harmonic of an FM transmission contain intelligible modulation? (19-13)

42. What is the difference between frequency and phase modulation? (19-15)

43. Describe briefly the operation of the Armstrong and the phasitron methods of obtaining phase modulation. (19-15, 19-16) [4]

44. Explain briefly what occurs in a waveform if it is phase-modulated. (19-15)

45. Draw a diagram of a phase-modulation system and explain its operation. (19-15, 19-17)

46. Discuss wide- and narrowband FM with respect to deviation and bandwidth. (19-17)

47. Draw a diagram of an amplifier with a pre-emphasis circuit. (19-17)

48. Draw a block diagram of an FM broadcast station and explain briefly. (19-18) [4]

49. By what methods may operating power at FM broadcast stations be computed? (19-18) [4]

50. Why are high-gain antennas used at FM broadcast stations? (19-18) [4]

51. What is the frequency tolerance of an FM broadcast station? (19-18) [4]

52. What might be the effect if a tripler stage in an FM transmitter were slightly detuned? (19-18)

53. What is meant by *frequency tolerance?* (19-18)

54. Draw a block diagram of an FM deviation (modulation) meter, including mixer, IF amplifier, limiter, discriminator, and peak-reading voltmeter. Explain its operation. (19-18)

55. Would an FM deviation meter respond if a transmitter is phase-modulated? (19-18)

56. If an FM transmitter without modulation is within frequency tolerance but with modulation is outside frequency tolerance, what are some possible causes? (19-18)

57. Exclusive of monitors, what indicating instruments are required in an FM broadcast station? (19-18) [4]

58. What is the required accuracy of meters indicating the plate current and plate voltage of the final stage and the transmission-line current or voltage at an FM broadcast station? (19-18) [4]

59. What is the required frequency range of the

indicating device on the frequency monitor at an FM broadcast station? (19-18) [4]

60. What should an operator do if (a) the frequency-monitor meter or (b) the modulation monitor becomes defective? (19-18) [9]

61. Draw the face of the following meters and show how to read each: (a) frequency monitor, (b) percent of modulation (VU). (19-18) [9]

62. What is the maximum permissible percent of modulation for FM BC stations? (19-18) [9]

63. What stations may be operated by a Third Class broadcast operator? (19-18) [9]

64. What are the operator license requirements for FM broadcast stations? (19-18) [4]

65. What is the best antenna site for an FM broadcast station? (19-18)

66. Why is narrowband rather than wideband FM used in communication systems? (19-19)

67. What sideband harmonic attenuation is required of FM transmitters in the Public Safety Radio Service? How can it be determined? (19-19)

68. What are the authorized bandwidth and frequency deviation of Public Safety stations operating at (a) about 30 MHz and (b) about 160 MHz? (19-19)

69. The carrier frequency of a Public Safety transmitter operating at 160 MHz with a license power of 50 W must be maintained within what percent of the licensed value? (19-19)

70. What legal requirements must be met before installing and operating a radio station in services such as the Public Safety? (19-19)

71. Where may standard forms applicable to the Public Safety Radio Service be obtained? (19-19)

72. What notification must be forwarded to the Engineer-in-Charge of the Commission's district office prior to testing a new radio transmitter in the Public Safety Radio Service? (19-19)

73. Define the terms *authorized bandwidth*, *bandwidth occupied by an emission*, and *station authorization*. (19-19)

ANSWERS TO CHECKUP QUIZ ON PAGE 494

1. (*1 μV or better*) 2. (*Spark plugs, generator, loose body parts, metering systems, wheel static*) 3. (*Adder must be linear to prevent heterodyning*) 4. (*Yes*) (*L + R*) 5. (*Doubles to 38 kHz, 4 kHz above highest AF and 4 kHz below highest L − R SB frequency*) 6. (*AF*) (*DSSC*) 7. (*45%*) (*45%*) (*10%*) 8. (*L + R AF, L − R DSSC, 38-kHz carrier*) (*L, R*) 9. (*In matrix*) 10. (*AF and DSSC AM*) (*FM*) (*AM or FM*) 11. (*Yes*) (*No*) 12. (*So high above 15 kHz that no signal would be left*) 13. (*105*) (*360*) (*Seven-eighths*) (*High*)

74. May Public Safety stations be operated for short periods of time without a station authorization issued by the Commission? (19-19)

75. Define *control point* as the term refers to Public Safety Radio Service transmitters. (19-19)

76. What are the general requirements for transmitting the identification announcements for stations in the Public Safety Radio Service? (19-19)

77. What are the Commission's general requirements for the records or logs which must be kept by stations in the Public Safety Radio Service? (19-19)

78. When a radio operator makes transmitter measurements required for a station in the Public Safety Radio Service, what information should be transcribed into the station's records? (19-19)

79. Outline the transmitter measurements required by the Commission's rules for stations in the Public Safety Radio Service. (19-19)

80. What percentages of modulation are allowed AM stations in the Public Safety Radio Service? (19-19)

81. What is the minimum equipment necessary for satisfactory installation and maintenance of a mobile transmitter in a station employing FM in the Maritime Mobile Radio Service? (19-19) [6]

82. What specific equipment-performance measurements must be made at all FM broadcast stations annually? (19-19) [4]

83. During what time period may FM or AM broadcast stations transmit signals for testing and maintenance purposes? (19-18) [4]

84. Draw and label a block diagram of a voice-modulated (press-to-talk microphone) indirect (PM) FM transmitter having a crystal multiplication of 12 and an output of 155.46 MHz. (19-20)

85. Consider the transmitter in question 84 as completely detuned. Explain a proper tuning procedure for stages up to the final amplifier. (19-20)

86. Describe a procedure for tuning the power amplifier of the transmitter in question 84 to an antenna if an adjustable coupling from power amplifier to antenna is assumed. (19-20)

87. Under what usual conditions of maintenance and/or repair should a transmitter be retuned? (19-20)

88. How could the frequency of operation of a transmitter operating on 155.45 MHz be measured with a heterodyne frequency meter tunable to any frequency between each 1-MHz interval over a range of 20 to 40 MHz if the harmonics were usable up to 640 MHz? (19-20)

89. If a transmitter has an assigned frequency of

155.46 MHz and a 0.0005% required tolerance, what are the minimum and maximum frequencies as read on a frequency meter with a 0.0002% accuracy which will assure transmission within tolerance? (19-20)

90. If an FM transmitter is overmodulating and has a limiting adjustment, what adjustments can be made to remedy the fault? (19-20)

91. In an FM transmitter what would be the effect on antenna current if the grid bias on the final power amplifier were varied? (19-20)

92. When checking frequency deviation with a modulation meter, would the greatest deviation be produced by whistling or by speaking in a low voice into the microphone? (19-20)

93. How can it be determined if a power-amplifier (PA) stage in an FM transmitter is self-oscillating, and what adjustments could be made? (19-20)

94. If the PA of an FM transmitter is a pentode, how is the power input determined? (19-20)

95. What is the basic difference between *type approval* and *type acceptance* of equipment? (19-20)

96. In general, what type of changes in authorized stations must be approved by the FCC and what type does not require FCC approval? (19-20)

97. Discuss the cause and prevention of interference to radio receivers in motor vehicles. (19-21)

98. Draw a block diagram and explain briefly the overall operation of a multiplex FM broadcast transmitter. (19-22, 19-23, 19-25) [4]

99. What are the stereophonic transmission standards provided by FCC rules? (19-23) [4]

100. What is *SCA,* and what are some possible uses of it? (19-23) [4]

101. What are the transmission standards of *SCA* multiplex operations? (19-23) [4]

102. What items must be included in an SCA operating log? (19-23) [4]

103. Define the following terms as they apply to FM broadcast stations: *center frequency, FM broadcast band, FM broadcast channel, FM broadcast station, frequency modulation, frequency swing, multiplex transmission, percent of modulation, cross talk, left signal, left stereophonic channel, main channel, pilot subcarrier, stereophonic separation, stereophonic subcarrier, stereophonic subchannel.* [4]

20 ANTENNAS

The objective of this chapter is to discuss fundamentals of energy radiation through space and the various types of antennas used in radio transmission and reception. Transmission lines and their uses in high-frequency applications and in broadcast systems are explained.

20-1 RADIO WAVES

An antenna, or aerial, in one form consists of a piece of wire or other conductor, with insulators at both ends, suspended well above the ground, as in Fig. 20-1. If the wire is cut to the equivalent of a half wavelength, it acts as an oscillating LC circuit. When excited by some source of RF alternating voltage, the free electrons in it will oscillate back and forth from one end to the other. If the wire were a perfect conductor and had a high Q, once shock-excited the electrons might be expected to continue to oscillate back and forth along it indefinitely. However, some of the energy

imparted to the electrons is lost in heating whatever resistance is in the wire. A much greater amount of the energy is radiated into space from the wire. It is this energy lost by the antenna into space that is the *radio wave*.

If an antenna is shock-excited, free electrons in it will be driven to one end, pile up there, and produce a highly negative charge. At the same time the other end of the wire is left with a deficiency of electrons and is positively charged. An electrostatic strain is developed from one end to the other, and an electrostatic field is formed along the antenna, as shown by the dashed lines in the illustration.

With nothing to hold the electrons at one end, they reverse their previous motion and start toward the opposite end. As they move, they produce a magnetic field, outward *around* the wire, as shown. The energy that had been stored in the electrostatic field is now transferred to an electromagnetic field. As the potential difference be-

Fig. 20-1   A basic antenna, or aerial, develops electromagnetic fields around it and electrostatic fields end to end.

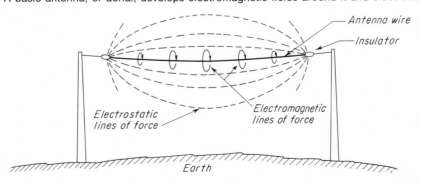

tween the ends of the antenna drops to zero, the magnetic field reaches maximum and then begins to contract back into the wire, driving electrons toward the once-positive end of the antenna. This end now has an excess of electrons on it, charging it negative, the other end, having lost electrons, becoming positive. The energy is again stored in an electrostatic field.

This oscillating of electrons and energy in the antenna produces expanding and contracting electrostatic and electromagnetic fields. Parts of these fields travel out so far that they cannot return to the antenna when the fields alternate. They are lost to the antenna and are radiated into space.

Actually, neither an alternating static nor a magnetic field can exist alone. If one is present, the other will be also, but 90° from the first.

If the frequency of the ac exciting the antenna is low, the antenna must be long to prevent the electric impulse from reaching the end and returning before the driving source potential reverses. As a result, low-frequency antennas are long and high-frequency antennas are short. The lowest radio frequency is 10 kHz and requires a half-wave antenna 15,000 meters (nearly 9 miles) long. An antenna for the middle of the standard broadcast band, 1,000 kHz, is 150 m (468 ft). For a frequency of 10 MHz, a wire 15 m (46.8 ft) is required; for a frequency of 100 MHz, 1.5 m (4.68 ft); and for 10,000 MHz (10 GHz), 0.015 m (0.0468 ft). By necessity, antennas of less than a half wavelength are used for low-frequency transmissions. They are *loaded* (tuned) with inductors until they will tune properly to the desired frequency.

Radiations of all the lower-frequency (longer-wavelength) radio waves tend to travel along the surface of the earth without attenuation. The *ground* wave, that portion traveling just above the surface of the ground (Fig. 20-2), is usable for hundreds to thousands of miles, day or night.

With higher frequencies, the ground wave weakens as energy from it is more effectively absorbed by the surface over which it is moving. Frequencies in the 500- to 1,500-kHz region may have usable ground-wave signals for only 600 to 50 miles.

As the frequency is increased above 5 or 10 MHz, the usable ground-wave signal exists for

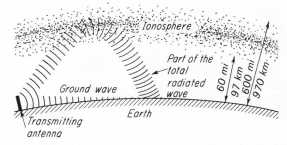

Fig. 20-2 The ionosphere, a layer of ionized air (gas) above the earth, can refract radio waves back to earth.

only a few miles. At these frequencies it is possible to transmit farther by using the *direct* wave. When the receiving and transmitting antennas are within sight of each other, the signal is considered a direct wave. At high frequencies, where the ground wave is insignificant, aircraft several thousand feet in the air may receive direct waves 100 miles or more away; satellites, thousands of miles away.

If the electrostatic and electromagnetic fields of energy traveling through the air cross a conductor, an emf at the frequency of the energy wave will be induced in the conductor. The explanations that follow will consider the antenna as a transmitting device. If an antenna transmits maximum signals east and west, it will receive best from east and west also. The radiation resistance of an antenna is the same for transmitting as for receiving. The only differences are the insulation requirements and wire diameter, since transmitting antennas may carry large currents.

20-2 THE IONOSPHERE

Near the earth the air is rather dense, but from about 60 to 600 miles above the earth the air is quite thin and radiated energy from the sun can ionize the widely spaced air molecules. The different degrees of ionization produced form into several recognizable layers. The ionized atmosphere allows the radiated wave to travel faster through it than in the more dense, un-ionized lower air. As a result, the top part of a wave front moving into the *ionosphere* speeds up and forges ahead of the lower part of the wave front and eventually may turn, or *refract,* downward.

The lower the frequency of the waves, the less

penetrating effect they have and the greater the proportion of them that may be turned back toward earth. The higher the frequency the more penetrating energy the radio wave contains. Signals with frequencies of 2 to 50 MHz may be deflected (refracted) or they may penetrate the ionosphere, depending on the time of day, the angle at which the wave strikes the ionosphere, and the degree of ionization present. With weak ionization they penetrate; with stronger ionization they may be refracted. With still stronger ionization the wave energy may be totally absorbed and dissipated in the ionosphere. During times of sunspot activity and while the aurora borealis is active, ionization is considerably increased and long-distance transmission may be interrupted because of almost complete absorption of all waves by the ionosphere. Only short-range communications may be dependable, except for some unexpected long-haul openings at times. Aurora-reflected signals have a rapid fluttering sound.

The maximum usable frequency (MUF) is the highest-frequency at which ionospheric-refracted signals return to earth with usable strength. MUF varies greatly with time of day, distance, direction, and season.

Besides the frequency factor, the angle at which the radio wave enters the ionosphere determines the penetration or refraction of the wave. While there may be some reflection of lower-frequency signals traveling directly upward, almost all higher-frequency waves that are transmitted at an angle of nearly 90° above the surface of the earth either penetrate or are absorbed by the ionosphere, as indicated in Fig. 20-3. As the angle becomes less than 90°, there is more chance of refraction. The higher the frequency, the greater the penetration and the lower the angle required to produce refraction. At high frequencies, there may be long distances between the end of the usable ground-wave signal and the reappearance of the reflected *sky* wave. At lower frequencies the sky wave often returns to earth in the ground-wave region.

If the sky wave returns to earth and strikes a good conducting surface such as salt water, it will be reflected back upward and take a double hop. A double hop may carry a signal a very long distance.

Active satellites parked in orbit over the equator receive and transmit at frequencies in the gigahertz range. Signals from and to earth are refracted almost not at all and remain the same night and day. The signals that are received at the satellite are translated to another frequency and relayed back to some other point on earth.

20-3 FADING

The fading of signals stems from two major causes. If a sky wave is being received, variations in the ionosphere may refract more or less energy to any given receiving point at different instants, producing a varying amplitude or fading signal.

When the receiving antenna is within both ground- and sky-wave range, the addition of the

Fig. 20-3 Paths of radio waves in the ionosphere. Low and medium frequencies are affected by the ionosphere more than high frequencies are.

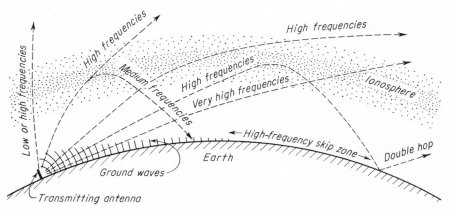

two waves may be in or out of phase. When arriving in phase, the waves add to each other, producing a strong signal. When out of phase 180°, they tend to cancel each other. Variations in the ionosphere can change the sky-wave travel distance and therefore the phase relationship of the ground and sky waves. Two sky waves refracted to the same point by two different areas of the ionosphere also may arrive in or out of phase.

In the VHF or UHF regions, where the direct wave is used, as in television, an airplane flying overhead will act as a reflector of the direct wave. A receiving antenna below the plane will receive the direct, or ground, wave from the TV station plus the wave reflected by the airplane. Since the plane is moving, the transmitter-reflector-receiver distance constantly changes and the relationship of the direct and the reflected waves at the receiving point is alternately in and out of phase, producing a continually fading signal.

20-4 NIGHT AND DAY TRANSMISSIONS

The ground wave remains the same night and day. Only the sky wave changes.

As night approaches, the sun can no longer ionize the atmosphere above the darkened part of the earth, and the ionized layers become thinner. The thinly ionized layers turn the waves back to earth over a wide arc, and sky waves return to earth much farther away than during the daytime.

At low frequencies there is not too much difference between night and day transmissions, although at distances of several thousand miles the nighttime signals will be stronger. This is mainly ground-wave communication.

At medium frequencies, out past the ground-wave range, the signals improve materially at night. It is possible to pick up signals several thousand miles away, in comparison with several hundred miles during the daytime.

At high frequencies, from 3 to 10 MHz, daytime refracted signals may return to earth 20 to 500 miles or more away. At nighttime the ionosphere lifts and refracted signals return to earth 200 to many thousand miles away.

At still higher frequencies, 10 to 30 MHz, the daytime refracted signals may return 200 to 5,000 miles or more away. At nighttime the signals may pierce the ionosphere and not return to earth at all.

A high-frequency communication system between two points 2,000 or 3,000 miles apart may have to shift from one frequency to another during the day and night to keep the signals strong enough to be usable.

Signals of 30 to 100 MHz may sometimes be refracted during the daytime, but never at night. These frequencies are considered unreliable for long-distance communication. They are used for ground-wave, or direct-wave, communications.

Signals above 100 MHz are rarely subject to refraction by the ionosphere, acting more like light waves. They are used for ground-wave, direct-wave, or extraterrestrial transmissions.

Beaming strong UHF or VHF signals toward the horizon causes induction of the signal into conductive areas or objects on the horizon. The signals may be refracted or reradiated from them. These signals out past the horizon form a *scatter*-type transmission which greatly increases the relatively reliable range of such frequencies.

Normally, the greater the altitude the colder the air. If for any reason a layer of warmer air forms above a colder stratum, a *heat inversion* is present. The two layers have different densities of air and can affect VHF or UHF radio waves enough at times to refract them back to earth at distances of a few hundred miles or less. When the layer is thin, it may act as a *duct,* or pipeline, for UHF and SHF (superhigh-frequency) signals. Ducted signals may travel hundreds of miles before leaving the duct and returning to earth. These ducts usually form over water areas.

20-5 EFFECT OF LIGHTNING ON RADIO RECEPTION

A bolt of lightning produces RF energy across almost the whole usable radio spectrum, but the percent of energy decreases as the frequency increases. At lower frequencies there is considerable energy, and with good ground-wave transmission, storms hundreds or thousands of miles away can produce considerable interference. The higher frequencies, particularly between 5 and 15 MHz, are subject to local storms, but storms in the skip zones may not be heard. As a result, the higher frequencies are much less subject to lightning-produced radio noise (static).

20-6 POLARIZATION

The polarization of the radiated wave is considered to be in the direction of the electrostatic field of the antenna. Therefore the antenna in Fig. 20-1 transmits a horizontally polarized radio wave. An antenna erected vertically will radiate a vertically polarized wave as in Fig. 20-2.

TV and FM broadcasts are made with horizontally polarized antennas. A receiving antenna to pick up a maximum signal voltage from the transmitted wave must have the same polarization. As a result, all TV and FM receiving antennas have horizontal elements. If a TV receiving antenna is erected vertically, almost no signal will be received from the transmitting station.

Most transmissions below 2 MHz will be vertically polarized. Standard-broadcast-band transmitters (535 to 1,605 kHz) use vertically polarized antennas. Transmitters on frequencies lower than 500 kHz may use horizontal wires, but the vertical radiation component may be greater than the horizontal.

In the frequency ranges from 3 to 30 MHz, under certain conditions, and for a given transmission distance, vertical polarization sometimes operates better than horizontal. Under other conditions, horizontal polarization may produce the better-received signal.

For frequencies above 30 MHz, most communication transmissions, with the exception of TV and FM, are vertically polarized.

Because of reflection from nonvertical objects or because of ionosphere contour, the polarization of a ground, direct, or sky wave may be twisted. Thus at times a vertically polarized wave may be received best on a horizontal antenna some distance away.

20-7 THE HALF-WAVE ANTENNA

The velocity of a radio wave is 300,000,000 m/s (actually 299,792,462 m/s). This is a little more than 186,291 mi/s or 983,616,670 ft/s.

If a 1,000,000-Hz RF ac transmitter is exciting an antenna, the radio wave from it will travel 300 m in the time it takes the ac to complete one

cycle. A frequency of 1 MHz has a wavelength of 300 m.

The wavelength of a radio wave, a sound wave, or any other wave varies inversely as the frequency and is determined by the basic formula

$$\lambda = \frac{v}{f}$$

where λ = wavelength in whatever unit of length is used
v = velocity of wave in the same unit of length
f = frequency in Hz

From this are derived three formulas often used in radio:

$$\lambda = \frac{300,000,000}{f} = \frac{300,000}{f_{kHz}} = \frac{300}{f_{MHz}}$$

where
λ = wavelength in m
f = frequency in Hz
f_{kHz} = frequency in kHz
f_{MHz} = frequency in MHz
300,000,000 = velocity of radio waves in m/s

If used with a 1-MHz radio transmitter, the basic length of the antenna wire will be equal to one-half of the full 300-m wavelength, or 150 m (492 ft). When an antenna of this length is excited by a 1-MHz RF ac, electrons should have just enough time to reach the end of the wire as the source RF ac reverses polarity. At no time should the natural period of electron oscillation be ahead of or behind the phase of the exciting RF.

While this theory is basically correct, it neglects *end effect*. End effect must be considered with any antenna. It may be regarded as a dielectric effect of the air at the end of the antenna that effectively lengthens it. The result of end effect is to make a half-wave-antenna wire act as if it were about 5% longer than it actually is. This will produce an interference between the exciting and the oscillating currents and a lessening of the oscillation amplitude, with a corresponding lessening of the radiated field. To overcome end effect and make the antenna resonate properly, it is necessary to cut the wire to a physical length approximately 95% of the electrically computed half wavelength. For 1 MHz the half-wave antenna will be about 468 ft. The ratio of length in feet, end effect, and frequency holds close enough to set up a general formula to determine the length in feet of any half-wave antenna supported at the end by insulators. The formula is

$$\text{Half wavelength in feet} = \frac{468}{f_{\text{MHz}}}$$

The length of a 7-MHz antenna is

$$\text{Length} = \frac{468}{f_{\text{MHz}}} = \frac{468}{7} = 66.9 \text{ ft}$$

If the half-wave-antenna element is self-supporting at the middle and is without end insulators, the end effect is less. The factor 478 may be used in the above formula.

If an antenna wire is close to ground, to a building, to trees, etc., the end-to-end capacitance increases and a greater end effect must be considered.

When an antenna is a full wave in length, it is composed of two half waves, but there is still only one pair of end effects to consider. The total length of such an antenna is equal to a half wave with end effect plus a half wave without end effect. The factor 492 is used in the formula to compute the half wave without end effect.

It is sometimes necessary to compute the length, in wavelengths, of an antenna.

EXAMPLE: A 405-ft antenna is to be operated at 1,250 kHz. What is its wavelength at this frequency? If 1 m = 3.28 ft, the length of the antenna in meters is 405/3.28, or 123.4 m. The wavelength of 1.25 MHz is 300/1.25, or 240 m. The ratio of 240 to 123.4 m gives the decimal fraction of the wavelength of the antenna, or 123.4/240 = 0.514 wavelength.

20-8 HERTZ AND MARCONI ANTENNAS

Any antenna complete in itself and capable of self-oscillation, such as a half or full wavelength, is known as a *Hertz* antenna.

When an antenna utilizes the ground (earth) as part of its resonant circuit, it is a *Marconi* antenna. An example of a Marconi antenna is a quarter-wave antenna, where the ground operates as the missing quarter wavelength. Most low- and medium-frequency antennas are Marconi types.

Test your understanding; answer these checkup questions.

1. What is the relation in degrees of the electrostatic and electromagnetic fields of an antenna? _____ Is this their relation in space? _____
2. What HF waves are attenuated within a few miles? _____
3. Does wave velocity increase, decrease, or remain the same as it passes from air to ionosphere? _____
4. What wavelength radiations tend to be transmitted entirely between ionosphere and earth? _____
5. What effect do sunspots have on the ionosphere? _____
6. What is a double-hop signal? _____
7. What are two major causes of fading? _____
8. Where is the skip zone? _____
9. What wave is the same day or night? _____ Why might it appear weaker at night? _____
10. Why do HF communication systems shift frequencies at different times of day? _____
11. Scatter transmission is used at what frequencies? _____
12. Over what areas do ducts usually form? _____
13. Is polarization named for the static or magnetic component of the wave? _____
14. What polarization is used in AM broadcasting? _____ TV and FM broadcasting? _____ Mobile communications? _____ Satellite communications? _____
15. What is the length in feet of an antenna wire for 4 MHz if the antenna is three half waves in length? _____
16. What is the name of antennas that are grounded at one end? _____ That do not use ground? _____

20-9 CURRENT AND VOLTAGE IN A HALF-WAVE ANTENNA

A half-wave antenna excited by an RF ac source produces an oscillation of the free electrons in the wire. Since electrons pile up at the ends, the maximum charge (voltage) always occurs at the far ends of the antenna.

When electrons move from one end of a half-wave antenna to the other, the greatest number must move past the midpoint of the wire. Therefore, the center has the maximum current flowing through it. Because the electrons that pile up at the end of the antenna do not move past the end, there is zero current at (through) the far end.

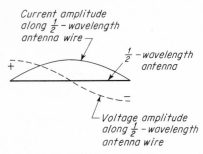

Fig. 20-4 Voltage and current distribution on a half-wave antenna.

Figure 20-4 illustrates the voltage and current relations on a half-wave antenna.

The current distribution in a half-wave wire, if measured by inserting an ammeter at different points in the antenna, would be very nearly sinusoidal. If the voltage at different points could be conveniently measured, it would also vary in a sine-wave manner.

When the electrons pile up at the far ends, the voltage at the ends is at a maximum and the current is zero. When the current at the center reaches a maximum, the voltage difference between the ends is zero. This is the same 90° voltage and current phase relation that exists in any freely oscillating LC circuit.

20-10 THE RADIATION RESISTANCE

When an antenna is excited into oscillation, it radiates energy in the form of electrostatic and electromagnetic waves. Insofar as space is concerned, the antenna is acting as a source of power. Any source must have an internal resistance or impedance. If radiated power and current maximum in an antenna are known, the power formula $P = I^2R$ can be rearranged to $R = P/I^2$. The resistance value computed is known as the

radiation resistance of the antenna. It is the ratio of radiated power to the square of the antenna center current. Radiation resistance is also the ratio of the voltage at any point on an antenna to the current flowing at that point ($R = E/I$). Since current decreases toward the ends of a half-wave antenna and the voltage increases, the minimum resistance point will be the center of any half-wavelength antenna.

Theoretically, the radiation resistance of a half-wave dipole a quarter-wave high is 73 Ω, but only if constructed of infinitely thin wire and away from any reflecting surfaces. As it is lowered toward the ground, the resistance decreases almost linearly to zero. Elevating it to one-third of a wave above ground raises the resistance to about 95 Ω, but the resistance returns to 73 Ω at a half-wave height. At two-thirds of a wave high the resistance drops to about 58 Ω, returning to 73 Ω when three-quarters of a wave high and at every multiple of a quarter wave higher. The higher above earth, the less the resistance deviates from the 73-Ω value when not a quarter-wave multiple in height.

Antennas are not made of infinitely thin but of reasonably thick wire, in some cases rods or, in broadcasting, metal towers. The practical radiation resistance when a quarter wave high ranges from about 65 Ω for wire antennas to perhaps 50 Ω for towers. The ohmic and skin-effect resistance of the antenna is probably no more than 1 Ω.

If a half-wave antenna (some multiple of a quarter wave high) is cut in two at the center, the two severed points will appear to any transmission line connected to them as a load of approximately 65 Ω. If the antenna is not cut but two points equally spaced from the center are selected, there will be a resistive value between the two points. If the points are close together, the resistance is low; if far apart, the resistance is high, as in Fig. 20-5.

Besides the resistive component, an antenna may be reactive: capacitive if it is less than a half wave in length and inductive if more than a half wave.

20-11 LOADING OR TUNING AN ANTENNA

To produce optimum operation, an antenna may be tuned to resonate at the frequency on

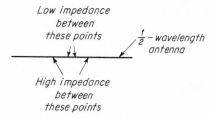

Fig. 20-5 The farther from the center of a half-wave antenna the higher the impedance.

which it is to operate. This tuning is accomplished, basically, by cutting the wire to the proper length, usually a half wave.

If an antenna is shorter than desired, it can be cut apart and a *loading coil* inserted in series with it, as in Fig. 20-6. The wire of the coil can be

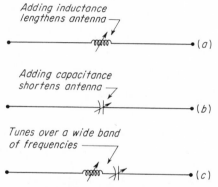

Fig. 20-6 (*a*) Adding inductance in an antenna lowers the resonant frequency. (*b*) Adding capacitance in series raises the resonant frequency. (*c*) Variable coil and capacitor allow the antenna to be tuned over a wide band of frequencies.

considered as being part of the required length of the antenna but coiled into a confined space. A short antenna may also be thought of as having too little inductance and, as a result, too little inductive reactance. Since $X_L = X_C$ at resonance, a short antenna must have some capacitive reactance canceled by inductive reactance. To cancel X_C and attain resonance, X_L (a coil) is needed.

When the antenna is longer than desired, the antenna can be cut apart and a capacitor inserted in series with it. An antenna longer than a half wavelength appears as an inductive reactance to the frequency of operation. The addition of the proper-value capacitor in series with it cancels the excess inductive reactance, making the antenna

resonant. It then acts as a pure resistance to the feed line.

If both the inductance and the capacitance inserted in the antenna are variable, it is possible to tune the antenna to resonance over a wide band of frequencies.

20-12 TRANSMISSION LINES

As in all phases of electricity, to produce maximum RF transference of energy from one circuit to another the impedance of the two circuits must match. An example of impedance matching to produce maximum antenna current and radiation is illustrated in Fig. 20-7.

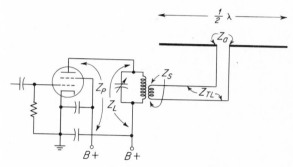

Fig. 20-7 Impedances that must be matched for maximum energy transfer to the antenna from the transmitter.

The impedance of the antenna Z_a (65 Ω) must be matched by the *surge impedance* of the two-wire transmission line Z_{TL}, which must be matched by the impedance of the secondary of the output transformer Z_s, which must couple tight enough to reflect an impedance across the primary Z_L that will equal the plate impedance of the tube Z_p. A variation of any part in this chain may prevent maximum output.

The two parallel wires are indicated as having an impedance. Any two parallel wires held apart a constant distance will have a characteristic, or surge, impedance value. This impedance is a function of the series inductance and shunt capacitance of the parallel wires, the diameter of the wires, and their distance of separation:

$$Z = 276 \log \frac{d}{r}$$

where Z = impedance in Ω

d = center-to-center distance of separation of conductors

r = radius of conductors (using same unit of measurement as with d)

From above, two No. 14 wires (diameter, 0.064 in.; radius, 0.032 in.), held 2 in. apart center to center, form a transmission line with an impedance of 495 Ω. If the wires are thicker, the impedance is lower. If the distance of separation is increased, the impedance increases as the log of the ratio of d/r.

If a 500-Ω line is infinitely long and is connected across a source of ac having 500-Ω impedance, maximum power will be taken from the source and dissipated along the line in the form of heat. No energy will return. If the line is a few yards long and is terminated with a 500-Ω resistor, essentially all the power from the source will be delivered to the resistor. Such a transmission line is an efficient means of transmitting energy from a source to a remote load, provided the transmission line has the proper impedance. A 300-Ω transmission line connected between a 500-Ω source and a 500-Ω load is a mismatch. The load will not draw maximum power from the source.

When the transmission line does not match the load impedance, not all the energy fed down the line flows into the load. Some is reflected back, forming *standing waves* on the line. Every half wave along the line, high-voltage and low-current points appear. Halfway between the high-voltage points will be low-voltage high-current points. The ratio of voltage across the line at the high-voltage points to that at the low-voltage points is known as the *standing-wave ratio,* or SWR. The SWR is also the ratio of the current values at the high and low points on the line. The standing-wave formula is

$$\text{SWR} = \frac{E_{\max}}{E_{\min}} \quad \text{or} \quad \text{SWR} = \frac{I_{\max}}{I_{\min}}$$

When the impedance at the far end matches the line impedance, there will be no standing waves. The current at all points along the line is the same, the SWR is 1:1, and the line is said to be *flat.*

If standing waves appear on a line that should be flat, it is necessary either to change the transmission-line impedance until it matches the antenna or to change the antenna impedance until it matches the line. This is important if optimum transmission and reception are desired from an antenna.

The ratio of current (or voltage) delivered to an antenna to that reflected back the line, called the reflection coefficient, ρ, is equivalent to

$$\rho = \frac{\text{SWR} - 1}{\text{SWR} + 1}$$

An impedance mismatch of 70 Ω to 35 Ω produces a SWR of 2:1 and a reflection coefficient of $(2 - 1)/(2 + 1)$, or $\frac{1}{3}$. Since power is proportional to I^2 (or E^2), the power reflected will be the square of ρ, or in this case, $(\frac{1}{3})^2$, or $\frac{1}{9}$. This means that eight-ninths of the power indicated by meters in the transmission line would actually be delivered to the antenna. The remaining one-ninth is reactive power (voltamperes), which is not actually a loss.

Mismatching a transmission line to an antenna results in the line at the transmitter end appearing to have either X_L or X_C, which will detune an LC circuit to which it is coupled. A matched antenna system should not detune a final amplifier appreciably.

Practical open-wire lines spaced with insulators, called spreaders, every few feet can be constructed to have impedances from about 150 to 800 Ω.

It is interesting to note that while standing waves are not wanted on transmission lines, they are developed on resonant antennas as evidenced by high-voltage points at the ends and low-voltage points at the center of a *dipole* (half-wavelength antenna).

Transmission lines for low-power transmitters and receivers may use two spaced wires held apart by a ribbon of plastic material. An example of this is the 300-Ω twin lead used in many TV receiving antennas.

When lower-impedance lines are required, coaxial, or concentric, cables can be used. These may consist of a copper tube, as the outer conductor, and a copper wire, centered inside the tube with insulating beads, as the other conductor. A coaxial cable may consist of a copper wire covered with plastic insulation, which in turn is covered with a braided-copper outer conductor. Coaxial cables are manufactured with various impedances, such as 50, 52, 72, 75, 93, and 125 Ω.

Hollow coaxial cables have the advantage of having their working surfaces (the inner surface of the outer conductor and the outer surface of the inner conductor) protected from the weather. This prevents oxidation of the surfaces, which would change the skin resistance as well as the dielectric constant between them and alter the impedance of the line.

The surge impedance of air-dielectric coaxial transmission lines is approximately

$$Z = 138 \log \frac{d_i}{d}$$

where Z = surge impedance in Ω
d_i = inside diameter of hollow tubing
d = diameter of center conductor

Although it may not be apparent in short, flat lines, when RF ac transmission over long lines is required, an attenuation of energy due to skin-effect, radiation, conductor resistance and dielectric losses occurs. As a result, the value of current flowing into a long, flat transmission line at the transmitter end may be significantly more than that flowing into the antenna at the far end.

20-13 DIRECTIVITY OF ANTENNAS

The half-wave antenna radiates energy in a direction at right angles to the wire itself. A horizontal antenna running north and south radiates maximum energy east and west, up into the sky, and down toward the earth. If the antenna is suspended in free space, well above the ground, the radiation can be represented by vector arrows as in Fig. 20-8. If the antenna is close to ground,

Fig. 20-8 Amplitude of radiation from a horizontal half-wave antenna in free space (end view of the wire).

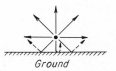
Fig. 20-9 When near the ground, energy is reflected, changing the radiation pattern of the antenna.

the part of the wave striking the ground reflects back upward and outward, as in Fig. 20-9.

When the antenna is a half wave above ground, the reflected wave travels to ground, reverses phase, is reflected upward, and reaches the antenna the equivalent of one wave later but 180° out of phase because of the reflection. As a result, the upward radiation is almost entirely canceled by the reflected wave. When all the vectors of all the angles from such an antenna are combined, the vertical radiation pattern of this antenna is as shown in Fig. 20-10. The angle of maximum radi-

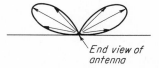
Fig. 20-10 Vertical radiation pattern of a horizontal antenna a half wave above ground as seen from a distance.

ation is about 30° above the horizon. There is a relatively strong wave being transmitted from a few degrees above the horizon up to more than 50°.

With the antenna only a quarter wave above ground, the reflected wave returns just in time and phase to add to the vertical radiation. The result is a raising of the angle of maximum radiation, as shown in Fig. 20-11.

With the antenna three-quarters of a wave above ground, the reflected waves return in time and phase to add to the upward radiation and also to the outward radiation. The radiation pattern is

Fig. 20-11 Vertical radiation pattern of a horizontal antenna a quarter wave above ground as seen from a distance.

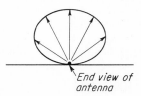

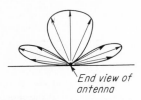

End view of
antenna

Fig. 20-12 Vertical radiation pattern of a horizontal antenna three-quarters of a wave above ground as seen from a distance.

then made up of three lobes as in Fig. 20-12. The higher the antenna, the more lobes that will be developed—one lobe for each quarter wave.

When looking down on a horizontal half-wave antenna from above, the radiation pattern in the horizontal direction forms a figure eight. Figure 20-13 shows maximum radiation at right angles

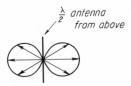

$\frac{\lambda}{2}$ antenna
from above

Fig. 20-13 Radiation from a half-wave horizontal antenna as seen from above (or radiation from a vertical antenna in free space).

to the wire and no radiation in the direction of the wire. The radiated-energy vectors form a doughnut around the wire if the antenna is suspended in free space. (If the antenna is brought near a reflecting surface, the doughnut shape is altered.)

A vertical half-wavelength antenna will radiate equally well in all horizontal directions, provided there are no nearby objects to alter its field. The effect of bringing the antenna near ground (by making it a quarter wave in length) is to raise the angle of maximum radiation, as in Fig. 20-14, but there is no effect on the circular horizontal radiation pattern (maximum in all directions). A vertical

Fig. 20-14 Comparative vertical radiation patterns for vertical half-wave and quarter-wave antennas.

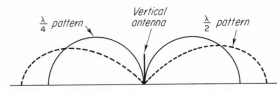

$\frac{\lambda}{4}$ pattern

Vertical
antenna

$\frac{\lambda}{2}$ pattern

antenna is said to be *omnidirectional* (all-directional).

It will be noted that directly above a vertical antenna there will always be a *cone of silence*. An airplane flying above a vertical transmitting antenna does not hear the station when directly over it. While there is a theoretical cone of silence at the ends of a horizontal half-wave antenna, in actual practice the reflected wave from the earth or nearby surfaces closes the cone and signals may be received directly off the ends.

Test your understanding; answer these checkup questions.

1. Where is (are) the maximum E point(s) on a half-wave antenna?_____ Maximum I point(s)?_____
2. Does Fig. 20-4 represent E and I at any one instant?_____
3. If current maximum and power radiated from an antenna are known, what two other facts can be determined?_____
4. What is the theoretical radiation resistance of a horizontal half-wave dipole one half wave high?_____
5. Which reactance would a dipole have if 0.6 wave long?_____ If 0.4 wave long?_____
6. How is X_C compensated in an antenna?_____ X_L?_____
7. What is the Z_o of a No. 14 wire transmission line if the wires are 4 in. apart?_____
8. Under what condition will a short transmission line have no reflected power?_____ What would the SWR be?_____
9. Is a dipole a center-fed half-wave antenna only?_____
10. What is the impedance of twin lead?_____
11. Why might more power flow into a transmission line than out of it?_____
12. For best long-distance communication why should a horizontal antenna be a half wave rather than a quarter wave high?_____
13. Would received signals off the end of a horizontal antenna be horizontally or vertically polarized?_____
14. Would vertical half-wave or quarter-wave antennas have the greater ground-wave range?_____
15. Is the radiation pattern of an antenna the same for transmitting and receiving?_____

20-14 THE QUARTER-WAVE ANTENNA

The antenna used with most MF (medium-frequency) and LF (low-frequency) transmitters is a quarter-wave vertical. If a quarter-wave vertical wire or mast is connected to a large conducting surface, such as the earth, as in Fig. 20-15a, the conducting surface will operate as the missing

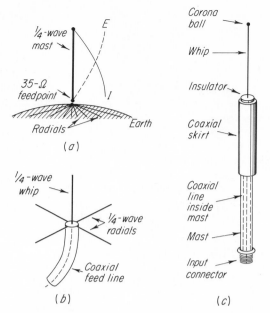

Fig. 20-15 (*a*) Quarter-wave vertical for LF and MF. (*b*) VHF quarter-wave coaxial-fed ground-plane whip antenna. (*c*) Coaxial half-wave vertical antenna.

quarter wave (or a great many multiples of quarter waves) to make the antenna resonate as a half-wave device. The radiation resistance of the quarter-wave antenna, between ground and the lower end of the wire, is just half of that of a half-wave antenna, theoretically 36.5 Ω. The antenna is often fed with 50-Ω coaxial transmission line without excessive SWR, but shunt feed may be more desirable (Sec. 20-24). Usually, quarter-wave copper-wire radials will be run out over (or under) the ground every 15° from the base to make the ground surface more constant electrically and to serve as one terminal for the coaxial feed line. When a quarter-wave whip is installed on top of an automobile, the roof of the car constitutes the *ground plane*. If the whip is mounted near the rear bumper, the back of the car and capacitance to the roadway act as the other required quarter wave. (When a wire-ground system is used beneath a horizontal antenna, it is called a *counterpoise*.)

In the VHF and UHF range, vertical antennas often employ four or more horizontal quarter-wave radials as the ground plane, as in Fig. 20-15*b*. If the radials point downward about 45°, the an-

tenna is called a *drooping* ground plane. This has the effect of raising the feed-point resistance to about 50 Ω. If the ground-plane radials are bent straight down and connected together or a cylinder or sleeve is used, as in Fig. 20-15*c*, a (half-wave) *coaxial* antenna is formed. These may be seen on tops of buildings or on the back bumpers of automobiles. Sometimes a second skirt is added to the support mast a quarter wave below the bottom of the coaxial skirt; this acts as a trap to prevent standing waves from forming on the mast if it is metal.

The small metal or plastic ball often seen at the top of whip antennas is used to prevent corona discharge (ionization of the air). Transmitting antennas develop thousands of volts at their ends. This alternately drives electrons out into the surrounding air molecules and attracts them from the molecules, producing a corona glow and crackling noises, particularly if the antenna tip is sharp.

A quarter-wave antenna, like any other antenna, has a high-voltage point at the far end and a low-voltage high-current point a quarter wave down the antenna.

20-15 THE FULL-WAVE ANTENNA

When the half-wave antenna is fed an ac to which it is resonant, electrons oscillate back and forth along it. If the antenna is a full wavelength, electrons will move from both ends toward the middle during one-half of the excitation cycle, as indicated in Fig. 20-16*a*. When the cycle reverses, the electrons travel from the middle toward the ends. In this way the maximum voltage points appear at the middle and at the ends of a full wave.

Fig. 20-16 (*a*) Current flows in opposite directions in two halves of a full-wave antenna at the same instant. (*b*) *E* and *I* in an antenna three-halves or three half waves long.

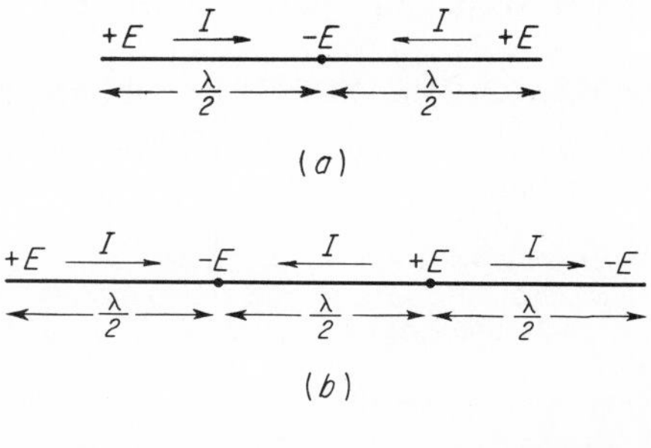

A 1-, 1½-, 2-, etc., wave wire will have maximum voltage points every half wave. The electrons in adjacent half waves are always traveling in opposite directions at any one instant, as shown in Fig. 20-16*b*.

The development of voltage maximums every half wave is actually the development of standing waves on the antenna wire. This produces maximum current and maximum energy radiation. If the wires are not an exact multiple of a half wave, lower-amplitude standing waves will be present and the antenna may be a less efficient radiator.

Since the full-wave antenna has equal current flowing in opposite directions at any one instant, the radiation from one of the half waves exactly equals the radiation from the other. With the polarity of the two being 180° out of phase, the result is zero effective radiation at right angles to the wire. The horizontal radiation pattern of a horizontal full-wave antenna is shown in Fig. 20-17. The angle of maximum radiation is approximately 45° from the direction of the wire.

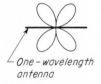

Fig. 20-17 Radiation pattern of a full-wave antenna in free space.

With a 1½-wave antenna, two of the half waves cancel the radiation of each other at right angles to the wire, leaving the third half wave to radiate in this direction. As a result, the major lobes are depressed toward the wire, and the direction of maximum radiation tends to follow the wire, although there is a lobe at right angles to it. The more half waves used, the more the maximum-radiation lobes are formed in the direction of the wire. When about four waves long, an antenna

---

ANSWERS TO CHECKUP QUIZ ON PAGE 508

1. (*Ends*)(*Middle of half wave*) 2. (*No, 90° out of phase*)
3. (*Radiation resistance and maximum voltage*) 4. (*73 Ω*)
5. (X_L)(X_C) 6. (*Add series L*)(*Add series C*) 7. (*580 Ω*)
8. (*Terminated by resistance equal to surge Z*)(*1:1*) 9. (*No, half-wave or less*) 10. (*300 Ω*) 11. (*Line losses*) 12. (*No wasted vertical radiation*) 13. (*Vertically*) 14. (*Half-wave*)
15. (*Yes*)

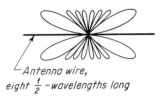

Antenna wire,
eight ½ -wavelengths long

Fig. 20-18 Radiation pattern of a four-wavelength antenna in free space.

becomes directional *in line with the wire* (Fig. 20-18). If the same antenna wire is used at different frequencies, it will appear as having a different number of wavelengths and therefore different directional characteristics on each frequency. Note that there is a lobe for each half wavelength.

20-16 FEEDING THE ANTENNA

There are many methods of feeding energy to an antenna. Cutting a half-wave antenna in the middle and feeding it with a 50- to 73-Ω transmission line is a satisfactory method. This is known as *center-feeding,* or *current-feeding* the antenna, since maximum current is present at the feed point.

If the antenna is a full wave in length and is cut in the middle, two points of high impedance are developed, usually having between 2,000- and 3,000-Ω impedance. Since a two-wire transmission line of more than 700 or 800 Ω is not very practical, such an antenna cannot be fed by matching it with a flat transmission line. Instead, a high impedance is developed on a transmission line by deliberately producing standing waves and thereby high- and low-impedance points on it. Maximum standing waves are obtained when the transmission line is some multiple of a quarter wave. If the transmission line feeding the center of a full-wave antenna is a multiple of a half wave in length, it may be fed as in Fig. 20-19.

If cut to a multiple of a half wave, a transmission line repeats its terminal impedance. In Fig. 20-19, by starting at the far end of one of the half-wave radiators (always a high-impedance or high-voltage point) and progressing toward the feed point, low impedance appears at the center of each half-wave wire and another high impedance appears at the feed point. By progressing down the transmission line a half wave, another point of high impedance is reached. By connecting this

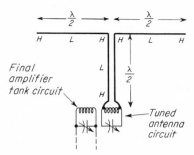

Fig. 20-19 Two half-wave antennas fed with a half-wave tuned feeder. (H and L indicate high- and low-impedance points.)

high-impedance end of the transmission line across the relatively high impedance of a tuned antenna LC circuit, a satisfactory impedance match is produced and maximum current flows in the tuned transmission line and in the two half-wave antenna wires.

By starting at the center of the two half-wave radiators but progressing down the transmission line only a quarter wave (or any odd multiple of a quarter wave), a point of low impedance is reached. The low-impedance-point terminals can be connected to a few turns of wire (a low-impedance coil) inductively coupled to the tuned LC circuit of a transmitter as in Fig. 20-20. The

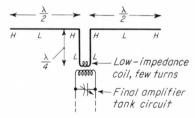

Fig. 20-20 Two half-wave-antenna wires fed with a quarter-wave tuned feeder.

addition of the few turns to the transmission line adds inductive reactance to the circuit. A variable capacitor, not shown, may be added in series with the coil to balance out this inductive reactance.

Any resonant antenna may be end-fed by using a tuned feeder system as above, leaving one end of the feed line unconnected, as in Fig. 20-21. This *Zepp antenna* (first used on zeppelins) is somewhat unbalanced.

Another method used to feed an antenna is known as the *delta match* (Δ match). A flat trans-

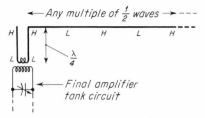

Fig. 20-21 Zepp antenna using a quarter-wave tuned feeder.

mission line, usually of 600-Ω impedance, is brought up to the center of a half-wave wire. To match the feed line to two points representing 600 Ω on the antenna, it is necessary to spread out the feeders at the antenna end as in Fig. 20-22. The approximate ratio of lengths to match a 600-Ω line to a half-wave antenna is also shown.

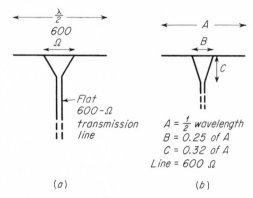

Fig. 20-22 (a) Δ-fed half-wave antenna and (b) ratio of dimensions to match a 600-Ω line.

Similar to Δ matching, a low-impedance coaxial line, 73 Ω, for example, will couple to an antenna by connecting the outer sheath of the cable to the center of the antenna and extending the inner conductor to a point on the antenna equivalent to 73 Ω. This results in a slightly unbalanced *gamma-match* (Γ-match) feed system. A balanced system can be produced by using two parallel coaxial lines or a special coaxial cable having two inner conductors parallel to each other. This type of coupling is commonly termed a *T match* (Fig. 20-23). Capacitors should be added in series with the extended inner conductors to cancel their inductive reactance.

A shorted or an open *stub* is sometimes inserted in the middle of a half-wave antenna to act

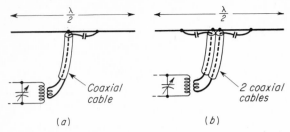

Fig. 20-23 (a) Coaxial transmission line terminated with a gamma match. (b) Two coaxial lines forming a T match.

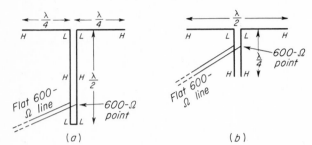

Fig. 20-24 Half-wave dipoles center-fed with (a) a half-wave shorted stub and (b) a quarter-wave open stub with flat lines coupled to the stubs.

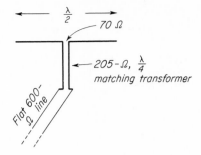

Fig. 20-25 A quarter-wave matching transformer.

wave section is found by the formula

$$Z_o = \sqrt{Z_1 Z_2} = \sqrt{70(600)} = \sqrt{42,000} = 205\ \Omega$$

where Z_o = quarter-wave line impedance
Z_1 = impedance connected across one end of transformer
Z_2 = impedance at other end

If a quarter-wave matching transformer is inserted between two *equal* impedances, it must have the same impedance and the whole line becomes flat with a 1:1 SWR.

Since there are no free ends on a quarter-wave matching transformer, there is no end effect and the length of the line is computed by the formula

$$L = \frac{246\ V}{F}$$

where L = length in ft
F = frequency in MHz
V = *velocity factor* of transmission line

When transmission lines are constructed of parallel wires with polyethylene insulation or any other types of dielectric material between the conductors, the velocity of the wave traveling down the line is less than in free air. As a result, tuned transmission lines as well as matching transformer lengths must be multiplied by a velocity factor to produce the proper working lengths. Table 20-1 lists some velocity factors. Note that the factor does not depend on frequency.

A transition from a higher- to a lower-impedance line can be produced by using a tapered line about four wavelengths long between two unequally spaced lines.

as an impedance-matching device for a flat transmission line. Figure 20-24 shows a half-wave shorted stub connected to a half-wave antenna. The low impedance of the center of the antenna is repeated as a low-impedance point at the shorted end of the half-wave stub. The whole antenna can be tuned by adjusting the shorting point. By connecting a 600-Ω flat transmission line at two points on the stub at which an impedance of 600 Ω appears, an excellent impedance match between transmission line and antenna system can be produced. A quarter-wave open stub can be utilized in a similar manner. While a stub might be any multiple of quarter waves, the fewer quarter waves used, the less radiation from the feeder system.

The quarter-wave transmission line has another useful property. It can act as an impedance-matching device between a high- and a low-impedance circuit if the line has the proper intermediate impedance. This is illustrated in Fig. 20-25, where the 70-Ω center impedance of a half-wave antenna is coupled to a flat 600-Ω transmission line through a *quarter-wave transformer*. The required impedance of the quarter-

Table 20-1 VELOCITY FACTORS

Dielectric material between wires	Velocity factor
Air-insulated parallel line	0.975
Air-insulated coaxial cable	0.85
Polyethylene parallel line (twin lead)	0.82
Polyethylene coaxial cable	0.66

20-17 COLLINEAR BEAM ANTENNAS

A shorted quarter-wave stub is actually a half-wave length of wire. The voltages at the open ends of a properly excited stub will be 180° out of phase. If the open ends of the stub are connected to two half-wave horizontal radiators as shown in Fig. 20-26, the array is not a full-wave

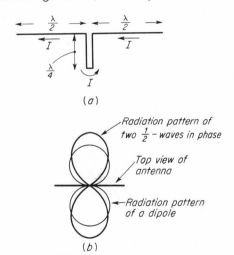

(a)

(b)

Fig. 20-26 (a) Collinear array composed of two half waves in phase. (b) Comparison of its radiation pattern with that of a simple dipole.

antenna but *two half-waves in phase,* with the currents in both radiating half-wave elements in the same direction at any given instant. This forms a two-element beam antenna. It has a maximum radiation in the horizontal direction at right angles to the wire, as does a half-wave dipole, but the lobes are narrower and longer. A gain of a little more than twice the power, about 4 dB, is produced in the direction of maximum radiation with the same power input to the antenna. Twice the power is not radiated, but energy radiation is less than with a single half-wave radiator in directions

more than 30° from right angles to the antenna.

Two half waves in phase can be fed with a flat transmission line tapped up on the stub or with tuned quarter- or half-wave feeders. Also, the stub can act as a matching transformer.

Greater energy radiation in the desired direction can be obtained by adding more half-wave elements, each separated from adjacent elements by a quarter-wave shorted stub, by half-wave open stubs, or by any other means of changing the phase 180°, such as an LC circuit tuned to the frequency of operation. This makes up a *collinear,* or *Franklin,* antenna. It gives a gain whether used horizontally or vertically. When used vertically, it has the advantage of holding the angle of radiation down, toward the horizon.

When collinear elements are connected in phase, the center impedance of the center element is several hundred ohms with five or more elements.

20-18 DRIVEN ARRAYS

The narrowing, or beaming, of the radiation lobe of a multielement array, with its increased energy radiated at right angles to the antenna, is highly effective in producing strong signals at long distances. For reliable point-to-point communication, a beam antenna is very desirable. For broadcasting, an omnidirectional vertical antenna may be best to reach listeners in all directions. However, many broadcast stations are located on one side of a population concentration and require an antenna that will beam its signals toward the nearby city and possibly put a minimum signal in some other direction.

Several types of beam antennas are used in broadcasting. Most of them consist of two or three vertical antennas so placed and *driven* that their radiated signals reinforce in certain directions and cancel each other in others. An example of a two-element vertical beam is shown in Fig. 20-27.

The two half-wave vertical radiators are fed by a flat transmission line tapped into a shorted quarter-wave stub, with a half-wave open-wire tuned line between the bases of the two radiators. With the two radiators being fed 180° out of phase, the signal approaching the reader from one antenna would cancel the signal from the other, resulting in zero radiation at right angles

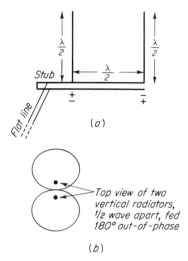

Fig. 20-27 (*a*) Two driven dipoles fed 180° out of phase. (*b*) Radiation pattern.

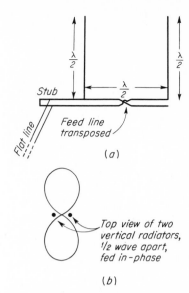

Fig. 20-28 (*a*) Two driven radiators fed in phase. (*b*) Radiation pattern.

to the plane of the page. The radiated wave from the first antenna expands and follows the wave along the transmission line. The radiated wave from the first antenna arrives in phase with the driven wave in the second antenna, and maximum signal is radiated in a line running through the two radiators. The radiation pattern, looking down on the vertical elements, is shown.

When the same two antennas are fed by a *transposed* feed line, as in Fig. 20-28, they are being fed in-phase signals. The signal approaching the reader is the sum of the two, or a stronger signal than would be radiated by a single antenna element. By the time the radiated wave from the first antenna reaches the second, it finds that the second antenna is being driven out of phase and zero signal is transmitted in the plane of the page. The radiation pattern, looking down on the vertical elements, is shown.

It is also possible to feed quarter-wave or other-length radiators in a somewhat similar manner. The more radiators fed in phase, or *stacked,* the greater the gain of the system in the maximum signal directions and the less radiation in all other directions.

It is common to speak of the separation of two antennas in degrees. Two antennas separated by a full wavelength are said to be 360° apart. If separated by a half wave, they are 180° apart. Two 950-kHz antennas separated by 120° are

one-third of a wave apart. The wavelength of 950 kHz is $\lambda = V/f_{\text{kHz}}$ or 300,000/950, or 316 m. One-third of 316 m is 105.3 m. Since a meter is equal to 3.28 ft, the two towers are separated by approximately 345 ft.

In broadcast stations two or more antennas may be fed different current values at different phase angles to control directivity of the radiation. Phase differences can be produced by using different-length transmission lines or by using *L*- or *π*-type phasing networks. Current variations can be controlled with the coupling networks. Changing either the current amplitude or the phase angle in any of the antenna elements shifts the directional pattern of the array. Such arrays are usually tuned by variable coils in the coupling or phasing networks.

Test your understanding; answer these checkup questions.

1. How many ground radials should a quarter-wave vertical antenna use? _____
2. What is the advantage of drooping ground planes? _____
3. Why are small balls attached to whip antennas? _____
4. What is the feed-point *Z* at the base of a quarter-wave vertical? _____ At the center of a half-wave?

_____ At the center of a full-wave? _____
A quarter-wave from the end of a full-wave? _____

5. Which radiates N and S, a $\frac{3}{2}$-wave or a $\frac{4}{2}$-wave horizontal E-W wire? _____

6. What length resonant line would couple an antiresonant circuit to the end of a half-wave antenna? _____ To the center of the same antenna? _____

7. Why does a delta feed line fan out at the antenna? _____

8. What type of transmission line is used in a gamma match? _____

9. How long would a stub usually be? _____

10. What is the Z_o of a quarter-wave transformer to match 500 Ω to 50 Ω? _____

11. Does frequency affect the velocity factor of a transmission line? _____

12. Would a radio wave travel faster in an air-insulated or in a solid-dielectric coaxial cable? _____

13. What are beams called if all elements are in line? _____ What is the phase difference between elements? _____ What is used between elements? _____

14. Why do broadcast stations usually use verticals? _____

15. What is a transmission line said to be if twisted 180°? _____

16. What is the phase separation of two antennas $\frac{3}{8}$ wave apart? _____

20-19 PHASE MONITORS

When there are two or more elements in a broadcast antenna array, it is necessary that the currents in the elements be maintained within 5% of the licensed values to maintain directivity.

To determine the phase of the currents in the elements, a *phase monitor* is used. Basically, a phase monitor consists of coaxial lines of equal length bringing in a sampling of the voltages developed in RF pickup coils at the base of the elements, as shown in Fig. 20-29. If the ampli-

Fig. 20-29 A basic phase-monitor circuit.

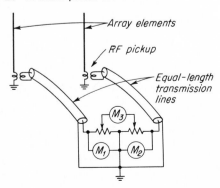

tudes of the two signals fed to RF voltmeters M_1 and M_2 are adjusted to be equal, meter M_3 will read zero when the voltages from the antennas are in phase. The farther the voltages are out of phase, the greater the difference of voltage across M_3 and the greater its deflection. This results in a direct indication of the phase difference between the two antennas. These sampling voltages can also determine if the antenna current amplitudes are within the licensed limits.

20-20 PARASITIC ARRAYS

If a driven half-wave element is a half wave from another similar *undriven* element, as in Fig. 20-30,

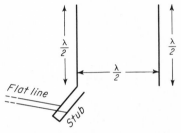

Fig. 20-30 A half-wave parasitic element spaced a half wavelength from a driven element.

the second element has a voltage induced in it by the radiated field from the first and is said to be *parasitically excited*. The current that is induced in the parasitically excited element is 180° out of phase with the wave from the first element that produces it (Lenz's law). The parasitic-element current produces a radiated wave that tends to cancel the originating wave, reducing the radiation in the plane of the page, but is in phase with the wave from the first antenna insofar as the reader is concerned. Such an array will radiate at right angles to the plane of the page. Since the reradiated wave cannot be as strong as the driven wave, there is no complete null in any direction.

If the parasitically excited element is moved up to within a quarter wave (90°) of the driven element, as in Fig. 20-31, it intercepts more energy and reradiates a stronger wave of its own. With quarter-wave separation, the radiated wave travels 90° to the parasitic element and reverses phase 180°. Some of it returns 90° to the driven element, arriving exactly in phase with the next half

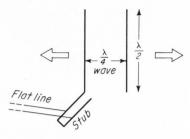

Fig. 20-31 A half-wave parasitic element spaced a quarter wavelength from a driven element. Maximum radiation direction is indicated by arrows.

cycle. Maximum signal is produced toward the left and right, as indicated by the arrows in Fig. 20-31.

The parasitic element is acting as a *reflector and director,* and the array has a gain of nearly 5 dB in both forward and backward directions. The radiation to the sides (toward the reader) is materially reduced.

A half-wave parasitic reflector gives greatest forward gain when spaced about 0.2 wavelength from the driven element. The radiation resistance at the center of the driven element drops to about 40 Ω. The more elements used in parasitic arrays, the *lower* the center impedance of the driven element.

If a parasitic element is a half wave long and is placed within about 0.1 wavelength of the driven element, the induced voltage and current relation is such that it acts as a director more than as a reflector and produces nearly 6 dB gain in the forward direction.

When the reflector is tuned for maximum forward gain (made about 5% longer than a half wave) and a director is also tuned for maximum forward gain (made about 5% shorter than a half wave), a three-element Yagi beam is produced. Such a beam antenna can have more than 8 dB gain in the forward direction with a 20-dB differ-

---

---

ence between the forward and backward radiation strength. A delta-fed beam of this type, shown in Fig. 20-32, may be recognized as a popular TV

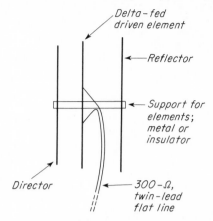

Fig. 20-32 A three-element Yagi consisting of a driven element, a reflector, and a director.

receiving antenna. For more forward gain, additional director elements are added.

The radiation resistance of close-spaced three-element Yagi antennas is approximately 10 Ω. This is somewhat difficult to center-feed, although it can be fed with a quarter-wave matching transformer or with the delta-feed system shown. The driven element may be made into a *folded dipole,* which will raise its radiation resistance. Figure 20-33 illustrates a dipole and a

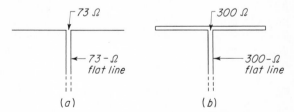

Fig. 20-33 (*a*) Center-fed dipole. (*b*) Center-fed folded dipole.

folded dipole. The impedance at the open point of the folded dipole is four times what it would be if the upper wire were not added. With three wires (or if the second wire has twice the surface area of the first), the impedance will be nine times the impedance of the dipole. By formula,

$$Z = 73\eta^2$$

where η = number of wires used

Plane-surface-reflector, corner-reflector, and parabolic-reflector antennas are parasitic types. A plane-reflector antenna uses a plane surface about one wave square behind the driven element, as in Fig. 20-34a. It has a gain of about

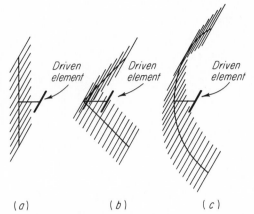

Fig. 20-34 Beam antennas. (a) Plane-reflector type. (b) 90° corner-reflector antenna. (c) Parabolic-reflector type.

7 dB in the forward direction. Spacing the driven element 0.2 wavelength from the reflector surface produces a center impedance of about 70 Ω for this element.

If the plane surface is bent into a 90° corner, with each side about two wavelengths long as in Fig. 20-34b, the antenna is more directive and has about 12-dB forward gain. Spacing the driven element 0.35 wavelength from the corner produces a center impedance of about 70 Ω for the driven element.

If the corner-reflector antenna is carefully engineered into a parabolic shape and the reflecting surface is sufficiently large, as in Fig. 20-34c, a beam width of 2° or 3° can be produced, with gains of well over 30 dB being possible. Such beams are used in radar.

20-21 LONG-WIRE BEAMS

There are several types of long-wire beam antennas. Since a wire several waves long exhibits directional properties in the line of the wire, it is possible to utilize the lobes of such an antenna to form a V beam, as in Fig. 20-35.

Simplified lobes have been indicated on the two

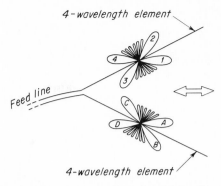

Fig. 20-35 Unterminated V-beam antenna four wavelengths on a leg.

legs of the antenna. Lobes 1 and A are in the same direction, as are 4 and D, resulting in a maximum radiation to the right and left on the page. Lobes 2 and B tend to cancel each other, as do lobes 3 and C, resulting in little radiation upward and downward on the page. Such a V beam has high gain in the forward and backward directions. The longer the legs, in number of wavelengths, the greater the gain. With twice the number of wavelengths there is roughly twice the power, or 3 dB gain, in the direction of maximum radiation. This holds fairly true for the number of elements in most arrays.

If desired, the backward radiation of a V beam can be effectively reduced without affecting the forward radiation by terminating the ends of the antenna with noninductive resistors to ground. These resistors must have approximately the impedance of the far end of the antenna, usually about 800 Ω. When terminated with resistance, the antenna becomes nonresonant, therefore not frequency-selective, and has no standing waves on it! It acts as a properly terminated but improperly spaced transmission line, and therefore radiates energy. The V beam should not be confused with an inverted-V antenna, a half-wave dipole supported at the center and with both ends dropping toward ground. This antenna radiates at right angles to the plane of the wires.

Another long-wire beam, called the *rhombic,* is composed of four legs, as in Fig. 20-36, instead of the two of the V beam but operates on the same general theory of addition and cancellation of lobes. It usually has more gain; it cancels side radiation more effectively, and the resistance ter-

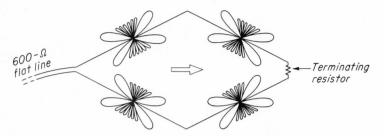

Fig. 20-36   Terminated rhombic-beam antenna four wavelengths on a leg.

mination at the far end tends to balance more evenly. A gain of 15 dB or more is possible. A rhombic antenna is not resonant to any particular frequency when terminated, as driven or parasitic arrays are. A single rhombic antenna may operate satisfactorily in one direction over a frequency range from 6 to 30 MHz, although the gain increases at higher frequencies. It is bidirectional when unterminated. Rhombics are used extensively in high-frequency point-to-point communication systems.

20-22 LOOP ANTENNAS

When speaking of loop antennas, the loop is usually considered as being a 1- to 3-ft vertically wound coil of wire having a few closely wound turns. It may be round or square in shape. The loop coil forms the inductance of a resonant LC circuit, as shown in Fig. 20-37a. Such a loop antenna is rarely used for transmitting, as its radiating efficiency is rather low, but it finds use as a receiving antenna because of the sharp *nulls* (zero signals) produced.

Radio waves passing across the loop induce a voltage in it. As the loop is rotated 360°, there will be two points where little or no voltage is induced in it and intermediate points where a maximum signal voltage is produced. The horizontal-plane reception pattern for a loop antenna is shown in Fig. 20-37b. When the loop is viewed from above, zero signal is received from a station at the top and bottom of the page and maximum signal from a station at the left or right of the loop. This is the opposite field-strength pattern of a half-wavelength wire. The figure eight pattern of the loop can be nearly perfect if the loop is balanced electrically (see Chap. 29).

If the loop antenna is held in a horizontal plane,

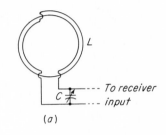

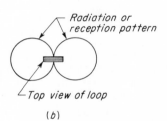

Fig. 20-37   (a) A small loop antenna forms the L of a tuned LC circuit. (b) Radiation or reception pattern of a loop antenna as seen from above.

it becomes omnidirectional, receiving equally well in all directions.

When a loop antenna is used for transmitting, its circumference should be a major portion of a wavelength of the frequency used. As the diameter approaches one wavelength, the maximum lobes are no longer in the line of the loop but break up into four lobes, as shown in Fig. 20-38. As the diameter is further increased, the loop becomes many-lobed with maximum response at nearly right angles to it, which is quite different from the reception pattern of a small-diameter loop.

When the sides of a square loop are a quarter wave each, as in Fig. 20-39, the loop acts as two quarter waves in phase, radiating and receiving maximum at right angles to the wires of the loop and with the vertical currents canceling each

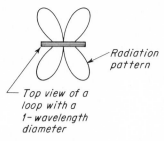

Fig. 20-38 Radiation pattern of a loop antenna having a diameter of approximately one wavelength.

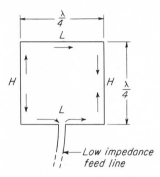

Fig. 20-39 Current flow at a given instant in a center-fed loop having quarter-wavelength sides.

other. The loop is shown as center-fed with a low-impedance line. If fed with a high-impedance line and opened at the top, it will result in a vertically polarized radiation with the horizontal currents canceling. Such a loop is bidirectional but may be made more unidirectional by using parasitic reflectors behind it or directors in front of it.

20-23 TOP LOADING

Low-frequency (30- to 300-kHz) antennas are usually quarter-wave Marconi types with ground radials. In the upper MF range (500 to 3,000 kHz) it becomes practical to use quarter- to half-wave vertical antennas. At the low end of the broadcast band the antennas become quite long, and it is often desirable to *top-load* them. Top loading may take the form of a metal wheel-like *hat* structure attached to the top of the antenna. This hat increases the length of the antenna to the edge of the hat, but more importantly, it materially increases the capacitance between the top of the antenna and ground. As a result of the increased length, or inductance, and the increased capaci-

tance, a short antenna can be made to resonate at a relatively low frequency.

Top loading may also be produced by using the top portion of the antenna guy wires as the top hat, as in Fig. 20-40. The antenna strain in-

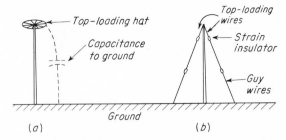

Fig. 20-40 Top-loading a vertical antenna by (a) using a hat and (b) using the top of the guy wires.

sulators are preferably a glazed-surface porcelain. A hard, glossy surface tends to discourage the accumulation of dirt and reduces losses across the insulators. In areas near salt water, any encrustation may have relatively high conductance and produce unwanted leakage as well as corona effect on insulators.

Guy wires are usually broken up into lengths that are not harmonically related to the frequency of transmission to prevent pickup of energy and reradiation. The insulators used are always egg or strain insulators. If these insulators break, the two wire loops are still joined and the guy wire does not part. If the usual antenna end insulators, as in Fig. 20-1, supporting a horizontal antenna wire break, the antenna wire comes down.

20-24 FEEDING ANTENNAS

There are many methods of coupling an antenna to a transmitter (or receiver). The simplest is by connecting the bottom of the antenna directly to a point on the tank coil as in Fig. 20-41. The end of any length of wire has a resistance and a reactance component to any frequency. If a point can be found on the tank circuit where both antenna and tank R and X values equal or cancel, the antenna will take power. The closer the antenna is connected to the ground end of the coil, the lower the degree of coupling. If the antenna has considerable reactance (is not resonant), an inductor or capacitor can be added in

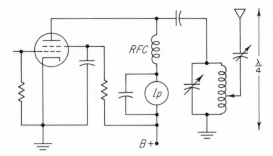

Fig. 20-41   Direct-coupled quarter-wave antenna.

series with the antenna, as shown, to tune out the reactance. This coupling has no means of reducing harmonic radiation.

A circuit that is superior to direct coupling is the π network shown in Fig. 20-42. L, C_1, and

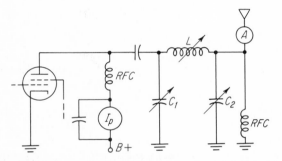

Fig. 20-42   A π-network coupling system to match the plate circuit to the bottom of the antenna.

C_2 act as a tank-tuning as well as antenna-coupling network and must resonate at the transmitting frequency. The base impedance of the antenna must match the impedance across C_2 to take maximum power. Also, the impedance of the plate circuit of the tube must match the impedance across C_1. Coil L must be the correct inductance to produce resonance in the circuit when C_1 and C_2 are adjusted to the proper capacitance values. If C_2 has a large capacitance value, its reactance will be small, resulting in a low voltage-drop across it. The antenna will have little voltage excitation; only a small current will flow into it; and it will take little power. Decreasing C_2 increases the coupling to the antenna.

The radio-frequency choke (RFC) coil between antenna and ground discharges any static charges picked up by the antenna. It has no effect on the tuning of the network.

To tune a π network, start with maximum C_2 value (minimum coupling). Tune C_1 to resonance, shown by minimum plate current. Note the antenna current. Decrease C_2 somewhat; then redip the plate current. The antenna current and I_p minimum should both rise. Keep decreasing C_2 and retuning C_1 until the I_p minimum is the desired value for the tube used according to a tube manual. Note the antenna current. Repeat this tuning procedure, but use less inductance L. Then try with more inductance. When maximum antenna current has been obtained with rated I_p, the stage can be considered as tuned.

The advantage of a π network is the output capacitor. At the fundamental frequency it helps to tune the circuit to resonance, but at the second-harmonic frequency it has half the reactance, reducing the possible harmonic frequency voltage fed to the antenna. Higher-order harmonics are bypassed still more.

If the base of a vertical quarter-wave antenna is insulated from ground, the resistance to be matched may be about 25 Ω. This can be matched by using two 50-Ω coaxial lines in parallel, connecting the center wires to the base of the antenna and the outer conductors to the central point of the ground-radial system. This is *series feeding*.

To match the low impedance at the base of a quarter-wave vertical to a higher-impedance feed line, an *L*-type impedance-matching network, such as in Fig. 20-43, can be employed. To obtain the step-down ratio of impedances the *LC* configuration must be as shown. This network can

Fig. 20-43   Step-down-ratio *L* network between line and antenna including arcing points and a possible quarter-wave shorted stub.

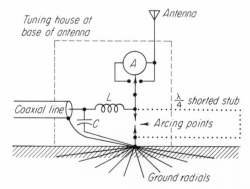

present a pure resistance at input and output, but there is always a phase shift in an L network. This network can also be designed to match a flat transmission line (pure resistance) to the base of an antenna which is reactive (not exactly a quarter wave).

Note the antenna-ammeter circuit. When not in use, the switch is thrown to the right, taking the ammeter out of the circuit to prevent burnout due to lightning striking near or directly on the antenna. The meter may also be removed if a jack and plug arrangement is used. Between the base of the antenna and ground a lightning gap may be installed. This gap must be just wide enough not to arc with the voltages built up by the transmitter but close enough to break down and arc with the higher voltages of a lightning strike.

A quarter-wave shorted stub, shown in dotted lines, can be connected between the base of an ungrounded antenna and ground. The stub presents a very high impedance for RF ac at its resonant frequency. It forms a low-resistance path to ground, discharging electrostatic charges, but has no effect on the RF being transmitted. It also affords a low-impedance path to ground for all *even-order* harmonics, preventing them from being transmitted. It acts as a wave trap.

To match the high impedance at the bottom of an insulated half-wave vertical to a lower-impedance transmission line, the capacitor C of the L network is moved to the other end of the inductor L. This produces the required step-up ratio of impedances.

A π network can produce either step-up or step-down impedance ratios or controlled phase shifts, as can the T-type network of Fig. 20-44. A series wave trap from antenna to ground can

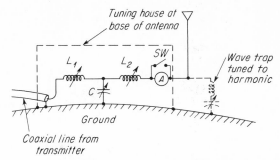

Fig. 20-44 A T-network coupling circuit to match the feed line to the antenna. A wave trap cancels a harmonic.

be used to decrease radiation of any harmonic to which it is tuned. Coil L_1 is tuned for minimum I_p, coil L_2 is tuned for maximum antenna current.

If the base of a vertical quarter- ($\frac{3}{4}$-, $\frac{5}{4}$-, etc.) wave antenna is connected directly to a ground-radial system, it can be shunt-fed by a half-delta feed system similar to that in Fig. 20-45. Whatever the transmission-line impedance, there is some point on the antenna with the same impedance value. However, the length of the lead up to the desired point on the antenna represents inductance or inductive reactance. To balance out the added X_L, a capacitor must be added to the feed line as shown. While the point on the tower can be computed, a series of cut and try tests is usually required to establish the optimum point and lowest SWR on the feed line.

Coaxial transmission lines have been shown between transmitter and antenna. They may be of the solid-dielectric or the gas-filled type. Coaxial lines have many advantages: all fields are confined inside the outer conductor, radiation is nonexistent if properly terminated, and the lines may even be buried if desired. To allow for expansion

Fig. 20-45 Shunt-fed antenna coupled to a push-pull, plate-neutralized RF amplifier.

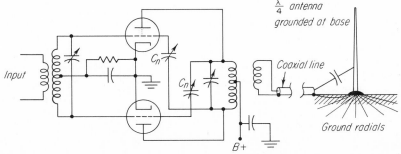

and contraction, long runs are usually constructed to have a slow bend or two. To prevent damp air from being drawn into the cable when cooled, air-dielectric lines are usually sealed and put under about 15 psi of nitrogen or dry air. Any loss of pressure indicates trouble. Location of leaks can be determined by watching for bubbles when soapy water is painted on the line or joints. Sections of hollow coaxial line can be butted together, and a sleeve sweated (soldered) over the joint, or the sleeve can be clamped with rings. Screw-type end couplings can be used on solid-dielectric lines. A shorted point in a line can be determined by taking several frequency readings with a grid-dip meter, which will give indications of quarter-wave, plus all other odd-quarter-wave shorted-stub lengths. Sharp bends should be avoided with any concentric lines to prevent mechanical pressures and breakage, as well as change of impedance where the line is deformed. Care must be taken that no sharp points are developed inside a cable. An internal arcing point may be indicated by heating at that point under loaded conditions.

Open-line feeders are usually run on top of closely spaced wooden poles about 8 ft above ground level from transmitter to the tuning house at the base of the antenna. Any excessive lumps of wire or solder form discontinuities and can reflect energy and produce standing waves. Kinks in the wire will do this too, as well as weaken the wire. Transmission-line wires must be maintained at the same spacing distance to prevent change of impedance and resulting reflections.

Test your understanding; answer these checkup questions.

1. What is used to determine phase difference between two broadcast antennas? _____
2. Which gives more forward gain, a director or reflector? _____ Which is longer? _____ Which is closer to the driven element? _____ How much do they differ in length from a half wave? _____
3. How much more feed-point Z does a folded dipole have than a normal dipole? _____
4. Would it be possible to produce 70-Ω twin lead? _____
5. With similar-size parabolic, plane-reflector, and corner-reflector beams, which has greatest gain? _____ Narrowest lobes? _____
6. Why would the wires of a V beam be closer together

when used on higher frequencies? _____ How is the beam made unidirectional? _____
7. Why might a rhombic be better than a V-beam? _____ _____
8. Are the beams in Fig. 20-34 vertically or horizontally polarized? _____
9. Why is the usual rhombic not frequency-sensitive? _____
10. In what direction is the null of a small-diameter loop in relation to the plane of the loop? _____
11. When is a loop omnidirectional? _____
12. Why are verticals sometimes top-loaded? _____
13. Why is π-network superior to direct coupling to an antenna? _____
14. Why might L networks be used between an antenna and transmission line? _____ _____
15. Why are antenna ammeters often shorted out or disconnected from the antenna except when readings are desired? _____

20-25 DETERMINING IMPEDANCE OF AN ANTENNA

Within limits, any length of antenna will take power at any frequency. For example, a 200-ft-high mast can be made to radiate on any frequency in the standard broadcast band. At 560 kHz, it would be only about one-eighth of a wave high, but at 1,600 kHz it would represent about one-third of a wave. It would be capacitively reactive at the lower frequency and inductively reactive at the higher. What feed-line impedance should be used?

To bring the antenna into resonance at 560 kHz, the base connection must be broken and a loading coil inserted to cancel the capacitive reactance of the antenna. A sensitive RF thermocouple ammeter is then added, as shown in Fig. 20-46a. A calibrated variable-frequency oscillator of several watts can be loosely coupled to the inductor and set accurately to the desired frequency of operation. The loading inductor is varied until the antenna ammeter indicates a peak of current. The antenna is now resonant. The oscillator is turned off. The antenna circuit is broken at the desired feed point *F,* and an RF impedance bridge is inserted and adjusted to a balance. The impedance indicated should be the required feed-line impedance. (Being shorter than a quarter wave, the base resistance will be quite low, 10 to 20 Ω.) The power into the antenna can be determined by

$$P = I^2 R$$

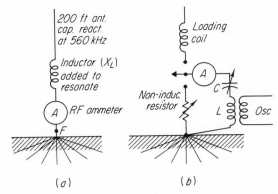

Fig. 20-46 Antenna arrangements to measure base resistance by (a) impedance bridge and (b) substitution.

where I = antenna current when transmitter is coupled to it

R = measured impedance or resistance at base or feed point

The ammeter can be inserted anywhere in the antenna when tuning for resonance; but when computing antenna power, it must be at the point where resistance is measured.

Another method of determining antenna resistance is by substitution. As with the impedance bridge, the antenna must first be brought to resonance by using the same procedure as above. The circuit between the loading coil and ground is broken, and components are connected as shown in Fig. 20-46b. The oscillator is coupled to the series L and C, and the switch is thrown to the antenna position. The capacitor tunes out the reactance of L and brings the antenna back into resonance. The ammeter reading is noted. The switch is then thrown to the dummy-antenna position, and the noninductive resistor is varied until the ammeter reads the same value as when the antenna was connected. Tle resistor value should be equal to the antenna resistance. The antenna resistance is usually measured at intervals up to 20 kHz above and below the operating frequency.

For 1,600-kHz operation the antenna is longer than a quarter wave, and it is necessary to insert a loading capacitor to tune out the inductive reactance of the antenna. A similar method of impedance measurement and substitution can be used.

In all these measurements care must be taken that lead lengths are as short as possible and that stray coupling paths do not exist.

20-26 FIELD INTENSITY

To measure the received strength of a radio signal at a certain point in space, the voltage developed in a 1-m-long wire is considered as the standard. *Field intensity* is therefore measured in received volts per meter. Since only a small fraction of a volt will be induced in a remote wire, it is more usual to express the field intensity in either millivolts or microvolts per meter. Thus, if a wire 3 m long has 0.001 V induced in it by a certain signal, the field intensity is 0.001/3, or 0.000333 V/m, or 0.333 mV/m, or 333 μV/m.

The *effective height* of an antenna can be expressed in meters if the voltage induced in it is known and if the field intensity in which it is erected is known. If the field intensity is known to be 25 mV/m and 2.7 V is induced in the antenna, the effective height is 2.7/0.025, or 108 m.

The field strength of a transmitted signal is dependent on ground, direct, reflected, and refracted sky waves, as well as the directivity of the antenna. As a result field strength will vary at different times of the day unless the path is very short.

The ground wave over seawater decreases almost inversely as the distance from the station, particularly in the low- and medium-frequency ranges. If the field strength at 10 mi is 50 mV/m, at 20 mi, it will be only slightly less than 25 mV/m. As the distance from the station increases to more than 100 mi, the signal decreases more than in a simple inverse proportion. With higher frequencies the signal decreases much more than in a simple inverse proportion.

Because power is proportional to both current squared and voltage squared ($P = I^2R$ and also $= E^2/R$), twice the current in an antenna represents four times the power. Similarly, twice the voltage represents four times the power. If 1,000 W of RF power in an antenna produces 10 mV/m in a remote antenna, 4,000 W will produce 20 mV/m. If the field strength doubles at a remote point, it indicates that the power in the transmitting antenna has quadrupled (increased 6 dB).

If the power in an antenna is doubled, the voltage in it (and the field intensity produced by it) is increased by $\sqrt{2}$, or by 1.414. If 1 kW in a transmitting antenna can produce 3 mV/m in an antenna, 2 kW in the same antenna will produce 3(1.414), or 4.24 mV/m.

20-27 FIELD INTENSITY OF HARMONICS

The field intensity of harmonic radiations from a transmitter can be measured in microvolts per meter. The difference between the field intensity of the fundamental and the harmonic is usually expressed in decibels (Sec. 8-6). A ratio of either 10 times, or one-tenth of the voltage equals 20 dB. Twice, or half of the voltage is a ratio represented by 6 dB. If the harmonic-field intensity is one-tenth of the intensity of the fundamental, it is 20 dB down. If the harmonic is one two-hundredth of the fundamental, how many dB down is it? One-tenth of 200, or 20, equals −20 dB. One-tenth of 20, or 2, equals another −20 dB. One-half of 2 equals 1 (or unity), which equals −6 dB (approximately). Therefore, the harmonic is −20, −20, and −6 dB, or −46 dB below the fundamental.

If the fundamental is 147 mV/m and the harmonic is measured as 405 μV/m, the voltage ratio between the two is 0.147/0.000405, or 363:1. One-tenth of this is 36.3 and represents a loss of −20 dB? One-tenth of 36.3 is 3.63, which represents another loss of −20 dB. One-half of 3.63 is 1.815, which represents −6 dB. One-half of 1.815 is 0.908 and another −6 dB. However, this has brought the ratio below unity, or 1:1. The total loss represented is −20, −20, −6, and −6, or −52 dB. Since this has gone down too far, the approximate loss must be about −51 dB. The

exact number of decibels is

$$dB = 20 \log \frac{E_1}{E_2} = 20 \log 363$$

$$= 20(2.56) = 51.2 \text{ dB}$$

Because of the different transmission characteristics of different frequencies at the same time of day, a harmonic signal from a transmitter, if not sufficiently attenuated, may have a greater field intensity at a distant location than does the fundamental and may interfere with other radio services when the fundamental cannot be heard at all by these other services. Thus, a 5-MHz transmitter may not be heard 2,000 miles away during the daytime, but an unattenuated 10-MHz harmonic may be quite appreciable at that distance.

There are many methods of decreasing harmonic radiation (Sec. 16-18). Figure 20-47 illus-

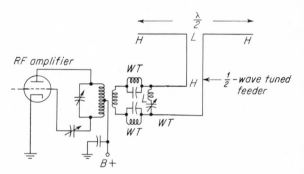

Fig. 20-47 Coupling a Hertz antenna to an RF amplifier. Wave traps are used in the transmission lines to attenuate harmonic radiation.

trates parallel wave traps connected in an antenna transmission line to attenuate any harmonics to which they are tuned. A series wave trap across the transmission line at any point will effectively decrease radiation of the frequency to which it is tuned.

20-28 FIELD GAIN

Multielement transmitting antennas may be rated in *field gain*. If a multielement transmitting antenna produces a 500-mV/m signal in a remote receiving antenna, whereas a simple dipole transmitting antenna will produce only a 250-mV/m signal, the field gain of the multielement antenna is 2, or 6 dB.

The *effective radiated power,* abbreviated erp, of an FM or TV transmitter considers the field (voltage) gain of the antenna.

EXAMPLE: A transmitter has a 370-W input to a class C final amplifier operating at 65% efficiency. Therefore, the output power fed to the antenna feed line is 240.5 W. If the transmission line delivering the RF power to the antenna radiating elements is 75% efficient, there are 240.5 × 0.75 or 180.4 W, being radiated. Since the field gain is expressed in voltage and power is proportional to voltage squared, with an antenna having a field gain of 1.3, the erp is 180.4 × 1.3^2, or 304.9 W.

A beam antenna may be rated in *power gain.*

EXAMPLE: The output of a transmitter is 1,000 W. The antenna transmission-line loss is 50 W. With a power gain of 3 (or 4.77 dB), the antenna has an erp of 3(950), or 2,850 W.

20-29 THE GROUND

The surface of the earth directly below and surrounding an antenna is known as the *ground.* With Marconi antennas the ground is used as one-half of the antenna. With self-resonant Hertz antennas operated high above the earth, the ground may play a slightly less important role. With both types of antennas the ground acts as a reflector of transmitted and received waves and is responsible for some portion of the total transmitted or received signal.

The best ground would be a silver sheet extending several wavelengths in all directions under the antenna. While this is not practical, it is possible to use copper ground radials. Each wire is separated from the next by 5° to 15° of arc and is at least one-quarter of a wavelength long. If some of these ground radials are broken off, the resistance of the antenna will change, as will the directivity of the radiation pattern. One of the best practical grounds is a salt marsh.

Without ground radials, the actual or virtual ground may vary from a few inches below the surface of a salt marsh to several feet below the surface of a dry, sandy soil. Because an antenna

is erected a quarter wave above ground physically does not mean that it is operating electrically at that height. Losses result from poor grounds.

20-30 COMPUTING THE POWER IN AN ANTENNA

When an antenna is tuned properly, it may be considered to be a resistance. It exhibits neither capacitive nor inductive reactance to the feed line. As a result, the power, voltage, current, and resistance in a tuned antenna can be computed by Ohm's-law and power formulas. For example, if resistance and current at the base of a Marconi antenna are known, the power radiated by the antenna can be found by the formula

$$P = I^2R$$

where P = power in W
R = radiation resistance
I = current in A, at the point in the antenna where radiation or feed-point resistance is measured

Some power is lost as heat in the antenna conductor itself, but most is radiated.

This and other power formulas can be rearranged to find the unknown as in the following examples.

The surge impedance of a flat transmission line is 500 Ω, and the line has a current of 3 A flowing in it. The power being fed to the antenna is

$$P = I^2R = 3^2 × 500 = 4,500 \text{ W}$$

If the daytime power in a broadcast station antenna is 2,000 W and the antenna has a resistance of 20 Ω, the current is

$$I = \sqrt{\frac{P}{R}} = \sqrt{\frac{2,000}{20}} = 10 \text{ A}$$

If the antenna current of this antenna is cut in half for nighttime operation, the transmitter has an antenna power of

$$P = I^2R = 5^2 × 20 = 500 \text{ W}$$

If the daytime transmission-line current of a 10,000-W transmitter is 12 A, the line impedance is

$$R = \frac{P}{I^2} = \frac{10,000}{144} = 69.4 \ \Omega$$

If it is required to reduce to 5,000 W at sunset, the new value of transmission-line current is

$$I = \sqrt{\frac{P}{R}} = \sqrt{\frac{5,000}{69.4}} = 8.49 \ A$$

A 72-Ω concentric transmission line is carrying 5,000 W. The rms voltage between the inner conductor and the sheath is

$$E = \sqrt{PR} = \sqrt{5,000(72)} = \sqrt{360,000} = 600 \ V$$

Since the radiated power of an antenna is directly proportional to the antenna current squared, if the antenna current is doubled, the power is increased 2^2, or 4 times. If the current is increased 2.77 times, the power increases 2.77^2, or 7.67 times.

A long, flat transmission line delivers 10,000 W at 4.8 A to an antenna. The impedance of the line is

$$R = \frac{P}{I^2} = \frac{10,000}{4.8^2} = 434 \ \Omega$$

If the current fed into this transmission line is 5 A, the power fed into it is

$$P = I^2R = 5^2(435) = 10,850 \ W$$

With 10,850 W delivered into the transmission line and 10,000-W output, 850 W must be lost in the line itself.

When antenna resistance and current are known, the *output* power of a transmitter can be determined. When the plate current and voltage are known, the *input* power can be determined. From these two values the efficiency of the amplifier can be found.

EXAMPLE: If the dc input is 1,500 V and 0.7 A, the power input to the final amplifier is 1,050 W.

If the antenna current is 9 A and the antenna resistance is 8.2 Ω, the antenna power is

$$P = I^2R = 9^2(8.2) = 664 \ W$$

The efficiency is determined by output/input, or 664/1,050, or 0.632, or 63.2%.

20-31 OMNIDIRECTIONAL ANTENNAS

It has been mentioned that the vertical antenna is omnidirectional in the horizontal plane, with vertical polarization.

TV and FM transmissions are required to be horizontally polarized. Transmitting antennas for these services should transmit equally well in all directions and still have horizontal polarization.

The horizontal half-wave dipole transmits a two-lobed horizontal pattern with nulls off the ends of the antenna. By bending a dipole into a horizontal loop, the antenna exhibits less directional properties but is not a true omnidirectional antenna.

If two half-wave dipoles are arranged to form a horizontal X, as shown in Fig. 20-48, and the

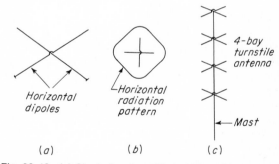

Fig. 20-48 (a) Single-bay turnstile antenna. (b) Horizontal radiation pattern of a turnstile antenna. (c) A four-bay turnstile (feed lines not shown).

two antennas are fed 90° out of phase through a quarter-wave section of transmission line, the resultant horizontally polarized radiation pattern approximates a circle. This forms a basic turnstile-type antenna. For greater gain, turnstile antennas are usually stacked with two or more *bays,* one above the other, as shown. Two dipoles can also be fed 90° out of phase by using a transmission line 90° (a quarter wave) longer for one than is used for the other.

Another omnidirectional antenna is the slotted cylinder. It consists of a metal cylinder longer than a half wavelength and with a vertical and half-wave slot cut out of it. The slot is fed with a coaxial transmission line at the points marked F in Fig. 20-49. When a cylinder about three-eighths of a

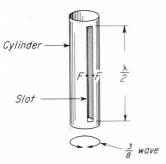

Fig. 20-49 A vertical slotted-cylinder antenna produces a horizontally polarized radiation.

wavelength in circumference is used, current flows from one side of the slot to the other, around the outside of the cylinder, producing an omnidirectional pattern. If the circumference is greater than a half wave, the antenna becomes directional with maximum radiation from the slotted side of the cylinder.

Other omnidirectional antennas are forms of horizontal loop antennas in which the loop is broken up into three or four segments. Each segment is formed from a major portion of a dipole. Each dipole is fed so that the current flow is in the same direction in all segments at the same time.

Since most of these omnidirectional antennas are used for VHF and UHF transmissions (FM and TV), which are essentially line-of-sight transmissions, they are usually found on tops of hills or peaks to allow maximum range of transmission or reception.

Test your understanding; answer these checkup questions.

1. What were the two methods explained to determine the feed-point Z of a vertical antenna? _____
2. In what unit is field intensity measured? _____
3. What formula is used to determine effective height of an antenna if field intensity and induced voltage are known? _____
4. How much does low-frequency field strength attenuate? _____ High-frequency? _____
5. If 4 kW in an antenna produces 50 μV/m in a receiving antenna, what will 16 kW produce? _____ 8 kW? _____ 1 kW? _____
6. Under what condition might a harmonic of a transmission be heard at a distant point when the fundamental can not be? _____
7. Would field gain be measured in voltage or dB? _____
8. What is involved in the erp of an antenna? _____
9. Why do ground radials assure constant feed-point Z? _____
10. What feed requirements must be met to produce a circular radiation pattern when using two crossed dipoles? _____
11. When speaking of antennas, what is a bay? _____
12. In what range of frequencies are most omnidirectional horizontally polarized antennas used? _____ Rhombics? _____ Parabolic reflectors? _____
13. A flat transmission line delivers 5 kW at 10 A. What is its Z? _____ The E across it? _____
14. If the antenna current increases 3.3 times, how much does the radiated power increase? _____ The field strength? _____

COMMERCIAL LICENSE QUESTIONS

Amateur license questions will be found in the Addendum.

FCC Elements 3, 4, and 6 require an ability to answer questions similar to those below. Sections in which questions are answered are shown in parentheses. A question followed by a bracketed number is required for that element alone.

1. What kinds of fields emanate from a transmitting antenna, and what relationship do they have with each other? (20-1)
2. Can either of the two fields from an antenna produce an emf in a receiving antenna? (20-1)
3. What are the lowest frequencies useful in radio communications? (20-1)
4. What bearing do the angle of radiation, density of the ionosphere, and frequency of emission have on the length of the skip zone? (20-2)
5. What effect do sunspots and the aurora borealis have on radio communications? (20-2)

6. What effects will the ionosphere have on different radio frequencies? (20-2)
7. Why is it possible for a sky wave to "meet" a ground wave 180° out of phase? (20-3)
8. What frequencies have line-of-sight characteristics and are unaffected by the ionosphere? (20-4)
9. What advantages may be expected from the use of HF in radio communications? (20-4, 20-5)
10. What frequencies are useful for continuous long-distance communications? (20-4)
11. What is the relation between operating frequency and ground-wave coverage? (20-4)
12. Explain polarization with respect to antennas for transmission or reception. (10-6)
13. What is the velocity of propagation of RF waves in space? (20-7)
14. What is the relation between the electrical and physical lengths of a Hertz antenna? (20-7)
15. What will be the effect upon the resonant frequency if the physical length of an antenna is reduced? (20-7)
16. If a vertical antenna is 400 ft high and is operated at 1,300 kHz, what is its physical height in wavelengths? (1 m = 3.28 ft.)(20-7) [4]
17. What is the difference between Hertz and Marconi antennas? (20-8)
18. Explain the voltage and current relations in a half-wave dipole antenna. (20-9)
19. What is meant by *radiation resistance*? (20-10)
20. What effect do the magnitudes of the E and I at a point on a half-wave dipole in free space have on the impedance at that point? (20-10)
21. Does radiation resistance vary in different types of antennas? (20-10, 20-17, 20-20, 20-21)
22. What is the effect on the resonant frequency of an antenna if (*a*) a capacitor and (*b*) an inductor is connected in series with the antenna? (20-11)
23. Why is a loading coil sometimes used with an antenna? Would absence of the coil mean a capacitive antenna impedance? (20-11)
24. How is it possible to operate on a lower frequency than the resonant frequency of the antenna? (20-11)
25. What is meant by the *characteristic,* or *surge,* impedance of a transmission line? To what physical characteristics is it proportional? (20-12)
26. Why is the impedance of a transmission line an important factor with respect to matching a transmitter into an antenna? (20-12)
27. If the conductors in a two-wire RF transmission line are replaced by larger conductors, how is the surge impedance affected? (20-12) [4]
28. If the spacing of the conductors in a two-wire transmission line is doubled, what change takes place in the surge impedance of the line? (20-12) [4]
29. Why might a solid-dielectric cable be preferable to a hollow pressurized cable for use as a transmission line? (20-12, 20-24)
30. Describe the procedure of installing transmission lines between broadcast transmitters and antennas. Include information on impedance, bends, kinks, connections, and gas-filled lines. (20-12, 20-24) [4]
31. What is meant by *standing waves* and *standing-wave ratio* (SWR) and how can they be minimized? (20-12)
32. What is the ratio of currents at opposite ends of a quarter-wave line if terminated in an impedance equal to the surge impedance? (20-12) [4]
33. What is the primary reason for terminating a transmission line in an impedance equal to the surge impedance of the line? (20-12)
34. Why is dry air or an inert gas often used in transmission lines in broadcast systems? (20-24) [4]
35. If standing waves are desirable on a transmitting antenna, why are they undesirable on some transmission lines? (20-12)
36. Describe the directional characteristics of horizontal and vertical antennas. (20-13)
37. Draw a sketch of the horizontal and vertical radiation patterns of a quarter-wave vertical antenna and discuss them. Would these patterns apply to a similar receiving antenna? (20-13)
38. Draw the radiation pattern of a half-wave horizontal antenna $\frac{3}{4}$ wave above ground. (20-13)
39. What antenna has minimum horizontal directional characteristics? (20-13)
40. What is the reception pattern of a vertical antenna? (20-13)
41. Explain the voltage and current relations in a quarter-wave grounded antenna. (20-14)
42. What would constitute the ground plane if a quarter-wave grounded whip antenna, 1 m in length, were mounted (*a*) on the metal roof of an automo-

---

ANSWERS TO CHECKUP QUIZ ON PAGE 527

1. (*Z bridge, R substitution*)   2. (*mV/m or μV/m*)
3. (*Induced E/field intensity*)   4. (*Directly proportional to distance*)(*Greater than simple ratio*)   5. (*100 μV/m*)(*70.7 μV/m*)(*25 μV/m*)   6. (*When fundamental frequency skipping over but harmonic refracting in*)   7. (*In both*)   8. (*P_o times field gain*)   9. (*Virtual ground constant in any weather*)
10. (*Feed 90° apart*)   11. (*A section which would be a complete antenna by itself*)   12. (*VHF, UHF*)(*HF, VHF*)(*UHF, SHF, EHF*)   13. (*50 Ω*)(*500 V*)   14. (*10.89 times*)(*3.3 times*)

bile and (*b*) near the rear bumper of an automobile? (20-14)

43. Draw a sketch of a coaxial whip antenna, identifying and explaining the whip, insulator, skirt, trap, support mast, coaxial line, corona and the corona ball, and input. (20-14)

44. Explain the voltage and current relations in a full-wave antenna. (20-15)

45. Diagram a two-wire RF transmission line feeding a Hertz antenna. (20-16, 20-27)

46. What is meant by *stub tuning*? (20-16)

47. Explain a quarter-wave matching section in an RF transmission line. (20-16)   [4]

48. What should be the surge impedance of a quarter-wave matching line used to match a 600-Ω feeder to a 70-Ω antenna? (20-16)   [6]

49. Will the velocity of signal propagation differ in different materials? Would wavelength or frequency have any effect? (20-16)

50. How does a directional-antenna array at an AM broadcast station reduce radiation in some directions and increase it in others? (20-18)   [4]

51. What is the direction of maximum radiation from two vertical antennas spaced 180° having equal currents in phase? (20-18)   [4]

52. Discuss the directivity and physical characteristics of stacked arrays. (20-18)

53. If the two towers of a 900-kHz directional antenna are separated by 120 electrical degrees, what is the tower separation in feet? (20-18)   [4]

54. What might cause the directional-antenna pattern of an AM station to change? (20-18)   [4]

55. What adjustable controls are normally provided at an AM broadcast station to maintain the directional pattern? (20-18)   [4]

56. What is the function of a phase monitor? (20-19)   [4]

57. The current in the elements of a directive broadcast antenna must be held to what percent of the licensed value? (20-19)   [4]

58. Discuss the directivity and characteristics of parasitic arrays. (20-20)   [4]

59. Discuss the directivity and characteristics of a corner-reflector antenna. (20-20)

60. Discuss the directivity and characteristics of a V-beam antenna. (20-21)

61. Describe the directional characteristics of horizontal and vertical loop antennas. (20-22)

62. Why do some broadcast stations use top-loaded antennas? (20-23)   [4]

63. What material is best suited for use as an antenna strain insulator? (20-23)   [6]

64. What is the effect upon a transmitter of dirty or salt-encrusted antenna insulation? (20-23)   [6]

65. Why are insulators sometimes placed in antenna guy wires? (20-23)

66. Draw a diagram of coupling a final stage to a quarter-wave Marconi antenna other than by link or transmission line. (20-24)

67. Discuss series- and shunt-feeding of quarter-wave antennas with respect to impedance matching. (20-24)

68. How is the degree of coupling varied in a pi network between a plate circuit and an antenna? (20-24)   [6]

69. Diagram some methods of coupling broadcast transmitters to antennas, impedance matching, attenuating harmonics, and guarding against lightning damage. (20-24)   [4]

70. Draw a circuit diagram of a push-pull triode final power amplifier, with transmission-line feed to a shunt-fed quarter-wave antenna, and indicate a method of plate neutralization. (20-24)

71. Draw a diagram of a T-type coupling and impedance-matching network coupling a coaxial line to an antenna. Include means for harmonic attenuation. How is the network tuned? (20-24)   [4]

72. How may a standard-broadcast-antenna ammeter be protected from lightning? (20-24)   [4]

73. If the antenna resistance and the current at the base of a Marconi antenna are known, what formula could be used to determine the power in the antenna? (20-25, 20-30)   [6]

74. Describe how to tune a broadcast antenna by (*a*) the RF bridge method and (*b*) the substitution method. (20-25)   [4]

75. Define *field intensity*. Explain how it is measured. (20-26)   [4]

76. If the field intensity of 25 mV/m develops 2.0 V in a certain antenna, what is its effective height? (20-26)   [4]

77. How does the field strength of a standard broadcast station vary with distance from the antenna? (20-26)   [4]

78. If a 500-kHz transmitter produces 100 mV/m at 50 mi, what field strength would it produce at 150 mi? (20-26)   [6]

79. If the power of a transmitter is doubled, what would be the percent change in (*a*) field intensity and (*b*) decibels? (20-26)   [6]

80. What is the relation between antenna current and radiated power? (20-26)   [6]

81. If the power output of a transmitter is doubled, what effect will this have on the field intensity? (20-26)   [4]

82. If the power output is increased so that the field intensity is doubled, what increase has taken place in antenna current? (20-26)   [4]

83. If the antenna current at a 500-kHz transmitter is reduced by 50%, what percent of field-intensity change is produced? (20-26) [6]

84. Why might harmonic radiation from a transmitter sometimes cause interference at distances from a transmitter where the fundamental signal could not be heard? (20-27) [6]

85. Draw a diagram of an RF amplifier coupled to a Hertz antenna. (20-27)

86. A 2,738-kHz transmitter has a field strength of 100 mV/m and a second harmonic of 400 μV/m at a distant point. How much is the harmonic attenuated in decibels? (20-27)

87. What is *effective radiated power* and how is it calculated? (20-28) [4]

88. What is the erp if the output of a transmitter is 1 kW, the antenna transmission-line loss is 50 W, and the antenna power gain is 3? (20-28) [4]

89. What is the importance of ground radials associated with broadcast antennas? What may result if many of such radials break? (20-29) [4]

90. If the daytime input power to an antenna having a resistance of 20 Ω is 5 kW, what would be the nighttime input power if the antenna current were cut in half? (20-30) [4]

91. If the daytime transmission-line current of a 10-kW transmitter is 15 A and the transmitter is required to reduce current to 5 kW at sunset, what is the new value of line current? (20-30) [4]

92. The power input to a 72-Ω concentric line is 10 kW. What is the rms voltage between inner conductor and sheath? (20-30) [4]

93. The ammeter at the base of a Marconi antenna has a certain reading. It is increased 3.5 times. What is the increase in output power? (20-30) [4]

94. A long transmission line delivers 10 kW to an antenna. At the transmitter the line current is 6 A, and at the coupling house it is 5.7 A. If proper line termination and negligible coupling losses are assumed, what power is lost in the line? (20-30) [4]

95. The dc input power to the final-amplifier stage is 2,500 V and 500 mA. The antenna center resistance is 10 Ω and the antenna current is 9 A. What is the plate-circuit efficiency of the final amplifier? (20-30) [4]

96. Explain the operation of a turnstile TV antenna. (20-31) [4]

97. What type of antenna site is technically best for an FM broadcast station? (20-31) [4]

21
MEASURING FREQUENCY

The objective of this chapter is to outline some of the methods used to determine the operating frequency of transmitters, receivers, and tuned circuits by using such devices as wavemeters, heterodyne frequency meters, and counters.

21-1 MEANS OF MEASURING FREQUENCIES

It is imperative that a transmitter be operated on its assigned frequency, or reasonably close to its assigned frequency, to prevent interference with other stations. In order to comply with this requirement, commercial transmitter oscillators are crystal-controlled or a frequency synthesizer is used to generate the desired frequency. However, crystals can vary in frequency if mistuned or overheated, and synthesizers can malfunction. To make sure the transmitter is on frequency, some external measuring device must be used periodically. When frequency tolerance was less stringent, absorption wavemeters or lecher wires were used. Now that frequency counters have been developed (Sec. 12-25), they are used to give direct frequency readouts. Another measuring method is to compare the frequency being measured with some known frequency in a heterodyne frequency meter. In this case the known frequency must first be checked against a primary or secondary frequency standard.

To find the resonant frequency of an LC circuit, a grid-dip meter is used. To calibrate the dial indications of a receiver, a secondary frequency standard or a frequency synthesizer can be used.

21-2 FREQUENCY TOLERANCE

How close to its assigned frequency a transmitter must be is determined by FCC rules and regulations. In the AM broadcast band (about 1 MHz) the frequency tolerance is ±20 Hz. In the FM broadcast band (about 100 MHz) the carrier must not drift more than 2,000 Hz from its assigned frequency. Commercial SSB (A3J) emissions must be held to within 10, 20, or 50 Hz of the licensed frequency, depending on whether they are land-based or mobile.

In other services the frequency tolerance may be stated in either percent of the assigned frequency or in *parts per million* (ppm). For example, a tolerance of 0.01% for a 9-MHz carrier is 0.0001 times 9,000,000, or 900 Hz. The carrier must always be within 900 Hz of 9 MHz. This could also be expressed in parts per million, which is the percent coefficient value of 0.000100, or 100 ppm. Since the carrier is assigned as 9,000,000 Hz, its tolerance is 9 $\times$ 100 ppm, or 900 Hz. The transmitter may operate on any frequency between 9,000,900 and 8,999,100 Hz, but preferably as close to 9 MHz as possible.

When doublers are used in a transmitter, the oscillator stage will have the same frequency tolerance requirements as has the carrier emitted by the final amplifier. If a 12-MHz carrier has an assigned frequency tolerance of 0.0001, it may operate on any frequency 1,200 Hz above and below 12 MHz. If its oscillator is followed by two doubler stages, the oscillator must be on 12 $\div$ 2 $\div$ 2, or on 3 MHz. It must not drift more than 300 Hz from its desired 3-MHz frequency to prevent the final amplifier from drifting more than 1,200 Hz.

Amateur transmitters must be operated in given bands of frequencies, and no part of their emissions may extend past the band limits. (The 40-m band, for instance, is from 7 to 7.3 MHz.) No frequency tolerance is required for amateur emissions in this case.

When measuring frequencies of transmitters, the frequency tolerance of the measuring device itself must be at least one-half that of the transmitter. If the transmitter has a frequency tolerance of 200 ppm, the frequency meter must have a frequency tolerance of 100 ppm or less.

21-3 ABSORPTION WAVEMETERS

The simplest and least accurate RF measuring device is the absorption wavemeter. It consists of a coil and a variable capacitor in parallel and some form of indicator, as in Fig. 21-1. The varia-

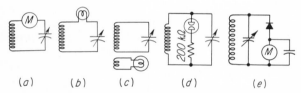

Fig. 21-1 Some possible indicators used with wavemeters. (a) Sensitive thermocouple milliammeter. (b) Flashlight lamp. (c) Inductively coupled lamp or meter. (d) Neon lamp. (e) Dc microammeter and rectifier.

ble capacitor may have a calibrated dial, or scale, attached to it.

When a wavemeter is held close to the LC tank circuit of an oscillator or low-powered amplifier, energy is absorbed by the meter if it is tuned to the same frequency as the tank. The plate or collector current of the stage will increase when the wavemeter absorbs energy, or the indicator will provide a maximum reading. The wavemeter should be held as far from the tank circuit as possible and still obtain an indication of resonance by the rise in plate current. The looser the coupling, the more accurate the indication given and the less the meter loads or detunes the circuit being measured. When a self-excited oscillator is

measured, a shift of frequency may be considerable if the coupling is tight. With the wavemeter adjusted to the resonant point, indicated by the peak plate current, the frequency of the transmitter and the wavemeter are indicated on the dial. Figure 21-2 illustrates a possible wavemeter dial calibrated to read from 3 to 5 MHz.

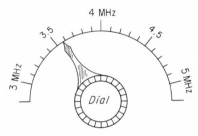

Fig. 21-2 Wavemeter dial and a scale calibrated directly in frequency.

When taking absorption-wavemeter measurements, it is important to remember that transmitter tank coils may have dangerously high dc or RF ac voltages on them.

Hand capacitance (detuning due to the presence of a hand changing the dielectric constant of the space surrounding a capacitor or inductor) will cause less detuning of a circuit being measured if the wavemeter is coupled to the cold (close-to-ground) end of the tank circuit.

Absorption wavemeters seldom have a frequency accuracy better than 0.05%. They find application in determining the approximate fundamental frequency, being relatively unresponsive to harmonic output.

A wavemeter may have several plug-in coils in order to cover a wide band of frequencies. In this case an arbitrary scale from 0 to 100° is used on the dial. A calibration chart, or graph, is used with each plug-in coil. With the graph, arbitrary dial indications can be converted to frequencies. Figure 21-3 shows an arbitrarily marked dial scale and a frequency-versus-dial-division calibration chart. The curve on the chart is used to convert dial-division readings to frequency.

According to the indication shown on the dial (fifty-seventh division), the frequency of resonance of the wavemeter from the graph is approximately 5.1 MHz. Note that the line on this graph is not linear. With special straight-line-frequency

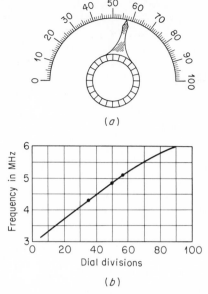

(a)

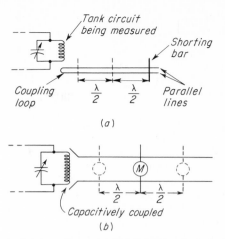

Fig. 21-4 Lecher wires: (a) inductively coupled and (b) capacitively coupled.

Fig. 21-3 (a) Dial and uncalibrated or arbitrarily marked scale and (b) a calibration chart that might be used with it.

variable capacitors, a nearly straight curve can be produced.

If a wavemeter has an error proportional to frequency and is accurate within 200 Hz at a frequency of 1,000 kHz, it will be accurate within 400 Hz at a frequency of 2,000 kHz. If a frequency-measuring device is accurate to 20 Hz when set at 1,000 kHz, its error when set at 1,250 kHz can be determined by setting up the ratio, or proportion, of 1,250 is to 1,000 as X is to 20 Hz:

$$\frac{1,250}{1,000} = \frac{X}{20}$$

$$1,000X = 25,000$$

$$X = 25 \text{ Hz possible error}$$

21-4 LECHER WIRES

A parallel-conductor transmission line can be used to determine the frequency of a transmitter by taking measurements of the standing waves developed on it. One end of a one- to three-wavelength transmission line is formed into a loop, which is loosely coupled to the tuned circuit of the amplifier or oscillator being measured, as shown in Fig. 21-4a. Linear tank wavemeters of this type are known as *lecher wires*.

The lecher wires can be constructed of two No. 12 bare copper wires, or $\frac{1}{8}$-in. copper tubing, held apart about 1 in., stretched tightly to prevent movement, and supported by insulators at the far ends only. A knife-edged metal shorting bar is held at exact right angles across the two parallel lines. By starting near the coupling loop and slowly moving the shorting bar along the line, a point will be reached where the plate current of the transmitter will rise to a sharply defined peak. This point is marked on the lecher wires. The shorting bar is moved farther down the lines until a second current peak is found. The distance between the two points represents a half wavelength. Moving farther down the line, the next peak represents another half wave, or a full wave from the first. By measuring the full-wave distance with a meter stick (marked in decimeters, centimeters, and millimeters) the operating wavelength is determined. The frequency can be found by using the basic frequency formula (Sec. 20-7):

$$f = \frac{v}{\lambda} = \frac{300,000,000}{\text{wavelength in meters}}$$

If a meter stick is not available, the length of the full wave can be measured in inches and the frequency in megahertz can be determined by the formula

$$f_{\text{MHz}} = \frac{11{,}810}{\text{length in inches}}$$

If the lecher wires are not long enough for a full-wave measurement, a half wave can be used with little loss of accuracy.

An alternative method of coupling to lecher wires is by capacitance, as in Fig. 21-4b. The two wires at one end of the transmission line are spread slightly and held near the two ends of the tuned circuit being measured.

A sensitive thermogalvanometer with contact bars can be moved along lecher wires instead of a shorting bar. Visual indications of current loops (high values) and nodes (low values) will be discernible.

With loose coupling and considerable care, measurements of frequency to about 0.1% can be made with lecher wires.

21-5 GRID-DIP METERS

To measure the frequency of a resonant circuit when not in operation, a *grid-dip meter* can be coupled to it. A grid-dip meter consists of a vacuum-tube oscillator, usually a Colpitts circuit. A sensitive milliammeter is connected in series with the grid leak resistor, as in Fig. 21-5. When the

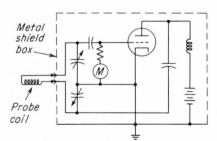

Fig. 21-5 Grid-dip-meter circuit.

stage is oscillating, grid current flows. When the oscillating tank probe coil is coupled to any other external LC circuit tuned to the same frequency, the external circuit will absorb energy from the oscillating tank, weaken its amplitude of oscillation, and lessen the grid current. If the grid-dip oscillator is coupled to a circuit having an un-

known frequency of resonance, the grid current dips as the oscillator is tuned past the resonant frequency of the LC circuit being tested.

The lower the degree of coupling between the grid-dip meter and the circuit being tested, the more accurate the results. When the meter is accurately calibrated and loosely coupled to an external circuit, its indications are fairly reliable, but it is not designed to have the accuracy of true frequency meters. To cover a wide band of frequencies, grid-dip meters usually have plug-in probe coils. There is a separate calibration scale for each plug-in coil used. The scales are often calibrated directly in frequency. A grid-dip meter is valuable when constructing, revamping, or testing receiving and transmitting LC circuits and resonance of antennas.

When BJTs or FETs are used in dip meters, the RF ac generated by the meter itself may be rectified and displayed on the meter. When an external LC circuit absorbs energy from the probe, the rectified ac decreases, or dips. For solid-state meters the term *dip oscillator* may be used.

Most dip meters act as wavemeters when the active device in them is turned off.

21-6 PRIMARY STANDARDS OF FREQUENCY

The National Bureau of Standards (NBS) in Fort Collins, Colorado, has three transmitters, WWVL, WWVB, and WWV. WWVL transmits intermittent experimental programs on 20 kHz. WWVB transmits continuously with an exact 60-kHz carrier which decreases 10 dB in amplitude each second and resumes normal amplitude according to a special time code.

The main station, WWV, transmits carriers continuously on 2.5, 5, 10, 15, 20, and 25 MHz (each accurate within a small fraction of one cycle). These carriers are amplitude-modulated with a pulse of five cycles of 1,000-Hz ac (producing a tick sound) every second, excepting the 29th and 59th, of every minute. During the 1st and 31st minutes of each hour there is a station identification announcement made by a man's voice, followed by 7 s of silence, and then a UTC (universal time coordinated, previously GMT, or Greenwich mean time) hour-and-minute time announcement. This is followed by a 0.8-s 1,000-Hz tone and then a standard 600-Hz tone for 45 s. After 7 s of silence there is another UTC hour-and-minute time announcement. This program continues, voice announcements on odd minutes and tones on even minutes, except that there is no modulation other than hour-and-minute announcements transmitted during a semisilent period between the 45th and 50th minutes of each hour. The tones on the even minutes change from 600 Hz on the 2d minute to 440 Hz (standard musical A above middle C) on the 4th minute, back to 600 Hz on the 6th minute, and so on. The 45-s voice announcements consist of propagation forecasts, storm warnings, geophysical alerts, or subscribed announcements from government or other agencies. If there are no voice announcements scheduled, a standard 500-Hz tone is transmitted instead. As each tick is produced, a special 100-Hz coded tone is started and lasts for a significant portion of the interval between ticks. During each hour, therefore, standard frequencies of 100, 440, 500, 600, and 1,000 Hz are available. (NOTE: NBS periodically changes its format.)

Another NBS station, WWVH, in Kekaha, Kauai, Hawaii, transmits on 2.5, 5, 10, 15, and 20 MHz. It makes its voice announcements on the 59th and 29th minutes. WWVH alternates its 600- and 440-Hz tones on the minutes when WWV has voice announcements and starts each hour with a 60-Hz tone. The WWVH 0.8-s minute-starting tone is at a frequency of 1,200 Hz. Its semisilent period is from 15 to 20 min after each hour. Voice announcements are made using a woman's voice.

The National Bureau of Standards uses a cesium-beam atomic standard. By international agreement "one second is the duration of 9,192,-631,770 periods of the radiation corresponding to the transition between the two hyperfine levels of the ground state of the cesium-133 atom." A cesium-beam-controlled oscillator is known as a primary frequency standard. By means of frequency dividers, oscillators, and heterodyning, a cesium-beam oscillator can be made to synchronize a 2.5-MHz oscillator, for example, and hold it at the required frequency to about 1 cycle in 10^{14}. NBS RF broadcasts are held to about 2 parts in 10^{11} or better.

The three major United States television networks, NBC, CBS, and ABC, use rubidium gas cell frequency standards to maintain their color subcarrier emissions at 3,579,545 Hz above their assigned carriers. Thus, the oscillation frequency of the crystal oscillator in a color TV receiver will provide a frequency standard of a few parts in 10^{11}.

By comparing a local 10-MHz oscillator with one of the carrier frequencies of the NBS stations, it is quite possible to check the local-oscillator frequency to 1 part in 10^6. To do this, one of the NBS carriers is tuned in on a radio receiver. At the same time an approximately equal-amplitude signal from the local oscillator is fed to the receiver. The difference between the two frequencies appears as an output beat signal that wavers in amplitude a certain number of times per second. This waver frequency is the difference in frequency between the local oscillator and the NBS carrier frequencies.

Test your understanding; answer these checkup questions.

1. What are the two sources of stable oscillation frequencies for receivers or transmitters? _____ _____
2. How do the tolerance percentages of AM and FM broadcast stations compare? _____
3. Why are A3J transmitters subject to tighter frequency tolerances than AM or FM? _____ In what units are their tolerances specified? _____
4. How many ppm is represented by a tolerance of 0.005%? _____
5. An oscillator with a tolerance of 0.001% has three doublers following it. What is the output frequency tolerance? _____
6. What is the hot end of an LC circuit? _____
7. In Fig. 21-2, what frequency is indicated? _____
8. In Fig. 21-3, what frequency corresponds to a dial reading of 50? _____ Of 35? _____
9. Lecher wires are what type of wavemeter? _____
10. What would be the formula to use to solve for f in a Lecher wire wavemeter if measurements are made in centimeters? _____
11. In what frequency bands would Lecher wires be practical? _____ Why not in the HF band? _____
12. What device is used to measure the resonant frequency of an LC tank circuit? _____
13. What are the carrier frequencies of WWVB? _____ WWV? _____ WWVH? _____ Which station uses a woman's voice? _____
14. List five standard frequencies that can be obtained from WWV besides the carrier frequencies. _____ _____ _____ _____ _____ What four other types of information can be obtained from WWV? _____ _____ _____
15. What is used as the primary frequency standard at WWV? _____ What is another possible primary standard? _____

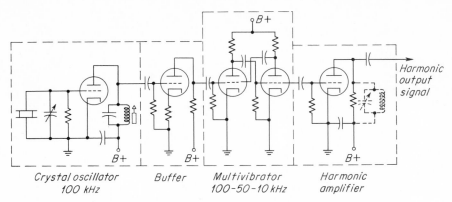

Fig. 21-6 Essential circuits of a secondary frequency standard.

21-7 SECONDARY FREQUENCY STANDARDS

A secondary frequency standard is relatively simple compared to a primary standard. It consists of two oscillators; one is a nonsinusoidal wave type; the other is a stable low-frequency synchronizing crystal oscillator. The harmonics of the first oscillator are the signals that are used.

A secondary standard might consist of the circuits shown in Fig. 21-6:

1. A multivibrator oscillator to generate a low-frequency fundamental sawtooth or square wave with many harmonics
2. A resistance-coupled amplifier stage to amplify the multivibrator harmonics
3. A 100-kHz crystal oscillator, usually temperature-controlled, to synchronize or lock the multivibrator into a desired frequency of oscillation
4. A resistance-coupled buffer stage to isolate the crystal-oscillator stage from the multivibrator
5. A voltage-regulated power supply (not shown in the illustration)

The source of usable signals from a secondary frequency standard is the multivibrator oscillator, not the crystal oscillator as is sometimes believed. The multivibrator oscillator, a two-tube RC oscillator, generates a steep-sided ac waveform, rich in harmonics. In practice its harmonics may be usable up to the three-thousandth or higher.

A 100-kHz multivibrator can produce a harmonic signal every 100 kHz throughout the radio spectrum. By comparing the relative spacing of an unknown signal between two known 100-kHz harmonics, it is possible to determine the frequency of the unknown. If a station is tuned in on a receiver and is found to be exactly halfway between the sixth and seventh 100-kHz harmonics, the station must be transmitting on 650 kHz. A calibrated receiver is necessary in order to identify which harmonics are being heard.

The calibration of a receiver can be checked by feeding the signal from a 100-kHz secondary standard to it. A signal from the standard should appear on the receiver dial at every point representing a multiple of 100 kHz. Signals should be received where the dial is marked 600 kHz, 900 kHz, 1.5 MHz, 2.7 MHz, etc.

By incorporating multiple-contact switches (not shown) in the secondary-standard multivibrator circuit, it is possible to change the grid resistors and coupling capacitors and thereby shift the frequency of oscillation from 100 to 50 or to 10 kHz.

To increase the output strength of the multivibrator-oscillator harmonics, a tuned circuit may be added in the output stage (dashed lines). The harmonics falling on or near the frequency of this tuned circuit are amplified more than they would be otherwise. This extends the usable high-frequency range of the secondary standard.

The stability of the multivibrator oscillator is improved by feeding a synchronizing signal into one of its grid circuits from the 100-kHz crystal oscillator. The sine-wave output of the crystal oscillator triggers the multivibrator circuit, forcing it to start its oscillations at the proper instant, thereby assuring proper output frequency from the multivibrator. The sinusoidal crystal output can trigger the sawtooth multivibrator at the fundamental or any *submultiple* of the crystal frequency. Thus, a 100-kHz crystal can hold a multivibrator oscillator at 100, 50, 33.3, 25, or 20 kHz down to about the tenth submultiple, which is 10 kHz. If the multivibrator normally oscillates at 49.9 kHz by itself, a 100-kHz synchronizing oscillator will force it to start a little early, at exactly 50-kHz oscillations. Regardless of its natural frequency, the multivibrator will be synchronized to the closest higher-frequency submultiple of the 100-kHz frequency.

To check the frequency of oscillation of the crystal oscillator, the secondary frequency standard should be turned on and allowed a $\frac{1}{2}$- to 1-h warm-up period to permit the circuits to stabilize. A shortwave receiver is then tuned to WWV. Some of the signal from the secondary standard set to 100-kHz harmonic output is also fed to the receiver. If the harmonics of the multivibrator produce a zero beat (no audible difference of frequency between the unmodulated WWV signal and the harmonic of the multivibrator), the crystal must be oscillating on 100 kHz. If there is a difference frequency, the trimmer capacitor across the crystal can be varied until a zero beat is produced. When the 100-kHz harmonics are correctly adjusted, there should be no need to readjust the crystal if the multivibrator is switched to 50- or 10-kHz harmonic output.

21-8 FREQUENCY MEASUREMENTS WITH A SECONDARY STANDARD

It is possible to determine the approximate frequency of any transmitter by tuning it in on a receiver and identifying the two 10-kHz multivibrator harmonics between which the signal is received.

Assume that a signal of 4,673 kHz is to be measured. After warm-up and calibration, the secondary standard is switched to 100-kHz harmonic output and fed into a communication receiver with the BFO (beat-frequency oscillator) turned on. The signal to be measured is also tuned in. From the calibration of the receiver dial it is found that the signal falls between the 4,600- and 4,700-kHz harmonics (Fig. 21-7).

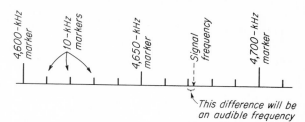

Fig. 21-7 Marker-signal positions on a receiver dial and the signal to be measured.

The secondary standard is next switched to 50-kHz harmonic output, and the signal is now observed to be between the 4,650- and 4,700-kHz markers. The receiver is set to 4,650 kHz, and the standard is switched to 10-kHz harmonic generation. The receiver is carefully tuned toward the signal, and each time the receiver zero-beats against a 10-kHz harmonic, another 10-kHz is added to the 4,650-kHz starting frequency. When the receiver comes to zero beat with the 4,670-kHz marker, a 3-kHz difference frequency will also be heard. This is produced by the beating of the 4,670-kHz harmonic and the received 4,673-kHz signal. It is impossible to tell by ear that the difference is 3 kHz. Some means must be used to determine accurately how far above 4,670 kHz the signal frequency is. There are three methods by which this difference can be found: (1) interpolation, (2) audio comparison, and (3) counter.

The interpolation method of determining the displacement of the signal from a marker signal requires a variable RF oscillator of good stability, incorporating a straight-line frequency capacitor to produce a linear frequency-versus-dial relationship. This oscillator is calibrated not in frequency, but in arbitrary units only.

The receiver is left tuned to the 4,670-kHz marker, and the BFO in the receiver is turned off.

The variable oscillator is fed into the receiver and tuned until a zero beat is produced against the 4,670-kHz marker.

The dial reading of the variable oscillator, when at zero beat with 4,670 kHz, is noted on a piece of paper. The receiver is tuned to the received signal; the oscillator is tuned to a zero beat with this; and the oscillator dial reading is noted again. The receiver is finally tuned to the next higher, or 4,680-kHz, marker, and the variable oscillator is brought to a zero beat at this frequency. The oscillator dial reading is again noted. The ratio of the received signal to the 4,670-kHz marker in comparison with the separation between markers will indicate the received frequency.

EXAMPLE: If

$$4,670 \text{ kHz} = 40° \text{ on oscillator dial}$$
$$\text{Received signal} = 49° \text{ on oscillator dial}$$
$$4,680 \text{ kHz} = 70° \text{ on oscillator dial}$$

Then

- Call the difference between 4,670 kHz and the received signal a.
- Call the difference between 4,670 and 4,680 kHz b.
- Call the difference between 40 and 49° c.
- Call the difference between 40 and 70° d.

By using these factors, it is possible to set up the ratio needed to solve for the unknown, the difference between 4,670 kHz and the signal:

$$\frac{a}{b} = \frac{c}{d}$$

By substituting known values and solving for a,

$$\frac{a}{10 \text{ kHz}} = \frac{9°}{30°} \qquad a = \frac{9(10)}{30} = 3 \text{ kHz}$$

The signal has now been computed to be 3 kHz above 4,670 kHz, making it 4,673 kHz.

The accuracy of this method depends on the accuracy of the three zero-beat dial readings and the linearity of the dial on the oscillator. To improve the accuracy of the results, it is possible to zero-beat the oscillator against the marker or signal frequency and then switch on the BFO of the receiver. The BFO is adjusted until a tone of about 1,000 Hz is heard. The tone produced by the BFO beating against the marker and oscillator zero beat will be heard to waver in amplitude. As the oscillator is brought closer to true zero beat, the waver of the audible signal will become slower. When the wavering is down to a fraction of a hertz, the oscillator can be assumed to be zero beat with the marker or signal frequency.

When the signal being measured is within a few kilohertz of a marker signal, the audio comparison method can be used. With the receiver tuned to 4,670 kHz and with the BFO off, a beat tone between the 4,670-kHz marker and the signal frequency will be produced in the receiver. A second loudspeaker can be connected to the output of an accurately calibrated audio signal generator. The signal generator is varied until the tone from it has exactly the same pitch as the beat signal from the receiver. If the two tones have the same pitch with the audio signal generator set to 3,000 Hz, the frequency of the received signal must be 3 kHz from the nearest marker. A better method is to apply the receiver signal to the vertical input of an oscilloscope and the signal generator to the horizontal. The signal generator is tuned until a circle or ellipse (Llssajous figure) is shown. The two signals are then the same frequency.

When the signal is closer to a higher frequency marker, 4,680 kHz in the example, the audible difference between 4,680 kHz and the signal is found and is subtracted from 4,680 kHz to give the correct frequency. If a frequency counter is available, the beat tone frequency will register directly on the readout scale of the counter. This value can be added to or subtracted from the nearer 10-kHz marker frequency.

21-9 HETERODYNE FREQUENCY METERS

The simplest form of heterodyne frequency meter might consist of only a well-shielded Hartley-type ECO with earphones in the plate circuit, forming a one-tube receiver with a calibrated dial. However, strong input signals will pull the frequency of oscillation of the LC circuit toward the incoming frequency and inaccurate zero-beat indications would result. This tendency of two oscillations of slightly different frequencies to synchronize must be guarded against in frequency measuring.

To overcome any pulling effect, heterodyne frequency meters employ a system as in Fig. 21-8,

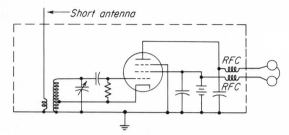

Fig. 21-8 Heterodyne-frequency-meter circuit with isolated oscillator not subject to frequency pulling.

in which the calibrated VFO is coupled to one grid of a mixer tube. The signal to be measured is fed to the other mixer grid. The heterodyne of these two signals produces an AF component in the plate circuit of the mixer. Assume that the oscillator tunes from 1 to 1.5 MHz and that its dial is straight-line frequency and is graduated in 100 equal divisions. Each division would be equivalent to 5 kHz. If the antenna of the meter is held close to the source to be measured and the dial is rotated, a zero beat should be heard in the earphones if the source is producing an output between 1 and 1.5 MHz. Suppose a zero beat is heard at 28 on the dial. The frequency must be 28(5), or 140 kHz above 1 MHz, or 1,140,000 Hz. This could also have been determined by proportion by $(28/100)(500,000) + 1$ MHz, where 500,000 is the frequency difference represented by 100 divisions on the dial.

If the source had been at half the frequency (570,000 Hz), the second harmonic would have heterodyned against the oscillator and a zero beat would have been received at the same spot on the dial. Furthermore, had the source been oscillating at twice the frequency (2,280,000 Hz) the second harmonic of the *meter's* oscillator would have heterodyned against the source and produced a zero beat at the same point on the dial. Thus, these meters can be used to measure frequencies much higher and much lower than those on the calibrated scale. While this is quite an

advantage, care must be exercised not to be confused by harmonic beats. A wavemeter could be used to determine the actual fundamental frequency of a source.

If the dial had read somewhere between 28 and 29, it would be necessary to guess at the actual frequency. To provide more accuracy a vernier scale, as shown in Fig. 21-9, could be added to

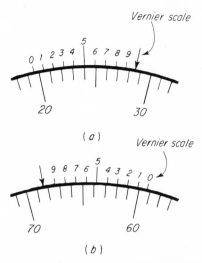

Fig. 21-9 (*a*) Vernier scale on a dial reading 28.5. (*b*) Vernier dial reading 68.3.

the dial. With this vernier the graduations on the main dial (20, 21, 22, etc.) can be accurately read to one-tenth of a division. For example, in Fig. 21-9*a*, the indicator reads between 28 and 29. The number 5 on the vernier scale lines up with one of the main dial markings. No other vernier mark lines up with any other main dial marking. This dial is indicating 28.5. If the source had produced a zero beat at 28.5, the frequency would have been $(28.5/100)(500,000) + 1$ MHz, or 1,142,500 Hz. In Fig. 21-9*b* the dial indicates 68.3, which would be equivalent to a frequency of $(68.3/100)(500,000) + 1$ MHz, or 1,341,500 Hz. It can be seen that with this vernier there are 1,000 usable markings and that reading to at least 500 Hz in the 1-MHz region is possible. To allow more accuracy, heterodyne-frequency-meter dials may have 50 main dial markings, with each main marking broken into 100 parts by a second geared dial; each of these parts is read to 10 parts by a vernier, making a total of 50,000 usable markings.

Most heterodyne frequency meters have equally spaced dial divisions. When a frequency indication is obtained, the dial reading is compared with a calibration chart or book to determine the actual frequency The following is an example.

An absorption-type wavemeter indicates that the approximate frequency of a transmitter is 500 kHz. At the same time the transmitter signal produces a zero beat on an accurately calibrated heterodyne frequency meter at a dial reading of 374.1. The frequency-meter calibration chart indicates the following dial readings versus frequencies:

Dial	Frequency
$367.0°$	$= 499.4$–998.8 kHz
$371.5°$	$= 499.6$–999.2 kHz
$376.0°$	$= 499.8$–999.6 kHz

Since the absorption wavemeter indicates the approximate frequency to be 500 kHz, the dial reading of 374.1 must indicate a frequency between 499.6 and 499.8 kHz, rather than one between two of the 999-kHz harmonic frequencies. To find the exact frequency from the given information, it is necessary to interpolate, as explained previously:

$$b\left\{a\begin{cases}371.5° = 499.6\text{ kHz}\\374.1° = \text{ the signal}\\376.0° = 499.8\text{ kHz}\end{cases}c\right\}d$$

a = difference between 371.5 and 374.1
    = $2.6°$
b = difference between 371.5 and 376.0
    = $4.5°$
c = difference between 499.6 and signal
    = unknown
d = difference between 499.6 and 499.8
    = 0.2 kHz

By setting up the ratio and solving for the unknown difference:

$$\frac{a}{b} = \frac{c}{d}$$

$$\frac{2.6}{4.5} = \frac{c}{0.2}$$

$$4.5c = 0.2(2.6)$$

$$c = \frac{0.52}{4.5} = 0.1156 \text{ kHz}$$

Therefore, the unknown frequency is 0.1156 kHz greater than 499.6 kHz, or 499.7156 kHz.

The following is a somewhat similar problem:

A heterodyne frequency meter has a dial reading of 31.7 for a frequency of 1,390 kHz and a dial reading of 44.5 for a frequency of 1,400 kHz. What is the frequency of the ninth harmonic of the frequency corresponding to a scale reading of 41.2?

By interpolation, the frequency corresponding to the scale reading of 41.2 is

$$b\left\{a\begin{cases}31.7° = 1,390\text{ kHz}\\41.2° = \text{unknown signal}\\44.5° = 1,400\text{ kHz}\end{cases}c\right\}d$$

$$\frac{a}{b} = \frac{c}{d}$$

$$\frac{9.5}{12.8} = \frac{c}{10\text{ kHz}}$$

$$12.8c = 9.5(10)$$

$$c = \frac{95}{12.8} = 7.422 \text{ kHz}$$

The unknown frequency is 7.422 kHz above 1,390, or 1,397.422 kHz. The ninth harmonic is 9 × 1,397,422 Hz, or 12,576.798 kHz.

Because of mechanical play, called *backlash,* in most dials, frequency-meter dials should always be brought to zero beat while tuning in the same direction (clockwise, for instance).

Commercial heterodyne frequency meters usually have special crystal checkpoint oscillators, and often the variable-frequency oscillator can be modulated by an AF tone. A block diagram of such a meter is shown in Fig. 21-10. The basic

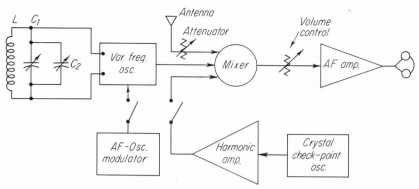

Fig. 21-10 Block diagram of a commercial type of heterodyne frequency meter with CW or modulated-signal output capabilities.

operation is similar to that explained above, but all variable-frequency oscillators tend to drift in frequency. A crystal oscillator, oscillating perhaps at 500 kHz, will maintain its frequency quite well. Before using the heterodyne frequency meter, it should be warmed up for about 30 min to allow all components to come to a stable operating temperature. Then the crystal oscillator is turned on, and the variable-frequency dial is set to some harmonic frequency of the crystal, called a *checkpoint*. If the LC circuit of the variable-frequency oscillator covers 1 to 1.5 MHz, either of these frequencies could be used as the checkpoint, as both are harmonics of 500 kHz. With the dial set for exactly 1 MHz, the crystal oscillator should produce a zero beat when it is turned on. If some audible beat tone is heard, the trimmer capacitor C_2 should be adjusted to zero beat. Now the meter has been calibrated by comparison with the checkpoint frequency, and all dial readings should be correct when the main tuning capacitor is varied.

To make sure that the crystal is operating at the proper frequency, station WWV can be tuned in on a receiver at perhaps 5 MHz. The antenna connection of the frequency meter is also coupled to the receiver. When the crystal oscillator of the meter is turned on, one of its harmonics (the tenth) should fall on WWV's frequency and produce a zero beat in the receiver. If an audible tone is heard, the trimmer capacitor across the crystal can be adjusted to bring the crystal on frequency. When approaching zero beat, a weak rising and falling hissing sound indicates how far off fre-

quency the crystal is from WWV. If the hiss rises and falls twice per second, the harmonic of the crystal is 2 Hz off at WWV's frequency (0.2 Hz at the crystal's fundamental frequency). It is a good idea to check the crystal frequency periodically, and particularly if tubes have been changed in the oscillator circuit. These meters are delicate instruments and must be handled with care. Should a tuning capacitor or coil become damaged, many hours of work will be required to recalibrate the equipment. When an oscillator tube is replaced, the equipment should be allowed several hours warm-up; then the meter should be tested at several points on the dial to see if it is indicating accurately. More than one tube may have to be tried before one with the same characteristics as the original is found.

A heterodyne frequency meter can be used as a calibrated source of RF energy. Some of the variable-frequency oscillator energy feeds into the antenna circuit through the mixer and is radiated. If a signal is tuned in on a radio receiver, it is possible to measure the frequency of a signal being received on a nearby receiver. The meter's dial is tuned until a zero beat between the variable-frequency oscillator (or its harmonic) and the received signal produces a zero beat in the receiver. The frequency is read off the meter's calibrated dial (or read and multiplied by the harmonic being used). It is necessary that the receiver dial indicate the received signal's approximate frequency to assure proper harmonic identification.

The radiated signal (or the harmonics) from a

heterodyne frequency meter can be used as an RF signal generator to line up the tuned circuits of a receiver (Chap. 18). For easy identification of the signal on the receiver, the AF modulator in the frequency meter is turned on. This stage is an AF oscillator that produces an ac of perhaps 400 Hz. Adding this ac to the suppressor grid of the variable-frequency oscillator modulates the output signal, and a 400-Hz tone is heard when the receiver is tuned to the signal generator frequency. The AF modulator is not used while measuring frequency.

Test your understanding; answer these checkup questions.

1. How do the two oscillators in a secondary frequency standard differ? _____ The output of which is used when measuring? _____
2. What frequencies are generated by the multivibrator in the usual secondary frequency standard? _____ What holds them constant? _____
3. Name the stages used in a secondary frequency standard. _____ _____ _____ _____
4. What must be used in conjunction with a secondary frequency standard when measuring frequency? _____
5. Could a signal frequency be measured more accurately with a secondary frequency standard or with a receiver having a synthesizer local oscillator? _____
6. What ratio of signal strengths gives a maximum beat-signal output? _____
7. What are the three methods of determining the AF beat difference between a signal and a marker? _____ _____ _____
8. If 8.5 MHz = 25° on a dial, a signal is on 32°, and 8.6 MHz is on 42°, what is the signal frequency? _____
9. Why should a heterodyne frequency meter antenna not be coupled to the VFO? _____

10. Over how many main-dial divisions does a vernier scale extend? _____
11. How is dial backlash prevented from producing an erroneous reading? _____
12. When would AF modulation be used with heterodyne frequency meters? _____
13. How long should heterodyne frequency meters be warmed up before use? _____ Against what is the crystal checked? _____
14. Can frequencies above the oscillation range of a heterodyne frequency meter be measured? _____ Below? _____

21-10 A CONSTANT FREQUENCY INDICATOR

When a constant indication of the frequency of a transmitter is required, as in a broadcast station, other types of meters are used. In standard AM broadcast stations, the transmitter must maintain its assigned frequency at all times within ±20 Hz. This rigid frequency tolerance is required to prevent two stations on the same frequency but in different parts of the country from producing an audible beat tone in a receiver. With a 20-Hz tolerance, the maximum difference of frequency between any two stations is 40 Hz, a tone normally inaudible in receivers. In most cases a station maintains its assigned frequency within about 5 Hz, resulting in a maximum beat tone of only about 10 Hz.

Maintaining this degree of frequency stability is possible by use of temperature-controlled crystals in the transmitters. A frequency monitor for such a station also uses a temperature-controlled crystal oscillator. A monitor might use a crystal oscillator on the assigned frequency of the transmitter and indicate the number of hertz's

Fig. 21-11 Simplified circuit of a broadcast-station frequency monitor that presents a constant frequency indication.

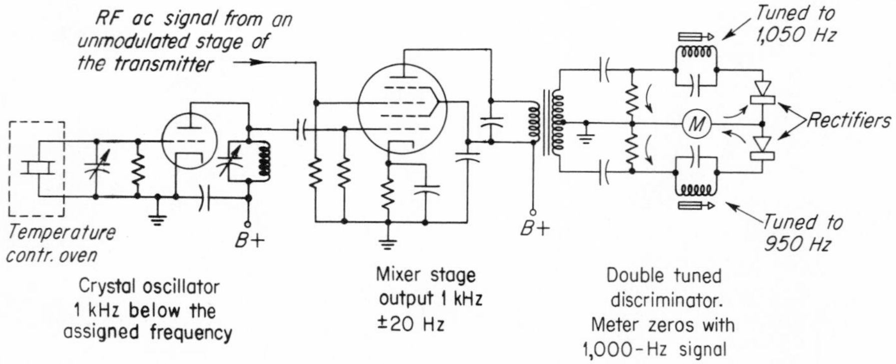

difference between the transmitter and the monitor. However, producing a frequency-indicating device that will readout in the 1- or 2-Hz range is difficult. Instead, the monitor crystal is usually a kilohertz above or below the assigned frequency, and the difference frequency is read. Figure 21-11 is a simplified diagram of a possible frequency-monitoring circuit.

In this circuit, an RF signal of constant amplitude from one of the unmodulated stages in the transmitter is fed to one of the grids of the mixer tube. The output of the monitor crystal oscillating at a frequency 1,000 Hz lower than the assigned frequency of the transmitter is fed into the other mixer grid. In the plate circuit of this tube a difference frequency of 1,000 Hz appears. This is fed to a discriminator type of circuit utilizing two tuned circuits, one tuned to 1,050 Hz and the other to 950 Hz. With the transmitter on frequency and the monitor 1,000 Hz lower in frequency, the two resonant circuits produce equal and opposite rectified dc voltages across the zero-center meter M and a resultant of zero current through it. The meter reads center scale, which is marked 0 Hz.

If the transmitter shifts up in frequency by 10 Hz, the beat tone becomes 1,010 Hz, the circuit tuned to the higher frequency develops more voltage across it, and current flows through the meter in the diagram from left to right. The needle moves up to a point on the meter scale which would be marked +10 Hz.

If the transmitter shifts down in frequency 5 Hz, the beat tone becomes 995 Hz and more voltage is developed across the circuit tuned to 950 Hz. Current from this circuit flows from right to left through the meter and moves the indicator to a point which would be marked −5 Hz.

Frequency monitors in broadcast stations may drift and give erroneous readings. They must be checked periodically for accuracy. Commercial radio companies with primary standards can check the frequency of the transmitter at a specified time. As an example, a station makes arrangements for a frequency check at 1 A.M. on a certain day. At that time the operator notes the frequency monitor as reading +4 Hz. The commercial radio company reports the frequency of the station to be −2 Hz. The difference between the two readings, 6 Hz, is the error in the monitor. In this case the station monitor is 6 Hz high. The monitor crystal can be adjusted to indicate a frequency 6 Hz lower, and the meter can be assumed to be correct. The transmitter frequency may also be trimmed, although it is expected that the carrier will drift back and forth a few hertz under normal operation.

21-11 MEASURING FREQUENCY WITH COUNTERS

The development of digital frequency counters (Sec. 12-25) has made accurate frequency measurements relatively simple. The block diagram of Fig. 21-12 represents a simplified counter system

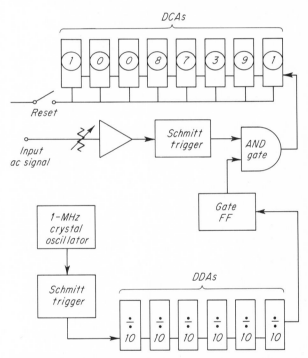

Fig. 21-12 Counter circuits required to measure frequency.

used to indicate frequency. An input signal is amplified and fed to a Schmitt trigger circuit which converts each cycle of input ac to a square-wave cycle to assure proper operation of the decimal counting assemblies (DCAs). The AND-gate circuit requires a simultaneous positive signal at both of its inputs before it will allow any signal to pass through it. If the gate flip-flop (FF) circuit applies a continuous positive signal to the AND gate, the DCAs will totalize (register) all input cycles. If the reset switch is closed, all DCAs are biased to zero readings. When the reset is opened, the DCAs start totalizing or counting again.

When operating as a frequency meter, harmonics of the 1-MHz crystal oscillator are compared with a received WWV signal and the crystal is corrected until it is exactly on frequency. The crystal output is fed through a Schmitt trigger circuit to produce square waves to properly operate the decimal divider (divide-by-10) assemblies (DDAs). With an input of 1 MHz and six DDAs the result is a precise 1-s output. This 1-s pulse is fed to the flip-flop gate, which in turn enables the AND gate for one second. During this 1-s period the DCAs count every cycle of input signal. If there are 10,087,391 cycles (as shown) in this time, the frequency of the input signal must be 10,087,391 cycles per second (Hz), plus or minus one count. Activation of the reset switch, either manually or automatically after a desired period of time, clears the DCAs and readies the counter for the next 1-s enabling pulse from the DDAs and the FF-gate circuit.

To measure the frequency of an oscillator or a transmitter a small link-coupling coil can be connected to the counter input. The counter is turned on and the link is slowly coupled to the ac source until just enough signal (usually less than one volt) is developed to produce uninterrupted counting. Too much voltage due to excessively close coupling can damage the input circuitry of a counter. For measuring at a distance of a few yards or meters, a tuned circuit with a small antenna can be connected to the input of the counter. However, if interference pulses of any kind are picked up, they will either be added to the RF being measured or may blot out some of the cycles that should be counted, resulting in erroneous readings.

When measuring the frequency of a received signal, a transfer oscillator is used. This is a stable oscillator that can be tuned to the frequency of the signal to be measured. Provision is made to couple the transfer oscillator output to the counter input. The signal is tuned in; the transfer oscillator is adjusted to an exact zero beat as heard on the receiver; and the frequency of the transfer oscillator is read off on the counter while the signal is still at zero beat.

Counters may have top frequencies of 100 kHz, 1 MHz, 10 MHz, or 50 MHz or, with special heterodyning circuits, can read up into the thousands of gigahertz. Assume a 14-MHz signal is to be measured but the only available counter has a 10-MHz top frequency. If a stable 10-MHz oscillation (from the master 1-MHz crystal signal multiplied by 10) is heterodyned against the 14-MHz signal, the resultant 4-MHz difference frequency can be read on the counter. In this case the signal frequency is 10 + 4, or 14 MHz. Thus, a 20-MHz transfer oscillation would allow measurements between 20 and 30 MHz, a 30-MHz oscillation would produce measurements between 30 and 40 MHz, and so on.

Transmitters and receivers using frequency synthesizers as their oscillators may have better frequency tolerances than most measuring devices (other than counters) provided their master crystal oscillators are on frequency. However, if a synthesizer malfunctions, it could develop an improper output frequency. This would normally be considerably off the desired frequency and counters or other frequency-measuring meters would show that a malfunction was present.

21-12 FREQUENCY CONSIDERATIONS FOR AMATEURS

Commercial radio systems are assigned one or more specific frequencies of operation. Frequency measurements are made to determine if the transmitting equipment is on frequency within FCC tolerances.

Amateur transmitting stations are not assigned any specific frequencies but bands of frequencies. For example, an Extra Class licensed amateur station can operate on any desired frequency

---

ANSWERS TO CHECKUP QUIZ ON PAGE 542

1. (*Type of ac generated*)(*Nonsinusoidal*) 2. (*100, 50, and 10 kHz*)(*Crystal*) 3. (*Crystal oscillator, buffer, multivibrator, RF amplifier*) 4. (*Receiver*) 5. (*With synthesizer local oscillator if it generates in 1-Hz steps*) 6. (*1:1*) 7. (*Ear comparison*) (*Scope ellipse*)(*Counter*) 8. (*8,541,177 Hz*) 9. (*Signal may pull oscillator*) 10. (*9*) 11. (*Always approach zero beat with same turning direction*) 12. (*To identify signal or when used as signal generator*) 13. (*At least ½ h*)(*WWV*) 14. (*Yes*)(*If they contain harmonics*)

in the 7,000- to 7,300-kHz band. It is necessary only that the amateur be assured that his transmitter is not outside these frequencies. When operating near the band limits, accurate frequency determination becomes necessary and frequency tolerance of equipment being used must be considered.

If an amateur has a frequency meter (wavemeter) that has a possible error of 1% and he wishes to operate as close to 7,000 kHz as possible, on what frequency, according to his meter, can he operate and know that he is in the band? This can be stated:

$$f_x = 7,000 \text{ kHz} + 1\% (f_x)$$

where f_x = unknown frequency

To solve for f_x:

$$f_x = 7,000 + 0.01(f_x)$$
$$f_x - 0.01(f_x) = 7,000$$
$$f_x(1 - 0.01) = 7,000$$

$$f_x = \frac{7,000}{1 - 0.01}$$

$$= \frac{7,000}{0.99} = 7,070.7 \text{ kHz}$$

To assure operation inside the band, the amateur must set his transmitter to a frequency not lower than 7,071 kHz according to his frequency meter.

An amateur is to purchase a crystal (guaranteed within 0.05% of the frequency shown on its holder) to operate as close to the 3,500-kHz end of the 3.5- to 4-MHz amateur band as possible. What is the lowest safe crystal frequency? From above:

$$f_x = 3,500 + 0.0005(f_x)$$
$$f_x - 0.0005(f_x) = 3,500$$
$$f_x(1 - 0.0005) = 3,500$$

$$f_x = \frac{3,500}{1 - 0.0005}$$

$$= \frac{3,500}{0.9995} = 3,501.7 \text{ kHz}$$

To assure operation inside the band, the lowest frequency to order would be 3,502 kHz. However,

the same crystal may operate on one frequency in one circuit but several hundred hertz higher or lower in another type of circuit. To allow for this possibility, an extra kilohertz might well be added, making 3,503 kHz the crystal frequency to order.

In place of the mathematical steps to determine the frequency, the formula for the problems given above is

$$f_x = \frac{\text{lower frequency limit}}{1 - \text{the percent}}$$
$$+ \text{ any additional safety allowances}$$

When computing the same type of problem to find what frequency crystal to order close to the *high*-frequency end of the band, the formula is

$$f_x = \frac{\text{upper frequency limit}}{1 + \text{the percent}}$$
$$- \text{ any additional safety allowances}$$

When an amateur transmitter is amplitude-modulated, allowance must be made for the sideband frequencies developed on both sides of the carrier during modulation as well as the frequency tolerance of the carrier.

EXAMPLE: If the frequency tolerance of the oscillator dial of an amateur transmitter is 0.1% and a maximum AF in the modulation system is 3,000 Hz, what is the closest frequency to the upper end of the 3.5- to 4-MHz band to which the carrier can be safely set without any sidebands appearing outside the band limits? This may be computed by the last formula above, using the 3 kHz of the sidebands as the additional safety allowance. Thus

$$f_x = \frac{4,000}{1 + 0.001} - 3$$

$$= \frac{4,000}{1.001} - 3 = 3,996 - 3 = 3,993 \text{ kHz}$$

Test your understanding; answer these checkup questions.

1. In the constant-frequency AM station monitor, why is RF taken from an unmodulated stage? _____ Is the monitor an AM or FM detector? _____
2. Where is the zero-hertz indication on the constant-

frequency monitor? _____ If the monitor is off calibration, what is adjusted to bring it back on frequency? _____

3. In a digital frequency counter what determines the accuracy of the count? _____ How is it maintained? _____

4. What does an AND gate require to make it pass a signal? _____

5. What is meant by a decimal divider? _____

6. Why is a digital counter always said to be ±1 count? _____

7. What does a Schmitt trigger circuit do? _____

8. Why can the frequency of a received signal at the second detector not be measured by a counter? _____

9. Why might a TRF receiver not give accurate frequency indications if the signal at its detector is measured by a counter? _____

10. Would a receiver using a frequency synthesizer having 1-Hz step capabilities make a good frequency-measuring device? _____

11. An amateur operating CW has a heterodyne frequency meter accurate to 0.02%. How close can he operate to 3.5 MHz according to his meter in the 3.5–4-MHz amateur band? _____ To 4 MHz? _____

COMMERCIAL LICENSE QUESTIONS

Amateur license questions will be found in the Addendum.

FCC Elements 3, 4, and 6 require an ability to answer questions similar to those below. Sections in which questions are answered are shown in parentheses. A question followed by a bracketed number is required for that element alone.

1. If a ship telephone station is assigned the frequency of 2,700 kHz and the maximum frequency tolerance is 200 ppm, what are the highest and lowest frequencies within tolerance? (21-2)

2. If an aircraft station is assigned the frequency of 3,200 kHz and the maximum tolerance is 0.01%, what are the highest and lowest frequencies within the tolerance limits? (21-2)

3. A ship A3J station has an assigned frequency of 12.5 MHz and a tolerance of ±50 ppm. If the oscillator operates at one-eighth of the output frequency, what is its maximum permitted deviation which will not exceed the tolerance? (21-2)

4. What should be the accuracy of a frequency-measuring device with respect to the desired accuracy of the frequency being measured? (21-2)

5. Draw a diagram of an absorption wavemeter with a galvanometer, explaining its operation and applications. (21-3, 21-9)

6. If a wavemeter has an error proportional to frequency and is accurate to 20 Hz when set at 1 MHz, what is its error when set at 1.5 MHz? (21-3)

7. What are the advantages and disadvantages of using an absorption-type wavemeter in comparison with other types of frequency meters? (21-3)

8. Draw a diagram of an absorption-type wavemeter. (21-3)

9. What precautions should be observed in using an absorption-type frequency meter to measure the frequency of a self-excited oscillator? (21-3)

10. Discuss lecher wires and their properties and uses. (21-4)

11. Draw a diagram of a grid-dip meter, explaining its operation and applications. (21-5)

12. With measuring equipment readily available, is it possible to measure a frequency of 10 MHz to within 1 Hz of the exact frequency? (21-6)

13. In accordance with the Commission's rules, what is the primary standard for RF measurements of radio stations in the various services? (21-6)

14. Draw a block diagram showing the stages which would illustrate the principle of operation of a secondary frequency standard and explain the functions of each stage. (21-7)

15. What determines the fundamental operating frequency of a multivibrator oscillator? (21-7)

16. What is the name of a device which derives a standard frequency of 10 kHz from a standard-frequency oscillator operating on 100 kHz? (21-7)

17. What is the meaning of *zero beat* as used in connection with frequency-measuring equipment? (21-7, 21-8)

18. What is a *multivibrator,* and what are its uses? (21-7)

19. Explain, step by step, how a secondary frequency standard could be calibrated against a WWV signal. (21-7)

20. Describe the technique used in frequency measurements employing a 100-kHz oscillator, a 10-kHz multivibrator, a heterodyne frequency meter of known accuracy, a suitable receiver, and a standard-frequency transmission. (21-8)

21. In frequency measurements using the heterodyne

zero-beat method, what is the best ratio of signal emf to heterodyne oscillator emf? (21-8)

22. Explain how the operating frequency of a transmitter can be determined by the use of a secondary frequency standard. (21-8)

23. Describe the operation of a heterodyne frequency meter. (21-9)

24. Do oscillators operating on adjacent frequencies have a tendency to synchronize oscillation or to drift apart? (21-9)

25. Describe how the unknown frequency of a transmitter could be determined by using a heterodyne frequency meter with headphones. (21-9)

26. Explain how to read the vernier on a frequency meter employing a vernier scale. (21-9)

27. If in measuring a frequency with a heterodyne frequency meter the tuning dial should show an indication between two dial-frequency relationships in the calibration book, how could the frequency value be determined? (21-9)

28. Describe how the unknown frequency of a transmitter could be determined by using a heterodyne frequency meter and a receiver. (21-9)

29. An absorption-type wavemeter indicates that the approximate frequency of a ship transmitter is 500 kHz. At the same time the transmitter signal produces a zero beat on an accurately calibrated heterodyne frequency meter at a dial reading of 370. The frequency-meter calibration book indicates dial readings of 367.0, 371.5, and 376 for frequencies of 499.4, 499.6, and 499.8 kHz, respectively. What is the frequency of the ship transmitter? (21-9) [6]

30. Draw a block diagram of a heterodyne frequency meter, including a variable-frequency oscillator, mixer detector, AF amplifier, crystal oscillator and harmonic amplifier, and AF modulator. (21-9)

31. What is meant by calibration *checkpoints* in a heterodyne frequency meter, and when should they be used? (21-9)

32. Describe how to check the crystal oscillator of a heterodyne frequency meter against WWV with a receiver. (21-9)

33. Under what conditions would it be necessary to recalibrate the crystal oscillator of a heterodyne frequency meter? (21-9)

34. What procedure should be adopted if it is found necessary to replace a tube in a heterodyne frequency meter? (21-9)

35. What precautions should be taken before using a heterodyne-type frequency meter? (21-9)

36. How can a heterodyne frequency meter be used as an RF generator? (21-9)

37. Under what conditions would the AF modulator be used in a heterodyne frequency meter? (21-9)

38. What is the purpose of using a frequency standard or service independent of the transmitter frequency monitor or control? (21-10) [4]

39. What is the reason why certain broadcast-station frequency monitors must receive their energy from an unmodulated stage of the transmitter? (21-10) [4]

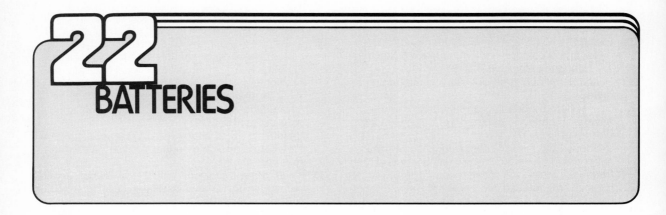

22 BATTERIES

The objective of this chapter is to outline some of the many types of primary and secondary cells used in batteries and to explain the charging methods employed with secondary cells.

22-1 A, B, AND C BATTERIES

An electric *battery* is a combination of two or more electrochemical cells. A cell is either not rechargeable and called a *primary* cell or is rechargeable and called a *secondary* cell. Cells and batteries store energy in chemical form in such a way that they can produce electric energy when required to do so.

Whenever a dc voltage is required, a cell or battery can be used as the power supply. In vacuum-tube radios, batteries used to heat a tube filament are called A batteries; those used to supply plate potential, B batteries; and those that supply grid bias voltage, C batteries. Either primary or secondary cells may be used as A, B, or C batteries.

ANSWERS TO CHECKUP QUIZ ON PAGE 545

1. (*Prevent SBs from affecting reading*)(*FM*)   2. (*Center scale*)(*Crystal; unlikely LC circuits would detune*)   3. (*Crystal frequency*)(*Checked with WWV*)   4. (*Both inputs enabled at same time*)   5. (*Divides by 10*)   6. (*Always possibility it will start counting too late in first cycle*)   7. (*Square-wave pulse output regardless of input waveshape*)   8. (*Signal at second detector is the IF*)   9. (*Received static affects count*)   10. (*Yes*)   11. (*3,500.7 kHz*)(*3,999.2 kHz*)

22-2 PRIMARY CELLS

Almost any two dissimilar metals immersed in a dilute acid or alkaline solution, as in Fig. 22-1,

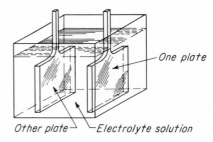

Fig. 22-1 Basic electric cell consisting of two dissimilar plates and an acidic or alkaline electrolyte.

will produce a difference of potential between them. The chemical action that takes place in such a cell produces a negative charge on one plate and a positive charge on the other. The solution into which the plates are immersed is known as an *electrolyte*. Such a cell may produce a few tenths of a volt to nearly 3 V, depending upon the metals and chemicals used. Unless the materials employed are selected properly, a continuous chemical action takes place and the cell wears itself out in a short time.

When carbon is the positive electrode or pole (cathode), zinc is the negative electrode (anode), and a solution of sal ammoniac and zinc chloride is the electrolyte, a 1.5-V cell known as a *Leclanché* cell is produced. This cell will slow or reduce its chemical action when not in use to

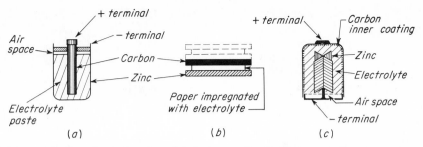

Fig. 22-2 Different types of dry-cell construction. (a) Carbon-rod cell. (b) Layer-built square-type cell. (c) Inside-out leakproof cell.

such an extent that it may retain a *shelf life* (nearly full charge) from 1 to about 3 years.

When the Leclanché cell is under load, the chemical action produces hydrogen atoms at the positive electrode. This forms a nonconductive hydrogen-gas sheath around the surface of the carbon; current flow through the cell is hindered; and the cell is said to be polarized.

The hydrogen gas can be made to combine with oxygen to form water by adding manganese dioxide, which is rich in oxygen, to the electrolyte. In this way, the cell is depolarized and a relatively heavier current can flow through it. This is the basic theory of operation of the common *dry cell,* which is actually quite damp inside.

Dry cells are made in three basic forms. The oldest form uses a cylindrical zinc can as the negative pole, as in Fig. 22-2a. A mixture of ground carbon, manganese dioxide, sal ammoniac, and zinc chloride forms the thick black paste electrolyte nearly filling the can. A carbon-rod positive pole is held in the center of the electrolyte. A layer of sealing wax seals the top of the cell. A small air space is left between the sealing wax and the electrolyte. The cell may be covered with cardboard or other insulating covering.

When several cylindrical cells of this type are packaged to form a battery, the space between cylinders is wasted. To reduce this waste, the cells may be constructed in square layers, as indicated in Fig. 22-2b.

Another form is the *inside-out* dry cell. It uses an outer cylinder with a coating of carbon on its inner surface. A connection from the carbon coating is brought to the top of the cell as the positive connection. The same electrolyte is employed. The negative plate consists of a sheet-zinc structure extending up into the electrolyte, as in Fig. 22-2c. A lead from the zinc is connected to the metal base of the cell to form the negative pole.

As the chemical action of any dry cell continues, the zinc is eaten away by chemical interaction with the electrolyte. In cells of the zinc-can type this produces holes in the can and the electrolyte oozes out, or the sides of the cell may swell. The inside-out cell has the carbon, which is not affected by the chemical reaction, on the outside and the chemically active zinc inside. Thus, the cell does not puncture and leak. Other leakproof cells use a steel outer protective jacket.

The condition of charge of a dry cell is determined by measuring the voltage across it, preferably with a load on the cell. A new cell has a voltage of 1.55 V. As a cell ages, the voltage across it decreases. A 30-cell battery should read at least 45 V. When the voltage without load drops 15%, to about 1.27 V, the cell is usually considered as being near the end of its chemical life. Under a heavy load the cell voltage drops more, however, and a cell may not be considered as nearing the end of its chemical life until it drops below 0.9 V. When a primary cell reaches the end of its chemical life, it is discarded. Recharging a zinc-carbon cell by forcing a current backward through it may depolarize it somewhat but does not actually recharge it by returning the chemicals to their original form. (This cell must not be confused with the newer completely sealed nickel-cadmium *rechargeable dry cell.*)

Batteries in receivers may produce noise when they polarize because of small variations in current through the cell.

The shelf life of a dry cell is increased by storing

the cell in a dry, cool place, usually near the floor. If a cell is allowed to heat, internal chemical reactions increase and the cell life decreases. Although dry cells are not damaged if exposed to low temperatures, they will not produce their normal current while they are extremely cold.

A large No. 6 dry cell ($2\frac{5}{8}$ by $6\frac{1}{2}$ in.) has a short-circuit current of approximately 32 A. The largest size of flashlight cell, type D, has a short-circuit current of about 6.5 A. The small penlite size produces about 4.5 A. A short-circuit test is not a true indication of the life expectancy of a dry cell and is not recommended.

22-3 THE LEAD-ACID BATTERY

The secondary cell in most general use has a lead dioxide (peroxide) (PbO_2) positive plate, a pure, spongy lead (Pb) negative plate, and an electrolyte of dilute sulfuric acid (H_2SO_4), as in Fig. 22-3. A group of these cells in series forms

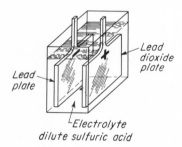

Fig. 22-3 Basic lead-acid cell consisting of a negative plate, a positive plate, the electrolyte, and a container or case.

a *lead-acid* battery. These batteries are used in automobiles and as power supplies for receivers and radio equipment of many types.

In practice, several positive plates and several negative plates are interleaved as in Fig. 22-4. Since a greater negative than positive plate area is required, there is always one more negative plate.

To prevent the positive and negative plates from contacting each other, thin wooden, glass, rubber, or plastic separator sheets (not shown) are inserted between adjacent plates. The separators are vertically grooved to allow gas bubbles that form on the plates during the charging process to float to the surface and escape from the cell.

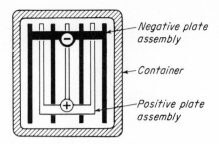

Fig. 22-4 Interleaved lead-acid cell plates in a container (top view). Separators not shown.

The plates and electrolyte are enclosed in a hard-rubber, plastic, or glass container. A screw-type cap in the top cover of each cell allows access to the electrolyte, which must be $\frac{1}{4}$ to $\frac{1}{2}$ in. above the plates at all times.

When a load is connected across a charged lead-acid cell, the chemical action that takes place internally as the cell discharges can be represented by

$$
\begin{array}{ll}
(+\text{plate}) & PbO_2 \rightarrow PbSO_4 \\
(\text{Electrolyte}) & H_2SO_4 \rightarrow H_2O \\
(-\text{plate}) & Pb \rightarrow PbSO_4
\end{array}
$$

During the time the cell is discharging, the sulfuric acid is combining chemically with both the lead dioxide and the pure lead plates, changing them to lead sulfate. When both plates have been changed to lead sulfate, they are no longer different materials, no potential difference is produced between them, and the battery is completely dead. The sulfuric acid, having combined with the plates, leaves only water as the electrolyte.

A lead-acid cell should never be allowed to become completely discharged. During discharge, lead sulfate forms as crystals on the plates, causing them to swell, which will bend, or buckle, the plates, possibly splitting the separators or even the battery case itself.

To recharge a lead-acid cell, a source of dc emf greater than that of the cell must be used. This forces a current backward through the cell, reversing the chemical action of discharging (explained above). The sulfate in both plates is chemically broken down, reappearing as sulfuric acid in the electrolyte, leaving the positive plate lead dioxide, the negative plate pure lead, and the cell again charged.

Lead-acid cells must be recharged as soon as possible after being discharged. The lead sulfate first forms small, soft crystals on the plates. If allowed to age, the crystals grow and turn hard. Soft crystals respond to recharging and easily change back into the sulfuric acid, lead, and lead dioxide. Hard crystals require a long, slow recharging to complete the chemical reversal. *Sulfated* batteries that are allowed to remain in a semicharged or discharged state for long periods may never completely recharge or may become completely destroyed.

Jarring a discharged battery knocks crystals off the plates, and they fall to the bottom of the cell. This active material is lost to the plates. An old battery will have a thick layer of this sludge in the pockets built into the case beneath the plates. If the sludge builds up to the extent that it covers the bottom edges of any two positive and negative plates, the cell becomes shorted and useless. The cell also becomes shorted if a separator cracks and the sludge can form a conducting path through the separator.

A battery becomes warm if charged or discharged rapidly. This heat is caused by the internal chemical action and the I^2R loss of the current that flows through the lead-antimony (or lead-calcium) grids that hold the active materials of the plates. Some battery manufacturers add silver and nickel to the grids to decrease their resistance and thereby allow a greater current flow and less heat loss. A battery should not be discharged or recharged at a rate that will heat it to more than 110°F. Except for its heating, a lead-acid battery can be discharged at as high a rate as the load demands.

22-4 SPECIFIC GRAVITY OF A LEAD-ACID CELL

A voltage reading of a lead-acid cell under load gives a fair indication of state of charge, but measurement of the *specific gravity* of the electrolyte is better.

The specific gravity (sp gr) of any liquid or substance is a comparison of the weight of the liquid or substance with the weight of an equal volume of water. Water has 1.000 sp gr. Chemically pure sulfuric acid has 1.835 sp gr, since it weighs 1.835 times as much as water per unit volume. The specific gravity of the electrolyte solution in a lead-acid cell ranges from 1.210 to 1.300 for new, fully charged batteries. This is about 20% acid and 80% water. The higher the specific gravity, the less internal resistance of the cell and the higher the possible output current. To start an automobile motor, 150 A at 6 V may be needed. As a result, automobile batteries have a fully charged specific gravity of 1.280 to 1.300. However, the higher the specific gravity, the more active the chemical action and the faster the battery will discharge itself. A 12-V automobile may require only 75 A to start.

When used in stationary services, as standby or emergency power supplies, or in any service where a relatively low-current drain over a long period of time is required, lead-acid batteries usually have a 1.210 to 1.220 sp gr, resulting in a longer life but a higher internal resistance and less current.

The voltage of a lead-acid cell varies somewhat with the specific gravity. For example,

Sp gr	No-load voltage
1.300	2.2
1.280	2.1
1.220	2.05

In most cases, a battery should be recharged before its specific gravity drops 100 points. A 1.280 sp gr battery should be recharged when it reaches a 1.180 sp gr.

When the specific gravity is 1.280, the freezing temperature of the electrolyte is −90°F. With a 1.180 sp gr, the freezing point is −6°F. With a 1.100 sp gr, the electrolyte will freeze at about 18°F. If the cell does freeze, it may split its case and be ruined. A charging battery will not freeze because of the internal heat developed by the charging process. While cold prevents the chemical action from occurring rapidly, it does not necessarily damage a battery.

22-5 THE HYDROMETER

Specific gravity is measured with a *hydrometer* of the syringe type, shown in Fig. 22-5, having a compressible rubber bulb at the top, a glass barrel, and a rubber hose at the bottom of the

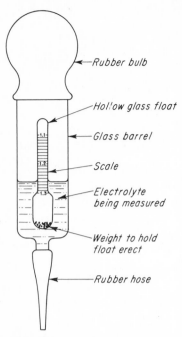

- Rubber bulb
- Hollow glass float
- Glass barrel
- Scale
- Electrolyte being measured
- Weight to hold float erect
- Rubber hose

Fig. 22-5 Hydrometer used to measure specific gravity of the electrolyte in a lead-acid cell.

barrel. A bottom-weighted, calibrated, thin, hollow glass float is inside the barrel.

To measure the specific gravity, the cap is unscrewed from the top of the cell and the hydrometer hose is dipped into the electrolyte above the plates. The bulb is compressed *slowly* and then released, drawing some of the electrolyte into the barrel. The calibrated float rides in the solution as shown. The heavier the electrolyte, the higher the float rides. The specific gravity is indicated by the highest calibration that is visible below the surface of the electrolyte (1.260 in the illustration).

Many hydrometers have a thermometer in the base of the barrel that indicates the temperature of the electrolyte being measured. Since heating the electrolyte causes it to expand, warm electrolyte is lighter per unit volume. Therefore, the warmer the solution (above 70°) the greater the correction to add to the specific gravity reading to express what it will be at the reference temperature of 70°F (3°F = sp gr change of 0.001).

If the temperature of the electrolyte is lower than 70°, the solution is more dense and the correction must be subtracted from the hydrometer indication. The proper corrections are marked on the thermometer scale.

When hydrometer readings are taken, care must be exercised to prevent drops of the electrolyte from spilling on the top of the cells or touching clothing or hands. The sulfuric acid is very corrosive and will eat holes in fabrics and burn human skin. When an acid solution is spilled on flesh or clothing, immediate flushing with water is recommended. The acid can then be counteracted by applying a weak base or alkali solution such as dilute ammonia or baking soda.

Batteries that are frequently tested for specific gravity may lose enough electrolyte by the drop or two taken out in the hydrometer each time the battery is measured to require additional sulfuric acid after a period of time. When the fully charged specfic-gravity reading drops 20 or 30 points, it may be necessary to add sulfuric acid to the electrolyte. (CAUTION: When mixing electrolyte, always pour the acid into the water. Never pour water into concentrated acid, or the heat produced by ionization will cause a corrosive steam and acid explosion.) It is rarely necessary to replace electrolyte in batteries.

22-6 WATER FOR BATTERIES

The cap on a storage cell usually has a tiny hole in it to relieve internal gas that develops, but the hole is small enough to prevent dirt and foreign objects from falling into the electrolyte. There is little evaporation of water from a battery unless it overheats. Whenever a battery is charging, the chemical action produces hydrogen gas on one plate surface and oxygen gas on the other. These gases bubble to the surface and escape through the vent hole in the cap. Thus, water (H_2O) is lost to the cell when the gases leave.

The water that escapes must be replaced to maintain the proper electrolyte level. Any impurity in the added water will combine chemically with the sulfuric acid or the plates and form a stable compound that will not enter into the charge or discharge action of the battery. The presence of impurities on a cell plate, either developed there during manufacture or added later, produces *local action* at that point. The part of the plate involved in the local action is effectively lost, and the cell decreases its capacity to some extent. To be safe, only distilled water should be added to a cell.

22-7 CAPACITY OF A BATTERY

The *capacity* of a battery is rated in ampere-hours, abbreviated Ah. The number of ampere-hours produced in an 8-h period to bring a lead-acid cell down to 1.75 V is one standard of measurement used. If the battery is forced to discharge faster than this, the number of ampere-hours produced will be somewhat less. If discharged slower, the number of ampere-hours obtainable will be considerably greater because of lessened internal heat.

The capacity of a storage battery determines how long it will operate at a given discharge rate. An 80-Ah battery must be recharged after 8 h of an average 10-A discharge. When fully charged, it may be capable of 1,000-A discharge for a short period without damage. When the electrolyte rises to more than 110°F, the discharge must be stopped.

The capacity required of a storage battery to operate an emergency radiotelegraph transmitter for 6 h under a continuous transmitter load of 70% of the key-locked demand of 40 A (with the telegraph key held down) plus a continuous emergency-light load of 1.5 A can be computed:

$$\text{Capacity} = \text{amperes} \times \text{hours}$$
$$= (0.7 \times 40 \times 6) + (1.5 \times 6)$$
$$= 168 + 9 = 177 \text{ Ah}$$

When a lead-acid cell is constructed, pink lead peroxide is molded into the grids of the positive plates. The battery is then *formed* by *cycling*, which is an alternate charging and discharging until the pink oxide has been changed to brown lead dioxide. With each cycle the ampere-hour output of the battery increases. After a user purchases the battery, its capacity may continue to increase for a few discharge-charge cycles. With sufficient cycling, all the lead oxide is finally converted to lead dioxide and the battery reaches its maximum capacity. *Dry-charge* batteries are manufactured with brown lead dioxide in the plates. To activate such cells it is necessary only to add electrolyte.

Test your understanding; answer these checkup questions.

1. Why would transistor circuits not use A batteries? _____ C batteries? _____ Do they use B batteries? _____
2. Why does touching a fork to a tooth filling sometimes produce an electric shock? _____
3. What are the basic constituents of a Leclanché cell? _____
4. What forms around the + electrode that hinders current in a Leclanché cell? _____ What is this effect called? _____
5. Why does an inside-out cell not leak? _____
6. What is the fully charged voltage of a dry cell? _____ At what voltage is the cell discharged? _____
7. What is the polarity of the outer plate of a lead-acid cell? _____ Why? _____
8. Why do lead-acid cell caps always have vent holes?
9. What is the constituent of a charged lead-acid cell's + plate? _____ − plate? _____ Electrolyte? _____
10. What is the constituent of a discharged lead-acid cell + plate? _____ − plate? _____ Electrolyte? _____
11. What causes plates to buckle? _____
12. Why should cells be recharged as soon as possible? _____
13. What are two causes of shorting in a lead-acid cell? _____ _____
14. What is affected if a lead-acid cell is overcharged? _____
15. If a lead-acid cell reads 1.220 sp gr, what should be done? _____
16. Does specific gravity read high or low on a cold day? _____ If the battery is charging? _____
17. What will counteract spilled battery acid? _____
18. What kind of water must be used in batteries? _____ What does this prevent? _____
19. In what unit is the capacity of a battery expressed? _____

22-8 CHARGING BATTERIES

When a current is forced backward through a lead-acid battery, the chemical action of discharge is reversed and the battery is recharged.

A *constant-voltage* charging circuit is shown in Fig. 22-6. Note the positive-to-positive and negative-to-negative connection of the battery and charging source. The dc generator has an output voltage between 5% and 25% greater than the full-charge voltage of the battery, depending on the rate of charge desired. With a discharged battery, the differential between battery and

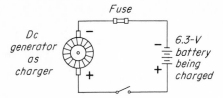

Fig. 22-6 A constant-voltage battery-charging circuit.

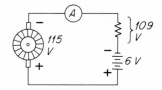

Fig. 22-7 A constant-current battery-charging circuit.

charging voltages is great and relatively high current flows. As the battery charges, the differential between battery and charging voltages is smaller and less charging current is produced.

EXAMPLE: A discharged storage battery of three cells has an open-circuit voltage of 1.8 V per cell and an internal resistance of 0.1 Ω in each of the cells. The potential needed to produce an initial charging rate of 10 A for this 5.4-V 0.3-Ω internal-resistance battery will be the voltage in excess of 5.4 that will produce 10 A through the 0.3 Ω, or

$$E_r = IR = 10(0.3) = 3 \text{ V}$$
$$E_{\text{bat}} = 5.4 \text{ V}$$
$$E_{\text{chg}} = 3 + 5.4 = 8.4 \text{ V}$$

A *constant-current* charging circuit is shown in Fig. 22-7. In this circuit the charging source has a potential several times that of the battery. The charging resistor limits the value of current that will flow. With low resistance the charging current will be high. With high resistance the charging current will be low. Since the difference of battery voltage between full charge and discharge is relatively small in comparison with the charging volt-

age, the charging current will decrease very little as the battery nears full charge. If the charging current is one-hundredth of the ampere-hour rating of the battery, it is known as a *trickle* charge. Batteries used only intermittently may be given a constant trickle charge.

EXAMPLE: If a 3-A charging rate is required, the necessary charging resistance value is computed by Ohm's law. If the battery has approximately 6 V and the source 115 V, the charging resistance must produce 115 V less 6 V, or a 109-V drop across it when 3 A flows:

$$R = \frac{E}{I} = \frac{109}{3} = 36.3 \ \Omega$$

The minimum power rating of the charging resistor will be

$$P = EI = 109(3) = 327 \text{ W}$$

A 500-W resistor would probably be used to allow a margin of safety for the resistor.

EXAMPLE: A 12.5-V battery requires a 0.5-A trickle charge. If the source voltage is 110 V, what is the value of the charging resistor?

$$110 - 12.5 = 97.5 \text{ V across resistor}$$
$$R = \frac{E}{I} = \frac{97.5}{0.5} = 195 \ \Omega$$

In an emergency, a 100-W 120-V light in series with a 6-V battery across a 120-V dc line will produce approximately a 1-A charging rate. If the polarity of a charger were reversed (or the battery were reverse-connected in the circuit), the battery would discharge very rapidly.

An electronic battery-charging circuit, as in Fig. 22-8, uses a multitapped transformer and a half-wave glass-enveloped, argon-gas, tungsten-

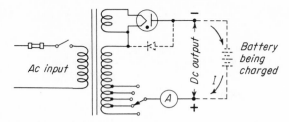

Fig. 22-8 Electronic-tube (solid-state diode in dashed lines) battery-charging circuit.

filament, carbon-plate diode tube. When the voltage differential between the ac voltage of the transformer secondary and the voltage of the battery exceeds the ionization potential of the tube, current flows through the circuit, charging the battery. Two tubes can be used in a full-wave center-tap circuit and give twice the average charging current of the half-wave circuit with the same ac voltage per tube. A capacitive filter should not be used with gaseous tubes. Diode rectifiers employed as battery chargers cannot discharge the battery if the output voltage falls below the battery voltage, since current cannot flow backward through a diode. (A dc generator used to charge a battery will operate as a motor if its output voltage drops below the battery voltage and will discharge the battery.)

Quick-charge automobile battery chargers may use a starting current of 30 to 60 A or more. These chargers employ high-current dc generators or low-resistance solid-state or gaseous tube rectifiers in a full-wave center-tap or bridge-rectifier circuit.

Different values of charging current are used with batteries. A trickle charge is usually about one-hundredth of the ampere-hour rating. Thus, a 120-Ah battery may use a 1.2-A trickle charge. A low charge may be about one-tenth of the ampere-hour rating of the battery, while a high charge may be as much as one-half of the ampere-hour rating.

As charging progresses, the charging rate must be tapered off to prevent excessive heat and electrolysis (loss of water due to electrolytic action). When the battery becomes fully charged, there is no further chemical action between the plates and electrolyte but the water of the electrolyte continues to be driven off in the form of hydrogen and oxygen gas. A constant watch should be maintained to determine the amount of gassing produced. If no hydrometer is available, an indication of probable full charge is given by rapid gassing with a low charging rate. A trickle charge of several hours is usually beneficial at the end of a rapid charge.

Frequently it is possible to operate a receiver or a transmitter from a storage battery and charge the battery at the same time. In this case, the battery acts as a voltage-regulating device. The energy to operate the equipment is coming from the charging source, but if the charging source drops off, the battery feeds energy to the equipment. This is essentially what happens in mobile radio equipment. Unshielded leads from a battery to a transmitter should be kept as short as possible to prevent pickup of RF ac. Bypassing to ground at battery and transmitter reduces such pickup.

When a constant radio watch is maintained, using batteries, it is possible to use one battery and charge another at the same time. When one becomes discharged, the other is switched in and the first is placed on charge. Figure 22-9 shows such a circuit.

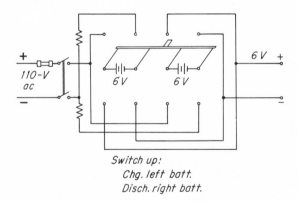

Switch up:
Chg. left batt.
Disch. right batt.

Fig. 22-9 Circuit to charge one battery while the other is being discharged.

If the polarity of a charging source is inadvertently reversed, the battery will discharge and the voltage across the battery will read less than normal, whereas the voltage across a charging battery will always be higher than its fully charged voltage.

With a constant-voltage charger, incorrect polarity across the battery will produce an extremely high current and any overload relay in the circuit will refuse to stay closed or a fuse will burn out.

If one of the cells of a battery becomes faulty, shorts out, and discharges itself, the cell then acts as a series resistance. When the battery is under a load, the voltage-drop across this cell will be found to be the reverse of its original polarity. This can be understood by referring to Fig. 22-10. The

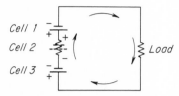

Fig. 22-10 Cell polarities in a discharging circuit with one cell dead.

dead cell is represented by a resistor. If it is remembered that current flows through a resistor from its negative terminal to its positive, the polarity will be as shown. A battery can produce very little output current with one dead cell.

A test for a dead cell is the gas produced in it by electrolysis when the battery is under a heavy load. If a copper bar is held across the terminals of a 6-V storage battery that is known to be operating improperly and one of the cells gasses profusely, that cell is faulty.

There are several methods of determining the polarity of a charging line. When a voltmeter reads correctly across a line, the terminal of the meter marked positive will be connected to the positive line. When wires from the positive and negative terminals of a source of dc are held in a glass of salt water, the wire having the greater number of bubbles developed around it by electrolysis is the negative terminal. If two copper wires are stuck into a raw potato, a blue color will develop around the positively charged wire.

22-9 MAINTAINING LEAD-ACID CELLS

There are several important points regarding the maintenance of lead-acid batteries:

1. Keep flames and sparks away from a charging or recently charged battery. The mixture of hydrogen and oxygen gases given off during charge is highly explosive.

2. Be careful when using a hydrometer. Avoid spilling acid.
3. Keep the tops of the cells clean and free from moisture to prevent leakage across the surface of the cell top and to prevent dust and dirt from falling into the electrolyte.
4. Keep the terminals of cells coated with petroleum jelly to prevent corrosion.
5. Keep cell tops on while charging to prevent electrolyte droplets from spraying out of the cell.
6. Maintain the proper electrolyte level.
7. Use only distilled water to replace lost water.
8. Take a hydrometer reading at least once a week.
9. Test-operate the battery at least once a month if possible.
10. Trickle-charge unused batteries at least one full day a month.
11. Always bring batteries up to full charge after a discharge.
12. Use only chemically pure sulfuric acid (diluted) if necessary to add new electrolyte.
13. Provide adequate ventilation while charging.
14. If a battery box is used, clean the inside surfaces once a year to remove sulfuric acid droplets that may form during charging.
15. Do not overcharge a battery, as this affects the grids.
16. Batteries may be stored several months if first fully charged and then kept refrigerated (not frozen).
17. If necessary to store for a year or more, batteries should be fully charged, the electrolyte removed, and the cells flushed with clear water and then filled with distilled water.
18. Plates and wooden separators must never be allowed to dry.
19. When removing caps, do not turn them over or place them on an unclean surface. This precaution will prevent the transportation of foreign materials into the cells when the caps are replaced.

22-10 EDISON BATTERIES

A lighter, more rugged secondary cell is the *Edison,* or *nickel-iron-alkaline* cell. When fully charged, it has a positive plate of nickel and nickel hydrate in small perforated nickel-plated steel tubes. It has a negative plate with iron as the active material in small pockets in a nickel-plated steel plate. The electrolyte is potassium hydroxide, a little lithium hydrate, and water.

While discharging, the electrolyte transfers oxygen, chemically, from the nickel-oxide positive plate to the iron negative plate, producing iron oxide (rust). When an Edison cell is recharged, the iron oxide is reduced to iron and the oxygen appears in the positive plate as a higher oxide of nickel.

The electrolyte has about 1.220 sp gr but does not change during charge and discharge, making a hydrometer useless. The voltage of the cell, measured under load, gives the best indication of state of charge. An ampere-hour meter may also be used in the circuit to determine state of charge.

The Edison cell has a fully charged voltage of approximately 1.4 V with no load. When a load is applied, the voltage decreases rapidly to about 1.3 V. As the discharge progresses, the voltage drops off more or less linearly to about 1.1 V. When the voltage reaches 1.0 or 0.9 V, the cell should be recharged.

The Edison cell can be completely discharged without damaging it. In fact, it can be recharged to the opposite polarity and will return to normal operation when charged properly. It may be stored for any length of time in a completely discharged condition without damage.

A lead-acid battery uses a highly corrosive acid electrolyte. The Edison cell has a somewhat less corrosive alkaline electrolyte. The lead-acid battery can produce high current with its low internal resistance. The Edison cell, with approximately three times the internal resistance of a lead-acid cell of the same capacity, produces less current under load.

While charging, the Edison cell also produces hydrogen and oxygen gases, and as a result the electrolyte requires replenishing with distilled water. The electrolyte should not drop below the level of the plates.

Loss of capacity of an Edison cell will result if an excessively high discharge rate is used. If the battery is trickle-charged, it will not reach full charge. If the electrolyte becomes contaminated, the capacity will be lowered. If the battery becomes hot, the capacity will be less, although the current output will be greater.

To forestall absorption of carbon dioxide from the air by the electrolyte, Edison cells have capped vents, preventing outside air from entering the cell.

22-11 OTHER CHEMICAL CELLS

Many other types of cells are available. Most of these have been developed in an attempt to improve the older Leclanché cell.

Manganese-alkaline-zinc primary cells have a manganese-dioxide cathode ($+$), a zinc anode ($-$), and a potassium hydroxide electrolyte. The nominal output is 1.5 V. These cells have low internal resistance, relatively high current capabilities, and a long shelf life, and they operate well at low temperatures. They find use in such applications as photoflash or small-motor power sources.

Mercury (or Ruben) cells have a cathode of mercuric oxide, an anode of zinc, and an electrolyte of potassium hydroxide and zincate. They are highly efficient and have a high capacity-volume ratio, a low internal resistance, a constant voltage under load, a long shelf life, and good high-temperature characteristics. They are made in 1.35- and 1.40-V no-load voltage ratings, operating at approximately 1.25 to 1.31 V under load.

Silver-oxide-alkaline-zinc primary cells have a silver-oxide cathode, a zinc anode, and a potassium (or sodium) hydroxide electrolyte. They have a flat 1.55-V characteristic under light load, are useful as a reference voltage source, and have a low internal resistance. They are used in hearing aids and, with the sodium-hydroxide electrolyte, in watches that operate continuously for a year or more.

Chargeable alkaline secondary cells are hermetically sealed units with manganese dioxide cathodes, zinc anodes, and potassium hydroxide as the electrolyte. They are relatively inexpensive and can be recharged many times. These cells have a nominal 1.5 V unloaded. They may be overcharged without damage but should not be discharged below 0.9 V. The charging rate is about $2\frac{1}{2}$ times the ampere-hours of discharge. They are called rechargeable dry cells.

Nickel-cadmium secondary cells sometimes are hermetically sealed systems with a nickel hydroxide cathode, a cadmium anode, and a potassium hydroxide electrolyte. They have an operating voltage of about 1.25 V and a low internal resistance, holding their voltage relatively constant until nearly discharged. They should be charged at 1.4 times the ampere-hours of discharge. They also are called *rechargeable dry cells*.

22-12 NONCHEMICAL CELLS

There are a number of cells that do not operate on the principle of chemical reactions and are therefore neither primary nor secondary types. When light strikes the junction between a strip of selenium (a semiconductor) and iron, as in Fig. 22-11, an emf of about 0.4 V is developed be-

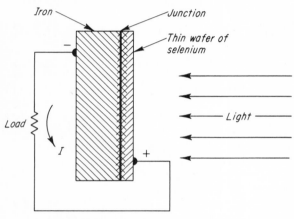

Fig. 22-11 A basic selenium or photovoltaic cell. The selenium is so thin that light energy can penetrate to the junction.

tween the two strips. This is the basis on which *photovoltaic*, or *selenium, cells* operate. When photons (light energy) strike the selenium, electron-hole (− and +) pairs are developed at the junction, resulting in a drift of electrons across to the iron. A single selenium cell may be useful as a photographic exposure meter. Many cells in series form a battery to deliver a small amount of power as long as they are illuminated, as from the sun. The efficiency is only about 1% for these cells.

One of the most practical of the *solar batteries* is made of special silicon cells in series. Each cell consists of two wafer-thin silicon strips, with the face of each infused with an opposite-polarity impurity. When the strips are pressed together,

a *PN* junction is formed. When they are illuminated by sunlight, these junctions produce an emf, as do the selenium cells, but they have an efficiency of nearly 10%. A square meter of these solar cells make up a battery that can produce more than 50 W of electric power.

Atomic, or *nuclear, cells* are somewhat similar to solar cells. A *PN* junction is bombarded by beta particles (electrons) from radioactive material painted on one side of the semiconductor wafers. Strontium-90 gives off an almost constant stream of high-velocity beta particles. In 20 years the intensity of the beta emission decreases by half. These cells produce about 0.3 V.

As previously explained, current flowing through an electrolyte chemically disassociates the hydrogen and oxygen of the water in the electrolyte. Hydrogen gas is given off at the negative electrode and oxygen gas is given off at the positive electrode. This is called electrolysis. The reverse chemical process is also possible. That is, hydrogen and oxygen gases can be made to combine, producing a current of electrons and water. This is the basis of operation of the *fuel cell*.

A basic fuel-cell system is illustrated in Fig. 22-12. Hydrogen gas under pressure is forced

Fig. 22-12 A basic fuel cell using hydrogen and oxygen gases as fuels.

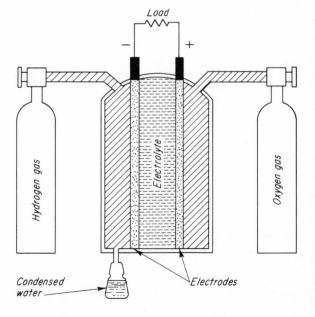

into the left chamber, which has a porous conductive electrode partitioning off the right side of its area. On the other side of this negative-pole electrode is a liquid electrolyte of potassium hydroxide. Oxygen gas under pressure is forced into the right chamber, which also has a porous conductive electrode wall between the oxygen and the electrolyte.

The potassium hydroxide electrolyte ($KOH + H_2O$) breaks up into positive potassium ions, K^+ (lacking an electron), and negative hydroxyl ions, OH^- (having one excess electron). These ions filter into the porous electrodes.

Because of the catalysts embedded in the electrodes, wherever the hydrogen gas contacts the surface of its electrode, it combines with the negative hydroxyl ion to form water, H_2O, plus one free electron (e^-).

Where the oxygen gas contacts the surface of its electrode, it combines with water and electrons returning from the load to form hydroxyl ions, OH^-, which travel into the electrolyte and over to the negative (hydrogen) electrode. This completes the electrical system. Current continues to flow as long as hydrogen and oxygen are fed as fuel to the cell. A residue of pure condensed water is produced in the hydrogen system. The catalysts in the electrodes are usually platinum and palladium. Increasing the temperature of these cells increases their current output. At about 250°C, cells must be maintained under pressure but will produce about six times as much current as they do at room temperature. Other fuels, such as alcohol, gasoline, methane, kerosene, ammonia, and hydrazine, are also used but must be operated at high temperatures.

These cells may produce approximately 1,000 A/m^2 of electrode area (100 A/ft^2) at about 0.85 V, with efficiencies of about 75%. At one-third this load value the output of the cell is about 1 V.

Test your understanding; answer these checkup questions.

1. Which has the higher voltage, a constant-voltage or a constant-current charger? _____
2. How much current produces a trickle charge? _____ Low charge? _____ High charge? _____
3. What does rapid gassing of a charging cell indicate? _____ Of a cell in a heavily loaded battery? _____
4. What is wrong if the voltage-drop across a cell under load reads reversed? _____
5. What is another name for an Edison cell? _____
6. What is used to determine the state of charge of an Edison cell? _____ _____
7. What is the fully charged voltage of an Edison cell? _____ Discharged voltage? _____
8. What can be used to counteract spilled electrolyte from an Edison cell? _____
9. Why do Edison cell caps have no vent holes? _____
10. What cell is used in wristwatches? _____
11. Name two "rechargeable dry cells." _____
12. Which two cells mentioned could be classified as solar cells? _____ _____
13. What are cells that depend on beta-particle bombardment called? _____
14. What two fuels are used in the fuel cell that was explained? _____ _____ What is the advantage of using these fuels? _____
15. List as many of the 19 points on maintaining lead-acid cells as you can. _____

COMMERCIAL LICENSE QUESTIONS

Amateur license questions will be found in the Addendum.

FCC Elements 3, 4, and 6 require an ability to answer questions similar to those below. Sections in which questions are answered are shown in parentheses. A question followed by a bracketed number is required for that element alone.

1. What form of energy is stored in lead-type storage batteries? (22-1)

2. How does a primary cell differ from a secondary cell? (22-1, 22-3)
3. What are *A, B,* and *C batteries?* (22-1) [6]
4. Describe an electrolyte. (22-2, 22-4, 22-10)
5. What are two types of radio equipment in which an electrolyte is used? (22-2)
6. What material is used in the electrodes of a common dry cell? (22-2)

7. How may a dry cell be tested to determine its condition? (22-2)
8. What is polarization of a primary cell and how may its effect be counteracted? (22-2)
9. How can the condition of charge of dry B batteries be determined? (22-2) [6]
10. What precaution should be observed in storing spare B batteries? (22-2) [6]
11. Draw a sketch showing the construction of a storage cell. (22-3)
12. What is the chemical composition of the (a) positive plate of a lead-acid cell, (b) negative plate, (c) electrolyte? (22-3, 22-4)
13. What may cause sulfation of a lead-acid storage cell? (22-3)
14. What are the effects of sulfation? (22-3)
15. What is the cause of the heat developed within a storage cell under charge or discharge conditions? (22-3)
16. What may cause the plates of a lead-acid cell to buckle? (22-3)
17. What will be the result of discharging a lead-acid storage cell at an excessively high rate? (22-3)
18. What is the approximate fully charged voltage of a lead-acid cell? (22-4)
19. Why is a low internal resistance desirable in a storage cell? (22-4)
20. Define *specific gravity* as used in reference to storage batteries. (22-4)
21. What is the effect of low temperatures upon the operation of a lead-acid battery? (22-4) [6]
22. A lead-acid storage battery has 1.120 sp gr. What should be done? (22-4) [6]
23. How is the state of charge of a lead-acid storage cell determined? (22-5)
24. What chemical will neutralize a storage-cell acid electrolyte? (22-5)
25. What special precautions should be taken when lead-acid storage cells are subject to low temperatures? (22-5) [6]
26. What should be done if the electrolyte level in a lead-acid cell is low? (22-6, 22-9) [6]

27. When and why should distilled water be added to a lead-acid cell? (22-6)
28. What is the effect of local action in a lead-acid cell and how may it be prevented? (22-6)
29. How is the capacity of a battery rated? (22-7)
30. What battery capacity is required to operate a 50-W emergency radiotelegraph transmitter for 5 h if a continuous transmitter load of 75% of the key-locked demand of 30 A is assumed? (The emergency-light load is 1 A.) (22-7) [6]
31. If the charging current through a storage battery is maintained at the normal rate but its polarity is reversed, what will result? (22-8)
32. What value of resistance should be connected in series with a 12-V battery that is to be charged at a 3-A rate from a 125-V dc line? (22-8)
33. Describe the construction and characteristics of a battery-charging rectifier tube. (22-8)
34. Why should unshielded leads from a battery to a transmitter be kept short? (22-8)
35. How may the polarity of the charging source to be used to charge a storage battery be determined? (22-8)
36. Without a hydrometer, how may the state of charge of a storage battery be determined? (22-8) [6]
37. What is indicated if, in testing a storage battery, the voltage polarity of some of the cells is found to be reversed? (22-8) [6]
38. Why does the charging rate to a storage cell, being charged from a fixed voltage source, decrease as charging progresses? (22-8) [6]
39. If the emergency batteries are placed on charge and the overload circuit breakers refuse to stay closed, what is the trouble? (22-8) [6]
40. If it were impossible to keep the receiver storage A battery charged and at the same time maintain the required watch period, what remedy might be found? (22-8) [6]
41. Draw a diagram of the charging circuit of two batteries using a four-pole double-throw switch such that while one battery is on charge the other is on discharge. Indicate dc power source, voltage-dropping resistors, and connections to the battery load. (22-8) [6]
42. Draw a diagram of a battery-charging circuit using a battery-charging rectifier tube. (22-8) [6]
43. A storage battery with a terminal voltage of 12.5 V is to be trickle-charged at a 1.5-A rate. What value of resistance should be connected in series if the source is a 120-V dc line? (22-8) [6]
44. A discharged three-cell storage battery has an open circuit voltage of 1.9 V and an internal resistance of 0.1 Ω per cell. What potential is neces-

ANSWERS TO CHECKUP QUIZ ON PAGE 559

1. (*Constant I*)   2. (*0.01 of ampere-hours*)(*0.1 of ampere-hours*)(*0.5 of ampere-hours*)   3. (*Completely charged*)(*Shorted*)   4. (*Shorted cell*)   5. (*Nickel-iron-alkaline*)   6. (*Voltmeter*)(*Ampere-hour meter*)   7. (*Usually 1.37 V*)(*1 or 0.9 V*)   8. (*Weak acid solution, vinegar, etc.*)   9. (*Prevent carbon dioxide being drawn into cell*)   10. (*Silver-oxide-alkaline-zinc is one*)   11. (*Chargeable alkaline*)(*Nickel-cadmium*)   12. (*Selenium*)(*Silicon*)   13. (*Atomic or nuclear*)   14. (*H and O*)(*Operate at lower temperature*)   15. (*See page 556*)

sary to produce an initial charging rate of 12 A? (22-8) **[6]**

45. If an auxiliary storage battery has 12.4 V on open circuit and 12.2 V when the charging switch is closed, what is the difficulty? (22-8) **[6]**

46. Describe the care which should be given a group of storage cells to maintain them in good operating condition. (22-9)

47. What steps may be taken to prevent corrosion of lead-acid storage-cell terminals? (22-9)

48. Why should adequate ventilation be provided in the room housing a large group of storage cells? (22-9)

49. Why should the tops of lead-acid storage batteries be kept clean and free from moisture? (22-9)

50. What are the differences between Edison and lead-acid types of storage batteries? (22-10) **[6]**

51. What is the chemical composition of the (*a*) positive plate, (*b*) negative plate, (*c*) electrolyte, of an Edison cell? (22-10) **[6]**

52. How is the condition of charge of an Edison cell best determined? (22-10) **[6]**

53. Why should an Edison storage battery not be charged at less than normal rate specified by the manufacturer? (22-10) **[6]**

54. What are three causes of a decrease in capacity of an Edison-type storage cell? (22-10) **[6]**

23 MOTORS AND GENERATORS

It is the objective of this chapter to outline the significant points regarding ac generators (called alternators), dc generators, dc motors, and ac motors used in communications. Servicing and maintenance suggestions are included at the end of the chapter.

23-1 ELECTRIC MACHINES

A motor converts energy of one form into mechanical rotational or twisting power, called *torque.* Examples of motors are gasoline or diesel engines, which convert the expansion of gas by heating into torque, a steam engine, which converts the expansion of hot steam into torque, and an electric motor, which converts electricity into twisting effort by the interaction of magnetic fields.

A generator converts mechanical rotational power into electric energy and may be called a *prime source* of emf. The two basic forms are the dc generator and the ac generator, or *alternator.* All generators require a *prime mover* (motor) of some type to produce the rotational effort by which a conductor can be made to cut through magnetic lines of force and produce an emf. The simplest electric machine is the alternator.

23-2 ALTERNATORS

An alternator consists basically of (1) a magnetic field, (2) one or more rotating conductors, and (3) a mechanical means of making a continuous connection to the rotating conductors (Fig. 23-1).

The magnets, called the *field poles,* produce

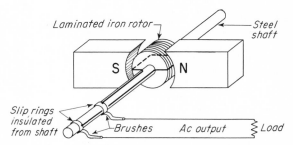

Fig. 23-1 Essential parts of a simple rotating-armature alternator.

lines of force. The laminated soft-iron rotor, also called the *armature,* presents a highly permeable path, resulting in fewer leakage lines and a greater magnetic-field strength in the gaps between the field-pole faces and the rotor. One end of the rotor conductor terminates at a brass *slip ring* on the insulated shaft to which the rotor is fixed. The other end of the conductor terminates at a second slip ring. Held against the slip rings by spring tension are two brushes made of copper, brass, or carbon.

If the shaft is rotated by a motor, the iron rotor and conductor turn. The conductor has an emf induced in it as it passes a field pole. As a wire rotates down past the north pole, the emf induced will be in one direction. As the conductor continues on upward across the face of the south pole, the same lines of force are being cut but in the opposite direction, resulting in an opposite-polarity induced emf. The continuous rotation of the conductor produces an alternating emf at the slip rings and brushes.

To produce a greater emf, the rotor may be

wound with a many-turn coil. The emf induced in each turn adds to that of all the other turns, and a higher-amplitude ac emf is produced. The number of magnetic lines of force produced by a permanent magnet is rather limited. A much more intense field can be produced by making the field poles of iron and wrapping them with several hundred turns of wire. When the field coils are excited by dc they form strong electromagnets. The greater the current flowing through the field-pole coils the greater the number of lines of force produced and the higher the output voltage of the rotor coil. Thus, a practical means of controlling the output emf of an alternator is to vary the dc excitation to the field coils. Another means of varying the output voltage is to vary the speed of rotation.

A diagram of an externally excited field alternator is shown in Fig. 23-2. The interlocked cir-

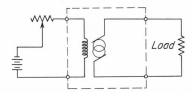

Fig. 23-2 Alternator having an externally excited field.

cles represent slip rings with brushes pressing against them. The coil represents the field windings. The rheostat varies the excitation current to the field and controls the output voltage. No prime mover is shown.

If the rotating field coil is excited with dc, as in Fig. 23-3, the rotor part becomes a strong electromagnet. In the position shown, a strong magnetic field is induced in the *stator* poles. The magnetic-field path is completed through the frame of the alternator. As the rotor moves to a point 90° from that shown, the stator poles are left without any magnetic field. Continued rotation of the rotor alternately produces and stops magnetic flux in the stator poles. The expanding and contracting alternating magnetic fields in the stator poles induce an ac emf into the turns wound around them, which forms the output of the machine.

In actual practice, the rotor is usually made into four-pole form, with opposite poles in parallel. Only one pair of slip rings is required. The station-

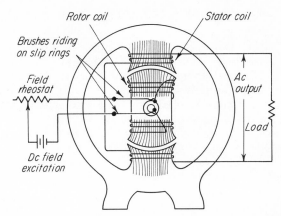

Fig. 23-3 Essential components of a rotating-field alternator.

ary part of the machine also has four poles, with all stationary armature windings in series for maximum voltage output.

A third type of ac generator is the *inductor* alternator, shown in Fig. 23-4. With the toothed

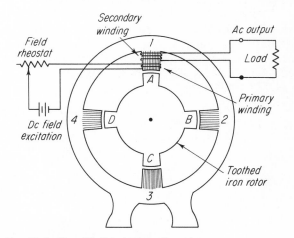

Fig. 23-4 Simplified inductor alternator.

soft-iron rotor in the position shown, the dc excitation on the stator-pole piece produces a magnetic flux that takes the path from pole 1 into tooth *A,* out of teeth *B* and *D,* and back to pole 1. In this position a maximum number of lines of force are in the iron pole piece. When the toothed rotor moves 45°, the magnetic path between pole 1 and the rotor teeth lengthens, greatly lessening the number of lines of force in the field pole. As the rotor continues to rotate, the alternately stronger and weaker magnetism in the pole induces an ac into the secondary winding.

Inductor alternators have primary and secondary windings on all pole pieces, of course. The windings are series-connected to add to the effect of the single pole illustrated.

In practical applications, inductor alternators have many pairs of pole pieces, with a tooth on the rotor for each pole piece. They are used when 500-Hz or higher-frequency ac is required.

Iron field poles of all motors and generators, both ac and dc, are laminated.

23-3 VOLTAGE OUTPUT OF AN ALTERNATOR

An alternator operates as an inductance. With a resistive load, the output voltage may be somewhat less than the no-load output, but if the load is inductive, its inductive reactance adds to the reactance of the alternator and the output voltage sags considerably. If the load is capacitive, however, the capacitive reactance counteracts the inductive reactance inherent in the alternator, and a voltage output in excess of the no-load value may result.

23-4 FIELD EXCITATION OF AN ALTERNATOR

All alternators require dc field excitation. This can be from a battery, rectified ac, or, in some cases, a dc generator attached to the same prime mover that rotates the alternator. As soon as the prime mover rotates, the generator (exciter) produces dc, which excites the alternator field coils, allowing ac to be generated.

23-5 PARALLELING ALTERNATORS

When it is necessary to connect a second alternator across a line already being fed by one alternator, either to aid the first in carrying the load or to allow the first to be removed from the line without interruption of service, the second machine must be synchronized and cut in at an instant when the *voltage* and *phase* of the two are equal.

The voltage of the second can be controlled by the field rheostat and the voltage phase compared

with the present line voltage. There are several methods by which the phase can be compared. One is by using a 240-V electric light connected as in Fig. 23-5.

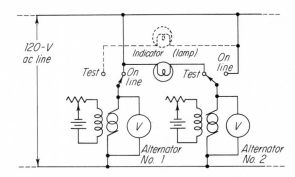

Fig. 23-5 Circuit to determine when two alternators are in phase and may be paralleled.

Alternator 1 is supplying the line. Alternator 2 is to be coupled to the line also. Alternator 2 is first brought to 120-V output. With the switch as shown, the voltage across the lamp will be something between 240 V (alternators 180° out of phase) and zero (machines in phase). The speed of the prime mover of alternator 2 must be adjusted until the light varies in intensity slowly. This indicates a slow shift of phase between the two machines. The switch is thrown to the line position while the lamp is dark (indicating that the two ac voltages are in phase). An ac voltmeter, or an oscilloscope might also be used as the indicator. Division of the load between the alternators is adjusted by the speed of the respective prime movers.

23-6 DC GENERATORS

The basic dc generator is similar to the basic alternator except for the substitution of a *commutator* for the slip rings, as shown in Fig. 23-6. When the shaft is rotated (by some type of motor), the rotor and conductor turn and ac is induced in the armature coil. With the conductors rotating, and in the position shown, a maximum emf will be produced between the brushes.

When the armature rotates to a point 90° from that shown, the conductors will be moving with the magnetic lines of force, not across them, and

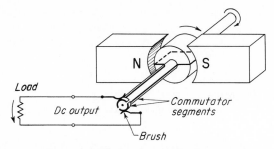

Fig. 23-6 Essential details of a simple dc generator.

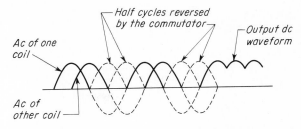

Fig. 23-8 Output pulses of a two-pole two-coil four-segment-commutator dc generator.

no voltage will be induced in the conductors. At this instant, the commutator segments have rotated, and each brush is making contact with both segments. The brushes are now shorting the segments together but at a time when zero voltage is being induced in the armature. If the armature windings and the commutator segments are not properly oriented, the brushes may short the segments when there is an emf between them, and sparking at the brushes will occur. Such sparking is one of the difficulties experienced with dc motors and generators, particularly in radio applications.

As the armature rotates another 90°, the emf induced in the conductors is again at a maximum and again in a direction out of the top segment and brush and into the bottom brush and segment.

With continual rotation of the armature, the ac emf induced in the armature conductors is made to enter and leave the brushes in the same direction at all times. The action of commutator and coil results in a full-wave-rectified output, as seen in Fig. 23-7. This is a form of mechanical rectification of the armature ac.

Fig. 23-7 Output pulses of a two-pole two-segment-commutator dc generator.

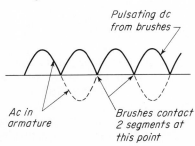

When a four-segment commutator with two armature coils wound at right angles to each other is used, the dc produced does not drop to zero at any time, as shown in Fig. 23-8. The more pairs of commutator segments and armature coils, the less variation in the output dc.

Test your understanding; answer these checkup questions.

1. What is an electrical prime mover called? _____
2. What is an alternator? _____
3. What are the names of four basic components of an alternator? _____ _____ _____ _____
4. In Fig. 23-1, if the conductor is moving down past the N pole, what is the current direction in the load? _____
5. What are two methods of increasing the output voltage of an alternator? _____ _____ Why is only one used? _____
6. Why would a rotating-field alternator be better than a rotating-armature type? _____
7. What type of alternator operates on a transformer principle? _____
8. What type of alternator is used for high-frequency-ac generation? _____
9. Why are field poles always laminated? _____
10. What kind of load tends to increase the output voltage of an alternator? _____
11. What are the three ways of exciting alternator fields? _____ _____ _____
12. In Fig. 23-5, should alternator 2 be switched "on line" when the lamp is dark or bright? _____
13. If alternator 1 ac is applied to the horizontal plates of an oscilloscope and alternator 2 ac is applied to the vertical plates, when should alternator 2 be switched on line? _____
14. What are the four basic components of a dc generator? _____ _____ _____ _____
15. Is ac or dc developed in the armature of a dc generator? _____

23-7 THE EXTERNALLY EXCITED DC GENERATOR

As with the alternator, the magnetic field of a dc generator is normally supplied by an electromagnet rather than by a permanent magnet. The diagrammatic symbol of a dc generator with a separately excited field is shown in Fig. 23-9. The

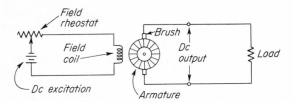

Fig. 23-9 An externally excited dc generator circuit.

external excitation is from a battery and is controlled by a field rheostat. The brushes and the multisegment commutator of the armature are indicated in symbol form.

The voltage output depends on the speed of armature rotation and the magnetic-field strength. The greater the field-coil current, the greater the output voltage of the generator, up to the point of field-pole saturation. Increasing field-coil current past field-pole saturation will not increase the output voltage.

The output voltage is also subject to an internal voltage-drop because of resistance in the windings of the armature. As the output current increases, the internal voltage-drop increases. As a result, the separately excited generator has a drooping voltage output with increase of load.

23-8 THE SERIES DC GENERATOR

When the field coils have a few turns of heavy wire and are connected in series with the arma-

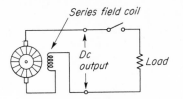

Fig. 23-10 A series-field dc generator circuit.

ture, as in Fig. 23-10, a *series generator* results. With no load it has almost no voltage output. With the load disconnected, no current flows through the field coils. The rotating conductors cut only the residual lines of force of the field poles, producing very low output voltage.

When a load is connected across the output, current begins to flow in the circuit. The field poles build up a stronger magnetism, and the voltage output of the generator increases. The heavier the load, the greater the field current and the higher the output voltage of a series generator up to the point where saturation of the field poles occurs. Because of its varying voltage output with a varying load, such a generator finds little application.

23-9 THE SHUNT DC GENERATOR

A more practical dc generator is produced by connecting the field coils across the armature, as in Fig. 23-11, to form a *shunt generator*. The

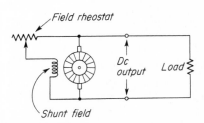

Fig. 23-11 A shunt-field dc generator circuit.

shunt field coils have many turns of fine wire to produce a maximum number of ampere-turns of magnetic force with as little drain on the output current as possible.

The rheostat in the field of the shunt generator controls the current flow through the field coils. This determines the magnetic-field strength and controls the voltage output. Increased field current produces increased voltage output.

The shunt-type generator has a drooping voltage characteristic underload due to the voltage-drop in the armature when current flows through it, which in turn decreases the voltage excitation to the shunt field.

23-10 THE COMPOUND DC GENERATOR

In some applications the load varies over a relatively wide range. For example, a radio-telegraph transmitter requires little power when the key is up and full power each time the key is pressed. To maintain a constant-voltage output from the source, neither the series generator with its increasing output voltage under load nor the shunt generator with its decreasing output voltage under load is suitable, but it is possible to use both series and shunt windings on each field pole, as in Fig. 23-12. By selection of the proper pro-

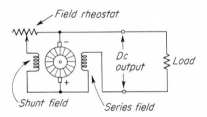

Fig. 23-12 A compound-field dc generator circuit.

portions of each winding, a *flat-compounded* generator can be produced. If the series field coils have too many turns, an *overcompounded* machine (rising voltage with increase of load) results. If the series field coils have too few turns, an *undercompounded* machine results. If the power lines carrying the current to a distant load are long, a slightly overcompounded generator may be desirable to compensate for voltage-drop across the resistance in the line. The voltage output of a compound generator is controlled by the shunt-field excitation current.

If the series- and shunt-field effects are additive, the machine is said to be *cumulatively* compounded. If the field effects oppose, a *differentially* compounded machine results, which has a sharply decreasing output voltage under load, making it unsuitable in normal applications.

23-11 THE THIRD-BRUSH GENERATOR

In automobiles a dc generator was used to charge the battery. The output voltage of such generators and therefore the charging current would vary with the speed of the engine. To prevent overcharging and overcurrent flow in the generator, the shunt field was connected to a third brush on the commutator, as in Fig. 23-13. The

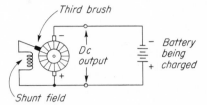

Fig. 23-13 A third-brush generator used as a battery charger.

farther down the commutator the third brush was placed the lower the field excitation and maximum current output from the generator. The charging current might be a few amperes at 20 miles per hour, 20 A at 25 mph, and 30 A at 30 mph and remain at or near 30 A at all speeds above 30 mph. These generators are rarely seen now.

23-12 COMMUTATING POLES, OR INTERPOLES

In the description of the simple dc generator (Sec. 23-6), it was mentioned that the brushes must move from one segment of the commutator to the next during a time when no voltage difference is being produced between them to prevent sparking at the brushes.

For any given armature-rotation speed and output current, the brushes can be adjusted to a *neutral point* and there will be negligible sparking. The neutral point is determined by the interaction of the magnetic field of the field poles and the magnetic field of the armature. If the output current (or the rotation speed) of the generator changes, the magnetic field of the armature shifts and the neutral position changes. To reduce sparking, the brushes must be rotated forward or backward around the commutator to the neutral point. When the load is changing rapidly, a physical shifting of the brushes is not feasible.

With no load, the mechanical neutral (position

of the brushes) and the magnetic neutral are exactly halfway between field poles. With a load, the magnetic neutral shifts a few degrees in the direction of armature rotation in a generator (backward in a motor). An electromagnet *commutating pole,* or *interpole,* can be placed between field poles at the mechanical neutral, as shown in Fig. 23-14.

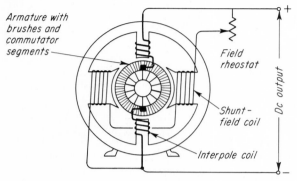

Fig. 23-14 Interpole connections on a shunt-wound dc generator (or motor).

Its winding is connected in series with the armature. When the armature (output) current increases, the interpole current increases, producing an opposite field to that of the armature. This cancels the effect of the armature field and holds the magnetic neutral at the mechanical-neutral point. The brushes can now be set at the mechanical neutral, and minimum sparking will occur at all load values. Interpoles are used in almost all larger dc generators and motors.

23-13 BRUSH SPARKING

In radio, any nearby electric sparking can produce interference to received signals. Sparking at a commutator or slip ring pits it and wears out the brushes at an abnormal rate. It is necessary to lessen all sparking as much as possible. Causes of sparking at the brushes are worn or dirty commutators, worn or improperly fitting brushes, overloading the machine, an open armature coil, a shorted interpole coil, or improper neutral positioning.

To prevent interference to local reception, bypass capacitors can be connected between brushes, or from each brush of a generator or motor to ground. A low-pass filter composed of

an RF choke in series with the ungrounded line and bypass capacitors from each side of the choke to ground may be used. A similar low-pass filter may be used in the field-coil lines. This also prevents RF ac from a transmitter from entering a generator or motor and possibly breaking down insulation in the machine.

23-14 DC MOTORS

A dc motor is the same machine as a dc generator. A series dc generator will operate as a series dc motor. A shunt generator will operate as a shunt motor, and a compound generator will operate as a compound motor.

23-15 SERIES DC MOTORS

Whereas series-wound generators are rarely used, many series-wound motors are in use.

When current is fed to a series motor, as in Fig. 23-15, the low resistance of the field and armature

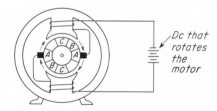

Fig. 23-15 Series motor. As the armature rotates, the commutator segments slide under the brushes.

windings allows a high current to flow through them and strong magnetic fields to form around them. The armature and the stator fields oppose each other and produce a pushing effort on the armature conductors, forcing them to turn away from the stator field poles. The twisting effort experienced by the armature is called torque.

In the figure, the brushes are in contact with the two commutator segments marked *A.* When current flows, the armature coils (not shown) connected to these segments are pushed away from the field poles, perhaps in a clockwise direction, and segments marked *B* are moved into position under the brushes. Now current flows through the *B* segments and their coil, producing further clockwise torque, moving the *C* segments under the brushes, and so on. Continuous rotation of the armature results.

If the current direction through the motor is reversed, the polarities of the field poles *and* the armature conductors are both reversed, resulting in a pushing effort between them in the same direction as before. Because of this, a series motor will rotate when excited by an alternating current, its direction of rotation being the same for both halves of each cycle. The lower the frequency of the ac, the greater the torque that can be produced. If the frequency is high, the inductive reactance of the coils limits the current that can flow and the torque of the motor.

When dc is applied to the series motor, the current flowing has nothing to oppose it except the resistance in the circuit. This results in high starting current and torque.

As the armature starts to rotate, the movement of the armature conductors through the magnetic field of the field poles induces a counter emf in the armature wires. If the counter emf could rise to the value of the source emf, the motor could no longer increase in speed. However, the counter emf can never equal the source emf in a series motor. The motor may accelerate until it flies apart by centrifugal force *if operated without a load*.

When starting a series motor, a rheostat may be included in series with it, as shown in Fig. 23-16. This adds resistance to the circuit and

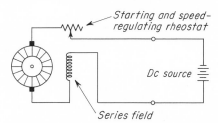

Fig. 23-16 A series dc motor circuit with a rheostat for speed regulation and starting.

reduces the starting current. As the motor picks up speed, counter emf is developed and the rheostat resistance can be lessened. The rheostat can adjust the speed of the motor to a certain extent. Small motors may require no starting resistors, but larger motors do require them.

Small series motors are used in fans, in blowers, in some less expensive electric drills, or in any application in which high-speed rotation is re-

quired but in which good speed regulation is not important. The high starting torque makes larger series-type motors useful for such heavy-duty jobs as starting streetcars. The poor speed regulation of the series motor makes it undesirable for use in a motor-generator set in radio applications.

The direction of rotation of a series motor can be reversed by reversing the direction of the current through either the armature or the field coil but not through both. Figure 23-17 illustrates a double-pole double-throw switch used as a rotation-reversing switch.

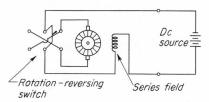

Fig. 23-17 Polarity-reversing switch in armature circuit to reverse motor rotation.

23-16 SHUNT DC MOTORS

The shunt-type motor has good speed regulation and is used in many applications.

A shunt-type motor with a three-terminal starter is shown in Fig. 23-18. To start the motor the field

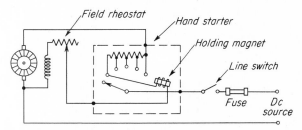

Fig. 23-18 A three-terminal (no-field release) manual starter connected to a shunt motor.

rheostat is first adjusted to zero resistance (low running speed) to produce maximum counter emf in the armature. The line switch is then closed, and the starter-rheostat arm is moved to the second contact. Current begins to flow through the armature and field, through the starter resistor and the holding magnet. Without the starting resistance the current through the armature would be excessive. The magnetic fields of the armature and the shunt fields repel each other, and the armature starts to turn. As the armature conductors move, a counter emf is induced in them. As the speed increases, the counter emf increases and the current through the armature begins to decrease. The starter-control arm is moved from contact to contact slowly enough to allow the counter emf at each point to build up as the resistance of the starter is being decreased. One or two seconds on each starting contact will prevent excessive current flow through the machine and starter resistor.

If the starter rheostat is advanced too rapidly, counter emf will not develop fast enough, current through the machine may be excessive, and the starter rheostat, armature, or fuses in the line may burn out. If the starter rheostat is moved too slowly, it may become overly heated, particularly if there is any load on the motor. In general, a motor should be started with as little load on it as possible.

When the iron rheostat arm is brought up to the full ON position, the holding magnet holds it in the running position against the tension of a spring on the arm.

When the line switch is opened and the motor is disconnected from the source, the motor becomes a dc generator, with the same polarity as it had when it was operating as a motor. As a generator, it supplies current to the holding magnet until it slows down to the point where its emf will not produce enough current to energize the magnet coil. The rheostat arm then snaps back to the OFF position by the action of the arm spring.

A shunt motor has less starting torque than a series motor but builds up a counter emf equal to the source voltage minus the voltage losses in the armature. This limits the speed to a definite value. To increase the speed of a shunt motor, the generation of the counter emf in the armature must be reduced. This can be done by decreasing the field-coil current. When the resistance in series with the shunt field is increased, the field current is less and the counter emf decreases. With less counter emf, more current flows through the armature, which picks up speed until it can again produce a counter emf almost equal to the source voltage.

If a shunt motor, running without load, has its shunt field suddenly opened or burned out, the counter emf will fall to nearly zero and the armature will continually increase its speed of rotation until the machine flies apart. If the motor were loaded, loss of the shunt-field current would result in stopping of the motor, burnout of the armature or fuse, or opening of any overload relay in the line. With the starter shown, an open field circuit de-energizes the holding magnet and the starter arm immediately snaps to the OFF position.

In larger motors, commutating poles are used to assure a constant neutral position for the brushes (not shown).

The shunt motor is usually reversed by reversing the current through the armature (and interpoles, if any).

The inductive reactance in the shunt winding limits current through this field so much that a shunt motor will not usually run on ac.

23-17 COMPOUND DC MOTORS

The compound dc motor has a series and a shunt field and, usually, interpoles. A simple schematic diagram with a motor starter is shown in Fig. 23-19.

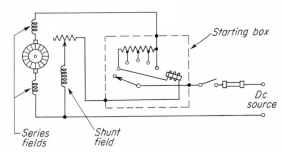

Fig. 23-19 A three-terminal manual starter connected to a compound motor.

Speed control of a compound motor is the same as in a shunt motor. A rheostat in series with the shunt field controls the counter emf generated in the armature and therefore the speed of the machine. Maximum resistance results in maximum speed. Starting and reversing direction are similar to those of a shunt motor.

It was mentioned that the differentially compounded machine produced a generator with a badly drooping voltage output under load. However, a differentially compounded motor has good speed regulation.

Test your understanding; answer these checkup questions.

1. What is the advantage of multisegment over two-segment commutators? _____
2. In Fig. 23-9, what makes the machine rotate? _____
3. What are two methods of increasing output E of a dc generator? _____ _____ Are both practical? _____
4. What is the main disadvantage of a separately excited dc generator? _____ A series generator? _____ A shunt generator? _____
5. In Fig. 23-11, is maximum output produced with maximum or minimum rheostat resistance? _____
6. Which has the larger field-coil wire, a series or a shunt generator? _____
7. What are the three types of compound dc generators? _____ _____ _____ Which would be best to feed a local load? _____ A distant load? _____
8. What do commutating poles reduce? _____
9. What value bypass capacitors do you think might be best to reduce radio noise from a dc generator? _____
10. In Fig. 23-15, how many coils would be wound on the armature? _____ Why would this not be a good dc generator? _____
11. Why might a series motor operate on dc or ac? _____
12. Why might a shunt motor not operate on ac? _____
13. If the motor in Fig. 23-18 is running unloaded, will the starter arm snap off as soon as the line switch is opened? _____ Why? _____
14. Will an unloaded series motor reach a reasonable running speed? _____ A shunt motor? _____ A compound motor? _____
15. Basically, what determines the running speed of a series motor? _____ Shunt motor? _____ Compound motor? _____

23-18 AC MOTORS

In a few cases, as where dc is the primary source of available power or where a wide range of speed control is desired, dc motors may be employed. However, most modern motors operate from an ac source. While there are many different types of ac motors, only three basic types will be discussed, the *universal,* the *synchronous,* and the *squirrel-cage* motors.

23-19 UNIVERSAL MOTORS

The series-type dc motor will rotate when dc or low-frequency ac is applied to it. Such a *universal motor* is used in fans, blowers, food mixers, portable electric drills, and other applications in which a high speed under light load or a slow speed with high torque is required.

One of the difficulties with universal motors is the radio interference or noise caused by commutator sparking. This noise may be reduced by bypassing the two brushes to the frame of the motor and grounding the frame.

23-20 SYNCHRONOUS MOTORS

A single-phase alternator works as a motor under certain circumstances. If the field is excited by dc and ac is fed to the slip rings and rotor coil, the machine will *not* start rotating. While the rotor-coil field may be alternating magnetically, during one half cycle it will try to move in one direction but during the other half cycle it will try to move in the opposite direction. The net result is no movement. The machine will only heat and may burn out.

The rotor of a two-pole alternator must make one complete rotation to produce one cycle of ac. It must rotate 60 times per second, or at 3,600 rpm, to produce 60-Hz ac. If such an alternator can be rotated at 3,600 rpm by some outside mechanical device, such as a dc motor, and the armature is then excited with a 60-Hz ac, it will continue to rotate as a synchronous motor at 3,600 rpm. As long as the load is not too heavy, a synchronous motor will run at its synchronous speed and at this speed only. If the load becomes too great, the motor will slow down, lose synchronism, and come to a halt.

One example of a synchronous motor is an electric clock. As long as the ac is maintained at the correct *frequency,* the clock keeps correct time. A precise voltage amplitude is not important.

23-21 SQUIRREL-CAGE MOTORS

Most of the motors that operate from single-phase ac have a *squirrel-cage* type of rotor. A simple form is illustrated in Fig. 23-20. A practical squirrel-cage rotor is considerably more massive than the one pictured and has a laminated iron core.

The conductors running the length of the squirrel cage are copper or aluminum and are welded to the metal end pieces. Each conductor forms a shorted turn with the conductor on the

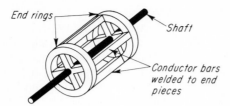

Fig. 23-20 The basic form of a rotor in a squirrel-cage type of motor.

opposite side of the cage. When this cage is between two electromagnetic field poles that are being magnetized by an alternating current, an alternating emf is induced in the shorted turns, a heavy current flows in them, and a strong counter field is produced which bucks the field that produced the current (Lenz's law). Although the rotor may buck the field of the stationary poles, there is no reason for it to move in either one direction or the other, and so it remains stationary. This is similar to the synchronous motor, which is not self-starting either. What is needed is a *rotating field* rather than an alternating one.

How the field is made to have a rotary effect names the type of squirrel-cage motor. A *split-phase* motor uses additional pairs of field poles that are fed out-of-phase currents, allowing the two sets of poles to develop maximum current and magnetic fields at slightly different times. The out-of-phase windings on the out-of-phase field poles could be fed by two-phase ac and produce a rotating magnetic field, but for single-phase operation the second phase is usually developed by connecting a capacitor (or resistor) in series with the out-of-phase winding. This can shift the phase by more than 20° and produces a maximum magnetic field in the phased winding that leads the magnetic field in the main winding. The

effectively moving maximum strength of the magnetic field passing from one pole to the next one attracts the squirrel-cage rotor with its induced currents and fields, rotating it. This makes the motor self-starting. The split-phase winding can be left in the circuit, or it can be cut out by use of a centrifugal switch that disconnects it when the motor reaches a predetermined speed. Once the motor starts rotating, it operates better without the split-phase winding. Not being a synchronous motor, it does not have to maintain a particular synchronous speed. In fact, the rotor of a split-phase induction motor always slips behind in speed by a small percent of what would be the synchronous speed. If the synchronous speed would be 1,800 rpm, the squirrel-cage rotor with a certain load may rotate at 1,750 rpm. The heavier the load on the motor, the more the rotor slips. If the rotor is constructed with two flat sides, it will maintain a synchronous speed. Under optimum operating conditions a split-phase motor with the phased poles disconnected may operate at approximately 75% efficiency.

Another method of producing a rotary field in a motor is to *shade* the field poles. This is accomplished by slotting the field poles and connecting a copper ring around one part of the slotted pole as shown in Fig. 23-21.

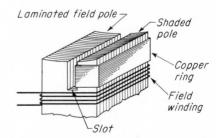

Fig. 23-21 Details of the end of a field pole of a shaded-pole ac motor.

As an alternation is increasing in amplitude in the field coil, the magnetic field expands and induces an emf and current in the copper ring. This produces a magnetic field around the ring that bucks the magnetism in the part of the pole surrounded by the ring. Maximum magnetic field is developed at this time in the unshaded part of the pole, and minimum in the shaded part. As the

---

field-current cycle reaches a maximum, the magnetic field no longer moves, the copper ring has no current induced in it, and the ring has no effect. Maximum magnetic field is now developed across the whole pole face. As the alternation decreases in amplitude, the field collapses, induces an emf and a current in the ring in the opposite direction, and produces a maximum field in the shaded part of the pole face. Thus the maximum magnetic field moves from the unshaded over to the shaded part of the field pole as the cycle progresses. This maximum-field movement produces the necessary rotating field in the motor to make the squirrel-cage rotor self-starting. The efficiency of shaded-pole induction motors is not high, ranging from 30 to 50%.

One of the main advantages of all squirrel-cage motors, particularly in radio applications, is the lack of commutator and brushes, or slip rings and brushes, which results in interference-free operation.

23-22 POLYPHASE MOTORS

When three-phase ac (Sec. 11-32) is available, a three-phase motor is usually desirable. Each of the three phases is fed to separate field coils wound around field poles in the motor. Since each phase is 120° from all other phases, current maximums appear in one field pole, then 120° later in the next pole, 120° later in the third pole, and so on. This action produces a true revolving field. All such motors are self-starting, are quite efficient, are usually squirrel-cage rotor types, and operate at a nearly constant speed, although they have some slip.

At the instant of starting, a very high current flows. Either series resistances must be used in two of the three power lines feeding the motor to limit the starting current or the line voltage must be stepped down until the machine approaches running speed.

When the machine is rotating at normal speed, the induced currents and fields in the rotor occur at such a time and in such phase that they do not cancel the field-coil inductance. With effective inductance in the field coils, their reactance limits the line current to a low value. When a load is applied, the rotor begins to fall behind the rotating fields. This increases the current induced in the rotor, which in turn reduces the inductance (and X_L) of the field windings, demanding more line current. The result is greater torque and horsepower from the motor as it tries to maintain a nearly synchronous speed.

A two-pole three-phase motor (two poles per phase) operating on 60-Hz ac will rotate at a speed of 3,600 rpm; a four-pole motor will rotate at 1,800 rpm; and so on.

23-23 MOTOR-GENERATORS

When it is desired to convert one form of electricity to another, a motor and generator are often coupled together. The motor may be either ac or dc, and the generator may be either ac or dc. For example, 1,500-V dc may be required, but the power lines carry 120-V ac. A 120-V ac motor coupled to a 1,500-V dc generator will provide the necessary high-voltage dc. A diagram of such a motor-generator set is shown schematically in Fig. 23-22.

A motor-generator set is quite heavy and bulky and makes a certain amount of audible noise. It is rather costly to purchase, although maintenance is not great. Being partly mechanical, it requires lubrication; commutator or slip rings need cleaning; and brushes must be replaced

Fig. 23-22 An ac motor and dc generator system circuit diagram.

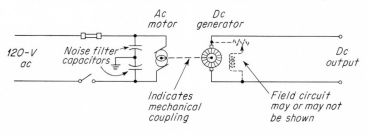

periodically. The set has fairly good voltage regulation, although the output voltage may be somewhat limited because of difficulty of insulating the armature and commutator components in dc machines. Sparking at brushes may cause radio interference. In comparison, a vacuum-tube or solid-state rectifier is lighter and less expensive, has almost unlimited voltage possibilities, operates without noise or vibration, is more efficient, and produces no radio interference. It may have poorer voltage regulation and require new tubes periodically, but in most cases it may be operated in any position.

23-24 DYNAMOTORS

A *dynamotor* is a form of a dc motor-generator but with a common field coil for both motor and generator. Dynamotors are normally used for low-power radio equipment, as in mobile stations. The motor part normally operates from a 6- to 28-V battery. The output dc voltage ranges from about 225 V when the dynamotor is used for a receiver to 1,000 V or more when it is used for transmitters.

Advantages of a dynamotor are light weight and compactness when compared with a motor-generator, made possible by a common field winding. One end of the common armature core carries the commutator segments for the motor. The other end carries the commutator segments for the generator. The two sets of armature coils are separate but are wound into the same slots in the armature. Figure 23-23 shows a schematic and pictorial diagram of a dynamotor.

The output voltage of a dynamotor cannot be changed efficiently. If the primary source voltage is changed, the output will change, but best efficiency and voltage regulation are produced with a specific input voltage. The overall efficiency ranges from 50 to 60%, somewhat less than that of a motor-generator. The voltage regulation of most dynamotors is fair but does not compare with that of a properly designed motor-generator. The dynamotor's high output power for its weight is produced by its high speed-of-rotation characteristic.

In general, the dc-to-dc application of dynamotors has been superseded by use of the tran-

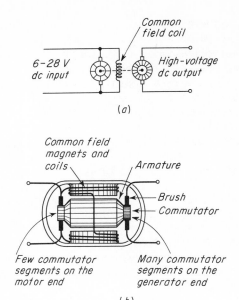

Fig. 23-23 (*a*) Schematic diagram of a dynamotor. (*b*) Component placement in the device.

sistor power oscillator, whose output ac is stepped up, rectified, filtered, and often regulated electronically.

23-25 RATING GENERATORS AND MOTORS

Dc generators are rated by watts output, but when an alternator feeds power to a circuit, it is not known what power factor will be involved. If the load is completely resistive (*E* and *I* in phase), the power output will be equal to $E \times I$ and will be the true power. Should the load be completely reactive (*E* and *I* out of phase by 90°), the current drawn from the machine may be the maximum that the wires of the alternator can stand without overheating but the true power delivered to and used by the load is zero ($EI \cos \theta = E \times I \times 0$) watts. Current is flowing and producing a field, but the field collapses and returns the energy to the circuit. The wires are carrying current, but no power is being lost. In such a case, an alternator normally thought of as capable of 2-kW output when used on electric lights alone will be delivering no true power to a reactive load but may still be running at a maximum current value. For this reason, alternators are always rated in apparent power (volt-amperes, or VA). Large alternators are rated in kilovolt-amperes, or kVA.

It is customary to rate motors in output horse-power: 746 W is equivalent to 1 hp. A 3-hp dc motor delivers the equivalent of 2,238 W of turn-ing power. However, the machine is not 100% efficient and therefore must be drawing more than this number of watts from the line. If the machine is 85% efficient, 2,238 W must be 85% of the input power, or

$$0.85(P_i) = 2,238 \text{ W}$$

$$P_i = \frac{2,238}{0.85} = 2,633 \text{ W}$$

From this, the formula to determine input power is

$$P_i = \frac{P_o}{\%}$$

where P_i = input power in W
P_o = output power in W
$\%$ = percent of efficiency
expressed as decimal

If the motor above is operating from a 110-V dc line, from the power formula $P = EI$ the line current would be

$$I = \frac{P}{E} = \frac{2,633}{110} = 23.9 \text{ A}$$

Alternating-current motors are also rated in horsepower, but besides their efficiency the power factor must be considered.

A 7-hp ac motor operating at full load, with a power factor of 0.8 and 95% efficiency, will have what power input? Power factor can be consid-ered a type of ac efficiency rating. The power output developed is equivalent to 7(746), or 5,222 W. The input power is 5,222/0.8(0.95), or 6,871 W. If the motor is operating from a 120-V line, the line current is $I = P/E$, or 6,871/120, or 57.3 A.

23-26 MAINTENANCE OF MOTORS AND GENERATORS

It is usually an assignment of an operator at a radio installation to see that the motors and gen-erators are in proper operating condition at all times. The following are some suggestions re-garding such maintenance.

Oil on rubber insulation causes the insulation to soften and weaken. Care must be exercised when oiling motors and generators that no oil is allowed to drip onto wiring.

The bearings of rotary machines must be oiled periodically to prevent overheating and freezing. If a bearing overheats because of lack of oil, it should be flushed, while still running, with light oil until cool.

Sparking at the brushes indicates trouble. The brushes may be worn and be making poor con-tact. The slip rings or commutator may be dirty. An armature coil may be shorted or open. The machine may be running with too heavy a load. The interpoles may be inoperative or incorrectly wired.

If slip rings or commutators become dirty, they can be cleaned with a piece of heavy canvas and a liquid solvent, or they may be sanded smooth with a very fine grade of sandpaper. Emery paper and steel wool must never be used to smooth a commutator. Metal particles from them may lodge between segments. The mica insulation between commutator segments may wear more slowly than the copper or brass segments, leaving the brushes to ride on insulated ridges instead of on the segments. The mica insulation should be *undercut* below the level of the surface of the commutator segments.

Motors and generators should be cleaned and dusted regularly and given a close visual exami-nation. The length of the brushes should be checked. If worn, replace with brushes of the correct size and type. In an emergency, a larger brush may be sanded down to fit the machine. The spring tension holding the brush against the commutator should produce adequate pressure.

When the commutator segments show signs of excessive wear, the armature may have to be taken out of the machine and the commutator turned down on a lathe. The mica insulation must be undercut before turning the commutator. After turning a commutator, remove all metal particles that may be deposited in the mica insulation between segments. If an armature coil becomes shorted, the armature will heat and the brushes will spark.

If a motor or motor-generator switch is closed but the machine does not start, the operator should check the line voltage, the fuses, and the overload relay. If the overload relay will not hold in when the motor is switched on, the bearings of the machine may be frozen, the armature shaft may be locked, the armature may have a shorted turn, or the field-coil circuit may be open or burned out. If the fuses or overload relays are not open and the line voltage is normal, a brush may not be making contact with the commutator or slip ring because of dirt under the brush or improperly undercut segments. There may be a poor electrical connection leading into the motor.

Test your understanding; answer these checkup questions.

1. What are the three basic ac motor types? _____ _____ _____

2. What is a universal motor? _____
3. If an alternator is to be used as a synchronous ac motor, what must be fed to it and under what conditions? _____ _____ _____
4. What is the synchronous speed of a two-pole alternator being fed 50-Hz ac? _____
5. Will an electric clock keep accurate time if the line voltage drops 20%? _____ Rises 20%? _____
6. How is a squirrel-cage motor made synchronous? _____ What is it said to have if not synchronous? _____
7. What are the two types of squirrel-cage motors discussed? _____ _____
8. Why do squirrel-cage motors produce no radio noise?
9. How many field poles does a 2-pole 1-ϕ motor have? _____ A 2-pole 3-ϕ motor? _____
10. Are 3-ϕ motors normally synchronous? _____
11. How does a dynamotor differ from a motor-generator?
12. In what units are large alternators rated? _____ Dc generators? _____
13. Does a 1-hp dc motor use more, less, or exactly 746 W? _____ Why? _____
14. A 2-hp ac motor when operating has a pf of 0.9 and is 80% efficient. What is the power input? _____ Power output? _____
15. Why is emery cloth not used to clean electric machines? _____
16. What is it called when the mica insulation between commutator segments is reduced in height? _____

COMMERCIAL LICENSE QUESTIONS

Amateur license questions will be found in the Addendum.

FCC Elements 3, 4, and 6 require an ability to answer questions similar to those below. Sections in which questions are answered are shown in parentheses. A question followed by a bracketed number is required for that element alone.

1. How may the output voltage of a separately excited ac generator, at constant output frequency, be varied? (23-2)
2. What conditions must be met before two ac generators can be operated in parallel? (23-5)   [6]
3. What is the purpose of a commutator on (*a*) a dc generator, (*b*) a dc motor? (23-6, 23-15)
4. Why is laminated iron generally used in the construction of the field and armature cores of motors and generators instead of solid metal? (23-6, 23-15)
5. Describe the characteristics of a series dc generator. (23-8)

6. Explain the principle of operation and the operating characteristics of dc shunt and compound generators. Draw diagrams of each and show how voltage is controlled. (23-9, 23-10)
7. What are *commutating poles* (*interpoles*) in a dc motor? (23-12)
8. What are four causes of excessive sparking at the brushes of a dc motor or generator? (23-13)
9. Why are bypass capacitors often connected across the brushes of a high-voltage dc generator? (23-13)
10. Why are RF chokes sometimes placed in the power leads between a motor-generator power supply and a high-powered radio transmitter? (23-13, 23-23)
11. How may RF interference, often caused by sparking at the brushes of a high-voltage generator, be minimized? (23-13)
12. Describe the action and characteristics of a series dc motor. (23-15)

13. What is the danger of operating a dc series motor without a load? (23-15)
14. Why is it sometimes necessary to use a starting resistance when starting a dc motor? (23-15)
15. What determines the speed of (a) a dc series motor, (b) a synchronous motor, (c) an induction motor? (23-15, 23-20, 23-21)
16. Why is a series motor not used in radio power-supply motor-generators? (23-15)   [6]
17. Draw a diagram of three kinds of dc motors, including a starting device. (23-15–23-17)
18. Describe the action and characteristics of a shunt-type dc motor. (23-16)
19. Draw a diagram of a shunt-wound dc motor. (23-16)
20. What is meant by counter emf in a dc motor? (23-16)
21. If the field of a shunt-wound dc motor were opened while the machine was running under no load, what would be the probable result? (23-16)
22. When a large dc motor-generator set is started, what adjustment should be made to the motor field rheostat? (23-16)   [6]
23. What might be the result of starting a motor too slowly when using a hand starter? (23-16)
24. Explain the principle of operation and characteristics of a compound-wound dc motor and show how speed is controlled. (23-17)
25. Explain the principles of induction motors and how they are started. (23-21)   [6]
26. What is the approximate speed of a 220-V, 50-Hz, 4-pole, 3-φ induction motor? (23-22)
27. To increase output voltage of a motor-generator, what is the usual procedure? (23-23)   [6]

28. List the comparative advantages and disadvantages of motor-generator and transformer-rectifier power supplies. (23-23)
29. Describe construction and characteristics of a dynamotor. (23-24)
30. How may the output voltage of a dynamotor be regulated? (23-24)
31. What are the advantages in the use of a dynamotor, rather than a motor-generator, to furnish power to a mobile transmitter? (23-24)
32. In what units is an alternator output ordinarily rated? (23-25)   [6]
33. What is power factor? Give an example of its use. (23-25)
34. What is the line current of a 1-φ 5-hp ac motor when operating from a 120-V line at full rated load and at 0.9 pf and 85% efficiency? (23-25)   [6]
35. What is the effect of an inductive load on the output voltage of an alternator? (23-25)   [6]
36. If a 6-hp 110-V dc motor is 80% efficient when developing its rated output, what will be the line current? (23-25)   [6]
37. What materials should be used to clean the commutator of a motor or generator? (23-26)
38. Why should emery cloth never be used to clean a commutator? (23-26)
39. What will be the effects of a short circuit in an armature coil of a dc motor? (23-26)   [6]
40. What may be the trouble if a motor-generator fails to start when the start button is depressed? (23-26)   [6]
41. What may cause a motor-generator bearing to overheat? (23-26)

24

BROADCAST STATIONS

The objective of this chapter is to outline some of the definitions and requirements relating to transmitters and operators of AM broadcast stations, particularly for FCC Elements 4 and 9. Some FM broadcast station information that is similar to AM broadcast stations is also included. Technical theory and circuits that apply to both AM and FM broadcast stations are discussed in preceding chapters such as those on AF amplifiers, basic transmitters, amplitude modulation, frequency modulation, and antennas.

24-1 STANDARD BROADCAST STATIONS

Standard broadcast stations operate in the standard broadcast band, which is between 535 and 1,605 kHz. This band is divided into 106 *channels* of 10 kHz each. The center frequency of each channel, beginning at 540 kHz, is assigned to one or more stations in the country as its assigned, or carrier, frequency. Broadcast transmitters must operate on the assigned frequency with a tolerance of ±20 Hz. (Given an

assigned carrier frequency of 1,260 kHz, a station carrier must operate between 1,259,980 and 1,260,020 Hz.) The type of emission is always amplitude modulation (A3) with an assigned unmodulated operating carrier power between 100 and 50,000 W, depending on the geographical area the station is expected to serve. Since the channels are only 10 kHz wide, transmission of audio frequencies in excess of 5 kHz produces sidebands in adjacent channels and may interfere with reception of stations assigned to these channels.

The service area of a broadcast station is described as *primary* if there is no fading of the signal, *secondary* if there is fading but no objectionable interference, and *intermittent* if the signal is subject to some interference and fading.

For broadcasting there are three parts of each day: the experimental period, from midnight to local sunrise; daytime, from local sunrise to local sunset; and nighttime, between local sunset and local sunrise. Some stations broadcast in daytime only; others, in nighttime only; and still others, for the full broadcast day.

Standard broadcast stations may be located in the central portion of cities, but the tendency is to locate the transmitter and antenna adjacent to populated areas in order that reception may be good but not so strong that local receivers are overdriven. The transmitters and their antennas are often located on marshy land to take advantage of the excellent ground-wave signals such an area provides or on a slight rise overlooking a populated area. However, VHF FM broadcast sta-

---

ANSWERS TO CHECKUP QUIZ ON PAGE 576

1. (*Universal, synchronous, squirrel-cage*)   2. (*Series dc motor that runs on ac or dc*)   3. (*Ac, dc, at synchronous speed*)   4. (*3,000 rpm*)   5. (*Yes; rotation determined by frequency*)(*Yes*)   6. (*Two flat sides of rotor*)(*Slip*)   7. (*Split-phase, shaded-pole*)   8. (*No commutators or slip rings*)   9. (*2*)(*6*)   10. (*No*)   11. (*Single field*)   12. (*E, frequency, φ, kVA, speed*)(*E, I, W, hp*)   13. (*More*)(*Hp delivered, not input*)   14. (*2,072 W*)(*1,492 W*)   15. (*Metal particles in it*)   16. (*Undercutting*)

---

tions (88 to 108 MHz) and VHF and UHF TV stations (54 to 890 MHz) are usually located atop the highest nearby peak for best line-of-sight transmission.

24-2 COMPONENTS OF A BROADCAST SYSTEM

Many of the circuits of equipment used in standard broadcast stations have been discussed in earlier chapters. A block diagram of a possible broadcast transmitter and associated equipment is shown in Fig. 24-1. The complete system includes the main transmitter with small local studios and facilities, a radio link (studio-transmitter link, or STL) from remote downtown main-studio facilities, and leased lines from transmitter and main studios either for telephone communication or for transmission of programs if the STL equipment is not used. A remote-pickup transmitter capable of transmitting an athletic event or other program may also be used from a remote location to a receiver at the main transmitter (or to the main studio).

The main transmitter consists of a crystal-controlled oscillator, one or more RF buffer amplifiers, a driver amplifier, and a radio-frequency final or power amplifier, plate-modulated for high efficiency, feeding the antenna. The program to be transmitted may originate in the local studios or be played from the local magnetic tape recorders or from the local turntables (TT). Announcements may be made in the local studio, in the announce booth, or from the operator's microphone. The operator sitting at the *console,* or control board, can switch in any of these circuits as desired; and with the *gain controls,* also known as *attenuators, faders,* or *pots* (potentiometers), he can vary the amplitude of the program to keep the highest peaks of modulation above the required 85% and below 100% on negative peaks. He is aided in this by the limiter (Sec. 24-7).

A monitor receiver audibly indicates the operation of the transmitter. The required modulation monitor and, possibly, an oscilloscope give a constant visual indication of the transmitted signal. A required frequency monitor indicates, in hertz, any deviation of the carrier from the assigned frequency.

This station has an *auxiliary transmitter* that is maintained for transmitting the regular programs of the station in case of failure of the main transmitter or for emergency broadcasts. There may be auxiliary power equipment such as a gasoline engine driving an alternator to supply power in case the public utility lines fail. An auxiliary transmitter may operate with the same power as the main transmitter (or less), on the same frequency, and with the same technical requirements of operation. It must be tested at least once a week. This is not the same as an *alternate transmitter,* which is a duplicate of the main transmitter, used by stations operating on a 24-h schedule to allow servicing and maintenance without interruption of programs.

Most stations have a workshop or equipment-repair room where microphones, amplifiers, etc., can be serviced by the operators.

If programs originate downtown, the studios may be considerably more elaborate than those at the transmitter. The largest studio (A) may have a special control board of its own at which an operator mixes several studio microphones into the line feeding the program to the master studio control board. The operator at the studio control board may do the major mixing and program switching while the operator at the transmitter makes sure that the output of the STL receiver or telephone line is producing the desired percent of modulation of the transmitter.

To produce a disk-jockey record program properly it is necessary to have three turntables, three tape machines, or both, and a microphone.

It is not unusual to have a program originating in studio A being broadcast while another program in studio B is being tape- or disk-recorded for replay at a later time. Perhaps a program originating at some other city is also being taped for later use. The control-board operators must be familiar with their equipment in order not to switch the wrong program on the air at any time. In smaller stations the radio station operator may also spin the records and make announcements. He is known as a combination, or *combo,* operator.

When remote programs, football, races, etc., are broadcast, leased lines may be used or a remote-pickup base station (possibly having mo-

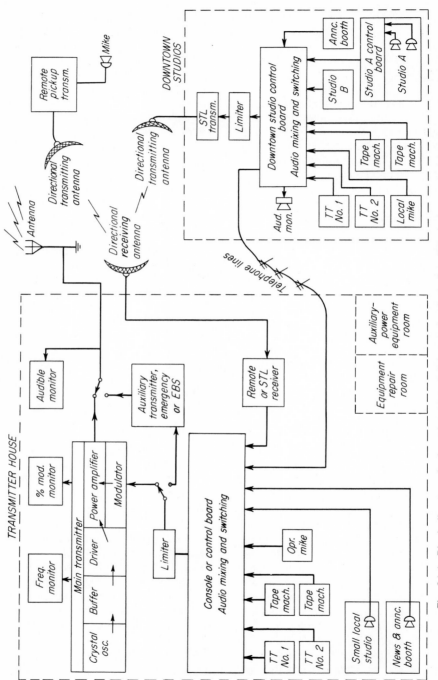

Fig. 24-1 Block diagram of a broadcast transmitting station including STL and remote-pickup transmitter.

bile units operating with it) may be established to communicate back to the main transmitter. These stations use frequencies from 1,606 kHz to 456 MHz in various bands, employing A3 or F3, with usually less than 200 W, and frequency tolerances of the base stations of 0.01 to 0.005% and mobiles of 0.02 to 0.005%.

STL (studio-transmitter link) fixed, FM stations operate in the 947- to 952-MHz range, employ directive (usually parabolic) antennas, have a frequency tolerance of 0.005%, and use only sufficient power output to assure satisfactory service. *Intercity relay* stations may be used to link two or more broadcast stations and operate with the same standards as STL stations. Licensed operators (any commercial license) are required to set up and tune all these stations. Permit holders may operate them.

24-3 THE BROADCAST CONSOLE

The nerve center of the broadcast station is its console, or control board. It is a complicated series of audio amplifiers, switches, relays, and gain controls. The operator switches the desired studio, turntable, microphone, or remote program through the console into the limiter and modulator stage of the transmitter.

The block diagram of Fig. 24-2 illustrates a simplified console circuit that will demonstrate some of the requirements of broadcast operating. Included are the console, microphone, turntable, and speaker for the console operator and a studio with a microphone and loudspeaker.

For a program originating in the nearby studio, the studio microphone switch S_1 is thrown to the program, or P, position. The output of the microphone, through its preamplifier and gain control G_1, is connected to the input of the program amplifier. The output of the program amplifier is fed through the master gain control to the transmitter. At the same time the console operator hears the signal output from the program amplifier, via the monitor amplifier, provided his local microphone switch S_2 is in the central or OFF position. The operator also has a visual indication of the signal strength by the VU meter in the output line of the program amplifier. Note that when S_1 is in the P position, the studio speaker is automatically disconnected from the monitor

Fig. 24-2 Functional block diagram of a simple broadcast console and studio.

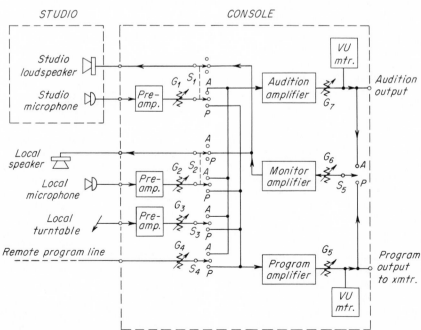

amplifier output, preventing an audio feedback howl from being produced in the studio.

While the studio is on the air (being fed into the program line), the operator can be recording an announcement or program to be used later by switching his microphone, local turntable, or the remote-line switches to the audition, or A, position. This feeds his program into the audition amplifier, the output of which can be fed to a tape- or disk-recording machine. The local speaker and microphone are interconnected by S_2 so that the speaker is disconnected at any time the microphone is in use, preventing a feedback howl.

When the studio program has been completed, the operator switches off the studio circuit S_1 and, by throwing his local microphone switch S_2 to P, can make a *station break* (station identification) or any other desired announcement. In many stations there is an announce booth (not shown), in which a staff announcer may make the station break.

If the next program is coming in from a distant network station, the operator usually checks the remote-line signal strength a few minutes before the program is scheduled to start by throwing S_4 and S_5 to the A position. This feeds the remote signal through the audition and monitor amplifier and to his local speaker. When it is time for the remote program to start, S_4 is connected to its program position and the program is fed through the program amplifier to the transmitter.

On many occasions it is necessary to provide *talk-back* facilities between the console operator and the people in the studio before they go on the air. This can be done with the audition channel of this console. The operator first switches S_5 to the audition position. When he throws his local-microphone switch to audition, he can be heard in the studio. When he throws his microphone switch to the central or OFF position and the studio microphone to the audition position, he can hear anyone speaking in the studio. Talk-back may be used to notify a studio that it is about to be switched on the air. On-the-air is usually indicated by a red light glowing in the studio.

When the program is originating at the console, as in a disk-jockey show, the turntable (or tape machine) playing the announced record is in the program position. The next record is set up on another turntable (not shown), and its output is fed into the audition channel. By switching the monitor amplifier switch S_5 to the A position, the operator can hear the output of the second turntable. He sets the pickup needle in the outer groove of the record and then spins the turntable by hand until he hears the modulation on the record. By backing the turntable past the first sign of modulation he has *cued* the record, and it is ready to be played. (Tapes can be manually cued, or they can be automatically cued by special cuing tones recorded on them.) This turntable is switched to the program line. A fraction of a second after the motor switch for the turntable has been turned on, the music of the record will begin. In this way there are no long pauses after the announcement of the record as the needle moves over the first unmodulated grooves.

The console described here is far simpler than those in actual use, although its basic functions are similar. With this console, the studio is fed the output of the program amplifier, allowing those in the studio to hear what is being broadcast at all times when their studio is not on the air. When the operator is cuing a record, however, the studio is fed the sounds of the cuing rather than the program being broadcast. To prevent this a separate cuing amplifier could be used.

The attenuators G_1, G_2, G_3, and G_4 may be termed *channel* gain controls. Attenuator G_5 is known as the *program master* gain since it controls any console signal. The *audition master* is G_7. The monitor gain is G_6.

24-4 AUDIO LEVELS

There are three important requirements during a broadcast. (1) Timing. The programs must start and stop when they should. Announcements and station breaks must be squeezed into the time allotted for them. (2) Fidelity. All amplifiers must be operating in such a manner that they do not add distortion to the program. This means constant maintenance and testing of the console and associated circuits. The input and output impedance of the lines of all equipment must be properly matched and terminated. (3) Maintaining the proper audio signal amplitude that modulates the transmitter.

Audio levels are usually expressed in decibels (dB), volume units (VU), or dBm (Sec. 8-5). In broadcast work, zero dB or VU represents a reference level of 0.001 W in a 600-Ω line. The abbreviation dBm indicates decibels using a reference of 1 mW, which is, for most purposes, the same as VU. In any of these, a gain of 10 dB, VU, or dBm equals a gain of 10 times the power, or 3.16 times the voltage. A gain of 20 dB equals 100 times the power, or 10 times the voltage.

With only these two facts in mind,

10 dB = power change of 10 times

20 dB = voltage change of 10 times

the following problems can be solved easily:

If the power output of a modulator is decreased from 1,000 to 10 W, how is the power reduction expressed in decibels? A loss of 10 dB reduces the power to 100 W; another 10 dB (−20 dB) reduces the power to 10 W. Therefore, the reduction from 1,000 to 10 W equals a 20-dB loss of power.

An AF amplifier has an overall gain of 40 dB, and the output is 6 W. What is the input power? Starting with the output of 6 W and working back, 10 dB gives a power of 0.6 W. Another 10-dB loss (−20 dB) gives a power of 0.06 W. Another 10-dB loss (−30 dB) gives a power of 0.006 W. Another 10-dB loss (−40 dB) gives a power of 0.0006 W. If 0.0006 W is fed into the amplifier, the output will be 6 W (proper input and output impedances are assumed).

The output of a voltage amplifier is −40 dB. This is fed into a mixer with a loss of 10 dB. How much voltage amplification is required to produce a +10-dB output power? The total decibel loss is −40 and −10, or −50 dB. The total number of decibels between −50 and +10 is 60 dB. Since 20 dB equals a voltage gain of 10 times, 40 dB equals a gain of 10 × 10, or 100 times the voltage, and 60 dB equals a gain of 10 × 10 × 10, or 1,000 times the voltage.

A microphone feeding into a console may have a rating of −85 dBm with a given value of sound striking it. Therefore, the console must be capable of amplifying the signal at least 85 dB to produce a zero-VU-level output. Weaker signals striking the microphone will require more gain in the am-

plifiers, possibly 100 dB. At a normal setting, there is usually a loss of at least 6 to 10 dB in the gain control of a microphone channel amplifier.

Another input, a turntable, may have an output signal of −45 dBm and need an amplifier of only 45-dB gain to bring its output up to the required zero VU used to feed the transmitter. Between these two input signals there is a difference of about 40 dB, or of two vacuum-tube or transistor amplifier stages.

A third input signal, a remote program, may be coming in at a zero-VU level on telephone lines. (The nominal signal level on commerical telephone lines is zero dBm to prevent cross talk between lines running in the same cable.)

To bring up very low levels to a value that can be adjusted with a gain control without introducing excessive amounts of noise, a preamplifier, a one- or two-stage amplifier before the gain control, is used. If the mixing (switching and gain controlling) is accomplished at a −10 dBm level, the amplifier that follows the mixing must be capable of a gain of at least 10 dB and preferably 20 to 30 dB. The preamplifier must be capable of 50 to 80 dB for the microphone. The preamplifier for the turntable must be capable of 35-dB gain or more. The remote program must lose 10 dB before being fed into the mixer (Sec. 24-5).

Note that it is desirable to control the volume of a microphone or other input signal in the zero-VU range rather than at the microphone level. All potentiometers introduce some noise as the arm is moved along the resistance. The more amplifiers that follow the gain control, the more this noise is amplified. Broadcast gain controls are usually multiple-contact (20 to 30) rotary step-type attenuator switches with the required amount of resistance between switch points to produce a 1- or 2-dB change of volume. If the contacts become dirty, they should be cleaned with soft cloth, or burnished, and then coated with a contact cleaner or petroleum jelly.

Test your understanding; answer these checkup questions.

1. What are the frequency limits of the standard AM BC band? _____ What is the width of a channel? _____ What emission is used? _____

2. What are the designations of the three service areas for AM BC stations? _____ _____ _____

3. What are the three parts of the BC day? _____ _____ _____

4. What type BC service(s) might have their antennas on top of hills? _____ On marshy land? _____

5. What does STL mean? _____ What is another name for a console pot? _____

6. The highest peaks of modulation should be held between what two percentages? _____

7. What is the name of the second transmitter used when a station operates 24 h a day? _____

8. Why are three turntables required for recorded programs? _____

9. What is a combo operator? _____

10. In Fig. 24-2, if a studio program is to be tape-recorded, to what is the tape input connected? _____

11. What term is used to indicate intercom operation in a BC station? _____

12. What does a red light glowing in a studio indicate? _____

13. What is the attenuator which controls the output of the program amplifier called? _____

14. What are the three important operational requirements during a broadcast? _____ _____

15. If 3 dB equals twice the power, what is the power gain of 43 dB? _____ Of 26 dB? _____

16. About how many dB gain would a single-stage voltage amplifier produce? _____

24-5 ATTENUATOR PADS

It is possible to use the same type of high-gain preamplifiers for all inputs to a console but intentionally lose signal level by inserting *H-* or *T-pad* attenuators between inputs and preamplifiers. These pads are groups of resistors that act as voltage dividers, at the same time holding the output and input impedances of the circuits being coupled at a constant value, often 600 Ω.

The T pad, shown in Fig. 24-3, is used with

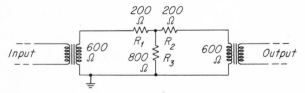

Fig. 24-3 An unbalanced T-type attenuator pad for a 600-Ω line.

unbalanced lines (one of the lines grounded). If it is desired to drop the voltage to one-half (6 dB) and maintain an impedance match at both transformers, the values shown can be used. The 600-Ω secondary of the left-hand transformer

looks at a 200-Ω resistance in series with two parallel 800-Ω impedances, or 600 Ω. The other transformer sees the same thing. If 1 V is delivered by the input transformer, the ratio of 200:400 Ω in series produces a voltage-drop across the 400-Ω resistance of $\frac{2}{3}$ V. Since the 200-Ω R_2 and the 600-Ω primary are in series across $\frac{2}{3}$ V, the primary sees $\frac{600}{800}$ or $\frac{2}{3}$, or $\frac{6}{12}$, or $\frac{1}{2}$ V.

If a loss of approximately 10 dB is desired, R_1 and R_2 may be 300 Ω each and R_3 400 Ω. For a 20-dB loss, R_1 and R_2 may be 500 Ω and R_3 120 Ω. (See an engineering handbook for formulas to develop other losses.)

Most long transmission lines are balanced to ground, as in Fig. 24-4. An H pad is used in this

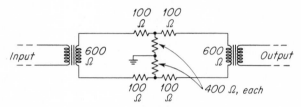

Fig. 24-4 A balanced H-type attenuator pad.

case. The series resistors are one-half the value of those in the T pad above. The resistance across the line is the same as in the T pad except that it is center-tapped.

Pads are also used to terminate amplifiers that are feeding long transmission lines to the transmitter equipment. Such lines may have differing transmission characteristics for different frequencies. If coupled to a line by a −6- to −10-dB pad, the characteristics of the line and the eventual load will have little effect on the amplifier and less distortion will be produced by the amplifier. The matching of impedances in all audio equipment is important to reduce distortion.

One application of a T pad is the variable attenuator, or gain control, of a mixer system, as in Fig. 24-5. The arms of all three resistors vary at the same time, holding the input and output impedance essentially constant regardless of the setting of the control arm. This attenuator is superior to a simple potentiometer type of gain control which always changes both input- and output-impedance values as the moving arm is varied. With 800-Ω resistors used on the three arms, at full gain the 600-Ω transformer sees an 800-Ω and

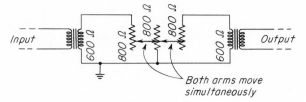

Fig. 24-5 A variable unbalanced T-type attenuator, or gain control.

a 600-Ω impedance in parallel, which is lower than its own impedance. At minimum gain the transformer sees an 800-Ω resistance, which is higher than its own impedance. Near midgain the impedances match satisfactorily, but at high and low settings they are somewhat mismatched.

When coupling two audio circuits having unequal impedances, either a coupling transformer or an L or U pad (named by configuration) may be used. The L pad shown in Fig. 24-6 matches

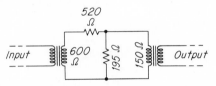

Fig. 24-6 An L-type pad to match a 600-Ω output to a 150-Ω input circuit.

a 600-Ω output to a 150-Ω input. There is no selection of loss in these pads. To match these two impedances only the two resistors shown can be used. The loss is about 10 dB.

For balanced lines, U pads, as in Fig. 24-7, are

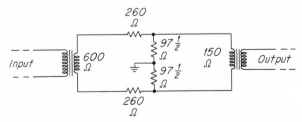

Fig. 24-7 A U-type balanced pad to match dissimilar impedances.

used. As with H pads, the series-resistance value of the unbalanced L pad is halved and the shunt resistor is center-tapped.

At the receiving end of a transmission line a center-tapped primary transformer may be used.

The center tap is grounded. This tends to balance out hum or noises picked up by the lines. The center-tapped shunt resistor of an H or U pad may be a potentiometer to allow a more accurate balance of the lines. A center-tapped transformer is not used in this case.

To reduce capacitive coupling of noise impulses picked up by transmission lines, the transformer used at the receiving end often has an electrostatic shield between primary and secondary windings. This shield is connected to the core, which is grounded.

24-6 LINE EQUALIZERS

A short 600-Ω line carrying an audio signal produces very little attenuation of any of the audio frequencies. If a mismatched line exceeds a few hundred meters in length, the inductance of the lines and the capacitance between the two wires do affect the transmission of the signal, attenuating the high frequencies more than the low. If the line and transformers are 150 Ω, the high-frequency losses will be less.

The use of a line pad at the output of the amplifier feeding the line helps to flatten the frequency response of the final-amplifier stage, but for high-fidelity transmissions the lines should transmit all audio frequencies equally well. For standard broadcast, this is usually considered to be from 50 to 7,500 Hz (±2dB, 100 to 5,000 Hz); for FM and TV, 50 to 15,000 Hz.

If a line is found to attenuate the higher frequencies, a *line equalizer* can be placed across the receiving end of the line to decrease the amplitude of all lower frequencies. This can be accomplished, as shown in Fig. 24-8, by using a parallel-resonant circuit in series with a variable resistance across the line. The circuit is made resonant at some high frequency—5,000 Hz, for

Fig. 24-8 A parallel-resonant type of line equalizer.

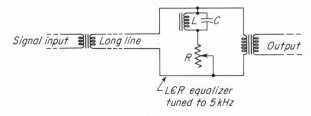

example. The high impedance of the circuit to frequencies near 5,000 Hz allows these frequencies to pass without attenuation. Lower frequencies (as well as higher) find that the equalizer presents a lower impedance, and they are attenuated somewhat. By adjustment of the resistance, it is possible to control the attenuation of the lower frequencies and equalize the frequency characteristics of the line from 50 to 5,000 Hz. More than one equalizer may be required.

24-7 PEAK-LIMITER AMPLIFIERS

Speech and music produce relatively high peaks of audio voltage at times. If the operator adjusts the gain controls to allow the average signal level of a musical program to produce 75 to 85% modulation, the peaks that occur may produce more than 100% modulation, distortion, and excessive sidebands, may actuate overload relays, or may damage tubes and associated equipment. In speech communication systems, peak limiters, or clippers, are usually used (Sec. 19-17). The distortion that is produced with this type of limiting is too great for broadcast standards, and special higher-fidelity *limiting amplifiers* are used. The basic idea of a limiting amplifier is shown in simplified form in Fig. 24-9.

The program signal is fed to the grid of an amplifier stage and, at the same time, to a second amplifier coupled to a biased full-wave diode rectifier. When the signal level exceeds the positive bias on the diode cathodes, current flows through the tubes and produces a negative voltage across the capacitor *C*. This negative voltage is used as a bias on one of the grids in the amplifier tube, decreasing its amplification and reducing the amplitude of the output peak signal. The value of the

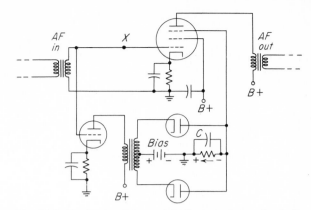

Fig. 24-9 Basic circuit to explain the principles of a limiting (or automatic gain control, AGC) amplifier.

bias voltage on the diodes determines the value at which limiting will occur.

Once charged, capacitor *C* discharges relatively slowly. If music rises in volume slowly, the bias voltage across *C* tends to rise also, attenuating the signal and acting as an automatic gain control. If a sharp, single peak of sound occurs, however, capacitor *C* charges and desensitizes the amplifier for a period of time. All signals that follow the short peak are held for a time at a low level by the bias voltage of *C*. This results in an unpleasant "hole" punched in the program following high-amplitude sounds.

A steep-sided audio wave can beat the buildup of the bias voltage and may produce an instantaneous overmodulation peak before gain reduction occurs. To prevent this, a limiting amplifier may incorporate a series-inductance shunt-capacitance *delay line* in the circuit at point *X* in the diagram. This results in the bias voltage being developed before the signal has arrived at the limiting-amplifier grid. Limiters allow a *higher average* percent of modulation of a transmitter.

Another useful circuit is the *compression amplifier*. The greater the amplitude of a signal the less proportional gain produced. At the other end of the system an *expander amplifier* is used. The greater the signal fed to this amplifier the greater proportional gain it produces. With matched compression and expander amplifiers a system may transmit a 40-dB variation of sound with only a 30-dB signal variation. Some records have compressed loud passages. An expander amplifier may produce more normal playback signals.

ANSWERS TO CHECKUP QUIZ ON PAGE 583

1. (535 to 1,605 kHz)(10 kHz)(A3) 2. (Primary)(Secondary) (Intermittent) 3. (Daytime)(Nighttime)(Experimental period)
4. (FM and TV)(AM BC) 5. (Studio-transmitter-link)(Attenuator, gain control, or fader) 6. (85–100%) 7. (Alternate)
8. (While one plays, cue second, cue recorded commercial)
9. (Combination of announcer, disk jockey, and engineer)
10. (Audition output) 11. (Talk-back) 12. (On the air)
13. (Master gain control) 14. (Timing)(Fidelity)(AF amplitude)
15. (20,000 times)(400 times) 16. (15–30 dB)

24-8 CLASSIFICATIONS OF POWERS

When a broadcast station is licensed for operation, the instrument of authorization will specify a certain *licensed power,* or *authorized operating power.* Operating power is that fed to the antenna by the transmitter. The tolerance of the operating power of the station must not exceed the licensed power by more than 5% nor fall more than 10% below the licensed power except in emergencies, when lower power may be used for a limited time. The operating power may also be known as the *carrier power* and is always the unmodulated value.

The *maximum rated carrier power* is the maximum power at which that model of transmitter can be operated satisfactorily. It is determined primarily by the tubes and plate voltages used in the final amplifier. The maximum rated carrier power capabilities of a transmitter must always be equal to, or exceed, the licensed power. For example, a station licensed for 500 W may emit 500 W from a transmitter that has a capability of 1,000 W.

The *plate input power* of a broadcast station transmitter is determined by the product of plate voltage and plate current ($P = E_p I_p$) of the final-amplifier tube or tubes during a time when no modulation is being applied.

The operating (output) power of an AM broadcast station is determined by the direct method. This is computed by

$$P = I^2 R$$

where I = antenna current with no modulation
R = impedance of antenna at the point where current is measured

EXAMPLE: An antenna has a feed-point resistance of 50 Ω, and the antenna ammeter at the feed point reads 4 A. The power being fed to the antenna is $4^2(50)$, or 800 W.

The indirect method, which is used to determine the output power of FM broadcast stations and TV aural transmitters, is employed in standard broadcasting only in an emergency or temporarily when authorized installation changes, changes of equipment, or antenna changes are being made. In these cases, the indirect power is determined by

$$P = E_p I_p F$$

where E_p = plate voltage of final amplifier
I_p = plate current of final amplifier
F = factor between 0.35 and 0.8 (see Table 24-1)

EXAMPLE: A class C final amplifier is plate-modulated and has a maximum rated carrier power of 5 kW. It has a plate voltage of 4,000 V and a plate current of 1.563 A. The plate power input is 4,000(1.563), or 6,250 W. With an efficiency factor of 0.8 (from the table) the power output is 5,000 W.

24-9 BROADCAST STATION TESTS

Before a broadcast station is constructed, it is first necessary to obtain a *construction permit* from the FCC. During the construction period, after the transmitter has been installed, it is permissible to test the transmitting equipment for short periods between midnight and local sunrise. These are known as *equipment* tests.

After construction has been completed and

Table 24-1

Factor (F)	Method of modulation	Maximum rated carrier power of the final amplifier	Class of final amplifier
0.80	Plate	5,000 W and over	C
0.70	Plate	100–1,000 W	C
0.65	Low level	100 W and over	BC
0.35	Low level	100 W and over	B
0.35	Grid	100 W and over	C

until the station license has been issued by the FCC, application can be made to test the station on the air. These are known as *service,* or *program,* tests.

The period of the day between midnight and local sunrise is used as the *experimental period.* During this time, broadcast stations may transmit for testing and maintaining equipment, provided that such tests do not interfere with other stations broadcasting on the frequency at that time.

All broadcast stations must be capable of transmitting 85 to 95% modulation with a total AF harmonic distortion not to exceed 7.5%. With less than 85% modulation, the total harmonic distortion must not exceed 5%. The distortion is measured by using modulating frequencies of 50, 100, 400, 1,000, 5,000, and 7,500 Hz, plus any intermediate or higher frequencies found necessary.

When a station is in operation, the percent of modulation must be maintained at as high a level as possible consistent with good quality of transmission but in no case less than 85% on peaks (no more than 100% on negative, or 125% on positive, peaks of modulation).

Another test that is required of operating stations is the Emergency Broadcast System test described in Sec. 24-18.

On an annual basis all broadcast stations must make on all its audio equipment a proof-of-performance test. The response of the transmitter to audio frequencies from 30 to 7,500 Hz must be made at 25, 50, 85, and 100% (if possible) and produced in response-curve form. Harmonic content, carrier shift, hum, and spurious radiations must also be measured. These measurements must be made during the last four months preceding filing for renewal of the station license.

Periodically a proof-of-performance test may be required on the antenna radiation pattern of any directional array.

24-10 BROADCAST STATION METERS

Ammeters and voltmeters associated with the final radio amplifier stage plate circuit must have an accuracy of at least 2% of the full-scale reading. For example, a 1,000-mA meter must be accurate to within 20 mA or less at any point on its scale. The full-scale reading of these meters shall not be greater than five times the minimum indication when the transmitter is in normal operation. Antenna-current meters are usually thermocouple types and have the same accuracy requirements as the dc meters.

A remote-reading antenna ammeter may be used to indicate the antenna current if its thermocouple is installed next to the main ammeter at the base of the antenna, if it has the same scale accuracy as the main meter, if its calibration is checked at least once a week against the main ammeter, and if its wiring is properly shielded.

When directive antennas are used, currents in the elements must be held within 5% of the authorized values.

If a required plate-circuit or antenna meter burns out and no substitute conforming to required specifications is available, an appropriate entry must be made in the operating log indicating when the meter was removed from and restored to service. The FCC Engineer-in-Charge of the radio district must be notified immediately after the meter has been found to be defective and again as soon as it has been replaced. The meter must be repaired or replaced within 60 days. Since remote antenna ammeters are not normally required, if such a meter becomes defective, it is necessary only that the ammeter at the base of the antenna be read and logged once daily until the remote meter has been returned to service.

Test your understanding; answer these checkup questions.

1. In Fig. 24-4, what is the dB loss of this pad? _____
2. What are the names of two unbalanced pads? _____ _____ Of two balanced pads? _____
3. What is the advantage of using balanced lines in audio systems? _____ T-pad attenuators over potentiometers? _____
4. What device is used to make up for the high-frequency losses in long AF lines? _____
5. Why does a peak-limiter increase the average output signal? _____
6. How does a limiter-amplifier differ from a peak-clipper? _____
7. What would a "compander" be? _____
8. What is another term meaning "operating power"? _____ Is it measured with or without modulation? _____
9. What does maximum rated carrier power refer to? _____

10. In what two ways can operating power be measured in BC stations? _____ _____ Which may be used in emergencies only? _____
11. According to Table 24-1, what type of modulation is the most efficient? _____ The least? _____
12. When are equipment tests made? _____
13. What is the maximum allowable distortion at 80% modulation? _____ At 90%? _____
14. On what are proof-of-performance tests made? _____ _____
15. What meters are required in the output of a BC transmitter and what is their required accuracy? _____ _____ _____
16. Within what limits must the antenna current of a directional antenna be held? _____

24-11 FREQUENCY MONITORING

Broadcast transmitters use low-temperature-coefficient-type crystals in the oscillator circuit. A station may also have a second standby oscillator stage ready to switch into service if the first one fails. The variation of the temperature inside the temperature-controlled crystal chamber (Sec. 13-14) must not exceed $\pm 1\,°C$ for the usual crystals ($\pm 0.1\,°C$ for X or Y crystals).

A frequency monitor operating independently of the frequency-control circuit of the transmitter and of a type approved by the FCC must be in operation at all times that the transmitter is on the air, at either the transmitting station or the remote-control point. Such a frequency monitor must give a constant visual indication of the operating frequency in hertz (Sec. 21-10).

The accuracy of the frequency-monitor indications can be checked by comparing a frequency report from a commercial measuring service with the indication of the monitor. If the frequency-measuring report indicates the station is 15 Hz low and the monitor shows the frequency to be 5 Hz high at the same time, the monitor must be indicating 20 Hz too high and should be serviced and recalibrated.

If the frequency monitor becomes defective, the station may use some other suitable measurement device at least once a week and log the results until the monitor has been restored to working order (60 days are allowed). Entries regarding removal and restoration of the monitor must be made in the maintenance log, and the FCC Engineer-in-Charge of the radio district must be notified of this information.

24-12 MODULATION MONITORS

Each station is required to have in operation at the transmitter or at the place the transmitter is controlled a modulation monitor that will give a visual indication of the percent of modulation at all times.

The modulation monitor must have (1) a dc meter to read the average rectified carrier value, which will also read carrier shift during modulation, (2) a rapidly moving type of meter to indicate the percent of the negative and positive peaks of modulation selected by a switch, (3) a peak-indicating light or alarm that can be set at any value from 50 to 120% modulation of positive peaks or from 50 to 100% of negative peaks, or both. Figure 24-10 is a simplified circuit of such a monitor.

Fig. 24-10 Modulation-percent indicator, including a carrier-shift meter and an overmodulation-warning flashing circuit.

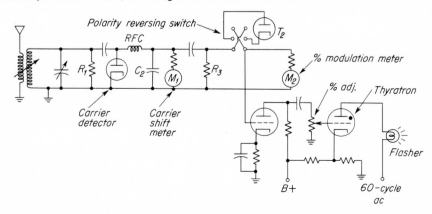

A small amount of modulated RF alternating current is coupled into the tuned circuit from the transmitter. Across this tuned circuit is a shunt-type half-wave diode carrier-detector circuit with the meter M_1 and its resistor as a load. The RF choke and C_2 remove the RF ac component, leaving only a steady current flowing through M_1 when there is no modulation and a varying direct current when the carrier is modulated. The input circuit is tuned to a peak indication on the carrier meter, and the input coupling is adjusted until the carrier meter indicates half scale. When no modulation is applied to the transmitter, the carrier meter, which is acting as a carrier-shift indicator, does not move. If the indicator moves when modulation is applied, carrier-shift distortion is present in the emission. The tuned circuit is shunted by R_1 to lower its Q and prevent it from peaking too sharply, allowing equal response of all sidebands generated.

The output of the carrier detector is capacitively coupled to the resistor R_3. With no modulation, no current flows through R_3; but when modulation occurs, an AF ac flows through it. A sensitive dc meter and a diode T_2 are in series across R_3. With the polarity-reversing switch in the downward position, whenever the top of R_3 becomes positive, current flows through the meter M_2. On the negative half cycles of modulation, no current flows. When the switch is reversed, however, the negative half cycles produce current flow through the meter. Depending on the position of the switch, either the extent of the positive peaks of modulation or the negative peaks can be made to register current and cause deflection of meter M_2, which is calibrated in percent of modulation. It is necessary to maintain the M_1 carrier level at the same value as when M_2 was calibrated, usually to center scale.

ANSWERS TO CHECKUP QUIZ ON PAGE 588

1. *(6 dB)* 2. *(T, L)(H, U)* 3. *(Reduce noise pickup)(Match Z's better)* 4. *(Line equalizer)* 5. *(Decreases peaks and allows gain to be raised higher)* 6. *(Limiter reduces gain of amplifier)* 7. *(Compresser-expander system)* 8. *(Licensed power, carrier power)(Without)* 9. *(Final-amplifier power rating)* 10. *(Direct, indirect)(Indirect)* 11. *(Plate modulation)(Low-level or grid types)* 12. *(Experimental period)* 13. *(5%)(7.5%)* 14. *(Directive antennas, AF system)* 15. *(I_p, 2%; E_p, 2%; antenna ammeter, 2%)* 16. *(5%)*

The overmodulation-indicator flasher circuit can utilize the polarized signal voltage fed to the percent-of-modulation meter circuit. These half-wave pulses are amplified and actuate a thyratron-tube circuit with 60-Hz ac as its plate-voltage supply. The thyratron-tube cathode is connected across a voltage-divider circuit, making it positive in respect to ground. The grid of the thyratron is returned to ground and is thereby made more negative than the cathode. When the amplified signal voltage fed to the thyratron grid exceeds the bias sufficiently, the thyratron ionizes and conducts current, flashing the lamp in its plate circuit. The tube de-ionizes on the negative half cycle of the 60-Hz ac. The potentiometer dial is calibrated in percent of modulation required to produce a flash of the lamp.

Many stations also use an oscilloscope displaying either an envelope or a trapezoidal indication of the percent of modulation of the transmitter. This can give a more accurate check on instantaneous peaks of modulation than any modulation monitor can.

If the modulation monitor becomes defective, the station may use an oscilloscope until the monitor has been repaired or replaced (60 days are allowed). Entries regarding removal and restoration of the monitor must be made in the maintenance log. The FCC Engineer-in-Charge of the radio district must be notified of this information.

24-13 INTERNATIONAL BROADCAST STATIONS

Some licensed broadcast stations in the United States beam their transmissions to the general public in foreign countries. These are shortwave stations and operate in one or more of seven shortwave bands, commonly known as:

Band A, 5.95–6.20 MHz
Band B, 9.50–9.775 MHz
Band C, 11.70–11.975 MHz
Band D, 15.10–15.45 MHz
Band E, 17.70–17.90 MHz
Band F, 21.45–21.75 MHz
Band G, 25.60–26.10 MHz

These stations must maintain a carrier frequency tolerance of ±0.003% on all frequencies of operation and have a licensed power output of not less than 50,000 W. Because of the variations of transmission paths during different times of the day and year, these stations shift from one band to another during the day to operate on the band that will have the best chance of reaching the audience desired. Directional antennas having a minimum power gain of 10 (10 dB) in the desired direction are required for these stations. Frequency monitors, modulation monitors, auxiliary or alternate transmitters, and other technical items are similar to those of standard broadcast stations. Radiotelephone First Class licenses are required of all operators at the transmitters of these stations.

24-14 LOGS

Three types of logs are required in broadcast stations: A *program log,* an *operating log,* and a *maintenance log.*

A program log contains:

1. An entry of the time each station identification announcement (call letters and location) is made
2. An entry briefly describing each program (music, drama, speech), the program name, sponsor, beginning and ending times, any mechanical record used, time announced as a mechanical record, and political affiliation of political-candidate speakers
3. An entry showing that each sponsored-program broadcast has been announced as sponsored, paid for, or furnished by the sponsor
4. An entry showing, for each program of network origin, the name of the network originating the program

An operating log contains:

1. An entry of the time the station begins to supply power to the antenna and the time it stops
2. An entry of the time each program begins and ends
3. An entry of each interruption to the carrier wave, its cause, duration, and remedy required
4. If required, an entry regarding antenna-structure illumination

5. An entry of the following, every 30 min, before making any necessary minor adjustments: (*a*) plate-circuit voltage and current of the final RF amplifier stage, (*b*) antenna current, (*c*) frequency-monitor reading

A maintenance log contains:

1. Weekly entries of such things as antenna-ammeter base current, time and results of auxiliary transmitter tests, results of any frequency checks made, and results of calibration checks on automatic recording devices
2. An entry showing date and time of removal from service or restoration of the modulation monitor, frequency monitor, final-stage plate voltmeter or ammeter, antenna ammeter, or common-point ammeter
3. A record of special tower-light inspections.
4. Descriptions of experimental operation during experimental periods

Logs are retained for a period of two years unless they contain information regarding claims, complaints, or distress communications. They are kept by persons competent to do so (licensed operators for maintenance and operating logs), who have actual knowledge of the facts required and who sign the logs when starting duty and again when going off duty. There is no prescribed form of a log other than that it be suitable and orderly. Key letters or abbreviations may be used in a station log if proper meanings of the letters are shown elsewhere in the log. (See also Sec. 31-9.)

The licensee of a remote-pickup broadcast station must keep logs or records of hours of operation, programs transmitted, frequency, point of program origination, receiver location, and antenna-structure information.

24-15 BROADCAST OPERATOR LICENSE REQUIREMENTS

It would be desirable to have all operators in a broadcast station hold Radiotelephone First Class licenses so that any operator could adjust any piece of equipment at any time. However, stations have found it difficult to find a sufficient number of qualified operators. Under the conditions listed below, holders of lesser licenses or permits may operate certain types of broadcast station controls.

At nondirectional-antenna stations of 10,000 W or less, for routine operations, holders of Radio-telephone Second Class licenses, broadcast-endorsed (Element 9) Radiotelephone Third Class permits, or Restricted Radiotelephone Operator permits may make certain routine adjustments: (1) turn the transmitter on and off, (2) adjust external controls to maintain proper power-supply voltages, (3) adjust external controls to maintain proper modulation percent, (4) adjust external controls to correct operating-power requirements, and (5) adjust external controls to effect operation during an Emergency Action Condition (Sec. 24-18). All other adjustments must be made by a holder of a Radiotelephone First Class license. At all stations a complete inspection of all transmitting equipment in use must be made at least once each day, 5 days each week, at intervals of no less than 12 h, by a First Class licensed operator, regularly employed by the station or under signed contract and on call.

Broadcast stations with nondirectional antennas with an excess of 10 kW output may use Second Class licensees, or permittees with broadcast endorsements, for routine operations, provided there is at least one full-time First Class licensee employed.

Stations with directional antennas, if required to maintain antenna current within 5% of a given value, must have First Class licensees on duty at all times that the directional antenna is in use, unless an approved antenna phase monitor is used and a First Class licensee is employed full time and takes daily readings of the antenna current. Second Class licensees and permittees may be used for routine operations.

All stations operating with an excess of 10 kW output or with directional antenna must have a First Class licensee designated as Chief Operator, who is responsible for station operation.

24-16 REMOTE CONTROL

When the transmitter is remote from the operating position, it is possible to use remote control. The following are among the significant items about remote control: (1) An indication must be given that any directional antenna system is in proper adjustment and is stable. (2) The transmitter and operating equipment cannot be operated by unauthorized persons. (3) Open circuits or grounds in control circuits automatically turn off the transmitter. (4) If improper meter readings occur, transmission must be stopped and a licensed operator must man the transmitter. (5) Licensed operators at the remote control must be able to perform all FCC required functions. (6) Antenna currents must be logged within 2 h of commencement of operation each day. (7) Telemetry of desired remote information from the transmitter may be by direct wire or by tones on a telephone line, or be transmitted on the broadcast station carrier at a modulating frequency lower than 30 Hz with a modulation percent of not over 6% whenever commanded. Commands may be made via telephone lines, STL transmitters, or microwave links.

24-17 SOME OTHER FCC REQUIREMENTS

Since all radio stations are under the jurisdiction of the FCC, representatives from the Commission may request certain information from a station at any reasonable hour. Some of the items that they may wish to examine are the program, operation and maintenance logs, results of equipment-performance tests, or recent antenna and field-intensity measurements.

All broadcast radio stations are required to identify themselves by transmitting their call letters and location at the beginning and end of each broadcast day, as well as on the hour. They may also identify on the half or the 15- and 45-min periods of each hour. While it is not necessary to interrupt a continuous program for such an announcement, the identification (ID) announcement should be made within 5 min of the required time.

Any program over 1 min in length that contains recorded portions must be announced as being partly or wholly recorded at either the beginning or at the end of the program.

Whenever a radio station is reimbursed in any way for any program or material being broadcast, it must make an announcement naming the sponsor.

The station license must be posted in a conspicuous place at the control point of the transmitter. The licenses of the operators of the station must be posted at the place where the operators are on duty.

Normally, broadcast stations may not direct information to any specific person. However, during emergencies in which safety of life and property are involved (hurricanes, floods, earthquakes, etc.), they may address information to dispatch aid. Whenever this is done, the FCC in Washington, D.C., and the Engineer-in-Charge of the radio district must be notified immediately.

No changes may be made to the final amplifier, its tube types, the number of tubes, or the system of modulation without prior approval of the FCC, but changes of audio and RF circuits other than those of the final amplifier and modulator can be made without approval.

Test your understanding; answer these checkup questions.

1. What are the two required monitors at the operating position in a BC station? _____ _____
2. Which monitor usually gives a carrier-shift indication? _____
3. What device can give the most accurate instantaneous peak-of-modulation indications? _____
4. In how many bands do U.S. International Broadcast Stations operate? _____ What frequency tolerance must they maintain? _____
5. List four items required in a program log. _____ _____ _____ _____
6. List five items in an operating log. _____ _____ _____ _____ _____
7. List four items in a maintenance log. _____ _____ _____ _____
8. What is the minimum time BC station logs must be held? _____
9. What are the five things considered routine operations in BC stations? _____ _____ _____ _____ _____
10. What class operator has no restrictions on technical work in a BC station? _____
11. What classes of operators are restricted? _____ _____ _____
12. If improper transmitter indications are given at the remote-control point, what should be done? _____
13. What can the FCC request to see at any reasonable time? _____ _____ _____
14. How often must IDs be made in a BC station? _____
15. Where must station licenses be posted? _____ Operator licenses? _____
16. Can final-amplifier tubes be replaced without notifying the FCC? _____

24-18 EMERGENCY BROADCAST SYSTEM (EBS)

In case of an *Emergency Action Condition* caused by a war, threat of war, state of public disaster, or other national emergency, some AM, FM, and TV broadcast stations will go into action under the *National Defense Emergency Authorization* (NDEA), broadcasting information to the public in their area according to an *Emergency Broadcast System Plan*. All other stations will make a short announcement notifying listeners that an *Emergency Action Condition* exists and will immediately shut down, observe radio silence, and monitor an NDEA station in their area.

NDEA stations are specifically exempt from operating-power limitations. At receipt of an *Emergency Action Notification* they will make the announcement: "We interrupt this program for a nationwide alert." They will then transmit the *Emergency Attention Signal* by cutting their transmitter carrier for 5 s (sound only for TV stations), return the carrier for 5 s, cut the carrier for 5 s, return the carrier to the air, and broadcast a 1,000-Hz tone for 15 s. This will be followed by a prescribed message explaining that an alert exists and how the station is going to operate.

Operation under an Emergency Action Condition will continue until the station receives an *Emergency Action Condition Termination* from the Federal government, at which time the following termination message will be sent: "This concludes operations under the Emergency Broadcast System. All broadcast stations may now resume normal broadcast operations."

All broadcast stations must make at least one unscheduled EBS test each week between the hours of 8:30 A.M. and local sunset. Results are forwarded to the FCC.

At each broadcast transmitter control point there must be equipment that will receive any Emergency Action Notification from a known NDEA station in the area. The frequency of the NDEA station must be monitored continuously, or some automatic alarm device must be included in the monitoring device.

Previously, an emergency broadcast system called *Conelrad* was used. It has been supplanted by the EBS, in which stations will indicate their operational areas but will not transmit their station call signs.

24-19 DISK RECORDING

Program material to be broadcast may be recorded on disk recordings (records), magnetic tape, or magnetic wire.

The recording of audible sounds on a disk involves the cutting of a continuous spiral groove inward from the outside of the record blank or outward from near the center of the record blank. The cutting instrument, a needle, or stylus, is sharpened to an 89° angle to cut a 2.5-mil (0.0025-in.) to 3-mil groove, or to a 76° angle to cut a 1-mil groove. The 2.5- or 3-mil groove is used on the older type of 78-rpm 8-, 10-, or 12-in.-diameter records and for the 33-rpm 16-in.-diameter transcriptions. The 1-mil groove is used on modern 45-rpm and 33-rpm LP (long-play) records and the 16-rpm book records. Whereas a 78-rpm 3-mil-groove record may have about 100 grooves to the inch, a 45-rpm 1-mil-groove LP has about 250 grooves/in. A pickup stylus is slightly narrower than the groove.

The grooves are modulated by vibrating a cutting stylus sideways by magnetic means as it cuts a groove in a blank disk. A 500-Hz modulation produces 500 lateral wavers in the groove in 1 s. When a playback needle rides along the groove, it is made to waver back and forth 500 times per second, which produces a 500-Hz ac in the playback pickup head.

The uncut area between the grooves is known as the *land area*. When recording, the operator must check the land and groove areas and keep them approximately equal. If the stylus and the recording head exert too much downward pressure, the cut becomes too deep, the groove too wide, and the land area too narrow, and cross talk results.

If the stylus is not modulated sufficiently, it does not cut a wide enough waver in the grooves and the playback mechanism reproduces a weak signal. If the stylus is modulated too much, one groove may waver over into, or too close to, adjacent grooves and distortion will be produced. The extent of the groove modulation is checked by making a test cut and observing it through a microscope.

A cutting stylus that is not properly sharpened, that becomes dull, or that has a chip in it produces a rough-surfaced groove that results in noise on playback. Old recording blanks become dry, and the groove tends to chip out instead of cutting out smoothly, again resulting in noise.

When the stylus has the proper pressure and angle and when the record blanks have a live surface, a thread or chip of recording material is produced as the stylus cuts the groove. This thread must be pushed in toward the center of the record by the operator during outside-in recordings but will accumulate near the center of the record with inside-out recordings. Usually a vacuum device with its opening riding along with the stylus point sucks up the recording thread and deposits it in a container of water. (The thread cut from acetate-surfaced record blanks is explosive.)

The recording stylus requires about 1 W of AF, usually supplied by a 10- to 15-W amplifier to reduce distortion. The cutting stylus may be steel (3 to 10 min of cutting), sapphire (30 to 60 min), or diamond (many hours).

The turntables used in broadcast recording are usually 16 in. in diameter, are made of cast aluminum, and are driven by a relatively heavy-duty, mechanically quiet, synchronous motor. Any mechanical *rumble* or vibration produced by the motor is transferred to the recorded grooves. If the motor does not produce absolutely constant rotational speed or if the turntable is not perfectly flat, musical tones will *wow* (vary in tone).

A strong low-frequency sound produces a large groove-stylus swing and groove deviation. An equally strong high-frequency sound produces very little groove swing. To allow a maximum number of grooves, the output of the recording

ANSWERS TO CHECKUP QUIZ ON PAGE 593

1. (*Frequency*) (*Modulation*) 2. (*Modulation*) 3. (*CRT oscilloscope*) 4. (*7*) (*0.003%*) 5. (*ID, program description, sponsors, network name*) 6. (*Antenna power on-off*) (*Programs on-off*) (*Carrier interruptions*) (*Antenna lights*) (*Meter readings*) 7. (*Antenna I checks*) (*Meter replacements*) (*Light inspections*) (*Experimental-period operation*) 8. (*Two years*) 9. (*Transmitter on-off*) (*Power-supply E*) (*Modulation %*) (*Operating power*) (*Operating during EAC*) 10. (*First phone*) 11. (*Second phone*) (*Restricted phone permits*) (*Third phone permits endorsed*) 12. (*First phone operator to transmitter*) 13. (*Logs*) (*Tests*) (*Measurements*) 14. (*On, off, hour, half-hour*) 15. (*Transmitter*) (*Operating position*) 16. (*Yes*)

amplifier is run through a filter that lowers the amplitude of the modulating signal below about 500 Hz and pre-emphasizes signals above about 2,000 Hz in order to keep the recorded signal well above the surface noise of the disk. As a result, pickup heads must be followed by filters that emphasize the low frequencies and de-emphasize the high frequencies for flat reproduction of the recorded signal.

24-20 PLAYBACK EQUIPMENT

The playback turntables used in broadcasting are usually 16 in. in diameter, are made of cast aluminum, and are belt-, rim-, or directly driven by a synchronous motor. The motor and turntable are usually spring-suspended to prevent rumble in the pickup.

Most pickup heads are magnetic or reluctance types with a diamond stylus for long playing life and protection of the disks. It is necessary to blow the dust from the stylus periodically.

The rotational speed of turntables can be checked by placing a stroboscopic disk on the turntable and illuminating it with a fluorescent light. The group of printed areas that appear to stand still indicates the speed at which the turntable is rotating. If the indication does not stop but slides slowly forward, the turntable is rotating too fast.

The frequency response of a pickup head can be checked with a test record on which are recorded at constant amplitude, frequencies of 50, 100, 400, 1,000, 3,000, 5,000, 10,000, and 15,000 Hz. If the pickup head, its filter, and the amplifiers of the system are operating correctly, the VU meter on the amplifier should read the same amplitude for all frequencies. An audio amplifier must have high fidelity to produce the desired results as well as satisfy the requirements of the FCC regarding purity of transmissions.

24-21 TAPE RECORDERS

The broadcasting industry uses a great many tape recorders today. It is standard practice for network stations to record programs on tape and hold them for a few minutes to several hours before transmitting them. Both the picture and the sound of a TV program can be tape-recorded and transmitted when desired.

Audio-frequency recording tape is made of plastic a few thousandths of an inch thick and approximately $\frac{1}{4}$ in. wide. One surface is painted with a finely ground magnetic material such as iron oxides or chromium dioxide, which has high retentivity. The tape is wound onto thin spools made of plastic or aluminum. The basic operation is indicated in Fig. 24-11.

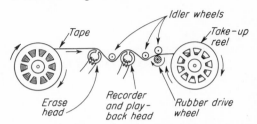

Fig. 24-11 Essential parts of a tape recorder-playback system.

When the constant-speed rotating drive wheel is forced up against the tape and the idler wheel, the tape is pulled along toward the take-up reel. The take-up reel has a light torque that allows it to reel in the tape that is fed to it. The other reel has a very slight drag to keep the tape tight against the heads.

When recording, AF ac fed to the recorder-head electromagnet produces magnetic flux at the gap, magnetizing the passing tape. The erase head is fed a supersonic-frequency ac that erases the tape before it reaches the recording head.

To rewind the tape, the drive-wheel pressure is released, the take-up reel has a slight drag placed on it, and the other reel is driven backward at high speed.

During playback, the drive wheel is engaged and the erase head is disconnected electrically. The AF ac produced in the playback coil as the tape passes over the gap is amplified and corresponds to the original AF ac that magnetized the tape during the recording process.

For high fidelity and best high-frequency response, the speed of the tape moving over the gap is 15 in./s. For slightly less fidelity, a speed of $7\frac{1}{2}$ in./s is used, allowing twice the playing time for the same reel of tape. For speech only, a speed of $3\frac{3}{4}$ in./s may be considered adequate

This speed may be halved for a low-fidelity system that will operate for a relatively long period of time. Frequency response of a tape pickup head can be checked with special test tapes recorded at several audio frequencies, all at the same amplitudes.

Some recorders record on the upper half of the tape; then by interchanging reels the other half of the tape can be recorded. This results in twice as much recording time on the same tape length. Stereo tape recorders record simultaneously on both halves (tracks) of the tape. Many such recorders split the tape into four tracks recording two-channel stereo on two of the tracks; then by interchanging reels they record stereo on the other tracks. If the tape is recorded with four tracks, it can also be used as single tape of quadrasonic sound. If recorded with eight tracks, it can be used as eight single-tracks, four 2-channel stereos, or two quadrasonic programs.

The magnetic gap of the record-playback head is filled with a nonmagnetic material, but this area may clog with some of the magnetic material from the tape that passes over it. Such clogging results in a loss of high frequencies and distortion during recording as well as playback. For this reason heads should be cleaned with a brush and iso-propyl alcohol or special cleaning solutions after every few hours of use. If the tape does not press firmly against the gap, high frequencies will also be lost. If the tape is inadvertently turned over (shiny side toward the head), the modulation will be quite muffled.

Originally all recorded commercials were cut as disk recordings with ungrooved spaces between each separate recording so that the operator could find the desired commercial easily. Visual cuing with magnetic tape on reels is not so easily accomplished, but some broadcast tape machines are made to stop automatically when a specially-recorded cuing tone is picked up from the tape as it moves forward or backward. In many stations plastic plug-in tape cartridges wound for continuous-loop operation are used. This allows immediate playback of taped commercials, no loss of time in threading the tape into the machine, and a minimum of cuing time. Such cartridges operate from a cue tone and come in various lengths, ranging from 40 s to 31 min of playing time. Cassettes—two-reel plug-in units—are also used.

24-22 BROADCAST MICROPHONES

These may be of the dynamic, velocity, or condenser types discussed in Secs. 17-28 to 17-30. Many broadcast microphones have tiny AF transformers in them to convert their output impedance to 150 or 600 Ω (or whatever is used in the particular station) so that all microphones can be interchanged without worrying about impedance matching. Generally, the lower the impedance of microphone lines, the less electrical noise they are likely to pick up.

When two or more microphones are used in a studio, sometimes the *phase* of their connections may have to be reversed, or their positions changed, because two microphones placed with one a half wavelength closer to a sound source will produce signals that tend to cancel in the amplifier.

24-23 RADIO STATION ANTENNA TOWERS

Radio towers are required to be painted and lighted if there is a possibility that they may constitute a menace to air navigation or, generally, if they exceed 170 ft in height.

Towers must be painted with equal-width alternate bands of aviation surface orange (TT-P-59) and outside white (TT-P-102), terminating with orange bands at top and bottom. (Paint samples are obtainable at the Federal Supply Service Center, Seventh and D Streets, SW, Washington, D.C., 20407.) While there is no required repainting period, a tower should be repainted as often as necessary to maintain good visibility.

For any compulsorily lighted tower above 150 ft in height there must be an electric code beacon equipped with two 500- or 620-W lamps installed at the top of the tower, as well as one or more lights down the tower. All lights must burn continuously from sunset to sunrise or be controlled by a light-sensitive device adjusted so that the lights turn on at a north-sky light intensity of 35 fc (35 foot-candles) and turn off at an intensity of 58 fc. A sufficient supply of spare lamps must be available at all times. Replacement and repair of lights and systems should be accomplished as soon as practicable.

The following light information must be included in the station's maintenance log: (1) time lights are turned on and off each day, (2) time when required daily light check was made, (3) time and nature of any failure of lights, (4) time nearest Federal Aviation Agency (FAA) establishment is notified of light failure, if not corrected within 30 min, (5) time lights are replaced, and (6) time FAA is notified that lights are back in service. At the completion of the required 3-month periodic inspection, the log must include date and time of inspection, condition of lights and lighting-control devices, indicators, and alarms, and any adjustments or repairs made.

It is not necessary to protect equipment in the antenna tuning house and the base of the antenna with screens and interlocks, provided the doors to the tuning house and the antenna base are fenced and locked at all times and the keys are in the possession of the operator on duty at the transmitter. All nearby fencing or wires should be grounded.

Test your understanding; answer these checkup questions.

1. What does EBS stand for? _____ EAC? _____ NDEA? _____ EAN? _____ EAS? _____ EACT? _____

2. How often must EBS tests be made? _____ What is done with the results? _____
3. What must all non-NDEA stations have? _____
4. What is the advantage of inside-out disk recording? _____ Disadvantage? _____ What groove diameters are used today? _____
5. What is the area between grooves on a disk called? _____
6. What stylus material is used in BC stations? _____
7. What advantages are there to having heavy turntables? _____ Light pickup heads? _____
8. What is the width of recording tapes? _____ The thickness? _____ What magnetic material is used? _____
9. What frequency is used to erase tapes? _____
10. What tape speed produces highest fidelity? _____
11. What should be used to clean recording heads? _____
12. How does a tape cassette differ from a cartridge?
13. What impedances are usually used with BC microphones? _____
14. Above what height must broadcast antenna towers be painted? _____ What colors must be used? _____
15. Above what height must BC antenna towers have beacon lights mounted on them? _____ What power lamps are used? _____
16. In what log is antenna light information entered? _____
17. Who should be notified if antenna lights fail? _____

COMMERCIAL LICENSE QUESTIONS

Applicants for Element 4 should know answers to all questions. Applicants for Element 9 need answer only questions followed by **[9]**.

1. Define the following: *standard broadcast station, standard broadcast band, FM station, FM band, daytime, nighttime, broadcast day.* (24-1)   **[9]**
2. Define the following: *standard broadcast channel, experimental period.* (24-1)
3. What is the permissible frequency tolerance of (*a*) standard broadcast stations, (*b*) FM stations? (24-1, 19-18)   **[9]**
4. If a broadcast station is licensed to operate on 1,100 kHz, what are the minimum and maximum frequencies at which it may operate and still be within frequency limits? (24-1)
5. What is the effect of 10,000-Hz modulation of a standard broadcast station on adjacent channel reception? (24-1)

6. What type of antenna site is technically best for an AM broadcast station? (24-1)
7. Draw a block diagram of a standard AM broadcast transmitter and explain briefly. (24-2)
8. What is the maximum permissible percent of modulation for AM and FM stations? (24-2) **[9]**
9. Why is plate modulation more desirable than grid modulation for use in AM broadcast transmitters? (24-2)
10. What is an *STL system?* (24-2)
11. What is the basic difference between STL and intercity relay broadcast stations? (24-2)
12. What type of antenna must be used with STL and intercity relay broadcast stations? (24-2)
13. What is the uppermost power limitation for remote-pickup broadcast stations, STL stations, and intercity relay broadcast stations? (24-2)
14. What is the frequency tolerance for an STL station? (24-2)

15. What is an *auxiliary broadcast transmitter?* (24-2)
16. How frequently must an auxiliary broadcast transmitter be tested? (24-2)
17. If the power output of a modulator is decreased from 10 kW to 10 W, how many dB is this? (24-4)
18. If an amplifier has an overall gain of 50 dB and the output is 1 W, what is the input? (24-4)
19. An amplifier with a 600-Ω output is connected to a microphone. The output is −60 dB. If the mixer is assumed to have a loss of 20 dB, what must be the voltage amplification in the amplifier to feed +10 dB to the output line? (24-4)
20. Why are preamplifiers used ahead of mixing systems? (24-4)
21. Why is it important to keep contact points on attenuator pads in a console clean? How are they cleaned? (24-4)
22. Why should impedances be matched in speech-input equipment? (24-5)
23. What are the purposes of H- or T-pad attenuators? (24-5)
24. Why is an audio amplifier generally isolated from the line by a pad? (24-5)
25. Why are grounded center-tap transformers used to terminate program wire lines? (24-5)
26. Why are electrostatic shields used between windings in coupling transformers? (24-5)
27. What are *line equalizers,* why are they used, and where in the transmission line are they normally placed? (24-6)
28. Draw a diagram of an audio line equalizer. (24-6)
29. What are *limiting amplifiers,* why are they used in broadcast stations, and where are they normally placed in the program circuit? (24-7)
30. Explain the operation of limiting amplifiers. (24-7)
31. What are the uses of peak-limiting amplifiers? (24-7)
32. Explain the operation and uses of compression amplifiers. (24-7)

33. Define the following terms: *operating power, plate input power, antenna power, antenna current.* (24-8)
34. Explain how operating power is computed by direct measurement and by indirect measurement. Under what conditions at a standard broadcast station is the indirect method used? (24-8)
35. What is the FCC requirement regarding maintenance of operating power? (24-8)
36. Under what conditions may a broadcast station operate at a power lower than is specified in its license? (24-8)
37. Are the antenna current and plate current and voltages of radiotelephone transmitters modulated or unmodulated values in computations of power output? (24-8)
38. What are the power limitations on broadcast stations? (24-9)   [9]
39. What is meant by *equipment, program,* and *service* tests? (24-9)
40. May a standard broadcast station licensed "daytime only" operate during the experimental period without specific authorization? (24-9)
41. What percentage-of-modulation capability is required of a broadcast station? (24-9)
42. When a broadcast station is operated at 85% modulation, what is the maximum permissible combined audio harmonic output? (24-9)
43. What is the Commission's requirement for maintenance of percent of modulation? (24-9)
44. What is the required full-scale accuracy of the ammeters and voltmeters associated with the final radio stage of a broadcast transmitter? (24-10)
45. What should an operator do if the remote-control devices at a station so equipped malfunction? (24-10, 24-16)   [9]
46. What should an operator do if a remote antenna ammeter becomes defective? (24-10, 24-16) [9]
47. If the plate ammeter in the last stage of a broadcast transmitter burns out, what should be done? (24-10)
48. Under what conditions may remote-reading antenna ammeters be used to indicate antenna current? (24-10)
49. A broadcast station receives a frequency-measurement report indicating the station was 15 Hz low at a certain time. The transmitter log shows the frequency to be 5 Hz high. What is the error in the frequency monitor? (24-11)
50. What should be done if a station's frequency monitor becomes defective? (24-11)
51. What should be done if a station's modulation monitor becomes defective? (24-12)

ANSWERS TO CHECKUP QUIZ ON PAGE 597

1. (*Emergency BC System*) (*Emergency Action Condition*) (*National Defense Emergency Authority*) (*Emergency Action Notification*) (*Emergency Attention Signal*) (*Emergency Action Condition Term*)   2. (*Weekly*) (*Forwarded to the FCC*)   3. (*NDEA station monitor*)   4. (*Chips fall toward center*) (*Most records are outside-in*) (*1 mil*)   5. (*Land*)   6. (*Diamond*)   7. (*Constant speed, less vibration and rumble*) (*Follow grooves well*)   8. (*0.25 in.*) (*103 mils*) (*Iron oxides, chromium dioxide*)   9. (*Supersonic, usually above 30 kHz*)   10. (*15 in./s*)   11. (*Isopropyl alcohol*)   12. (*Cassette has two separate reels*)   13. (*150 or 600 Ω mostly*)   14. (*170 ft*) (*Orange and white*)   15. (*150 ft*) (*500 to 620 W*)   16. (*Maintenance*)   17. (*FAA*)

52. What is the frequency tolerance allowed an international broadcast station? (24-13)
53. Who keeps broadcast station logs? (24-14) **[9]**
54. How and by whom may station logs be corrected? (24-14, 31.9) **[9]**
55. What logs must be kept by broadcast stations? (24-14) **[9]**
56. When may key letters or abbreviations be used in a station's logs? (24-14) **[9]**
57. How long must a station log be retained? (24-14) **[9]**
58. Should minor corrections to a transmitter be made before or after logging the meter readings? (24-14) **[9]**
59. What entries are made in the (a) program log and (b) operating log? (24-14) **[9]**
60. What entries shall be made in a station's maintenance log? (24-14)
61. How many times and when must the operating log be signed by an operator who goes on duty at 10 A.M. and off duty at 6 P.M.? (24-14)
62. What records of operation must be maintained for each licensed remote pickup broadcast station? (24-14)
63. What are the operator requirements at (a) a nondirectional 5-kW standard broadcast station, (b) a directional 5-kW broadcast station, (c) a nondirectional 20-kW standard broadcast station? (24-15, 24-16)
64. What are the requirements concerning stations which operate their transmitters by remote control? (24-16)
65. What information must be given an FCC inspector at any reasonable hour? (24-17) **[9]**
66. At what place must (a) a station license and (b) operator's licenses be posted? (24-17)
67. Should the sponsor's name ever be omitted when reading commercials on the air? (24-17) **[9]**
68. When should an operator announce a program as having been recorded? (24-17) **[9]**
69. Under what conditions may a broadcast station use its facilities for communication directly with individuals or other stations? What notice shall be given when a station is operating during a local emergency? (24-17)
70. Changes to the broadcast transmitter of what general nature require FCC approval? (24-17)
71. What type of changes to a broadcast transmitter do not require FCC approval? (24-17)
72. What specific equipment-performance measurements must be made at all broadcast stations on an annual basis? (24-9)
73. During what period of time preceding the date of filing for renewal of a station license should its equipment-performance measurements be made? (24-9)
74. What is a *proof-of-performance* test and how does it differ from annual equipment-performance measurements? (24-9)
75. What is *EBS?* What is an *Emergency Action Condition?* (24-18) **[9]**
76. During a period of Emergency Action Condition what should all nonparticipating stations do? (24-18) **[9]**
77. How often and during what time period should EBS test transmissions be sent? (24-18) **[9]**
78. What equipment must be installed in broadcast stations in regard to reception of an Emergency Action Notification? (24-18) **[9]**
79. Define the following terms: *Emergency Broadcast System (EBS), National Defense Emergency Authorization (NDEA), Emergency Action Notification, Emergency Action Condition Termination, Emergency Action Condition, Emergency Broadcast System Plan.* (24-18) **[9]**
80. Must stations operate in accordance with Sec. 73-57 *(authorized power)* of the FCC rules during an Emergency Action Condition? (24-18)
81. Describe the Emergency Action Notification Attention Signal. (24-18) **[9]**
82. Under normal conditions all standard, FM, and TV broadcast stations must make what provisions for receiving Emergency Action Notification and Termination? (24-18)
83. What type of station identification shall be given during an Emergency Action Condition? (24-18) **[9]**
84. What are *wow* and *rumble* with reference to turntables? How can they be prevented? (24-19)
85. Explain the use of a stroboscopic disk in checking turntable speed. (24-20)
86. What type of playback stylus is generally used in broadcast station turntables and why? (24-20)
87. Why is good fidelity an important consideration when replacing amplifiers in a broadcast station? (24-20)
88. How can frequency response of the pickup unit of a turntable be tested? (24-20)
89. How can frequency response of the pickup unit of a tape recorder be tested? (24-21)
90. What factors can cause a serious loss of high frequencies in tape recordings? (24-21)
91. How does dirt on the playback head of a tape recorder affect the audio output and how is it cleaned? (24-21)
92. What is meant by *phasing* two microphones and when is it necessary? (31-9, 17-31)
93. Under what two general conditions must antenna structures be painted and lighted? (24-23)

94. What color or colors should antenna structures be painted and where can paint samples be obtained? (24-23)
95. Generally speaking, how often should an antenna tower be painted? (24-23)
96. What items regarding the operation of antenna-tower lighting should be included in a maintenance log? (24-23)
97. What action should be taken if the tower lights at a station malfunction and cannot be immediately repaired? (24-23)
98. Is it necessary to have available replacement lamps for a station's antenna-tower lights? (24-23)
99. If a tower is required to be lighted, the lights are controlled by a light-sensitive device, and the device malfunctions, when should the tower lights be on? (24-23)
100. How often should automatic control devices and alarm circuits associated with antenna-tower lights be checked for proper operation? (24-23)
101. Light-sensitive devices used to control tower lights should face which direction? (24-23)
102. If the operation of a station's tower lights is not continuously monitored by an alarm device, how often should the lights be visually checked? (24-23)
103. Generally speaking, how soon after a defect in the antenna-tower lights has been noted should the defect be corrected? (24-23)
104. Who keeps the keys to the fence which surrounds the antenna base at a broadcast station and where are the keys usually kept? (24-23)

25 TELEVISION

The objective of this chapter is first to discuss the makeup of the basic monochrome (black-and-white) television transmitting system and describe some of the specialized circuits and devices that might be found in it, plus the blanking and synchronizing pulses used, and modulation requirements. The basic monochrome TV receiving system circuits are briefly developed. The fundamentals of some of the additional circuits required in both transmitters and receivers of color telecasts as well as general color TV systems operation are discussed. It is assumed the reader is familiar with preceding chapters on transmitters and receivers.

25-1 A TV BROADCAST SYSTEM

A television transmitting system involves many of the basic radio circuits and systems already discussed. A monochrome TV transmitter is actually two separate transmitters coupled to a single antenna, as shown in Fig. 25-1.

The *aural,* or sound, transmitter is essentially the same as an FM broadcast station (Chap. 19), except that 100% modulation is represented by only a 25-kHz frequency swing of the carrier instead of 75-kHz deviation. The same 30- to 15,000-Hz audio capability of the amplifiers is required. The microphone signals are amplified, controlled by an operator, and fed to the FM modulator in the transmitter. The FM transmitter signal (F3) is monitored and fed to the antenna through a *diplexer,* which prevents any FM signal from being coupled into the visual transmitter.

The *video,* or visual signal, transmitter receives its modulating voltages from a photosensitive tube in a camera instead of from a microphone. These video signals range in frequency from a few hertz to more than 4.5 MHz. Since no transformer can handle such a wide span of frequencies, RC-coupled grid modulation is used in the visual transmitter. The video signals from the camera are monitored by a control-room operator, synchronizing pulses are added to them, and they are fed to the visual transmitter, where they amplitude-modulate the carrier. The visual RF signal from the transmitter is fed to a vestigial sideband filter in which some of the below-carrier-frequency sidebands are attenuated. These vestigial sidebands, the carrier, and all the above-carrier-frequency sidebands (type A5C emission) are monitored by a TV receiver and are fed to the antenna through a diplexer, which prevents any of them from entering the FM transmitter.

It will be assumed that a TV receiver has been observed and that it has been noted that the picture is composed of many illuminated horizontal lines across the face of the picture tube. The illumination of the different parts of each line changes in intensity as the televised scene changes.

It is the task of the TV camera and transmitter to scan, or slice, the scene to be televised into about 495 separate horizontal lines, each with varying intensities, transmitting a stronger signal where the lines are dark and a weaker signal where the lines are light. (This is called negative-type transmission. Some countries use positive-type transmission, in which dark scenes produce the weakest output.)

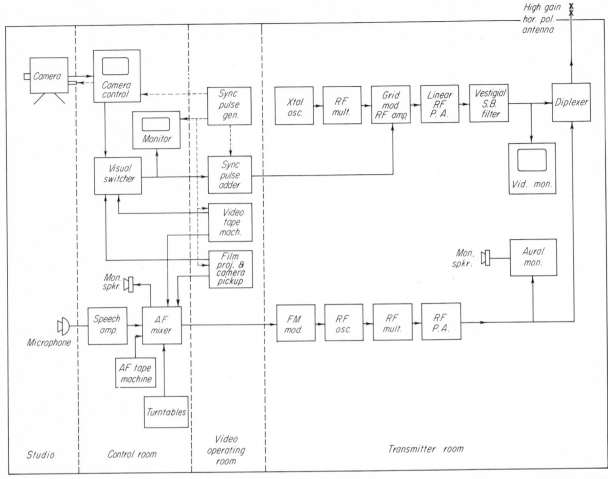

Fig. 25-1 Block diagram of a simple TV transmitting station including studio, control room, video operating room, and transmitter room. Camera operator intercom not indicated.

The *synchronizing-pulse generator* generates the necessary pulses and adds them to the visual transmitted signal to enable TV receivers to synchronize their sawtooth-waveform oscillators with the sawtooth scanning voltages applied to the camera tube at the transmitter.

The signals radiated from the two transmitters are received by viewers on superheterodyne TV receivers, shown in block form in Fig. 25-2. The received TV signal is amplified, converted to a 42-MHz IF, and detected. This detected signal contains (1) varying dc components that represent changes in illumination for the lines of the picture (0–4.2-MHz video frequencies), (2) a

group of varying dc frequencies centered around 4.5 MHz, which is the frequency-modulated aural information, and (3) horizontal and vertical sync pulses added to the signal at the transmitter. All this information is amplified by the video amplifier. The frequencies centered around 4.5 MHz are accepted by a tuned IF amplifier, detected, amplified, and fed to a loudspeaker. The low-frequency sync pulses are used to synchronize a 60-Hz sawtooth oscillator to sweep the lines down the picture tube at this rate. The high-frequency sync pulses are used to synchronize a 15,750-Hz sawtooth oscillator to sweep lines across the picture tube at this rate. The video information signals are

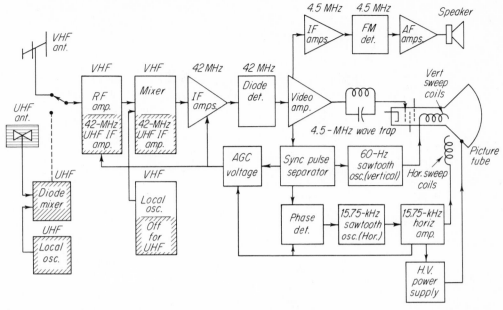

Fig. 25-2 Block diagram of a monochrome VHF TV receiver. Shaded blocks represent a possible arrangement of front-end circuits when receiving UHF TV signals.

passed to the grid of the picture tube to change the intensity of the illumination as the lines are being written across the face of the tube.

To receive UHF stations, a special UHF diode mixer and local oscillator are used. The VHF RF amplifier and mixer (tuned to 42 MHz) may now act as additional IF amplifiers. Functioning of the various other blocks in the complete system will be discussed later in greater detail.

25-2 THE TV TRANSMITTER ANTENNA SYSTEM

The transmitted signal from a TV station is essentially 6 MHz wide, requiring a broadband antenna. Both aural and visual TV transmissions must be horizontally polarized (by FCC rule). Either a turnstile or a slotted-cylinder type of antenna can be used (Chap. 20). Transmission lines can be balanced, shielded two-wire types, unbalanced coaxial, or, for UHF, waveguide (Chap. 26). The standing-wave ratio (SWR) can be determined by SWR bridges or meters, or a pair of directional couplers can be used. A directional coupler samples a small known fraction of the

power that is traveling along a transmission line *in one direction only*. The current (or voltage) traveling up toward the antenna shows on one directional coupler. Reflected current shows on the other coupler connected to the line in reverse. The ratio of the two is called the *reflection coefficient, ρ*. The two directional couplers used in this manner make up a *reflectometer*. The standing-wave ratio is found by.

$$SWR = \frac{1 + \rho}{1 - \rho}$$

The diplexer must pass both the visual and the aural signals to the same antenna and prevent either from coupling to the other transmitter. This is accomplished with a balanced bridge-type circuit, shown in Fig. 25-3.

The visual signal is coupled directly to the two center feed points of the antenna through a shielded two-wire feed line. The shield acts as a center tap of the two wires. Two similar inductors, L_1 and L_2 (or two capacitors), across the line provide a center tap also. The aural signal is fed between the two center taps. Because of the bal-

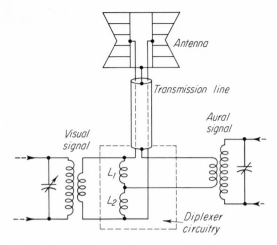

Fig. 25-3 Essentials of a diplexer.

ance of the system there can be no transmission from transmitter to transmitter, but both transmitters excite the antenna and radiate signals.

25-3 MAGNETIC DEFLECTION AND FOCUSING

In Sec. 12-29 a cathode-ray tube (CRT) that used an electrostatic (positive and negative) means of deflecting the electron beam was discussed. Electrostatic deflection can be used in TV cathode-ray tubes, but practically all cathode-ray devices employed in TV systems are magnetically deflected.

A beam of moving electrons represents an electric current. The direction of lines of magnetic force around a current is shown by the left-hand rule, in which the thumb points to the current direction and the fingers point in the direction of

Fig. 25-4 An electron beam moving into a magnetic field is deflected at right angles to the field.

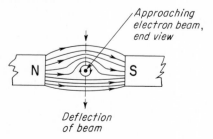

the lines of force. In Fig. 25-4 an electron beam is moving toward the reader through a magnetic field.

The part of the line of force surrounding the beam at the top has a direction similar to that of the lines of force of the stationary field. Since like lines of force repel, the beam tends to deflect downward. Similarly, the line of force of the beam that is below it is found to be in the opposite direction to the stationary lines. Since unlike lines attract, the beam is attracted downward still more. Note that the electron-beam deflection is at *right angles* to the magnetic field through which it is moving, not toward either magnetic pole. By sending a beam of electrons through the area between two poles of an electromagnet, it is possible to control the amount of deflection of the beam by controlling the strength of the current that produces the magnetic field and the direction of deflection (up or down) by controlling the direction of the current flow through the electromagnets and therefore the magnetic line direction.

Figure 25-5 illustrates placement of vertical-

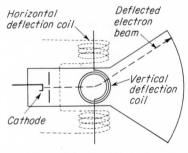

Fig. 25-5 Positions of horizontal- and vertical-deflection coils around the neck of a TV picture tube.

deflection coils at the neck of a CRT. When current flows through the coils, a horizontal magnetic field is produced across the neck of the CRT. The amount of vertical deflection will depend on the strength of the magnetic field. The direction of deflection (up or down) will depend on the polarity of the field.

When current flows through the horizontal-deflection coils (shown dotted), a vertical magnetic field is produced through the tube. The extent of horizontal deflection depends on the strength of the field, and direction of deflection depends on polarity of the field.

By use of both horizontal- and vertical-deflection coils, the beam can be moved to any position on the tube face by applying the proper polarity and strength of current to the two sets of coils. Two horizontal and two vertical coils form a *deflection yoke*.

The electron beam in a CRT may be focused by electrostatic means (Sec. 12-29) or by electromagnetic means. If a wire carrying a steady current is coiled around the neck of a CRT, as in Fig. 25-6, the lines of force of the magnetic field

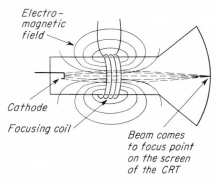

Fig. 25-6 Position of an electromagnetic focusing coil around a CRT.

produced are parallel to the axis of the CRT neck. Electrons in the beam tend to converge as they pass through this field of parallel magnetic lines. The current in the coil and therefore the strength of the field determine where the electrons of the beam will come to a point of focus. Television picture tubes have magnetic deflection yokes and magnetic focusing coils, or magnetic deflection and electrostatic focusing.

25-4 INTERLACED SCANNING

The current fed to the horizontal-deflection coils has a sawtooth waveform to produce a relatively slow movement of the electron beam across the face of the tube during the writing of a horizontal line, followed by such a rapid retrace that essentially nothing would be written on the tube face. While it would be simpler to scan from top to bottom of the picture with about 495 sawtooth-produced lines, it is better to interlace the scanning lines. Consider the following.

If a motion-picture film is projected with a *frame* (single-picture) frequency of about 15 per second, an illusion of movement is created, but the motions seem jerky and flicker of bright areas is apparent. Increasing the frame frequency to 24 or 30 per second will make motion appear smooth, but a slight flicker can still be noted on more brilliantly lit areas of the picture. Doubling the frame frequency would eliminate the flicker, but too much film would be required. Instead, it is possible to flash each picture on the screen twice by the use of a shutter that opens and closes 48 or 60 times per second, and the flicker disappears. Interlaced scanning of the TV picture has a similar effect.

The TV picture always has an *aspect ratio* of 4:3; that is, the picture is always four units wide to three units high. The size of the tube is measured from one corner to the opposite corner, which has no effect on the number of lines received. On the larger tubes each line is merely thicker and stretched out longer.

Satisfactory vertical definition can be produced by breaking a picture into about 495 horizontal lines. (Any number of lines between 482 and 495 may be employed at the discretion of the transmitting station.) By allowing time for 525 lines per picture, an allowance of 30 to 40 lines is included during which the scanning circuits can be brought into proper synchronism to assure that both transmitters and receivers start scanning at the top of the picture at the same time.

To produce satisfactory horizontal picture definition (sufficient dark and light units per line) requires a bandwidth of about 4.2 MHz.

To prevent flicker and still not have to show a complete picture more than 30 times per second, interlaced scanning, as in Fig. 25-7, is employed. The first time the picture is scanned, only the even-numbered lines are used. This produces a picture with only 262.5 lines, known as a *field*. The field by itself would be a rather poorly defined picture vertically. However, the picture is immediately scanned again, this time using the odd lines. This second field is interlaced between the lines of the first. Since the two are presented within $\frac{1}{30}$ s, the eye accepts them as one *frame,* with

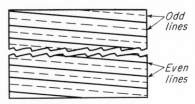

Fig. 25-7 Interlaced scanning. The first-field lines are shown broken; the second field is shown solid.

satisfactory vertical and horizontal definition and showing no flicker on brightly lit areas. Because no more lines have been added, the bandwidth (4.2 MHz) required to transmit the picture signal has not been increased. In a TV transmission the frame frequency is 30 per second and the field frequency is 60. Accurately timed synchronizing pulses must be utilized to enable all horizontal lines to start at the correct time to produce straight vertical edges of the picture, as well as to start the half line at the top of the picture properly.

Since each horizontal line is below the preceding one, a gradually changing current must be fed to the vertical-deflection coils to move successive lines continually farther down the picture. This is known as the *vertical sweep* current. The sweep current must return rapidly to the starting value as soon as a field has been scanned all the way to the bottom. A sawtooth waveform can produce the required relatively slow vertical movement downward, as well as the rapid return upward at the end of the cycle. This sawtooth wave must have a frequency of 60 Hz.

There are about 24 sweep lines above the TV picture that carry no picture modulation. Lines 17 through 20 may carry test signals indicating modulation reference values, cue, and control signals. Lines 21 through 23 may carry coded information regarding local or network origination, or other identification.

With 525 lines per frame and 30 frames per second, the horizontal sawtooth ac required to sweep the beam across the camera and picture tubes is 525(30), or 15,750 Hz.

25-5 THE ICONOSCOPE

One of the first successful TV camera tubes was the iconoscope. It consists of an electron gun, deflection coils, a mosaic picture screen on an insulating sheet, a conductive plate, and a collector anode, as shown in Fig. 25-8.

Microscopic islands of photoelectric silver are deposited on one side of a thin mica insulating sheet. Each island is electrically insulated from all those adjacent to it. The other side of the mica sheet is coated with a conductive material. Thus, each island forms a tiny capacitor between itself and the back plate, with the mica sheet as the dielectric. When a scene is focused on the mosaic screen, *photons* (light energy) strike the photoelectric islands, releasing electrons and leaving the light areas positive and the darker areas less positive. The electrons released from the islands are drawn to the collector-anode ring. This leaves the mosaic screen with areas of electric charge

Fig. 25-8 Iconoscope camera tube and basic circuit.

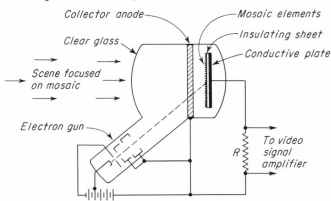

that conform to the light and dark areas of the scene focused on it, each tiny capacitor being charged in proportion to the light striking it. The positively charged capacitor plates pull a charging current of electrons upward through the load resistor R.

When an electron beam is made to scan across the mosaic, it feeds electrons to the positive islands, discharging them. As they are discharged, they no longer tend to hold electrons in the circuit connected to the top of the resistor, and a discharging electron current flows down through the resistor. As scanning continues, a signal current flows up and down the resistor in accordance with the light striking the part of the scene being scanned and discharged at the time. These currents produce voltage-drops across the load resistor, forming the video signal that is amplified and transmitted.

25-6 THE IMAGE ORTHICON (IO)

This camera tube gives good definition and is quite sensitive to light of low intensity. The fundamental components of the tube are shown in Fig. 25-9. The scene to be televised is focused on the front of a negatively charged, thin, semitransparent photocathode through a series of glass lenses. Lighter portions of the scene liberate many electrons from the back of the photosensitive cathode, whereas dark portions liberate none. The freed electrons are attracted toward the less negatively charged target plate, made of a thin sheet of special glass having a carefully controlled conductivity from one flat surface to the opposite.

To strike the glass plate the electrons must pass through the spaces of a very fine wire screen. If each electron hitting the target glass also produces a secondary emission of two electrons from the front face, the glass becomes positive at any point where electrons strike it. All secondary-emission electrons from the target plate are attracted to the positive metal target screen and are led from the tube to the power supply. This leaves the glass target plate with positive charges corresponding to the light areas focused on the photocathode and with no charge where dark areas fall. The visible scene has now been transferred to the glass target plate as a scene composed of areas of various degrees of electric charge instead of areas of light and dark as on the photocathode.

An electron gun at the far end of the tube emits electrons that are focused and made to scan the back of the conductive glass target plate. Wherever this beam strikes the back of a positively charged area on the target, it loses some of its electrons. When the beam electrons hit an area of zero charge, they rebound from the glass and travel back down the tube toward a positively charged electron-collecting plate surrounding the outlet of the cathode-ray beam. The outer surface of this plate is coated with a substance that produces considerable secondary emission. Such a plate may be called a *dynode* (dynatron-type anode).

As the electron beam sweeps across the electrical scene, from light to dark areas, the number of returning electrons will vary. In this way, it is possible to produce an electron current back down the tube that varies in accordance with the

Fig. 25-9 Essential components of an image orthicon camera tube.

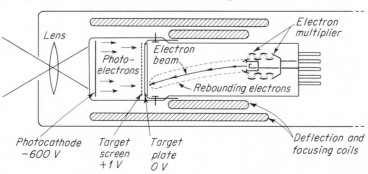

target-plate scene. To reproduce the whole visible scene for its counterpart of electric impulses, it is necessary to scan it. The beam is made to sweep across the top of the scene from left to right, blank out, move down the width of the beam, retrace to the left, unblank, sweep a second line across the scene, and so on until the whole scene has been scanned from top to bottom and electric impulses have been collected from all parts of the scene. These electric video impulses are amplified and used to amplitude-modulate the TV transmitter. At the receiver the amplitude of these impulses determines the illumination produced by the scanning beam as it sweeps across the cathode-ray picture tube.

The returning beam in the image orthicon is a relatively weak current of electrons. To increase the output current, a series of dynodes are used to form an *electron multiplier*. As returning electrons strike the first dynode plate, each electron may liberate two secondary electrons. These liberated electrons are attracted to the next dynode plate, which has a higher positive potential. Each of these electrons liberates two or more electrons, and so on. By the use of five dynode elements it is possible to produce a current amplification of more than 100 times, resulting in relatively high sensitivity for the image-orthicon camera tube.

One of the difficulties experienced in IO operation is the requirement of maintaining the target plate between 95 and 140°F and at approximately the same temperature as the outer glass envelope of the tube. This requires a warm-up period of 30 min or more before the tube can be used. Normally, heat from the amplifier tubes and the deflection coils wrapped around the tube keep the tube at an operating temperature. During cold weather, a special internal heating system may be required if the camera is used out of doors.

If a bright scene is focused on the tube for a few minutes, the scene will continue to be observed for several minutes after the camera has been focused on another scene. The operator must be careful not to allow a bright scene to remain stationary on an IO screen. A system has been developed whereby the passage of photoelectrons is interrupted during blanking times, reducing the average time the scene is applied to the target plate and decreasing image "sticking" or "burn-in."

25-7 THE VIDICON

The most modern broadcast or closed-circuit TV camera tube is the *vidicon*. It is much smaller, lighter, less expensive, and simpler in design than the IO; it has comparable sensitivity without an electron multiplier; and optical lenses for it cost only a fraction of those for larger tubes. Its definition is slightly inferior.

A simplified representation of one type of vidicon is shown in Fig. 25-10. The light-active area

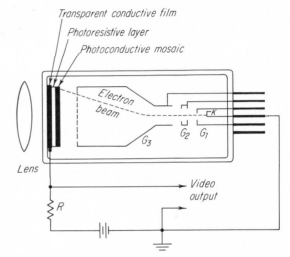

Fig. 25-10 Essential elements of one of several types of vidicon camera tubes.

at the front of the tube consists of three very thin films on the inside surface of the flat end of the tube. The first is a transparent conductive film. Next is a semiconductor photoresistive layer that is deposited on the conductive film. Lastly, a photoconductive mosaic layer is developed on the semiconductor layer. The middle semiconductor layer has the property of extremely high resistance (in the region of 50 MΩ) when in the dark but reduces its resistance radically when photons of light strike it. Thus, areas between the photoconductive mosaic islands and the conductive film act as tiny capacitors whose dielectric *leakage* is directly dependent on the light intensity at these areas.

The cathode *K* and the three grids develop an electron beam but limit its velocity to a low level, so that as the beam scans the mosaic areas, it merely charges these million or so tiny capacitors.

Light on the mosaic areas discharges the capacitors through the load resistor *R,* producing a current through it. The scanning beam recharges the mosaic capacitors and produces a video signal voltage-drop across *R* that is proportional to the light intensity at the areas being scanned.

Focusing can be accomplished electrostatically by controlling the voltages between grids or magnetically by wrapping a focusing coil around the midsection of the tube. Scanning is produced electromagnetically with horizontal and vertical yokes. Centering can be adjusted either by placing permanent magnets around the tube or by dc-biasing the deflection yoke coils.

Test your understanding; answer these checkup questions.

1. What is considered 100% modulation for the aural TV transmitter? _____ The video transmitter? _____
2. What device prevents aural RF from entering the video transmitter and vice versa? _____
3. About how many lines make up a TV picture? _____
4. In a TV receiver, what is the video IF? _____ Aural IF? _____ Horizontal sync frequency? _____ Vertical sync frequency? _____
5. What is the width of a TV channel? _____
6. What is the polarization of TV video signals? _____ Aural? _____
7. If a reflectometer shows 50 V radiated and 25 V reflected, what is the SWR? _____
8. What type deflection is used in oscilloscope CRTs? _____ In TV CRTs? _____
9. What direction deflection is produced by a deflection coil above a CRT? _____ What is the unit that is composed of horizontal- and vertical-deflection coils called? _____
10. What is eliminated by using interlaced scanning? _____
11. What is the aspect ratio of a TV picture? _____
12. What is the field frequency? _____ Frame frequency? _____
13. What was the name of the first TV camera tube? _____ Two others now in use? _____ Which is the smallest? _____
14. Which camera tube requires electron multipliers? _____ Which is used in closed-circuit TV? _____

25-8 THE SYNCHRONIZING-PULSE GENERATOR

The timing of the TV transmitter is developed in the synchronizing-pulse generator, or sync-pulse generator. This complex piece of equipment may require 60 tubes or transistors to generate the pulses shown in Fig. 25-11.

1. 31,500-Hz *equalizing* pulses having a duration of approximately 2.7 μs each, the leading edges separated by 31.75 μs
2. 15,750-Hz *horizontal sync* pulses, having a duration of about 5.4 μs, the leading edges separated by 63.5 μs
3. 15,750-Hz *horizontal blanking* pulses, having a duration of about 10 μs each
4. 60-Hz *vertical sync* pulses, having a total duration of 190 μs but slotted, or serrated, by 4.4-μs spacings every 27.3 μs
5. 60-Hz *vertical blanking* pulses, having a duration of 830 to 1,330 μs, depending on how

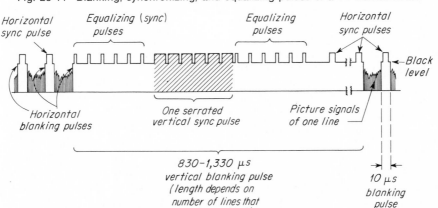

Fig. 25-11 Blanking, synchronizing, and equalizing pulses of a TV transmission.

many lines are used in the complete picture
6. Horizontal and vertical *driving* pulses similar to the sync pulses that are transmitted to receivers but slightly out of time to properly synchronize studio cameras and equipment

Note that the horizontal sync pulses ride on top of the narrow horizontal blanking pulses, except for a few (the number is determined by how many lines are used in the picture) that ride on top of the wide vertical blanking pulse, following the equalizing pulses. The visual modulation is inserted between the horizontal blanking pulses.

The wide vertical blanking pulse carries one serrated vertical pulse plus equalizing and horizontal sync pulse spaces.

Basically, the sync generator consists of an oscillator having a frequency of twice the horizontal sweep, or 31,500 Hz. Each cycle synchronizes a circuit to generate a narrow pulse at this frequency. When shaped properly, these are the equalizing pulses used before and after the vertical sync pulse. These equalizing pulses are also used to serrate, or slot, the vertical sync pulse into six parts. Serrations keep the horizontal sweep oscillators in receivers in synchronism during the long-duration vertical sync pulse. The vertical sync pulse fed through a low-pass filter triggers the vertical sweep oscillator in receivers. The equalizing and horizontal sync pulses fed through a high-pass filter trigger the horizontal sweep oscillator in receivers. The equalizing pulses allow the receiver horizontal oscillator to hold sync during the periods between the end of the last line of a field and the beginning of the next field line at the top of the picture. The equalizing pulses are used to produce the horizontal sync pulses at the transmitter by dropping out every other one, giving the required frequency of 15,750 horizontal sync pulses per second.

ANSWERS TO CHECKUP QUIZ ON PAGE 609

1. (*+25-kHz deviation*) (*Peak of sync pulse*) 2. (*Diplexer*)
3. (*495*) 4. (*41–47 MHz*) (*4.5 MHz*) (*15,750 Hz*) (*60 Hz*)
5. (*6 MHz*) 6. (*Horizontal*) (*Horizontal*) 7. (*3:1*) 8. (*Electrostatic*)   (*Electromagnetic*)   9. (*Horizontal*)   (*Yoke*)
10. (*Flicker*) 11. (*4:3*) 12. (*60*)(*30*) 13. (*Iconoscope*)(*IO and vidicon*) (*Vidicon*) 14. (*IO*) (*Vidicon*)

Just before the horizontal sync pulse is produced between lines, a blanking pulse must appear to cut off the electron beam in the camera tube in the transmitter and in the picture tube in the receivers. During the interval of the blanking pulse, the horizontal oscillators return the electron beam to the starting condition for the beginning of the next line. In receivers the sweep current is generated by a local sawtooth oscillator. The sync pulses transmitted as part of the complete TV signal are used to make any necessary correction to this oscillator frequency in order to assure the proper starting time of each line after the blanking pulse drops off. Thus, the equalizing, the horizontal, and the vertical sync pulses are added to the transmitted TV signal only as an aid to receiver oscillator operation; they are not used in picture pickup in the camera.

At the transmitter each piece of equipment has its own local sawtooth sweep oscillators. The driving pulses of the sync generator are used to keep these oscillators in proper synchronism. These pulses are similar to the transmitted horizontal and vertical sync pulses, except that they are timed to produce the required time difference for the transmitting equipment.

It is desirable to lock the sync generator into synchronism with the local public utility 60-Hz source. This prevents a vertical waver of the picture tube if there is ripple in the power-supply voltage. Ripple may also cause dark horizontal *hum bars* across the picture.

The sync generator also has a crystal oscillator capable of producing an accurate 60-Hz frequency if local power is not available, as is often the case when the equipment is operating in the field. With field equipment in operation, the remote sync generator develops the pulses for the main transmitter. The sync generator at the main station has an input circuit to allow it to lock in on signals originating from remote sync generators, from network programs, or from field equipment.

Transmission to the main transmitter from a studio or from a field pickup is usually made by a microwave transmitter using high-gain parabolic directional transmitting antennas to beam the signals in the 6.875- to 7.125-GHz (gigahertz) or 12.7- to 13.25-GHz bands.

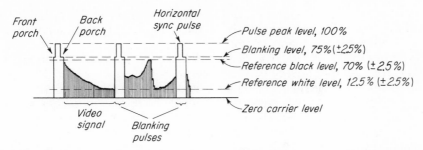

Fig. 25-12 Detail of monochrome horizontal blanking and sync pulses with two lines of picture information, or line writing.

25-9 MODULATION PERCENTAGES OF THE VISUAL CARRIER

There is never a constant-amplitude carrier in a visual transmission. The only parts of the emission that remain constant are the peaks of the sync pulses and the blanking level, shown in Fig. 25-12.

The top of any sync pulse is the maximum value of the emitter carrier and may be considered the 100% level.

The *blanking,* or *pedestal,* level, on which are found the *front porch* and the *back porch,* is 75% of the peak value (±2.5%).

The *reference black* level, which will produce the darkest black in the transmitted picture, is approximately 70% of the peak level.

The *reference white* level, which will produce the whitest white in the picture, is 12.5% of the peak value. It is undesirable to transmit a whiter white than this, as the power of the emission is then so low that noise at the receiving location may interfere with picture quality. More important, there will be insufficient carrier to beat against the aural FM carrier to create the 4.5-MHz beat frequency that produces audible signals in intercarrier receivers. This results in a buzzing sound in the receiver, often noticeable on weak signals.

The illustration shows two complete lines of video (picture), three blanking pulses with front and back porches, and three horizontal sync pulses on top of the horizontal blanking pedestals. The first line illustrates the line writing of a scene that is black at the left-hand side and white at the right-hand side. The second line is gray at the left-hand side, changes to black at the center, abruptly changes to white, and then to a light gray.

Another "IEEE standard scale" of modulation uses the blanking level as 0 and the reference white as 100. Zero carrier is then 120 and sync pulse peaks −40.

25-10 CAMERA CHAINS

A TV camera with an electronic viewfinder mounted on top of it and a camera-control unit with the necessary power supplies make up a *camera chain.* A block diagram of two camera chains and other video equipment necessary for the pickup of a live studio program is shown in Fig. 25-13.

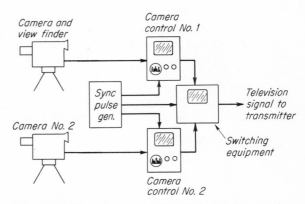

Fig. 25-13 Block diagram of two camera chains, synchronizing-pulse generator, and switching and monitoring equipment.

The camera usually has an electronic viewfinder with a 5- or 7-in. picture tube mounted directly above the camera itself and appearing to be part of the camera. The viewfinder operates much the same as a TV-receiver picture tube, except that it receives its picture information directly from the output of the video-frequency amplifiers in the camera and its blanking and driving pulses from the sync generator. The output of the camera is also fed to a nearby control position by low-impedance coaxial lines and to a camera-control unit. This unit has two cathode-ray tubes, one a 7- to 10-in. monitor tube similar to the viewfinder in the camera and the other a 3- to 5-in. oscilloscope with a horizontal sweep frequency that synchronizes the video and blanking signals in such a manner that a composite display of pulses and modulation is continually shown on the screen. The operator watches the video signal amplitude on the oscilloscope, controlling it to prevent video modulation peaks from exceeding the reference black and reference white levels.

A camera chain might be composed of one or two film projectors and a slide projector, all aimed at an iconoscope or vidicon tube, working through a *film multiplexer,* with a camera-control unit, as indicated in Fig. 25-14.

25-11 LOGS AND PERSONNEL REQUIRED

As might be expected, TV broadcast station log requirements are similar to those of standard broadcast and FM broadcast stations (Sec. 24-14). Program, operating, and maintenance logs are required. For the operating log it is necessary to take the transmission-line readings for both aural and video transmitters of the TV station. The maintenance log must contain information regarding removal and restoration to service of visual and aural modulation- and frequency-monitoring equipment.

Many people are involved in the production of even a simple live TV studio program. On the studio floor is a boom-microphone operator who keeps the microphone near the actors but out of sight of the cameras, one or more camera operators, and a floor director.

At the operational console in the control room, one or more camera-control operators are continually watching the technical condition of the camera signals. A program director watches the camera pictures and advises which camera to use for the program. Seated in this same room is an audio operator, controlling the microphone output that is fed to the aural FM transmitter. There is a telephone (chest mike and earphone) intercom system between control room and operating personnel in the studio by which directions are received and transmitted.

The motion-picture and slide-projection operator and equipment may be located in a separate room. The output of the film cameras also terminates at the switching position in the control room, with its separate monitor screen.

From the switching position at the visual console the TV signal is fed to the TV transmitter, while the aural signal is fed to the FM transmitter. In the transmitter room one or more operators monitor the visual and aural signals being transmitted and also keep a constant check on the operation of the transmitting equipment. These operators keep the operating and maintenance logs, and in small stations they may also keep the

Fig. 25-14 Block diagram of a single film camera with a multiplexer to allow it to pick up pictures from three projectors.

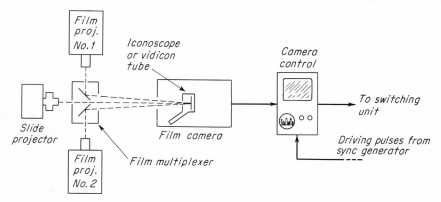

program log, although this job is usually detailed to other personnel. Operators at the TV station transmitter or remote-control point must have Radiotelephone First Class licenses.

25-12 MOTION-PICTURE PROJECTION FOR TV

The TV frame frequency was selected as 30 per second to allow transmitters and receivers to lock into synchronization and prevent waver of the received picture. Unfortunately, both 35- and 16-mm motion pictures have a frame frequency of only 24 per second. To reduce flicker, each frame must be produced as two or more fields for both motion pictures and TV. A standard 16-mm projector can be used for TV projection by using either a xenon-gas, electronically flashed light or an incandescent projection lamp with a special shutter to allow the following sequence of operations with a storage-type pickup tube (iconoscope, vidicon, etc.):

1. The film is pulled down into position in the projector. No flashed or shuttered light shines through the film as yet.
2. During the TV vertical blanking-pulse interval, the camera tube scanning beam is blanked out and is returning to the starting point. At this time a flash of light is allowed to shine through the film, projecting the picture on the mosaic.
3. The mosaic retains the electrically charged image of the picture.
4. The mosaic is scanned for one field, and the video signals obtained are transmitted.
5. At the end of the field, during the next blanking pulse and scanning retrace period, the projector light is again flashed, once again projecting the same picture on the mosaic.
6. The second (interlaced) field is scanned, and the video signals obtained are transmitted.
7. During the scanning time of the second field the next frame of the film is pulled down into position. During this frame, because of the difference between 24 and 30 frames per second, there is enough time to flash the projector light three times during blanking pulses before the next frame is pulled down.

In this way, the first film frame is scanned twice, producing 2 fields for the TV picture, but the second frame is scanned as 3 fields. This allows the correction that is necessary to change 48 fields per second for the film to 60 fields per second for the TV picture.

Accurate synchronization is required between blanking pulses and the operation of the shutter, or the flashing of the xenon light. This necessitates synchronous motors for the shutters, or a multivibrator-keyed light source synchronized by driving pulses from the sync-pulse generator.

25-13 TELEVISION TRANSMISSION REQUIREMENTS

A TV station has one antenna (a turnstile or similar horizontally polarized array) common to two transmitters, one the frequency-modulated audio signal and the other the amplitude-modulated picture signal. These two transmitters always operate with the center frequency of the aural, or sound, transmitter 4.5 MHz higher than the carrier frequency of the picture transmitter, ± 1 kHz. The whole TV transmission for one station must be contained in a 6-MHz-wide channel, as shown in Fig. 25-15. The visual carrier is

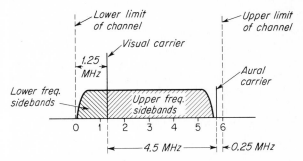

Fig. 25-15 Location of carriers and sidebands in a 6-MHz monaural TV channel.

1.25 MHz above the lower channel limit, ± 1 kHz.

The higher-frequency visual sidebands extend out more than 4 MHz before they are attenuated as the sound carrier frequency is approached. The lower-frequency sidebands are allowed to exist for only about 0.75 MHz before they are attenuated by a vestigial sideband filter in the output circuit of the transmitter (or by other means). Since only a vestige of the lower-frequency sidebands is transmitted, the emission is termed *vestigial sideband*. This emission (A5C) requires considerably less bandwidth than a true double sideband TV (A5) emission would.

The operating power of the aural transmitter is determined by either the direct or the indirect method (Sec. 24-8) with no modulation applied to the carrier.

The operating power of the visual transmitter is determined at the output of the vestigial sideband filter if one is used; otherwise, at the transmitter output terminal. The average power is measured while operating the transmitter into a dummy load of substantially zero reactance, with a resistance equal to the transmission-line surge impedance, and while transmitting a standard black picture. The average power is determined by direct measurements. When the antenna is coupled to the transmitter, the final-amplifier E_p and I_p values must be substantially the same as with the dummy antenna. Television transmitters are rated in peak power output, which is determined by multiplying the average power above by the factor 1.68.

The effective radiated power (erp) of a TV transmitter is the peak power times the antenna field gain (voltage gain over a simple dipole) squared (Sec. 20-28). Thus, if a transmitter has a peak power of 10 kW and an antenna has a field gain of 3, the erp is 10,000 $\times$ 3^2, or 90,000 W.

Since the transmitted sync pulse peaks always remain at the same amplitude regardless of percent of visual modulation, it is the peaks that are monitored to determine whether the transmitter output is within licensed limits. The peak power must be held within 10% above and 20% below the authorized power, except in emergencies. These same power limits are required of the aural FM transmitter, although the carrier power of the aural transmitter must be held to a value less than 70% of the peak power of the visual transmitter. This produces nearly equally effective visual and aural signals in receivers.

Test your understanding; answer these checkup questions.

1. What are the three frequencies produced by a sync-pulse generator? _____ _____ _____
2. What are the names of the six signals from a sync-pulse generator? _____ _____ _____ _____ _____ _____

3. Which pulse(s) is (are) serrated? _____ The longest? _____ The shortest? _____ The highest amplitude? _____
4. Which pulses hold the CRT beam cut off? _____ _____ Which impede the start of oscillators? _____
5. What is the waveform of the sweep voltages? _____
6. If the pulse peak is 100% modulation, what is the blanking level? _____ Black level? _____ White level? _____
7. What units are involved in a camera chain? _____
8. What three types of logs are required in TV? _____ _____ _____
9. What device allows one camera to accept pictures from three or more projectors? _____
10. Why would an IO not be used to pick up motion-picture signals? _____
11. What is the order of scanning four successive motion-picture frames for TV? _____
12. From the low-frequency end of a channel, what is the visual carrier frequency? _____ The aural carrier frequency? _____
13. What method of power-output measurement is used for the aural transmitter? _____ The visual? _____
14. A 10-kW transmitter feeds an antenna with a field gain of 3.5. What is the erp? _____
15. How much more is the peak power of a TV transmitter than the average power? _____

25-14 THE VISUAL TRANSMITTER

The visual transmitter is required to pass the carrier plus about 4.2 MHz of upper sidebands and about 0.75 MHz of lower sidebands. This is a total bandwidth of about 4.95 MHz. Past these limits the sidebands must be attenuated, at the high-frequency end to prevent visual sidebands from occurring at the aural carrier frequency and at the low-frequency end to prevent sidebands from appearing in the adjacent channel.

A common means of producing the desired transmitted band configuration is to grid-modulate the final amplifier and add a vestigial sideband filter between the output circuit and the antenna diplexer. This filter consists of tuned cavity wave traps with added LC circuits to present a constant impedance to the amplifier for all frequencies but still attenuate the sideband signals that must not be fed to the antenna.

Another way to develop the desired bandpass is to grid-modulate a low-level RF amplifier with the video signal. This amplifier must have a flat tuning characteristic over at least 8.4 MHz to produce acceptably the RF carrier and video sidebands 4.2 MHz on each side of the carrier. The class B linear amplifiers that follow the modulated stage can be overcoupled and stagger-tuned to produce the required 4.95-MHz vestigial sideband transmission. Series-circuit wave traps are also used in the amplifiers to assure proper attenuation of both sidebands.

To tune a TV transmitter properly, a *sideband analyzer* is used. This consists essentially of a narrowband receiver that is swept (reactance-tube-tuned) across the 6-MHz channel of the transmitter while a complex signal that consists of the carrier and all the video sidebands is being transmitted. The received signal amplitude is displayed on an oscilloscope CRT. The oscilloscope has a horizontal sweep in step with the receiver tuning sweep. The output of the receiver is fed to the vertical-deflection plates of the oscilloscope. The amplitude of the received signal as the receiver sweeps across the channel is displayed on the scope tube. Figure 25-16 illustrates two sideband-analyzer displays, one of a properly tuned amplifier producing both upper and lower sidebands properly and the other of an improperly

tuned amplifier in which the upper sidebands are attenuated excessively. This is a narrowband form of *spectrum analyzer*.

25-15 THE TV RECEIVER FRONT END

Frequency allocations for commercial TV are shown in Table 25-1. Each channel is 6 MHz wide. The lowest frequency of all channels is given. Only every tenth UHF channel is listed, since these channels are all contiguous. (Note that channel 1 has been deleted. Frequencies originally allotted to this channel are being used for other services.) There are three TV bands, lower VHF, upper VHF, and UHF.

It has been indicated (Sec. 25-1, Fig. 25-2) that the duty of the TV superheterodyne receiver is to amplify the received TV signal and separate it into aural and video components. The video signal is applied to the grid-cathode circuit of the picture tube. Horizontal and vertical sync pulses are separated from the visual signal and are used to synchronize the frequency of oscillation of horizontal and vertical sawtooth oscillators that produce the sweep currents for the picture tube. The aural signal is detected, amplified, and fed to the loudspeaker.

The *front end* of a superheterodyne receiver includes the antenna input, the RF amplifier stage or stages, the local-oscillator circuit, and the mixer stage.

VHF TV receiving antennas may consist of two Yagi beams. One is used to pick up low-band and the other high-band stations. Broad-banding techniques, such as the use of large-diameter antenna elements or the use of a folded-dipole

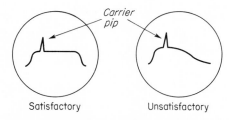

Fig. 25-16 Examples of a satisfactory and an unsatisfactory presentation on a sideband analyzer of a TV test transmission.

Table 25-1

Lower VHF band		Upper VHF band		UHF band	
Channel	Lowest frequency, MHz	Channel	Lowest frequency, MHz	Channel	Lowest frequency, MHz
2	54	7	174	14	470
3	60	8	180	24	530
4	66	9	186	34	590
	(4 MHz skipped)	10	192	44	650
5	76	11	198	54	710
6	82	12	204	64	770
		13	210	74	830
				83 (highest)	884

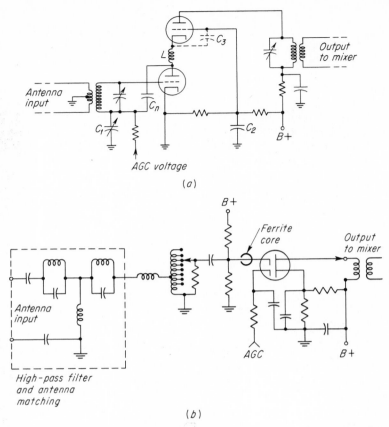

Fig. 25-17 RF amplifiers used in TV receivers. (*a*) VT cascode. (*b*) MOSFET.

main element, may be employed. Many V-beam-driven arrays are in use also. In the UHF band, broadband bow-tie, or conical antennas with sheet reflectors, are commonly used because of their ability to accept all frequencies above their fundamental with almost equal response. In weak-signal areas it may be necessary to have

a special beam antenna for each frequency to be received. Actually, this is desirable whenever optimum reception is necessary. Practically all modern TV receivers have a 300-Ω antenna input-impedance circuit. This allows use of 300-Ω twin lead as the transmission line, which is relatively inexpensive and simple to install.

The RF amplifier in a TV receiver must be capable of producing amplification and have low inherent noise. Any noise generated in this stage develops white dots, termed *snow*, in the picture. Pentodes or BJTs may be used as the RF amplifier, but triodes and FETs have a lower noise figure. A low-noise RF amplifier circuit found in many VT TV receivers is the *cascode* (Sec. 14-44). The circuit in Fig. 25-17*a* has two triodes in series, the first stage directly coupled to a grounded-grid second stage. Capacitor C_n forms a grid-neutralizing circuit for the first triode. Capacitor C_1 acts as a means of controlling neutralization and bypassing the grid coil to ground.

The amplified signal is fed directly to the cathode of the second tube through the small inductance L. The grid of this tube is bypassed to ground with C_2 and biased with a voltage divider. Being a grounded-grid amplifier, the second triode requires no neutralization.

The inductance L is in series with the grid-cathode interelectrode capacitance of the second tube, C_3, and with C_2 is made series-resonant to the middle of the upper VHF TV band. Signals near the resonant frequency produce relatively heavy current flow and therefore relatively high voltage across C_3 (cathode to grid). This provides added gain at these frequencies, which are normally subject to losses. The cascode RF amplifier also acts as a satisfactory isolation stage to prevent radiation of local-oscillator signals.

The diagram shows standard coil and capacitor tuning. In TV receivers each channel may have its own separate slug-tuned coil, utilizing distributed capacitance to complete the tuned circuit. When the channel selector in the receiver is changed from one channel to the next, a separate set of coils is connected into the RF amplifier-grid circuit, into the mixer-grid circuit, and into the oscillator circuit. Each of these coils may be separately peaked to give optimum signal for its own channel.

Figure 25-17b illustrates a MOSFET RF amplifier. After passing through a high-pass filter, the antenna signal is fed to a switch-selected coil for the desired channel and then to one of the MOSFET gates. The ferrite core on the gate lead stabilizes the circuit. The other gate is fed an AVC (called AGC, for automatic gain control, in TV). Note that B+ is used instead of V_{CC} in TV diagrams. The output signal is fed to the mixer stage.

The mixer stage of the VHF portion of a TV receiver is similar to the mixer stage of any superheterodyne. The local oscillator is usually a separate circuit coupled to the mixer and is often an ultraudion (sometimes termed a Pierce) circuit. It is possible to incorporate automatic frequency control (AFC) by adding a 45.75-MHz IF amplifier coupled to a reactance tube, reactance transistor, or Varactor diode across the local oscillator LC

circuit (Sec. 19-13) and feeding it the voltage developed at the discriminator. Such a circuit makes a fine-tuning, or trimmer, adjustment unnecessary. Vacuum tubes, BJTs, FETs, and diodes are used as mixers.

The UHF front end of a TV receiver usually has no RF amplifier, consisting instead of tuned tank circuits fed by the antenna, a triode or BJT oscillator, and a diode mixer. The circuit in Fig. 25-18

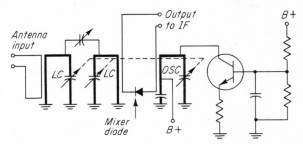

Fig. 25-18 UHF mixer circuit.

shows a hairpin antenna loop inductively coupled to a tuned hairpin tank capacitively coupled to a second tuned tank to improve selectivity. The second tank is coupled to a pickup hairpin with a diode mixer in series with it. The BJT oscillator is also coupled to the diode hairpin. The three tuned circuits are ganged.

25-16 THE IF STRIP

The output of the mixers is an intermediate frequency, or IF (Sec. 18-13). TV receivers now use a 6-MHz-wide IF, from 41 to 47 MHz. In this band of frequencies the sound carrier usually is made to fall on 41.25 MHz and the video carrier on 45.75 MHz.

Modern TV receivers amplify both the video carrier and sidebands as well as the sound carrier and sidebands with the same IF amplifiers. This is known as an *intercarrier* receiver. The transmitted signal and the desirable bandpass characteristics of the TV-receiver IF section are shown in Fig. 25-19. (In older, so-called *conventional* TV receivers, the sound IF was taken from the mixer stage separately. Slight variations of the local-oscillator frequency due to drift, detuning, etc., detuned the sound noticeably. This does not occur with the intercarrier receiver, because the

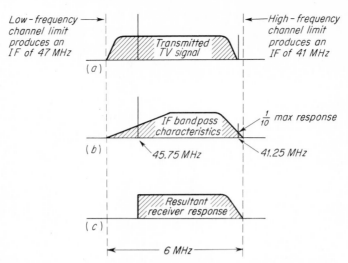

Fig. 25-19 (*a*) Transmitted carriers and sidebands of a TV signal. (*b*) Idealized IF response curve in a receiver. (*c*) Idealized resultant IF receiver response.

4.5-MHz IF is produced by the beating of the video and aural carriers, which are always transmitted 4.5 MHz apart.)

At the transmitter it would be difficult to cut off all the lower-frequency sidebands and leave just the carrier and the higher-frequency sidebands. Instead, a vestigial lower sideband, as shown in Fig. 25-19*a,* is transmitted. To utilize the lower-frequency sideband, it is desirable to slope the receiver bandpass characteristics, as illustrated in Fig. 25-19*b*. In this way the addition of the upper and lower sidebands near the carrier, those carrying the low video modulating frequencies, produces a sum equal to the maximum of the higher-modulating-frequency sidebands. The result is an idealized response to the transmitted signal, as shown by Fig. 25-19*c*. The output from the IF amplifiers contains all the sidebands starting at the carrier out to more than 4 MHz with equal amplitude. Signals displaced farther than this are attenuated.

The slope of the bandpass curve must drop off in such a way that only 10% of the maximum amplification is given to the sound carrier. This prevents the sound signal from forcing its way through to the picture tube and producing dark and light bars across the screen. It also makes the sound-carrier amplitude similar to the whitest white-signal amplitude, which allows the sound and video carriers to beat together efficiently to

produce a 4.5-MHz sound-signal IF in the receiver.

Different types of coupling have been used in TV IF stages to produce the required bandpass characteristics. Some are impedance-coupling, single-tuned transformers, double-tuned transformers, autotransformers, and the *bifilar* transformer. A bifilar transformer in Fig. 25-20 has the

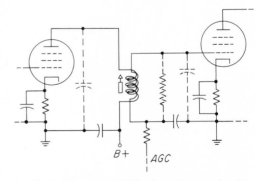

Fig. 25-20 An interstage bifilar transformer circuit.

primary and secondary turns interwound in a form of unity coupling. The lumped plate-ground and grid-ground distributed capacitance of the two pentodes shown acts as the capacitance of the tuned circuit. A single powdered-iron-core slug broadly tunes the bifilar-transformer circuits. The resistance shown in broken lines is sometimes used to lower the Q, broaden the tuning, and decrease the possibility of oscillation in the stages. Bifilar-transformer coupling, besides being economical to manufacture, has the advantage of reducing noise impulses on the picture-tube screen that result when capacitive-coupling circuits are employed. A high-amplitude noise pulse charges the coupling capacitor of an impedance-coupling circuit. The discharge of the capacitor through high-resistance coupling resistors represents a relatively long-duration signal that produces white tails on any black noise spots on the screen, making the noise much more noticeable.

The IF strip in a TV receiver has two to four amplifier stages, with AGC applied to them. To produce the desired bandpass characteristics the stages are stagger-tuned. The first transformer may be peaked at 43.1 MHz, the second at 45.3 MHz, and the third at 43.1 MHz. This arrangement should result in a bandpass characteristic similar to the curve in Fig. 25-19b.

There are several frequencies other than the desired TV channel signals that may be able to force their way through the front end and the IF strip and cause visual or audible interference. To prevent this interference, wave traps are used in TV receivers. To prevent blanketing or overpowering of the front end by local commercial or amateur radio signals, a high-pass filter is incorporated in the antenna input circuit, designed to attenuate all frequencies lower than 54 MHz but pass all higher frequencies without loss.

Because of the broadness of the IF strip, one source of interference may be the sound carrier of a station operating on the adjacent lower channel. Another source of interference may be the visual carrier of a station operating on the adjacent higher channel. Either series-resonant or parallel-resonant traps tuned to the IF frequencies produced by such interfering signals, namely, 47.25 and 39.75 MHz, may be included in the IF amplifiers. Such traps not only attenuate the undesired responses but also tend to narrow the skirts of the IF bandpass out past the channel limits.

An important wave trap, connected in the first video amplifier, is tuned to 4.5 MHz to prevent the sound signals from reaching the picture tube.

When operating from higher-voltage power supplies, transistor stages may be connected in series across the supply as in Fig. 25-21. The input signal feeds a bifilar transformer with two wave traps (47.25 and 41.75 MHz) between it and ground. Note the collector current path through the two BJTs in series and also the series biasing network. If a positive AGC voltage is applied across the first emitter resistor, collector current in both BJTs will decrease, lowering the gain of the stages. The 47-Ω resistors help to prevent oscillation of the relatively low-Q stages. Ferrite cores might be used instead.

Fig. 25-21 Series-connected two-stage IF amplifier.

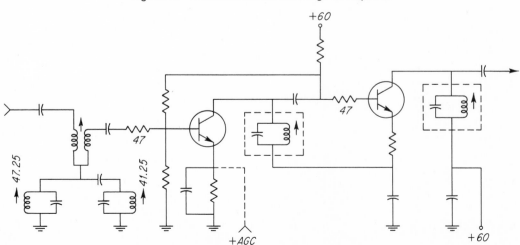

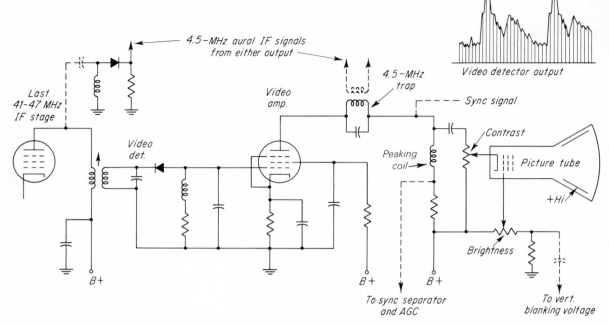

Fig. 25-22 Last IF stage to picture tube of a monochrome TV receiver.

Although shown here as discrete VT, BJT, or FET stages, modern TV receivers may have several ICs. One IC may be used for the whole IF section except for the transformers, another for the whole aural section, another for the video amplification, and so on.

25-17 DETECTOR, AURAL, AND VIDEO CIRCUITS

Signals from the last IF stage are fed to a diode detector, which rectifies them to a varying dc having a waveform that includes the sync and blanking pulses as well as the video information, as indicated in Fig. 25-22. Since both the video carrier (45.75 MHz) and the aural carrier (41.25 MHz) are mixed in the diode, their difference frequency, 4.5 MHz, is also available. This 4.5-MHz beat frequency is modulated by the FM produced at the aural transmitter and is also amplitude-modulated by the visual-carrier sidebands, both of which would be amplified by the video amplifier that follows the detector. To prevent the aural signals from producing black sound bars on the picture tube, a 4.5-MHz wave trap can be included in the video amplifier as shown. If the wave trap is wound with a secondary coupled to it, 4.5-MHz aural carrier signals for the aural IF stages can be extracted at this point.

Another method of obtaining the 4.5-MHz aural IF is by capacitively coupling from the output of the last IF stage to a separate mixer diode, the output of which is fed to the aural IF circuit.

The aural circuits consist of a 4.5-MHz IF amplifier stage feeding a 4.5-MHz ratio detector, discriminator, or with VT receivers, a quadrature FM detector, similar to circuits discussed in Chap. 19. The quadrature detector requires only a single beam-power amplifier to produce loudspeaker volume. With either solid-state or VT ratio detectors or discriminators, one or two low-level AF amplifiers are required to drive the AF output amplifier. In many TV receivers all the aural circuitry up to the power amplifier may be contained in a single IC.

The video-blanking-sync signals from the second detector diode of monochrome receivers are fed to a single, direct-coupled, wideband (4.2-MHz) video amplifier and then directly to the grid-cathode circuit of the picture tube, somewhat as illustrated. These voltages control the intensity of the beam striking the face of the cathode-ray tube. To extend the high-frequency capabilities of the video amplifier circuit past about 2 MHz, a special resonant or peaking circuit may be added in the plate circuit. It forms a parallel-resonant circuit with distributed capacitance at a frequency of about 4 MHz. This results in a nearly flat amplifier response out to the required frequency limits.

If the video amplifier is not directly coupled to the picture tube, it is necessary to add a bias that remains just at the blanking level but which will follow any variation of the amplitude of the peak of the sync pulses due to fading. In this way variations of signal voltages of light and dark scenes will not affect the bias to which the video signals are added. Thus, light scenes remain light and dark scenes remain dark. Such a biasing circuit is known as *dc reinsertion,* and it is sometimes applied to signals being fed to the picture tube grid. The amount of CRT cathode-grid bias controls the *brightness* of the picture displayed. The video amplifier signal gain circuit, shown highly simplified, controls the *contrast* between light and dark signals on the screen. The grids indicated past the control grid on the CRT are the accelerator and focusing electrodes. The aquadag coating on the inside of the CRT near the screen, sometimes called the *ultor,* is supplied a positive potential of 9 to 12 kV greater than the cathode in monochrome tubes and 22 to 25 kV with color tubes.

In transistorized TV receivers the video amplification section is quite similar to vacuum-tube circuitry except that there are usually two amplifier stages. Later models may use a single IC for the whole section.

Test your understanding; answer these checkup questions.

1. What device may be used to hold video sidebands within required limits? _____ To indicate if all required sidebands are being transmitted? _____
2. What are the three TV bands called? _____ _____ _____
3. What circuits make up the front end of a TV receiver? _____
4. What is the impedance at the antenna input terminals of a TV receiver? _____
5. Why are triodes and MOSFETs best for TV RF amplifiers?

_____ In what circuit configuration are they used? _____
6. What type of LC tank is used in the UHF portion of a TV receiver? _____
7. When receiving channel 2, on what frequency must the mixer oscillator be operating? _____
8. Why are TV IF stages stagger-tuned? _____
9. What are two wave-trap frequencies that might be used in a 42-MHz TV IF strip? _____ _____ What wave trap is used in the video amplifier? _____
10. Why might transistorized IF stages not use bifilar transformers? _____
11. From what two points might aural IF information be taken in a TV receiver? _____
12. What four signals might a video amplifier amplify? _____ _____ _____ _____
13. What does a peaking coil do in a video amplifier? _____
14. What controls brightness of a CRT picture? _____ The contrast? _____
15. When is dc reinsertion required? _____

25-18 SYNC-PULSE SEPARATION

Thus far, the TV signal has been received, converted to an IF, amplified, and detected. The aural signal has been separated, amplified, detected, and fed to a loudspeaker. The video signal, consisting of blanking and sync pulses plus the modulation of each line of picture information, has been amplified and fed to the cathode-grid circuit of the picture tube. The intensity-modulated CRT electron beam determines how many electrons strike the fluorescent inner coating on the CRT face during horizontal beam traces and the amount of light developed on the screen.

The narrow sync pulses, perched on top of the blanking pulses, are also fed to the CRT cathode-grid circuit, but since they represent a "blacker than black" voltage, they produce no effect on the screen.

The video signal is also fed to a tube or transistor that biases itself to pass only the sync pulses having amplitudes greater than the blanking-voltage level. Such a circuit, known as a *sync separator,* may employ diodes, BJTs, triodes, or pentodes. Basic operation of such a stage can be explained by the circuit in Fig. 25-23. With no signal, the 10-MΩ resistor holds the bias on the triode grid to a fraction of a volt positive, resulting in a relatively high I_p. Incoming video signals charge C enough to produce an average class C bias (zero I_p) for the triode. The only parts of

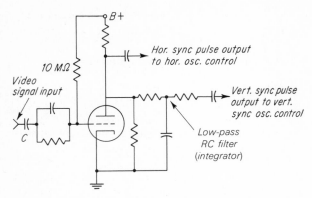

Fig. 25-23 Sync-separator circuit using a triode.

the video signal that can appear above I_p cutoff and produce I_p are the peaks of the horizontal sync pulses, equalizing pulses, and serrated vertical sync pulses. These pulses, which will syn-

chronize the oscillator frequency of the horizontal oscillator at 15,750 Hz, are taken off through a small series capacitor. A high-frequency pulse passing circuit such as this is known as a *differentiator* circuit. To synchronize the vertical oscillator at 60 Hz, the sync-separator amplifier output is also passed through a low-pass RC filter, in this application called an *integrator* circuit. The only pulses that can pass through the integrator are the low-frequency (60 pulses/s) vertical sync pulses. The narrow serrations on the vertical sync pulses represent high frequencies and do not appear at the output of the integrator.

25-19 VERTICAL-DEFLECTION CIRCUITS

To develop a TV picture on the screen it is necessary to drive the electron beam horizontally across the face of the tube 15,750 times a second and, at the same time, move it down the screen relatively slowly and then back up rapidly 60 times a second. The sawtooth ac that produces the downward vertical deflection and the flyback is the simpler of the two deflection systems and will be described first.

A two-tube multivibrator-type circuit used to develop a 60-Hz sawtooth ac is shown in Fig. 25-24. Two triodes or two BJTs might also be

Fig. 25-24 Unbalanced power-multivibrator circuit used to produce sawtooth sweep voltages for vertical deflection.

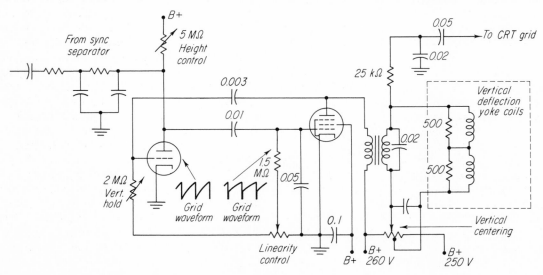

used. This particular circuit consists of a triode coupled to a beam-power tetrode with capacitive feedback to the triode grid. The selection of proper R and C values results in a relatively slow, linear voltage buildup on the tetrode grid, even though the voltage waveform at the grid of the triode may have a nonlinear RC time-constant curve as indicated. The 0.003-μF capacitor feeds back (in phase) some of the amplified signal, maintaining oscillation of the two-tube circuit. The grid-to-grid dc circuit feedback via the three resistors provides a method of controlling linearity of the waveform applied to the tetrode grid. Controlling the plate-load resistance of the triode determines the amplitude of the sawtooth sweep voltage, thereby furnishing a height control for the *raster* (unmodulated lines on the screen).

The 2-MΩ rheostat in the triode grid circuit, in conjunction with the 0.003-μF capacitor, determines the frequency of oscillation and enables the oscillator to be synchronized by the received vertical sync pulses to produce the correct vertical sweep frequency. This is called the *vertical-hold* control.

For vertical centering of the picture on the screen, a dc bias can be introduced into the yoke coils by using a center-tapped potentiometer as shown. Centering can also be accomplished mechanically by positioning of the yoke coils.

If the phase of the synchronizing pulses fed to the oscillator through the RC integrator circuit is incorrect, a single amplifier stage may be used to invert the phase.

Note that each coil of the vertical yoke has a 500-Ω damping resistor shunted across it to decrease inductive effects during the rapid-retrace or flyback periods. During the flyback time a voltage from the vertical yoke is fed to the CRT grid (or cathode) to blank the screen, preventing any retrace line from the bottom of the screen to the top from showing. Although shown in series, vertical yoke coils are often connected in parallel.

Transistorized vertical-sweep circuits may use a multivibrator or a blocking-type oscillator that develops narrow pulses at the sweep frequency. The pulses are amplified by a driver stage, and the required sawtooth waveform is developed by an RC circuit at the input of the output amplifier stage. An IC may be used for all this circuitry

except for the larger capacitors and the vertical-linearity, hold, and height-control potentiometers.

25-20 HORIZONTAL-DEFLECTION CIRCUITS

The horizontal section of a TV receiver includes many functions. It must generate the 15,750-Hz sawtooth current that sweeps the beam across the screen in exact synchronism with the sweep of the beam in the camera tube at the transmitter. The same ac that accomplishes the horizontal sweep is also stepped up and rectified to produce the required 10-kV ultor voltage for monochrome picture tubes or the 23 kV for color tubes. A small portion of the horizontal flyback voltage is used to key an automatic gain control (AGC) circuit into operation. Some of the output voltage may be shaped and added to the vertical yoke to decrease *pincushion* distortion of the raster (pulling in at the center of the top, bottom, and right and left sides of the picture), or it may be shaped and used for convergence correction in color TV circuits. A basic horizontal-deflection circuit is shown in Fig. 25-25. The 15,750-Hz multivibrator oscillator is in the center of the system.

The multivibrator could be synchronized by feeding a positive or negative pulse (from the sync separator) to one of its grid circuits. However, any received static or noise pulses could upset the proper synchronization and result in tearing out of portions of the picture. A better system is the one shown, using a phase inverter feeding a phase detector to form an automatic frequency control (AFC) circuit for the multivibrator. The AFC circuit feeds either a negative or positive biasing voltage to the grid of one of the oscillator tubes or transistors. The additional bias forces the oscillator to increase or decrease its frequency until it agrees with that of the pulses coming from the sync separator.

The 15,750-Hz ac generated by the multivibrator appears across the 0.0005-μF capacitor. The basic frequency of oscillation is controlled by adjusting the grid-leak resistor in the second tube, the *horizontal-hold* control. The parallel LC tank in the plate circuit of one of the multivibrator tubes is resonant to 15,750 Hz and stabilizes oscillations. The frequency of oscillation

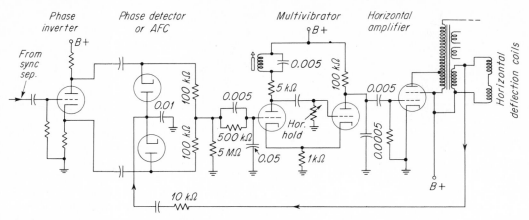

Fig. 25-25 Horizontal-deflection system with AFC, sawtooth multivibrator oscillator, and horizontal amplifier feeding the deflection coils.

can be varied slightly by variation of any grid or plate component of this cathode-coupled type of oscillator, however.

The sawtooth ac output is fed to a grid-leak-biased class C beam-power-tetrode (BJT or SCR) amplifier, which is transformer-coupled to the horizontal-deflection coils. The buildup of current in these coils deflects the electron beam across the face of the picture tube 15,750 times per second.

A flyback pulse of this sawtooth wave is fed back to the AFC circuit to be compared with horizontal sync pulses from the sync-separator via a phase-inverter circuit. Each sync pulse is fed by the phase inverter to one diode as a positive voltage and at the same time as an equal-value negative voltage to the other diode. Thus, both diodes try to charge the 0.01-μF capacitor, but since the charging currents are 180° out of phase and assumably equal, the capacitor receives no charge. The center tap between the two 100-kΩ diode resistors would also be at zero potential insofar as the pulses are concerned. However, if the deflection circuit pulses are ahead of or behind the sync pulses, the sum of deflection pulses and sync pulses will result in one diode drawing more current than the other, and the 0.01-μF capacitor will charge either positively or negatively. This will result in the center tap of the 100-kΩ resistors having either a positive or a negative potential. Changing the bias potential of the multivibrator grid circuit in this way will change the time of conduction of one grid of the oscillator

circuit and correct the frequency of oscillation. The RC network between the AFC and multivibrator circuits has a long time constant to prevent rapid changes in the frequency of oscillation.

25-21 HORIZONTAL OUTPUT CIRCUITS

The horizontal amplifier output circuit feeds the deflection current to the horizontal-deflection coils, produces the high voltage for the picture tube, and also develops a boosted $B+$ voltage for the horizontal amplifier and for the first anode of the picture tube. A simplified circuit is shown in Fig. 25-26.

As the sawtooth ac builds up slowly on the grid of the horizontal amplifier, the plate current builds up through the transformer primary and through the damper tube. This linear current change produces a constantly increasing deflection-coil current and magnetic-field strength, resulting in the electron beam's being deflected relatively slowly but evenly across the picture-tube screen. The current flows in a downward direction through the coils during the buildup of the magnetic field.

When the sawtooth ac on the grid drops suddenly to a high negative value, plate current in the amplifier drops to zero. The magnetic field around the deflection coils collapses rapidly inducing a high-amplitude flyback emf across the transformer secondary. This produces a sharp pulse of high-amplitude flyback current flow upward in the secondary of the transformer and a

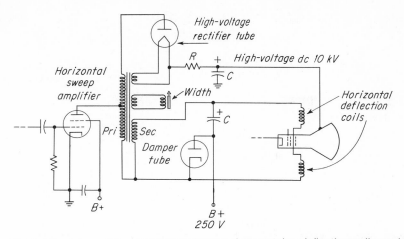

Fig. 25-26 Details of the horizontal amplifier, output transformer, damper tube, deflection coils, and high-voltage circuit.

high induced flyback emf downward in the primary winding, making the top of the primary positive. The many thousand volts induced into the primary by this action is rectified by the high-voltage rectifier, filtered with an RC network, and connected to the high-voltage anode of the cathode-ray picture tube.

During the flyback period, the energy stored in the magnetic field around the deflection coils would tend to produce damped oscillations in the LC circuit consisting of these coils and any distributed capacitance that existed. This would interfere with production of the next sawtooth wave. However, when the collapsing magnetic field induces a current flow downward in the coils, the damper tube passes current, charging the capacitor C. The energy that would have produced oscillations is now contained in the capacitor. The polarity of this charge is such that it is in series with the plate supply voltage. This results in a source of "boosted" voltage for the horizontal amplifier plate circuit of about twice the power-supply voltage.

The width of the picture can be controlled by shunting a small variable inductance across a few turns of the output transformer, as shown.

The high voltage developed by horizontal flyback and connected to the picture tube ultor often exceeds 10,000 V. When a TV set is serviced, be sure to discharge this circuit to ground after the TV receiver has been turned off before attempting to make any internal adjustments to the set. Care

must also be taken not to strike or jar the picture tube, since it can implode violently and throw broken glass and parts many feet.

25-22 AUTOMATIC GAIN CONTROL

AGC in a TV receiver is similar to automatic volume control, or AVC, in AM receivers. A received AM carrier is rectified, filtered with a long-time-constant RC filter, and fed as a negative bias to the grid circuits of the RF and IF amplifier tubes to produce nearly constant average signal at the second detector regardless of the strength of the signal received. AVC is possible in an AM (or FM) receiver because the average voltage of the carrier is constant, with or without modulation.

An AVC circuit operating in a similar manner from the second detector of a TV receiver would develop a low bias voltage for light scenes and a high bias voltage for dark scenes, which would not produce the desired pictures. However, if the AGC voltage is developed from the amplitude of only the sync pulses, which do not change when picture amplitudes change, a satisfactory bias voltage will result. The first AGC system employed a sync-separator circuit and a low-pass RC filter. The horizontal and vertical pulses were filtered to a smooth dc and used as the AGC voltage. However, airplanes flying overhead reflected rapidly changing in-phase and out-of-phase signals to the receiver and produced wildly changing AGC amplitudes that sometimes threw the receiver

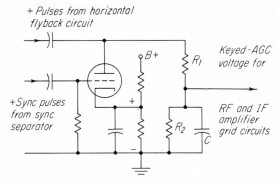

Fig. 25-27 A method of producing keyed AGC voltage.

completely out of synchronization. This was overcome by the use of *keyed AGC.*

The basic theory of a keyed AGC system can be explained by the simplified diagram of Fig. 25-27. Two sets of pulses are fed to the AGC tube. Positive pulses developed by flyback in the horizontal output transformer are fed to the triode plate. These pulses last for only a few microseconds. Positive horizontal sync pulses from the sync-separator stage are fed to the negatively biased grid. If both grid and plate are keyed positively at the same instant, the tube conducts and electrons flow from cathode to plate. When the pulses drop off, the electrons on the plate flow down through resistors R_1 and R_2, charging capacitor C negative at the top. This negative voltage is fed to the RF and IF grid circuits as the AGC voltage. If noise pulses are fed to the grid through the sync separator, they cannot produce plate current unless the plate is also positive. Thus, the keyed AGC discriminates against noise impulses that occur between sync pulses (during the reception of the visual modulation). The RC time constant can be relatively short, allowing the AGC voltage to follow reasonably rapid fading, such as that produced by airplanes, making this fading only slightly noticeable. In some cases a dc amplifier is used to increase the AGC effectiveness.

It should be noted that this and the other circuit diagrams shown may not be exactly the same as those found in actual TV receivers. For each simplified basic circuit described here, there are dozens of variations in practical use. For more detailed explanations of TV circuits, refer to TV-receiver texts.

Test your understanding; answer these checkup questions.

1. When coupling from a sync separator, what is used to obtain the horizontal sync pulses? _____ Vertical? _____
2. About how many lines would make up a raster? _____ A TV picture? _____
3. What does a vertical-hold control do? _____
4. What is controlled by varying a bias current in yoke coils? _____
5. What is the purpose of a voltage fed from the vertical sweep to the CRT grid? _____ When must it occur? _____
6. What are two ways of obtaining sawtooth ac used in TV sweep circuits? _____ _____
7. Where does pincushion distortion occur? _____
8. What holds the horizontal oscillator on frequency? _____
9. What are four uses of the signal from the horizontal output transformer? _____ _____ _____ _____
10. What does a horizontal-hold control do? _____
11. Carrier amplitude determines AVC voltage. What determines AGC voltage? _____
12. Why is keyed AGC better than simple pulse-voltage AGC? _____
13. What is the result of using a damper tube? _____
14. What are the three types of active devices used in horizontal output stages? _____ _____ _____
15. If solid-state diodes have a maximum inverse voltage rating of only about 500 V, how could they be used as the high-voltage rectifier in a TV receiver? _____

25-23 COLOR

Radio waves are a form of radiant energy that will travel through space. Light differs from radio waves in frequency or wavelength only. Red light has a wavelength of about 700 nm (700 nanometers, or billionths of a meter; also measured in millimicrons, mμ), green light about 530 nm, and blue light about 450 nm. If the correct proportions of only these three pure (saturated) colors (hues) are added together, any desired color can be produced. For example, a blue, a green, and no red produces a blue-green, called cyan. A green, a red, and no blue results in yellow. A blue, a red, and no green produces a purple color called magenta. With proper levels of green, blue, and red, all color perception in the eye is canceled and white is seen. Thus, to transmit a picture in full color it will only be necessary to scan the picture simultaneously for its blue content, for its green content, and for its red content. When the same percent of each color is projected on a screen at the same time and in the correct places, the picture appears as the original. Bright white areas in the picture will be projected as strong pure green, strong pure red, and strong pure blue colors. Gray areas will be projected as weak green, weak red, and weak blue colors. Black (dark) areas will be projected with no green, no red, and no blue. By proper combinations of hues and intensities, any desired color can be reproduced as well as white, grays, and black.

The color triangle in Fig. 25-28 represents a

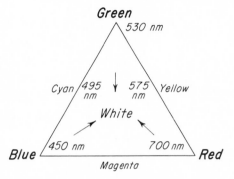

Fig. 25-28 Color triangle with wavelengths of hues.

rough approximation of the distribution of colors it is possible to reproduce by striking picture-tube phosphors with an electron beam. None of the colors shown can actually be reproduced in a 100% saturated form, but the possible percent is equal to, or better than, that obtained with printing inks. Color TV can reproduce very lifelike colors if both transmitting and receiving systems are operating properly.

25-24 TRANSMITTING COLOR SIGNALS

The TV color camera consists of a lens system that focuses the picture to be televised onto and through special *dichroic* glass mirrors. A dichroic mirror will reflect the color for which it is made and pass all other colors through it. The three reflected color scenes are picked up by three separate camera tubes (vidicons), as shown in Fig. 25-29. In this way the red, green, and blue color content of a picture is separated into three different signals. The three camera tubes scan their respective color scenes in unison, developing modulated-line information much as a monochrome camera tube does. Eventually, these three electrical signals will be made to illuminate red, green, and blue phosphor dots or lines on the face of a color-picture tube and the original scene will be reproduced in color.

At the transmitter, the outputs of the camera tubes are amplified by a series of video amplifiers and are *gamma*-corrected. Gamma (a measurement of contrast) correction is required because the camera tubes produce a brightness output that does not correspond to the brightness recognition of the human eye. The three gamma-corrected video signals are then fed to a transmitter *matrix*. In the matrix the three color signals are combined into one composite signal with a ratio of 59% green, 30% red, and 11% blue. This is done to produce a signal that will discriminate better against noise and also produce a good rendition of whites, grays, and blacks when the signal is viewed on a monochrome receiver. The output from this part of the matrix is known as the *Y*, or *luminance*, signal. It contains video frequencies up to 4.2 MHz and produces a black-

Fig. 25-29 Block diagram of color camera and circuits used to produce the *Y*, or luminance, signal.

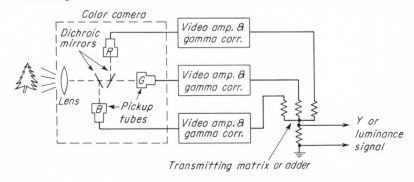

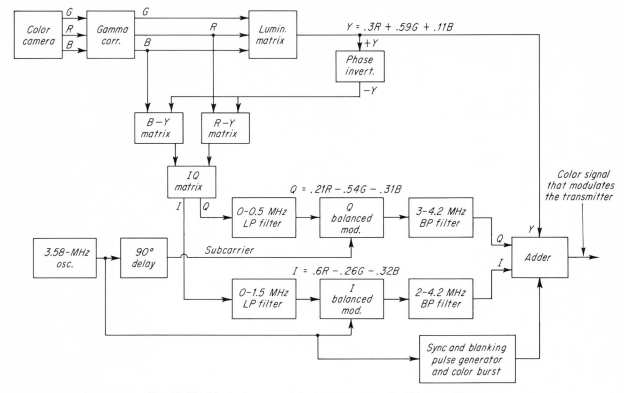

Fig. 25-30 Block diagram of the color system of a TV transmitter.

and-white picture on any monochrome TV receiver. Thus, the color transmission will be *compatible* (receivable on either a color or a monochrome receiver).

A simplified block diagram of the color section of a TV transmitter is shown in Fig. 25-30. The Y signal is properly proportioned in the luminance matrix and is fed in two directions. One is to the adder, in which luminance, color, sync, and blanking pulses, and a sample of the color subcarrier frequency, called a color burst, are added

to form the modulating signal for the color TV transmitter.

The positive potential luminance ($+Y$) signal is also fed through a phase inverter, becoming a $-Y$ signal. The blue signal is added to the $-Y$ signal in a $B - Y$ matrix. Similarly, a $R - Y$ signal is developed. These are further added into in-phase (I) and quadrature (Q), meaning 90°-out-of-phase, components, each having red, green, and blue voltages with the polarities and relative amplitudes indicated.

The $+Y$ signal controls the brightness of the color picture and also produces the visible signal on monochrome receivers. The I and Q signal modulation is used to carry the color information from the transmitter to color receivers, where it will be added in proper proportions of red, green, and blue to the Y signals and then applied to the three-color TV tube screens. The amplitude of the I and Q signals will determine the saturation (purity) of the colors. The phase developed by the difference in amplitude between the I and Q sig-

---

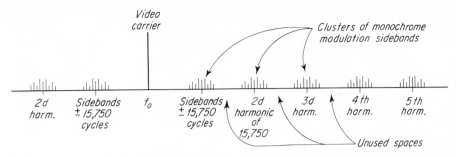

Fig. 25-31 A monochrome emission groups its sidebands around harmonics of 15,750 Hz, leaving space for color information in between.

nals will determine the *hue* (actual color) produced on the receiver screen.

The problem of transmitting the regular black-and-white luminance signal, requiring 4.2 MHz of sidebands, and at the same time the chrominance information, without increasing the required bandwidth, was ingeniously solved. Sidebands that are produced by monochrome modulation were found to cluster around frequencies that are harmonics of the line frequency of 15,750. As a result, there are spaces between these clusters that are not used, as shown in Fig. 25-31.

By careful selection of available frequencies, it was found that a subcarrier with a frequency of 3,579,545 Hz above the video carrier would produce its sidebands in the spaces between the *Y* sidebands. In determining the desired subcarrier frequency it was found necessary to change the line frequency slightly. A monochrome transmitter uses a line frequency of 15,750 Hz and a field frequency of 60 Hz. A color transmitter uses a line frequency of 15,734.26 Hz and a field frequency of 59.94 Hz. The difference between these frequencies is so slight that the TV-receiver oscillators will hold synchronism without adjustment of the sync controls.

If the subcarrier frequency is amplitude-modulated by *I* and *Q* signals, the sidebands that are produced will fall in the spaces between the sideband clusters produced by the luminance modulation. In this way the luminance and the chrominance modulation can be transmitted within the same 4.2-MHz bandwidth. The positions of the video carrier and its sidebands, the chrominance subcarrier and its *I* and *Q,* sidebands, and the aural-carrier center frequency are shown in Fig. 25-32.

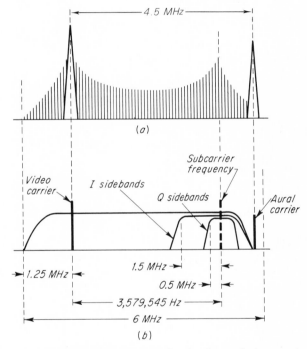

Fig. 25-32 (a) Appearance of color transmission on a spectrum analyzer. (b) Placement of color signal subcarrier and sidebands in a 6-MHz TV channel.

A significant difference between the luminance and the chrominance bandwidths is indicated. It has been found that the fine detail made possible by the highest-frequency sidebands of a TV signal is not necessary in transmission of color. The human eye sees color in large areas and to a certain extent in moderately small areas, but color is not apparent in fine detail. Therefore the fine detail in a color transmission can be carried by the *Y* signal alone, and a narrow chrominance bandwidth may be used. When the chrominance information is inserted where it appears to be high-frequency sidebands to the video carrier, it shows on the monochrome screen as very-small-detail patterns. (In this respect the chrominance signals act as weak interference signals to the luminance carrier and information.) If the interlaced color fields are transmitted properly, a near cancellation of the chrominance signals occurs on the picture-tube screen, making black-and-white renditions of color signals quite acceptable.

The subcarrier frequency in Fig. 25-32b is shown by a broken line because it is suppressed and is not actually transmitted. If it were transmitted, it would produce an objectionable pattern on the receiving picture-tube screen.

At the transmitter, the *I*-chrominance signal from the matrix is first fed through a 0–0.5-MHz low-pass filter to remove all higher frequencies. (It is still a video-frequency camera signal, not sideband, because it has not yet been used to modulate a carrier.) This filtered signal is fed to a balanced modulator that is being fed the 3,579,545-Hz subcarrier (3.58 MHz for simplicity) at the same time. The balanced modulator cancels the carrier but leaves the upper and lower sidebands produced by the *I* signal. These sidebands are ac signals ranging in frequency from $3.58 - 0.5 = 3.08\,\text{MHz}$ to $3.58 + 0.5 = 4.08\,\text{MHz}$. They are fed through a 3–4.2-MHz bandpass filter to remove any spurious products of modulation and are added to the *Y* signal before it modulates the video carrier of the transmitter (Fig. 25-30).

The *Q* signal is also used to modulate the 3.58-MHz subcarrier in a balanced modulator. It is first passed through a 0–1.5-MHz low-pass filter to remove any higher (finer-detail) frequencies. The carrier that the *Q* signal modulates is also 3,579,545 Hz, but the *Q* and *I* carriers are out of phase by 90°. When the *I* and *Q* signals are added to the *Y* signal, the two sets of amplitude-modulated chrominance sidebands are 90° out of phase. Together they form a resultant which acts as phase modulation. The phase angle is determined by the relative amplitude of the two signals. In the receiver the resultant of these out-of-phase signals will produce the hue, or color, of the signal to be displayed, while the amplitudes

of the signals will determine the saturation, or depth, of the color.

Two other modulating signals are required in the transmission of color TV signals. One is a short burst of the subcarrier frequency. Eight cycles of this frequency are added to the back porch of the horizontal blanking signal, as shown in Fig. 25-33. The color bursts are used to synchronize

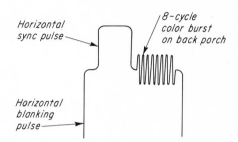

Fig. 25-33 Placement of color burst on back porch of the horizontal blanking pulse.

a crystal oscillator in the receiver and hold it to the exact frequency and phase of the missing subcarrier. The received sidebands must be mixed with this carrier frequency in the receiver in order that they may be detected properly. Lastly, vertical and horizontal synchronizing and blanking pulses must be added.

It would appear that the green signal has been forgotten entirely. However, since the *Y*, *Q*, and *I* signals all contain components of three colors, in the receiver it is possible to combine the three components and produce a green-chrominance signal even though none is transmitted as such.

25-25 COLOR-TV RECEIVERS

The transmitted VHF or UHF color-TV signal contains three separate systems of sidebands, all within a band of 4.2 MHz. The luminance signal sidebands extend the full 4.2 MHz. The two separate 90°-out-of-phase sets of chrominance sidebands occupy the spaces between the luminance sideband clusters in the upper 2.2 MHz of the luminance band. A color burst of about eight cycles of the suppressed subcarrier frequency of 3.58 MHz amplitude-modulates the back porch of the blanking pulses.

The requirement for the receiver is to develop signals equal to the three original color signals picked up in the camera tubes and reassemble them in their own component colors and intensities in their proper places on the screen of the color-picture tube.

The antenna, front end, IF amplifiers, AGC, detector, video, sweep circuits, and sound systems are very similar to corresponding systems in monochrome receivers except that the color receiver requires more accurate alignment of RF and IF channels to produce a flat response curve. A falling off of response of the higher-frequency sidebands in a monochrome receiver will produce a less-well-defined picture. Since all chrominance information is in the higher-frequency end of the IF bandpass, such a loss of response will mean

loss of proper color rendition on the picture tube.

The block diagram in Fig. 25-34 shows the section of the color TV receiver that separates the three color signals. It starts with the composite detected signal from the video (second) detector.

The whole composite signal, including all video frequencies that lie between zero and 4.2 MHz, is passed through a 1-μs delay circuit and is then fed to the luminance or Y amplifier. This delay is required because the chrominance signals passing through narrow-passband LC filters will be delayed about 1 μs.

When the receiver is receiving a monochrome transmission, the Y signal produces the black-and-white images. The Y signal is fed to the red, green, and blue matrices and their adders. The 0–4.2-MHz Y-signal output is fed with equal am-

Fig. 25-34 Block diagram of the color signals from the receiver detector to the three signals fed to the color grids in the picture tube.

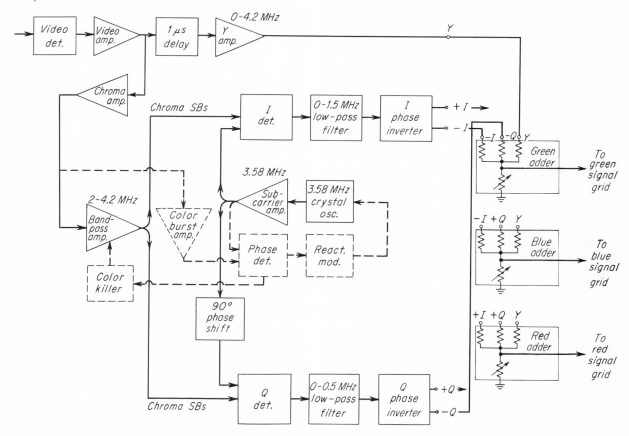

plitude to all three color grids. (If red is strong, so are green and blue. If red is weak, so are green and blue.) This results in varying degrees of brightness on the screen (white, light gray, dark gray, and, when the Y signal is zero, no illumination, which appears black).

When the receiver is receiving a color transmission, the whole composite detected signal (0 to 4.2 MHz plus the color burst) passes through the chroma amplifier and is fed to the 2–4.2-MHz bandpass filter-amplifier. This stage passes on the frequencies of the I and Q sidebands that must beat with the missing subcarriers to produce the color in the picture tube.

A 3.58-MHz crystal oscillator is forced into exact synchronization with the color-burst signal through the use of a phase detector and a reactance modulator. If the oscillator is not in exact phase with the color burst, this is sensed by the phase detector, which adds a bias to the reactance device to force the crystal into synchronization. The accuracy of the color-burst signals transmitted by the ABC, CBS, and NBC TV network stations allows the output of any properly operating color TV crystal oscillator to be used as a highly accurate signal source.

The amplified 3.58-MHz signal is fed directly to the I detector. Since this injected carrier has the same frequency and phase as the subcarrier that originally produced the I sidebands in the balanced modulator, the output of the I detector or mixer is the same as the I modulating signals. (All Q sidebands were produced by a 90°-out-of-phase subcarrier and will therefore not appear in the output of the I detector.)

The signal from the I detector is now in the form of video frequencies of 0–1.5 MHz. These pass through a 1.5-MHz low-pass filter and to a phase inverter or splitter to produce equal but out-of-phase I signals to feed the three sections of the matrix, $+I$ to $+I$ points, $-I$ to $-I$ points.

The same 3.58-MHz carrier is shifted 90° and is fed to the Q detector, which is receiving the same sidebands as is the I detector. The phase of this carrier allows the Q detector to produce a Q output signal. Since the Q signals are 90° out of phase with the I signals, no I information can appear in the output of the Q detector. The Q signal passes through a 0.5-MHz low-pass filter

and to its phase inverter. The output of the phase inverter is fed to the three sections of the matrix as indicated.

The Q, I, and Y voltages are mixed in the top section of the matrix in proper proportions to produce a resultant which is equivalent to the green signal picked up by the green camera tube. In the center matrix the proportions of the I, Q, and Y voltages produce the blue signal. In the lower matrix the red signal is developed. The three signals that have now been separated are relatively the same as the three signals produced in the color-camera tubes at the transmitter. It is now necessary to use these three signals to key three scanning beams in such a manner that the proper color proportions will be laid down where they belong on the face of the picture-tube screen. (Although shown as a resistance matrix, diodes, BJTs, or VTs may be used.)

When a TV signal is transmitted in monochrome, no color burst is transmitted. This means that no color-burst signals are being fed to the phase detector by the color-burst amplifier. As a result, the color-killer circuit receives no bias from the phase-detector circuit, and a positive flyback pulse from the horizontal yoke produces current flow in the color-killer tube. This current in turn is made to bias the bandpass-filter amplifier past cutoff, so that no signals are fed through it to the I and Q color-signal systems.

The scanning circuits of the color receiver are basically the same as for monochrome receivers. For example, the vertical-deflection system consists of a 59.95-Hz oscillator, an amplifier, and the deflection coils or yoke. If a three-electron-gun color tube is used, a vertical parabolic voltage is developed in the vertical output and used in the convergence system. Convergence is necessary to make the three separate electron beams converge to a point at the shadow mask just behind the picture screen (discussed in Sec. 25-26).

The horizontal-deflection system consists of an AFC circuit, a horizontal oscillator, amplifier, damper, flyback high-voltage rectifier circuit, and high-voltage regulator. When a three-gun tube is used, a horizontal parabolic convergence voltage, or magnetic field, is developed from the horizontal amplifier stage, and a 4,500-V focusing voltage is required. A block diagram of the vertical and

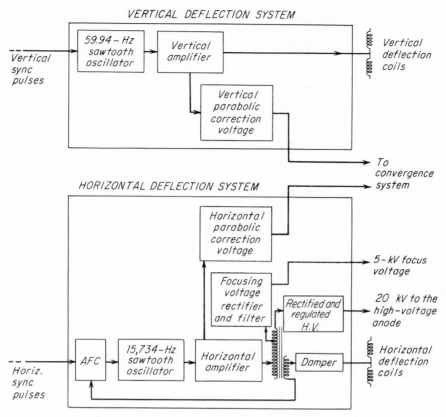

VERTICAL DEFLECTION SYSTEM

HORIZONTAL DEFLECTION SYSTEM

Fig. 25-35 Block diagrams for vertical- and horizontal-deflection systems of a three-gun color tube.

horizontal systems is shown in Fig. 25-35. Note that the *HV* power supply must be regulated, usually with a shunt regulator.

Test your understanding; answer these checkup questions.

1. What would be produced by weak blue, weak green, and weak red? _____
2. What is the name of a mirror that reflects one color but passes all others? _____ How many such mirrors are required in a color TV camera? _____
3. What is a combining circuit called in TV? _____
4. What is the letter symbol for the luminance signal? _____
5. What is the total $R + G + B$ for a Y signal? _____ For a Q signal? _____ For an I signal? _____
6. If a white scene is picked up, what is the Y value? _____ Q value? _____ I value? _____
7. If a black scene is being picked up, what is the Y value? _____ Q? _____ I? _____
8. What determines the purity of received colors? _____ The actual color? _____

9. Which leads by 90°, I or Q? _____
10. What is the color horizontal sweep frequency? _____ Vertical sweep? _____ Subcarrier frequency? _____
11. Why can chrominance sidebands have narrower bandwidths than luminance? _____ What bandwidths are used? _____
12. Why are chrominance sidebands used without a carrier? _____
13. Where is a color burst found? _____ What is its frequency? _____ Duration? _____
14. Why is the luminance signal delayed? _____
15. What circuit does a color-killer kill? _____
16. List the color TV receiver systems that are similar to those in a monochrome receiver. _____
17. List the color TV receiver systems that differ from those in a monochrome receiver. _____

25-26 COLOR-PICTURE TUBES

With the three color signals and a suitable deflection system, the color picture can be displayed

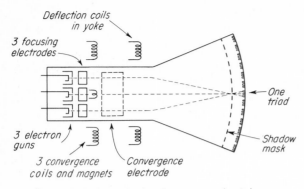

Fig. 25-36 Components of a three-gun color tube.

on a color tube. One tube in use is the three-gun shadow-mask tricolor type illustrated in Fig. 25-36.

The three electron guns when viewed from the front are actually oriented to be equidistant and 120° from each other rather than being stacked one on top as they are drawn. Each projects an electron beam toward the screen end of the tube. Three focusing electrodes are excited with separate potentials (+4,500 V) to focus their own beams to a tiny spot on the screen. Three small permanent magnets are spaced around the neck of the tube to aid convergence of the beams so that they will all pass through the same tiny holes in the metal *shadow mask* mounted a short distance from the fluorescent screen. The three beams cross in the holes and deconverge, each striking its own phosphor color spot on the screen, as shown. Each trio of color dots, called a *triad,* is arranged as shown in Fig. 25-37. The

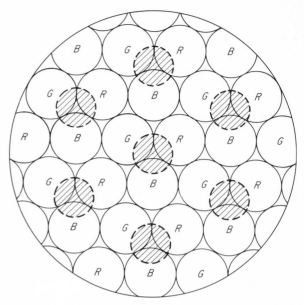

Fig. 25-37 Triads of color dots on the inner face of a color tube. Shaded areas are holes in the shadow mask behind the color dots.

dashed circle represents one hole in the shadow mask behind the dots painted on the inside surface of the screen.

The metal shadow mask of a medium-size tube may have about 250,000 tiny holes in it. For each hole there is a green, a red, and a blue phosphor dot on the screen, or a total of 750,000 color dots. When three beams are all passing through one hole, the red-signal beam strikes a red phosphor, the blue-signal beam a blue phosphor, and the green-signal beam a green phosphor. If all beams are of equal intensity, the triad area appears white to the viewer. When only the green-signal and the red-signal electron-gun grids are receiving color signals, only these two phosphors are illuminated and the viewer sees a yellow color. By scanning the picture tube in synchronism with the camera-tube scanning, the original scene can be reproduced in full color on the tricolor picture tube.

25-27 ONE-GUN PICTURE TUBES

Another picture tube in general use is the Trinitron. Whereas the three-gun shadow-mask tubes have three guns, three focusing lenses, and a relatively complicated convergence system, the

Trinitron has only one gun, requires only one focusing lens, and has a simpler convergence system. Although it has only one gun, it has three separate cathodes, all in parallel, as indicated in Fig. 25-38, to which blue, green, and red signals

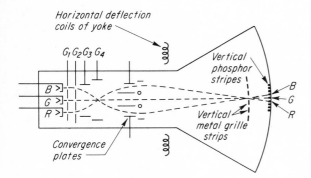

Fig. 25-38   One-gun Trinitron color TV picture tube.

are fed. The first grid is a single sheet of metal with three holes in it through which electrons from the three cathodes can pass. The second grid is an accelerating anode, pulling electrons out past the three G_1 openings. Once past G_2, the three beams of electrons fall under the influence of focusing electrodes G_3 and G_4, acting similar to focusing electrodes in oscilloscope tubes. The three beams are made to converge or focus to a point and then diverge as they continue on. The green beam moves straight through the two zero-charged convergence plates, bending only when passing under the influence of the deflection yokes. The blue and red beams approach the outer negatively charged convergence plates and are repelled back toward each other, meeting the green beam at an opening in the metal grille mounted near the face of the screen, provided the bias on the plates is correct. The blue and red beams deconverge after passing through the grille openings and strike phosphers painted on the inner surface of the screen.

Unlike the shadow-mask tube with its three color dot triads, the Trinitron has its phosphors painted in vertical stripes on the inside of the screen. Mounted near the stripes is a vertical metal grille. If only the green cathode is driven negative, only the green beam travels through the tube and strikes a green phosphor. If the red and blue cathodes are driven negative, these two colors will be activated and the area will appear magenta in color.

Proper horizontal convergence to produce the correct landing of the beams out at the edges of the screen requires a small parabolic convergence voltage added to the horizontal sweep. This can be aided by proper positioning of the deflecton yoke forward or backward. In actual use, the beams are wide enough to strike two adjacent grille openings, producing relatively bright illumination of the color screen. No vertical convergence circuitry is required.

Test your understanding; answer these checkup questions.

1. Why does the subcarrier signal to the Q detector have to be shifted 90° in phase? _____
2. For what is the 23-kV potential used? _____ The 4.5-kV? _____ Which must be regulated? _____
3. If the blue dot is always at the bottom of a triad, what position must the blue gun always have? _____
4. How many cathodes in a shadow-mask picture tube? _____ How many guns? _____ How many first grids? _____
5. How many cathodes in a grille-type picture tube? _____ Guns _____ First grids? _____
6. In which picture tube(s) is a horizontal parabolic correction voltage needed? _____ Vertical? _____
7. Can electrostatic, electromagnetic, or both, types of convergence be used with three-gun tubes? _____ With one-gun tubes? _____
8. In which tube is one color beam not affected by the convergence field? _____ Which beam is it? _____
9. In what form are the color phosphors laid down in the three-gun tube? _____ In the one-gun tube? _____
10. What would happen if a permanent magnet were laid on top of a color TV receiver? _____

FCC Element 4 requires an ability to answer questions similar to those below. Sections in which questions are answered are shown in parentheses.

1. Draw a block diagram of a monochrome TV transmitter and explain it briefly. (25-1)
2. What is the range of AF that the aural transmitter of a TV broadcast station is required to be capable of transmitting? (25-1)
3. What is meant by *100% modulation* of the aural transmitter at a TV station? (25-1)
4. Does the video transmitter at a TV station employ FM or AM? What does the sound transmitter employ? (25-1)
5. Why is grid modulation desirable in television video transmitters? (25-1)
6. What is meant by *vestigial sideband transmission* of a TV broadcast station? (25-1, 25-13)
7. Where, in TV transmitters, are sawtooth waves employed and why? (25-1)
8. Besides the camera signal, what signals and pulses are included in a complete TV broadcast signal? (25-1, 25-8)
9. Describe scanning as used by TV stations and the way in which the scanning beam moves across the picture in a receiver. (25-1, 25-4)
10. In TV, what do these terms mean: *aural transmitter, frequency swing, monochrome transmission, negative transmission?* (25-1)
11. What polarization is used in the transmission of the (a) aural and (b) visual TV signal? (25-2)
12. Why is a diplexer a necessary stage in TV transmitters? (25-2)
13. Draw a circuit diagram of a typical bridge-type diplexer used to transmit both video and audio from a turnstile antenna. (25-2)
14. For what purpose are reflectometers or directional couplers used in TV transmission systems? (25-2)
15. Describe the scanning technique used in TV transmissions. Why is interlacing used? (25-4)
16. Does the size of a TV picture tube have any effect on the number of scanning lines per frame? (25-4)
17. What is the meaning of *aspect ratio* in TV? What aspect ratio is used? (25-4)

18. How many frames are transmitted per second in TV? How many fields? (25-4)
19. In TV, what is meant by *field, frame, scanning, scanning lines, interlaced scanning?* (25-4)
20. What is a *mosaic* plate in a TV camera tube? (25-5)
21. Explain the operation of an image-orthicon camera tube. Include a schematic diagram of the tube, showing focusing and scanning details. (25-6)
22. What are advantages and disadvantages of vidicon TV camera tubes compared with image orthicons? (25-7)
23. What are (a) *blanking pulses* and (b) *synchronizing pulses* in a TV broadcasting and receiving system? (25-8, 25-9, 25-21)
24. Show by a sketch equalizing, blanking, and synchronizing pulses of a standard monochrome TV transmission. (25-8)
25. In TV broadcasting, why is the field frequency made equal to the frequency of the commercial power supply? (25-8)
26. In TV, what is meant by *blanking level, reference black level, reference white level?* (25-9)
27. In a TV signal, what is the relation between peak carrier level and blanking level? (25-9)
28. Sketch the amplitude characteristics of an idealized picture transmission of a monochrome TV station. (25-9)
29. What is a monitor picture tube at a TV broadcast station? (25-10)
30. What items must be included in a TV station's (a) operating log and (b) maintenance log? (25-11)
31. What are the operator requirements for TV stations? (25-11)
32. How wide is a TV broadcast channel? (25-13)
33. If a TV station transmits in channel 6 (82 to 88 MHz), what is the center frequency of the (a) aural transmitter and (b) video transmitter? (25-13)
34. How is the operating power determined for the (a) visual transmitter and (b) aural transmitter at a TV station? (25-13)
35. Under what conditions should the indicating instruments of a TV visual transmitter be read in order to determine the operating power? (25-13)
36. Within what limits is the operating power of a TV aural or visual transmitter required to be maintained? (25-13)
37. What is meant by *antenna field gain* of a TV broadcast antenna? (25-13)
38. What is the frequency tolerance for TV stations? (25-13)

ANSWERS TO CHECKUP QUIZ ON PAGE 635

1. (*To bring it in quadrature to demodulate properly*)
2. (*Ultor*) (*Focusing anode*) (*23-kV*) 3. (*Top*) 4. (*3*) (*3*) (*3*)
5. (*3*) (*1*) (*1*) 6. (*One- and three-gun tubes*) (*Three-gun*)
7. (*Both*) (*Electrostatic*) 8. (*One-gun*) (G) 9. (*Triads*) (*Vertical stripes*) 10. (*Deconvergence, colors wrong*)

39. In TV, what is meant by *aural center frequency, effective radiated power, peak power, standard TV signal, TV transmission standards, vestigial sideband transmission, visual transmission power?* (25-13)
40. Describe the procedure and adjustments necessary to couple a VHF visual transmitter to its load circuits. (25-14)
41. What is meant by (*a*) *TV channel* and (*b*) *TV broadcast band?* (25-15)
42. For what is a voltage of sawtooth waveform used in a TV receiver? (25-15, 25-19, 25-20)
43. What safety precautions should be observed to avoid personal injury when making internal adjustments to a TV receiver? (25-2)
44. What are *AGC amplifiers* and why are they used? (25-22)
45. Describe the scanning process employed in connection with color-TV transmissions. (25-24)

46. Describe the composition of the chrominance subcarrier used in the authorized system of color TV. (25-24)
47. In TV what is meant by *chrominance, chrominance subcarrier, color transmission, luminance?* (25-24)
48. Draw a block diagram of the color portion of a color-TV broadcast transmitter and explain basic operation of the circuits. (25-24)
49. Make a sketch showing the difference between blanking and sync pulses used for color and for monochrome. (25-24)
50. Where on a TV synchronizing waveform would the color burst for color transmissions appear? (25-24)
51. Explain the operation of a turnstile antenna used in TV. (See Sec. 20-31.)

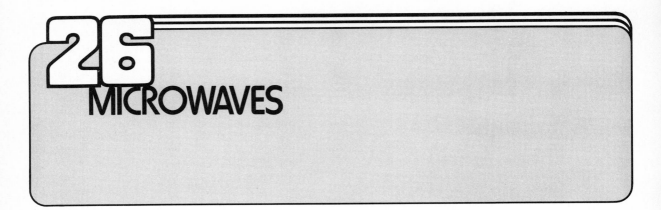

26 MICROWAVES

The objective of this chapter is to outline some of the fundamental microwave techniques required of technicians working in the microwave area, to describe some of the hardware used in microwave systems, and to lay a foundation for the chapter on marine radar (and FCC Element 8) which follows this chapter.

26-1 MICROWAVE RANGE

As communication knowledge has increased, the usable spectrum has expanded to higher and higher frequencies. Common miniature vacuum tubes and high-frequency transistors reach a practical limit in the range of 1,000 MHz (1 GHz). This can be considered the beginning of the microwave region. Specially designed UHF and lighthouse triodes and some transistors operate up to about 3 GHz. Specialized microwave vacuum tubes—magnetrons, klystrons, traveling-wave tubes (TWTs), backward-wave oscillators (BWOs), and many solid-state devices now operate well above 40 GHz, some at 100,000 GHz and higher. Above this comes the infrared or heat region, usually considered to be 300,000 to 375,000 GHz. Light visible to the human eye ranges from about 275,000 (red) to 790,000 (violet) GHz. In the light region, laser oscillators are becoming highly important communication devices.

Most microwave activity is from 1 to 50 GHz. Electronics and radio technicians of today should be familiar with the basic concepts of equipment operating in this region. Besides being employed in radar systems and point-to-point radio commu-nications, microwaves are used extensively in research laboratories, and many companies are engaged in building and servicing microwave test equipment and components. Microwave ovens operate at 2.45 GHz.

26-2 MICROWAVE TRANSMISSION LINES

Low-frequency ac can be carried effectively by a pair of wires, such as a lamp cord. In the RF spectrum special constant-impedance two-wire transmission lines or coaxial cables are efficient devices for carrying energy. When the frequency is higher than 3 GHz (0.1-m, or 10-cm wavelength), a hollow pipe, slightly larger in diameter than a half wavelength, can act as an acceptable confined space down which energy of this or higher frequencies can be propagated. This process is similar in some respects to the propagation of radio waves into space from an antenna. Such a hollow metal pipe, called a *waveguide* (WG), may be round or rectangular. Polarization may shift in the round, however. Most rectangular WG has a height about one-half of the width for the usual voltage and current oscillation mode employed. (There are several possible modes of oscillation in waveguides.) The width of the guide must be slightly greater than the half wavelength of the ac to be transmitted. A common band of operation is known as the 3-cm, or X band, with lower and upper frequencies of 8.2 and 12.4 GHz, respectively. X-band waveguide is about 3 cm wide and $1\frac{1}{2}$ cm high.

Other bands of microwave frequencies have been given labels. Some popular designations are

Table 26-1

Band	Frequency range, GHz	Waveguide size, in.
L	1.12–2.7	6.5 × 3.25
S	2.6–3.95	3.0 × 1.50
G	3.95–5.85	1.87 × 0.87
C	4.9–7.05	1.59 × 0.795
J	5.85–8.2	1.37 × 0.62
H	7.05–10.0	1.12 × 0.497
X	8.2–12.4	0.9 × 0.4
M	10.0–15.0	0.75 × 0.375
P	12.4–18.0	0.62 × 0.31
N	15.0–22.0	0.51 × 0.255
K	18.0–26.5	0.42 × 0.17
R	26.5–40.0	0.28 × 0.14

presented in Table 26-1 (there is no one standardized letter designation of bands). Frequencies above the operational frequency (half wavelength) of a waveguide will be propagated, but all lower frequencies will be sharply attenuated. Thus, waveguides act as high-pass filters. They have a surge impedance of 50 Ω to the band for which they are fabricated.

Waveguides may be constructed of brass, copper, or aluminum. Since currents on the walls of waveguides oscillate only on the inner skin, low-loss sections will be silver-plated on the inside to reduce skin effect. While currents do flow on the waveguide walls and voltages are developed between the upper and lower sides, it is not possible to measure them with the usual meters. Instead, the electrostatic and magnetic fields developed by the currents are sampled by inserting a pickup probe (antenna) into the waveguide. The indica-

tions obtained can be converted to the desired values.

At radio frequencies it is customary to think of current as traveling on the surface of the wire and energy as being radiated when the antenna impedance matches the impedance of space. In the microwave case, currents and voltages may be relegated to a secondary role and the radiated electrostatic and magnetic waves, always at right angles to each other, are considered to be carrying the energy from source to load and matching the load impedance.

Waveguides are made into various-length sections. They may be straight, as in Fig. 26-1, be bent to some desired direction, be twisted to some desired angle, or even be made flexible. At each end of a waveguide section there is a precisely machined flat metal flange allowing one section to be coupled to another by bolting the flanges together.

Flat flanges may be butted together, but for minimum losses and reflections from the joint one of the flanges should be a *choke* flange. In a choke flange, part is machined away so that a half-wavelength-long cavity is developed, as indicated in Fig. 26-2. From antenna theory, a half wavelength from a low-impedance point on a stub or transmission line is another low-impedance point. The half-wave cavity in the flange dead-ends (0-Ω impedance) in the flange metal, reflecting 0-Ω impedance back to the spacing between waveguide sections. This represents perfect continuity for the wave between sections and therefore no reflection of energy at the joint.

Fig. 26-1 (a) Round waveguide. (b) Rectangular-waveguide section. (c) Straight section coupled to a 90° elbow. (d) Section in (c) in schematic diagram form.

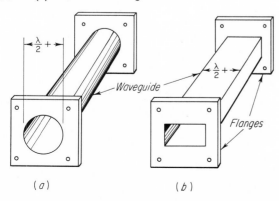

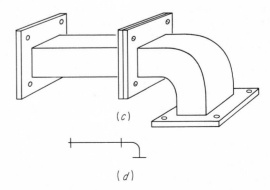

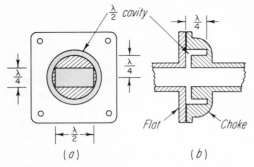

Fig. 26-2 Choke joint. (a) Flange-face view. (b) Cross section of flat and choke flanges mating.

Coaxial cables also may be used for microwave transmission lines. Solid dielectric coax may have rather high losses at higher microwave frequencies. In addition, coax coupling devices tend to present discontinuities and undesired reflections of energy back up the cable. Reflections produce standing waves on the line and prevent full-power transfer from source to load, as well as high- and low-voltage points along the cable, which is important in high-power applications. If only a few decimeters long, however, coaxial cable is often employed up to X-band frequencies.

Microstrip is a microwave transmission line consisting of a flat metal base on which is laid an insulator, or dielectric material. A thin metal strip is laid on the dielectric, as in Fig. 26-3a. Strip width and dielectric constant and thickness determine the impedance of this type of transmission line. This is a handy printed-circuit transmission line. Two other planar transmission lines are stripline and co-planar line, shown in (c) and (d).

26-3 SOME WAVEGUIDE DEVICES

As with other transmission lines, terminating a waveguide with a load resistance equal to the surge impedance of the guide produces no reflected energy and maximum transfer of power to the load. If the impedance match is not correct, the reflection may appear to be either capacitive or inductive. To cancel an inductive-reactance effect a projection down into the waveguide, as in Fig. 26-4a and b, has a capacitive effect. (If it

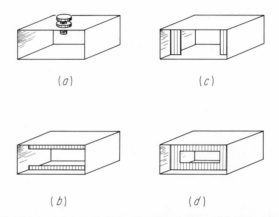

Fig. 26-4 (a) Adjustable capacitive probe. (b) Capacitive iris. (c) Inductive iris. (d) Resonant or decoupling window.

projects down more than a quarter wave, it becomes inductive.) A projection into the side of the waveguide, as in Fig. 26-4c, has an inductive effect. A metal window can be made resonant to pass certain frequencies. The optimum placement of a capacitive tuning screw can be determined

Fig. 26-3 (a) Microstrip pictured. (b) Microstrip cross-sectional view. (c) Stripline cross-sectional view. (d) Co-planar transmission line.

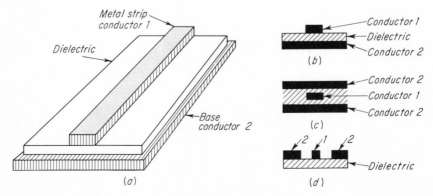

by replacing a waveguide section by a *slide-screw tuner*, which has a slot down the length of it and is fitted with an adjustable retractable probe that can be moved along the slot. With this device the optimum distance for the reactive screw from the reactive load and the best probe depth can be determined by lowest SWR (Sec. 20-12).

One dummy load used in waveguides is a long pyramid of carbonized material with a sharp point to prevent reflections from it, as shown in Fig. 26-5*a*. The tapered pyramid absorbs energy,

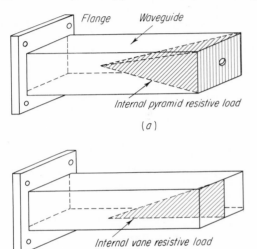

Fig. 26-5   Low-power waveguide loads. (*a*) Pyramid type. (*b*) Resistive vane type.

leaving none to be reflected. Another form of low-power resistive load is a tapered resistance-coated strip at the end of the waveguide, as in Fig. 26-5*b*. Higher-power dummy loads are air-, oil-, or water-cooled.

When it is desired to lose some fraction of the power flowing along a waveguide, either a flap or a vane *attenuator* may be used (Fig. 26-6). Attenuation is often required when making microwave measurements because most test equipment operates in the milliwatt range, whereas many practical applications operate with several watts of average power, which is enough to burn out the test equipment. These attenuators decrease the signal from 0 dB (no decrease) to more than 30 dB (one-thousandth of the power).

Directional couplers are produced by welding

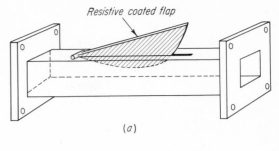

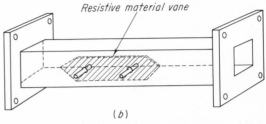

Fig. 26-6   Waveguide attenuators. (*a*) Flap type; maximum attenuation with flap lowered. (*b*) Vane type; maximum attenuation as vane nears center of guide.

two pieces of waveguide together and opening one or more holes between them, as indicated in Fig. 26-7. The larger the holes or the greater the number of them the greater the power transfer to the secondary waveguide section. If two holes are

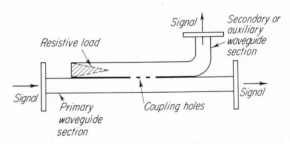

Fig. 26-7   Directional waveguide coupler.

a quarter wave apart, the propagation is such that most of the energy induced in the secondary or coupled section is in the forward direction. A dummy load in the backward-direction end absorbs any reflected transmission, making the coupler a true forward coupler. These devices are rated in decibels. A 3-dB coupler transmits one-half of the power to the secondary section; a 10-dB coupler transmits one-tenth of the power; a 20-dB coupler, one-hundreth of the power. A 1-kW input to a 30-dB coupler couples one-thousandth of the power (1 W) to the secondary section, while 999 W passes through the primary section. As discussed in Sec. 25-2, two directional couplers back to back can sample the energy moving forward and that reflected by the load. From this the reflection coefficient and the SWR can be found.

One method of determining reflection coefficient and the SWR caused by a mismatched load is illustrated in Fig. 26-8. If the oscillator has a

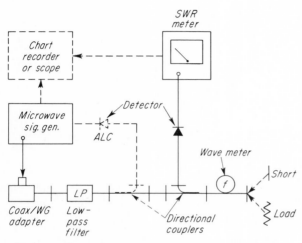

Fig. 26-8 Reflectometer to measure SWR.

coaxial outlet, it is fed to a coax/WG adapter and then to a low-pass filter to remove the possibility of any harmonic energy being read. A forward-direction coupler can be used to produce a feedback voltage to form an automatic leveling control (ALC) if the oscillator is being swept over a band of frequencies. A WG-short is substituted in place of the load, reflecting all transmitted energy back up the waveguide. A reversed direction coupler picks up part of the energy being reflected, which is attenuated a known amount and is displayed as a voltage on the meter. The load is then connected in place of the short and the attenuator is reduced to zero. The indication on the meter can represent the reflection coefficient, or it can be calibrated directly in SWR. (The calibrations on the meter must match the curve of the diode detector used.) If the load accepts all the energy, no power will be returned and the meter will not read. Reflection coefficient is then infinite and SWR is $(1 + \infty)/(1 - \infty)$, or $1:1$. When sweep-

ing a band of frequencies, it is often desirable to chart the SWR along the band. The dashed lines indicate the circuits that would be used.

26-4 COUPLING TO WAVEGUIDES

There are two common methods of coupling energy into (or out of) a waveguide other than the hole method used between waveguide sections in directional couplers. One method, shown in Fig. 26-9a, is similar to link coupling. Energy

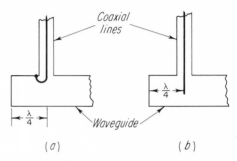

Fig. 26-9 Coupling coaxial line to waveguide. (a) Loop or hook method. (b) Antenna-probe method.

from a coaxial line terminates in a single-turn loop connected to a wall of the waveguide. Magnetic fields due to the current in this loop induce voltages in the waveguide space and currents in the walls, allowing energy to radiate down the waveguide. An alternative method is shown in Fig. 26-9b, in which the coaxial line terminates in essentially a quarter-wave vertical antenna projecting into the waveguide space. Energy radiation from this probe is transmitted down the waveguide. In both cases a reinforcement of the energy transfer occurs if the coupling devices are some odd quarter wavelength from the sealed end of the waveguide. The sealed end then acts as a parasitic reflector. In waveguides containing coupling devices, the near end will often be adjustable or tunable to assure maximum reflection from the end.

26-5 DETECTING DEVICES

To sense the amplitude of the SHF (superhigh-frequency) ac energy in waveguides, either solid-state diodes or bolometers are inserted across a waveguide section. By using a crystal

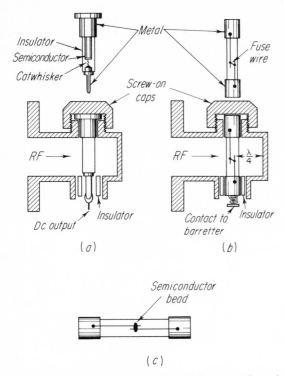

Fig. 26-10 (a) Microwave crystal and placement in a detector mount. (b) Barretter and mount. (c) Thermistor.

cies. The crystal detectors are used when pulse or modulation envelope displays on an oscilloscope are desired.

26-6 RESONANT CAVITIES

To produce a resonant LC circuit for microwaves, the number of turns of the coil is reduced to one and the capacitance across the circuit is only that between the two ends of the coil wire, shown in Fig. 26-11a. This forms a quarter-

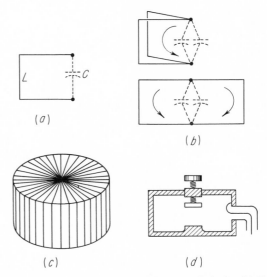

Fig. 26-11 (a) L and C of a quarter-wave hairpin. (b) Two paralleled hairpins. (c) Cavity of infinite paralleled hairpins. (d) Capacitive tuning of cavity and low-impedance coupling of a coaxial cable to the cavity.

diode in a waveguide detector mount across the waveguide, as in Fig. 26-10a, ac energy is picked up, rectified, and made to operate dc meters, indicate modulation, etc.

Bolometers are devices that change their resistance when heated. Those with a positive temperature coefficient are called *barretters* and are actually only fine resistance wires in a waveguide mount, as in Fig. 26-10b. Another type of bolometer is the *thermistor,* a small bead of semiconductor substance between two connecting wires, as shown in Fig. 26-10c. It has a negative temperature coefficient. Ac energy passing along a waveguide will heat a bolometer element. If the bolometer is used as the fourth arm of a bridge circuit that is balanced when the bolometer is cold, the amount of bridge unbalance indicates the relative value of ac power in the waveguide. The barretter is used for higher-power levels. The thermistor, being quite sensitive, is used for low-power measurements. Bolometers react slowly and are limited to indicating only very low modulating frequen-

wavelength *hairpin* tank. However, the resonant frequency will be somewhat lower than the measured wavelength value. An attempt to reduce the inductance further by paralleling another similar hairpin, as in Fig. 26-11b, does lower the L value but increases the C value. The resonant frequency does not change, but Q increases. Note the distribution of current at a given instant if the two resonant hairpins are fanned out 180°. They appear as a cross section of a rectangular waveguide.

If many hairpins were added to a central point, a cavity similar to a round tuna can, Fig. 26-11c, would be developed. A tuna can has a radius and height of about 4 cm and should have a natural resonant frequency a little lower than 2 GHz if excited. A larger cavity would resonate at a lower frequency; a smaller, at a higher frequency. The frequency could be varied by a small percent by installing a variable capacitor in the center of the cavity (Fig. 26-11*d*). This cavity is shown with an inductive coupling loop in it.

If a resonant cavity is made variable by installing a movable plunger in it and is also hole-coupled to a section of waveguide, as in Fig. 26-12, a wavemeter, or microwave frequency

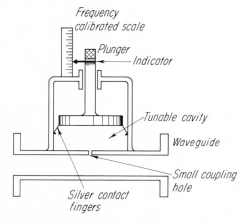

Fig. 26-12 Cavity wavemeter (frequency meter) coupled to a waveguide section.

meter, results. Energy in the waveguide passing the hole induces a voltage and current into the cavity. If the cavity is resonant to the frequency, it will absorb some of the energy. This can be noted in a detector coupled to the waveguide system, or a detector coupled to the wavemeter will show when resonance occurs. If the cavity is not resonant, it will not absorb any energy. An indicator on the plunger handle operating along a calibrated scale indicates the frequency of resonance.

Test your understanding; answer these checkup questions.

1. What are the limits of microwaves today? _____
2. What are the two transmission lines used for microwave frequencies? _____ _____ Which is more efficient? _____
3. How wide must a WG (waveguide) be? _____ How high? _____

4. What is the designation of the lowest-frequency microwave band? _____ The highest? _____
5. What is reduced by using choke flanges? _____
6. What is the name of the microwave transmission line that is used with printed circuits? _____
7. Why are attenuators so often used with microwave test equipment? _____
8. Where is a resistive load placed in a directional coupler? _____
9. If the transmission-line $z = 50\ \Omega$ and the load is 100 Ω, what is the value of ρ? _____ SWR? _____
10. Why are detectors used in SWR reflectometers? _____
11. What are the three listed methods of coupling into or out of a WG? _____ _____ _____
12. What are the names of two types of bolometers? _____ _____ Which is a semiconductor? _____
13. Why are bolometers not used to detect modulation? _____
14. What would be two advantages of a resonant cavity over a hairpin tank? _____ _____
15. What type of coupling is used between a WG and a WG wavemeter? _____
16. Would a thermistor have a negative or positive coefficient of resistance? _____ Why? _____

26-7 KLYSTRONS

Four important types of vacuum tubes are used to generate or amplify microwave ac. These are the klystrons, the magnetrons, the traveling-wave tube (TWT), and the backward-wave oscillator (BWO).

There are two basic types of klystrons: reflex (oscillators) and multicavity (amplifiers). The low-power reflex klystrons produce a stream of electrons from a hot cathode that is drawn to a cylindrical cavity with grids on top and bottom, as shown in Fig. 26-13. If the cavity is oscillating, at one instant the top grid will be going positive while the bottom grid is going negative. This bunches any electrons traveling through the cavity and toward the repeller plate at the top. Since the repeller is negative, it returns the now well-bunched electrons to the cavity in such a phase as to increase the strength of cavity oscillation. If the repeller has the wrong voltage, the bunched electrons will return out of phase and no oscillations can be produced. Several repeller voltages will produce oscillation of the cavity. Some of these modes of oscillation will be stronger than others.

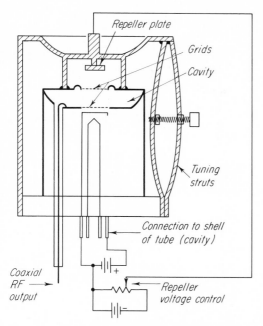

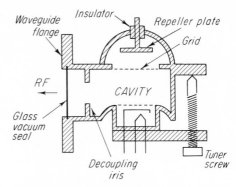

Fig. 26-14 Reflex klystron with waveguide mount.

Fig. 26-13 Reflex klystron. The distance the coaxial probe projects into the waveguide determines coupling. Screwing struts together stretches the cavity apart, raising the frequency.

The frequency of oscillation can be changed by a few hundred megahertz (in X band) by stretching the grids apart physically by screwing in the tuning strut screw and deforming the cavity at the center. This reduces center-cavity capacitance and raises the frequency. The frequency is also electronically variable by a few megahertz by changing the repeller voltage slightly from the value that produces highest output. Modulating the repeller voltage by 1 or 2 V is a practical and simple method of producing wideband FM in these klystrons.

The original reflex klystrons produced only a few milliwatts of output power and were very inefficient. Newer tubes, as in Fig. 26-14, can produce considerably more power. When high power is required, however, two-, three-, or four-cavity klystrons are used. These range in efficiency from about 20 to 50%.

A three-cavity gridless power klystron is shown in Fig. 26-15. Electrons are emitted from the cathode and are attracted by the collector and modulating anode. However, a magnetic focusing coil

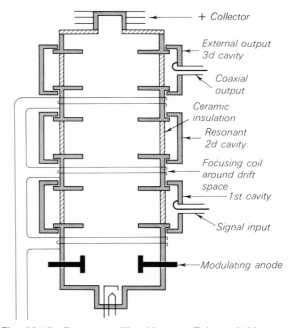

Fig. 26-15 Power-amplifier klystron. Tubes of this type range in height to more than 10 ft.

around the tube forces the electrons up toward the collector, past the externally connected cavities. As the electrons stream past the first cavity, input-signal ac oscillations in it begin to bunch them. The bunched electrons move through the first drift space, where they continue to improve their bunching, to the second cavity, which is resonant to the same frequency and produces still greater bunching. The well-bunched electrons pass the third, or output, cavity, giving up most of their energy to this tank circuit. The electrons then pass on to strike the air- or water-cooled anode. By varying the dimensions of the external cavities, the frequency of operation of these tubes can be changed over a wide range by substituting different-size cavities.

Reflex klystrons are used as the mixer oscillator in microwave superheterodynes or in signal generators. Power-amplifier klystrons can produce 10 to 50 kW of SHF power. They find use in TV transmitters, radar (producing megawatts of peak pulse power), and *tropospheric* beyond-the-horizon *scatter* communications. (Waves striking atmospheric discontinuities refract back to earth some distance away.)

26-8 MAGNETRONS

The magnetron was developed to produce high-power microwave pulses for radar. A 25-W plate dissipation magnetron may be visualized as being a cylindrical brass block about $2\frac{1}{2}$ in. in diameter and about $1\frac{1}{2}$ in. thick. A large hole is drilled down the center, and eight smaller holes are drilled between the center and outer edges, as illustrated in Fig. 26-16. Slots interconnect the small and large holes. Both ends are sealed with end plates. The smaller holes form cavities. When the magnetron is operating, electrons take a back-and-forth path along the walls of the cavities, as indicated by the vector arrow in one of the cavities.

A cylindrical cathode with an internal heater wire is located in the center of the central hole. One heater lead connects to the near side of the heater; the other lead connects to the far side.

A small hook in one of the cavities acts as a pickup loop for this particular tube, extracting microwave energy from this cavity (and all others thereby) when it oscillates. This energy is fed to a short concentric transmission line, which is ter-

Fig. 26-16 Essentials of a multicavity magnetron.

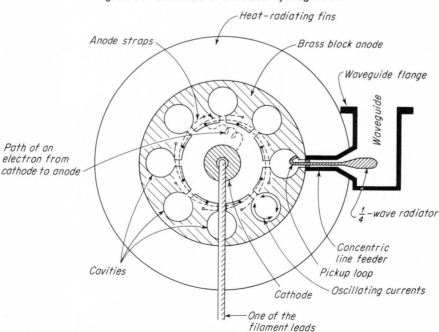

minated by a quarter-wave radiator protruding into the end of a waveguide.

The magnetron block acts as the anode or plate and is connected to ground. When the tube is pulsed, the cathode is driven negative by perhaps 10 to 20 kV. This makes the plate relatively positive, and electrons from the hot cathode start moving toward it. However, a strong external horseshoe-shaped permanent magnet, with its north pole at one end of the cathode and its south pole at the other end, produces an intense magnetic field down the central hole. According to the right-hand motor rule (Sec. 3-17), the electrons will be deflected at right angles to the lines of force through which they are passing. This results in an elliptical path for the electrons, as shown, as they progress toward the anode areas.

The positive potential of the anode accelerates the electrons toward it. This is the same as saying the electrons pick up energy from the difference of potential. As the electrons whirl past the slots between anode areas, they induce voltages between the slot faces which drive currents into oscillation along the surfaces of the cavity walls. In this way the energy of the cathode electrons is transferred to the oscillating current in the cavities. All the cavities are of the same size and oscillate at the same frequency. However, adjacent cavities have opposite-direction currents in them at any one time. Strapping every other anode face together, as shown in broken lines in Fig. 26-16, increases the efficiency to about 50% from an unstrapped 35%.

A magnetron may have an average power output of only 20 W, but when it is used in pulsed circuits, it can produce high-peak-power pulses. For example, if a 20-W magnetron is pulsed 1,000 times per second and each pulse is only 1 μs long, the total time it is operating is only $\frac{1}{1,000}$ s. Each pulse may therefore have a peak power of 20,000 W. If pulsed with 1-μs pulses only 500 times per second, each pulse could have 40,000 W and not exceed the 20-W average power-dissipation value of the tube.

The frequency of resonance of a cavity can be raised by making the volume of the cavity smaller by pushing a plug into it. Some magnetrons are made variable in frequency by moving slugs into all the cavities simultaneously. Other magnetrons

are electronically tuned by varying the anode voltage on them. For example, one such low-power magnetron can change from 400 to 1,200 MHz by changing the anode voltage from 700 to 1,900 V.

26-9 TWTs AND BWOs

A microwave tube that can tune over a relatively wide band of frequencies electronically is the traveling-wave tube (TWT), or traveling-wave amplifier (TWA). It consists of electron-focusing gun to deliver a beam of electrons down the center of a spiral (helical) coil to the anode at the far end, as in Fig. 26-17. The helix, anode, and elec-

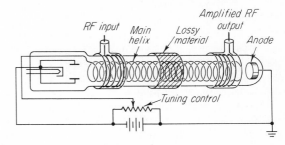

Fig. 26-17 Traveling-wave tube (TWT).

tron gun are all in a vacuum. The anode and helix are made highly positive in respect to the cathode to pull electrons down the helix center. Outside the evacuated glass tubing are slipped two additional helixes which are 50-Ω-impedance input and output coupling devices.

When a microwave signal is induced into the left end of the main helix, it travels along the surface of this wire at essentially the speed of light. Since the helix is a coil, however, the actual velocity of propagation down the tube is considerably less than the speed of light. If electrons are fed down the core of the main helix at a slightly higher velocity than the signal wave is traveling on the helix, electrons and wave interact and the electrons inside the main helix become bunched as they progress down the length of the helix. This results in some of the electrons slowing down and losing much of their energy to the induced wave on the main helix. As the energized helix waves pass the output coupling device, they induce energy into it. The gain of a TWT ranges from 30 to 60 dB. Available power outputs range from milliwatts to 10 W or more.

There is no cavity in a TWA or TWT. The only tuning required is to maintain the helix potential at an optimum synchronous value. To prevent the electron beam from being attracted to the positive helix, a strong axial permanent or electromagnetic field is developed down the center of the helix by external magnets (not shown). This magnetic field focuses the electrons and holds them in the center of the main helix area.

To prevent amplified energy from working back the helix from the output to the input device, which would produce oscillations, a lossy attenuator is wrapped around the glass envelope that surrounds the main helix. Without such an attenuator, and with only one coupling device, the TWT would be essentially the same as a backward-wave oscillator tube. A BWO can operate with relatively high efficiency and is voltage-tunable by varying the helix voltage. For example, one tube can tune over an octave and a half (from 1 to 3 GHz) by varying the helix voltage from 300 to 2,000 V. Such a variation of voltage results in an unequal amplitude output over the operating range. It is necessary to use a *leveling* or limiting circuit to maintain equal amplitude output at all frequencies.

The BWO shown in Fig. 26-18 uses a bifilar (two-wire) helix and has a balanced output. To couple to an unbalanced line, such as a coaxial cable, and to convert to some other impedance,

if desired, a *balun* (a balanced-to-unbalanced coupling transformer) must be employed. The balun may be part of the equipment that comes with the BWO tube. Basically, a balun is a center-tapped autotransformer, as in Fig. 26-19a. Since

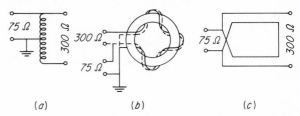

Fig. 26-19 (a) Simple coil balun. (b) Toroidal coil balun. (c) Linear balun.

a 1:2 turns ratio has a 1:4 impedance ratio in a transformer, the balun can change a 75-Ω unbalanced coaxial line to a 300-Ω balanced line. Figure 26-19b shows the same circuit wound on a toroidal form. Figure 26-19c illustrates a linear balun, in which the 75-Ω input is across one turn of a two-turn coil. A cavity-type balun is shown in Fig. 26-18.

26-10 MICROWAVE TRANSISTORS AND DIODES

Lighthouse-type triodes are limited in frequency by interelectrode capacitances, inductance of leads, and transit time. The same is true of transistors. But by improving the geometry and using overlay epitaxial planar types, transistors can be made to amplify or oscillate in the 6- to 10-GHz range. Monolithic integrated circuity is used at lower microwave frequencies.

Tunnel diodes were mentioned in Sec. 10-3. A highly stable oscillator is produced by loosely coupling a tunnel diode to a high-Q cavity, as in Fig. 26-20. By using a short antenna probe protruding into the cavity and by feeding off-center of the cavity, loose coupling is attained. The power output of such oscillators is a few hundred μW, but this is sufficient to act as local oscillators for microwave superhets.

One application of a varactor diode is to produce a multiplication of a given ac frequency. It will be remembered that any distortion of a sine wave indicates the presence of harmonics of the sine-wave frequency. The circuit in Fig. 26-21

Fig. 26-18 Bifilar-helix backward-wave oscillator (BWO) with balun to convert from balanced to unbalanced output.

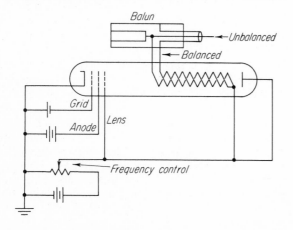

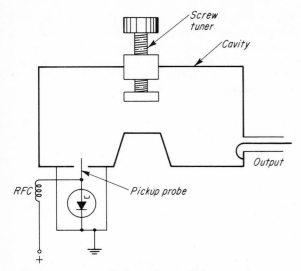

Fig. 26.20 Tunnel-diode cavity oscillator.

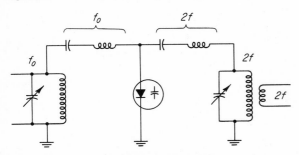

Fig. 26.21 Varactor frequency multiplier.

indicates the *LC* circuit of an oscillator producing sine-wave ac at frequency f_o. The ac voltage across the *LC* tank circuit is essentially sinusoidal. This ac is coupled through a high-Q series resonant circuit to the varactor diode. The ac waveform of the voltage fed to the diode is sinusoidal; but because of the very nonlinear current-voltage relationship in a diode of this type, the voltage-drop across the diode will be very nonlinear (contain considerable harmonic content). A second series resonant circuit, tuned to the second harmonic frequency (2*f*), is used to couple second harmonic energy to the output tank circuit, which is also tuned to 2*f*. Any diode can be used in this kind of a circuit, but the rectification by a varactor, because of its nonlinear junction charge storage, is particularly distorted. Efficiency of such a multiplier circuit can be as high as 75%. Had the 2*f* resonant circuit and the 2*f* antiresonant tank been

tuned to 3*f*, the output circuit would have picked up third-harmonic energy; and the output would have been three times the fundamental frequency but at somewhat reduced efficiency. As doublers, varactors can deliver as much as 25 W at 1 GHz and 100 mW at about 35 GHz.

Another diode that can be used in this same type of frequency-multiplier circuit is the *step-recovery* diode. By special doping of its junction, charges store in the junction during forward-bias current-flow time. As the forward-biased pulse starts to fall off, the charges start flowing out of the junction. By the proper doping density, all the charges stored by the high-frequency pulse stop flowing at approximately the same instant. The result is a very fast cessation of current flow some time after the pulse drops off and reverse bias starts. The sharp current break, or step, is rich in harmonics, which is the reason for the step-recovery diode's use. As much as 10 times multiplication can be produced with step-recovery diodes.

The impact avalanche and transit time diode (IMPATT) is a special microwave diode that utilizes the delay time of attaining an avalanche condition plus transmit time to produce the 180° voltage-current condition of negative resistance. As with other microwave diodes, it may use silicon, germanium, or gallium arsenide. It can oscillate and produce up to 1 W at 10 GHz and about 0.1 W at 100 GHz, but at an efficiency below 10% at the latter frequency.

The hot-carrier, or Schottky, diode is an excellent low-noise detector for use in the higher microwave frequencies.

Microwave PIN diodes can be used in waveguides as switches or modulators. They operate as diodes up to a few megahertz, and then the intrinsic layer produces too much transit time. However, they have a useful high resistance with reverse bias and low resistance with forward bias. A PIN diode mounted across a 50-Ω waveguide and reverse-biased has almost no effect on energy being transmitted down the line. If the junction is forward-biased, the low resistance acts as a wall to the energy, and almost all transmission can be stopped by it. If the bias of the diode is modulated, its resistance varies, and an amplitude modulation can be obtained at the output of the waveguide. While one diode cannot switch high powers, several in parallel can be used to pulse several thousand watts of peak power in a waveguide.

26-11 GUNN AND LSA DIODES

These semiconductor devices are called diodes only for the lack of a better term. They do not have junctions but instead depend on other peculiarities of semiconductors to produce negative-resistance effects. They can be called *active-area* or *bulk-effect* devices.

The Gunn diode consists of a thin slice of N-type gallium arsenide between two metal conductors, as shown in Fig. 26-22a, assembled in

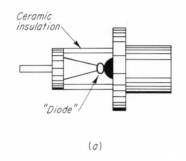

(a)

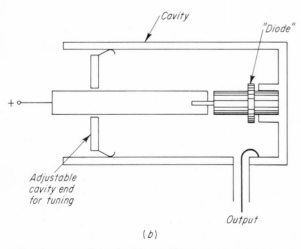

(b)

Fig. 26-22   (a) Gunn diode. (b) Gunn-diode oscillator.

a cylindrical metal and ceramic body. The diode is fitted into a holder inside a cavity and is fed the required dc to make it oscillate. If the voltage

across the diode is increased from 300 V/mm to 400 V/mm, at some voltage the electrons from the outer, partly filled energy ring of the atoms of the semiconductor crystal jump across the very narrow forbidden energy gap of gallium arsenide and actually decrease their mobility in the crystal. This produces a negative-resistance effect in the semiconductor. If the voltage is increased further, the current will begin to increase in proportion to the applied dc voltage, and an *EI* curve similar to that of a tunnel diode is produced.

Figure 26-22b represents the placement of a Gunn diode in a cavity that is made tunable over several GHz by making one end wall adjustable.

The other active-area device is the *limited space-charge accumulation* (LSA) diode. It is somewhat similar to a Gunn diode except that it is considerably thicker and produces its negative-resistance effect at a dc voltage in the range of 800 V/mm.

Both the Gunn and LSA diodes operate considerably above the 25-GHz region, but the LSA produces several times the power output, in the range of 3 W at 50 GHz. At lower microwave frequencies, peak pulsed powers of 100 kW are possible.

26-12 ISOLATORS AND CIRCULATORS

Ferrites (compounds of oxygen, iron, and several other metals) are playing an important role in modern microwaves. They have two important properties. One is *ferromagnetic resonance absorption*. With the magnetic field magnetizing the ferrite rod in the waveguide, as shown in Fig. 26-23, energy entering the waveguide at the resonant frequency of the ferrite molecules will be absorbed by the ferrite. Energy entering from the opposite direction will not be affected by the molecular resonance of the ferrite rod. As a result, an on-off waveguide switch can be produced by reversing the magnetism of the electromagnets. A ferrite rod of this type is called an isolator. It can act as a one-way buffer between a microwave oscillator coupled to the end of a waveguide and variations of the loads further down the line. The resonant frequency of the ferrite is controlled by the external-magnetic-field strength.

A second useful property of a ferrite rod is

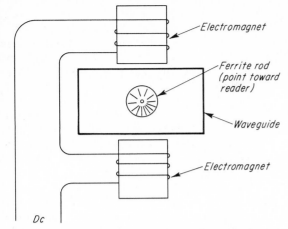

Fig. 26-23 Ferromagnetic resonance absorption switch.

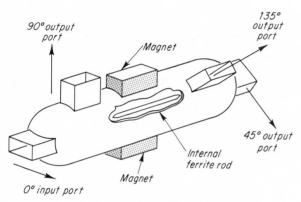

Fig. 26-24 Faraday rotation circulator.

known as *Faraday rotation*. In this case, the frequency of the energy entering the 0° port, in Fig. 26-24, must not be near the resonant frequency of the ferrite molecules. As the signal transfers from rectangular waveguide to the circular waveguide, the energy wave rotates 45° as it flows past the ferrite rod that is in the field of two magnets. The output-port waveguide must be at 45° from the input-port waveguide. If the magnets were removed, no rotation would result and almost no energy would be transmitted out the 45° port, because of improper field polarization or orientation. If the 45° port has an efficient reflecting terminal, energy is reflected back, past the ferrite rod again, is rotated 45° more, and now emerges from the 90° port. If this energy strikes

a reflecting load, it will re-enter the circular waveguide and emerge from the 135° port. If this is reflected and returned to the circular waveguide, it will emerge from the 0° port, either in phase or out of phase, depending on the total travel distance of the wave through the *circulator*. A purely resistive load, matching the impedance of the waveguide, coupled to the 90° port, produces an isolator. Energy entering the 0° port can emerge only from the 45° port, since energy reflected from any load coupled to the 45° port will be completely dissipated by the resistive load at the 90° port.

Figure 26-25 illustrates one use of a circulator.

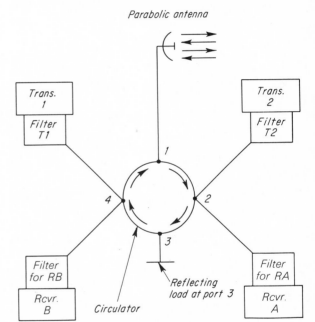

Fig. 26-25 Circulator to multiplex two transmitters and two receivers to same antenna.

It allows one parabolic-reflector, focusing-type, high-gain (±30-dB), broadband antenna to feed two separate microwave transmitters and receivers at the same time. Each bandpass filter is tuned to its own receiver or transmitter frequency, reflecting all other frequencies. A signal for receiver B is picked up by the antenna, feeds to port 2, is reflected, is reflected again at port 3; but, at port 4, the signal is accepted by the RB filter and passes to receiver B. Transmitter 2 energy is reflected from filter RA, from port 3, from the filters of port 4, feeding its signal to the antenna. Transmitter 1 and receiver A make only one 45° transition to and from the antenna.

26-13 TUNNEL-DIODE AMPLIFIERS

Operation of a tunnel diode, biased to its negative-resistance center point and coupled to a tuned circuit, produces an oscillator at the tuned-circuit frequency. If the input to the tunnel diode can be tuned but the feedback due to the tunnel diode being connected across the tuned circuit cannot reflect energy back to the tuned circuit, the tunnel diode cannot oscillate but will amplify. A circulator can be employed, as shown in Fig. 26-26, with a tunnel diode to produce an

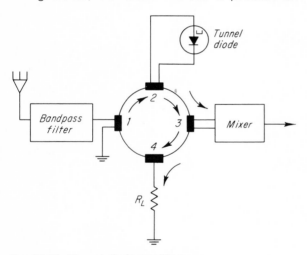

Fig. 26-26 Tunnel-diode amplifier

RF amplifier ahead of a microwave mixer stage. The bandpass filter feeds the desired-frequency received signals into port 1 and to the tunnel diode in port 2. The amplified signals are fed from port 2 to port 3 and to the mixer input. If any signal is reflected from port 3, it passes to port 4, where it is absorbed by the matched load resistance. There is no way for the tunnel diode to produce feedback of energy, so it cannot oscillate.

Gunn diodes can be used in similar amplifier circuitry. They have more gain than the tunnel diode, but they have more noise also. LSA diodes can be used as microwave amplifiers using similar circulator circuitry.

26-14 YIG RESONATORS

A device that can have very high Q at microwave frequencies is a tiny, grown-crystal, highly polished ball of ferrimagnetic yttrium-iron-garnet (YIG). Such a sphere, smaller than the head of a pin, if exposed to a fixed magnetic field, will rotate until its crystalline domains align with the external field. If the sphere is then anchored and a small loop of wire is wound around it in a direction that allows dc flowing through the loop to produce a magnetic field at right angles to the fixed field, the resultant poles of the YIG sphere shift from the original fixed-field direction. If ac is fed to the loop, the resultant magnetic YIG poles will attempt to *precess,* or rotate around the axis of the fixed field at the frequency of the applied ac. At one particular frequency, determined by the geometry of the YIG sphere and the strength of the fixed field, the precession finds little molecular opposition and the ac sees the YIG sphere as presenting almost no load (high Q) to it.

If the strength of the fixed field is decreased, ease of precession of the poles of the YIG sphere decreases, and the original ac frequency sees the YIG as having a lower Q and being a more lossy device. Now, by lowering the fixed field strength, a value of magnetization that will permit a free resonant precession at the lower frequency can be found, and the YIG sphere again appears as having a high Q. Conversely, a YIG sphere produces a resonant effect at a higher frequency if the dc magnetic field strength is increased. Thus, the YIG acts as a tank circuit whose resonant frequency is tunable by varying the dc excitation to an electromagnet between whose poles the sphere is mounted.

Figure 26-27a represents a YIG sphere mounted between the north and south poles of an electromagnet. There will be some value of dc excitation which will allow the ac input to see the YIG as a high-Q parallel resonant circuit.

In Fig. 26-27b, the YIG sphere is mounted in a field but is surrounded by two loops at right angles to each other. If the dc excitation produces the proper magnetic field strength, the YIG will allow maximum transformer effect between primary and secondary loops and maximum output at its resonant frequency. At all other input frequencies there will be no magnetic resonant pre-

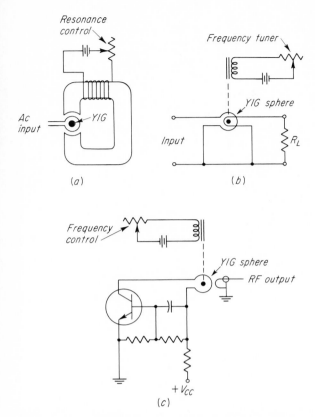

Fig. 26-27 (*a*) Magnetic poles of YIG precess with ac excitation. (*b*) Basic YIG voltage-variable narrow BP filter. (*c*) Basic voltage-variable Hartley-type YIG-tuned oscillator.

cession of the YIG and there can be no induced emf from primary to secondary because they are at right angles to each other. The device operates as a narrow-bandwidth tunable bandpass filter.

In Fig. 26-27*c* a YIG sphere forms the tank circuit of a Hartley-type oscillator circuit. Base and collector are coupled to opposite ends of the single-turn inductor, with the ground emitter connected to the center. Actually, at microwave frequencies, distributed and interelement capacitances make simple Colpitts or ultraudion oscillator circuits possible, with the oscillator output frequency determined entirely by the field strength of the dc-controlled electromagnet. In the SHF range, transistors may be supplanted by tunnel, IMPATT, Gunn, or LSA diodes as the active devices in oscillator circuits. YIG-tuned devices can operate up into the tens-of-gigahertz region.

26-15 MICROWAVE MEASUREMENTS

Technicians working in microwaves will be expected to work with:

- *Signal generators*. These use klystron or BWO tubes, transistors, YIGs, or Gunn diodes, having pulse, CW (continuous-wave), or sine- or square-wave modulated-signal outputs. The generators may be hand-variable or be set to sweep a selected band of frequencies. They generate signals to more than 40 GHz.
- *Calorimetric power meters*. These are oil-cooled bridges with bolometer arms reading directly in watts or in decibels. Their power capabilities range from 10 mW to 10 W.
- *Power meters*. These indicate the resistance change due to microwave energy striking a thermistor mount at the end of a waveguide or coaxial cable, reading directly in watts or in decibels. If the power in the system is greater than can be read directly, accurately calibrated waveguide or coaxial attenuators are used ahead of the thermistor mount.
- *Standing-wave indicators*. These read SWR directly as a carriage and its detector probe are moved along a slotted section of waveguide.
- *Ratio meter*. This device reads the ratio between the two directional couplers of a reflectometer setup, indicating SWR directly, regardless of the frequency being swept by the signal generator.
- *Frequency meters*. These are either the calibrated wavemeter type or heterodyne oscillators beating against the unknown signal and reading out on a frequency counter.
- *Sampling oscilloscopes* can display microwave signals up to about 20 GHz. They may use variable-storage CRTs, and often cameras are fastened over the CRT to take pictures of waveforms.
- *Spectrum analyzers* are swept receivers with a CRT display of the amplitude of received signals versus the frequency band across which the receiver is being tuned. The swept bands may be relatively narrow or broad, but top frequencies are in the 40-GHz range.
- *Electronic counters* can count directly up to 18 GHz; but by using prescalers and heterodyne converters, counting can be extended to above 40 GHz.

- *Function generators* are used to modulate signal generators with a variety of waveshapes and pulse widths. PIN diodes across coaxial lines or waveguides produce absorption-type modulation of microwave signals.
- *X-Y recorders* are used to plot relatively slow variations of circuit voltages or other parameters by moving a stylus up and down a sheet as the stylus is moved along horizontally. *Strip-chart recorders* have a stylus that is horizontally stationary but moves up and down a sheet of paper which is being drawn slowly along beneath the stylus.
- *Connectors* used to and from microwave equipment are of the N, BNC, SMA, GR, APC-7, and banana types. Most of these afford a relatively small impedance discontinuity which prevents reflection of signals in the line.
- *Time domain reflectometry* (TDM) utilizes a sampling oscilloscope, sending a very narrow pulse down a line and measuring the time it takes for a return reflected signal from any discontinuities in the line to appear. Time is then converted to distance to locate the discontinuity.

Some of the measurements that are made are:

- Absolute power in a system, using bolometer terminations to a power meter.
- Relative power levels in systems. As a system is tuned or the loading is changed, measurements are taken on a power meter.
- Attenuation produced by a device added in a waveguide system. A reading is taken at the load without the device and then again with the device coupled into the system. The difference between readings is the attenuation produced by the device.
- VSWR measurements. Standing-wave ratio indicates the reflection produced by a mismatched load. Either a detector probe in a slotted section or a reflectometer may be used to produce the voltages that are read.
- Frequency of signals in a system.
- Impedance. Since a matched load produces a 1:1 SWR, analysis of the SWR in a waveguide allows determination of the impedance of a load if the waveguide impedance is known.

26-16 INSTALLING WAVEGUIDES

When waveguides are installed permanently, as in radar, it is important that there be no long horizontal runs which might result in an accumulation of dust and moisture droplets from condensation on the inside walls of the sections. These would attenuate energy transmitted down the line.

Since condensation does occur in exterior waveguides, it is usual to drill a small hole in the elbow of a waveguide section at the lowest point to allow an escape vent for the water. Choke flanges must always be mounted with their half-wave cavities pointed upward so that they will not fill with condensation.

Waveguides must be handled with care. The slightest dent in a wall of a section produces a discontinuity and increases the SWR and loss of power transfer along the system.

To prevent radiation of undesired signals and interference to nearby receivers, waveguides should be firmly bolted to walls and grounded as often as possible.

Test your understanding; answer these checkup questions.

1. Name five microwave VTs. _____ _____ _____ _____ _____
2. Which microwave tube(s) has (have) a repeller? _____ A series of cavities? _____ A helix? _____ A horseshoe magnet? _____
3. What is the advantage of gridless klystrons? _____
4. How are magnetrons tunable? _____ _____ _____
5. What does TWT mean? _____ BWO? _____
6. Which VT do you think will oscillate at the highest frequency? _____
7. How would a balun be used to change 150 Ω to 50 Ω? _____ How could a balun change 100 Ω to 900 Ω? _____
8. How high a frequency can transistors generate? _____
9. What is the microwave application of a Varactor? _____ What diode does even better? _____
10. Name four diodes that generate microwave ac. _____ _____ _____ _____
11. What does PIN mean? _____ For what are PIN diodes used in microwaves? _____
12. What are two other names for Gunn and LSA diodes? _____ _____ What does LSA stand for? _____ What is its advantage over a Gunn diode? _____
13. What is wrong with the term "Gunn diode"? _____

14. What two important properties do ferrites have at microwave frequencies? _____ _____ Which is used as a switch? _____ Which is used in circulators? _____
15. What is the advantage of using a circulator with two transmitters and two receivers? _____
16. To what must the fourth port be connected in a tunnel-diode amplifier? _____ The third port in a multiplex circulator? _____
17. List 12 items microwave technicians might use. _____
18. Why should outdoor WG not be installed horizontally? _____

COMMERCIAL LICENSE QUESTIONS

Applicants for Element 8 (radar endorsement) should know answers to all questions. Applicants for Element 3 should know answers to questions followed by [3]. Sections in which questions are answered are shown in parentheses.

1. What is a *lighthouse triode* and for what frequency range was it designed? (26-1) [3]
2. Why are waveguides used in preference to coaxial lines for the transmission of microwave energy in most shipboard radar installations? (26-2)
3. Why are rectangular-cross-section waveguides used in preference to round waveguides? (26-2)
4. Describe briefly the construction and purpose of a waveguide. (26-2) [3]
5. Discuss with respect to waveguides the relation between frequency and size, modes of operation, coupling of energy into the waveguide, general principles of operation. (26-2, 26-4) [3]
6. Why are choke joints often used in preference to flange joints to join sections of waveguides together? (26-2)
7. Draw a longitudinal section of a waveguide choke joint. Explain briefly its principle of operation. Indicate the two quarter-wave slot dimensions. (26-2)
8. Explain the principles of operation of a cavity resonator. (26-6) [3]
9. Explain briefly the principle of operation of the reflex klystron. (26-7)

10. Describe the physical structure of a klystron tube and explain how it operates as an oscillator. (26-7) [3]
11. Describe the physical structure of a multicavity magnetron. Explain its operation. (26-8) [3]
12. Draw a simple cross-sectional diagram of a magnetron and show the anode, the cathode, and the direction of electron movement under the influence of a magnetic field. (26-8)
13. Draw a diagram showing the construction and explain the principles of operation of a traveling-wave tube. (26-9) [3]
14. Why are long horizontal waveguides not desired? (26-16) [3]
15. Why should the interior of a waveguide be clean, smooth, and dry? (26-16)
16. Why is a 1/8-in.-diameter hole sometimes drilled on the underside of a waveguide elbow near the point where the waveguide enters the radar transmitter? (26-16)
17. What precautions should be taken in the installation and maintenance of a waveguide to ensure proper operation? (26-16)
18. What precautions should be taken when installing vertical sections of waveguides with choke coupling flanges to prevent moisture from entering the waveguide? (26-16) [3]

27 RADAR

The objective of this chapter is first to describe the general purpose of a radar set and then to discuss separately some possible basic radar transmitter, receiver, and antenna circuits. Finally, the component circuits are explained as a complete PPI transmit-receive-display system.

27-1 PRINCIPLES OF RADAR

Radar is an electronic system of determining the direction and range of anything that will reflect microwave radio waves, representing one application of the microwave theory of Chap. 26.

The use of echoes as an aid to navigation is not new. When running in fog near a rugged shoreline, ships have sounded a short blast on their whistles, fired a shot, or struck a bell. The time between the origination of the sound and the returning echo indicated how far the ship was from the cliffs or shore. Sound is known to travel approximately 1,100 ft/s. If an echo is heard after 2 s, it indicates a total sound-travel distance, outward and return, of 2,200 ft. The ship must be half this distance, or 1,100 ft, from shore. The direction from which the echo approaches indicates the bearing of the shore.

Today, ships transmit a short pulse (type P∅ emission) of microwave energy and receive the echo produced when the radio wave is reflected from any object. By use of an antenna with a narrow radiation beam, the direction of the reflecting target can be accurately determined. By indicating electronically the time between pulse transmission and reception, the range of the target is known. This is radar, from RAdio Direction And Range.

Besides being employed as a marine navigational aid, radar is used by aircraft, by airfields, and by the armed services as a means of locating enemy targets and aiming guns. Even the police can determine accurately the speed of an unwise and unsuspecting motorist.

The indicators of U.S. ship radar sets are cathode-ray tubes (CRT), usually calibrated in U.S. nautical miles of 6,080 ft (the international nautical mile is 6,076 ft). Land-based radar may be calibrated in statute miles, of 5,280 ft. International equipment may be calibrated in kilometers.

Radio waves travel 162,000 nautical miles per second (186,000 statute miles per second, or 300,000 kilometers per second). Thus, in 1 μs (1 microsecond) they travel 0.162 mi. To travel 1 nautical mile a radio signal requires 1/0.162, or 6.17 μs. A *radar mile* is considered to be 12.3 μs, since the wave must travel for this period of time, to and from the target, if the target is 1 mile away.

If a target is 10 mi away, it takes 123 μs from the time of transmission before the echo signal returns and is displayed on the CRT of the radar.

The frequencies generally used for marine radar are in the SHF (superhigh-frequency) part of the radio spectrum, either in the 3,000- to 3,246-MHz band (10-cm, or S, band) or in the 9,320- to 9,500-MHz band (3-cm, or X, band). A third band, 5,460- to 5,650-MHz, is also available.

For the 10-cm band, a half-wave antenna is only 5 cm (2 in.) long. An antenna reflector 2 m (2 meters) wide can form a radiated beam with a horizontal width of 2° or 3°. In the 3-cm band, the antenna is only 1½ cm in length, and a 2-m reflector can form a 1° or 2° beam. At these frequencies the effective range is only slightly more than line of sight.

Marine radar is usually made to operate with a maximum range of 20 to 40 mi and with a minimum range of less than 100 yards from the antenna.

The number of pulses transmitted per second, called the *pulse repetition rate* (PRR), varies between about 800 and 2,000 per second. The lower PRR is used for longer ranges. The pulses on long-range equipment have a longer duration, representing a greater power output from the transmitter to produce stronger echo signals from distant objects.

The pulse width is about 0.25 μs for short-range indications. A pulse width of more than 2 μs is used for long-distance operation.

27-2 A BASIC RADAR SYSTEM

Before the operation of one of the more complex radar systems is explained, the functioning of the simplified one in Fig. 27-1 will be described.

The heart of the system is in the modulator, keyer, or timer. In this section, a 0.2-μs pulse is formed and fed to a magnetron oscillator, resulting in the transmission of a 0.2-μs pulse of SHF radio energy. The timer pulse is also fed to the indicator, starting a dot moving horizontally, tracing from the center of the CRT face to one edge. If the maximum range of the radar set is to be 10 mi, the time for the barely visible dot to move to the edge of the scope will be 123 μs. The timer pulse is also fed to the grid (or cathode) of the CRT, producing a bright spot at the center of the tube at the start of each trace.

After each RF pulse has been transmitted, the

Fig. 27-1 Block diagram of the basic elements of a simple radar system.

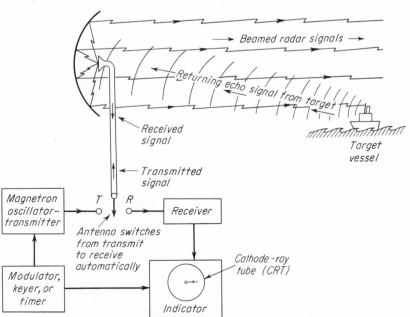

antenna automatically switches to the receiver. The receiver waits for the return of the echo signal. If a target happens to be 5 mi away and in the beam of the antenna, after 61.5 μs a weak echo signal is received by the antenna, fed to the receiver, amplified, and fed to the grid of the indicator scope. This signal increases the brightness of the moving dot wherever it happens to be on the trace line, and a bright spot appears. The distance from the center spot to the echo spot is an indication of the range of the target. In this case the target *blip* will appear halfway across the 10-mi trace. The direction of the target is indicated by the direction to which the antenna must be rotated to pick up the echo signal.

As soon as the barely visible dot has traveled for 123 μs, the indicator tube is desensitized and the dot returns to the center without producing any visible indication.

With a PRR of 800, every 1,250 μs a new pulse is produced and the target signal is registered at the same point on the trace. If the target is approaching the radar set, the distance between blip and center spot (main bang) on the trace shortens. In this way a constant check can be maintained on the range of the target vessel.

The parabolic antenna reflector should have a radiated beam of about 2° horizontal width for good *bearing resolution*. (It has a vertical beam height of 15° to 20°.) Bearing resolution is the ability to separate adjacent targets the same distance away. *Range resolution* is the ability to distinguish two or more targets in the same direction but at different distances.

If the antenna rotation were controlled manually, it might be difficult to keep the target in the 2° horizontal beam if either the target or the radar ship were moving. It is necessary to improve this basic radar system to make it a practical aid to navigation.

If the antenna is made to rotate horizontally at a constant speed, 10 times per minute, it will make one rotation in 6 s. With a PRR of 1,000, it will fire 6 × 1,000, or 6,000 pulses per single rotation. This is about seventeen pulses per degree of rotation. If horizontal-deflection coils are rotated physically around the neck of the CRT in exact synchronism with the rotation of the antenna, targets can be shown on the indicator face in exact relation with their range and bearing. This

is known as a plan-position-indication (PPI) type of presentation.

CRTs used in radar employ electromagnetic deflection and vary in diameter from 7 to 16 in. They differ from television tubes in screen persistence. The radar-tube faces are coated with a little fluorescent and considerable phosphorescent material. The phosphors retain a latent image for a period of 10 s or more, which is longer than is required to make one antenna rotation. As a result, a PPI presentation forms a constant, well-illuminated plan, map, or chart of the targets in all horizontal directions from the ship. A block diagram of the component parts of the system is shown in Fig. 27-2. The motor that rotates the

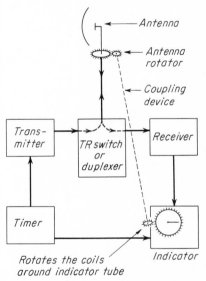

Fig. 27-2 Block diagram of the elements of a PPI navigational radar system.

antenna is shown mechanically coupled to the deflection coils around the neck of the indicator CRT.

27-3 A MARINE RADAR SYSTEM

A step-by-step explanation of the important circuits in a marine radar system will be given. Most of the circuits in the receiver and indicator are similar to those found in other types of equipment, explained previously. New circuits will be explained. The circuitry has been simplified as much as possible to give a general understanding, since

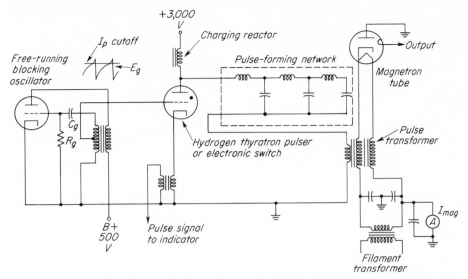

Fig. 27-3 Diagram of transmitter portion of a simplified radar system.

it is impossible to cover all the various types and models of radar.

27-4 THE TRANSMITTER SECTION

The radar transmitter will be considered as the blocking oscillator, the pulse-forming circuit, and the magnetron oscillator, shown in Fig. 27-3. The blocking-oscillator circuit depends on the C_g and R_g values to determine how many times per second it will operate. The circuit shown resembles an Armstrong oscillator, but any oscillator circuit will block if high values of grid-leak resistance and capacitance are used. As the circuit starts to oscillate, plate current flows for an instant, inducing a charge in C_g well beyond plate-current cutoff. During the I_p-off time, C_g discharges slowly through R_g and the bias voltage decreases until finally plate current can begin to flow again. Circuit regeneration develops a rapid rise in I_p. These short-duration plate-current pulses induce high-amplitude pulses in the grid winding of the transformer. The number of pulses of current per second in this stage determines the PRR of the whole radar system.

The grid of the hydrogen thyratron is triggered by the positive pulse, and the tube ionizes. (Hydrogen is used in radar thyratrons because it ionizes and de-ionizes more rapidly than argon or mercury.) Ionization discharges the capacitors of the pulse-forming network that have charged to a 3,000-V or higher potential from the power supply through the charging reactor during the nonoperating period of the oscillator. This charge produces a narrow *square-wave* pulse of current through the primary of the pulse transformer, inducing a high voltage in the two secondaries. These two voltages are of equal value and are in phase, raising the filament and cathode of the magnetron to a negative potential of several thousand volts above the magnetron plate without changing the potential between the two filament terminals. The plate of a magnetron (Sec. 26-8) consists of a series of tuned metal cavities surrounding the cathode. As a safety measure, the plate cavities of the magnetron with their metal cooling fins are always grounded and the high-voltage pulses are fed to the cathode. The average value of magnetron plate current, perhaps 2 to 5 mA, is indicated by the meter between ground and the secondary of the filament transformer. The square-wave pulse of current from the pulse transformer excites electrons into oscillation in the cavities of the magnetron. These oscillations are coupled to the antenna and form the RF output burst.

A pulse-forming network, as shown, may also be called an *artificial transmission line*. A transmission line has the ability to produce a square-wave output when triggered with a short burst of energy, provided its impedance is matched to the load impedance. The length of the pulse will be a function of the values of inductance and capacitance used in the pulse-forming network. (The pulse length = $2\sqrt{LC}$, with L in henrys and C in farads.) An artificial transmission line can be called a *delay line* because a pulse voltage fed across the input end will appear a few microseconds later across the output end (delay time = $\sqrt{LC}$). If an electric impulse travels 300 m in 1 μs, a 300-m transmission line will delay the voltage 1 μs. An artificial line with similar values of series inductance and shunt capacitance will delay the same time.

The transformer in the cathode of the thyratron pulser tube has a primary of two or three turns, which induces a voltage into the secondary. This voltage is fed to the indicator system to trigger the circuit that starts the trace moving across the CRT screen.

One of the earlier radar systems employed a rotating spark gap to produce its pulses. Capacitor C in Fig. 27-4 charges to 5,000 V while the spark gap is open. When the gap rotates closed, the capacitor discharges through the pulse line and the primary of the pulse transformer, developing a square-wave pulse to fire the magnetron.

Test your understanding; answer these checkup questions.

1. What does "radar" mean? _____
2. In what distance units are marine radars calibrated? _____

3. What time unit is the equivalent of a radar mile? _____
4. In what frequency bands does marine radar operate? _____
5. What does PRR mean? _____ What values are used? _____
6. Why must short-range radar have narrow pulse widths? _____
7. To what two circuits would a radar timer circuit feed signals? _____ _____
8. What is a target display on a CRT called? _____
9. Why must a radar antenna be pointed directly at a target? _____
10. What is the main bang on a PPI radar set? _____
11. What is the ability to separate adjacent equidistant targets called? _____
12. What is the approximate rotational rate of a radar antenna? _____ About how many pulses are transmitted per degree? _____
13. In what two ways do radar CRTs differ from TV? _____ _____
14. How would sector scanning differ from PPI presentations? _____
15. What circuits make up the radar transmitter? _____
16. What type of oscillator determines the PRR? _____ What other circuit might be used? _____
17. Why are hydrogen-gas thyratrons used in radar? _____
18. Why is a magnetron anode grounded? _____
19. What are two other names for a pulse-forming network? _____ _____
20. What determined the PRR in old-time radar sets? _____

27-5 AVERAGE POWER AND DUTY CYCLE

At the marine radar frequencies, 3 and 9 GHz, the most practical transmitting oscillator tube is the magnetron. Voltage pulses fed to it may range from 5 to more than 20 kV, with output pulses of

Fig. 27-4 Diagram of old type rotating-spark-gap radar transmitter.

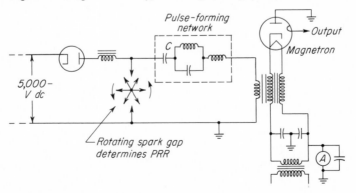

perhaps 15 kW (megawatts for military radar). What power rating must such tubes have?

Suppose a radar transmitter has a PRR of 900 pps, with each pulse having a duration of 0.5 μs (0.000 000 5 s) and a peak pulse power of 15 kW. The total emission duration must be 900 × 0.0000005, or 0.00045 s. The transmitter is on 0.00045 s each second. It has a *duty cycle* of 0.00045. The average power output of the transmitter is the peak power (15 kW) times the fraction of a second it operates, or 15,000 × 0.00045 = 6.75 W. The tube is transmitting pulses with peaks of 15 kW but is transmitting an average power of only 6.75 W. If the magnetron is 50% efficient, the average dc power input is only 13.5 W. From this, formulas for average power output would be

$$P_{av} = P_{peak} \times \text{PRR} \times \text{pulse width}$$
$$= P_{peak} \times \text{duty cycle}$$

From this information, two formulas for duty cycle must be

$$\text{Duty cycle} = \frac{P_{av}}{P_{peak}}$$
$$= \text{PRR} \times \text{pulse width}$$

The duty cycle is also the ratio of the pulse width to the time between the beginning of two pulses (called the *pulse repetition time,* or PRT). With a PRR of 900, the PRT is $\frac{1}{900}$ s, or 0.00111 s. For the transmitter above, the duty cycle must be 0.000 000 5/0.00111, or 0.00045.

EXAMPLE: What are the peak power and duty cycle of a radar transmitter with a pulse width of 1 μs, a PRR of 900, and an average power of 18 W? By solving for P_{peak} from the P_{av} formula,

$$P_{peak} = \frac{P_{av}}{\text{PRR} \times \text{pulse width}}$$
$$= \frac{18}{900(0.000001)} = 20,000 \text{ W}$$
$$\text{Duty cycle} = \frac{P_{av}}{P_{peak}} = \frac{18}{20,000} = 0.0009$$

27-6 THE ANTENNA SYSTEM

The radar antenna system consists of waveguide from the magnetron to the antenna and a *duplexer* near the magnetron, as shown in block form in Fig. 27-5. The end of the conductor from

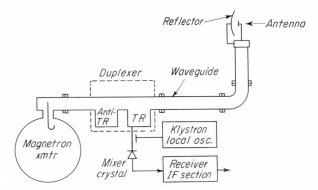

Fig. 27-5 Block diagram of duplexer and components coupled to it.

the magnetron acts as a quarter-wavelength antenna, transmitting RF energy along the waveguide. At the far end, the walls of the waveguide may be expanded to form a horn, and the RF energy is radiated from the horn into a parabolic antenna reflector. In some equipment the waveguide carries the signal up to the antenna, where the inner conductor of a coaxial cable projects into the waveguide, acting as a receiving antenna. The signal is transferred a short distance by coaxial cable to a tiny vertical half-wavelength dipole which radiates RF energy into the reflector.

Radio waves of 3 or 9 GHz propagate in much the same way as light rays. As in a flashlight, these radio waves can be focused into a narrow beam with a metal parabolic reflecting surface by placing the radiator at the focal point, as in Fig. 27-6. By shaping the reflector properly, the emitted wave can be formed into the desired 2° horizontal beam width and 15° vertical beam height.

Radiator and reflector are rotated constantly, at about 10 rpm, in synchronism with the sweep coils rotating around the neck of the CRT in the indicator.

The horn, or dipole radiator, is usually covered with polystyrene or some other plastic to protect it from weather. This plastic must not be painted. An excessive amount of soot or dirt on either the polystyrene cover or the active surfaces of the

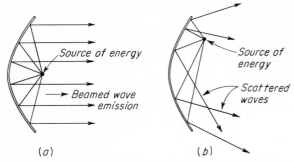

(a)   (b)

Fig. 27-6 Parabolic reflectors. (*a*) With source at the focal point, all reflected waves are radiated parallel, forming a narrow beam. (*b*) Waves scatter when the source is not at the focal point.

reflector may decrease the transmission and reception to a degree. The radiated power is considerable with some radar equipment. People standing in the beam of high-powered radar an-

tennas for a few minutes can be cooked internally and may die within a few days. Radar beams have been known to ignite a shipment of photograph flashbulbs and to start fires.

The placement of the radar antenna is important; it should be as high and as much in the clear as possible. The higher the antenna, the farther it can "see" targets. If the radiated beam is above masts, booms, and stacks, these objects will not shadow, or reflect signals and produce false blips on the indicator.

27-7 TR BOXES

The radar transmitter and receiver use the same antenna system. Some means must be provided to prevent the powerful transmitter signal from feeding directly into the receiver and burning out the input circuit. This is accomplished by using a resonant transmit-receive (TR) cavity with a special gas-filled spark-gap TR tube in it. The cavity is coupled to an opening in the antenna waveguide. Signals in the waveguide excite the tuned cavity into oscillation. A coupling loop in the cavity feeds these signals to the crystal-diode mixer, shown in Fig. 27-7. The local oscillator is also coupled to the mixer circuit.

When the transmitter emits a pulse from the magnetron, these high-powered waves induce enough voltage across the TR cavity to ionize the TR tube, and it conducts. The tube then effectively shorts out the cavity, detunes it, and prevents

---

ANSWERS TO CHECKUP QUIZ ON PAGE 660

1. (*Radio direction and range*)   2. (*Nautical miles*)   3. (*12.3 μs*)   4. (*3 GHz, 10 cm, S; 9.5 GHz, 3 cm, X*)   5. (*Pulse repetition rate*) (*800–2,000*)   6. (*Transmission pulse must stop before first of echo returns*)   7. (*Modulator, indicator*)   8. (*Blip*)   9. (*Narrow beam*)   10. (*Center spot*)   11. (*Bearing resolution*)   12. (*10 rpm*)(*15–20*)   13. (*Round*) (*Persistence*)   14. (*Antenna moves back and forth over a sector*)   15. (*Blocking oscillator, pulse-forming circuit, magnetron*)   16. (*Blocking or pulse*) (*Unbalanced multivibrator*)   17. (*Fast ionize, de-ionize*)   18. (*Coupled to waveguide, which must be grounded*)   19. (*Artificial transmission line*) (*Delay line*)   20. (*Spark gap speed*)

Fig. 27-7 Details of the TR, ATR, and crystal mixer.

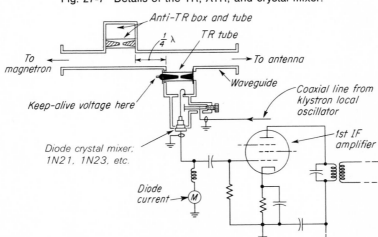

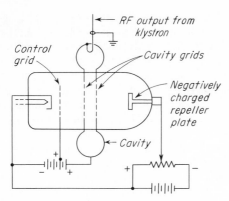

Fig. 27-8 SHF reflex klystron oscillator circuit.

high-amplitude signals from forming in it. At the conclusion of the transmitted pulse, the TR tube de-ionizes and the cavity is ready to receive any returning echo signals. To make the TR tube more sensitive, a dc *keep-alive* voltage is applied across it at all times. This voltage is not quite high enough to support ionization, but a small increase in voltage across the gap will ionize it. When the TR tube ages, it requires a greater signal voltage to produce ionization, it does not protect the crystal diode in the mixer circuit of the receiver, and the diode burns out. Generally, when a diode must be replaced, the TR tube is also replaced.

The distance from the point where the magnetron is coupled to the waveguide and where the TR box is installed is very critical. A less critical means of coupling the TR box to the waveguide is to use a second tuned cavity, placing it exactly a quarter wavelength from the TR box opening. This second cavity is called an *anti-TR* (ATR) *box*.

The anti-TR box has a TR tube in it also, but since it does not have to protect any circuits, it has no keep-alive voltage. The ATR ionizes during each transmitted RF pulse, presenting a low impedance across its opening in the waveguide. This allows the pulse moving up the waveguide to pass unattenuated. When the pulse ceases, the ATR tube de-ionizes and presents a high-impedance point to the waveguide. Now echo signals coming down the waveguide can no longer pass to the magnetron but are reflected at the ATR opening and dispose of their energy in the TR cavity, which is the input circuit of the receiver.

27-8 THE RADAR RECEIVER

Radar receivers are always superheterodynes, with an IF of approximately 30 MHz. This requires an oscillator operating at a frequency 30 MHz or more from the transmitting frequency (3 to 9 GHz). A reflex klystron (Sec. 26-7) is usually the local-oscillator circuit. The basic klystron oscillator is shown in Fig. 27-8.

A radar receiver consists of the crystal-diode mixer stage, a klystron local oscillator, six or more 30-MHz IF amplifiers, a diode second detector, and two or more stages of video amplification capable of amplifying signals up to several megahertz (Fig. 27-9). This wide range is required because a pulse of 0.5-μs duration represents a frequency of 1 MHz. The square waveshape of the pulse requires many harmonics of the 1-MHz frequency to reproduce the square waveshape. A

Fig. 27-9 Block diagram of radar receiver with AFC.

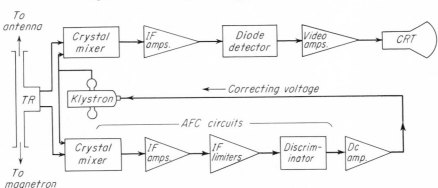

bandwidth of 10 MHz can be attained with a 30-MHz or higher IF strip.

The AFC (automatic-frequency-control) system keeps the klystron circuit oscillating on the correct frequency. A second crystal diode is coupled to the duplexer-mixer cavity. The output of this crystal mixer is similar to that of the receiver mixer, feeding an IF amplifier, limiters, and a 30-MHz discriminator circuit. When the transmitter and klystron local oscillator are separated by the proper IF difference, no output is delivered to the klystron repeller by the discriminator. Should either the transmitter or the klystron drift in frequency, a dc output voltage is produced by the discriminator, is amplified, and is made to vary the repeller voltage of the klystron, forcing it to oscillate at a frequency 30 MHz from the transmitter frequency. When a mixer crystal must be replaced, the AFC crystal should be changed at the same time.

The IF stages of a radar receiver are basically similar to other broadband superheterodynes. However, to prevent strong returning echoes from nearby sea waves, known as *sea return,* from being shown on the indicator, it is necessary to desensitize the receiver immediately after the transmitted pulse. The IF stages must then return to maximum sensitivity in 10 to 15 μs. Sea-return elimination may be accomplished by using a thyratron tube in a *sensitivity-time-control* (STC) circuit. A simplified STC circuit is shown in Fig. 27-10. When a positive trigger impulse from the

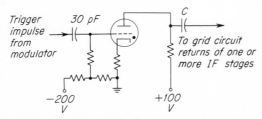

Fig. 27-10 Thyratron STC circuit to develop a decaying bias for IF amplifiers.

transmitter modulator is applied to the thyratron grid circuit, it overcomes the bias. The tube ionizes, conducts heavily, and discharges capacitor C. This drives electrons from the right-hand plate of the capacitor to the grid circuits of the IF stages, biasing them negatively and desensitizing

them. At the completion of the trigger pulse, the thyratron de-ionizes and capacitor C begins charging, reducing the negative charge on the grids to normal in a few microseconds. Thus, as the transmitted pulse is produced, the IF stages are highly biased and insensitive. As time progresses, the bias falls off and in 10 to 15 μs normal sensitivity returns. STC is controlled by a switch on the panel.

The signal from the IF stages is detected by a diode and fed to the video stages, where it is limited to reduce *blooming* (excessively expanded blips on the CRT screen).

The output impulses of the video amplifiers are fed to either the grid or the cathode of the CRT, producing the visible echo signals on the screen. A manual gain control in the video stages acts as the brilliance control for the scope presentation. A manual gain control on the IF amplifiers acts as the sensitivity control. These controls are adjusted to produce a just-visible trace on areas of the CRT screen where no targets are displayed.

Test your understanding; answer these checkup questions.

1. With a PRR of 800 and a 1-μs pulse width, if the peak power is 100 kW, what is the average power? _____ The duty cycle? _____
2. What is the PRT with a PRR of 800? _____ What does the answer mean? _____
3. What is used to carry RF energy from magnetron to antenna in a radar set? _____
4. What are two methods of illuminating a parabolic reflector with RF? _____ _____
5. Would a radar reflector be parabolic-shaped vertically or horizontally? _____ Why? _____
6. In what way are radar emissions similar to microwave ovens? _____
7. What method of coupling is used between klystron local oscillator and diode mixer in Fig. 27-7? _____
8. To what tube(s) is a dc keep-alive voltage applied? _____
9. If a mixer diode burns out, what are replaced? _____ _____
10. What is the cavity between magnetron and mixer cavity called? _____
11. Radar receivers use what IFs? _____ Why? _____ What kind of LO? _____ What solid-state types might be used? _____
12. Does an ATR tube aid transmitting or receiving? _____

13. To what is the AFC voltage applied in a klystron? _____

14. What is the control that desensitizes a radar receiver for 10–15 μs called? _____

15. How is blooming prevented? _____ How is brilliance controlled? _____ Sensitivity? _____

27-9 CIRCUITS OF THE INDICATOR

The radar indicator consists of a CRT and the circuits necessary to produce a sawtooth sweep current for the deflection coils that rotate around the CRT neck, plus range-marker pulses. At first the CRT grid will be considered biased so that no electrons can strike the face of the tube.

The originating signal for the deflection system comes from the hydrogen thyratron (Fig. 27-3). The pulse transformer signal, in negative phase, is fed to a one-shot multivibrator *gating* circuit, as in Fig. 27-11. Tube V_1 normally conducts heav-

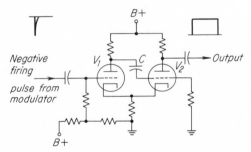

Fig. 27-11 One-shot multivibrator oscillator to develop a square-wave intensifier pulse.

ily because of the positive bias on it. The resulting cathode bias of this stage holds V_2 in nonconduction. A negative pulse from the thyratron drives V_1 into nonconduction for an instant charging C, and V_2 conducts for a period of time. When C discharges far enough, V_1 starts conducting again and V_2 cuts off once more. For each pulse from the modulator a negative-going square-wave intensifying output pulse is produced from V_2. This pulse is reversed in phase through an amplifier and fed to the grid of the CRT, through an adder circuit, overcoming the bias on the grid and allowing a few electrons to strike the face of the tube.

The leading edge of the intensifying pulse triggers a sawtooth circuit into one cycle of oscillation. The sawtooth current is amplified and fed through slip rings and brushes to the deflection coils that rotate around the CRT neck, as shown in Fig. 27-12. With the intensifying pulse operating, deflection coils rotating, and the sawtooth current deflecting the beam, a weak glow appears all over the face of the tube as the coils rotate. The glow is somewhat brighter at the center, where all the sweep lines originate.

Now, when echo signals are received and fed in a positive phase to the CRT grid, through the adder circuit, they produce much brighter spots than are produced by the intensifying pulse. These target signal blips are readily visible.

The negative intensifying pulse is also fed to a *range-marker* circuit. Figure 27-13 illustrates a possible circuit. Tube V_1 in this case has no bias and conducts heavily, producing a strong stationary magnetic field around the cathode coil L. As the negative intensifying pulse suddenly biases V_1 into nonconduction, the magnetic field collapses and the high-Q LC circuit is driven into damped oscillations. This ac is fed to amplifier V_2. The amplified output voltage is fed through resistor R_1 to a zener diode, which acts essentially as a short circuit to the negative half of the ac and limits the positive excursion to its zener-voltage rating. This produces the half-wave-rectified, constant-amplitude square-wave pulses indicated. These are differentiated (coupled through a short-time-constant R_2C_1 circuit), a process which develops a narrow positive pip on the rise of the square wave and a narrow negative pip on the fall of the square wave. The negative pip is shorted to ground by diode D_2, leaving only the positive pip to be fed to the adder circuit and to the CRT grid.

An LC oscillation frequency of 80.7 kHz produces one complete cycle every 12.3 μs. It will be remembered that this is the time equivalent to 1 radar mile. If some 80.7 kHz pips are added to each outward moving sweep on the CRT, they will produce concentric rings of illumination, each separated from the next by the equivalent of 1 mi. By counting range markers, the range of any observed target blip can be accurately estimated. A gain control on the output of the range-marker circuit controls the intensity of the rings. On longer-range presentations, the markers may be generated every 3, 5, or 10 mi by lowering the LC circuit oscillation frequency.

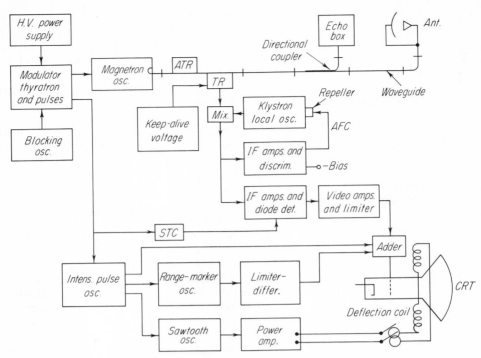

Fig. 27-12 Block diagram of a simplified PPI radar system.

The CRT used in radar has 10-s persistence instead of the few hundredths of a second of TV tubes. A TV receiver employs both vertical- and horizontal-deflection coils, but in a radar indicator there is only one set of coils. These are gear-driven around the neck of the tube 10 times per minute, in synchronism with the rotation of the antenna.

With the exception of the synchronizing system between antenna and sweep coils on the CRT, a complete radar system has been outlined.

Summarizing the operation, in Fig. 27-12, the sequence began with the blocking oscillator generating a series of pulses at approximately 1,000 pps. These pulses were shaped and used to fire a magnetron, as well as being used to trigger the indicator circuits. The magnetron emitted a strong RF burst from the slowly rotating antenna. The TR and ATR tubes protected the receiver and allowed only received echo signals to enter the mixer cavity of the receiver. The received signals were amplified, detected, and fed to the cathode-ray tube. At the same time, the trigger impulse started an intensifying pulse that was fed to the grid of the CRT, enabling received signals to produce indications on the screen. The intensifying pulse also started a sawtooth wave that produced the sweep trace from the center of the screen out to the edge, 1,000 times per second as the sweep coils were rotating. This resulted in a presentation of all radar targets in their relative position around the ship. The direction of the targets is determined by their angle from the top of the screen; their range is determined by how far they are from the center. As an aid to determining distance, range markers can also be turned on. To reduce sea-return echoes, the STC circuit can be adjusted to the lowest

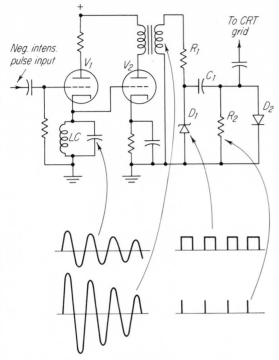

Fig. 27-13 Possible circuit to develop range-marker signals and the waveforms developed.

value that does not produce blurred light areas near the center of the screen.

27-10 ANTENNA SYNCHRONIZATION

If the antenna is mounted on or just above the indicator, as in small-boat radars, it is possible to use a single vertical drive shaft with gears at both ends to rotate the antenna and the deflection coils in perfect synchronization. In larger vessels this

is not feasible. Some sort of synchro system is needed. A basic form consists of a selsyn generator and a selsyn motor. The two units are similar; each has a rotor and three stator windings interconnected as shown in Fig. 27-14.

The connections of the stator windings make this appear to be a three-phase system, but this is not true. There is only the single phase of the 115-V 60-Hz power line ac. This emf is fed to both motor and generator rotors. The magnetic fields from these rotors induce voltages in the stators. As long as the two rotors are resting at the same relative angle between similar field coils, the voltages induced in these stator coils will equal each other and a condition of balance occurs. If the motor is held in position and the generator rotor is moved by hand in a clockwise direction, the voltages induced in the two sets of stators will no longer be similar. This results in a magnetic pulling, counterclockwise by the generator rotor and clockwise by the motor. When the generator is turned, the motor will respond to the proportionate magnetic changes produced in its stator fields and will follow the angular rotation of the generator rotor.

Mechanically coupling the rotating radar antenna to a selsyn generator and the selsyn motor to the rotating mechanism that drives the deflection coils around the neck of the CRT provides a possible means of synchronizing antenna and deflection-coil rotation. However, in operation, rotation of the selsyn motor must always lag that of the generator by a few degrees. The lag angle may change with variations of friction. The possibility of change of lag angle and an inherent lack of sensitivity of selsyns make other forms of synchros more desirable.

Fig. 27-14 A basic selsyn system. Dashed lines represent mechanical coupling.

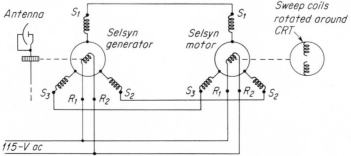

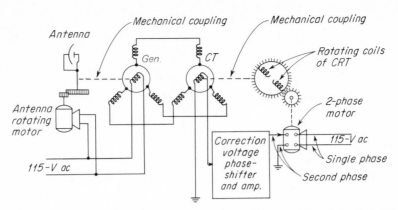

Fig. 27-15 Servomechanism to synchronize rotation of the radar antenna and the deflection coils around the CRT.

The system in Fig. 27-15 uses the emf induced in the unexcited *control transformer* (CT) rotor winding as a correcting voltage. If the two rotors are in the same angular position, no voltage will be induced in the CT rotor. As the rotors are varied in angular placement, the CT-rotor-induced voltage will change, but the rotor itself does not try to turn. However, any 60-Hz ac induced in the CT is shifted in phase 90°, is amplified, and is fed to one winding of a two-phase ac motor. The power-circuit ac is fed to the other winding of this motor. With both phases applied, the motor rotates, turning the CT rotor and the deflection coils of the CRT. If the antenna tends to rotate faster than the deflection coils, a greater voltage is induced in the CT rotor. This correction voltage increases the speed of the two-phase motor, and the deflection coils speed up. The CT rotor must always lag the generator somewhat, but in this system the amplifiers reduce the lag or variation in lag to a small value, resulting in satisfactory synchronization for the radar system. The amplifier and driving motor are known as a *servomechanism*. Note that the rotation of the CRT coils is completely dependent on the rotation of the antenna. If the antenna motor stops, the selsyn generator is no longer rotated by the antenna rotation, and the correcting voltage is no longer developed. Without both phases the two-phase motor stops and the CRT coils stop.

27-11 HEADING FLASH

As the radar antenna turns toward the bow of the ship, it trips a microswitch, which feeds a short positive pulse of voltage to the grid of the CRT. This results in a trace being made from the center of the screen to the edge. Such a *heading flash* indicates, on the CRT, the direction the ship is taking on the chartlike presentation of the PPI screen. The circuit can be turned on or off by a control on the indicator panel.

27-12 ECHO BOX

When at sea, with no targets available, or whenever it is desired to test the overall sensitivity and operation of the radar set, an *echo box* can be used. This is a high-Q cavity resonant to the transmitter frequency coupled into the waveguide with a directional coupler, as in Fig. 27-12. Each transmitted pulse shock-excites the cavity into oscillation, and it responds with a damped-wave output. The coupling to the echo box is usually adjusted to ring for about 12 μs. As long as it is active, it produces a tapering-off signal for the receiver and will result in illumination of the screen, outward from the center, for a distance equal to about a mile. If tubes or crystals become weakened or the system is not operating properly, the distance indicated is less, or no echo-box signal will be seen at all.

Since an echo box will blot out all target signals within a 1-mi radius, it is necessary either to decouple it by some mechanical means or to detune it far enough that it will not ring. One method utilizes a plunger that tunes the box through resonance as it is pushed down. This results in a flash on the radial traces, but only during the time that the box is being tuned through resonance.

27-13 OPERATING THE RADAR SET

The master of the vessel or any person designated by him may operate the radar set. No radio license is required. Furthermore, such a person may replace fuses or receiver-type tubes in the set, although this duty usually falls to one of the radio operators. However, whenever the equipment requires maintenance other than this, only persons holding First or Second Class licenses with radar endorsements or persons working under direct supervision of such a license holder may make adjustments to, service, or install radar equipment.

Each radar installation must have an installation and maintenance record, kept at the radar station. This record will include the date and location of installation and the name and license number of the person installing it. All subsequent maintenance, tubes, fuses, oiling, interference reports, tuning, etc., must be noted, with date and action taken, and signed by the person responsible. The station licensee, usually through the master, is jointly responsible with the operator concerned for the faithful and accurate making of such entries.

It is required that at least one set of instructions for the use and operation of the particular type of radar being used, as well as the FCC publication "Part 83—Stations on Shipboard in the Maritime Services," (from Volume IV, "Rules and Regulations") be on board the vessel.

A radar transmitter is one of the few RF emissions that require no specific identifying emissions or call letters.

27-14 RADAR INTERFERENCE

In most cases, the only interfering signals received on a radar set are due to other radar transmitters that are operating in the same area. This interference may take the form of curved dotted lines across the screen.

The radar transmitter can produce interference to radio receiving or electronic devices in its vicinity if they are not properly shielded and grounded. Improper bonding or grounding of radar equipment, connecting cables and waveguides, or inadequate bypassing of the input power lines may result in interference.

Interference with a radio receiver by radar is characterized by a harsh tone having a frequency of the PRR, about 1,000 Hz. The noise may increase and decrease as the radar antenna rotates or be steady if originating at the radar set itself. If grounding and bypassing power lines and other circuits do not help, it may be necessary to change the position of the receiving antenna. Rotation of the RDF (radio-direction-finder) loop may indicate nulls on interference produced by a radar transmitter.

Any motors or generators in the radar set having slip rings, or armatures and brushes may cause a constant scratching sound. Such interference may appear to peak at certain frequencies but may be picked up on all frequencies used at sea (100 kHz to 150 MHz).

On the loran screen, radar interference appears as either *grass* or *spikes*. Impulse noises appear as many vertical pips across the traces of the loran screen and actually look like grass. Interfering signals due to the constant-rate pulsed emission of the radar transmitter appear as spikes on the loran screen. These spikes may appear to drift in one direction or the other but may seem to synchronize for short periods of time.

On an auto-alarm receiver, radar interference will sound the same as on any other receiver if earphones are used. It presents a constant signal which may activate and hold the first dash counter circuit. The red light on the bridge and in the radio station will glow, indicating trouble.

Intercommunication, motion-picture, or public address systems on the vessel may also pick up radar impulses if they are not properly shielded and grounded or if they have poor connections at some points in them.

27-15 BASIC RADAR MAINTENANCE

Although cabinet enclosures may be protected by interlocks that remove high voltages when opened, some circuits with up to 200 or 300 V may not be interlocked. Interlocks should never be jumpered or shorted to operate the high-voltage systems with the enclosure doors open.

In most cases, faulty operation of a radar set is the result of weak tubes, faulty TR tubes or diodes, or open fuses. Tubes may be checked with a tube tester, or similar tubes may be substituted, one by one. When a TR tube weakens, the mixer diode usually fails also, and its current drops. When mixer diodes are replaced, replace the TR tube at the same time. The diodes may be quite sensitive to mechanical shock, magnetic fields, and electric current. Under certain conditions, the operator may attain a static charge. When he pushes the cartridgelike diode (about the size of a 22-caliber shell) into its socket, he may discharge through it, burning it out. To prevent this, the operator should always touch the mixer cavity with one hand to ground himself while inserting the diode with the other hand. When a diode is handed from one person to another, it should be kept in its metal-foil capsule to prevent static discharge and burnout.

The magnetron current should be checked periodically. No plate current may indicate an open filament or no modulator pulses. If the current is abnormally high, it may mean a gassy magnetron or a high PRR. The permanent magnet used in conjunction with the magnetron is quite strong. There is danger, when a magnetron is being removed or installed, that iron or steel tools may be grabbed by the magnet and cause damage to the tube. The filament leads and the output circuit have long glass seals that may be fractured by mechanical jarring.

If the permanent magnet weakens, the magnetron current will increase, the output power will lessen, and the frequency of operation may change so much that the AFC will not hold the receiver in tune.

If the AFC circuit, the adder, the intensifying pulse circuit, or the magnetron is functioning improperly, bright pie sections may appear on the screen.

Remember that the filament leads to the magnetron have several thousand volts on them when the set is in operation.

Most equipment has a series of jacks into which a test meter can be inserted to test the operation of the different circuits. It will be necessary to check the instruction booklets that accompany the equipment to determine what the readings should be.

If the sensitivity control is turned to maximum but little or no grass or signals appear and the diode current is low, the diode may be removed and checked by measuring its resistance in both directions with a sensitive ohmmeter adjusted to read "R times 1,000." If the front-to-back ratio is less than about 10:1, the crystal should be replaced. No crystal current may also indicate a defective klystron. Turn off the equipment before changing this tube, which may have several hundred volts on the shell of its cavity. When klystrons or TR tubes are replaced, it is usually necessary to retune the screws in the associated cavities to bring the set up to optimum performance.

Cathode-ray tubes are dangerous to service because of high voltages applied to them plus the possibility of implosion. Heavy gloves and a face mask should be worn when changing such tubes.

Motors, generators, and synchros should be checked every 200 to 300 h of operation. They should be cleaned, and any brushes should be checked and replaced if necessary. Oil or grease should be applied where necessary. Remember that oil left on rubber and other electrical insulations may cause them to deteriorate.

Before a ship leaves the dock, the radar set should be turned on and tested. At this time it should be dusted thoroughly and observed carefully. If any signs of overheating of any component or improper functioning of mechanical parts are noted, corrective steps should be taken immediately.

27-16 NONMARINE RADAR

Although there are many different types of radar (gunfire control, landing approach, distance measuring, etc.), only two others not used in the merchant marine will be mentioned.

One type of radar used near airports for surveillance of all flying aircraft in the area is basically similar to the PPI system discussed. To prevent the radar from displaying trees, buildings, bridges, and hills in its range, it can use a discriminator as the second detector in its receiver. All echo signals returning at exactly the same frequency as transmitted produce zero output from the discriminator. However, an approaching target will compress any radio waves striking it. The returned wave will be slightly higher in frequency. Any departing target returns waves that have a longer wavelength and lower frequency. This is known as *Doppler effect*. The discriminator might give a positive signal for all approaching targets and a negative one for departing targets. Thus, on Doppler radar, moving targets produce blips, but all stationary targets produce no signal on the screen. It is necessary to produce electronically a series of chart signals of the area and feed this to the CRT to enable the watcher to know the position of the craft in relation to ground points. If any aircraft is carrying a *transponder,* when the radar signal strikes an antenna on the aircraft, this signal is detected and keys a small transmitter. It returns a short burst of RF to the radar receiver which is displayed as an identifying blip next to the target indication on the CRT (usually to several similar CRTs at different positions). By switching to an amplitude detector, the CRT will display both moving and stationary targets.

Automobile speed-check radar operates on a different principle. It consists of a low-power CW emission from a directive antenna. An adjacent antenna is used to pick up the returning echo signal. The faster a target is approaching the greater the phase shift between transmitted and received signals. A phase-shift detector produces a direct readout in miles per hour on a meter and can also print out a speed indication on a paper tape.

Test your understanding; answer these checkup questions.

1. How does a one-shot multivibrator differ from the common circuit? _____
2. How much intensifying pulse is used? _____
3. If echo signals are developed as negative pulses, to what part of the CRT would they be fed? _____
4. What starts the range-marker circuit ringing? _____
5. Why is a selsyn pair not too successful for radar antenna synchronization? _____
6. What is fed to the rotor of a selsyn motor? _____ Generator? _____
7. What is fed to a control transformer rotor? _____ To a synchro-system coil-turning motor? _____
8. List six components of the servo system described. _____ _____ _____ _____ _____ _____
9. What is used to indicate the bow of the ship on the CRT? _____
10. When is an echo box used? _____
11. What qualification must a person have before he is eligible to make repairs to a radar set? _____
12. List seven types of electronic equipment aboard ship that may be interfered with by radar. _____ _____ _____ _____ _____ _____ _____
13. What simple device can be used to test diodes? _____ What ratio should be obtained?
14. What is the advantage of using Doppler radar? _____ Why is it not used on ships? _____
15. How could Doppler radar give speed indications? _____ What are other applications of Doppler radar? _____ _____

COMMERCIAL LICENSE QUESTIONS

Applicants for Element 8 (radar endorsement) should know answers to all questions. Applicants for Element 6 should know answers to questions followed by [6]. Sections in which questions are answered are shown in parentheses.

1. How many nautical miles does a radar pulse travel in 1 μs? (27-1)   [6]
2. In what part of the RF spectrum do marine radar systems operate? (27-1)   [6]
3. Within what frequency bands do ship radar transmitters operate? (27-1)
4. What is the distance to a target if it takes 246 μs for a radar pulse to travel to and from the target and be displayed on the PPI scope? (27-1)
5. Explain briefly the principle of operation of a radar system. (27-1, 27-2)
6. Draw a block diagram of a radar system and label the antenna, duplexer, transmitter, receiver, modulator, timer, and indicator. (27-2)   [6]
7. Explain the principle of operation of the cathode-ray PPI tubes used in radar. (27-2)
8. At what approximate speed does the antenna of a marine radar rotate? (27-2)   [6]
9. What is meant by *bearing resolution* of a radar set? Range resolution? (27-2)
10. What component in a radar set determines PRR? (27-4)
11. What circuit elements determine the operating frequency of a self-blocking oscillator? (27-4)
12. Draw a simple diagram of an artificial transmission line showing inductance and capacitance, source of power, the load, and the electronic switch. (27-4)

13. What is the purpose of an artificial transmission line in a radar set? (27-4)
14. What is the purpose of the rotary spark gap used in some radar sets? (27-4)
15. Why is the anode of a magnetron in a radar transmitter normally maintained at ground potential? (27-4)
16. What is the average plate-power input to a radar transmitter if the peak pulse power is 25 kW, the pulse length is 1 μs, and the pulse repetition frequency is 1,000 pps? (27-5)   [6]
17. What is the peak power of a radar pulse if the pulse width is 1.0 μs, PRR is 800, and the P_{av} is 40 W? What is the *duty cycle?* (27-5)
18. Describe how a radar beam is formed by a parabolic reflector. (27-6)
19. How are waveguides terminated at the radar antenna reflector? (27-6)
20. What is the best location of the radar antenna assembly aboard ship? (27-6)
21. What effect, if any, does the accumulation of soot or dirt on the antenna reflector have on the operation of a ship radar? (27-6)
22. Is there any danger in testing or operating radar equipment aboard ship when explosive or inflammable cargo is being handled? (27-6)
23. Draw a simple block diagram of a radar duplexer system and label the waveguide, TR box, ATR box, receiver, and transmitter. (27-6)
24. Describe briefly the construction and operation of radar TR and ATR boxes. What is the purpose of a keep-alive voltage? (27-7)
25. Draw a simple frequency-converter or mixer circuit such as is often used in radar superheterodyne receivers and indicate the crystal stage. (27-8)
26. What is the purpose of the klystron tube in a radar set? (27-8)
27. Draw a simple block diagram of a radar receiver and label the signal crystal, local oscillator, AFC, IF amplifier, and discriminator. (27-8)
28. What is the nominal IF commonly found in radar receivers? (27-8)
29. What type of detector is used in radar receivers? (27-8)
30. What is the purpose of the discriminator stage in a radar receiver? (27-8)
31. What is *sea return* on a radarscope? (27-8)
32. Explain briefly the purpose of the STC circuit in a radar set. (27-8)
33. What adjustment is made to a radar set by the operator to reduce sea return? (27-8)   [6]
34. Describe how heading flash and range-marker circles are produced on a radar PPI scope. (27-9, 27-11)
35. Draw a simple diagram showing how a synchro generator located in the radar antenna assembly is connected to a synchro motor located in the indicator to drive the deflection coils. Show proper designation of leads, ac voltages, etc. (27-10)
36. What is an *echo box?* Explain its purpose, principle of operation, and indications given by it on a radarscope when the radar is operating correctly and incorrectly. (27-12)
37. Who may operate a ship radar station? (27-13)
38. Under what conditions may a person who does not hold a radio operator license operate a ship radar station? (27-13)
39. May fuses and receiving-type tubes be replaced in ship radar equipment by a person whose operator license does not have a ship radar endorsement? (27-13)
40. What are the FCC license requirements for the operator who is responsible for the installation, servicing, and maintenance of ship radar equipment? (27-13)
41. Who has the responsibility for making and who may make entries in the installation and maintenance record of a ship radar station? (27-13)
42. What entries are required in the installation and maintenance record of a ship radar station? (27-13)
43. Describe how various types of interference from a radar installation may be apparent to a person listening to a radio communications receiver. (27-14)
44. Why is radar interference to a radiotelephone receiver characterized by a steady tone? (27-14)
45. Why is it important that all units of a radar installation be thoroughly bonded to the ship's electrical ground? (27-14)
46. On what frequencies does radar interfere with communications receivers on ships? (27-14)

ANSWERS TO CHECKUP QUIZ ON PAGE 671

1. (*Not free-running; produces output only when triggered*)
2. (*Just enough to produce light flicker on screen*)
3. (*Cathode*)   4. (*Intensifying pulse*)   5. (*Variable lag angle*)   6. (*Power-line ac*) (*Same*)   7. (*Nothing; emf is induced into it*) (*Power ac 90° shifted*)   8. (*Generator, CT, phase shifter, amplifier, 2-φ motor, coil assembly*)   9. (*Heading flash*)   10. (*Testing only*)   11. (*Radar endorsement*)   12. (*Commercial receivers, loran, RDF, auto-alarm, PA system, motion pictures, intercoms*)   13. (*Ohmmeter*) (*Better than 10:1*)   14. (*Shows only moving targets*) (*Navigators must see everything*)   15. (*Calibrate discriminator output*) (*Ground-speed indicator, missile velocity, etc.*)

47. In checking a radio direction finder for interference caused by radar equipment, would it be a good policy to check for interference while the direction-finder loop is being rotated? (27-14)

48. Is there any likelihood of a radar installation causing interference to radio receivers if long connecting lines are used between the radar transmitter and the radar modulator? (27-14)

49. What steps might be taken by a radar serviceman to eliminate a steady-tone type of interference to radio communications receivers or interference to loran receivers evidenced by spikes? (27-14)

50. List at least two types of indications on a loran scope that signify that a radar installation is causing interference to the loran. (27-14)

51. What steps might be taken by a serviceman to reduce grass on a loran scope or motor-generator noise in communications receivers? (27-14)

52. How are the various types of radar interference recognized in (a) auto-alarm equipment and (b) RDF equipment? (27-14)

53. Name four pieces of radio or electronic equipment aboard ship that might suffer interference from the radar. (27-14)

54. What precautions should the service or maintenance operator observe when replacing the CRT in a radar set? (27-15)

55. What precautions should a radar serviceman observe when making repairs or adjustments to a radar set to prevent injury to himself or others? (27-15)

56. What precautions should a radar serviceman take when working with or handling a magnetron to prevent weakening of or damage to the magnetron? (27-15)

57. What care should be taken when handling silicon crystal rectifier cartridges for replacement in radar receivers? (27-15)

58. In a radar set, what indicates (a) a defective magnetron, (b) a weak magnet in the magnetron, (c) a defective crystal in the receiver converter stage? (27-15)

59. What tests may a radar serviceman make to determine whether the radar receiver mixer crystal is defective? (27-15)

60. What may cause bright, flashing pie sections to appear on a radar PPI scope? (27-15)

28
SHIPBOARD RADIO EQUIPMENT

The objective of this chapter is to describe radio communication equipment of both telegraphic and telephonic types which may be found aboard compulsorily equipped or other seagoing vessels. This includes medium-, high-, and very-high-frequency transmitters and receivers, as well as auto-alarms. The duties of radio operators are outlined briefly. More detailed equipment requirements can be obtained from the "FCC Rules and Regulations," Part 83.

28-1 RADIO ABOARD SHIP

Small vessels may not carry radio equipment, but ships carrying passengers or those over certain tonnage ratings may be required to have radio equipment aboard for the safety of life at sea. Determining which ships must carry radio equipment is rather complex and is set by international agreement. Usually, however, ships of more than 500 gross tons or those carrying more than 12 passengers on deep-sea voyages are compulsorily radio-equipped.

Rather sophisticated communication equipment may be found aboard larger ships carrying one, two, or three radio operators. Smaller vessels, commercial fishing boats, tugs, boats on inland waters, etc., may carry a low-powered radiotelephone transmitter and receiver for communication with other ships, nearby coast stations, or the Coast Guard. This equipment may operate in the 1600–2850-kHz band or in the 156–162-MHz band, is fixed-tuned, and requires no radio operator but only a radiotelephone permittee to operate it.

Ships usually have 115-V 60-Hz ac but may have 115-V dc generated on board. As a result, shipboard radio equipment may be manufactured to operate on 115-V ac or dc and in some cases from either 12- or 24-V batteries.

Unlike land-type power systems, in which one line is at ground potential, neither of the dc power lines aboard ship may be connected to ground (the metal hull of the vessel). This requires that all motors have both brushes insulated from ground, with bypass capacitors to ground. If a voltmeter between ground and either line reads the full line voltage, it indicates that the other line has become grounded at some place on the ship.

As in other radio services which require the sharing of frequencies, the minimum transmitting power for satisfactory communication should be used at sea. For most distress communications the maximum power output may be desirable, except when conservation of batteries may be a factor.

28-2 COMPULSORY RADIOTELEGRAPH INSTALLATIONS

When larger ships, or those sailing internationally, are required to have radio aboard by the communications act or by the radio safety convention, the minimum *radiotelegraph* requirements are usually (1) a main 405–535-kHz transmitter of at least 160-W A1 and 200-W A2, (2) a main receiver capable of receiving A1 and A2 on 100–200- and 405–535-kHz bands, (3) a main power supply, usually the electrical system of the ship, and (4) a main antenna with a safety link.

In addition, it must have (1) a reserve 405–535-kHz 25-W A2 or A2H transmitter, (2) a reserve receiver capable of receiving A1 and A2 on the 100–200- and 405–535-kHz bands, (3) a 6-h minimum reserve power supply, usually a motor-generator set or batteries, (4) a reserve antenna, (5) an emergency-light system for the operating area, and (6) an automatic radiotelegraph alarm signal keyer.

Besides the basic requirements, an efficient two-way intercom system is required between the radio room and the ship's bridge. A reliable clock of 5 in. or larger diameter marked off in silence-period segments must be mounted within sight of the radio operator.

If there are an insufficient number of operators to stand 24 h of watch daily, an approved auto-alarm (Sec. 28-16) will be used.

28-3 COMPULSORY RADIOTELEPHONE INSTALLATIONS

When a ship is required to have radiotelephone because of communications act or safety convention requirements, it must carry on the bridge a radiotelephone transmitter capable of at least a 25-W A3 carrier (until January 1, 1977), or a 50-W PEP A3H (or A3A or A3J as of January 1, 1977) emission on 2182 kHz (distress and calling frequency) and 2638 kHz (safety), and at least two other frequencies in the 1605–2850-kHz ship-to-shore and intership band. It must have a device that will transmit the international radiotelephone alarm signals, a 30–60-s warbling 2.2- and 1.3-kHz 0.25-s series of tones. It must have two receivers, one capable of being switched to any of the assigned transmitter frequencies, and one manually tuned for A3 (as of January 1, 1977, A3H, A3A, A3J) emissions on any frequency between 1605 and 3500 kHz. This equipment uses a vertical antenna on the bridge. A reserve power source (motor-generator or batteries) that will operate all the radiotelephone equipment plus an emergency-light system for the operating area for a period of 6 h must be provided. A 5 in. minimum-diameter clock must be provided at the operating position, and a spare antenna must be carried.

All compulsorily equipped U.S. ships after January 1, 1977 will carry a VHF bridge-to-bridge transmitter capable of transmitting 8 to 25 W of 15-kHz-deviation F3 on 156.3 MHz (safety frequency) and 156.8 MHz (distress and calling), plus other frequencies in the 156–162-MHz band used for ship-to-shore communications in the area in which the ship is to be navigated. It must carry a receiver preset to and capable of accurate and convenient selections of the frequencies of 156.3 and 156.8 MHz, as well as the frequencies of its associated transmitter. The reserve power must be capable of supplying the equipment for 3 h. See also Sec. 28-10.

28-4 A MAIN TRANSMITTER

A main radiotelegraph transmitter aboard a compulsorily equipped ship must be capable of a minimum of 160 W of A1 power output to the main antenna, a minimum of 200 W of A2 output to the main antenna, and break-in operation.

The A2 (or A2H) emission required when transmitting an SOS on 500 kHz must be at least 70%, but not over 100%, modulated. The frequency of the modulating tone must be between 300 and 1,250 Hz.

The transmitter must be capable of transmitting on the international calling and distress frequency of 500 kHz, on the direction-finding frequency of 410 kHz, and on at least two other working frequencies between these two. The frequency tolerance is 1,000 Hz/MHz. The transmitter must have an antenna ammeter, a final-amplifier plate voltmeter, and a final-amplifier plate milliammeter. It must have some means of reducing the plate input power to 200 W or less for tuning and short-range operation, usually by changing the plate voltage to the final tubes.

The antenna power is determined by the formula

$$P = I^2R$$

where P = power in W
I = antenna current in A
R = resistive component of impedance of antenna at point where antenna current is measured

A commercial marine radiotelegraph main transmitter in simplified block form is shown in Fig. 28-1.

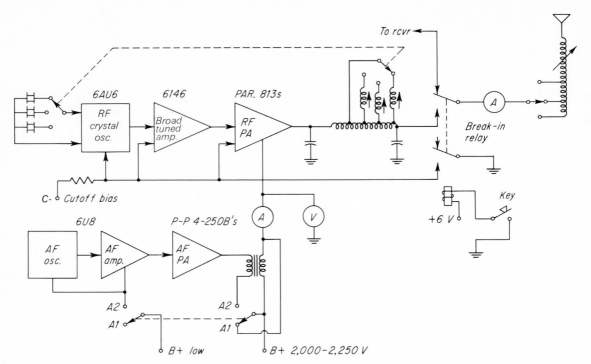

Fig. 28-1 Block diagram of a representative marine main transmitter (from ITT Mackay Marine type 2012 with 472 modulator).

The RF oscillator stage is a Pierce oscillator using a 6AU6 pentode tube. The operating frequency is selected by switching to the desired crystal. The oscillator drives a broadly tuned (410–535-kHz) 6146B-type RF pentode stage, which in turn drives the parallel 813 pentodes in the power amplifier. This stage is pi-network-coupled (Sec. 20-24) to the antenna through the keying relay. With the transmitting key up, the relay connects the antenna to the receiver and allows all RF stages to be biased beyond cutoff, preventing any output from the transmitter. When the key is pressed, the relay arms move down, disconnecting the antenna from the receiver and connecting it to the PA stage. At the same time the high bias is grounded and all RF stages start operating. Note that the pretuned powdered-iron PA inductors are ganged to the crystal switch, so that no tuning is required by the operator when he switches from one frequency to another. The antenna circuit has tapped inductances plus a *variometer* (variable inductor) to allow resonating regardless of the length of the antenna used on

the particular ship. Variometers tune over a much wider range of frequencies in the MF range than do similar-sized variable capacitors.

When it is desired to transmit using A2 (modulated CW, or MCW), the A1-A2 switch is moved to the A2 position. This turns on a sinusoidal phase-shift AF oscillator and AF amplifier using a type 6U8 triode-pentode tube and an AF power amplifier having two 4-250B power tetrodes in push-pull. The AF output is added in series with the B+ to the RF PA through the modulation transformer, modulating both plate and screen-grid circuits of the output stage with a constant AF tone.

28-5 A RESERVE TRANSMITTER

Reserve transmitters must produce four frequencies with at least 25 W of A2 type RF output in the 410–535-kHz band. An interesting solid-state reserve transmitter is shown in semiblock form in Fig. 28-2. The RF oscillator is a *Butler* circuit having the crystal between the emitters of

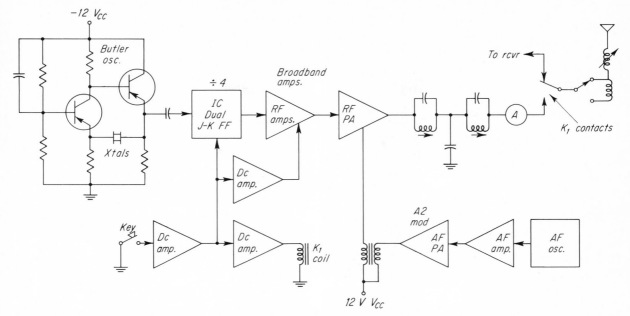

Fig. 28-2 A solid-state reserve transmitter in block form (from ITT Mackay Marine type 2017).

an amplifier-type stage and a direct-coupled emitter follower, as shown. The feature of this system is the frequency of the oscillator, which is four times the eventual output frequency. In this way the oscillator can run continuously but cannot be picked up by the local receiver (as it would be if the oscillator operated on the transmitting frequency or on some subharmonic frequency).

The keying circuit activates a dual J-K flip-flop IC circuit that divides the input frequency by 2 twice, resulting in a 500-kHz output from a 2-MHz crystal. However, the FF output is square-wave and must be sinusoidally shaped by the following RF amplifiers. The final RF power amplifier consists of three parallel transistors feeding a bandpass filter circuit (the parallel LC circuits and capacitor to ground) and the antenna tuning circuit. In this case the antenna is the LC circuit for the PA stage, but since part of the circuit is a bandpass filter, no harmonics can be transmitted, which would be the case if an amplifier were directly coupled to an antenna circuit. The keying circuit also activates dc amplifiers that key the driver amplifier and feeds the antenna relay coil with dc. The modulation is produced by a 500-Hz AF oscillator. This ac is amplified and is coupled to

the collectors of the parallel transistors in the RF PA. The AF stages run continuously. The power supply is a 12-V battery, which must be capable of 6 h of continuous operation, and the output is a nominal 40 W of A2 RF.

28-6 THE MAIN ANTENNA

The main antenna must be as efficient as practicable under the prevailing physical limitations on board ship. It usually consists of a single wire hung from the peak of the foremast to the peak of the mainmast, with a lead-in to the radio station (Fig. 28-3). On smaller vessels it is made as long and kept as high as possible.

A *safety link* must be incorporated in the antenna. It is a planned weak section in the halyard at one end of the antenna. A short weak wire and a longer heavy wire are fastened across one another and are inserted in the halyard as illustrated. If the ship is subjected to a sudden stress, because of collision, grounding, torpedoing, etc., the weak part of the link breaks as the masts are forced apart and the antenna takes up the slack in the longer, stronger part of the link. This causes the antenna to sag but prevents its falling, and

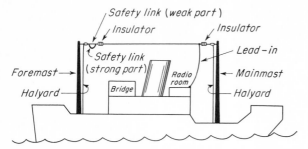

Fig. 28-3 Essentials of a shipboard main-antenna installation.

it remains usable. The sag can be corrected by tightening one of the halyards.

Unless care is taken in hoisting an antenna, insulators may fracture if they strike a metal object. The wire must not be allowed to kink, since it weakens at such a point and may break under sudden strain.

When the main antenna is used on the 410- to 515-kHz band, it is operating as a base-loaded quarter-wavelength wire. The far end is the maximum voltage point, and the part leading into the transmitter is the high-current part. When used on high-frequency bands, the far end is always a high-voltage point, with other high-voltage points every half wave along the wire and high-current points halfway between the voltage points. Ships using high frequencies may use shortwave dipole antennas for receiving and the main antenna for transmitting.

Test your understanding; answer these checkup questions.

1. If a ship has only dc aboard, how could 60-Hz ac be developed for radio equipment? _____
2. What is the minimum power requirement for a main 500-kHz transmitter? _____ A reserve? _____
3. With what part of the ship is the radio station connected by intercom? _____
4. After January 1, 1977, what radiotelephone equipment must all U.S. ships carry? _____
5. What is the required frequency stability in percent, of a 500-kHz radio transmitter? _____
6. What is the required modulating frequency for A2 SOS transmissions? _____ Modulation %? _____
7. Is shipboard antenna power determined by a direct or by an indirect method? _____
8. In Fig. 28-1, why would neither crystal connection be grounded? _____ What circuits must be tuned when changing frequency? _____ What type of tuning is used in the PA? _____ Why is break-in

used? _____ What are the three required meters? _____ _____ _____
9. What might the 4-250B designation on the power tetrode tubes mean? _____
10. What is a Butler circuit? _____
11. In Fig. 28-2, what does a single J-K FF do? _____ How many circuits must be tuned when the frequency is changed? _____ Why can the oscillator run but not be heard in the receiver? _____
12. For what is a safety link used? _____

28-7 MEDIUM- VERSUS HIGH-FREQUENCY COMMUNICATIONS

During daylight hours the medium frequencies between 405 and 515 kHz provide reliable communication for a 200-W transmitter over a range of 300 to possibly 600 mi under normal conditions. During nighttime the range may extend to more than 2,000 mi. However, medium frequencies are subject to high-amplitude static, making long-range communication difficult at times. Ships on long voyages find that frequencies between 4 and 24 MHz provide more satisfactory operation over long distances (Sec. 20-1, 20-2). Most ships engaged in international trade will have high-frequency as well as medium-frequency main transmitters.

28-8 HIGH-FREQUENCY TRANSMITTERS

The HF marine bands are in the 2–23-MHz part of the spectrum, commonly known as the 2-, 4-, 6-, 8-, 12-, 16-, and 22-MHz bands. Since the first six bands are harmonically related, marine HF CW (A1) transmitters may be relatively simple types using crystal oscillators for the assigned frequencies and frequency multipliers to drive the final-power-amplifier stage. The 22-MHz band, not being harmonically related to the other bands, requires its own separate crystals. Heterodyne-type transmitters (Sec. 16-23) can also be used to cover the marine bands.

Since marine communications have expanded from simple radiotelegraph to SSB radiotelephone communications with shore-based telephone companies, as well as radioteleprinter communications, the modern marine HF transmitter may be quite sophisticated. The transmitter shown in block form in Fig. 28-4 represents a greatly simplified modern marine HF transmitter capable of

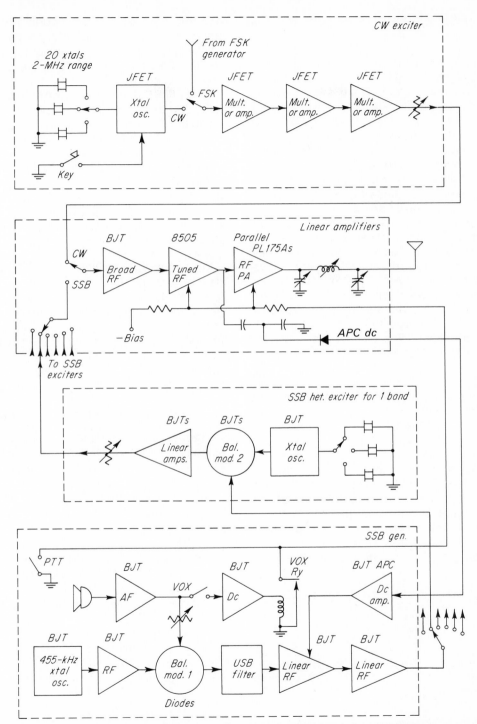

Fig. 28-4 Simplified block diagram of a modern marine HF transmitter (from ITT Mackay Marine type MRU-27).

transmitting A1, F1 (FSK), A3A, A3H, and A3J signals.

To transmit A1 signals, the CW-SSB switch is thrown to CW and the desired assigned frequency crystal is selected. Band switches (not shown) determine whether the CW exciter stages operate as multipliers or amplifiers and to which band the linear amplifier LC circuits are tuned. When the key is pressed, the transistor crystal oscillator is keyed on. The transistor amplifiers (or multipliers) amplify (or multiply) the signal that has been generated. Only the last two linear amplifiers are power vacuum tubes. Since there is no break-in relay shown in the transmitter, a transmitting and a receiving antenna are required. The drive to the broadband transistor amplifier in the linear amplifier section is controlled at the output of the last CW exciter multiplier or amplifier. High power (1,000 W) and low power (150 W) can be selected by a switch which changes the bias on the power amplifiers. If the plates of the tubes dissipate too much heat due to overloading, a light-sensitive cadmium sulfide cell senses it and the overload relay is tripped, turning off the linear amplifiers.

To transmit F1 teleprinter signals, the output from a separate FSK generator (not shown) with its own crystals, mixers, and amplifiers is fed into the FSK input and the CW-FSK switch is thrown to FSK.

To transmit A3A, A3H, or A3J signals, the CW-SSB switch on the linear amplifier is thrown to the SSB position. Either push-to-talk (PTT) or voice-operated transmission (VOX) can be selected on the SSB generator panel to key the transmitter on. The microphone AF ac is amplified and fed to balanced modulator 1 to produce a double-sideband-without-carrier signal based on

455 kHz. By passing this through a filter only the upper sidebands of the 455 kHz carrier remain. These are amplified by linear amplifiers and are fed to balanced modulator 2 in one of six heterodyne exciters (one for each marine band). The 455-kHz USB signals are mixed in balanced modulator 2 with the proper crystal frequency to develop the desired frequency on the desired marine band. The heterodyned USB signals are amplified by the linear-amplifier section and are fed to the antenna.

To prevent overmodulation and also to allow the average microphone gain to be raised, an automatic power control (APC) circuit is used. This is shown as a capacitively coupled USB RF output to a diode. The diode is biased so that it will develop dc only when the SB peak signal exceeds a level equal to the maximum desired power. The dc that is developed is fed back as a bias voltage to one of the 455-kHz linear amplifiers in the SSB generator, reducing the gain in this stage and thereby in all following stages. The result of this feedback is to allow no more than the maximum desired peaks but to produce an increase in the average modulation by about 15 dB—a very great increase in volume for receivers.

28-9 RADIOTELEGRAPH FREQUENCIES

Frequency bands used by *ship radiotelegraph* stations for calling, working, and distress traffic are:

o LF or 90–160-kHz band, in kHz
   143 (calling, A1 only)
   152–158
o MF or 405–535-kHz band, in kHz
   500 (calling and distress)
   512 (calling when 500 occupied by distress)
   410 (RDF), 425, 444, 454, 468, 480
o HF or 2–23-MHz band
   Calling, all ships, in kHz
   2089.75–2092.25
   4178.5–4186.5
   6267.75–6279.75
   8357–8373
      8364 safety of life only
   12535.5–12559.5
   16714–16746
   22225–22262.5

Working, high-traffic ships, in kHz
  4172.5–4176
  6258.75–6264
  8342–8355
  12505.5–12532.5
  16662–16710
  22189–22219
Working, low-traffic ships, in kHz
  4188–4229
  6282–6343.5
  8376–8458
  12564–12687
  16752–16916
  22272.5–22345

Frequency bands in which *coastal stations* communicating with ship radiotelegraph stations will be assigned:

o 90–160-kHz band, in kHz
   112.85–153
o 415–525-kHz band, in kHz
   416–512
o 2–23-MHz band, in kHz
   2036–2063
   4238–4316
   6351.5–6512.5
   8502–8726
   12745.5–13119
   16933.2–17242.4
   22407–22617

28-10 RADIOTELEPHONE FREQUENCIES

The 2–23-MHz or HF bands in which *ship radiotelephone* stations may be assigned for calling, working, and distress traffic are:

o 2000–2850 kHz, A3 or SSB (A3A, A3H, A3J) until January 1, 1977. After that only SSB.
o **2182** kHz. International distress frequency
o 4069.2–4434-kHz SSB
o 6210.4–6455-kHz SSB
o 8201.2–8780-kHz SSB
o 12358–12428-kHz SSB
o 16474–16572-kHz SSB
o 22028–22108-kHz SSB

The 2–23-MHz or HF bands in which *coastal stations* may be assigned to communicate by SSB with ships are:

o 2009–2782 kHz
o 4071–4371 kHz
o 6147.5–6455 kHz
o 8207–8783.2 kHz
o 12379–13158 kHz
o 16488–17283 kHz

The 156–162-MHz or VHF band is divided into 25-kHz channels in which narrowband F3 emissions are used. Some representative channels are listed in Table 28-1.

Table 28-1

Channel	Frequency for		Type of communication
	Ship	Coast	
16	156.8	156.8	Distress, safety, calling
6	156.3	156.3	Intership safety
7	156.35	156.35	Commercial
9	156.45	156.45	Noncommercial
12	156.6	156.6	Port operations
*13	156.65	156.65	Navigational, bridge-to-bridge
15		156.75	Environmental
24	157.2	161.8	Public correspondence
68	156.425	156.425	Yachts
WE$_1$		162.55	Weather broadcasts
WE$_2$		162.4	Weather broadcasts

* A continuous listening watch must be maintained without interruption on this channel when vessels are in heavy-traffic areas and on inland waters.

28-11 MARINE MAIN RECEIVERS

The history of radio receivers aboard ships started with coherer detectors (encapsulated iron filings which magnetized and adhered to each other when subjected to modulated RF currents and thereby became alternately better and worse conductors) and then galena or other metal crystal diode detectors. With the vacuum tube came regenerative detectors and TRF receivers. In the 1930s the first superheterodynes appeared aboard ships. They were single-conversion receivers with a BFO for A1 detection. To improve image rejection, double- and triple-conversion receivers were developed. Superheterodynes were designed to cover the low frequencies also, allowing them to qualify for main receivers. Today a ship installation may use an all-wave (15 kHz to 30 MHz) superheterodyne capable of receiving the distress frequencies of 500 and 8364 kHz, plus a 15–560-kHz reserve or main receiver in case the all-wave receiver malfunctions or to watch 500 kHz while listening to traffic on HF or distress on 8364 kHz.

An example of a modern-day reserve receiver that also qualifies as a main receiver covering the two required bands of 100–200 and 400–535 kHz is shown in block form in Fig. 28-5. It is a solid-state regenerative-detector-type TRF receiver with four printed-circuit boards. It has ganged tuning capacitors and band switches to cover 15 to 560 kHz in four overlapping bands. The two protect-circuit series diodes are reverse-biased to 6 V. If an ac signal of more than 3 V is received, the diodes conduct and protect the front-end coils and transistor. The first stage is a tuned dual-gate MOSFET RF amplifier followed by an untuned dual-gate MOSFET amplifier. The RF gain is controlled by varying the bias on the second gates of the MOSFETs. The regenerative detector is a Hartley-type JFET with a potentiometer across the tickler part of the coil as a regeneration control. The detector is followed by a BJT AF preamplifier and an IC AF amplifier feeding earphones or a loudspeaker. The power supply is either a 12-V battery or a 12-V regulated rectifier-filter supply. The sensitivity and selectivity of the TRF main receiver may not be quite that of a superheterodyne, but it is very adequate for LF and MF reception. If the circuit were used in the HF region, it would be inferior to a superheterodyne.

Test your understanding; answer these checkup questions.

1. List the HF marine bands. _____ The MF. _____
2. In Fig. 28-4, what would control the A1 P_o? _____ The A3J P_o? _____ What does APC stand for? _____ Why would the carrier oscillator not block the transmitting frequency? _____
3. In Fig. 28-4, what does PTT mean? _____ VOX? _____ In what band are the A1 crystals? _____ The SSB carrier oscillator? _____ The SSB channel crystals? _____ What has to be changed in the SSB generator when shifting from band to band? _____ What does closing the PTT or VOX switches do? _____
4. What is the result of using APC? _____
5. On which side of the calling frequencies are the low-

Fig. 28-5   Solid-state 15–560-kHz marine main or reserve TRF receiver (from ITT Mackay Marine type 3018).

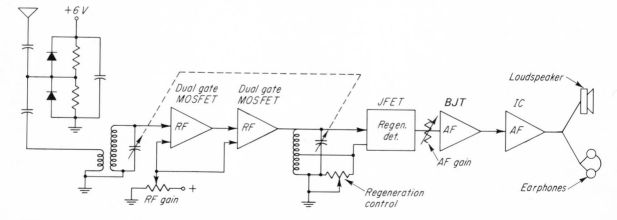

traffic working bands? _____ The high-traffic bands? _____

6. When is 512 kHz used? _____ For what is 410 kHz used? _____
7. Are coastal stations found above or below ship working frequencies? _____
8. Are ship HF radiotelephone bands in the same general bands as ship radiotelegraph? _____
9. What VHF channel is used for intership safety? _____ For distress, safety, and calling? _____ For bridge-to-bridge communication? _____ On which channel must a continuous watch be maintained? _____
10. What are the two radiotelegraph distress frequencies? _____

11. In Fig. 28-5, what basic receiver circuit is this? _____ What frequency bands must it cover? _____ What does it cover? _____ Why are MOSFETs particularly useful in RF amplifiers? _____ What is the maximum rms RF voltage that will be fed to the receiver circuits? _____

28-12 MARINE SUPERHETERODYNES

The IF frequencies of the original marine superheterodynes were usually in the 455-kHz region to produce a narrow IF passband. Unfortunately, signals near any IF frequency that follows the first mixer tend to force their way through to the detector and interfere with other signals being received. Double conversion, with the first IF in the 2–8-MHz region, reduced this effect for low-frequency signals and also resulted in better image signal rejection, but with two oscillators in the receiver, unexpected beats, called spurious responses or birdies, appeared at various places on the receiver dial. Another difficulty in older HF superheterodynes was the requirement of ganging and tracking one or two RF amplifier tuned circuits with the mixer and local oscillator circuits. A 15-kHz to 30-MHz "third generation" superheterodyne circuit that overcame most of these objections is shown in block form in Fig. 28-6.

The signal passes by the protect-circuit diodes; and if it is too strong because of proximity to the ship, it is attenuated and fed to the RF-tuned circuit and amplifier as in the TRF above.

After being amplified and fed through a 30-MHz low-pass filter, it is mixed with frequencies produced by crystals ranging in frequency from 39 to 67 MHz in 2-MHz steps. A 1-MHz signal would mix with the 39-MHz crystal to produce a 36-MHz difference or IF frequency. A 29-MHz signal would mix with the 67-MHz crystal to produce a 38-MHz

difference and also an IF frequency. From this it can be deduced that, to receive signals from 15 kHz (relatively zero MHz) to 30 MHz in 2-MHz bands, the first IF must be from 37–39 MHz. For this reason the first IF strip consists of a tuned amplifier and a 2-MHz-wide bandpass filter, passing any 37–39-MHz signals from the mixer. Note that the front-end oscillator frequencies are well above any received signal frequencies, and except for a trimmer control to peak the RF signal, there are no circuits that must be gang-tuned.

The received-signal frequency-determining circuit, a variable-frequency oscillator (VFO) tunable over the frequency band of 3–5 MHz is mixed with a 47.94-MHz crystal oscillator to form a 42.94–44.94-MHz injection signal. This signal and the IF_1 signal beat in the second mixer to form 5.94-MHz second-IF signals. These are amplified and fed to a third mixer, where they are heterodyned with a 5.485 crystal to form a 455-kHz third IF frequency. The 455-kHz frequency is selected because of the availability of excellent narrow-bandpass filters for this frequency, plus the natural selectivity of circuits at this frequency. As shown, either one of two bandpass filters, perhaps 400 Hz or 2.8 kHz, can be selected for CW or SSB signals, or with no filter the bandpass should be wide enough for A3 emissions. A cathode-follower amplifier feeds the IF signal to outside auxiliary equipment.

The detector circuit shown is a diode detector for A3 or A3H signals with a variable BFO for CW or A3J reception. One of two crystals can also be used to produce the proper beat frequency for USB or LSB A3A or A3J reception. A shunt-type noise clipper can be switched on in the circuit between the detector and the first AF amplifier. A delayed AVC (DAVC) feeds both the RF amplifier and the IF_3 amplifier. A sensitive dc meter reads the AVC voltage as an indication of RF signal strength, or the AF voltage can be rectified to indicate the AF signal strength.

The advantages of this receiver are all-wave reception, excellent sensitivity and selectivity, practically no image signals or birdies, and only one tuning circuit plus a trimmer to peak signals (not always required). Points of difficulty might be possible drifting of the VFO and the variable BFO, plus the requirement of band switching and tuning of the manual dial.

One of several types of fourth generation, and what appears to be nearly the ultimate in receivers, is shown in block form in Fig. 28-7. As in previous receivers, the received signal passes the protect circuit, is attenuated if necessary, is trimmed to a peak, and is fed through a 30-MHz low-pass filter to the RF amplifier(s). The signal is then mixed with a *frequency synthesizer* acting as a first local oscillator.

A frequency synthesizer does what its name indicates—it puts together different frequencies produced by a stable source by multiplying, dividing, and/or mixing the source frequency to produce desired output frequencies. This particular synthesizer has a temperature-controlled 8-MHz crystal oscillator as a standard which shifts less than a hertz in frequency once it reaches operating temperature. The desired frequency is dialed on the digital frequency selector indicators (12.3456 MHz shown). This operates six binary-coded decimal circuits that cause the major and minor loop circuits to synthesize the desired frequency. Since this frequency synthesizer is being used not as a calibrated signal generator, but as a local oscillator for a superheterodyne, the frequency output is offset enough to beat against the incoming signal and produce the desired first IF frequency, in this case in the 92–122-MHz range for a received signal on 12.3456 MHz.

After conversion to the desired 92-MHz IF, the received signal is mixed in a second mixer with an 84-MHz crystal oscillator from the frequency synthesizer that also is feeding into the major loop circuit. Should the 84-MHz crystal drift a few hertz high, this will increase the major loop output frequency a like amount so that there is no effective change in the 92-MHz signal. Thus, the 84-MHz oscillator does not have to be tempera-ture-controlled. The output of the second mixer is an 8-MHz IF, which is chosen because of general availability of 8-MHz crystal bandpass filters and because the 8-MHz standard can be used as a BFO. The desired bandpass is selected from five different filters plus a wide-bandpass circuit. After further amplification the 8-MHz signal is fed to an emitter-follower amplifier for external equipment and to either a diode detector to detect A3 or A3H signals or a product detector for A3A, A3J, A1, or F1 detection. The product detector is actually a mixer, and it is fed either a fixed BFO signal from the 8-MHz standard or the output from an 8-MHz crystal oscillator which can be tuned 100 Hz or so.

The frequency synthesizer can be set to within 100 Hz of any desired frequency. If a transmitter radiates a 12.3456-MHz signal, the receiver frequency selector is set to that frequency and the product detector will produce a zero-beat output. This is desirable for A3A, A3H, and A3J. However, to receive a CW station on this frequency the frequency select would have to be changed. If the operator prefers to listen to code at a 400-Hz pitch, the frequency select could be set to either 12.3460 or 12.3452. If he wants to vary the tone a little, he can switch to the variable BFO and vary the tone by 100 Hz.

To provide a medium-fast AVC or AGC voltage for A3 or A3H signals and a slow-decay AGC for CW and A3J signals, the 8-MHz IF is amplified, rectified, properly filtered, and fed to the first IF strip amplifiers as a delayed AGC and to the second IF strip amplifiers as an undelayed AGC. This particular receiver features all-solid-state devices for minimum power drain and fault-free operation.

If frequency synthesizers can be used as local oscillators in receivers, it is easy to see that a frequency synthesizer capable of generating frequencies from perhaps 5 to 7 MHz in 1-Hz steps could be used in a heterodyne circuit, such as Fig. 16-28, to produce an exciter for a *transmitter* on any frequency desired. It would have an accuracy of ± 1 Hz, and provided the crystal oscillator is supplanted by the internal synthesizer frequency standard, no drift would occur. The exact frequency for such a transmitter to the last cycle per second could be digitally dialed. Such will undoubtedly be the type of equipment used in the future.

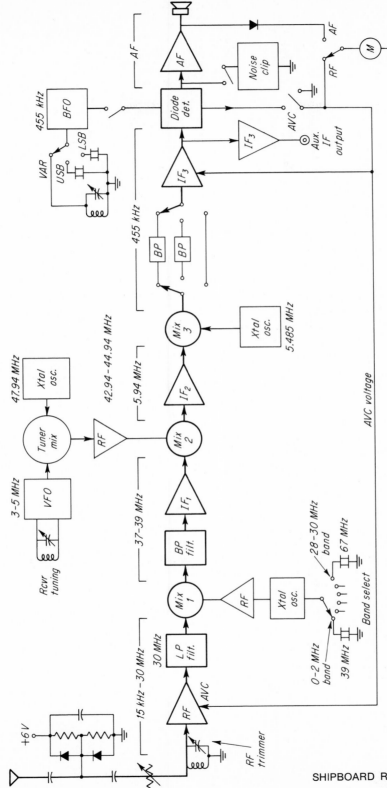

Fig. 28-6 Triple conversion marine superheterodyne (from ITT Mackay Marine type 3010).

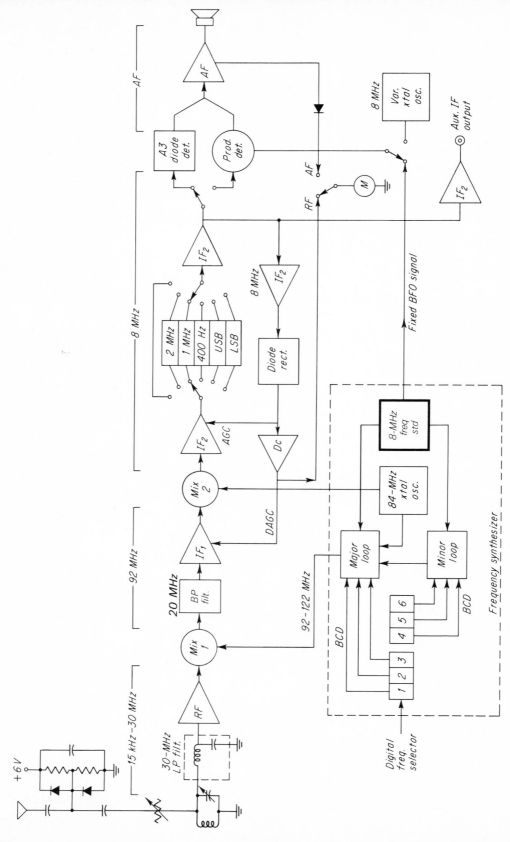

Fig. 28-7 Simplified modern solid-state superheterodyne using a digital frequency synthesizer for the local oscillator (from ITT Mackay Marine type 3020).

28-13 BRIDGE-TO-BRIDGE RADIOTELEPHONE

In 1973 it was made mandatory for all U.S.-flag power-driven vessels of over 300 gross tons, or of 100 gross tons and carrying one or more passengers, or any towing vessel 26 ft or over in length operating in navigable waters of the United States to have an 8- to 25-W 16F3 (16-kHz bandwidth FM) transmitter and accompanying receiver operating on channel 13 in the 156–162-MHz band. The transmitter must also be capable of operating at a reduced power level of 0.75 to 1 W output. Such type-accepted equipment must be mounted on or operated from the navigational bridge. The transmitters have a pre-emphasis response curve of 6 dB per octave from 300 to 3,000 Hz, and the receivers must have a similar de-emphasis response curve. Only the master or the person in charge of navigating the vessel is to operate the equipment. A continuous watch must be maintained on the simplex (transmit and receive on the same frequency) channel 13 (156.65 MHz) as long as the vessel is within 100 miles of U.S. shores. Foreign vessels may have portable equipment brought aboard by pilots.

Although the basic requirements seem relatively simple, the fact that a continuous watch must be maintained means that a bridge-to-bridge radiotelephone system must have two receivers if it is desired to communicate on other channels regarding port operations, commercial traffic, public correspondence, or international distress or to listen to weather broadcasts on 162.4 or 162.55 MHz.

An example of the sophistication that may be found in some equipment is the ITT Mackay Marine type 222 VHF radiotelephone system. This is an all-channel solid-state transmitter, two receivers, a synthesizer, a scanning system, and a ringer. It features automatic shift from simplex operation (used on 156.3 and 156.375 to 156.875 MHz) to duplex operation (used on all other channels, with the receiving frequency always 4.6 MHz higher than the transmitting frequency) at any time that the receiving channel is switched. The system has one transmitter but two separate receivers feeding a single audio circuit. One receiver uses the transmitting antenna, and the other uses its own separate antenna. The synthesizer supplies both the fundamental frequency for the transmitter and the local oscillator signal for the receivers. The system may also incorporate a channel-scanner circuit that continually samples up to six channels to see if there are any calls coming in on them. A coded tone transmission from a calling station can activate a ringer circuit on the called vessel only.

The bridge-to-bridge systems aboard all vessels navigating in the same area provide instant intercommunication, as well as afford assured response to distress traffic. In general, the VHF channels should carry much of the traffic that was previously carried by the 1600–2850-kHz radiotelephones, since port authorities, telephone company facilities, etc., are available on VHF. For long-haul communications the 4–24-MHz bands are still to be used.

28-14 SURVIVAL RADIO EQUIPMENT

One form of survival-craft radio equipment that may be required by law is the portable type consisting of a radiotelegraph transmitter with an attached telegraph key and capable of emitting not less than 2 W of A2-type emission on 500 kHz and 4 W of A2 on 8364 kHz to a self-contained antenna rod (or wire) and ground connector. It must operate manually on 500 and 8364 kHz and be capable of automatically transmitting the auto-alarm signal followed by SOS three times followed by a long dash of 30 s or more. The power supply is a manually operated electric generator operated by crank handles rotated at 70 rpm or less.

To ensure operation within the frequency tolerance of 5,000 Hz/MHz, such equipment is usually a crystal MOPA transmitter modulated with a single-tone AF oscillator. The frequency of operation is selected by switch.

The receiver is fixed-tuned on 500 kHz and is broad enough to receive A2 signals on any frequency from 492 to 508 kHz. The receiver must tune from 8266 to 8745 kHz and be capable of receiving A1 and A2 signals. The equipment must be capable of floating in seawater and of withstanding a drop of at least 20 ft into the water.

Nonportable lifeboat equipment is permanently installed aboard a lifeboat in a housing large enough to hold the equipment and the operator.

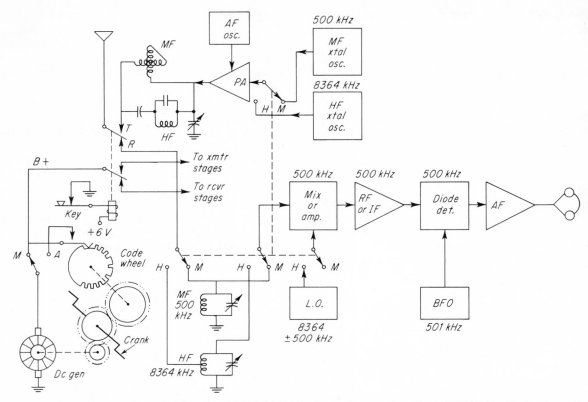

Fig. 28-8 Block diagram of a possible portable 500–8364-kHz survival transceiver (from ITT Mackay Marine type 401).

The transmitter must have not less than 30 W of RF output on 500 kHz and 40-W output on 8364 kHz. It is powered by batteries (6 or 12 V) of sufficient capacity to operate the equipment for at least 6 h. It must be capable of manual operation on either frequency and have the same automatic transmissions as the portable types.

A simplified portable survival transceiver is shown in block form in Fig. 28-8. The frequency selection switch is shown in the MF (500-kHz) position. With the key up, the relay coil is energized and the antenna is connected to the MF tuned circuit of the receiver, which feeds 500-kHz signals to the mixer or amplifier input. Since the local oscillator is not connected, the mixer stage acts as a 500-kHz amplifier, as does the next stage. This feeds 500-kHz signals to the diode detector, which beats against the BFO signal to produce the audio frequency to be amplified and fed to the earphones. The receiver operates as a TRF on 500 kHz.

When the key is pressed, the relay coil is de-energized and the TR relay arms move up, shifting high-voltage dc from the receiver to the transmitter and also shifting the antenna from receiver to transmitter. A 500-kHz signal is now generated by the MF crystal oscillator, is amplified, and is fed to the antenna. A 550-Hz AF oscillator modulates the PA to produce an A2 emission. The variometer in the antenna circuit tunes the power amplifier to resonance for maximum power output.

Power-supply voltage (and filament with vacuum tubes) is generated by hand-cranking a dc generator through rotation-increasing gears. If the M–A switch is moved from manual to automatic, a geared-down code wheel is rotated, which makes and breaks electric contacts as a wiper moves over projections along the wheel edge to produce automatic SOS signals followed by a long dash to allow nearby vessels to take direction-finder bearings on the lifeboat.

When the master switch is moved to HF, the local oscillator in the receiver section is turned on and the antenna is connected to an 8364-kHz tuned circuit which is coupled to the mixer input. The LO is tunable over several hundred kilohertz, allowing the operator to locate answering signals even if they are not exactly on 8364 kHz. On HF the receiver operates as a superheterodyne.

It will be noted that the MF and HF antenna tuning circuits are in parallel. For HF ac the variometer has so much inductance it looks like an RFC and has little effect on circuit operation. For MF ac the HF tuned circuit and its series capacitor appear as a small capacitor across the variometer and have little effect on circuit operation. The variable capacitor acts as a means of tuning the antenna to 8364 KHz but has little effect on the MF circuit. The variometer trims the antenna on MF.

Test your understanding; answer these checkup questions.

1. In Fig. 28-6, why does this receiver have high sensitivity? _____ Selectivity? _____ Image rejection? _____ What would be the image frequency when tuned to 30 MHz? _____ 2 MHz? _____ How many band selector crystals are needed? _____

2. In Fig. 28-7, over what band of frequencies must the frequency synthesizer generate frequencies to be able to receive signals from nearly zero frequency to 20 MHz? _____

3. Why might Fig. 28-7 be better for A3J than Fig. 28-6? _____

4. Why might DAGC be better at the front end of a receiver than AGC? _____

5. What is the purpose of the 20-kHz BP filter in Fig. 28-7? _____ The 400-Hz filter? _____ The 2-MHz filter? _____

6. How can a crystal oscillator be made variable? _____

7. If the dial in Fig. 28-7 is adjusted to 07.5412, to what frequency should the receiver then be tuned? _____ Within how many hertz of this value will it be? _____

8. What would be the only LC circuit in Fig. 28-7 that would have to be band-switched? _____ Why? _____

9. What would 2.8A3J mean? _____ 6A3? _____ 0.2A1? _____ 0.6F1? _____

10. What is meant by simplex? _____ Duplex? _____

11. Why is a synthesizer used with VHF radiotelephone sets? _____

12. What is the P_o of a portable survival transmitter on 500 kHz? _____ On 8364 kHz? _____ Of a nonportable lifeboat transmitter on 500 kHz? _____ On 8364 kHz? _____

13. In Fig. 28-8, what CW letters are shown on the code wheel? _____ Can the BFO be turned off? _____ Why is the dc generator geared up? _____

28-15 AUTO-ALARM KEYERS

When no radio operator is standing watch on a compulsorily equipped ship at sea, a device known as an auto-alarm must be in operation.

By international agreement, prior to sending an SOS on 500 kHz, all ships will transmit (if time allows) an auto-alarm signal consisting of alternate 4-s dashes and 1-s spaces for a period of 1 min. Whenever possible, the auto-alarm signal will be followed by a 2-min silence before the SOS message is transmitted.

An auto-alarm receiver is designed to register the dash-space auto-alarm transmissions, rejecting them if they are not timed correctly but accepting them if they are reasonably close to the required lengths. After registering three or four successive, properly made dashes and spaces, the auto-alarm rings a bell in the radio station, on the navigation bridge, and in the radio operator's living quarters. (Note that the auto-alarm does not stand watch for an SOS but for the auto-alarm signal that should precede any SOS message.)

While the auto-alarm signal may be transmitted manually by an operator watching the sweep second hand of a clock, an alarm can be transmitted by automatic means.

The keyer shown schematically in Fig. 28-9 is operated from an emergency battery. When the switch is turned on, a reed-type 60-Hz vibrator is set into vibration. When the reed is in the UP position, current flows downward through the motor; when it is in the DOWN position, current flows upward through the motor. Thus, the motor is fed an alternating-type current.

The synchronous motor rotates at 3,600 rpm, which is reduced through a series of gears to a speed of 12 rpm (5 s per revolution) to rotate a circular cam. One-fifth of the outer rim of the cam is raised, as indicated. When the raised portion strikes the microswitch (light-pressure switch), the keying-relay circuit is opened and the relay con-

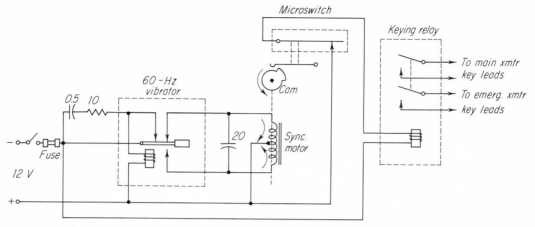

Fig. 28-9 An auto-alarm signal keyer used to transmit an auto-alarm signal automatically (RCA AR8651).

tacts open. Thus for 4 s the keying-relay circuit remains closed, followed by a 1-s open period. One of the two sets of contacts on the keying relay is permanently connected across the key leads of the main transmitter, and the other across the key leads of the reserve transmitter.

A simplification of an electronic timer used in one AA keyer is shown in Fig. 28-10. Transistors Q_1 and Q_2 form an asymmetrical astable multivibrator. Since the ratio of resistances in the two circuits is about 5:1, one transistor conducts for four times as long as the other. The dc amplifier, Q_3, coupled to the collector of Q_2, operates a sensitive relay, closing it for 4 s and opening it for 1 s as the two transistors conduct alternately. The relay could key a transmitter directly or operate a second relay to key either a main or a re-

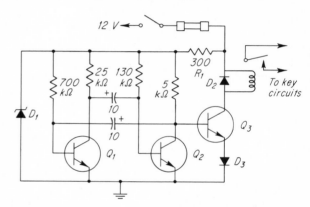

Fig. 28-10 Electronic AA keyer (from ITT Mackay Marine type 5103).

serve transmitter, at the same time keying an indicator light on and off.

Auto-alarm keyers must transmit dashes 3.8 to 4.2 s long with spaces of 0.8 to 1.2 s duration.

28-16 AUTO-ALARMS

Auto-alarm receiving equipment must allow for imperfectly made dashes and spaces, as well as interference signals adding to the received dash length. Consequently, they must accept dashes from 3.5 to 6 s in length and spaces from 0.1 to 1.5 s duration.

The auto-alarm receiver is turned on by throwing the main antenna switch to the AA position. (This also disables all transmitters.) The receiver must detect any A2 or type B emission that is modulated by frequencies between 300 and 1,350 Hz, must have 100-μV or better sensitivity, and must not overload with less than 1 V RF input. The receiver is fixed-tuned to 500 kHz ($\pm$8 kHz bandwidth) and is either a superheterodyne or a TRF type. An oscillator that generates approximately 100 μV on 500 kHz must be incorporated in the input of the receiver to be used for testing.

After three or four properly received auto-alarm dashes and spaces, an audible alarm (usually bells) must be sounded in three places: (1) radio room, (2) bridge, and (3) radio operator's quarters. The only way that the bells can be silenced is by pushing a reset switch button on the auto-alarm panel in the radio room. After silencing the bells, the radio operator listens on the main receiver for distress traffic on 500 kHz. If none is heard after a period of several minutes, it is assumed that the alarm was false. This is noted in the radio log and the auto-alarm is placed back in service.

Audible warnings will be sounded and a fault light will glow on the panel if the AA receiver circuitry malfunctions (relay falls out due to decreased current, as from a burned-out filament) or if the line voltage varies excessively. If the bells do not stop when the reset button is pressed, there is a fault in the equipment and the auto-alarm must be turned off and serviced.

Any time that an AA receives a constant dash of 3.5 s or longer, the counting circuits activate a red light on the AA panel. When the dash exceeds 6 s, the light goes out. The light indicates a proper-length dash is being received.

Prior to signing off watch, a radio operator turns on the auto-alarm, tests it to see if it is operating properly, and makes an entry as to this fact in the radio log. While the ship is at sea, the auto-alarm and all other emergency equipment, batteries, etc. must be tested once a day. Prior to leaving port they must also be tested. To test the alarm, the operator pushes the bridge-bell-off switch with one hand and the test signal button with the other, observing the dash-present light to indicate proper operation of the equipment. Alternatively, the bells can be rung for a short period before stopping them.

Several methods have been used to time received AA signals—the charging of RC circuits operating stepping relays which ring the bells after receiving four properly made dashes and spaces, dc motors operating electromagnetic drums that are engaged to slowly rotating axles which when rotated do the timing, received dashes and spaces gating on tone generators with counter circuits to count the number of cycles of ac produced to determine the dash and space times, etc. In all cases the receiver and timing circuits normally operate from the ship's main power, but the alarm bells operate from storage batteries.

Early AA equipment had manual sensitivity controls. When static noise increased, there were innumerable false alarms. Later models used slow-rise and fast-fall AGC voltages which improved operation considerably and made manual sensitivity adjustments unnecessary.

One modern auto-alarm system (ITT Mackay Marine type 5003) is a solid-state TRF receiver with a 16-kHz-bandpass 500-kHz filter feeding five tuned RF amplifiers, a diode detector, and a 1,350-Hz low-pass AF filter feeding a slow-rise and fast-fall AGC voltage for the first three RF amplifiers. Incoming dashes, converted to dc by the diode detector, key a 2-Hz pulse oscillator into operation. A counter consisting of a series of flip-flop stages begins to react to the pulses. After the seventh pulse (3.5 s), the short-dash circuit drops out. After the twelfth pulse (6 s), the long-dash circuit stops the oscillator and the logic circuits wait for another beginning dash to start. If the dash is between 3.5 and 6 s long, it is accepted in a dash-storage flip-flop. If the space is proper, the next dash is counted; and if it is correct, it is stored in dash-storage also. When four proper dashes are accepted, the alarm circuits are activated, and the bells ring.

Should the power supply fail, the line fuse burn out, or the line voltage be interrupted or drop below about 65 V, a relay coil across part of the power supply receives insufficient voltage excitation, its arm is pulled up, and the alarm bells are connected across the storage battery. If the power-line voltage varies above or below 65 V, the bells ring intermittently.

28-17 THE MAIN-ANTENNA SWITCH

Besides the main antenna, a ship may have an auxiliary antenna. Either of these antennas may be connected to the main-antenna switch, which will have several different positions, as follows:

o *Ground*. The main antenna is switched to ground when the radio watch is secured or when the ship is in an electrical storm.

o *Main transmitter*. When the switch is moved to this position, the main antenna is connected to the main receiver and to the main transmitter through a break-in relay.

o *Emergency transmitter*. The main antenna is connected to the emergency receiver and to the emergency transmitter through a break-in relay.

o *AA*. The main antenna is connected to the auto-alarm receiver, and power is connected to the AA equipment. All transmitters are disabled by interconnecting circuits.

o *DF* (*direction finder*). The main antenna is open-circuited or an auxiliary antenna is connected to the main receiver, depending on the antenna conditions when the radio direction finder was calibrated. The transmitters are disabled.

o *AA-DF*. (This may be labeled AA only.) Main or auxiliary antenna is connected to the auto-alarm, as when the radio direction finder was calibrated. Power is applied to the AA equipment.

o *HF* (*high-frequency*) *transmitter*. Main antenna is connected to the HF transmitter. (It may be connected to the HF receiver and to the HF transmitter through a break-in relay if an HF doublet antenna is not used on the HF receiver.) The auxiliary antenna may be connected to the main receiver when the main antenna is used on the HF transmitter.

28-18 TELEPRINTERS

A piece of communication equipment that is now being found aboard some ships is the tele-printer. It has a keyboard similar to that of a type-writer on which messages or news can be typed. When a letter key is pressed, the machine develops a series of open-closed circuit conditions that form a code for that particular letter or function (punctuation, spacing, line feed, etc.). The open-closed type code can be used to key a transmitter, normally as an FSK (F1) emission. At the receiving end the open-closed (called *space* and *mark*) conditions are received and converted

to off-on signals. These are mechanically set up by the receiving machine as a particular letter, actuating the proper type-bar to print the letter on the paper.

Since the timing of the mark-space code is done mechanically, the timing devices must run at the same speed at both transmitting and receiving ends. For this reason teleprinter internal parts are moved by either synchronous or carefully governed constant-speed motors.

Each letter consists of seven parts: a *start* signal (always a space), followed by five mark- or space-coded periods, and finally a *stop* signal. For 60 word per minute (wpm) machines the mark-space code segments are formed at a rate of 22 ms (22 milliseconds) each, with a 31-ms stop signal. For 100-wpm machines the code segments are at a rate of 13.5 ms and the stop is 19 ms.

A teleprinter, also known as a Teletype (trade name) machine, is either a page printer or a tape printer. In the latter case, the letters are printed on narrow paper tape. In the former case, after about 65 letters, it is necessary to send *carriage return* and *line feed* function signals to make the printer return to the left-hand edge of the page and also drop one line down. These functions do not cause a print-out, so that they are not indicated on tape copy.

To transmit, the mark-space signals generated by an operating teleprinter machine can be processed into the standard 170-Hz frequency shift at the desired transmitting frequency. This is then fed to amplifier stages of a transmitter (CW exciter, Fig. 28-4), resulting in an F1 emission. Methods of converting on-off to FSK might be by biased Varactors or biased diodes across LC or crystal oscillator circuits, as in Fig. 16-18.

To receive, either a separate RTTY (Radio Tele-TYpe) receiver is used or the IF output from a station receiver such as those shown in Figs. 28-6 and 28-7 may be used. The F1 received signal is converted to off-on dc currents by one of several methods. One of the original methods employed a *polar relay*. This relay has separate coils on both sides of its magnetized contactor arm, or armature (Fig. 28-11), along with separate permanent biasing magnets on each side of it. If the magnets are properly positioned, the armature is magnetically biased halfway between the coils and the contacts on both sides of the armature

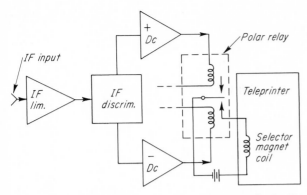

Fig. 28-11 Simple polar-relay RTTY IF-type terminal unit.

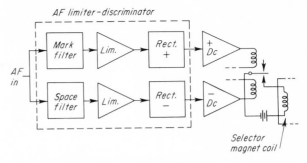

Fig. 28-12 Simple polar-relay RTTY AF-type terminal unit.

are open. A small current in either coil will cause the contacts to close according to the direction of the current.

A possible converter for F1 to machine-operating dc is the *terminal unit* (TU) shown in Fig. 28-11. Assuming a 455-kHz IF from the receiver, a 455-kHz limiter might feed a 455-kHz discriminator which would develop + and − signals from the received FSK signal. These voltages are amplified by dc amplifiers and might actuate the field coils of a polar relay. The contacts of the polar relay might key mark currents into the selector magnet coil in the teleprinter and no current for the space part of each letter. The machine mechanically selects the letter from the coded pulse input and causes the proper type-bar to strike the paper. Instead of activating the polar relay, the amplified + and − signals can act as on-off bias for transistor or vacuum-type amplifying devices in which the selector magnet coil forms the output circuit load.

Another form of TU operates from two-tone audio signals. If an F1 signal is being received on an A1 detector, the mark and space signals will produce two separate tones. For example, if the mark signal is tuned in to produce a 300-Hz tone, the space signal will produce 300 ± 170 Hz, or either a 470- or a 130-Hz tone. Assume the mark signal is 300 Hz and the space is 470 Hz. (Space-high mark-low is standard for marine teleprinters.) In Fig. 28-12 after passing through tone filters (470 and 300 Hz), the two signals are limited to prevent fading from affecting the operation of the system. Then both tones are rectified, one to a positive voltage and the other

to a negative one. Both voltages are amplified and can be made to either actuate a polar relay or key an electronic circuit to feed on-off dc pulses to the selector magnet coil in the teleprinter. To prevent the machine from chattering constantly when no signal is present, some form of timed bias must be used to switch the TU to a mark condition if the last signal was a space. The first space-frequency signal starts the machine's timing circuits.

A block diagram of a complete teleprinter station is shown in Fig. 28-13. Some radioteleprinter

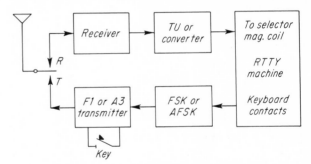

Fig. 28-13 Basic components in a teleprinter station.

systems transmit two-tone A3 emissions (AFSK) instead of FSK.

By necessity this description has been highly simplified. A wide variety of transmitting and receiving systems are in actual use in radioteleprinter stations.

Test your understanding; answer these checkup questions.

1. What makes up an AA signal? _____ When should the SOS be sent? _____ Why? _____
2. In Fig. 28-9, what would the 20-μF capacitor do?

_____ The 0.5 µF *C* and 10-Ω *R*? _____
If the cam has the microswitch closed when the keyer is turned off, might this leave a transmitter on? _____

3. In Fig. 28-10, what is the function of D_1? _____ D_2? _____ D_3? _____ R_1? _____ Is the 12 V positive or negative? _____

4. What tolerance do AA keyers have for their dash and dot lengths? _____ What tolerance do auto-alarms have? _____

5. When is an AA used? _____ How is an AA turned on? _____ What is its required bandwidth? _____

6. Where are audible AA warnings sounded? _____ How are they shut off? _____ What does a red light indicate? _____

7. What might cause constant audible alarms from an AA? _____ Intermittent alarms? _____

8. What powers an AA set? _____ Its bell system? _____

9. What circuit in a modern AA system makes manual setting of the RF gain control unnecessary? _____ What are its general characteristics? _____

10. List the seven positions of a radio-room antenna switch.
_____ _____ _____ _____
_____ _____ _____ Would all ships have all of these? _____ Why? _____

11. What are three possible uses of RTTY aboard ships? _____ _____ _____

12. Are all coded letters and functions of a teleprinter of equal length? _____

13. What is fed to a teleprinter to make it print? _____ What comes from it when transmitting? _____ Does it also print when transmitting? _____

14. What are the two types of discriminators used in RTTY TUs? _____ _____

28-19 SHIPBOARD RADIO OPERATORS

Freighters and tankers usually employ one radio operator who stands 8-h watch daily, using the auto-alarm for the other 16 h. The operator is a ship's officer and eats with the other ship's officers. A Radiotelegraph First Class license or a Second Class license with an endorsement of 6 months' satisfactory service is required. The minimum age is twenty-one years.

Passenger vessels usually employ three watch-standing operators. The chief operator usually has passenger privileges. On larger ships there may be a chief and three watch-standing operators. When radar is used aboard the ship, one or more of the operators must have radar endorsements (FCC Element 8) on their licenses to allow them to service the equipment.

The minimum age for any radiotelegraph operator license is eighteen years. The minimum age for a Radiotelegraph First Class license is twenty-one. One year of experience aboard ship is required of any applicant for a Radiotelegraph First Class license. The Radiotelegraph First and Second Class licenses authorize the holder to operate any radiotelephone equipment aboard the vessel.

28-20 STANDING WATCH

The radio operator is aboard ship to enable the ship to communicate in case of emergency, to receive distress messages from other ships, to transmit weather reports, to handle messages pertaining to shipping business, and to send and receive routine messages for passengers or crew members.

The main duty of the operator is to keep an efficient watch by earphones or loudspeaker on the international calling and distress frequency, 500 kHz. He must log any signals heard on this frequency during the two 3-min silence periods from 15 to 18 min and from 45 to 48 min after each hour and also log at least one other call heard on this frequency every 15 min. If no signals are heard, he makes a log entry to this effect. All log entries are in 24-h Greenwich mean time (GMT), which is more properly called universal coordinated time (UTC). See also Sec. 33-14.

The radio-room clock is an 8-day windup type marked with 12 h and with additional markings up to 24 h as shown in Fig. 28-14. It has a sweep second hand to allow reading time accurately and to aid the operator if he must transmit an auto-alarm signal by hand. It has the two 3-min silence

Fig. 28-14 Shipboard radio-room clock marked for 24-h time, with 3-min silence periods shown, and 4-s timing for an auto-alarm signal.

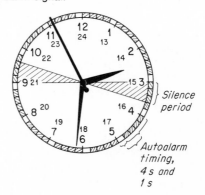

periods marked off in red, and around the minute or second scale it has twelve 4-s red arcs separated by 1-s spaces representing the auto-alarm signal. On compulsorily equipped vessels this clock is compared daily with a standard time transmission from such stations as WWV, Bureau of Standards station, near Boulder, Colorado, which transmits constant time signals on 2.5, 5, 10, 15, 20, and 25 MHz; similar transmissions from WWVH in Honolulu; or time ticks from NSS in Washington, D.C., or NPG in San Francisco.

The operator must abide by all Federal as well as international telecommunication laws, rules, and regulations, as applicable. If he observes a ship station flagrantly violating international radio regulations, he may make a report of the violation on a form supplied by the radio company servicing his ship and submit the report to the FCC, Washington, D.C.

On single-operator ships the operator stands an aggregate of 8 h daily, using the auto-alarm for the off hours. Watch hours are determined by the geographical area in which the ship is operating. Ship stations whose service is not continuous should not close before finishing all operations resulting from a distress call or an urgency or safety signal or before exchanging, insofar as possible, all traffic for the ship to or from any coastal station within its range.

A ship station which has no fixed working hours normally advises the nearest coastal station with which it is in communication the hours of its closing and reopening, unless the ship is in foreign waters, in which case such transmissions are prohibited.

Radio operators should know that New York City and San Francisco are the principal Atlantic and Pacific Coast ports or shipping terminals. They are also the major centers of telecommunication for communication with Europe or the far Pacific, although the greatest traffic is handled out of New York, both shipping and by radio. The greatest number of telecommunication channels are between New York and Europe.

28-21 POSITIONS AND TIME

A shipboard radio operator is expected to know a reasonable amount of world geography, approximate locations of the continents and well-known ports, and the meaning of latitude, longitude, international date line, and GMT.

The earth's being a nearly round sphere rather than having a flat, plane surface presents some difficulties to a navigator who must sail his vessel from one point on the sphere to another. To help him, the earth is divided into the Northern Hemisphere, with the North Pole at the "top," and a Southern Hemisphere with the South Pole at the "bottom." An imaginary line running from the North to the South Pole is known as a *line of longitude,* or a *meridian*. The sphere is divided into 360°, with a starting point of 0° being the imaginary longitudinal line running through Greenwich, a borough of London, England. Progressing westward from 0°, the meridians are termed *west*. New York is located on a meridian about 75° west of Greenwich, the Panama Canal about 80° W, Los Angeles 120° W, and Honolulu about 157° W. Moving to the eastward of the 0° meridian, Cairo is 30° E, Bombay 73° E, Shanghai and Manila 120° E, and Tokyo about 142° E. At the 180th, the east and west meridians meet in the *international date line*. Here, halfway around the world from Greenwich, the day begins. When it is noon tomorrow in Greenwich, it is midnight on the date line and today is considered to be starting. (The date line runs through no populated areas, being mostly over the ocean.) When a ship passes over this line from west to east, it skips a day, possibly going from 6 P.M. Wednesday to 6 P.M. Thursday. When traveling in the opposite direction, it will have two identical days, perhaps two Fridays.

Since there are 360° of longitude and there are 24 h in a complete rotation of the earth, $\frac{1}{24}$ of the total rotation, or 15°, represents how far the sun travels in 1 h. Therefore, each 15° of longitude represents 1 h that the traveler must turn his clock back if going westward or forward if progressing eastward. A ship keeps its clocks on *local,* or sun, time. The local time of a ship traveling east or west may change a few minutes per day near the equator to more than half an hour in the high-latitude regions where the lines of longitude are closer together.

Since Shanghai is 120°, or 8 h, east of Greenwich, when it is 1700 GMT (12 h plus 5 h, or 5 P.M.) Monday in London, it is 17 h plus 8 h, which would be 25 o'clock, or 1 A.M., in Shanghai on the next day, Tuesday. Since New York is 75° W, or 5 h behind GMT, when it is 1 A.M. Tuesday in Shanghai, it is 12 noon Monday in New York (and 9 A.M. in San Francisco).

The *equator* is an imaginary line girdling the world in the middle. From the equator to either pole is one-fourth of a circle, or 90°. Imaginary lines parallel to the equator are known as lines of *latitude,* or *parallels.* Since lines of latitude do not converge, the distance between 5° of latitude anywhere in the world may be considered the same. A degree is often considered to be composed of 60 equal parts called minutes. A minute of latitude is equal to 1 nautical mile, approximately $1\frac{1}{8}$ land miles. Latitude lines above or below the equator are said to be so many degrees north or south. The position of a point on the earth, Colón, Panama, for example, can be indicated accurately by its longitude and latitude. Thus, "Colón is at 80° W and 9° N" specifies only one place in the world where it can be found. The island of Guam, located in the Pacific Ocean, is at approximately 14° N and 146° E; Sydney, Australia, is at 33° S and 151° E; and so on.

All radio-room clocks are kept on GMT. One-operator ships usually have the operator standing watch according to GMT for the area or region of the world in which the ship is traveling. Three-operator ships usually have the operators stand-ing watch by local time to prevent confusion at mealtimes, but the log is still kept in GMT. The three radio regions of the world are (1) Europe, Russia, and Africa, (2) North, Central, and South America, and (3) Australia and Asia.

Test your understanding; answer these checkup questions.

1. How much shipboard experience is required for a First Class license? _____ To be sole operator on a ship? _____
2. Who must have a radar-endorsed license? _____
3. What is the minimum number of log entries in 1 h? _____
4. What is meant by GMT? _____ UTC? _____ Are they the same? _____
5. In Fig. 28-14, what time(s) is (are) shown? _____
6. During what time(s) during any hour is routine traffic and calling not allowed on 500 kHz? _____ Why? _____
7. How are watch hours determined on one-operator ships? _____
8. Is the equator a line of latitude or longitude?
9. Where does 0° longitude start? _____ What is another name for 180° longitude? _____
10. How many hours difference between San Francisco and Greenwich? _____ Which is ahead? _____
11. According to what time is a radio log kept? _____ Does (do) the operator(s) eat? _____
12. What are the three radio regions of the world? _____ _____ _____

COMMERCIAL LICENSE QUESTIONS

The following questions refer to Element 6 only. Sections in which questions are answered are shown in parentheses.

1. What is indicated if a voltmeter connected between the negative side of a ship's dc line and ground reads the full line voltage? (28-1)
2. In all cases other than those in which the trans-mitter output must be maintained at a fixed value, what amount of power should be employed for routine communication? (28-1)
3. Between what points on a ship, compulsorily equipped with a radiotelegraph installation, is a reliable intercommunication system required? (28-2)
4. With what type or types of emission and upon

ANSWERS TO CHECKUP QUIZ ON PAGE 693

1. (*12 dashes, 11 spaces*) (*2 min after AA signal*) (*Give alerted operators time to get to receiver*) 2. (*Shape ac*) (*Reduce contact sparking*) (*No*) 3. (*Hold E constant*) (*Damp out inductive E of coil*) (*Bias Q3 to allow cutoff*) (*E drop so D_1 can operate*) (*+ for NPNs*) 4. (*± 0.2 s*) (*3.5–6 s, 0.01–1.5 s*) 5. (*When operator is off watch*) (*Throw antenna switch*) (*16 kHz*) 6. (*Bridge, radio room, operator quarters*) (*At AA*) (*Trouble, or long dash*) 7. (*AA signal combinations of static and signals, faults, low line E*) (*Line-E variations*) 8. (*Ship's line*) (*Battery*) 9. (*AGC E*) (*Slow rise, rapid fall*) 10. (*Ground, main transmitter, emergency transmitter, AA, DF, AA-DF, HF transmitter*) (*No*) (*Some have no HF*) 11. (*Weather, news, messages*) 12. (*Yes*) 13. (*Coded pulses*) (*Open-closed circuit*) (*Yes*) 14. (*IF*) (*AF*)

what frequency should a radiotelegraph transmitter be adjusted to transmit a distress call? (28-2, 28-4)

5. For how long a period of continuous operation should the emergency power supply of a compulsorily equipped ship station be capable of energizing the emergency radiotelegraph installation? (28-2, 28-5)

6. How is the power output of a marine vacuum-tube radiotelegraph transmitter ordinarily adjusted? (28-4)

7. What is the purpose of the iron-compound cylinders which are found in the inductances of certain marine radiotelegraph transmitters? (28-4)

8. Why might marine transmitters employ variometers rather than variable capacitors as the tuning elements? (28-4)

9. What care should be taken in hoisting the antenna of a shipboard radiotelegraph station to avoid damage to the antenna wire and insulators? (28-6)

10. Draw a sketch of a typical shipboard antenna for transmitting on 500 kHz showing the supporting insulators, the safety link, and the lead-in wire. How does voltage vary along the length of the lead-in and along the antenna? (28-6)

11. At what point on a shipboard antenna system will the maximum potential be found? (28-6)

12. Upon what band, in addition to the 350- to 515-kHz band, must a main receiver on a United States ship be capable of operation? What is the purpose of this additional band? (28-11)

13. If broadcast signals interfered with your reception of signals on 500 kHz, how would you reduce or eliminate such interference? (28-12)

14. What is the purpose of an auto-alarm signal keying device on a compulsorily equipped ship? (28-15)

15. During the time that a vessel is at sea, how frequently must the auto-alarm be tested? (28-16)

16. What signal will cause an approved auto-alarm receiver to ring the warning bell? (28-16)

17. If a vacuum-tube heater burns out in an approved auto-alarm, what causes the warning bells to ring? (28-16)

18. Describe the number of dashes, or dots, and spaces which compose the international auto-alarm signal and indicate the time intervals involved. (28-16)

19. To what frequency or band of frequencies is an approved auto-alarm receiver tuned? (28-16)

20. While in port, how frequently should the emergency equipment be tested? (28-16)

21. How frequently must the quantity of fuel in the supply tank for use with an oil- or gas-driven emergency generator be checked while the vessel is in the open sea? (28-16)

22. While the vessel is in the open sea, how frequently must the specific gravity of the emergency battery be taken? (28-16)

23. While the vessel is in the open sea, how frequently must the emergency equipment be tested? (28-16)

24. If an auto-alarm bell rings and upon pressing the release button it stops, what could be the cause or causes? (28-16)

25. If you were a radio operator on a vessel equipped with an aproved type of auto-alarm, describe what would happen upon failure of a vacuum-tube filament. (28-16)

26. If an auto-alarm bell rings and upon pressing the release button it does not stop what could be the cause or causes? (28-16)

27. On a vessel equipped with an approved auto-alarm, where is the control button which silences the warning bells located? (28-16)

28. When the auto-alarm bell rings, what should the operator do? (28-16)

29. What factor or factors determine the setting of the sensitivity control of an auto-alarm receiver approved for installation on a vessel of the United States? (28-16)

30. On a vessel equippped with an approved type of auto-alarm, what factors cause (a) the bell to sound, (b) the warning light to operate, (c) intermittent ringing of the bells? (28-16)

31. What means usually are provided to prevent the operation of the ship's transmitter when the auto-alarm receiver is in use? (28-16, 28-17)

32. What is the purpose of an auxiliary receiving antenna installed on a vessel which is also fitted with a direction finder? (28-17)

33. What experience is the holder of a First or Second Class Radiotelegraph Operator license required to have before he is permitted to act as chief or sole operator on a compulsorily radio-equipped cargo ship? (28-19)

34. Are there any age requirements that a person must meet before he can be issued a radiotelegraph operator license? (28-19)

35. How frequently should an entry be made in a ship radiotelegraph log while a radio watch is being maintained? (28-20)

36. Upon compulsorily equipped vessels, which are required to have an accurate clock in the radio room, how frequently must this clock be adjusted and compared with standard time? (28-20)

37. Why is the clock in a compulsorily equipped ship radiotelegraph station required to have a sweep second hand? (28-20)

38. What action, if any, may a radio operator take when he observes a ship station which is flagrantly violating the international radio regulations and

causing harmful interference to other stations? (28-20)

39. Under what conditions may a ship station close if its service is not required to be continuous? (28-20)

40. What exceptions are permitted to the regulation which states that a ship station which has no fixed working hours must advise the coast station with which it is in communication of the closing and reopening time of its service? (28-20)

41. What is the principal port of the United States on the Pacific Coast at which navigation lines terminate? (28-20)

42. In what city is the major telecommunication center of the United States located? (28-20)

43. What is the principal Atlantic Coast port of the United States at which navigation lines terminate (28-20)

44. To what continent do the greatest number of telecommunication channels from the United States extend? (28-20)

45. What is the GMT time and the day of the week in Shanghai when it is Wednesday noon in New York City? (28-21)

46. What is the approximate latitude of Colón, Republic of Panama? (28-21)

47. In what ocean is the island of Guam located? (28-21)

ANSWERS TO CHECKUP QUIZ ON PAGE 696

1. (*1 yr*) (*6 months*)   2. (*One operator if radar aboard*) 3. (*Four*) 4. (*Greenwich mean time*) (*Universal coordinated time*) (*Yes*) 5. (*0231:56 or 1431:56*) 6. (*Silence periods*) (*All listen for SOSs*) 7. (*By geographical position*) 8. (*Latitude*) 9. (*Greenwich*) (*Date line*) 10. (*8 h*) (*Greenwich*) 11. (*GMT, or UTC*) (*Local ship's*) 12. (*1 = Europe-Africa*) (*2 = Americas*) (*3 = Australia-Asia*)

---

29 RADIO DIRECTION FINDERS

The objective of this chapter is to explain the fundamentals of devices capable of determining the direction of approaching radio waves, particularly equipment which might be found aboard ships. Such radio direction finders may have rotating loop antennas or two stationary fixed loops. A block diagram of an automatic direction finder is explained briefly, as is the basic theory of position determination.

29-1 BASIS OF RADIO DIRECTION FINDING

A simple horizontal dipole antenna capable of being rotated, coupled to a radio receiver, forms a rudimentary radio direction finder (RDF or DF). When an antenna wire is pointing toward a transmitting station, a minimum signal is received, but if parallel to the wave front of an approaching radio wave from the station, a maximum signal is received. The indication of the maximum signal is very broad—the antenna can be swung many degrees before any change in signal amplitude is discernible. On the other hand, the signal will drop to a minimum and back up again in a very few degrees of rotation. For this reason the minimum signal gives a more definite line of direction, or bearing, and is the one normally used in DF work.

A short rotatable dipole has a fair signal-reception capability. For a given size, a small loop antenna is better, but its pickup response differs from that of the dipole.

29-2 THE DF LOOP

The usual DF antenna is a loop constructed in either circular or square form. Figure 29-1 can be used to illustrate the basic properties of a loop.

A vertically polarized signal, such as most LF and MF signals, striking the loop wires induces currents in the vertical sides S_1 and S_2. These currents are equal in amplitude and polarity if the transmitter is an equal distance from the two sides, as when the radio waves are traveling

Fig. 29-1 (*a*) Basic balanced RDF loop circuit. (*b*) Reception pattern of the loop.

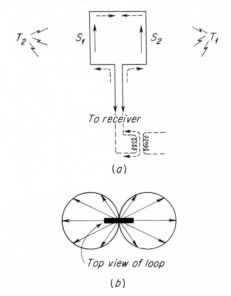

(*a*)

Top view of loop

(*b*)

toward the loop from the position of the reader or striking it from behind the page. Two equal induced currents flowing upward, as shown, would cancel at the top of the loop and in the receiver, leaving a zero resultant received signal. On the opposite half cycle cancellation also occurs. Rotating the loop 180° places the wires in a similar relative position, and again zero signal is received.

With a transmitter at either position T_1 or T_2, the currents induced in S_1 and S_2 will not be of equal amplitude during the major portion of any cycle because they are different distances from the transmitter, and a current equal to the difference between the two induced currents will flow in the receiver. While this *difference current* may be weak, multiturn loops with a diameter of about 3 ft produce satisfactory pickup for the sensitive receivers used.

When a loop is rotated 360°, a figure-eight reception pattern will be produced. Looking down on the loop, signals approaching at a 90° angle from the plane of the loop will be received as nulls. Signals approaching in the direction of the plane of the loop will be maximums. Signals from other directions will be intermediate in amplitude. The loop is said to have a *bilateral,* or *bidirectional* (two-way), response, giving a good null in two directions 180° apart and two maximums 180° apart.

Note that the horizontal portions of the loop are considered to be picking up no signal at all from approaching radio waves. Any signal induced in the top portion should be canceled by that induced in the bottom portion as long as the wave is traveling parallel to the earth, regardless of how the loop is positioned. If a signal is approaching the loop from above the horizon, however, a horizontal difference current will be induced in these portions of the loop. This appears as a residual signal, shifting the null or making a complete null impossible, or both. This occurs with signals reflected from the ionosphere (sky waves) but is nonexistent for ground-wave signals.

Practical loop antennas are 10- to 15-turn coils of 3-ft diameter encased in a hollow brass or aluminum case. This shield is broken only at the top, at which point an insulating segment is inserted as shown in Fig. 29-2. Without the insulating segment no signal would be able to penetrate

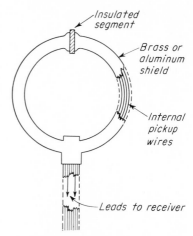

Fig. 29-2 A shielded RDF loop antenna.

to the pickup wires inside. The insulation allows oscillating currents to be induced in the metallic shield, inducing voltages into the internal pickup wires.

The loop is rotated by hand. The loop shaft terminates below deck at the operating position and is coupled mechanically to a wheel. Rotating the wheel 360° rotates the loop 360°.

29-3 AN UNBALANCED LOOP

A loop antenna that is properly balanced electrically produces a perfect figure-eight reception pattern. Small differences in length of wire between the two sides, or differences of distributed capacitance between the two sides and ground (or shield), will unbalance the loop. Unbalance results in one of the reception lobes being greater than the other, in the lack of complete nulls, and in a shifting of the nulls from their normal 180° relative positions. Figure 29-3 illustrates S_1 as being shorter electrically and picking up less signal than S_2, resulting in noncancellation. A possible reception pattern is shown. Note the lack of complete nulls and also that the best nulls are no longer 180° apart. In the unbalanced condition neither minimum is suitable for determining direction. The lack of a deep, sharp null on stations within 50 or 100 mi usually indicates loop unbalance and erroneous bearings.

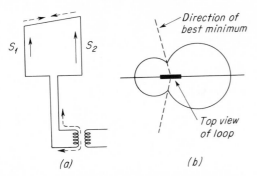

Fig. 29-3 (a) Unbalanced loop results in a difference current in the receiver. (b) Reception pattern of the unbalanced loop.

29-4 METHODS OF BALANCING A LOOP

By proper manufacturing precautions a loop that has little electrical unbalance can be constructed. If such a loop is placed in a clear space or in the central portion of an airplane, for example, where nearby portions of the body on both sides are symmetrical, errors tend to balance out and no electrical balancing may be required.

On a ship there may be no place where such a physically symmetrical condition is available. As a result, a means of electrical balance must be used. There are three basic means. The first loop in Fig. 29-4a uses a differential variable capacitor to balance both sides of the loop to ground. Rota-

ting this capacitor increases the capacitance to ground from the left side as the capacitance to the right side is decreased, and vice versa. This can make up for electrical unbalance due to greater capacitance to ground from one side of the loop than from the other. The center of the loop coil is connected to the shield (ground) to aid in balancing both sides of the loop. A push-pull input circuit aids balancing.

The second loop, shown in Fig. 29-4b, uses a differential capacitor and a short vertical antenna. The reception pattern of the vertical antenna is circular (equal reception in all horizontal directions). Signals from this *sense antenna* can be fed by the differential capacitor to the side of the loop that is not picking up sufficient signal because of improper electrical balance.

The third loop, Fig. 29-4c, employs a rotatable coil to induce the required amplitude and phase signal voltage into one side of the loop to produce balance.

It is necessary to balance the loop on each bearing taken, as a single balance adjustment may not hold for more than a few degrees of loop rotation.

The usual procedure for balancing a loop is to tune in the signal, rotate the loop for the best minimum obtainable, and then rotate the balancing control to the weakest response possible. If the null is not a zero signal, the loop must be

Fig. 29-4 Three means of balancing a loop. (a) Balancing to ground with a differential capacitor. (b) Balancing with a vertical antenna and a differential capacitor. (c) Inductive balancing with a vertical antenna.

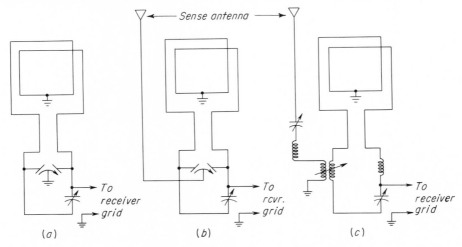

rotated to a still greater minimum and the balancing control adjusted until a sharp null results. If the DF has been compensated correctly, the sharp null should indicate an accurate bearing or direction of the station being received.

29-5 UNIDIRECTIONAL BEARINGS

The loop alone gives bidirectional or bilateral bearings (two nulls for 360° rotation). If the loop is intentionally unbalanced, a distorted figure-eight pattern results, and if the distortion is sufficient, the amplitudes of the two maximums are noticeably different. The stronger maximum is used as the indication of the direction of the station. This is known as a *unidirectional,* or *unilateral,* bearing. Since it is a maximum-signal indication, it is not accurate by itself. It is necessary either to sense the direction of the signal first (take a unidirectional bearing) and then take a balanced (bidirectional) null bearing on the station or to reverse this procedure.

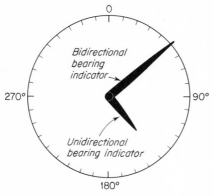

Fig. 29-5 An RDF bearing indicator rotates with the loop. Bidirectional and unidirectional indicators are at right angles.

The indicator on a DF is directly coupled to the rotator wheel and normally has two pointers, as in Fig. 29-5. One is long and points out the bearing of the null on a compass ring marked in degrees. The other is the unidirectional indicator, is short, and points at right angles to the bearing indicator.

The unbalancing of the loop is accomplished by pressing a sense switch that resonates and couples the sense antenna to one side of the loop. (The sense antenna is usually about 20 to 30 ft long, is erected as close to the loop as practical, and must be vertical.)

The amount of sense-antenna signal, determined by the antenna length, tuning, and degree of coupling to the loop, affects the shape of the reception pattern, as indicated in Fig. 29-6. A small value of sense signal decreases one maximum a little and increases the other. With the correct value of signal, one of the maximums can be canceled entirely and the other increased materially. This results in a *cardioid,* or heart-shaped, reception pattern. Still greater sense signal produces a pattern with no useful minimums.

While the cardioid would seem to be a good general-purpose bearing indicator, its null is not sharply enough defined to make it practical. Furthermore, if the received frequency is changed, the cardioid shape is altered.

29-6 DF ERRORS

There are several errors to which a DF is subject. All may cause significant changes in bearings.

COASTLINE REFRACTION. This is also known as *land effect.* Radio waves crossing from a land to a water area at any angle other than a right angle will refract, or bend, toward the coastline.

Fig. 29-6 Reception patterns with four values of sense signal. (*a*) Balanced loop. (*b*) Small value of sense signal. (*c*) A cardioid with still more sense signal. (*d*) Excessive sense signal.

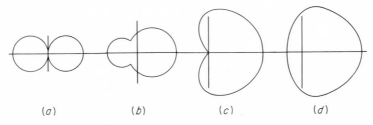

(*a*)   (*b*)   (*c*)   (*d*)

The resulting error may not be appreciable if the transmitting station is within a few hundred yards of the coast but may be significant if it is situated inland several miles.

NONOPPOSITE MINIMUMS. This is also known as *antenna effect*. It is a form of loop unbalance due to stray signal pickup through earphone cords or power lines or improper shielding or grounding of the loop. To lessen this effect, earphones are bypassed and low-pass filters are installed in any power lines feeding the DF. The shield constructed around the loop coil materially decreases the effect, reducing nonopposite minimums to less than a degree.

RERADIATION. Signals striking the loop may also strike nearby metal objects, inducing currents in them that produce reradiated signals. In this case, the loop receives the same signal from two directions, at two different amplitudes, and displaced in phase. The loop responds to the vector sum of the two signals, which represents a deviation error when compared with the true direction of the approaching signal. Such signals also require an adjustment of the balancing control to feed in an opposite-phase signal to the loop circuit to attain a sharp null. Bearings taken within $\frac{1}{2}$ mi of large bridges, ships, or other metal objects may be subject to considerable error because of reradiation and should be used with caution.

GREAT-CIRCLE ERROR. Radio signals approach the receiver on a great-circle track. Since both Tokyo and San Francisco are at approximately 37° N, it would be assumed by looking at the usual map that the shortest route between the two would be to travel directly east or west along the 37th parallel. Since the world is a sphere, however, the shortest track between the two cities takes a route northward up to almost the 50th parallel. Thus, for all normally used maps or charts (except great-circle charts), radio waves, which are received strongest via the most direct route, appear to travel a curved course rather than a straight line. The only time they conform to the usual maps is when they are traveling either north or south or along the equator. When the transmission distance is only a few miles, great-circle curvature can be ignored, but for distances in excess of 100 mi it may have to be considered in navigation.

POLARIZATION ERRORS. As mentioned previously, ground-wave signals striking the loop induce currents in the vertical portions, with any induced currents in the horizontal portions balancing out. However, signals approaching the loop from above induce nonbalancing currents in the horizontal portions that add vectorially with the vertical signals and result in loss of null as well as deviation. With medium frequencies used in marine DF work, usually 200 to 535 kHz, the ground wave is so much stronger than the sky wave for 50 mi or more that any downward-approaching sky waves may be too weak to be significant. Furthermore, during the daytime the sky wave tends to be attenuated. During nighttime hours the sky wave refracts from the ionosphere and is receivable for long distances. When the sky-wave amplitude approaches the ground-wave amplitude, the null of a loop antenna may no longer be reliable. This is known as *night effect*. Loop reception is considered unreliable for navigation with sky-wave signals alone.

The variation of the ionosphere at night produces a fading sky-wave condition that may make the null appear to vary back and forth over several degrees in a short period of time. During the periods of $\frac{1}{2}$ h before and $\frac{1}{2}$ h after sunrise and sunset the ionosphere is varying so wildly that bearings at even 50 mi may be affected.

29-7 CALIBRATING A DF

A shipboard DF will normally be quite accurate for signals approaching from dead ahead, from directly astern, or from either beam. However, between these points it may deviate several degrees. These are known as *quadrantal* errors. By running the vessel in a circle within sight of a radio transmitting antenna, it is possible to take simultaneous visual and radio bearings every 5° or 10° as the ship turns. Comparison of the two bearings will indicate how far the DF bearings are in error. This is known as *calibrating* the DF. By building into the DF indicator system a mechanical means of producing opposite errors at desired points, it is possible to produce a resultant DF indicator with no appreciable error. Adding the mechanical opposite errors is known as *compensating*.

Another form of compensation is sometimes

used. The masts, deck, and smokestack, plus guy and other wires aboard ship, form a semiresonant circuit that may introduce considerable quadrantal error in a DF that is located in this area. This error can often be reduced by closing the mast-deck-stack circuit with a wire connected between the top of the mast and the top of the stack. (Sometimes this increases the error.)

Because the DF is sensitive to signals reflected from nearby objects, it is important that the calibration be carried on when the ship is in its seagoing condition insofar as halyards, guy wires, masts, and booms are concerned. Since the main radio antenna usually is installed over the top of the loop, the calibration of the DF may shift if the resonant frequency of the radio antenna is changed. To prevent this, the DF is interlocked with the main-antenna switch in the radio station. When the bridge officers turn on the DF (located on the bridge), a red light is turned on in the radio room. The operator throws his main-antenna switch to the DF position. This disconnects his main antenna, closes a circuit that allows the DF to start operating on the bridge, and connects his watch-standing receiver to an auxiliary antenna that has no effect on the DF. (It is in this condition that the DF is calibrated.)

29-8 FREQUENCIES USED FOR MARINE DF

Marine radio direction finders must be capable of operating ón 500 kHz to take bearings on ships in distress, on the international DF frequency of 410 kHz when requesting DF bearings of other stations, and on the frequencies used by the marine radio-beacon stations operating between 285 and 320 kHz. Most DF bearings are taken on radio-beacon stations. As a result the DF is usually calibrated at a frequency near 315 kHz. The farther from the frequency at which it is calibrated, the greater the quadrantal error that may be present in it.

Test your understanding; answer these checkup questions.

1. Why are signal nulls used in RDF work? _____ How many are there in one loop rotation? _____
2. Seen from above, what horizontal reception pattern does a small vertical loop have? _____

3. Under what condition do the horizontal portions of a loop pick up difference currents? _____ What effect does this have? _____
4. Why must there be a break in loop shielding? _____
5. What are two results of an improperly balanced loop? _____ _____
6. What are the three methods of balancing a loop? _____ _____ _____
7. For what are unidirectional bearings used? _____ How can a loop be made to give them? _____
8. Why might it be better to take unidirectional bearings before taking bidirectional ones? _____ Are they always required? _____
9. What kind of an antenna is a sense antenna? _____ Where should it be located? _____
10. What happens to a radio wave path as it moves outward across a coastline at less than 90°? _____
11. What is the result of antenna effect? _____ Of reradiation of signals? _____
12. What path is always taken by radio waves? _____ When would bearings on such waves plot correctly on charts? _____
13. Does a ground wave fade? _____ Does it vary in strength from day to night? _____
14. What causes night effect? _____ At what times of the day are DF bearings least accurate? _____
15. At what angles are quadrantal errors maximum? _____ How are they corrected? _____
16. Why must the main switch in the radio room be thrown to DF or AA-DF before the DF can be used? _____

29-9 GONIOMETER-TYPE DIRECTION FINDERS

Rotating-loop types of direction finders have been used aboard ships for many years. Such equipment has some disadvantages. The loop and receiver must be relatively close together, with a mechanical connection between the loop and the operating position in the cabin below the loop. The moving parts of such equipment require maintenance to keep them in proper working condition.

A nonrotating, double-loop *goniometer* type of DF is also used. The nonrotating loop can be 100 ft from the DF equipment and be coupled to it by a transmission line. The theory of operation can be explained by reference to the simplified diagram of Fig. 29-7.

Two separate fixed loops are set at right angles to each other, usually on top of a mast, and centered above the navigation bridge. Such a centered position eliminates most of the quadrantal errors to which lower rotating-loop installations

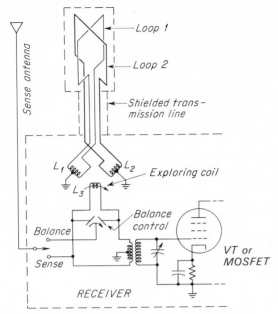

Fig. 29-7 Goniometer loops coupled to a receiver.

received. An indicator arrow is attached to the shaft of the exploring coil and sweeps over a circular compass scale calibrated in 360°.

The goniometer-type DF may be balanced and sensed by methods similar to those used with rotating-loop direction finders. The goniometer coil provides the same function as does a rotating loop. The two types of direction finders are operated similarly when taking bearings.

29-10 RDF RECEIVERS

The receiver used in conjunction with the DF antenna is either a superheterodyne or a TRF. To enable the receiver to operate successfully on weak signals, or signals having no modulation, a BFO is used. Since an aural null is the indication of direction, an AVC circuit is not desirable. An RF gain control is included in the circuit, but no audio volume control would be used.

To attain a better balance, earlier direction finders employed a push-pull input stage in the receiver. Later models use a single-ended input stage.

Some receivers use a form of tuning-eye tube as a visual indicator of the amplitude of the received signal, although earphone or loudspeaker operation is generally considered most suitable. More elaborate land-based or aircraft-type direction finders may utilize a cathode-ray-tube display as a bearing indicator.

On ships having only 110-V dc for power, a 6-V storage battery may be used for the filament supply and either dry batteries or a dynamotor operating from the storage battery as the plate-voltage supply. On ac-powered ships the filaments operate from ac and the plate supply is a rectifier and filter circuit. Transistorized DFs may use self-contained batteries or power from the ship's lines.

29-11 MAINTENANCE OF DIRECTION FINDERS

Besides the usual electrical maintenance of vacuum tubes, batteries, etc., in VT equipment, a DF of the rotating-loop type requires periodic checks for freedom of rotation and noisy slip rings. The rotating-loop mechanism requires greasing or oiling once or twice a year, and the silver-plated slip rings and brushes that connect

are subject. Each loop has its own transmission line that terminates in a pair of fixed coils in the receiver. The two sets of coils are installed at right angles to each other, as are the loops. A rotary exploring, or goniometer, coil L_3 can be adjusted to pick up signals from either L_1 or L_2 or from both simultaneously.

If loop 1 happens to be receiving a null from a station, then loop 2, at right angles, will be receiving a maximum. If the exploring coil is rotated for maximum coupling to L_1, with no signal in loop 1, no signal is picked up by the exploring coil. Since L_3 is 90° from coil L_2, it is completely decoupled and can pick up no signal from this coil either. Therefore, no signal is heard in the receiver. When the exploring coil is rotated in either direction, the coupling between it and L_2 increases and signals are received.

If the transmitted wave approaches at an angle of 45° from both loops, the exploring coil will receive equal-amplitude and opposite-phase signals from L_1 and L_2 when it is midway between them. This results in a zero signal output when the exploring coil is at 45° from L_1 and L_2. Thus, the exploring coil is pointing in the direction of the approaching signal whenever a null is being the rotating loop to the receiver require cleaning with a dry cloth from time to time. The insulating segment between the two tubular sections at the top of the loop must not be painted, or loss of signals will result.

29-12 AN AUTOMATIC DIRECTION FINDER

One of the latest additions to shipboard radio equipment is the solid-state automatic direction finder (ADF). Within 2 s of the time it is turned on and the frequency of a desired station is selected, the indicator swings to the correct bearing of the station. This movement is accomplished by the use of a servomotor rotating a goniometer or exploring coil until a minimum signal is reached. At this point the error signal developed by the received signal drops to a minimum and the indicator stops on the correct bearing.

A block diagram of such an ADF is shown in Fig. 29-8. Assume that a bearing is desired on a station transmitting on 400 kHz. The 0.1 kHz-step-frequency synthesizer local oscillator is set by thumbwheel dials to a reading of 400.0 kHz as indicated. The frequency synthesizer develops a local oscillator frequency of 8 MHz (the IF) + 400 kHz, or 8.4 MHz for the dialed 400-kHz signal. This also electronically switches in the proper 300–550-kHz bandpass filter. Note that all stages up to the mixer are broad-tuned types. Among many signals picked up by the loop antenna is the 400-kHz signal. This is amplified (along with all others) and is fed to a ring-type balanced modulator. The modulating frequency being fed to this circuit is 75 Hz. The result is a 75-Hz-modulated 400-kHz carrier signal, but with the carrier canceled. However, the two 75-Hz sidebands, 400,075 and 399,925 Hz, are fed to the adder circuit. If the exploring coil is picking up a strong signal from the station, strong 75-Hz sidebands of 400 kHz are developed. If the goniometer coil happens to be on the correct bearing, no signal or 75-Hz sidebands will be sent to the adder.

The same 400-kHz signal (plus many others) is also picked up by the sense antenna. This signal is always present, although it is shifted in phase by 90° before being amplified and fed to the adder in order to prevent 180° ambiguity of bearing indications.

The same 75-Hz ac that was fed to the balanced modulator is amplified and excites one of the windings of a two-phase ac servomotor coupled to the goniometer coil. Only when 75-Hz ac is fed to the second phase winding will the servomotor turn. When the goniometer coil is rotated to zero signal by the servo, the servo receives zero 75-Hz ac and the indicator and servo stop. The zero signal bearing is indicated on the compass ring in this manner.

The 400-kHz sense antenna signal may be considered as a carrier that is fed through the whole RF system. It picks up the 75-Hz sidebands as modulation in the adder. The local oscillator converts this carrier and sidebands to 8 MHz, which is then fed through a narrow bandpass filter. In this filter all other signals in the 300–550-kHz band that were amplified all along the system are now rejected. Only the 400-kHz carrier and its sidebands pass through the 8-MHz filter and are amplified further. The 8-MHz IF signal is rectified by a diode detector. The 75-Hz output component is amplified and fed to the second-phase or control winding of the servomotor, producing motor and goniometer coil rotation.

If the signal being received is A3, its modulation is also detected and fed through the AF amplifier to a loudspeaker or to earphones. If the signal is A1, an 8.001-MHz crystal oscillator can be turned on to produce a 1-kHz beat tone which can be heard in the loudspeaker or earphones.

The ADF can be used as a simple receiver by disabling the loop circuits. The sense antenna and its amplifiers feed signals to the mixer. Any station to which the synthesizer frequency-set is dialed can then be heard on the loudspeaker.

ANSWERS TO CHECKUP QUIZ ON PAGE 704

1. (Sharper than maximums) (2) 2. (Figure 8) 3. (Sky waves) (Residual signal, poor or shifting nulls) 4. (Allow E to be induced in wires) 5. (Nulls shift, not complete nulls) 6. (C to ground, C to sense antenna, sense antenna to one side of loop) 7. (Resolve 180° ambiguity) (Unbalance) 8. (Possible 180° ambiguity) (Not if approximate direction is known) 9. (Vertical) (Near loop) 10. (Bends toward shore) 11. (Nulls shift) (Nulls shifted) 12. (Great circle) (If N-S, along equator, or if less than 50 mi) 13. (No) (No, see Chap. 20, Antennas) 14. (Ground plus sky waves) (Sunset, sunrise) 15. (45°, 135°, 225°, 315°) (Calibration, compensation) 16. (Interlocked so radio antennas are in same condition as when DF was calibrated)

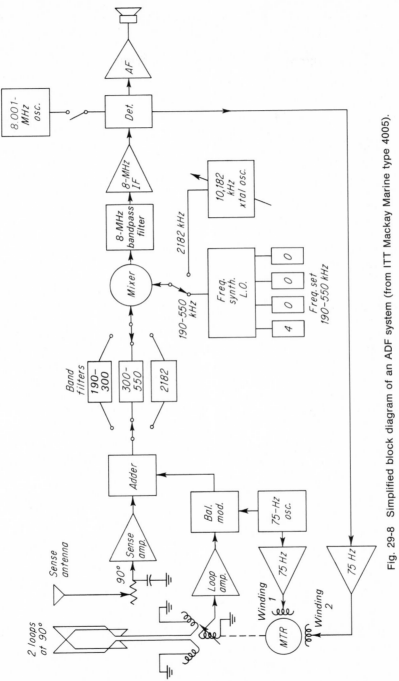

Fig. 29-8 Simplified block diagram of an ADF system (from ITT Mackay Marine type 4005).

If manual control is required (due to static or noises which can confuse the automatic circuits) after determining the approximate bearing automatically, one phase of the 75-Hz ac can be reduced to zero by disabling the sense antenna. The goniometer coil can then be rotated by hand to an exact audible null. The human ear can identify a null through noise far better than an automatic circuit can.

An additional feature of this equipment is its ability to take RDF bearings on the calling and distress frequency of 2182 kHz. This is accomplished by switching from the frequency synthesizer to a 2182-kHz variable crystal oscillator for the local oscillator. This crystal oscillator can be tuned ±15 kHz to allow trimming in case the transmitting station is not exactly on frequency.

29-13 ADCOCK ANTENNAS

The loop-type direction finders are satisfactory for shipboard work because the frequencies used are low and a strong ground wave exists for a considerable distance. At higher frequencies the ground-wave range decreases and a loop becomes inaccurate at distances over 10 to 30 mi because of loss of ground wave, which means that a preponderance of the sky-wave signal is being received.

For shortwave use or for long-distance low-frequency use, an *Adcock* antenna can be used with a DF receiver. It consists of two vertical antennas separated by perhaps 10 to 20 ft. From the centers of the two verticals the signal is coupled to a two-wire horizontal transmission line that is coupled in turn to the receiver, as shown in Fig. 29-9. The transmission line balances out any signals striking it, leaving the vertical portions of the antenna as the only signal-accepting elements. Waves approaching the antenna, either in a direction parallel to the earth or downward from the ionosphere, induce currents in the vertical elements. The sky and ground waves can be out of phase, but accurate bearings can be taken as long as the sky wave has not been changed in travel direction by the ionosphere. By the use of sky waves alone, past the ground-wave range, reasonably accurate bearings can be obtained as long as the ionosphere is relatively constant in density. Because of their large size, Adcock di-

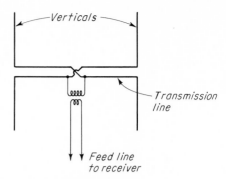

Fig. 29-9 Basic Adcock antenna.

rection finders are usually found on shore stations only.

Rather than rotate such a large antenna, direction finders may be constructed by using two pairs of Adcock verticals installed at right angles and fed to a set of fixed coils similar to a goniometer DF. The exploring coil is the only part that rotates.

29-14 DETERMINING A POSITION BY RDF

A vessel can determine its position by taking a bearing on two shore transmitters whose locations are known and then plotting the bearings on a map, or chart, of the local area. The point on the chart where the two bearings cross is the position of the vessel. For example, a vessel takes a bearing on two shore transmitter stations S_1 and S_2 whose positions are shown on the chart in Fig. 29-10. The bearing of S_1 happens to be due east, or 90° (90° from true north). The bearing of S_2 happens to be due south, or 180° from true north. On the chart, a pencil line is drawn through the position of S_1 at an angle of 90° (east and west). The angle is taken from the *compass rose* on the chart and transferred to the position of S_1 by a device known as *parallel rules*. A line is drawn through S_2 at an angle of 180°. The position indicated by the crossing of the two lines of position is known as a *fix*.

Many lighthouses and light vessels along coastlines have radio-beacon transmitters to allow navigators to take DF bearings on them. During clear weather many of them do not operate, but during foggy weather they are all in operation, allowing navigators to check their positions by this means. These beacon stations transmit an identi-

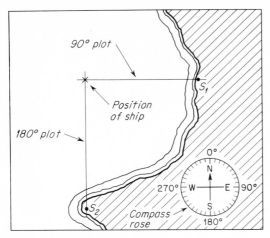

Fig. 29-10 A fix of a ship's position by RDF bearings on two shore stations or by plotting bearings of the ship from the two stations.

fying signal (listed in the publication "Radio Aids to Navigation," aboard all ships) such as dash-dash-dot, or perhaps dot-dot-dash-dot, to enable the navigator to identify the station on which he is taking a bearing.

While a bearing on a single station gives only a *line of position,* it is possible to cruise a few minutes and then take a second bearing on the same station. When these two bearings through the position of the beacon station are properly plotted on a chart and the ship's course and speed and the elapsed time are known, the position of the ship at the time of both bearings can be determined.

The position of a ship can be determined by two RDF stations ashore taking bearings on radio transmissions from the ship. Plotting the two bearings on a chart, as in Fig. 29-10, indicates the position of the ship. Such a service was previously available to ships through the United States Coast Guard but has been discontinued. Many foreign countries still operate RDF stations, however.

Test your understanding; answer these checkup questions.

1. Where is the rotating loop in a goniometer-type RDF system? _____ Must it be balanced? _____
2. Why might goniometer DFs have less quadrantal error than rotating loops? _____
3. Name four methods by which aural nulls may be detected by the operator. _____ _____ _____ _____
4. What are three important points regarding rotating-loop RDF system maintenance? _____ _____ _____
5. How is it that none of the 120–550-kHz ADF circuits are tuned to the desired station? _____ What is the only hand-tuned circuit in the ADF? _____
6. What forms the error signal that actuates the ADF servomotor? _____ What stops the servomotor? _____
7. When would the 8.001-MHz ADF oscillator be used? _____ Is it required during ADF operation? _____
8. What is the local oscillator frequency to receive 500 kHz? _____ To receive 2182 kHz? _____
9. Which ADF antenna(s) is (are) used for manual operation? _____ For simple receiver operation? _____ For ADF operation? _____
10. Does the ADF adder ever receive any loop signals? _____
11. What is heard by earphones in the ADF? _____ Why is an AF signal desirable? _____
12. Why are Adcock DFs better than loops? _____
13. What is the point where two bearings cross on a chart called? _____ How are bearings laid out on a chart? _____ _____
14. If two bearings are determined from a radio beacon or station over a period of a few minutes, what else must be known to determine the ship's position? _____

COMMERCIAL LICENSE QUESTIONS

The following questions refer to Element 6 only. Sections in which questions are answered are shown in parentheses.

1. Describe the construction and operation of a shielded loop antenna as used with a DF. (29-2)
2. What is the directional reception pattern of a loop antenna? (29-2, 29-3)
3. Why are loop antennas, associated with radio direction finders, metallically shielded? (29-2, 29-6)
4. What is the purpose of an auxiliary receiving antenna installed on a vessel which is also fitted with a DF? (29-4)
5. What is the function of the balancing capacitor in a DF? (29-4)

6. What is the principal function of a vertical antenna associated with a unilateral RDF? (29-5)
7. What is the principal function of a vertical antenna associated with a bilateral RDF? (29-5)
8. What figure represents the reception pattern of a properly adjusted unilateral RDF? (29-5)
9. From how many simultaneous directions is a DF capable of receiving signals if adjusted to take unilateral bearings through 360°? (29-5)
10. How is the unilateral effect in a DF obtained? (29-5)
11. What is indicated by the bearing obtained by the use of a unilateral RDF? (29-5)
12. On shipboard, what factors may affect the accuracy of a DF after it has been properly installed, calibrated, and compensated? (29-6, 29-7)
13. What is a *compensator* as used with radio direction finders? What is its purpose? (29-7)
14. Within what frequency-band limits do all United States marine radio-beacon stations operate? (29-8)
15. What is indicated by the bearing obtained by the use of a bilateral RDF? (29-14)
16. Draw a sketch showing how a fix on a ship station can be obtained by taking DF bearings. (29-14)

ANSWERS TO CHECKUP QUIZ ON PAGE 709

1. (*Exploring coil between four loop coils*) (*Yes*) 2. (*Above reflecting objects*) 3. (*Earphones, loudspeaker, CRT, tuning eye*) 4. (*Insulation at top of loop; clean moving contacts; oil bearings of loop*) 5. (*Broadband amplifiers; narrow-pass-band filter selects desired station*) (*2182-kHz signal LO*) 6. (*75-Hz SBs*) (*No 75-Hz SB signals*) 7. (*If A1 signals received*) (*No*) 8. (*8.5 MHz*) (*10.182 MHz*) 9. (*Loop*) (*Sense*) (*Both*) 10. (*No*) 11. (*A3 modulator or 1-kHz beat*) (*Station identification and manual operation*) 12. (*Angle of signal from sky does not matter*) 13. (*Fix*) (*Parallel rules, from compass rose*) 14. (*Speed, elapsed time, and ship's course*)

LORAN, OMEGA, AND FACSIMILE

The objective of this chapter is to discuss (1) the fundamentals of the loran and Omega systems of navigation by reception of radio signals on specialized receivers and (2) a facsimile system that can provide up-to-date maps or charts showing the cloud cover and storm centers, or can transmit photographs or whole pages of text.

30-1 THE LORAN A SYSTEM

Radio direction finders (RDF) have been in general use as a means of navigating by radio since the early 1920s. During the Second World War, a new, entirely different method of radio navigation called *loran* (LOng-RAnge Navigation), was introduced. Whereas the maximum reliable

range of RDF is only 100 or 200 mi, loran is usable for approximately 700 mi during the daytime and 1,400 mi at night.

The loran A system aboard a ship (or airplane) consists of a short vertical antenna, a superheterodyne or TRF receiver, a cathode-ray tube (CRT), a comparing and timing device called an *indicator,* and loran charts. No shipboard transmitter or special rotating antenna is required.

The loran A receiver picks up signals from a *pair* of transmitters located on shore an average of about 300 miles apart. One of these stations is known as the *master* and the other as the *slave.* Both transmit 40-μs pulses of RF energy on the same frequency. Figure 30-1 illustrates a pair of loran transmitters. If both stations transmit their

Fig. 30-1 Basic loran chart showing center and base lines and two 123.6-μs hyperbolic curves if master and slave were transmitted at the same time.

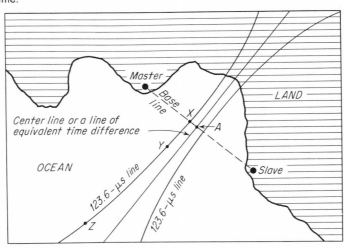

pulses simultaneously, a ship on the base line, exactly midway between them, at point A, will receive the two pulses at the same time. In fact, anywhere on the center line, the pulses will be received simultaneously. Thus, the center line is a line of equal time difference.

Radio waves travel 1 mi in 6.18 μs. If the ship were 10 mi closer to the master than to the center line, at point X, the pulse of the master would arrive 123.6 μs (20 × 6.18) ahead of the slave pulse. At points Y and Z the same difference occurs. If a sufficient number of such similar time-difference points were located and a line intersecting them were drawn, a curve known as a hyperbola would be developed. The center line and such other hyperbolic curves are known as loran lines of position. If a loran receiver indicates a difference in arrival time of 61.8 μs, the ship would be somewhere on the hyperbola of this time difference, although it might be on another, similar hyperbolic curve on the opposite side of the center line. To prevent this last possible ambiguity, the master transmits its pulse first. The slave receives the pulse by the constant-velocity ground wave, delays a fraction of a second, and then transmits its pulse a carefully controlled time later. Each major hyperbolic curve on a loran chart is numbered. This number is the actual number of microseconds time difference between master and slave signals that will be received when on that particular curve.

The pulses of loran A master stations are transmitted at the basic H (high) rate of 33⅓ pps (pulses per second), the L (low) rate of 25 pps, or the S (slow) rate of 20 pps. If the slave is made to transmit its pulse a specific time after the mid-time between master pulses, as shown in Fig. 30-2, it is possible to produce on the loran receiver CRT a display of two horizontal lines. The

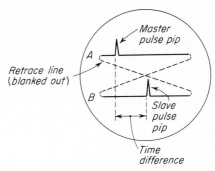

Fig. 30-3 CRT display showing the master pip on the A trace and the slave pip on the B trace.

master pulse is made to appear as a pip on the top line and the slave pulse as a pip on the bottom line (Fig. 30-3). A measurement of the horizontal displacement distance between the leading edges of the two pulses indicates the time difference between the two signals and on which hyperbolic line of position the ship is located.

How the master and slave pulses are properly set on the two CRT traces and what the time difference between them is are the items that must be determined when finding a loran line of position.

A single line of position may be of little use. It is usually necessary to employ another pair of loran transmitters for a second line of position to obtain a fix. Where the two hyperbolic lines of position cross on the special loran chart is the location of the receiver.

In many cases the master is double-pulsed and is operated in conjunction with a second slave station. In this way, only three stations are required for two loran pairs. Figure 30-4 illustrates a few hyperbolic curves for the second slave station S_2. A navigator first locates his ship as being on the curve marked 2H4-3300, for example, then takes another loran reading on the second slave and master, and finds he is on curve 2H3-1600. By reference to his chart he knows that he is at the latitude (marked along the vertical sides of the chart) and longitude (marked along the horizontal sides of the chart) of the point where these curves cross (point Z). When the received time difference falls between two hyperbolic curves on the chart, the navigator must interpolate by penciling in lines to allow a more accurate fix.

Fig. 30-2 Delaying the slave pulse past the midtime causes it always to appear to the right of the master on the CRT.

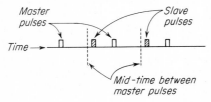

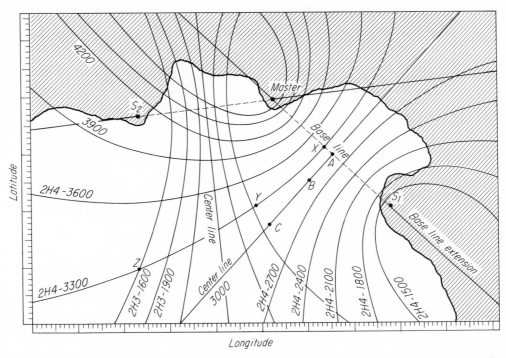

Fig. 30-4 A loran chart using one master to double-pulse two slave stations and showing station identification and time differences on the curves.

30-2 LORAN A STATION DESIGNATIONS

Each pair of loran stations is designated by a three-character identification symbol, such as 2H4. The first number represents the channel on which the transmitters operate:

Channel 1 = 1,950 kHz
Channel 2 = 1,850 kHz
Channel 3 = 1,900 kHz

The letter indicates the *basic pulse recurrence rate,* such as

H = 33⅓ pps   L = 25 pps   S = 20 pps

The last number indicates the *specific pulse recurrence rate* and will be between 0 and 7. When the specific pulse recurrence rate (PRR) is designated as 0, the specific and basic pulse rates are the same. As the numbers increase, the specific rate becomes higher in frequency than the basic rate. For example,

H0 = 33³⁄₉ pps	L0 = 25 pps	S0 = 20 pps
H1 = 33⁴⁄₉ pps	L1 = 25¹⁄₁₆ pps	S1 = 20¹⁄₂₅ pps
H2 = 33⁵⁄₉ pps	L2 = 25²⁄₁₆ pps	S2 = 20²⁄₂₅ pps
H3 = 33⁶⁄₉ pps	L3 = 25³⁄₁₆ pps	S3 = 20³⁄₂₅ pps
H4 = 33⁷⁄₉ pps	L4 = 25⁴⁄₁₆ pps	S4 = 20⁴⁄₂₅ pps
H5 = 33⁸⁄₉ pps	L5 = 25⁵⁄₁₆ pps	S5 = 20⁵⁄₂₅ pps
H6 = 34 pps	L6 = 25⁶⁄₁₆ pps	S6 = 20⁶⁄₂₅ pps
H7 = 34¹⁄₉ pps	L7 = 25⁷⁄₁₆ pps	S7 = 20⁷⁄₂₅ pps

Each curve on a loran chart carries not only the time-difference indication but the master-slave identification symbol, as shown in Fig. 30-4.

The identity of any received loran signal can be determined by adjusting the channel, the basic PRR, and the specific PRR switches until the desired master-slave signals synchronize and stand still on the screen. If the channel switch is set to 1, the basic pulse rate is set to L, and the specific PRR is 5, the station pair being received is 1L5. This is the only type of identification of such a station. On an ordinary receiver, loran emissions sound like a continuously firing machine gun that changes tone slowly.

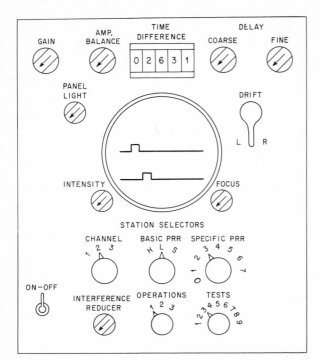

Fig. 30-5 Front-panel controls of a loran receiver indicator.

30-3 DETERMINING TIME DIFFERENCE

Assume that it is desired to determine a line of position when within range of the loran pair 2L3. An example of the front panel of a loran receiver is shown in Fig. 30-5.

The receiver tuning control, a three-position switch, is turned to channel 2. The receiver in this position is fixed-tuned to 1,850 kHz.

The basic-pulse-recurrence-rate panel switch, marked L-H in earlier sets, or H-L-S in newer models, is set to L. This selects a sweep frequency of twice 25 Hz which will produce the two horizontal sweep lines on the CRT. A series of signals will be seen drifting across the two CRT traces.

The seven-position specific-PRR switch is set to 3. This changes the CRT sweep frequency to twice $25\frac{3}{16}$ Hz, and the two signals from the 2L3 loran pair, which are now in synchronism with the sweep frequency, appear to stand still on the screen. Any other loran station signals received will not be in synchronism and will continue to drift. These are ignored by the operator. (In some

equipment the basic and specific PRR controls are combined into one multitapped switch.)

At this point, with the function, or operations, switch in the first position, the screen display may appear as shown in Fig. 30-6a. On the upper, or A, trace, one signal is stopped and a fixed rectangular *pedestal* is seen near the left end of the trace. If both signals are on one trace, the *left-right*, or *drift*, control is pressed until one of the signals is moved to the left-hand end of the A pedestal. If the other signal is now on the B trace but to the left of the signal on the A trace, the master and slave are reversed, and the left-right control must be operated until the other signal, the master, is on the A pedestal. The slave will then be on the B trace, under and to the right of the master.

The left-right control increases or decreases the sweep rate slightly, depending on the direction in which it is pressed. This shifts the position of the received signal pips along the traces.

The B-trace pedestal must now be moved to a position under the slave signal by operating the *coarse delay* control one way or the other. This should result in a display as in Fig. 30-6b, with the slave pip at the left end of the B pedestal. If the signals continue to drift slowly, they can be stopped by adjusting the drift control, which slightly changes the basic oscillator frequency from which the sweep voltages are developed. (If automatic frequency control is included in the equipment, the AFC circuit can be switched on to stop the drift.)

The function, or operation, switch is now set to position 2. This displays only the tops of the two pedestals with their two signal pips. The remainder of the A and B traces are blanked out with special internal circuits. The pip display now appears magnified, as shown in Fig. 30-6c. If the two pulses are not directly above one another, further adjustment of the fine delay control is required. The balance or master and slave gain controls are adjusted until both signals are as nearly the same height as possible. The RF gain control determines the amplitude of the signals fed to the indicator section. A balance control varies the relative signal strengths fed to the two pedestals.

When the function switch is set to position 3, only one trace is displayed. Both signals now

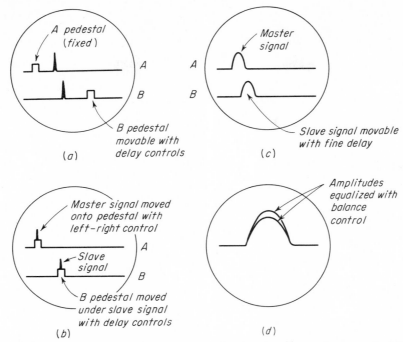

Fig. 30-6 Loran displays. (a) When station is first tuned in. (b) Master signal moved onto pedestal and pedestal moved under slave. (c) Display of pedestals only. (d) Both signals displayed on the same trace and expanded.

appear on one trace. By careful manipulation of the delay control, the leading, or left-hand, edges of the two signals can be made to coincide. Again, the balance and gain are adjusted for equal amplitude of both signals. The cranking of the delay control to position the two pips properly has resulted in the operation of a mechanically coupled (or electronic timing) device. How far the delay control has been moved to make the master and slave signals coincide or superimpose perfectly represents the time difference between the arrivals of the two signals. This time difference is shown by the numbers on the counter (or electronic display). If the number shown is 2100, it indicates that the position of the ship is somewhere on the hyperbolic curve marked 2L3-2100 on the loran chart.

In early loran sets, the basic frequency of oscillation was 100 kHz. Frequency-divider circuits were used to develop the twice 33⅓-Hz and the twice 25-Hz sweep frequencies, as well as 10-, 50-, 500-, and 2,500-μs marker signals. After the master and slave signals were brought to coincidence, a fourth function position was switched

on, which displayed the A and B traces again. This time, along with the pedestals and signal pips, a series of these marker signals was superimposed on the two sweep lines. It was then necessary to count the number of time-marker signals between master and slave pips to determine the time difference. Some of these models may still be in use.

30-4 USING SKY-WAVE SIGNALS

Whenever possible, only ground-wave signals should be used for loran lines of position. Ground-wave signals can be received about 800 mi by day and 600 mi by night. (The noise level increases at night, because of better long-distance reception, although the ground-wave signal remains constant.)

Sky waves travel upward, are refracted from the ionosphere, and return to earth. Because of their longer travel distance, sky waves are received a little later than the ground-wave signals are. Furthermore, the sky-wave signals fade up and down in amplitude, whereas the ground-wave signals remain constant. As another means of identification, sky-wave signals may be the resultant of two or more signals striking the receiving antenna from slightly different points of the ionosphere, and the received pip may vary in width and height on the screen.

Since the sky wave is received some time after the ground wave, it appears as a separate signal to the right of the ground-wave pip. A third signal, a two-hop sky-wave pip, is often seen to the right of the first sky-wave signal, as shown in Fig. 30-7.

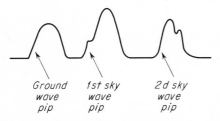

Fig. 30-7 Ground-wave, first sky-wave, and second sky-wave pips, as seen on the CRT.

Only the ground and first sky-wave signals are usable.

When out of ground-wave range, the first sky wave is used. It is important that only the base of the left edges of the two pips be brought into coincidence. When fading is occurring, it may be necessary to watch for a while to determine whether the true left-hand edges are being compared.

Because the sky-wave signal travels a greater distance than does the ground wave, a correction must be added to any reading taken on sky-wave signals. The required corrections are indicated on the loran charts or tables. Sky-wave signals should not be used when within 250 mi of either station, as the corrections are no longer reliable at this range. However, at this distance the ground wave should be satisfactory.

30-5 LORAN TRANSMITTERS

Loran transmitters emit 40-μs pulses. For H-rate pairs the basic recurrence interval is every 30,000 μs; for L-rate pairs, 40,000 μs; and for S-rate pairs, 50,000 μs. Each pulse may have a peak power in excess of 200,000 W, but since the duty cycle (the ratio of time the transmitter is on duty) for an L-rate pair, for example, is only 40/40,000, or 0.001, of the time, the transmitter has an average power output of only about 200 W (200,000 × 0.001).

If any trouble occurs at either the master or the slave station that might impair the accuracy of the pulse timing, the transmitters will be operated ON for a period of about 2 s and then OFF for a like time. This appears to the receiving operator as a blinking signal. Blinking signals must not be used for navigational purposes.

30-6 LORAN RECEIVERS

The loran receiver may consist of a fixed-tuned superheterodyne with an RF amplifier, a mixer, two or more 50-kHz bandwidth IF stages (wide bandwidth is required to accept the sharp-cornered pulse wave plus its many harmonic-frequency sidebands), a detector, and two or more audio or video amplifiers. The output signals from the video amplifiers are coupled to the vertical-deflection plates of a 3-in. CRT, producing the visible pips on the horizontal traces.

The indicator section is quite complicated. It consists of a horizontal sweep generator, a pedestal generator, delay multivibrators to delay the beginning of the pedestal on the trace, a basic oscillator from which the desired sweep rates are developed with frequency-divider circuits or flip-flop counter circuits, sweep speed controls, and amplitude-balance circuits, as well as gating, amplifying, clamping, inverting, and other circuits.

The CRT requires an anode potential of 1,500 to 2,200 V. Caution must be observed by an operator when servicing such equipment to disable the high-voltage circuit, shut off the equipment, then touch a grounded wire or screwdriver to the high-voltage lead. Removal of the type 2X2 high-voltage rectifier will prevent the dc high voltage from being developed but will not de-energize the high-voltage transformer lead to the rectifier. Without high-voltage dc, no display will be visible on the screen but all other circuits should remain operative.

30-7 LORAN C

The newest type of loran is known as loran C. The operation of the receiving equipment is much the same as with loran A, with some exceptions. The frequency of a loran C master and its two or more slaves is 100 kHz. A master transmits nine pulses, eight of which are equally spaced and the ninth doubly spaced. If any transmitter develops difficulties, the ninth pulse is made to blink on and off. It is the first pulse which is moved on to the pedestals and is made to coincide or superimpose to produce the time differential readout marked on the lines of position on loran charts.

Modern loran receivers have a function switch that can select the usual loran A channels 1, 2, and 3, plus a C position for loran C, plus a C-CM position for loran C with cycle-matching. After a loran C master and its desired slave envelopes are superimposed on the CRT, by switching from position C to C-CM, internal circuitry brings the 100-kHz RF cycles of the master and slave pulses into exact phase, resulting in a CRT display of about 15 superimposed cycles. This allows determination of time differential to about 0.1 μs rather than about 1 μs possible by manual alignment of master and slave pulse envelopes. A meter can also be used as an aid. The fluctuations of its needle indicates how far from being cycle-matched two signals are.

Loran C stations have four basic rates: SH, SL, SS, and SC. The first three are half of the H, L, and S rates of loran A, or 16⅔, 12½, and 10 Hz, and the SC rate is 20 Hz. Since only one frequency is used in loran C, identification of a line of position for a given master-slave pair will carry no channel number but might be similar to SSO-Y-35,000, indicating basic rate SS, specific rate O, slave station Y, and 35,000-μs time differential between arrival of master and slave signals.

Whereas loran A operates with signals between 50 and 9,950 μs and requires a four-digit readout, loran C operates with time differentials between 10,050 and 98,000 μs and requires a five-digit readout (often by neon-gas Nixie tubes).

The basic rates available on the modern loran receiver panel will be the H, L, and S of loran A, plus SC, SH, SL, and SS for loran C. The specific rates which may be selected for both types of loran are from 0 to 7.

The function switch may have the same three positions by which master and slave pulses are moved into superposition. With loran C a mode switch is added; by means of it, the operator can select manual operation. AGC is applied to the master signal to hold at least one signal steady, or achieve automatic frequency control.

Whereas loran A slaves operate within about 300 mi of their masters, loran C slaves are 500 to 700 mi from their masters, extending the useful range of loran C. Since there are always two or more slaves with loran C, it is possible to obtain the crossing of two or more lines of position to produce a better fix accuracy. Because lower frequencies have longer ground-wave ranges, loran C can be useful at greater distances than loran A.

Test your understanding; answer these checkup questions.

1. What does loran A mean? _____ What is its range at night? _____ During the day? _____
2. What is the minimum number of loran stations required for a navigational line of position? _____ For a fix? _____
3. What is the distance between loran A masters and slaves? _____ For loran C? _____
4. Besides receiving equipment, what is required to use loran navigation? _____
5. What shape does a loran line of position have? _____
6. What letters indicate the three basic pulse rates? _____ How many loran A specific pulse rates are there? _____
7. In respect to master pulses, when do slaves transmit? _____ When are masters double-pulsed? _____ When do they blink? _____
8. What does 1S5-1600 indicate? _____
9. What does a loran CRT show if the function switch is set to 1? _____ To 2? _____ To 3? _____
10. What does the balance control balance? _____
11. What parts of pulses must coincide? _____
12. Are ground or sky waves preferable? _____ Which extend farther? _____ Which do not change? _____
13. If a pulse appears twice on the CRT, what does it indicate? _____ If three times? _____ If the pulse is deformed? _____
14. What is the frequency of loran C? _____ What are two advantages of loran C over A? _____
15. What letters indicate the basic rates of loran C? _____ What would SL2-W-26,000 mean? _____
16. How many slaves does a loran C master have? _____

30-8 OMEGA RADIO NAVIGATION

The lowest-frequency application of radio transmissions is the Omega navigation system operating on 10.2 kHz. With eight transmitters dispersed around the world, complete global navigation is possible. Each Omega station transmits eight dashes approximately 1 s long in 10 s on 10.2, 11.3, or 13.6 kHz. The first station transmits a 0.9-s 10.2-kHz dash. After 0.2 s wait, the second station transmits a 1.0-s 10.2-kHz dash while the first station is transmitting on one of the other frequencies. After another 0.2-s pause the third station transmits a 1.1-s 10.2-kHz dash while all other stations are transmitting dashes on other frequencies. In this way only one of the eight stations is transmitting the 10.2-kHz signal at any one time. The various other frequencies help identify the stations. With transmissions at 10.2 kHz the usable ground-wave range is thousands of miles.

Omega navigation relies on the fact that 10.2 kHz has a wavelength of 16 nautical miles. By having a 10.2-kHz oscillator in the receiver and comparing this with received dashes from Omega transmitters, each time the ship moves 8 mi nearer to a transmitter the phase of signals being compared changes 180°. By keeping track of how many phase changes occur over a period of hours or days, the receiver will be able to indicate how much closer (or farther away) the ship is to the transmitter. By comparing phases on two Omega stations, a line of position (LOP) can be determined at any time, provided that the proper LOP information was fed into the receiver at the beginning of the voyage and provided that the receiver operates continuously.

The receiver must be turned on at least 30 min before sailing. After 15 min of warm-up the circuits are tested to see that the internal oscillator is synchronized with received dashes, so that the receiver circuits shift from station A to station B signals as those stations transmit their 10.2-kHz signals. The proper hyperbolic LOP of constant phase difference for two receivable stations (A and B, for example) according to the ship's present location is determined. This LOP information is fed into the equipment in numerical form. The hyperbolic LOP for the present ship's location, according to two other stations (C and D, or perhaps B and C), is found and is fed into the receiver. As long as the ship stays at this position, the readout of the LOP of either pair that was fed into the receiver will show on the indicator. It will be some value, such as 756.42. If the readout device is switched to the other station pair, the readout might be something like 321.87. However, as soon as the ship sails, phase changes begin to occur for all stations (all eight are received and information regarding them is continually stored in the receiver memory). After the ship has moved a few miles, the LOP of 756.42 may change to a readout of 758.01. By finding on a chart the LOP for the new number, the navigator can tell on what new LOP the ship is. To get a fix it is necessary to switch the readout to the other pair of stations being used. Where the second LOP crosses the first on the chart is the present position of the ship. If a ship moves out of the usable area of one station, it will be necessary to shift to some other more readable pair.

30-9 FACSIMILE TRANSMISSIONS

A radio-operated device now being used aboard some ships to aid in weather prognostication is a facsimile (fax) recorder. With it, weather charts or maps transmitted by about 50 shore-based stations of more than 20 countries, or relayed down from geostationary-orbit applied technology satellites (ATS) can be printed out aboard ships at sea. Only HF SSB and 135-MHz satellite fax emissions will be discussed.

A facsimile (A4 or F4 emission) picture is roughly comparable to a single frame of a TV picture, except that dark and light portions of each line are laid down across a sheet of special

---

ANSWERS TO CHECKUP QUIZ ON PAGE 717

1. (*MF long range navigation*) (*1,400 mi*) (*700 mi*) 2. (*2*) (*3*) 3. (*About 300 mi*) (*500–700 mi*) 4. (*Loran charts*) 5. (*Hyperbolic*) 6. (*H, L, S*) (*8 or 24*) 7. (*After mid-pulse time*) (*If two slaves*) (*If timing off*) 8. (*Channel 1, slow, specific PRR 5, 1,600 μs between master and slave*) 9. (*Two lines with pips and pedestals*) (*Two lines with tops of pedestals and pips*) (*Pips superimposed*) 10. (*Pulse heights*) 11. (*Left base edges*) 12. (*Ground*) (*Sky*) (*Ground*) 13. (*Ground and sky, or two sky*) (*Ground and two sky*) (*Multipath signals*) 14. (*100 kHz*) (*Longer range and more accurate*) 15. (*SH, SL, SS, SC*) (*Basic SL, specific PRR 2, slave W, 26,0000 μs*) 16. (*Two or more*)

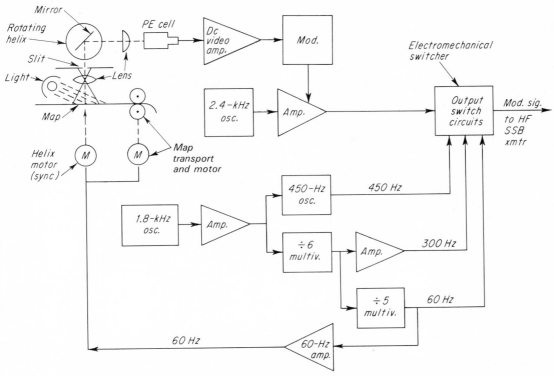

Fig. 30-8 Basic block diagram of facsimile scanner (from Alden model 9165KJL/AN-GXT-2).

paper instead of across a CRT. The dark portions are developed electrochemically by applying a dc through the paper as the line is being scanned. Instead of the one-thirtieth of a second required for a TV frame, it takes about 10 min to produce a fax picture or map.

A basic flat-copy scanner used to develop the video signals that are transmitted is shown in simplified block diagram form in Fig. 30-8. In this system it will be assumed that the standard 120-rpm line scanning rate and the 96 lines per inch (LPI) being traced on the paper are being used. The equipment can be automatically started by radio with the transmission of a 5-s 300-Hz tone and stopped with a 5-s 450-Hz tone.

The picture information is obtained through the use of an ingenious rotating black-coated 19-in.-long glass drum with a single-turn helical scratch from one end to the other. Only through the thin scratched line can light penetrate to the center of the drum. Through a lens system an image of the map is projected on a flat metal surface with

a very narrow slit across it, as shown in the illustration. Any light that passes through the slit falls on the rotating helix. As it turns once, at 120 rpm, the helix allows light-dark variations to pass through the rotating scratch on its surface. These are the video signals for one horizontal line of the map. With mirrors and lenses, the light impulses passing through the helix opening are fed to a photoelectric (PE) cell and are amplified. These video signals are used to modulate a 2.4-kHz carrier, with signals from black areas producing maximum carrier strength and signals from lighter areas developing less carrier strength. During picture transmission times, the video-modulated 2.4-kHz carrier can be fed as AF-range signals to the AF input of a HF SSB transmitter. This is the signal that can be picked up by fax receivers aboard ships.

A 1.8-kHz crystal (or tuning fork) oscillator acts as a frequency standard. Its stabilized output is amplified and used to hold a divide-by-6 multivibrator in synchronism to produce the ac for the

300-Hz start signal. It also locks in a 450-Hz phase-shift oscillator to produce the ac for the stop signal. The 300-Hz ac synchronizes a divide-by-5 multivibrator to produce an exact 60-Hz ac output. The 60-Hz ac is amplified and operates the synchronous helix motor as well as the map transport rollers that pull the map slowly under the video-pickup PE-cell system.

An electromechanical switcher acts as a timing device through cams and contacts to transmit first the 5-s 300-Hz start signal followed by 25 s of phasing signal (95% of each line in black, then 5% of white), then a 1-s burst of 60 Hz, and then the map signals. At the end-of-map transmission the stop signal is switched in for 5 s and the equipment can then shut down.

There can be several choices of transport speeds (LPI), helix rotation speeds (rpm), and carrier frequencies other than the standards mentioned. As an example, if the starting signal is 852 Hz instead of 300 Hz, the helix motor will be switched to rotate at 240 rpm instead of 120.

The video-modulated 2.4-kHz carrier modulates a HF SSB transmitter. This results in a form of FSK, with black signals producing lower frequency sidebands than white signals. (This is reversed with some types of facsimile transmissions.) As a result, a receiver tuned to the transmitter carrier demodulates the video as tones that range from about 2,300 Hz down to about 1,500 Hz, which sounds to a listener as a warbling tone.

30-10 FACSIMILE RECEPTION

The facsimile equipment aboard ships usually has its own HF receiver for SSB map transmissions from shore stations and/or a 135.6-MHz down-link FM receiver to pick up the transmissions being relayed from stationary-orbit satellites parked 22,300 mi above the equator. The satellite transmissions are narrowband FM (10 kHz) and are relatively noise-free. The receiver feeds its signal to a facsimile recorder, such as the Alden* model 519/4A shown in highly simplified block form in Fig. 30-9.

To reproduce maps or pictures, a moist, electrosensitive paper is pulled slowly between a 120-rpm

* Alden Research Center, Westborough, MA 01581.

rotating drum and an essentially stationary ruler-straight steel electrode blade that extends across the width of the 19-in.-wide moving paper. On the nonconductive revolving drum is mounted a single spiral or helical turn of wire. The whole rotating assembly is called a helix, with the wire turn functioning as a second electrode. When paper is between the two electrodes and 60 V is applied, the paper is electrochemically activated, marking the paper at one point (between recording blade and helical wire) with a dark spot. The darkness of the marking is directly proportional to the voltage between electrodes.

The rotation of the helix results in a left-to-right traveling point of contact between helix wire and blade. If a constant 60 V is applied between helix and blade, a dark line is developed across the paper. If the voltage is modulated, the horizontal line will consist of darker and lighter spots. As the right end of the helix wire passes beneath the upper electrode, the left end of the single-turn helix is back at the left side of the paper again, ready to start the next line. Since the paper is constantly being pulled ahead slowly, one line is never laid down over the preceding one.

The detected 1,500- to 2,300-Hz signals from the receiver are first limited, are then passed through a linear-slope filter that passes 1,500 Hz 10 times stronger than 2,300 Hz, and are amplified. This amplified signal has three functions: It is the marking voltage that marks the paper. It can be fed to a speaker for aural monitoring. It feeds control signals to the recorder control system. Through an amplifier it can actuate either 300- or 450-Hz resonant relays to produce start and stop signals for the recorder control. The 25-s phasing signals that follow a start signal can increase the divide-ratio of a logic circuit that normally divides the 2,400-Hz crystal or tuning fork oscillator by 40 to produce 60-Hz ac. The 60-Hz ac drives the synchronous helix motor in exact synchronism with the scanner helix at the transmitter. The phasing signals should reduce the 60-Hz frequency and slow the helix to move the left margin of the map to the left side of the recording paper. If for some reason they do not, a manual control can be actuated a few times to reduce the 60-Hz frequency and move the margin of the map toward the left edge of the paper.

The 1-s 60-Hz synchronizing signal that follows

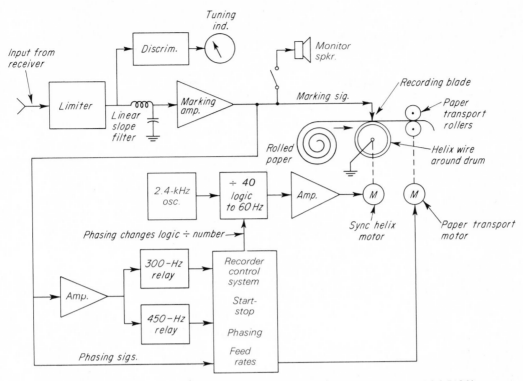

Fig. 30-9 Simplified block diagram of a facsimile recorder (from Alden model 519A).

the phasing signals is not used in the system being discussed, but it can produce map alignment and start-signaling for other systems.

When the map is completed, the 450-Hz stop signal is transmitted, shutting down the transmitter and, if on automatic mode, the recorder circuitry also.

Inasmuch as the received map video signals may be considered as either FM or FSK, with signals varying from about 2,300 to 1,500 Hz, the signals, after limiting, can be fed to a 1,900-Hz discriminator and to a zero-center tuning indicator meter. The signals are tuned in correctly when the meter needle wavers least during map reception.

To prevent rapid wearing due to the helix wire wiping against it through the paper, the blade is made in the form of an endless steel loop moving along at about $\frac{1}{8}$ in. per minute. It should last three to six months with relatively constant use.

The recorder paper feed speed must be set to the same rate as that of the map feed rate of the scanner. This can be determined by published information in U.S. Government Radio Facsimile Publications HO 118A and HO 118B. The rolls of 19-in.-wide electrosensitive paper are packaged in 170-ft lengths.

Transmissions from parked satellites are relayed signals fed up to the satellites from ground stations. These are cloud-cover pictures picked up from polar-orbiting weather satellites overlaid on charts of that portion of the earth's surface. The satellite converts the received map information to 135.6-MHz FM signals and transmits this down to earth. Such a transmission from a single satellite can cover nearly half of the surface of the globe.

Test your understanding: answer these checkup questions.

1. On what frequency are Omega signals measured? _____ How many dashes are sent in 10 s? _____ Omega receivers keep track of how many stations simultaneously? _____
2. What is measured in the Omega system? _____

3. If the Omega value 546.72 is shown, what does this mean? _____
4. Where is a fax scanner found? _____ What are the standard scanning and LPI rates? _____
5. What tone is used to start fax? _____ To stop it? _____
6. What tones carry map information? _____
7. What is 1.8 kHz used for in a scanner? _____ Could it be used as the video-modulated carrier? _____
8. To what is the output of a scanner fed? _____
9. Basically, what is a scanner helix? _____ A recorder helix? _____
10. What does a scanner switcher switch in? _____ _____ _____ _____

11. With what type transmitter is fax transmitted on HF? _____ On VHF? _____
12. The helix wire forms one end of a circuit; what forms the other? _____ What emf is required across them to form a dark spot? _____
13. In a recorder, what is used to make 1,500-Hz ac stronger than 2,300 Hz? _____ To generate 60-Hz ac? _____
14. If the tuning meter read full scale, what would this indicate? _____ Mid-scale? _____ Low-scale? _____
15. Why is the blade an endless loop? _____
16. What should bring the left margin of a map to the left side of the paper in a recorder? _____
17. What is the makeup of the phasing signal? _____

COMMERCIAL LICENSE QUESTIONS

The following questions refer to Element 6 only. Sections in which questions are answered are shown in parentheses.

1. In determining a fix, or position, by a marine loran system, what is the minimum number of land transmitters involved? (30-1)
2. Draw a simple sketch showing relative positions of pairs of master and slave stations of a loran system and indicate lines of position of each pair of stations. (30-1)
3. What is the relation between a master and a slave station in reference to loran systems? (30-1)
4. Explain why pulse emission rather than continuous waves is used by loran transmitters. Approximately what pulse repetition frequency, pulse duration, and operating frequency are used in loran systems? (30-1, 30-2, 30-7)

5. How can the operator of a loran receiver on shipboard identify the transmitting stations that are being received? (30-2)
6. When several pairs of loran transmitting stations are operating on the same frequency, how does the operator at the loran receiver select the desired pair of transmitting stations? (30-2, 30-3)
7. During daytime hours approximately what is the maximum distance in nautical miles from loran transmitting stations that loran lines of position can be determined? (30-4)
8. What is the purpose of blinking in a loran system and how is blinking recognized at the receiver? (30-5)
9. What precautions should an operator or serviceman observe when working with cathode-ray tubes and the associated circuits of loran receivers? (30-6)

31

RADIO LICENSES AND LAWS
FCC ELEMENT 1

The objective of this chapter is to outline some of the basic laws, rules, and regulations that are involved in radio operator licensing, the material required to answer Element 1 of all FCC commercial licenses.

31-1 RADIO LAWS AND REGULATIONS

To prevent intolerable interference, radio stations throughout the world operate according to agreements set up at international communication meetings. In 1934, the United States combined laws of many previous acts and agreements into the Communications Act of 1934. The Communications Act set up general laws to be followed in the United States that would coincide with communication agreements with other countries. To execute and enforce the provisions of this act, the Federal Communications Commission was constituted. The FCC has developed a series of rules and regulations for different types of communications services. The first seven volumes contain information of general radio communication interest and are outlined in Appendix F. Each part is subject to change as the art of radio progresses in its particular field. Thus, a rule regarding broadcast station operation last year may not exist or may be changed this year. It is important that owners and operators of radio systems keep themselves informed about the rules and regulations referring to their particular service.

Any person who willfully does anything prohibited by the Communications Act or knowingly omits to do anything required by the act is subject, upon conviction, to a fine of not more than $10,000 or imprisonment for a period of 1 yr on the first offense and 2 yr on the second offense, or both.

Anyone willfully violating any *rule, regulation, restriction,* or *condition* set up by the FCC by authority of the Communications Act or by international treaty to which the United States is a party is subject, upon conviction, to a fine of not more than $500 for each day during which such offense occurs.

One of the many functions of the FCC is the issuance of operator and station licenses to those qualifying for them. This requires the administration of operator examinations. The FCC also has authority to and does make inspections of licensed United States radio stations of all types whenever necessary to assure operation in accordance with FCC rules and regulations.

Messages transmitted by radio are subject to secrecy provisions of law. No persons receiving such messages (amateur, distress, broadcast, and messages preceded by CQ—"attention all stations"—excepted) may divulge their content to anyone but the legal addressee, his agent, or attorney, nor may they use information so gained to their advantage.

31-2 LICENSE ELEMENTS

All commercial radio operator license or permit examinations given by the FCC will be composed of one or more of the following nine elements:

o *Element 1: Basic Law.* Twenty questions on provisions of laws, treaties, and regulations with which every radio operator should be familiar (covered in this chapter).

- Element 2: Basic Operating Practice. Twenty questions on radio operating procedures and practices in communicating by means of radio-telephone stations (Chap. 32). Element 2 is subdivided so that a candidate may elect to answer questions in the maritime field or questions of a general nature.
- Element 3: Basic Radiotelephone. One hundred questions on technical, legal, and other matters applicable to the operation of radiotelephone stations other than broadcast stations (most chapters up to Chap. 27).
- Element 4: Advanced Radiotelephone. Fifty questions on advanced technical, legal, and other matters applicable to the operation of broadcast stations (most chapters, particularly Chaps. 17 through 25).
- Element 5: Radiotelegraph Operating Practice. Fifty questions on operating procedures and practices for radiotelegraph stations (Chap. 33).
- Element 6: Advanced Radiotelegraph. One hundred questions on technical, legal, and other matters applicable to the operation and practices of all radiotelegraph stations, radio navigational aids, message traffic routing, accounting, etc. (most chapters, particularly Chaps. 16, 28, 29, 30, and 33).
- Element 7: Aircraft Radiotelegraph. (Not covered in this book.)
- Element 8: Ship Radar Techniques. Fifty questions on specialized theory and practice applicable to the installation, servicing, and maintenance of ship radar equipment (Chaps. 26 and 27).
- Element 9: Basic Broadcast. Twenty questions

on specialized elementary theory and practice in operation of standard AM and FM broadcast stations. (Chaps. 19 and 24)

Test questions are of the multiple-choice type.

31-3 LICENSES AND PERMITS

All United States commercial and amateur radio stations must be licensed by the FCC and must be operated by FCC-licensed operators or permittees. (Exclusions: armed-service and government stations and some citizen-band equipment of 100 mW or less.)

All applications for licenses or permits must be accompanied by a small fee. Application for renewals, for endorsements, for duplicates, or for replacement licenses also require a fee.

Four operator licenses are issued by the FCC. The license examination requirements for these are:

- Radiotelephone, First Class. Elements 1, 2, 3, 4. (No code.)
- Radiotelephone, Second Class. Elements 1, 2, 3. (No code.)
- Radiotelegraph, First Class. Elements 1, 2, 5, 6. Code test: 20 code groups per minute and 25 wpm (words per minute) plain language.
- Radiotelegraph, Second Class. Elements 1, 2, 5, 6. Code test: 16 code groups per minute and 20 wpm plain language.

Two lower-grade operating permits are issued. Holders are not allowed to make any adjustments that might result in improper transmitter operation unless these are made under the direct supervision of a licensed operator. These permits are:

- Radiotelephone, Third Class Operator Permit. Elements 1, 2. (No code.)
- Radiotelegraph, Third Class Operator Permit. Elements 1, 2, 5. Code test: 16 code groups per minute and 20 wpm plain language.

The lowest-grade permit is the Restricted Radio-Telephone Operator Permit, requiring no test but only a certification in writing that the applicant requires the permit, can receive and transmit spoken messages in English, is familiar with rules and regulations, and can keep a log. Such permits

---

ANSWERS TO CHECKUP QUIZ ON PAGE 721

1. (10.2 kHz) (8) (8)   2. (Phase and phase reversals) 3. (546.72 phase reversals to station)   4. (At origin of fax signals)(120.96)   5. (300 Hz)(450 Hz)   6. (1,500–2,300 Hz) 7. (Develop 300, 450, and 60 Hz) (Yes)   8. (Transmitter) 9. (Opaque-sided drum with one-turn scratch) (Insulated drum with one-turn wire)   10. (300Hz, phasing signal, 450 Hz, 60 Hz, map signals)   11. (SSB) (FM)   12. (Blade) (60 V)   13. (Linear filter)(2,4-kHz oscillator and logic dividers) 14. (1,500 signal or mistuned) (1,900 or average) (2,300 signal or mistuned)   15. (Wears slower)   16. (Phasing signals, or manual switch)   17. (25 s of solid black lines 95% across paper)

are used by taxi drivers, policemen, etc. Chapters 31 and 32 are adequate for such operations.

Element 8 is a radar *endorsement* applicable to any license. If the *endorsement* is taken with a license, there is no fee. Element 9 is an endorsement applicable to a Radiotelephone Third Class Operator permit.

All licenses and permits require an ability to transmit and receive spoken messages in English and are issued only to citizens of the United States. An applicant must be twenty-one years of age to be eligible for a Radiotelegraph First Class license examination and at least fourteen years old for a Restricted Radiotelephone Operator permit.

To obtain a commercial license it is necessary to take a test at one of the field engineering offices of the FCC, located in many of the larger cities of the country. An application form must be filled out before the license examination will be given. Most offices set specific days for certain types of license examinations. This information should be obtained by phone or mail before appearing at the nearest office for an examination. An FCC form 756 is required to take an examination.

If when taking any commercial license examination an element is failed (grade less than 75%), no test containing that element may be taken for 2 months (60 days).

Licenses are normally issued for a period of 5 yr. To renew a license, a renewal application may be obtained by mail and filled out. The original license and the renewal application may then be presented to the nearest office in person or by mail, within the last year of the license term, or within a 1-yr grace period after its expiration. Temporarily, renewal applications need not be supported by documentary evidence describing in detail the service performed and showing that the applicant actually operated in a satisfactory manner. During the period when the license and the renewal application are in the mail or at the FCC office, the operator may continue operating by posting an exact, signed copy of his renewal application in lieu of the original license or permit.

The license of an operator normally must be posted at the place where he is on duty. If he is working at two different stations, he may post his license or permit at one station and a duly issued *verified statement* (FCC Form 759) at the other.

A license or permit holder may apply for a *verification card* (FCC Form 758-F). This card may be carried on the person of the operator in lieu of the original license or permit when operating any station at which posting of an operator license is not required, provided the license or permit itself is reasonably accessible if needed.

If a license or permit is lost, mutilated, or destroyed, the FCC should be notified and a duplicate license or permit requested. If the license is lost, the application for duplicate should state that reasonable search has been made and that, if the license is found, either the original or duplicate will be returned for cancellation. The applicant should also submit documentary evidence of the service (experience) that has been obtained under the original license or permit or a statement under oath or affirmation embodying that information. A signed copy of the application for duplicate license must be used in lieu of the license until the duplicate is issued.

When a licensee qualifies for a higher-grade FCC license (or permit), the lower-grade license must be turned in to the FCC to be marked "canceled." The canceled license is then returned to the licensee.

31-4 SUSPENDED LICENSES

The FCC has authority to suspend the license of any operator upon proof that he has violated a provision of acts, treaties, or FCC rules; failed to carry out a lawful order of the master or person lawfully in charge of the ship or aircraft on which he is employed; willfully damaged or permitted radio apparatus to be damaged; transmitted superfluous radio communications or signals, obscene language, false or deceptive signals, or call signals not assigned by proper authority to the station he is operating; willfully or maliciously interfered with any other radio communications; or assisted another to obtain or himself obtained an operator's license by fraudulent means.

Such a suspension becomes effective 15 days after the licensee receives the notice of suspension. In this 15-day period the licensee may make application for a hearing on the suspension order. The suspension will be held in abeyance until the conclusion of the hearing, and the licensee may continue to operate until that time at least.

31-5 NOTICES OF VIOLATIONS

The FCC maintains several monitoring stations throughout the country. The operators at these stations spend their time listening to and checking all receivable signals. If a radio station appears to have violated any provision of the Communications Act or FCC rules and regulations, it will be served with a *notice of violation* calling the facts to the station's attention and requesting a written statement concerning the matter within 10 days (unless another period is specified). The answer must be addressed to the office of the FCC originating the notice and be a full explanation of what occurred, what steps have been taken to prevent future violations or what new apparatus has been or will be installed, and the name and license number of the operator in charge if the notice of violation relates to lack of attention or to improper operation of the transmitter. If an answer cannot be sent or an acknowledgment made within the required period by reason of illness or other unavoidable circumstances, acknowledgment and answer must be made at the earliest practical time, along with a satisfactory explanation of the delay.

31-6 WHO MAY OPERATE TRANSMITTERS

Any transmitter employing radiotelegraph (International Morse code) must be operated by a person having a suitable radiotelegraph license or permit.

Transmitters employing a microphone for communication are normally expected to be operated by persons holding a suitable radiotelephone and, in many cases, radiotelegraph license or permit. However, a nonlicensed person may speak over a microphone in some services, provided that a licensed operator is in control of the transmitting equipment. In the Public Safety Radio Service (police, fire, etc.), an unlicensed person, such as a patrolman, may operate a mobile station during the course of normal rendition of service, after having been authorized to do so by the station licensee.

Radiotelegraph licenses and permits are used by operators on ship radiotelegraph and radiotelephone stations, coastal radiotelegraph stations communicating with ships, aircraft radiotelegraph stations, aeronautical radiotelegraph stations communicating with aircraft stations, police radiotelegraph zone or interzone point-to-point stations, or any other services employing radiotelegraph.

Radiotelephone First Class licenses are required in broadcast, FM, and TV stations. Almost all other radiotelephone communications, such as experimental TV, police, fire, forestry, highway-maintenance, special-emergency, aircraft, aeronautical, power, petroleum, forest-products, motion-picture, relay-press, industrial, motor-carrier, railroad, taxicab, and automobile radio services, as well as experimental broadcasts, require only a Radiotelephone Second Class license to operate and, in most cases, only a Third Class Operator permit.

31-7 WHO MAY SERVICE OR ADJUST TRANSMITTERS

Except for standard broadcast, FM, and TV stations, which require Radiotelephone First Class license holders, radiotelephone transmitters in other services, listed above, may be serviced, tuned, or adjusted by holders of Radiotelephone and, in some cases, Radiotelegraph Second Class *licenses*. Permit holders are not authorized to make technical adjustments to any transmitters.

Radiotelegraph transmitters (such as shipboard, coastal, aircraft, aeronautical, zone, etc.) cannot be serviced, tuned, or adjusted by radiotelephone license holders. Technical adjustments must be made by radiotelegraph license holders.

No person holding only an amateur license is authorized to operate or make technical adjustments to commercial radio transmitting equipment (except limited operation of disaster communication services stations).

No person holding only a commercial license or permit is authorized to operate any amateur radio transmitting equipment.

31-8 CLASSIFICATION OF COMMUNICATIONS

The normal transmissions from radio stations may be classified as *routine*. Transmissions made during times of emergency or to prevent possible disasters, by reason of their importance, demand a higher priority. The order of high-priority radio transmissions or messages is (1) *distress,* (2) *urgency,* (3) *safety.*

A mobile station in *distress* is in need of immediate assistance. By radiotelegraphy the distress signal is the transmission of the signal SOS (... — — — ...) sent as one character with no spacing between letters. By radiotelephone the distress signal is the word MAYDAY (from the French *m'aider*), usually transmitted three or more times to attract attention. A distress message should contain the name of the station, particulars of its position, nature of the distress, kind of assistance desired, and any other information which might facilitate rescue. The mobile station in distress is responsible for the control of distress-message traffic. However, if the station in distress is not itself in a position to transmit the distress message or if the master of a station observing the one in distress believes that further help is necessary, then the observing station can send a distress message. Distress messages are not subject to the secrecy provisions of law, as are most other radio communications. Any false or fraudulent signals of distress are prohibited by law and punishable by a fine of not more than $10,000 and not more than 1 yr in prison.

Radio messages with an *urgency* classification refer to a situation that requires immediate attention and might conceivably become distress in nature. By radiotelegraphy the urgency signal is the transmission of the three letters XXX. By radiotelephone the urgency signal is the spoken word PAN repeated three or more times.

Radio communications with a *safety* classification refer to meteorological information, particularly information about storms, hurricanes, etc., or to other navigational warnings. By radiotelegraphy the safety signal is the transmission of the three letters TTT. By radiotelephone the safety signal is the word SECURITY (from the French *sécurité*).

The 1959 Geneva Treaty designates any emission which endangers the functioning of a radio navigation service or other safety service or degrades or obstructs or interrupts a radio communication service as *harmful interference* and subject to legal action.

31-9 LOGS

All communication systems are required to keep logs of their transmissions. Log entries normally show the date, time, operator on duty, station with which the communication was carried on, and an indication of what communications occurred or what traffic was handled.

Logs are made out by those legally competent to do so. If an error is made in a log, it may not be erased. The error should be struck out, the correct entry made above the error, and then the correction initialed and dated by the operator who made the original entry. Logs are usually required to be kept for at least 1 yr. If they contain information pertaining to distress traffic, they should be kept for at least 3 yr. If they contain information pertaining to an investigation being made by the FCC, they must be kept until authority to destroy them is received in writing.

Logs are also discussed in Chaps. 19, 24, 25, 32, 33, and 34.

Test your understanding; answer these checkup questions.

1. Is the Communications Act of 1934 a national or international set of laws? _____
2. What is the penalty for violating a rule, regulation, restriction, or condition set up by the FCC? _____ Of willfully doing anything prohibited by the Communications Act? _____
3. What four types of radio messages are not subject to secrecy provisions of law? _____ _____
4. What FCC test element is not covered in this book? _____
5. What are the four operator licenses issued by the FCC? _____ _____ _____ _____ The two permits? _____ _____ The lowest-grade permit? _____
6. What is the lowest passing grade for any FCC element? _____
7. Which FCC elements are endorsements for licenses? _____
8. If an FCC commercial license test is failed, how long must the applicant wait before retaking? _____
9. Is it possible to hold two grades of radiotelephone licenses simultaneously? _____ A telephone and telegraph simultaneously? _____
10. If an operator is served with a suspension, when is the suspension effective? _____
11. Can a First Class telegraph license holder tune or operate an amateur station? _____
12. What is the order of high-priority radio transmissions or messages? _____ _____ _____

13. What are the telegraph indicators of the three highest priority types of messages? _____ _____ _____ The telephone? _____ _____ _____

14. Under what condition may incorrect entries in a radio log be erased and corrected? _____

15. What is an FCC form 400? _____

COMMERCIAL LICENSE QUESTIONS: ELEMENT 1

Applicants for all commercial licenses should know the answers to all these Element 1 questions. Sections in which questions are answered are shown in parentheses.

1. What government agency inspects radio stations in the United States? (31-1)
2. When may an operator divulge the contents of an intercepted message? (31-1, 31-8)
3. Under what conditions may messages be rebroadcast? (31-1)
4. What are the penalties provided for violating (a) a provision of the Communications Act of 1934 and (b) a rule of the FCC? (31-1)
5. Where and how are FCC licenses and permits obtained? (31-3)
6. Who may apply for an FCC license? (31-3)
7. If a license or permit is lost, what action must be taken by the operator? (31-3)
8. What is the usual license term for radio operator licenses? (31-3)
9. When may a license be renewed? (31-3)
10. When a licensee qualifies for a higher grade of FCC license or permit, what happens to the lesser-grade license? (31-3)
11. What messages and signals may not be transmitted? (31-4)
12. May an operator deliberately interfere with any radio communication or signal? (31-4, 31-8)
13. If a licensee receives a notice of suspension of his license, what must he do? (31-4)
14. What are the grounds for suspension of operator licenses? (31-4)
15. If a licensee is notified that he has violated an FCC rule or a provision of the Communications Act of 1934, what must he do? (31-5)
16. What type of communication has top priority in the mobile service? (31-8)
17. What is meant by *harmful interference?* (31-8)
18. Who keeps the station logs? (31-9)
19. Who corrects station logs? (31-9)
20. How may errors in the station logs be corrected? (31-9)

32 VOICE COMMUNICATIONS
FCC ELEMENT II

It is the objective of this chapter to present the basic radiotelephone operating practices and procedures included in Element 2 of all FCC commercial license or permit tests. Candidates for licenses can elect to answer (1) some basic questions plus questions in the marine field or (2) more detailed questions of a general nature. Sections containing general information carry a (G) following the section heading.

32-1 RADIOTELEPHONE OPERATION (G)

Radiotelephone communications take place from ship to ship, ship to shore, land station to land station, mobile to mobile, mobile to land station, aircraft to ground, and so on.

Police, fire, and taxi services represent one type of communication. Mobiles talk to their base station or to other mobile units of the same system. The base station may be on one frequency, the mobiles on another, or each station may have two or more frequencies on which it may operate. The base station has an assigned call, such as KMA539, but the mobile units may be known only as car 216 or some such designation. Logs of communications are kept at the base station only.

Many commercial vessels, fishing boats, and pleasure craft use the 2,000–3,000-kHz or the 156–162-MHz VHF bands for radiotelephonic communications from ship to ship or from ship to shore (Sec. 28-10). These services should use the more formal marine procedures outlined below, each station maintaining watch on the calling frequency, 2,182 kHz or 156.8 MHz.

Every licensed station is assigned a *call* to distinguish it from all other stations. The call may consist of letters only or a combination of letters and numbers. When signing a call, the operator should do so clearly and distinctly to prevent the possibility of confusing one station with some other station.

Since there are actually hundreds of thousands of mobile and fixed stations in communication every day, it is important that some form of logical procedure be followed to reduce confusion and interference. Rules, regulations, and accepted procedures of good operating are outlined in this chapter.

32-2 MESSAGE PRIORITIES

Urgent types of messages take preference over routine traffic. Wherever applicable, message priority is

1. Distress calls, messages, and traffic
2. Communications preceded by the urgency signal
3. Communications preceded by the safety signal
4. Communications relative to radio direction finding
5. Messages relative to the navigation of aircraft
6. Messages relative to navigation, movements, and needs of ships and official weather-observation messages
7. Government messages for which priority has been claimed
8. Service messages relating to previously transmitted messages or to the operation of the communication service
9. All other routine communications

32-3 DISTRESS

A distress call has absolute priority over other transmissions. Stations hearing it must immediately cease any transmissions capable of interfering with it and must listen on the frequency used for the distress call. A *distress* call sent by radiotelephony consists of

1. The distress signal MAYDAY spoken three times
2. The words THIS IS, followed by the identification of the mobile station in distress, the whole being repeated three times

The *distress* message must follow the distress call as soon as possible and should contain

1. The distress call
2. The name of the ship, aircraft, or vehicle in distress
3. Position, nature of distress, and kind of assistance desired
4. Any other information which might facilitate rescue

The position is given in latitude and longitude or by bearing in degrees from true north and distance in miles from a known geographical point.

An operator hearing a distress call must listen on the distress frequency for the distress message and must immediately acknowledge receipt of the distress message if the mobile station in distress is in the vicinity. An acknowledgment by radiotelephone might be (1) call of the station in distress three times, (2) the words THIS IS, followed by the call of the acknowledging station three times, (3) the words ROGER YOUR MAYDAY MESSAGE, (4) the word OUT.

Every mobile station which acknowledges receipt of a distress message must, on the order of the person responsible for the ship, aircraft, or other vehicle, transmit as soon as possible (1) its name, (2) its position at the present time, and (3) the speed at which it is proceeding toward the station in distress.

A station must ensure that it will not interfere with emissions of other stations better situated to render assistance.

Any operator in the mobile service who has knowledge of distress traffic must follow such traffic, even if he does not take part in it. For the duration of the distress traffic no station must use the distress frequency for other types of calls or traffic. If a transmitting operator is told that he is interfering with distress traffic, he must cease transmitting immediately and listen for distress signals.

An operator situated in such a position that he cannot assist in a distress message must take all possible steps to attract the attention of stations which might be in a position to render assistance.

Stations not directly involved in distress traffic may continue normal service on frequencies that will not interfere with the distress traffic after the distress traffic has been observed to be well established.

Any land station receiving a distress message must, without delay, advise authorities who might participate or be participating in rescue operations. Otherwise, the land station must maintain silence on the distress frequency unless involved in the distress traffic. If it appears that the distress call and message have not been acknowledged, all steps should be taken to attract the attention of stations in position to render assistance.

If a radio watch is required on a distress frequency, it is desirable that the receiver be tuned to this frequency as soon as traffic has been completed on other frequencies.

When distress traffic is ended, an announcement should be made on the distress frequency, such as (1) the words MAYDAY ALL STATIONS, three times, (2) the words THIS IS, followed by the call letters of the station transmitting the message, (3) the name of the station in distress, (4) the words DISTRESS TRAFFIC ENDED. OUT.

32-4 URGENCY SIGNALS

The urgency signal by radiotelephone is PAN, spoken three times, followed by the words THIS is and the call of the transmitting station. It indicates an urgent message, one which concerns the safety of a ship or person but is not quite of distress priority, is to follow.

Mobile stations hearing the urgency signal must continue to listen for at least 3 min. If no urgency message is heard, they may then resume their normal service. Operations under urgency conditions are similar to distress traffic.

32-5 SAFETY MESSAGES

The *safety signal* by radiotelephone is SECURITY, spoken three times, followed by the words THIS IS and the call of the transmitting station three times. It indicates a storm warning, danger to navigation, or other navigational-aid message is to follow.

All stations hearing the safety signal must continue to listen on this frequency until they are satisfied that the message is of no importance to them. They must make no transmissions likely to interfere with the message.

A *safety communication* is not necessarily the same as a message following the safety signal. A safety communication pertains to any distress, urgency, or safety messages which, if delayed in transmission or reception, might adversely affect the safety of life or property. Stations handling paid radio messages cannot charge for forwarding any safety-communication messages.

32-6 INTELLIGIBILITY (G)

Communication by radiotelephone may be hampered by static, fading signals, interference due to other stations, noise in the receiving room, noise picked up by the transmitting microphone, unusual voice accents, colloquialisms, improper enunciation or pronounciation of words, and by speaking too fast. To improve intelligibility at the receiving end, the transmitting operator should speak slowly and clearly, using well-known words and phrases and simple language. Unusual or important words should be repeated or spelled out if it is known that the receiving operator is experiencing any difficulty in reception.

Speaking too far from the microphone may result in weak, hard-to-hear signals. Shouting into the microphone produces a distorted output signal that may be difficult to understand even with perfect reception conditions. Most communication microphones are constructed for close talking but in a normal tone of voice. If there is considerable talking or local noise in the area of the microphone, it may help intelligibility to cup the hands around the microphone and speak directly into the microphone in a moderate voice. Directing the front of the microphone away from noise sources may also help.

Distortion of the voice is produced by fading signals and by improper functioning of the transmitting circuits. In the latter case, the fault must be found by a licensed operator or serviceman. In many cases, a distorted transmission is readable if the operator speaks slowly and distinctly.

32-7 PHONETIC ALPHABET (G)

When words are spelled out in a radiotelephone communication, considerable confusion can result because many letters of the alphabet sound similar unless clearly heard. For example, the letters B, C, D, E, G, P, T, V, and Z all have the same *ee* ending sound. The word *get,* when spelled out, might be copied as b-e-t, p-e-t, b-e-d, etc. To prevent this confusion, each letter of the alphabet may be represented by a well-known word. Thus, *golf* for G, *echo* for E, and *tango* for T might be used. When the word *get* is spelled out, using this phonetic alphabet, it is spoken "golf echo tango." The receiving operator writes down the first letter of each word and receives "get."

There have been many phonetic alphabets in the past, using names, cities, and other words. Unfortunately, non-English-speaking people mispronounced English words so badly that confusion still resulted. An international phonetic alphabet using words that are familiar to most languages (Table 32-1) has been selected.

Table 32-1 *INTERNATIONAL PHONETIC ALPHABET*

A	Alpha	J	Juliet	S	Sierra
B	Bravo	K	Kilo	T	Tango
C	Coca	L	Lima	U	Union
D	Delta	M	Metro	V	Victor
E	Echo	N	Nectar	W	Whiskey
F	Foxtrot	O	Oscar	X	Extra
G	Golf	P	Papa	Y	Yankee
H	Hotel	Q	Quebec	Z	Zulu
I	India	R	Romeo		

32-8 OPERATIONAL WORDS (G)

In the course of radiotelephone operation many special words or phrases have a definite meaning and are desirable to use. A list of some of these follows:

Words	Meaning
Roger	I received your message.
Over	I have completed transmitting and await your reply.
Go ahead	Same as over.
Out	I have completed my communication and do not expect to transmit again.
Clear	I have no further traffic. (Sometimes used in place of out.)
Stand by	Wait for another call or further instructions.
Break	I am changing from one part of the message to another [address to text . . .]. (Also used to request the receiving operator to indicate if he has received the portion of the message transmitted thus far.)
Words twice	Transmit each word or phrase twice, or, I will transmit each word or phrase twice.
Read back	Read the message back to me.
I spell	I will spell [usually phonetically] the word I just said.
Say again all after	Repeat all words transmitted after . . . [give last correctly received word].
Say again all before	Repeat all words transmitted before . . . [give first correctly received word].
Say again . . . to . . .	Repeat all words transmitted from . . . [word before missing portion] to . . . [word after missing portion].

32-9 CALLING AND WORKING

Station KBBB wishes to communicate with station KAAA. KBBB transmits on a frequency known to be monitored by KAAA, saying,

KAAA KAAA KAAA   THIS IS   KBBB KBBB KBBB   OVER

The called station answers, saying,

KBBB   THIS IS   KAAA OVER (or GO AHEAD)

Note that only on the first call is it necessary to call three times and sign three times. The fewer calls the better.

A formal type of message from KBBB to KAAA might be transmitted as follows:

KAAA   THIS IS   KBBB. MESSAGE NUMBER ONE FROM SAN FRANCISCO FEBRUARY TWELFTH NINETHIRTY AM. BREAK. MACPHERSON UNIT TWENTYTHREE. BREAK. ADVISE EXPECTED ARRIVAL SAN FRANCISCO. BREAK. SIGNED WILLIAMS. BREAK (or END OF MESSAGE). THIS IS KBBB. OUT.

The operator at KAAA should copy the message in a form similar to:

KBBB NR 1 SAN FRANCISCO ⟋Preamble
                FEBRUARY 12 1975 9:30 AM
MACPHERSON ←Address   Text
     UNIT 23   ↙
ADVISE EXPECTED ARRIVAL SANFRANCISCO
              Signature →WILLIAMS
              Service →9:50A LC

Note the terminology for the different parts of the message. The service indicates the time the operator (initials) received the message.

The receiving operator acknowledges the message and also signs out:

KBBB   THIS IS   KAAA.   ROGER YOUR MESSAGE NUMBER ONE.   OUT.

After contact has been established, continuous two-way communication is usually desirable without identification on each transmission until termination of the contact (unless of 15 min duration or more).

A mobile station calling a particular station by radiotelephone must not continue for a period of more than 30 s in each instance. If the called station is not heard to reply, it should not be called again until after an interval of 1 min (emergencies excepted). A coast station must not call for more than 1 min and must wait 3 min between calls.

If it is desired to call a vessel within sight when its identity is not known, an operator may call on 2,182 kHz, saying, in effect, ''Calling the green, two-masted yacht passing Point Conception. [Three times.] This is WMBD WMBD WMBD. Over.'' If the yacht hears, it should answer on the same frequency.

In heavy radio traffic, the duration of communications between two stations should not exceed 5 min (excluding distress or emergency communications). Calls should be spaced to allow other stations to communicate.

32-10 RADIOTELEPHONE STATION IDENTIFICATION

Both mobile and fixed stations should identify themselves with their FCC assigned call letters in English, with modulated code transmitted by a licensed telegraph operator, or with an approved automatic device.

If a mobile station has no assigned call letters, it must use its full name, transmitted in English, as "Yacht Cleopatra." Coast stations may identify themselves by location, as "Washington marine operator," if this has been approved.

Mobile or coastal stations should identify themselves at the beginning and end of communications as well as at the beginning and end of test or other transmissions. Mobiles must transmit their call letters at intervals not exceeding 15 min whenever transmission is sustained for a long period. Stations in the Public Safety Radio Service (fire, police) transmit identifying calls at the end of each exchange of transmissions, or at 30-min intervals, as the licensee prefers.

32-11 GOOD OPERATING PRACTICES (G)

Before making a call, or testing, an operator should monitor the frequency on which he is going to transmit to ensure that interference will not be caused to communications which may be already in progress. Under normal conditions an operator should never transmit on a frequency he cannot monitor with a receiver, as he may interrupt or interfere with important communications on that frequency.

To allow maximum use of a frequency or channel and to prevent possible interference to emergency messages, all communications should be as brief as possible and all unnecessary calls and transmissions should be avoided. Transmitters and receivers should be capable of changing frequency rapidly and should be in constant readiness to make or answer a call.

With routine-type messages, if receiving and transmitting conditions are poor and difficulty is experienced in communicating because of static, fading, and interference, it may be better to wait for improved conditions rather than tie up a frequency with slow-moving nonemergency traffic.

A radiotelephone transmitter should be tested at least once a day to assure proper operation of the equipment. If no other communications are made during a day, a suitable test may consist of turning on the transmitter briefly, saying, "This is [call letters] testing." If all meters indicate normal values, it is assumed the transmitter is operating properly. However, an operator should not press the push-to-talk switch on a transmitter except when he intends to speak into the microphone. Radiation from a transmitter may cause interference even when voice signals are not being transmitted.

An operator must always operate his station according to the provisions of his station license except during distress or emergency. Even then, the station should not use unauthorized frequencies or higher power than has been authorized.

When two or more groups of stations are sharing the use of one frequency or channel, it is good practice to leave an interval between transmissions in case one of the other sharing stations desires to transmit emergency traffic.

It must be remembered that radiotelephone transmissions may be received by many unauthorized persons and are not confidential. It is sometimes necessary to choose the phrasing of messages carefully to attain a desired secrecy of meaning.

Anyone may speak over the microphone, provided a licensed operator exercises control and ensures the proper operation of the station. A licensed operator must remember that he is responsible for any transmissions made by a station under his control. If anyone uses obscene language, it is the operator's duty to stop such transmission immediately.

Operators must take all steps necessary to prevent unauthorized use of transmitting equipment. Transmitters in mobile stations in a public place must not be left unattended. The transmitter should be turned off when an operator leaves a car.

The license of the operator of a radiotelephone station should be posted in plain view in the operating room. The 1959 Geneva Treaty states that this license will be available to authorities of any country in which the ship or vehicle is operating.

As in other communication services, an accurate log must be kept of all transmissions made, tests or maintenance performed, messages handled, and distress, urgency, and safety messages heard. Each sheet is to be serially numbered and signed by the operator or person authorized to do so. Logs for international voyages are kept in GMT or universal coordinated time (UTC); others may be in local time or local 24-h time (1:30 P.M. being 1330 h).

32-12 CALLING AND WORKING FREQUENCIES

In the Maritime Mobile Radio Service *calling* and *working* frequencies are used. Original contacts between two stations are made on a calling frequency, such as the international radiotelephone distress and general calling frequency of 2,182 kHz. This frequency is reserved for short calls and answers or for distress, urgency, or safety communications. However, it may be used for short tests of equipment or to broadcast short lists of stations for which a coastal station has traffic.

Normally, as soon as contact between two stations has been accomplished on the calling frequency, operations are shifted to a correspondence channel (or "working frequency") on which all routine-type messages must be transmitted. It is good practice to listen first on the working frequency to see if it is in use before making contact with a station on the calling frequency. Communications in the 2,000- to 3,000-kHz range must consist of safety or maritime traffic, never social or personal messages.

The frequency 156.8 MHz is known as the short-range international radiotelephone frequency for calling, safety, intership, and harbor-control purposes for the maritime mobiles in the 156–162-MHz band.

To contact a coast station (KSA) for routine traffic, the operator (of WPNB) would first check the working frequency of the coast station, next the calling frequency, and then call on 2,182 kHz, saying,

KSA KSA KSA THIS IS WPNB WPNB WPNB ANSWER ON . . . . . (Frequency) OVER.

As soon as KSA has acknowledged the call, WPNB shifts his transmitter to the frequency he has designated as his working frequency and completes the communication on that frequency.

The United States Coast Guard usually monitors the maritime calling frequencies (2,182 kHz and 156.8 MHz for radiotelephone, 500 and 8,364 kHz for radiotelegraph). The three letters NCU signify "Calling all Coast Guard stations." For example, the call "NCU NCU NCU THIS IS WPNB WPNB WPNB OVER" on 2,182 kHz, is a general call to any Coast Guard station to answer on that calling frequency. For distress traffic only the Coast Guard can be called on the government frequency, 2,670 kHz.

Ship stations licensed for radiotelephone operation only in the 1,600–3,500-kHz band must maintain an efficient watch on 2,182 kHz when not in operation on another frequency.

Use of the calling frequencies (2,182 kHz and 156.8 MHz) should be as brief as possible, in no case over 3 min duration for one exchange of communication (emergency traffic excepted).

32-13 ANTENNA-TOWER LIGHTS (G)

Radio transmitting or receiving towers on land, high enough above surrounding terrain to be considered dangerous to aerial navigation, are required to carry one or more warning lights on them. These lights may burn steadily or be rotating beams. For further information, see Sec. 24-23 and "FCC Rules and Regulations," Volume I, Part 17.

32-14 COAST STATIONS

Coast stations must maintain an accurate log during their hours of service of the following items: an "on duty" and "off duty" entry of the licensed operator or operators on watch; an indication of the call signs of all stations worked and the time in UCT (except on inland waters, where a local standard time may be used); any interruptions of the watch, including reasons and when the watch was resumed; all distress, urgency, and safety messages intercepted or transmitted copied in full into the log; all tests made; a daily comparison of the required station clock or clocks with standard time signals; any measurements of transmitter frequency; all service or maintenance work performed on the transmitter; and entries on antenna-tower lights (if any). Each sheet of the log is to be numbered in sequence and dated and is to carry the call sign of the station and the signature of the licensed operator or operators performing operating duties. (*Signature* indicates a minimum of first initial and last name, written, not printed.)

Similar logs are also kept by compulsorily radio-equipped ships.

A coast station communicates with ship or aircraft stations. It communicates with other coast stations only to facilitate communications with ship stations. Except for safety communications and short calls on calling frequencies, a coast station operates on its assigned working frequency as much as possible.

Test your understanding; answer these checkup questions.

1. List the three top priority messages in order. _____ _____ _____ What is the lowest-priority message? _____

2. When should a radiotelephone distress message be sent? _____
3. What should a distress message contain? _____
4. What should an operator do if he hears a distress message in his area? _____ If out of his area? _____
5. How would KAAA conclude distress traffic? _____
6. Which of these indicates a safety communication to follow: PAN, SECURITY, MAYDAY? _____
7. For best communication, how close should an operator's mouth be from the microphone? _____
8. How would the name "Dean" be spelled phonetically? _____
9. In formal messages, what is the preamble? _____ Text? _____ Address? _____ Service? _____ Signature? _____
10. Basically, when should stations identify? _____ _____
11. When and how often should a radiotelephone transmitter be tested? _____ _____
12. Where should operator licenses be posted? _____
13. After communication is established on a calling frequency, to what frequency should the stations move? _____
14. What are the two international radiotelephone calling frequencies? _____ _____
15. What is a coast station? _____
16. Where is complete information regarding antenna lighting found? _____
17. Where might an operator listen for a coast station for routine traffic? _____ For emergency traffic? _____

COMMERCIAL LICENSE QUESTIONS ELEMENT 2

An applicant for any license has the choice of answering 20 questions in (1) the general category or (2) the marine category. Sections in which questions are answered are shown in parentheses.

GENERAL

1. Why is the station's call sign transmitted? (32-1)
2. How should a microphone be treated when used in noisy locations? (32-6)
3. What may happen to the received signal if an operator shouts into a microphone? (32-6)
4. Why should an operator use well-known words and phrases? (32-6, 32-7)
5. What is meant by a *phonetic alphabet* in radiotelephone communications? (32-7)
6. What are the meanings of *clear, out, over, roger, words twice, repeat,* and *break*? (32-8)

7. When may an operator use his station without regard to certain provisions of his station license? (32-11)
8. What should an operator do if he hears profanity being used at his station? (32-11)
9. Who bears the responsibility if an operator permits an unlicensed person to speak over his station? (32-11)
10. What precautions should be observed in testing a station on the air? (32-11)
11. What should an operator do when he leaves a transmitter unattended? (32-11)
12. Why should radio transmitters be off when signals are not being transmitted? (32-11)
13. Where does an operator find specifications for obstruction marking and lighting (where required) of the antenna towers of a particular radio station? (32-13)

MARINE

1. In the case of a mobile radio station in distress, what station is responsible for the control of distress-message traffic? (32-3)
2. What information must be contained in distress messages? What procedure should be followed by a radio operator in sending a distress message? What is a good choice of words to be used in sending a distress message? (32-3)
3. What do *distress, safety,* and *urgency* signals indicate? What are the international distress, safety, and urgency signals? (32-3–32-5)
4. What actions should be taken by a radio operator who hears (*a*) a distress message and (*b*) a safety message? (32-3, 32-5)
5. If a radio operator is required to stand watch on an international distress frequency, when may he stop listening? (32-3)
6. When may a coast station not charge for messages it is requested to handle? (32-5)
7. What are the meanings of *clear, out, over, roger, words twice, repeat,* and *break*? (32-8)
8. In regions of heavy traffic, why should an interval be left between radiotelephone calls? (32-9)
9. How long may a radio operator in the mobile service continue attempting to make contact with a station which does not answer? (32-9)
10. How often should the station's call sign be sent? (32-10)
11. Why are test transmissions sent? How often should they be sent? What is the proper way to send a test message? (32-11)
12. In the mobile service, why should radiotelephone messages be as brief as possible? (32-11)
13. Why should a radio operator listen before transmitting on a shared channel? (32-11)
14. How does the licensed operator of a ship station exhibit his authority to operate a station? (32-11)
15. Does the 1959 Geneva Treaty give other countries the authority to inspect United States vessels? (32-11)
16. What is the importance of the frequency 2,182 kHz? (32-12)
17. What are the requirements for keeping watch on 2,182 kHz? (32-12)
18. What is the difference between *calling* and *working* frequencies? (32-12)
19. Under what circumstances may a coast station make contact with a land station by radio? (32-14)

ANSWERS TO CHECKUP QUIZ ON PAGE 735

1. (*Distress, urgent, safety*) (*Routine*) 2. (*Immediately after distress signal*) 3. (*Signal, call, name of craft, position, nature of distress, assistance required, etc.*) 4. (*Acknowledge*) (*Listen, help if needed*) 5. (MAYDAY *all stations this is KAAA (3) SS Blank distress traffic ended. Out*) 6. (*All*) 7. (*Within an inch usually*) 8. (*Delta-echo-alpha-nectar*) 9. (*Origin, date, number*) (*Body of message*) (*Where going*) (*Received time and operator*) (*Sender's name*) 10. (*Calling, conclusion, every 15 min*) 11. (*When channel clear, daily*) 12. (*In view of operating position*) 13. (*Working*) 14. (*2182 kHz and 156.8 MHz*) 15. (*Communicates with ships and aircraft*) 16. (*FCC Rules, Vol. 1, Part 17*) 17. (*Coast working frequency*) (*International calling frequency*)

33

COMMUNICATING BY RADIOTELEGRAPH

The objective of this chapter is to answer FCC Element 5 type license test questions except those covered in Chaps. 16 and 18. Information regarding radiotelegraph messages for stations open to public correspondence (paid messages), message construction, charges, etc., of particular interest to marine operators is included. Amateur messages are similar to those discussed except that words in the address and signature may not be counted.

33-1 FUNDAMENTALS OF OPERATING

There is a basic similarity of radiotelephone and radiotelegraph regulations. For instance, the list of message priorities is the same for both (Sec. 32-2), radiotelegraph stations must not be operated by unlicensed persons, and operator licenses should be posted at the operating position.

When an operator exercises good common sense in communications, he will usually be conforming with regulations. Such regulations have been developed to enable the greatest number of stations to handle the most traffic, in the least amount of time possible, with the least confusion. Long calls, failure to listen on a frequency before transmitting, transmitting on a frequency already in use, employing either more or less transmitting power than necessary (the minimum necessary should always be used), sending faster than the receiving operator can copy or faster than an operator can legibly telegraph, using improperly functioning equipment, and using abbreviated procedures with stations not understanding them are common operating faults that can needlessly delay the communications of all stations concerned.

33-2 RADIOTELEGRAPH LICENSES

There are three types of radiotelegraph licenses or permits: First Class license, Second Class license, and Third Class permit (Chap. 31). All require passing at least FCC Elements 1, 2, and 5, plus code tests. Holders of the First and Second Class licenses are authorized to make any technical adjustments to any radiotelegraph transmitters, as well as any radiotelephone transmitters in the mobile service. Radiotelegraph permit holders are not authorized to make technical adjustments, but they can operate any equipment that can be operated by a Radiotelephone Third Class permit holder and also operate radiotelegraph stations or transmit radiotelegraph call signs at certain stations having an RF power output of 250 W or less. The radiotelegraph permit holders cannot operate any broadcast, TV, or FM stations (except noncommercial educational FM of 10 W or less), ship stations required to have a radiotelegraph installation, coastal radiotelephone stations of more than 250 W, or aircraft stations employing radiotelegraphy.

With radiotelegraphy, unlike radiotelephone, there is no time when an unlicensed person can transmit with a telegraph key, with the exception of a dire emergency such as in a lifeboat with no operator aboard and adrift at sea. Anyone may operate radiotelephone transmitters, provided he is authorized by the person in charge or is under the direct supervision of a licensed operator.

33-3 THE MORSE CODE

The telegraphic code used for normal commercial and amateur radiotelegraphic communication is the International, or Continental, Morse Code, consisting of dots and dashes. (This differs from the American Morse code, consisting of dots, dashes, and spaces.) In International Morse code a *dot* is made by pressing the telegraph key down and allowing it to spring back up again rapidly. The length of a dot is the basic time unit. A *dash* is made by pressing the key down and holding it for a period of three basic time units. The spacing between two dots or between a dot and dash in the same letter is equal to one time unit. The spacing between two letters in one word is equal to three units. The spacing between two words is equal to seven units.

THE INTERNATIONAL, OR CONTINENTAL, MORSE CODE

A ·—	N —·	1 ·————
B —···	O ———	2 ··———
C —·—·	P ·——·	3 ···——
D —··	Q ——·—	4 ····—
E ·	R ·—·	5 ·····
F ··—·	S ···	6 —····
G ——·	T —	7 ——···
H ····	U ··—	8 ———··
I ··	V ···—	9 ————·
J ·———	W ·——	0 —————
K —·—	X —··—	
L ·—··	Y —·——	
M ——	Z ——··	

. (period)	·—·—·—
, (comma)	——··——
? (question mark)($\overline{IMI}$)	··——··
/ (fraction bar)	—··—·
: (colon)	———···
; (semicolon)	—·—·—·
((parenthesis)	—·——·
) (parenthesis)	—·——·—
' (apostrophe)	·————·
- (hyphen or dash)	—····—
$ (dollar sign)[1]	···—··—
" (quotation marks)	·—··—·
or,	·—··—·[1]

[1] These characters are not listed internationally but may be heard in domestic communications.

Error sign (8 dots)	········
Separation indicator also known as $\overline{BT}$	—···—
End of transmission of a message ($\overline{AR}$)	·—·—·
Invitation to transmit	—·—
Wait ($\overline{AS}$)	·—···
End of work ($\overline{SK}$ or $\overline{VA}$)	···—·—
Starting signal	—·—·—

A number and a fraction are transmitted as number, hyphen, and fraction. For example, $45\frac{1}{2}$ is transmitted: 45-1/2. (Sometimes ·—··— is used instead of the hyphen.)

Four dashes (————) may be used as the two letters *ch,* or an an end-of-paragraph indicator.

One or two question marks are sometimes used instead of the eight-dot error signal.

There are several other codes used in communications. One is the American Morse code used on telegraph lines (*landlines*). Another code is used in teletype circuits. It is a five-character on-off, or mark-space, code, in which the length of time of each letter, number, or function is the same. Some languages such as Japanese and Russian have their own special radiotelegraph codes.

33-4 FREQUENCIES USED

Radiotelegraph-equipped ships on the high seas maintain either a constant or a specified number of hours watch on the international calling and distress frequency of 500 kHz (previously known as 600 m). All coastal radiotelegraph stations maintain a constant watch on 500 kHz. Most of these stations also keep watch on one or more HF (high-frequency, 4-, 6-, 8-, 12-, 16-, and 22-MHz) bands (Sec. 28-9).

There is a less used international maritime mobile band between 90 and 160 kHz, having a calling frequency on 143 kHz.

The calling frequency of 8,364 kHz is reserved for aircraft, lifeboats, and other survival craft for communication with stations of the Maritime Mobile Service. This frequency is monitored by the United States Coast Guard and Navy and, being the center of the ship calling-frequency band, is constantly scanned by coastal stations listening for calls by ships.

33-5 CALLING BY RADIOTELEGRAPH

An original call is made by radiotelegraph by transmitting the call sign of the station wanted not more than three times, followed by the letters DE (meaning "from"), the call sign of the calling station, and the letter K (meaning "go ahead"). For example,

KAAA KAAA KAAA   DE   KBBB KBBB KBBB   K

In the bands between 4 and 23 MHz, when the conditions of establishing contact are difficult, the call signs may be transmitted more than three times but not more than eight times.

Calls are made on a frequency that is known to be monitored by the station to be called. If within a few hundred miles, the frequency of 500 kHz is used for ship-to-ship or ship-to-shore calls. If farther away, high-frequency calling frequencies are used.

Except for distress traffic, all stations handle traffic on working frequencies, not on calling frequencies. A ship station calling a coastal station listens for the coastal station on its assigned working frequency and notifies the coast station to listen for the ship on one of its assigned working frequencies. Thus, "KFS DE KDMW QSW 463" means "KFS from KDMW, I am going to send on 463 kHz. Listen for me there." If KFS hears the call, it acknowledges on its working frequency and no other transmission is made on the calling frequency.

In the 405- to 535-kHz band, coastal stations may shift to the calling frequency to call a ship station or to advise all stations after sending the call CQ ("attention all stations") that it is going to transmit a list of traffic on hand by using its working frequency. In the HF bands the coastal stations always remain on their working frequencies.

The general call to all stations is CQ. If a coast station has traffic for one or two ships, it may transmit their call signs on the calling frequency. If the list of traffic is for three or more ships, the coast station must send a traffic list on its working frequency. It will first send CQ on the calling frequency and the Q signal (Appendix G) to listen on its working frequency. For example,

CQ CQ CQ   DE   WSV WSV WSV   TFC QSY   408 $\overline{AS}$

This indicates a traffic list will be transmitted by WSV on 408 kHz as soon as the station can switch to that frequency.

The letters CP followed by two or more call signs on a calling frequency are a general call to those stations only and indicate that no reply is expected from them. This call precedes a general broadcast or information to a special group of receiving stations, such as press, on a working frequency.

33-6 ANSWERING BY RADIOTELEGRAPH

When the operator of a ship station hears another ship or a coast station calling him on a calling frequency, he should answer on the same frequency (or on any frequency designated by the calling station) as soon as possible. As an example, KAAA hears KBBB calling on 500 kHz and responds with

KBBB   DE   KAAA   QSY   468   K

This notifies KBBB to change to 468 kHz and assumes that KAAA will also shift to the same frequency. KBBB usually responds with an R, indicating "I received your message (and concur with its meaning)."

A ship operator responding to a call of a coastal station would normally respond with Q signals. (The word "up" is sometimes used to notify the other station to move up to its working frequency or to indicate that both stations are to shift up to their usual working frequencies.)

If a station hears another station calling but cannot make out the call sign, it should transmit the Q signal meaning "By whom am I being called?" followed by its call signs, as

QRZ?   DE   KAAA   K

If a station hears another calling but is busy at the time, if possible it should answer the calling station and transmit $\overline{AS}$ ($. - ...$) or the Q signal QRX, followed by a number indicating how many minutes to wait.

33-7 TUNING AND TESTING

When it becomes necessary to test or tune a transmitter on the air, a time should be chosen when the frequency to be used is idle. The tuning should be accomplished as rapidly as possible. At the conclusion the call sign of the station must be transmitted. A radiotelegraph station usually transmits a series of VVV as a test signal, followed by DE and its call sign.

33-8 STATION IDENTIFICATION

Besides the normal station identification that occurs during calling and answering, radiotelegraph stations usually transmit a station identification by their call signs in Morse code at the completion of each transmission, at the conclusion of an exchange of transmission, or every 15 min (10 min for amateurs) if a transmission is sustained for a period exceeding this time. In the Public Safety Radio Service a station may sign after each transmission, after each exchange, or every 30 min, as the licensee desires.

33-9 AUTO-ALARM SIGNAL

A signal of 1 min duration, composed of twelve 4-s dashes with 1-s spacing between them, is known as an auto-alarm signal. Any four consecutive alarm dashes will activate an auto-alarm receiver and sound an alarm aboard any ship on which an operator is off watch. The auto-alarm signal is used only before distress calls from ships or aircraft or before urgent cyclone warnings transmitted by coastal stations. A period of 2 min should elapse before a message follows an auto-alarm signal to allow time for operators who have been summoned by the alarm to get to the radio room to stand watch (see Sec. 28-15).

33-10 SILENCE PERIODS

All ships and coastal stations operating on 500 kHz must listen on, but not transmit on, 500 kHz during the 15–18- and 45–48-min periods of every hour of their operation. These 3-min periods are known as *silence periods*. Ships in distress should repeat their distress messages during these periods to assure reception.

During silence periods (also during distress, urgency, and safety traffic on 500 kHz) routine transmissions are forbidden in the band from 485 to 515 kHz.

33-11 DISTRESS

One of the main reasons why a ship carries radio equipment is to enable it to signal its condition and position in case of emergencies or distress.

The distress signal by radiotelegraph is ... — — — ... transmitted three times. This is usually referred to as an SOS but is actually a single character selected for its easy identification. It is sent only on the authority of the master or person responsible for the ship, aircraft, or other vehicle and only when the distressed station is in grave and imminent danger and requests immediate assistance.

In a distress situation, if time allows, the auto-alarm signal should be transmitted for 1 min on 500 kHz. After a 2-min interval, to allow operators summoned by the auto-alarm signal to assume a watch, the distress call is transmitted. This consists of the distress signal three times, DE, and the call sign of the station three times.

The distress call must be followed as soon as possible by the distress message transmitted at no more than 16 wpm (16 words per minute). The distress message consists of (1) the distress call again, (2) the name of the ship, aircraft, or vehicle in distress, (3) particulars of its position, the nature of the distress, the kind of assistance desired, and speed and course if underway, and (4) any other information which might facilitate the rescue.

After the distress message, the mobile station should transmit two 10-s dashes, followed by its call sign, to permit direction-finding stations to determine its position.

The distress message should be repeated again in the next silence period.

All subsequent distress messages must contain the SOS indicator in their preamble.

Distress signals, calls, and messages should be transmitted by using A2 (modulated-code) if possible.

Stations that cannot use 500 kHz should use their normal calling frequency for distress transmissions.

The station in distress is in complete control of distress traffic unless it delegates control to another station.

An operator hearing a distress call must cease transmitting and listen on the frequency of the distress call. If the station in distress is without doubt in the vicinity, after the completion of the distress message, the operator must acknowledge receipt of the message by transmitting (1) the call of the distressed station (three times), (2) DE followed by the call of the receipting station (three times), and (3) R R R $\overline{SOS}$.

Stations receiving a distress message not in their vicinity must allow a short interval of time before acknowledging receipt to permit closer stations to acknowledge.

When distress traffic is well established, stations not involved and in no position to help may continue normal service on other frequencies, provided they do not interfere with any distress traffic.

When all distress traffic has ceased or when silence is no longer necessary, a station which has controlled the distress traffic must send a message to terminate the distress condition. Such a message will take the form (1) $\overline{SOS}$ CQ CQ CQ DE followed by the call of the station sending, (2) time of the message, (3) name and call of distressed station, and (4) QUM $\overline{SK}$.

33-12 THE URGENCY SIGNAL

In radiotelegraphy, the urgency signal is the group XXX sent slowly three times, usually on 500 kHz or another calling frequency. It indicates that the calling station has a very urgent message to transmit concerning the safety of a ship, aircraft, or other vehicle or of some person on board or within sight. It must be authorized by the person responsible for the transmitting ship.

The urgency signal transmitted by a ship may be addressed to a specific station. The urgency signal, with the approval of the responsible authority, may be transmitted from a coast station and addressed to all ships.

Mobile stations hearing the urgency signal must continue to listen for at least 3 min. If no urgency message has been heard by then, they may resume their normal service.

An auto-alarm signal may be transmitted before an urgent cyclone warning by a coastal station authorized to do so by its government. A period of 2 min must elapse between transmission of the auto-alarm signal and the cyclone-warning message.

33-13 THE SAFETY SIGNAL

In radiotelegraphy, the safety signal consists of three repetitions of the group TTT sent slowly, followed by the station call (3 times). It indicates that the station is about to transmit a message concerning the safety of navigation or important meteorological warnings and is sent on the distress frequency. The safety signal is usually transmitted during the last minute of the first available silence period. The safety message is then transmitted at the conclusion of the silence period.

Operators hearing a safety signal must continue to listen to the message until they are satisfied that it is of no importance to them. They may then resume normal service on frequencies that will not interfere with the safety transmission.

33-14 RADIOTELEGRAPH LOGS

All ship stations authorized to use telegraphy must maintain an accurate radiotelegraph log. The first page is a *title page*. At the completion of the voyage the following information is placed on it: the name of the ship, the call letters, the period of time covered by the log, the number of pages, a statement whether any distress message entries are contained in it and on what pages, the operator's signature and mailing address, and his license number, class, and date of issuance.

Each page is numbered serially for the voyage and contains the name of the ship, call letters, and name of the operator on watch. The entry "on watch" must be made by the operator beginning a watch, followed by his signature. The entry "off watch" must be made by the operator being relieved or terminating a watch, followed by his signature.

The log is kept in GMT or UTC (EST for ships in the Great Lakes) and must contain all calls or tests transmitted by the ship, stations contacted, and serial numbers of messages handled, stating times and frequencies. A positive entry with respect to reception on 500 kHz should be made at least once in each 15 min. Entries stating whether or not the international silence period was observed shall be made twice per hour, noting any signals heard during these periods. All distress, urgency, and safety signals heard must be entered, with complete text of distress messages if possible. Any harmful interference noted should be logged. Once a day the position of the ship and a comparison of the radio station clock with standard time, including errors noted, must be entered. Times of arrival and departure from ports are logged. Failures of equipment and corrections taken should be noted. Results of emergency-equipment tests, the battery specific gravity when placed on and taken off charge, and the quantity of fuel for emergency generators must be entered. On cargo vessels, the time the auto-alarm was placed in service and when out of service, as well as the setting of the sensitivity control (if any), results of tests, alarms, and false alarms must be logged.

Logs are kept by the licensee of the station for a period of 1 yr. If the logs contain distress or disaster traffic, they must be kept 3 yr. Station logs which include entries of communications incident to or involved in an investigation by the FCC and concerning which the station licensee has been notified are retained until authorization to destroy them is received in writing.

Test your understanding; answer these checkup questions.

1. What words are counted in commercial radio messages? _____ In amateur messages? _____
2. When only might an unlicensed person transmit radiotelegraph code? _____
3. What is the basic unit of Morse code? _____ How long is a dash? _____ Space between letters? _____ Between words? _____
4. Must a calling station always call at least three times and sign three times? _____
5. What are the three marine CW calling frequencies? _____ _____ _____
6. What does CQ mean? _____ QSW? _____ QSY? _____ QRZ? _____
7. On HF bands, why might a ship call KPH KPH KPH DE WAAA WAAA WAAA KPH K? _____
8. If a coast station tells a ship on 500 kHz "up," what does it mean? _____
9. What is the test signal for CW? _____
10. A ship is in distress at 0952 UTC. When should the operator send the SOS message? _____ At what speed? _____ What should be sent after the message? _____

11. What does QUM mean? _____ XXX? _____ TTT? _____
12. Besides prefacing an SOS message, when may an AA signal be sent? _____
13. What is considered a minimum signature on a log sheet? _____
14. How often must log entries be made while at sea? _____
15. How long must logs be kept if they contain distress messages? _____

33-15 COMMERCIAL RADIOTELEGRAPH MESSAGES

Radiotelegraph messages from ship to ship or from ship to shore are composed of a preamble, address, text, and signature, as explained in Chap. 32. An example of a standard paid ship-to-ship message, as transmitted from KBBB to KAAA, might be as follows:

P 1 KBBB CK 13 SS GOLDEN ARROW 12 $\overline{}$
                          2145Z $\overline{BT}$

FRED MANGELSDORF
SS GOLDENSPEAR (or SS KAAA) $\overline{BT}$

MAKING ARRANGEMENTS MEET YOU HILO TWENTYFIRST $\overline{BT}$
                    STEVE AND FREDA $\overline{AR}$
1 KAAA ES 2213 12

In the preamble, P 1 KBBB indicates a paid message No. 1 from the station of origin, KBBB, the SS *Golden Arrow*. The check (CK) is the number of words paid for in the address, text, and signature. The 12 indicates the twelfth day of the month. The 2145Z indicates the filing time of 2145 UTC. The serial numbering of messages begins at midnight (0000 h) UTC and continues to the next midnight (2400 h). (Stations on inland waters may use local time.) The date and time may be sent as a date-time group, in this case, 122145Z. (The word "*date*" may be sent to indicate the filing date is the same as the sending date.) The same preamble might be transmitted in one of the following forms:

P 1 13 SS GOLDEN ARROW DATE 2145 UTC

SS GOLDEN ARROW NR1 CK 13 122145Z

NR 1 KBBB CK 13 P SS GOLDEN ARROW
                          2145 GMT 12

An internationally acceptable preamble may be: ship name, call, message number, check, date, and filing time.

The address includes the name of the person to whom the message is being sent and sufficient address to deliver it.

The text is the body of the message.

The signature, if any, is the person sending the message. If no signature is to be sent, the words *no sig* are transmitted in place of the signature. The sign $\overline{AR}$ indicates "end of message."

The servicing on the message to Fred Mangelsdorf is on the line below the signature; it is placed on the message by the transmitting operator. It indicates the message was sent as No. 1 to KAAA by operator Emery Simpson at 2213 on the twelfth. Somewhere on every message the full date (month, day, and year) must be shown. If it is not in the preamble, it must be in the servicing.

The receiving operator should copy the message in a form similar to the following:

P 1 KBBB CK 13 SS GOLDEN ARROW
          FEBRUARY 12 1975 2145 UTC

FRED MANGELSDORF
SS GOLDENSPEAR

MAKING ARRANGEMENTS MEET YOU HILO
TWENTYFIRST

          STEVE AND FREDA
          12 2213 BW

Note that the operating signals $\overline{BT}$ and $\overline{AR}$ are not shown on the received message. Also, the receiving operator has typed in the complete date in the preamble.

A ship-to-shore message is addressed to some destination on land. It is transmitted to a coastal station and has a form similar to that of the ship-to-ship message.

A message filed at a telegraph station ashore for a ship is relayed by telegraph to a coastal station and transmitted to the ship as soon as it can be reached. An example of a shore-to-ship message as received by a ship operator might be as follows:

P 1 RENO NEV CK 14 FEB 12 1975 8:45 AM

SASSER
SS GOLDENARROW
SANFRANCISCORADIOKPH

MEET YOU ON ARRIVAL EUREKA FRIDAY
AFTERNOON WITH RONADLO

        JANE
        2301 12 BW

Note that the name of the coastal station transmitting the message is counted as one word. Actually, the call letters of the coastal station are all that may be transmitted; the receiving operator fills in the complete name of the station. Also note that the receiving operator double-spaces after the fifth word (*meet*) and drops to the next line after the tenth word (and after every multiple of 10 words in long messages) to facilitate counting the check while he is copying the message.

The punctuation in the filing time may be transmitted as a period, although the letter R also may be used. It may be copied as a period or as a colon.

When a message contains words that appear to the receiving (or transmitting) operator as possibly improperly spelled or copied, he should confirm (collate) any such odd words, figures, or symbols. For example, the last word in the text of the shore-to-ship message above appears suspicious. At the completion of the transmission of the message the receiving (or transmitting) operator may transmit CFM RONADLO K, meaning "Confirm the spelling of the word Ronadlo." If the questioned spelling is correct, the transmitting operator responds with R, or preferably C, meaning "yes," or "correct." If the receiving operator confirms a word incorrectly, the transmitting operator must correct him, as N RONALDO (if this happens to be the actual spelling). The N indicates negation, or "no."

33-16 COUNTING WORDS IN MESSAGES

In any of the acceptable languages, English, French, German, etc., any word or name in the text up to 15 letters long counts as only one word. If it has 16 or more letters, it is charged for as two words (31 letters as three words).

When pronounceable but meaningless code or cipher words are used in the text in such a manner as to hide the meaning of the message, the same number of letters per word is counted as in plain-language texts.

When unpronounceable or code words made up of letters, numbers, or letters and numbers are used, one word is counted for each five letters. CBS3 is one word. RCBS3 is one word. RCBX32 is two words. RCBX321465L is three words. The 45-$\frac{1}{2}$ has six transmitted characters, but since the fraction to follow sign, or hyphen, is not counted, it is one word. The number 125-$\frac{3}{4}$ would be counted as two words.

Names in the address may be longer than 15 letters and still be counted as one word. The words *street, avenue, boulevard,* etc., may be added to the street name and counted as one word if the total of all letters does not exceed 15, as ELMSTREET. If transmitted as two words, it is counted as two words. The name of a ship is always one word in the address, as PRESIDENTHOOVER. If used, the SS (steamship) always counts as a word. The state (and country, if necessary) may be sent attached to the city name, as SANFRANCISCOCALIFORNIA, or the state may be enclosed in parentheses and may not be counted, as SANFRANCISCO (CALIFORNIA). Sometimes a fraction bar is used to tie city and state names together.

When requested to be transmitted, punctuation marks such as period, comma, colon, question mark, parentheses, and quotation marks are counted as separate words. Parentheses and quotation marks require two transmitted characters for each single word count.

ANSWERS TO CHECKUP QUIZ ON PAGE 742

1. (*Address, text, signature*) (*Text*)   2. (*Dire emergency*)   3. (*Dot length*) (*3 units*) (*3 units*) (*7 units*)   4. (*Not if sure other station can copy*)   5. (*143, 500, 8364 kHz*)   6. (*All stations*) (*Will send on ... frequency*) (*You send on ... frequency*) (*By whom am I being called?*)   7. (*Coast stations scan band, may not hear first part of call*)   8. (*Shift to working frequencies*)   9. (*VVV*)   10. (*0955*) (*16 wpm*) (*2 long dashes*)   11. (*Distress traffic ended*) (*Urgency signal*) (*Safety signal*)   12. (*Before cyclone warning*)   13. (*Initial and last name written*)   14. (*Every 15 min*)   15. (*3 yr*)

33-17 MESSAGE CHARGES

Each word in the address, text, and signature of a radiogram is charged for. (This differs from the domestic or landline telegraph messages, which charge for the text only.) Between United States ships and shore stations the rates are charged for in dollars and cents. When a United States ship has traffic with a foreign station, the rates (other than its own ship charges) are charged for in gold francs, with 2.5 to 3 gf equaling $1. (Subject to change.)

For a regular, full-rate type radiogram (P) the charges per word are 8 cents to the transmitting ship, 8 cents (or 0.40 gf) to any receiving ship, $17\frac{1}{2}$ cents to the coastal station (varying rates in gold francs to foreign-country coastal stations), and $7\frac{3}{4}$ cents landline charge to any place in the United States. Thus, a ship-to-ship message between United States ships costs 16 cents per word, and a ship-to-shore message costs $33\frac{1}{4}$ cents per word. (Subject to change.)

33-18 DIFFERENT TYPES OF MESSAGES

- CDE. *Code* messages consisting of a text having five-letter coded words only, but carry regular P charges.
- CODH, or DHCO. *Company deadhead* messages refer to traffic, licenses, repairs, supplies, etc., exchanged by radio between ships and offices of the same radio service. There are no charges.
- DH or PDH. These complimentary franked messages may require payment of only the landline charges, the ship and coastal charges being franked (transmitted free). The sender must possess a frank card.
- DH MEDICO. *Deadhead medical* messages are free (deadhead) messages pertaining to medical or surgical advice for persons aboard a ship and are normally addressed to MARINE HOSPITAL in a city served by the coastal station receiving them.
- DH OPR. *Operator deadhead* messages are personal messages of radio officers and may be forwarded through the coast stations of the same radio service subject only to landline charges.

- GOVT. *Government radiograms* by accredited officials of the United States government on official government business are charged full rates.
- HYDRO. *Hydrographic* messages report menaces to navigation and are addressed to HYDRO WASHINGTON. There are no charges insofar as the ship operator is concerned. These are sent to U.S.C.G. coastal stations.
- MSG. *Master's* messages are from masters of vessels and pertain to ship's business. Only coastal and landline charges are collected.
- NRT. *Night radio telegrams* (minimum of 22 words charged) must be filed before midnight (ship time) and are transmitted to coastal stations, telegraphed to destination city, and delivered sometime the next morning. Ship charge is 8 cents; coast, $8\frac{3}{4}$ cents; landline, $3\frac{7}{8}$ cents.
- OBS. *Observer* messages are meteorological reports from ships at sea to the U.S. Weather Bureau addressed OBSERVER in the city where the report is to be sent. There are no charges insofar as the ship operator is concerned.
- PC. *Acknowledged delivery of message.* Coastal station notifies sender via telegraph of time and date message was delivered to the ship destination. J1PC means advise sender if undelivered after day 1. J12hrPC means after 12 h, etc.
- PRESSE, or PX. *Press* messages transmitted by authorized members of the press when addressed to newspapers, magazines, or broadcast stations are charged for at a rate 50% of the standard radiogram rate. The word *presse* is also transmitted as the first word of the address and is charged for. The minimum charge is for 14 words.
- RP. *Reply-prepaid* messages carry the amount prepaid as the first word of the address, as RP$3.99 (transmitted RPDOLS3.99 to U.S. stations and RPGF11.97 to foreign stations). Any message in reply to an RP is known as an *ANSTORP* (answer to RP) and is prepaid, in this case, up to $3.99. If the charges of the answering message are greater than this, the sender must pay the excess. Full rates are charged for an RP message, plus the prepaid amount.
- SLT. *Ship letter telegrams* are mailed to desti-

nation by the receiving coastal station. For the first 22 words or less, ship charge is 88 cents and coast is 1.92\frac{1}{2}$. For additional words, ship charge is 4 cents and coast is $8\frac{3}{4}$ cents.
- ST. *Paid service* messages are made out in the more formal preamble, address, text, and signature form. This type of service adds words to or corrects a message because of errors made by the originator of the message. The sender must pay standard rates for only the words in the text that are required to make the correction.
- SVC, or A. *Service* messages refer to previous messages that have been handled, the operation of the communication system, or the non-delivery of messages. Since SVC messages refer to previous traffic, they have priority over any other routine messages. They carry no charges. They may or may not carry a serial number. An example of a service message might be:

SVC 3 (or A 3)
   KAAA

RE OUR NR 1 12 CORRECT FIRST WD
TXT READ MAILING

       KBBB 13

This message advises the operator at KAAA to change the No. 1 message of KBBB of the 12th of this month to make the first word of the text read "mailing" instead of the way it was originally transmitted. The signature of the SVC is the call letters of the sending ship and date of the service message.
- TR. *Position reports* may be requested by land stations or mobile stations. The message indicates approximate position and next place of call when furnished by the master of the ship or vehicle.

33-19 TRANSMITTING SPEED

The average speed of radiotelegraphic communications is probably about 20 wpm. This is a comfortable telegraphic hand-key speed. Sending 25 to 30 wpm with a hand key is possible. However, with a mechanical key, called a *bug,* or an electronic keyer, good operators may work easily at speeds of 35 to 40 wpm under fair to good conditions. The same operators may not be able to average 15 wpm when static, fading, and interference are bad. The limiting factor is always how well the receiving operator can read the transmitted signal. The transmitting operator must gauge his sending to that speed which the receiving operator can copy without breaking too often. When receiving conditions are poor, an operator should not hesitate to send QRS in order to attain accuracy, the most important factor in communications.

In distress, urgency, and safety traffic, the speed should not exceed 16 wpm to assure a maximum number of operators' copying correctly.

33-20 TRANSMITTING RADIOTELEGRAPH

There are two basic types of mechanisms by which Morse code is transmitted by hand. The first is the *telegraph key,* as shown in Fig. 33-1. The

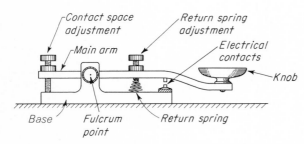

Fig. 33-1   Telegraph, or hand, key. Electrical connections are made by pushing the top contact down onto the lower contact.

operator places his first finger on the top of the knob and his thumb at the side of the knob and, with a downward pressure of his first finger, closes the contacts. A quick make and break produces a dot. A dash is produced by holding the contacts closed for three times as long as for a dot. For speeds over about 18 wpm, two fingers are used on the top of the knob. The contacts are adjusted for a maximum opening of about $\frac{1}{16}$ in. The spring is adjusted to the desired, comfortable upward pressure. For higher speeds a

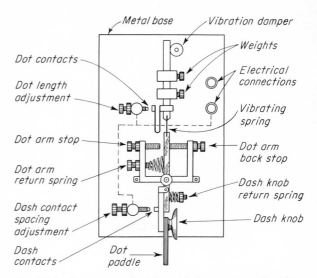

Fig. 33-2   Semiautomatic telegraph key, or bug. Pushing the dot paddle causes the weighted bar to vibrate and make a series of dots. Placement of weights determines speed of dots.

slightly lighter spring tension and less contact spacing may be better.

The second mechanism is a semiautomatic key, called a bug, shown in Fig. 33-2. When the knob is pressed to the left with the first finger, the contacts close, as with a telegraph key. When the paddle is pressed to the right with the tip of the thumb, the vibrating arm makes a series of dots. The speed of the dots is controlled by the placement of the weights on the arm on the bug. The dots are made automatically, but the operator must make each dash separately. The dot contact moves about $\frac{1}{8}$ in. and the dash contact about $\frac{1}{32}$ in. before making contact, although these adjustments are made at the pleasure of the operator. (An *electronic keyer* not only makes the dots automatically but also makes a series of dashes when the dash knob is pressed. It can produce very precise sending.)

When an error in sending has been made, the transmitting operator should immediately send a series of eight dots, stop, and start at the beginning of the word in which the error was made. If the error occurs on the first letter of a word, he should transmit the error signal and start at the beginning of the word before the word in which the error was made.

33-21 CALL LETTERS

The nationality of a station is indicated by the first or first two letters (or number and letter) of its call sign. All calls beginning with K, W, N, and AA to AL are those of United States stations.

G and 2A calls are assigned to Great Britain, F calls to France and its colonies and protectorates, R and U calls to the Union of Soviet Socialist Republics, V calls to British colonies, 4U to United Nations, and so on.

In general, three-letter calls indicate coastal stations in the maritime service or land stations in other services. Four-letter calls are ship stations or land stations. Five-letter calls are aeronautical mobile stations. Radiotelephone land or coast stations are assigned three-letter and two- or three-number calls, as KMA539. Mobile radiotelephones are assigned two-letter and three- or four-number calls, as KH7603.

33-22 OPERATING SIGNALS

To facilitate communications between operators speaking the same or different languages, a group of Q signals which have the same meanings internationally have been developed (Appendix G). Radiotelegraph abbreviations or signals that are in more or less common use by operators are listed in Appendix H.

Test your understanding; answer these checkup questions.

1. What items make up a preamble of a ship-to-shore message? _____ Of a shore-to-ship message? _____
2. What time is used in ship messages? _____ Land messages? _____
3. What is included in the servicing of a transmitted message? _____ Received message? _____
4. Why should a receiving operator double-space after each fifth word? _____
5. What punctuation in the text of a paid message is not counted? _____
6. What three charges are made for paid ship-to-shore messages? _____ _____ _____
7. What is another designation for a SVC message? _____ Paid service? _____ Franked message? _____ Press? _____ Company deadhead? _____ A Master's message? _____
8. What would J2PC mean on a message? _____ J24hrPC? _____
9. Besides his own abilities, list what limits how fast an operator should transmit a message. _____
10. If an error is made in the first letter of a word, how is it corrected? _____ If in the fourth letter? _____
11. What type of stations might have calls such as WSL? _____ KH4267? _____ KDOZ? _____ KNL123? _____ WNGPL? _____
12. What is meant by the operating signal K? _____ C? _____ AA? _____ AB? _____ CFM? _____ TU? _____ WA? _____ ETA? _____

COMMERCIAL LICENSE QUESTIONS

An ability to answer questions similar to the following is required for FCC Elements 5 and 6. Sections in which questions are answered are shown in parentheses.

ELEMENT 5

1. What is meant by the statement, "A station is open to public correspondence"? (33-1)
2. Where should the operator on duty at a manually operated radiotelegraph station normally post his operator license or permit? (33-1)
3. Is the holder of a Radiotelegraph Third Class operator permit authorized to make technical adjustments to (a) a radiotelephone transmitter and (b) a radiotelegraph transmitter? (33-2)
4. List three classes of stations which may not be operated by the holder of a Radiotelegraph Third Class operator permit. (33-2)
5. In order to avoid confusion in transmitting numbers involving a fraction, how should such numbers be transmitted? Given an example of such a number showing how it should be transmitted. (33-3)
6. If a radiotelegraph operator makes an error in transmitting message text, how does he indicate that an error has been made? (33-3, 33-20)
7. What radiotelegraph signal is generally employed in a call to all stations? (33-5)
8. Describe a procedure of radiotelegraph transmission in which one station calls another. Give an example. (33-5)

9. Describe a procedure of radiotelegraph transmission in which one station answers the call of another. Give an example. (33-6)
10. If, upon being called by another station, a called station is busy with other traffic, what should the operator of the called station do? (33-6)
11. When testing a radiotelegraph transmitter, what signals are generally transmitted? (33-7)
12. What are the requirements for station identification at radiotelegraph stations in the Public Safety Radio Service? (33-8)
13. Radiotelegraph code transmission in cases of distress, urgency, or safety must not in general exceed what speed? (33-10, 33-19)
14. What is the radiotelegraph (a) distress signal, (b) urgency signal, (c) safety signal? (33-11–33-13)
15. What is meant by the *preamble* in a radiotelegraph message? What information is usually given in the preamble? (33-15)
16. In addition to the preamble, what parts does a radiotelegraph message contain? (33-15)
17. What is meant by a *service prefix* or *indicator* in a radiotelegraph message? (33-15)
18. What is the meaning of *word count,* or *check,* in a radiotelegraph message? (33-15)
19. At what time or times does the serial numbering of radio messages begin? Does the period of numbering vary in some services? (33-15)
20. Immediately following the transmission of a radiotelegraph message containing figures or odd symbols, why are such figures sometimes collated? (33-15, 33-18)
21. Code or cipher groups are often used in radiotelegraph messages for what purpose? (33-16)
22. In general, what is the purpose of a service message in radiotelegraph communication? (33-18)
23. Should the speed of transmission of radiotelegraph signals be in accordance with the desire of the transmitting or receiving operator? (33-19)

ANSWERS TO CHECKUP QUIZ ON PAGE 747

1. (*Type of message indicator, number, ship name, ship call, date, time*)(*Type of message, number, originating city, date, time*) 2. (*UTC or GMT*)(*Local of sender*) 3. (*Number, call of station sent to, operator initials, date*) 4. (*To facilitate count of total check*) 5. (*Fraction-to-follow*) 6. (*Ship, coastal, landline*) 7. (*A*)(*ST*)(*DH or PDH*)(*PX*)(*DHCO*)(*MSG*) 8. (*Advise if not delivered in 2 days*)(*If not delivered in 1 day*) 9. (*Receiving conditions, receiving operator ability, type keyer used*) 10. (*Error sign, repeat last correctly sent word*)(*Error sign, repeat from first letter of the word*) 11. (*Land*)(*Mobilephone*)(*Ship or land*)(*Base phone*)(*Aeronautical*) 12. (*Go ahead*)(*Yes*)(*All after*)(*All before*)(*Confirm*)(*Thank you*)(*Word after*)(*Estimated time of arrival*))

24. How should a manual radiotelegraph transmitting key be adjusted for good operation? Is the adjustment always the same for slow as it is for high speed? (33-20)
25. Describe how an automatic key, or bug, should be properly adjusted to send good readable radiotelegraph signals. (33-20)
26. Why are Q signals or other arbitrarily selected procedure signals used in radiotelegraph communications? (33-22)
27. What is meant by the following signals: QRA, QRM, QRN, QRT, QRZ, QSA, QSV, QUM, QRL? (Appendix G)
28. If receiving conditions are bad and you desire that the transmitting station send each word or group twice to facilitate reception, what operating signal would be appropriate to use? (Appendix G)
29. If the signal strength of a radiotelegraph signal is reported on a scale of 1, 2, 3, 4, 5, what scale number would indicate (a) a very strong signal and (b) a very weak signal? (QSA, Appendix G)
30. What is meant by the following radiotelegraph operating signals: R, $\overline{AS}$, $\overline{IMI}$, C, $\overline{BT}$, K, $\overline{AR}$, $\overline{VA}$, DE? (Appendix H)

ELEMENT 6

1. In all cases other than those in which the transmitter output must be maintained at a fixed value, what amount of power should be employed for routine communication? (33-1)
2. Indicate the order of priority of the various types of radio communications. (33-2)
3. Under what circumstances may a station be operated by an unlicensed person? (33-2)
4. In the transmission of the International Morse code, what are the relative time lengths of dashes, dots, and spaces? (33-3)
5. You intercept CQ CQ WSV TFC QSY 408 $\overline{AS}$. What does this message mean? (33-5)
6. If, upon being called by another station, a called station is unable to proceed with the acceptance of traffic, what should the operator of the called station do? (33-6)
7. At what time or times must the international silence period be observed? (33-10)
8. At what time or times are routine transmissions in the band from 485 to 515 kHz forbidden? (33-10)
9. Describe how a distress call should be made. (33-11)
10. What station shall be in control of distress traffic? (33-11)
11. What is the international radiotelegraph distress

frequency for stations in the mobile service? (33-11)

12. With what type or types of emission and upon what frequency should a radiotelegraph transmitter be adjusted to transmit a distress call? (33-11)

13. What transmission should precede the sending of a distress call? (33-11)

14. What space of time should elapse between the transmission of the international auto-alarm signal and the distress call? (33-11)

15. Upon hearing an SOS, what should an operator do? (33-11)

16. After a distress call has been transmitted, every distress-traffic radiotelegram shall contain what symbol in the preamble? (33-11)

17. Under what circumstances is a station in the mobile service not required to listen to distress traffic? (33-11)

18. During what periods must a distress message be repeated following the initial transmission? (33-11)

19. How long must mobile stations listen after they have heard an urgency signal? (33-12)

20. What interval of time must elapse between the end of the auto-alarm signal and an urgent cyclone warning? (33-12)

21. Under what circumstances and by whom may the international auto-alarm signal be transmitted to announce an urgent cyclone warning? (33-12)

22. During what periods must the safety signal be transmitted? (33-13)

23. Upon hearing a safety signal, what should the operator at the receiving station do? (33-13)

24. Under what circumstances must log entries regarding observance of the international silence period be made? (33-14)

25. What time system shall be used in making log entries with respect to the observance of the international silence period? (33-14)

26. For what period of time must a station log which contains entries incident to a disaster be retained? (33-14)

27. Explain cable count and the use of standard service abbreviations and show the difference between cable count and domestic word count. (33-15, 33-17)

28. Construct a plain-language radiotelegram and indicate what portions comprise (a) the preamble, (b) the address, (c) the text, (d) the signature. (33-15)

29. Explain the use and meaning of the following indicators, or prefixes, on radiotelegrams and describe the difference in handling of the various types of radiotelegrams: RP, PC, TR, MSG, CDE, OPS, PDH, CODH. (33-18)

30. If you receive a distress call signed by a call signal composed of five letters, what type of craft transmitted the signal? (33-21)

34

AMATEUR RADIO

The objective of this chapter is to outline information about amateur radio operating and the various grades of available licenses. References are made to sample test questions for each class or license. (See new rules, page 757.)

34-1 THE AMATEUR RADIO SERVICE

Since the beginning of radio many persons have been interested in experimenting with radio equipment without any pecuniary interest. Such persons are known as amateurs solely because they do not accept money for their on-the-air operations and communications. Many of the foremost radio and electronics engineers operate as amateurs during times when they are not occupied professionally. When properly licensed, such operators may communicate on amateur bands with their own transmitting and receiving equipment.

Rules governing the Amateur Radio Service are given in their entirety in the "FCC Rules and Regulations," Volume VI, Part 97. Since this chapter can discuss only a relatively few of the many amateur rules and regulations, a copy of Part 97 should be on hand at every amateur station. For information on how to obtain Volume VI, see Appendix F.

Amateurs are subject to the same penalties for violating FCC rules as commercial operators are (Sec. 31-1). FCC rules have been developed to (1) recognize the value of amateurs as a voluntary noncommercial service, particularly for providing emergency communications, (2) provide an outlet for the extension of the amateur's ability to contribute to the radio art, (3) encourage advancement in the communication and technical phases of radio, (4) develop a reservoir of trained operators and technicians, and (5) continue the amateur's unique ability to enhance international goodwill.

The beginning of amateur radio operations dates back to the turn of the century. In 1914 the Amateur Radio Relay League (ARRL) was formed, and since then it has acted as one of the strongest spokesmen for the radio amateur. Its monthly magazine, *QST*, carries information regarding amateur activities, experimental circuits, and other items of interest to amateurs. Among other activities, the ARRL HQ station WIAW transmits code practice daily (see "WIAW Schedules" in *QST*). As the FCC adds new questions to amateur licenses they will appear in *QST*. Other magazines carrying amateur information are *73, Ham Radio,* and *Worldradio News*.

34-2 AMATEUR COMMUNICATIONS

Amateurs may communicate with other U.S. amateurs or with amateurs of most foreign countries. Foreign contacts may be limited to technical and general-interest items, provided communication with that particular country is not forbidden (lists of forbidden countries are published periodically in amateur magazines). Amateurs may also communicate with certain civil defense stations on a temporary basis.

Most amateur communications are between

one amateur and another, but it is possible to have third-party contacts. These may be in the form of messages from one amateur to another that are to be delivered to a third party. Also, a telephone patch, in which an amateur receiver and transmitter are used in conjunction with a telephone line to relay communications from some other amateur station, involves a third party. Third-party communications must not involve payments of any type and must not include commercial business information, nor may they involve amateur communications from countries not having third-party agreements with the United States. A third party at an amateur station may speak into the microphone, but the amateur must control the transmitter.

While no broadcasting of information of interest to the general public from amateurs is allowed, some one-way broadcasts are permitted. These include code-practice broadcasts, emergency-practice drills, and some transmissions of information of interest to radio amateurs operating in groups or "nets."

An amateur station must transmit its call sign at least every 10 min of operation and at the beginning and end of each transmission, using either radiotelephone or radiotelegraph, plus station identification with the method of communication in use (such as radioteleprinter). When a series of communications, each of less than 3-min duration, occurs between two or more stations on the same frequency, in general it is only necessary to identify every 10 min of operation.

As in all other services, distress signals and messages have priority over all other types of communications.

Except for short tests, all amateur emissions on frequencies below 51 MHz should be carrying information of some type. Above 51 MHz it is possible to transmit control signals for model aircraft, boats, etc., and to turn on remote transmitters. Such transmissions require no identification and are considered to carry no intelligence.

Retransmission of signals other than those received on amateur bands are forbidden. Each amateur station must have a control point from which the station is controlled. The licensee of an amateur radio station is responsible for all emissions from that station. If another amateur is des-ignated by the licensee as the control operator, that amateur also is responsible for the proper operation of the station. The station may be operated only to the extent permitted by the control operator's license. If the control operator has a higher-grade license than the station licensee, the station may be operated according to the control operator's license, provided proper station identification procedures are performed. For example, a General licensee operating a Novice station identifies as WN6XXX/W6XX.

Anyone operating an amateur station must have his amateur operator's license with him. Any amateur station being operated must have its station license at the operating position.

Calling and working procedures, Q signals, and the phonetic alphabet in the amateur service generally follow those outlined in Chaps. 32 and 33. If the calling station is on or near the frequency of the called station, a short call (W6ECU W6ECU W6ECU DE W6BNB W6BNB W6BNB K) is all that may be necessary. If the two stations are removed somewhat in frequency, the called station's call letters may be repeated more than three times. When calling CQ (general call to all stations), a satisfactory procedure is to repeat CQ five times and then transmit DE and the call letters of the transmitting station two or three times, repeating CQs and call signs for a period of perhaps a minute.

Amateur operation may be *fixed* if the station is in a home, *mobile* if in an automobile, aircraft, or boat, or *portable* if carried to some location other than the licensed address and operated from that point. In the latter case, if the operation will exceed 15 days, the Engineer-in-Charge of the nearest FCC field office (Appendix F) must be notified regarding the portable location and dates of expected operation. Mobile stations should notify the Engineer-in-Charge of any district outside the home district in which the mobile station will be operating.

Each amateur should keep a log of all his transmissions showing date, starting time, frequency band, type of emission, signal strengths, remarks, ending time, and the signature of the control operator for each entry. However, he is only required to enter (1) his station call sign and his signature (or photocopy of station license); (2) locations and dates either fixed or portable operation was initiated and terminated; (3) dates, times, call signs, and signatures of other licensed amateurs operating the station; and (4) dates and names of third parties and brief description of traffic content.

To operate control transmitters for model craft (no more than 1 W) a Transmitter Identification Card (FCC Form 452-C) or a plate indicating station call sign and licensee's name and address must be affixed to the transmitter.

Amateurs are prohibited from the following: broadcasting; some forms of third-party traffic; transmitting music or codes or ciphers; obscenity, indecency, or profanity; false signals; unidentified signals or communications; intentional interference with other stations; damage to any licensed radio apparatus; obtaining any license fraudulently; traffic which is contrary to Federal, state, or local law.

If amateur stations cause interference to local radio reception on receivers of good engineering design, the amateur must not operate daily 8:00 to 10:30 P.M. local time and in addition, on Sunday from 10:30 A.M. to 1 P.M. The amateur should take such steps as may be necessary to minimize or eliminate any such interference. If a second violation is received within 12 months, the silent periods will be increased.

What frequency to use for communications depends on which band will provide best communication at that time of day and year, whether local or distant communications are desired, whether to answer a CQ or to call CQ (answer CQs on the frequency of the other station; select a clear frequency for calling CQ), and selecting a carrier frequency so that any sidebands generated will be within band limits.

Amateurs report each other's radiotelegraph signals by using an RST system. R stands for readability (1 for unreadable to 5 for perfectly readable). S stands for strength (1 for extremely weak to 9 for 54 dB above an S-1 signal, using 6 dB per S unit). T stands for tone (1 for extremely poor to 9 for very pure). For example, RST 579 indicates perfectly readable, reasonably loud, very pure tone signal.

Radiotelephone reports are given using only the RS portion of the RST system. For example, "4

and 9" indicates somewhat unreadable for some reason, but quite strong.

Amateur stations should always be operated in accordance with good engineering and good amateur practices, even if the activity is not covered by FCC rules.

34-3 AMATEUR STATIONS

A simple amateur station may consist of only a radiotelegraph transmitter and key, a receiver with earphones or loudspeaker, a dipole antenna with a transmit-receive switch, as shown in block form in Fig. 34-1, and some separate means of

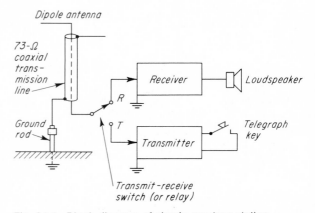

Fig. 34-1 Block diagram of simple amateur station.

checking the frequency of the transmitter. A more advanced station may include SSB, FM or AM transmitters, radioteleprinter equipment, or perhaps slow-scan or VHF TV equipment.

Many amateur-maintained repeater stations are located on mountain peaks to receive VHF signals from mobile amateur stations, detect the signals, have them modulate a transmitter on another frequency, and radiate the signal in all directions, thus extending the normally limited distances of low-antenna VHF transmissions.

Some amateurs like to communicate by radiotelegraph; others prefer radiotelephone; some prefer radioteletype. Some like to relay or handle messages ("traffic"). Some are interested in testing a variety of new circuits or different antennas on the air. Many like to see if they can contact other amateurs a long distance (DX) away. Some communicate via Oscar satellites. Voice communications may be by AM (6-kHz bandwidth), SSB (2.8-kHz), narrowband FM (6-kHz), or wideband FM (150-kHz). Wideband FM can be used on all amateur bands above 52.5 MHz; narrowband from 3.775 MHz to the highest-frequency band.

An amateur station must make regular measurements of its emitted carriers with frequency-measuring equipment (Chap. 21) other than its receiver and transmitter to assure that its transmissions are always within band limits.

Amateur antennas may be simple single wires, dipoles, verticals, or fixed or rotatable beams. If the antenna impedance matches the transmitter and receiver, they may be intercoupled directly. If the impedances do not match, a balun is used in some cases. A *transmatch,* or antenna tuner, can be used to convert impedances by the use of tuned LC circuits. Amateur antennas must conform to any local building codes. Basically, if more than 200 ft above ground, they may have to be lighted. They must not be more than 1 ft in height for every 100 ft from any airfield runway (approximately 52 ft/mi). If in the path of an airfield runway or if required to be lighted, they must meet specifications listed in ''FCC Rules,'' Part 17. Since high voltages are developed on antenna systems when high-powered transmitters are used, care must be taken that no one will inadvertently be able to touch such points. Towers should be so engineered that they will withstand the highest known wind velocities in the area. To protect equipment from lightning storms, antenna lead-in wires should be connected by a heavy-duty switch to ground when necessary. Lightning arresters may be installed on towers in some cases. Arcing contacts may be installed between transmission lines and ground.

34-4 AMATEUR CALL SIGNS

An amateur station normally identifies at the start of a transmission and at the conclusion, or every 10 min if a transmission exceeds this period. Identification is made by use of the assigned station call letters, either in English or in International Morse code. If a radiotelephone station is mobile or portable, the operator must add the word ''mobile'' or ''portable'' after the call sign. When using CW, a fraction bar is sent after the

call sign of a portable station followed by the number of the call area in which the station is operating. If mobile, the letter M and the call area number follow the fraction bar. MM is used for mobile marine.

The call sign area numbers of the first 48 states and the District of Columbia are:

1 (ME, CT, MA, NH, RI, VT)
2 (NJ, NY)
3 (DE, MD, PA, DC)
4 (AL, FL, GA, KY, NC, SC, TN, VA)
5 (AR, LA, MS, NM, OK, TX)
6 (CA)
7 (AZ, ID, MT, NV, OR, UT, WA, WY)
8 (MI, OH, WV)
9 (IL, IN, WI)
0 (CO, IA, KS, MN, MO, NE, ND, SD)

An an example, W6BNB is the author's California licensed call sign. W2, K2, WA2, WB2, and WN2 are examples of second-area prefixes. The WN indicates a novice station. The call sign prefix KH6 is used for Hawaii, KL7 for Alaska, KP4 for Puerto Rico, KM6 for Midway, etc. There will be either two or three letters following the area-number indicator.

34-5 AMATEUR EXAMINATIONS

While many amateur examinations are given at a field office of the FCC, licenses such as the Novice, Conditional, and Technician classes may be given by volunteer licensed amateurs as explained in later sections. Sometimes the FCC orders such license holders to appear for an FCC supervised reexamination. If the operator fails the examination or fails to take the examination, his license is canceled.

Any U.S. citizen or national is eligible to take amateur radio license examinations. If a license is obtained, it should be kept at the control point or on the person of the operator.

All written amateur license examinations include questions on rules and regulations, radio phenomena, operating procedures, emission characteristics, electrical principles, practical circuits, circuit components, antennas and transmission lines, and radio communication practices.

Amateur license examinations consist of two or more test elements from the following groups:

INTERNATIONAL MORSE CODE SENDING AND RECEIVING TEST

- Element 1(A). 5 wpm
- Element 1(B). 13 wpm
- Element 1(C). 20 wpm

THEORY TEST QUESTIONS

- Element 2. Basic laws essential to beginner operation, plus elementary radio theory.
- Element 3. General amateur practices, operation, apparatus and rules.
- Element 4(A). Intermediate radio theory and operation, including radio telephony and telegraphy.
- Element 4(B). Advanced radio theory, operation, and techniques, including radio telephony, telegraphy, and control of remote objects.

All amateur licenses except Novice class are renewable by application during the last 60 days of the license term. The license and a completed form 610-B are sent to the FCC at Gettysburg, PA 17325. If the license expires, there is a grace period of 1 yr in which it may be renewed, but no operation of transmitting equipment is allowed until the renewed license is received. If the license is not renewed during the 1-yr grace period, the examinations must be taken over again.

The amateur radio operator license and the amateur station license are two separate things. Since both may be applied for at the same time, it is usual to apply for a station license (and call letters) when applying for an operator's license. Amateur station licenses will be issued only to licensed amateur operators.

34-6 AMATEUR BANDS

Licensed amateurs may operate on a series of possible bands. Each band may be broken up into segments. Extra class licensees can operate on any amateur frequency. Advanced class licensees are restricted from operating on a few frequencies. General and Conditional class licensees are still further limited. Technical class licensees can not operate on any frequencies below 50.1 MHz

but, except for 1 MHz in the 2-m (2-meter) band, have full amateur privileges on all higher-frequency bands. Novice class amateurs have only narrow segments on four of the HF bands, and they must operate using A1 emissions only. The amateur bands are:

1.8–2.0 MHz (160-m band)
3.5–4.0 MHz (80-m)
7.0–7.3 MHz (40-m)
14.0–14.35 MHz (20-m)
21.0–21.45 MHz (15-m)
28.0–29.7 MHz (10-m)
50.0–54.0 MHz (6-m)
144–148 MHz (2-m)
220–225 MHz (1-m)
420–450 MHz
1.215–1.3 GHz
2.3–2.45 GHz
3.3–3.5 GHz
5.65–5.925 GHz
10–10.5 GHz
21–22 GHz
Above 50 GHz

Amateur bands are restricted to certain types of emissions. The graph of Fig. 34-2 illustrates the possible frequencies and emissions that may be used by the various classes of licensees for the bands from 160 to 2 m.

There is one *non-amateur* band of frequencies open to experimental or other communications (FCC Part 15.203) which requires no license at all. This is the 160–190-kHz band. The three transmitter requirements for this band are as follows: The power input to the final RF stage must not exceed 1 W. Any emissions outside the band must be suppressed 20 dB or more below the unmodulated carrier. The total length of the transmission line plus the antenna must not exceed 50 ft, which greatly restricts radiation efficiency.

34-7 THE NOVICE CLASS LICENSE

The test for the Novice class license consists of Elements 1(A) and 2.

The applicant first sends for an FCC application form 610 from the closest FCC field office. After filling out the form, he chooses a volunteer General, Advanced, or Extra class amateur 21 years

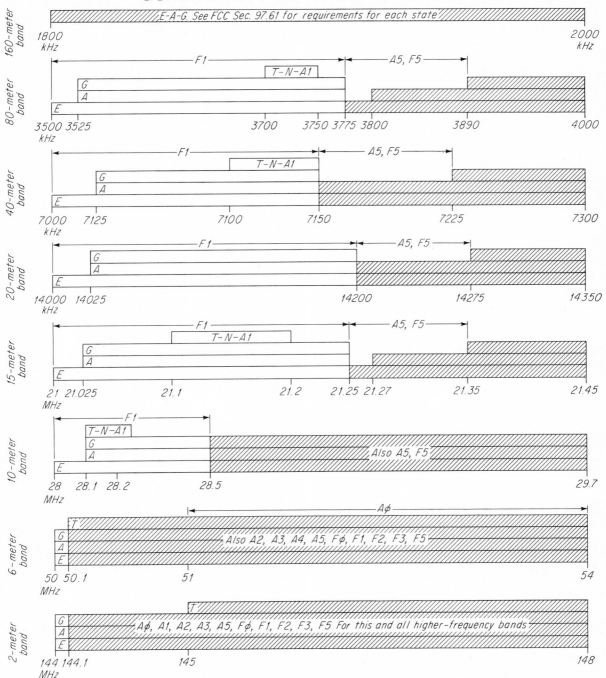

Fig. 34-2 Graph of possible frequencies and emissions that may be used by various classes of licensees for bands from 160 to 2 m.

of age or older (or a commercial radiotelegraph license holder or the operator of a radiotelegraph station operated by the United States). The volunteer gives the code test. If passed, the applicant must then submit the completed Form 610 to FCC, Gettysburg, PA 17325. The application must also include any fee required and a written request from the volunteer examiner for the Novice class license examination papers. The examiner's request must include: (1) names and permanent addresses of both examiner and applicant, (2) description of examiner's qualifications, (3) examiner's statement that the applicant passed the code test for the class license involved within 10 days prior to the submission of the request, and (4) examiner's signature. Within a few weeks the written examination papers are sent to the volunteer examiner, who then administers the theory test to the applicant. The examiner must return the completed test papers within 30 days of the mailing date from Gettysburg.

The test is graded by the FCC, and either the license and call letters (such as WN5XXX) are mailed to the applicant within a few weeks if the test was passed with a grade of 74% or higher or the applicant is notified that the test was failed. If failed, after 30 days the applicant may reinitiate the licensing procedure.

The Novice license is issued for a period of 2 yr but is not renewable. After that period the applicant can retake the examination, however. It is hoped that in this 2 yr he has advanced and can now take the General class license test.

Novice transmitters are limited to a maximum of 75 W of dc input to the final RF stage. Since this is meant to be a learner's license, only a few bands of frequencies are open to Novices, as indicated in Fig. 34-2, and all communications must be by A1 radiotelegraph.

Some representative study questions for Novice class license examinations are listed in the Addendum on page 773.

34-8 THE TECHNICIAN CLASS LICENSE

The test consists of Elements 1(A) and 3. As with the Novice license, volunteer examiners give the code and theory tests (Sec. 34-7). The Technician class license is renewable after its 5-yr term.

Technician licensees have full amateur privileges above 50.1 MHz on the 6-m band and all privileges above 145 MHz, as indicated in Fig. 34-2. For examples of representative study questions for a Technician license examination, see General class questions in the Addendum on page 774.

34-9 THE GENERAL CLASS LICENSE

This is the fundamental all-band amateur license. The test consists of Elements 1(B) and 3.

The General test must be given by a representative of the FCC at either an FCC field office or some designated location at which tests are given periodically. If both code and theory tests are passed, a license will be issued from Gettysburg within a period of a few weeks. If failed, the applicant will be so notified, but after 30 days another test may be taken again. There is a small fee for any amateur test.

General class licensees may operate on the major portions of all amateur bands, as shown in Fig. 34-2. The legal limit of dc input power to the final RF stage of a home or mobile transmitter is 1,000 W except for the 160-m band (FCC Sec. 97.61). Amateur repeater stations are legal only above 52 MHz, and they have special power limitations (FCC Sec. 97.67).

Examples of some representative study questions for a General class license examination are listed in the Addendum on page 774.

34-10 THE CONDITIONAL CLASS LICENSE

This is similar to the General class license except that it is given if the applicant lives more than 175 mi from any FCC testing point, has a disability, or is in the armed forces and unavailable to an FCC testing point. It is given by a volunteer examiner, essentially the same as in Sec. 34-7, except that the test elements will consist of Elements 1(B) and 3.

The Conditional licensee operates with the same restrictions as a General class licensee. Reexamination is not required if the licensee moves within 175 mi of an FCC test point or if a test point is established later within that distance of his home.

See General class study questions for this license in the Addendum on page 774.

34-11 THE ADVANCED CLASS LICENSE

The test consists of Elements 1(B), 3, and 4(A). Credit may be given for Elements 1(B) and 3 if the applicant already has a valid General class license. The test must be taken from an FCC representative at some FCC testing point.

The added privileges of the Advanced license over the General are indicated in Fig. 34-2.

Examples of some representative study questions for an Advanced class license examination are listed in the Addendum on page 776.

34-12 THE EXTRA CLASS LICENSE

The Extra class license requires at least 1 yr of operation with a Conditional, General, or Advanced class license or proof of amateur operation prior to May 1917. The test consists of Elements 1(C), 3, 4(A), and 4(B). Credit for Elements 3 and 4(A) may be granted if the applicant holds a valid General or Advanced class license, as applicable. Credit for Element 1(C) may be granted if the applicant holds a First Class commercial radiotelegraph license or a license with an aircraft radiotelegraph endorsement. The test must be taken at an FCC testing point.

The added privileges of the Extra class license over the other amateur licenses are indicated in Fig. 34-2.

Examples of some representative study questions for an Extra class license examination are listed in the Addendum on page 780.

NEW AMATEUR RULES AND REGULATIONS

By FCC order, effective July 23, 1976, the following changes in Amateur rules and regulations were made. Consider them when reading this chapter.

ALL EXAMINATIONS. Will include element 2, the Novice theory examination.

EXAMINATION CREDIT. Will be given for all elements which were previously passed without Commission supervision.

TESTS BY MAIL. Only Novice license examinations and those for applicants who show by physician's certification that they are unable to appear at an FCC examination point will be given by mail. The FCC will select the volunteer examiner for the latter tests. Present holders of licenses obtained by mail will be able to renew as if the license examination had been Commission conducted.

NOVICES. Maximum input power to the last RF stage is increased to 250 W dc. Transmitter oscillators may be VFO types. *All stations* operating in Novice subbands are limited to a maximum dc power input to the final RF stage of 250 W.

TECHNICIANS. Will have full Novice privileges, including use of the Novice HF CW subbands. They will become General class by passing the 13 wpm code test 1(B) at an FCC examination point.

CONDITIONALS. When renewed become General class automatically, resulting in the phasing out of the Conditional licenses.

appendixes

GREEK ALPHABET

Upper case	Lower case	Name	English equivalent	Upper case	Lower case	Name	English equivalent
A	α	Alpha	a	N	ν	Nu	n
B	β	Beta	b	Ξ	ξ	Xi	x
Γ	γ	Gamma	g	O	o	Omicron	ŏ
Δ	δ	Delta	d	Π	π	Pi	p
E	ϵ	Epsilon	ĕ	P	ρ	Rho	r
Z	ζ	Zeta	z	Σ	σ, s	Sigma	s
H	η	Eta	ē	T	τ	Tau	t
Θ	θ	Theta	th	Υ	υ	Upsilon	u
I	ι	Iota	i	Φ	ϕ, φ	Phi	ph, f
K	κ	Kappa	k	X	χ	Chi	ch
Λ	λ	Lambda	l	Ψ	ψ	Psi	ps
M	μ	Mu	m	Ω	ω	Omega	ō

STANDARD COMPONENT VALUES

Carbon resistors, 0.1- to 3-W types, are available with ±5% tolerance in all of the values listed below. They are also available in ±10% tolerance, but only in the values shown in bold figures. Capacitors, in picofarads (pF), are also generally available in the 10% values shown.

Ohms	Ohms	Ohms	Ohms	Ohms	Ohms	Ohms	Ohms	Ohms	Ohms
2.7	13	**68**	360	**1,800**	9,100	**47,000**	0.24 M	**1.2 M**	6.2 M
3.0	**15**	75	**390**	2,000	**10,000**	51,000	**0.27 M**	1.3 M	**6.8 M**
3.3	16	**82**	430	**2,200**	11,000	**56,000**	0.30 M	**1.5 M**	7.5 M
3.6	**18**	91	**470**	2,400	**12,000**	62,000	**0.33 M**	1.6 M	**8.2 M**
3.9	20	**100**	510	**2,700**	13,000	**68,000**	0.36 M	**1.8 M**	9.1 M
4.3	**22**	110	**560**	3,000	**15,000**	75,000	**0.39 M**	2.0 M	**10.0 M**
4.7	24	**120**	620	**3,300**	16,000	**82,000**	0.43 M	**2.2 M**	11.0 M
5.1	**27**	130	**680**	3,600	**18,000**	91,000	**0.47 M**	2.4 M	**12.0 M**
5.6	30	**150**	750	**3,900**	20,000	0.1 M	0.51 M	**2.7 M**	13.0 M
6.2	**33**	160	820	4,300	**22,000**	0.11 M	**0.56 M**	3.0 M	**15.0 M**
6.8	36	**180**	910	**4,700**	24,000	**0.12 M**	0.62 M	**3.3 M**	16.0 M
7.5	**39**	200	**1,000**	5,100	**27,000**	0.13 M	**0.68 M**	3.6 M	**18.0 M**
8.2	43	**220**	1,100	**5,600**	30,000	**0.15 M**	0.75 M	**3.9 M**	20.0 M
9.1	**47**	240	**1,200**	6,200	**33,000**	0.16 M	**0.82 M**	4.3 M	**22.0 M**
10	51	**270**	1,300	**6,800**	36,000	**0.18 M**	0.91 M	**4.7 M**	
11	**56**	300	**1,500**	7,500	**39,000**	0.20 M	**1.0 M**	5.1 M	
12	62	**330**	1,600	**8,200**	43,000	**0.22 M**	1.1 M	**5.6 M**	

MILITARY PRECISION STANDARD VALUES

Carbon or wire-wound precision (1%) resistors are available from $\frac{1}{8}$ to 10 W in the values indicated below. Only one decade is shown. For other decade values, multiply by 0.1, 10, 100, 1,000, etc. Preferred values are shown in bold figures.

1.00	**1.47**	**2.15**	**3.16**	**4.64**	**6.81**
1.02	1.50	2.21	3.24	4.75	6.98
1.05	1.54	2.26	3.32	4.87	7.15
1.07	1.58	2.32	3.40	4.99	7.32
1.10	**1.62**	**2.37**	**3.48**	**5.11**	**7.50**
1.13	1.65	2.43	3.57	5.23	7.68
1.15	1.69	2.49	3.65	5.36	7.87
1.18	1.74	2.55	3.74	5.49	8.06
1.21	**1.78**	**2.61**	**3.83**	**5.62**	**8.25**
1.24	1.82	2.67	3.92	5.76	8.45
1.27	1.87	2.74	4.02	5.90	8.66
1.30	1.91	2.80	4.12	6.04	8.87
1.33	**1.96**	**2.87**	**4.22**	**6.19**	**9.09**
1.37	2.00	2.94	4.32	6.34	9.31
1.40	2.05	3.01	4.42	6.49	9.53
1.43	2.10	3.09	4.53	6.65	9.76

TABLE OF NATURAL
TRIGONOMETRIC FUNCTIONS

Angle, °	sin	tan	cot	cos		Angle, °	sin	tan	cot	cos	
0.0	.00000	.00000	∞	1.00000	**90.0**	**6.0**	.10453	.10510	9.5144	.99452	**84.0**
.1	.00175	.00175	572.96	1.00000	.9	.1	.10626	.10687	9.3572	.99434	.9
.2	.00349	.00349	286.48	0.99999	.8	.2	.10800	.10863	9.2052	.99415	.8
.3	.00524	.00524	190.98	.99999	.7	.3	.10973	.11040	9.0579	.99396	.7
.4	.00698	.00698	143.24	.99998	.6	.4	.11147	.11217	8.9152	.99377	.6
.5	.00873	.00873	114.59	.99996	.5	.5	.11320	.11394	8.7769	.99357	.5
.6	.01047	.01047	95.489	.99995	.4	.6	.11494	.11570	8.6427	.99337	.4
.7	.01222	.01222	81.847	.99993	.3	.7	.11667	.11747	8.5126	.99317	.3
.8	.01396	.01396	71.615	.99990	.2	.8	.11840	.11924	8.3863	.99297	.2
.9	.01571	.01571	63.657	.99988	.1	.9	.12014	.12101	8.2636	.99276	.1
1.0	.01745	.01746	57.290	.99985	**89.0**	**7.0**	.12187	.12278	8.1443	.99255	**83.0**
.1	.01920	.01920	52.081	.99982	.9	.1	.12360	.12456	8.0285	.99233	.9
.2	.02094	.02095	47.740	.99978	.8	.2	.12533	.12633	7.9158	.99211	.8
.3	.02269	.02269	44.066	.99974	.7	.3	.12706	.12810	7.8062	.99189	.7
.4	.02443	.02444	40.917	.99970	.6	.4	.12880	.12988	7.6996	.99167	.6
.5	.02618	.02619	38.188	.99966	.5	.5	.13053	.13165	7.5958	.99144	.5
.6	.02792	.02793	35.801	.99961	.4	.6	.13226	.13343	7.4947	.99122	.4
.7	.02967	.02968	33.694	.99956	.3	.7	.13399	.13521	7.3962	.99098	.3
.8	.03141	.03143	31.821	.99951	.2	.8	.13572	.13698	7.3002	.99075	.2
.9	.03316	.03317	30.145	.99945	.1	.9	.13744	.13876	7.2066	.99051	.1
2.0	.03490	.03492	28.636	.99939	**88.0**	**8.0**	.13917	.14054	7.1154	.99027	**82.0**
.1	.03664	.03667	27.271	.99933	.9	.1	.14090	.14232	7.0264	.99002	.9
.2	.03839	.03842	26.031	.99926	.8	.2	.14263	.14410	6.9395	.98978	.8
.3	.04013	.04016	24.898	.99919	.7	.3	.14436	.14588	6.8548	.98953	.7
.4	.04188	.04191	23.859	.99912	.6	.4	.14608	.14767	6.7720	.98927	.6
.5	.04362	.04366	22.904	.99905	.5	.5	.14781	.14945	6.6912	.98902	.5
.6	.04536	.04541	22.022	.99897	.4	.6	.14954	.15124	6.6122	.98876	.4
.7	.04711	.04716	21.205	.99889	.3	.7	.15126	.15302	6.5350	.98849	.3
.8	.04885	.04891	20.446	.99881	.2	.8	.15299	.15481	6.4596	.98823	.2
.9	.05059	.05066	19.740	.99872	.1	.9	.15471	.15660	6.3859	.98796	.1
3.0	.05234	.05241	19.081	.99863	**87.0**	**9.0**	.15643	.15838	6.3138	.98769	**81.0**
.1	.05408	.05416	18.464	.99854	.9	.1	.15816	.16017	6.2432	.98741	.9
.2	.05582	.05591	17.886	.99844	.8	.2	.15988	.16196	6.1742	.98714	.8
.3	.05756	.05766	17.343	.99834	.7	.3	.16160	.16376	6.1066	.98686	.7
.4	.05931	.05941	16.832	.99824	.6	.4	.16333	.16555	6.0405	.98657	.6
.5	.06105	.06116	16.350	.99813	.5	.5	.16505	.16734	5.9758	.98629	.5
.6	.06279	.06291	15.895	.99803	.4	.6	.16677	.16914	5.9124	.98600	.4
.7	.06453	.06467	15.464	.99792	.3	.7	.16849	.17093	5.8502	.98570	.3
.8	.06627	.06642	15.056	.99780	.2	.8	.17021	.17273	5.7894	.98541	.2
.9	.06802	.06817	14.669	.99768	.1	.9	.17193	.17453	5.7297	.98511	.1
4.0	.06976	.06993	14.301	.99756	**86.0**	**10.0**	.17365	.17633	5.6713	.98481	**80.0**
.1	.07150	.07168	13.951	.99744	.9	.1	.17537	.17813	5.6140	.98450	.9
.2	.07324	.07344	13.617	.99731	.8	.2	.17708	.17993	5.5578	.98420	.8
.3	.07498	.07519	13.300	.99719	.7	.3	.17880	.18173	5.5026	.98389	.7
.4	.07672	.07695	12.996	.99705	.6	.4	.18052	.18353	5.4486	.98357	.6
.5	.07846	.07870	12.706	.99692	.5	.5	.18224	.18534	5.3955	.98325	.5
.6	.08020	.08046	12.429	.99678	.4	.6	.18395	.18714	5.3435	.98294	.4
.7	.08194	.08221	12.163	.99664	.3	.7	.18567	.18895	5.2924	.98261	.3
.8	.08368	.08397	11.909	.99649	.2	.8	.18738	.19076	5.2422	.98229	.2
.9	.08542	.08573	11.664	.99635	.1	.9	.18910	.19257	5.1929	.98196	.1
5.0	.08716	.08749	11.430	.99619	**85.0**	**11.0**	.19081	.19438	5.1446	.98163	**79.0**
.1	.08889	.08925	11.205	.99604	.9	.1	.19252	.19619	5.0970	.98129	.9
.2	.09063	.09101	10.988	.99588	.8	.2	.19423	.19801	5.0504	.98096	.8
.3	.09237	.09277	10.780	.99572	.7	.3	.19595	.19982	5.0045	.98061	.7
.4	.09411	.09453	10.579	.99556	.6	.4	.19766	.20164	4.9594	.98027	.6
.5	.09585	.09629	10.385	.99540	.5	.5	.19937	.20345	4.9152	.97992	.5
.6	.09758	.09805	10.199	.99523	.4	.6	.20108	.20527	4.8716	.97958	.4
.7	.09932	.09981	10.019	.99506	.3	.7	.20279	.20709	4.8288	.97922	.3
.8	.10106	.10158	9.8448	.99488	.2	.8	.20450	.20891	4.7867	.97887	.2
.9	.10279	.10334	9.6768	.99470	.1	.9	.20620	.21073	4.7453	.97851	.1
6.0	.10453	.10510	9.5144	.99452	**84.0**	**12.0**	.20791	.21256	4.7046	.97815	**78.0**
	cos	**cot**	**tan**	**sin**	**Angle, °**		**cos**	**cot**	**tan**	**sin**	**Angle, °**

Angle, °	sin	tan	cot	cos		Angle, °	sin	tan	cot	cos	
12.0	.20791	.21256	4.7046	.97815	78.0	18.0	.30902	.32492	3.0777	.95106	72.0
.1	.20962	.21438	4.6646	.97778	.9	.1	.31068	.32685	3.0595	.95052	.9
.2	.21132	.21621	4.6252	.97742	.8	.2	.31233	.32878	3.0415	.94997	.8
.3	.21303	.21804	4.5864	.97705	.7	.3	.31399	.33072	3.0237	.94943	.7
.4	.21474	.21986	4.5483	.97667	.6	.4	.31565	.33266	3.0061	.94888	.6
.5	.21644	.22169	4.5107	.97630	.5	.5	.31730	.33460	2.9887	.94832	.5
.6	.21814	.22353	4.4737	.97592	.4	.6	.31896	.33654	2.9714	.94777	.4
.7	.21985	.22536	4.4373	.97553	.3	.7	.32061	.33848	2.9544	.94721	.3
.8	.22155	.22719	4.4015	.97515	.2	.8	.32227	.34043	2.9375	.94665	.2
.9	.22325	.22903	4.3662	.97476	.1	.9	.32392	.34238	2.9208	.94609	.1
13.0	.22495	.23087	4.3315	.97437	77.0	19.0	.32557	.34433	2.9042	.94552	71.0
.1	.22665	.23271	4.2972	.97398	.9	.1	.32722	.34628	2.8878	.94495	.9
.2	.22835	.23455	4.2635	.97358	.8	.2	.32887	.34824	2.8716	.94438	.8
.3	.23005	.23639	4.2303	.97318	.7	.3	.33051	.35020	2.8556	.94380	.7
.4	.23175	.23823	4.1976	.97278	.6	.4	.33216	.35216	2.8397	.94322	.6
.5	.23345	.24008	4.1653	.97237	.5	.5	.33381	.35412	2.8239	.94264	.5
.6	.23514	.24193	4.1335	.97196	.4	.6	.33545	.35608	2.8083	.94206	.4
.7	.23684	.24377	4.1022	.97155	.3	.7	.33710	.35805	2.7929	.94147	.3
.8	.23853	.24562	4.0713	.97113	.2	.8	.33874	.36002	2.7776	.94088	.2
.9	.24023	.24747	4.0408	.97072	.1	.9	.34038	.36199	2.7625	.94029	.1
14.0	.24192	.24933	4.0108	.97030	76.0	20.0	.34202	.36397	2.7475	.93969	70.0
.1	.24362	.25118	3.9812	.96987	.9	.1	.34366	.36595	2.7326	.93909	.9
.2	.24531	.25304	3.9520	.96945	.8	.2	.34530	.36793	2.7179	.93849	.8
.3	.24700	.25490	3.9232	.96902	.7	.3	.34694	.36991	2.7034	.93789	.7
.4	.24869	.25676	3.8947	.96858	.6	.4	.34857	.37190	2.6889	.93728	.6
.5	.25038	.25862	3.8667	.96815	.5	.5	.35021	.37388	2.6746	.93667	.5
.6	.25207	.26048	3.8391	.96771	.4	.6	.35184	.37588	2.6605	.93606	.4
.7	.25376	.26235	3.8118	.96727	.3	.7	.35347	.37787	2.6464	.93544	.3
.8	.25545	.26421	3.7848	.96682	.2	.8	.35511	.37986	2.6325	.93483	.2
.9	.25713	.26608	3.7583	.96638	.1	.9	.35674	.38186	2.6187	.93420	.1
15.0	.25882	.26795	3.7321	.96593	75.0	21.0	.35837	.38386	2.6051	.93358	69.0
.1	.26050	.26982	3.7062	.96547	.9	.1	.36000	.38587	2.5916	.93295	.9
.2	.26219	.27169	3.6806	.96502	.8	.2	.36162	.38787	2.5782	.93232	.8
.3	.26387	.27357	3.6554	.96456	.7	.3	.36325	.38988	2.5649	.93169	.7
.4	.26556	.27545	3.6305	.96410	.6	.4	.36488	.39190	2.5517	.93106	.6
.5	.26724	.27732	3.6059	.96363	.5	.5	.36650	.39391	2.5386	.93042	.5
.6	.26892	.27921	3.5816	.96316	.4	.6	.36812	.39593	2.5257	.92978	.4
.7	.27060	.28109	3.5576	.96269	.3	.7	.36975	.39795	2.5129	.92913	.3
.8	.27228	.28297	3.5339	.96222	.2	.8	.37137	.39997	2.5002	.92849	.2
.9	.27396	.28486	3.5105	.96174	.1	.9	.37299	.40200	2.4876	.92784	.1
16.0	.27564	.28675	3.4874	.96126	74.0	22.0	.37461	.40403	2.4751	.92718	68.0
.1	.27731	.28864	3.4646	.96078	.9	.1	.37622	.40606	2.4627	.92653	.9
.2	.27899	.29053	3.4420	.96029	.8	.2	.37784	.40809	2.4504	.92587	.8
.3	.28067	.29242	3.4197	.95981	.7	.3	.37946	.41013	2.4383	.92521	.7
.4	.28234	.29432	3.3977	.95931	.6	.4	.38107	.41217	2.4262	.92455	.6
.5	.28402	.29621	3.3759	.95882	.5	.5	.38268	.41421	2.4142	.92388	.5
.6	.28569	.29811	3.3544	.95832	.4	.6	.38430	.41626	2.4023	.92321	.4
.7	.28736	.30001	3.3332	.95782	.8	.7	.38591	.41831	2.3906	.92254	.3
.8	.28903	.30192	3.3122	.95732	.2	.8	.38752	.42036	2.3789	.92186	.2
.9	.29070	.30382	3.2914	.95681	.1	.9	.38912	.42242	2.3673	.92119	.1
17.0	.29237	.30573	3.2709	.95630	73.0	23.0	.39073	.42447	2.3559	.92050	67.0
.1	.29404	.30764	3.2506	.95579	.9	.1	.39234	.42654	2.3445	.91982	.9
.2	.29571	.30955	3.2305	.95528	.8	.2	.39394	.42860	2.3332	.91914	.8
.3	.29737	.31147	3.2106	.95476	.7	.3	.39555	.43067	2.3220	.91845	.7
.4	.29904	.31338	3.1910	.95424	.6	.4	.39715	.43274	2.3109	.91775	.6
.5	.30071	.31530	3.1716	.95372	.5	.5	.39875	.43481	2.2998	.91706	.5
.6	.30237	.31722	3.1524	.95319	.4	.6	.40035	.43689	2.2889	.91636	.4
.7	.30403	.31914	3.1334	.95266	.3	.7	.40195	.43897	2.2781	.91566	.3
.8	.30570	.32106	3.1146	.95213	.2	.8	.40355	.44105	2.2673	.91496	.2
.9	.30736	.32299	3.0961	.95159	.1	.9	.40514	.44314	2.2566	.91425	.1
18.0	.30902	.32492	3.0777	.95106	72.0	24.0	.40674	.44523	2.2460	.91355	66.0
	cos	cot	tan	sin	Angle, °		cos	cot	tan	sin	Angle, °

TABLE OF NATURAL TRIGONOMETRIC FUNCTIONS (*continued*)

Angle, °	sin	tan	cot	cos		Angle, °	sin	tan	cot	cos	
24.0▼	.40674	.44523	2.2460	.91355	66.0	30.0▼	.50000	.57735	1.7321	.86603	60.0
.1	.40833	.44732	2.2355	.91283	.9	.1	.50151	.57968	1.7251	.86515	.9
.2	.40992	.44942	2.2251	.91212	.8	.2	.50302	.58201	1.7182	.86427	.8
.3	.41151	.45152	2.2148	.91140	.7	.3	.50453	.58435	1.7113	.86340	.7
.4	.41310	.45362	2.2045	.91068	.6	.4	.50603	.58670	1.7045	.86251	.6
.5	.41469	.45573	2.1943	.90996	.5	.5	.50754	.58905	1.6977	.86163	.5
.6	.41628	.45784	2.1842	.90924	.4	.6	.50904	.59140	1.6909	.86074	.4
.7	.41787	.45995	2.1742	.90851	.3	.7	.51054	.59376	1.6842	.85985	.3
.8	.41945	.46206	2.1642	.90778	.2	.8	.51204	.59612	1.6775	.85896	.2
.9	.42104	.46418	2.1543	.90704	.1	.9	.51354	.59849	1.6709	.85806	.1
25.0	.42262	.46631	2.1445	.90631	65.0	31.0	.51504	.60086	1.6643	.85717	59.0
.1	.42420	.46843	2.1348	.90557	.9	.1	.51653	.60324	1.6577	.85627	.9
.2	.42578	.47056	2.1251	.90483	.8	.2	.51803	.60562	1.6512	.85536	.8
.3	.42736	.47270	2.1155	.90408	.7	.3	.51952	.60801	1.6447	.85446	.7
.4	.42894	.47483	2.1060	.90334	.6	.4	.52101	.61040	1.6383	.85355	.6
.5	.43051	.47698	2.0965	.90259	.5	.5	.52250	.61280	1.6319	.85264	.5
.6	.43209	.47912	2.0872	.90183	.4	.6	.52399	.61520	1.6255	.85173	.4
.7	.43366	.48127	2.0778	.90108	.3	.7	.52547	.61761	1.6191	.85081	.3
.8	.43523	.48342	2.0686	.90032	.2	.8	.52696	.62003	1.6128	.84989	.2
.9	.43680	.48557	2.0594	.89956	.1	.9	.52844	.62245	1.6066	.84897	.1
26.0	.43837	.48773	2.0503	.89879	64.0	32.0	.52992	.62487	1.6003	.84805	58.0
.1	.43994	.48989	2.0413	.89803	.9	.1	.53140	.62730	1.5941	.84712	.9
.2	.44151	.49206	2.0323	.89726	.8	.2	.53288	.62973	1.5880	.84619	.8
.3	.44307	.49423	2.0233	.89649	.7	.3	.53435	.63217	1.5818	.84526	.7
.4	.44464	.49640	2.0145	.89571	.6	.4	.53583	.63462	1.5757	.84433	.6
.5	.44620	.49858	2.0057	.89493	.5	.5	.53730	.63707	1.5697	.84339	.5
.6	.44776	.50076	1.9970	.89415	.4	.6	.53877	.63953	1.5637	.84245	.4
.7	.44932	.50295	1.9883	.89337	.3	.7	.54024	.64199	1.5577	.84151	.3
.8	.45088	.50514	1.9797	.89259	.2	.8	.54171	.64446	1.5517	.84057	.2
.9	.45243	.50733	1.9711	.89180	.1	.9	.54317	.64693	1.5458	.83962	.1
27.0	.45399	.50953	1.9626	.89101	63.0	33.0	.54464	.64941	1.5399	.83867	57.0
.1	.45554	.51173	1.9542	.89021	.9	.1	.54610	.65189	1.5340	.83772	.9
.2	.45710	.51393	1.9458	.88942	.8	.2	.54756	.65438	1.5282	.83676	.8
.3	.45865	.51614	1.9375	.88862	.7	.3	.54902	.65688	1.5224	.83581	.7
.4	.46020	.51835	1.9292	.88782	.6	.4	.55048	.65938	1.5166	.83485	.6
.5	.46175	.52057	1.9210	.88701	.5	.5	.55194	.66189	1.5108	.83389	.5
.6	.46330	.52279	1.9128	.88620	.4	.6	.55339	.66440	1.5051	.83292	.4
.7	.46484	.52501	1.9047	.88539	.3	.7	.55484	.66692	1.4994	.83195	.3
.8	.46639	.52724	1.8967	.88458	.2	.8	.55630	.66944	1.4938	.83098	.2
.9	.46793	.52947	1.8887	.88377	.1	.9	.55775	.67197	1.4882	.83001	.1
28.0	.46947	.53171	1.8807	.88295	62.0	34.0	.55919	.67451	1.4826	.82904	56.0
.1	.47101	.53395	1.8728	.88213	.9	.1	.56064	.67705	1.4770	.82806	.9
.2	.47255	.53620	1.8650	.88130	.8	.2	.56208	.67960	1.4715	.82708	.8
.3	.47409	.53844	1.8572	.88048	.7	.3	.56353	.68215	1.4659	.82610	.7
.4	.47562	.54070	1.8495	.87965	.6	.4	.56497	.68471	1.4605	.82511	.6
.5	.47716	.54296	1.8418	.87882	.5	.5	.56641	.68728	1.4550	.82413	5
.6	.47869	.54522	1.8341	.87798	.4	.6	.56784	.68985	1.4496	.82314	.4
.7	.48022	.54748	1.8265	.87715	.3	.7	.56928	.69243	1.4442	.82214	.3
.8	.48175	.54975	1.8190	.87631	.2	.8	.57071	.69502	1.4388	.82115	.2
.9	.48328	.55203	1.8115	.87546	.1	.9	.57215	.69761	1.4335	.82015	.1
29.0	.48481	.55431	1.8040	.87462	61.0	35.0	.57358	.70021	1.4281	.81915	55.0
.1	.48634	.55659	1.7966	.87377	.9	.1	.57501	.70281	1.4229	.81815	.9
.2	.48786	.55888	1.7893	.87292	.8	.2	.57643	.70542	1.4176	.81714	.8
.3	.48938	.56117	1.7820	.87207	.7	.3	.57786	.70804	1.4124	.81614	.7
.4	.49090	.56347	1.7747	.87121	.6	.4	.57928	.71066	1.4071	.81513	.6
.5	.49242	.56577	1.7675	.87036	.5	.5	.58070	.71329	1.4019	.81412	.5
.6	.49394	.56808	1.7603	.86949	.4	.6	.58212	.71593	1.3968	.81310	.4
.7	.49546	.57039	1.7532	.86863	.3	.7	.58354	.71857	1.3916	.81208	.3
.8	.49697	.57271	1.7461	.86777	.2	.8	.58496	.72122	1.3865	.81106	.2
.9	.49849	.57503	1.7391	.86690	.1	.9	.58637	.72388	1.3814	.81004	.1
30.0	.50000	.57735	1.7321	.86603	60.0▲	36.0	.58779	.72654	1.3764	.80902	54.0▲
	cos	cot	tan	sin	Angle, °		cos	cot	tan	sin	Angle, °

TABLE OF NATURAL TRIGONOMETRIC FUNCTIONS (*continued*)

Angle, °	sin	tan	cot	cos		Angle, °	sin	tan	cot	cos	
36.0	.58779	.72654	1.3764	.80902	54.0	40.5	.64945	.85408	1.1708	.76041	49.5
.1	.58920	.72921	1.3713	.80799	.9	.6	.65077	.85710	1.1667	.75927	.4
.2	.59061	.73189	1.3663	.80696	.8	.7	.65210	.86014	1.1626	.75813	.3
.3	.59201	.73457	1.3613	.80593	.7	.8	.65342	.86318	1.1585	.75700	.2
.4	.59342	.73726	1.3564	.80489	.6	.9	.65474	.86623	1.1544	.75585	.1
.5	.59482	.73996	1.3514	.80386	.5	41.0	.65606	.86929	1.1504	.75471	49.0
.6	.59622	.74267	1.3465	.80282	.4	.1	.65738	.87236	1.1463	.75356	.9
.7	.59763	.74538	1.3416	.80178	.3	.2	.65869	.87543	1.1423	.75241	.8
.8	.59902	.74810	1.3367	.80073	.2	.3	.66000	.87852	1.1383	.75126	.7
.9	.60042	.75082	1.3319	.79968	.1	.4	.66131	.88162	1.1343	.75011	.6
37.0	.60182	.75355	1.3270	.79864	53.0	.5	.66262	.88473	1.1303	.74896	.5
.1	.60321	.75629	1.3222	.79758	.9	.6	.66393	.88784	1.1263	.74780	.4
.2	.60460	.75904	1.3175	.79653	.8	.7	.66523	.89097	1.1224	.74664	.3
.3	.60599	.76180	1.3127	.79547	.7	.8	.66653	.89410	1.1184	.74548	.2
.4	.60738	.76456	1.3079	.79441	.6	.9	.66783	.89725	1.1145	.74431	.1
.5	.60876	.76733	1.3032	.79335	.5	42.0	.66913	.90040	1.1106	.74314	48.0
.6	.61015	.77010	1.2985	.79229	.4	.1	.67043	.90357	1.1067	.74198	.9
.7	.61153	.77289	1.2938	.79122	.3	.2	.67172	.90674	1.1028	.74080	.8
.8	.61291	.77568	1.2892	.79016	.2	.3	.67301	.90993	1.0990	.73963	.7
.9	.61429	.77848	1.2846	.78908	.1	.4	.67430	.91313	1.0951	.73846	.6
38.0	.61566	.78129	1.2799	.78801	52.0	.5	.67559	.91633	1.0913	.73728	.5
.1	.61704	.78410	1.2753	.78694	.9	.6	.67688	.91955	1.0875	.73610	.4
.2	.61841	.78692	1.2708	.78586	.8	.7	.67816	.92277	1.0837	.73491	.3
.3	.61978	.78975	1.2662	.78478	.7	.8	.67944	.92601	1.0799	.73373	.2
.4	.62115	.79259	1.2617	.78369	.6	.9	.68072	.92926	1.0761	.73254	.1
.5	.62251	.79544	1.2572	.78261	.5	43.0	.68200	.93252	1.0724	.73135	47.0
.6	.62388	.79829	1.2527	.78152	.4	.1	.68327	.93578	1.0686	.73016	.9
.7	.62524	.80115	1.2482	.78043	.3	.2	.68455	.93906	1.0649	.72897	.8
.8	.62660	.80402	1.2437	.77934	.2	.3	.68582	.94235	1.0612	.72777	.7
.9	.62796	.80690	1.2393	.77824	.1	.4	.68709	.94565	1.0575	.72657	.6
39.0	.62932	.80978	1.2349	.77715	51.0	.5	.68835	.94896	1.0538	.72537	.5
.1	.63068	.81268	1.2305	.77605	.9	.6	.68962	.95229	1.0501	.72417	.4
.2	.63203	.81558	1.2261	.77494	.8	.7	.69088	.95562	1.0464	.72297	.3
.3	.63338	.81849	1.2218	.77384	.7	.8	.69214	.95897	1.0428	.72176	.2
.4	.63473	.82141	1.2174	.77273	.6	.9	.69340	.96232	1.0392	.72055	.1
.5	.63608	.82434	1.2131	.77162	.5	44.0	.69466	.96569	1.0355	.71934	46.0
.6	.63742	.82727	1.2088	.77051	.4	.1	.69591	.96907	1.0319	.71813	.9
.7	.63877	.83022	1.2045	.76940	.3	.2	.69717	.97246	1.0283	.71691	.8
.8	.64011	.83317	1.2002	.76828	.2	.3	.69842	.97586	1.0247	.71569	.7
.9	.64145	.83613	1.1960	.76717	.1	.4	.69966	.97927	1.0212	.71447	.6
40.0	.64279	.83910	1.1918	.76604	50.0	.5	.70091	.98270	1.0176	.71325	.5
.1	.64412	.84208	1.1875	.76492	.9	.6	.70215	.98613	1.0141	.71203	.4
.2	.64546	.84507	1.1833	.76380	.8	.7	.70339	.98958	1.0105	.71080	.3
.3	.64679	.84806	1.1792	.76267	.7	.8	.70463	.99304	1.0070	.70957	.2
.4	.64812	.85107	1.1750	.76154	.6	.9	.70587	.99652	1.0035	.70834	.1
40.5	.64945	.85408	1.1708	.76041	49.5	45.0	.70711	1.00000	1.0000	.70711	45.0
	cos	cot	tan	sin	Angle, °		cos	cot	tan	sin	Angle, °

TABLE OF LOGARITHMS
(FOUR-PLACE MANTISSAS)

No.	0	1	2	3	4	5	6	7	8	9
10	0000	0043	0086	0128	0170	0212	0253	0294	0334	0374
11	0414	0453	0492	0531	0569	0607	0645	0682	0719	0755
12	0792	0828	0864	0899	0934	0969	1004	1038	1072	1106
13	1139	1173	1206	1239	1271	1303	1335	1367	1399	1430
14	1461	1492	1523	1553	1584	1614	1644	1673	1703	1732
15	1761	1790	1818	1847	1875	1903	1931	1959	1987	2014
16	2041	2068	2095	2122	2148	2175	2201	2227	2253	2279
17	2304	2330	2355	2380	2405	2430	2455	2480	2504	2529
18	2553	2577	2601	2625	2648	2672	2695	2718	2742	2765
19	2788	2810	2833	2856	2878	2900	2923	2945	2967	2989
20	3010	3032	3054	3075	3096	3118	3139	3160	3181	3201
21	3222	3243	3263	3284	3304	3324	3345	3365	3385	3404
22	3424	3444	3464	3483	3502	3522	3541	3560	3579	3598
23	3617	3636	3655	3674	3692	3711	3729	3747	3766	3784
24	3802	3820	3838	3856	3874	3892	3909	3927	3945	3962
25	3979	3997	4014	4031	4048	4065	4082	4099	4116	4133
26	4150	4166	4183	4200	4216	4232	4249	4265	4281	4298
27	4314	4330	4346	4362	4378	4393	4409	4425	4440	4456
28	4472	4487	4502	4518	4533	4548	4564	4579	4594	4609
29	4624	4639	4654	4669	4683	4698	4713	4728	4742	4757
30	4771	4786	4800	4814	4829	4843	4857	4871	4886	4900
31	4914	4928	4942	4955	4969	4983	4997	5011	5024	5038
32	5051	5065	5079	5092	5105	5119	5132	5145	5159	5172
33	5185	5198	5211	5224	5237	5250	5263	5276	5289	5302
34	5315	5328	5340	5353	5366	5378	5391	5403	5416	5428
35	5441	5453	5465	5478	5490	5502	5514	5527	5539	5551
36	5563	5575	5587	5599	5611	5623	5635	5647	5658	5670
37	5682	5694	5705	5717	5729	5740	5752	5763	5775	5786
38	5798	5809	5821	5832	5843	5855	5866	5877	5888	5899
39	5911	5922	5933	5944	5955	5966	5977	5988	5999	6010
40	6021	6031	6042	6053	6064	6075	6085	6096	6107	6117
41	6128	6138	6149	6160	6170	6180	6191	6201	6212	6222
42	6232	6243	6253	6263	6274	6284	6294	6304	6314	6325
43	6335	6345	6355	6365	6375	6385	6395	6405	6415	6425
44	6435	6444	6454	6464	6474	6484	6493	6503	6513	6522
45	6532	6542	6551	6561	6571	6580	6590	6599	6609	6618
46	6628	6637	6646	6656	6665	6675	6684	6693	6702	6712
47	6721	6730	6739	6749	6758	6767	6776	6785	6794	6803
48	6812	6821	6830	6839	6848	6857	6866	6875	6884	6893
49	6902	6911	6920	6928	6937	6946	6955	6964	6972	6981
50	6990	6998	7007	7016	7024	7033	7042	7050	7059	7067
51	7076	7084	7093	7101	7110	7118	7126	7135	7143	7152
52	7160	7168	7177	7185	7193	7202	7210	7218	7226	7235
53	7243	7251	7259	7267	7275	7284	7292	7300	7308	7316
54	7324	7332	7340	7348	7356	7364	7372	7380	7388	7396
No.	0	1	2	3	4	5	6	7	8	9

TABLE OF LOGARITHMS (*continued*)

No.	0	1	2	3	4	5	6	7	8	9
55	7404	7412	7419	7427	7435	7443	7451	7459	7466	7474
56	7482	7490	7497	7505	7513	7520	7528	7536	7543	7551
57	7559	7566	7574	7582	7589	7597	7604	7612	7619	7627
58	7634	7642	7649	7657	7664	7672	7679	7686	7694	7701
59	7709	7716	7723	7731	7738	7745	7752	7760	7767	7774
60	7782	7789	7796	7803	7810	7818	7825	7832	7839	7846
61	7853	7860	7868	7875	7882	7889	7896	7903	7910	7917
62	7924	7931	7938	7945	7952	7959	7966	7973	7980	7987
63	7993	8000	8007	8014	8021	8028	8035	8041	8048	8055
64	8062	8069	8075	8082	8089	8096	8102	8109	8116	8122
65	8129	8136	8142	8149	8156	8162	8169	8176	8182	8189
66	8195	8202	8209	8215	8222	8228	8235	8241	8248	8254
67	8261	8267	8274	8280	8287	8293	8299	8306	8312	8319
68	8325	8331	8338	8344	8351	8357	8363	8370	8376	8382
69	8388	8395	8401	8407	8414	8420	8426	8432	8439	8445
70	8451	8457	8463	8470	8476	8482	8488	8494	8500	8506
71	8513	8519	8525	8531	8537	8543	8549	8555	8561	8567
72	8573	8579	8585	8591	8597	8603	8609	8615	8621	8627
73	8633	8639	8645	8651	8657	8663	8669	8675	8681	8686
74	8692	8698	8704	8710	8716	8722	8727	8733	8739	8745
75	8751	8756	8762	8768	8774	8779	8785	8791	8797	8802
76	8808	8814	8820	8825	8831	8837	8842	8848	8854	8859
77	8865	8871	8876	8882	8887	8893	8899	8904	8910	8915
78	8921	8927	8932	8938	8943	8949	8954	8960	8965	8971
79	8976	8982	8987	8993	8998	9004	9009	9015	9020	9025
80	9031	9036	9042	9047	9053	9058	9063	9069	9074	9079
81	9085	9090	9096	9101	9106	9112	9117	9122	9128	9133
82	9138	9143	9149	9154	9159	9165	9170	9175	9180	9186
83	9191	9196	9201	9206	9212	9217	9222	9227	9232	9238
84	9243	9248	9253	9258	9263	9269	9274	9279	9284	9289
85	9294	9299	9304	9309	9315	9320	9325	9330	9335	9340
86	9345	9350	9355	9360	9365	9370	9375	9380	9385	9390
87	9395	9400	9405	9410	9415	9420	9425	9430	9435	9440
88	9445	9450	9455	9460	9465	9469	9474	9479	9484	9489
89	9494	9499	9504	9509	9513	9518	9523	9528	9533	9538
90	9542	9547	9552	9557	9562	9566	9571	9576	9581	9586
91	9590	9595	9600	9605	9609	9614	9619	9624	9628	9633
92	9638	9643	9647	9652	9657	9661	9666	9671	9675	9680
93	9685	9689	9694	9699	9703	9708	9713	9717	9722	9727
94	9731	9736	9741	9745	9750	9754	9759	9763	9768	9773
95	9777	9782	9786	9791	9795	9800	9805	9809	9814	9818
96	9823	9827	9832	9836	9841	9845	9850	9854	9859	9863
97	9868	9872	9877	9881	9886	9890	9894	9899	9903	9908
98	9912	9917	9921	9926	9930	9934	9939	9943	9948	9952
99	9956	9961	9965	9969	9974	9978	9983	9987	9991	9996
No.	0	1	2	3	4	5	6	7	8	9

EMISSIONS

AMPLITUDE-MODULATED

A0   Carrier carrying no modulation or information

A1   Telegraphy; on-off; no other modulation

A2   Telegraphy; on-off; amplitude-modulated tone

A3   Telephony; carrier with double sideband

A3A Telephony; reduced carrier with single sideband

A3J Telephony, suppressed carrier with single sideband

A3H Telephony; full carrier with single sideband

A3B Telephony with two independent sidebands

A4   Facsimile (slow-scan TV)

A5C Television with vestigial sideband

A9B Telephony or telegraphy with independent sidebands

FREQUENCY- OR PHASE-MODULATED

F1   Telegraphy; frequency-shift-keyed

F2   Telegraphy; on-off; frequency-modulated tone

F3   Telephony; frequency- or phase-modulated

F4   Facsimile

F5   Television

F6   Telegraphy; four-frequency diplex

PULSE-MODULATED

P0   Radar (pulsed carrier without information)

P1D Telegraphy; on-off keying of pulsed carrier

P2D Telegraphy; pulsed-carrier tone-modulated

P2E Telegraphy; pulse-width tone-modulated

P2F Telegraphy; phase or position tone-modulated

P3D Telephony; amplitude-modulated pulses

P3E Telephony; pulse-width-modulated

P3F Telephony; pulses phase- or position-modulated

NOTE: When a number precedes the emission designation, it indicates the allowable bandwidth in kilohertz. Thus, 0.1A1 indicates a slow-speed radiotelegraph on-off emission; 2.1A2, a slow-speed, 1,000-Hz tone-modulated telegraph emission; 6A3, a 3,000-Hz voice-modulated double-sideband emission; 3A3J, a suppressed-carrier single-sideband 3,000-Hz voice modulation; 10A3, sound broadcasting of 5,000-Hz music and speech; and 15750A5C, the broadcast TV video emission.

FCC RULES AND REGULATIONS AND FIELD OFFICES

The "FCC Rules and Regulations" are sold in volumes by the Superintendent of Documents, Government Printing Office, Washington, D.C. The price of a volume entitles the purchaser to receive its amended pages for an indefinite period. The seven most useful volumes comprise individual parts, as follows:

Volume I
    Part 0: Commission Organization
    Part 1: Practice and Procedure
    Part 13: Commercial Radio Operators
    Part 17: Construction, Marking, and Lighting of Antenna Structures
Volume II
    Part 2: Frequency Allocations and Radio Treaty Matters; General Rules and Regulations
    Part 5: Experimental Radio Services (Other than Broadcast)
    Part 15: Radio Frequency Devices
    Part 18: Industrial, Scientific, and Medical Equipment
Volume III
    Part 73: Radio Broadcast Services
    Part 74: Experimental, Auxiliary, and Special Broadcast Services
Volume IV
    Part 81: Stations on Land in the Maritime Services
    Part 83: Stations on Shipboard in the Maritime Services
    Part 85: Public Fixed Stations and Stations of the Maritime Services in Alaska
Volume V
    Part 87: Aviation Services
    Part 91: Industrial Radio Services
    Part 93: Land Transportation Radio Services
Volume VI
    Part 95: Citizens Radio Service
    Part 97: Amateur Radio Service
    Part 99: Disaster Communications Service
Volume VII
    Part 21: Domestic Public Radio Services (Other than Maritime Mobile)
    Part 23: International Fixed Public Radiocommunication Services
    Part 25: Satellite Communications

The following are the mailing addresses for Federal Communications Commission field offices. Street addresses can be found in local directories under the heading "United States government." Address all communications to *Engineer-in-Charge, Federal Communications Commission.*

Mobile	Alabama 36602
Anchorage	Alaska 99501
(P.O. Box 644)	
Long Beach	California 90807
San Diego	California 92101
San Francisco	California 94111
San Pedro	California 90731
Denver	Colorado 80202
Washington	D.C. 20554
Miami	Florida 33130
(P.O. Box 150)	
Tampa	Florida 33602
Atlanta	Georgia 30303
Savannah	Georgia 31402
(P.O. Box 8004)	
Honolulu	Hawaii 96808
Chicago	Illinois 60604
New Orleans	Louisiana 70130
Baltimore	Maryland 21202
Boston	Massachusetts 02109
Detroit	Michigan 48226
St. Paul	Minnesota 55101
Kansas City	Missouri 64106
Buffalo	New York 14203
New York	New York 10014
Portland	Oregon 97204
Philadelphia	Pennsylvania 19106
San Juan	Puerto Rico 00903
(P.O. Box 2987)	
Beaumont	Texas 77701
(P.O. Box 1527)	
Dallas	Texas 75202
Houston	Texas 77002
Norfolk	Virginia 23510
Seattle	Washington 98104

Q SIGNALS

Signal	Meaning	Signal	Meaning
QRA?	What is the name of your station?	QRA	The name of my station is . . .
QRB?	How far approximately are you from my station?	QRB	The approximate distance between our stations is . . . miles.
QRC?	By what private enterprise (or, state administration) are charges for your station settled?	QRC	The accounts for charges of my station are settled by . . .
QRD?	Where are you bound, and where are you from?	QRD	I am bound for . . . from . . . .
QRE?	What is your estimated time of arrival at . . . ?	QRE	ETA at . . . is . . . hours.
QRF?	Are you returning to . . . ?	QRF	I am returning to . . . .
QRG?	Will you tell me my exact frequency (or, that of . . .)?	QRG	Your exact frequency (or, that of . . .) is . . . kHz (or MHz).
QRH?	Does my frequency vary?	QRH	Your frequency varies.
QRI?	How is the tone of my transmission?	QRI	The tone of your transmission is: (1) Good. (2) Variable. (3) Bad.
QRK?	What is the readability of my signals (or, those of . . .)?	QRK	Readability is: (1) Unreadable. (2) Readable now and then. (3) Readable with difficulty. (4) Readable. (5) Perfectly readable.
QRL?	Are you busy?	QRL	I am busy (or, busy with . . .). Please do not interfere.
QRM?	Are you being interfered with?	QRM	I am being interfered with.
QRN?	Are you troubled by static?	QRN	I am troubled by static.
QRO?	Shall I increase power?	QRO	Increase power.
QRP?	Shall I decrease power?	QRP	Decrease power.
QRQ?	Shall I send faster?	QRQ	Send faster (. . . wpm).
QRR?	Are you ready for automatic operation?	QRR	I am ready for automatic operation. Send at . . . wpm.
QRS?	Shall I send more slowly?	QRS	Send more slowly.
QRT?	Shall I stop sending?	QRT	Stop sending.
QRU?	Have you anything for me?	QRU	I have nothing for you.
QRV?	Are you ready?	QRV	I am ready.
QRW?	Shall I inform . . . that you are calling him on . . . kHz?	QRW	Please inform . . . that I am calling him on . . . kHz.
QRX?	When will you call me again?	QRX	I will call you again at . . . hours.
QRY?	What is my turn?	QRY	Your turn is number . . . .
QRZ?	Who is calling me?	QRZ	You are being called by . . . .
QSA?	What is the strength of my signals (or, those of . . .)?	QSA	Your signals are: (1) Scarcely perceptible. (2) Weak. (3) Fairly good. (4) Good. (5) Very good.
QSB?	Are my signals fading?	QSB	Your signals are fading.
QSC?	Are you a cargo vessel?	QSC	I am a cargo vessel.
QSD?	Is my keying defective?	QSD	Your keying is defective.
QSG?	Shall I send . . . messages at a time?	QSG	Send . . . messages at a time.

Signal	Meaning	Signal	Meaning
QSJ?	What is the charge to be collected per word to . . . including your internal telegraph charge?	QSJ	The charge to be collected per word to . . . including my internal charge is . . . francs.
QSK?	Can you hear me between your signals?	QSK	I can hear you between my signals.
QSL?	Can you acknowledge receipt?	QSL	I am acknowledging receipt.
QSM?	Shall I repeat the last telegram which I sent you, or some previous telegram?	QSM	Repeat the last telegram which you sent me [or, telegram(s) number(s) . . .].
QSN?	Did you hear me (or, on . . . kHz)?	QSN	I did hear you (or, on . . . kHz).
QSO?	Can you communicate with . . . direct or by relay?	QSO	I can communicate with . . . direct or by relay through . . . .
QSP?	Will you relay to . . . free of charge?	QSP	I will relay to . . . free of charge.
QSQ?	Have you a doctor on board or, is . . . (name of person) on board?	QSQ	I have a doctor on board or, . . . (name of person) is on board.
QSU?	Shall I send or reply on this frequency (or, on . . . kHz) (with emissions of class . . .)?	QSU	Send or reply on this frequency (or, on . . . kHz) (with emissions of class . . .).
QSV?	Shall I send a series of V's on this frequency (or, . . . kHz)?	QSV	Send a series of V's on this frequency (or, . . . kHz).
QSW?	Will you send on this frequency (or, . . . kHz) (with emissions of class . . .)?	QSW	I am going to send on this frequency (or, . . . kHz) (with emissions of class . . .).
QSX?	Will you listen to . . . (call sign) on . . . kHz?	QSX	I am listening to . . . (call sign) on . . . kHz.
QSY?	Shall I change to transmission on another frequency?	QSY	Change to transmission on another frequency (or, on . . . kHz).
QSZ?	Shall I send each word or group more than once?	QSZ	Send each word or group twice (or, . . . times).
QTA?	Shall I cancel telegram number . . . as if it had not been sent?	QTA	Cancel telegram number . . . as if it had not been sent.
QTB?	Do you agree with my counting of words?	QTB	I do not agree, I will repeat the first letter of each word.
QTC?	How many telegrams have you to send?	QTC	I have . . . telegrams for you (or, for . . .).
QTE?	What is my TRUE bearing from you (or, What is my TRUE bearing from . . . ; or, What is the TRUE bearing of . . . from . . .)?	QTE	Your TRUE bearing from me is . . . degrees (at . . . hours) (or, Your TRUE bearing from . . . was . . . degrees at . . . hours; or, The TRUE bearing of . . . from . . . was . . . degrees at . . . hours).
QTF?	Will you give me the position of my station according to the bearings taken by the direction-finding stations which you control?	QTF	The position of your station according to the bearings taken by the direction-finding stations which I control was . . . latitude . . . longitude.
QTG?	Will you send two dashes of 10 s each followed by your call sign (repeated . . . times)?	QTG	I am going to send two dashes of 10 s each followed by my call sign (repeated . . . times).

Signal	Meaning	Signal	Meaning
QTH?	What is your position in latitude and longitude (or, according to any other indication)?	QTH	My position is . . . latitude . . . longitude (or, according to any other indication).
QTI?	What is your TRUE track?	QTI	My TRUE track is . . . degrees.
QTJ?	What is your speed?	QTJ	My speed is . . . knots.
QTK?	What is the speed of your aircraft in relation to the surface of the earth?	QTK	The speed of my aircraft is . . . knots.
QTL?	What is your TRUE heading (TRUE course with no wind)?	QTL	My TRUE heading is . . . degrees.
QTN?	At what time did you depart from . . . (place)?	QTN	I departed from . . . (place) at . . . hours.
QTO?	Have you left dock (or, Are you airborne)?	QTO	I have left dock (or, I am airborne).
QTP?	Are you going to enter dock (or, Are you going to alight)?	QTP	I am going to enter dock (or, I am going to alight).
QTQ?	Can you communicate with my station by means of the International code of signals?	QTQ	I am going to communicate with your station by means of the International code of signals.
QTR?	What is the correct time?	QTR	The correct time is . . . hours.
QTS?	Will you send your call sign for . . . minute(s) now (or, at . . . hours) (on . . . kHz) so that your frequency may be measured?	QTS	I will send my call sign for . . . minute(s) now (or, at . . . hours) (on . . . kHz) so that my frequency may be measured.
QTU?	What are the hours during which your station is open?	QTU	My station is open from . . . to . . . hours.
QTV?	Shall I stand guard for you on the frequency of . . . kHz (from . . . to . . . hours)?	QTV	Stand guard for me on the frequency of . . . kHz (from . . . to . . . hours).
QTX?	Will you keep your station open for further communication with me until further notice (or, until . . . hours)?	QTX	I will keep my station open for further communication with you until further notice (or, until . . . hours).
QUA?	Have you news of . . . (call)?	QUA	Here is news of . . . (call).
QUB?	Can you give me, in the following order, information concerning visibility, height of clouds, and direction and velocity of ground wind at . . . (place)?	QUB	Here is the information requested . . . .
QUC?	What is the number (or other indication) of the last message you received from me [or, from . . . (call sign)]?	QUC	The number (or other indication) of the last message I received from you [or from . . . (call sign)] is . . . .
QUD?	Have you received the urgency signal sent by . . . (call sign)?	QUD	I have received the urgency signal sent by . . . (call sign).
QUF?	Have you received the distress signal sent by . . . (call sign)?	QUF	I have received the distress signal sent by . . . (call sign).
QUG?	Will you be forced to land?	QUG	I am forced to land immediately [or, I shall be forced to land at . . . (position or place)].
QUH?	Will you give me the present barometric pressure at sea level?	QUH	The present barometric pressure at sea level is . . . (units).

Signal	Meaning	Signal	Meaning
QUI?	Are your navigation lights working?	QUI	My navigation lights are working.
QUJ?	Will you indicate the TRUE course for me to steer toward you (or, . . .) with no wind?	QUJ	The TRUE course for you to steer toward me (or, . . .) with no wind is . . . degrees at . . . hours.
QUK?	Can you tell me the condition of the sea observed at . . . (place)?	QUK	The sea at . . . (place) is . . . .
QUL?	Can you tell me the swell observed at . . . (place)?	QUL	The swell at . . . (place) is . . . .
QUM?	Is the distress traffic ended?	QUM	The distress traffic is ended.
QUN?	Will vessels in my immediate vicinity (or in the vicinity of . . . latitude . . . longitude) (or, of . . .) please indicate their position, TRUE course, and speed?	QUN	My position, TRUE course, and speed are . . .
QUO?	Shall I search for . . . [(1) aircraft, (2) ship, (3) survival craft] in the vicinity of . . . latitude . . . longitude (or according to any other indication)?	QUO	Please search for . . . [(1) aircraft, (2) ship, (3) survival craft]) in the vicinity of . . . latitude . . . longitude (or according to any other indication).
QUP?	Will you indicate your position by . . . [(1) searchlight, (2) black-smoke trail, (3) pyrotechnic lights]?	QUP	My position is indicated by . . . [(1) searchlight, (2) black-smoke trail, (3) pyrotechnic lights].
QUQ?	Shall I train my searchlight nearly vertical on a cloud, occulting if possible, and if your aircraft is seen, deflect the beam upwind and on the water (or land) to facilitate your landing?	QUQ	Please train your searchlight on a cloud, occulting if possible, and if my aircraft is seen or heard, deflect the beam upwind and on the water (or land) to facilitate my landing.
QUR?	Have survivors . . . [(1) received survival equipment, (2) been picked up by rescue vessel, (3) been reached by ground rescue party]?	QUR	Survivors . . . [(1) are in possession of survival equipment dropped by . . . , (2) have been picked up by rescue vessel, (3) have been reached by ground rescue party].
QUS?	Have you sighted survivors or wreckage? If so, in what position?	QUS	Have sighted . . . [(1) survivors in water, (2) survivors on rafts, (3) wreckage] in position . . . latitude . . . longitude (or, according to any other indication).
QUT?	Is position of incident marked?	QUT	Position of incident is marked (by . . .).
QUU?	Shall I home ship or aircraft to my position?	QUU	Home ship or aircraft [(1) . . . (call sign) to your position by transmitting your call sign and long dashes on . . . kHz, (2) . . . (call sign) by transmitting on . . . kHz] courses to steer to reach you.

RADIOTELEGRAPH OPERATING SIGNALS

Signal	Meaning	Signal	Meaning
AA (?AA)	All after . . .	K	Invitation to transmit
AB (?AB)	All before . . .	MN or MIN	Minute(s)
ABV	Repeat figures in abbreviated form	N	No
ADS	Address	NIL	I have nothing for you
AS	Wait	NW	Now
BK	Used to interrupt a transmission in progress	OK	We agree
		PBL	Preamble
BN	All between . . . and . . .	R	Received
BQ	A reply to an RQ	REF	Reference to . . .
C	Yes	RPT	Repeat
CFM	Confirm	RQ	Indication of a request
CL	I am closing my station	SIG	Signature
COL	Collate, or I collate	SYS	See your service message
CP	Call to two or more stations	TFC	Traffic
CQ	General call to all stations	TU	Thank you
CS?	Call sign?	TXT	Text
ER or HR	Here	$\overline{VA}$ or $\overline{SK}$	End of work
ETA	Estimated time of arrival	W or WD	Word
ITP	The punctuation counts	WA	Word after . . .
JM	Make a series of dashes if I may transmit, dots if not to transmit	WB	Word before . . .

addendum

AMATEUR LICENSE SAMPLE QUESTIONS

NOVICE, Technician, Conditional, General, Advanced, Extra. Applicants for all these license examinations should know the answers to the following questions.

N1. What is the Amateur Radio Service? (34-1)

N2. What part of the Federal Communications Commission's Rules govern the Amateur Radio Service and what are the penalties for violating these rules? (34-1, 31-1)

N3. The Rules encourage and improve the Amateur Radio Service by providing for advancing skills in what two phases of the radio art? (34-1)

N4. What is the definition of an amateur radio operator? Of an amateur radio station? (34-1, 34-2, 34-3)

N5. For how long is a Novice class license valid? May it be renewed? (34-7)

N6. May a transmitting station be operated in the Amateur Radio Service without being licensed by the FCC? (No)

N7. Who may hold an amateur radio station license? (34-5)

N8. Where must an amateur radio operator license be retained? The radio station license? (34-2)

N9. Who is responsible for the proper operation of an amateur radio station? (34-2)

N10. What is the definition of a control operator and who may be the control operator of an amateur station? (34-2)

N11. What is the log of an amateur station, what information must it contain, and how long should it be preserved? (34-2)

N12. What are the frequency privileges authorized to Novice class licensees? (34-6)

N13. What are the emission privileges authorized to Novices? (34-6)

N14. What is the maximum transmitter power privilege for Novices? (34-7)

N15. What are the rules regarding the measurement of the frequency of emissions from an amateur station? (34-3)

N16. How fast, in meters per second, do radio waves travel in free space? (20-7)

N17. What is the relation between the frequency and the wavelength of a radio wave? (20-7)

N18. What are the approximate wavelengths for the frequency bands available to Novice class licensees? (Fig. 34-2)

N19. How are radio signals transmitted across great distances? (20-1 to 20-5)

N20. Which Novice bands are most likely to result in long-distance communication during daylight hours? (15, 28 MHz) At night? (3.7, 7 MHz)

N21. When transmitted by telegraphy, what is the meaning of DE, K, $\overline{AR}$, $\overline{SK}$ (33-3), and CQ? (33-5)

N22. What is the meaning of "RST 579"? (34-2)

N23. What are Q signals? What do QRM, QRS, QRU, QRZ, QTH, and QSL mean? (Appendix G)

N24. How should a transmitting frequency be selected for an amateur station, particularly when near one end of the authorized frequency band? (34-2)

N25. What is an A1 emission? (Appendix E)

N26. What are the characteristics of a good-quality A1 emission? (16-1, 16-2)

N27. What are current, electromotive force, and electric power and what are their units of measurement? (1-7, 1-8, 2-5)

N28. What are direct and alternating currents? (1-11)

N29. How can ac be converted into dc? (11-1 to 11-4)

N30. What is meant by a cycle, kilocycle, megacycle, hertz, kilohertz, and megahertz? (4-6)

N31. What is r.f. or RF? (Radio-frequency ac)

N32. What is the relation between a fundamental frequency and its second and third harmonics? (15-21)

N33. What are resistance, inductance, and capacitance and what is the unit of value of each? (1-12, 5-2, 6-1)

N34. Draw a schematic diagram of a circuit having the following: (*a*) battery with internal resistance,

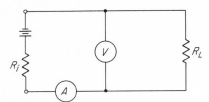

Fig. A-1 Circuit for question N34. Internal resistance is usually shown external of a source in a diagram.

(*b*) resistive load, (*c*) voltmeter, (*d*) ammeter. (See Fig. A-1)

N35. From values indicated by the meters in question N24, how can the resistance of the load be determined? (2-1) The power consumed by the load? (2-5)

N36. In question N34, what must the value of the load be in order for maximum power to be delivered by the battery? (2-18)

N37. Draw a schematic diagram of an RF power amplifier having (*a*) triode vacuum tube, (*b*) pi-network output tank, (*c*) high-voltage source, (*d*) plate current meter, (*e*) plate voltage meter, (*f*) RF choke, (*g*) bypass capacitors, (*h*) coupling capacitor. (See Fig. A-2)

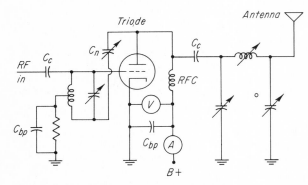

Fig. A-2 A possible circuit for question N37.

N38. What is the proper tuneup procedure for the circuit of N37? (20-24)

N39. What is an insulator, a conductor, and a semiconductor? Give an example of each. (1-7, 1-12, 10-2)

N40. Draw schematic symbols of a resistor (Fig. 1-17), a capacitor (Fig. 6-3), an inductor (Fig. 5-7), a transformer (Figs. 5-20, 5-21), and a choke (Fig. 11-10).

N41. Draw the schematic symbol of a diode (Fig. 11-1), a transistor (Fig. 10-12), and a triode vacuum tube (Fig. 9-21).

N42. What is a dipole antenna? (20-10)

N43. What are the approximate lengths in feet for half-wave antennas for the Novice bands? (From Sec. 20-7: 125, 66, 21.2, 16.7 ft)

N44. What is a transmission line and what are some commonly used types? (20-12)

N45. What are some advantages and disadvantages of a multiband antenna? (Only one antenna needed for several bands; will radiate harmonics if any are fed to it.)

N46. What precautions can be taken to reduce the possibility of shock hazard in amateur radio stations? (16-19, 11-23, 27-15. Do not use metal ladders near power lines. Never run antennas under or over power lines.)

N47. Draw a schematic block diagram of an amateur radio station with (*a*) a receiver, (*b*) a speaker, (*c*) a transmitter, (*d*) a telegraph key, (*e*) a transmission line, (*f*) an antenna, (*g*) a ground rod, (*h*) a transmit-receive antenna switch. (Fig. 34-1)

N48. Explain the purpose of each component in the diagram of question N47. (34-3)

N49. What is the power input to a vacuum tube for the following operating conditions: (*a*) driving power 0.5 W, (*b*) plate voltage 600 V, (*c*) plate current 140 mA, (*d*) screen voltage 175 V, (*e*) screen current 10 mA, (*f*) filament voltage 6.3 V, (*g*) filament current 0.8 A? (From Sec. 15-6, $P = E_p I_p$, or 84 W)

N50. What methods are used by amateur radio licensees for determining that an emission from a transmitter is within an authorized frequency band? (Receivers, frequency meters, spectrum analyzers)

N51. What methods are most often used by amateurs to determine the quality of emissions from their stations? (Monitor receivers, frequency meters, oscilloscopes.)

N52. What is a transmatch and what are the advantages of using one? (34-3)

GENERAL, CONDITIONAL, TECHNICIAN, Advanced, Extra.

Applicants for all these license tests should know the answers to the following questions as well as all those listed for prior tests.

G1. What are the five principles expressing the fundamental purposes of the Amateur Radio Service? (34-1)

G2. What is the definition of amateur radiocommunications? (34-1)

G3. What is the definition of fixed, portable, and mobile operation? (34-2)

G4. With what stations may an amateur radio station communicate? (34-2)

G5. What types of transmissions may be made by U.S. amateurs to amateurs in foreign countries? (34-2)

G6. What is third-party traffic and what types of it are prohibited? (34-2)

G7. Under what limitations may a third party participate in an amateur radiocommunication? (34-2)

G8. What types of one-way amateur transmissions are permitted and what types are prohibited? (34-2)

G9. May an amateur station automatically restransmit programs or signals emanating from any class of station other than amateur? (No)

G10. When there is a violation of the rules at an amateur station, how is (are) the responsible person(s) determined? (34-2)

G11. In regard to amateur station logs, what is the minimum information to be contained in them?

G12. What determines the operator privileges at an amateur station if the control operator is other than the station licensee? How should the station be identified? (34-2)

G13. What is a control point and what types of amateur stations must have one? (34-2)

G14. In what manner and at what intervals must an amateur station be identified by the transmission of its call sign? (34-2)

G15. What are the rules regarding the transmission of music, interference, codes and ciphers, obscenity, indecency, profanity, false signals, and unidentified signals? (34-2)

G16. When is a notice of operation away from the authorized location of an amateur station required and where must the notice be sent? (34-2)

G17. What are the requirements to qualify for the special provisions in the Rules for stations used only for radio control or remote craft? (34-2)

G18. What are the consequences if the holder of an amateur license obtained by mail examination is ordered to appear for a Commission-supervised examination and either fails to appear or fails to pass the examination? (34-5)

G19. What are the propagation characteristics of the HF (3–30 MHz) and VHF (30–300 MHz) amateur frequency bands? (20-1 to 20-6)

G20. What are some of the propagation factors that influence transmission and reception of amateur bands? (20-1 to 20-6)

G21. What are some good operating procedures that can be employed to minimize interference and congestion of amateur bands? (32-11, 33-19, 33-22. Call locals on their frequency, DX slightly off their calling frequency.)

G22. In what manner should an amateur station be operated in respects not specifically covered by the rules? (34-2)

G23. What system is used in the rules for classifying and designating emissions from amateur radio stations? (Appendix E)

G24. What are the characteristics and standards of good-quality telephony emissions from amateur stations? (16-13, 16-14, 17-1, 17-32, 19-3, Appendix E)

G25. What range of audio frequencies is usually adequate for excellent voice intelligibility in communication systems? (200–3000 Hz)

G26. How is voice information conveyed in amplitude-modulated emissions (17-1, 17-2, 17-7, 17-9), frequency-modulated emissions, (19-1, 19-3), and phase-modulated emissions? (19-15 to 19-17)

G27. What is meant by the occupied bandwidth of an emission? (Number of hertz between highest and lowest significant sidebands produced by the emission. 8-6)

G28. What is wideband F3, what is narrowband F3, and on what amateur frequencies may each be used? (34-3)

G29. What is the maximum percent of modulation permitted by the rules and what does the term mean? (100%, 17-10)

G30. What is peak-envelope power (PEP) in an emission from an RF amplifier? (17-39)

G31. How can PEP be determined? (17-39)

G32. What is average power of an emission from an RF amplifier and how can it be determined? (P_{avg} = power in carrier if unmodulated = $\pm 70\%$ of $E_p I_p$ of amplifier.)

G33. How can the average power input to the final amplifying stage supplying power to the antenna be determined? (15-6)

G34. In a single-sideband suppressed-carrier A3J emission (SSB or SSSC) transmitter, what determines the PEP-to-average power ratio? (Modulating waveform) What are some typical values? (2-1 to 4-1)

G35. What does the term S/D ratio in a SSB transmitter mean and how can the ratio be determined? (Sideband distortion is a ratio of radiated desired RF signal to third and fifth harmonic energy also being developed and radiated.)

G36. What are some common RFI (radio-frequency interference) problems encountered by amateurs and what are some of the solutions? (For harmonic radiation, 16-18; for front-end overload or heterodyning of two local signals, 18-22)

G37. How do resistors combine in parallel and in series? (2-10, 2-11, 2-14, 2-15)

G38. How do capacitors combine in parallel and in series? (6-14, 6-15)

G39. How do inductors combine in parallel and in series? (5-9, 5-10)

G40. What is inductive reactance and how do such reactances combine in parallel and in series? (5-12)

G41. What is capacitive reactance and how do such reactances combine in parallel and in series? (6-13, 6-15)

G42. What is impedance? (7-2)

G43. What is Ohm's law and how does it relate to resistive and reactive impedances? (2-1, 7-2)

G44. What is impedance matching and why is it important? (2-18, 20-12)

G45. What is a decibel? (8-5)

G46. What is an S unit? (18-18)

G47. What are the distinguishing features between series and parallel resonant circuits, and how is the resonant frequency determined? (8-1 to 8-3)

G48. What is the Q of a series or a parallel resonant circuit? (8-4)

G49. What is the operating principle of a transformer? (5-15)

G50. How can a transformer provide a desired voltage? (5-21) A desired impedance? (14-33)

G51. Why do circuits oscillate? (13-1 to 13-4)

G52. What are some ways of minimizing harmonic generation im amplifiers? (15-21, 16-8)

G53. What are some ways of minimizing harmonic radiation from transmitters, transmission lines, and antennas? (16-8)

G54. What are the main classes of amplifier operation? (Figs. 14-23, 14-21) For what uses is each class suited? (14-22, 14-23, 14-28, 14-29, 15-3) What are typical efficiencies associated with each? (9-24, 13-5, 14-22, 15-3)

G55. What is neutralization and how does it contribute to proper amplifier operation? (15-12 to 15-14)

G56. What procedure should be followed to neutralize an RF amplifier properly? (15-15, 15-16)

G57. Compare the operating characteristics of grounded-grid (15-24, 15-25), grounded-cathode (14-1, 14-2), and grounded-plate or cathode-follower amplifiers. (14-30)

G58. Why are resonant circuits used in RF amplifiers? (15-1)

G59. What is the principle of a semiconductor diode? (10-3)

G60. What is the principle of a transistor? (10-5)

G61. Compare the elements of a transistor with those of a triode vacuum tube. (Compare 9-22 and 10-18)

G62. What is the principle of an electrolytic capacitor and why are such capacitors widely used? (6-7)

G63. What are the advantages and disadvantages of a toroidal inductor compared to a helical inductor? (5-3)

G64. How is the length of a half-wave dipole related to its resonant frequency? (20-7)

G65. How are the length and resonant frequency of a quarter-wave antenna related? (20-14)

G66. Compare the characteristics of a horizontal half-wave dipole with those of a quarter-wave ground-plane antenna. (20-13)

G67. How may a center-fed nonresonant horizontal antenna be used on several HF bands? (Fig. 15-18 or 20-24, or by using a transmatch)

G68. What is SWR and how can it be determined on a transmission line from the incident and reflected voltages? (20-12)

G69. What are the characteristics that determine the impedance of a parallel-conductor transmission line? (20-12)

G70. What are the factors that determine the characteristics of a coaxial transmission line? (20-12)

G71. What one instrument will give more information, more accurately, for the proper adjustment of a radiotelephony transmitter than will any other collection of instruments generally available to amateur operators? (Oscilloscope) How is it used? (17-21, 17-39)

G72. How is the input power to the plate circuit of a final amplifying stage of a SSB transmitter determined? ($P = E_p I_p$ with modulation) If power is also supplied to the antenna by the driver stage, how is the power determined? (With grounded-grid final amplifiers the total input power producing RF output is determined by the sum of the final and driver stages input dc power.)

G73. Outline the basic stages of a filter system SSB transmitter and explain the purpose of each stage. (17-35)

G74. Outline the basic stages of an F3 transmitter and explain the purpose of each stage. (19-20)

G75. What precautions should be taken in the construction and operation of amateur radio equipment to avoid the danger of electric shock? (N46. Also, during bench testing, stand on an insulated surface and use insulated-handle tools.)

ADVANCED, Extra. Applicants for these license tests should know the answers to the following questions, as well as all those listed for prior tests.

A1. A resistor, capacitor, and inductor each has 100 Ω of resistance or reactance. What is the equivalent series impedance of these three elements? (From 7-8 and 8-2, 100 Ω)

A2. Compare transistors and tubes. What are the advantages and disadvantages of each? (10-1)

A3. What is the vacuum-tube counterpart of a grounded emitter circuit? (Grounded cathode, 10-5) Of a grounded base circuit? (10-9) Of a grounded collector circuit? (10-10)

A4. Power dissipation in what part of a transistor warrants careful observance of power ratings? (Collector junction, 10-8)

A5. Why does a type 6146 tube have three prongs connected to the cathode? (15-4. Three parallel inductors have one-third the inductance of one.)

A6. Should a voltmeter have high- or low-internal circuit resistance? Explain. (12-10)

A7. What are lissajous figures? (12-31) How are lissajous loop patterns related to the frequencies applied to the scope plates? (12-31, Fig. A-3)

2X   3X   4X

Fig. A-3 Lissajous figures for question A7, when the horizontal to vertical sinusoidal frequency ratio is exactly 2:1, 3:1, and 4:1.

A8. How does a full-wave bridge rectifier operate? What is the schematic diagram of this rectifier circuit? (11-4, 11-29)

A9. What are some common types of oscillators, how can they be identified in diagrams, and how is feedback produced in each? (13-4, 13-6, 13-8, 13-9, 13-10, 13-11, 13-12, 13-16, 13-20, 13-21)

A10. How do parasitic oscillations affect circuits and what can be done to prevent or eliminate them? (13-5, 13-22, 15-8)

A11. When is an amplifier operating class A? (14-21, 14-22) Class B? (14-21, 14-29) Class C? (15-3, 15-4) In which amplifier stages of an amateur transceiver are these classes normally used? (Receiver and AF circuits usually class A, transmitter RF circuits class A, AB, or C.)

A12. A transformer with 115 V applied across the primary has a primary/secondary turns ratio of 10:1. If a 5-Ω load is connected across the secondary, what is the impedance reflected back into the primary? ($10^2 \times 5 = 500\ \Omega$.

14-33) How much voltage appears across one-half of the turns of the primary? (57.5 V, 5-26)

A13. How are bypass capacitors used? (9-26, 13-6, 13-8, 14-7, 14-14, 15-5, 16-8) How should the impedance of a bypass capacitor compare to the element it shunts? (The X_c should be one-tenth to one-hundredth of the bypassed circuitry.)

A14. Why is neutralization important in amplifiers? (15-12) What points in an amplifier circuit should be coupled to provide good neutralization? (Figs. 15-5 to 15-20)

A15. Compare the pentode, tetrode, and triode for use in an RF amplifier stage. (9-14, 9-15, 9-22, 9-26, 9-28) Give advantages and disadvantages of each. (15-2, 15-3, 15-12, 15-24)

A16. What is backwave ("leak-through") radiation? (16-4) How can it be eliminated? (By neutralizing the amplifier and shielding both oscillator and amplifier.)

A17. How is the power output of a 100% sine-wave-modulated AM signal related to the carrier power? (100-W carrier + 50-W sidebands = total power. Carrier power is 66% of total power output. 17-13, 17-14)

A18. What factors affect the peak envelope power of a transmitter? (The envelope is the pattern seen on the oscilloscope, Fig. 17-23. The output power is affected by plate voltage, grid bias voltage, grid driving signal, screen voltage, coupling to antenna, and condition of tube. 17-8, 17-9, 17-11, 17-17)

A19. What do oscilloscope patterns showing 25, 50, and 100% sine-wave-modulated signals without distortion look like? (Figs. 17-23, 17-25, 17-26, 25% is midway between 0 and 50%, 75% is midway between 50 and 100%.)

A20. What methods are most commonly used to generate single-sideband signals? Draw a block diagram of the filter method showing essential stages to convert a low-frequency SSB signal to the desired transmitting frequency. (17-35, 17-38)

A21. How does the peak-envelope-power *input* of an amplifier used for CW compare to the PEP input of a SSB amplifier with 1 kW input? (The modulated peaks would be assumed to be reaching 2 kW to show an average of 1 kW on the E_p and I_p meters.)

A22. The ratio of peak envelope power to the average power in an SSB signal is primarily dependent on what? (On the waveshape of the modulating signal; there is no envelope power with no modulation in SSB.)

A23. How can SSB signals be amplified with little or

no distortion? (Use linear class A, AB, or B RF amplifiers. 17-22)

A24. How should a linear amplifier be adjusted for linear operation? [17-23. For SSB, (1) bias to about one-third normal I_p with no signal input, (2) introduce weak signal and tune amplifier and antenna, coupling for maximum output, and (3) modulate with voice and watch envelope patterns on scope. Operate so no peaks of modulation, "flat top."]

A25. What happens to even-order products in RF linear amplifiers? (In well-balanced push-pull stages, even-order products should be nearly canceled. Not true for single-ended stages.)

A26. What types of emissions can be received with selectable sideband receivers? (Without BFO, A2, A3, and F2. With BFO on, A0, A1, A2, A3, A3J, F1, F2.)

A27. How can receiver sensitivity and selectivity be improved? (Greater RF and/or IF amplification. Narrower IF bandpass. 18-9, 18-11, 18-13, 18-21.)

A28. A superheterodyne receiver having an intermediate frequency of 455 kHz is to be adjusted to receive a signal on 3,900 kHz. To what frequencies can the high-frequency oscillator be set in order to give a beat signal at the intermediate frequency? (From 18-23, 4,355 or 3,445 kHz)

A29. What function does a variable-mu tube perform in an RF or IF amplifier stage of a receiver? (9-28, 18-14)

A30. How does automatic gain control (AGC or AVC) operate? (18-14) When can it be used for SSB operation? (18-26) For CW operation? (18-14, 18-26)

A31. Define the shape factor and selectivity of a crystal lattice bandpass filter. (8-8, 17-37)

A32. How do noise limiters operate? (18-19)

A33. What are the advantages and disadvantages of using the same antenna for receiving and transmitting? (16-3. Same antenna has similar transmitting and receiving directivity and polarization. Separate antennas are better for break-in and duplex operation.)

A34. What is meant by describing a radio wave as horizontally or vertically polarized? Which type is most suitable for sky- and ground-wave propagation? (20-6)

A35. What factors affect the state of ionization of the atmosphere? (20-2)

A36. Define maximum usable frequency, or MUF. (20-2)

A37. How does the sunspot cycle affect wave propagation? What are the best frequencies to use for day and night, short- and long-distance communications? (20-2)

A38. Which amateur band is the most suitable for daytime communication over a distance of about 200 mi? (7 MHz is always under the MUF.)

A39. How can the resonant frequency of an antenna be increased and decreased? (20-11)

A40. When can a low-pass filter be installed in a coaxial cable without causing a large power loss? (By matching filter-to-transmission line surge or characteristic-impedance values, minimal SWR is developed, and most power can be coupled to the load. The cutoff frequency of the filter must be higher than the signal frequency.)

A41. A 70-Ω half-wave antenna operating on a frequency of 7,300 kHz is to be matched to a 50-Ω transmission line. What is the physical length of the antenna? (From 20-7, 61.4 ft) What is the characteristic impedance of a quarter-wave matching section? (From 20-16, 59.2 Ω) What is the SWR between the antenna and transmission line without a matching section? (SWR $= Z_1/Z_2 = 1.4:1$)

A42. A transmission line that feeds an antenna has a power loss of 10 dB. If 10 W is delivered to the input, how much power is delivered to the antenna? (From 8-5, 1 W) List possible causes of power loss. How can the SWR of the line be made as low as possible? (20-12)

A43. What happens to the voltage, current and impedance along a transmission line with an SWR of 1? (Short line: E, I, and Z are constant. Long line: Z is constant, but E and I decrease.)

A44. What is a good indication that a high standing-wave ratio is present on a transmission line? (Transmission line will not accept power from amplifier, acts as reactance instead of resistance when coupled to amplifier. High and low E and I points on line. SWR bridge reads high ratio.) Where is the best point on a long transmission line to measure SWR? (Usually, for convenience, measured at input end, but more accurate value obtained if measured at antenna end.)

A45. Define frequency deviation in FM transmissions. (19-3)

A46. How close to the edges of a certain amateur band can you safely operate a VFO CW transmitter if using a frequency meter having a maximum possible error of 0.01%? (From 21-11, 7299.27 kHz)

A47. What is a third-party agreement? What countries have such agreements with the United States? (34-2. Some countries having third-party agreements with the United States are Argentina,

Bolivia, Brazil, Canada, Chile, Colombia, Costa Rica, Cuba, Dominican Republic, Ecuador, El Salvadore, Haiti, Honduras, Israel, Liberia, Mexico, Nicaragua, Panama, Paraguay, Peru, Uruaguay, and Venezuela. Messages must not be handled with any other countries.)

A48. How can a low-frequency SSB signal be converted to a HF transmitting frequency? (17-35)

A49. On what frequencies do SSB transmissions become more difficult? (As VHF oscillator frequency stability lessens.)

A50. List some of the advantages SSB provides over double sideband with carrier operation. (17-34)

A51. How can the generation of excessive harmonics be avoided? (16-18)

A52. Which class of amplifier operation is most favorable to the generation of harmonics? (15-21)

A53. How should the addition of a reactive element to a purely resistive circuit affect the sum of the voltage-drops around a closed loop in the circuit? (Increase them.)

A54. What effect would a reactive load have on the output frequency of an oscillator? (16-2)

A55. What can the value of the dc voltage across the grid-leak resistor of an oscillator reveal about the oscillator's performance? (13-5)

A56. How can TVI caused by cross-modulation, or by heterodyning of two signals in the receiver RF amplifier, be remedied? (Trap out one or both signals. 8-8)

A57. Where in a TV receiver should a TVI filter be mounted to best reduce television interference? (Between antenna and TV antenna terminals.)

A58. What formula is used to determine the resonant frequency of an antenna? (20-7)

A59. What is the center impedance of a half-wave dipole? (20-10) Of a folded dipole? (20-20) Of an inverted-V (feed point at apex; radiators drop toward ground) dipole? (Less than 73 Ω)

A60. What amateur frequencies between 7 and 148 MHz are most affected by weather conditions? (Lightning QRN decreases at higher frequencies.)

A61. What factors determine whether or not a transformer having a center-tapped high-voltage winding can be used in a bridge rectifier circuit? (Wire size of windings due to possible increase in load current when dc voltage is doubled.)

A62. During the application of a single-tone test to a SSB linear amplifier, how does the average power input to the amplifier relate to the PEP produced? (The dc plate power input times the efficiency of the amplifier equals the PEP. As RF input increases, PEP increases.)

A63. How should a linear amplifier affect the directional pattern of a beam antenna? (20-17, 20-18, 20-20, 20-21)

A64. What parameters affect the directional pattern of a beam antenna? (20-17, 20-18, 20-20, 20-21)

A65. What are some precautionary measures that should be taken before replacing faulty circuit elements? (11-34)

A66. Compare the operating characteristics of wire-wound and carbon-type resistors. (1-12)

A67. How can amateur equipment be protected from damage induced by electrical storms? (34-3)

A68. Define single and double conversion in a receiver. (18-24)

A69. What is an intermediate frequency (IF) in a receiver? (18-10, 18-13)

A70. How does an image frequency relate to the desired signal frequency in a receiver? (18-23)

A71. Why should the grid wiring in an RF transmitter be as far removed as possible from the plate circuitry? (Stage may oscillate. 13-6)

A72. What is a dummy antenna and how can it be of use to amateur operators? (16-7)

A73. What is meant by percent of modulation? (17-10)

A74. What determines if a carrier wave is under- or over-modulated? (17-10)

A75. What effect would a self-oscillating buffer stage have on the output frequency of a transmitter? (Amplifier may transmit frequency generated by buffer instead of frequency of oscillator.)

A76. What is meant by the effective value of a voltage? (4-4)

A77. What is meant by the peak-to-peak value of a voltage? (4-4)

A78. What is a wave trap? (8-8, 18-22)

A79. Draw some common wave-trap configurations. (Fig. 18-32)

A80. What circuit conditions may be indicated by a high current reading in the grid meter in the final class C amplifier stage of a transmitter? (Excessive RF drive, low bias, no plate current.)

A81. Briefly discuss the advantages and disadvantages of paper, mica, air, and ceramic capacitors. (6-9)

A82. What happens to a circuit when a capacitor develops a leakage resistance? (Malfunctions; oscillators cease oscillating or change frequency; amplifiers lose gain; high currents may develop. 6-10)

A83. Discuss the characteristics of series and parallel resonant circuits. (8-2, 8-3)

A84. Be prepared to answer any other questions on subjects covered by Part 97 of the FCC Rules.

EXTRA. Applicants for this license test should know the answers to the following questions, as well as all those listed for prior tests.

E1. How do mica and paper dielectric bypass and filter capacitors compare at different frequencies? (6-9, 6-17)

E2. What does the term *power factor* mean in reference to electric power circuits? (6-10, 7-6, 12-23)

E3. What are inductive and capacitive reactance and how are the phase angles and power dissipation related? (5-12, 6-13, 7-8)

E4. How is the decibel used for voltage and power calculations? (8-5)

E5. How do NPN-type transistors differ from PNP types and how does their bias differ? (10-6)

E6. How are transistors biased for amplifier operation? (10-6) How are they biased for cutoff, or open circuit? (Reverse biased) For saturation, or short circuit? (Heavily forward-biased)

E7. Define the alpha cutoff frequency of a transistor. How is this parameter of use in circuit design? (10-6)

E8. What is the phase relation between the input and output signals in transistor common-emitter circuits? (180°) Common-base circuits? (In phase) Common-collector circuits? (In phase)

E9. How does a cathode-ray tube operate? What voltages should be applied to the plates of a CRT? (12-29)

E10. An oscilloscope is used to study the relation between the input and output of an amplifier produced by a voice signal. How would the scope pattern display a linear relation? (Use trapezoidal display. 17-21)

E11. What precaution(s) should be taken when measuring the rectified grid voltage in an oscillator with a dc voltmeter? (Use a meter with at least 10,000 Ω/V sensitivity with an RF choke or 50-kΩ resistor attached to probe to prevent disturbing grid circuit.)

E12. How does the positioning of a powdered-iron tuning slug affect the frequency of the oscillator it is tuning? (Iron core out decreases the inductance and raises the frequency of the oscillator.)

E13. What factors determine the frequency at which a quartz crystal will oscillate? What are some advantages of using crystal oscillators? (13-12, 13-13)

E14. For maximum stability, to what frequency should a crystal oscillator circuit be tuned? (13-12)

E15. How can the safe power input to a crystal oscillator circuit be determined? (Excessive power heats the crystal; it may expand and drift in frequency. If a crystal oscillator drifts after a few seconds of operation, the power of the oscillator should be reduced.)

E16. How can parasitic oscillations be prevented? (13-22)

E17. What determines the fundamental operating range of a multivibrator? (13-21)

E18. List several advantages and disadvantages for class A, B, and C amplifier operations. (14-22, 14-29, 15-4. Class A: low drive, simple to bias, low distortion, may use only one tube, but low efficiency. Class B: must be push-pull for audio, high drive required, requires regulated bias, some distortion, high efficiency. Class C: high bias, high drive, distorts, not usable for audio, highest efficiency.)

E19. What improper operating conditions are indicated by the upward or downward fluctuation of a class A amplifier plate current when a signal voltage is applied to the grid? (14-22, 17-31. improper bias, excessive signal input, weak tube, low heater voltage, nonlinear load.)

E20. What is meant by frequency-shift keying and how is it accomplished? (16-13)

E21. What is the proper way of identifying an RTTY (radio-teletype) transmission? (At the beginning and end of each transmission, call signs should be transmitted using RTTY, followed by identification in A1 or F1 if in a CW band, or in A3 or F3 if in a phone band.) Draw a block diagram of an RTTY system and indicate the function of each stage. (28-18)

E22. How are ground-grid amplifiers used in electronic circuits? List some advantages and disadvantages of their use. (15-24. Will not plate-modulate satisfactorily; has low input impedance.)

E23. How can a transmitter be tested for self-oscillation? (Prevent oscillator from operating by pulling out tube. If grid current or RF output is shown, transmitter is self-oscillating.) What precautions should be observed during this test? (If grid-leak-biased, stages will draw excessive plate current and may damage the tubes if not oscillating.)

E24. How is the output circuit of a transmitter adjusted to increase or decrease its coupling to the antenna system? (15-6, 16-5, 16-7, 20-24)

E25. What are some causes of the excessive production of harmonics in RF amplifiers and how can they be remedied? (16-18)

E26. How do filters attenuate harmonic emissions? (8-8, 16-18)

E27. What are sideband frequencies? During 100% sinusoidal-amplitude modulation, what percent of the average power is in the sidebands?

(17-14. 100-W carrier + 50-W sidebands = total power. Therefore, sideband power = 33% of total.)

E28. What is a grid-bias-modulated amplifier? Should the source of fixed bias have a high or low internal resistance? (17-17. The bias supply must have low internal resistance, good regulation.)

E29. What would happen if the grid-bias supply of a class C modulated amplifier were suddenly short-circuited? (High I_p, no positive peaks of modulation, decreased carrier power, possible loss of RF amplifier tube unless some grid-leak resistance were in the circuit.)

E30. What radiotelephone transmitter-operating deficiencies may be indicated by a decreasing antenna RF current during modulation of the final RF amplifier? (17-24, 17-25, 17-27)

E31. What may be the cause of a decrease in antenna current when a modulating signal is applied to a class B RF amplifier? (17-25)

E32. How may an amateur check his transmitter for spurious sidebands? (Remove antenna from receiver and listen to adjacent frequencies for spurious signals out past 6 kHz for a receiver with a bandpass of 3 kHz and out past 9 kHz for a receiver with a bandpass of 6 kHz.)

E33. What useful function does a balanced modulator perform in a radio transmitter? (17-33)

E34. When designing SSB transmitters, why is the lowest audio-frequency response usually 250 to 300 Hz? (Allows 500- to 600-Hz skirts for filters which must cut off other sideband signals. In phase systems it is difficult to retain linear 90° AF phase shift below these frequencies. Very little speech sounds below 250 Hz.)

E35. How can the two-tone test output of a linear amplifier be used to tell if an SSB transmitter is working properly? (A single AF tone into an SSB transmitter produces an oscilloscope presentation as a simple RF carrier. If two AF tones are fed simultaneously into the transmitter, an oscilloscope presentation of proper modulating conditions is as shown in Fig. 17-52a. If the linear amplifier is overdriven, the result will be as shown in b.)

E36. In what section of a properly operating SSB transmitting system is distortion most likely to originate? (In the linear amplifier because coupling to the antenna is critical and grid may be overdriven.)

E37. How should a wave trap be connected to a receiving antenna circuit to attenuate an interfering signal? (18-22)

E38. How do trimmer and padder capacitors affect receiver tuning? (18-29)

E39. Define the conversion efficiency of a mixer tube. (The conversion gain of a mixer tube is a comparison of the RF signal voltage amplitude fed into the grid of the tube with the IF signal voltage developed across the plate-circuit load.)

E40. Of what importance is the signal-to-noise ratio of a receiver and at what frequencies is it most important? (Receiver noise is developed in the RF amplifier or mixer stages usually. The S/N ratio indicates how weak a signal voltage can be received over the receiver noise level. Low S/N is most important in VHF and UHF receivers.)

E41. Where in a receiver circuit should a limiter-blanker stage be placed to provide maximum utility? (18-19)

E42. Why is there a practical limit to the number of stages that can be cascaded to amplify a signal? (18-13)

E43. How are phasing condensers used in crystal filters? (18-21)

E44. How can you distinguish between a product and an envelope detector? (Product detectors use a BFO or a local oscillator and mixer. 18-12, 18-15. Envelope detectors are usually diode-type detectors with no oscillator circuit. 18-2, 18-4, 18-6, 18-14)

E45. How will a long- and a short-time-constant AVC circuit affect reception? (18-14)

E46. How does a squelch circuit operate? Draw a commonly used squelch circuit. (18-16)

E47. What is the image response of a receiver and how can it be reduced? (18-23)

E48. How do receivers for remote control of objects and regular communications receivers differ in basic operation? (Remote-control receivers may have a fixed frequency and operate a relay in addition to a loudspeaker. They may be portable, lighter, and battery-operated.)

E49. How does the beat-frequency oscillator affect the tuning of a single-sideband signal? (18-26)

E50. How can a receiver be adjusted for SSB reception when the receiver does not have a product detector? (18-26. If no BFO is incorporated in the receiver, an RF signal generator set at the carrier frequency of the SSB signal to be detected can be fed into the antenna circuit to insert the required carrier.)

E51. How will the reception of a single-sideband signal be affected if the carrier is not completely suppressed? (When properly tuned in, it will not affect reception at all. If mistuned, a beat tone may be heard.)

E52. How may a limiter be employed in an FM receiver? (19-9)

E53. How are reactance tubes used? (19-13, Fig. 19-10)

E54. What type of signal will be produced when the output of a reactance modulator is coupled to a Hartley oscillator and multiplied in frequency? (19-13)

E55. What are some different types or sources of noise voltages in reception and how are they generated? (19-21, 20-5)

E56. What determines the skip distance of radio waves? (20-1, 20-2, 20-3, 20-4)

E57. What are aurora-reflected VHF signals and what do they sound like? (20-2)

E58. A 70-Ω transmission line is connected to a 35-Ω antenna. What is the standing-wave ratio? If 10 A is flowing at the antenna terminals, what is the current value in the transmission line at the current node or minimum point? What is the reflection coefficient? What is the percent reflected power? (20-12)

E59. What are the current and voltage characteristics along a transmission line when it is matched and mismatched? (20-12)

E60. Explain the properties of a quarter-wave section of RF transmission line. (20-16)

E61. Can a lossy transmission line be used to transmit signals? (Open-wire transmission lines constructed of resistance wire maintain a constant impedance value but lose all power heating the wires and none is radiated, acting as a constant-Z dummy load. Open-wire lines spaced more than one-hundredth of a wavelength apart radiate considerable power similarly to V-beam and rhombic antenna theory. 20-21)

E62. What effect does an untuned antenna and transmission line have on a transmitter? (Reflects inductive or capacitive reactance into the output tank circuit and refuses to accept power.)

E63. What constitutes a parasitic antenna element? (20-20)

E64. How does the directivity of an unterminated V antenna and a parasitic beam antenna compare? (Unterminated V beams are bidirectional, whereas three- or more element parasitic beams are usually unidirectional. 20-20, 20-21)

E65. List some different types of beam antennas. (20-16, 20-17, 20-18, 20-20, 20-21, 20-22) Which antennas are not resonant and therefore do not reduce harmonic radiation? (Terminated V beams, rhombics, long-wires.)

E66. What means may be employed to measure low frequencies? (12-25, 21-9) High frequencies? (21-3, 21-5, 21-9, 21-7) VHF and UHF? (21-4, 21-9)

E67. What are A5 and F5 emissions? On what amateur frequencies can these emissions be transmitted? (Amplitude-modulated television, frequency-modulated TV, 420 MHz and above. 34-6)

E68. Why are synchronizing pulses transmitted with television signals? (25-1)

E69. How does amateur TVI usually affect television reception? (Visual bars, "herring bone," loss of picture, weakening of signal, audible response from speaker.)

E70. How can unwanted VHF resonances in a transmitter amplifier be moved from TV channel frequencies? (Parasitic chokes, lead dress, parts shielding, losser resistors, shielding of transmitting equipment.)

E71. What must the value of an inductor be to cancel a capacitive reactance of 12.6 kΩ at an operating frequency of 2 MHz? (From 8-1, 0.001 H)

E72. Discuss advantages and disadvantages of electrolytic versus paper filter capacitors. (6-7, 6-9)

E73. Compare silicon and vacuum-tube diodes. What is meant by the *forward voltage-drop* of a conducting diode? (9-9, 9-11, 10-3, 11-15. The barrier voltage that must be overcome before a solid-state diode conducts is known as the forward voltage-drop of the device.)

E74. Describe briefly how an ac power supply can produce an ac output voltage. (11-1, 11-3, 11-4, 11-6, 11-7)

E75. Discuss the merits of using choke-input versus capacitor-input filters in power supplies. (11-8, 11-9, 11-13, 11-15)

E76. How does leakage resistance of power-supply filter capacitors affect the output voltage? (If excessive, it may reduce the voltage somewhat, especially in high-voltage low-current supplies.)

E77. What is voltage regulation as related to power supplies? (11-25, 11-26, 11-27)

E78. What visual observation within an operating vacuum-tube envelope would indicate that the tube is gaseous? (11-34)

E79. What are the bandwidths normally used for A1 emissions? (50 to 400 Hz depending on keying speed.) A1? (6 kHz) A3J? (2.8 kHz) Narrowband FM? (6 kHz) Wideband FM? (Up to 150 kHz)

E80. What is push-pull amplifier operation? (14-23, 15-9)

E81. How can the final amplifier of a transmitter be tested for self-oscillation? (With no RF drive, there should be no RF output. If there is one, the stage is oscillating.)

E82. If a crystal lattice bandpass filter has bandwidths of 3 kHz at the 60-dB points and 1.5 kHz at the 6-dB points, what is the shape factor? (From 8-8, 2 : 1)

E83. What effect will extending the low-frequency

audio response of a signal have on the design of an SSB transmitter? (Requires sharper cutoff filters or more elaborate phasing circuits, otherwise some of other sideband will be transmitted.)

E84. Describe briefly the basic circuits of a single-sideband (SSB) transmitter. (17-35, 17-38)

E85. How might spurious signals in the output of a mixer stage of an SSB transmitter be suppressed? (17-35. Reduce amplitude of signals being mixed; raise Q of mixer output circuit.)

E86. How does a frequency converter (mixer) operate? (18-10, 18-12)

E87. What is a Q-multiplier and how is it used in amateur equipment? (A separate IF amplifier which feeds back a controlled amount of *regenerative* signal to bring one of the IF amplifiers up to near-oscillation, thereby raising the IF circuit Q and thus narrowing the bandpass.)

E88. What improper operating conditions are indicated by grid current flow in a class A amplifier? (14-22)

E89. Define the deviation ratio in a frequency-modulated signal. (19-3)

E90. What is meant by "end effects" of an antenna and how can they be compensated for in half-wave antennas? (20-7)

index

A

Absorption wavemeter, 532
Active devices, 146
ADF (automatic direction finder), 706
Admittance, 116
Aerial, 479
AF (audio-frequency) amplifiers, 290
  BJT, 320
  cascade, 323
  cascode, 322
  classes of, 303–308
  complementary symmetry, 323
  compression, 586
  Darlington, 321
  differential, 322
  expander, 586
  operational, 323
  power, 290
  push-pull, 304
  receiver, 446
  vacuum tube, 319
  voltage, 290
AFC (see Automatic frequency control)
AGC (see Automatic gain control)
ALC (automatic leveling control), 642
Alexanderson alternator, 264, 356
Alignment, receiver, 455
Alloyed transistor, 179
Alpha, BJT, 188
Alpha cutoff, 181
Alternating current (ac), 11, 63
Alternation, 63
Alternators, 562
  Alexanderson, 264, 356
  brushes for, 562
  externally excited, 563

Alternators:
  field excitation of, 564
  Goldsmith, 264, 356
  inductor, 563
  paralleling, 564
  three-phase, 224
  voltage output, 564
AM or A3 emission (see Amplitude modulation)
Amateur radio, 750
  Advanced Class license, 757, 774
  bands, 753, 755
  call signs, 752
  calling and working procedures, 751
  Conditional Class license, 756, 774
  Extra Class license, 757, 780
  General Class license, 756, 774
  logs, 751
  mobile or portable, 751
  Novice Class license, 754, 773
  RST system, 752
  stations, 752
  Technician Class license, 756, 774
  tests, 752
  third-party communications, 751, 778
  transmitters, 373, 375
Ammeters, 31
  dc, 235
  electrodynamometer, 248
  hot-wire, 248
  repulsion, 249
  thermocouple, 247
Ampere, 7
Ampere-hour meter, 252
Ampere-turn, 55
Amplification, 155
  factor, μ, 158, 190

Amplifiers:
  classes of: A, 157, 161, 187
    AB_1, 306
    AF, 303–308
    B, 161, 307
    BC, 404
    RF, 332
    [See also AF (audio-frequency) amplifiers; RF (radio-frequency) amplifiers]
Amplitude, ac, 67
Amplitude distortion, 313
Amplitude modulation, 382
  absorption, 385
  AF power, 390
  ammeter indications, 409
  asymmetrical, 407
  bandwidth, 395
  carrier shift, 407
  checking, 401
  code emission, 366
  downward, 407
  dynamic instability, 386
  envelope, 387
  grid, 398, 401
  Heising system, 397
  high-level, 401
  loop, 385
  low-level, 401
  negative peak, 385
  operating power, 395
  overmodulation, 389
  pentode tubes, 393
  percent of, 389
  plate, 388, 390
  positive peak, 388
  screen-grid, 400
  series, 387
    simple, 386
  sidebands, 396
  SSB (see Single sideband)

Amplitude modulation:
  suppressor-grid, 400
  tetrode tubes, 393
  transformer, 393
  transistors, 415
  transmitter tuning, 413
Analog signal, 178
AND gate, 543
Angular velocity, 77
Anode, vacuum tube, 148
  cooling, 168
Antennas, 498
  Adcock, 708
  beam (see Beam antenna)
  broadcast towers, 596
  coaxial, 509
  computing power, 525
  current in long, 510
  DF loop, 700
  dipole, 506
  directivity, 507
  director, 516
  driven arrays, 513
  E and I in, 503
  effective height, 523
  end effect, 502
  feeding, 510, 519
  field gain, 524
  folded dipole, 516
  full-wave, 509
  ground area, 525
  ground plane, 509
  half-wave, 502, 503
  Hertz, 503
  impedance, 522
  inverted-V, 517
  loading, 504
  lobes, 508
  loop, 518
  Marconi, 503
  omnidirectional, 508, 526
  polarization, 502
  quarter-wave, 508
  radar, 661
  radiation resistance, 504
  reflector, 516
  safety link, 677
  sense, 701
  switch, ship, 691
  top loading, 519
  vertical, 508
  Zepp, 511
Antiresonance, 130
Apparent power, 109

Aquadag coating, 254
Arc, 10
  inductive, 70
Arc-back, 210
Armature, 562
Armstrong oscillator, 266
  PM circuit, 478
Artificial antenna, 360
Aspect ratio, 605
Atom, 2, 4
Atomic cell, 558
Atomic theory:
  of magnetism, 51
  of matter, 3
ATR (antitransmit-receive) box,
    663
Attenuator, 641
Attenuator pads, 584
Audio frequencies, 290
  [See also AF (audio frequency)
    amplifiers]
Audio oscillator, 282
Aural signal, 601
  amplifier, 620
Aurora signals, 500
Auto-alarm, 690
Auto-alarm keyer, 689
Autodyne, 433
Automatic direction finder (ADF),
    706
Automatic frequency control
    (AFC), 477
  radar, 664
Automatic gain control (AGC),
    441
  keyed, 626
  TV, 625
Automatic leveling control (ALC),
    642
Automatic volume control (AVC),
    437, 441
  delayed, 442
Avalanche effect, 175
AVC (automatic volume control),
    437, 441
Average value:
  ac, 66
  pulsating dc, 200

B

Back porch, TV, 630
Backlash, 540
Backward-wave oscillator, 647

Backwave, 358
Balanced modulator, 417, 419
Balun, 648
Bands:
  AM broadcast, 578, 590
  amateur, 753, 755
  FM broadcast, 463
  loran, 713, 717
  microwave, 639
  radar, 657
  ship, 680
  TV, 615
Bandwidth, 137, 396
  emissions, 329
  receiver, 440
  transformers, 138
Barretters, 643
Barrier voltage, 174
Base, 179
Batteries, 548
  A, 169, 548
  B and C, 156, 169
  capacity of, 553
  charging, 550, 553
  dry-charge, 553
  Edison, 556
  lead-acid, 550
  maintenance of, 556
  storing of, 550
  water for, 552
  (See also Cells)
Beam antenna:
  bay, 526
  collinear, 513
  corner reflector, 517
  director, 516
  driven array, 513
  element of, 514
  erp, 525
  Franklin, 513
  parabolic, 662
  parasitic, 515
  reflector, 516
  rhombic, 517
  stacked array, 514
  turnstile, 526
  V-beam, 517
  Yagi, 516
Beam-power tetrodes, 163
Bearing resolution, 658
Bearings, DF: bidirectional, 700
    unidirectional, 702
Beating, 396
Bel, 135, 383

Beta, BJT, 181
BFO (beat-frequency oscillator), 437, 443
BH curve, 51
Bias:
 BJT: fixed, 181
  self-, 182
  voltage-divider, 182
 VT: AF amplifier, 296
  automatic, 297
  cathode resistor, 297
  classes of, 303
  contact potential, 298
  cutoff value, 157
  filament tube, 299
  grid-leak, 267, 337, 368
  oscillator, 267
  self-, 297
  voltage, 156
Bifilar-helix backward-wave oscillator, 648
Bifilar transformer, 618
Bimetallic element, 277
Bipolar junction transistor (BJT), 179
BJT (bipolar junction transistor), 179
Blanking pulse, 609
Bleeder resistor, 214
Blip, 658
Block diagrams, 256
Blocked-grid keying, 364
Blooming, 664
Bolometer, 643
Boosted B+, 625
Break-in keying, 358
Bridge rectifier, 200
Bridges, R and L, 253
Broadcast (BC) bands:
 AM, 578, 590
 FM, 463
Broadcast stations:
 AF levels, 582
 alternate transmitter, 579
 AM, 578
 antenna towers, 596
 areas, 578
 attenuator pads, 584
 auxiliary transmitter, 579
 components of, 579
 console, 581
 cueing, 582
 disc recording, 594
 EBS, 593
 experimental period, 588

Broadcast stations:
 international, 590
 licenses, 591
 logs, 591
 meters, 588
 microphones, 596
 monitoring, 589
 peak limiters, 586
 power classifications of, 587
 remote control, 592
 station break, 582
 STL, 579
 talk-back, 582
 tests, 587
 tolerance, 578
 TV, 579, 601, 613–615
Brush discharge, 10
Brushes, 562
 sparking, 568
Buckshot, 389
Buffer amplifier, 367
 oscillating, 779
Buffer capacitor, 222
Bug key, 746
Butler oscillator, 676
BWO (backward-wave oscillator), 647

C

Calibration chart, 532
Call letters, 729, 733, 740, 747, 753
Capacitance, 89
 distributed, 134
 hand, 271, 371
 junction, 184
Capacitor(s), 89
 buffer, 222
 bypass, 163, 777
 charge, 93
 color code, 102
 dielectric, 91
 electrolytic, 93
 energy in, 93
 in parallel, 98
 in series, 99
 time constant, 90
 types of, 95
 variable, 93
Capacitor analyzer, 229
Carrier, 382
 pilot or vestigial, 418

Carrier shift, 407
 meter, 407
Cascade amplifier, 323
Cascode amplifier, 322, 616
Cat whisker, 431, 643
Cathode, 146
 directly heated, 147
Cathode-ray tube (CRT), 254
 deflection (see Deflection)
 storage of, 259
Cavities, 643
Cavity oscillator, 281
Cells, 9
 dry, 549
  rechargeable, 558
 forming, 553
 Leclanché, 548
 manganese-alkaline-zinc, 557
 mercury, 557
 nonchemical, 558
 primary, 548
 Ruben, 557
 secondary, 550
 shelf life of, 549
 silver-oxide-alkaline-zinc, 557
 specific gravity, 551
 sulfated, 551
 (See also Batteries)
Center frequency, 463
Center-tapped transformers, 200
Center-tapping of filaments, 348
Cesium-beam standard, 535
Characteristic, log, 135
Characteristic curve:
 BJT, V_{EC}/I_C, 184, 186
 MOSFET, V_{DS}/I_D, 192
 VT, 149, 156, 158, 165, 291, 306
Checkpoint, 541
Chirping sounds, 357
Choke:
 parasitic, 286, 337
 RF, 334
 smoothing, 205
 swinging, 206
Choke coil, 72
Choke flange, 639
Chrominance circuits, 631
Circuit breakers, 372
Circular mil, 12
Circulator, 650
Clamp circuit, 368
Clipping, peak, 291
Clock, shipboard, 694
Clock pulse, 424

C-MOS, 191
Coastal station, 681, 734
Coaxial tank oscillator, 281
Code practice oscillator, 282
Coefficient of coupling, 74
Coercive force, 53
Coherer detectors, 682
Coiling wires, 70
   formulas, 71
Cold-cathode rectifier, 217
Collector, 179
Color, TV, 626
      burst, 630
         circuits, 627–633
         picture tube, 633
Colpitts oscillator, 272
Common-base circuit, 187
Common-collector circuit, 188
Common-emitter circuit, 180
Communications:
   classification of, 726
   voice, 729, 731, 733
Commutator, 564
   undercutting, 576
Compass, magnetic, 56
Complementary symmetry, 195,
   323
Complex dc circuits, 41
Complex number, 122
Compound dc generator, 567
Compound dc motors, 570
Condenser, 89
   microphone, 411
Conductance, 37
   mutual, 159
Conduction bands, 4, 173
Constant-current modulation, 398
Construction permit, 486, 587
Contact potential, 148
   VTVM probe, 244
Control grid, 154
Control transformer, 668
Converters, 437
   dc-to-dc, 221
Copper loss, 83
Copper-oxide rectifier, 220
Core, 48, 60, 71
   gap in, 73
Corona, 10
   antenna, 509
Cosine, 107
COS/MOS, 191
Coulomb, 7, 90
Counter, frequency, 543, 653
Counter emf, 69, 570

Coupling, 293
   coefficient, 74
   critical, 140
   direct, dc, 296
   inductive, 293
   impedance, 295
   loose and tight, 74
   π-network, 351
   resistance, 294
   transformer, 293
   unity, 138
Crosby circuit, 476
Crystal oscillators, 274, 278
Crystals:
   cuts, 276
   holders, 278
   IF filter, 447
   materials used, 274
   microphone, 411
   oscillation modes, 275
   ovens, 277
   temperature coefficients, 276
Cueing, 582
Current, 6
   hole, 174
Current gain, BJT, 181
Current transformer, 249
Curves:
   $E_g/_p$, 149, 156, 165
   $E_p/_p$, 149, 158, 165
   load line, 158
   static and dynamic, 156
   $V_{DS}/_D$, 192
   $V_{EC}/_C$, 184, 186
Cut-off frequency, filter, 141
Cycle of ac, 63
Cycles per second, 66

D

Damped ac, 11
Damping, meter, 236
Darlington circuit, 321
D'Arsonval meter, 223
DAVC (delayed AVC) circuit, 442
DCA (decimal counting
     assemblies), 543
DCU (decimal counting units),
   252
DDA (decimal divider assemblies)
     circuit, 544
Decade box, 253

Decibel (dB), 135
   in broadcasting, 583
Decimal counting units (DCUs),
   252
Decoupling, 302
De-emphasis circuit, 472
Deflection, CRT: electrostatic,
   255
      magnetic, 604
      yoke, 605
Degeneration, 182, 189, 267
Delay line, 586, 660
Demagnetizing, 54
Demodulator, 429
Desoldering tool, 18
Detection, 430
Detectors:
   autodyne, 433
   crystal, 430
   diode, 429
   envelope, 441, 781
   first and second, 437
   gated-beam, 470
   grid-leak, 432
   microwave, 642
   plate, 431
   power, 431
   product, 443, 781
   ratio, 469
   regenerative, 433
   square-law, 432
   superregenerative, 434
Deviation ratio, 465
Diac, 194
Diagrams, block, 256
Diamagnetic, 52
Dichroic mirror, 627
Dielectric constants, 91
Dielectric losses, 92
Dielectric strength, 92
Differential amplifier, 322
Differentiator, 665
Digital panel meter, 259
Digital signal, 178
Diodes, 151, 174
   copper-oxide, 220
   Gunn, 650
   hot-carrier, 176
   light-emittings 176
   photo, 177
   PIN, 177
   Schottky, 176
   selenium, 220
   solid-state, 174
   solid-state lamp, 176

Diodes:
   tunnel, 176
   varactor, 176
   xenon, 209
   zener, 175
Dip oscillator, 534
Diplexer, 601, 603
Dipole, 506
   folded, 516
Direct current (dc), 10
Direct wave, 499
Direction finders, 699
   Adcock, 708
   automatic, 706
   balancing, 701
   calibrating, 703
   compensating, 703
   errors, 702, 703
   frequencies, 704
   goniometer, 704
   indicator, 702
   loop antenna, 699
   maintenance of, 705
   navigation, 708
   night effect of, 703
   receivers, 705
   sense antenna, 701
   unidirectional, 702
Directional coupler, 603, 641
Discriminator:
   Foster-Seeley, 467
   stagger-tuned, 466
Distortion, 158, 312
   AM, 405
   carrier shift, 407
   crossover, 321
Distress frequencies, 680
   priority of, 729
   radiotelegraph, 740
   radiotelephone, 727, 730
   transmitting speed, 746
Distributed capacitance, 134
Diversity reception, 452
Doherty circuit, 406
Domains, 51
Doping, 174
Doppler effect, 670
Double conversion, 448
Doubler:
   frequency, 345
   voltage, 213
Dummy antenna, 360
Dummy load, microwave, 641
Duplexer, radar, 661
Duty cycle, 660

Dynamic drain resistance, 190
Dynamic electricity, 3
Dynamic instability, 386
Dynamotors, 574
Dynatron oscillator, 282
Dynode, 607

E

Earphones, 308
Echo box, 668
Eddy currents, 82
Edison cell, 556
Effective ac values, 65
Effective radiated power (erp), 525, 614
Efficiency:
   circuit, 44
   modulation, 399
   RF amplifier, 336, 406
Electrodynamometer, 248
Electrolyte, 9, 548
Electrolytic capacitor, 93
Electromagnet, 48
Electromotive force, 8
Electron-coupled oscillator, 273
Electron gun, 254
Electron-hole pair, 178
Electron multiplier, 608
Electronic voltmeter (EVM), 243
Electrons, 1
   free, 4, 173
   valence, 4
Electroscope, 4
Electrostatic field, 3
Electrostatic voltmeter, 237
Elements:
   atomic, 4
   beam, 514
Emergency Broadcast System (EBS), 593
Emergency repairs, 459
Emission, cathode, 147
   secondary, 162
Emissions:
   AØ, 361
   A1, 356, 358
   A2, 365
   B, 356
   F1 (FSK), 365
   (See also Appendix E)
Emitter, 179
End effect, antenna, 502

Energy, 27
   capacitor, 93
   inductor, 72
Envelope, modulation, 387
Equalizer, line, 585
Equalizing pulse, 609
Equalizing resistors, 100, 209
Erp (effective radiated power) of TV transmitter, 525, 614
EVM (electronic voltmeter), 243
Excitation control, 374
Experimental period, 485

F

Facsimile, 718
   FM station, 493
   recorder, 720
   scanner, 719
Fading, 500, 542
   A3 versus A3J, 416
Faraday rotation, 651
Faraday shield, 293, 370
FDM (frequency division), 423
Feedback:
   degenerative, 302
   factor, 315
   regenerative, 302
Feeder (see Transmission line)
Ferrite core, 71, 616
Ferromagnetism, 52
FET:
   drain resistance, 190
   junction, 189
   MOSFET, 191
   μ, 190
Fidelity, 300
Field, TV, 605
Field gain, antenna, 524
Field intensity:
   antenna, 523, 524
   magnetic, 50
Field poles, 562
Filament, 6, 146
   center-tapping, 150, 348
   metals used, 148
Filters:
   frequency, 140
      balanced, 142
      bandpass, 140
      bandstop, 140
      constant-k, 141
      crystal, 420
      half-full lattice, 420

Filters, frequency:
   high-pass, 141
   low-pass, 141
   m-derived, 142
   mechanical, 421
   shape factor, 420
   SSB, 420
  power supply, 200
   capacitive, 200
   capacitive input, 203
   choke, 205
   configurations, 205
   inductive, 202
   inductive input, 203
   π-type, 203
   RC, 205
Fixed bias, transistor, 181
Flux, 49
Flux density, 49
Flywheel effect, 131, 264
FM (*see* Frequency modulation)
Focusing:
  electrostatic, 254
  magnetic, 604
Folded dipole, 516
F1 (*see* Frequency-shift keying)
Forming current, 180
Forward biasing, 174
Foster-Seeley circuit, 467
Frame, TV, 605
Free electron, 4, 173
Frequencies, calling and working, 734
  (*See also* Bands)
Frequency:
  ac, 66
  conversion, 438
  distortion, 313
  image, 448
  pulling, 434
  resonance, 126
  spectrum, 67
  stability, 359
  tolerance, 360, 531
Frequency meters, 251
  digital counter, 543
  heterodyne, 538
  induction, 251
  vibrating reed, 251
  wavemeter, 532
  waveguide, 644
Frequency modulation (FM), 463
  AFC, 477
  broadcast band, 463
  broadcast station, 482

Frequency modulation (FM):
  channel, 465
  Crosby circuit, 476
  de-emphasis, 472
  direct method, 476
  discriminators, 466, 467
  educational, 484
  emissions, 466
  experimental period, 485
  FAX, 493
  four fields, 463
  gated-beam detector, 470
  licenses required, 484
  licensing, 486, 488
  limiters, 471
  operating power, 483
  pre-emphasis, 472
  public safety services, 484
  ratio detector, 469
  receiver alignment, 475
  receivers, 472
  SCA, 493
  slope detection, 466
  stereo multiplex, 489
  tolerances, 486
Frequency multipliers, 345, 368
  push-pull, 346
  push-push, 345
Frequency-shift keying (FSK), 365, 452
Frequency standards:
  cesium-beam, 535
  primary, 534
  rubidium, 535
  secondary, 536
FSK (frequency-shift keying), 365, 452
Fuel cell, 558
Full-wave rectification, 200
Function generator, 654
Fuses, 30

G

Gain, stage, 292
Gain control (*see* Volume control)
Galvanometer, 223, 253
Gamma correction, 627
Gaseous tubes, 152
Gated-beam circuit, 470
Gauss, 55
Generator, dc, 564
  commutating poles, 567

Generator, dc:
  commutator, 564
  compound, 567
  externally excited, 566
  maintenance of, 575
  ratings of, 574
  series, 566
  shunt, 566
  sparking, 568
  third brush, 567
Generator rule, 58
Getter, 152
Gilbert, 55
Gimmick, 459
Goldsmith transmitter, 264, 356
Goniometer, 704
Grid, 154
  gaseous tube, 154
Grid current, 156, 267, 308, 333
Grid-leak bias, 267, 337, 368
Grid-leak detectors, 432
Grid modulation, 398
Ground area, 525
Ground plane, 509
Ground wave, 499
Grounded-base circuit, 187
Grounded-emitter circuit, 187
Grown transistor, 179
Gunn diode oscillator, 650

H

Hairpin tank, 643
Hairpin-tank oscillator, 281
Half-wave rectification, 199
Hand capacitance, 271, 371
Harmonics, 67, 346
  attenuation, 369, 524
  field intensity, 524
  generation, 305
  oscillator, 273
  radiation, 369
Hartley oscillator, 270
Hash interference, 208
Hazeltine balance, 341
HCD (hot-carrier diodes), 176
Heading flash, 668
Hearing, 384
Heat sink, 186
Heated cathodes, 147
Heising modulation, 397
Helium, 4
Helix, 647
Henry, 70

Hertz, 66
Heterodyne, 396
  frequency meter, 538
  transmitter, 375
Hole current, 174
Homotaxial diffusion, 179
Horizontal deflection, TV, 623
Horsepower, electric, 27
Hot-carrier diodes (HCD), 176
Hum in amplifier, 317
Hydrogen, 4
Hydrometer, 551
Hysteresis loop, 52
Hysteretic loss, 53, 82

I

IC (integrated circuits), 195
  digital and linear, 196
Iconoscope, 606
IF (intermediate frequency)
  amplifier, 437, 439
IGFET, 191
Image frequency, 448
Image orthicon (IO), 607
Impedance, 106
  antenna, 522
  with dc, 44
  pentode, 165
  plate, 159
  surge, 505
Impedance matching, 187
  networks, 520
Inductance, 69
  air-core formula, 71
  in capacitors, 96
  definition of, 70
  energy, 72
  parallel, 75
  series, 75
Inductive reactance, 76, 105
Inductor alternator, 563
Instantaneous values, 64
Insulator, 7, 66
Integrated circuits (IC), 195
Integrator circuit, 260
Interference, harmful, 727
Interlock switch, 371
Intermediate frequency (IF), 437,
  439
Intermittent operation, 318
Intermodulation:
  distortion, 313
  transmitter, 425

International broadcast stations,
  590
Interpoles, 567
Interrupted direct current, 11
Inverse feedback:
  AF, 313
  RF, 341
Inverse peak voltage, 199, 209
Inversion layer, 501
IO (image orthicon), 607
Ion, 10
Ionization, 10, 152
  double, 208
Ionosphere, 499
Isolator, 650

J

j operator, 115
JFET, 189
Joule, 27
Junction capacitance, 184
Junction FET, 189
Junction transistor, BJT, 179

K

Kc, 66
Key click, 361
Key-click filter, 362
Keying:
  BJTs, 364
  break-in, 358
  cathode, 361
  duplex, 358
  frequency shift, 365, 452
  grid-block, 364
  hard and soft, 362
  plate, 361
  primary, 363
  relay, 357
  shaping, 361
  simplex, 357
  vacuum-tube, 363
Keys, telegraph, 746
KHz, 66
Kirchhoff's current law, 33
Kirchhoff's voltage law, 33
Klystron, 644

L

Laminated core, 82
LASCR (light-activated SCR),
  195
Latitude, 695
Lattice, semiconductor, 173
Laws, Federal Communications
  Commission (FCC), 723
LC product, 127
Leakage flux, 83
Lecher wires, 533
LED (light-emitting diodes), 176
Left-hand generator rule, 58
Licenses, Federal Communica-
  tions Commission (FCC):
  amateur (see Amateur
    radio)
  broadcast, 591
  commercial, 724
  elements for, 723
  endorsements, 725
  posting, 733, 737
  radiotelegraph, 737
  shipboard operator, 694
  station, 486
  suspension of, 725
Light-activated SCR (LASCR),
  195
Light-emitting diode (LED), 176
Lighthouse tube, 169
Lightning, 501
Lights, antenna, 734
Limiters:
  AF, 390, 480, 586
  antenna circuit, 682
  IF, 471
Linear radio-frequency amplifiers,
  403–407
  A3, 403
  A3J, 422, 778
  clamp tube, 423
  Doherty, 406
Linear scale, 234
Lissajous figures, 257, 777
Litzendraht wire, 133
Load, 6
  plate circuit, 149, 156
Load line:
  BJT, 184
  triode, 158
Loading antennas, 504
Local oscillator, 437
Loftin-White circuit, 296

Logarithms, 135
Logs:
    AM broadcast stations, 591
    amateur, 751
    marine, 727
    radiotelegraph, 741
    TV stations, 612
Longitude, 695
Loop antenna, 518
Loop stick, 450
Loran, 711
    A-type, 713
    blinking signals, 716
    C-type, 716
    master-slave, 712
    receivers, 716
    sky-waves, 715
    time difference, 714
    transmitters, 716
Losser resistor, 341
Loudspeaker, 310
LSA (limited space-charge ac-
        cumulation) diode, 650
Luminance signal, 627

M

Magnetic field, 47
    left-hand rules for, 48
Magnetic units, 55
Magnetism, 47
    earth's, 56
Magnetizing, 54
Magnetohydrodynamic, 9, 62
Magnetomotive force (mmf), 49
Magnetostriction, 59
Magnetron, 646
Mantissa, 135
Mark and space keying, 365
Mark and space signals, 692
Matching impedances, 43
Mathematic basics, 23
Matrix, 627
Maxwell, 55
MAYDAY, 727
Mc, 66
Measuring frequency, 531
    amateurs, 544
    broadcast station monitor, 542
    with counters, 543
    grid-dip meter, 534
    with heterodyne meters, 538
    by interpolation, 537
    lecher wires, 533

Measuring frequency:
    with secondary standard, 537
    with wavemeter, 532
Mercury-vapor diode, 199, 207
    paralleled, 208
Mesa transistor, 179
Messages:
    charges for, 744
    check count of, 743
    radiotelegraph, 739, 742
    radiotelephone, 732
    types of, 744
Meters, 233
    ac, 245
    ammeter, 31, 235
    ampere-hour, 31, 252
    broadcast station, 588
    bypassing, 359
    dc, 233
    decibel, 246
    digital panel, 259
    electrodynamometer, 248
    electrostatic, 237
    EVM, 243
    frequency, 251
    grid-dip, 534
    heterodyne frequency, 538
    I² scales, 252
    inclined coil, 250
    inclined vane, 250
    iron vane, 250
    moving vane, 249
    ohmmeter, 31, 239
    peak reading, 246
    power, 246
    rectifier, 245
    repulsion, 249
    sensitivity, 236
    taut band, 234
    thermocouple, 247
    voltmeter, 30, 238
    VOM, 240
    VTVM, 240, 243
    VU, 246
    watthour, 31, 251
    wattmeter, 250
    wavemeter, 532
Mho, 37
Microphones:
    broadcast, 596
    carbon, 384
    condenser, 411
    crystal, 411
    directivity, 412
    dynamic, 410

Microphones:
    magnetic-induction, 410
    phasing, 413
    ribbon, 411
    single-button, 384
    sound-powered, 411
    variable reluctance, 411
    velocity, 411
Microstrip, 640
Microwave measurements, 653
Microwave transistors, 648
Microwave transmission lines,
        638
Microwaves, 638
    backward-wave oscillator tube,
        647
    bands, 639
    detectors, 642
    devices, 640
    diodes, 648
    klystrons, 644
    magnetrons, 646
    traveling-wave tube, 647
Mil, 12
Mile:
    nautical, 656
    radar, 656
Miller effect, 301
Minority carriers, 175
Mixer, 437
    conversion efficiency, 781
Mixing, 396
Mmf (magnetomotive force), 49
Modulated amplifier, 390
Modulation:
    phase (PM), 478
    visual, 614
    (See also Amplitude modu-
        lation; Frequency
        modulation)
Modulation index, 465
Modulators:
    AM, 387
    balanced, 417, 419
Molecule, 2
Monitors:
    frequency, 542, 589
    modulation, 589
Monochrome, 628
MOPA (master oscillator power
        amplifier), 358
Morse code, 738
Mosaic, 606
MOSFETs, 191
Motor-generators, 573

Motor rule, 58
Motorboating, 293
Motors:
  ac, 571
    maintenance of, 575
    polyphase, 573
    ratings of, 574
    shaded-pole, 572
    split-phase, 572
    squirrel-cage, 571
    synchronous, 571
    universal, 571
  dc, 568
    compound, 570
    maintenance of, 575
    ratings of, 574
    series, 568
    shunt, 568
    starters, 569
MUF (maximum usable
    frequency), 500
Multipath signals, 475
Multiplexer, film, 612
Multiplexing:
  by frequency division, 423
  by time division, 423
Multiplier:
  frequency, 345, 368
  meter resistor, 238
  voltage, 213
Multivibrator, 284
  one-shot, 665
Mutual conductance, 159
Mutual inductance, 74

N

Navigation, radio: line of position,
    709, 712
  loran, 711
  omega, 718
  RDF, 708
NCU call, 734
Negative lines of force, 2
Negative resistance, 176, 282
Negative transmission, 601
Neutralization:
  direct, 343
  grid, 342
  with I_g meter, 343
  inductive, 344
  pentode, 344
  plate, 341
  push-pull stage, 343

Neutralization:
  with RF indicator, 342
Neutron, 3
Night effect, 703
Nixie tube, 259
NPN-transistor circuit, 179
Noise:
  blanker, 446
  figure, 438
  limiter, 445
  reduction of, 489
  white, 375
Nonlinear scale, 234
Nonsynchronous vibrator, 221
North Pole, geographical, 56
North pole, magnetic, 48
Nucleus of the atom, 4

O

Octave, 346
Oersted, 55
Ohm, 12
Ohmmeter, 31, 239
  caution, 243
  multirange, 240
Ohm's law, 21, 106
  complex circuits, 33
  magnetic circuits, 54
Omega navigation, 718
Operating power, 395
  AM, 587
  FM, 483
Operational amplifier, 260, 323,
    325
Opto-isolator, 178
Oscillate, 265
Oscillation, indications of, 286
Oscillators, 264
  Armstrong, 266
  audio, 282
  beat-frequency, 437
  Butler, 676
  carrier, 443
  Colpitts, 272
  crystal, 274
  dip, 534
  dynatron, 282
  electron-coupled, 273
  grid-voltage, 780
  Hartley, 270
  high-frequency, 280
  multivibrator, 284
  overtone, 279

Oscillators:
  pentode, 280
  Pierce, 278
  RC, 283
  relaxation, 283
  stability, 287
  TPTG, 268
  tunnel-diode, 283
Oscilloscopes, 254
  with A3, 401
  with A3J, 422
  envelope display, 401
  free-running, 256
  Lissajous figures, 257, 777
  sampling, 259, 653
  trapezoidal display, 402
  triggered, 258
Oven, temperature control, 277
Overdriving, 291
Overlay transistor, 179
Overload relay, 372
Overtone, 383
  oscillator, 279

P

PAN (urgency signal), 727
Parabolic reflector, 517, 662
Parallel operation:
  ac circuits, 115
  AF amplifiers, 312
  batteries, 40
  resistors, 38
  RF amplifiers, 337
  tubes, 312
Paramagnetic, 52
Paraphase circuit, 316
Parasitic array, 515
Parasitic choke, 286, 337
Parasitic oscillation:
  high-frequency, 286
  low-frequency, 268, 285
Passivated planar BJT, 179
Passivation, 177
Passive devices, 146
PC (printed circuit) boards, 18,
    460
Peak envelope power (PEP), 422
Peak limiter, 390, 471, 480, 586
Peak value, ac, 65
Pentagrid converter, 438
Pentode, 164
  AF amplifier, 317
  oscillator, 273, 279

Pentode:
    remote-cutoff, 165
    RF amplifier, 330
    variable-μ, 165
PEP (peak envelope power), 422
Percent of modulation, 389
Permanent magnet, 53
Permeability, 50
Permeance, 51
Permits (*see* Licenses)
Phantom antenna, 360
Phase:
    ac, 67
    angle, 107
    capacitive circuit, 100
    distortion, 313
    inductive circuit, 78
    inverter, 315
Phase modulation (PM), 478
Phase monitors, 515
Phase-shift networks, 421
Phasitron, 479
Phasor, 122
Phonetic alphabet, 731
Photodiode, 177
Photoetching, 18
Photojunction, 177
Photons, 173
Photoresistive device, 177
Phototransistor, 180
Photovoltaic cell, 177, 558
Pi-network (π-network), 351
    antenna, 520
    resistors, 43
Picture tubes:
    color, 633
    monochrome, 604
    shadow-mask, 634
    single-gun, 635
Pierce oscillator, 278
Piezoelectric, 9, 274
    earphones, 310
Pilot carrier, 418
PIN diode, 177, 649
Pinch-off bias value, 189
Pincushion distortion, TV, 623
Pitch of sound, 383
Planar epitaxial BJT, 179
Plasma, 208
Plate, vacuum tube, 148
    cooling, 168
    impedance, 159
    improper current, 168
    load, 149
    voltage, 149

PN junction, 179
PNP transistor, 180
Point contact, BJT, 179
Polar notation, 122
Polyphase alternating current, 224
Positive electric field, 2
Poulsen arc, 264
Power, 26
    amplifier, 291
    apparent, 109
    true, 109, 250
Power factor, 110
    capacitor, 96
    compensation, 213
Power output, triode, 160
    maximum, 161, 294
    RF amplifier, 336
    undistorted, 161, 294
Power supply, 198
    dc converter, 221
    failure, 228
    full-wave, 200, 212
    half-wave, 199
    motor-generator, 573
    overload protection, 219
    regulated, 217
    three-phase, 227
    vibrator, 222
PPI (plan-position-indication)
    radar, 658
Preamplifier, 319, 414
Preemphasis circuit, 472
Printed circuit (PC) boards, 18, 460
Priority of messages, 729
Probes, test, 261
Proof-of-performance test, 588
Proton, 2
PRR (pulse repetition rate), 657
PTT (push-to-talk), 680
Public safety radio, 485
Pulsating direct current, 10, 199
Pulse-width demodulation, 435, 471
Push-pull:
    AF amplifier, 304–308, 320
    RF amplifier, 338, 350
Push-push circuit, 345
Pythagorean theorem, 106

Q

Q of circuit, 133, 443
Q-multiplier, 783

Q signals, 768–771
Quadrantal counterweights, 234
Quadrantal errors, 703
Quadrature, 420
Quantisizing, 425
Quarternary winding, 81
Quiescent value, 187

R

Radar, 656
    antenna, 661
    ATR, 663
    duty cycle, 660
    echo box, 668
    heading flash, 668
    indicator, 665
    interference, 669
    maintenance of, 669
    mile, 656
    nonmarine, 670
    operating, 669
    PPI, 658
    PRR, 657
    range marker, 665
    receiver, 663
    servo, 668
    STC, 664
    TR, 662
    transmitter, 659
Radian, 65
Radiation resistance, 504
Radio, amateur (*see* Amateur radio)
Radio direction finders (*see* Direction finders)
Radio-frequency amplifiers [*see* RF (radio-frequency) amplifiers]
Radio-frequency chokes, 334
Radio-frequency interference (RFI), 775
Radio waves, 498
    direct, 499
    ground, 499
    polarization, 502
    scatter, 501
    skipping, 500
    sky, 500
    velocity, 502
Radiotelegraph, 737
    calling, 739
    frequencies, 738
    keys, 746

Radiotelegraph:
  licenses, 737
  messages, 739
Radiotelephone, 729
  calling and working, 732
  frequencies, 734
  messages, 729
  operating, 733
  phonetic alphabet, 731
  single-sideband, 415
Ramp voltage, 258
Range marker, 665
Range resolution, 658
Raster (unmodulated lines on TV
    screen), 623
RC circuit, 89
RC filter, 205
RC oscillator, 283
Receivers, 429
  aligning, 455
  all-wave, 451
  automobile, 450
  broadcast, 451
  color TV, 630
  communication, 451
  crystal, 430
  DF, 705
  double-conversion, 448
  FM, 472
  front end, 440
  FSK, 452
  loran, 716
  low-frequency, 451, 682
  operating, 451
  radar, 663
  ship, 682
  shortwave, 451
  superheterodyne, 436–452
  transistor, 449
  TRF, 435, 682
  TV, 615, 630
Recorder:
  disc, 594
  tape, 595
  turntable, 594
  X-Y, 654
Rectangular notation, 122
Rectification, ac bridge, 151, 200
    full-wave, 200
    half-wave, 199
Rectifiers, 198
  cold-cathode, 217
  copper-oxide, 220
  high-vacuum, 206
  mercury vapor, 207

Rectifiers:
  selenium, 220
  xenon, 209
Reference levels, TV, 611
Reflection coefficient, 506, 603
Reflectometer, 603, 642
Regeneration, 267, 433
Regulation, voltage: resonant
    power transformer, 218
  series, 217
  shunt, 217
  VR tube, 216
Relaxation oscillator, 283
Relays, 59
  circuit breaker, 372
  contacts, 358
  keying, 357
  overload, 337, 372, 379
  polar, 692
  recycling, 379
  time delay, 379
Reluctance, 50
Remanence, 52
Remote control, 592, 781
Remote-cutoff tube, 165
Repairing equipment, 376
Residual magnetism, 52
Resistance, 11
  negative, 176, 282
  power in, 29
  radiation, 509
Resistors, 13
Resonance, 113, 126
  C and L values, 347
  ferromagnetic, 650
  parallel, 130
  series, 127
Retentivity, 52
Reverse breakdown voltage, 193
RF (radio-frequency) amplifiers,
    329, 340
  buffer, 367
  classes of, 161, 332, 403
  coupling, 335, 339
  grounded-grid, 349
  linear A3, 403
  linear A3J, 403
  low-level, 330
  output, 336
  pentode, 330, 344
  power, 331
  receiver, 437
  tetrode, 332, 344
  triode, 340
  tuning, 335

RF (radio-frequency) amplifiers:
  VHF and UHF, 350
RF (radio-frequency) chokes, 334
RFI (radio-frequency inter-
    ference), 775
Rhombic array, 517
Right-hand motor rule, 58
Ripple, frequency, 210
  plate current, 150
Rms (root-mean-square) value,
    65
Root-mean-square (rms) value,
    65
RTTY (see Teleprinter)
Rubidium gas standard, 535

S

S meters, 444
Safety link, 677
Safety messages, 727, 731
Safety signal, 741
Safety tuning transmitters, 414
Satellites, 500
Saturation, plate, 149, 161
Sawtooth alternating current, 63
Scanning, interlaced, 605
Schottky diode, 176
SCR (silicon-controlled rectifier),
    193
Screen grid, 163
  AF amplifier, 317, 320
  modulation, 400
  RF amplifier, 330, 332
S/D (sideband distortion) ratio,
    775
Sea return, 664
Secondary emission, 162
SECURITY, 727
Selectivity:
  detector, 431, 435
  IF, 440
Selenium cell, 558
Selenium rectifier, 220
Self-bias, BJT, 182
Self-induction, 70
Selsyn, 667
Semiconductors, 173
Sensitivity:
  control, 437
  detector, 431
  meter, 236
Servomechanism, 668
Shadow-mask tube, 634

Shape factor, 420
Shaping continuous wave
   emissions, 361
Shelf life of cells, 549
SHF (superhigh radio
   frequencies), 67
Shielding:
   choke, 205
   RF fields, 371, 436
Shielding tubes, 166
Shipboard radio, 674
   AA, 689
   antenna switch, 691
   bridge-to-bridge, 687
   compulsorily equipped, 674
   frequencies, 680
   geography, 695
   HF transmitters, 678
   main antenna, 677
   main transmitter, 675
   operators, 694
   radiotelephone, 681
   receivers, 682–686
   reserve transmitter, 676
   safety link, 677
   standing watch, 694
   survival radio, 687
Shock, electric, 774, 776
Shock excitation, 264
Shorted turn, 76
Shot effect, 190
Shunt, feed, 270, 339
   meter, 235
Shunt dc generator, 566
Shunt dc motors, 569
SI (Système International
   d'Unités) units, 54
Sideband:
   A1 keying, 362
   A2 keying, 366
   single (see Single sideband)
   spark, 356
   vestigial, TV, 614
Sideband analyzer, 615
Sideband filters, 420
Siemens, 37
Signal generator, microwave,
   653
Signal-to-noise (S/N) ratio, 438,
   781
Silence periods, 740
Silicon, doped, 174
Silicon-controlled rectifiers
   (SCRs), 193
Sine, 107

Single sideband (SSB), special
   uses of, 423
   multiplexing, 423
   VOX, 425
Single-sideband radiotelephone,
   415
   fading signals, 416
Single-sideband transmitters:
   balanced modulators, 417, 419
   filter-type, 417
   input power, 776
   linear amplifiers, 422
     PEP, 422, 779
     two-tone modulation tests,
      423, 781
   phase-type, 421
   pilot or vestigial carrier, 418
Sinusoidal alternating current, 63
Skin effect, 133
Skip zone, 500
Sky wave, 500
Slide-screw tuner, 641
Slip rings, 562
Slug, ferrite, 71, 780
   brass, 76
S/N (signal-to-noise) ratio, 438,
   781
Snow on TV screen, 616
Soft tube, 152
Solar cells, 558
Soldering:
   gun, 460
   iron, 17
Solid-state devices, 173
Solid-state lamp, 176
SOS (distress signal), 740
   keyer, 688
Sound, 382
   pitch, 383
   threshold of, 383
Source of electric circuit, 6
South Pole, geographical, 56
South pole, magnetic, 48
Space charge, 147
Spark gap, radar, 660
Spark transmitter, 265
Specific gravity, 551
Specific inductive capacity, 91
Specific resistance, 12
Spectrum, 67
Spectrum analyzer, 615
Speech amplifier, 387, 414
Splatter, 389
Split-load circuit, 316
Split tuning, 140

SPST relay, 59
Spurious emissions, 389
   checking, 781
Square root, 24
Square-wave alternating current,
   63
Squelch circuit, 443
   differential, 473
SSB (see Single sideband)
SSL (solid-state lamp), 176
Stabilizing circuit, NPN, 267
Stage, single, 317
Standing-wave ratio (SWR), 506
   measurement of, 642
   presence of, 778
Star connection, 224
Static, 3
STC (sensitivity-time-control)
   circuit, 664
Stereo multiplex, 489
STL (studio-transmitter link), 579
Stub, 511
Subcarrier, TV, 630
Sulfation, 551
Summing junction (SJ), 325
Supercontrol tube, 165
Superheterodyne, 436–443
   fourth generation, 684
   third generation, 683
Superregeneration, 434
Suppressor grid, 164
   modulation, 400
Surge impedance, 505
Susceptance, 116
Swinging choke, 206
Switches, interlock, 371
Switching transistors, 285
SWR (see Standing-wave ratio)
Symbols:
   semiconductor, 196
   tube, 166
Sync-pulse separator, 621
Synchronizing-pulse generator,
   602, 609
Synchronous vibrator, 222
Synthesizer, 684

T

Tangent, 107
Taper:
   audio, logarithmic, 300
   linear, 299
Taut-band meter, 234

TDM (time division), 423
Telegraph (*see* Radiotelegraph)
Telephone:
  simple one-way circuit, 384
  use of frequency division
    multiplexing, 423
  (*See also* Radiotelephone)
Teleprinter, 692
  identifying, 780
Television (TV), 601
  AGC, 625
  aspect ratio, 605
  broadcast system, 601
  camera chain, 611
  channels, 613
  chrominance circuits, 631
  color (*see* Color, TV)
  color receivers, 630
  color tubes, 633
  deflection, 605
  deflection circuits, 622
  frame, 605
  lines, 605
  logs, 612
  luminance, 627
  modulation, 611
  multiplexer, 612
  power, 614
  projectors, 613
  pulses, 609
  receiver, 615
  scanning, 605
  subcarrier, 630
  synchronizing-pulse generator,
    602, 609
  visual transmitter, 614
Temperature coefficient, 369
  of capacitor, 102
  of crystal, 276
Temporary magnet, 53
Tertiary winding, 81
Tesla, 55
Testing transmitters, 740
Tetrode, VT, beam power, 163
Tetrode transistor, 180
Thermal runaway, 187, 267
Thermal stability, 182
Thermistor, 183, 321, 643
Thermocouple, 277
Thermocouple ammeter, 247
Thermostat, 277
Three-phase alternating current,
  224
  delta, 224
  neutral, 225

Three-phase alternating current:
  star or wye, 224
  supplies, 227
  wattmeters, 226
Thyratron, 153, 664
  solid-state, 193
Thyristor, 194
Tickler coil, 266, 433
Time constant:
  RC, 90
  RL, 72
Time domain reflectometry, 654
Time line, 64
Time ticks, 695
Tone control, 300
Toroid, 71
Torque, 27, 562
TPTG (tuned-plate tuned-grid)
  oscillator circuit, 268
TR (transmit-receive) box, 662
  anti-, 663
  keep-alive voltage, 663
Transadmittance, forward, 190
Transceiver, 452, 688
Transconductance:
  conversion, 439
  JFET, 190
  VT, 159
Transducer, 66, 308
Transfer oscillator, 544
Transformer, 74, 80
  AF, 86
  bifilar IF, 618
  control, 668
  current, 249
  output, 311
  phase in, 467
  power, 86
  quarter-wave matching, 512
  RF, 87
Transformer losses:
  copper, 83
  eddy current, 82
  external induction, 83
  hysteretic, 82
Transformer ratios:
  current, 84
  power, 84
  voltage, 83
Transient variation of power-
  supply voltage, 362
Transistors:
  BJT, 179
  CW transmitter, 372
  FET, 189

Transistors:
  IGFET, 189
  JFET, 189
  junction, 179
  MOSFET, 189
  NPN, 179
  photo, 180
  PNP, 181
  point contact, 179
  RF power amplifier, 350
  switching, 285
  tetrode, 180
  unijunction, 192
Transit time, 169, 350
Translate, 419
Transmatch, 753
Transmission line:
  air-filled, 522
  artificial, 659
  delta match, 511
  flat, 506
  gamma match, 511
  lossy, 782
  phase lengths, 514
  reflection coefficient, 506
  standing waves, 506
  stub, 511
  surge impedance, 505
  SWR, 506
  transposed, 514
  velocity factor, 513
  waveguide, 638
Transmitters:
  Alexanderson, 264, 356
  amateur band, 373, 375
  arc, 357
  aural TV, 601
  basic, 356
  Goldsmith, 264, 356
  heterodyne, 375
  intermodulation, 425
  loran, 716
  modern, 372
  operation, 726
  radar, 659
  servicing, 726
  shipboard, 676
  spark, 356
  transistor CW, 372
  VFO, 373
  visual TV, 614
Transponder, 671
Traveling-wave tube (TWT), 647
TRF (tuned-radio-frequency)
  receiver, 435

Triac, 194
Trickle charge, 555
Trigger voltage, 258
Triodes:
  amplification factor, μ, 155
  bias voltage, 156
  dc plate resistance, 155
  gaseous, 153
  lighthouse, 169
  load resistance, 156
  mutual conductance, g_m, 159
  plate dissipation, 161
  plate saturation, 161
  power output, 160
  RF amplifier, 340
  vacuum, 154
Tripler, frequency, 345
Tropospheric scatter, 646
Troubleshooting:
  AF amplifiers, 317
  receivers, 457
  RF amplifiers, 352
  transistor circuits, 459
  transmitters, 376
Tuned-plate tuned-grid (TPTG)
       oscillator circuit, 268
Tungar rectifier, 153
Tungsten filaments, 148
Tuning:
  BJT RF stages, 350
  transmitters, 359, 367, 373
  VT RF stages, 350
Tuning indicators:
  eye tube, 443
  S meter, 444
Tunnel diode, 176
Tunnel-diode amplifier, 652
Tunnel-diode oscillator, 283, 648
TV (see Television)
TWT (traveling-wave tube), 647
Type acceptance of licensed
       transmitting equipment,
       488
Type approval of licensed trans-
       mitting equipment,
       488

U

UHF (ultra high radio frequen-
       cies), 67
  amplifiers for, 350
  antennas, 509
UJT (unijunction transistor), 192

Ultor, 621
Ultraudion oscillator, 272
Unblanking, 258
Unijunction transistor (UJT), 192
Unity coupling, 74, 138
Universal resonant curve, 139
Universal winding, 334
Up-converted, 437
Urgency signal, 727, 730, 741

V

Vacuum capacitor, 95
Vacuum tubes (VT), 146
  acorn, 169
  bias (see Bias, VT)
  CRT, 254
  diode, 151
  hard, 152
  high-frequency, 169
  lighthouse, 169
  receiving, 167
  soft, 152
  transmitting, 167
  water-cooled, 168
Valence electron, 4
Varactor, 176
  multiplier, 649
Variable-μ tube, 165
Variometer, 676
Varying direct current, 10
Vector, 64
Vector addition, 108
Vertical deflection, 622
Vestigial carrier, 418
Vestigial sideband, 614
VFO (variable frequency
       oscillator), 373
VHF (very high radio frequency),
       67
  amplifiers for, 350
  antennas, 509
  oscillators, 280
  transmitter, 486
  (See also Microwaves; Radar;
       Television)
Vibrator:
  nonsynchronous, 221
  synchronous, 222
Video signal, 601
  amplifier, 620
Vidicon, 608
Volt (V), 8
  by induction, 74

Volt-amperes, 84
Volt-coulombs, 28
Volt-ohm-milliammeter (VOM),
       240
Voltage, 8
  versus power amplifiers, 290
Voltage divider, 220
Voltage-doubler circuit, 213
Voltage-drop, forward, 782
Voltage-multiplier circuits, 213
Voltage regulation, 214
  resonant transformer, 218
  series, 217
  shunt, 217
  VR tube, 216
Voltage-to-current converter, 260
Voltage-to-frequency circuit, 260
Voltage variable capacitor, 176
Voltmeter, 30
  loading, 238
  vacuum-tube, 240
Volume control, 299
  AF, 441
  RF-IF, 437, 441
VOM (volt-ohm-milliammeter),
       240
VOX (voice-operated trans-
       mission), 425, 680
VR (voltage-regulator) tube, 216
VT (see Vacuum tubes)
VTVM (vacuum-tube voltmeter),
       240
VU (volume unit), 136
VU (volume-unit) meter, 246

W

Watt, 26
Watthour meter, 31, 251
Wattmeter, 31, 250
Wattsecond, 27
Wave-soldering, 18
Wave trap, 140
  antenna, 521, 524
  receiver, 447
Waveguides, 638
  attenuator, 641
  coupling, 642
  devices, 640
  installation, 654
  loads, 641
  measurements, 642
Wavelength, 67, 347

Wavemeter:
    absorption, 532
    cavity, 644
Weber, 55
Wire sizes, 16
WWV transmitters, National
        Bureau of Standards
        (NBS), 534
Wye connection, 224

X

Xenon rectifier, 209

Y

Yagi antenna, 516
YIG (yttrium-iron-garnet)
        resonator, 652

Yoke, deflection, 605

Z

Zener diodes, 175
Zener effect, 175
Zepp antenna, 511